T0155493

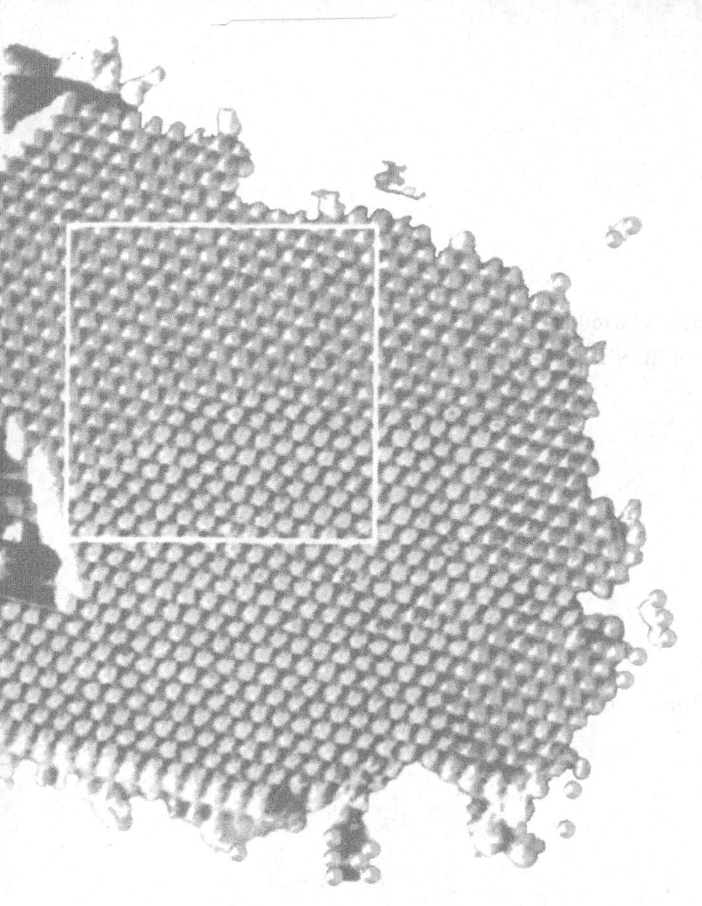

Introduction to Condensed Matter Physics

Volume 1

Feng Duan & Jin Guojun

Nanjing University

NEW JERSEY • LONDON • SINGAPORE • BEIJING • SHANGHAI • HONG KONG • TAIPEI • CHENNAI

Published by

World Scientific Publishing Co. Pte. Ltd.
5 Toh Tuck Link, Singapore 596224
USA office: 27 Warren Street, Suite 401-402, Hackensack, NJ 07601
UK office: 57 Shelton Street, Covent Garden, London WC2H 9HE

British Library Cataloguing-in-Publication Data
A catalogue record for this book is available from the British Library.

First publihsed 2005
Reprinted 2007

INTRODUCTION TO CONDENSED MATTER PHYSICS
Volume 1

ISBN-13 978-981-238-711-0
ISBN-10 981-238-711-0
ISBN-13 978-981-256-070-4 (pbk)
ISBN-10 981-256-070-X (pbk)

Printed in Singapore

Preface

Condensed matter physics is one of the most important, as well as most fertile, branches of contemporary physics. It is characterized by a multitude of research workers, a bewildering variety of research results, widespread influences on technical developments and rapid infiltrations into interdisciplinary areas. Historically, condensed matter physics has gradually evolved from solid state physics. However, due to the lack of a clear recognition of their inter-relationship (and the unfamiliarity with the new conceptual systems within the unifying theoretical framework introduced by condensed matter physics), though there are numerous excellent textbooks on solid state physics, a comprehensive introductory textbook on condensed matter physics is still waiting to appear. This unsatisfactory state of affairs is most clearly shown by the enormous gap between traditional textbooks on solid state physics and the frontier of present day research in condensed matter physics. It is a familiar sight to see students who have already taken courses in solid state physics (and even solid state theory) on approaching the frontier of condensed matter physics, for instance when browsing current issues of journals such as *Physical Review Letters*, who generally feel perplexed and alienated: They find the literature hard to understand. These difficulties in understanding do not stem principally from the derivation of the formulas or things related to experimental situations, the crux of the matter is that the mind is unprepared, so it is hard to comprehend why certain topics are chosen and whence the fundamental ideas descend. This situation has been realized by many eminent physicists who expressed the need to establish a graduate course to bridge the gap between traditional solid state physics and the research frontier. This course should be situated between two extremes: on the one hand, a traditional course which includes such long-standing subjects as the periodic structure, energy bands, and lattice dynamics, etc., comprehending these subjects and mastering their language prove to be indispensable to the students who wish to communicate scientifically. On the other hand, there are current topics, such as high T_c superconductivity, localization, quantum Hall effect, giant and colossal magnetoresistance, quantum dots, fullerenes and carbon nanotubes, photonic bandgap crystals, etc. Without touching these latter subjects they can hardly begin their research. Our book is just such an attempt to fill the gap between traditional solid state physics texts and the research frontier, and bears the name *Introduction to Condensed Matter Physics* to stress its introductory approach. There is bewildering variety and apparent complexity in contemporary condensed matter physics. However, due to the conceptual unity in the structure of matter, condensed matter physics can be organized into a coherent logical structure which is ripe for a systematic exposition in textbook form.

The basic concepts of condensed matter physics have been already penetratingly analyzed by P. W. Anderson in his monograph, "*Basic Notions of Condensed Matter Physics*", however, this book is for *cognoscenti*, perhaps too difficult for uninitiated graduate students. We acknowledge our indebtedness to this inspiring book, full of creative ideas, and try to make concrete some of these basic ideas, illustrating them with examples and placing them into suitable contexts. We were faced with the formidable task of assimilating and arranging this enormous mass of material into a satisfactory logical framework guided by a unifying conceptual framework, and finally incorporating it into a readable text. This text was originally intended for first year students of our graduate school, and some previous exposure to an undergraduate course in solid state physics is desirable but not necessary. In our institution, there is another course on condensed matter theory or solid state theory after this one, so in this course the physical concepts are stressed, simplified theoretical derivations are given to facilitate understanding, while cumbersome details of mathematics are avoided. Since condensed matter physics is closely linked to technological developments and interdisciplinary fields, we hope this book may also serve as a reference book for researchers in condensed matter physics, materials science, chemistry and engineering.

This book is organized into eight parts.

Part I gives the structural foundation of condensed matter physics. It starts from a brief resumé of classical crystallography. Its basic concepts such as symmetry and lattice are introduced and prepared for later generalization. Although the importance of symmetry is emphasized, complementary notions stemming from differential geometry and topology are also introduced to give a more rounded picture of structural crystallography and related topics. What follows may be called generalized crystallography; it means that the basic concepts of crystallography are generalized to cases in which the strict periodicity is absent, such as quasiperiodic structures, homogeneously disordered structures, supramolecular structures and inhomogeneous structures. Of course, statistical concepts such as order parameter, distribution functions and correlation functions are introduced to facilitate the treatment of various types of disordered structure. Mathematical aspects are introduced through Fourier transforms and other topics. Liquid crystals, self-assembled membranes, polymers and biopolymers are also discussed in order to help readers to explore the new field of soft condensed matter physics.

Part II embodies the results of carrying out the program based on the enlargement of the original paradigm of solid state physics. Starting from wave propagation in a periodic structure, besides briefly summarizing the basics of energy bands and lattice dynamics, an introduction of the photonic bandgaps is added; and wave behavior in quasiperiodic structures is also introduced. Then the dynamics of Bloch electrons and elementary treatment of transport properties including spin transport are given. It further leads to wave scattering by impurities and alloys, and after that, to wave localization due to disorder. More recent topics of quantum transport in mesoscopic systems have also been introduced. The unity of wave behaviors of different types of waves are emphasized through parallel treatments.

Part III may be taken as a duet between bond approach and band approach as well as a step towards many-body physics. These are based on the paradigms of quantum chemistry and solid state physics respectively. It is found that these two approaches are complementary to one another. Topics on electron correlation are introduced gradually, starting from Heitler and London's treatment of H_2 molecule, and finally reaching strongly correlated electronic systems with anomalous physical properties. This shows there are many important problems which lie beyond these two conventional approaches, some of them are still waiting to be solved. Finally current topics such as quantum confined nanostructures are discussed to illustrate the usefulness of these approaches in present day investigations. These first three parts may be regarded as the stage of transition from solid state physics to condensed matter physics.

Part IV deals with phase transitions and ordered phases. The Landau theory of second order phase transitions is introduced concurrently with the concepts of broken symmetry and order parameter. Then a lot of systems are discussed within this framework, such as crystals, quasicrystals and liquid crystals, which are the results of broken translational or rotational symmetry; ferromagnets or antiferromagnets, which are results of broken time reversal (or spin rotation) symmetry; and superfluids and superconductors, which are results of broken gauge symmetry. Finally, the concept of broken symmetry is generalized to include broken ergodicity for introducing the gas-liquid, wetting, glass, spin glass and metal-insulator transitions.

Part V deals with critical phenomena. After an introduction on fluctuations and related topics such as correlation and dissipation, the concepts of scaling and universality are formulated, and the renormalization group method is used to elucidate critical behaviors. Then the renormalization group method is applied to various other phenomena, such as percolation, localization of electrons, etc. Quantum critical phenomena are also discussed.

Part VI deals with elementary excitations. A general introduction, together with the scheme for classification of elementary excitations, is given. Then, more detailed discussions on vibrational excitations, magnetic excitations and electronic excitations follow, emphasizing those aspects which lie beyond conventional treatments. The theories of Fermi liquids, quantum Hall effect and Luttinger liquids are introduced and discussed.

Part VII deals with topological defects. These are nonlinear or topological excitations. Beginning from generalized elasticity and hydrodynamics, topological properties of defects are followed, then the structure and the energetics of defects, as well as the phases and the phase transitions associated with the defect assemblies, are described. Furthermore, the concept of generalized rigidity is introduced and used to elucidate those physical properties that are structure-sensitive.

It is reasonable to consider that the material from Part IV to Part VII forms the main body of condensed matter physics, unified by the concepts of broken symmetry and order parameter, and stratified according to the energy scale into different levels i.e., the ground states, the elementary excitations, the topological defects and the critical phenomena near T_c.

Part VIII may be regarded as an extension of condensed matter physics. It deals with physical kinetics and nonlinear phenomena in nonequilibrium states. The kinetics of phase transitions are introduced first, covering both second-order and first-order phase transitions; then the growth and form of crystalline materials follows, treating various facets of crystal growth; these provide the theoretical foundation for some crucial topics of materials science as well as an introduction to nonlinear phenomena. Then we treat nonlinear phenomena far from equilibrium in more detail: thermal convection is used as a classical example to illustrate the onset of hydrodynamic instability; then strange attractors and the routes to chaos are explained; finally, diverse phenomena involving spatio-temporal instabilities and pattern formation, such as avalanche and turbulence as well as some problems related to biology and technology are discussed. It shows that there is still a large number of problems in the neighborhood of condensed matter physics related to complexity, waiting to be explored and solved; it testifies that condensed matter physics today is still a flourishing subject.

This book will be published in two volumes: Vol. 1 contains Part I to Part IV, while Vol. 2 contains Part V to Part VIII. In general we assume that readers are already familiar with the basics of quantum mechanics and statistical physics. However different parts have different prerequisites, for instance, Part I, VI, VII and VIII as well as the first two chapters of Part IV can be read through even by those without any training in quantum mechanics. It is expected that this book may be used at different levels and for people with different areas of major interest. Each part may be read quite independently despite their interconnectedness. The whole book is intended as a text for a course in two semesters, while a sequence of selected parts may serve for a one semester course.

We would like to make some comparisons with existing textbooks in this field. Since the publication of *Modern Theory of Solids* by F. Seitz in 1940, many excellent textbooks on solid state physics have appeared. Among them: Kittel's *Introduction to Solid State Physics* is most popular, it has run into numerous editions; while Ashcroft and Mermin's *Solid State Physics* is noted for its painstaking effort in clarifying basic concepts. However, due to the limitation of the conceptual framework, these texts do not communicate many important basic concepts as well as as catch up to the flowering richness of contemporary condensed matter physics. In recent years, a few graduate texts bearing the title condensed matter physics have appeared. We admire these pioneering attempts and have read them with much interest and profit. However, in our opinion or prejudice, they are not entirely satisfactory because they are either not comprehensive enough to cover this broad field, or not systematic enough to give a precise idea about basic concepts threading through them. In this book we try to give a pedagogically understandable exposition on the basic concepts of condensed matter physics illustrated with many concrete physical problems, as well as to give a comprehensive and coherent picture of the contemporary scene, thereby bridging the gap between the traditional texts of solid state physics and the current literature scattered throughout physics journals. Condensed matter physics gradually unfolds in this book from more traditional parts of solid state physics, its basic concepts are emphasized and carefully explained, most equations are derived with not too complicated mathematics, in order to be accessible to first year graduate students and research workers, especially for experimental workers.

Since 1983, one of the authors, Duan Feng, had been seriously concerned with the clarification of the conceptual framework of condensed matter physics and had written a number of articles to expound it. Since 1989, this new approach had been thoroughly discussed and illustrated with various current topics at the frontier of this field in a series of lectures. Then, in collaboration with the other author, Guojun Jin, a book "*New Perspective on Condensed Matter Physics*" (in Chinese) was written and published in 1992. Beginning in 1990 a graduate course called *Introduction to Condensed*

Matter Physics was established in the Physics Department of Nanjing University jointly by the authors. More recently, the teaching of this course in Nanjing University has been done singly by Guojun Jin. Courses for the *Summer School on Condensed Matter Physics* in 1996 (organized by Tsinhua and Peking University, Beijing) and in 1998 (organized by Nanjing University, Nanjing) have been jointly given by the authors. Based on the lecture notes for these courses used over many years, the manuscript for this book gradually evolved. Everyone recognizes that the field of condensed matter physics has exploded in recent years and we have chosen an impossible task: New and significant research results emerge every day. Thus, we can only apologize for overlooking some topics and trust that the reader will understand the difficult task we have set for ourselves. We invite criticism and advice on how to improve this book.

The authors thank the National Laboratory of Solid State Microstructures and Department of Physics of Nanjing University for the encouragement, help and support for this course and the writing of this book. We thank the Education Reform Project of Nanjing University, National Science Foundation of China, and Nanjing University-Grinnell College Exchange Program for their financial support. We are much indebted to various colleagues, friends and graduate students for their discussion, criticism and help. We thank Profs. Changde Gong, Hongru Zhe, Zhengzhong Li, Yuansheng Jiang, Dingyu Xing, Jinming Dong, Weijiang Yeh and Brian Borovksy for reading through some chapters or sections of our earlier manuscript and giving their valuable comments and suggestions for improvement. We especially thank Prof. Charles Cunningham for his painstaking efforts to read through our whole book just before its publication, detecting and correcting the errors and mistakes in it page by page, and improving the English conspicuously, making this book more readable and comprehensible. One of the authors (Guojun Jin) would like to express his thanks to Mr. Jiaoyu Hu for providing a personal computer at the beginning of writing this book; Prof. Zidan Wang for inviting him twice (1996, 1998) to visit the University of Hong Kong, besides to engage in cooperative researches, also to get the opportunity to collect useful materials; Mr. Dafei Jin for solving some technical difficulties in editing the book in LaTeX format. Finally we wish to express our sincere thanks to Prof. K. K. Phua for his persistent concern for the publication of book, to the editors of this book for the patience and carefulness in their editing.

Duan Feng and Guojun Jin
Mar. 2005

Acknowledgments

The authors would like to acknowledge the following publishers, as well as the authors, for their kind permission to reproduce the figures found in this book:

Prof. Jianguo Hou for the STM micrograph on our book cover and Fig. 12.4.19;

Prof. Kunji Chen for Fig. 2.3.4;

Academic Press for granting permission to reproduce Figs. 8.3.5, 9.2.6, 13.1.4, 19.4.1, 19.4.2 and 19.4.3;

American Association for Advancement of Sciences for Fig. 14.5.6;

American Institute of Physics for Figs. 5.3.2, 8.3.4, 10.4.2, 10.4.3, 10.4.4, 10.4.5 and 14.4.2;

American Physical Society for Figs. 2.2.7, 2.3.11, 5.3.3, 5.4.3, 5.4.4, 5.5.2, 5.5.4, 6.2.2, 7.4.2, 8.3.9, 9.1.1, 9.1.2, 9.2.4, 9.2.9, 9.3.6, 9.3.7, 9.3.8, 10.1.2, 10.3.2, 10.3.4, 10.3.5, 10.3.7, 10.4.6, 12.4.13, 13.1.13, 13.2.1, 13.2.2, 13.2.7, 13.2.10, 13.2.11, 13.2.12, 13.3.8, 13.4.1, 14.2.6, 14.2.7, 14.3.5, 14.4.7, 14.4.1, 14.5.5, 14.5.7, 16.3.6, 18.1.3, 18.1.4, 18.1.8, 18.2.5, 18.3.5, 18.3.11, 18.4.2, 18.4.5, 18.4.6, 19.3.6, 19.3.7, 19.4.4, 19.4.7, 19.4.8, 19.4.9, 19.3.3 and 19.3.4;

Cambridge University Press for Figs. 4.3.6 and 8.3.2;

Elsevier for Figs. 8.2.2, 10.3.3 and 10.3.6;

Elsevier Sequoia S. A. for Fig. 13.4.2;

European Physical Society for Fig. 19.4.5;

Institute of Physics Publishing, Bristol, UK for Figs. 5.5.5, 14.3.1, 14.3.2, 14.3.3 and 19.3.8;

John Wiley & Sons for Figs. 4.2.5, 4.2.6, 8.2.4 and 8.2.6;

McGraw-Hill for Figs. 8.2.3, 11.1.4 and 11.1.5;

Nature Publishing Group for Figs. 3.2.5, 13.3.6, 13.3.7, 14.5.2 and 14.5.4;

North-Holland for Figs. 3.2.6, 3.2.7 and 3.2.8;

Oxford University Press for Figs. 17.1.7 and 17.1.9;

Pergamon for Fig. 4.1.8;

Princeton University Press for Figs. 3.4.11, 3.4.12, 4.3.3 and 7.4.3;

The Royal Society, London for Fig. 12.4.17;

The Royal Society of Chemistry for Fig. 13.1.14;

Science Press for Figs. 17.3.3 and 17.3.4;

Springer for Fig. 14.4.5;

Taylor & Francis for Figs. 8.2.7, 8.2.8, 14.2.1, 14.2.2, 14.2.4, 14.2.5, 14.3.4 and 19.3.2;

The Yukawa Institute for Theoretical Physics and the Physical Society of Japan for Fig. 13.3.5.

Contents

Preface v

Acknowledgments ix

Overview 1

0.1 Stratification of the Physical World 1
0.1.1 Physics of the 20th Century 1
0.1.2 Simplicity versus Complexity and Unity versus Diversity 4
0.1.3 Emergent Phenomena 5
0.2 The Terrain of Condensed Matter Physics 8
0.2.1 Theoretical Descriptions: Quantum versus Classical 8
0.2.2 Condensation Phenomena 9
0.2.3 Ordering 11
0.3 Historical Perspective and Conceptual Framework 12
0.3.1 From Solid State Physics to Condensed Matter Physics 12
0.3.2 The Paradigm for Solid State Physics and Its Extension 13
0.3.3 Bond Approach versus Band Approach 15
0.3.4 The Paradigm for Condensed Matter Physics 16
Bibliography 22

Part I Structure of Condensed Matter 25

Chapter 1. Symmetry of Structure 27

1.1 Basic Concepts of Symmetry 27
1.1.1 Symmetry and Symmetry Operations 28
1.1.2 Some Theorems for the Combinations of Symmetry Elements 29
1.1.3 Symmetry Group 30
1.1.4 Representations of Symmetry Groups 32
1.2 Finite Structures and Point Groups 34
1.2.1 Combination Rules for Symmetry Axes 34
1.2.2 Cyclic and Dihedral Groups 35
1.2.3 Platonic Solids and Cubic Groups 35
1.3 Periodic Structures and Space Groups 37
1.3.1 Periodic Structure and Lattice 37
1.3.2 Bravais Lattices 39
1.3.3 Space Groups 41
1.3.4 The Description of Crystal Structure 42
1.4 Structures and Their Fourier Transforms 42
1.4.1 The General Case 42
1.4.2 The Reciprocal Lattice 43
1.4.3 Fourier Transform of Periodic Structure 44
1.5 Generalized Symmetry 45

1.5.1 High-Dimensional Space Groups 45
1.5.2 Color Groups 45
1.5.3 Symmetry of Reciprocal Space 47
1.5.4 Other Extensions of Symmetry 47
Bibliography 47

Chapter 2. Organization of the Crystalline State **49**
2.1 Geometrical Constraints 49
2.1.1 Topological Constraints 49
2.1.2 Curvature — Curves and Surfaces 50
2.1.3 Tiling of Space 52
2.2 Packing Structures and Linkage Structures 53
2.2.1 Sphere Packings and Coverings 54
2.2.2 The Voids in Packing Structures 56
2.2.3 Linkage Structures 58
2.2.4 Fullerenes and Carbon Nanotubes 60
2.2.5 The Structure of Perovskites 61
2.3 Quasiperiodic Structures 63
2.3.1 Irrational Numbers and Quasiperiodic Functions 63
2.3.2 1D Quasiperiodic Structure 64
2.3.3 The Cut and Projection from a 2D Periodic Lattice 65
2.3.4 2D Quasiperiodic Structures 67
2.3.5 3D Quasicrystals 68
2.3.6 Discussions about Some Basic Notions 70
Bibliography 72

Chapter 3. Beyond the Crystalline State **73**
3.1 Alloys and Substitutional Disorder 73
3.1.1 Ordered and Disordered Alloys 73
3.1.2 Distribution Functions and Correlation Functions 75
3.2 Liquids and Glasses 76
3.2.1 Overview 76
3.2.2 Statistical Description 77
3.2.3 Structural Models for the Amorphous State 79
3.3 The Liquid-Crystalline State 82
3.3.1 Overview 82
3.3.2 Nematic Phase and Cholesteric Phase 82
3.3.3 Smectic Phase and Columnar Phase 84
3.3.4 Lyotropics 86
3.4 Polymers 87
3.4.1 Structure and Constitution 87
3.4.2 Random Coils and Swollen Coils 88
3.4.3 The Correlation Function of Single Chain and Experimental Results 91
3.4.4 Ordered and Partially Ordered Structure 92
3.5 Biopolymers 94
3.5.1 The Structure of Nucleic Acid 94
3.5.2 The Structure of Protein 94
3.5.3 Information and the Structure 96
Bibliography 97

Chapter 4. Inhomogeneous Structure 99
4.1 Multi-Phased Structure . . . 99
4.1.1 Structural Hierarchies . . . 99
4.1.2 Microstructural Characteristics of Heterogeneous Material . . . 101
4.1.3 Effective Medium Approximation: The Microstructure and Physical Properties of Two-Phase Alloys . . . 105
4.2 Geometric Phase Transition: Percolation . . . 106
4.2.1 Bond Percolation and Site Percolation . . . 106
4.2.2 Overview of Percolation Theory . . . 107
4.2.3 Examples of Percolation . . . 109
4.3 Fractal Structures . . . 110
4.3.1 Regular Fractals and Fractal Dimension . . . 110
4.3.2 Irregular Fractal Objects . . . 112
4.3.3 Self-Affine Fractals . . . 113
4.3.4 The Basic Concept of the Multifractal . . . 114
Bibliography . . . 117

Part II Wave Behavior in Various Structures 119

Chapter 5. Wave Propagation in Periodic and Quasiperiodic Structures 121
5.1 Unity of Concepts for Wave Propagation . . . 121
5.1.1 Wave Equations and Periodic Potentials . . . 121
5.1.2 Bloch Waves . . . 122
5.1.3 Revival of the Study of Classical Waves . . . 124
5.2 Electrons in Crystals . . . 125
5.2.1 Free Electron Gas Model . . . 125
5.2.2 Nearly-Free Electron Model . . . 126
5.2.3 Tight-Binding Electron Model . . . 128
5.2.4 Kronig–Penney Model for Superlattices . . . 129
5.2.5 Density of States and Dimensionality . . . 131
5.3 Lattice Waves and Elastic Waves . . . 132
5.3.1 Dispersion Relation of Lattice Waves . . . 132
5.3.2 Frequency Spectrum of Lattice Waves . . . 134
5.3.3 Elastic Waves in Periodic Composites: Phononic Crystals . . . 136
5.4 Electromagnetic Waves in Periodic Structures . . . 137
5.4.1 Photonic Bandgaps in Layered Periodic Media . . . 137
5.4.2 Dynamical Theory of X-Ray Diffraction . . . 139
5.4.3 Bandgaps in Three-Dimensional Photonic Crystals . . . 142
5.4.4 Quasi-Phase-Matching in Nonlinear Optical Crystals . . . 145
5.5 Waves in Quasiperiodic Structures . . . 146
5.5.1 Electronic Spectra in a One-Dimensional Quasilattice . . . 146
5.5.2 Wave Transmission Through Artificial Fibonacci Structures . . . 149
5.5.3 Pseudogaps in Real Quasicrystals . . . 150
Bibliography . . . 152

Chapter 6. Dynamics of Bloch Electrons 155
6.1 Basic Properties of Electrons in Bands . . . 155
6.1.1 Electronic Velocity and Effective Mass . . . 155
6.1.2 Metals and Nonmetals . . . 157
6.1.3 Hole . . . 158
6.1.4 Electronic Specific Heat in Metals . . . 159
6.2 Electronic Motion in Electric Fields . . . 160
6.2.1 Bloch Oscillations . . . 161

6.2.2 Negative Differential Resistance . . . 162
6.2.3 Wannier–Stark Ladders . . . 163
6.3 Electronic Motion in Magnetic Fields . . . 165
6.3.1 Cyclotron Resonance . . . 165
6.3.2 Landau Quantization . . . 167
6.3.3 de Haas–van Alphen Effect . . . 171
6.3.4 Susceptibility of Conduction Electrons . . . 173
Bibliography . . . 176

Chapter 7. Surface and Impurity Effects **177**
7.1 Electronic Surface States . . . 177
7.1.1 Metal Surface . . . 177
7.1.2 Semiconductor Surface States . . . 181
7.2 Electronic Impurity States . . . 183
7.2.1 Shielding Effect of Charged Center . . . 183
7.2.2 Localized Modes of Electrons . . . 185
7.2.3 Electron Spin Density Oscillation around a Magnetic Impurity . . . 187
7.3 Vibrations Related to Surface and Impurity . . . 188
7.3.1 Surface Vibrations . . . 189
7.3.2 Impurity Vibration Modes . . . 190
7.4 Defect Modes in Photonic Crystals . . . 192
7.4.1 Electromagnetic Surface Modes in Layered Periodic Structures . . . 193
7.4.2 Point Defect . . . 194
7.4.3 Line Defects . . . 196
Bibliography . . . 197

Chapter 8. Transport Properties **199**
8.1 Normal Transport . . . 199
8.1.1 Boltzmann Equation . . . 199
8.1.2 DC and AC Conductivities . . . 200
8.1.3 Microscopic Mechanism of Metallic Conductivity . . . 203
8.1.4 Electric Transport in Semiconductors . . . 205
8.1.5 Other Transport Coefficients . . . 207
8.2 Charge Transport and Spin Transport in Magnetic Fields . . . 208
8.2.1 Classical Hall Effect . . . 208
8.2.2 Shubnikov–de Haas Effect . . . 210
8.2.3 Ordinary Magnetoresistance and Its Anisotropy . . . 212
8.2.4 Spin Polarization and Spin Transport . . . 215
8.2.5 Resistivity and Magnetoresistance of Ferromagnetic Metals . . . 217
8.3 Tunneling Phenomena . . . 218
8.3.1 Barrier Transmission . . . 218
8.3.2 Resonant Tunneling through Semiconductor Superlattices . . . 221
8.3.3 Zener Electric Breakdown and Magnetic Breakdown . . . 223
8.3.4 Tunneling Magnetoresistance . . . 224
8.3.5 Scanning Tunneling Microscope . . . 226
Bibliography . . . 229

Chapter 9. Wave Localization in Disordered Systems **231**
9.1 Physical Picture of Localization . . . 231
9.1.1 A Simple Demonstration of Wave Localization . . . 231
9.1.2 Characteristic Lengths and Characteristic Times . . . 232
9.1.3 Particle Diffusion and Localization . . . 233
9.2 Weak Localization . . . 234
9.2.1 Enhanced Backscattering . . . 235

9.2.2 Size-Dependent Diffusion Coefficient 237
9.2.3 Interference Correction to Conductivity 238
9.3 Strong Localization 240
9.3.1 Continuum Percolation Model 240
9.3.2 Anderson Model 242
9.3.3 Mobility Edges 244
9.3.4 Edwards Model 245
9.3.5 Hopping Conductivity 246
9.3.6 Strong Localization of Light 247
Bibliography 250

Chapter 10. Mesoscopic Quantum Transport **251**
10.1 The Characteristics of Mesoscopic Systems 251
10.1.1 Prescription of the Mesoscopic Structures 251
10.1.2 Different Transport Regimes 252
10.1.3 Quantum Channels 253
10.2 Landauer–Büttiker Conductance 255
10.2.1 Landauer Formula 255
10.2.2 Two-Terminal Single-Channel Conductance 256
10.2.3 Two-Terminal Multichannel Conductance 258
10.3 Conductance Oscillation in Circuits 259
10.3.1 Gauge Transformation of Electronic Wavefunctions 259
10.3.2 Aharonov–Bohm Effect in Metal Rings 260
10.3.3 Persistent Currents 263
10.3.4 Altshuler–Aronov–Spivak Effect 265
10.3.5 Electrostatic Aharonov–Bohm Effect 265
10.4 Conductance Fluctuations 266
10.4.1 Nonlocality of Conductance 267
10.4.2 Reciprocity in Reversed Magnetic Fields 269
10.4.3 Universal Conductance Fluctuations 270
Bibliography 272

Part III Bonds and Bands with Things Between and Beyond **273**

Chapter 11. Bond Approach **275**
11.1 Atoms and Ions 275
11.1.1 A Hydrogen Atom 275
11.1.2 Single-Electron Approximation for Many-Electron Atoms 276
11.1.3 Intraatomic Exchange 278
11.1.4 Hund's Rules and Magnetic Moments in Ions 279
11.2 Diatomic Molecules 279
11.2.1 The Exact Solution for the Hydrogen Molecular Ion H_2^+ 279
11.2.2 The Molecular Orbital Method 283
11.2.3 Heitler and London's Treatment of Hydrogen Molecule 286
11.2.4 The Spin Hamiltonian and the Heisenberg Model 288
11.3 Polyatomic Molecules 289
11.3.1 The Molecular Orbital Method for Polyatomic Molecules 289
11.3.2 Valence Bond Orbitals 290
11.3.3 The Hückel Approximation for the Molecular Orbital Method 291
11.3.4 Electronic Structure of Some Molecules 293
11.4 Ions in Anisotropic Environments 295
11.4.1 Three Types of Crystal Fields 295
11.4.2 ***3d*** Transition Metal Ions in Crystal Fields 296

11.4.3 Jahn–Teller Effect 297
11.4.4 Ions in Ligand Fields 298
Bibliography 300

Chapter 12. Band Approach **301**
12.1 Different Ways to Calculate the Energy Bands 301
12.1.1 Orthogonized Plane Waves 301
12.1.2 Pseudopotential 303
12.1.3 The Muffin-Tin Potential and Augmented Plane Waves 304
12.1.4 The Symmetry of the Energy Bands and the $\boldsymbol{k} \cdot \boldsymbol{p}$ Method 305
12.2 From Many-Particle Hamiltonian to Self-Consistent Field Approach 307
12.2.1 Many-Particle Hamiltonians 307
12.2.2 Valence Electrons and the Adiabatic Approximations 308
12.2.3 The Hartree Approximation 309
12.2.4 The Hartree–Fock Approximation 310
12.3 Electronic Structure via Density Functionals 312
12.3.1 From Wavefunctions to Density Functionals 312
12.3.2 Hohenberg–Kohn Theorems 313
12.3.3 The Self-Consistent Kohn–Sham Equations 314
12.3.4 Local Density Approximation and Beyond 315
12.3.5 Car–Parrinello Method 317
12.4 Electronic Structure of Selected Materials 318
12.4.1 Metals 318
12.4.2 Semiconductors 322
12.4.3 Semimetals 324
12.4.4 Molecular Crystals 325
12.4.5 Surfaces and Interfaces 328
Bibliography 331

Chapter 13. Correlated Electronic States **333**
13.1 Mott Insulators 333
13.1.1 Idealized Mott Transition 333
13.1.2 Hubbard Model 335
13.1.3 Kinetic Exchange and Superexchange 337
13.1.4 Orbital Ordering versus Spin Ordering 339
13.1.5 Classification of Mott Insulators 341
13.2 Doped Mott Insulators 343
13.2.1 Doping of Mott Insulators 343
13.2.2 Cuprates 343
13.2.3 Manganites and Double Exchange 347
13.2.4 Charge-Ordering and Electronic Phase Separation 349
13.3 Magnetic Impurities, Kondo Effect and Related Problems 351
13.3.1 Anderson Model and Local Magnetic Moment 351
13.3.2 Indirect Exchange 353
13.3.3 Kondo Effect 354
13.3.4 Heavy-Electron Metals and Related Materials 357
13.4 Outlook 359
13.4.1 Some Empirical Rules 359
13.4.2 Theoretical Methods 361
Bibliography 362

Chapter 14. Quantum Confined Nanostructures **363**
14.1 Semiconductor Quantum Wells . . . 363
14.1.1 Electron Subbands . . . 363
14.1.2 Hole Subbands . . . 366
14.1.3 Optical Absorption . . . 368
14.1.4 Coupled Quantum Wells . . . 369
14.2 Magnetic Quantum Wells . . . 370
14.2.1 Spin Polarization in Metal Quantum Wells . . . 370
14.2.2 Oscillatory Magnetic Coupling . . . 372
14.2.3 Giant Magnetoresistance . . . 374
14.3 Quantum Wires . . . 377
14.3.1 Semiconductor Quantum Wires . . . 378
14.3.2 Carbon Nanotubes . . . 380
14.3.3 Metal Steps and Stripes . . . 381
14.4 Quantum Dots . . . 383
14.4.1 Magic Numbers in Metal Clusters . . . 383
14.4.2 Semiconductor Quantum Dots . . . 386
14.4.3 Fock–Darwin Levels . . . 388
14.4.4 Coulomb Blockade . . . 390
14.4.5 Kondo Effect . . . 393
14.5 Coupled Quantum Dot Systems . . . 395
14.5.1 Double Quantum Dots . . . 395
14.5.2 Semiconductor Quantum Dot Superlattices . . . 398
14.5.3 Metal Quantum Dot Arrays . . . 399
Bibliography . . . 400

Part IV Broken Symmetry and Ordered Phases **403**

Chapter 15. Landau Theory of Phase Transitions **405**
15.1 Two Important Concepts . . . 405
15.1.1 Broken Symmetry . . . 405
15.1.2 Order Parameter . . . 406
15.1.3 Statistical Models . . . 409
15.2 Second-Order Phase Transitions . . . 411
15.2.1 Series Expansion of Free Energy . . . 411
15.2.2 Thermodynamic Quantities . . . 412
15.2.3 System with a Complex Order Parameter . . . 413
15.3 Weak First-Order Phase Transitions . . . 415
15.3.1 Influence of External Field . . . 415
15.3.2 Landau–Devonshire Model . . . 416
15.3.3 Landau–de Gennes Model . . . 418
15.3.4 Coupling of Order Parameter with Strain . . . 419
15.4 Change of Symmetry in Structural Phase Transitions . . . 420
15.4.1 Density Function and Representation Theory . . . 421
15.4.2 Free Energy Functional . . . 422
15.4.3 Landau Criteria . . . 423
15.4.4 Lifshitz Criterion . . . 424
Bibliography . . . 426

Chapter 16. Crystals, Quasicrystals and Liquid Crystals 427
16.1 Liquid-Solid Transitions 427
16.1.1 Free Energy Expansion Based on Density Waves 427
16.1.2 Crystallization 429
16.1.3 Quasicrystals 431
16.2 Phase Transitions in Solids 433
16.2.1 Order-Disorder Transition 433
16.2.2 Paraelectric-Ferroelectric Transition 435
16.2.3 Incommensurate-Commensurate Transitions 437
16.3 Phase Transitions in Soft Matter 440
16.3.1 Maier–Saupe Theory for Isotropic-Nematic Transition 440
16.3.2 Onsager Theory for Isotropic-Nematic Transition 443
16.3.3 Phase Separation in Hard-Sphere Systems 445
Bibliography 447

Chapter 17. Ferromagnets, Antiferromagnets and Ferrimagnets 449
17.1 Basic Features of Magnetism 449
17.1.1 Main Types of Magnetism 449
17.1.2 Spatial Pictures of Magnetic Structures 452
17.1.3 Band Pictures of Magnetic Structures 456
17.1.4 Hamiltonians with Time Reversal Symmetry 459
17.2 Theory Based on Local Magnetic Moments 461
17.2.1 Mean-Field Approximation for Heisenberg Hamiltonian 462
17.2.2 Ferromagnetic Transition 463
17.2.3 Antiferromagnetic Transition 465
17.2.4 Ferrimagnetic Transition 466
17.2.5 Ferromagnetic and Antiferromagnetic Ground States 468
17.3 Theory Based on Itinerant Electrons 471
17.3.1 Mean-Field Approximation of Hubbard Hamiltonian 471
17.3.2 Stoner Theory of Ferromagnetism 472
17.3.3 Weak Itinerant Ferromagnetism 476
17.3.4 Spin Density Waves and Antiferromagnetism 478
Bibliography 479

Chapter 18. Superconductors and Superfluids 481
18.1 Macroscopic Quantum Phenomena 481
18.1.1 The Concept of Bose–Einstein Condensation 481
18.1.2 Bose–Einstein Condensation of Dilute Gases 484
18.1.3 The Superfluidity of Liquid Helium 486
18.1.4 Superconductivity of Various Substances 490
18.2 Ginzburg–Landau Theory 495
18.2.1 Ginzburg–Landau Equations and Broken Gauge Symmetry 495
18.2.2 Penetration Depth and Coherence Length 497
18.2.3 Magnetic Properties of Vortex States 499
18.2.4 Anisotropic Behavior of Superconductors 500
18.3 Pairing States 502
18.3.1 Generalized Cooper Pairs 502
18.3.2 Conventional Pairing of Spin-Singlet ***s***-Wave 505
18.3.3 Exotic Pairing for Spin-Singlet ***d***-Wave 507
18.3.4 Pseudogaps and Associated Symmetry 508
18.3.5 Exotic Pairing for Spin-Triplet ***p***-Wave 510
18.4 Josephson Effects 513
18.4.1 Josephson Equations 513
18.4.2 The Josephson Effects in Superconductors 514

18.4.3 Phase-Sensitive Tests of Pairing Symmetry 516
18.4.4 Josephson Effect in Superfluids 517
Bibliography 519

Chapter 19. Broken Ergodicity **521**
19.1 Implication of Ergodicity 521
19.1.1 Ergodicity Hypothesis 521
19.1.2 Involvement of Time Scale 523
19.1.3 Internal Ergodicity 525
19.2 From Vapor to Amorphous Solid 526
19.2.1 Vapor-Liquid Transition 526
19.2.2 Wetting Transition 529
19.2.3 Glass Transition 532
19.3 Spin Glass Transition 534
19.3.1 Spin Glass State 535
19.3.2 Frustration and Order Parameter 537
19.3.3 Theoretical Models 539
19.4 Metal-Nonmetal Phase Transitions 542
19.4.1 Semi-Empirical Criteria 543
19.4.2 Wigner Crystallization 545
19.4.3 Gutzwiller Variation Method and Phenomenological Treatment of Mott Transition 547
19.4.4 Electron Glass 549
Bibliography 552

Appendices **555**

Appendix A. Units and Their Conversion **557**
A.1 Maxwell Equations and Related Formulas in Three Systems of Units 558
A.2 Schrödinger Equation for a Many-Electron Atom in Three Systems of Units 558
A.3 SI Prefixes 558
A.4 Conversion of Units Between SI and Gaussian Systems 559
A.5 Fundamental Physical Constants 560

Appendix B. List of Notations and Symbols **561**

Index **565**

Overview

The accumulation of so enormous a mass of substantial truth is not possible without organization. The faculty for order is just as much a creative one as the faculty for representation. Or they are simply different aspects of one and the same faculty. Out of the truth of countless isolated phenomena there arise the truth of the relationship existing among them: in this way a world is produced.

— Hugo von Hofmannstahl

In this age of increasing specification it is comforting to realize that basic physical concepts apply to a wide range of seemingly diverse problems. Progress made in understanding one area may often be applied in many other fields. This is truc not only for various fields of materials science but for structure of matter in general. As examples we illustrate how concepts to develop to understand magnetism, superfluid helium and superconductivity have been extended and applied to such diverse fields as nuclear matter, weak and electromagnetic interactions, quark structure of the particles of high energy physics and phases of liquid crystals.

— John Bardeen

In this introductory chapter, we shall give a brief overview of condensed matter physics. It begins with a brief résumé of the development of physics in the 20th century, so that we may view condensed matter physics in the context of other branches of physics, and note the important consequences due to the stratification of the physical world into its different layers, each with its quasistable constituent particles and elementary interactions. These make us realize why condensed matter physics is still a subject so full of vitality even though the interactions of the constituent particles (nuclei and electrons) are thoroughly known in principle as a result of the formulation of quantum mechanics. Next, we shall outline the scope of research on condensed matter physics, involving theoretical treatment by quantum mechanics and classical physics, and the fertile and multitudinous phenomena brought about by condensation and ordering. Then we will discuss the relationship between solid state physics and condensed matter physics, noting the enlarged range of phenomena studied and analyzing the change in their respective unifying conceptual framework.

§0.1 Stratification of the Physical World

0.1.1 Physics of the 20th Century

Here we briefly outline the development of physics in the 20th century. Two outstanding theoretical breakthroughs occurred at the beginning of that century: relativity and quantum theory.

Relativity may be regarded as the crowning achievement of classical physics, for special relativity was found to be a continuation and extension of classical electrodynamics and brought with it a new mechanics for fast moving bodies and the revision of our basic concepts of space-time and matter-energy. Meanwhile general relativity extended classical mechanics into the realm of strong gravitational fields and also constructed the geometrical theory of gravitation, which profoundly influenced modern astronomy. Quantum theory heralded the new era of microscopic physics, culminating in the formulation of quantum mechanics in 1925–1928. With the establishment of quantum mechanics, the enigmatic microscopic dynamics of atomic systems was elucidated at one stroke, to the delight of physicists. What next?

One route is from atomic physics downwards, physicists pushed forward to probe more and more microscopically, penetrating into the new worlds of nuclear physics and the physics of subnuclear particles. In quest of elusive 'elementary' particles, several generations of high energy accelerators were built, and leptons, quarks, gluons and intermediate bosons were discovered or inferred, advancing the frontier of physics on the microscopic side.

Obviously, there is another frontier of physics on the cosmic scale: the exploration of the universe. General relativity gave a curved space description of the universe, which marked the birth of modern cosmology. Modern astrophysics brought forth its tremendous observational data with far-reaching consequences, these pose intellectually challenging problems for theorists.

It is surprising that the twin frontiers of physics, apparently lying in opposite directions, the one facing the world at smallest scale, the other facing the world at largest scale, are actually not disparate, but miraculously brought together by a circuitous route and merged into one frontier with Janus-like faces confronting the smallest as well as the biggest, i.e., the microscopic-cosmic. Nowadays high-energy physics provides archaeological information for the early stage of the universe, while stars and the cosmos serve as gigantic laboratories to test various theories of fundamental physics. There are two 'standard models', one for particle physics, the other for cosmology. They stand side by side and interpenetrate each other, as two landmarks for microscopic physics and cosmic physics in the latter half of the 20th century.

There was another route for development of physics after the establishment of quantum mechanics, i.e., from atomic physics upward, applying quantum mechanics to molecules, quantum chemistry was born; applying quantum mechanics in conjunction with statistical physics to the crystalline solids, the foundation of solid state physics was laid. Later confronted with various cooperative phenomena as well as more complicated condensed matter, the scope of solid state physics was enormously enlarged. At the same time, its basic concepts were thoroughly revised and deepened, and gradually and almost imperceptibly it was transformed into condensed matter physics in the 1970s. Moreover, linking the microscopic world described by quantum mechanics and macroscopic world of complicated condensed matter is by no means a easy task. It poses unforeseen intellectual challenges, though it reaps unexpected practical rewards, as exemplified by discoveries of high temperature superconductivity, quantum transport in mesoscopic systems, C_{60} molecules and solids, giant and colossal magnetoresistance, and the realization of Bose–Einstein condensation. The exciting thing is that there are immense possibilities waiting to be explored and utilized. So the physics of condensed matter bordering with complexity naturally forms another frontier for the development of physics. This new frontier for physics, besides its own interests, is closely linked with development of high technology and is profusely penetrating into interdisciplinary fields. Therefore it is attracting more attention and assuming more importance with the opening of the 21th century.[a] Figure 0.1.1 shows the stratification of the physical world according to length scale and energy scale, while Fig. 0.1.2 shows the different branches of physics coordinated according to the length scales of the objects studied.

Formerly physicists were accustomed to deal with simple matter. The initial successes of solid state physics were achieved in the realm of simple solids, such as copper, silicon, etc. However, when solid state physics was transformed into condensed matter physics, its scope was enlarged to include various types of soft matter, such as liquid crystals, self-assembled membranes, granular

[a]For problems connected with the development of physics of the 20th century, one may consult L. M. Brown, A. Pais, and B. Pippard (eds.), *Twentieth Century Physics*, Vols. I–III, Bristol and Philadelphia, Institute of Physics Publishing Ltd. (1995).

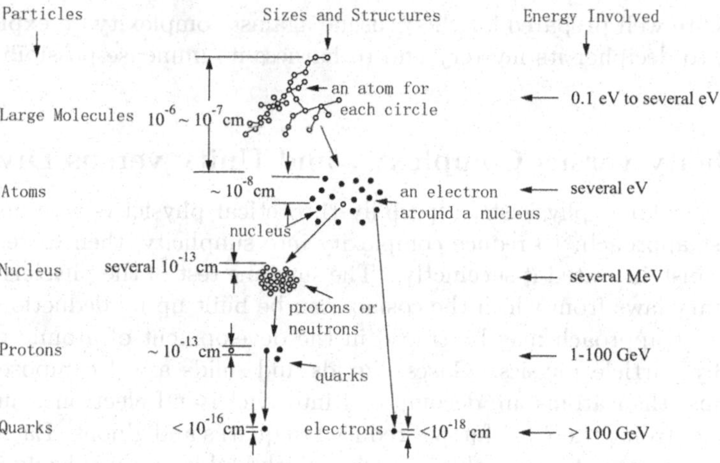

Figure 0.1.1 The stratification of physical world according to length scale and energy scale.

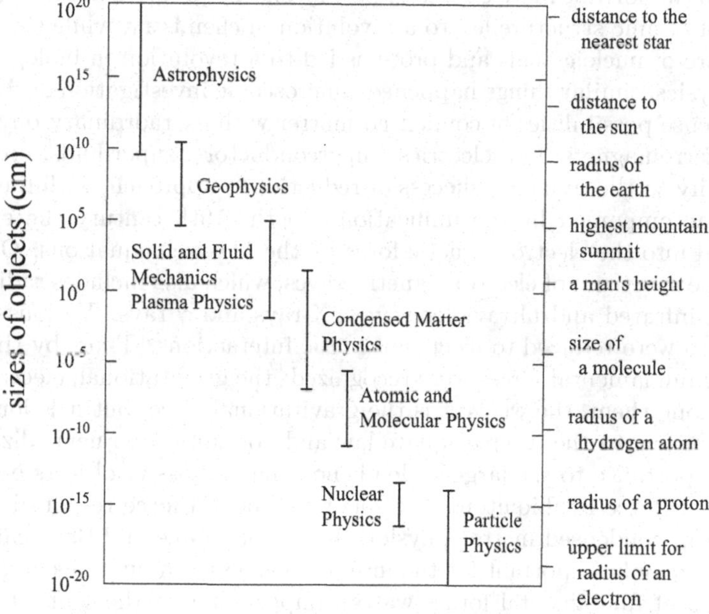

Figure 0.1.2 Different branches of physics coordinated according to their length scales.

materials, polymers and biopolymers. Thus, matter with increasing complexity became the object of study for condensed matter physicists. The concept of broken symmetry which emerged from the study of cooperative phenomena became the key to uncover the properties of complicated matter. The development of nonlinear science fashioned new theoretical tools for dealing with complexity, and the development of materials science and technology provided the multitudes of new techniques for fabrication and characterization of complicated matter. Chemists devised ever-new ways to synthesize and organize new materials. Advances in computational techniques brought with them *ab initio* calculations of the electronic structure of matter as well as simulations of various transformation processes. So with this changed climate of research, condensed matter physicists have become more familiar with either simple materials within the contexts of added complexity, such as artificial microstructures and nanostructures, or chemically complex materials, such as complex oxides, organic materials, polymers and biopolymers. The cooperation between physicists, materials scientists, chemists and biologists in interdisciplinary fields has become more frequent and fruitful. So, associated with physicists of other branches, as well as scientists of other disciplines, condensed matter physicists are well prepared for this crusade against complexity, to explore the real material world around us, to decipher its mystery and to harness its immense possibilities to the service of human kind.

0.1.2 Simplicity versus Complexity and Unity versus Diversity

Traditionally speaking, physicists, especially theoretical physicists, are accustomed to the so-called reductionist approach: to reduce complexity into simplicity, then to reconstruct complexity from simplicity. Einstein stated it succinctly, "The supreme test of the physicist is to arrive at those universal elementary laws from which the cosmos can be built up by deduction."

This reductionist approach may be traced in the development of atomic physics, then nuclear physics and finally particle physics. Gases, liquids and solids are decomposed into aggregates of molecules or atoms; then atoms are decomposed into nuclei and electrons; nuclei are decomposed into protons and neutrons; these are further reduced to quarks and gluons. Each stage of reduction is marked by the appearance of quasistable particles which had been mistakenly identified as 'elementary' particles. With each stage of reduction, the length scale drastically decreases, while the energy scale drastically increases (see Fig. 0.1.1). This coupling between nearby strata in structure of matter proved to be an important motive force in promoting the progress of science in the 20th century. The elucidation of atomic structure led to a revolution in chemistry, while the determination of the molecular structure of nucleic acids and proteins led to a revolution in biology. In our field of condensed matter physics, similar things happened: microscopic investigation of the structure of matter revealed the immense possibilities of condensed matter with extraordinary physical properties, e.g., semiconductors, ferromagnets, ferroelectrics, superconductors, superfluids, etc. These tremendous achievements testify to the immense success of reductionist approach, within certain limits.

Reduction is accompanied by the unification. In the 19th century the electric and magnetic forces were unified into the electromagnetic force by the Maxwell equations. Optical waves were incorporated into the spectrum of electromagnetic waves, which also includes radio waves, microwaves, millimeter waves, infrared and ultraviolet waves, X-rays and γ rays. Various interactions between atoms or molecules were reduced to electromagnetic interactions. Thus, by the middle of the 20th century, only four fundamental forces were recognized: the gravitational, electromagnetic, weak and strong forces. Among them, the weakest is the gravitational force, but it is long-ranged, decreasing with distance according to the inverse square law and not subject to neutralization or shielding, so it is of extreme importance to the large-scale phenomena of massive objects observed in astronomy. The electromagnetic force is ubiquitous, the sole fundamental force responsible for various complicated phenomena in condensed matter physics. The strong force and the weak force are extremely short-ranged and are only important for the interactions between nucleons and subnuclear particles. Further unification of fundamental forces was an important ingredient of 20th century theoretical physics. The later years of Einstein were spent in an unsuccessful endeavor to unify the gravitational force and the electromagnetic force through classical field theory. A more fruitful pursuit of unification, based on quantum field theory, was taken up later by other scientists. The electromagnetic interaction was incorporated into quantum field theory with the foundation of quantum electrodynamics in the 1930s and 1940s. Further, quantum field theory, treating strong interactions, was realized as quantum chromodynamics. In the 1960s, Glashow, Weinberg and Salam succeeded in unifying the weak force and the electromagnetic force into the electro-weak force. Together with the quark model of nucleons, these are the main ingredients of the standard model for particle physics, which has been thoroughly verified in detail by experiments. Furthermore, grand unified theories have attempted to enlarge the unifying scheme to include the strong force, and further attempts at unification to include the gravitational force at ultramicroscopic level are on the way, for example the superstring theory or other theories related to supersymmetry. These are sometimes misnamed as 'theory of everything' in particle theorist circles, indicating that the tantalizing hope that the supreme unification at an ultramicroscopic level may be attained.[b] Further progress in this direction certainly will cause a profound impact on particle physics and cosmology; however, its value for the rest of the physical sciences should be regarded with healthy skepticism for the reason that there is

[b] S. Weinberg, *Dreams of a Final Theory*, Random House, New York (1972).

decoupling as well as coupling between different strata in the physical world. This decoupling will interrupt the influence of high energy physics on other branches of physics with much lower energy.

In order to realize why the decoupling occurs, we may take the renormalization group theory for critical phenomena as an example.[c] In the 1960s Widom and Kadanoff found evidence for scaling and universality of critical phenomena, indicating that the structural details have no influence on the critical phenomena. The renormalization technique was previously used with enormous success to eliminate the divergences in quantum field theory, leading to the flowering of quantum electrodynamics in the late 1940s. In the early 1970s K. G. Wilson had the insight to perceive that the renormalization was not only a technical trick to eliminate the divergence, but also a very useful expression for the variation of the rule of interactions with change in scale of the phenomena studied. So he developed the renormalization group method which took seriously the energy cut-off Λ, i.e., in formulating an 'effective' field theory for a phenomenon observed at accessible energies (of the order E), the terms of the order $(E/\Lambda)^2$ may be safely neglected. It explained nicely why the physics of matter at microscopic length scale and high energy are irrelevant to critical phenomena. It may be extended to explain why the important recent advance of particle physics such as the discovery of top quark has practically no effect on condensed matter physics at all. So, in spite of the immense success it achieved, the reductionist approach has its limitations as well; it cannot be pushed too far.

0.1.3 Emergent Phenomena

The structure of matter is stratified into a hierarchy of layers, each with its own quasistable constituent particles and distinctive length and energy scales. Consequently, there is decoupling between different levels, which makes the construction from simplicity and unity to complexity and diversity not so easy as the reductionists once thought. P. W. Anderson in 1972 questioned the validity of the pure reductionist approach: "The ability to reduce everything to simple fundamental laws does not imply the ability to start from those laws and reconstruct the universe. ... The constructionist hypothesis breaks down when confronted by the twin difficulties of scale and complexity. The behavior of large and complex aggregates of elementary particles, it turns out, is not to be understood in terms of a simple extrapolation of the properties of a few particles. Instead, at each level of complexity entirely new properties appear, and the understanding of the new behaviors require research which I think is as fundamental in its nature as any other." So with each stage of aggregation, entirely new properties emerge. These are called emergent properties, and they lie outside the realm of the physics of their constituent particles. The reductionist approach is being gradually superseded by the study of emergent phenomena. As an apology for this viewpoint, we may further quote an eloquent passage due to Kadanoff: "Here I wish to argue against the reductionist prejudice. It seems to me that considerable experience has been developed to show that there are levels of aggregation that represent natural subject areas of different groups of scientists. Thus, one group may study quarks (a variety of subnuclear particle), another, atomic nuclei, another, atoms, another, molecular biology, and another, genetics. In this list, each succeeding part is made up of objects from the preceding levels. Each level may be considered to be less fundamental than the one preceding it in the list. But at each level there are new and exciting valid generalizations which could not in any natural way have been deduced from any more 'basic' sciences. Starting from the 'least fundamental' and going backward on the list, we can enumerate, in succession, representative and important conclusions from each of these sciences, as Mendelian inheritance, the double helix, quantum mechanics, and nuclear fission. Which is the most fundamental, the most basic? Which was derived from which? From this example, it seems rather foolish to think about a hierarchy of scientific knowledge. Rather, it would appear that grand ideas appear at any levels of generalization."

It is by no means an accident that these advocates for the study of emergent properties are prominent condensed matter theorists, for the study of emergent properties occupies a foremost place in condensed matter physics. Dirac declared, immediately after quantum mechanics achieved enormous successes: "the general theory of quantum mechanics is now almost complete. ... The underlying physical laws necessary for a large part of physics and the whole of chemistry are thus

[c] S. S. Schweber, *Phys. Today* **46**, 34 (1993).

completely known, and the difficulty is only that the exact application of these laws leads to equations much too complicated to be soluble." Surely condensed matter physics is found among 'a large part of physics' in above quotation, and though Dirac's statement is correct, i.e., condensed matter physicists have already 'the theory of everything' in their hands, he certainly underestimated the difficulties, as well as the intellectual challenges and the creative opportunities, associated with the subtle interplay of experimental discoveries, theoretical insights and practical realizations in the study of emergent properties of condensed matter. It should be noted that the argument against the reductionist approach as a philosophical doctrine is for its abuse, certainly it is not to deny the importance and effectiveness of its legitimate use. Otherwise, we would fall into the trap of another dubious philosophical doctrine at the opposite pole, i.e., the so-called positivist approach advocated by E. Mach and his associates in the late 19th century, to limit scientific investigations to phenomenology. Actually, it is in the subtle interplay of the coupling and the decoupling of different strata in structure of matter that the gist of emergent phenomena lies. Genetics has its roots in the molecular structure of DNA without bothering about the shell structure of atomic nucleus; the electrical conductivity of semiconductors has its roots in the energy bands of the electronic structure of solids without bothering about the quark structure of nucleons.[d]

Here we may take superconductivity as an example to illustrate some peculiarities associated with the study of emergent properties of condensed matter. Superconductivity was first discovered in 1911 by K. Onnes. Previously he had already liquefied helium, then a laboratory was set up to study physical properties of matter at low temperatures. He found experimentally that mercury lost all trace of electrical resistance below a well-defined critical temperature $T_c = 4.15$ K. Subsequently, it was found that a large number of metallic elements and alloys are superconductors at low temperature. However, superconductivity remained a mystery to physicists for many decades. After the successful application of quantum mechanics to explain the electrical conductivity in metals by F. Bloch in his seminal paper in 1928. Pauli, the famous theoretical physicist, thought the time was ripe for a similar attack on superconductivity, and entrusted Bloch, his assistant then, to carry it out. Bloch failed in this mission, to the annoyance of Pauli, who underestimated the difficulty of this problem. However, as a by-product of this research, Bloch proved a theorem showing that a current-carrying state cannot be the ground state of the system, which later became a criterion for identifying wrong theories of superconductivity. In 1933, the Meissner effect was discovered, showing that the superconducting state is a thermodynamic equilibrium state and that superconductors are perfect diamagnets which expel all magnetic flux from their interiors. This may be taken as a firm experimental basis for the construction of a theory, so, in 1934, the London brothers developed a phenomenological theory of electrodynamic properties of superconductors. The first intuitive recognition of its real nature was recorded in F. London's monograph "*Superfluids*", published in 1950, in which he regarded superconductivity as a manifestation of macroscopic quantum phenomena to be described by a macroscopic wavefunction, whose gauge symmetry (the arbitrariness of phase in its wave function) is broken. In the same year V. I. Ginzburg and L. D. Landau identified a macroscopic wavefunction-like order parameter to construct a comprehensive phenomenological theory for superconductors, based on an effective field approach. Prior to 1950, there had been many attempts to construct microscopic theories of superconductivity by physicists, among them, some eminent theoretical physicists. Most of these attempts were complete failures. The first approach that appeared hopeful was due to H. Fröhlich in 1950, who proposed a mechanism for superconductivity based on the interaction of electrons as a result of their coupling to lattice vibrations. The concurrent experimental discovery of the isotope effects on T_c of superconductors gave plausibility to this mechanism. However Fröhlich's treatment, based on perturbation methods; was found inadequate. It remained for J. Bardeen and his young associates to persevere in this direction and to borrow some theoretical techniques from quantum field theory to treat this intricate many-body problem, until finally in 1957 Bardeen, Cooper and Schrieffer proposed the famous Bardeen–Cooper–Schrieffer (BCS) theory which is able to account for many key properties of superconductors, in particular the existence of an energy gap with the right order of magnitude, BCS theory and it passed the scrutiny of experimental tests. Later Bogoliubov, Eliashberg, Nambu and others improved the theoretical

[d] For discussions about emergent phenomena one may consult P. W. Anderson, *Science* **177**, 393 (1972); L. P. Kadanoff, *From Order to Chaos*, World Scientific, Singapore (1993), p. 339; and P. Coleman, *Ann. Henri Poincaré* **4**, 1 (2003).

techniques, facilitating the applications of Bardeen–Cooper–Schrieffer (BCS) theory; while Gor'kov derived Ginzburg–Landau equations from BCS theory, linking together two levels of description for superconductivity. In the late 1930s L. V. Shubnikov had discovered experimentally the type II superconductors in which the Meissner effect is not complete. In 1957 A. A. Abrikosov derived a vortex lattice model from the Ginzburg–Landau equations to account for the magnetic properties of the type II kind superconductors theoretically. In 1965 neutron diffraction experiment verified the existence of the flux-line lattice. Since the 1960s practical superconducting cables for solenoids producing strong magnetic field have been developed. In 1962 B. Josephson predicted the tunneling of paired electrons through an insulating barrier based on the concept of broken symmetry; later it was verified in the laboratory and known as the Josephson effect. Further, it was used to fabricate Josephson junctions which became key devices for the new technical field known as superconducting electronics. These facts demonstrate clearly the close connection between theoretical conception, experimental verification and practical utilization in the study of emergent phenomena.

Into the 1980s, some exotic superconductors such as organic and heavy fermionic superconductors were discovered. Then a breakthrough occurred in 1986 with Bednorz and Müller's discovery of high temperature superconductivity in doped lanthanum cuprate; it broke the T_c barrier which had stood for decades. It triggered a unprecedented worldwide research activity in this field. Within a few years the record of T_c soared to 134 K at ambient pressure and 163 K under high pressure. Even after intense research for more than 15 years, no consensus was reached for the explanation of its anomalous properties and pairing mechanism and it is still a hot topic alive with possibilities. Early in 1964, high T_c organic superconductors were conjectured theoretically by W. Little with pairing mechanism of excitons in polymers with side branches. In 1980, Jerome first discovered superconductivity in charge-transfer salt with T_c less than 1 K, then T_c was pushed to 11.2 K in $(BEDT\text{-}TTF)_2Cu(NCS)_2$ in 1987. In 1990–1991, the superconductivity in solid C_{60} chemically doped with alkali metals was discovered, the highest T_c reached 30 K at ambient pressure. In 2000 a superconductor of metallic alloy MgB_2 with $T_c = 39$ K was discovered, and its practical importance demonstrated. On the fundamental side, evidence for exotic pairing states involving the coexistence of superconductivity and weak ferromagnetism is accumulating. So a new wave of research on superconductivity begins. Ten scientists engaged in research on superconductors have already been Nobel prize winners. More is to be expected. This demonstrates the complexity, unpredictability and fertility of research on emergent properties in condensed matter.

There is another aspect of emergent phenomena that should be noted: the emergence of diverse complex structures out of very simple interactions. Since the details of elementary interactions are generally irrelevant to the complex structures they produce, so extremely simplified models for these interactions may be devised to account for these complex structures. Just as Kadanoff noted, "all richness of structure observed in the natural world is not a consequence of the complexity of physical laws, but instead arises from the many-times repeated application of quite simple laws." Here we take the Ising model for cooperative phenomena as an example of a simple model that can go very far to account for complexity.

The Ising model was first proposed in a doctorate thesis in 1920 as a simplified model for a ferromagnetic phase transition. It considers a collection of classical spins on lattice sites, each spin pointing either upward or downward. Thus the value of σ_i at arbitrary lattice point i is a constant with a plus or a minus sign, and J_{ij} is the interaction energy between i and j spins, which may be further simplified by setting $J_{ij} \neq 0$ only for nearest neighbors. Then the Hamiltonian for the Ising model may be expressed by the following equation,

$$\mathcal{H} = -\sum_{i<j} J\sigma_i\sigma_j. \tag{0.1.1}$$

This model can be used to illustrate the ferromagnetic transition, although it is not very realistic for it is extremely anisotropic, and it completely ignores the quantum nature of magnetic interaction. It may be used also to describe the atomic configuration of binary alloys and the lattice gas. It is an extremely useful model for the second order phase transition despite its apparent simplicity. In the mean-field approximation, it may be easily solved, showing there is a critical temperature T_c. Below T_c is the ordered phase with lower symmetry, while above T_c is the disordered phase with higher symmetry, in agreement with the phenomenological Landau theory of the second-order phase transition.

However, the Ising model with extreme simplicity permits more precise handling. The exact solution for the one-dimensional case was preserved in Ising's original thesis, showing no phase transition to occur down to $T = 0$ K. The exact solution for the two-dimensional case is more difficult but in 1944 Onsager published his famous result; it is regarded as a landmark for the development of statistical physics in the 20th century. The exact solution for the two-dimensional Ising model shows its behavior in the critical region (specific heat at T_c with a logarithmic singularity, the critical indices different from mean-field values, etc.), to be quite different from that of Landau theory. It raised the question: Which theory is the correct one for the treatment of critical phenomena? Though the exact solution of the three-dimensional Ising model is still missing, numerical methods of series expansion have yielded quite accurate results. In the 1960s accurate experimental results for the determination of critical indices became available and discrimination between rival theories became possible, with the verdict in favor of the Ising model. Kadanoff examined the theoretical consequences of the exact solution of Ising model and formulated the basic ideas of scaling and universality for the modern theory of critical phenomena. So the complex structure of domain droplets with scaling invariance observed in the critical region may be understood as the direct consequence of the Ising model. Besides, the complex structure of spin glasses may be understood in the framework of a modified Ising model, with the interaction is extended to a long-range oscillatory type, while spins are distributed on the sites of a lattice gas. Further traces of the Ising model have also been found in some problems in biophysics, e.g., the theory for helix-coil transition of biopolymers and the modeling of neural networks. The Ising model in a magnetic field was among the first ventures into the theory of quantum phase transitions, i.e., the phase transition at $T = 0$ K by changing pressure, magnetic field, composition, or other parameters.

§0.2 The Terrain of Condensed Matter Physics

We shall cut out a slice from the hierarchy of strictures in physical world to be the object of our study and focus our attention on condensed matter, which lies in the middle span of the whole hierarchy. Its length scale is in the range from several meters to several 0.1 nanometers; the time scale is in the range from several tens of years to several femtoseconds; the energy scale expressed in thermodynamic temperature in the range from several thousands of kelvins to nanokelvins; the number of particles is generally in the range from 10^{27} to 10^{21}, which may be considered as approaching the thermodynamic limit $N \to \infty$; however, cases of much smaller number of particles are also encountered. Some parts of this slice of the physical world are directly accessible to our senses, while the details may be observed with the help of various types of microscopy. So it is no wonder that this branch of physics has proved to be most relevant to our daily life as well as high technology. In our study, nuclei and electrons are mostly regarded as 'elementary' particles in the sense that they could not be further decomposed in the energy range we are interested. Interactions between them are governed by quantum mechanics. By common sense, condensed matter means solids and liquids, as well as some mesophases between them. Now we shall examine the range and the substance of condensed matter physics in more details from a fundamental viewpoint.

0.2.1 Theoretical Descriptions: Quantum versus Classical

Here we consider systems of many identical particles. According to quantum mechanics, all particles manifest wave-particle duality, so all material particles have wave character as well, this is what makes quantum mechanics different from classical physics. The de Broglie wavelength λ is inversely proportional to its momentum p, $\lambda = h/p = h/mv$, where h is Planck's constant, m is the mass and v the speed of the particle. In thermal equilibrium, $v \sim T^{1/2}$ and the thermal de Broglie wavelength is inversely proportional to the square root of temperature.

We may imagine a swarm of flying particles with tentacles about one-half wavelength long stretched out. When the chance of these tentacles touching each other amounts to certainty, the wave character of these particles will manifest itself conspicuously. So we may use the condition of the thermal de Broglie wavelength λ of the particle equal, to or larger than, the average inter-particle spacing a as a criterion for the employment of a quantum description. Let the mean kinetic energy of particles $mv^2/2$ be equal to thermal energy $3k_BT_2$ (here k_B is the Boltzmann constant), setting $\lambda = a$, we obtain the quantum degeneracy temperature,

$$T_0 = \frac{h^2}{3mk_Ba^2}. \tag{0.2.1}$$

At $T \gg T_0$, the wave aspect can be neglected and we may safely treat these problems with classical physics, while at $T \leq T_0$, the wave aspect predominates and we must treat these problems with quantum mechanics and quantum statistics. Thus T_0 or k_BT_0 sets the temperature (or energy) scale for a system that requires quantum description. For a collection of particles, (0.2.1) may serve as the criterion which determines what type of theory, quantum vs. classical, is the right one for the problem in question.

In solids and liquids, the value of a is about 0.2–0.3 nm. For electrons, $m \approx 10^{-27}$ g, $T_0 \approx 10^5$ K, so the quantum degeneracy condition is always fulfilled for electrons in solids or liquids. For atoms, m is about A (the atomic mass number) times the proton mass (1.6×10^{-24} g), so roughly $T_0 \simeq (50/A)$ K in solids or liquids. So systems of atoms or ions in condensed matter will show quantum mechanical effects only in the low temperature region, and apparently only with very light atoms such as H or He. However, the case for H is often complicated by associating into molecules H_2 and it cannot maintain the liquid state to low temperatures, so only He liquids under 5 K display their quantum nature most conspicuously. Generally speaking, for solids, delocalized electrons must be treated as waves, while heavier particles such as atoms or ions are usually treated as particles. This dichotomy is quite justified in view of the enormous difference in the degeneracy temperatures for these two kinds of particles. So in condensed matter physics, the classical picture of particles and quantum mechanical picture of waves are generally used simultaneously for the description of various phenomena. For example, in the explanation of the electrical conductivity of metals, ions arranged in the lattice are regarded as classical particles, while electrons must be treated as de Broglie waves. However, in treating lattice vibrations a classical description is usually adopted, supplemented by quantization of the vibrational energy. Full quantum mechanical treatment is required for the theory of liquid He, so these liquids are known as quantum liquids. It should be noted that most interactions between atoms and ions are mediated by electrons, or speaking more figuratively, electrons serve as the glue between atoms or ions. To treat these interactions adequately and in detail, quantum mechanics is indispensable, though sometimes classical approximations such as empirical or semi-empirical potentials (e.g., Lennard–Jones, Morse, Stilliner–Weber, etc.) may be effective for some approximate estimates. For instance, in the Car–Parrinello method, the electronic structure is calculated quantum mechanically, while the behavior of atoms is described by classical molecular dynamics. The above statements are also valid for the case of molecules, which are mixed systems of atoms (or ions) and electrons.

Inter-particle spacing also plays an important role in T_0 for its inverse square relationship in (0.2.1). In the case of gases, due to the large values of a, the values of T_0 are extremely low; in the submicrokelvin range. To reach there is a challenge for experimentalists. Experimental difficulties were surmounted only with laser cooling by the end of the 20th century. On the other hand, the estimated T_0 for protons and neutrons in atomic nuclei is about 10^{10} K due to the fact that the inter-particle spacing is extremely small ($\sim 10^{12}$ cm); neutron stars are in the same situation.

0.2.2 Condensation Phenomena

Condensation phenomena are typical collective phenomena which are well-known in everyday experience. Gases condense into liquids or solids. Liquid differs from solid by its fluidity which is shown macroscopically by the fact that its shear modulus is equal to zero, while microscopically their atoms or ions may readily change places within the liquid. Liquids differ from gases by the fact that there is a free surface dividing the substance into two parts with different densities. At the critical point, the density difference and the free surface disappear together, so liquid and gas phases merge into one.

For organic substances, sometimes it is hard to determine the demarcation line between liquids and solids and a plethora of mesophases such as liquid crystals, self-assembled membranes or gels may be formed. Polymers and biopolymers also may exist in the solid state or the liquid state, as well as some of the mesophases between these.

From the viewpoint of statistical physics, the gist of condensation phenomena lies in the compartmentalization of phase space. First we shall consider the case of position space. When gases condense into liquids, compartmentalization of position space is realized through the appearance of the liquid surface, which divides the configuration space into two parts. Near the surface there is a potential barrier that maintains the density difference on the two sides by making the net flux of particles zero at thermal equilibrium. When liquids are frozen into solids, further compartmentalization of position space occurs by subdivision of position space into a huge number of cells so that each particle is nearly confined within its cell, as exemplified by atoms or ions in crystals or glasses.

However, position space is only one part of the phase space for a collection of particles envisaged by statistical physics. We may anticipate that there are some phenomena related to compartmentalization in momentum space as well. These may be called condensation phenomena in momentum space and will be important especially for systems of particles in which the wave aspect dominates, i.e., at temperatures below the quantum degeneracy temperature T_0.

Indistinguishibility is a characteristic of identical particles in quantum mechanics. There are two types of particles: particles with integer (including zero) spin, which have symmetric wave functions and particles with half integer spin, which have antisymmetric wave functions. The former are bosons, which obey Bose–Einstein statistics; the latter are fermions, which obey Fermi–Dirac statistics. The electron, proton and neutron all have spin 1/2 and are therefore fermions, while the photon has spin 1 and is a boson. A complex particle is made up of many elementary particles; their spins are added together to form the total spin, which determines the nature of the complex particle. Two isotopes of He may be taken as examples: the ^{3}He atom has two protons; one neutron and two electrons, so it is a fermion; while a ^{4}He atom has two protons, two neutrons and two electrons and so it is a boson.

For a system of bosons, in which more than one particle may occupy the same quantum state, at $T = 0$ K all particles occupy the state with lowest energy as well as momentum, i.e. the ground state. This is a condensate in momentum (or wavevector $\boldsymbol{k}$) space. In 1924 Einstein predicted that, for a system of ideal bosons, when decreasing the temperature to a definite value, a macroscopic number of particles will occupy the ground state, i.e. a Bose–Einstein condensation (BEC) occurs. In 1950 F. London conjectured that the transition to superfluidity at 2.17 K in ^{4}He is the Bose–Einstein condensation of a system of interacting bosons. In 1995 the Bose–Einstein condensation of ^{87}Rb gas in a magnetic trap at a temperature of about 200 nK was first demonstrated with laser cooling plus evaporation cooling. Since then, similar condensation of gases of ^{23}Na, ^{7}Li as well as spin-polarized ^{1}H have been reported. Due to the very weak interactions between atoms, it is commonly said that these experiments verify the Bose–Einstein condensation as predicted by Einstein's theory; however, some reservations should be noted: the small number of atoms (10^3–10^6) in the trap deviates from the thermodynamic limit, the trap exerts a force field on the atoms, and the influence of weak interactions between atoms in some cases are non-negligible.

For a system of fermions, the Pauli exclusion principle prevents more than one particle from occupying the same quantum state. So, for an ideal gas of fermions at $T = 0$ K, the particles will occupy all momentum states up to some maximum value, the Fermi momentum. The locus of the Fermi momentum is called the Fermi surface, which divides occupied states from unoccupied ones in momentum space. This is a close analog for a liquid drop in position space, so it may be understood also as Fermi condensate versus Bose condensate in momentum space, although no sharp critical temperature like Bose–Einstein condensation temperature is found for a system of fermions. Nearly free electrons in metals are the most commonly encountered case of a Fermi gas. In 1999 the dilute atomic gas of ^{40}K in a magnetic trap was found to be quantum degenerate at microkelvin temperatures, providing another example of a Fermi gas. ^{3}He liquid below 5 K and electrons in metals with moderate interactions are Fermi liquids. If there are suitable interactions between a pair of fermions, then pairs of fermions may be formed and condensed into macroscopic quantum states. Examples include superconductors with paired electrons and ^{3}He superfluids with paired ^{3}He atoms. Whether there are superfluids with paired fermionic atoms in dilute gases at extremely low temperature is a hot topic of contemporary research.[e]

0.2.3 Ordering

According to thermodynamics, each equilibrium phase is situated at the minimum of the Helmholtz free energy $F = U - TS$ or the Gibbs free energy $G = U - TS + PV$, in which, U is the internal energy, T is the thermodynamic temperature, S is the entropy, P is the pressure, V is the volume. Different phases appearing in phase diagrams are due to the competition of the internal energy term U and the entropy term TS; usually the former term favors order, while the latter term favors disorder.[f] In general, condensation phenomena are closely connected with ordering processes, various condensates are phases with some kind of ordered structures.

For atoms considered as classical particles, the order is most conspicuously manifested as the positional order which signifies the correlations of atomic positions at different places. If the range of correlations tends to infinity, the system has long-range order; if the range of correlations is limited to nearby atoms, then the system has short-range order; if there is no correlation at all, the distribution of atoms is fully random, so the system is in perfect disorder. Crystalline solids and substitutionally-ordered alloys are examples of long-range order with periodicity; incommensurate phases and quasicrystals are examples of long-range order with quasiperiodicity; glasses and liquids are examples of short-range order without periodicity; while gases are phases with nearly perfect disorder. For the phases consisting of anisotropic molecules, the orientational order becomes important; for instance, liquid crystals and plastic crystals. Both the positional order and the orientational order are shown clearly only in configuration space; these may be identified as order in configuration space or particle order. It should be noticed that order in position space is not limited to a system of classical particles; degenerate electron gases or liquids may manifest order in position space due to interactions as exemplified by charge density waves (CDW), spin density waves (SDW) and the Wigner crystal, i.e., electrons periodically arranged in space.

For a system of particles in the quantum regime, i.e., below T_0, the important thing is ordering in momentum space. Order in momentum space, or wave order, is manifested by the occupation of states in momentum space. For a system of bosons, perfect order is achieved when all particles occupy the ground state or a state with definite momentum. For a system of fermions, perfect order means every state below the Fermi surface has been occupied. This is quite analogous to the liquid state limited by a sharp surface in configuration space. These are wave orders for large systems, i.e., where the number of particles, $N \to \infty$. Wave order will also manifest in a system composed of a finite number of fermions, especially in a central field or a spherical trap. This kind of order is most clearly shown by the display of a shell structure: in the case of electrons in atoms it provides the scientific basis of the periodic table of chemical elements; in the case of atomic nuclei (mixed fermions) magic numbers and shell structure occur; in the case of electrons in metallic clusters or quantum dots, magic numbers and shell structure also occur. Wave order in a system of fermions is a direct consequence of the Pauli exclusion principle.

The physical origin of ordering for a system of classical particles is mostly due to the interactions between particles though, for a system of hard spheres, ordering of entropic origin is sometimes found. However wave order is a purely quantum mechanical effect; perfect order may develop even in a system of particles without any interactions at all, i.e., an ideal gas. Interactions actually may diminish, instead of enhancing wave order as indicated by the lowering of the Bose–Einstein condensation temperature for a system of interacting bosons. The same situation holds for a system

[e] It has been demonstrated that the dilute gas of fermionic ^{40}K atoms, after cooling under T_0 to form quantum Fermi gas, then further cooling will form weakly bound localized atom pairs, in which the Bose–Einstein condensate has been detected, see M. Greiner, C. A. Regal, and D. S. Jin, *Nature* **426**, 537 (2003).

[f] There are exceptional cases, such as entropy-driven ordering. For instance, in systems composed of hard rods or hard balls, when internal energy or temperature is kept constant, increase of entropy may promote ordering. This mechanism was first proposed by L. Onsager for phase transition in liquid crystal in the 1940s. Recently it has been widely used in the self-assembly of soft matter in which the change of internal energy may be neglected.

of fermions. Perfect order is shown by the non-interacting Fermi gas. By switching on interactions between particles adiabatically, the Fermi gas is changed into a Fermi liquid; its wave order is somewhat diminished, though it still retains the sharp Fermi surface and its elementary excitations are still analogous to the Fermi gas. Increasing the interactions between particles further, the Fermi liquid may be changed into a strongly-correlated fermionic system displaying many anomalous properties which are still perplexing topics for current investigations. On the further end, strong interactions may induce Wigner crystallization in an electronic system, as well as ordinary crystallization in an atomic system.

Uncertainties in position and momentum of a particle in quantum mechanics are coupled together by the Heisenberg uncertainty principle. So it is expected that positional order and momentum order in a quantum system appear to be mutually exclusive: Bose–Einstein condensation (BEC) does not occur in crystalline solids and Wigner crystallization develops at the expense of momentum order of electrons.

It may be mentioned in passing that the concept of momentum order is not necessarily limited to quantum systems. We may extend this concept to nonequilibrium classical systems such as hydrodynamic systems. In hydrodynamic flow, the highly correlative velocity field is analogous to a kind of momentum order in classical system. So it is no wonder that some hydrodynamic problems, such as thermal convection, pattern formation and turbulence, have been studied with profit by the methods of condensed matter physics.

There is also another sort of condensation (or ordering) phenomenon associated with particles with spins. For localized particles with spins, we may imagine spins attached to each lattice site. Ordering of spins occurs when the disordered spin state (paramagnet) transforms into the ordered spin state (ferromagnet or antiferromagnet). This is analogous to the classical description of crystallization in position space, while transition into a spin glass is analogous to glass transition from the liquid state. For spins of itinerant electrons, the quantum description of ordering in momentum space is more appropriate: The Fermi energy E_F may be different for up and down spins in the spin-ordered phase. Also there may be two kinds of pair formation for fermions: Pairs in configuration space are known as Schafroth pairs, while pairs in momentum space are Cooper pairs.

From the above discussions, we may conclude that condensed matter means condensates in phase space which, in general, can be subdivided into position space and momentum space. Hence what is known as condensed matter should enlarge its scope by including both condensates in position space and momentum space, some of these even may be gases, so compared with solid state physics condensed matter physics surely has a more rich and variegated collection of objects for its study.

The phase transitions discussed above are all thermodynamic ones. Phase transition temperatures are always higher than 0 K. Even if the interactions responsible for a phase transition are of quantum nature, the appearance of the phase transition is induced by a classical thermal fluctuation. There is another kind of transition, the so-called quantum phase transition which appears at 0 K, through adjustment of pressure or composition or by application of a magnetic field. At 0 K, thermal fluctuation is absent, so the quantum fluctuation implied by Heisenberg uncertainty principle plays the dominant role. Because the 0 K condition cannot be realized in a laboratory, what happens at 0 K can only be inferred from experimental results near 0 K. However, exact theoretical results may be obtained to compare with experimental results near 0 K. Such results even may be valid in quite a wide range of temperature. The study of quantum phase transitions may help us to understand some puzzling problems in condensed matter physics, such as strongly correlated electronic states.[g]

§0.3 Historical Perspective and Conceptual Framework

0.3.1 From Solid State Physics to Condensed Matter Physics

The foundation of solid state physics was laid in the 1930s, and by 1940 substantial parts of solid state physics had been established and the field was ready for a comprehensive survey: Structural crystallography was established by X-ray and electron diffraction; lattice dynamics was formulated

[g]One may consult S. Sachdev, *Quantum Phase Transitions*, Cambridge University Press, Cambridge (1999).

by the application of quantum theory and statistical physics to the study of thermal properties of solids; electron band theory arose from the application of quantum mechanics and statistical physics to the study of electrical conductivity of solids; the theory of magnetic properties of solids was based on the quantum mechanical treatment of exchange interactions. Thus the publication of the influential monograph "*Modern Theory of Solids*" by F. Seitz in 1940 marked the coming-of-age of solid state physics. Subsequently many textbooks on solid state physics have been published.

By the end of the 1940s, explosive growth of solid state physics occurred in the wake of the invention of the transistor, so the stage for theoretical treatments was set and the practical utility of the research activities had been demonstrated. The range of research topics was much enlarged, so in the mid-1950s, Seitz and Turnbull found that the field had become too wide to be included in a single monograph, so the multi-volumed *Solid State Physics—Advances in Research and Applications*, which published comprehensive review articles on recent progress in some subfields at the rate of about one volume per year, was launched and has continued to this day. It testifies to the diversity and the fertility of this field and remains an excellent set of reference books for many topics developed since the 1950s.

After the 1970s and the 1980s, the term 'solid state physics' has often been superseded by 'condensed matter physics': Why this change? What are the meanings hidden behind these two different terms? In this section, we shall try to answer these questions. In our analysis, the reasons for this change are twofold: One is more obvious, it concerns the enlarged scope of objects studied; the other is more subtle, it concerns a change or shift in conceptual frameworks.

Besides the enlargement of scope, there is also the change or the shift in conceptual framework associated with the transformation from solid state physics to condensed matter physics. A noted historian of science, T. S. Kuhn emphasized the important role played by the paradigm in the development of science. Before the establishment of the paradigm, a given branch of science was still immature. Although much knowledge had been amassed, divergent viewpoints may be raised, but a coherent conceptual framework was still lacking, so it remained in the stage of pre-science. After the paradigm was established, it went into the stage of normal science, and quick and stable growth resulted. At this stage the discipline became mature, many monographs and textbooks were written to expound it. After quite a number of anomalies appeared leading to a crisis, a scientific revolution took place, forming a new paradigm to replace the old one. Although the actual development of science is too complex to be fitted into such simple pigeonholes, undeniably the paradigm can be seen to play a substantial role in the history of science. In this and later sections we shall try to decipher the paradigms involved in solid state physics and condensed matter physics.[h]

0.3.2 The Paradigm for Solid State Physics and Its Extension

In retrospect, although solid state physics was characterized by variety of subject matter and diversity of theoretical treatments, we may still recognize a common conceptual framework unifying a large part of this branch of science, i.e., a paradigm in Kuhnian sense. As a paradigm for solid state physics, we may take the title of a book by L. Brillouin, "*Wave Propagation in Periodic Structures*". In this book, fundamental problems of solid state physics were illustrated by a unifying viewpoint. He began his discussions with a historical perspective, starting from Newton's derivation of the formula for sound speed with a lattice model, then discussing the mechanical model of the dispersion of light as envisaged by Cauchy, Kelvin and others in the 19th century, finally emphasizing the importance of the concept of pass-bands and cut-off frequencies, widely used by engineers in their treatment of electric filters with one-dimensional periodic structures. Brillouin emphasized that there are common characteristics of wave propagation for different types of waves: It does not matter whether the wave is de Broglie or classical, elastic or electromagnetic, transverse or longitudinal. Solid state physics is mainly concerned with structures and properties of the crystalline state which is characterized by the existence of periodic structures. Propagation of elastic waves or

[h]T. Kuhn, *The Structure of Scientific Revolution*, 2nd ed., The University of Chicago Press, Chicago (1971). Although we acknowledge the importance of the concept of 'paradigm' introduced by Kuhn, however, it does not imply that we endorse without reservation all the ramifications of the Kuhnian philosophy of science.

lattice waves in periodic structures lead to lattice dynamics mainly formulated by M. Born and his school; propagation of short wavelength electromagnetic waves led to the theory of X-ray diffraction in crystals, the dynamical theory of which was formulated by P. P. Ewald, C. G. Darwin, and M. von Laue; propagation of de Broglie waves (electrons) in crystals led to the band theory of the electronic structure of solids, formulated by F. Bloch, A. C. Wilson, L. Brillouin and others. There are common features of these theories: In the treatment of wave propagation, Bloch formalism is adopted to take advantage of simplicity introduced by translational symmetry (periodicity), and the crucial role played by dispersion and its visualization in wavevector (or reciprocal) space. Subsequent consolidations and applications of these fields became the main tasks accomplished by solid state physicists, both experimentalists and theorists. It should be noted that, even today, the vitality of this paradigm is by no means exhausted, a new lease on life was obtained through investigations of photonic crystals with band gaps in the late 1980s and after.

The paradigm once established faced extension and modification with the progress of science. After the immense success achieved by band theory and lattice dynamics in the treatment of periodic systems, the challenge of aperiodic systems became acute. For crystals with dilute impurities, the case of Bloch waves scattered by a single impurity atom is of prime importance. This lead to the formulation of Friedel oscillations and related topics. For concentrated alloys, multiple scattering from different sites becomes important and some averaging schemes such as average t-matrix (ATM), coherent potential approximation (CPA), and effective medium approximation (EMA) have been proposed to give a picture of the average band. These methods were also extended to the cases of classical waves. In addition, these approximations have been applied to weakly disordered systems with considerable success. For more strongly disordered systems, P. W. Anderson, in his seminal paper in 1958, introduced the concept of localization of de Broglie waves due to strong disorder. About 10 years later, Mott gave a physical interpretation of Anderson localization to explain the behavior of amorphous semiconductors and the metal-insulator transition induced by disorder with considerable success. Thus, the physical picture for localization of electrons was accepted by the scientific community. Just as Brillouin emphasized that there is a common trait for all type of waves, so localization phenomena are not limited to de Broglie waves; classical waves also show its manifestations. There has been a flurry of research activity concerned with the localization of classical waves both theoretically and experimentally in recent years. Wave behaviors in quasiperiodic structures, such as incommensurate phases and quasicrystals, situated midway between periodic structure and a homogeneously-disordered one, is another topic worthy of attention. The energy spectra of these structures are shown to be characteristic of a critical state which is intermediate between delocalized and localized states, displaying a self-similar structure. Inhomogeneous structures are another type of disordered system. The traditional method of obtaining physical properties of phase mixtures is to use EMA, but with the development of percolation theory, we may go beyond it. Fractal structure with scale invariance has been transformed from mathematical curiosity to a model applicable to many real problems studied by physicists. The concept of fracton emerged from the study of its vibrational and electronic spectra.

Periodic potentials that satisfy translational invariance should be infinite in extent and without any boundaries, but actual crystals have surfaces on which the periodicities of atomic arrays are interrupted. A new subfield called surface physics, which studies crystallography, lattice dynamics and the electronic structure of surfaces, has been established. Low-dimensional physics is another new subfield full of vitality. Although theoretical physicists rigorously proved that fluctuations should destroy the long-range periodicity in one-dimensional (1D) and two-dimensional (2D) structures, it is found that the existence of inter-chain or inter-sheet coupling, as well as its finite extent, may stabilize the periodic structure in certain quasi-2D or quasi-1D materials. It should be noted that effects due to disorder are particularly conspicuous in these low-dimensional materials. The scaling theory of localization shows that electrons in 1D and 2D metals are generally localized. Not only low-dimensional structures are related to disorder, but structures with dimensions higher than 3 are also related to disorder. 1D incommensurately modulated structure corresponds to the projection from periodic structure in a four-dimensional (4D) hyperspace; three-dimensional (3D) quasicrystal corresponds to the projections from periodic structures in six-dimensional (6D) hyperspace; while structure of amorphous materials may be modeled as the projections from periodic structure in a 4D curved space. Fractal structures may have fractional dimensions. Thus, the study of wave behavior in these aperiodic and non-3D structures becomes an important extension of the original paradigm and is found in active areas of contemporary research.

Another active field is related to coherence effects of de Broglie waves. Anderson localization and the subsequently-discovered weak localization of electrons are really due to interference effects. With the fabrication of artificial nanostructures, such as superlattices, quantum wells, quantum wires, quantum dots and small rings, in which transport phenomena with interference of coherent de Broglie waves are studied, ballistic transport as well as tunneling transport may be readily observed, especially at low temperatures. This has given birth to a new exciting subfield, mesoscopic physics.

Electrons have spins as well as charges, but to incorporate spins into this paradigm is not an easy task. One may introduce spin-dependent scattering or one may consider propagation of spin waves in a periodic array of localized spins phenomenologically modelled on that of lattice waves. Only partial success has been achieved, for the subject of ferromagnetism is at heart a many-body problem which lies outside this paradigm. This also applies to superconductivity.

0.3.3 Bond Approach versus Band Approach

When band theory was first proposed and theoretical treatments of the hydrogen molecule based on quantum mechanics appeared, this marked the first success of quantum chemistry. So a paradigm for quantum chemistry was established alongside that of solid state physics. Although both band theory and quantum chemistry are based on quantum mechanics, the difference in viewpoints should be noted. Band theory adopts a global view, stressing the delocalized valence electrons and the dispersion relation in $\boldsymbol{k}$ space, while quantum chemistry, in general, adopts a local view, stressing the atomic configurations, bond formation and charge transfers in real space. Of course, both band theory and quantum chemistry have their merits as well as their disadvantages: Band theory is most successful for transport properties of solids; however, in the treatment of binding of solids, it is not so intuitively transparent as quantum chemistry. It should be noted that quantum chemistry also has its shortcomings; the calculations are done by summing up contributions by different atoms, the difficulty in calculation scales up drastically with the addition of more atoms and the results cannot extrapolated to the case of infinite number of atoms, which is the most interesting case for solid state physics. However, since each theory has its limitations, actually the band approach and the bond approach are complementary to each other, so it is unwise to see things only from one point of view. Formerly physicists have been too partial on the side of the band approach but nowadays condensed matter physics is increasingly concerned with complex oxides and organic materials and it is quite desirable for condensed matter physicists to be familiar with the language of quantum chemistry.

Even in the heyday of solid state physics, the shortcomings of band theory became apparent when confronted with the fact that NiO, CoO, MnO etc., are transparent insulators, whereas, according to band theory, they should be metals. These oxides were later called Mott insulators, in recognition of Mott's contributions to this problem. Taking NiO as an example, the current passing through it depends on configurations Ni^{3+} and Ni^{+} which could move, and on-site correlation energy which could prevent their formation. The energy required to create these configurations, if the ions are a long way from each other, is called the Hubbard energy U equal to

$$U = I - E , \tag{0.3.1}$$

where I is the energy required to remove an electron from Ni^{2+} to form Ni^{3+} and E is the energy gained when a free electron at rest is added to Ni^{+}. This distinction between a metal and a insulator has more 'chemical' flavor compared with Wilson's formulation in band theory. Later Hubbard incorporated this idea in combination with band theory into the Hubbard model to treat the Mott transition, i.e., the metal-insulator transitions due to electron correlation. It should be noted that the on-site correlation effect has been already considered in Heitler and London's treatment of the valence bond in hydrogen molecule, and its generalization to the case of magnetic interactions is the starting point of the Heisenberg model for ferromagnetism. Parenthetically, the mainstream of quantum chemistry, the molecular orbital approach, neglected the on-site correlations just like band theory does.

In the 1950s Anderson proposed a theory of superexchange for antiferromagnetic or ferrimagnetic oxides; the on-site correlation energy also plays an important role in this subject. About the same time Zener developed a theory of double exchange to account for the ferromagnetic metallic state of certain calcium manganites. About 40 years later in the 1990s, it was found that these oxides show colossal magnetoresistance (CMR), i.e., with application of magnetic field there is about a 10^2-fold to 10^6-fold change of resistance.

In the study of dilute magnetic impurities in normal nonmagnetic metals, a series of interesting physical effects was discovered showing anomalous resistivity versus temperature as well as anomalies in magnetic properties. It is certainly related to the mixing and interaction of s and d electrons. Anderson in his theoretical treatment gave a simplified Hamiltonian in which the on-site correlation energy appeared, this was later called Anderson model. The Anderson Hamiltonian may be transformed into the Kondo Hamiltonian which is crucial for the elucidation of the Kondo effect, i.e., the appearance of a resistance minimum in alloys of noble metals with magnetic impurities. The Anderson model may be generalized to the periodic Anderson model just as the Kondo model can be generalized into the Kondo lattice model for the treatment of more complex problems such as heavy electrons and Kondo insulators. After 1986, a series of high T_c superconductors were discovered. The prototypes of these superconducting oxides are mostly Mott insulators, so it is expected that strong correlations between electrons also play a decisive role in the mechanism of the normal state conductivity as well as superconductivity in these doped Mott insulators. Surely these problems indicate that the development of theories going beyond a simple band approach or simple bond approach, with appropriate treatment of electron correlation, is much needed.

0.3.4 The Paradigm for Condensed Matter Physics

Cooperative phenomena such as phase transitions are important examples of many-body physics. Historically, mean-field theories have been repeatedly proposed for various physical systems by different scientists. The first one was van der Waals theory of gas-liquid transitions in 1873, then Weiss theory for paramagnetic-ferromagnetic transition in 1907, later still the Bragg and Williams theory for order-disorder transition in alloys in 1934. In 1937 Landau formulated a phenomenological theory of second-order phase transitions with sufficient generality that it contained all the essence of these mean-field theories. Heisenberg's theory of ferromagnetism and the Bardeen–Cooper–Schrieffer (BCS) theory of superconductivity were highlights of the quantum theory of cooperative phenomena. The discovery of antiferromagnetism by Néel, the discovery of superfluidity in ^{4}He by Kapitza and Allen and its explanation by Landau and London; the discovery of superfluidity in ^{3}He by Lee, Richardson, and Osheroff, explained theoretically by Legget, all added new members to the repertoire of ordered phases with broken symmetry.

Researches were extended to the excited states of ordered phases. First among them was the Debye theory of phonons which gives a rough idea of elementary excitations. Bloch introduced the idea of spin waves or magnons. The Bohm and Pines theory of plasmons and the Landau theory of Fermi liquids marked important advances in many-body theory.

In 1934 the theory of dislocations was proposed by Taylor, Orowan , and Polanyi, independently, to explain why metallic crystals are easily plastically deformed; about the same time, the domain theory for ferromagnets was proposed by Landau and Lifshitz. While the latter was immediately verified by experiments, the former had to wait 20 years for clear-cut experimental verifications. In 1957 Abrikosov predicted the vortex lattice in type II superconductors, which also waited many years to be verified. In the 1970s defects (including disclinations) in liquid crystals were intensively studied. In 1976 Toulouse and Klémen proposed a scheme of topological classification of defects and gave a unified treatment of defects for the first time. Topological defects are recognized as singularities in ordered media, in which topological stability is related to the dimensions of media and the number of components of the order parameters.

Though critical phenomena in the region near T_c were observed very early in the 19th century by T. Andrews, precise experimental measurements of critical indices in the 1950s and 1960s found that the Landau theory was not valid in the critical region, so the modern theory of critical phenomena, with strong and long-range fluctuations, was born. Its basic concepts and theoretical methods incorporating scaling, universality, and the renormalization group technique not only neatly solved the problem of critical phenomena, but also found important applications in other fields.

In the 1970s and after, these concepts of cooperative phenomena were used fruitfully by P. G. de Gennes, S. F. Edwards and others to explore new territories, such as the physics of liquid crystals and polymers.

In 1976 J. A. Hertz extended the theory of critical phenomena from the classical regime to the quantum one by treating time as a new dimension, and so broke the ground for the study of quantum critical phenomena. The importance of this kind of study for condensed matter physics was fully recognized only in the 1990s. From the works discussed above, condensed matter physics acquired a new physiognomy compared with more traditional solid state physics. The shortcomings of the original paradigm, due to its inadequate treatment of interactions between particles, became apparent and the time was ripe for the formulation of a new paradigm for condensed matter physics.

During the transformation from solid state physics to condensed matter physics, two outstanding scientists, Landau and Anderson played important roles in it not only by their creative theoretical contributions but also by distilling and clarifying some basic concepts that serve as the foundation for the edifice of condensed matter physics. Landau in his theory of the second-order phase transitions formulated the concepts of broken symmetry and generalized the concept of the order parameter. In his theories of superfluidity of ^{4}He and Fermi liquids, he introduced the general idea of elementary excitations. Anderson in his book "*Concepts in Solids*" recognized the importance of broken symmetry and elementary excitations; in his later book "*Basic Notions of Condensed Matter Physics*", he made systematic and insightful expositions on basic concepts, such as broken symmetry, elementary excitations, generalized rigidity, topological defects, adiabatic continuity and renormalization groups.

Based on the endeavors of Landau, Anderson and others, we shall try to formulate explicitly a new paradigm for condensed matter physics. In contrast with the former one, this new paradigm emphasizes many-body effects and its central place is reserved for broken symmetry.

Landau's idea of broken symmetry may be summarized as follows: The presence or the absence of a certain symmetry element in a certain state of matter is never ambiguous, it is either there or not there. The sudden disappearance of a certain symmetry element in the high-symmetry phase corresponds to the occurrence of a phase transition with the concurrent appearance of the low-symmetry phase. Broken symmetry signifies the appearance of an ordered phase with the value of certain order parameter to be different from zero. The order parameter is the average value of a certain physical quantity; it may be a scalar, a vector, a complex number or a more complicated quantity with many components. It is zero in the high temperature phase, it has a finite value in the low temperature phase. The critical temperature T_c marks the temperature at which the second-order phase transition occurs. This order parameter is used to describe the qualitative as well as the quantitative aspects of the loss of symmetry in the low symmetry phase.

As we know, matter at sufficiently high temperatures is in the gaseous state. It is homogeneous and isotropic, and maintains full translational and rotational symmetry, which are compatible with the symmetry of the controlling equations. Few spectacular physical properties are manifested by matter in the gaseous state, while matter in the solid state is entirely different: Solids have rigidity, can conduct electricity, and can manifest a full spectrum of interesting properties. The difference may be traced to the different symmetries that characterize the gaseous state and the solid state. Most solids are crystalline materials in which full translational symmetry and rotational symmetry are broken. The only remaining symmetries are invariance with the displacement of lattice vectors and a specific set of discrete rotations. So broken symmetry is closely connected with a change in structure and the emergence of new physical properties.

Though Landau theory was precisely formulated only for second-order phase transitions, its basic concepts may be extended to some first-order phase transitions as well.

The importance of the Landau theory of broken symmetry can never be overestimated. It gives insight into how the important events in condensed matter happen, and it is very comprehensive in scope and very flexible in handling, so it may be utilized as a new framework for various theoretical

Table 0.3.1 Symmetry in condensed matter physics.

Transformations	Unobservables	Conservation laws and selection rules
Translation in space $\boldsymbol{r} \to \boldsymbol{r} + \Delta$	Absolute position in space	Momentum
Translation in time $t \to t + \tau$	Absolute time	Energy
Rotation $\boldsymbol{r} \to \boldsymbol{r}'$	Absolute direction in space	Angular momentum
Space inversion $\boldsymbol{r} \to -\boldsymbol{r}$	Absolute left or right	Parity
Time reversal $t \to -t$	Absolute sign of time	Kramers degeneracy
Sign reversion of charge $e \to -e$	Absolute sign of electric charge	Charge conjugation
Particle substitution	Distinguishibility of identical particles	Bose or Fermi statistics
Gauge transformation $\psi \to e^{iN\theta}\psi$	Relative phase between different normal states	Particle number

treatments. Strictly speaking, Landau's original formulation of broken symmetry is only valid within classical systems, but it was successfully extended to quantum mechanical many-body systems by Heisenberg in the 1950s.

Symmetry is of prime importance for the whole realm of physics, for it is closely connected with the hypothesis that certain physical quantities are unobservable, and its direct consequences are the conservation laws or selection rules. We are going to examine a few examples: The invariance for arbitrary translation in space implies that the absolute position in space is unobservable, and its direct consequence is the conservation of momentum. The invariance of arbitrary translation in time implies that the absolute time is unobservable and its direct consequence is the conservation of energy. The invariance for arbitrary rotation implies that the absolute direction in space is unobservable, it leads to the conservation of angular momentum. The invariance for inversion in space is connected with parity conservation, and that inversion in time, i.e., time reversal, is connected with the energy degeneracy for reversed spins, while reversion of the sign of electric charge is connected with charge conjugation. There is a more subtle symmetry, i.e., the gauge symmetry which implies that the relative phase between different normal states is unobservable, with conservation of particle number as its direct consequence. The different symmetries discussed above are tabulated in Table 0.3.1. These symmetries may be divided into two classes according to the locality of the symmetry operations: Global symmetry which means that the symmetry operation in question affects every point in space or time in the same manner indiscriminately; while local symmetry, whose symmetry operations affect each point in question independently. For instance, gauge symmetry may be either a global or a local symmetry, the latter is exemplified by gauge field theory. However, in condensed matter physics we shall meet both cases. According to the continuity of symmetry operations, the symmetries may be also divided into two classes, the continuous vs. the discrete. In Table 0.3.1 we can find discrete as well as continuous symmetries.

Now we shall discuss some typical cases of broken symmetry in condensed matter. The liquid state and the crystalline state differ in translational symmetry and rotational symmetry, the former retains full translational and rotational symmetry in the statistical sense, while the latter adopts a periodically ordered structure with the symmetry described by one of 230 space groups. This is a case of broken translational as well as rotational symmetry, though liquid-solid transitions are always first-order. Another case of broken translational symmetry is the quasicrystals with its symmetry described by quasiperiodical space groups. Nematic liquid crystals are examples of pure broken orientational symmetry, while a smectic liquid crystal is superposed with broken translational symmetry in one dimension. Broken space inversion symmetry is related to ferroelectrics and antiferroelectrics;

while broken time reversal symmetry is related to ferromagnets and antiferromagnets. Broken global gauge symmetry leads to macroscopic wave functions in which phase coherence is maintained at the macroscopic length scale, leading to superfluids or superconductors. It is no exaggeration to say that the infinite variety of condensed matter is just the manifestation of broken symmetry.

A fundamental problem for various ordered phases is what are their ground states, i.e., the states at $T = 0$ K? For an ideal gas of fermions and bosons, this problem but has been answered. Now we are going to examine this problem for some real systems. To do this, we must take account of the interactions between particles, so, in general, it is a complicated many-body problem of quantum mechanics. We shall take the magnetic ordered phase as the simplest example. For a generalization of the Ising model, let us start from the Heisenberg Hamiltonian

$$\mathcal{H} = -\sum_{i<j} J \boldsymbol{s}_i \cdot \boldsymbol{s}_j \,. \tag{0.3.2}$$

If $J > 0$, it is easy to prove that the state of lowest energy has all spins aligned parallel; this is the ferromagnetic ground state. If $J < 0$, it is expected that neighboring spins will be arranged antiparallel. However, the situation is not so simple as it seems, and the determination of the antiferromagnetic ground state proved to be a rather difficult problem. In the 1D case, this problem was exactly solved by the Bethe ansatz, and it is found that it is not a fully ordered phase. Anderson treated the 2D case, obtained the resonant valence bond (RVB) state, somewhat akin to a spin liquid. The Heisenberg Hamiltonian is for localized spins, so the ferromagnetic ground state of itinerant electrons is also a quite difficult problem.

Moreover, there is a problem about the compatibility of the different types of order, for instance, spin order and lattice order. Ferromagnetic spin order is always compatible with lattice order, while the situation is different for antiferromagnetic spin order, e.g., antiferromagnetic spin order is not compatible with a triangular lattice. Incompatibility of these two different types of order may lead to frustration, the inability to satisfy simultaneously the decrees issued by different ordering schemes.

Small attractive interactions between electrons may lead to the formation of Cooper pairs, so a Fermi liquid can be transformed into the superconducting ground state, which breaks gauge symmetry. For conventional superconductors, the pairing mechanism is due to the electron-phonon interaction, which has been superbly described by the Bardeen–Cooper–Schrieffer (BCS) theory. For unconventional superconductors the situation is quite different: There may be different symmetries for the order parameters of different superconductors, s-wave, p-wave and d-wave; also there may be different mechanisms, the electron-phonon interaction, electron-electron interaction and others. So after more than 15 years of intense research on high T_c superconductivity, even though the symmetry of the order parameter has been identified as d-wave, the mechanism for it still has not been uniquely identified. The pairing of ^{3}He superfluid is found to be in the triplet state and is associated with magnetic effects, the mechanism being identified as p-wave pairing.

To find the ground state of a certain ordered phase is a quite intricate problem of many body theory; to find corresponding excited states is certainly a tremendously laborious task. However, some schemes to simplify the theoretical treatment have been devised and found to be very effective.

For low lying excited states, which are responsible for a number of interesting physical properties such as specific heat, magnetic susceptibility, electrical and thermal conductivities, Landau's concept of elementary excitations plays the crucial role. Theories of phonons, magnons, quasi-electrons, plasmons, excitons, polarons and polaritons were developed and form an important branch of condensed matter theory. Utilizing the concept of elementary excitations, we may regard the low lying excited states of an ordered phase as a collection of quasi-particles in which the interactions between the particles may be neglected in most cases. So, statistics for the ideal gas may be used in deriving the relevant physical properties, and thus a great simplification is achieved.

Anderson emphasized that broken symmetry is related to the appearance of generalized rigidity, just as a crystal breaks the translational symmetry, so that each atom is locked to a particular position, from which it acquires rigidity and can transmit force without dissipation. This phenomenon can be generalized to other cases of broken symmetry. For instance, in a superconductor, the 'rigidity' is the phase coherence of Cooper pairs, it may transmit persistent current without dissipation.

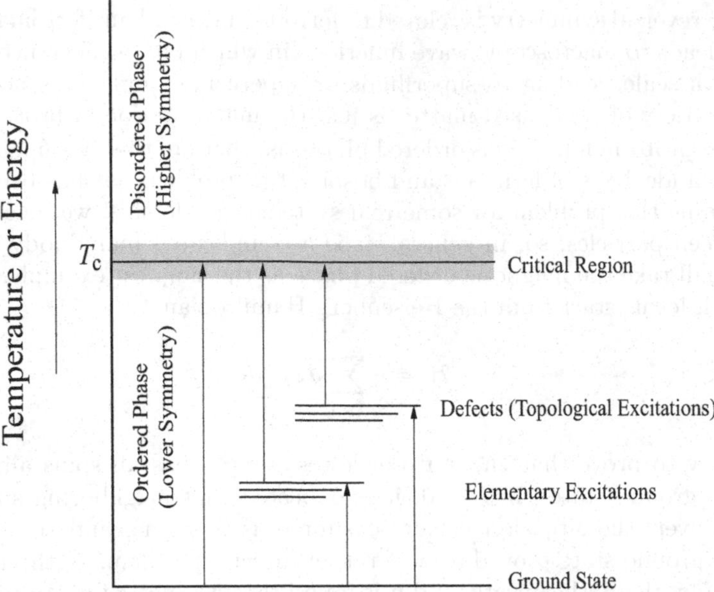

Figure 0.3.1 Schematic diagram for the energy states of a typical many particle system, as an illustration of broken symmetry in condensed matter.

In general, elementary excitations are too gentle to have much influence on generalized rigidity, however, the presence of topological defects may partially break the 'rigidity' of broken-symmetry phases. This is precisely the role played by topological defects in certain physical properties, for instances dislocations in crystal plasticity, magnetic domain walls in magnetic coercivity, and Abrikosov vortex lattice in critical currents of type II superconductors.

Both elementary excitations and topological defects show a tendency to regain the symmetry that has been broken, so it is quite natural that these may play important roles in phase transitions. The influence of the former ones is most clearly shown by some modes going soft in certain phase transitions, while the influence of the latter is exemplified by some defect-mediated phase transitions. For 3D systems the difference between elementary excitations and topological defects is clear-cut, while for the low-dimensional case the demarcation is somewhat blurred.

We have already mentioned that the interactions between electrons changes the Fermi gas into Fermi liquid. Other ground states may be formed due to the electron-electron and electron-phonon interactions. In general there is Coulomb repulsion between electrons. For an electronic system at high density, the Coulomb interaction between electrons is strongly shielded, and the ground state is the ordinary Fermi liquid. On the other hand, for a system in the low density limit, where the potential energy due to the Coulomb interaction dominates, it is predicted that a lattice of electrons, known as a Wigner crystal will be formed. In low-dimensional systems the effects of interactions become particularly striking. For 1D or quasi-1D metals, electron-lattice interaction may lead to new ground states, the charge density wave (CDW) and the spin density wave; interactions between electrons may lead to a Luttinger liquid. Exact solutions are available for these 1D problems. Many interesting phenomena are observed in systems with intermediate density and dimensionality, known as systems of strongly correlated electrons. These include the fractional quantum Hall effect, anomalous properties of heavy electrons, anomalous transport properties and exotic superconducting state of high T_c superconductors, colossal magnetoresistance (CMR), etc. What are the ground states, elementary excitations and topological defects of these systems? These are hot topics for current research. Full elucidation of these problems is a major challenge for present day investigations, and the answers are expected to be further breakthroughs for condensed matter physics.

Near the T_c of second-order phase transitions there are critical regions in which critical phenomena are manifested. Due to strong fluctuations, it is found that mean-field theory (in consequence, Landau theory) is not valid in this region. There are empirical relationships such as scaling laws

Table 0.3.2 Systems in condensed matter showing broken symmetry phenomena.

Phenomena	Broken Symmetry	Ordered Phase	Order Parameter	Elementary Excitations	Topological Defects
Ferroelectricity	Space Inversion	Ferroelectric	$\boldsymbol{P}$	Optical phonon	Domain wall
Antiferroelectricity		Antiferroelectric	$\boldsymbol{P}_{\text{sublattice}}$		
Ferromagnetism	Time Reversal	Ferromagnet	$\boldsymbol{M}$	Spin wave	Domain wall
Antiferromagnetism		Antiferromagnet	$\boldsymbol{M}_{\text{sublattice}}$	Spin wave	
Superconductivity	Gauge Invariance	Superconductor	$\langle\psi\rangle = \rho^{1/2}\exp(-i\theta)$	Electron	Vortex line
Superfluidity	Gauge Invariance	^{4}He Superfluid	$\langle\psi\rangle = \rho^{1/2}\exp(-i\theta)$	Phonon, Roton	Vortex line
	Gauge Invariance + Time Reversal	^{3}He Superfluid	$d_{ij} = \langle\psi\psi\rangle_{M_L,M_S}$		Vortex line, Disgyration Singular point
Liquid-Crystalline Phenomena	Rotation	Nematics	$\bar{n}$		Disclination
	Rotation+	Cholesterics	$\bar{n}$+1D $\rho(Q)$		Disclination,
	Translation	Smetics	$\bar{n}$+1D ρ_G		Dislocation
Crystallization	Translation	Crystal	ρ_G	Phonon	Dislocation
	Translation	Quasicrystal	$\rho_{G'}$	Phonon Phason	Dislocation
Electronic crystallization	Translation	Wigner crystal	3D (2D) ρ_G	Ripplon	
	Translation	Charge density wave	1D ρ_G	Phason, Amplitudon	Discommensuration
	Translation	Spin density wave	1D ρ_G^+	Phason, Amplitudon	Discommensuration

and universality in the critical region; these are signs that the correlation length of the fluctuations reaches a macroscopic scale, so that the microscopic details of physical interactions do not matter. Renormalization group theory plays a crucial role in this regime. By changing the chemical constitution or applying pressure or a magnetic field, T_c may be suppressed all the way to absolute zero. At 0 K classical thermal fluctuations are also fully suppressed, so instead quantum fluctuations due to the uncertainty principle play the decisive role, leading to quantum critical phenomena.

Now we are ready for a graphical display of the paradigm for condensed matter physics. The central role is played by the concept of broken symmetry. Broken symmetry leads to the ordered phase. The ground state is the fully ordered phase. The excited states show tendencies to regain the original symmetry, so various types of elementary excitations and topological defects appear. In the critical region, the correlation length of fluctuations reaches a macroscopic scale until, at T_c, the ordered phase merges with the disordered phase. These relationships are displayed diagrammatically in Fig. 0.3.1. In Table 0.3.2 are listed some important systems in condensed matter that show broken symmetry phenomena.

However, it should be noted that some important phase transitions are not related directly to broken symmetry, for instance, gas-liquid and metal-nonmetal transitions, as well as liquid-glass and paramagnet-spin glass transitions. Ergodicity was introduced by Boltzmann in the 19th century as the foundation for statistical mechanics. Ergodicity in a system may be broken by the compartmentalization of phase space, which includes configuration space as well as momentum space, so that some parts of these spaces are inaccessible or hardly accessible to constituent particles.

Liquids, glasses and spin glasses are examples of broken ergodic phases in configuration space; while metal-insulator transitions are related to ergodicity-breaking in momentum space. Though broken symmetry is always accompanied by broken ergodicity, the reverse is not always true. We can find in the above-cited cases in which the ergodicity is broken while the symmetry still remains intact. Moreover, the concept of broken symmetry may be generalized to include that of broken ergodicity, i.e., the asymmetric distribution of particles in phase space. Spin glass is the most intensively studied system which exhibits broken ergodicity. Its theory may be applied to wide-ranging problems such as neural networks and protein folding. However, compared with more conventional broken symmetry, broken ergodicity is still in a more primitive stage of research.

In some systems, far from equilibrium, phase-transition-like behavior may be observed, for instance, the onset of thermal convection at the critical Rayleigh number. Sometimes, an analogy to broken symmetry may be used to illustrate these transitions to some advantage. Both Prigogine, in his theory of dissipative structure, and Haken, in his synergy ingenuously used this analogy to construct some far-reaching generalizations to embrace a variety of topics drawn from different disciplines. Thom in his theory of catastrophe found some mathematical links between phase transitions and transitions in nonequilibrium states. These theories are mostly based on analogies and similitudes. However, there are other alternatives to these. Take the onset of a convection state as an example: It may be interpreted as a transition into the ordered state from a disordered state. However, the state with Benard cells lacks the rigidity which characterizes the conventional broken symmetry phase in equilibrium; it is also much influenced by the boundary conditions. Further, with increasing Rayleigh number (or other characteristic number), the system may undergo a series of bifurcations, and finally reach some spatio-temporal chaotic state or turbulence. The situation is made clearer if we take an alternative interpretation: The quiescent state, or the state with laminar flow, is like the ground state of perfect order in momentum space, while the convection state is like one of the excited states with defect-assemblies. With the development of nonlinear science for dynamical systems, systems with few degrees of freedom have been thoroughly analyzed, and the theory of chaos emerges as a shining example. However, nonlinear phenomena in condensed matter far from equilibrium are generally accompanied by a large number of degrees of freedom. Some simplified dynamic models for dissipative many-body systems such as the sand-pile model for self-organized criticality proposed by P. Bak, *et al.* have found wide applications in different branches of science. Turbulence containing a large number of degrees of freedom remains a hard problem for physicists. So there are still a lot of problems to present challenges to condensed matter physicists confronting complexity.

It should be noted that the validity of many basic concepts of this new paradigm is not limited to the realm of condensed matter physics; it may be extended and generalized to other branches of physics. For instance, broken symmetry has already played outstanding roles in particle physics and cosmology: scientists envisaged that there are superfluidity in the interior of neutron stars and liquid metallic hydrogen in the interior of Jupiter, the supershell structure was first proposed as a possible structure for nuclei of super-heavy elements and subsequently was actually found in some metallic clusters. These substantiate the truth of a statement by Bardeen in 1980: there is an unity of concepts in the structure of matter.

Bibliography

A. MONOGRAPHS AND TEXTS ON CONDENSED MATTER PHYSICS

[1] Anderson, P. W., *Basic Notions of Condensed Matter Physics*, Benjamin, Menlo Park (1984).

[2] Isihara, A., *Condensed Matter Physics*, Oxford University Press, Oxford (1991).

[3] Chaikin, P. M., and T. C. Lubensky, *Principles of Condensed Matter Physics*, Cambridge University Press, Cambridge (1995).

[4] Committee on Condensed Matter Physics, National Research Council, *Condensed Matter and Materials Physics*, National Academic Press, Washington D. C. (1999).

[5] Marder, M. P., *Condensed Matter Physics*, John Wiley, New York (2000).

[6] Taylor, L. T., and O. Heinonen, *A Quantum Approach to Condensed Matter Physics*, Cambridge University Press, Cambridge (2002).

B. SOME TEXTBOOKS ON SOLID STATE PHYSICS AND SOLID STATE THEORY

[7] Seitz, F., *Modern Theory of Solids*, McGraw-Hill, New York (1940).

[8] Peierls, R., *Quantum Theory of Solids*, Clarendon Press, Oxford (1955).

[9] Wannier, G. H., *Elements of Solid State Theory*, Cambridge University Press, Cambridge (1959).

[10] Anderson, P. W., *Concepts in Solids*, Benjamin, New York (1963).

[11] Ziman, J. M., *Principles of Solid State Theory*, Cambridge University Press, Cambridge (1972).

[12] Ashcroft, N. W., and N. D. Mermin, *Solid State Physics*, Holt, Rinehart and Winston, New York (1976).

[13] Madelung, O., *Solid State Theory*, Springer Verlag, Berlin (1986).

[14] Callaway, J., *Quantum Theory of the Solid State*, 2nd ed., Academic Press, New York (1991).

[15] Kittel, C., *Introduction to Solid State Physics*, 7th ed., John Wiley, New York (1995).

[16] Elliot, S. R., *The Physics and Chemistry of Solids*, Wiley, New York (1998).

C. SOME IMPORTANT MONOGRAPHS AND REFERENCES

[17] Mott, N. F., and H. Jones, *The Theory of the Properties of Metals and Alloys*, Clarendon Press, Oxford (1936).

[18] Brillouin, L., *Wave Propagation in Periodic Structures*, John Wiley, New York (1946).

[19] Born, M., and K. Huang, *Dynamical Theory of Crystal Lattices*, Clarendon Press, Oxford (1954).

[20] London, F., *Superfluid*, Vols. I, II., Wiley, New York (1950, 1954), revised 2nd ed, Dover, New York (1961, 1964).

[21] Goodstein, D. L., *States of Matter*, Prentice-Hall, Englewood Cliffs (1975).

[22] White, R. M., and T. H. Geballe, *Long Range Order in Solids*, Academic Press, New York (1978).

[23] Abrikosov, A. A., *Fundamentals of the Theory of Metals*, North Holland, Amsterdam (1988).

[24] Sheng, P., *Introduction to Wave Scattering, Localization and Mesoscopic Phenomena*, Academic Press, New York (1995).

[25] March, N. H., *Electronic Correlation in Molecules and Condensed Phases*, Plenum Press, New York (1996).

[26] Fazekas, P., *Lecture Notes on Electron Correlation and Magnetism*, World Scientific, Singapore (1999).

[27] Kadanoff, L. P., *Statistical Physics, Statics, Dynamics, and Renormalization*, World Scientific, Singapore (2000).

[28] Seitz, F., D. Turnbull, H. Ehrenreich, and F. Spaepen, *Solid State Physics — Advances in Research and Applications*, Vols.1-56, Academic Press, New York (1955-2002).

Part I
Structure of Condensed Matter

Countless laws of construction and constitution penetrate matter like secret flashes of mathematical lightning. To equal nature it is necessary to be mathematically and geometrically exact. Number and fantasy, law and quantity, these are living creative strengths of nature; not to sit under a green tree, but to create crystals and to form ideas, that is what it means to be one with nature.

— Karel Ĉapek (1924)

One of the great lessons of condensed matter physics is nature is more fertile than human imagination in devising ways for matter to organizc itself into coherent structures. Yet given the initial clue from nature, the human imagination has proved to be remarkably adept at eventually inventing simple theoretical models that display and illuminate strange new kinds of behavior.

— D. C. Wright and N. D. Mermin (1989)

Chapter 1

Symmetry of Structure

For aeons humans have been fascinated by the striking symmetry of many natural objects such as flowers, snowflakes and mineral crystals, and embodied the idea of symmetry in numerous artificial objects such as decoration patterns, handiworks, buildings and monuments. Here is a quote from a passage in a text of ancient China, *Peripheral Notes on Poetry by Han Scholars* (circa 200 B. C.), "while flowers from plants are mostly fivefold, only snowflakes are sixfold." Perhaps it indicated the first glimmering of consciousness of the subtle difference in symmetry between ordinary flowers and snowflakes. In 1952, Hermann Weyl, a mathematician, expressed this idea in more accurate modern scientific language: "While pentagonal symmetry is frequent in the organic world, one does not find it among the most perfectly symmetrical creations of inorganic nature, among the crystals." In 1611, J. Kepler, in his booklet *On a Hexagonal Snowflake*, speculated that hexagonal symmetry of a snowflake is a manifestation of the internal periodic structure of closed-packed particles postulated as identical balls. Later developments substantiated this bold hypothesis, so it stands in history of science as the first successful venture into the realm of condensed matter physics.

Scientific investigations about symmetry of the external shapes of mineral crystals flourished in the 17th and the 18th centuries and marked the beginning of the science of crystallography. It culminated in the 19th century with the formulation of the theory of crystal lattices and space groups, making crystallography an exact science, methodical just like astronomy, and it also stands as the first successful theory in condensed matter physics which is still valid today. This chapter is intended to give a brief introduction to this subject, emphasizing those aspects which are more closely associated with recent developments: since this book has been written after the discovery of the C_{60} molecule and quasicrystals, in discussing point groups, besides the crystalline point groups, we also introduce icosahedral groups as well as other groups incompatible with periodic structure; in discussing the space groups, we also treat space groups with dimension higher than 3 and complex groups in both real and reciprocal space.

§1.1 Basic Concepts of Symmetry

Symmetry is certainly one of the most important concepts in physics, especially in condensed matter physics. Different states and phases of matter have their characteristic symmetries, and phase transitions are mostly related to a change in symmetry. In its original sense, symmetry was used to describe the geometrical property of certain figures or patterns and found important applications in the science of crystallography. Later it was extended and deepened to cover the invariant aspects of physical properties, physical interactions and physical laws under certain transformations, and it played important roles in the physics of the 20th century. In this chapter we are mainly concerned with symmetry in the original and restricted sense, i.e. the symmetry of geometrical figures or structures, whether their extent is finite or infinite.

1.1.1 Symmetry and Symmetry Operations

Symmetry here means the invariance of a certain figure or structure with respect to coordinate transformations. Consider a certain physical quantity such as the density as a function of the position vector $\boldsymbol{r}$, i.e., $\rho(\boldsymbol{r})$. Define an operator g which leads to the transformation of coordinates,

$$\boldsymbol{r} \to g\boldsymbol{r} = \boldsymbol{r}', \tag{1.1.1}$$

if

$$\rho(\boldsymbol{r}') = \rho(g\boldsymbol{r}) = \rho(\boldsymbol{r}), \tag{1.1.2}$$

then g is a symmetry operation. Here symmetry means invariance of an object under certain transformation in space, while the corresponding coordinate transformation is called a symmetry operation for this object.

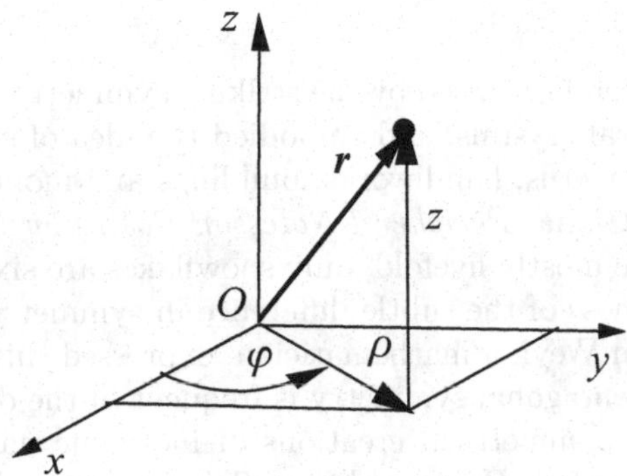

Figure 1.1.1 The Cartesian coordinate system (x, y, z) and cylindrical coordinate system (ρ, φ, z).

In the 3D Cartesian coordinate system (see Fig. 1.1.1), the transformation of coordinates may be expressed as

$$\boldsymbol{r}(x, y, z) \to \boldsymbol{r}'(x', y', z'). \tag{1.1.3}$$

In general, the transformation of coordinates may be decomposed into two parts: a matrix $\boldsymbol{M}$ denoting an operation without translation such as a rotation about a fixed axis or reflection by mirror, plus a translation vector $\boldsymbol{t}$, i.e.,

$$\boldsymbol{r}' = g\boldsymbol{r} = \boldsymbol{M}\boldsymbol{r} + \boldsymbol{t}, \ \boldsymbol{M} = (a_{ij}) = \begin{pmatrix} a_{11} & a_{12} & a_{13} \\ a_{21} & a_{22} & a_{23} \\ a_{31} & a_{32} & a_{33} \end{pmatrix}, \tag{1.1.4}$$

or, expressed in components,

$$x_i' = \sum_j a_{ij}x_j + t_i. \tag{1.1.5}$$

A restricted set of symmetry operations satisfies the condition for isometry, i.e., the constancy of the distance between any two points in the body under the symmetry transformation

$$|\boldsymbol{r} - \boldsymbol{r}'| = \sqrt{(x'-x)^2 + (y'-y)^2 + (z'-z)^2}, \tag{1.1.6}$$

and the preservation of the angles between lines and planes during the transformation. In this chapter we mainly deal with the symmetry operations that satisfy the condition of isometry, while in §1.5.3 we shall mention those symmetry operations that violate this condition, e.g., inflation or deflation related to the scale invariance, which will be treated more fully in Chap. 4.

For point symmetry, the determinant of the transformation is

$$|\boldsymbol{M}| = |a_{ij}| = \pm 1. \tag{1.1.7}$$

Consider a rotation about the z-axis from φ to φ'. If we let $\theta = \varphi' - \varphi$, the matrix of the transformation is

$$M = \begin{pmatrix} \cos\theta & -\sin\theta & 0 \\ \sin\theta & \cos\theta & 0 \\ 0 & 0 & 1 \end{pmatrix}, \tag{1.1.8}$$

which satisfies $|M| = +1$. And the mirror reflection corresponding to the xy plane can be expressed by

$$\boldsymbol{M} = \begin{pmatrix} 1 & 0 & 0 \\ 0 & 1 & 0 \\ 0 & 0 & -1 \end{pmatrix}, \tag{1.1.9}$$

where $|\boldsymbol{M}| = -1$. For inversion through the origin O,

$$\boldsymbol{M} = \begin{pmatrix} -1 & 0 & 0 \\ 0 & -1 & 0 \\ 0 & 0 & -1 \end{pmatrix}, \tag{1.1.10}$$

where also $|\boldsymbol{M}| = -1$.

So symmetry operations are divided into two kinds: g^{I} and g^{II} depending on whether the determinant of the transformation equals $+1$ or -1. We shall get the following rule easily

$$|\boldsymbol{M}(g_t)| = |\boldsymbol{M}(g_1^{\mathrm{I}} g_2^{\mathrm{I}} \cdots g_q^{\mathrm{I}})| = (+1)^q = +1. \tag{1.1.11}$$

It indicates that the successive operations of g^{I} for q times is still g^{I}; while successive operations of g^{II} for q times may be g^{II} or g^{I} depending on q is odd or even,

$$|M(g_t)| = |M(g_1^{\mathrm{II}} g_2^{\mathrm{II}} \cdots g_q^{\mathrm{II}})| = (-1)^q = \begin{cases} +1, & \text{when } q = \text{even, equal to } g^{\mathrm{I}}, \\ -1, & \text{when } q = \text{odd, equal to } g^{\mathrm{II}}. \end{cases} \tag{1.1.12}$$

It should be noted that the transformations for the symmetry operations of the I kind can be realized by the actual motion of the object, while that of the II kind, such as mirror reflection and space inversion, can never be realized by the actual motion of the original object, just as the mirror world is inaccessible by displacements from the ordinary world. For instance, a right-handed screw can be transformed into a left-handed screw by mirror-reflection, but this transformation cannot be realized by any actual motion of a screw in real space. So right-handed and left-handed screws are different objects with different chirality, i.e., handedness; examples for chiral molecules are shown in Fig. 1.1.2. In the 19th century, L. Pasteur discovered chiral molecules by observing the optical activity (rotation of the plane of polarization of plane-polarized light) of certain solutions. Chirality of molecules is an important topic for biology, for biologically important molecules are mostly chiral; for instance, proteins are mostly left-handed, while DNA is right-handed. The problem of life in living organism is closely connected with the chirality of these molecules.

However, successive operations of the II kind for even number of times is equivalent to an operation of the I kind. It reminds us of an episode in Lewis Carroll's fairy tale "*Through the Looking Glass*": Alice found in the mirror world a book containing some queer verse entitled "Jabberwocky" with mirror-inverted characters, she had the wit to reflect the book a second time, so restoring the original text of this 'nonsense poem' for reading (see Fig. 1.1.3). This nicely illustrates that two successive operations of a symmetry operation of the II kind is equivalent to a symmetry operation of the I kind.

1.1.2 Some Theorems for the Combinations of Symmetry Elements

Here we shall state some theorems for the combinations of symmetry elements; the readers may prove these by inspecting the figures accompanying the text.

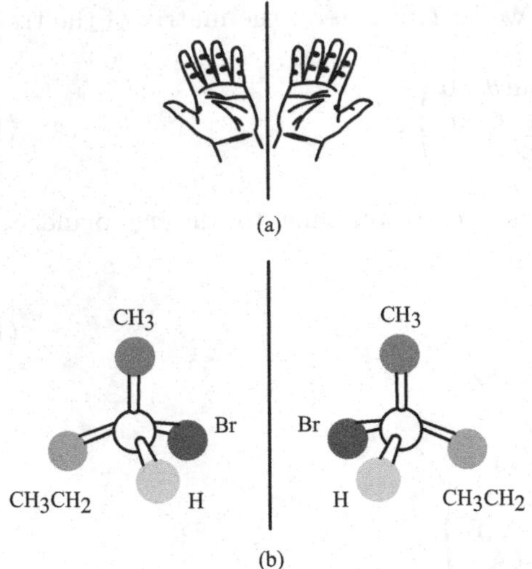

Figure 1.1.2 The schematic diagram of handedness. (a) The left hand and the right hand; (b) the mirror reflection of a chiral molecule.

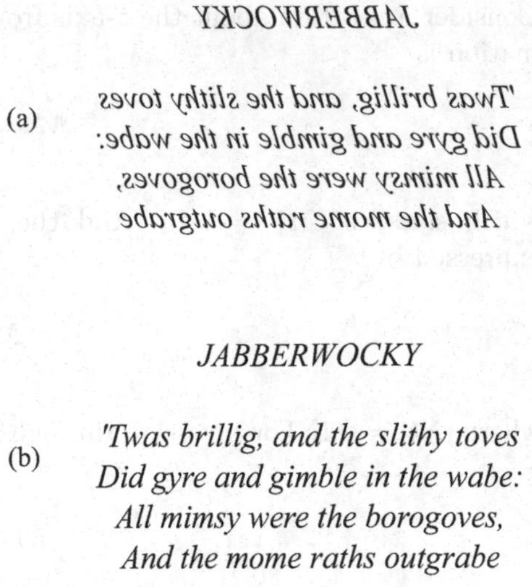

JABBERWOCKY

(a)

'Twas brillig, and the slithy toves
Did gyre and gimble in the wabe:
All mimsy were the borogoves,
And the mome raths outgrabe

JABBERWOCKY

(b)

'Twas brillig, and the slithy toves
Did gyre and gimble in the wabe:
All mimsy were the borogoves,
And the mome raths outgrabe

Figure 1.1.3 The mirror-reflected text of Jabberwocky and the restored text from a second mirror reflection.

(1) The equivalence of a mirror rotation and an inversion rotation

The mirror rotation $\tilde{N}_\alpha$ and the inversion rotation $\bar{N}_\beta$ ($\beta = \alpha - \pi$) are equivalent (see Fig. 1.1.4), i.e., $\tilde{N}_\alpha = \bar{N}_{\alpha-\pi}$.

(2) Kaleidoscope theorem

The line of intersection of two mirror planes m and m' at an angle $\alpha/2$ is equivalent to a rotation axis N_α (see Fig. 1.1.5). This is the theoretical foundation of the toy kaleidoscope. If the space between two mirrors at an angle of 30° is filled with multi-colored fragments of glass, various patterns of hexagonal symmetry are shown after two mirror reflections.

(3) Euler theorem

Successive rotations about two intersecting axes N_{α_1} and N_{α_2} are equivalent to a rotation about a third axis N_{α_3} (see Fig. 1.1.6), i.e., $N_{\alpha_1} + N_{\alpha_2} = N_{\alpha_3}$.

These theorems put severe restrictions on the possible combinations of symmetry elements, making the study of the geometrical theory of symmetry easier than one might otherwise expect.

1.1.3 Symmetry Group

Group theory is a necessary mathematical tool to understand the symmetry of crystals and related problems; here we briefly introduce some basic concepts of this theory. All distinct symmetry operations of a definite figure form the set of the elements of a symmetry group, $\mathcal{G}\{g_1, g_2, \ldots\}$. Mathematically speaking, a set of elements $(g_1, g_2, \ldots)$ forming a group should fulfill following 4 rules, known as group axioms:

(1) Closure rule

The multiplication of group elements, i.e., the successive operation of elements, $g_i \in \mathcal{G}$ and $g_j \in \mathcal{G}$ satisfies $g_i g_j = g_k \in \mathcal{G}$, it means the elements of a symmetry group form a closed set.

(2) Associativity rule

The multiplication satisfies the associative law $g_i(g_j g_l) = (g_i g_j) g_l$.

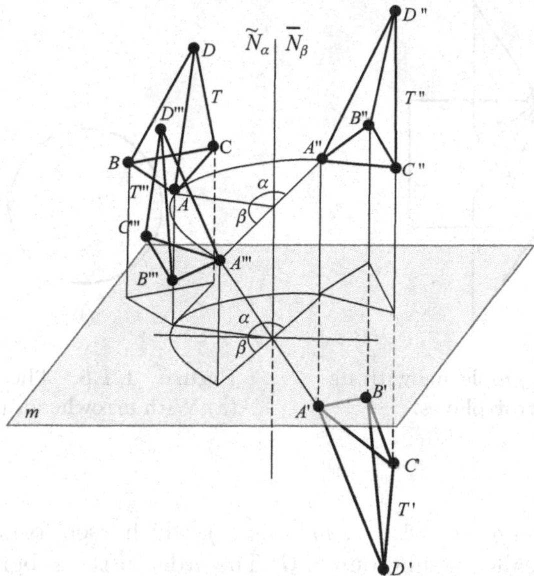

Figure 1.1.4 The demonstration of equivalence of mirror rotation $\tilde{N}_\alpha$ and inversion rotation $\bar{N}_\beta$ $(\beta = \alpha - \pi)$.

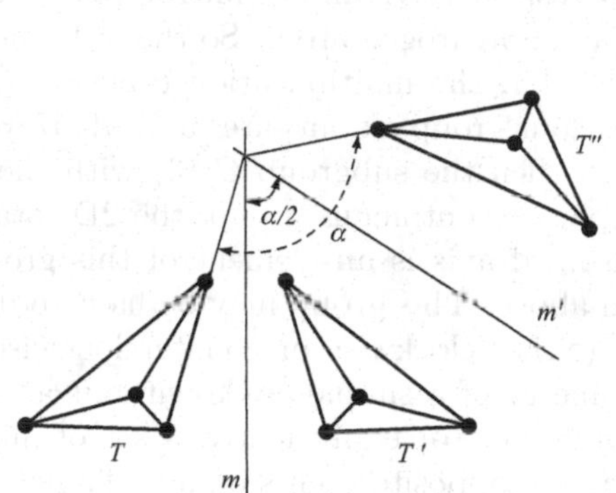

Figure 1.1.5 The demonstration of the kaleidoscope theorem.

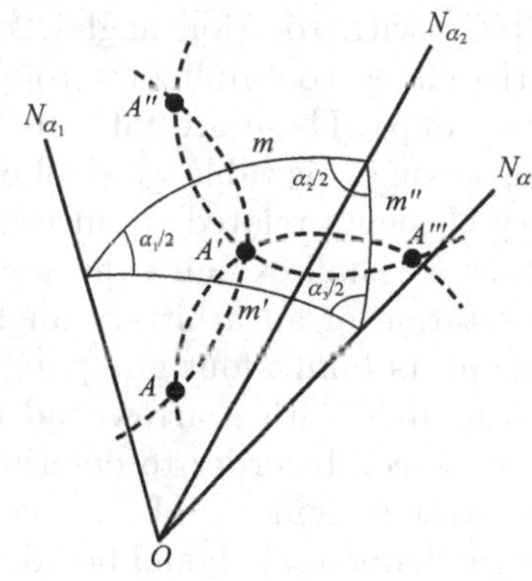

Figure 1.1.6 The demonstration of the Euler theorem.

(3) Identity element

There is an identity element $e \in \mathcal{G}$, such that for any $g_i \in \mathcal{G}$, $eg_i = g_i$.

(4) Inverse element

There is an inverse element g_i^{-1} for any $g_i \in \mathcal{G}$, such that $g_i g_i^{-1} = e$.

From the axioms above, we shall get the uniqueness of the identity element and the inverse element, i.e., $eg_i = g_i e$, $g_i g_i^{-1} = g_i^{-1} g_i$. In general, $g_i g_j \neq g_j g_i$, i.e., group elements are non-commutative; only in the special case of Abelian groups are every pair of elements commutative, i.e., $g_i g_j = g_j g_i$. The number of distinct elements of the group is called the order of the group. If the order of a group is finite, it is a finite or discrete group; if the order of a group is infinite, it is an infinite or continuous group. We may easily verify that symmetry elements of a figure satisfy these group axioms. Since successive rotations along axes pointing to different directions are non-commutative, so symmetry groups are in general non-Abelian; however, successive rotations along the same axis are commutative, i.e., the result does not depend on the sequence of rotations, so some symmetry groups may be Abelian. From the elements g_i $(i = i, \ldots, n)$ of group $\mathcal{G}$, we may

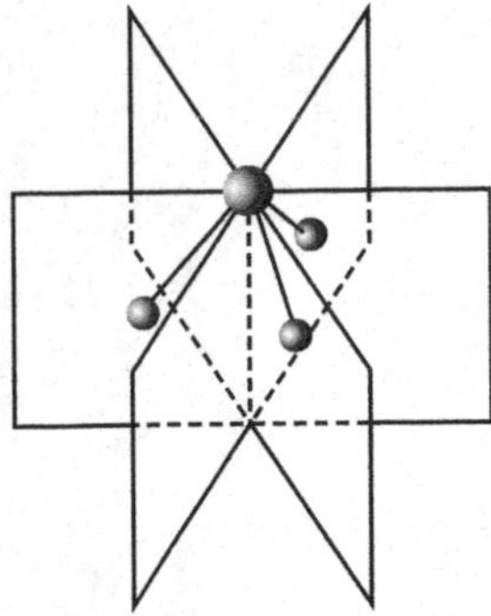

Figure 1.1.7 A NH_3 molecule with its rotation axis and mirror planes.

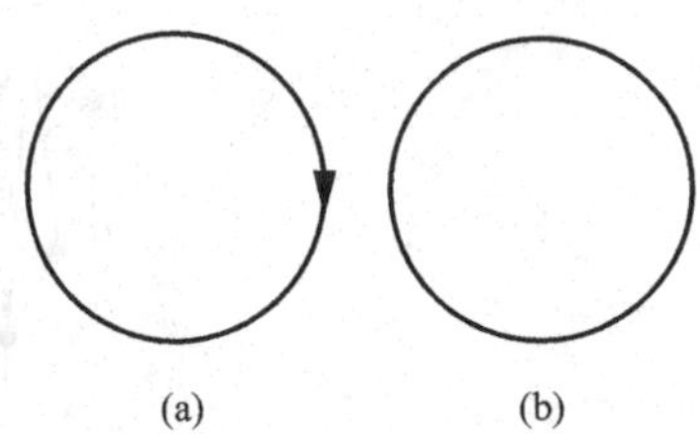

Figure 1.1.8 The symmetry of circles. (a) With arrowhead; (b) without arrowhead.

choose a subset of elements g_k ($k = 1, \ldots, m$, $m < n$), which itself forms a group $\mathcal{G}'$, i.e., it satisfies all group axioms, so it is called a subgroup of $\mathcal{G}$. The order of the subgroup $\mathcal{G}'$ is certainly less than that of $\mathcal{G}$.

Now we shall give two examples of a symmetry group. First consider the NH_3 molecule shown in Fig. 1.1.7. It includes following elements of symmetry: e, the identity element; 3-fold rotation elements C_3^1, C_3^2 with rotation angles $2\pi/3$, $4\pi/3$ respectively; and 3 mirror planes σ_a, σ_b, σ_c situated in the planes containing the rotation axis and a hydrogen atom. So the order of this group is 6, a finite group. These are tabulated in Table 1.1.1, the multiplication table for point group C_{3v}(3m). Inspecting this table we shall clearly seen four group axioms are fulfilled. If we eliminate all symmetry elements related to mirror reflections, then the subgroup C_3(3) with the order 3 is obtained. Now we shall examine the second example: A continuous group, the 2D rotation group $O(2)$. The rotation of an arbitrary angle along a fixed axis is an element of this group. Surely this set of elements fulfills four group axioms listed above. This group may be used to describe the symmetry of a circle with a arrowhead Fig. 1.1.8 (either clockwise or counterclockwise) attached to its circumference. In order to describe the symmetry of a simple circle without an arrowhead, besides the rotation elements of R, a center of symmetry (or equivalently, a set of mirror planes along arbitrary diameters) should be added, making the opposite points on any diameter identical.

Table 1.1.1 The multiplication table for a C_{3v}(3m) group.

$C_{3v}(3m)$	$E(1)$	$C_3^1(3^1)$	$C_3^2(3^2)$	$\sigma_a(m_a)$	$\sigma_b(m_b)$	$\sigma_c(m_c)$
$E(1)$	$E(1)$	$C_3^1(3^1)$	$C_3^2(3^2)$	$\sigma_a(m_a)$	$\sigma_b(m_b)$	$\sigma_c(m_c)$
$C_3^1(3^1)$	$C_3^1(3^1)$	$C_3^2(3^2)$	$E(1)$	$\sigma_c(m_c)$	$\sigma_a(m_a)$	$\sigma_b(m_b)$
$C_3^2(3^2)$	$C_3^2(3^2)$	$E(1)$	$C_3^1(3^1)$	$\sigma_b(m_b)$	$\sigma_c(m_c)$	$\sigma_a(m_a)$
$\sigma_a(m_a)$	$\sigma_a(m_a)$	$\sigma_b(m_b)$	$\sigma_c(m_c)$	$E(1)$	$C_3^1(3^1)$	$C_3^2(3^2)$
$\sigma_b(m_b)$	$\sigma_b(m_b)$	$\sigma_c(m_c)$	$\sigma_a(m_a)$	$C_3^2(3^2)$	$E(1)$	$C_3^1(3^1)$
$\sigma_c(m_c)$	$\sigma_c(m_c)$	$\sigma_a(m_a)$	$\sigma_b(m_b)$	$C_3^1(3^1)$	$C_3^2(3^2)$	$E(1)$

1.1.4 Representations of Symmetry Groups

Each group has a characteristic multiplication table of its elements. These elements may be expressed as numbers, symbols and functions with the same multiplication table. In the group $\mathcal{G}$, the elements of the subgroup $\mathcal{G}'$ can be expressed as e, g_2, $g_3, \ldots$, $g_{n'}$, and the products of all of them and some symmetry element g_x' in $\mathcal{G}$ but not belonging to $\mathcal{G}'$ are called the coset. The one that g'_x multiplied from right side is called the right coset, while from left is called the left coset.

$$\begin{aligned} \mathcal{G}'g'_x &= eg'_x,\ g_2g'_x, \ldots,\ g_{n'}g'_x, \quad \text{(right coset)};\\ g'_x\mathcal{G}' &= g'_xe,\ g'_xg_2, \ldots,\ g'_xg_{n'}, \quad \text{(left coset)}. \end{aligned} \tag{1.1.13}$$

According to the concept of the coset, we shall get the group table which is called the exact (or isomorphic) representation of the original group $\mathcal{G}$. Usually we take the square matrix

$$\boldsymbol{M}(\mathcal{G}) = \begin{pmatrix} a_{11} & a_{12} & \cdots & a_{1n} \\ a_{21} & a_{22} & \cdots & a_{2n} \\ \vdots & \vdots & \ddots & \vdots \\ a_{n1} & a_{n2} & \cdots & a_{nn} \end{pmatrix} = (a_{ij}) \tag{1.1.14}$$

to represent the symmetry operation of some vector, while the vector is the basis of the representation. So the symmetry operation as introduced in §1.1.1 can be expressed by these operation matrices as the representation of a group $\mathcal{G}$. Here the a_{ij} may be either a real number or a complex one. So we may use multiplication of matrices to represent the multiplication of symmetry elements of $\mathcal{G}$. It may be easily seen that these representations fulfill 4 group axioms introduced above, with the identity matrix $a_{ij} = 1\ (i = j)$, $a_{ij} = 0\ (i \neq j)$. Any square matrix may be expressed in the form

$$\begin{pmatrix} \boldsymbol{M}_1(A) & \boldsymbol{M}_2(B) \\ \boldsymbol{M}_1(B) & \boldsymbol{M}_2(A) \end{pmatrix}, \tag{1.1.15}$$

where $\boldsymbol{M}_1(A)$, $\boldsymbol{M}_1(B), \ldots$ are submatrices. If we can find a transformation which makes $\boldsymbol{M}_1(B) = \boldsymbol{M}_2(B) = 0$, then the original matrix can be reduced to matrices with lower rank, and it is called a reducible representation; on the other hand, if it is impossible to find a transformation which makes $\boldsymbol{M}_1(B) = \boldsymbol{M}_2(B) = 0$, then the original matrix is called an irreducible representation. Generally, we should reduce the group to its simplest form, i.e., the irreducible representation.

These rather abstract reasonings may be illustrated by some concrete examples: Take the group C_{3v}(3m) as an example of a finite group; take the 3D matrix introduced in §1.1.1 as 3D representations, and $\boldsymbol{r}(x, y, z)$ as a basic vector. Select certain basis functions to simplify representations into irreducible ones. Here we select z, (x, y) and R_z as basis functions, R_z indicates right-handed (or left-handed) screw along the z axis. We obtain two 1D representations for z and R_z, one 2D representation for (x, y) and they are all irreducible; they are tabulated in Table 1.1.2.

Table 1.1.2 The irreducible representation of $C_{3v}(3m)$.

$C_{3v}(3m)$	$E(1)$	$C_3^1(3^1)$	$C_3^2(3^2)$	$\sigma_a(m_a)$	$\sigma_b(m_b)$	$\sigma_c(m_c)$	Base
Γ_1	(1)	(1)	(1)	(1)	(1)	(1)	z
Γ_2	(1)	(1)	(1)	(−1)	(−1)	(−1)	R_z
Γ_3	$\begin{pmatrix} 1 & 0 \\ 0 & 1 \end{pmatrix}$	$\begin{pmatrix} -\frac{1}{2} & -\frac{\sqrt{3}}{2} \\ \frac{\sqrt{3}}{2} & \frac{1}{2} \end{pmatrix}$	$\begin{pmatrix} -\frac{1}{2} & \frac{\sqrt{3}}{2} \\ -\frac{\sqrt{3}}{2} & \frac{1}{2} \end{pmatrix}$	$\begin{pmatrix} 1 & 0 \\ 0 & -1 \end{pmatrix}$	$\begin{pmatrix} -\frac{1}{2} & \frac{\sqrt{3}}{2} \\ \frac{\sqrt{3}}{2} & \frac{1}{2} \end{pmatrix}$	$\begin{pmatrix} -\frac{1}{2} & -\frac{\sqrt{3}}{2} \\ -\frac{\sqrt{3}}{2} & \frac{1}{2} \end{pmatrix}$	(x, y)

Now take group $R(2)$ as an example of a continuous group. The matrix for the rotation can be changed from 3D to 2D, so the original group may be replaced by its representations $O(2)$ group, i.e., the 2D group of orthogonal transformations with determinants equal to +1. It may be further reduced to the 1D unitary group $U(1)$, which has matrix $\boldsymbol{M} = \exp(i\theta) = \cos\theta + i\sin\theta$. A matrix is unitary if its Hermitian conjugate, $\boldsymbol{M}^\dagger$, equals its inverse, i.e., $\boldsymbol{M}^\dagger = \boldsymbol{M}^{-1}$. If $\boldsymbol{M}$ is a matrix in 1D (i.e., with only one row and one column), then it is just a single number. This number must be a complex number of the form $\exp(i\theta)$ and its complex conjugate is $\exp(-i\theta)$. It should be noted that the irreducible representation plays an important part in dealing with the electronic structure of molecules and crystals using symmetry groups.

§1.2 Finite Structures and Point Groups

Point groups describe the symmetry of finite figures, excluding translation as a symmetry element. So all symmetry elements of a point group pass through a definite point, and the symmetries of most molecules belong to this kind.

1.2.1 Combination Rules for Symmetry Axes

We now discuss the combination rules for symmetry axes along different directions. If there are two axes N_1 and N_2, then a third axis N_3 is expected according to the Euler theorem. Their emergent points on a sphere will form a spherical triangle, and the entire sphere will be divided into such triangles (see Fig. 1.2.1). The angles at the vertices of these triangles are denoted by α_i equal to one half of the rotation angle of the corresponding axis, i.e., $2\pi/2N_i$. The sum of the inner angles of a spherical triangle must exceed π, therefore

$$\frac{1}{N_1}+\frac{1}{N_2}+\frac{1}{N_3}>1\,. \tag{1.2.1}$$

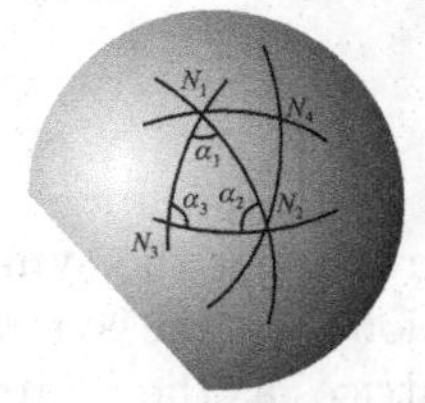

Figure 1.2.1 The combination of the symmetry axes at different directions (emergences of symmetry axes at the sphere surface form a spherical triangle).

I	II	IIIa	IIIb	IV	Va	Vb	VI	VIIa	VIIb
N-C_n	$N/2$-D_n	N-$S_n(C_{ni})$	N/m-C_{nh}	N_m-C_{nv}	Nm-D_{nd}	N/mm-D_{nh}	$N_1 N_2$	$\bar{N}_1 N_2$	$\bar{N}_1 N_2$
1 C_1	2 D_1	$\bar{1}$ C_i		m C_{1v}		$mm2$ C_{1h}			
2 C_2	222 D_2	m C_{2i}	$2/m$ C_{2h}	$mm2$ C_{2v}	$2/m$ C_{2h}	mmm D_{2h}			
3 C_3	32 D_3	$\bar{3}$ C_{3i}	$\bar{6}$ C_{3h}	$3m$ C_{3v}		$\bar{6}m2$ D_{3h}	23 T	$m\bar{3}$ T_h	$\bar{4}3m$ T_d
4 C_4	422 D_4	$\bar{4}$ S_4	$4/m$ C_{4h}	$4mm$ C_{4v}	$\bar{4}2m$ D_{2d}	$4/mmm$ D_{4h}	432 O	$m\bar{3}m$ O_h	
5 C_5	52 D_5	$\bar{5}$ C_{5i}	$\bar{10}$ C_{5h}	$5m$ C_{5v}	$\bar{5}m$ D_{5d}	$\bar{10}m2$ D_{5h}	235 Y	$m\bar{3}\bar{5}$ Y_h	
6 C_6	622 D_6		$6/m$ C_{6h}	$6mm$ C_{6v}	$\bar{3}m$ D_{3d}	$6/mmm$ D_{6h}			
↓	↓	↓		↓	↓			↓	
∞	$\infty 2$	∞/m		∞mm	∞/mm		$\infty\infty$	$\infty\infty m$	

Figure 1.2.2 The schematic diagram of point groups (showing the special arrangement of symmetry elements by stereograms).

From this inequality we may enumerate all the possible cases of combinations of rotation axes: (1) only one axis of high order $N_3 = N \geqslant 3$ then $N_1 = N_2 = 2$, and all the combinations $[N22]$ are possible; where N may be any integer; (2) two or more axes of high order, the only possibilities for $[N_1, N_2, N_3]$ are [233], [234], and [235]. This seemingly formidable problem of crystallography (including quasicrystallography) is resolved at one stroke with a little geometry. N-fold axes with $N = 5$ and $N \geqslant 6$, though excluded from ordinary crystallography, are basic ingredients of quasicrystallography. Thus, the consequences of these combination rules can help us understand why 3D quasicrystals only have icosahedral symmetry [235], and there is no restriction on the order of 2D quasicrystals from the theory of symmetry.

1.2.2 Cyclic and Dihedral Groups

We have already mastered enough knowledge to derive all kinds of point groups. We have shown that point symmetry elements may be divided into two types, type I contains only proper rotations (including the identity), while type II contains improper rotations as well as mirror reflections and inversion. Now we are only concerned with point groups with only one N-fold rotation axis. We begin with simple cyclic groups with N-fold axis without any type II symmetry elements, in International notation, denoted by N; in Schönflies notation, denoted by C_n. Next we should consider point groups related to the combinations of axes $[N22]$, i.e., dihedral groups. Further, the addition of type II symmetry elements may lead to more complex groups: groups (S_n) with a simple improper rotation axis $\bar{N}$; groups with an additional mirror plane perpendicular to the proper rotation axis, i.e., N/m or C_{nh}; groups with additional mirror planes bisecting two intersecting axes Nm and $\bar{N}m$, or C_{nv}; similar operations may be acting on dihedral groups, making $D_n \to D_{nh}$ and $D_n \to D_{nd}$. These results are all tabulated in Fig. 1.2.2. Groups with more than one high order rotational axes will be discussed in the next subsection.

1.2.3 Platonic Solids and Cubic Groups

To discuss point groups with more than one high order rotational axis, we shall review the problems of the solid geometry of the regular polyhedra. There are five regular polyhedra called Platonic solids, from the era of Plato. A regular polyhedron, since all its faces are identical, is characterized by a polygon with p faces; all vertices are also identical, each is connected to q polygons. So we may use the Schläffli notation $\{p, q\}$ to denote a regular polyhedron. Thus five Platonic solids are (see Fig. 1.2.3) tetrahedron (4 regular equilateral triangular faces), $\{3, 3\}$; cube (6 square faces), $\{4, 3\}$; octahedron (8 equilateral triangular faces), $\{3, 4\}$; dodecahedron (12 regular pentagon faces), $\{5, 3\}$; icosahedron (20 regular triangular faces), $\{3, 5\}$. It should be noted that some pairs of different regular polyhedra (for example octahedron and cube, icosahedron and dodecahedron) belong to the same symmetry group. These pairs are related by dualities, which means the exchange of the numbers of faces and vertices, i.e., the exchange of p, q values; the exchange of $\{3, 4\}$ and

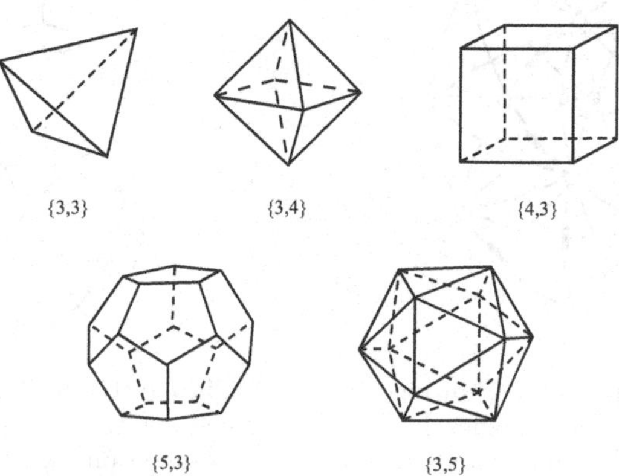

Figure 1.2.3 Five kinds of regular polyhedron.

$\{4,3\}$ or $\{3,5\}$ and $\{5,3\}$. Theatatus, a mathematician of ancient Greece, proved that there are only 5 regular polyhedra in existence. The proof may be stated simply: A regular polygon can be regarded as p vectors joined in a sequence but having the sum 0, and the angle between the two joined vectors is the external angle of a polygon, so the sum of all the angles equals to 2π. Every external angle is $2\pi/p$, while the internal angle of a polygon is equal to $(1-2/p)\pi$. Now we consider at each vertices of a polyhedron, the sum of internal angles of polygons connected to it must be less than 2π, since the polyhedron is convex, so an inequality is obtained

$$\frac{1}{p}+\frac{1}{q}>\frac{1}{2}, \tag{1.2.2}$$

or

$$(p-2)(q-2)<4. \tag{1.2.3}$$

It can be easily found that p and q have only combinations of {3, 3}, {3, 4} and {3, 5}, which correspond to {3, 3}, {3, 4}, {4, 3}, {3, 5}, and {5, 3} of the Platonic solids. This purely geometrical conclusion obtained 2000 years ago, was a prelude to the Euler theorem about the combination of high order rotational axes of the modern theory of symmetry.

To enumerate the symmetry groups of Platonic solids, we shall begin with their rotational subgroups $T(23)$, $O(432)$ and $Y(235)$. For a tetrahedron, there are 4 three-fold rotation axes passing through the center to vertices, and 3 two-fold rotation axes passing through the center to the centers of the edges; for a regular octahedron, there are 3 four-fold rotation axes passing through the center to the vertices, 4 three-fold rotation axes passing through the centers of the triangular faces, and 6 two-fold rotation axes passing through the center to the centers of the edges; for a regular icosahedron, there are 6 five-fold axes passing through the center to the vertices, 10 three-fold axes passing through the center to the centers of triangular faces, and 15 two-fold axes passing through the center to the centers of the edges. These rotational subgroups are all composed of the I kind of symmetry element. We shall introduce the II kind of symmetry element, reflection, to obtain higher order symmetry groups. For example from the T-group, 3 mirror planes perpendicular to the 2-fold axes may be added, thus $T(23)\to T_h(\frac{2}{m}\bar{3})$, and mirror planes bisecting the 2-fold axes normal to each other may be added, thus $T(23)\to T_d(\bar{4}3\frac{2}{m})$; for the octahedral group $O(432)$, mirror planes perpendicular to 3 4-fold axes and 6 2-fold ones may be added, thus $O(432)\to O_h(\frac{4}{m}\bar{3}\frac{2}{m})$; for $Y(235)$, mirror planes perpendicular to 15 2-fold axes may be added, making $Y(235)\to Y_h(\frac{2}{m}\bar{3}\bar{5})$. All vertices of a regular polyhedron are situated on the same sphere, so the angles extended by different axes (or faces) may be determined from the stereogram (see Fig. 1.2.4).

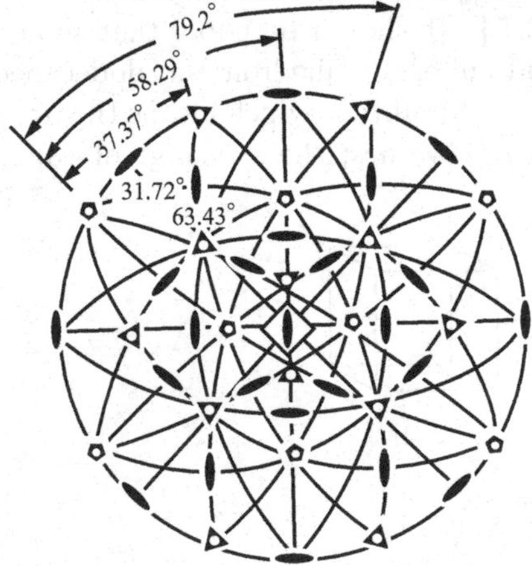

Figure 1.2.4 The stereogram showing the relation of the axial (face) angles between the cube and icosahedron symmetry groups.

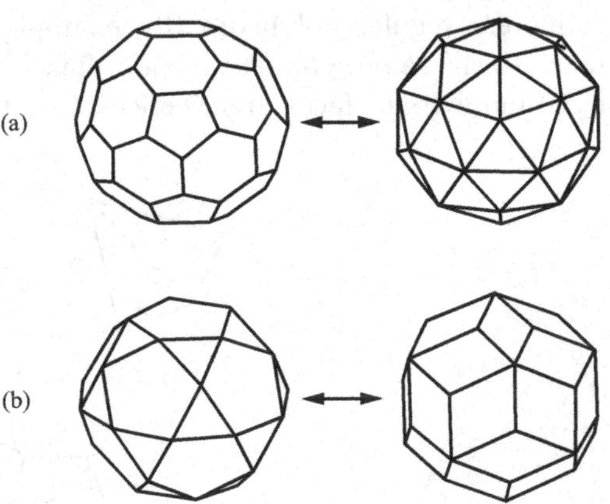

Figure 1.2.5 Two kinds of semi-regular polyhedra with icosahedral symmetry. (a) Truncated icosahedron and its dual, a polyhedron with 60 faces; (b) a dodeca-icosahedron and its dual, the rhombic triacontahedron.

Now we would like to explain the notation for point groups. Actually there are two systems in use: One is Schönflies notation, the other is the International notation.[a] In Schönflies notation, cyclic groups are denoted as C, dihedral groups are denoted as D, while tetrahedral, octahedral and icosahedral groups are denoted by T, O, and Y respectively. These may be further differentiated by the indices indicating the main order of rotation and the manner of adding mirror planes, such as D_{nd}, T_h, T_d, O_h and Y_h. In the International notation system, the order for the numerals are very important, for it gives a rough indication of the spatial arrangement of symmetry elements. If there is only one high order rotation axis, the first place numeral is reserved for this principal axis, while the other places are for two sets of axes perpendicular to it, e.g., 32, $\frac{4}{m}\frac{2}{m}\frac{2}{m}$; for the case of several high order axes, the first place is reserved for symmetry axes along cubic axes, while the second place is reserved for those symmetry axes along the body diagonals of the cube, while the third place is reserved for symmetry axes along the the other directions, e.g., 23, 432, $\frac{4}{m}\bar{3}\frac{2}{m}$, $\frac{2}{m}\bar{3}\bar{5}$. It should be noted that the appearance of 3 or $\bar{3}$ in the second place is the criterion for cubic or icosahedral symmetry. International notation may be written in abbreviated form, e.g., $\frac{2}{m} \rightarrow m$, or $\frac{4}{m} \rightarrow m$ in cubic symmetry, then $\frac{6}{m}\frac{2}{m}\frac{2}{m} \rightarrow \frac{6}{m}mm$, $\frac{4}{m}\bar{3}\frac{2}{m} \rightarrow m\bar{3}m$. The abbreviated notation is used in Fig. 1.2.2.

We may extend the notion of regular polyhedron to that of semi-regular polyhedron. Starting from regular polyhedra, through truncation, we may obtain 13 semi-regular polyhedra with identical vertices, but with faces which composed of two different types of polygons. These are called Archimedean solids (or polyhedra). For instance, starting from an icosahedron, by cutting out all vertices along the points at 1/3 of the edges, we obtain a truncated icosahedron with 32 faces (20 are regular hexagons, 12 are regular pentagons) and 60 vertices. Making the duality operation on this equi-vertex semi-regular polyhedron, i.e., exchanging vertices into faces and *vice versa*, a dual equi-face polyhedron is obtained. In the case of the truncated icosahedron, its dual is a polyhedron with 60 faces (each is an identical isosceles triangle); while the dual for the dodeca-icosahedron is a rhombic triacontahedron (in which each face is a rhombus with acute angle equal to 63.43°, obtuse angle equal to 116.57°) (see Fig. 1.2.5). Some of these polyhedra preserve icosahedral symmetry, though their faces are not regular polygons.

Icosahedral symmetry is important for the structure of molecules. The CH groups in $C_{20}H_{20}$ molecule are situated on the 20 vertices of the dodecahedron, while a cluster of 20 H_2O molecules has the same structure. In 1985, Kroto, Smalley and Curl studied the mass spectra of laser-evaporized carbon clusters, and detected a high peak of abundance at C_{60}, indicating its extreme stability. They conjectured that the existence of a cage-shaped carbon molecule with 60 atoms all occupying the positions of the vertices of a truncated icosahedron, and ultimately obtained the Nobel prize for chemistry of 1996. C_{60} is called buckminsterfullerene, later abbreviated to fullerene, in commemoration of the architect Buckminster Fuller who designed geodesic domes with a mixture of regular hexagons and pentagons. On the other side, in 1984 Shechtman *et al.* obtained the electron diffraction pattern of rapidly-quenched Al-Mn alloy showing icosahedral symmetry, discovered the quasicrystal with quasiperiodic structure and so enlarged the range of crystallography to include quasicrystallography. In the study of quasicrystals, icosahedral symmetry plays an important role, as will be shown in §2.3.

§1.3 Periodic Structures and Space Groups

The crystal, in which atoms are arranged periodically, is the major object for study in solid state physics, and it still maintains an important place in condensed matter physics.

1.3.1 Periodic Structure and Lattice

We start from any periodic structure, e.g., certain wallpaper (see Fig. 1.3.1), to deduce a lattice from it as the idealization of its periodicity. First we pick any point O (or O') of the structure as the

[a] At present both of the two notations are used. The Schönflies notation is used in electronic structure calculations (both the band theory of solids and quantum chemistry); while the International notation is used in crystallography and structure analysis. So, readers studying condensed matter physics should be familiar with both notations.

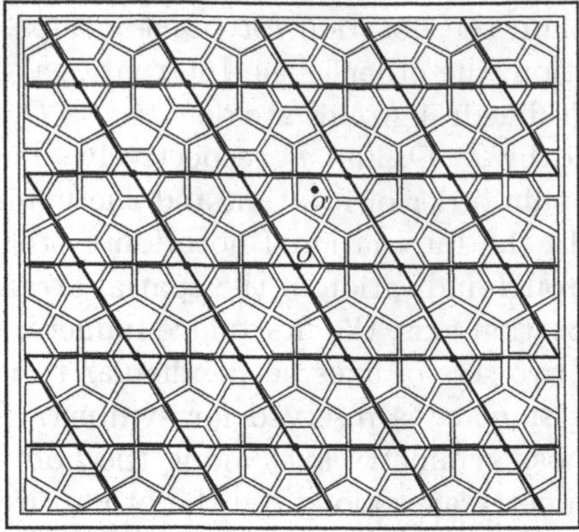

Figure 1.3.1 A 2D periodical figure and its lattice.

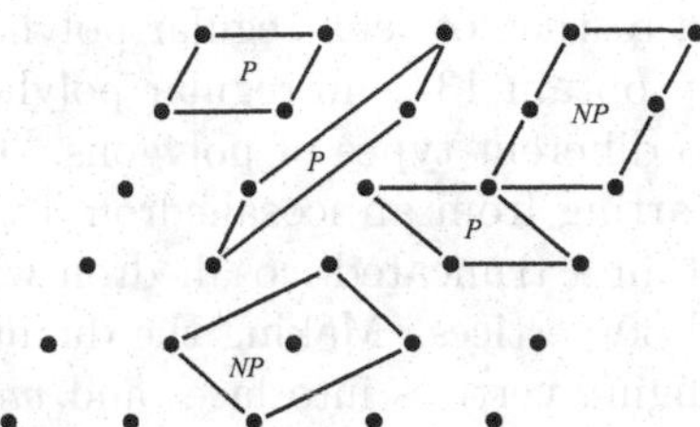

Figure 1.3.2 The different choices of unit cells in 2D lattice (P = primitive, NP = nonprimitive).

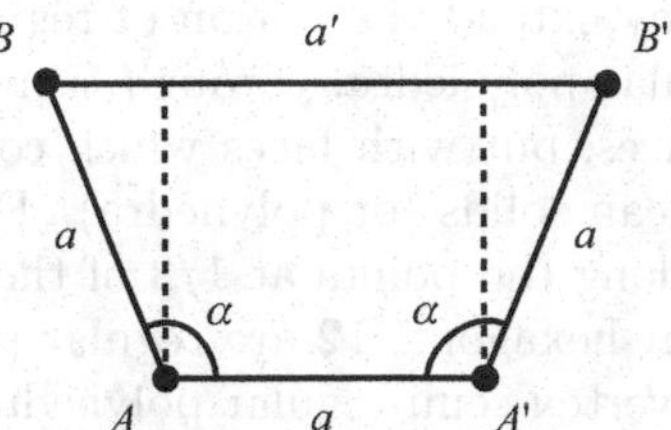

Figure 1.3.3 The compatibility of the N-fold rotation with periodicity.

origin, then we may find a series of points in the structure with the same environment as O. The full set of these points constitutes a lattice. We shall get the same figure by translation along any lattice vector. From the point of view of symmetry, the lattice stands for the translation symmetry of a figure; however the figure of translation symmetry cannot have any boundary and must extend to infinity. For 3D periodic structures, the lattice vector can be expressed as

$$\boldsymbol{l} = u\boldsymbol{a} + v\boldsymbol{b} + w\boldsymbol{c}, \tag{1.3.1}$$

where $\boldsymbol{a}$, $\boldsymbol{b}$, $\boldsymbol{c}$ are basic vectors, while u, v, w are integers, i.e., $0, \pm1, \pm2, \ldots, \pm\infty$.

It should be noted that for a given lattice the origin, the basic vectors and the unit cells are not unique. The primitive unit cell is defined as a unit cell that contains only one lattice point, while nonprimitive ones contain more than one lattice point (Fig. 1.3.2). Usually the chosen unit cell is highly symmetrical, but not always primitive. It should be also noted that the lattice is not equivalent to the original periodic structure. Something (the figure surrounding each lattice point or the content of the unit cell) is lost in the process of idealization. Only when the lost information (known as the decoration or basis) is added to the lattice, can we fully recover the original crystal structure, i.e.,

$$\text{Crystal Structure} = \text{Lattice} + \text{Decoration}.$$

In fact, as can be easily proven, the rotational symmetry is completely restricted by the periodicity. First we choose a lattice point A in the lattice (see Fig. 1.3.3), and let a N-order rotation axis, which is perpendicular to any lattice plane, pass through it. Then choose another lattice point A' next to point A, and make a vector $\overrightarrow{AA'}$ from A to A'. This vector has the minimum modulus of any other lattice vectors parallel to it, we then make an N-fold rotational operation on $\overrightarrow{AA'}$ to bring A' to B; then make the same operation along an axis passing through A' to $\overrightarrow{A'A}$, bring A' to B' (α is the rotational angle). The modulus of vector $\overrightarrow{BB'}$ should be an integer multiple of that of

$\overrightarrow{AA'}$, i.e.,

$$a + 2a\sin(\alpha - \frac{\pi}{2}) = a - 2a\cos\alpha = pa, \tag{1.3.2}$$

then

$$\cos\alpha = \frac{(1-p)}{2}. \tag{1.3.3}$$

Satisfying $-1 \leqslant \cos\alpha \leqslant 1$, we get $p = 3, 2, 1, 0$; and $\alpha = 2\pi/N$, where $N = 2, 3, 4, 6$. It is clear that only rotation axes 1, 2, 3, 4, 6 (including true and improper ones) are compatible with periodic structure. In other words, the point groups of crystal can only contain the compatible rotation axes derived above. Crystalline point groups only include the rotation axes compatible with periodicity, so there are 10 2D point groups, 32 3D point groups. Icosahedral point groups and other group containing 5-fold or more than 6-fold rotation axes are incompatible with the periodic structure, so these are excluded in traditional crystallography; they belong in quasicrystallography.

1.3.2 Bravais Lattices

Traditionally all the lattices in crystallography are called Bravais lattices. In 2D (Fig. 1.3.6), these are: The oblique, rectangular, centered rectangular (rhombic), square, and hexagonal (triangular). According to the different types of symmetry, it is easy to demonstrate that there are only five 2D Bravais lattices (translation groups), but for 3D, there are fourteen Bravais lattices. It should be noted that, for periodic structures, besides operations of point symmetry, other symmetry operations involving translations should be considered, they include: Pure translations, glide reflections and screw axes (rotations and inversion-rotations). Pure translations can be easily understood, while the other two, equivalent to a point symmetry operation plus some fractional lattice translation, are more complex. Glide reflections are usually related to translations of $\boldsymbol{t}/2$, or $\boldsymbol{t}/4$, where $\boldsymbol{t}$ is the lattice vector along the glide plane; while N-fold screw rotation is related to translation vectors $\boldsymbol{t}' = p\boldsymbol{l}/N$, where p is an integer number smaller than N, and $\boldsymbol{l}$ is the translation vector along the rotation axis (see Figs. 1.3.4 and 1.3.5).

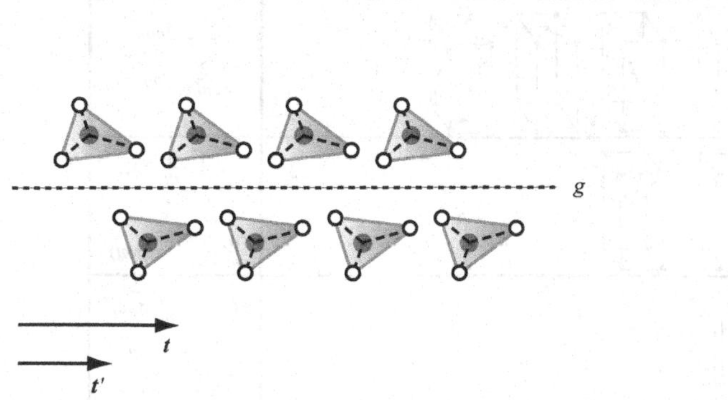

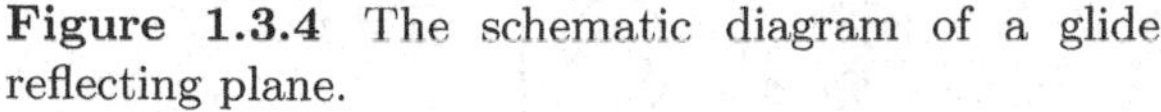

Figure 1.3.4 The schematic diagram of a glide reflecting plane.

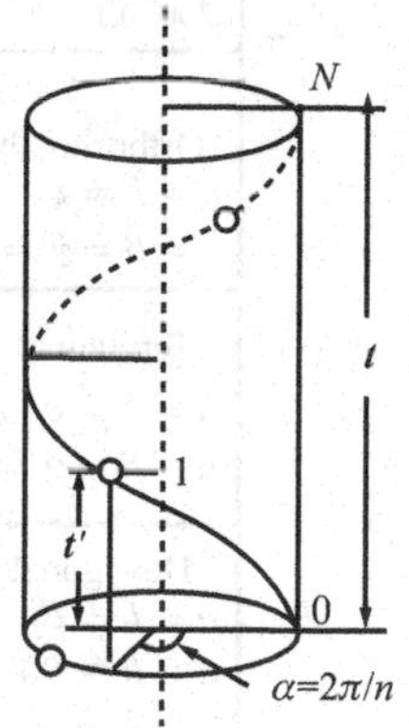

Figure 1.3.5 The schematic diagram of a screw axis.

Centered lattices are associated with added symmetry operations, such as glide reflections, screw axis operations, etc. In 2D space, the additional operation of the centered lattices is obvious: When compared with the noncentered rectangular lattice, the 2D centered one adds a set of parallel glide reflection lines (a plane changes into a line in the case of 2D), and the translation vector $\boldsymbol{l}$ equals $\boldsymbol{a}/2$, $\boldsymbol{b}/2$, so its symmetry is different from the noncentered lattice (see Fig. 1.3.6). Although in 3D space the symmetry of lattice is more complex than that of 2D space, the situation is quite similar. 14 Bravais lattices are divided into the 7 crystal systems: Triclinic, monoclinic, orthorhombic, tetragonal, rhombohedral, hexagonal and cubic. The rhombohedral point group can be considered as a special type of centered hexagonal lattice. If we choose this viewpoint, the 7 crystal systems are reduced to 6 (see Fig. 1.3.7). It should be noted that the classification of the Bravais lattices is

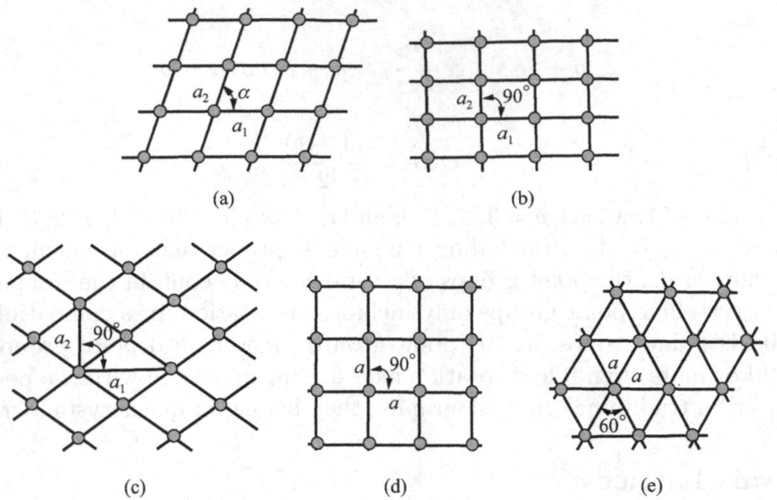

Figure 1.3.6 The 2D Bravais lattices: (a) Oblique; (b) rectangular; (c) rhombic; (d) square; (e) hexagonal.

Crystal System	Type of Lattices P	I	C	F	R	Related Point Group
Triclinic $a \neq b \neq c$ $\alpha \neq \beta \neq \gamma \neq 90°$	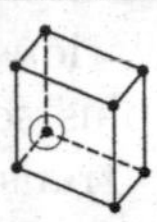					$1, \bar{1}$
Monoclinic $a \neq b \neq c$ $\alpha = \gamma = 90°$ $\beta \neq 90°$	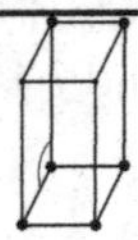		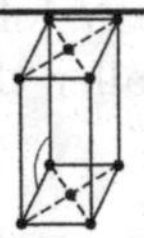			$2, m, \frac{2}{m}$
Orthorhombic $a \neq b \neq c$ $\alpha = \beta = \gamma = 90°$	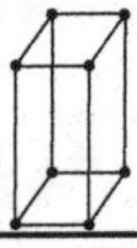	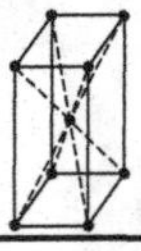	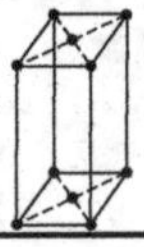	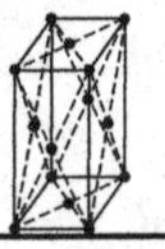		$222, 2mm,$ $\frac{2}{m}\frac{2}{m}\frac{2}{m}(mmm)$
Tetragonal $a = b \neq c$ $\alpha = \beta = \gamma = 90°$	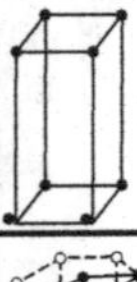	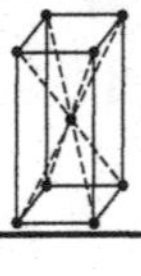				$4, \bar{4}, \frac{4}{m}\ 422,$ $4mm, \bar{4}2m,$ $\frac{4}{m}\frac{2}{m}\frac{2}{m}(4/mmm)$
Hexagonal $a = b \neq c$ $\alpha = \beta = 90°$ $\gamma = 120°$	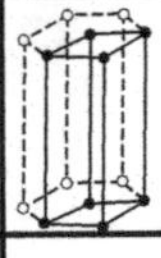					$6, \frac{3}{m}, \frac{6}{m}, 622,$ $6mm, \bar{6}2m,$ $\frac{6}{m}\frac{2}{m}\frac{2}{m}(6/mmm)$
Rhombohedral $a = b = c$ $\alpha = \beta = \gamma \neq 90°$					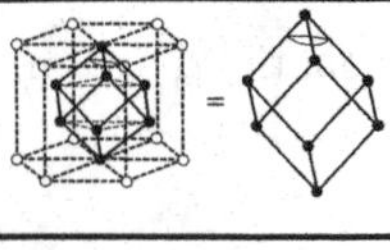 =	$3, \bar{3}, 32, 3m,$ $\bar{3}\frac{2}{m}\ (\bar{3}m)$
Cubic $a = b = c$ $\alpha = \beta = \gamma = 90°$	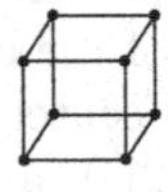	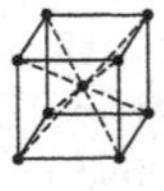		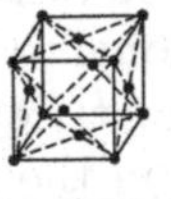		$23, \frac{2}{m}\bar{3}(\bar{3}m),$ $432, \bar{4}3m,$ $\frac{4}{m}\bar{3}\frac{2}{m}(m\bar{3}m)$

P (Primitive) , I (Body Center) , C (Bottom Center), F (Face Center), R (Rhombohedron)

Figure 1.3.7 14 kinds of Bravais lattices.

carried out by the differences in symmetry, so the types of the lattices coincide with the translation groups, i.e., the groups of pure translations.

1.3.3 Space Groups

In the early 1890s, Fedorov and Schönflies independently enumerated 230 space groups as a crowning achievement of classical crystallography. Fedorov's approach was more geometrical and intuitive, while Schönflies' approach was more mathematical and exact. However it is too tedious to repeat all these arguments leading to the final result. Here we shall take a far simpler case, 2D space groups or plane groups for a brief discussion.

As we know, there are 5 Bravais lattices. We may generate 13 symmorphic groups by combining each Bravais lattice with compatible point groups. The only nonsymmorphic operation permitted in 2D is the glide with reflection across a line in the ***ab*** plane, and a further 4 nonsymmorphic groups are generated, making a total of 17 in all. These may be used to classify surface structures (see Fig. 1.3.8).

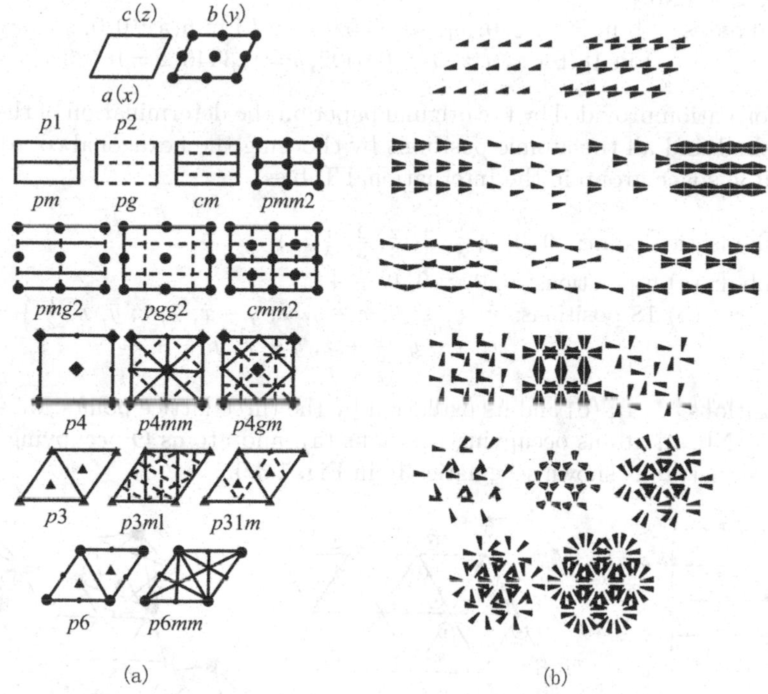

Figure 1.3.8 17 kinds of 2D space groups. (a) The arrangement of symmetry elements; (b) the typical patterns for these group.

In the case of 3D, the situation is the same, although more complicated. The nonsymmorphic symmetry operations include glide-reflections and screw-rotations, both involving suitable fractions of lattice translations. 73 symmorphic groups may be generated by combining Bravais lattices with compatible point groups, while 157 nonsymmorphic groups may be generated involving the nonsymmorphic symmetry operations. In fact, the nonsymmorphic symmetry operations play a more important role in 3D space, which makes the theory of space groups quite complex. Among space groups, there are 11 pairs of enantiomorphic space groups, i.e., one of them can be obtained by a mirror image of another, like left and right hands. For instance, $P6_222$ is the space group for right hand quartz, and while $P6_422$ is that for left hand quartz. Determination of the space group is a primary step in a full structure determination. Centered lattices and nonsymmorphic symmetry elements may be recognized by systematic extinctions in X-ray diffraction patterns. Whether an inversion center exists or not should be determined by supplementary physical tests, for instance, second harmonic generation of laser light, piezoelectricity or the anomalous dispersion of X-ray diffraction. The last method is based on the breakdown of the Friedel rule for the X-ray diffraction intensity of a pair of reflections, i.e., $I(h,k,l) \neq I(\bar{h},\bar{k},\bar{l})$.

1.3.4 The Description of Crystal Structure

In the study of specific substances, scientists usually need knowledge of the crystal structure. A complete description of the crystal structure includes lattice parameters and the type and coordinates of each atom in the unit cell. Often, familiar crystal structures can be found in reference books, but usually the data given in the literature are in shorthand form (such as the Wyckoff symbol). The space group table in the "International Tables for Crystallography" should be consulted for the full elucidation of the structure. Here we give an example of the use of the International Tables for Crystallography, Vol. I for $LiNbO_3$.

Chemical formula: $LiNbO_3$;
Lattice: Rhombohedrally-centered hexagonal with $a = 5.448$ Å, $c = 13.863$ Å;
Number of chemical formula in a unit cell: 6;
Space group: R3c (No.161);
Atomic coordinates: Nb in 6(a): $0, 0, w_1$ ($w_1 = 0.0186$); Li in 6(a): $0, 0, \frac{1}{3} - w_2$ ($w_2 = -0.0318$);
O in 18(b): x, y, z ($x = 0.0492, y = 0.3446, z = 0.0833$).

With the information provided by the original paper on the determination of the crystal structure of $LiNbO_3$, we shall find all the atomic positions by choosing the hexagonal coordinate on the page containing the R3c space group in the International Tables.

Position of the lattice point: $(0,\ 0,\ 0;\ \frac{1}{3},\ \frac{2}{3},\ \frac{2}{3};\ \frac{2}{3},\ \frac{1}{3},\ \frac{1}{3})$
Wyckoff symbol: (a) 6 positions: $0,\ 0,\ z;\ 0,\ 0,\ \frac{1}{2} + z;$
(b) 18 positions: $x,\ y,\ z;\ \bar{y},\ x-y,\ z;\ y-x,\ \bar{x},\ z;\ \bar{y},\ \bar{x},\ \frac{1}{2}+z;$
$x,\ x-y,\ \frac{1}{2}+z;\ y-x,\ y,\ \frac{1}{2}+z.$

From the positions of (a), (b) and its extention by the three lattice points, we shall derive all the atomic positions: Nb, Li atoms occupying positions (a), and atoms O occupying positions (b), the resulted crystal structure is shown schematically in Fig. 1.3.9.

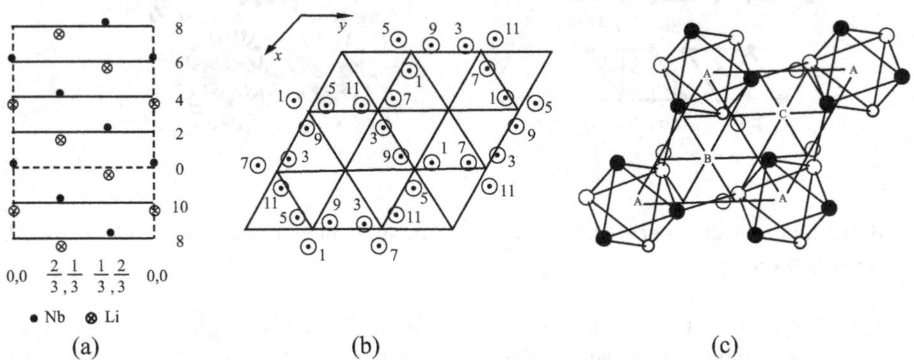

Figure 1.3.9 The structure of $LiNbO_3$. (a) Nb, Li atoms are placed on $(11\bar{2}0)$ cross section; (b) the projection of O atoms on (0001) plane, and the numbers are given in unit of $c/12$; (c) oxygen octahedra near the plane of $z = 0$, with the filled circles $z = c/12$ and the open circles $z = -c/12$.

§1.4 Structures and Their Fourier Transforms

1.4.1 The General Case

Structures may be determined by scattering, diffraction and imaging of waves such as X-rays, electrons, neutrons and visible light. The diffraction pattern, i.e., the distribution of scattering intensity of waves, is related to the distribution of scattering matter by way of its Fourier transform. For any well-behaved functions $f(x)$ we may define its Fourier transform by the following mathematical operation,

$$\varphi(k) = \int f(x)\exp(ikx)dx\,. \tag{1.4.1}$$

There is also an inverse Fourier transform to convert the Fourier transform back into the original function,

$$\varphi^{-1}[\varphi(k)] = \frac{1}{2\pi}\int \varphi(k)\exp(-ikx)dk = f(x)\,. \tag{1.4.2}$$

It may be readily extended to functions in 3D space, for example, the distribution of density of scattering centers in ordinary space,

$$\varphi(\boldsymbol{k}) = \int \rho(\boldsymbol{r})\exp(i\boldsymbol{k}\cdot\boldsymbol{r})d\boldsymbol{r}\,. \tag{1.4.3}$$

From the inverse transform we may recover the original

$$\varphi^{-1}[\varphi(\boldsymbol{k})] = \frac{1}{(2\pi)^3}\int \psi(\boldsymbol{k})\exp(-i\boldsymbol{k}\cdot\boldsymbol{r})d\boldsymbol{k} = \rho(\boldsymbol{r})\,. \tag{1.4.4}$$

Here $\boldsymbol{k}$ is an arbitrary vector with dimension $[length]^{-1}$ called the reciprocal lattice vector. Let vectors $\boldsymbol{r}$ and $\boldsymbol{k}$ be decomposed into their Cartesian components x, y, z and k_x, k_y, k_z, so the inner product of them is

$$\boldsymbol{k}\cdot\boldsymbol{r} = k_x x + k_y y + k_z z\,. \tag{1.4.5}$$

So, every structure has its reciprocal, i.e., its Fourier transform, which contains the full information of the original structure. If the Fourier transform is thoroughly known, then the original structure may be recovered easily by the inverse transform through calculation. The Fourier transform of a structure is connected to the experimental results of wave scattering through the following relations. We define the incident wave vector $\boldsymbol{K}_0$ and the scattering wave vector $\boldsymbol{K}$ with directions along the beams and with magnitudes equal to the reciprocal wavelength λ^{-1}, then

$$\boldsymbol{k} = \boldsymbol{K} - \boldsymbol{K}_0\,, \tag{1.4.6}$$

so the scattering amplitude is proportional to the Fourier transform associated with $\boldsymbol{k}$. In order to observe different regions of reciprocal space, the incident and scattering directions, as well as the wavelength of the incident radiation, may be changed. So the Fourier transform of a structure and its inverse are closely related to the physical process of observation and determination of structure. However, it should be noted that our knowledge of the Fourier transform of a structure is never complete, for the wavelength of radiation used has finite value, there are always regions in reciprocal space inaccessible to experimental probing, and most detectors cannot register the phase of radiation, so usually only the intensities instead of the amplitudes are recorded. These are fundamental difficulties for structure determination.

1.4.2 The Reciprocal Lattice

Next we shall show that the Fourier transform of a lattice gives the reciprocal lattice. Firstly let consider the simplest case, the 1D lattice with period a defined by Dirac's delta function $\delta(x - ua)$, $u = 0,\ \pm1,\ \pm2, \ldots,\ \pm\infty$, satisfying

$$\delta(x - ua) = \begin{cases} \infty, & \text{if } x = ua, \\ 0, & \text{otherwise}, \end{cases} \tag{1.4.7}$$

and

$$\int \delta(x - ua)dx = 1\,. \tag{1.4.8}$$

The Fourier transform of this delta function is also a delta function

$$\int \delta(x - ua)\exp(ikx)dx = \delta\left(k - \frac{2\pi u}{a}\right). \tag{1.4.9}$$

This delta function may be visualized as a series of planes perpendicular to the x axis with spacings equal to $2\pi/a$ ($u = 0, \pm1, \pm2, \ldots, \pm\infty$).

For a 3D lattice with basic vectors $\boldsymbol{a}$, $\boldsymbol{b}$ and $\boldsymbol{c}$,

$$\boldsymbol{l} = u\boldsymbol{a} + v\boldsymbol{b} + w\boldsymbol{c}, \qquad u,\ v,\ w = 0,\ \pm1,\ \pm2, \ldots,\ \pm\infty. \tag{1.4.10}$$

This may be written as a 3D delta function $\delta(\boldsymbol{r} - \boldsymbol{l})$, and its Fourier transform is readily found to be

$$\varphi(\boldsymbol{G}) = \delta(\boldsymbol{k} - \boldsymbol{G}). \tag{1.4.11}$$

This 3D delta function define a lattice, the reciprocal lattice, which may be visualized as the intersection points of 3 families of parallel planes which are parallel to the $\boldsymbol{bc}$-plane, $\boldsymbol{ac}$-plane and $\boldsymbol{ab}$-plane with spacings equal to $2\pi/a$, $2\pi/b$ and $2\pi/c$ respectively. Mathematically the reciprocal vector $\boldsymbol{G}$ is

$$\boldsymbol{G} = h\boldsymbol{a}^* + k\boldsymbol{b}^* + l\boldsymbol{c}^*, \tag{1.4.12}$$

where $h, k, l = 0,\ \pm 1,\ \pm 2, \ldots,\ \pm\infty$, and

$$\boldsymbol{a}^* \cdot \boldsymbol{a} = 2\pi,\ \boldsymbol{b}^* \cdot \boldsymbol{b} = 2\pi,\ \boldsymbol{c}^* \cdot \boldsymbol{c} = 2\pi. \tag{1.4.13}$$

It can be also rewritten as

$$\boldsymbol{a}^* = 2\pi\boldsymbol{b} \times \boldsymbol{c}/\Omega_0,\ \boldsymbol{b}^* = 2\pi\boldsymbol{c} \times \boldsymbol{a}/\Omega_0,\ \boldsymbol{c}^* = 2\pi\boldsymbol{a} \times \boldsymbol{b}/\Omega_0, \tag{1.4.14}$$

where Ω_0 is the volume of a unit cell,

$$\Omega_0 = \boldsymbol{a} \cdot \boldsymbol{b} \times \boldsymbol{c}. \tag{1.4.15}$$

If the basic vectors of the direct lattice are orthogonal to each other, then the directions of the basis vectors of the reciprocal lattice coincide with those of the direct lattice. In general, $\boldsymbol{a}^*$ may deviate from $\boldsymbol{a}$ but is always perpendicular to the plane defined by $\boldsymbol{b} \times \boldsymbol{c}$. It should be noted that the definition of the reciprocal lattice vector introduced here follows the convention adopted in texts on solid state physics; it is slightly different from that commonly used in texts on crystallography, in which the factors of 2π in (1.4.13) and (1.4.14) are replaced by unity, i.e., the relation appears to be more symmetrical.

Now we shall examine whether Fourier transformation has an effect on this. We know that there are 14 Bravais lattices and, besides primitive lattices, there are centered lattices, such as base-centered, body-centered and face-centered. The reciprocal lattices of primitive lattices retain the original lattice symmetry, but, the situation is different for centered lattices. It is easy to verify that a face-centered direct lattice is transformed into a body-centered reciprocal lattice and *vice versa*, while base-centered lattices remain unchanged. The reciprocal lattice is an important concept in condensed matter physics; it has played crucial roles in the physics of diffraction as well as condensed matter theories dealing with wave propagation in periodic structures.

1.4.3 Fourier Transform of Periodic Structure

The crystal structure may be understood as the content of the unit cell repeated by the lattice, i.e., each lattice point is associated with a unit cell. Let the density distribution of a unit cell be $f(\boldsymbol{r})$, it is repeated at each lattice point, in the language of Fourier transform this act is called convolution, denoted by $f(\boldsymbol{r})\delta(\boldsymbol{r} - \boldsymbol{l})$. According to the convolution theorem, the Fourier transform of the convoluted function is simply the product of their respective transforms, i.e.,

$$\int f(\boldsymbol{r})\delta(\boldsymbol{r} - \boldsymbol{l})\exp(i\boldsymbol{kr})d\boldsymbol{r} = F(\boldsymbol{k})\delta(\boldsymbol{k} - \boldsymbol{G}), \tag{1.4.16}$$

here $F(\boldsymbol{k})$ is the Fourier transform of $f(\boldsymbol{r})$ and is usually called the structure factor in diffraction physics; it is a continuous function in reciprocal space. In an actual diffraction experiment, the absolute values of $F(\boldsymbol{G})$ are sampled at the reciprocal lattice points $\boldsymbol{G}$. We also define the incident wave vector $\boldsymbol{K}_0$ and the scattering wave vector $\boldsymbol{K}$ ($|\boldsymbol{K}| = |\boldsymbol{K}_0| = 2\pi\lambda^{-1}$), where λ is the wavelength of the radiation), then

$$\boldsymbol{K} - \boldsymbol{K}_0 = \boldsymbol{k} = \boldsymbol{G}. \tag{1.4.17}$$

Because $\boldsymbol{G} = 2\pi d_{hkl}^{-1}$, the equation above is the generalized form of the Bragg equation ($2d_{hkl}\sin\theta = \lambda$). Now $F(\boldsymbol{k})$ can be expressed by the reciprocal lattice points h, k, l, in the form $F(h,k,l)$. Here we use the kinematical theory of diffraction, and ignore the interaction of the incident and the scattering wave (see Fig. 1.4.1). We shall discuss the dynamical theory of diffraction in §5.4.2.

Using the full spectrum of $F(hkl)$ values (complex numbers) we may reach the goal of crystal structure analysis, i.e., the distribution of density of scattering matter within the unit cell, by means of the inverse transform. Let (ξ, η, ζ) be the fractional coordinates of a point in an unit cell so that $x = \xi a$, $y = \eta b$, $z = \zeta c$, then

$$\rho(\xi,\eta,\zeta) = \frac{1}{\Omega_0}\sum_{hkl} F(h,k,l)\exp[2\pi i(\xi h + \eta k + \zeta l)]\,. \tag{1.4.18}$$

The problem is reduced to the summation of Fourier series, which can be done in a straightforward fashion. However, due to the fact that the phase of $F(hkl)$ cannot be determined directly from experiment, the phase problem still remains a crucial question in crystal structure analysis. Many ingenious methods have been developed to overcome this difficulty; in conjunction with enormous advances in the calculating capability of modern electronic computers, crystal structures of moderate complexity can be solved routinely. For problems related to crystal structure analysis, readers should consult the texts on this special subject, such as Bibs. [5, 6].

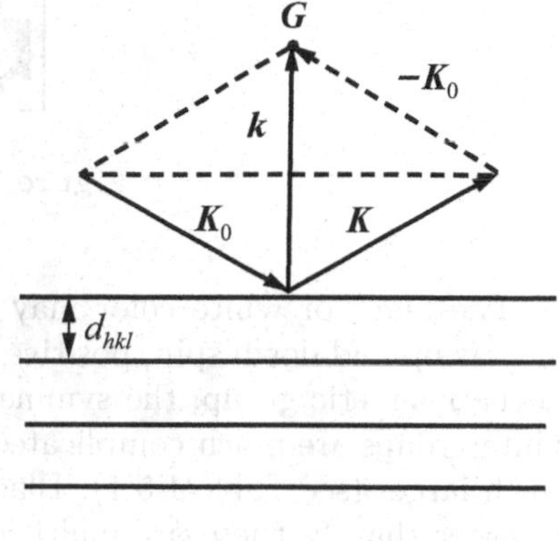

Figure 1.4.1 Bragg equation and reciprocal vector.

§1.5 Generalized Symmetry

Ordinary space has 3 dimensions and space groups give a complete description of its possible symmetries. In this section we shall discuss the generalized symmetries beyond ordinary space groups.

1.5.1 High-Dimensional Space Groups

Mathematically it is not difficult to generalize the symmetry of 3D Euclidean space to Euclidean spaces with dimensions higher than 3, the motivation for this generalization is to extract real physical consequences from some of these nD symmetries. For instance, 4D space groups are used in the study of incommensurate structures, while 6D space groups are used in the study of quasicrystals. These, together with the problems related to the projections from the high-dimensional spaces into lower-dimensional ones, will be treated in §2.3.

1.5.2 Color Groups

Another dimension with discrete values may be added into an ordinary 3D space. These discrete values may be expressed as different colors, so the symmetry groups of these spaces are called color groups. The simplest color group is the black-white group. The idea of black-white symmetry or antisymmetry may be introduced by inspecting Fig. 1.5.1 which shows a series of orthogonal mirror planes with which black patterns are transformed into white ones, and *vice versa*.

Figure 1.5.1 A figure of black-white symmetry.

The black or white color may acquire different physical meanings in actual crystal structure, such as up and down spin, positive and negative charges, etc., so the black-white group is equivalent to the magnetic group, the symmetry groups of magnetically ordered structures. Obviously black-white groups are much complicated than ordinary space groups, the number of symmetry groups is much larger (see Table 1.5.1). Black-white groups have only two colors, when the number of colors is larger than 2, these are multi-colored groups. Ions in crystals may have different orbitals with nearly the same energy so in order to describe detailed distribution of these orbitals on a crystalline lattice, multi-colored groups are required.

Table 1.5.1 Comparison of ordinary symmetry and black-white symmetry for fully periodic symmetry.

Kind of symmetry group	Number of dimensions	Number of groups				
		ordinary	1-color	gray	2-color	total
Transition group	1	1	1		1	2
	2	5	5		5	10
	3	14	14		22	36
Point group	1	2	2	2	1	5
	2	10	10	10	11	31
	3	32	32	32	58	122
Space group	1	2	2	2	3	7
	2	17	17	17	46	80
	3	230	230	230	1191	1651

We may introduce complex symmetry elements to represent color groups. A complex symmetry element Σ is a combination of an ordinary 3D symmetry element g and a compatible symmetry element on a complex plane σ, i.e., $\Sigma = (g, \sigma)$. If a complex symmetry element is to operate on a point $\boldsymbol{r}(x, y, z)$, then $F(\boldsymbol{r})$ will multiplied by a phase factor i.e.,

$$F(g\boldsymbol{r}) = F(\boldsymbol{r}) \exp[2\pi i \xi(\boldsymbol{r})], \tag{1.5.1}$$

where $\xi(\boldsymbol{r})$ is the phase factor, and is the function of $\boldsymbol{r}$.

Color groups with μ colors correspond to μ-fold rotation, i.e.,

$$\xi = \pm\frac{1}{\mu}. \tag{1.5.2}$$

When $\mu = 2$, $\sigma = \pm 1$, these are the black-white groups as a special example; when $\mu = 1$, $\sigma = 1$, they become the ordinary space groups. Thus, the introduction of complex symmetry elements gives a general way to derive color groups.

1.5.3 Symmetry of Reciprocal Space

Introduction of complex symmetry elements is indispensable to the study of the symmetry of reciprocal space. Consider the case of periodic structure in direct space: The distribution of its Fourier coefficients, i.e., the structure factor $F(h,k,l) = |F(h,k,l)|\exp(2\pi i\xi)$, is limited to the reciprocal lattice points (h,k,l). The distribution of matter in direct space is continuous, while the distribution of Fourier coefficients in reciprocal space is discontinuous, i.e., each lattice point is weighted with its corresponding structure factor. Due to the weighed structure factor, reciprocal space is no longer periodic. Thus, the space groups of direct space correspond to the complex point groups of index space. So it gives another way to derive 230 space groups, i.e., to derive it from complex point groups of index space.

In this case, symmetry operations of point group $\mathcal{G}$ will make $\rho(\boldsymbol{k}) \to \rho(g\boldsymbol{k})$, and

$$\rho(g\boldsymbol{k}) = \rho(\boldsymbol{k})\exp[2\pi i\varphi_g(\boldsymbol{k})], \tag{1.5.3}$$

where $\exp[2\pi i\varphi_g(\boldsymbol{k})]$ is called the phase function of the point group operation. With the introduction of the phase function, the point group is enlarged. Actually the nonsymmorphic symmetric elements of space group are implicitly contained in the phase function, for instance systematic extinction in centered lattices. It is customary to prepare tables showing the effects due to the symmetric elements of space group, now the way is reversed to derive space groups from complex point groups.

It should be noted that the δ-function-like Fourier transform is not limited to periodic structures. The study of symmetry in reciprocal space may be extended to derive space groups of quasiperiodic structure without invoking unphysical high-dimensional space.[b]

1.5.4 Other Extensions of Symmetry

According to the symmetry mentioned above, the structure should be the same after the symmetry operation. Another generalization of the concept of symmetry is to replace identity by approximate or statistical equality, i.e.,

$$F(g\boldsymbol{r}) = F(\boldsymbol{r}') \approx F(\boldsymbol{r}). \tag{1.5.4}$$

In this way we shall describe the statistical distribution of different kinds of atoms in alloy structure, atomic positions in glasses and liquids, or the orientation distribution of rod-like molecules along a given direction. This concept of statistical symmetry will be further discussed in Chap. 3.

Previously we have only considered symmetry operations that do not violate the condition of isometry, i.e., the lengths, angles, areas, and volumes are preserved in the symmetry operations. If we introduce a transformation by change of scale, such as deflation or inflation, a new type of symmetry, i.e., scale invariance is introduced. This will be discussed more fully in Chap. 4.

Bibliography

[1] Weyl, H., *Symmetry*, Princeton University Press, Princeton (1952).

[2] Vainshtein, B. K., *Modern Crystallography I*, 2nd ed., Springer-Verlag, Berlin (1994).

[3] Burns, G., and A. M. Glasser, *Space Groups for Solid State Scientists*, 2nd ed., Academic Press, New York (1990).

[b] Introducing the complex symmetry elements and deriving the ordinary space groups from reciprocal space was begun by A. Binenstock and P. P. Ewald, see A. Binenstock, P. P. Ewald, *Acta. Cryst.* **15**, 1253 (1962). Later N. D. Mermin derived the space groups of quasiperiodic structures, see D. A. Rabson, N. D. Mermin *et al.*, *Rev. Mod. Phys.* **63**, 699 (1991); N. D. Mermin, *Rev. Mod. Phys.* **64**, 3 (1992).

[4] Hahn, A. J., and A. J. Cochran, *International Tables for Crystallography, Vol. A, Space Group Symmetry*, 3rd ed., Kluwer Academic Publishers, Dordrecht (1992).

[5] Lipson, H., and W. Cochran, *The Determination of Crystal Structures*, 3rd ed., G. Bell and Sons, London (1966).

[6] Woolfson, M. M., and Fan Hai-fu, *Physical and Nonphysical Methods of Solving Crystal Structures*, Cambridge University Press, Cambridge (1995).

Chapter 2

Organization of the Crystalline State

The organization of the crystalline state is concerned with the aggregation of atoms, ions, and molecules into various structures. So it is expected that interactions, as well as entropy, will play outstanding roles in the structure of condensed matter. In the case of hard matter, interactions determine the main aspect of the structure of matter; while in soft matter, entropy plays the dominant role. We shall discuss packing and linkage structures that include most crystalline and quasiperiodic structures, as well as more exotic materials such as fullerenes and carbon nanotubes. Then the crystalline state will be generalized to include the quasiperiodic structures which also have long range order. These structures, which embody the organization of points, lines and faces, are subject to various geometrical constraints. So we shall begin with some mathematical preliminaries about topology and differential geometry of curves and surfaces.

§2.1 Geometrical Constraints

2.1.1 Topological Constraints

Topology is a kind of geometry that permits deformation of figures, while the metric aspects of geometry are totally ignored. A sphere, a cube, a dodecahedron and a cylinder have quite different metrical properties; however, intuitively we may imagine that by some rubber-like deformation process one object may be transformed into another. So from topological point of view, these objects are equivalent; however, this kind of object is inequivalent to a torus or a cup with a handle. The difference lies in their connectivity. On the surface of a sphere (or other objects which are equivalent to it topologically, e.g., a polyhedron) every closed loop on its surface can be shrunk continuously into a point, so it is called singly-connected. It is quite different in a torus, because there are some closed loops on its surface that cannot be shrunk to a point without meeting a insurmountable barrier, so it is called multiply-connected.

We may quantify the connectivity of a figure by the minimum number of cuts required for dissecting the figure into singly-connected regions, and this number of cuts is called the genus, denoted by g_t. For example, in a 1D topological space, such as a line or a loop, the minimum number of cuts required to make it singly-connected may be 0, 1 or $g_t > 1$. As to 2D topological space: On the surface of a sphere, any closed curve may shrink to a point, but it is not so for a torus, as in Fig. 2.1.1, the corresponding number of cuts is 0, 1 or 4.

The topological properties of geometrical figures may be characterized by topological invariants. The topological dimension is one such invariant, and the connectivity (or genus) is another.

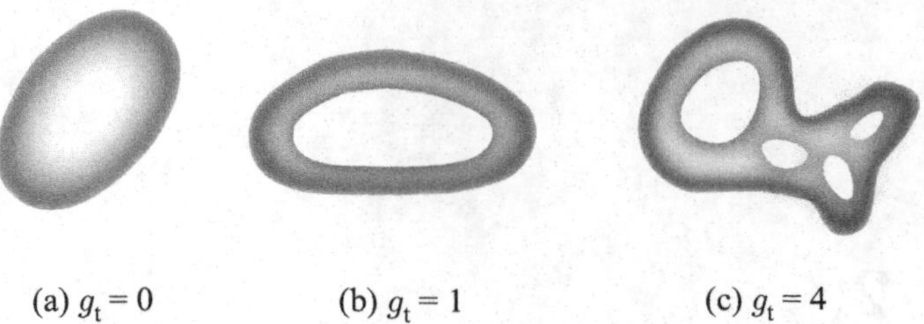

Figure 2.1.1 2D surfaces with different genus.

Another important topological invariant for geometric figures is the Euler–Poincaré characteristic χ, which is defined as

$$\chi = \sum_{p=0}^{n} (-1)^p N_p, \tag{2.1.1}$$

where N_0 is the number of vertices, N_1 the number of branches (or edges), N_2 the number of faces, N_3 the number of chambers (or cells). For any closed 2D surface of genus g_t, the Euler–Poincaré characteristic is found to be

$$\chi = 2(1 - g_t). \tag{2.1.2}$$

For $g_t = 0$, we get the Euler formula for any polyhedron,

$$\chi = N_0 - N_1 + N_2 = 2. \tag{2.1.3}$$

2.1.2 Curvature — Curves and Surfaces

Now we shall turn to another aspect of geometrical constraints, i.e., the differential geometry of curves and surfaces. First let us define curvature in the simplest case, a curve on a plane. As shown in Fig. 2.1.2, at any point P on the curve we can make a circle of best fit with radius r. The curvature at any point on the curve is defined by the ratio of the change in tangent vector $\Delta\psi$ to that of arc length Δs, i.e.,

$$\kappa = \frac{d\psi}{ds} = \lim \frac{\Delta\psi}{\Delta s} = \frac{1}{r}. \tag{2.1.4}$$

This shows that the reciprocal of the radius of the circle of best fit may serve as a measure of curvature. This definition of curvature may be readily extended to any curve in space. For a straight line, $r = \infty$, $\kappa = 0$; for a circle or helix, $r =$ const., $\kappa =$ const.

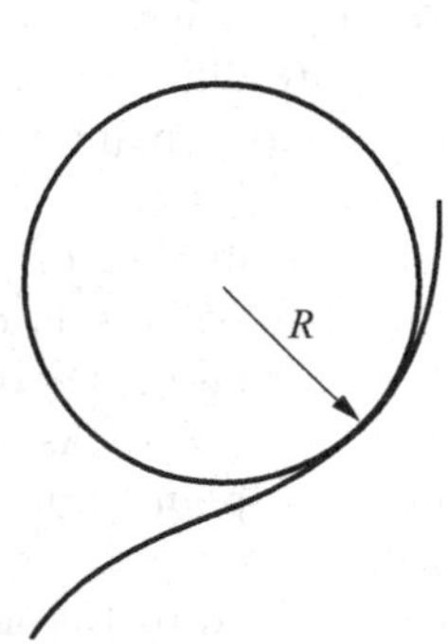

Figure 2.1.2 The curvature of a planar curve.

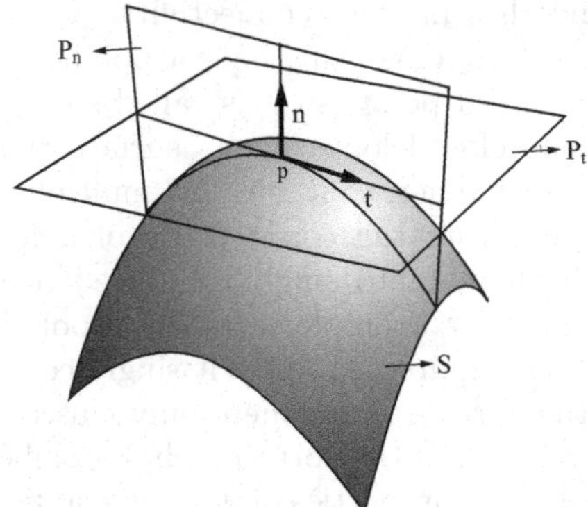

Figure 2.1.3 Definition of the normal curvature at a point P on a surface.

Many lines can be pleated into a thick line, like a pig tail; a long line may be folded back and forward into a sheet or a segment of lattice, or be rolled back and forward randomly to fill the space, or tightly bound together into a ball like woollen yarn. These possibilities are fully utilized in the organization of polymers and biopolymers.

The curvature of a curve may be readily extended to that of a surface. Imagine a point P on a surface with the normal vector $\boldsymbol{n}$ (see Fig. 2.1.3). Any plane containing $\boldsymbol{n}$ is normal to the surface at P and it intersects the surface on a curve with the normal curvature κ_n. Varying the orientation of the plane by rotating along $\boldsymbol{n}$ may give various values of κ_n, while its maximum and minimum values define the principal curvatures κ_1 and κ_2 corresponding to the principal directions. In general, the principal directions are mutually orthogonal. We define the mean curvature κ_{m} and the Gaussian curvature κ_{G} as

$$\kappa_{\mathrm{m}} = \frac{1}{2}[\kappa_1^{-1} + \kappa_2^{-1}], \ \kappa_{\mathrm{G}} = \frac{1}{\kappa_1 \kappa_2}. \tag{2.1.5}$$

Differential geometry gives a local description of the surfaces, while topology gives a global description. Now we shall give some rules connecting these two descriptions. First we may define the integral curvature as the area-weighted integral of the Gaussian curvature over the entire surface, i.e., $\oint \kappa_{\mathrm{G}} dS$. The quantitative relationship is summarized by the Gauss–Bonnet formula

$$2\pi\chi = \oint \kappa_{\mathrm{G}} dS, \tag{2.1.6}$$

where χ is the Euler–Poincaré characteristic which is related to the the genus g_{t} by $\chi = 2(1 - g_{\mathrm{t}})$. It means that all surfaces with the same genus are characterized by the same value of integral curvature.

In the 19th century, the absolute validity of the "parallel postulate" of Euclid had been questioned, so non-Euclidean geometries were formulated. Three different types of geometry may be distinguished by local values of the Gaussian curvature: $\kappa_{\mathrm{G}} = 0$, for Euclidean shapes; $\kappa_{\mathrm{G}} > 0$, for elliptic ones; $\kappa_{\mathrm{G}} < 0$, for hyperbolic ones (see Fig. 2.1.4). The sphere has constant positive Gaussian curvature, while the cylinder always has zero. In general, a surface may contain elliptic, Euclidean and hyperbolic regions, the 'average geometry' of a surface is characterized by the average value of its Gaussian curvature, i.e., $\langle \kappa_{\mathrm{G}} \rangle$. This is equal to the integrated curvature divided by the surface area S:

$$\langle \kappa_{\mathrm{G}} \rangle = \frac{\oint \kappa_{\mathrm{G}} dS}{\oint dS} = \frac{2(1 - g_{\mathrm{t}})}{S}. \tag{2.1.7}$$

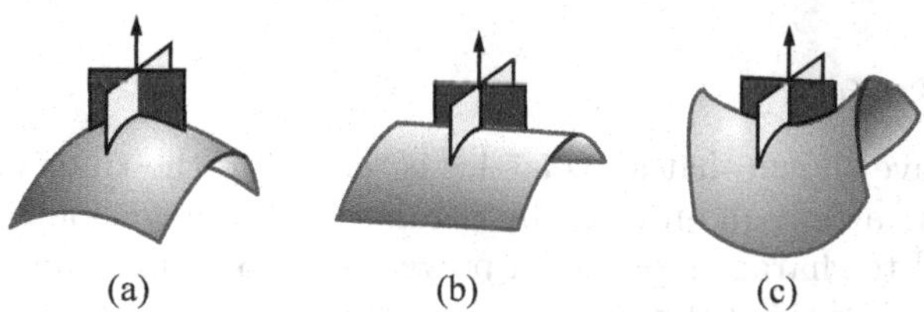

Figure 2.1.4 Curved surfaces of different types. (a) κ_1 and κ_2 with the same sign, $\kappa_{\mathrm{G}} > 0$; (b) $\kappa_1 = 0$, $\kappa_2 \neq 0$, $\kappa_{\mathrm{G}} = 0$; (c) κ_1 and κ_2 with different signs, $\kappa_{\mathrm{G}} < 0$.

Obviously it is impossible here to give a general survey of all curved surfaces. We can only give a rough introduction to some curved surfaces which may help us understand the organization of condensed matter.

Any curved surface is characterized by its intrinsic (metric and curvature) as well as extrinsic (embedded space) geometrical properties. Here we are concerned with 2D surfaces embedded in 3D space. 3D space is Euclidean, while a local region on 2D surface may be Euclidean (parabolic), or non-Euclidean, such as elliptic or hyperbolic. Near an elliptic point, the surface approximates an ellipsoid of the same curvature on the same side of the tangent plane, with principal curvatures of the same sign; near a parabolic point, the surface approximates a cylinder with radius equal to the reciprocal of the only non-zero principal curvature; near a hyperbolic point, the surface approximates a saddle surface, convex in one direction, concave in another direction and planar in a definite direction, with some part of the surface higher than the tangent plane, another part lower than it.

Surfaces with which we are mostly familiar are surfaces with identical Gaussian curvature (in value or sign), especially elliptic or parabolic, i.e., sphere, ellipsoid or cylinder. As to the hyperbolic case, it is impossible to form surfaces with identical curvature without introducing regions of singularity, so in general, hyperbolic surfaces contain regions of positive, negative and zero Gaussian curvature.

Surface tension induces a liquid film bounded by metallic frames to assume a shape with minimum surface area. One example of a minimum surface is a catenoid-shaped soap film extended between two identical metallic rings. This kind of minimum surface is characterized by two principal curvatures which have the same value but are opposite in sign, so that everywhere the mean curvature is zero, the Gaussian curvature is negative and it is a hyperbolic surface. Further, we are concerned with the minimum surfaces characterized by their curvature properties, not their minimum surface properties, for the latter is only significant for a finite region bounded by some definite boundaries.

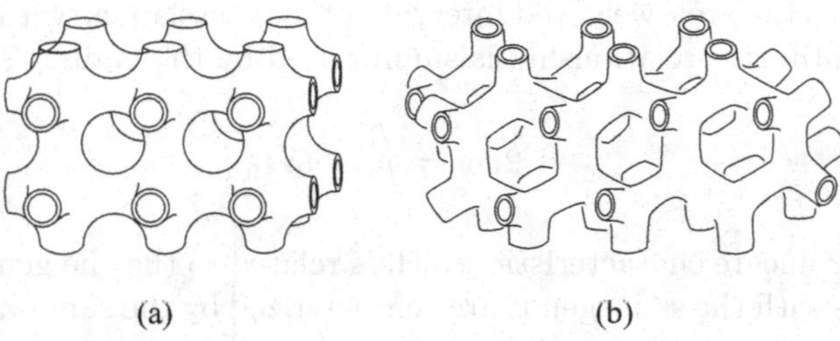

Figure 2.1.5 A schematic diagram for some IPMS. (a) P surface; (b) D surface.

For the study of structure of condensed matter, the infinite periodic minimum surfaces (IPMS) are the most significant. The two simplest examples of IPMS, P surface and D surface, were discovered by the 19th century mathematician H. Schwarz (see Fig. 2.1.5). These two surfaces may be visualized: P surfaces are formed by expanding rubber tubes connected as a simple cubic lattice, and D surfaces are formed by the corresponding diamond structure. The genus g_t of a single cell for these structures is 3, while in global sense, $g_t = \infty$. We may transform a P surface into a D surface or other IPMS with an isometric transformation. It should be noted that the space of these structures is divided into two regions, inside and outside the surface, so a bicontinuous space partition is formed.

2.1.3 Tiling of Space

In Fig. 2.1.6, we have shown that a 2D lattice is related to the tiling of identical parallelograms. Now we shall focus our attention on tiling by regular polygons without fissures or overlaps. Using Schläffli notation $\{p, q\}$ to characterize regular polygons, let α denote the angle between two adjacent sides. Since the external angle of the regular polygon $(\pi - \alpha)p = 2\pi$, we have

$$\alpha = \pi\left(1 - \frac{2}{p}\right) = \frac{2\pi}{q}, \tag{2.1.8}$$

or

$$(p-2)(q-2) = 4. \tag{2.1.9}$$

This equation determines the possible ways for the regular tiling of a plane: $\{3,6\}, \{4,4\}, \{6,3\}$, (see Fig. 2.1.6). Duality for these tilings should be noted, i.e., to interchange the number of sides with that of vertices, $\{3,6\} \leftrightarrow \{6,3\}$; while self-duality occurs for $\{4,4\}$. The impossibility of tiling an infinite plane by identical regular pentagons excludes the 5-fold symmetry in periodic structures. With a spherical surface there are increased possibilities for regular tilings. For tiling of S^2, at first we can use the Euler formula

$$N_0 - N_1 + N_2 = 2, \tag{2.1.10}$$

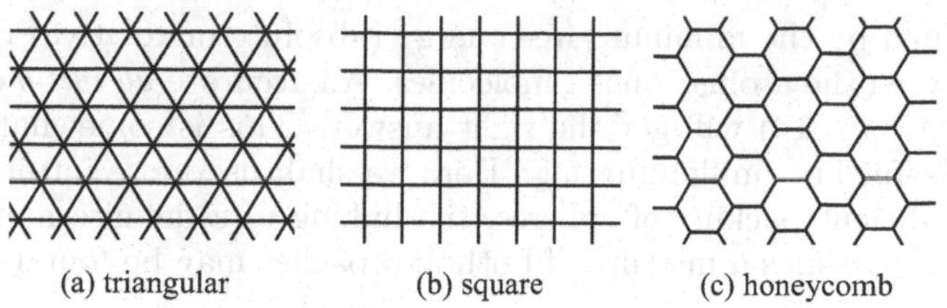

Figure 2.1.6 Regular tiling of plane.

then we calculate the number of edges, to get the equalities

$$qN_0 = 2N_1 = pN_2. \tag{2.1.11}$$

From (2.1.10) and (2.1.11), we find that

$$N_0 = \frac{4p}{4-(p-2)(q-2)},$$
$$N_1 = \frac{2pq}{4-(p-2)(q-2)}, \tag{2.1.12}$$
$$N_2 = \frac{4q}{4-(p-2)(q-2)}.$$

Noting that the angle around a point is less than 2π on a spherical surface, i.e., $q\alpha < 2\pi$, we have

$$(p-2)(q-2) < 4. \tag{2.1.13}$$

Therefore, only spherical polygons satisfy the regular tiling condition,

$\{3, 3\}$,	$\{3, 4\}$,	$\{4, 3\}$,	$\{3, 5\}$,	$\{5, 3\}$,
tetrahedron	octahedron	cube	icosahedron	dodecahedron

which correspond to Platonic polyhedra.

Tiling of the hyperbolic plane $\mathcal{H}^2$ requires

$$(p-2)(q-2) > 4, \tag{2.1.14}$$

and infinite sets of $\{p,\ q\}$ are possible.

Now, let us discuss the tiling of 3D Euclidean space. For tiling of 3D space with regular polyhedra, the Schläffli notation is enlarged to include r, which is the number of polyhedra around an edge, i.e., $\{p, q, r\}$. Since the dihedral angle of a Platonic solid may be expressed as $2\arcsin[\cos(\pi/q)/\sin(\pi/p)]$, it should be equal to $2\pi/r$ for a regular tiling, i.e.,

$$\cos\frac{\pi}{q} = \sin\frac{\pi}{p}\sin\frac{\pi}{r}. \tag{2.1.15}$$

The only solution with integer > 2 for this equation, is $\{4, 3, 4\}$, i.e., a regular stacking of cubes. It should be noted that with tetrahedra or octahedra only, one cannot tile the 3D space, we have recourse to mixed tiling called semiregular polyhedra.

If two types of regular polyhedra are allowed, then we have semiregular tiling. There are semiregular tilings, e.g., mixed tilings of tetrahedra and octahedra, which fill the 3D space. It is expected that for regular tiling of curved spaces $\mathcal{S}^3$ and $\mathcal{H}^3$, the equality of (2.1.15) will be changed into inequalities and that certainly will enlarge the possibilities for regular tiling.

§2.2 Packing Structures and Linkage Structures

In this section we shall be concerned with some geometrical principles for the architecture of hard condensed matter, which include most inorganic crystals and glasses. In principle, the actual structure is determined by the minimum free energy (absolute or relative) requirement dictated by the interactions between the atoms, ions or molecules. An accurate *ab initio* calculation of electronic band structure (see Chap. 12) will get the right answer. This is somewhat like an operation in a black-box — very useful but unilluminating. Here we shall use more intuitive approaches to treat this problem: The efficient packing of spheres, the linking of a definite number of chemical bonds between atoms and sometimes a mixture of both approaches may be found useful.

2.2.1 Sphere Packings and Coverings

In general, the efficiency of sphere packing in space is measured by the packing density f_p, which is defined as the ratio of the volume occupied by the spheres to the volume of space. For lattice-packing, i.e., the spheres occupy the sites of a lattice, f_{p} may be defined as the volume of one sphere to the volume of one primitive unit cell.

We may divide the whole of space into nearest neighbor regions around the center of every sphere, i.e., Wigner–Seitz (WS) cells. Each face of a WS cell is formed by perpendicular planes bisecting the straight lines linking neighboring sphere centers. The number of faces of a WS cell is equal to coordination number z, which is another important characteristic for sphere packing; it is equal to the number of spheres in contact with the central one, also called the kissing number by mathematicians.

Let P_i be the vertices of a WS cell, the distance from P_i to the center is a local maximum R_i. If we draw a series of spheres with radius equals to largest R_i from each center, these spheres just cover the space completely. We may define the covering density f_{c} as the ratio of the volume of the sphere (radius equals to largest R_i) to that of one primitive unit cell (or WS cell).

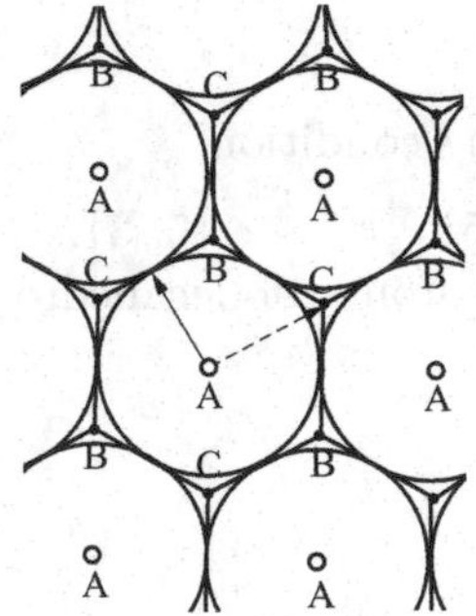

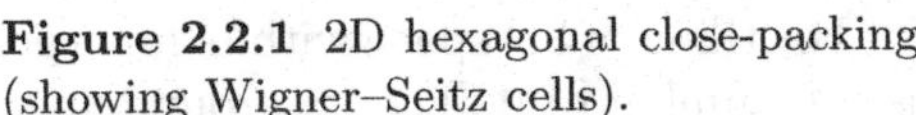

Figure 2.2.1 2D hexagonal close-packing (showing Wigner–Seitz cells).

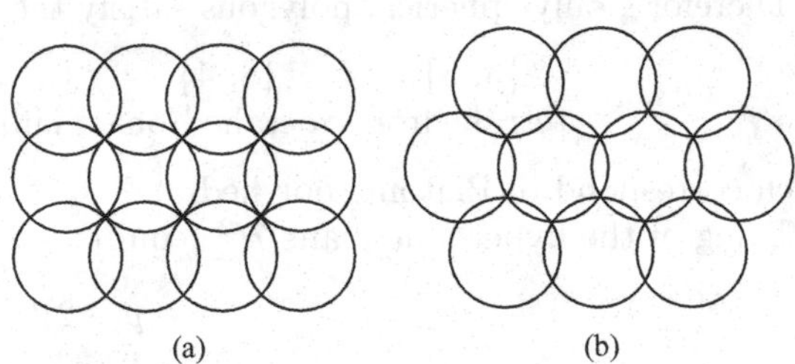

Figure 2.2.2 The lattice packing of a 2D plane. (a) Cubic; (b) hcp.

For the case of 2D space, the spheres degenerate into circular disks, and the lattice packing of the 2D plane is related to the packing of identical circular disks. Disks are placed at the vertices of triangular tiling with sides equal to a. Let the radii of the disks be equal to $a/2$. Then the corresponding WS cells are hexagonal honeycombs and the packing density f_{p} in this case, is

$$f_{\mathrm{p}} = \frac{\pi}{\sqrt{12}} = 0.9069\ldots, \tag{2.2.1}$$

so closest packing for a plane is achieved; in this case, $z = 6$, the highest one for 2D packing; while all R_i are identical, so the covering density

$$f_{\mathrm{c}} = \frac{2\pi}{3\sqrt{3}} = 1.2092\ldots, \tag{2.2.2}$$

the thinnest one, i.e., the least overlap between the disks, is achieved. It should be noted that $f_{\mathrm{c}} > 1 > f_{\mathrm{p}}$. We may apply the same procedures to examine the other types of packing, the disks sitting on vertices of other polygons, such as squares and hexagons: For square packing and covering,

$f_p = \pi/4 = 0.7854\ldots, z = 4, f_c = \pi/2 = 1.5708\ldots$; in the case of disks sitting on the vertices of a honeycomb, which is the dual for close-packing, with low packing density and definite coordination number $z = 3$. In this case, the formation of a definite number of bonds or links between neighboring disks will play the dominant role in structure-building (see Table 2.2.1).

Table 2.2.1 Coordination number, packing and covering density for some 2D structure.

Structure type	Coordination number	Packing density	Covering density
hcp	6	0.9069...	1.2092...
square	4	0.7854...	1.5708...
honeycomb	3	0.6046...	

It should be noted that in 2D structure there is an outstanding structure, i.e., hexagonal close-packing, in which the highest packing density, highest coordination number and least covering density are simultaneously achieved.

Kepler (1611) in his booklet "On Hexagonal Snow" gave the first scientific conjecture about closest sphere packings in 2D and 3D space.

For 3D packing, first hexagonal close-packed layers of spheres are formed, then the upper layer stacks on the voids of the lower layer, and so *ad infinitum.* Two main types of close-packed structures are formed:

(a) abcabc... →fcc, (b) ababab... →hcp.

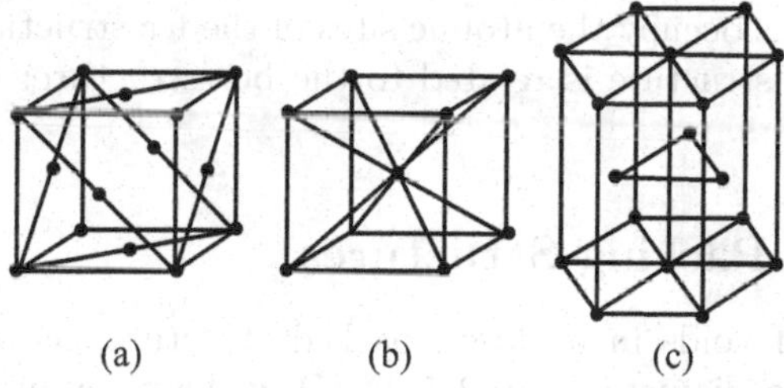

Figure 2.2.3 Unit cells of three kinds of packings. (a) fcc; (b) bcc; (c) hcp.

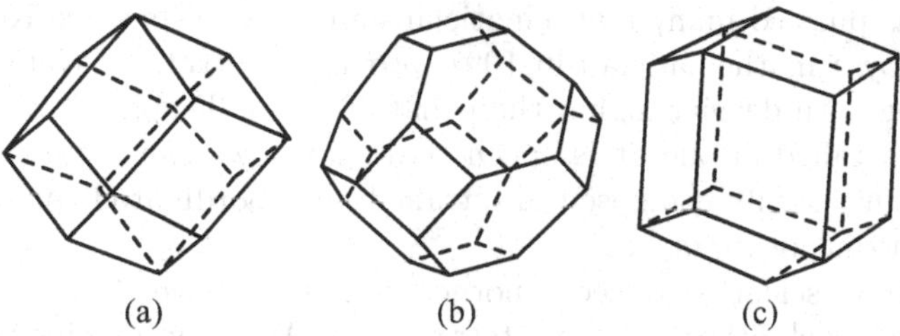

Figure 2.2.4 Wigner–Seitz cells of three kinds of packings. (a) fcc; (b) bcc; (c) hcp.

In the fcc (face-centered-cubic) structure, every sphere is situated on a lattice point, i.e., lattice packing; while in the hexagonal close-packed (hcp) structure, only one half of the spheres are situated on lattice points; besides hcp structure there are an infinite set of stacking sequences to fulfill the closed-packed condition. All these closed-packed structures have identical packing densities $f_p = 0.74048\ldots$, and the coordination number for closed packed structures is $z = 12$. The WS cell for the fcc structure is the rhombic dodecahedron, and the reciprocal lattice of fcc is bcc, whose WS cell is the truncated octahedron. Though bcc has lower packing density, it has thinner covering

Table 2.2.2 Coordination number, packing and covering density of some 3D structures.

Structure	Coordination number	Packing density	Covering density
fcc	12	0.74048	2.0944
hcp	12	0.74048	
bcc	8	0.68017	1.4635
diamond	4	0.3401	

density due to the fact that the WS cell for bcc is more sphere-like. For a system composed of hard (rigid) spheres, high packing density is the premium requirement for lowest energy; however, for a system of soft spheres which are deformable, low covering density may be more important.

Some types of crystal structures may be understood by sphere packing (or covering) requirements, these are called packing structures. In crystals, the structural units are atoms, ions and molecules and they are glued together by different kinds of chemical bonds. In Fig. 2.2.3 and Fig. 2.2.4, the unit cells and WS cells for fcc, bcc and hcp are shown respectively.

Van der Waals, metallic and ionic bonds favor densely packed structures with high packing density or low covering density. Now, let us examine the crystal structure of the elements: All crystalline structure of inert elements are fcc: Ne, Ar, Kr, Xe,...; many crystalline structure of metallic elements have closed-packed structures: Cu, Ag, Au, Al, Pd, Pt, Ir, Pb,... (fcc); Co, Ru, Os, Sc, Y, La, Ti, Zr, Hf,... (hcp); as well as bcc structure Li, Na, K, Rb, Cs, Fr, V, Nb, Ta, Cr, Mo, W, Fe,...; some exceptions are found to have more complicated structure such as Mn and U.

Structures of some ionic crystals may be understood as close-packing of unequal spheres: In the NaCl structure, the Cl^- ions occupy the atomic sites of the fcc structure, while the Na^+ occupy the octahedral voids. The CsI structure is related to the bcc structure: I^- occupy the cubic corners, Cs^+ occupy the cubic centers.

2.2.2 The Voids in Packing Structures

There are two kinds of voids in a closed packed structure, as shown in Fig. 2.2.5, one is a tetrahedral void, with hole diameter of 0.225 D (D is diameter of the sphere), the other is an octahedral void with diameter equal to 0.414 D. The number of tetrahedral voids is equal to twice the number of spheres, while the number of octahedral voids is equal to the number of spheres.

Whether there is any realization of 3D sphere packing with a packing density larger than 0.7405 is a problem that has puzzled many mathematicians and physicists since Kepler's conjecture. This problem was solved by a mathematician in 1997, proving the truth of Kepler conjecture, i.e. there is no possible packing with density higher than that of fcc in 3D space.[a] Since the tetrahedral void is smaller than the octahedral one, it is speculated that if we can eliminate all octahedral voids, i.e., the entire structure is only composed of tetrahedrally coordinated spheres, the packing density could be higher than that of fcc or hcp.

This concept inspires scientists to seek another route to achieve close packing. The starting unit is 4 spheres that touch each other to form a tetrahedron. We may imagine that the infinite tiling of these tetrahedra will produce an ideal structure with extraordinary high packing density since the larger octahedral voids will be absent. However, it can be easily demonstrated that this attempt is doomed to failure due to the impossibility of tiling 3D space with only regular tetrahedra. In a closed packed structure, every sphere is surrounded by 12 other spheres in contact with it, Fig. 2.2.6 shows the situation for fcc and hcp structure, note that the spheres are arranged in squares on surface layers, implying the half octahedra. We may imagine the 12 spheres are situated on the vertices of an icosahedron, forming 20 tetrahedra sharing a central sphere, with small gaps between spheres

[a] T. C. Hales, *Discr. Comput. Geom.* **17**, 1 (1997). Since this proof is extremely complicated, the validity of this proof is still under scrutiny.

on surface layer, forming the Mackay icosahedron. The existence of 5-fold symmetry excludes the possibility of building up of a periodic structure based solely on Mackay icosahedra. However, it does not hinder us from building up larger Mackay icosahedra shell by shell, with slightly deformed tetrahedra. The number of balls in the nth shell is found to be $10n^2 + 2$.

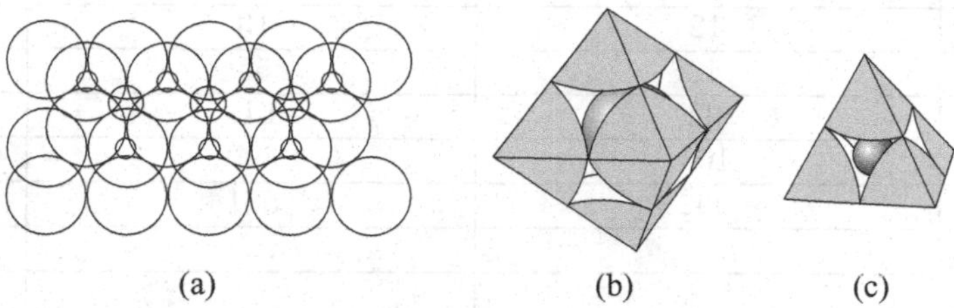

(a) (b) (c)

Figure 2.2.5 Two types of voids in close-packed structure. (a) The positions of the voids; (b) the octahedral void; (c) the tetrahedral void.

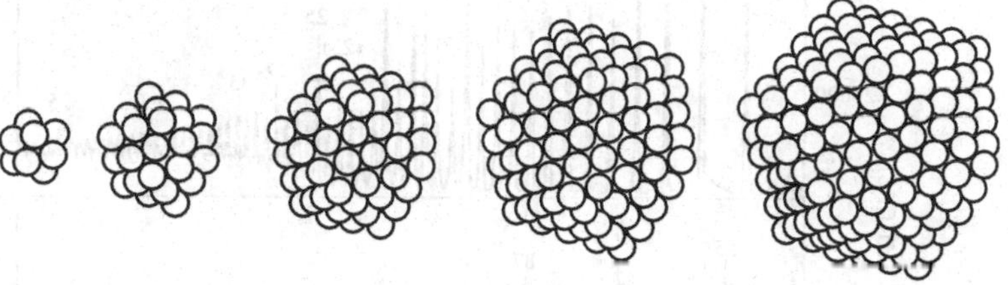

Figure 2.2.6 Mackay icosahedra.

Let d be the distance between the central ball and its neighbor, and l the distance between neighbors in the shell.

$$\frac{l}{d} = \frac{1}{3}(\sqrt{15} - \sqrt{3}) = 1.05146\dots. \tag{2.2.3}$$

Though the packing density for the first shell of Mackay icosahedra is slightly lower than fcc structure, the corresponding Voronoi cell is a regular dodecahedron which is certainly more sphere-like than the semiregular dodecahedron in the case of the fcc structure. So it is expected that the Mackay icosahedron has lower covering density compared with that of the fcc structure. Computer simulation using the Lenard-Jones potential confirmed that Mackay icosahedra are more stable than small *fcc* or *hcp* clusters. Experimental data on inert element clusters confirmed this, showing that Mackay icosahedra may persist to clusters with more than 1000 atoms, and it is also found experimentally that there are some magic numbers related to Mackay icosahedra, such as 13, 55, 145,... [b] (see Table 2.2.3 and Fig. 2.2.7).

Although it is impossible to fill space by a periodic close-packing of regular tetrahedra, it is possible to circumvent the difficulty if some distortion of tetrahedra is allowed. For example, a class of complex alloys have structures known as the tetrahedrally close-packed (tcp) structure (also known as Frank–Kaspar phase) in which local icosahedra, as well as other clusters dominated by tetrahedra, play important roles. In a basic Mackay icosahedron, the ratio of the length of edge to the distance from the center to vertex, i.e., $l/d = 1.05146\dots$, about 5% larger than 1.

If the central atom is slightly smaller than the surrounding ones, a more efficient packing may be realized. This is the cases of certain precipitation-hardened Al and Ni alloys, an outstanding example of which is $Mg_{32}(Al,Zn)_{49}$. According to L. Pauling, its unit cell may be decomposed into several shells of atoms with icosahedral symmetry, fitting one over the other, with the final shell slightly distorted to lower its symmetry from $m\bar{3}\bar{5}$ to $m\bar{3}$, suitable for building up a periodic structure (see Fig. 2.2.8). This structure is later found to be the crystalline approximant for a quasicrystal.

[b] In special cases, the length scale of Mackay icosahedra may reach several microns. For example, large clusters of BeO, see H. Hubert *et al.*, *Nature* **391**, 376 (1998). This kind of structure is certainly not a crystal, nor a quasicrystal, but may be regarded as an enormously large molecule, due to close-packing.

Table 2.2.3 Packing density of Mackay icosahedra.

No. of shells	No. of spheres in nth shell	No. of sphere in a cluster	Packing density
0	1	1	
1	12	13	0.72585
2	42	55	0.69760
3	92	147	0.69237
4	162	309	0.69053
5	252	561	0.68969
6	...	...	...
∞			0.68818

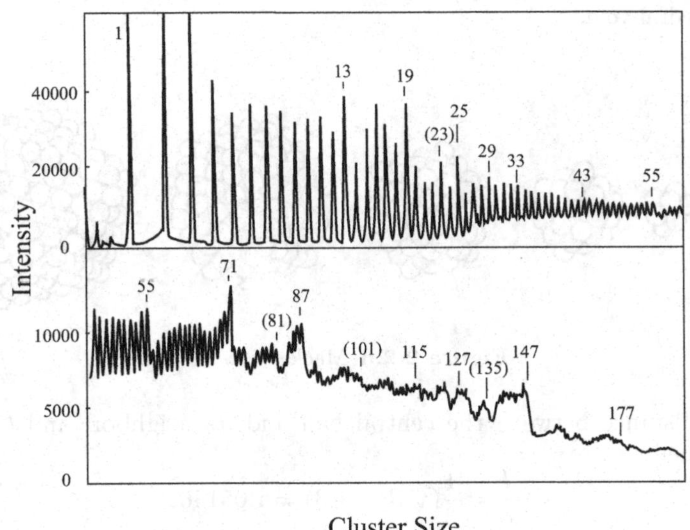

Figure 2.2.7 Mass spectrum of xenon clusters (showing high abundance peaks corresponding to closed shells of Mackay icosahedra). From O. Echt *et al.*, *Phys. Rev. Lett.* **47**, 1121 (1981).

The main characteristics of this type of structure are: (1) all interstices are tetrahedral, though the related atomic tetrahedra are somewhat distorted; (2) besides icosahedra X (with coordination number $z = 12$), the coordination types are limited to a set including P($z = 14$), Q($z = 15$) and R($z = 16$); these are shown in Fig. 2.2.9 and tabulated in Table 2.2.4. These 4 coordination types may be present in various combinations for different tcp structures. We may use some empirical combination formulas such as $X_xP_pQ_qR_r$ or $(PX_2)_i\ (Q_2R_2X_3)_j(R_3X)_k$ to designate tcp structures; here x, p, q, r are integers or rational fractions, while i, j, k are always integers. Some tcp structures are tabulated in Table 2.2.4. These are mostly alloys of transition metals with different atomic sizes. It will be shown later that these tcp structures have close relationships with some metallic glasses and quasicrystals.

2.2.3 Linkage Structures

Covalent bonding favors linkage structures, in which the premier structural requirement is the formation of definite number of bonds between each atom and its neighbors. So, here the importance of bonds is stressed.

We may take the diamond structure as the first example of a linkage structure, $z = 4$, all atoms are tetrahedrally bonded, see Fig. 2.2.10. Elemental semiconductors such as Si and Ge assume the diamond structure, as well as the namesake, the metastable phase of carbon. A variant of

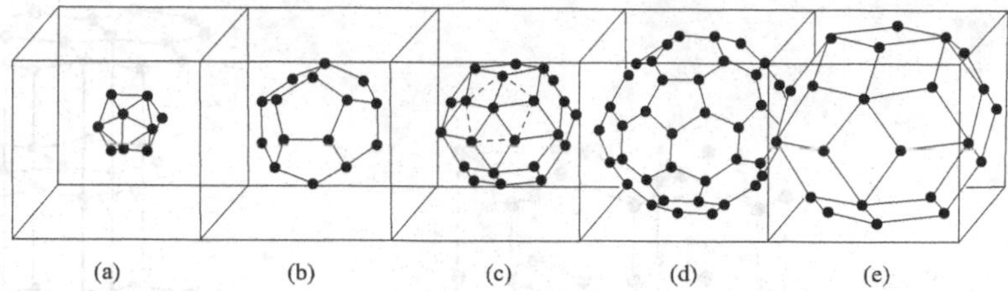

Figure 2.2.8 Pauling's model for tcp structure of $Mg_{32}(Al,Zn)_{49}$. (a) Icosahedron (icosahedra); (b) dodecahedron; (c) rhombic tricontahedron; (d) truncated icosahedron; (e) large rhombic tricontahedron.

Table 2.2.4 Structural parameters of tcp.

Type	Coordination number (CN)	Ideal point symmetry	No. of vertices (5-fold)	No. of vertices (6-fold)	No. of faces
P	16	T_d ($\bar{4}$3m)	12	4	28
Q	15	D_{3h} ($\bar{6}$m2)	12	3	26
R	14	D_{6h} (6/mmm)	12	2	24
X	12	Y_h (m$\bar{3}\bar{5}$)	12	0	20

the diamond structure is the zinblende structure, in which the two sites in the diamond structure are occupied by different atoms. Most compound semiconductors, such as GaAs, InP,... have this structure.

There is another variant of the diamond structure which is hexagonal diamond, also known as lonsdalite. It retains the tetrahedral coordination of each atom, while the symmetry is changed from cubic to hexagonal. A further variant of lonsdalite is wurzite with 2 types of atomic sites occupied by different atoms, just like the zincblende structure.

The graphite structure is composed of hexagonal sheets in which all atoms are triangularly bonded, among them 1/3 may be double bonds, while there are van der Waals bonds between hexagonal sheets. Since van der Waals bonds are much weaker than other types of bonds, the mixture of weak bonds with strong bonds may lead to structures with sheets or chains displaying quasi-low-dimensionality in physical properties.

Now we shall examine the crystal structure of ice at atmosphere pressure, i.e., the I_h phase of ice (see Fig. 2.2.12). The larger O atoms occupy the sites of hexagonal diamond, while the H atoms lie between them. If the H atoms lie midway between them, then a perfect crystal will be formed. Actually the situation is quite different. Each proton is strongly bonded to one O atom with a covalent bond and bonded to another O atom with a weaker hydrogen bond, so the position of the proton lies closer to one O atom, but farther from another. This creates a local configuration very close to a free H_2O molecule, but makes each unit cell of the crystal not identical. A special type of disorder, the so-called ice disorder, exists in the structure of ice even at zero temperature.

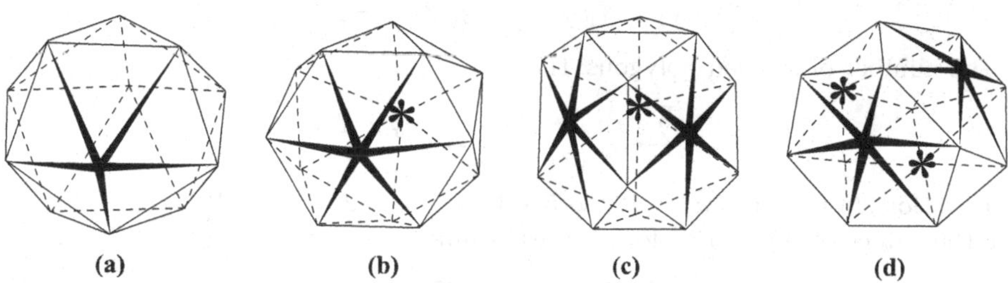

Figure 2.2.9 Kasper polyhedra. (a) X(CN = 12); (b) R(CN = 14); (c) Q(CN = 15); (d) P(CN = 16).

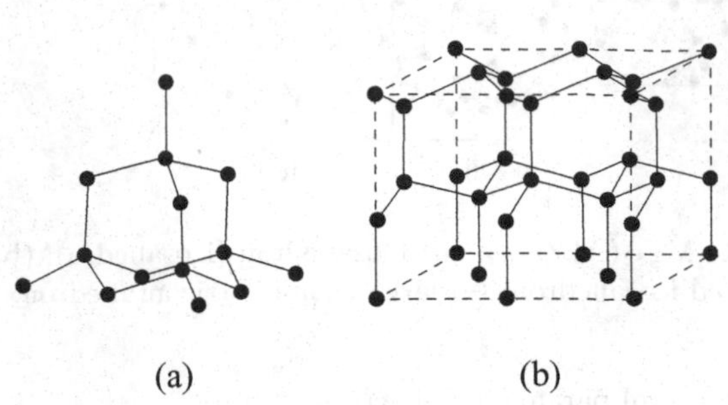

Figure 2.2.10 The structure of diamond (a) Cubic diamond; (b) hexagonal diamond (lonsdalite).

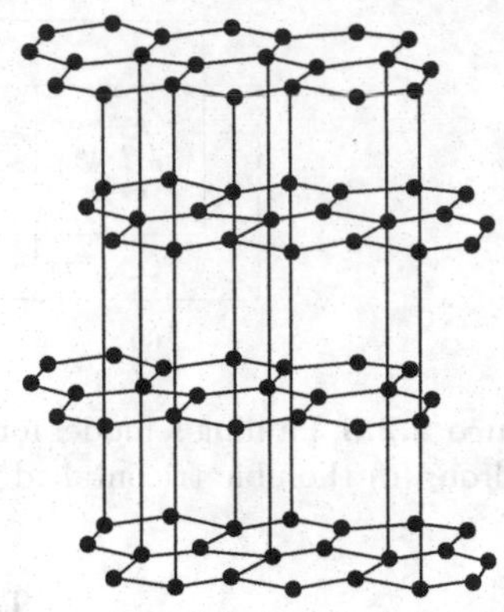

Figure 2.2.11 The structure of graphite.

In a crystal of ice with $2N$ bonds, there are 2^{2N} possible ways for arranging the protons. But not all these will satisfy the rules for making H_2O molecules. In fact, out of $2^4 = 16$ ways of arranging the H atoms on a tetrahedrally bounded O atom, only 6 satisfy the ice condition, i.e., 2 H atoms are near a O atom, similar to a H_2O molecule. The total number of allowed configurations in the whole crystal is about

$$\left(\frac{6}{16}\right)^N 2^{2N} = \left(\frac{3}{2}\right)^N, \tag{2.2.4}$$

so the residual entropy (i.e., the entropy at 0 K) is

$$S_0 = Nk_B \ln\frac{3}{2}. \tag{2.2.5}$$

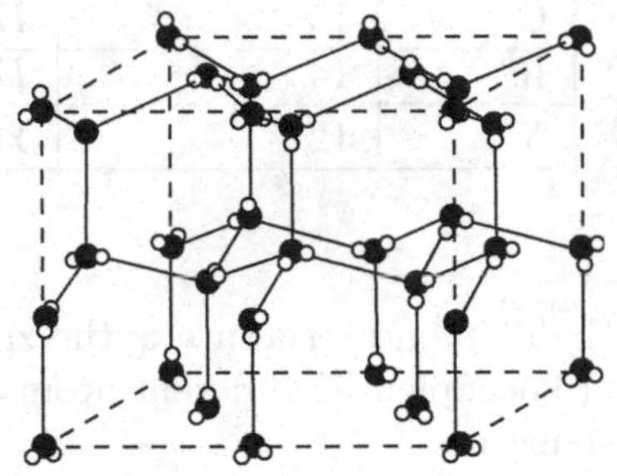

Figure 2.2.12 The structure of ice (I_h).

This is known as the Pauling formula and has been verified experimentally. The Pauling model for ice may be extended to other crystals with hydrogen bonds, such as the ionic compound KH_2PO_4 (potassium dihydrogen phosphate, 'KDP').

2.2.4 Fullerenes and Carbon Nanotubes

Let us start from a graphite sheet, i.e., graphene, in which carbon atoms occupy the vertices of a plane tiled by regular hexagons, this is a Euclidean plane with $\kappa_G = 0$. When we add pentagons to it, the sheet will curve up into an elliptic surface with $\langle\kappa_G\rangle > 0$; if we add heptagons into it, it will curve up into a hyperbolic surface with $\langle\kappa_G\rangle < 0$.

Fullerenes are closed cages of carbon atoms composed of hexagons and pentagons. Let N_0, N_1 and N_2 be the number of vertices, edges and faces respectively. According to the Euler formula, since each vertex has 3 branches, we have

$$N_0 - N_1 + N_2 = 2,\ 3N_0 = 2N_1. \tag{2.2.6}$$

Let n_x be the number of x-sided polygons, then

$$\sum_x xn_x = 2N_1,\ \sum_x n_x = N_2. \tag{2.2.7}$$

For $x = 5$ or 6, for closed cages $n_5 = 12$, so $5 \times 12 + 6n_6 = 2N_1$, i.e., $N_1 = 30 + 3n_6$. From this we may derive the values of N_0 and n_6 for a series of fullerenes:

$$n_6 = 0, 1, 4, 6, 12, 15, 19, 20, 25, 28, 31, 32, 37, 110, 260, \dots,$$
$$N_0 = 20, 22, 28, 32, 44, 50, 58, 60, 70, 81, 82, 84, 240, 560, \dots.$$

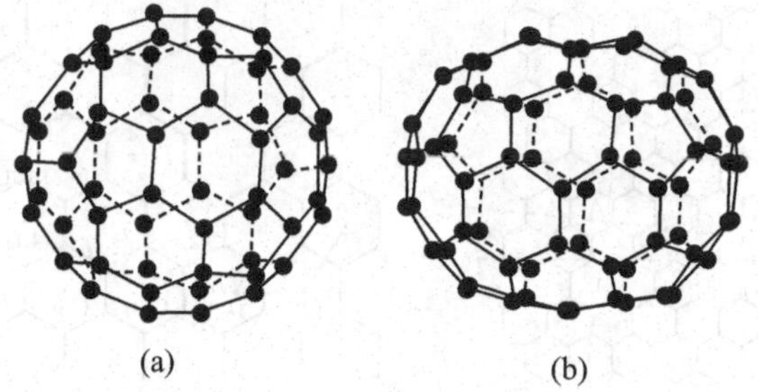

Figure 2.2.13 Typical fullerenes. (a) C_{60}; (b) C_{70}.

Graphene sheets may be rolled into carbon nanotubes, usually multiple-walled, in special cases, single-walled. Here we only consider the simplest case, single-walled nanotubes. For its main body, $\kappa_G = 0$; for its closed ends, $\kappa_G > 0$.

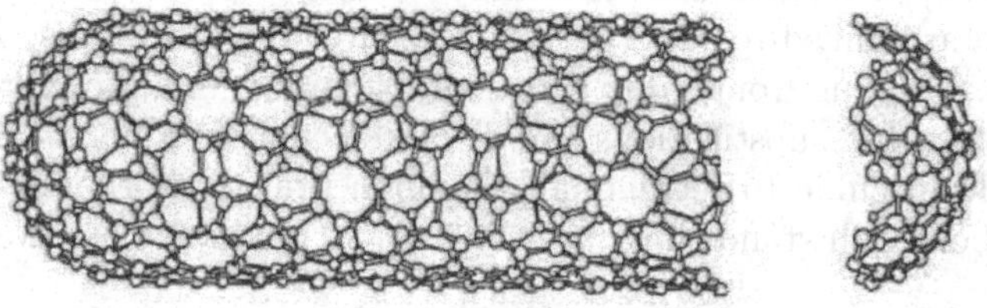

Figure 2.2.14 A single-walled carbon nanotube.

Carbon nanotubes may be indexed by a chiral vector $\boldsymbol{c} = m\boldsymbol{a}_1 + n\boldsymbol{a}_2$ with two integers m and n, where $\boldsymbol{a}_1$ and $\boldsymbol{a}_2$ are the unit vectors of graphite sheet. The geometrical nature of the chiral vector may be identified by inspecting Fig. 2.2.14. For $m = 0$, the configuration is called zigzag type; for $m = n$, armchair type; for $m \neq n$, chiral with the chiral angle $\theta < 30°$,

$$\theta = \arctan\left[\frac{(3)^{1/2}n}{2m+n}\right], \tag{2.2.8}$$

and the diameter

$$d_t = \sqrt{3}a_{\text{C-C}}\frac{m^2+mn+n^2}{\pi}. \tag{2.2.9}$$

The fullerites are periodic structures with fullerenes as building blocks. C_{60} solid is a crystalline substance formed of carbon atoms not arranged as diamond and graphite but rather as a close-packed structure built by C_{60} molecules; its structure and properties have been intensely studied. Another fullerite formed by C_{70} has been reported. Whether there are other types of periodic structure besides the fullerites is a problem still to be solved. There has been speculation about the possibility of the existence of a hyperbolic surface formed by carbon atoms. The graphene sheet may be modified by inserting heptagons periodically so that an IPMS with $\langle\kappa_G\rangle < 0$ is produced, it is tentatively called 'schwarzite'; however, it is still waiting for experimental confirmation.

2.2.5 The Structure of Perovskites

Many crystal structures may be understood as mixture of packing and linkage structures. We can take perovskites as an example. A large family of oxides is known as the perovskites. The name originally came from a mineral $CaTiO_3$, which itself actually is a distorted perovskite structure. The chemical formula for the perovskites is ABO_3 where A stands for metal with valence 1 or 2 such as K, Na, Ba, Sr, Ca, La,...; and B stands for a transition metal with valence 4 or 5, such as Ti, Nb, Zr, V,.... The structure of perovskite with the highest symmetry is cubic, with space group Pm3m, as shown in Fig. 2.2.16(a). The origin may be chosen at the site of a B atom [as

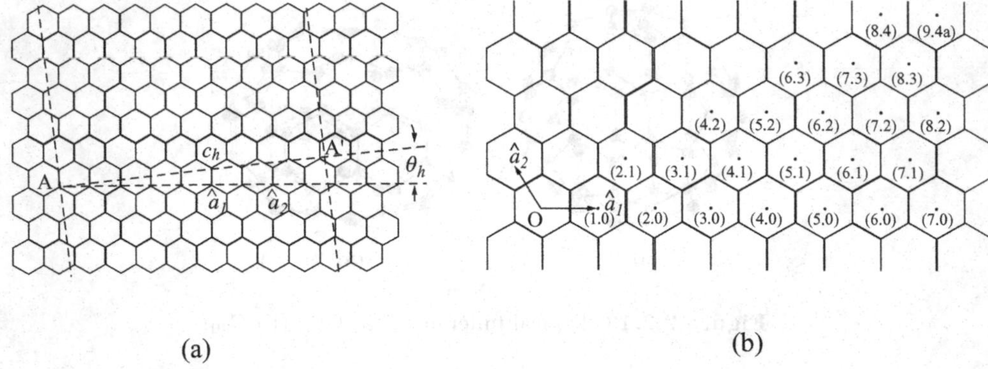

Figure 2.2.15 Chiral vector and chiral angle of a carbon nanotube. (a) The relationship of chiral vector c_h and chiral angle θ_h with basic vectors $\boldsymbol{a}_1$, $\boldsymbol{a}_2$; (b) the chiral vector in terms of coordinates h_1 and h_2.

shown in Fig. 2.2.16(b)] or an A atom. An easily visualized picture for the perovskite follows: A set of BO_6 octahedra are linked together by corner-shared oxygen atoms, with A atoms occupying the space in between. Starting from cubic perovskite, a large number of variants may be derived by some distortion of octahedra, substitutions and displacements of atoms, while the symmetry of these variants may range from cubic to tetragonal, rhombohedral, orthorhombic or monoclinic, though the topological backbone still stands intact.

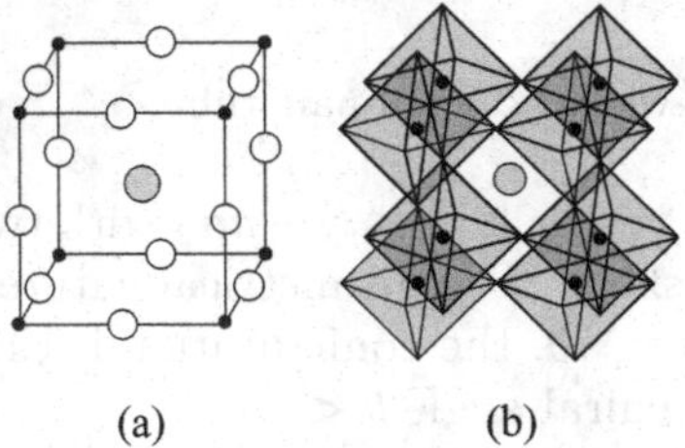

Figure 2.2.16 Structure of cubic perovskite. (a) Atomic positions, open circles — O atoms, shaded circles — B atoms, small full circles — A atoms; (b) showing BO_6 octahedra.

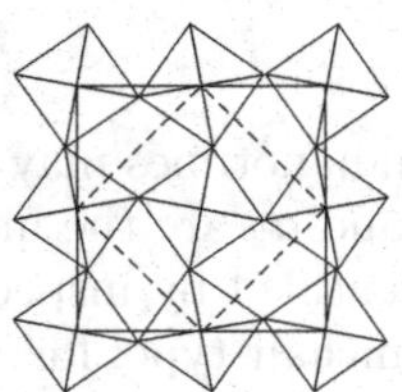

Figure 2.2.17 Structure of a distorted perovskites.

The oxides with perovskite structures are model materials for the study of structural phase transitions of the displacement type. A cubic-tetragonal phase transition occurs in $SrTiO_3$ at 103 K by tilting of neighboring TiO_6 octahedra in opposite directions. The tilt angle φ/φ_0 varies continuously, from zero starting at T_c.

In $BaTiO_3$, deviation of the axial ratio $(c/a) - 1$ starts from zero at T_c, and jumps to a finite value, i.e., it various discontinuously at the phase transition. In the ferroelectric regime, a further tetragonal-orthorhombic phase transition occurs at 273 K, while a orthorhomic-rhombohedral one occurs at 183 K. A similar series of structural phase transitions occurs in isostructural $KNbO_3$, with transition temperatures at 683 K, 480 K and 233 K.

$LaMnO_3$ is a orthorhombic perovskite with tilted MnO_6 octahedra. It attracts attention because of the discovery of colossal magnetoresistance (CMR) in doped $LaMnO_3$.

The perovskite family may be further enlarged to include the K_2NiF_4 structure, which may be regarded as a layered perovskite. Starting from the $BaTiO_3$ structure, we may stack 3 unit cells one over the other, interchanging the positions of Ti and Ba atoms in the middle one, yielding a tetragonal structure of the K_2NiF_4 type, shown in Fig. 2.2.18. The first high T_c superconductor $La_{2-x}Ba_xCuO_4$ discovered by Bednorz and Müller has a distorted K_2NiF_4 structure, which may be regarded as a variant of layered perovskite by substitution, stacking and distortion. Variants of this structure are found for other cuprate high T_c superconductors that were subsequently discovered.

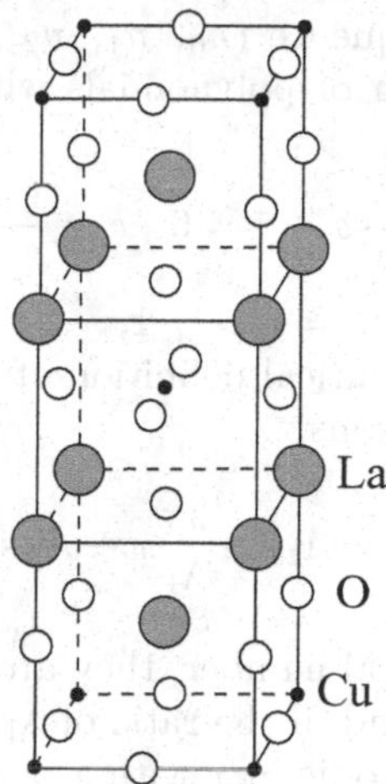

Figure 2.2.18 Structure of La_2CuO_4 (K_2NiF_4 type or layered perovskite).

§2.3 Quasiperiodic Structures

The discovery by Shechtman *et al.* (1984) of electron diffraction patterns showing the crystallographically forbidden icosahedral symmetry, and immediately interpreted by Levine and Steinhardt (1984) as a new state of matter called a quasicrystal, which lies between periodic crystals and amorphous glasses, startled the academic community of condensed matter physics. After a decade of intense research, quasicrystallography, as a new branch of generalized crystallography, has reached maturity. In this section we shall not follow strictly the historical sequence of the discoveries, but consider a wider perspective, starting from the simplest case, 1D structure, then going to more complicated cases, 2D and 3D structures, prefaced by some mathematical discussion on irrational numbers, and some discussion of the basic questions of quasicrystals following as a postscript.

As we know, there are two types of atomic structures for solids, one is crystal and the other glass. Crystal structures are highly ordered: (1) Long-range translational order characterized by a periodic repetition of unit cells; (2) long-range orientational order with a symmetry corresponding to specific crystallographic discrete subgroups of the rotation, as represented by the 5 2D and 14 3D Bravais lattices; (3) rotational point symmetry, also restricted to specific crystallographic discrete subgroups. Glass structures, by contrast, have no long-range correlation. For example, metallic glass can be modelled by spheres that are densely random-packed. However, there is another class of long-range ordered structure forbidden by traditional crystallography. They are quasicrystals, which have long-range quasiperiodic translational order and long-range orientational order.

2.3.1 Irrational Numbers and Quasiperiodic Functions

Real numbers are classified into two kinds: the rationals, which can be expressed as P/Q, where P, Q are integers, and the irrationals, which can be expressed as the the limit of a continued fraction,

$$x = n_0 + \cfrac{1}{n_1 + \cfrac{1}{n_2 + \cfrac{1}{n_3 + \dots}}}. \tag{2.3.1}$$

When one terminates the fraction at the nth stage, we shall get the nth approximation of the irrational number x, and the deviation of the nth approximation from its true value is bounded by the inequality

$$\left| \frac{P_n}{Q_n} - x \right| \leq \frac{1}{Q_n^2}. \tag{2.3.2}$$

According to the nature of the sequence ($n_0, n_1, n_2, \ldots$), these are algebraic numbers, if the sequence is periodic, they are solution of polynomials with integral coefficients, set to zero. For instance,

$$x^2 - x - 1 = 0, \ x = \frac{1+\sqrt{5}}{2}, \tag{2.3.3}$$

this corresponds to a special sequence $n_i = (1,\ 1,\ 1, \ldots)$.

Now we may introduce the mathematical definition of quasiperiodic functions. Let $f(x)$ denote the summation of two sinusoidal functions,

$$f(x) = A_1 \sin \frac{2\pi}{\lambda_1} x + A_2 \sin \frac{2\pi}{\lambda_2} x. \tag{2.3.4}$$

If the ratio of λ_1 and λ_2 is a rational number, they are commensurate and $f(x)$ is periodic in x with a longer period. On the other hand, if the ratio of λ_1 and λ_2 is an irrational number, they are incommensurate, and $f(x)$ is quasiperiodic in x with a period $\sim \infty$.

This may be further extended to the summation of n periodic functions

$$f(x) = \sum_n A_n \exp(2\pi i x/\lambda_n). \tag{2.3.5}$$

$\lambda_1/\lambda_2, \ldots, \lambda_{n-1}/\lambda_n, \ldots$ are all irrational numbers, then $f(x)$ is called almost periodic functions. H. Bohr (1935) proved mathematically that almost-periodic functions have δ function-like Fourier transforms.

2.3.2 1D Quasiperiodic Structure

Consider the problem of rabbit breeding which was first treated by the mathematician Fibonacci in 1202: The pairs of large and small rabbits, corresponding to A and B. After a generation, a pair of large rabbits give birth to a pair of small rabbits while a pair of small rabbits grow up to two large ones, according to the substitution rule

$$\mathrm{A} \to \mathrm{AB}, \mathrm{B} \to \mathrm{A}, \tag{2.3.6}$$

we can write down the matrix form as

$$M_{ij} = \begin{pmatrix} 1,1 \\ 1,0 \end{pmatrix}, \ M_{ij} \begin{pmatrix} \mathrm{A} \\ \mathrm{B} \end{pmatrix} = \begin{pmatrix} \mathrm{A,B} \\ \mathrm{A} \end{pmatrix} \to \mathrm{ABA}. \tag{2.3.7}$$

For the next generation

$$M_{ij} \begin{pmatrix} \mathrm{AB} \\ \mathrm{A} \end{pmatrix} = \begin{pmatrix} \mathrm{AB,A} \\ \mathrm{AB} \end{pmatrix} \to \mathrm{ABAAB} \ldots, \tag{2.3.8}$$

thus we have a sequence

$$\mathrm{B} \to \mathrm{A} \to \mathrm{AB} \to \mathrm{ABA} \to \mathrm{ABAAB} \to \mathrm{ABAABABA} \to \mathrm{ABAABABAABAAB} \ldots. \tag{2.3.9}$$

The total number of As and Bs for a generation is defined as the Fibonacci numbers: 1, 1, 2, 3, 5, 8, 13, 21, 34, ... Their recursion relation is

$$u_{n+1} = u_n + u_{n-1}, \tag{2.3.10}$$

so

$$\frac{u_{n+1}}{u_n} = \frac{u_n + u_{n-1}}{u_n} = 1 + \frac{u_{n-1}}{u_n} = 1 + \frac{1}{\dfrac{u_n}{u_{n-1}}} = 1 + \frac{1}{1 + \dfrac{u_{n-2}}{u_{n-1}}} = 1 + \cfrac{1}{1 + \cfrac{1}{1 + \cfrac{1}{1 + \cdots}}}. \tag{2.3.11}$$

A irrational number may be expressed by a infinite sequence of rational numbers. The limit of the above sequence can be written as the form

$$\tau = \lim_{n\to\infty} \frac{u_{n+1}}{u_n} = 1 + \frac{1}{1+\tau}, \quad \tau^2 - \tau - 1 = 0, \tag{2.3.12}$$

we shall get the limit, the so-called golden number

$$\tau = \frac{1+\sqrt{5}}{2} = 1.618\ldots. \tag{2.3.13}$$

Consider a set of atomic sites at distances from the origin according to the Fibonacci sequence

$$x_n = n + \frac{1}{\tau}\left\lfloor \frac{n+1}{\tau} \right\rfloor, \tag{2.3.14}$$

here $\lfloor y \rfloor$ means to take the integer part of y. The interval between neighboring sites

$$\Delta x \equiv x_n - x_{n-1}, \tag{2.3.15}$$

so

$$\Delta x = \begin{cases} 1, & \text{if } \left\lfloor \frac{n+1}{\tau} \right\rfloor - \left\lfloor \frac{n}{\tau} \right\rfloor = 0, \\ 1 + \frac{1}{\tau}, & \text{if } \left\lfloor \frac{n+1}{\tau} \right\rfloor - \left\lfloor \frac{n}{\tau} \right\rfloor = 1. \end{cases} \tag{2.3.16}$$

Thus there are two possible intervals between sites, the long one L and the short one S

$$L = 1 + \frac{1}{\tau}, S = 1, \tag{2.3.17}$$

which appear in a quasiperiodic sequence, where the ratio of the numbers of L's to that of S's is equal to τ. This is the Fibonacci lattice in which the parameter τ plays a crucial role.

The Fibonacci lattice shows a self-similar structure: The self-similarity implies that there are deflation and inflation rules to transform one Fibonacci lattice into another with a different scale. We may use the substitution rule: $L \to LS$ and $S \to L$ for the realization of deflation (see Fig. 2.3.1).

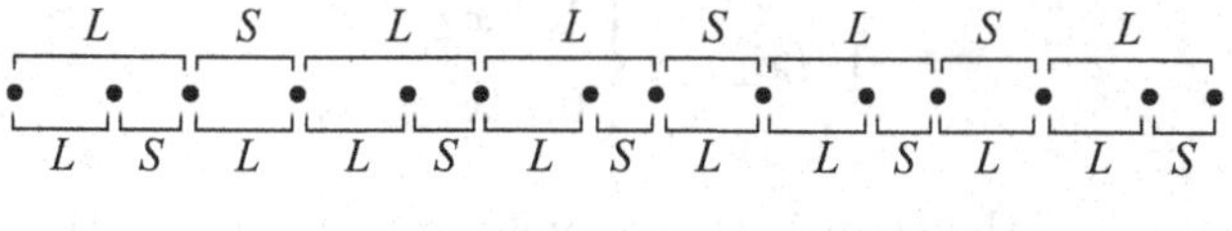

L S L L S L S L

L S L L S L S L L S L L S

Figure 2.3.1 The self-similarity of a Fibonacci lattice.

2.3.3 The Cut and Projection from a 2D Periodic Lattice

We may generate the 1D quasiperiodic lattice from a 2D periodic lattice. As shown in Fig. 2.3.2, starting from a 2D square lattice with lattice spacing a,

$$\rho(x, y) = \sum_{n,m} \delta(x - na)\delta(y - ma), \tag{2.3.18}$$

where n, m are integers.

We draw a straight line $R_{\|}$ for projection at an inclination angle α. If the slope of this line is irrational, then this line will not touch any lattice points of the square lattice. Then lattice points projected on line $R_{\|}$ will form a quasiperiodic lattice, provided that the projected lattice points are limited within a strip of certain width Δ along $R_{\perp}$ that is perpendicular to $R_{\|}$. Otherwise, the line $R_{\|}$ will be densely populated by the projected lattice points. Now we shall consider the special case

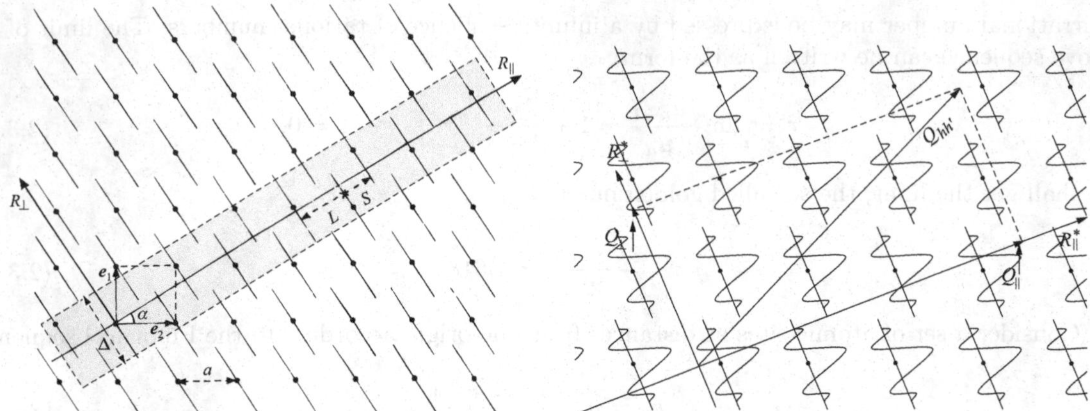

Figure 2.3.2 The cut and projection of 2D square lattice (projection line $\tan\alpha = 1/\tau$, stipe width Δ).

Figure 2.3.3 The cut and projection of 2D reciprocal lattice (showing the distribution of diffraction amplitude).

$\tan\alpha = \tau$. In this case the interval between neighboring projected points will be either $L = a\cos\alpha$ or $S = a\sin\alpha$, while the cut-out strip has the width $\Delta = a(\cos\alpha + \sin\alpha)$, so a Fibonacci lattice along $R_{\|}$ is generated by the cut and projection method.

Using the cut and projection method we can easily derive the Fourier transform of a Fibonacci lattice. For 2D lattice defined by (2.3.18), its Fourier transform is δ-functions centered on the reciprocal lattice, which is also a square lattice with lattice spacing $2\pi/a$ parallel to the original one, i.e.,

$$F_{hh'} = \sum_{h,h'} \delta(Q - Q_{hh'}), \tag{2.3.19}$$

where h and h' are integers (Miller indices).

Actually the Fourier components centered about the reciprocal lattice points have extensions in the reciprocal space as denoted by Fig. 2.3.3. Because we only consider a strip in 2D space, a window function $W(x_{\perp})$ for this strip is introduced, i.e.,

$$W(x_{\perp}) = \begin{cases} 1, & x_{\perp} \leq \Delta, \\ 0, & x_{\perp} > \Delta. \end{cases} \tag{2.3.20}$$

The Fourier transform of the strip is the convolution product of the Fourier transform of the infinite 2D square lattice with the Fourier transform of the window function. The latter turns out to be

$$G(Q_{\perp}) \sim \Delta \left[\frac{\sin Q_{\perp}\Delta}{2}\right] \left[\frac{Q_{\perp}\Delta}{2}\right]^{-1}, \tag{2.3.21}$$

where

$$Q_{\perp} = \frac{2\pi}{a} \frac{h - h'\tau}{(2+\tau)^{1/2}}. \tag{2.3.22}$$

In the 2D reciprocal space, though all the Fourier transforms are centered on the reciprocal lattice, only those touched on the irrational slope $R_{\|}^{*}$ that is parallel to $R_{\|}$ can be observed in the diffraction pattern corresponding to positions on $R_{\|}^{*}$ given by

$$Q_{\|} = \frac{2\pi}{a} \frac{h - h'\tau}{(2+\tau)^{1/2}}. \tag{2.3.23}$$

In Fig. 2.3.4 the electron diffraction pattern of an artificial Fibonacci superlattice is shown.

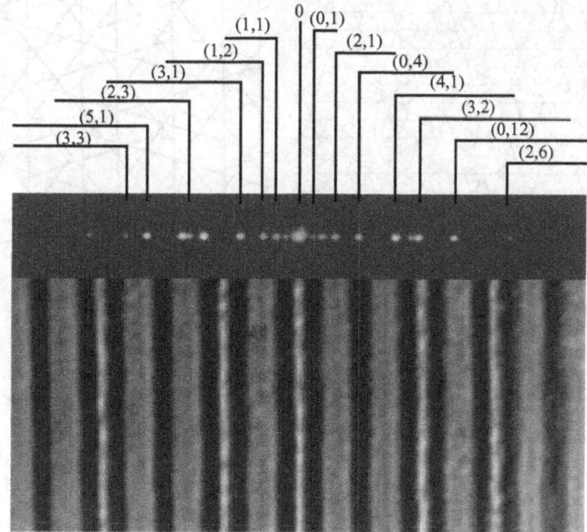

Figure 2.3.4 Electron micrograph and electron diffraction pattern of an artificial Fibonacci superlattice of a-SiH/SiN:H. Provided by Prof. K. J. Chen. Related paper: Chen Kunji *et al.*, *J. Noncrys. Solid.* **97** & **98**, 341 (1987).

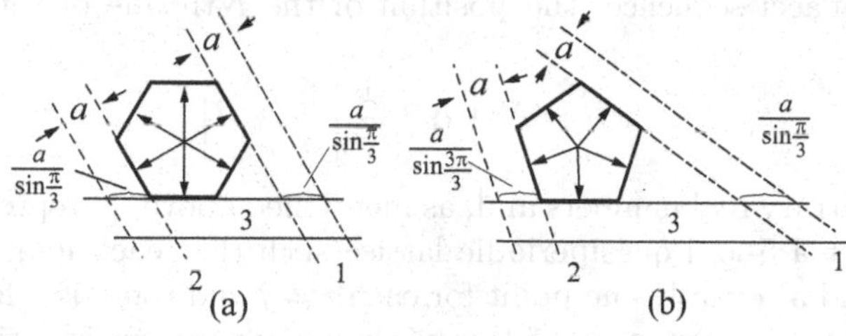

Figure 2.3.5 The sections of equal interval parallel lines perpendicular to symmetry axis. (a) regular hexagon; (b) regular pentagon.

2.3.4 2D Quasiperiodic Structures

In Chap. 1, we have already demonstrated that the rotational symmetries compatible with periodicity are limited to 2, 3, 4, 6; and 5-fold and all n-fold ($n > 6$) are forbidden symmetries in classical crystallography. We draw a star of unit vectors with n-fold forbidden symmetry in a plane, then a system of equidistant parallel lines with spacing a perpendicular to these vectors, forming an n-grid. We may then compare the case where n is crystallographic with the case where n is noncrystallographic, take the hexagrid and the pentagrid as example. In the case of the hexagrid, consider the intersections of sets 1 and 2 with set 3 (see Fig. 2.3.5), the spacing of intersections for each set is the same, so that a periodic lattice may be built up; while for a pentagrid, the spacings of intersections are different for sets 1 and 2 with set 3, the ratio of long and short sections equal to τ, an irrational number, the golden number

$$\frac{a\csc\pi/5}{a\csc 3\pi/5} = \tau = \frac{1+\sqrt{5}}{2} = 1.618\ldots, \tag{2.3.24}$$

for other noncrystallographic symmetries, corresponding irrational numbers are different, e.g., for the octagrid, the ratio is equal to $\sqrt{2}$,

$$\frac{a\csc\pi/5}{a} = \sqrt{2} = 1.414\ldots. \tag{2.3.25}$$

The foregoing statement may be summarized as follows: The quasiperiodicity is directly related to irrational numbers while the orientational symmetry constrains the quasiperiodicity (i.e., the

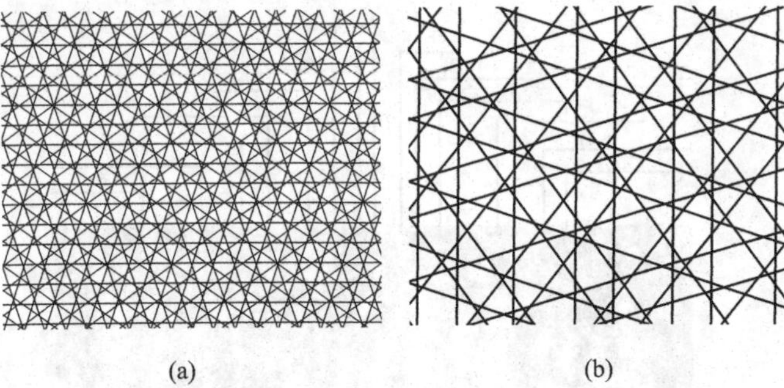

(a) (b)

Figure 2.3.6 The pentagrids. (a) Periodic; (b) quasiperiodic.

irrational length ratio). Thus, quasicrystals may be characterized by the corresponding irrational length ratios which determine the orientational symmetry and quasiperiodicity.

Consider a periodic pentagrid fully drawn: Since the 5-fold symmetry is incompatible with periodicity, infinite numbers of cells with different shapes are generated, some of the cells may be too small to satisfy the requirement for the minimum separation for an ordinary lattice. If the parallel lines are arranged quasiperiodically with the long and short spacings L and S $(L/S = \tau)$ according to the Fibonacci sequence, the position of the Nth line of the grid from the origin is given by

$$x_N = N + \alpha + \frac{1}{\tau}\left\lfloor \frac{N}{\tau} + \beta \right\rfloor, \tag{2.3.26}$$

where α and β are arbitrary real numbers and, as mentioned above, $\lfloor\ \rfloor$ represents the greatest integer function. Thus, we get a 5-fold quasiperiodic lattice such that each line in the ith grid intersects each line in the jth grid at exactly one point for each $i \neq j$ and there is a finite number (8 actually) of cell shapes so that the requirement of the minimum separation is satisfied. This is called the Ammann quasilattice. We may generate the dual lattice by placing the points at the mass centers of these cells. The resulting space-filling of 2 unit cells of the fat and the thin rhombs is called a Penrose tiling. Historically Penrose tilings were first realized by the nonperiodic tiling of space by two rhombs associated with the matching rule. A special decoration of rhombs with some line sections may produce the Ammann quasilattice. Deflation rules for rescaling are shown in Fig. 2.3.9; from this it may be demonstrated that the ratio of the numbers of the fat and the thin rhombs in a Penrose tiling is τ.

Penrose and related tilings are geometrical figures with quasiperiodicity. Similarly, atoms or ions may be packed together to form 2D quasicrystals. Various 2D quasicrystals (consisting of atoms and ions) that are quasiperiodic (with 8, 10 and 12-fold symmetries) in 2D and periodic in the direction normal to the plane have been discovered. Historically, the discovery of 2D quasicrystals followed that of 3D quasicrystal.

The arrangement of atoms for a decagonal phase is much like the Penrose tiling. However, Penrose tiling with its matching rules appears to be too artificial a model for a real quasicrystal. An alterative model for a 2D quasicrystal, the covering model, was proposed. This uses black-white decagons to cover the plane with overlaps, only requiring that overlapping parts should have the same structure (see Fig. 2.3.10). The covering model has proved to be geometrically equivalent to the Penrose tiling.[c] However, Penrose tiling is a purely geometrical theory, while the covering is somewhat connected to some energy consideration, high covering density means larger overlapping atomic clusters, and this may stabilize the quasicrystalline structure.

2.3.5 3D Quasicrystals

The signature for a 3D quasicrystal is the icosahedral symmetry shown in its diffraction patterns. The quasilattice displaying such a diffraction pattern may be deduced as follows: Starting from the

[c]P. Gummelt, *Geometric Dedicata* **62**, 1 (1996).

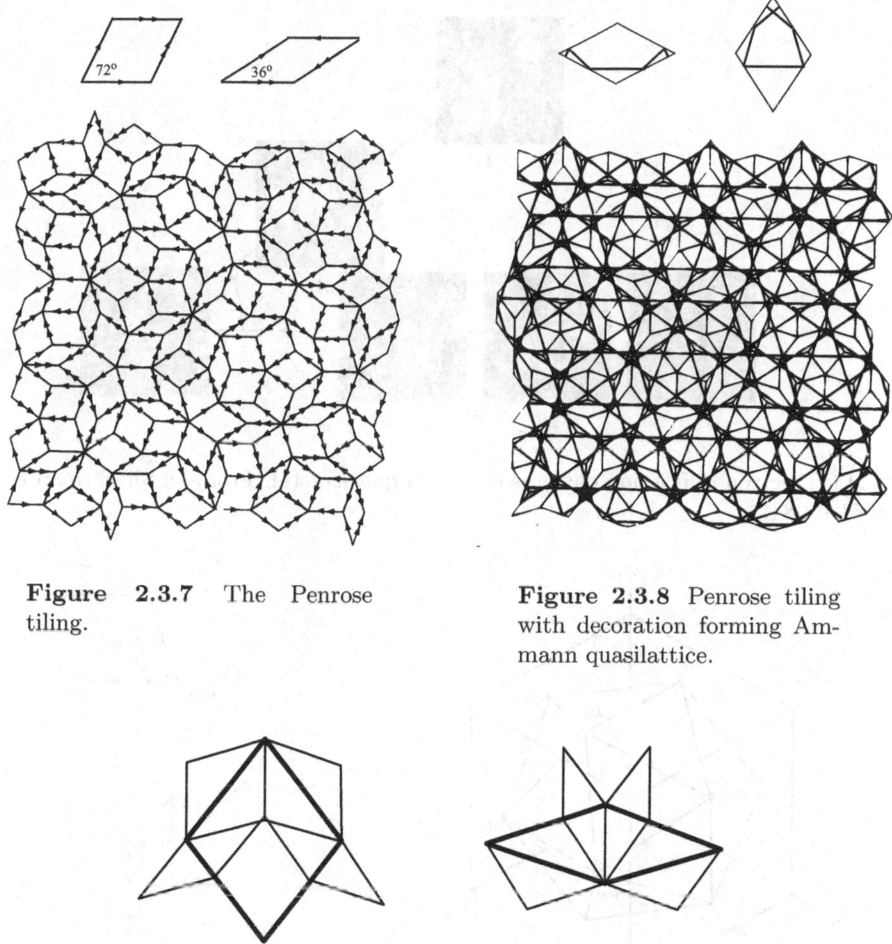

Figure 2.3.7 The Penrose tiling.

Figure 2.3.8 Penrose tiling with decoration forming Ammann quasilattice.

Figure 2.3.9 Inflation and deflation rules for Penrose tiles.

origin, a star of unit vectors $\boldsymbol{e}_i$ $(i = 1, 2, 3, 4, 5, 6)$ may be drawn parallel to the 5-fold symmetry axes of a regular icosahedron; a series of planes perpendicular to these vectors may be drawn at points whose distance from the origin is given by (2.3.16), as we have done in the 2D case: By the intersections of these planes, a quasilattice may be formed, provided that any triplet of planes in the ith, jth and kth grids does not intersect exactly at one point. A finite number of cells is found in this quasilattice, so the minimum separation requirement is satisfied. The duals for this quasilattice have 2 unit cells, the prolate and oblate rhombohedral, which are 3D analogs of the fat and the skinny rhombs of Penrose tiling. Larger rhombic dodecahedron, rhombic icosahedron and rhombic tricontahedron may be formed by tiling these 2 unit cells. These composite structural units (some of them displaying icosahedral symmetry) are frequently found in 3D quasicrystals.

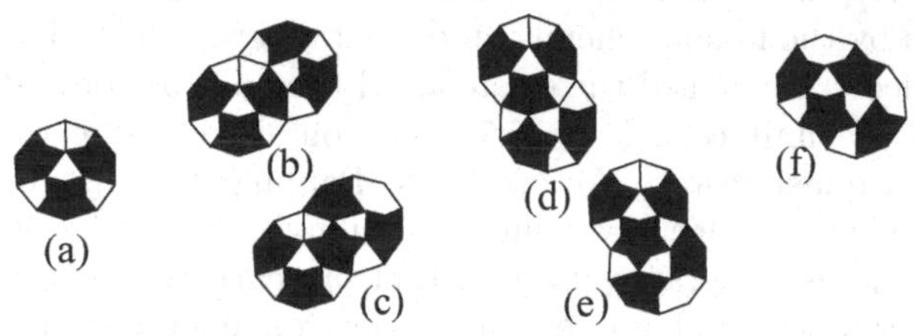

Figure 2.3.10 Schematic diagram for the covering of black-white decagons.

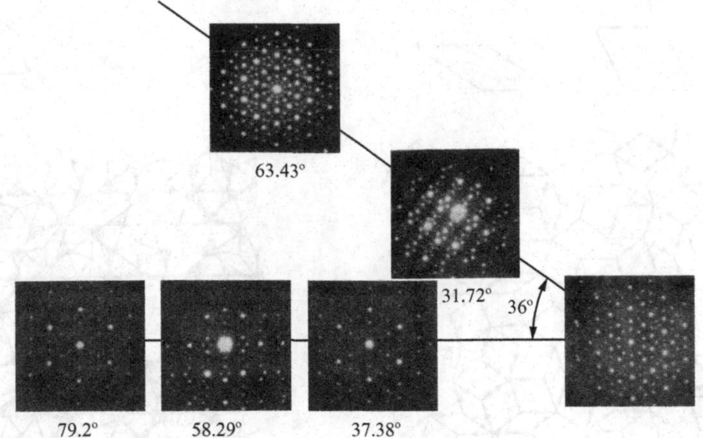

Figure 2.3.11 Electron diffraction patterns of Al-Mn quasicrystal. From D. Shechtman *et al.*, *Phys. Rev. Lett.* **53**, 1951 (1984).

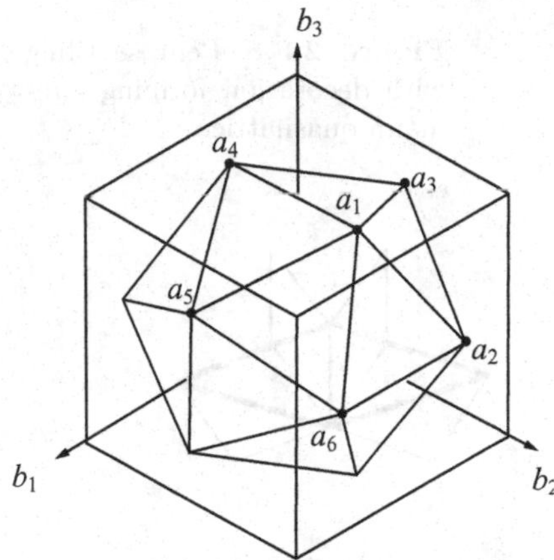

Figure 2.3.12 Regular icosahedron with Cartesian coordinates and 6 vertices which determine the directions of 5-fold axes.

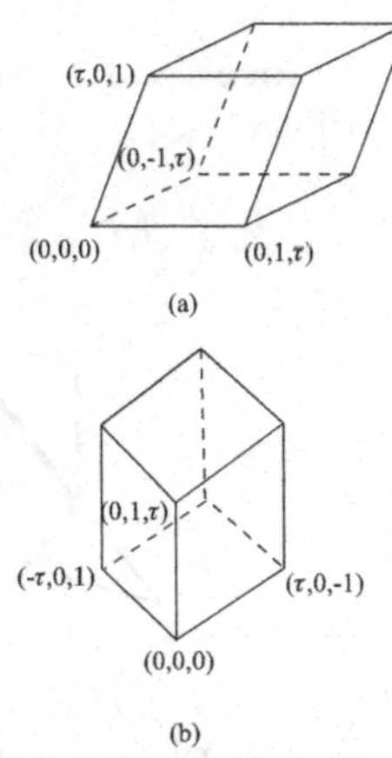

Figure 2.3.13 Unit cells for 3D Penrose tiling. (a) Oblate; (b) prolate.

Just as a 1D quasilattice may be formed by the cut and projection of a 2D periodic lattice, a 3D quasilattice may be formed by the cut and projection from a periodic lattice in 6D hyperspace, and 3D space may be divided into the parallel space and the perpendicular space just as we have done in §2.3.2.

2.3.6 Discussions about Some Basic Notions

After the discovery of quasicrystals in 1984, a quite different approach on icosahedral quasicrystal structure was proposed by the famous chemist and crystallographer, L. Pauling. Then a controversy over this proposal ensued.[d] He denied the existence of the quasiperiodic structure, and constructed a cubic crystal with large unit cell (a = 26.74 Å), containing more than 2000 atoms to fit the diffraction pattern of a quasicrystal. According to Pauling, the apparent icosahedral symmetry was due to multiple twins of cubic crystallites. Pauling's proposal was refuted by a number of experimentalists in quasicrystallography showing that presence or absence of multiple twinning may be experimentally determined by high resolution electron micrography. However, even without

[d]L. Pauling, *Nature* **317**, 512 (1985), *Phys. Rev. Lett.* **58**, 365 (1985); J. W. Cahn *et al.*, *Nature* **319**, 102 (1986).

multiple twinning, it cannot be denied that a sufficiently large unit cell may account for the diffraction pattern of quasicrystals to desired degree of precision.

It is found that though in some cases, 5-fold twins actually appear; while in most cases, no trace of multiple twin boundaries is found. Disregarding some details which proved to be questionable in Pauling's scheme, the controversy introduced by Pauling may be resolved in such a viewpoint that the quasicrystalline structures are an interpolation between two crystalline structures, just as an irrational number is an interpolation between two rational numbers. True quasicrystalline structures can be approached by a series of periodic structures (periodic approximants), the same way that irrational numbers are the limit of a series of rational numbers. For icosahedral phase or decagonal phase, replace τ by its rational approximants,

$$1, \frac{2}{1}, \frac{3}{2}, \frac{5}{3}, \ldots, \frac{611}{377}, \ldots.$$

Pauling's unit cell corresponds to the 5th approximant, which departs from the exact symmetry by 1/400.

Before the discovery of quasicrystals there were already many accounts of incommensurate phases, i.e., crystalline phases in which displacement of atomic sites, chemical concentration or spin are modulated by perturbations with independent incommensurate periodicity λ. The modulations were mostly 1D, e.g., $NaNO_3$, Na_2CO_3, $Ba_2NaNb_5O_{15}$, but sometimes they were 2D, e.g., quartz, or 3D, e.g., wüstite $Fe_{1-x}O$. This is another kind of quasiperiodic structure. In general, the modulation depths are small compared with the underlying crystalline structures, so the average structures characterized by crystalline symmetry are still discernable. We may take a 2D periodic structure with 1D incommensurate modulation as an example. Let the unmodulated atomic sites be determined by vectors

$$\boldsymbol{r} = a(n_1\boldsymbol{e}_1 + n_2\boldsymbol{e}_2), \tag{2.3.27}$$

where $\boldsymbol{e}_1$, $\boldsymbol{e}_2$ are orthogonal unit vectors.

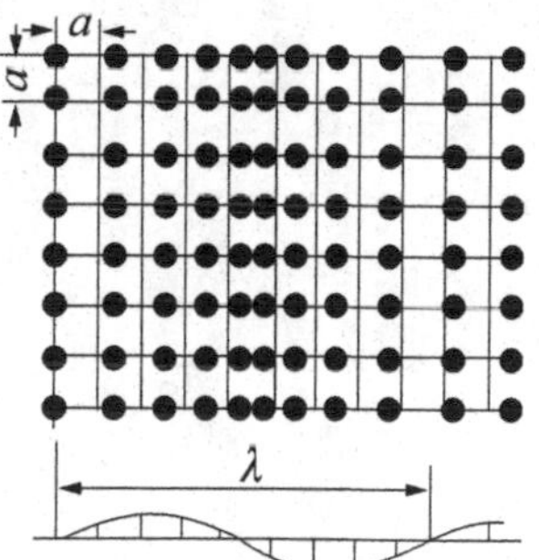

Figure 2.3.14 Schematic diagram showing incommensurately modulated structure (with modulated displacement wavelength λ).

A modulation along x axis is introduced to transform $\boldsymbol{r}$ into $\boldsymbol{r}'$, i.e.,

$$\boldsymbol{r}' = a[(n_1 + \epsilon \sin qn_1a)\boldsymbol{e}_1 + n_2\boldsymbol{e}_2] \tag{2.3.28}$$

for such 1D modulation, diffraction peaks occur at vector $\boldsymbol{Q}$ in the reciprocal space,

$$\boldsymbol{Q} = h\boldsymbol{a}_1^* + k\boldsymbol{a}_2^* + m\boldsymbol{q}. \tag{2.3.29}$$

where $\boldsymbol{a}_1^*$, $\boldsymbol{a}_2^*$ are reciprocal basic vectors of the unmodulated lattice, and h, k, m are integers. If $\boldsymbol{q}$ is incommensurate with $\boldsymbol{a}_1^*$, then three indices are required for indexing the diffraction pattern. The spots with $m = 0$ are called fundamental reflections, related to basic periodic structure, while the $m \neq 0$ spots are called mth order satellites. These may be easily extended to cases of n modulations ($n > 1$).

So the main distinction between quasicrystals and incommensurate phases lies in that the latter is unconnected with forbidden symmetry in classical crystallography. Surely, this is only conspicuous for the cases in which the modulation occurs in more than 1D; while for the 1D case, this distinction is somewhat blurred, a Fibonacci lattice may be considered to be equivalent to a periodic lattice with the spacing a modulated by a function with the period τa.

Bibliography

[1] Vainstein, B. K., V. M. Fridkin, and V. L. Indenbom, *Modern Crystallography*, Vol II. Spinger-Verlag, Berlin (1982).

[2] Megaw, H., *Crystal Structures: A Working Approach*, Saunders, Philadphia (1973).

[3] Hyde, S., S. Andersen, K. Larsson *et al.*, *The Language of Shape*, Elsevier, Amsterdam (1997).

[4] Ball, P., *Made to Measure-New Materials for the 21st Century*, Princeton University Press, Princeton (1997).

[5] Dresslhaus, M. S., G. Dresslhaus, and P. C. Eklund, *Science and Technology of Fullerenes and Nanotubes*, Academic Press, New York (1995).

[6] Janot, C., *Quasicristals: A Primer*, Clarendon Press, Oxford (1992).

[7] DiVincenzo, D. P., and P. J. Steinhardt (eds.), *Quasicrystals: The State of the Art*, World Scientific, Singapore (1991).

Chapter 3

Beyond the Crystalline State

In the crystalline state, long-range order for positions of atoms and orientations of atomic rows is apparent; in alloys, the positional order for atomic sites remains, but the occupation of atomic sites by different species of atoms is somewhat random. In liquids, both the long-range positional order and orientational order are lost; atomic positions show randomness as well as delocalization. If the liquid is rapidly quenched, then atomic sites are localized into a special noncrystalline state, i.e., glass. Thus, glasses are super-cooled liquids, characterized by short range order without long-range order. In addition, liquid crystal shows orientational order; polymers and biopolymers show more complex supramolecular structures. These mostly belong to the realm of soft condensed matter.

§3.1 Alloys and Substitutional Disorder

The simplest disordered system maintains the order of atomic sites but lets various species of atoms somewhat randomly occupy the sites, so the translational symmetry of the crystal lattice is destroyed. This is the structural characteristic of alloys with substitutional disorder.

3.1.1 Ordered and Disordered Alloys

Though substitutional disorder is prevalent in the structure of alloys, this characteristic is most spectacularly displayed in the order-disorder transitions in alloys. For instance, β-CuZn alloy has a critical temperature $T_c = 743$ K: It is in an ordered phase below T_c and a disordered bcc phase above T_c. The T_c of Cu_3Au is 665 K: Its high-temperature phase is a disordered fcc phase. The basic difference between the high and low temperature phases of this type of transition can be illustrated by the distribution of black and white pieces on the lattice shown in Fig. 3.1.1. The black and white pieces in the figure stand for two different kinds of atoms with identical concentrations, i.e., 50%. In figure (a) the two kinds of atomic sites a and b can be easily distinguished, where the a sites are occupied by white pieces and the b sites are occupied by black pieces, and the system stays in ordered state. On the other hand, in figure (b), the white and black pieces are distributed on the sites statistically, i.e., the a and b sites can be occupied by either kind of piece, so the system stays in a disordered state. Strictly speaking, in a disordered state the translation symmetry vanishes, but if the symmetry condition is relaxed by using the concept of statistical symmetry, the system may be regarded as a disordered square lattice as shown in figure (c). It should be noted that from (a) to (c), the transition from the ordered phase to a disordered one involves a change of lattice type: the former is a tilted and enlarged square, containing two kinds of lattice sites; while the latter is the primitive square cell with only one kind of lattice site. Also there is another way of ordering (or reversed ordering), as shown in figure (d): segregation into two distinct phases (white and black).

The perfectly ordered state shown in Fig. 3.1.1(a) is an ideal one ($T = 0$ K); in fact, there must be some wrong occupations at finite temperatures because of thermal fluctuations. The two kinds of sites a and b can be distinguished: It is right for a sites to be occupied by A atoms and wrong

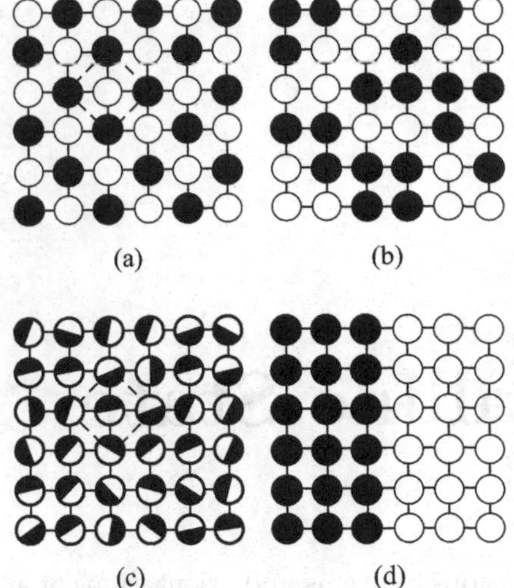

Figure 3.1.1 The illustration of 2D ordered and disordered phases: the distribution of black and white pieces. (a) the ordered phase; (b) the disordered phase; (c) the statistical description for disordered phase; (d) the reversed ordering: the separation into two distinct phases.

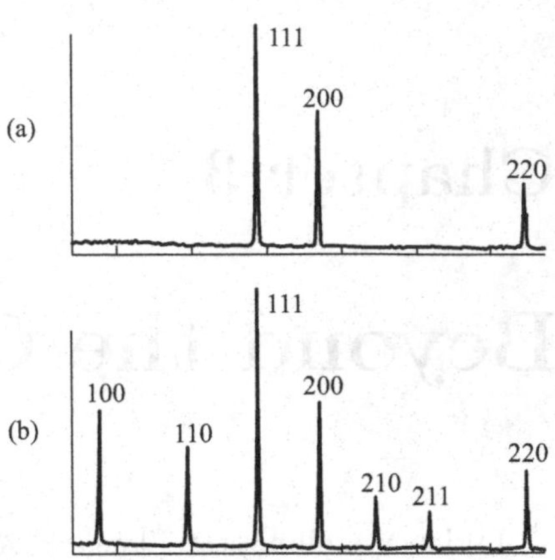

Figure 3.1.2 The powder diffraction patterns of Cu_3Au. (a) disordered phase; (b) ordered phase.

for them to be occupied by B atoms; conversely, it is right for b sites to be occupied by B atoms and wrong for them to be occupied by A atoms. Further, we can define the probability of proper occupation of a sites as r_a and the wrong occupation of them as w_a; similarly r_b and w_b can be defined. The order parameter η may be defined as

$$\eta = r_a - w_b = r_b - w_a = r_a + r_b - 1. \tag{3.1.1}$$

For the perfectly disordered state, the sites are occupied randomly, and the probabilities that the a sites are occupied by A atoms and b sites occupied by B atoms are c_A and c_B, which are the concentrations of A and B atoms respectively. Since $c_A + c_B = 1$, η may be also expressed as

$$\eta = \frac{r_a - c_A}{1 - c_A} = \frac{r_b - c_B}{1 - c_B}. \tag{3.1.2}$$

If $r_a = r_b = 1$, $\eta = 1$ indicating perfect order and if $r_a = c_A$, $r_b = c_B$, $\eta = 0$ indicating perfect disorder. The order parameter defined in this way can show the existence of an ordered lattice, so $\eta \neq 0$ is the condition for the presence of long-range order. The structure factor for X-ray or electron diffraction is

$$F = \sum_a (r_a f_A + w_a f_B) e^{2\pi i(hx_n + ky_n + lz_n)} + \sum_b (r_b f_B + w_b f_A) e^{2\pi i(hx_n + ky_n + lz_n)}, \tag{3.1.3}$$

where f_A and f_B are the structure factors of A and B atoms respectively, and x_n, y_n, z_n are the coordinates of the nth atom. In the ordered structure (based on the bcc) of the β-CuZn type: $\boldsymbol{b} = 000$; $\boldsymbol{a} = \frac{1}{2}\frac{1}{2}\frac{1}{2}$; $c_A = c_B = \frac{1}{2}$. If $h+k+l =$ even, $F = 2(c_B f_B + c_a f_A)$, is the structure amplitude of the main diffraction peaks shown in both the ordered and disordered phases; if $h + k + l =$ odd, $F = \eta(f_B - f_A)$ is the structure amplitude of the superstructure diffraction peaks, which are shown only in the ordered phase. In the ordered structure (based on fcc) of the Cu_3Au type: $\boldsymbol{b} = 000$; $\boldsymbol{a} = \frac{1}{2}\frac{1}{2}0, 0\frac{1}{2}\frac{1}{2}, \frac{1}{2}0\frac{1}{2}$; $c_A = \frac{3}{4}$, $c_B = \frac{1}{4}$; if the values of hkl are all even or all odd, $F = 4(c_B f_B + c_a f_A)$ for the main diffraction spots; if the values of hkl are even mixed with odd, $F = \eta(f_B - f_A)$ for the superstructure diffraction spots, only in the ordered phase. Figure 3.1.2 is the powder diffraction pattern of Cu_3Au, from which the difference between the ordered and disordered phase can be found: The lattice type of the disordered phase is fcc, and it has the systematic extinctions shown by this kind of lattice type. When it is ordered, the lattice type is changed into simple cubic, and the systematic extinctions for the fcc lattice are absent.

3.1.2 Distribution Functions and Correlation Functions

There are two extreme situations in binary alloys: Perfect order and perfect disorder. However the real situation will lie between these extremes, so the statistical problem of the occupation of the atomic sites should be considered more meticulously. For instance, in binary alloys, if the energy for an A-A bond or B-B bond is different from that of an A-B bond, the short-range order parameter for atomic distributions will appear in a disordered phase.

To describe this problem quantitatively, let the total number of closest A-B neighbors be $N_{\rm AB}$. If one atomic site has z closest neighbors, the total number of closest neighbors in the system with N atoms is $(1/2)zN$, so the probability of A-B as closest neighbors is

$$P_{\rm AB} = \lim \frac{N_{\rm AB}}{(1/2)zN}. \tag{3.1.4}$$

If the probability of a single site occupation is $c_{\rm A}$ or $c_{\rm B}$, and the probability for perfectly random occupation of sites is $2c_{\rm A}c_{\rm B}$, then the short-range order parameter that describes the closest neighbors is defined as

$$\Gamma_{\rm AB} = \frac{1}{2}P_{\rm AB} - c_{\rm A}c_{\rm B}, \tag{3.1.5}$$

$\Gamma_{\rm AB} \neq 0$ means the system has short-range order even if the system itself stays in the disordered state.

By the way, the short range order is not limited to closest neighbors; it can be extended to an arbitrary pair of sites. Suppose the distance between the two sites is R, and $\langle \alpha_{\rm AB}^{\rm ab} \rangle$ expresses the average probability of a and b sites with distance R between A and B, we get the pair correlation function

$$\Gamma_{\rm AB}^{\rm ab} = \langle \alpha_{\rm AB}^{\rm ab} \rangle - \langle \alpha_{\rm A} \rangle \langle \alpha_{\rm B} \rangle, \tag{3.1.6}$$

where $\langle \alpha_{\rm A} \rangle$ is the average probability of an A atom occupying any site, obviously $\langle \alpha_{\rm A} \rangle = c_{\rm A}$, similarly we shall get $\langle \alpha_{\rm B} \rangle = c_{\rm B}$. Generally, the correlation of a pair of atomic sites is defined for a definite $R_{\rm AB}$. In the disordered state, atomic correlation attenuates rapidly with increasing $R_{\rm AB}$, so take any lattice site as the origin, the correlation function can be expressed as

$$\Gamma(R) \sim R^{-n} \exp(-R/\xi), \tag{3.1.7}$$

where ξ is the correlation length and when $R \gg \xi$, $\Gamma(R) \to 0$. At a temperature near $T_{\rm c}$, $\xi \to \infty$ and the system enters into the critical region. Here n is a constant determined by the lattice dimension and the type of interaction; this problem will be discussed in Part V. At temperatures below $T_{\rm c}$, there will be long-range order. Since the sign of $\Gamma(R)$ alternates with the change of R, the definition of order parameter should use the absolute value of $\Gamma(R)$ and so the long-range parameter can be defined as

$$|\Gamma_\infty| = \lim_{R\to\infty} |\Gamma_R|. \tag{3.1.8}$$

This agrees with the definition of the long-range order parameter given in the previous subsection, however, long range and short range order parameters are unified by means of the correlation function. At some temperature above $T_{\rm c}$, $|\Gamma_\infty| = 0$, but $\Gamma(R) \neq 0$, so we could get the distribution of the intensity of diffuse scattering of X-rays (not concentrating on reciprocal lattice points) by Fourier transforming to $\Gamma(R)$. Figure 3.1.3 shows the measured values of the correlation function of Cu_3Au, which indicates that even above $T_{\rm c}$, $|\Gamma(R)|$ can extend about 10 atom layers.[a]

[a]The introduction to the experimental method and the result of X-ray diffraction and scattering studies on the ordered and disordered phase of binary alloys, see B. E. Warren, *X-Ray Diffraction*, Addison-Wesley, Reading (1969).

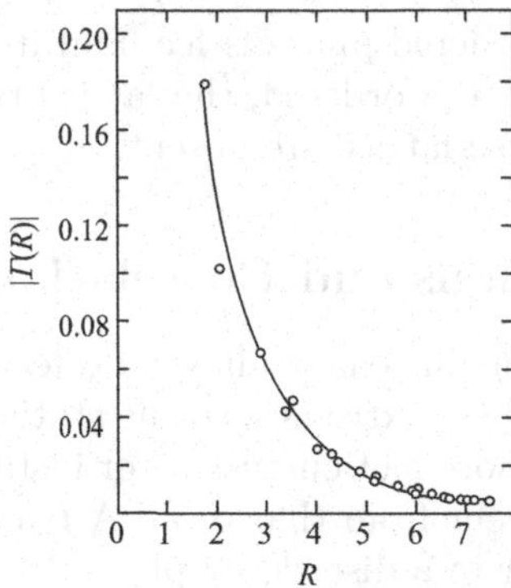

Figure 3.1.3 The measured values of the correlation function of Cu_3Au above T_c.

§3.2 Liquids and Glasses

3.2.1 Overview

First we give the basic physical picture for the melting of crystals: Below the melting temperature T_m, the thermal vibrations of atoms only causes irregular motion of atoms near lattice sites. When the temperature is raised above T_m, thermal fluctuations lead to the complete destruction of the lattice. This process can be simulated by computer, shown in Fig. 3.2.1. Figure 3.2.1(a) gives a vivid picture of the crystal at high temperature: Although there are anomalous thermal motions, the atoms stay near the sites of the lattice. For Fig. 3.2.1(b), the lattice is destroyed, and the positions of atoms are delocalized and no longer stay near any given lattice sites. Figure 3.2.1 shows the clear distinction between the structures of crystal and liquid indicating the disorder and delocalization of the atomic sites in the liquid state.

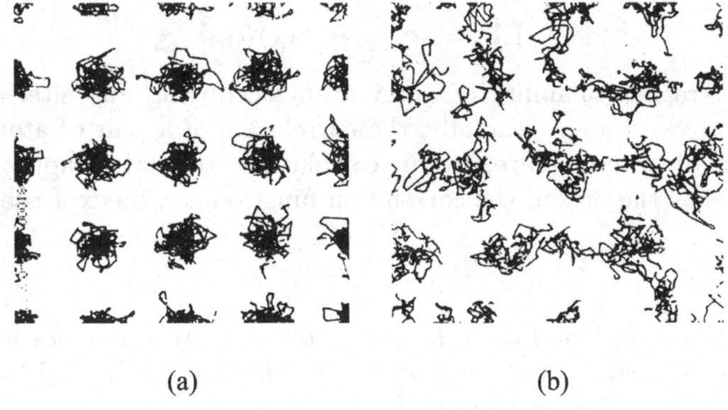

Figure 3.2.1 The computer simulation of the moving trajectories of 32 solid ball (by using periodic boundary condition): (a) a crystal at high temperature; (b) liquid state.

When the liquid is cooled to its melting point, it does not freeze or crystallize at once but exists as a supercooled liquid below the melting point. If the liquid is cooled quickly enough, the supercooled liquid does not crystallize, and a glass is formed. Ordinary glasses are mostly oxide glasses based on SiO_2. These oxides have complex crystal structures and exhibit strong viscosity in the liquid state, which makes the diffusion of atoms very difficult. Thus the formation and growth of the crystal nuclei is very slow, and the usual cooling speed ($10^{-4} - 10^{-1}$ K/s) is enough to prevent these liquid oxides from crystallizing so they instead form glasses.

The situation is quite different for metals or alloys. The diffusivity of atoms is very large, so a glass is not formed with the usual cooling speeds. In 1959, Duwez developed a kind of splat quenching technology, i.e., splashing the liquid drip onto a cool plate with a high thermal conductivity thus raising the cooling speed to 10^6 K/s. In this way he quenched an Au_3Si alloy into the glass state for the first time, and started the era of metallic glasses. Later people developed the melt spinning technique which made possible the production of metal glass on an industrial scale. In addition, a laser vitrification technology was developed, with a cooling speed of $10^{10} - 10^{12}$ K/s, but amorphous Si could not be formed even with this speed of cooling. At present, almost all elements and compounds can be quenched at high speed to the glass state from the melt, with the exception of only a few metallic elements. It can be assumed that with increasing cooling speed, eventually all materials can be prepared in the glass state. This assumption is shown to be viable by simulating quenching experiments with computers.

What is the essential feature of the glass transition? We have discussed the difference between crystal and liquid above. In a crystal atoms are both ordered and localized at lattice sites; whereas the liquid has fluidity, its structure is disordered, and the atoms are delocalized. The localization of atoms is the characteristic of a solid. The glass transition corresponds to the localization of atoms into a disordered structure. There are two types of transitions: Localization and ordering are coupled together in the crystallization of liquids; but in the glass transition, these two are decoupled, i.e., in this transition atomic localization is realized without concurrent ordering of structure.

3.2.2 Statistical Description

Because glass and liquid states are characterized by long-range disorder, a statistical description must be introduced to describe these two states.

In crystals we may form Wigner-Seitz cells by drawing perpendicular planes bisecting the lines linking atomic centers of nearest neighbors; the size and shape of these cells are identical for simple crystals; the same procedure may be applied to a glass, but we can get cells which are no longer identical. These are called Voronoi cells or Voronoi polyhedra (see Fig. 3.2.2). The number of faces of the Voronoi polyhedra corresponds to the coordination number z of the atom. In the topologically disordered system, z is not a constant, and the average value of z is a significant parameter with which to describe the structure.

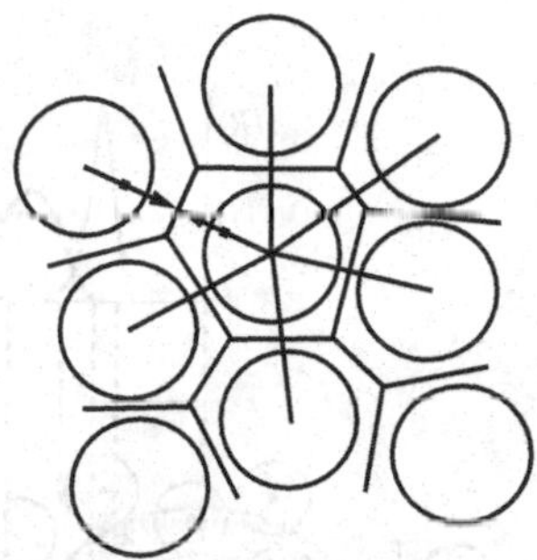

Figure 3.2.2 The 2D schematic diagram of Voronoi polyhedra with disordered system.

To describe the disordered structure quantitatively, we introduce the atomic distribution function: Define $n(\boldsymbol{r}_1)$, $n(\boldsymbol{r}_1, \boldsymbol{r}_2)$, $n(\boldsymbol{r}_1, \boldsymbol{r}_2, \boldsymbol{r}_3), \ldots$, as the statistical density of one-body, two-body, three-body, etc. Then the density distribution function

$$dP(\boldsymbol{r}_1, \boldsymbol{r}_2, \ldots, \boldsymbol{r}_s) = n(\boldsymbol{r}_1, \boldsymbol{r}_2, \ldots, \boldsymbol{r}_s) d\boldsymbol{r}_1 d\boldsymbol{r}_2 \ldots d\boldsymbol{r}_s \tag{3.2.1}$$

is the probability of finding an atom at position $d\boldsymbol{r}_1$ near $\boldsymbol{r}_1$, $d\boldsymbol{r}_2$ near $\boldsymbol{r}_2$, etc. The normalized distributed function is

$$g(\boldsymbol{r}_1, \boldsymbol{r}_2, \ldots, \boldsymbol{r}_s) = n(\boldsymbol{r}_1, \boldsymbol{r}_2, \ldots, \boldsymbol{r}_s)/n^s, \tag{3.2.2}$$

where n is the average density. If $s = 2$, this is the two-body (or pair) distribution function, and if $s = 3$, it is the three-body distribution function, etc. The two-body distribution function is the one most commonly used. Now, introduce a vector $\boldsymbol{R}_{12}$, and the probability of finding an atom in the unit volume in the tiny region around the end of the vector $g(\boldsymbol{r}_1, \boldsymbol{r}_2) \equiv g(\boldsymbol{R}_{12})$. Since liquids and glasses are isotropic, $\boldsymbol{R}_{12}$ may also be chosen in any direction, and the direction of the vector can be ignored, so we can rewrite the atomic distribution function as

$$g(R) = \frac{1}{\langle \rho \rangle} \frac{dn\,(R, R + dR)}{dv\,(R, R + dR)}. \tag{3.2.3}$$

This form of two-body distributed function is called as the radial distribution function (RDF). But, what is the physical meaning of $g(R)$? We may illustrate this problem with a series of examples. Starting from the central atom shown in Fig. 3.2.3, the average number of atoms in the spherical shell with radius from R to $R + dR$ is $g(R)4\pi R^2 dR$. Obviously, in the area of $R < R_0$ (R_0 is the radius of the atom), $g(R) = 0$. From R_0, $g(R)$ increases until the first peak value ($R = R_1$) and then begin to decrease. The first peak of $g(R)$ corresponds to the first coordination shell of the central atom, and the area under the first peak equals the coordination number z of this structure. Because of the existence of disorder, z is not always an integer. Analogously we can define the second nearest spherical shell but with wider peak width and lower peak value, and combined with other peaks gradually. Finally when $R \to \infty$, $g(R) = 1$. Here we introduce the correlation function

$$\Gamma(R) = g(R) - 1, \tag{3.2.4}$$

which can be used to express the deviation of the local region from statistical uniformity. We could also define the range of long-range order as L, i.e., if $R > L$, $\Gamma(R) \approx 0$. While crystal structures have long range order $L \to \infty$, the structures of liquid and glass have only short range order, i.e., L is limited to several atomic spacings. In order to compare with experimental results of X-ray, electron or neutron scattering, we may use $J(R) = 4\pi R^2 g(R)$ as another expression of the RDF and measure the relation between the scattering intensity $I(\theta)$ and scattering angle 2θ to get $J(R)$ or $g(R)$ through a Fourier transform.

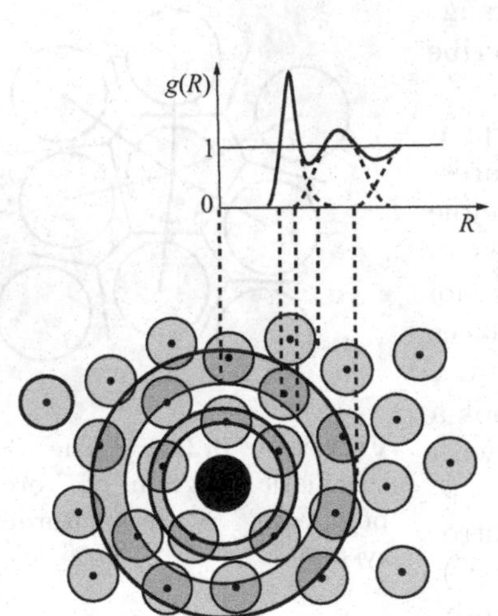

Figure 3.2.3 The 2D schematic diagram explaining the meaning of radial distribution function.

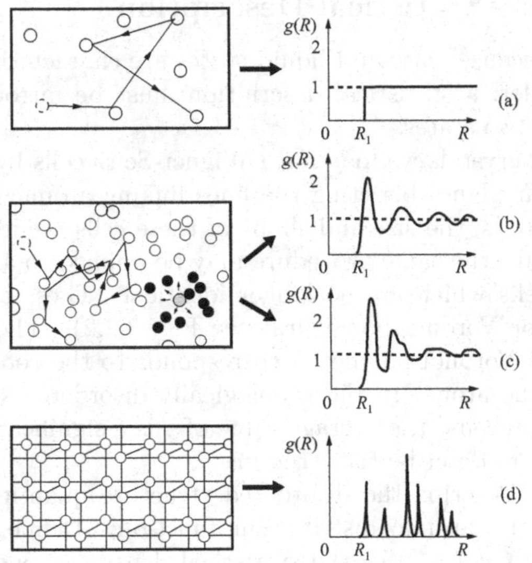

Figure 3.2.4 The RDF $g(R)$ and corresponding atom distributing state at some time of matter in four different states: (a) gas, (b) liquid, (c) glass, (d) crystal.

Figure 3.2.4 shows the RDF and the corresponding atomic distribution in gases, liquids, glasses and crystals schematically. We may conclude that the effect of the RDF in the description of structure comes from the the oscillating part of $4\pi R^2 g(R)$. By the measurement of the RDF, we shall have information about liquid and glass structures, such as short range order and chemical bonding; these are key points to test and distinguish different models. It is valuable to compare the $g(R)$ for the liquid and glass states. The liquid state has somewhat lower peak value, larger width and smoother variation, i.e., some details of the glass state are smoothed out.

The RDF is the average statistical result of all atoms and gives an average description of the surroundings of one atom in the solid. It cannot give the whole picture of the distribution of atoms in an amorphous structure, i.e., it ignores much information lost in the statistical procedure. For example, in a glass composed of different atoms, valuable information about chemical correlation and bonding properties may be missed. Therefore, there are some limitations in the description of amorphous structure by using RDF. In order to get more detailed description of the structure, we may use specific structure models. Some famous models will be introduced in the next subsection.

For liquids, the positions of atoms change with time. Unlike thermal vibrations in a crystal, atoms in a liquid are delocalized from lattice sites. Since the dynamic nature of liquid structure is more prominent, besides the space parameter $\boldsymbol{R}$, the time parameter t should be introduced for a more complete statistical description of structure. The density function in terms of the space and time parameters can be written as

$$n(\boldsymbol{R},t) = \frac{1}{V}\sum_i \delta\{\boldsymbol{R} - \boldsymbol{R}_i(t)\}, \qquad (3.2.5)$$

where V is the volume and δ is the Dirac δ function which has the value 1 if the atom stays at the point $\boldsymbol{R}_i(t)$, and 0 otherwise. Then we introduce the van Hove correlation function

$$\Gamma(\boldsymbol{R},t) = \langle n^*(\boldsymbol{R}',t')n(\boldsymbol{R}'+\boldsymbol{R},t'+t)\rangle. \qquad (3.2.6)$$

In principle, the van Hove correlation function gives a full description of the dynamic structure of condensed matter. Neutron inelastic scattering is the major experimental tool to probe the dynamic structure of condensed matter, and the Fourier transform of the van Hove correlation function is equal to the dynamic structure factor $S(\boldsymbol{k},\omega)$ of neutron inelastic scattering, where $\boldsymbol{k}$ is the wave vector and ω is the frequency. So the static structure factor can be expressed as

$$S(\boldsymbol{k}) = \int_{-\infty}^{\infty} S(\boldsymbol{k},\omega)d\omega. \qquad (3.2.7)$$

However, in order to determine dynamic structure from neutron scattering data, the construction of a suitable theoretical model is also required.

3.2.3 Structural Models for the Amorphous State

The random close packing model was originally introduced by British crystallographer J. D. Bernal in 1959 as a model of liquid structure, so perhaps it is an adequate model for glasses in which packing plays a prominent role, such as metallic glasses. The basic idea of the model is as follows: Consider a liquid as a homogeneous, coherent and essentially irregular assemblage of molecules containing no crystalline regions or holes large enough to admit another molecule. To avoid the complexity brought by the shape of the molecule, we only consider the packing problem of a monoatomic liquid. To develop the random close packing model, Bernal adopted an empirical approach, building models with plasticine balls, ball-bearings, as well as ball-and-spoke. He put many plasticine balls into a rubber container, at various pressures, and found that these balls became polyhedra of various shapes corresponding to Vororoi polyhedra in liquids and glasses. Some dodecahedra were found, while the majority were polyhedra containing many pentagons. Further he designed the experiments with a large assemblage of ball-bearings as well as computer simulation of the ball-and-spoke model. Based on these observations, Bernal proposed the random close-packing model for liquids and glasses. In this model, the space occupancy of balls is 63.66±0.004%, which is obviously lower than the corresponding value of crystalline closest packing 74.05%; the average number of faces on one Voronoi polyhedron is 14.251, and the average number of edges on one face is 5.158, approaching the number of a pentagon. Later the model had been further refined by other scientists and became more exact. If the interaction potential is introduced (so-called soft balls) to replace the initial hard balls, the model became more realistic, for instance, structure relaxation, a common behavior of glasses was also observed in this model, see Fig. 3.2.5(b).

The random close packing model can also be characterized by the distribution of void polyhedra (see Fig. 3.2.6). In the initial work of Bernal, the distribution of voids is continuous one with a high peak centered at tetrahedra then extended to octahedra and beyond. The shapes of the voids are not regular. However, with the addition of an interaction potential, structure relaxation results and the shapes of pores are adjusted so that only tetrahedra and octahedra are left.

The structure of metallic glass is close to the Bernal's random close packing model, and experimental results for the RDF by X-ray scattering match the theoretical calculation approximately.

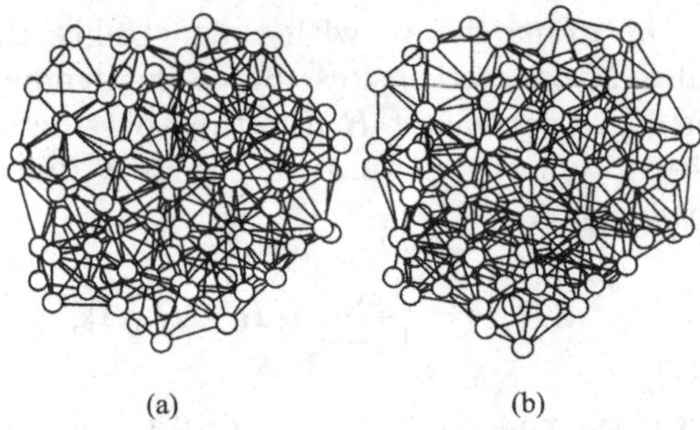

Figure 3.2.5 Ball-and-spoke model for random close packing (100 atoms). (a) hard balls; (b) soft balls (using the Lennard–Jones potential). From J. A. Barker *et al.*, *Nature* **257**, 120 (1975).

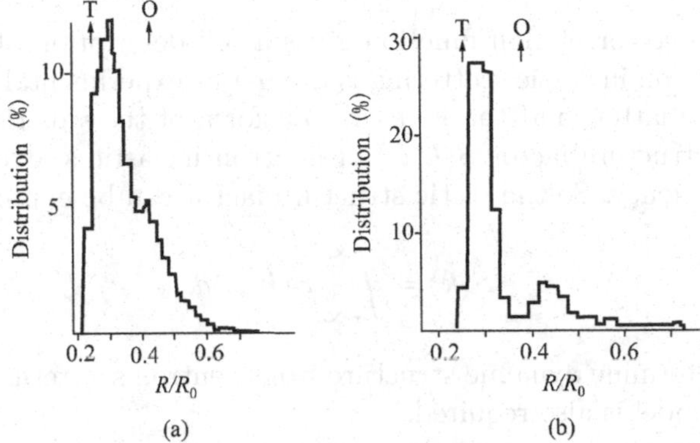

Figure 3.2.6 The distribution of voids in the random close packing model, R_0 is the radius of the sphere, T is the tetrahedral void, and O is the octahedral void. (a) hard balls; (b) soft balls (introducing the interaction potential) (using the Lennard–Jones potential). From J. L. Finney and J. Wallace, *J. Non-Cryst. Solids* **43**, 165 (1981).

The continuous random network model was proposed by W. H. Zachariasen in 1932 for the elucidation of the structure of SiO_2 glass. The basic idea is: A unit of the structure is a tetrahedron composed of 4 O atoms which are bonded with the central Si atom by four valence bonds. The adjacent tetrahedra share a common vertex, so with an infinite extension, they form SiO_2 glass. In this way, a random network is formed; however, the introduction of randomness may allow the Si-O-Si bond angle to deviate from the average value, and the bond length to be stretched. Even the azimuth of the tetrahedron can be varied by a small amount by rotation along the Si-O bond (see Fig. 3.2.7). Just like the random close packing model, we can deduce the coordinates of an atom, the density and the statistical number of the members forming a closed loop from the continuous random network model. Figure 3.2.8 shows the comparison of RDF derived from the continuous random network model and experimental results on Ge.

This model can describe not only the structure of glasses, but also qualitatively the structure of liquids with tetrahedral coordination, such as liquid Ge, Si, and even water. In Chap. 2, we discussed the crystal structure of ice, which also has tetrahedral coordination. The pair correlation function of O-O observed in water is shown in Fig. 3.2.9, and by taking the integral to the first shell we can get

$$n = 4\pi\rho \int g_{\text{oo}}(r) r^2 dr = 4. \tag{3.2.8}$$

By the way, after the second shell, g_{oo} tends to one asymptotically.

The statistical honeycomb model was proposed by mathematician H. S. M. Coxeter in 1958. The Voronoi polyhedron is denoted by Schläffli symbol $\{p, q, r\}$, where p is the number edges of polygons, q is the number of faces sharing a vertex, r is the number of polyhedra sharing an edge. For the random close packing model, the statistical distribution shows that $q = r = 3$, $5 \leqslant p \leqslant 6$. We also

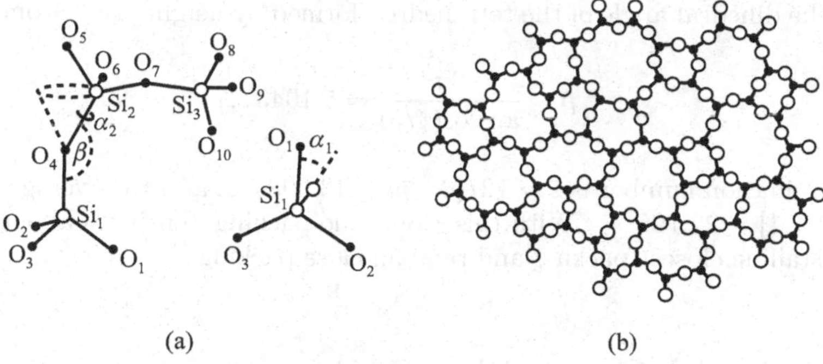

Figure 3.2.7 (a) Si-O tetrahedrons joined at common O atoms; (b) 2D schematic diagram of the continuous random network model proposed by Zachariasen. From A. C. Wright *et al.*, *J. Non-Cryst. Solids* **49**, 63 (1982).

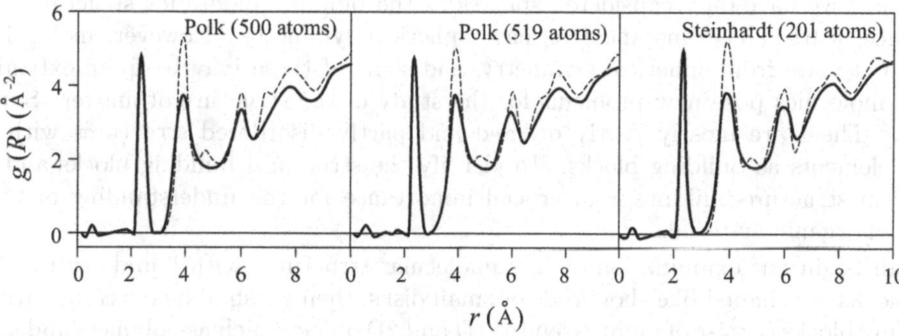

Figure 3.2.8 The RDF of Ge, the solid line is the experimental result while the dotted line is the calculated result according to the continuous random network model. From A. C. Wright *et al.*, *J. Non-Cryst. Solids* **49**, 63 (1982).

know that stacking regular polyhedra in 3D must satisfy the condition (see §2.1.3)

$$\cos\frac{\pi}{q} = \sin\frac{\pi}{p}\sin\frac{\pi}{r}. \tag{3.2.9}$$

In (3.2.8), no integer p satisfies $q = r = 3$, and the number p is a non-integer in $\{p, 3, 3\}$, which means the honeycomb can only exist in a statistical sense. In a Voronoi polyhedron, atoms stay at

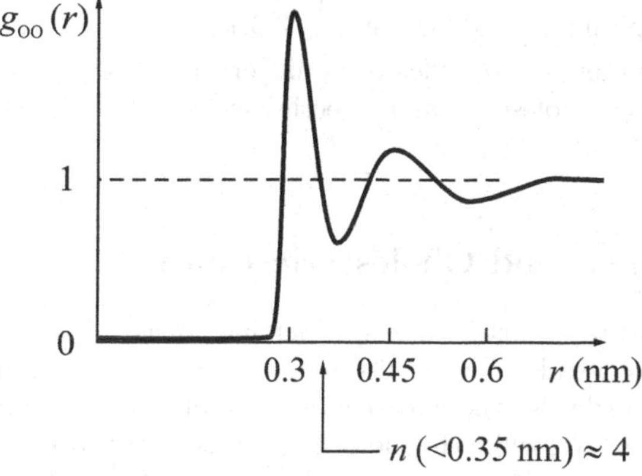

Figure 3.2.9 The pair correlation function of O-O in water.

the center, and the dihedral angle of the tetrahedron formed by neighboring atoms is arccos(1/3), so

$$\bar{p} = \frac{2\pi}{\arccos(1/3)} = 5.1043\ldots \tag{3.2.10}$$

The average coordination number is $\bar{z} = 12/(6-\bar{p}) = 13.398\ldots$, and the average number of vertices is $N_0 = 4/[(6/\bar{p}) - 1] = 22.796\ldots$. With this model the packing density is about 0.7071, which lies between the crystalline closest packing and random close packing.

§3.3 The Liquid-Crystalline State

3.3.1 Overview

Up to now we have only considered spheres as the building blocks for structures; this is quite natural since nearly all atoms and ions have spherical symmetry. However, molecules may have shapes that deviate from spherical symmetry, and some of these may be quite extended in 1D or 2D. These molecules pose new problems for the study of the structure of matter: Supramolecular structures. These are mostly partly ordered and partly disordered structures with 1D and 2D structural elements as building blocks. To identify the structural building blocks and to elucidate their roles in structure-building is of crucial importance for the understanding of these complex structures of organic materials.

We shall begin our examination of supramolecular structures with liquid crystals in which the building blocks are shaped like short rods or small disks, then we shall be concerned with structures with building blocks consist of more extended 1D and 2D objects such as polymers and self-assembled membranes. If structural building blocks are all identical or randomly mixed, statistics will be important for their structures, while several alternative building blocks are coded into sequence, then informatics shall play a crucial role in such structures, as exemplified by biopolymers.

Some of their properties are strongly anisotropic and exhibit a certain degree of fluidity. They may be called mesophases, i.e., phases with symmetries of structures as well as physical properties lying intermediate between that of crystals and liquids. Crystals have both long-range positional order and orientational order; common liquids have neither long-range positional order nor orientational order; while liquid crystals have long-range orientational order but lack positional order.

The transition into the mesophase may be simply made by lowering the temperature; the liquid crystals thus formed are called the thermotropics. Another way to form a liquid crystal is dissolving the organic molecules in a suitable solvent at a suitable concentration; these are called lyotropics. The building blocks forming the liquid crystalline state may be divided into four kinds: (a) rod-like molecules; (b) disc-like molecules; (c) (flexible) long chain polymers connected by rod-like or disc-like molecules; and (d) membrane composed by amphiphilic molecules. All these structural units and some representative compounds are shown in Fig. 3.3.1.

According to the symmetry and order of liquid crystal, thermotropics can be further divided into three phases: nematic, cholesteric and smectic (see Fig. 3.3.2). These will be discussed in the following subsection.

3.3.2 Nematic Phase and Cholesteric Phase

The nematic liquid crystal is a typical one, which has rather large fluidity. The basic characteristics of the nematic liquid crystals are the existence of long-range orientational order and the absence of long-range translation order for the mass-centers of molecules. The molecules are arranged along a special direction $\bar{n}$ (director), while the molecular mass-centers are randomly distributed in space. These account for the absence of long-range translational order, so the system shows the characteristics of liquid. Furthermore, even the molecular orientations have some variation. For a rod-like

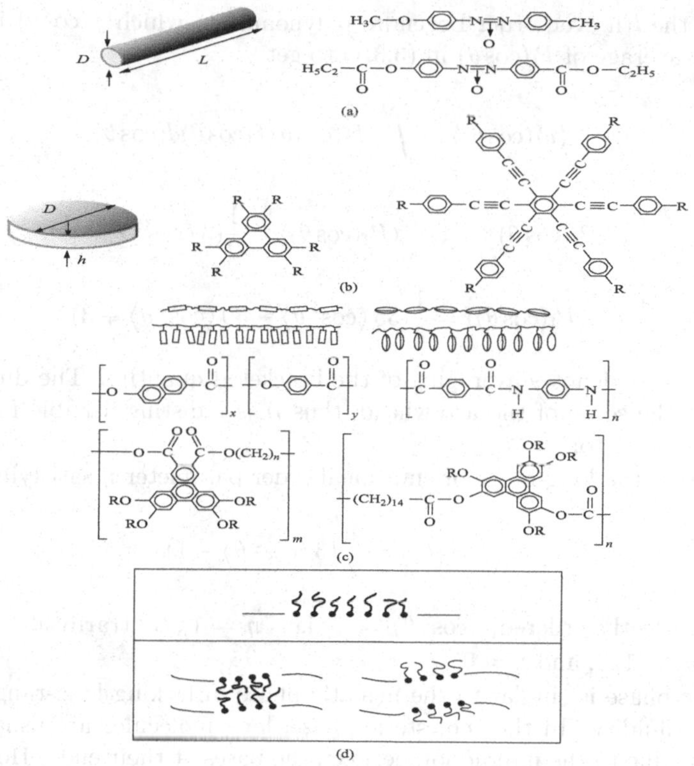

Figure 3.3.1 Four kinds of building blocks for the liquid crystalline state: (a) rod-like molecules; (b) disc-like molecules; (c) (flexible) long chain polymers connected by rod-like or disc-like molecules; (d) membranes composed by amphiphilic molecules.

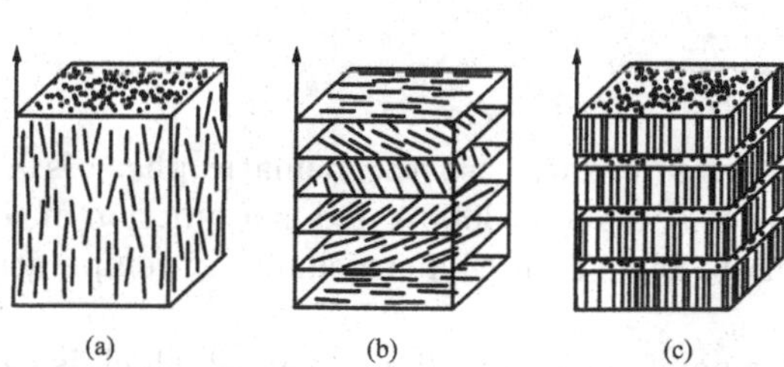

Figure 3.3.2 The schematic diagram of structures of liquid crystals. (a) nematic phase; (b) cholesteric phase; (c) smectic phase.

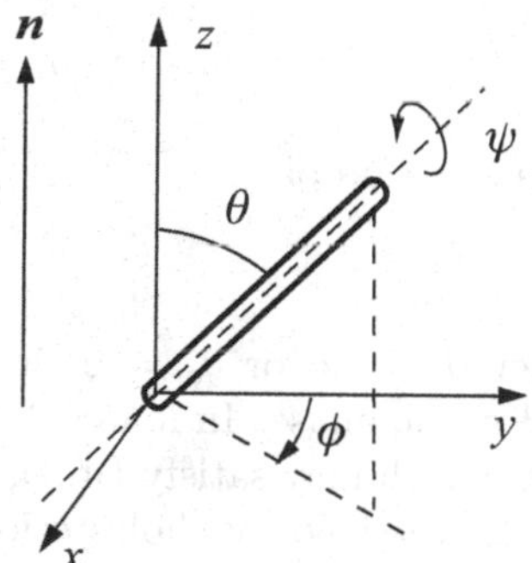

Figure 3.3.3 The Euler angles which describe the liquid crystal in the nematic phase.

molecule, we introduce three Euler angles θ, ϕ and ψ to describe the nematic liquid crystal (see Fig. 3.3.3), and the key point is the distribution with θ around the director $\bar{\boldsymbol{n}}$. So we shall introduce the distribution function

$$f(\cos\theta) = \sum_{l=0,\text{even}} \frac{2l+1}{2} \langle P_l(\cos\theta)\rangle \, P_l(\cos\theta), \tag{3.3.1}$$

where $P_l(\cos\theta)$ is the lth even-order Legendre polynomial in which $f(\cos\theta)$ is an even function of $\cos\theta$. We take the average of $P_l(\cos\theta)$ in (3.3.1) to get

$$\langle P_l(\cos\theta)\rangle = \int_{-1}^{1} P_l(\cos\theta)f(\cos\theta)d(\cos\theta),$$

$$\langle P_0(\cos\theta)\rangle = 1, \quad \langle P_2(\cos\theta)\rangle = \frac{1}{2}(3\left\langle\cos^2\theta\right\rangle - 1), \tag{3.3.2}$$

$$\langle P_4(\cos\theta)\rangle = \frac{1}{8}(35\left\langle\cos^4\theta\right\rangle - 30\left\langle\cos^2\theta\right\rangle + 3),$$

where the symbol $\langle\cdots\rangle$ denotes averaging of the bracketed quantity. The director $\bar{\boldsymbol{n}}$ has direction without specifying the sense of the arrowhead, thus $\bar{\boldsymbol{n}}$ is indistinguishable from $-\bar{\boldsymbol{n}}$, so it is quite different from a usual vector.

Now we introduce the long-range orientational order parameter η satisfying

$$\eta = \langle P_2\rangle = \frac{1}{2}\left(3\left\langle\cos^2\theta\right\rangle - 1\right). \tag{3.3.3}$$

If the system is perfectly ordered, $\left\langle\cos^2\theta\right\rangle = 1$, i.e., $\eta = 1$; contrarily if the system is perfectly disordered, $\left\langle\cos^2\theta\right\rangle = 1/3$, and $\eta = 0$.

The cholesteric phase is similar to the nematic phase in lacking long-range positional order, so it also has certain fluidity. In the cholesteric phase, long molecules are usually flat and arranged into lamellar layers due to the interaction between the bases at their ends. However, their long axes are lined up along the plane inside the layers, and the orientation of molecules inside a particular layer is similar to the situation in the nematic phase. Because of the left-right asymmetry of the molecular structure, the orientation of the director rotates at a steady rate as one moves normal to the layers, as shown in Fig. 3.3.2(b).

The screw axes superimposed on the director $\bar{\boldsymbol{n}}$ can be written as

$$n_x = \cos(q_c z + \phi), \; n_y = \sin(q_c z + \phi), \; n_z = 0, \tag{3.3.4}$$

with a period of

$$L = \frac{\pi}{|q_c|}. \tag{3.3.5}$$

When $L = \infty$ or $q_c = 0$, it changes into the nematic phase, so the nematic phase is a special cholesteric phase. In many cholesterics, the value of L is about 3000Å and is a function of temperature, which may satisfy Bragg law for diffraction of visible light. Thus a change of temperature may change the color of cholesterics.

It is interesting to note that there is special type of cholesteric liquid crystal which is called the liquid crystalline blue phase with a special complicated organization.[b]

3.3.3 Smectic Phase and Columnar Phase

In the common smectic phase, the mass-centers of rod-like molecules are arranged in parallel periodic layers with a definite spacing. Inside the layers, the molecules line up along a certain director $\bar{\boldsymbol{n}}$ which may coincide with the normal to the layer or at a definite angle to it, while the positions of the mass centers are disordered. The molecules can only move within the layers, while the layers may slip by each other. The smectic phase can be divided to two kinds, smectic A and C phases. In smectic A, the molecules are upright in each layer — see Fig. 3.3.2(c); while in smectic C, the molecules are inclined with respect to the layer normal.

[b]About the blue phase, see D. C. Wright and N. D. Mermin, *Rev. Mod. Phys.* **61**, 385 (1989); T. Seideman, *Rep. Prog. Phys.* **53**, 657 (1990).

The smectic phase is more ordered than the the nematic mentioned above, for it has not only 2D molecular orientational order but also 1D translational symmetry along the layer normal. Its distribution function

$$f(\cos\theta, z) = \sum_{l=0,\text{even}} \sum_{n=0} A_{ln} P_l(\cos\theta)\cos\left(\frac{2\pi nz}{d}\right), \tag{3.3.6}$$

satisfies the normalization condition

$$\int_{-1}^{1}\int_{0}^{d} f(\cos\theta, z)dzd(\cos\theta) = 1. \tag{3.3.7}$$

The results are

$$\begin{aligned} A_{00} &= \frac{1}{2d}, \\ A_{0n} &= \frac{1}{d}\left\langle \cos\left(\frac{2\pi nz}{d}\right)\right\rangle, (n \neq 0) \\ A_{l0} &= \frac{2l+1}{2d}\left\langle P_l(\cos\theta)\right\rangle, (l \neq 0) \\ A_{ln} &= \frac{2l+1}{2d}\left\langle P_l(\cos\theta)\cos\left(\frac{2\pi nz}{d}\right)\right\rangle, (l, n \neq 0). \end{aligned} \tag{3.3.8}$$

The order parameters are

$$\eta = \langle P_2(\cos\theta)\rangle, \ \tau = \langle \cos(2\pi z/d)\rangle, \ \sigma = \langle P_2(\cos\theta)\cos(2\pi z/d)\rangle, \tag{3.3.9}$$

where z is the coordinate of the molecular mass-center. It can be easily found that for the isotropic liquid phase $\eta = \tau = \sigma = 0$, for the nematic phase $\eta \neq 0$, $\tau = \sigma = 0$, and for the smectic phase $\eta \neq 0$, $\tau \neq 0$, $\sigma \neq 0$. Therefore, to describe the smectic phase, besides the orientational order parameter η, the translational order parameter τ should be introduced. If all of the molecular mass-centers are at $z = 0$, $\tau = 1$; if the mass-centers are uniformly distributed, $\tau = 0$. In the columnar liquid crystalline phase, dissimilar molecules are stacked into columns with a hexagonal structure, so it has 2D translational order as shown in Fig. 3.3.4.

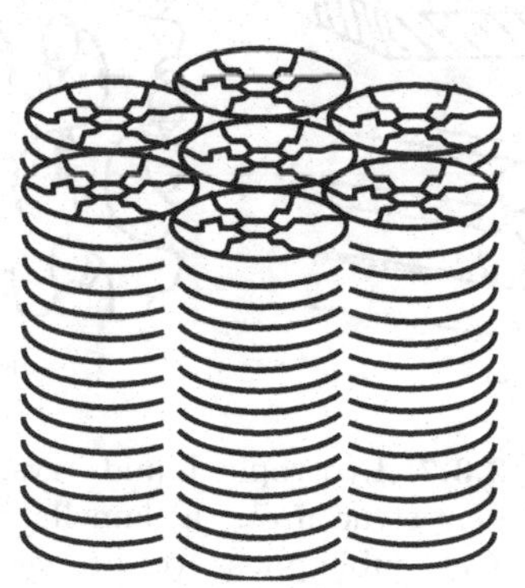

Figure 3.3.4 Columnar phase liquid crystal.

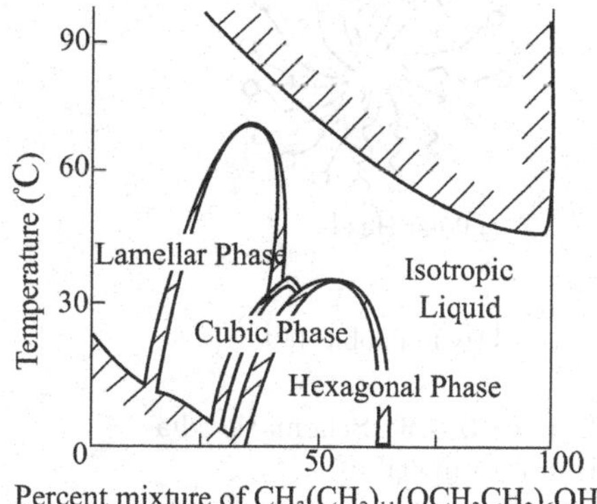

Figure 3.3.5 The phase diagram of soap-water.

When the temperature is increased, the translational order of the columnar phase disappears first, then the orientational order vanishes. With the disappearance of translational order, the smectic phase is first changed into the nematic phase, then the nematic phase is changed into an isotropic liquid. If the building blocks are disk-like molecules, this is called discotics. There are two distinct types of discotics: Columnar and nematic. For the columnar phase, discs are stacked one over the other aperiodically to form liquid-like columns, while different columns constitute a 2D lattice.

3.3.4 Lyotropics

The molecular order of lyotropic liquid crystals is quite different from that of the thermotropics in both structure and interaction.

Soap-water is a typical lyotropic liquid crystal, which shows birefringence. The constitution of this sort of liquid crystal is rather complex, and the interaction is very tortuous. The soap (lauric acid) molecule is a typical amphiphile characterized by the fact that the head and the tail have quite different affinities. Its head is a polar group $-CO_2$–K^+, which is soluble in water and is hydrophilic; while the tail of hydrocarbon chain $-(CH_2)_7CH_3$ is hardly soluble in water, and is hydrophobic. The amphiphilic molecules or the surfactant, i.e., in water solution they are inclined to swarm on the interface with the hydrophilic group touching the water, and the hydrophobic one pointing away from the water. So a layer of amphiphilic molecules at the interface constitute a membrane within liquid mixtures.

The building block of an ordinary crystal is the 0D (zero dimension) atom or ion, that of the thermotropic liquid crystal is the 1D rod-like molecule, while that of the lyotropic liquid crystal is the 2D liquid membrane. The structure of the liquid membrane itself does not have long-range order, however the lyotropic liquid crystal made up of these building blocks may acquire long range order. Figure 3.3.5 shows the phase diagram of soap-water, from which we can find that with the increasing concentration of soap, there are a series of lyotropic liquid crystalline phases with different structures.

In soap-water at low concentration, the liquid phase is isotropic, in which the amphiphilic molecules for sphere-like closed membranes, called micelles, are shown in Fig. 3.3.6. The sizes and the shapes of micelles are currently uncertain, but only keep statistical equilibrium with amphiphilic molecules dispersed in the liquid around them. If the solution is diluted by water, the micelles disappear rapidly; on the other hand, with increasing concentration, micelles with larger areas are formed and thus a series of lyotropic liquid crystalline phases would appear finally at different values of concentration.

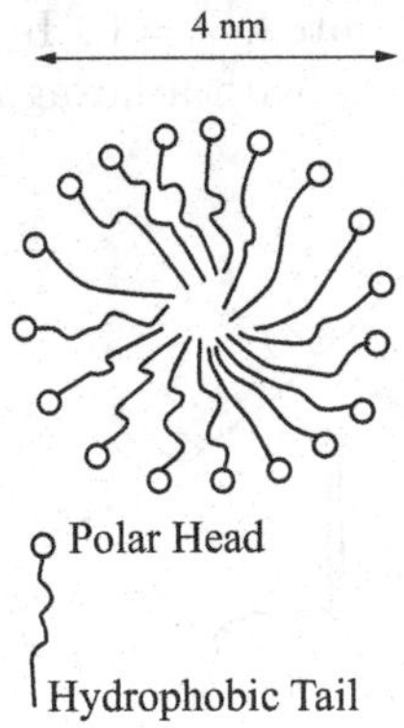

Figure 3.3.6 Schematic diagram of a micelle.

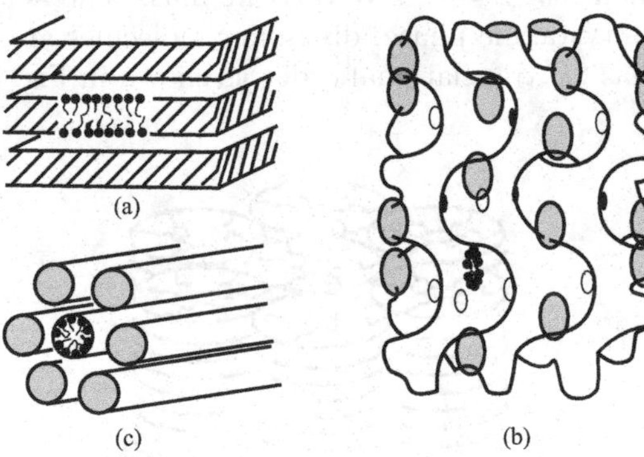

Figure 3.3.7 Lyotropic liquid crystal phases with different structures. (a) lamellar phase; (b) 'bicontinuous' cubic phase; (c) hexagonal phase.

Three types of lyotropic phases shown in Fig. 3.3.7 were obtained by X-ray diffraction experiments. The sharp diffraction lines in the small angle range indicate a clearly periodic structure in the range of several dozen Angstroms; but the diffraction peaks are fuzzy in the large angle range, which indicates that the liquid membranes themselves (in the range of several angstroms) have no long-range order. Hexagonal and lamellar phases are simple, while the cubic phase is complex. In the cubic phase, the membrane becomes bicontinuous, separating space into two distinct regions. The interface has minus Gaussian curvature κ_G and the genus g_t is infinite. This is just the IPMS mentioned in §2.1.2.

The study of lyotropics belongs to the growing field of self-assembly of soft condensed matter. This self-assembling process is also utilized in materials science to synthesize structural patterns on the mesoscale, i.e., midway between microscopic and macroscopic scale, e.g., to fabricate mesoporous materials, and templates for photonic crystals. This kind of structure is also encountered in biomembranes. Thus research on lyotropic liquid crystal not only helps us understand the physical phenomenon of surfactants but also has extensive applications in many fields, such as the food and pharmaceutical industry, cosmetics, petroleum reclamation, mineral separation, etc., and it is also important for biology.

§3.4 Polymers

3.4.1 Structure and Constitution

Polymers consisting of long chain molecules are also called macromolecules. The structural unit of a polymer molecule is the monomer, and the total number of monomers in a macromolecule ranges between 10^2 and 10^5. In Fig. 3.4.1, many structures of macromolecular monomers are shown. Figure 3.4.2(a) gives the space structure of $-CH_2-$ in polyethylene, and Fig. 3.4.2(b) gives the bonding configuration of a series of C–C bonds.

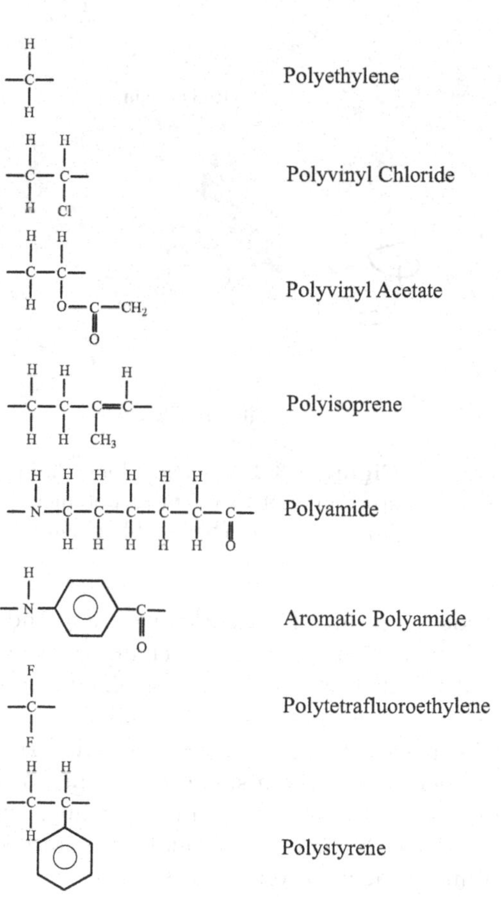

Figure 3.4.1 The structural formulas of monomers in some polymers.

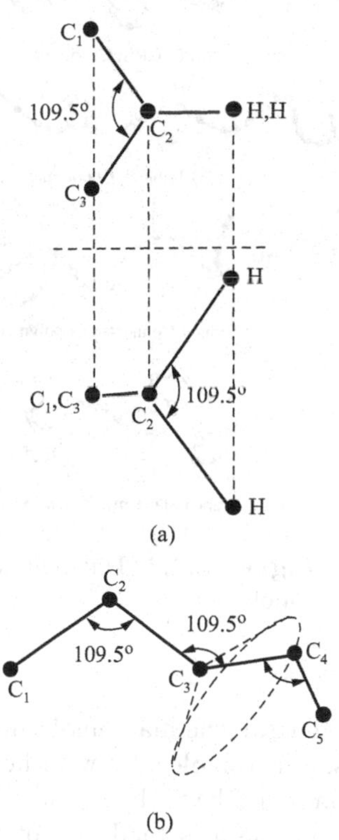

Figure 3.4.2 (a) CH_2 in the polyethylene (the upper figure is top view, and the lower one is a side view); (b) the bonding configuration of 5 C atoms.

Monomers can be repeated simply to constitute the macromolecule. For instance, let A stand for the repeated unit, then we will get

$$\text{X–A–A–A–A–} \cdots \text{–A–A–A–Y},$$

where X and Y are the initial and the terminal bases. In polyethylene, X=Y=H. The structures and types of monomers may be not completely the same in a polymer, and usually different variants may be formed. The arrangement of monomers in a polymer is also multifarious. For example,

$$\text{—A—B—B—A—B—A—A—A—B—A—},$$

where A and B are different monomers.

The copolymer is made up of two or more different monomers according to some modes of arrangement. According to different modes of arrangement, it has many types including random copolymer, block copolymer and so on, as shown in Fig. 3.4.3. In biopolymers, the monomers are no longer the same and this has important consequence for biological properties. However, besides the specific structures of the macromolecular chains, the configurations of polymers are also very complex.

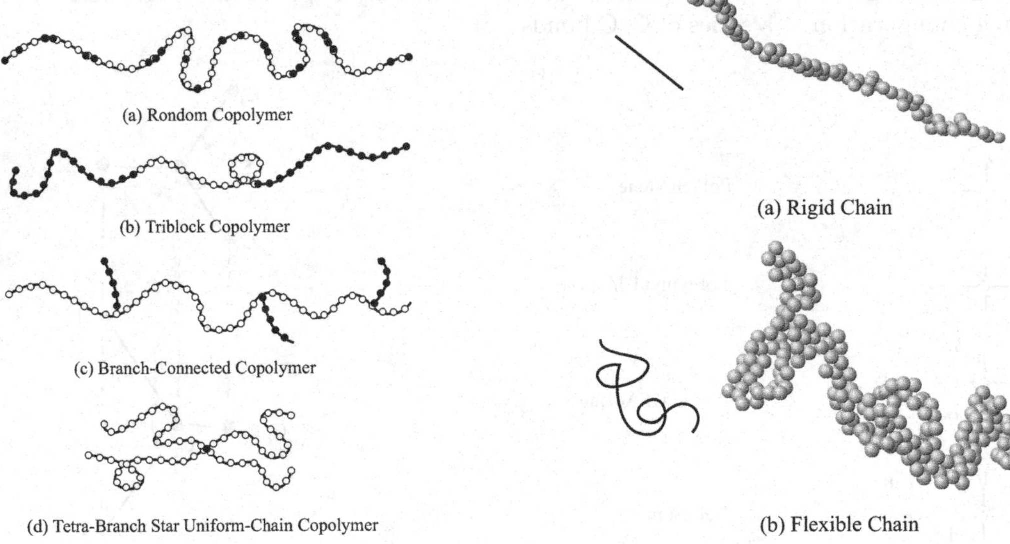

Figure 3.4.3 Different kinds of copolymer.

Figure 3.4.4 The long-chain structures of two typical polymers. (a) rigid chain; (b) flexible chain.

In general, the macromolecular chains can be divided into three kinds: Flexible, rigid and helical chains. Macromolecules with benzene rings or heterocyclic rings in the main chain are usually the rigid ones and have the rod-like shape shown in Fig. 3.4.4(a). For macromolecules like polyethylene, the main chain is made up of C–C bonds, and several conformations with similar energy value may attach to each other by rotations of bonds. This makes the molecular chain multifarious, like flexible coils, as shown in Fig. 3.3.4(b). There is another kind of chain shape which, in essence is a flexible one, and the potential barrier for internal rotation of the main chain is not high, with the interactions between different parts of the molecule, the stable helix conformation is formed. The α-helix in proteins and the double helix of DNA in nucleic acids belong to this type.

3.4.2 Random Coils and Swollen Coils

In this subsection we will introduce the structural model of disordered polymers with flexible long-chain molecules as the basic units — the random walk model. This model was proposed by

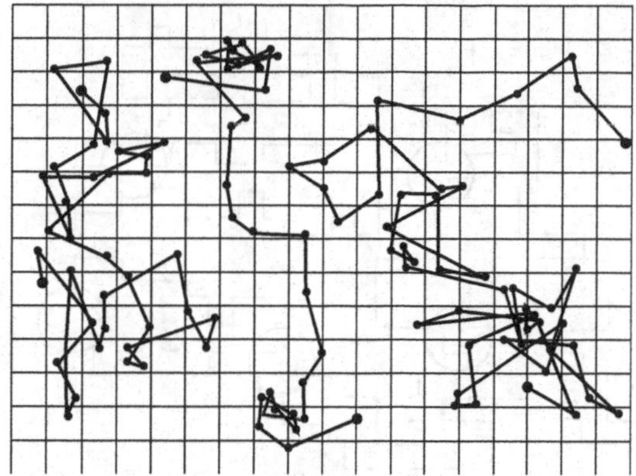

Figure 3.4.5 The loci of particles in Brownian motion (Perrin's result).

P. T. Flory in 1949 based on loci of the particle in Brownian motion. Figure 3.4.5 shows the observed result for the loci of particles in Brownian motion by J. Perrin in 1909. Following the random movement of the particles in the liquid, Flory divided the macromolecular long chain into many segments each with length a, which corresponds to a segment of the moving locus of a particle in Brownian motion. Starting from the origin, a flexible polymer changes its direction randomly and continuously in space, and moves randomly from the origin. After N steps ($\boldsymbol{a}_1$, $\boldsymbol{a}_2$, ..., $\boldsymbol{a}_N$), the distance away from the origin is

$$\boldsymbol{r} = \boldsymbol{a}_1 + \boldsymbol{a}_2 + \cdots + \boldsymbol{a}_N = \sum_{i=1}^{N} \boldsymbol{a}_i. \tag{3.4.1}$$

Since every step is arbitrary, the average of $\boldsymbol{r}$ is zero. More significant is the average value of r^2

$$\langle r^2 \rangle = \sum_{i=1}^{N} (\boldsymbol{a}_1 + \boldsymbol{a}_2 + \cdots + \boldsymbol{a}_N)^2 = \sum_{i=j} \boldsymbol{a}_i \cdot \boldsymbol{a}_j + \sum_{i \neq j} \boldsymbol{a}_i \cdot \boldsymbol{a}_j = Na^2 = R_0^2. \tag{3.4.2}$$

There are two sorts of terms in the formula above: One is the products of $\boldsymbol{a}_i$ and $\boldsymbol{a}_i$ with the same subscripts, which equal a^2 ; the other sort is the products of $\boldsymbol{a}_i$ and $\boldsymbol{a}_j$ which equal $a^2 \cos\theta_{ij}$, where θ_{ij} stands for the deviation angle between $\boldsymbol{a}_i$ and $\boldsymbol{a}_j$. Since the directions of $\boldsymbol{a}_i$ and $\boldsymbol{a}_j$ are random, so $\sum_{i,j} a^2 \cos\theta_{ij} = 0$ and we will get

$$R_0 = N^{1/2} a. \tag{3.4.3}$$

From (3.4.3) above, we find that the size R_0 of the random chain is proportional to $N^{1/2}$. The result is the same for any dimension of the system, so it is valid for 1D, 2D, 3D and even higher dimensions.

Let the probability by which the distance between the head and the tail of a flexible polymer is R be $P(R)$. For liquid phase, N is large, and the Gaussian distribution is a good approximation, i.e.,

$$P(R) = A \exp(-BR^2), \tag{3.4.4}$$

where $A = (2\pi/3)^{-3/2} R_0^{-3}$, $B = (3/2) R_0^{-2}$.

The random coil model of flexible macromolecules is the simplest one. However, the random coil can have many intersections with itself, which is impossible for real macromolecules because of steric repulsive interactions between intersecting monomers, making different parts of a flexible polymer avoid each other. So the self-avoiding walk (SAW) model is proposed. This model is based on the self-avoiding of different parts in a macromolecular chain and embodies the so-called impenetrability of the chain, as shown in Fig. 3.4.6. The areas inside the circles indicate the self-avoiding effect between the molecular monomers.

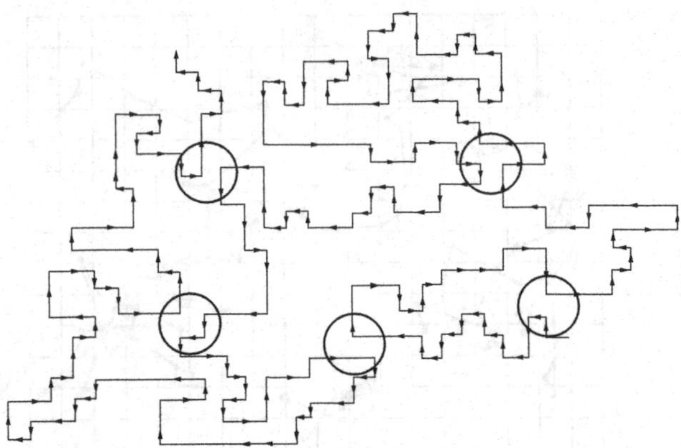

Figure 3.4.6 The self-avoiding walk in 2D square lattice (the areas inside the circles indicate the self-avoiding effect).

The self-avoiding walk is a difficult mathematical problem that has not been analytically solved yet. Of course, a computer simulation method can be used to solve the problem, and many different conformations are created, after which we can do an assembly average. The correlation between R and N can be expressed as

$$R_0 = aN^{\nu}. \tag{3.4.5}$$

Define

$$\mu_N = \langle R_{N+1}^2 \rangle / \langle R_N^2 \rangle, \ N = 1, 2, \ldots. \tag{3.4.6}$$

When $N \to \infty$, $\mu_N \to 1$, and

$$\lim_{N\to\infty} Nt(\mu_N - 1) = \lim_{N\to\infty} N\left[\left(1+\frac{1}{N}\right)^{2\nu} - 1\right] = 2\nu. \tag{3.4.7}$$

For a d-dimensional self-avoiding walk, the result by computer simulation is

$$\nu = \frac{3}{d+2}. \tag{3.4.8}$$

This formula indicates that the value of ν is related to the dimension of space d. If $d = 1$, since the long-chain molecule cannot intersect with itself and only move forward, $\nu = 1$; for a macromolecular chain in 3D space, $\nu = 3/5$. This result was first derived by Flory using a mean field approximation. Because $3/5 > 1/2$, for same N monomers the calculated size of the coil by self-avoiding walk is larger than that by random-walk as the coil expands or swells for the self-avoiding walk due to steric interaction.

The distribution of the distance R between the head and the tail can be expressed as

$$P(R) = R_0^{-d} f_p\left(\frac{R}{R_0}\right) = R_0^{-d} f_p(x). \tag{3.4.9}$$

The specific result for $d = 3$ is shown in Fig. 3.4.7. For large value of x, $f_p(x)$ decreases rapidly and can be expressed as

$$\lim_{x\to\infty} f_p(x) = x^k \exp(-x^{\delta}). \tag{3.4.10}$$

On the other hand, for small x, f_p drastically decreases to zero, which greatly reduces the probability of returning to the origin. Thus we have

$$\lim_{x\to 0} f_p(x) = C_0 \exp(-x^{\theta}), \tag{3.4.11}$$

where C_0 is a constant, and the exponentials of the two formulas k, δ, θ are constants related to the dimension d.

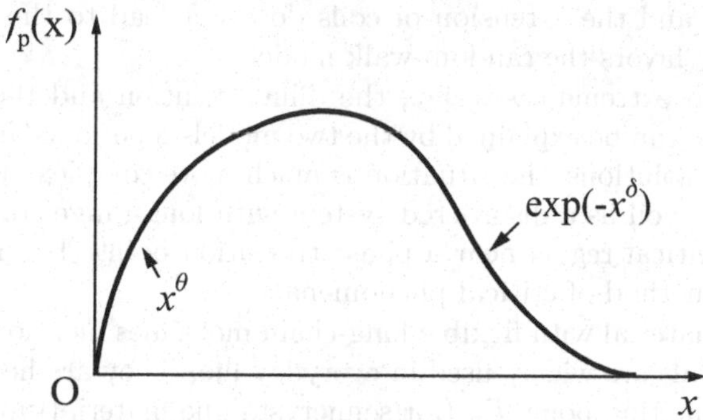

Figure 3.4.7 The distribution of the head-tail distance by self-avoiding chain ($x = R/R_0$).

3.4.3 The Correlation Function of Single Chain and Experimental Results

We introduce the correlation function $\Gamma(R)$ to express the distribution of polymer in space. Set a typical monomer as the origin O, and let the monomer density at a distance R from the origin be $c(R)$, then

$$\Gamma(R) = \langle c(R)c(O)\rangle - \langle c(R)\rangle\langle c(O)\rangle. \tag{3.4.12}$$

$\Gamma(R)$ stands for the correlation of the local density fluctuation between two points separated by a distance R. Generally, $\Gamma(R)$ decays with increasing R. In the following we shall estimate the specific forms of $\Gamma(R)$ for models of random and self-avoiding walks.

Obviously, the integral of $\Gamma(R)$ over the whole space is equal to the total number N of monomers in a chain, i.e.,

$$\int \Gamma(R) d^3R = N. \tag{3.4.13}$$

The power law of $\Gamma(R)$ can be deduced by the random walk or self-avoiding walk model. Assuming the number of monomers is n in a sphere with radius R and by using the random walk relation $R^2 = na^2$, we get the corresponding

$$\Gamma_r(R) \sim \frac{n}{R^3} = \frac{1}{a^2R}. \tag{3.4.14}$$

If the self-avoiding walk relation $R \sim an^\nu$ is used, and $\nu = 3/5$ in 3D space, we obtain

$$\Gamma_s(R) \sim \frac{n}{R^3} = \frac{1}{R^{4/3}a^{5/3}}. \tag{3.4.15}$$

$\Gamma(R)$ can be determined by the scattering of X-rays, neutrons or light, and its Fourier transform is

$$\varphi(k) = \int \Gamma(R)\exp(ikR) d^3R. \tag{3.4.16}$$

The Fourier transforms of the correlation functions of random walk and self-avoiding walk are respectively

$$\begin{aligned} \varphi_r(k) &\sim 1/a^2k^2, \qquad (kR_0 \gg 1); \\ \varphi_s(k) &\sim 1/(ka)^{5/3}, \quad (kR_0 \gg 1). \end{aligned} \tag{3.4.17}$$

If a deuterium doped long-chain molecule is introduced in the sample, the correlation functions of the polymer-chain can be measured by neutron scattering. For dilute solutions of the polymer, the experimental result agrees with the self-avoiding walk; whereas for a dense solution and even the melt, the result favors the random walk. The self-avoiding walk leads to a swollen chain because of the effect of the self-avoiding repulsive force, which makes the energy lower when the system is extended. In a dense solution of long-chain molecules, the macromolecule is excluded not only by the other monomers like itself, but also the monomers of other macromolecules. Thus the repulsive forces are compensated and the extension of coils does not lead to the lowering of energy, so the experimental result here favors the random walk model.

In general, these two extreme cases, i.e., the dilute solution and the dense one (including the melt) of macromolecules, can be explained by the two models mentioned above. For the intermediate case such as semi-dilute solutions, the situation is much more complex. P. G. de Gennes considered the long-chain molecular coil as a disordered system with long-range correlation, somewhat similar to fluctuations in the critical region near a phase transition point. He treated this problem by the renormalization group method of critical phenomena.

A macromolecular material with flexible long-chain molecules mentioned above is called thermoplastic, and such materials are widely used in everyday life. They are heated to the glass transition temperature T_g or the melting point T_m (for semicrystalline material) for processing.

3.4.4 Ordered and Partially Ordered Structure

Many macromolecules have structures with preferred orientation. The orientation of a polymer can be preferentially arranged so that the molecular chains and other structural units lie along some direction by some external action. The process of orientation is the ordering of molecules, and generally the model for preferred orientation of the macromolecules is proportional to its degree of crystallization. Molecular chain folding is one of the crystallization methods, and Fig. 3.4.8 shows the chain-folding model of a macromolecular monocrystalline thin film with a thickness of several microns. In large blocks of polymers, crystallized and amorphous zones are mixed together, but the ratio of the crystallized zone cannot exceed 40% (see Fig. 3.4.9).

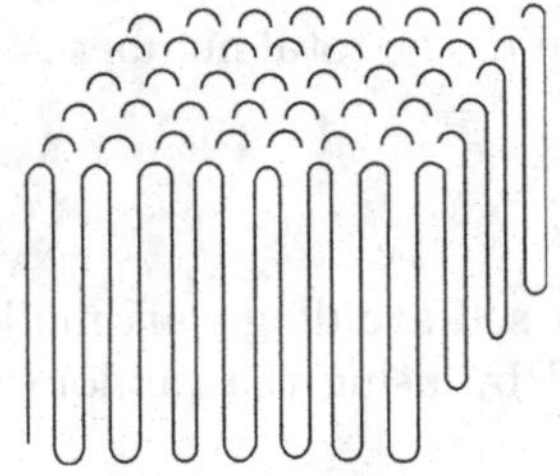

Figure 3.4.8 The chain-folding model for polymer crystallization of macromolecular monocrystalline thin film.

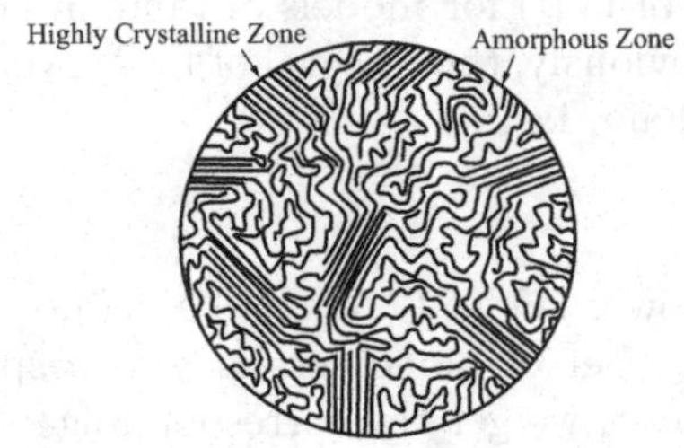

Figure 3.4.9 Schematic diagram of the structure for partially crystallized polyethylene.

Applying an external force can also make macromolecules with pronounced preferred orientation, such as directional crystallization, directional processing, precipitation from a macromolecular solution, etc. With the application of an external force, random macromolecules coils are preferentially oriented and arranged directionally as bundles of fibres along the long axis, crystallizing partly as in Fig. 3.4.10. After this kind of treatment, the macromolecular material acquires excellent mechanical and physical properties.

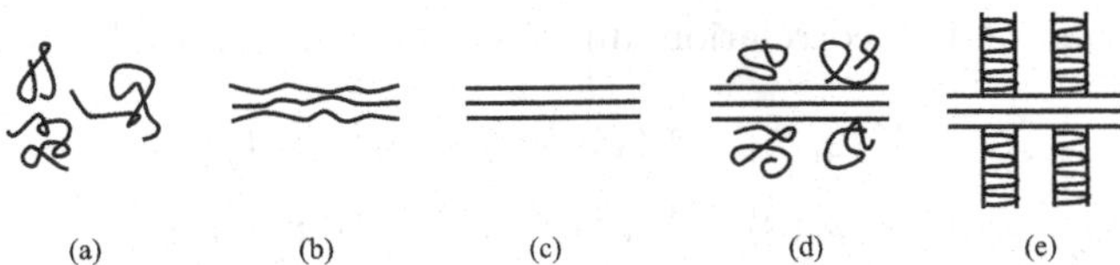

Figure 3.4.10 The schematic diagram of directional extrusion.

Another approach is to form the polymeric liquid crystals by precipitation from solution. It is an effective method of preparation of high-strength polymeric material. In the liquid crystalline state, the macromolecular chain basically arranges along a given direction. The C–C bonds in macromolecules are strong ones, and if the C–C bonds are arranged directionally, the material has high strength in this specific direction. For example, the solution of polyaramide molecules in concentrated sulfuric acid has orientational order, and the fibres precipitated from concentrated sulfuric acid (so-called Kevlar fibres) have extremely high strength, even exceeding that of the highest strength metal, i.e., piano wire; its strength density ratio is eight times higher than steel wire, and it is the material used in bullet-proof vests.

In §3.4.1, we mentioned block copolymers, which are composed of several blocks with different properties. Now let us consider the simplest situation, i.e., the di-block copolymer made up of A and B blocks. The chain blocks made up of homogeneous molecules form an ordered structure of mesoscopic dimension, as in Fig. 3.4.11. These structures are similar to the aforementioned lyotropics, but here the interface between A and B acts as the amphiphilic interface. The basic parameters controlling the structures are the relative lengths of A and B. With more kinds of blocks, the ordered structure becomes rich and colorful. The block copolymer provides the materials scientist with a new route of preparation for ordered structures with sizes ranging from microns to nanometers.

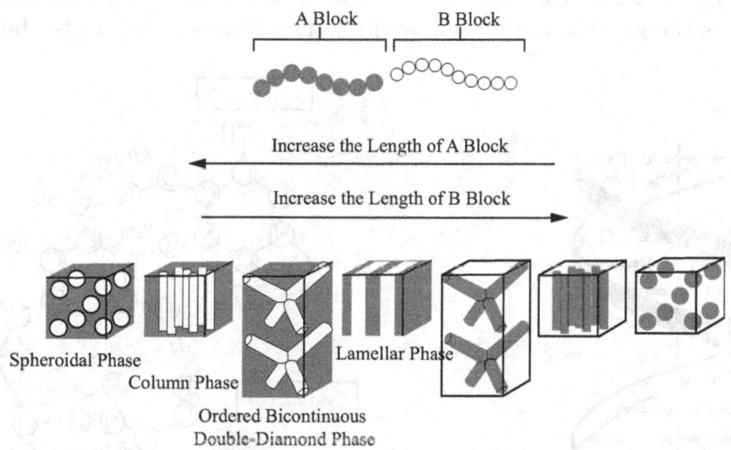

Figure 3.4.11 Different kinds of ordered structures of di-block copolymer. From P. Ball, *Made to Measure*, Princeton University Press (1997).

In §3.4.1, we also mentioned the branch-connected copolymer and the star copolymer; such dendrimers have been further developed, as shown in Fig. 3.4.12. Beginning from the center, the multiple-branched structure is formed. The composition and application of dendrimers is still in development.

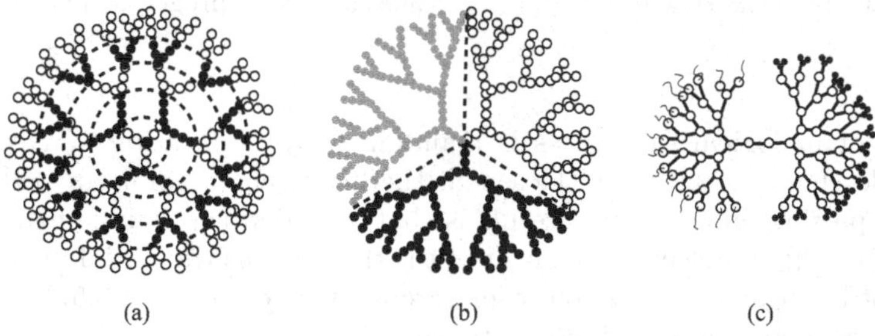

Figure 3.4.12 different kinds of the dendrimers. (a) a dendrimer with distinct concentric shells (b) a segmented dendrimer; (c) amphiphilic dendrimer (with the hydrophilicity and hydrophobicity in each two hemispheres). From P. Ball, *Made to Measure*, Princeton University Press (1997).

Although macromolecular materials have been widely used, studies of block copolymers and dendrimers indicate that tailoring of the mesoscopic structure of macromolecular materials may provide a new route to design and fabrication of new types of high performance macromolecular materials.

§3.5 Biopolymers

3.5.1 The Structure of Nucleic Acid

There are two important kinds of biopolymers: nucleic acids and proteins. Among nucleic acids, deoxyribonucleic acid (DNA) is the carrier which controls genetic processes. DNA in the nuclei of cells is the physical foundation of genetic matter, and the genetic information is contained in the structure of the DNA molecule.

In 1953, J. D. Watson and F. H. C. Crick established the double helix structure of the DNA molecules (see Fig. 3.5.1), and indicated that the genetic information, as well as the rule for its replication is encoded in its structure. The basic structural unit of DNA consists of a backbone of phosphate and deoxyribose molecular groups. Four different bases, i.e., adenine (A), guanine (G), cytosine (C) and thymine (T), are attached to it. The double helix structure of DNA is maintained by the matching of base pairs. The matching rule of the base groups is fixed: A matches T and G matches C (see Fig. 3.5.2), so the number of A and T, or G and C are the same. The sequence of the bases composes the genetic information and its arrangements constitute the genetic code.

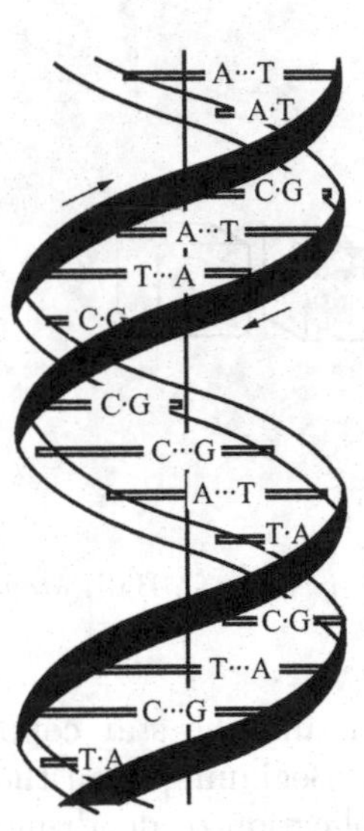

Figure 3.5.1 The schematic diagram of the double helix structure of DNA.

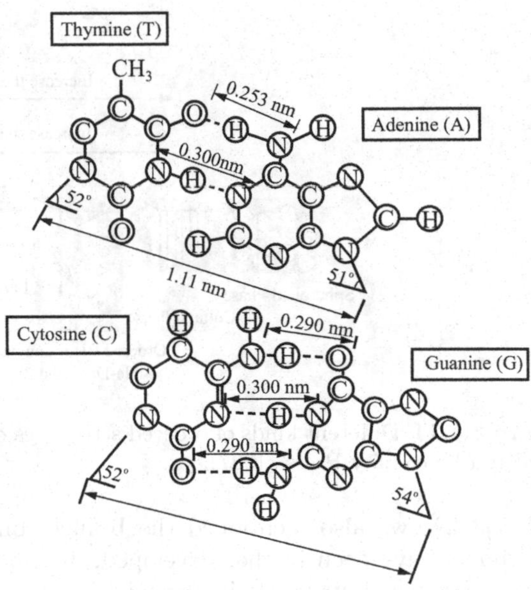

Figure 3.5.2 Two kinds of base pairing (dotted lines are the hydrogen bonds).

The heredity of biological substances is maintained by duplication of DNA at the molecular scale: The double chains of DNA are loosened, then each chain connects with a new one by the principle of base pairing, and the final result is that two identical double-helices derived from the initial one. Since the final molecular arrangement is the same as the initial one, this process realizes the duplication of biological information from parent to child (see Fig. 3.5.3), which indicates that this information sequence plays an important role in biology.

3.5.2 The Structure of Protein

Other important molecules in the living organism are proteins. Most of the biotic functions are realized by protein, and it has much a more complex structure than DNA molecule. The simplest protein molecule is myoglobin, in which strands of amino-acid chains fold to a globular shape (see Fig. 3.5.4). It is the tertiary structure of the protein, which is made up of α-helix chains or laminar β-sheets (secondary structure). By the way, these secondary structures are formed by lots of

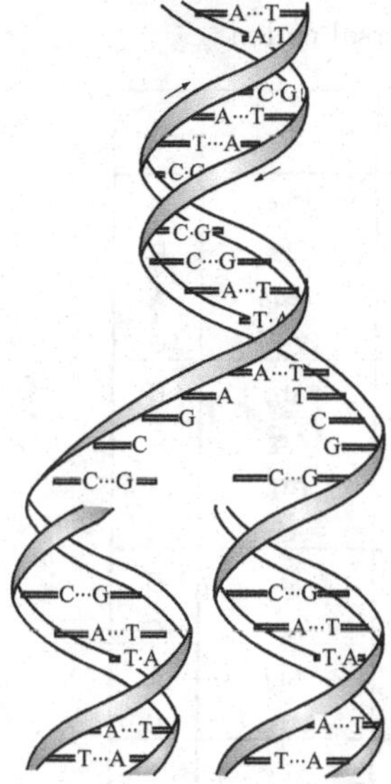

Figure 3.5.3 The schematic diagram of DNA duplicating mode with the precondition of complementarity of the base groups.

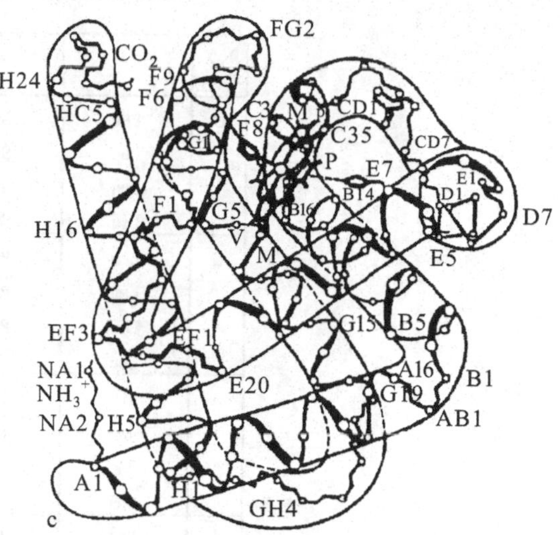

Figure 3.5.4 Structure of the myoglobin molecule.

molecules (primary structure). Different amino-acid chains lead to different folding states by which different protein functions are achieved.

The functions of proteins are controlled by 20 kinds of amino acids, while only 4 kinds of nucleic acids compose DNA. Since DNA should control the arrangement of amino acids, we must find the specific method of controlling the arrangement of 20 kinds of amino acids by 4 kinds of nucleic acids, i.e., encoding. If we use one nucleic acid as the code, we can only get 4 kinds of amino acids with each of the 4 kinds of nucleic acid; if we use two nucleic acids as the code, we will get 16 coden, which is still less than 20; and if we use so-called triplet code, and we shall get 64 coden, which is larger than 20. Shortly after the double helix structure of DNA was discovered, physicist R. I. Gamov first proposed this triplet code. In the 64 kinds of coden, 3 of them are termination coden and the other 61 kinds stand for 20 kinds of amino acids. So most of the amino acids have more than one coden, and this shows the degeneracy of the coden. Almost all life-forms use the same coden, and this indicates the universality of the coden. Table 3.5.1 shows the universal coden.

The first step of the expression of genetic information is transferring them onto messenger RNA (mRNA), but in mRNA base T is replaced by U (uracil). This process is called transcription. Information is further transcripted into transfer RNA (tRNA) and ribosomal RNA(rRNA). Then the information is translated into codens of amino acids of the proteins in the cytoplasm. The two catalytic functions of DNA and the flow direction of the genetic information is shown in Fig. 3.5.5.

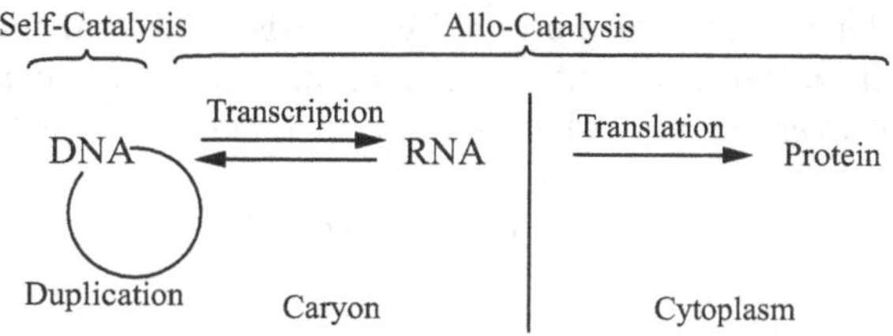

Figure 3.5.5 Two catalytic functions of DNA and the flow direction of genetic information.

Table 3.5.1 The table of universal coden.

x \ y	U	C	A	G	z
U	Phe	Ser	Tyr	Cys	U
	Phe	Ser	Tyr	Cys	C
	Leu	Ser	Term	Term	A
	Leu	Ser	Term	Trp	G
C	Leu	Pro	His	Arg	U
	Leu	Pro	His	Arg	C
	Leu	Pro	Gln	Arg	A
	Leu	Pro	Gln	Arg	G
A	Ile	Thr	Asn	Ser	U
	Ile	Thr	Asn	Ser	C
	Ile	Thr	Lys	Arg	A
	Met	Thr	Lys	Arg	G
G	Val	Ala	Asp	Gly	U
	Val	Ala	Asp	Gly	C
	Val	Ala	Glu	Gly	A
	Val	Ala	Glu	Gly	G

The molecules carrying information play a key role in the development of life. By the transcription of the genetic codes, forming the functional protein molecules, so information can be translated into function, this is the foundation of modern biology.

3.5.3 Information and the Structure

The structure of matter contains information at every level, such as the atomic level, the molecular level and even levels at larger scales. We have already discussed the order-disorder transition. In the ordered phase, the exact information about the positions of a small group of atoms suffices for the description of the whole structure. This is impossible for a disordered phase, because to specify at every atomic site needs too many figures ($\sim 10^{24}$), so the statistical method should be used. In early 1943, the famous physicist Schrödinger in his book *What is life?* indicated that the secret of life lies in the existence of genetic codes in an aperiodic crystal, and, if we wish to treat the information of aperiodic molecules in earnest, a quantitative theory for information is required.

How can we define information scientifically? In 1948, C. Shannon proposed his statistical theory of information: First consider there are P possible choices with the same probability, for instance, for Morse code, $P = 2$; for Latin letters, $P = 27$ (26 letters and one blank). If one in P is chosen, we get information. With larger P, there would be much more information from the choice. So the information content I is defined as

$$I = K \ln P, \tag{3.5.1}$$

where K is a proportionality constant.

Because mutually independent choice probabilities satisfy the multiplication theorem, the corresponding information content has additivity. Consider an information content as a series of mutually independent choices, and every choice is between 0 and 1. The total value of $P = 2^n$, so

$$I = K \ln P = nK \ln 2, \tag{3.5.2}$$

and let I equal n, then

$$K = \frac{1}{\ln 2} = \log_2 \mathrm{e}. \tag{3.5.3}$$

In this way we define the unit of information content, the bit, universally used in computer science. And if K is defined as the Boltzmann constant k_B, the information content can be measured in units of entropy. In Brillouin's theory, information is equated to the negative entropy.

There are A, T, G and C — 4 kinds of bases in the list of DNA structure. If we arrange by choosing two bases, there are $4^2 = 16$ kinds of different arrangements; if we choose 3, there are $4^3 = 64$ kinds. If there are 100 bases in a nucleic acid chain, there are 4^{100} kinds of arrangements, which is a huge number, larger than the total number of species in history (about 4×10^9), and larger than the total number of atoms in the solar system. The sequence of base-pairs in a human body is about 2.9×10^9. To identify all of them is an enormous task but an international program for the human genome project was undertaken between 1991 to 2001 and successfully completed. Thus, an enormous data bank of biological information is available to scientists for the construction of theoretical biology. It should be noted that there is abundant information stored in nucleic acids and proteins, these play a key role for understanding biology. Research on polymer structure with information is underway, and further clarification of the relationship between information and function will be an important topic for current research.

Bibliography

[1] Allen S. M., and E. L. Thomas, *The Structure of Materials*, Wiley, New York (1998).

[2] Zallen, R., *The Physics of Amorphous Solids*, John Wiley & Sons, New York (1983).

[3] Ziman, J. M., *Models of Disorder*, Cambridge University Press, Cambridge (1978).

[4] de Gennes, P. G., and J. Prost, *The Physics of Liquid Crystals*, 2nd ed., Oxford University Press, Oxford (1993).

[5] Chandrasekhar, S., *Liquid Crystals*, 2nd ed., Cambridge University Press, New York (1992).

[6] de Gennes, P. G., *Scaling Concepts in Polymer Physics*, Cornell University Press, Ithaca (1979).

[7] Strobl, G., *The Physics of Polymers*, Springer-Verlag, Berlin (1996).

[8] Ball, P., *Made to Measure, New Materials for 21 Century*, Princeton University Press, Princeton (1997).

[9] Hyde, S., S. Anderson, and K. Larsson, *et al.*, *The Language of Shape*, Elsevier, Amsterdam (1997).

[10] Venkataraman, G., D. Soho, and V. Balakrishnan, *Beyond the Crystalline State*, Springer-Verlag, Berlin (1989).

[11] Brillouin, L., *Science and Information Theory*, 2nd ed., Academic Press, New York (1962).

Chapter 4

Inhomogeneous Structure

In reality, condensed matter is both inhomogeneous and hierarchical, forming a world full of multiplicity and complexity. This chapter begins with a description of hierarchical systems with multi-phases, then introduces the basic concepts of percolation and fractal, which play important roles in inhomogeneous systems.

§4.1 Multi-Phased Structure

4.1.1 Structural Hierarchies

In physics, there are structures at various length scales. Roughly, they are divided into microscopic, the length scale at atomic dimension or less (~ 0.1 nm), and macroscopic, the length scale of every day life (0.1 mm $\sim$ 1 nm) and between these extremes there is the mesoscopic scale (10 μm $\sim$ 1 nm). Physical properties of condensed matter depend on the structural characteristics at these different length scales. Figure 4.1.1 gives a schematic diagram for length scales of various structures in condensed matter.

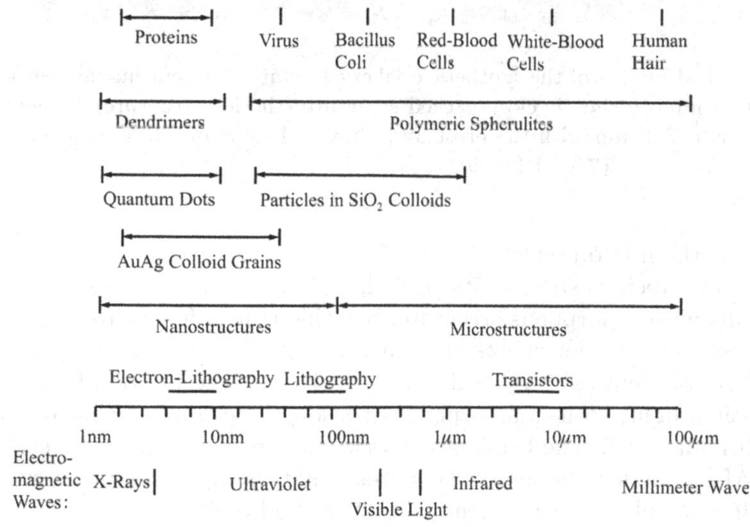

Figure 4.1.1 Length scales for various structures of condensed matters.

The structures discussed in previous chapters mostly appear to be homogeneous at the macroscopic scale, but, with the help of optical microscopy, electron microscopy as well as STEM, the inhomogeneity of materials, such as crystals, ceramics, glasses and polymers at the mesoscopic or microscopic scale are shown. We can divide inhomogeneities of materials roughly into the following three types:

1. Macroscopic inhomogeneities — these include mixtures of different phases as well some artificial composite materials.

2. Mesoscopic inhomogeneities — these include inhomogeneities due to some natural processes such as crystal growth, spinodal decomposition, co-precipitation as well some artificial fabrication process, such as sputtering, vapor deposition, etc. These static inhomogeneities are also known as microstructures, super-microstructures or nanostructures. In inorganic materials with multi-phases, the existence of these microstructures is the rule and may influence profoundly the physical properties of these materials. A big challenge for contemporary materials science lies in the design of various inhomogeneous materials at suitable length scales and to produce them by fabrication or self-assembly.

3. Microscopic inhomogeneities — the inhomogeneous distribution of different atomic species at lattice sites may produce microscopic inhomogeneities such as clustering or short range order in alloys.

Next we shall mainly concern ourselves with inhomogeneities at the mesoscopic scale. Certainly, there are some natural structures at this length scale, and with development of condensed matter physics, many mesoscopic structures are produced artificially. If the scale (or period) of the multi-phased structure is matched to some characteristic length in condensed matter physics, spectacular physical effects may appear. This becames the driving force which developed artificial meso-structures in the latter part of the 20th century. We shall give a brief sketch about the structures at micron and nanometer scales.

Figure 4.1.2 The SEM photos of the synthetic opal crystal with different magnification. Nearly isometrical colloidal SiO_2 balls compose the 3D close-packed structure (the fcc structure) autonomically. The average diameter of the balls is 300 nm with the error 5%. Provided by Prof. B. Y. Cheng. Related paper: B. Y. Cheng *et al.*, *Opt. Commun.* **170**, 41 (1999).

1. Structures at the micron scale

Natural opals are precious stones with brilliant colors. X-ray analysis shows that they are composed of SiO_2 balls with amorphous structure, but this is insufficient to explain why this precious stone has such spectacular color effects. In the 1970s, scientists using SEM showed that opal had a periodic structure of identical SiO_2 balls (see Fig. 4.1.2). The period of this structure is about a wavelength of visible light, so it is amorphous at a length scale of several tenths of a micron, but periodic at the 0.1 μm level. The brilliant colors are due to diffraction of visible light by this periodic structure. After the deciphering of the secret of natural opals, artificial opals were successfully fabricated by self-assembly of nearly identical SiO_2 colloid balls.

In the late 1990s, photonic crystals in the optical wavelength range were fabricated with a periodic structure in this length scale for their diffraction effect. Materials with high indices of refraction, such as C, Si, TiO_2 are substituted for SiO_2 in opals, or using opals as templates; the interstitial space is filled with material with a high index of refraction, then the SiO_2 ball etched out, forming inverted opal structures, shown in Fig. 4.1.3. The woodpile structure shown in Fig. 4.1.4 is also used for photonic crystals.

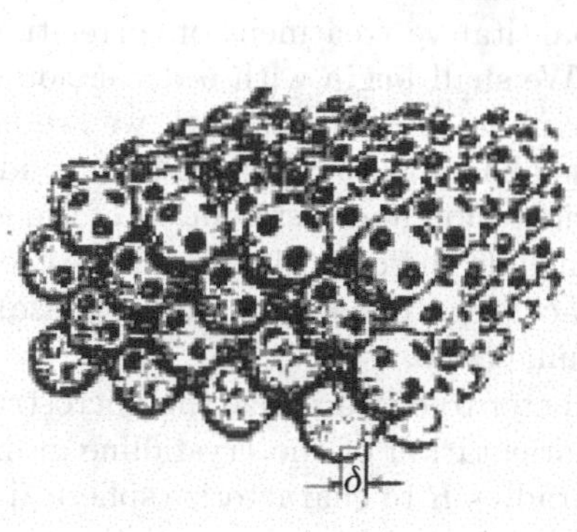

Figure 4.1.3 The inverted opal structure.

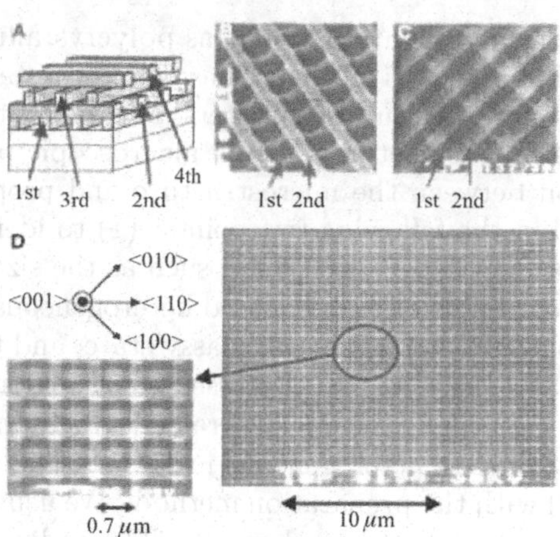

Figure 4.1.4 The woodpile structure.

Certainly, spectacular physical effects may be also achieved by fabricating micron structure 1D periodic multi-layers or 2D periodic multi-columns.

2. Nanostructures

Another class of artificial mesostructure is the nanostructure (1 nm to 100 nm). Nanometers are just at the scale of the de Broglie wavelength at the Fermi level for most solid state materials: For instance, it is about 50 nm for semiconductors; about 1 nm for metals. So it is expected that nanostructures will play outstanding roles in tailor-made electronic properties (see Chaps. 5 and 14). Especially as contemporary techniques for microelectronics are nearly approaching their physical limits, alternatives will be certainly derived from nanotechnology.

The self-assembling nanocrystal superlattices of metals, semiconductors and oxides (see Fig. 4.1.5) have long-range translational and orientational order.[a] The building blocks for the self-assembling unit are nanoscopic grains, spherical oxide grains or polyhedral crystal grains, and these nano-grains are coupled by a surfactant such as thiolate (this is different from the coupling of atoms by chemical bonds) with an adjustable structural period. This provides another route to the formation of 3D superlattices.

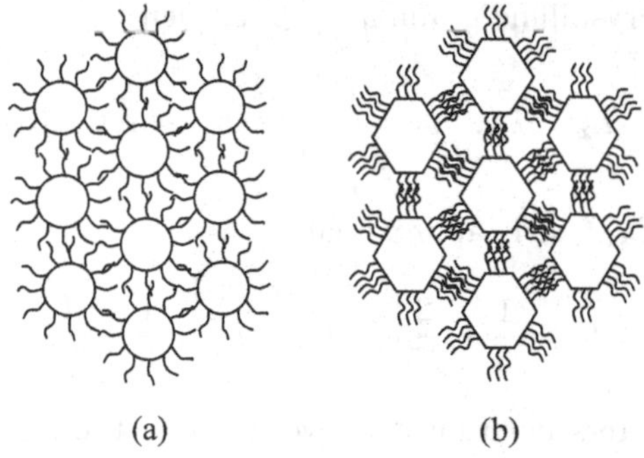

Figure 4.1.5 Self-assembling of nanocrystal superlattices with (a) spherical grains and (b) faceted grains.

4.1.2 Microstructural Characteristics of Heterogeneous Material

The multi-phased structures discussed above are mostly regular, and their geometrical relationship can be easily characterized; however, the irregular ones need further discussion.

[a] Z. L. Wang, *Adv. Mater.* **10**, 13 (1998).

Heterogeneous materials such as polycrystalline ceramics and multi-phased composites, etc. are composed of mixtures of single-phase microcrystalline grains of different sizes and characters. Knowing their microstructural characters is important for the quantitative treatment of correlations between the microstructure and the macroscopic properties. We shall begin with a discussion of the correlation between the microstructure and properties. For multi-phased material, we are usually interested in the following two points: (1) to identify various phases in the material; (2) to identify the microstructure of each phase, such as the size, shape, orientation and distribution of the grains.

Different kinds of multi-phased heterogeneous material include the crystalline state (isotropic or anisotropic), the amorphous or glassy state, and the pore space (regarded as the second phase). The ratio of each phase can be expressed as a mole, mass or volume ratio.

The mean size of the crystal grains is an important parameter to characterize the microstructure of a multi-phased heterogeneous material. The shape and orientation of the crystalline grains are correlated with the preparation method. We usually use the radius R to characterize spherical, near-spherical and square-shaped grains. The radius R is usually expressed as the logarithmic normal distribution

$$n(R) = \frac{1}{R\sqrt{2\pi\delta}} \exp\left\{-\left[\frac{\ln(R/R_0)}{\sqrt{2\delta}}\right]^2\right\}, \tag{4.1.1}$$

where R_0 is the geometric mean radius and δ is the deviation. For different materials, the effective radius R_0 ranges from several nanometers to tens of millimeter. For nonspherical (ellipsoidal) crystalline grains, the radii of main axes (a, b, c) can be used as parameters; the shape can be characterized by the eccentricity $e_{p(0)}$ and three geometrical parameters — depolarization factors L_x, L_y and L_z which satisfy the condition $L_x + L_y + L_z = 1$.

For the prolate spheroid, $a > b = c$, and the depolarization factors are

$$L_x = \frac{1 - {e_p}^2}{2{e_p}^3}\left(\ln\frac{1+e_p}{1-e_p} - 2e_p\right), \quad L_y = L_z = \frac{1}{2}(1 - L_x), \tag{4.1.2}$$

and the eccentricity

$$e_p = \sqrt{1-(b/a)^2}. \tag{4.1.3}$$

If the crystal grain approaches spherical shape, i.e., $e_p \ll 1$, approximately

$$L_x = \frac{1}{3} - \frac{2}{15}{e_p}^2, \quad L_y = L_z = \frac{1}{3} + \frac{1}{15}{e_p}^2. \tag{4.1.4}$$

For the oblate spheroid crystalline grain, $a = b > c$, then

$$L_x = L_y, \quad L_z = \frac{1 - {e_0}^2}{{e_0}^3}(e_0 - \arctan e_0), \tag{4.1.5}$$

for $e_0 = \sqrt{(a/b)^2 - 1}$, and if ${e_0}^2 \ll 1$, approximately

$$L_z = \frac{1}{3} + \frac{2}{15}e_0^2, \quad L_x = L_y = \frac{1}{3} - \frac{1}{15}e_0^2. \tag{4.1.6}$$

The microstructural parameters mentioned above, such as the size of the crystalline grains, the distribution of the size, the shape factor and the volume ratio, can be obtained quantitatively from the experimental 2D samplings.[b]

Another microstructural characteristic of heterogeneous material is the distribution of single-phase grains. The microstructure of multi-phased material would vary with the change of the volume ratio and three types of geometrical patterns may be distinguished — (1) isolated grains dispersed

[b]The depolarization factors indicate the influence of the geometrical shape on the polarization of the dielectric ellipsoid in an electric field and the magnetization of the magnetic ellipsoid in a magnetic field. The simplest case of a conducting ellipsoid in an electric field has been treated in L. D. Landau, E. I. Lifshitz, *Electrodynamics of Continuous Media*, 2nd ed. Pergamon Press (1984), §4.

in the matrix; (2) an aggregated grain structure; and (3) different grains connected together forming a large cluster. Consider the simpler two-phase systems, which can represent two-phase alloys, composite ceramics and porous materials (in which the pores may be considered as the second phase) in practical problems. With increasing percentage of the second phase, Fig. 4.1.6(a)–(d) shows the corresponding microstructures.

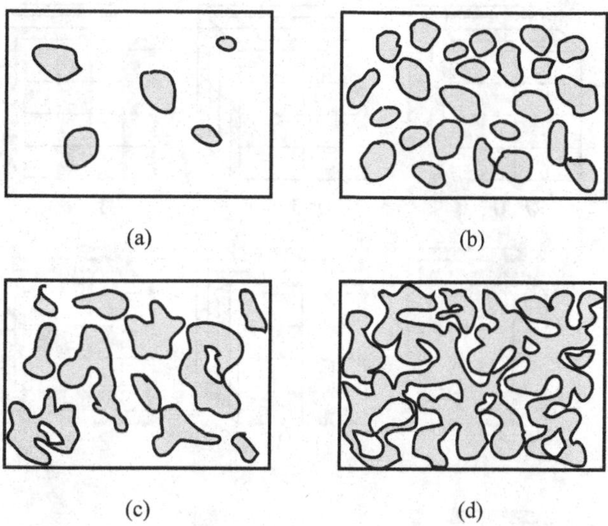

Figure 4.1.6 The schematic diagram of the microstructure of the two-phase alloy (the volume percentage of the second phase increases from (a) to (d) gradually).

(1) Dispersed grain structure. If the volume ratio of the minor phase is very small, the grains are dispersed in the matrix randomly [see in Fig. 4.1.6(a)]; if the volume ratio of the minor phase is somewhat larger, the grains of this phase become dispersed in the matrix quite uniformly [see Fig. 4.1.6(b)], and this situation corresponds to some granular metal films and the glass-ceramic controlled by nucleation and growth.

(2) Aggregated grain structure. With a bigger volume ratio, the minor-phase grains aggregate into grain clusters of a certain size [see Fig. 4.1.6(c)].

(3) Grains are connected into a network of large clusters. If the volume ratio exceeds some critical value, i.e., the percolation-like threshold value, most of the minor-phase grains are connected into a continuous network [see Fig. 4.1.6(d)].

The microstructures shown in Fig. 4.1.6 are basically isotropic, the shapes of the minor-phase grains in the material are spherical or sphere-like with random orientation and uniformly distributed in the matrix phase. For more complex cases, we can consider the obviously anisotropic shape of the minor phase grains, such as the rod-like, the plate-like or ellipsoidal. These are realized in some composite materials and the eutectic or eutectoid alloys. Also ellipsoidal grains may be oriented in their microstructures (as in Fig. 4.1.7).

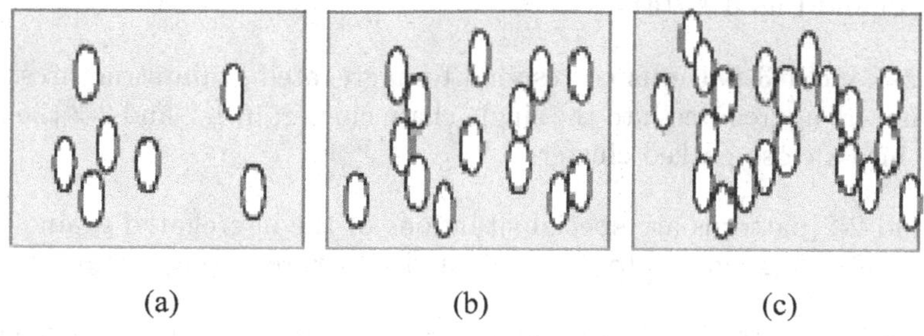

(a) (b) (c)

Figure 4.1.7 The schematic diagram of the distribution of oriented ellipsoidal grains is in a two-phase materials.

In the following, we shall discuss connectivity among the grains. We might as well use cubic grains instead of spherical ones as the building blocks of multi-phased materials, to show the different connectivity patterns. For the two-phase alloy, there are 10 kinds of connectivity patterns: 0-0, 0-1, 0-2, 0-3, 1-1, 1-2, 1-3, 2-2, 2-3 and 3-3, as shown in Fig. 4.1.8. Now we will give a brief explanation of these patterns:

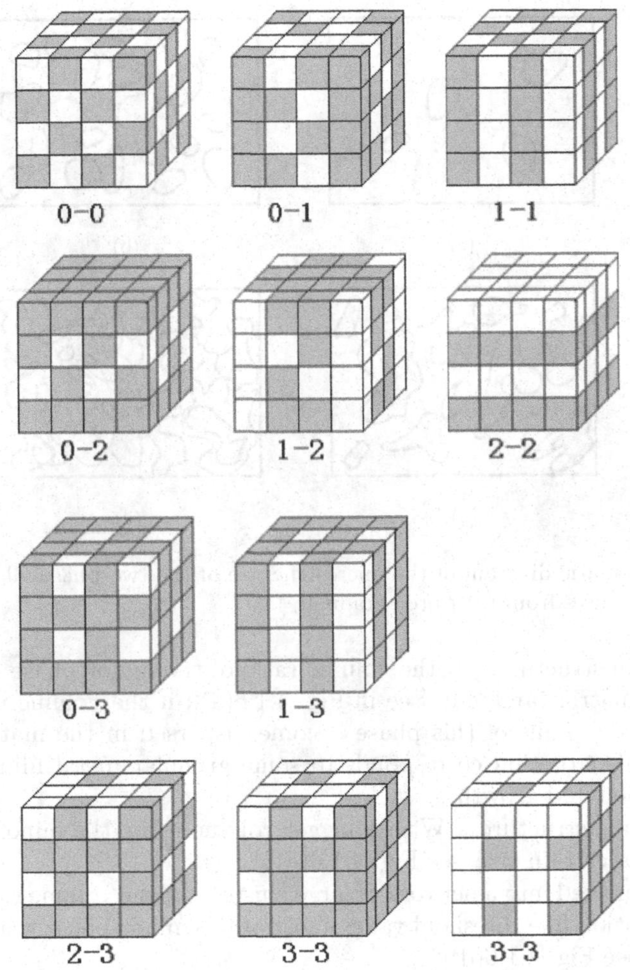

Figure 4.1.8 10 kinds of connectivity patterns for two-phase alloys, the last two are 3-3 mode seen from two different direction. From R. E. Newnham *et al.*, *Mat. Res. Bull.* **13**, 525 (1978).

1. The 0-0, 0-1, 0-2 and 0-3 patterns correspond to dispersed grain structures. The 0-0 pattern corresponds to the structure shown in Fig. 4.1.6(b); the 0-1, 0-2 and 0-3 ones are related to Fig. 4.1.6(a) and Fig. 4.1.7(a).

2. The 1-2, 2-3 and 1-3 patterns correspond to aggregated grain structures. In 1-3, the minor phase grains are aggregated into the single-chain cluster; in 1-2 and 2-3 the minor phase grains aggregate into a close-packed cluster.

3. The 1-1 and 2-2 patterns are special situations of the aggregated grain structures, in which the minor grains aggregate into the laminated structures along certain directions.

4. The 3-3 pattern corresponds to the structure with a large interconnected cluster as in Fig. 4.1.6(d) and Fig. 4.1.7(c), in which the two-phase material forms the interpenetrating 3D network (i.e., the percolation cluster).

From Figs. 4.1.6–4.1.8, we can find that the phase distribution of heterogeneous material is extremely complex. However, the phase distribution plays an important role in the correlation of microstructure and physical properties. The problems of completely dispersed laminated structures [Figs. 4.1.6(a) and (b), Fig. 4.1.7(a) and the 0-0, 0-1 and 0-3 patterns in Fig. 4.1.8] and the flaky aggregated structures (1-1 and 2-2 patterns in Fig. 4.1.8) have been solved. But, for most aggregated grain structures [Figs. 4.1.6(c) and (d), Figs. 4.1.7(b) and (c), 1-2, 1-3, 2-3 and 3-3 patterns in Fig. 4.1.8], such a complete description is almost impossible, for it requires the s-point grain correlation functions of all the grains (of all sizes). This is only available for high-ordered cases whereas the required experimental information is generally limited to the pair-correlation function of phase distribution, which can be expressed as

$$g_2(\boldsymbol{r}) = \frac{1}{N}\sum_{\boldsymbol{r}_1,\boldsymbol{r}_2} P_2(\boldsymbol{r}_1,\boldsymbol{r}_2)\delta(\boldsymbol{r}_1-\boldsymbol{r}_2-\boldsymbol{r}). \tag{4.1.7}$$

In this equation $g_2(\boldsymbol{r})$ is the probability of finding a grain at the position $\mathbf{r}$ and $P_s(\boldsymbol{r}_1,\boldsymbol{r}_2)$ is the probability of finding the s grains at the positions $\boldsymbol{r}_1,\boldsymbol{r}_2,\ldots,\boldsymbol{r}_s$ in the same grain cluster. $g_2(\boldsymbol{r})$ can be measured directly by light scattering, small angle neutron scattering or X-ray scattering. Generally, the pair correlation function $g_2(\boldsymbol{r})$ is in common use, while the high-order correlation function g_s $(s>2)$ is rarely used, apart from model building.

Figure 4.1.9 Micrograph of a multi-phased alloy.

4.1.3 Effective Medium Approximation: The Microstructure and Physical Properties of Two-Phase Alloys

Figure 4.1.9 shows a micrograph of a multi-phased alloy. We marvel at such complex structure with a somewhat frustrated feeling. In the past, physicists were accustomed to dealing with simple systems, but advanced research on condensed matter physics is leading to more complex systems. However, building models may somewhat simplify these problems.

Returning to the cases in which the minor phase grains are mainly isotropic (see Fig. 4.1.6), if the parameters of the two phases are known, what are the physical properties of the mixture of two phases? J. C. Maxwell gave a part of the answer in the 19th century; and now let us see how he solved this problem.

Assume the percentages of the two phase are c and $(1-c)$, and their dielectric constants are ϵ_1 and ϵ_2, then idealize a part of minor phase into a ball embedded in the uniform medium by a simple assumption that the effective dielectric constant is $\epsilon_{\rm m}$; this is the effective medium theory. If an external field is added to the effective medium in order to generate an electric-field $E_{\rm m}$, according to electrostatics, we find that the dipole moment is proportional to $E_{\rm m}(\epsilon_1-\epsilon_m)/(\epsilon_1+2\epsilon_{\rm m})$. Then we do a similar treatment for every volume in the original structure, and the total dipole moment per unit of volume (i.e., the polarization P) can be obtained:

$$P \propto cE_{\rm m}\frac{\epsilon_1-\epsilon_{\rm m}}{\epsilon_1+2\epsilon_{\rm m}} + (1-c)E_{\rm m}\frac{\epsilon_2-\epsilon_{\rm m}}{\epsilon_2+2\epsilon_{\rm m}}. \tag{4.1.8}$$

This equals the result of embedding a ball of a mixture into the effective medium with electric field $E_{\rm m}$.

The effective medium and the embedded ball with a mixture are indistinguishable in electrostatics, i.e., $P = 0$. According to this condition, we can get the dielectric constant of the effective medium $\epsilon_{\rm m}$. Because the steady-state current density and electrostatic field have the same form, similarly we can get the conductivity of the effective medium for the two-phase medium with conductivities σ_1 and σ_2,

$$\sigma_{\rm m} = \frac{1}{4}\left\{(3c-1)\sigma_1 - (3c-2)\sigma_2 + \sqrt{[(3c-1)\sigma_1 - (3c-2)\sigma_2]^2 + 8\sigma_1\sigma_2}\right\}. \tag{4.1.9}$$

For a mixture of metallic and insulating grains, let $\sigma_2 = 0$, then

$$\sigma_{\rm m} = \left(\frac{3}{2}c - \frac{1}{2}\right)\sigma_1, \ c > \frac{1}{3}. \tag{4.1.10}$$

This result indicates that $\sigma_{\rm m}$ depends on c linearly. If $c \leqslant 1/3$, $\sigma_{\rm m} = 0$. It is a reasonable result if the proportion of the conducting phases (the minor phases) is low because they cannot form an interconnecting path to conduct electricity. This also indicates the existence of the percolation-like threshold to be discussed in the next section. Obviously here we only give a rough sketch of the effective medium approximation (EMA). For more detail on later work where others made improvements please see Bib. [1]. Although this approximation gives useful results in certain areas, it also raises many new problems. This discussion would fail if the nonuniform area is very minute and the difference in electric conductivities between the two phases is very sharp. The validity of many conclusions near the conducting threshold value is doubtful. In a word, the effective medium approximation is the result of the averaging of properties of the heterogeneous material, and in the process of averaging, some characteristics of disordered systems may be averaged out; this is a fault of the method itself.

§4.2 Geometric Phase Transition: Percolation

4.2.1 Bond Percolation and Site Percolation

The percolation problem in condensed matter can be deduced from consideration of the conductivity of mixtures of metal and insulator grains. The electrical conductivity of the mixture changes with the proportion of metal, and the mixture conducts electricity only if the volume ratio of metal is larger than a certain threshold value, at which point the conductivity has a rapid increase from zero, as will be shown. This indicates that percolation in the metal-insulator transition in a disordered mixture is very complex. Aggregation, or the increase of concentration of metallic grains, would lead to a sudden increase in conductivity and a sudden appearance of long range connectivity of metallic grains. This process is a geometrical phase transition. Percolation in a lattice can be divided into two types: bond percolation and site percolation.

Consider the bond percolation problem in Fig. 4.2.1(a). In the network, the solid lines stand for the channels and the dashed lines stand for the blockages. Let the probability of blocking one

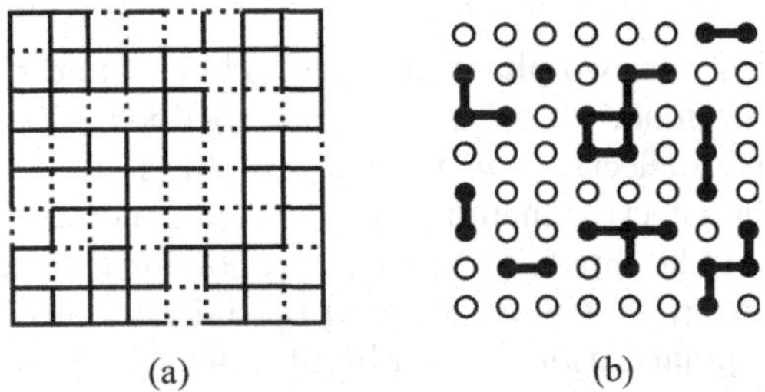

Figure 4.2.1 The schematic diagram of the percolation. (a) bond percolation; (b) site percolation.

channel be p, and if we imagine these form a conduit system for liquids, we would like to know whether the liquid can pass through from one side to another. Obviously, the basic condition is the existence of infinite penetrable channels, and the critical concentration of this channel is p_c, i.e., the percolation threshold value. This is bond percolation.

This randomness can be transformed from bonds to sites. Figure 4.2.1(b) shows a lattice whose sites are occupied by white or black pieces with occupation probabilities p and q respectively, which satisfy the condition $p+q=1$. Nearest neighbors with the same color are considered to be connected and the others disconnected; the critical concentration for the first appearance of an infinite cluster composed of black pieces is also defined as the percolation threshold value p_c.

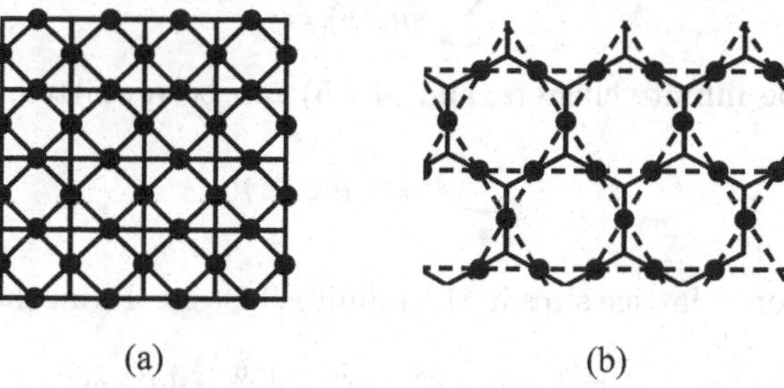

Figure 4.2.2 The transformation from bond percolation to site one. (a) the square lattice and its covering lattice; (b) honeycomb lattice and its covering lattice (the Kagomé lattice).

Bond percolation can be transformed to site percolation in another lattice (the so-called covering lattice). This is constructed by placing a site on each bond of the original lattice, and drawing connecting lines between the sites which are closest neighbors. A 2D square bond-percolation lattice is transformed into a covering lattice which is still a square one, but with half unit cell area and inclined at 45° to the original one. A similar example is provided by the honeycomb lattice for which the covering lattice, is the Kagomé lattice [Fig. 4.2.2(b)]. We can always convert bond percolation into site percolation, but the reverse is not always true, i.e., some site percolation lattices cannot be transformed into bond ones. So site percolation is a more fundamental problem.

4.2.2 Overview of Percolation Theory

The basic purpose of percolation theory is to calculate p_∞, the critical probability for the appearance of an infinite cluster, and P, the probability that each site (or channel) belongs to the infinite cluster. These are expressed as a function of p, i.e., they can be written as $p_\infty(p)$ and $P(p)$. It is difficult to calculate $P(p)$ accurately, and so far there are only some approximate results generally, while the percolation threshold values p_∞ have exact results.

First consider site percolation in a square lattice [see Fig. 4.2.2(a)]. Let p stand for the probability of a black piece occupying a lattice site; the corresponding probability for white one is $q=1-p$, and the total number of lattice sites $N\to\infty$. Now we shall discuss the probabilities of the existence of a singlet, a doublet and a triplet of black pieces denoted by $n_1(p)$, $n_2(p)$ and $n_3(p)$ respectively. If a single black piece occupies a site, there must be 4 white ones surrounding it, so

$$n_1(p) = pq^4; \tag{4.2.1}$$

so there must be 6 white pieces surrounding a doublet, however a doublet can have 2 orientations, one vertical and the other horizontal, so a weight factor 2 must be included, i.e.,

$$n_2(p) = 2p^2q^6; \tag{4.2.2}$$

for a triplet, it can be either collinear or rectangular with the corresponding weight factors 2 and 4, so

$$n_3(p) = 2p^3q^8 + 4p^3q^7. \tag{4.2.3}$$

In calculating $n_s(p)$, we must enumerate all different configurations of the s-connected sites, which is called the s-lattice animal. The key parameter is the number t of perimeter sites needed to isolate it, and the number of independent ways $g(s,t)$ to put it on the lattice. Then

$$n_s(p) = \sum_t g(s,t)p^s q^t = p^s D_s(q), \tag{4.2.4}$$

where $D_s(q) = \sum_t g(s,t)q^t$ and is called the perimeter polynomial.

For small p, infinite clusters do not exist, and according to the definition of s-cluster probability n_s,

$$\sum sn_s(p) = p. \tag{4.2.5}$$

For large p, there may be infinite clusters, and (4.2.5) can be rewritten as

$$\sum_s sn_s + p_\infty = p, \tag{4.2.6}$$

where Np_∞ is the number of lattice sites in the infinite clusters. From the above equation we can get

$$p_\infty = 1 - q - q^4 + q^5 - 4q^6 - 4q^7 \cdots . \tag{4.2.7}$$

The percolation probability $P(p)$ for a black piece belonging to an infinite cluster can be defined as

$$P(p) = \frac{p_\infty}{p} = 1 - \frac{1}{p}\sum_s sn_s(p), \tag{4.2.8}$$

here we shall get

$$P(p) = 1 - q^4 - 4q^6 - 8q^7 \cdots . \tag{4.2.9}$$

If we begin from the situation of low density, then with increasing p, the size of the cluster would grow, and diverge at the point p_c.

According to the theoretical treatment by Domb and Sykes,[c] the mean cluster size in this region is defined as

$$S(p) = \frac{1}{p}\sum_s s^2 n_s(p). \tag{4.2.10}$$

From the first few polynomials they derived the following series expansion

$$\begin{aligned} S(p) &= 1 + 4p + 12p^2 + 24p^3 + 52p^4 + 108p^5 \\ &\quad + 224p^6 + 412p^7 + 844p^8 + 1528p^9 + \cdots . \end{aligned} \tag{4.2.11}$$

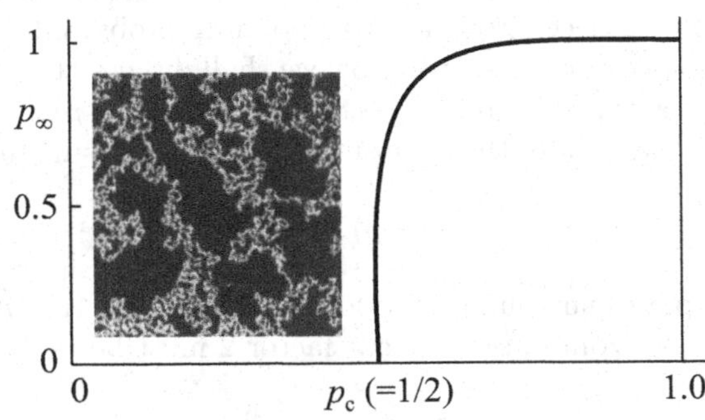

Figure 4.2.3 The probability for the appearance of infinite clusters versus percentage of occupation (inset shows the computer simulation figure of the infinite cluster (the result of triangular lattice).

[c] C. Domb and M. F. Sykes, *Phys. Rev.* **22**, 77 (1961).

Domb and Sykes defined the radius of convergence as the percolation threshold value p_c. It indicates that when $s \to \infty$, $n_s(p)$ has a dramatic discontinuity at p_c: When $p < p_c$, $P_\infty = 0$, and P_∞ has a sudden increase after p_c, as shown in Fig. 4.2.3.

The percolation threshold values of different lattices are shown in Fig. 4.2.1. It should be noted that 2D and 3D percolation threshold values have obvious differences, but if the threshold values are multiplied by the coordination number z or the packing ratio f_p, we shall get almost the same values for different 2D (or 3D) lattices. In this way the results of the lattice model may be extended to the continuum one.

Table 4.2.1 The threshold values of different lattices.

Dimension	Lattice	Bond threshold value p_c^b	Site threshold value p_c^s	Coordination number z	Packing density f_p	zp_c^b	$f_p p_c^s$
2	triangle	0.347	0.500	6	0.9069	2.08	0.45
2	square	0.500	0.593	4	0.7854	2.00	0.47
2	honeycomb	0.653	0.698	3	0.6046	1.96	0.42
3	fcc	0.119	0.198	12	0.7405	1.43	0.147
3	bcc	0.179	0.245	8	0.6802	1.43	0.167
3	sc	0.247	0.311	6	0.5236	1.48	0.164
3	diamond	0.388	0.428	4	0.3401	1.55	0.146

4.2.3 Examples of Percolation

When some small molecules are dissolved in a solvent, multiple reactions may take place with the appearance of giant molecules. If s monomers are combined together, a s-polymer is formed. When $s \to \infty$, the infinite macromolecule is called the gel; if s is finite, the s-polymer is called the sol. When sodium silicate (Na_2SiO_3) is dissolved in water, we get so-called water-glass, which is an example of the sol-gel transition. In 1941 (10 years before percolation was discovered), the chemist P. T. Flory proposed a theory of gelation. He analyzed it with a branching number z formed by bonds of the s-polymer and assumed that the branches were all tree-like without any closed path, which is called the Cayley tree. We can begin from a branch shown in Fig. 4.2.4, and reach the neighboring one, at which there are further $z-1$ branches starting from it. Assuming the probability of each channel is p, the tree-like branches could extend to infinity only if the condition $p(z-1) \geqslant 1$ is satisfied. The critical value for gelation $p_c = 1/(z-1)$, which is the bond percolation threshold value for the Bethe lattice (the infinite Cayley tree); this is actually the first theory of bond percolation on a Bethe lattice. This example shows that using simple and intuitive mathematical methods can get results that can be confirmed by more general theoretical methods.

Figure 4.2.5 shows an experimental set-up for a demonstration on percolation in the disordered structure. In the beaker there is a mixture of glass and metal balls of identical sizes, with aluminum foil used as electrodes at the bottom and the top, which are connected to a circuit. The percolation threshold value can be obtained by observing the first appearance of current by changing the percentage of the metal balls. Since the topologically random close-packed structure does not form a lattice, the threshold value for this disordered structure ϕ_c depends on the percentage (it equals to the product of p_c and the f_p packing density). Thus, percolation of the lattice is extended into percolation of the continuum (see Table 4.2.1)

To illustrate this, Fig. 4.2.6 is a map for a water reservoir with contours showing gravitational equipotentials, the white area is the region above the water surface, while the dark ones are the region below it. Figures 4.2.6(a)–(c) show different stages as the water level is raised: Fig. 4.2.6(a) is before the reservoir is constructed, the ground is connected and there are many mountains with ponds

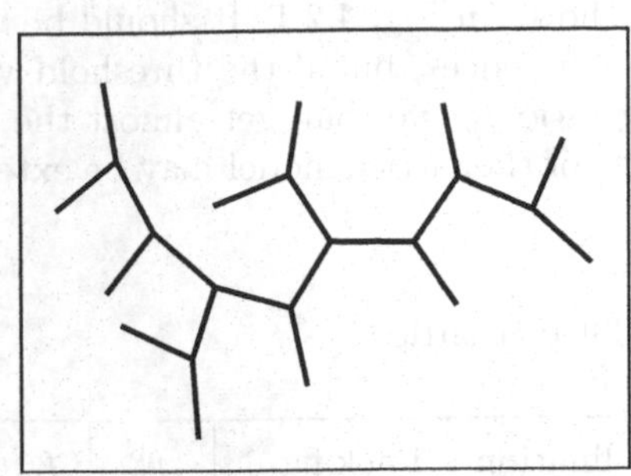

Figure 4.2.4 The schematic diagram of gelatinization (from Flory's original paper).

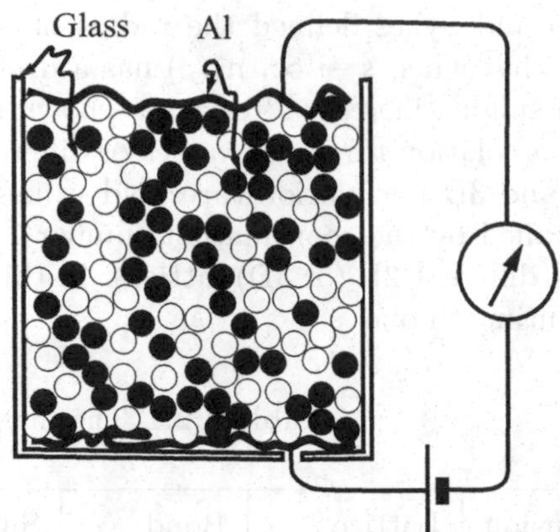

Figure 4.2.5 The percolation of the random close-packed two-phase mixture of Al balls and glass ones. From R. Zallen, *The Physics of Amorphous Solids*, John-Wiley & Sons, New York (1983).

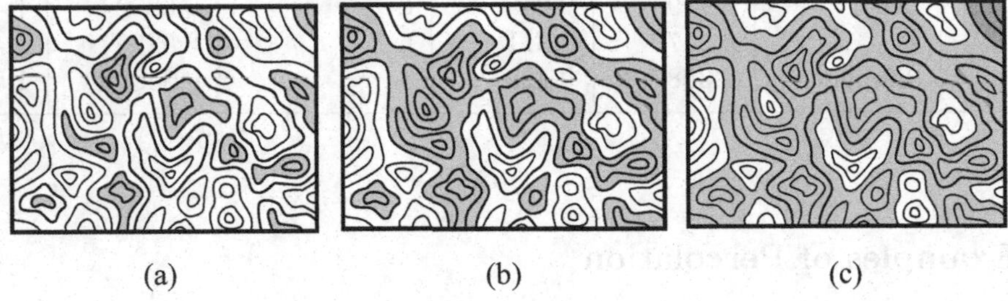

Figure 4.2.6 The percolation of continuum. From R. Zallen, *The Physics of Amorphous Solids*, John-Wiley & Sons, New York (1983).

dispersed in it; Fig. 4.2.6(b) corresponds with the situation when the ponds become interconnected into continuous channels; Fig. 4.2.6(c) is after the reservoir is constructed, most of the ground is submerged and only a few high peaks are left as isolated islands.

The percolation threshold value is approached in Fig. 4.2.6(b), for the inversion of shaded areas into unshaded ones just looks the same statistically as the initial pattern. The area of shaded areas just equals to that of unshaded ones. Figure 4.2.6(c) looks like the inversion of Fig. 4.2.6(a), so we conclude that the corresponding area at the percolation threshold value is about 1/2. This model can be used to mimic the behavior of a particle system in a random potential field, where the percolation threshold value corresponds to the transition from localized states to delocalized ones (see §9.3.1).

§4.3 Fractal Structures

B. Mandelbrot proposed the geometric concept of the fractal structure with self-similarity which can be used to describe certain fragmentary structures found in nature. After several decades of research, the importance of the concept of fractals is widely recognized in condensed matter physics.

4.3.1 Regular Fractals and Fractal Dimension

In 1883 the German mathematician G. Cantor conceived the following: Divide a bar into 3 equal parts and remove the middle one, and then apply the same procedure to the remainder, and so on *ad infinitum*; the imagined final figure is called Cantor bar (or set) (Fig. 4.3.1). The reason why Cantor thought about such an outlandish set is to illustrate the existence of some functions which are hard to treat with traditional methods. It is infinitely dividable on the one hand, while it is discontinuous on the other hand. Besides the the Cantor set, there are other similar structures with different rules, such as the Sierpinski carpet as shown in Fig. 4.3.2. These figures show self similarity and complex fine structure; their geometric properties cannot be characterized by traditional geometry, and they are named *fractal figures*.

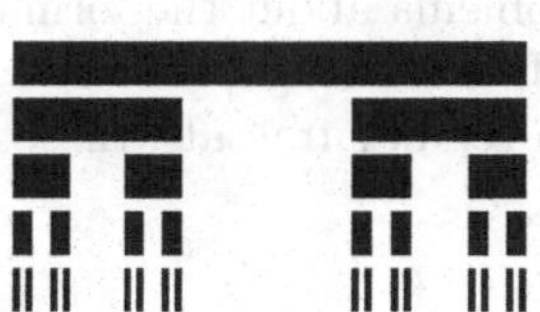

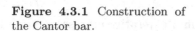

Figure 4.3.1 Construction of the Cantor bar.

Figure 4.3.2 Construction of the Sierpinski carpet.

The clue to explain these pathological figures can be obtained by consideration of their dimensionality. Spaces and figures have dimensions that are defined mathematically. According to Euclidean geometry, there are one-dimensional lines, two-dimensional planes, three-dimensional spaces, and even n-dimensional ($n > 3$) superspaces. We shall express the Euclidean dimension as d. According to topology, any figure can undergo a rubber-like deformation into another figure without rupture; their topological properties are invariant; and the topological dimension d_t can be introduced. We may take a line as an example: No matter how it is bent into any strange curve or folded back and forth, its topological dimension $d_t = 1$ and does not change. Analogously, for any curved surface, $d_t = 2$; these cases are not hard to explain. Besides, there is another dimension derived from self similarity. If the linear size is magnified L times, then for a D-dimensional geometric object, K original objects should be obtained, and

$$K = L^D. \tag{4.3.1}$$

For a square, if $L = 3$ and $K = 9$, we can get $D = 2$; for the cubic, if $L = 3$ and $K = 27$, then $D = 3$. These are the same as the intuitive Euclidean dimensions.

In 1919, the mathematician F. Hausdorff defined the dimension of a geometric object as

$$D = \frac{\ln K}{\ln L}, \tag{4.3.2}$$

which is called the Hausdorff dimension, D. From another viewpoint, if a D-dimensional object is divided into N units with volume r^D, then $r^D N = 1$. And we find an equivalent expression for D, i.e.,

$$D = \frac{\ln N}{\ln \frac{1}{r}}. \tag{4.3.3}$$

As defined by (4.3.2) or (4.3.3), D is not necessarily an integer. For the Cantor bar in Fig. 4.3.1, if the linear size is enlarged 3 times, the number of the units is twice as large as the initial value, so

$$D = \frac{\ln 2}{\ln 3} \approx 0.6309. \tag{4.3.4}$$

For the Sierpinski carpet in Fig. 4.3.2, if the linear size is increased 3 times, the number of units increases 8 times, and so

$$D = \frac{\ln 8}{\ln 3} \approx 1.8628. \tag{4.3.5}$$

Since B. Mandellbrot proposed the concept of the fractal, the Hausdorff dimension D is usually called the fractal dimension. It should be noted that, although many of these fractals have non-integer fractal dimension, some fractal objects may show integer dimension while the condition $D > d_t$ is satisfied, for D is usually less than d, thus, $D \leqslant d$.

The examples mentioned above are disordered, fragmentary, and have obvious discontinuity; but on the other hand, they are ordered with particular symmetries: the self similarity or scale invariance; the figure remains the same after inflation or deflation. This infinitely hierachical and self-similar structure is no stranger to scientists. For example the domain structure appearing in the critical region of a phase transition has this characteristic, and in problems about the formation of polymers as well as infinite clusters near the percolation threshold, self-similar figures can be found. However, fractal objects found in nature are somewhat different from regular fractals; these will be discussed in next subsection.

4.3.2 Irregular Fractal Objects

The regular fractal objects discussed above are described by infinite iterative processes according to mathematical rules, and these conditions can hardly be all satisfied in natural objects. First, the process cannot proceed infinitely. At some microscopic size ξ_1, there is a lower cut-off radius for the fractal due to the graininess of the object (either the atoms or colloid particles), while the maximum length of the figure ξ_2 gives the upper limit on the fractal. So self similarity can only be found in the region between the upper and lower limits ($\xi_1 < L < \xi_2$). Secondly, any physical process will be subject to fluctuations and perturbations; these may lead to randomness and stochasticity, leading to self-similarity only in the statistical sense.

We may introduce the density-density or pair correlation function

$$g(\boldsymbol{r}) = \frac{1}{V}\sum_{\boldsymbol{r}'} \rho(\boldsymbol{r}+\boldsymbol{r}')\rho(\boldsymbol{r}') \tag{4.3.6}$$

to describe fractal structures. The pair correlation function $g(\boldsymbol{r})$ gives the probability of finding a particle at position $(\boldsymbol{r}+\boldsymbol{r}')$ if there is one at $\boldsymbol{r}$. In (4.3.6) ρ is the local density, i.e., $\rho(\boldsymbol{r}) = 1$ if the point $\boldsymbol{r}$ belongs to the object, otherwise $\rho(\boldsymbol{r}) = 0$. Ordinary fractals are usually isotropic, which means that the density correlation depends only on the distance r; that is $g(\boldsymbol{r}) = g(r)$. An object is non-trivially scale-invariant if its correlation function determined by (4.3.6) is unchanged up to a constant under rescaling of length by an arbitrary factor b

$$g(br) \sim b^{-\alpha} g(r), \tag{4.3.7}$$

and the object has the scale-invariant with at least one nonzero variable, where α is number larger than zero and less than d. It can be shown that the only function that satisfies (4.3.7) is a power-law dependence of $g(r)$ on r

$$g(r) \sim r^{-\alpha}, \tag{4.3.8}$$

and we can find the fractal dimension through the exponent α. We calculate the number of particles $N(L)$ within a sphere of radius L from their radial density distribution

$$N(L) \sim \int_0^L g(r) d^d r \sim L^{d-\alpha}. \tag{4.3.9}$$

Comparing (4.3.1) with (4.3.9), we arrive at the desired relation

$$D = d - \alpha. \tag{4.3.10}$$

This equation is widely used for the determination of D from density correlations in a random fractal. The actual method is to determine the Fourier transform of the correlation function from small-angle scattering experiments using X-rays or neutrons.

Let us consider the locus of a particle undergoing random Brownian motion. Although the locus is very complex, it is a line, so the topological dimension $d_t = 1$ while the fractal dimension $D = 2$ which indicates that it would fill the plane finally. This result can also be used to describe the molecular configuration of the polymer melt. As for dilute solution of the polymer, self-avoiding walk provides a more realistic model. When $d = 3$, $R^{5/3} \propto N$, $D = 5/3$, and $d_t = 1$ (see §3.4.2).

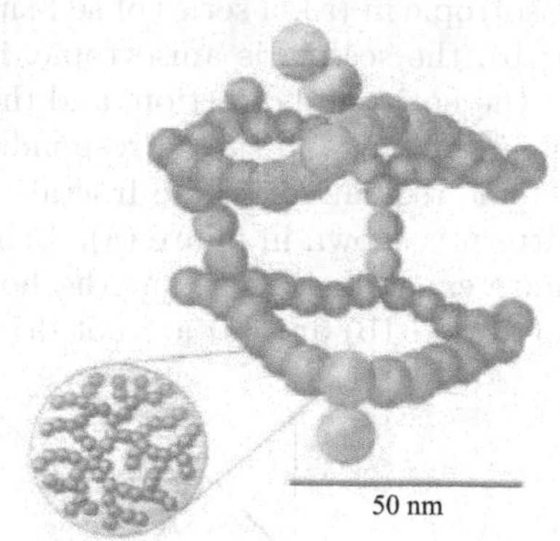

Figure 4.3.3 The schematic diagram for fractal structure of the silica aerogel. From P. Ball, *Made to Measure*, Princeton University Press, Princeton (1997).

Fractal geometry also provides an effective description of porous material, e.g., determination of a gel structure. Let us consider the case of silica aerogel (see Fig. 4.3.3), in which the solvent is replaced by air, thus forming a solid material with a large percentage of pore space. In this material, glassy SiO_2 takes up 1% of the space while the rest is air, so its density is extremely low (0.02 g/cm^3). On the nanometer scale, the system consists of glassy SiO_2 particles. In the range from nanometers to millimeters, these SiO_2 particles form a porous aggregation, and the pores have a fractal structure. It presents the appearance of a continuous solid when observed at a millimeter scale. The fractal dimension, determined by small-angle scattering of X-rays is found to be $D = 1.8$.

The fractal dimensions of common solid materials with pores such as rock and coal are between 2 and 3, the values of D for chemical catalysts with rough surfaces are always larger than 2; and the ones of the main trunks of protein molecules are between 1.2 to 1.8.

In the percolation model, the infinite clusters formed at the percolation threshold also have a fractal structure. In order to decipher the concept of fractal dimension in the context of the percolation model, we have to introduce another exponent τ, which describes the behavior at $p = p_c$ of the cluster-size distribution $n_s(p_c)$ in the asymptotic limit of large cluster size

$$n_s(p_c) \sim s^{-\tau}. \tag{4.3.11}$$

It can be demonstrated from (4.3.11) that the exponent τ must be a number between 2 and 3.

Computer experiments for large clusters show

$$s(p_c) \sim L^D(p_c), \tag{4.3.12}$$

which presents a relation between volume and linear size L of the large clusters. In two-dimensional space, $D \approx 1.9$, while in the three-dimensional space, $D \approx 2.6$.

4.3.3 Self-Affine Fractals

The scale invariance embodied by the self-similarity mentioned above is isotropic, i.e., it uses only one fractal dimension to describe the scale transformation in any direction. It can be anisotropic in other cases. Let us compare the scaling rules of the two dotted circles shown in Fig. 4.3.4. The scaling

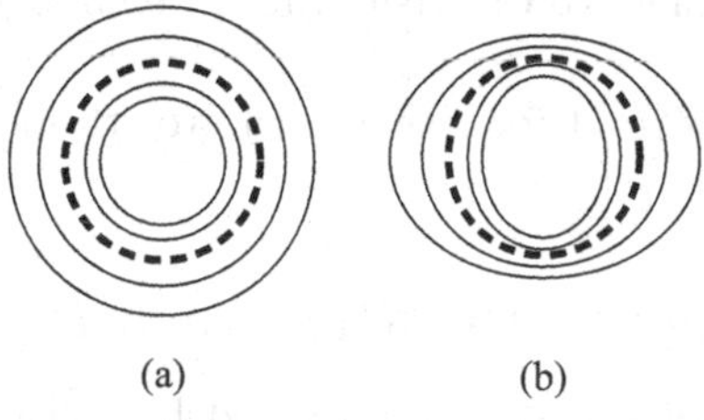

Figure 4.3.4 The scaling transformation to a circle. (a) self-similar transformation; (b) self-affine transformation.

is isotropic in (a), a series of self-similar circles is formed by inflation or deflation of the original one; in (b), the scaling is anisotropic, i.e., the scaling ratios are different for the vertical direction and for the horizontal direction, and the circle changes into ellipses. The second transformation is called the self-affine, and the corresponding fractal invariance is the self-affine one.

The regular self-affine fractal can be explained by the example shown in Fig. 4.3.5. The basic pattern is shown in figure (a). Using it as a template, (b) and (c) are generated. After n steps, the figure expands 4^n times at the horizontal direction and 2^n at the vertical direction. It should be noted that (b) and (c) are not drawn to scale, but on the axes the scales are indicated.

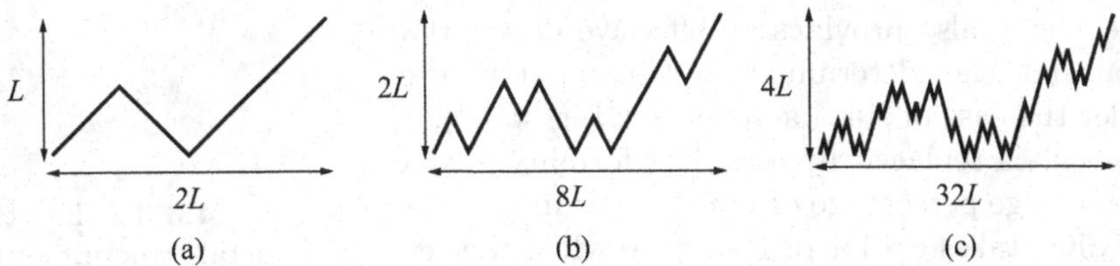

Figure 4.3.5 The regular self-affine fractal with Hurst exponent $H = 1/2$.

Just as in the case of the self-similar fractal, the irregular self-affine fractals are actually more important in natural processes. So we need statistical rescaling: Assume the transformation factor of the horizontal coordinate is b. The corresponding factor for the vertical coordinate is b^H, where H is the Hurst exponent. The result is the typical self-affine fractal, which can be expressed by the function $F(x)$ satisfying

$$F(x) \approx b^{-H} F(bx). \tag{4.3.13}$$

An important example of the random self-affine fractal is the distance $B(t)$ traversed by a particle in Brownian motion at the time interval t, for

$$\begin{gathered} \langle B(t) \rangle \approx t^{1/2}, \\ b^{1/2} B(t) \approx B(bt), \ \langle B(t) \rangle \approx b^{-1/2} B(bt), \end{gathered} \tag{4.3.14}$$

this situation is shown in Fig. 4.3.6.

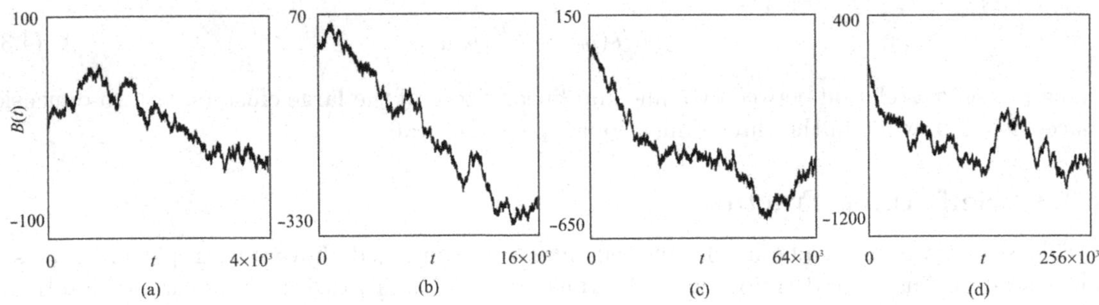

Figure 4.3.6 The patterns for particles in Brownian motion with different scales, while the scale of the vertical direction is the square root of that of the horizontal direction. From P. Meakin, *Fractal, Scaling and Growth Far from Equilibrium*, Cambridge University Press, Cambridge (1998).

The self-affine fractal is important for research on growth morphology of surfaces and it will be discussed in Part VIII.

4.3.4 The Basic Concept of the Multifractal

The concept of the multifractal was first proposed by Masndelbrot in 1972 in his research on turbulence. It is related to the distribution of physical quantities (such as density, intensity, flow velocity, growth probability, etc.) on its geometric support. All previous treatment of fractals was purely geometrical but now we introduce the distribution of physical quantities in fractal or non-fractal structures.

In order to explain the concept of the multifractal, we start with the simplest example, which involves the distribution of density on the Cantor set. We can imagine the process generating the Cantor set: Divide the original bar with unit length and mass into two equal parts, each with the mass $\mu_1 = 1/2$, then compress them into a length 1/3 of the initial one; obviously the density increases. Repeat this process n times and get $N = 2^n$ short bars each with length $l_n = (1/3)^n$ and mass $\mu_n = (1/2)^n$. Since the mass of the original bar is constant:

$$\sum_{i=1}^{n} \mu_i = 1, \tag{4.3.15}$$

then we will get

$$\mu_i = l_i^{\alpha}. \tag{4.3.16}$$

The exponent α here is defined as

$$\alpha = \frac{\ln 2}{\ln 3}, \tag{4.3.17}$$

with the corresponding density

$$\rho_i = \frac{\mu_i}{l_i} = l_i^{\alpha-1}. \tag{4.3.18}$$

For $\alpha < 1$, if $l_i \to 0$, the $\rho_i \to \infty$, and this scale exponent is named as the Hölder exponent which indicates the singularity of the physical quantity (the density here). Figure 4.3.7 shows the density change (expressed by the thicknesses of the bar) of the Cantor set and gives a physical explanation for the quantity α.

Figure 4.3.7 The density change of Cantor set.

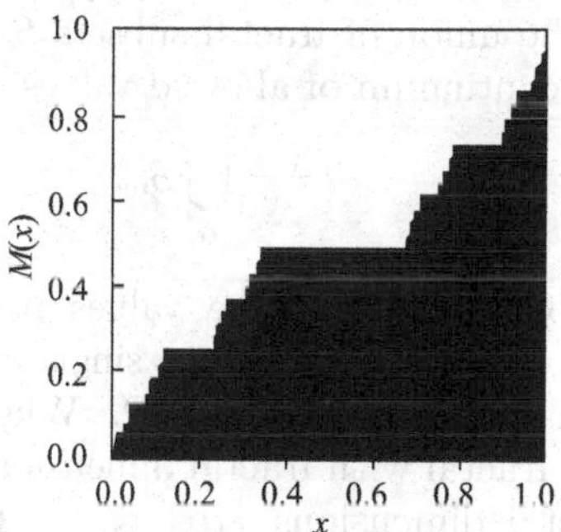

Figure 4.3.8 $M(x)$ versus x figure of Cantor set (showing the devils staircase).

From the above result of the Cantor set, we can get the total mass in the area of $[0, x]$

$$M(x) = \int_0^x \rho(x')dx' = \int_0^x d\mu(x'). \tag{4.3.19}$$

There are infinitely many points on the Cantor set, and $\rho(x') = \infty$. The total mass does not vary in the area corresponding to a gap, and with infinite recurrence, the total length of the gaps is 1 (in mathematical language, the Lesbeque measure of the Cantor set is 0) which equals the initial length of the bar. In these ranges, the value of $M(x)$ remains invariant, and it has a jump at a point of the Cantor set. Finally the total mass $M(1) = 1$. Figure 4.3.8 shows the relation between M and x, and the figure is named the devil's staircase.[d] Scientists have observed a lot of phenomena similar

[d] A popular introduction to the devil's staircase was written by P. Bak in *Phys. Today* **39**, Dec., 39 (1986).

to the devil's staircase, which are usually the result of competition among different periods in space or among different frequencies in the time domain; see §5.5.1.

The theory of measure can give a more complete description of multifractal objects. A set S consisting of N points will have N_i points in the ith cell. These points are sample points of an underlying measure. Let us use the mass or probability $\mu_i = N_i/N$ in the ith cell to construct the measure with q-moment

$$M_\delta(q,l) = \sum_{i=1}^{n} \mu_i^q l^\delta = N(q,l)l^\delta \xrightarrow{l\to 0} \begin{cases} 0, & \delta > \tau(q), \\ \infty, & \delta < \tau(q). \end{cases} \tag{4.3.20}$$

For mass exponent $\delta = \tau(q)$, the measure neither vanishes nor diverges as $l \to 0$. The measure is characterized by a whole sequence of exponents $\tau(q)$. The weighted number of boxes $N(q,l)$ has the form

$$N(q,l) = \sum_i \mu_i^q \sim l^{-\tau(q)}, \tag{4.3.21}$$

and the mass exponent is given by

$$\tau(q) = -\lim_{l\to 0} \frac{\ln N(q,l)}{\ln l}. \tag{4.3.22}$$

We first note that if we choose $q = 0$, then $N(q = 0, l) = N(l)$ is simply the number of boxes needed to cover the set, and $\tau(0) = D$ equals the fractal dimension of the set. The probabilities are normalized, i.e., $\sum_i \mu_i = 1$, and it follows from (4.3.21) that $\tau(1) = 0$.

The Hölder exponent describes local singularity, but it is easy to think that there are lots of boxes μ_i with same exponent α spreading out over the whole range of the set S. It may be helpful to consider that set S is the union of fractal subsets S_α with α chosen from the continuum of allowed values

$$S = \bigcup_\alpha S_\alpha, \tag{4.3.23}$$

The exponent α can take values from the interval $[\alpha_{+\infty}, \alpha_{-\infty}]$, $f(\alpha)$ is usually a single humped function with a maximum $f_{\max}(\alpha) = D$. When the complete set S is a fractal with fractal dimension D, the subsets have fractal dimensions $f(\alpha) \leqslant D$. For fractal subsets with fractal dimension $f(\alpha)$, the number $N(\alpha, l)$ of segments of length l need to cover the sets S_α is

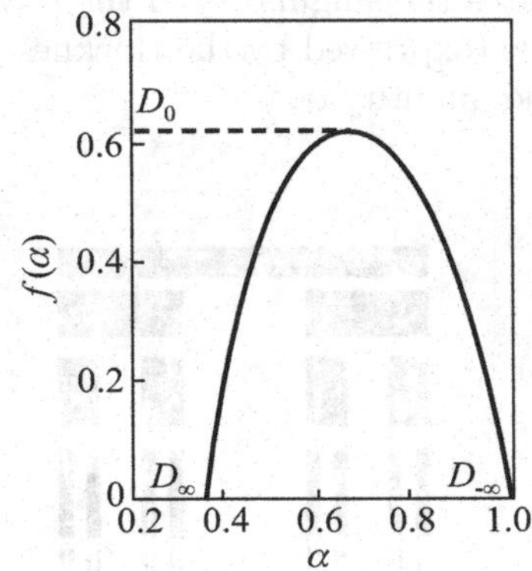

Figure 4.3.9 $f[\alpha(q)] \sim \alpha(q)$.

$$N(\alpha, l) = l^{-f(\alpha)}. \tag{4.3.24}$$

For these sets, we may write the measure $\mu_\alpha = l^\alpha$ in a cell of size l, and therefore the measure M for the set S may be written

$$M_\delta(q,l) = \int \rho(\alpha) l^{-f(\alpha)} l^{\alpha q} l^\delta d\alpha = \int \rho(\alpha) l^{q\alpha - f(\alpha) + \delta} d\alpha, \tag{4.3.25}$$

here the $\rho(\alpha)d\alpha$ is the number of sets from S_α to $S_{\alpha+d\alpha}$. The integral in the equation above is dominated by terms where the integrand has its maximum value, so

$$\frac{d}{d\alpha}(f\alpha - f(\alpha) + \delta) = 0, \tag{4.3.26}$$

in other words

$$\left.\frac{df(\alpha)}{d\alpha}\right|_{\alpha=\alpha(q)} = q. \tag{4.3.27}$$

The integral in (4.3.26) is therefore asymptotically given by

$$M_\delta(q,l) \sim l^{q\alpha(q)-f(\alpha)+\delta}. \tag{4.3.28}$$

In the limit $l \to 0$, M_δ remains finite if δ equals the mass exponent $\tau(q)$ and

$$\tau(q) = f(\alpha(q)) - q\alpha(q), \tag{4.3.29}$$

where $\alpha(q)$ is the solution of (4.3.27). Thus the mass exponent is given in terms of the Hölder exponent $\alpha(q)$ for the mass, and the fractal dimension $f(\alpha(q))$ of the set that supports this exponent. The relation between the both the above is shown in Fig. 4.3.9.

Multifractal analysis has been applied to a wide range of physical phenomena, including the distribution of dissipation in turbulent flows, nonequilibrium growth phenomena and electronic wavefunctions in disordered systems.

Bibliography

[1] Nan, C. W., Physics of inhomogeneous inorganic materials, *Progr. Mat. Sci.* **37**, 1 (1993).

[2] Deutcher, G., R. Zallen, and J. Adler (eds.), *Percolation Structures and Processes*, Adam Hilger, Bristol (1983).

[3] Zallen, R., *The Physics of Amorphous Solids*, John-Wiley & Sons, New York (1983).

[4] Mandelbrot, B. B., *Fractal Geometry of Nature*, Freeman, New York (1983).

[5] Vicsek, T., *Fractal Growth Phenomena*, 2nd ed., World Scientific, Singapore (1992).

[6] Bunde, A., and S. Haolin (eds.), *Fractals and Disordered Systems*, Springer-Verlag, Berlin (1991).

[7] Meakin, P., *Fractal, Scaling and Growth Far from Equilibrium*, Cambridge University Press, Cambridge (1998).

Part II

Wave Behavior in Various Structures

Like as the waves make towards the pebbled shore
So do our minutes hasten to their end,
Each changing place with that which goes before,
In sequent toil all forward to contend.

— William Shakespeare

Waves always behave in a similar way, whether they are longitudinal or transverse, elastic or electric. Scientists of last century always kept this idea in mind.... This general philosophy of wave propagation, forgotten for a time, has been strongly revived in the last decade...

— L. Brillouin (1946)

Chapter 5

Wave Propagation in Periodic and Quasiperiodic Structures

Since Newton deduced the formula for the speed of sound using a one-dimensional lattice model, wave propagation in periodic structures has been a topic of study by physicists. In the 19th century Kelvin *et al.* investigated the problem of dispersion of light also using a one-dimensional mechanical lattice model. At the beginning of the 20th century engineers considered a periodic structure composed of LC circuits, developed a theory of filters for electric waves and proposed such important concepts as the cut-off frequency, pass band and forbidden band. The behavior of electrons and lattice vibrations in periodic structures form the basis of solid state physics. In recent years, the study of classical waves, including electromagnetic and elastic waves, in periodic materials has expanded the domain of solid state physics. Since the discovery of quasicrystals in 1980s, wave propagation in quasiperiodic structures attracted much attention. Propagation of different kinds of waves in periodic and quasi-periodic structures can be treated in a unified manner.

§5.1 Unity of Concepts for Wave Propagation

In this section we shall discuss a unified picture for wave propagation in periodic structures. The formal analogy and similarity between three types of waves, i.e., electrons, lattice waves and electromagnetic waves, will be emphasized.

5.1.1 Wave Equations and Periodic Potentials

There are three types of wave equations, i.e., the Schrödinger equation for de Broglie waves; Newton's equation for lattice waves; and Maxwell equations for electromagnetic waves, with corresponding periodic potentials which influence their propagation.

As a scalar wave in a crystalline solid, an electron has a de Broglie wavelength $\lambda = 2\pi/k$ and hence momentum $\boldsymbol{p} = \hbar \boldsymbol{k}$, which is in fact the crystal momentum (or quasimomentum). Here $\boldsymbol{k}$ is the wavevector. The state of the electron is described by the Schrödinger equation

$$i\hbar\frac{\partial}{\partial t}\psi = \left[-\frac{\hbar^2}{2m}\nabla^2 + V(\boldsymbol{r})\right]\psi. \tag{5.1.1}$$

The first term on the right hand side is the kinetic energy, the second term the potential provided by the periodic array of atoms, such that

$$V(\boldsymbol{r}+\boldsymbol{l}) = V(\boldsymbol{r}), \tag{5.1.2}$$

where $\boldsymbol{l}$ is any lattice vector. After 1992, this treatment has been extended to the ultracold atoms as de Broglie waves in a periodic potential (See Bib. [11]).

The electron considered above is assumed to be moving in a rigid array of atoms. In reality, atoms vibrate about their equilibrium positions because of zero-point fluctuations and thermal fluctuations at finite temperature. Here we ignore the former, and consider only the thermal motion, which gives rise to lattice waves in periodic structures. Lattice waves have one longitudinal branch and two transverse branches for each wavevector. Let $\boldsymbol{u}$ be the atomic displacement and V the potential energy of the whole crystal, a function of the position of all atoms. Then in the harmonic approximation, one obtains the Newton equation

$$M_s \frac{\partial^2}{\partial t^2} u_{ls\alpha} = -\sum_{l's'\beta} V_{ls\alpha,l's'\beta} u_{l's'\beta}, \tag{5.1.3}$$

where $\alpha, \beta = x, y, z$; l and s denote the unit cell and the atom in the unit cell, respectively. $V_{ls\alpha,l's'\beta} \equiv \partial^2 V/\partial u_{ls\alpha}\partial u_{l's'\beta}|_0$ is the force constant defined at the atomic equilibrium positions. The force constant has periodicity

$$V_{ls\alpha,l's'\beta} = V_{0s\alpha,\bar{l}s'\beta}, \tag{5.1.4}$$

with $\bar{l} = l' - l$.

The theory of propagation of electromagnetic waves was perfectly summarized by the Maxwell equations

$$\nabla \times \boldsymbol{E} = -\frac{1}{c}\frac{\partial \boldsymbol{B}}{\partial t}, \ \nabla \cdot \boldsymbol{D} = 4\pi\rho,$$

$$\nabla \times \boldsymbol{H} = \frac{1}{c}\frac{\partial \boldsymbol{D}}{\partial t} + \frac{4\pi}{c}\boldsymbol{j}, \ \nabla \cdot \boldsymbol{B} = 0,$$

where $\boldsymbol{E}$ and $\boldsymbol{H}$ are the electric and magnetic fields, respectively, while $\boldsymbol{D}$ and $\boldsymbol{B}$ are the electric displacement and magnetic induction, respectively, and ρ is the charge density. In addition, there are several basic equations which describe the electromagnetic properties of linear media

$$\boldsymbol{D} = \epsilon \boldsymbol{E} = \boldsymbol{E} + 4\pi \boldsymbol{P} = (1 + 4\pi\chi)\boldsymbol{E},$$

$$\boldsymbol{B} = \mu \boldsymbol{H} = \boldsymbol{H} + 4\pi \boldsymbol{M} = (1 + 4\pi\chi_{\mathrm{m}})\boldsymbol{H},$$

where ϵ and μ denote the dielectric constant and magnetic permeability; χ and χ_{m} represent the electric and magnetic susceptibility. In general, ϵ, μ, χ and χ_{m} are all second-rank tensors. Under strong fields, there are nonlinear polarizations, and nonlinear polarization is a higher order tensor. In this chapter, inhomogeneous media are considered, and so ϵ, μ, χ and χ_{m} can all be expressed as functions of position.

Electromagnetic radiation passing through a solid will polarize the medium. For example, an external electromagnetic wave will cause a displacement of the negative charge with respect to the nuclei in a crystal, and there is an electric displacement vector $\boldsymbol{D}$, described by a wave equation

$$\frac{1}{c^2}\frac{\partial^2 \boldsymbol{D}}{\partial t^2} + \nabla \times \left(\frac{1}{\mu(\boldsymbol{r})}\nabla \times \frac{\boldsymbol{D}}{\epsilon(\boldsymbol{r})}\right) = 0. \tag{5.1.5}$$

Generally, $\mu(\boldsymbol{r})$ and $\epsilon(\boldsymbol{r})$ are tensors, but in many cases they can be taken as scalar quantities. In periodic structures with lattice vector $\boldsymbol{l}$,

$$\mu(\boldsymbol{r} + \boldsymbol{l}) = \mu(\boldsymbol{r}), \ \epsilon(\boldsymbol{r} + \boldsymbol{l}) = \epsilon(\boldsymbol{r}), \tag{5.1.6}$$

i.e., they have the lattice periodicity. It is noted that there are three other field quantities, $\boldsymbol{E}$, $\boldsymbol{B}$ and $\boldsymbol{H}$ and each of them can be used equivalently to write down a wave equation.

5.1.2 Bloch Waves

There are some common properties of wave propagation in periodic structures, although electrons are scalar waves, while lattice vibrations and electromagnetic radiation are vector waves. Before discussing these properties, it is important to notice that there is a tuning condition between periodic potentials and the wavelength of the propagating waves. The de Broglie wavelength of an electron in a solid can be obtained from $\lambda = (\hbar^2/2mE)^{1/2}$, which ranges from lattice spacing to the size of bulk materials, if the appropriate eigenenergy E is chosen. The lattice vibrations include both acoustic and optical branches. In general, their wavelengths also range from lattice spacing to macroscopic size. It is only for those waves which have a wavelength near the period of these structures that substantial effects in wave behavior will be expected. So, in the discussion of wave propagation in periodic structures, the tuning condition must be considered. It is implicitly fulfilled for most problems in solid state physics. On the other hand, the range of wavelengths of electromagnetic radiation may vary from γ-ray (< 0.4 Å), X-ray ($0.4 \sim 50$ Å), ultraviolet ($50 \sim 4000$ Å), visible light ($4000 \sim 7000$ Å), infrared ($0.76 \sim 600$ μm), to microwave and radio waves (> 0.1 mm), so that the lattice spacings of periodic structures for electromagnetic wave should be chosen correspondingly.

The waves propagating in periodic structures are not perfectly free. They are constrained to form Bloch waves. Take an electron in a crystalline solid as an example. The time-dependent equation (5.1.1) can be transformed into the stationary equation

$$\left[-\frac{\hbar^2}{2m}\nabla^2 + V(\boldsymbol{r})\right]\psi(\boldsymbol{r}) = E\psi(\boldsymbol{r}). \tag{5.1.7}$$

The electron will experience the periodic potential described by (5.1.2), so its wavefunction $\psi(\boldsymbol{r})$ and energy E determined by (5.1.7) will reflect the characteristics of the periodic potential.

It can be understood that the solution of (5.1.7) should be characterized by a wavevector $\boldsymbol{k}$, and it will be always possible to take the form

$$\psi_{\boldsymbol{k}}(\boldsymbol{r}) = u_{\boldsymbol{k}}(\boldsymbol{r})f_{\boldsymbol{k}}(\boldsymbol{r}), \tag{5.1.8}$$

where the function $u_{\boldsymbol{k}}(\boldsymbol{r})$ is assumed to have the same translational symmetry as the lattice, that is,

$$u_{\boldsymbol{k}}(\boldsymbol{r}+\boldsymbol{l}) = u_{\boldsymbol{k}}(\boldsymbol{r}). \tag{5.1.9}$$

We now have to determine the function $f_{\boldsymbol{k}}(\boldsymbol{r})$. Due to the periodic potential, one requires that the quantity $|\psi_{\boldsymbol{k}}(\boldsymbol{r})|^2$, which gives the electron probability, must also be periodic. This imposes the following condition on $f_{\boldsymbol{k}}(\boldsymbol{r})$:

$$|f_{\boldsymbol{k}}(\boldsymbol{r}+\boldsymbol{l})|^2 = |f_{\boldsymbol{k}}(\boldsymbol{r})|^2.$$

The choice which satisfies this requirement for all $\boldsymbol{l}$ is the exponential form $\exp(i\boldsymbol{k}\cdot\boldsymbol{r})$. Then we write the solution of (5.1.7) with periodic potential (5.1.2) in the form

$$\psi_{\boldsymbol{k}}(\boldsymbol{r}) = u_{\boldsymbol{k}}(\boldsymbol{r})\mathrm{e}^{i\boldsymbol{k}\cdot\boldsymbol{r}}. \tag{5.1.10}$$

This is the Bloch function from which we can establish the Bloch theorem

$$\psi_{\boldsymbol{k}}(\boldsymbol{r}+\boldsymbol{l}) = \psi_{\boldsymbol{k}}(\boldsymbol{r})\mathrm{e}^{i\boldsymbol{k}\cdot\boldsymbol{l}}. \tag{5.1.11}$$

The physical meaning of the Bloch theorem is that the wavefunctions at positions $\boldsymbol{r}+\boldsymbol{l}$ and $\boldsymbol{r}$ are almost the same, except for a phase factor $\exp(i\boldsymbol{k}\cdot\boldsymbol{l})$.

The Bloch theorem discussed above may be generalized to other cases of wave equations with periodic potentials. As stated in Chap. 1, if there is perfect periodicity in a direct lattice, Fourier transformation gives rise to a reciprocal lattice. Now each of the states is characterized by a wavevector to which an eigenenergy or eigenfrequency is related. We shall see later that a profound consequence of periodicity is that there are some ranges of energy or frequency, known as bandgaps, within which wave propagation is forbidden. Then, the dispersion relation can be divided into separated bands. A useful concept is the Brillouin zone (BZ), i.e., the Wigner–Seitz (WS) cell of the reciprocal lattice. A BZ may be regarded as a unit cell in $\boldsymbol{k}$ space including all the modes for propagating waves. All real solids are finite, but, if the solid is large enough, the Born–von Karman cyclic boundary conditions may be used to extend the solid to infinity in order to simplify the theoretical treatment.

5.1.3 Revival of the Study of Classical Waves

The propagation of classical waves, including elastic and electromagnetic waves, in media is an old scientific problem. Historically, X-ray diffraction in crystals demonstrated both the wave property of the X-rays and the periodic structures of crystals at one stroke. The Bragg equation

$$2d\sin\theta = n\lambda, \tag{5.1.12}$$

which combines the wavelength of X-ray λ and the lattice parameter d, is the cornerstone of the theory of diffraction of X-rays, as well as electrons and neutrons, in crystals. The elementary theory of X-ray diffraction, i.e., kinematical theory, in which the single-scattering approximation is adopted, is very useful and has reasonable validity for a wide range of diffraction experiments. More refined theory, like the dynamical theory of diffraction, takes into account the coherence of multiply scattered waves and proceeds by solving Maxwell equations to give a more correct picture of diffraction phenomena, especially in large and nearly perfect crystals.

To study the frequency-wavevector relation for electromagnetic waves in periodic structures, we at first transform (5.1.5) into a stationary equation for the electric displacement vector $\boldsymbol{D}$ as

$$-\nabla^2\boldsymbol{D} - \nabla\times\nabla\times[\chi(\boldsymbol{r})\boldsymbol{D}] = \frac{\omega^2}{c^2}\boldsymbol{D}, \tag{5.1.13}$$

where

$$\chi(\boldsymbol{r}) = 1 - \frac{1}{\epsilon(\boldsymbol{r})}, \tag{5.1.14}$$

the electric susceptibility, is also a periodic function of position.

According to the general characteristics of wave propagation in periodic structures, it is expected that electromagnetic waves can show bands and gaps as electrons do; these are known as photonic bandgaps. This is simply a window of frequencies, with finite width, in which electromagnetic wave propagation through a periodic structure cannot occur. From this point of view, we may expect there are bandgaps for X-rays in crystals. However, although the diffraction of X-rays has played a spectacular role in revealing the structure of crystalline solids, it is unfortunate that the perturbation of X-rays by the crystalline lattice is extremely small, so the energy gaps are too narrow to be generally noticed. On the other hand, diffraction of light waves by artificial structures, such as gratings, has a much longer history. Still, in most cases, the modulation depths are too low to show clear-cut energy gaps.

The crucial step in establishing propagation gaps lies in achieving a sufficiently large dielectric contrast for different media. These were first demonstrated with microwaves for three-dimensional dielectric structures which are called photonic crystals by Yablonovitch in 1989, and have been pursued in the late 1990s to light waves, which are crucial for application to photonics. The photonic crystals with photonic bandgaps for photonics are therefore the natural analog of semiconductors with electronic energy gaps for solid state electronics. The technological importance of this is much anticipated.

In the 1980s and the 1990s, there was renewed interest in the study of the propagation of classical waves in both periodic and aperiodic structures. It should be noted that the study of electronic wave started from band structure (Bloch, 1928) and then led to localization (Anderson, 1958; Edwards, 1958); however, the study of classical waves shows a reverse process, i.e., from localization (John, 1984; Anderson, 1985) to band structure (Yabolonovitch, 1987; John, 1987). Yabolonovitch and John studied the propagation of electromagnetic waves in a fcc structure by different methods. The former considered "inhibited spontaneous emission in solid-state physics and electronics"; while the latter investigated the band tail states, mobility edges and Anderson localization by introducing a known degree of disorder.

We should point out here that although electrons and photons both have characteristics of waves, there are some basic differences which influence their band structures. As a matter of fact, the underlying dispersion relation for electrons is parabolic, while for photons it is linear; the angular momentum of electrons is 1/2, so scalar wave treatment is always sufficient, while photons have spin 1, and vector-wave character plays a major role. In addition, the band theory of electrons is only an approximation due to the fact that there are always interactions among electrons, while photonic band theory is exact since interactions between photons are negligible. Thus the structure-property relationship for photonic crystals is essentially independent of length scale.

§5.2 Electrons in Crystals

There are a large number of electrons moving in a solid. If the interaction between electrons can be ignored, we arrive at the independent electron model, in which only the periodic ionic potential is felt by the electrons.

5.2.1 Free Electron Gas Model

If the potential is weak enough, we might take $V(\boldsymbol{r}) = 0$, then (5.1.7) has the plane wave solution

$$\psi_{\boldsymbol{k}}(\boldsymbol{r}) = \Omega^{-1/2} e^{i\boldsymbol{k}\cdot\boldsymbol{r}}, \tag{5.2.1}$$

where Ω is the volume of the crystal. The dispersion relation now takes the simplest form

$$E(\boldsymbol{k}) = \hbar^2 k^2/2m. \tag{5.2.2}$$

For a system with N electrons characterized by wavevectors $\boldsymbol{k}$, we can introduce the concept of the Fermi surface, which is the surface in $\boldsymbol{k}$ space within which all the states are occupied. We can define the Fermi energy at the Fermi surface

$$E_{\mathrm{F}} = E(\boldsymbol{k}_{\mathrm{F}}), \tag{5.2.3}$$

where k_{F} is the Fermi wavevector. It is strictly valid only at $T = 0$ K, but the effect of finite temperature on the Fermi surface is very small. It remains sharp even at room temperature. For alkali metals, such as Li, Na and K, with bcc structure, the reciprocal lattices are fcc, and the Brillouin zone (BZ) is a rhombic dodecahedron; their Fermi surfaces are almost spherical. For the noble metals Cu, Ag, and Au, having an fcc structure, the reciprocal lattice is bcc, the Brillouin zone is a truncated octahedron, so their Fermi surfaces are also almost spherical but distorted a little along the $\langle 111 \rangle$ direction. For these solids, the free electron gas model is a good starting point.

It is easy to find the Fermi wavevector and Fermi energy in free-electron approximation by writing

$$N = \sum_{\boldsymbol{k}} = \frac{2\Omega}{(2\pi)^3} \int d\boldsymbol{k}, \tag{5.2.4}$$

where the factor 2 comes from spin degeneracy. Noting that $\int d\boldsymbol{k} = 4\pi k_{\mathrm{F}}^3/3$, we get

$$k_{\mathrm{F}} = \left(3\pi^2 \frac{N}{\Omega}\right)^{1/3}, \tag{5.2.5}$$

and

$$E_{\mathrm{F}} = \frac{\hbar^2}{2m}\left(3\pi^2 \frac{N}{\Omega}\right)^{2/3}. \tag{5.2.6}$$

Equation (5.2.2) tells us that the electron energy depends on the wavevector $\boldsymbol{k}$. In fact, as periodicity exists, any state may be characterized by its reduced wavevector. Taking the empty lattice approach, Fig. 5.2.1 shows the energy as the function of wavevector for the one-dimensional free electron gas in the extended-zone scheme and the reduced-zone scheme. It is obvious that in the extended-zone scheme each state is represented by its real wavevector, but in the reduced-zone scheme there are many states with different energies corresponding to the same wavevector. To eliminate this uncertainty, the band index n is introduced to denote the electron energy as $E_n(\boldsymbol{k})$ and wavefunction as $\psi_{n\boldsymbol{k}}(\boldsymbol{r})$. It is shown that the number of states in each band inside the first zone $-(\pi/a) < k < (\pi/a)$ is equal to the number of unit cells in the crystal.

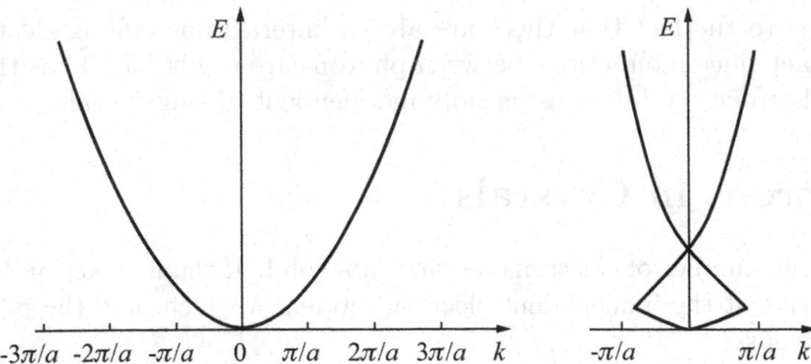

Figure 5.2.1 Dispersion curves of one-dimensional free electron gas for (a) extended-zone scheme, and (b) reduced-zone scheme.

5.2.2 Nearly-Free Electron Model

Strictly speaking, the crystal potential $V(\boldsymbol{r})$ cannot be ignored. However, it may be assumed that it is relatively weak and can be treated as a perturbation, so we can replace the free electron gas model by the nearly-free electron model. This model works well for simple metals such as Na, K, Al, and for narrow-gap semiconductors. The starting point for the nearly-free electron model are the wavefunctions $\psi_{n\boldsymbol{k}}^{(0)}$ and the eigenenergies $E_n^{(0)}(\boldsymbol{k})$ for the free electron model. By using perturbation theory, the dispersion relation and wavefunction can be obtained:

$$E_n(\boldsymbol{k}) = E_n^{(0)}(\boldsymbol{k}) + \left\langle \psi_{n\boldsymbol{k}}^{(0)} |V| \psi_{n\boldsymbol{k}}^{(0)} \right\rangle + \sum_{n'\boldsymbol{k}'}{}' \frac{\left| \left\langle \psi_{n'\boldsymbol{k}'}^{(0)} |V| \psi_{n\boldsymbol{k}}^{(0)} \right\rangle \right|^2}{E_n^{(0)}(\boldsymbol{k}) - E_{n'}^{(0)}(\boldsymbol{k}')}, \tag{5.2.7}$$

and

$$\psi_{n\boldsymbol{k}} = \psi_{n\boldsymbol{k}}^{(0)} + \sum_{n'\boldsymbol{k}'}{}' \frac{\left\langle \psi_{n'\boldsymbol{k}'}^{(0)} |V| \psi_{n\boldsymbol{k}}^{(0)} \right\rangle}{E_n^{(0)}(\boldsymbol{k}) - E_{n'}^{(0)}(\boldsymbol{k}')} \psi_{n'\boldsymbol{k}'}^{(0)}. \tag{5.2.8}$$

For simplicity and still without loss of generality, we take the one-dimensional case as example. In (5.2.7), the second term on the right hand side is a constant, and merely corresponds to a shift of the zero of energy. When we consider the lowest energy band $n = 1$, we see that the energy difference $E_1^{(0)}(\boldsymbol{k}) - E_{n'}^{(0)}(\boldsymbol{k}')$ rises rapidly for $n' \geq 3$, and so we only consider $n' = 2$:

$$E_1(k) \simeq E_1^{(0)}(k) + \frac{|V_{-2\pi/a}|^2}{E_1^{(0)}(k) - E_2^{(0)}(k)}, \tag{5.2.9}$$

with

$$V_{-2\pi/a} = \frac{1}{L} \int V(x) \mathrm{e}^{i2\pi x/a} dx.$$

There is only a slight modification for the eigenenergy of (5.2.9) from the parabolic dispersion curve of Fig. 5.2.1, if k is not at the Brillouin zone boundary. However, when $k \simeq \pi/a$, $E_1^{(0)} = \hbar^2 k^2/2m$, $E_2^{(0)} = \hbar^2(k - 2\pi/a)^2/2m$, then $E_1^{(0)} \simeq E_2^{(0)}$ and degenerate perturbation theory is needed. The result of applying this is:

$$E_\pm(k) = \frac{1}{2} \left\{ E_1^{(0)}(k) + E_2^{(0)}(k) \pm \left[\left(E_2^{(0)}(k) - E_1^{(0)}(k) \right)^2 + 4|V_{-2\pi/a}|^2 \right]^{1/2} \right\}, \tag{5.2.10}$$

and there is a energy gap

$$E_{\mathrm{g}} = E_+(k) - E_-(k) = 2|V_{-2\pi/a}| \tag{5.2.11}$$

at the boundary of the first Brillouin zone. The same procedure can be used to get the whole energy spectrum for the one-dimensional nearly-free electron model as shown in Fig. 5.2.2 where (a) is for the extended zone scheme and (b) for the reduced zone scheme. Thus, a periodic potential brings in energy bands and gaps. It should be noted that the value of the perturbation potential at the BZ boundary is crucial for the opening up of band gaps.

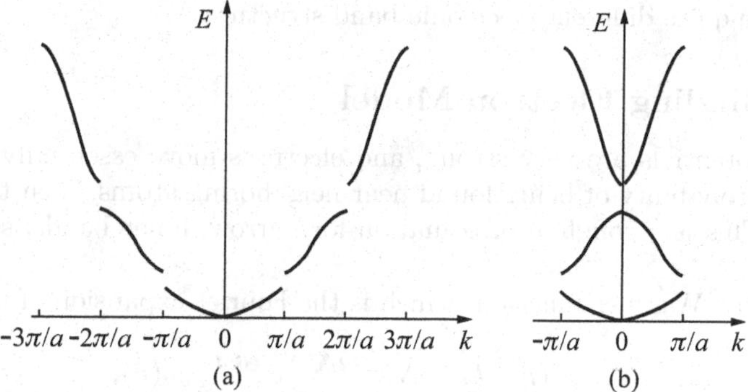

Figure 5.2.2 Bands and gaps in one-dimensional nearly-free electron model for (a) the extended-zone scheme, and (b) the reduced zone scheme.

The energy gap is related to the Bragg diffraction of electron waves. We can see this by observing that the first-order modified wavefunction of the first band for $-\pi/a < k < \pi/a$ is

$$\psi_{1k} = \psi_{1k}^{(0)} + \frac{V_{-2\pi/a}}{E_1^{(0)}(k) - E_2^{(0)}(k)}\psi_{2k}^{(0)}, \tag{5.2.12}$$

where the summation only involves the wavefunction of the second band. Here $\psi_{1k}^{(0)} = L^{-1/2}\exp(ikx)$ represents a wave travelling in the positive direction, whereas $\psi_{2k}^{(0)} = L^{-1/2}\exp[i(k-2\pi/a)x]$ a left-travelling wave. Near the zone edge, degenerate perturbation theory is still needed. The wavefunctions $\psi_{1k}^{(0)}$ and $\psi_{2k}^{(0)}$ are treated on an equal footing, then the wavefunctions at the BZ boundaries are

$$\psi_\pm(x) = \frac{1}{\sqrt{2L}}\left[\psi_{1,\pi/a}^{(0)}(x) \pm \psi_{2,\pi/a}^{(0)}(x)\right] = \frac{1}{\sqrt{2L}}\left(e^{i\pi x/a} \pm e^{-i\pi x/a}\right). \tag{5.2.13}$$

These two state functions $\psi_+ = (2/L)^{1/2}\cos(\pi x/a)$ and $\psi_- = (2/L)^{1/2}\sin(\pi x/a)$ represent standing waves. Their squared moduli are the electron probability distribution as shown in Fig. 5.2.3.

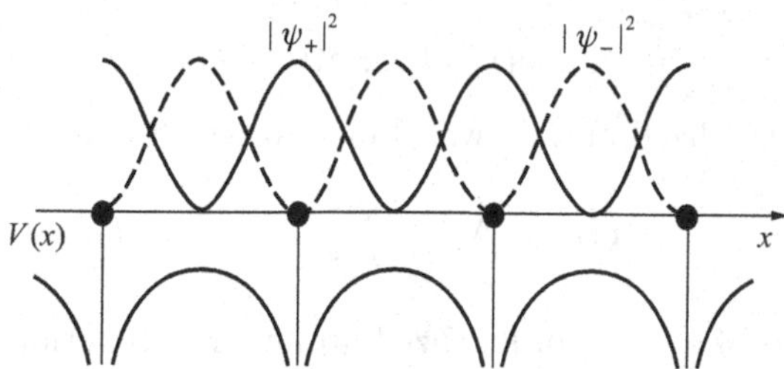

Figure 5.2.3 Bragg reflection of electron in a periodic structure.

From the viewpoint of scattering, at the zone edge $k = \pi/a$, the scattering is strong and the reflected wave has the same amplitude as the incident wave. This leads to Bragg diffraction at $\lambda = 2\pi/k = 2a$. This strong scattering is caused by the periodic potential. Bragg scattering at the zone boundary opens up energy gaps, that is, the interaction of electron waves with the atomic lattice results in destructive interference at certain wavelengths.

It is clear that Fig. 5.2.1 was changed into Fig. 5.2.2, when the crystalline potential was added. The continuous energy-wavevector dispersion relation characteristic of free space is therefore modified; this energy spectrum is referred to as the electronic band structure of the system. It is not only in one dimension that energy gaps are created at the BZ boundaries. In higher dimensional lattices, the Bragg condition is satisfied along all boundaries of the BZ, so energy gaps along those boundaries are also created. The quantitative details of a given electronic band structure depend on the specific form of the periodical potential. As a result, substances with different crystalline structures may have quite different electronic band structures.

5.2.3 Tight-Binding Electron Model

When atomic potentials are very strong, and electrons move essentially around a single atom, with only a small probability of being found near neighboring atoms, then the tight-binding model is more effective. This is a rough approximation for narrow, inner bands, such as the 3d bands in transition metals.

We begin with the Wannier function, which is the Fourier expansion of the Bloch functions, as

$$w_n(\boldsymbol{r}-\boldsymbol{l}) = N^{-1/2}\sum_{\boldsymbol{k}} \mathrm{e}^{-i\boldsymbol{k}\cdot\boldsymbol{l}}\psi_{n\boldsymbol{k}}(\boldsymbol{r}), \tag{5.2.14}$$

and conversely

$$\psi_{n\boldsymbol{k}}(\boldsymbol{r}) = N^{-1/2}\sum_{\boldsymbol{l}} \mathrm{e}^{i\boldsymbol{k}\cdot\boldsymbol{l}}w_n(\boldsymbol{r}-\boldsymbol{l}). \tag{5.2.15}$$

The Wannier functions have orthogonality for different n and $\boldsymbol{l}$, so $w_n(\boldsymbol{r}-\boldsymbol{l})$ is a localized function.

To illustrate the localized feature of the Wannier functions, we take a Bloch function

$$\psi_{\boldsymbol{k}}(\boldsymbol{r}) = N^{-1/2}u(\boldsymbol{r})\mathrm{e}^{i\boldsymbol{k}\cdot\boldsymbol{r}}, \tag{5.2.16}$$

with the assumption that the modulated amplitude $u(\boldsymbol{r})$ is independent of the wavevectors $\boldsymbol{k}$ in a band. We substitute (5.2.16) into (5.2.14), for a cubic lattice with lattice constant a, and the Wannier function at the origin is

$$w(\boldsymbol{r}) = \frac{\sin(\pi x/a)\sin(\pi y/a)\sin(\pi z/a)}{(\pi x/a)(\pi y/a)(\pi z/a)}u(\boldsymbol{r}). \tag{5.2.17}$$

This looks like $u(\boldsymbol{r})$ in the center of the cell, but it spreads out a long way with gradually decreasing oscillations.

When the separation a between atoms in solids is not small, it is reasonable to use the atomic wavefunction $\phi_{\mathrm{a}}(\boldsymbol{r}-\boldsymbol{l})$ instead of Wannier function $w(\boldsymbol{r}-\boldsymbol{l})$ to describe the behavior of electrons, that is

$$w(\boldsymbol{r}-\boldsymbol{l}) \simeq \phi_{\mathrm{a}}(\boldsymbol{r}-\boldsymbol{l}).$$

Here $\phi_{\mathrm{a}}(\boldsymbol{r})$ is a strongly localized atomic wavefunction which can be used to construct the Bloch function

$$\psi_{\boldsymbol{k}}(\boldsymbol{r}) = N^{-1/2}\sum_{\boldsymbol{l}} \mathrm{e}^{i\boldsymbol{k}\cdot\boldsymbol{l}}\phi_{\mathrm{a}}(\boldsymbol{r}-\boldsymbol{l}). \tag{5.2.18}$$

This function is composed of a series of localized atomic wavefunctions multiplied by phase factors $\exp(i\boldsymbol{k}\cdot\boldsymbol{l})$. The expectation value of the energy can be calculated from this wavefunction

$$E(\boldsymbol{k}) = \frac{\int \psi_{\boldsymbol{k}}^*\left\{-\frac{\hbar^2}{2m}\nabla^2 + V(\boldsymbol{r})\right\}\psi_{\boldsymbol{k}}d\boldsymbol{r}}{\int \psi_{\boldsymbol{k}}^*\psi_{\boldsymbol{k}}d\boldsymbol{r}} \simeq \sum_{\boldsymbol{h}} \mathrm{e}^{i\boldsymbol{k}\cdot\boldsymbol{h}}E_{\boldsymbol{h}}, \tag{5.2.19}$$

where

$$E_{\boldsymbol{h}} = \frac{1}{\Omega_{\mathrm{c}}}\int \phi_{\mathrm{a}}^*(\boldsymbol{r}+\boldsymbol{h})\left\{-\frac{\hbar^2}{2m}\nabla^2 + V(\boldsymbol{r})\right\}\phi_{\mathrm{a}}(\boldsymbol{r})d\boldsymbol{r}, \tag{5.2.20}$$

with Ω_{c} being the volume of a unit cell and $\boldsymbol{h} = \boldsymbol{l}-\boldsymbol{l}'$ as the relative position of the sites upon which the orbitals are centered.

The atomic wavefunctions ϕ_a can be obtained by solving the Schrödinger equation

$$\left[-\frac{\hbar^2}{2m}\nabla^2 + v_a(\boldsymbol{r})\right]\phi_a(\boldsymbol{r}) = E_a\phi_a(\boldsymbol{r}), \tag{5.2.21}$$

with $v_a(\boldsymbol{r})$ being the potential of an isolated atom. For a simple cubic lattice with $\boldsymbol{h} = (a,0,0)$, $(0,a,0)$ and $(0,0,a)$, the dispersion relation is

$$E(\boldsymbol{k}) \simeq E_a + 2E_{100}(\cos ak_x + \cos ak_y + \cos ak_z). \tag{5.2.22}$$

The BZ in this case is a cube, and the width of the band is $12|E_{100}|$, the lowest energy being $E_a + 6E_{100}$.

The tight-binding model demonstrates an important principle: If there are N atoms far apart and a is large, a single electron has N-fold degenerate states with a single energy. When a decreases, wavefunction overlap leads to delocalized states with energy bands, as shown in Fig. 5.2.4.

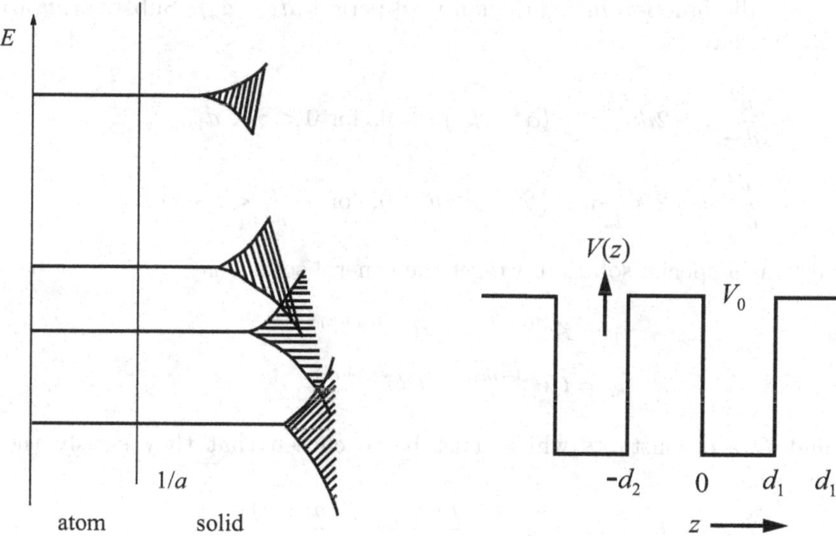

Figure 5.2.4 Atomic levels spreading into bands as lattice separation decreases.

Figure 5.2.5 One-dimensional superlattice potential.

5.2.4 Kronig–Penney Model for Superlattices

We consider a semiconductor superlattice GaAs-$Ga_{1-x}Al_xAs$. The longitudinal motion of an electron in the direction of the superlattice can be simplified as in the one-dimensional periodic rectangular potential well with lattice spacing $d = d_1 + d_2$, as shown in Fig. 5.2.5. The Kronig–Penney model gives energy solutions which form a series of minibands within the original conduction and valence bands.

The equation of motion for an electron can be written as

$$\frac{d^2}{dz^2}\psi + \frac{2m}{\hbar^2}E\psi = 0, \text{ for } 0 < z < d_1, \tag{5.2.23}$$

and

$$\frac{d^2}{dz^2}\psi + \frac{2m}{\hbar^2}[E - V_0]\psi = 0, \text{ for } -d_2 < z < 0. \tag{5.2.24}$$

Consider bound states, i.e., $E < V_0$, and define

$$\alpha^2 = \frac{2mE}{\hbar^2}, \ \beta^2 = \frac{2m(V_0 - E)}{\hbar^2},$$

where α, β are real. Equations (5.2.23) and (5.2.24) can be written as

$$\frac{d^2}{dz^2}\psi + \alpha^2\psi = 0, \ \text{for } 0 < z < d_1, \tag{5.2.25}$$

$$\frac{d^2}{dz^2}\psi - \beta^2\psi = 0, \ \text{for } -d_2 < z < 0. \tag{5.2.26}$$

To solve (5.2.25) and (5.2.26), we use the Bloch theorem

$$\psi(z) = u_k(z)\mathrm{e}^{ikz}, \tag{5.2.27}$$

where $u_k(z)$ is the periodic function in z direction with period $(d_1 + d_2)$. Substituting (5.2.27) into (5.2.25) and (5.2.26), we have

$$\frac{d^2}{dz^2}u + 2ik\frac{d}{dz}u + (\alpha^2 - k^2)u = 0, \ \text{for } 0 < z < d_1, \tag{5.2.28}$$

$$\frac{d^2}{dz^2}u + 2ik\frac{d}{dz}u - (\beta^2 + k^2)u = 0, \ \text{for } -d_2 < z < 0. \tag{5.2.29}$$

If we assume $u = \mathrm{e}^{mz}$ as a special solution, we get the general solutions

$$u_1 = A\mathrm{e}^{i(\alpha-k)z} + B\mathrm{e}^{-i(\alpha+k)z}, \tag{5.2.30}$$

$$u_2 = C\mathrm{e}^{(\beta-ik)z} + D\mathrm{e}^{-(\beta+ik)z}, \tag{5.2.31}$$

where A, B, C and D are constants which must be so chosen that they satisfy the boundary conditions, for example,

$$u_1(z)|_0 = u_2(z)|_0, \quad \left.\frac{du_1(z)}{dz}\right|_0 = \left.\frac{du_2(z)}{dz}\right|_0,$$

$$u_1(z)|_{d_1} = u_2(z)|_{-d_2}, \quad \left.\frac{du_1(z)}{dz}\right|_{d_1} = \left.\frac{du_2(z)}{dz}\right|_{-d_2}.$$

These lead to a 4×4 determinant, which gives a relation between the energy and the parameters

$$\frac{\beta^2 - \alpha^2}{2\alpha\beta}\sinh\beta d_2 \sin\alpha d_1 + \cosh\beta d_2 \cos\alpha d_1 = \cos k(d_1 + d_2). \tag{5.2.32}$$

The allowed energy bands in the superlattice can be calculated from the following expression

$$-1 \le \left(\frac{V_0}{2E} - 1\right)\left(\frac{V_0}{E} - 1\right)^{-1/2} \sin\left(\frac{d_1(2mE)^{1/2}}{\hbar}\right) \sinh\left(\frac{d_2[2m(V_0 - E)]^{1/2}}{\hbar}\right)$$

$$+ \cos\left(\frac{d_1(2mE)^{1/2}}{\hbar}\right) \cosh\left(\frac{d_2[2m(V_0 - E)]^{1/2}}{\hbar}\right) \le 1. \tag{5.2.33}$$

E is the electron energy in the superlattice direction. If, in the case of GaAs-$\mathrm{Al}_x\mathrm{Ga}_{1-x}\mathrm{As}$, where $V_0 \simeq 0.4$ eV, $m^* = 0.1m_0$ we assume $d_1 = d_2$, the lowest four allowed bands $E_1, \ldots, E_4$ calculated as a function of the width, are as shown in Fig. 5.2.6.[a]

[a]L. Esaki, in *Synthetic Modulated Structures*, eds. L. L. Chang and B. C. Giessen, Academic Press, Oriando (1985).

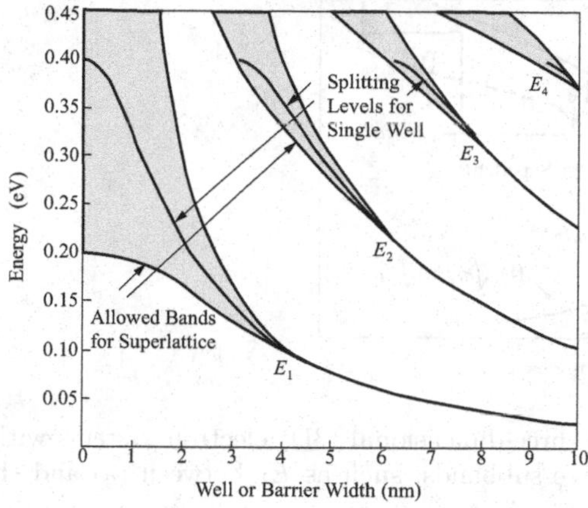

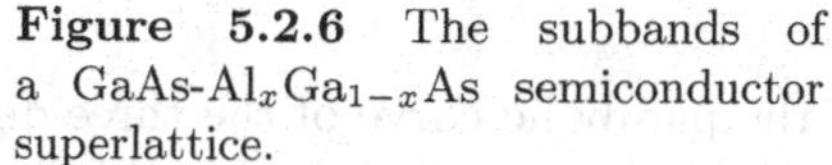
Figure 5.2.6 The subbands of a GaAs-Al$_x$Ga$_{1-x}$As semiconductor superlattice.

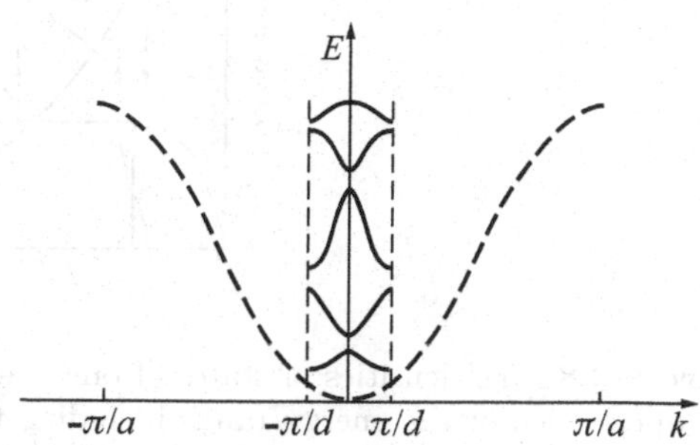

Figure 5.2.7 Folding of the Brillouin zone in a superlattice.

The superlattice period is usually much greater than the original lattice constant. The superlattice potential perturbs the band structure of the host materials and leads to the original Brillouin zone being divided into a series of minizones. These are narrow sub-bands separated by forbidden regions in the conduction band of the host crystal. For a periodic superlattice potential of period d, we shall find d/a sub-bands, as shown in Fig. 5.2.5. It may be said that the Brillouin zone is folded.

5.2.5 Density of States and Dimensionality

The density of states is the number of states per unit energy per unit volume. Consider a d-dimensional box with length L. Under periodic boundary conditions, the volume of each wavevector point is

$$\Delta \boldsymbol{k} = \left(\frac{2\pi}{L}\right)^d. \tag{5.2.34}$$

Then in reciprocal space, the number of states in the shell from k to $k + dk$ is

$$dN = 2\left(\frac{L}{2\pi}\right)^d \int_k^{k+dk} d\boldsymbol{k}. \tag{5.2.35}$$

For the parabolic dispersion relation $E(\boldsymbol{k}) = \hbar^2 k^2/2m$, we can get the density of states (DOS)

$$g(E) = \frac{1}{L^d}\frac{dN}{dE} = \begin{cases} \dfrac{1}{2\pi^2}\left(\dfrac{2m}{\hbar^2}\right)^{3/2} E^{1/2}, & \text{for } d = 3; \\ \dfrac{m}{\pi\hbar^2}, & \text{for } d = 2; \\ \left(\dfrac{2m}{\pi^2\hbar^2}\right)^{1/2} E^{-1/2}, & \text{for } d = 1. \end{cases} \tag{5.2.36}$$

The results are displayed in Fig. 5.2.8. It is obvious that the density of states is closely related to the dimensionality. The difference of the density of states for various dimensions will have a substantial influence on the behavior of the system.

We see that for a two-dimensional (2D) electron gas, the density of states is a constant $g(E_{xy}) = m/\pi\hbar^2$. However, a real system may often be quasi-two-dimensional (quasi-2D), which combines motion in the xy plane with a small but non-zero amount in the z direction, we can consider that for nth level, electrons could fill into n sub-bands, so that

$$g(E) = nm/\pi\hbar^2. \tag{5.2.37}$$

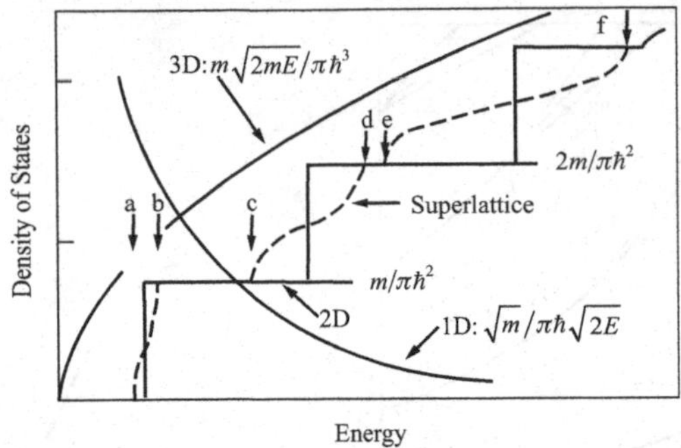

Figure 5.2.8 The densities of states of one-, two- and three-dimensional (3D) electron systems with that of a superlattice in the energy range including first three subbands, such as E_1 between (a) and (b); E_2 between (c) and (d); E_3 between (e) and (f).

The density of states of the superlattice is neither the parabolic curve of the three-dimensional crystal nor the step curve of the two-dimensional electron gas. It is the results of both effects, as also shown in Fig. 5.2.8. Here the electronic energy in the perpendicular direction is again not a discrete level, but extends into a narrow band. This certainly comes from the overlapping of electronic wave functions belonging to different wells. So we can write

$$g(E) = \left(\frac{m}{\pi\hbar^2}\right)\frac{1}{2\pi}k_z(E) \times 2. \tag{5.2.38}$$

where $k_z(E)$ comes from miniband in the dispersion curve, and the factor of 2 comes from the fact that for every E_z there are two values, corresponding to $\pm k_z$.

§5.3 Lattice Waves and Elastic Waves

The vibrations of atoms about their equilibrium positions due to thermal motion will form lattice waves in periodic structures. Many thermodynamic properties of crystals are a result of lattice waves. When the wavelength of the vibration is very long, the atomic details of the structure cannot be seen by the wave, and the medium may be regarded as an elastic continuum. However, in recent years, there is much interest in the study of elastic wave propagation in artificial periodic composites.

5.3.1 Dispersion Relation of Lattice Waves

From the time-dependent equation (5.1.3), we get an algebraic equation describing the stationary states of lattice vibrations

$$M_s\omega^2 u_{ls\alpha} = \sum_{l's'\beta} V_{ls\alpha,l's'\beta}u_{l's'\beta}, \tag{5.3.1}$$

from which the dispersion relation of lattice waves can be obtained.

Now let us consider the solution of (5.3.1). Periodicity suggests that the solution must be such that the displacements of equivalent atoms in different cells satisfy the Bloch theorem. Thus we try a special solution

$$u_{ls\alpha} = M_s^{-1/2}U_{s\alpha}(\boldsymbol{k})\mathrm{e}^{i\boldsymbol{k}\cdot\boldsymbol{l}}, \tag{5.3.2}$$

where $\boldsymbol{k}$ represents the wavevector, and $s = 1, \ldots, n$ if there are n atoms in a unit cell. Substitution of (5.3.2) into (5.3.1) leads to the following $3n$ simultaneous equations of wave amplitudes $U_{s\alpha}(\boldsymbol{k})$

$$\omega^2(\boldsymbol{k})U_{s\alpha}(\boldsymbol{k}) = \sum_{s'\beta} D_{s\alpha,s'\beta}(\boldsymbol{k})U_{s'\beta}(\boldsymbol{k}), \tag{5.3.3}$$

where

$$D_{s\alpha,s'\beta}(\boldsymbol{k}) = (M_s M_{s'})^{-1/2} \sum_{\boldsymbol{l}} V_{0s\alpha,l's'\beta} e^{i\boldsymbol{k}\cdot\boldsymbol{l}}. \tag{5.3.4}$$

Equation (5.3.3) may be expressed in a matrix form

$$\omega^2(\boldsymbol{k})U(\boldsymbol{k}) = D(\boldsymbol{k})U(\boldsymbol{k}), \tag{5.3.5}$$

where $D(\boldsymbol{k})$ is a $3n \times 3n$ square matrix called the dynamic matrix, and $U(\boldsymbol{k})$ is a $3n \times 1$ column matrix representing the eigenvector of the vibrational state.

As usual, the eigenfrequencies in (5.3.5) are determined by the secular equation

$$\|D(\boldsymbol{k}) - \omega^2(\boldsymbol{k})I\| = 0, \tag{5.3.6}$$

where I is a $3n \times 3n$ unit matrix. There are $3n$ eigenfrequencies $\omega_j^2(\boldsymbol{k})$, for $j = 1, \ldots, 3n$. Because the dynamic matrix $D(\boldsymbol{k})$ is Hermitian, i.e.,

$$D_{s\alpha,s'\beta}(\boldsymbol{k}) = D^*_{s'\beta,s\alpha}(\boldsymbol{k}),$$

and because there are some symmetric properties, such as $D(\boldsymbol{k}) = D(\boldsymbol{k}+\boldsymbol{G})$, $D(\boldsymbol{k}) = D(-\boldsymbol{k})$, $D(\boldsymbol{k}) = D(g\boldsymbol{k})$ (here g is a element of point group in real space) all the frequencies are real and satisfy the corresponding symmetric properties. After the eigenfrequencies $\omega_j^2(\boldsymbol{k})$ are known, the corresponding eigenvectors $U_{s\alpha,j}(\boldsymbol{k})$ can be extracted from (5.3.3).

Furthermore, taking into account that there are $3n$ wavelike solutions for each $\boldsymbol{k}$, and $3nN$ independent solutions in total, we can write the general solution of the atomic displacement as a superposition of the independent special solutions

$$u_{ls\alpha} = (NM_s)^{-1/2} \sum_{\boldsymbol{k}j} U_{s\alpha,j}(\boldsymbol{k}) \exp\{i[\boldsymbol{k}\cdot\boldsymbol{l} - \omega_j(\boldsymbol{k})t]\}. \tag{5.3.7}$$

This is the Fourier expansion for the displacement, where the time-dependent factor $\exp[-i\omega_j(\boldsymbol{k})t]$ is added for completeness.

In general, N is a large number and (5.3.1) is actually the infinite set of coupled equations which were reduced to $3n$ equations (5.3.3) by the periodicity. $3nN$ eigenfrequencies form continuous curves as a function of $\boldsymbol{k}$ and for $\omega = \omega_j(\boldsymbol{k})$ $(j = 1, \ldots, 3n)$ there are $3n$ branches of the dispersion curve.

To understand the physical properties of lattice waves, it is convenient to study a simple example: A linear diatomic chain. Two kinds of atoms with different masses, M_1 and M_2, are arranged alternately to compose a one-dimensional periodic lattice with the same atomic spacing a. There are two atoms in each primitive cell and the size of the cell is $2a$. It is assumed that there are only nearest-neighbor forces with a same force constant. The potential energy is of the form

$$V = \frac{1}{2} \sum_m \beta(u_m - u_{m+1})^2, \tag{5.3.8}$$

where β represents the force constant.

The equation of motion become

$$\begin{aligned} M_1 \frac{d^2}{dt^2} u_{2m+1} &= -\beta(2u_{2m+1} - u_{2m} - u_{2m+2}), \\ M_2 \frac{d^2}{dt^2} u_{2m+2} &= -\beta(2u_{2m+2} - u_{2m+1} - u_{2m+3}). \end{aligned} \tag{5.3.9}$$

We substitute a set of trial solutions

$$u_{2m+1} = U_1 e^{i[k(2m+1) - \omega t]}, \quad u_{2m+2} = U_2 e^{i[k(2m+2) - \omega t]} \tag{5.3.10}$$

into (5.3.9) and get the secular equations

$$\begin{aligned} -\omega^2 M_1 U_1 &= -2\beta U_1 + 2\beta\cos(ka)U_2, \\ -\omega^2 M_2 U_2 &= -2\beta U_2 + 2\beta\cos(ka)U_1. \end{aligned} \tag{5.3.11}$$

The eigenfrequencies can be obtained from the determinant

$$\begin{vmatrix} 2\beta - M_1\omega^2 & -2\beta\cos(ka) \\ -2\beta\cos(ka) & 2\beta - M_2\omega^2 \end{vmatrix} = 0. \tag{5.3.12}$$

There are two eigenfrequencies as functions of wavevector k

$$\omega_{\pm}^2 = \beta\left(\frac{1}{M_1} + \frac{1}{M_2}\right) \pm \beta\left\{\left(\frac{1}{M_1} + \frac{1}{M_2}\right)^2 - \frac{4\sin^2(ka)}{M_1M_2}\right\}^{\frac{1}{2}}. \tag{5.3.13}$$

The results of (5.3.13) are shown in Fig. 5.3.1. There are two branches, one is called the acoustic mode and the other the optical mode. These are two pass-bands. There are also two stop-bands, one lies between these two branches, the other lies above the optical branch. Within the pass-bands, k is real, and the lattice wave propagates without attenuation; within the stop-bands, $k = \alpha + i\beta$, is complex, and its imaginary part introduces attenuation. The frequency gap between acoustic and optical modes is surely due to the difference between M_1 and M_2: When the difference is reduced, the width of the gap decreases, disappearing when $M_1 = M_2$. However, the optical branch may remain in the case where the unit cell contains more than one atom of the same species at different positions. The high frequency cutoff is due to the discrete structure of crystal lattice.

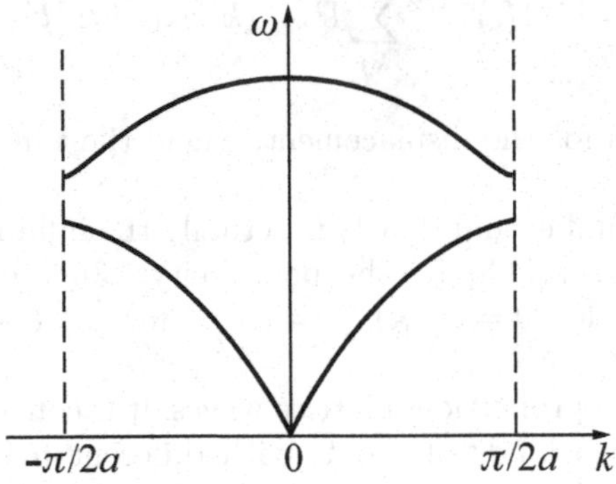

Figure 5.3.1 Pass-band and stop-band for a linear diatomic lattice.

5.3.2 Frequency Spectrum of Lattice Waves

In discussing the physical properties involved in the vibration of crystals, it is always necessary to know the frequency distribution function $g(\omega)$, which is defined as the fraction of vibrational frequencies ω per unit frequency interval. It can be written formally as

$$g(\omega) = \lim_{\Delta\omega\to 0}\frac{1}{\Delta\omega}\sum_j\sum_{\boldsymbol{k}}^{\omega\le\omega_j(\boldsymbol{k})\le\omega+\Delta\omega} = \sum_j\sum_{\boldsymbol{k}}\delta(\omega - \omega_j(\boldsymbol{k})). \tag{5.3.14}$$

Because there are $3nN$ eigenfrequencies in total, $g(\omega)$ satisfies the condition

$$\int_0^\infty g(\omega)d\omega = 3nN. \tag{5.3.15}$$

As the allowed wavevectors $\boldsymbol{k}$ are closely spaced, the summation over $\boldsymbol{k}$ in (5.3.14) may be replaced by an integration using the well-known relation

$$\sum_{\boldsymbol{k}} \to \frac{\Omega}{(2\pi)^3} \int_{\mathrm{BZ}} d\boldsymbol{k}, \tag{5.3.16}$$

where Ω is the volume of the crystal, and then (5.3.14) becomes

$$g(\omega) = \lim_{\Delta\omega \to 0} \frac{1}{\Delta\omega} \frac{\Omega}{(2\pi)^3} \sum_j \int_{\omega \le \omega_j(\boldsymbol{k}) \le \omega + \Delta\omega} d\boldsymbol{k}. \tag{5.3.17}$$

The volume integral over $\boldsymbol{k}$ may be converted to a surface integral. Let dS be an elementary area on a constant frequency surface S corresponding to frequency ω. Considering that $\nabla_{\boldsymbol{k}}\omega_j(\boldsymbol{k})$ is the gradient of the frequency function in $\boldsymbol{k}$-space, we may write

$$\Delta k = \Delta\omega / |\nabla\omega_j(\boldsymbol{k})|,$$

thus

$$\int_{\omega \le \omega_j(\boldsymbol{k}) \le \omega + \Delta\omega} d\boldsymbol{k} = \int_S dS \Delta k = \Delta\omega \int_S \frac{dS}{|\nabla\omega_j(\boldsymbol{k})|}. \tag{5.3.18}$$

With this result, (5.3.17) becomes

$$g(\omega) = \sum_j g_j(\omega) = \frac{\Omega}{(2\pi)^3} \sum_j \int_S \frac{dS}{|\nabla\omega_j(\boldsymbol{k})|}. \tag{5.3.19}$$

This formula is useful in the theory of spectra of lattice modes, as well as of electron states. At $|\nabla\omega_j(\boldsymbol{k})| = 0$, the integrand is divergent; $g(\omega)$ becomes singular at this point, which is known as a van Hove singularity.

The frequency distribution $g(\omega)$ is an important quantity, especially as it is related to several thermodynamic properties. However, it is difficult to get a general expression for the frequency distribution function, and many results are based upon numerical calculations. To do this we solve (5.3.5) for a suitably chosen mesh of values of $\boldsymbol{k}$, and find how many values of $\omega(\boldsymbol{k})$ fall into each range, $\Delta\omega$.

There are two approximate approaches which have proved to be effective in some contexts. The first is the Einstein model. Before the theory of lattice dynamics, one of the most elementary ideas to discuss the motion of atoms in a solid is the Einstein model. In this model each atom vibrates like a simple harmonic oscillator in the potential well of the force fields of its neighbors. The excitation spectrum of the crystal then consists of levels spaced a distance $\hbar\omega_{\mathrm{E}}$ apart, where ω_{E} is the Einstein frequency, i.e., the frequency of oscillation of each atom in its potential well. Then its frequency spectrum is

$$g(\omega) = 3nN\delta(\omega - \omega_{\mathrm{E}}). \tag{5.3.20}$$

This model is useful only in a few cases especially at relatively high temperature, when the assumption that the various atoms vibrate independently is justified.

The second is the Debye model for the specific heat of solids. To study this we shall go from a discrete lattice to a continuum. We may note that, of the $3n$ branches of the dispersion curves, the three acoustic branches approach zero frequency as $\boldsymbol{k}$ approaches zero. This corresponds to the case where the wavelength is very long and one may disregard the atomic details and treat the solid as an isotropic continuum. Such vibrations are referred to as elastic waves, which have dispersion relations:

$$\omega_{\mathrm{l}} = v_{\mathrm{l}} k, \ \omega_{\mathrm{t}} = v_{\mathrm{t}} k, \tag{5.3.21}$$

where subscripts l and t distinguish the longitudinal from the transversal modes. From (5.3.16), we can write for each mode

$$g(\omega) d\omega = \frac{\Omega}{(2\pi)^3} 4\pi k^2 dk = \frac{\Omega}{(2\pi)^3} 4\pi \left(\frac{\omega}{v}\right)^2 \frac{d\omega}{v},$$

and get the frequency distribution function

$$g(\omega) = \frac{\Omega}{2\pi^2}\frac{\omega^2}{v^3}. \tag{5.3.22}$$

Because there exist three different modes simultaneously, one longitudinal and two transverse, (5.3.22) must be modified into

$$g(\omega) = \frac{\Omega}{2\pi^2}\left(\frac{1}{v_{\mathrm{l}}^3} + \frac{2}{v_{\mathrm{t}}^3}\right)\omega^2. \tag{5.3.23}$$

The Debye model is effective for thermodynamic properties at low temperature, especially the T^3 temperature dependence of the low temperature specific heats of solids. In the model calculation, we need to know the range of frequency. The lower limit is evidently $\omega = 0$. The upper cut-off frequency can be determined by requiring that the total number of modes be equal to the number of degrees of freedom for the solid. From

$$\int_0^{\omega_{\mathrm{D}}} g(\omega)d\omega = 3nN, \tag{5.3.24}$$

where the cutoff frequency, denoted by ω_{D}, is called the Debye frequency, which can be determined by substituting (5.3.23) into (5.3.24). The result is

$$\omega_{\mathrm{D}} = \left(\frac{6\pi^2 nN}{\Omega}\right)^{1/3} v_a, \tag{5.3.25}$$

where v_a is defined as $3/v_{\mathrm{a}}^3 = (1/v_{\mathrm{l}}^3 + 2/v_{\mathrm{t}}^3)$. The Debye characteristic temperature θ_{D} is determined by the relation $\theta_{\mathrm{D}} = \hbar\omega_{\mathrm{D}}/k_{\mathrm{B}}$. The physical interpretation of θ_{D} is as follows: For $T > \theta_{\mathrm{D}}$, all lattice waves are excited, while for $T < \theta_{\mathrm{D}}$, only the low frequency lattice waves are excited. This explains the T^3 law for specific heats of solids at low temperatures.

5.3.3 Elastic Waves in Periodic Composites: Phononic Crystals

The propagation of elastic or acoustic waves in inhomogeneous media is an important problem in both geophysics and solid state physics. Attention is directed to the question of possible gaps (i.e., stop-bands) for elastic wave propagation in periodic media. It is difficult for elastic waves to have transmission gaps in inhomogeneous media, because there are both transverse and longitudinal modes, each of which must develop stop-bands overlapping with each other in order for a total spectral gap to appear. The overlapping of the different characteristic stop-bands is complicated by the different velocities for longitudinal and transverse waves. Thus it is a challenging problem to try to find spectral gaps for elastic waves.

The elastic wave equation for a medium that is locally isotropic is

$$\rho(\boldsymbol{r})\frac{\partial^2 u_\alpha}{\partial t^2} = \frac{\partial}{\partial x_\alpha}\left(\lambda(\boldsymbol{r})\frac{\partial u_\beta}{\partial x_\beta}\right) + \frac{\partial}{\partial x_\beta}\left[\mu(\boldsymbol{r})\left(\frac{\partial u_\alpha}{\partial x_\beta} + \frac{\partial u_\beta}{\partial x_\alpha}\right)\right], \ \alpha, \beta = 1, 2, 3, \tag{5.3.26}$$

where u_α are the Cartesian components of the displacement vector $\boldsymbol{u}(\boldsymbol{r})$, and x_α are the Cartesian components of the position vector $\boldsymbol{r}$; $\lambda(\boldsymbol{r})$ and $\mu(\boldsymbol{r})$ are the Lamé coefficients, and $\rho(\boldsymbol{r})$ is the density; summation over repeated indices is implied. The coefficients $\lambda(\boldsymbol{r})$, $\mu(\boldsymbol{r})$, and $\rho(\boldsymbol{r})$ are periodic functions of $\boldsymbol{r}$, such as

$$\lambda(\boldsymbol{r} + \boldsymbol{l}) = \lambda(\boldsymbol{r}), \ \mu(\boldsymbol{r} + \boldsymbol{l}) = \mu(\boldsymbol{r}), \ \rho(\boldsymbol{r} + \boldsymbol{l}) = \rho(\boldsymbol{r}), \tag{5.3.27}$$

and they can all be expanded in a Fourier series. As a result of the common periodicity of all the coefficients in (5.3.26), its eigensolutions, according to the Bloch theorem, can always be chosen so as to satisfy the relation $\boldsymbol{u}(\boldsymbol{r}) = \boldsymbol{U}_{\boldsymbol{k}}(\boldsymbol{r})\exp(\boldsymbol{k}\cdot\boldsymbol{r})$, where $\boldsymbol{k}$ is restricted to the first BZ and $\boldsymbol{U}_{\boldsymbol{k}}(\boldsymbol{r})$ is also a periodic function of $\boldsymbol{r}$, so it can be expanded in Fourier series. We then obtain

$$\omega^2 c u_{\alpha,\boldsymbol{k}+\boldsymbol{G}} = \sum_{\boldsymbol{G}'} \left\{ \sum_{\beta,\boldsymbol{G}''} \rho^{-1}_{\boldsymbol{G}-\boldsymbol{G}''} [\lambda_{\boldsymbol{G}''-\boldsymbol{G}'}(\boldsymbol{k}+\boldsymbol{G}')_\beta(\boldsymbol{k}+\boldsymbol{G}'')_\alpha \right.$$
$$+ \mu_{\boldsymbol{G}''-\boldsymbol{G}'}(\boldsymbol{k}+\boldsymbol{G}')_\alpha(\boldsymbol{k}+\boldsymbol{G}'')_\beta]\, u_{\beta,\boldsymbol{k}+\boldsymbol{G}'}$$
$$\left. + \sum_{\boldsymbol{G}''} \left(\rho^{-1}_{\boldsymbol{G}-\boldsymbol{G}''} \mu_{\boldsymbol{G}''-\boldsymbol{G}'} \sum_\beta (\boldsymbol{k}+\boldsymbol{G}')_\beta(\boldsymbol{k}+\boldsymbol{G}'')_\beta \right) u_{\alpha,\boldsymbol{k}+\boldsymbol{G}'} \right\}, \tag{5.3.28}$$

where c is the acoustic velocity and $\boldsymbol{G}$ are vectors of the reciprocal lattice. If the infinite series in (5.3.28) are approximated by a sum of N reciprocal vectors, (5.3.28) is reduced to a $3N \times 3N$ matrix eigenvalue equation for the $3N$ unknown coefficients $u_{\alpha,\boldsymbol{k}+\boldsymbol{G}}$. The number N is increased until the desired convergence is achieved.

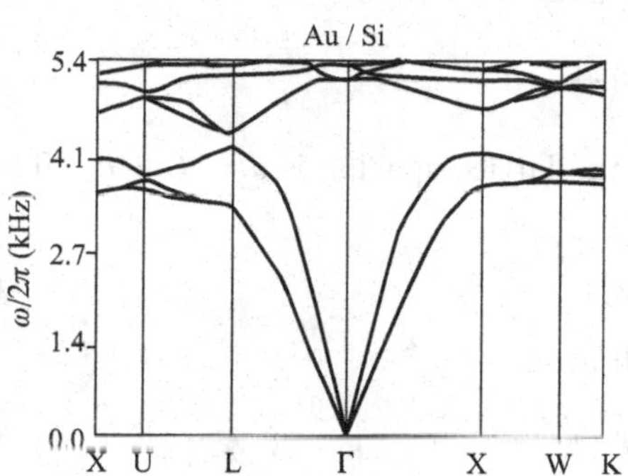

Figure 5.3.2 The dispersion relation of a periodic composite. From E. N. Economou and M. Sigalas, *J. Acoust. Soc. Am.* **95**, 1734 (1994).

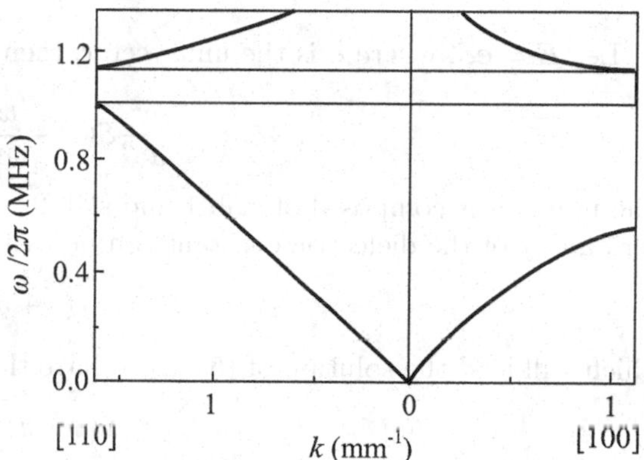

Figure 5.3.3 Experimentally observed dispersion relation of ultrasonic wave in a two-dimensional periodic composite. From F. R. Montero de Espinosa, E. Jimenez, and M. Torres, *Phys. Rev. Lett.* **80**, 1208 (1998).

Figure 5.3.2 shows the theoretically calculated bandgaps for the propagation of elastic waves in a three-dimensional periodic composite, Fig. 5.3.3 shows the bandgaps found experimentally in a two-dimensional periodic composite. These artificial structures with bandgaps for the propagation of acoustic waves are called phononic crystals.

§5.4 Electromagnetic Waves in Periodic Structures

We now consider electromagnetic waves propagating in a medium characterized by a spatially dependent dielectric constant $\epsilon(\boldsymbol{r})$. For simplicity, $\epsilon(\boldsymbol{r})$ is assumed to be a scalar function of position. Here we are only concerned with nonmagnetic media, so μ is a constant, simply $\mu = 1$. We shall also assume that there is a periodic variation in the dielectric constant satisfying equation (5.1.6).

5.4.1 Photonic Bandgaps in Layered Periodic Media

Before considering more complicated artificial periodic dielectric structures, it is instructive to investigate layered periodic media. We consider a situation where there are two kinds of slabs with thicknesses d_1 and d_2 arranged alternatively with dielectric constants ϵ_1 and ϵ_2, respectively as shown in Fig. 5.4.1(a). We assume that the stacking direction is z, and the x-y plane is infinite and homogeneous. Electromagnetic wave propagation in this system can be treated as a one-dimensional problem, because it is trivial for waves along the x and y directions.

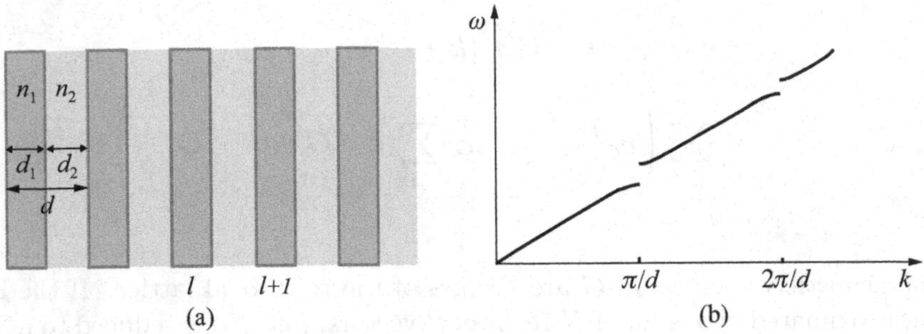

Figure 5.4.1 (a) Layered periodic medium composed with two kinds of slabs with thicknesses d_1 and d_2; (b) The dispersion relation for a one-dimensional periodic dielectric structure.

Let $\boldsymbol{E} = \boldsymbol{e}\mathcal{E}$, where $\boldsymbol{e}$ is the unit vector, then wave equation (5.1.13) can be reduced to

$$\frac{d^2}{dz^2}\mathcal{E}(z) + \frac{\omega^2}{c^2}\epsilon(z)\mathcal{E}(z) = 0. \tag{5.4.1}$$

The unit cell is composed of slab 1 and slab 2, so that the lattice spacing is $d = d_1 + d_2$. Then the periodicity of the dielectric constant can be written as

$$\epsilon(z + d) = \epsilon(z), \tag{5.4.2}$$

which will lead the solution of (5.4.1) to have the Bloch form

$$\mathcal{E}(z + d) = \mathcal{E}(z)\mathrm{e}^{ikd}. \tag{5.4.3}$$

Further, we write the solutions in slabs 1 and 2 of the lth unit cell and slab 1 of the $(l+1)$th unit cell as follows:

$$\begin{aligned} \mathcal{E}_l^{(1)}(z) &= A_l\mathrm{e}^{iq_1 z} + B_l\mathrm{e}^{-iq_1 z}, \\ \mathcal{E}_l^{(2)}(z) &= C_l\mathrm{e}^{iq_2 z} + D_l\mathrm{e}^{-iq_2 z}, \\ \mathcal{E}_{l+1}^{(1)}(z) &= A_{l+1}\mathrm{e}^{iq_1 z} + B_{l+1}\mathrm{e}^{-iq_1 z}, \end{aligned} \tag{5.4.4}$$

where we define $q_1 = \sqrt{\epsilon_1}\omega/c = n_1\omega/c$, $q_2 = \sqrt{\epsilon_2}\omega/c = n_2\omega/c$, with n_1 and n_2 the indices of refraction. A_l, B_l C_l and D_l are the oscillation amplitudes of the electric field.

For convenience, we take the interface at the left hand side of each slab as its local coordinate origin $z = 0$, then using the continuity conditions of $\mathcal{E}$ and $d\mathcal{E}/dz$ at boundaries, we can find that

$$\begin{pmatrix} A_{l+1} \\ B_{l+1} \end{pmatrix} = \boldsymbol{T}\begin{pmatrix} A_l \\ B_l \end{pmatrix}, \tag{5.4.5}$$

where $\boldsymbol{T}$ is a unimodular 2×2 transfer matrix with elements

$$\begin{aligned} T_{11} &= \mathrm{e}^{iq_1 d_1}\left[\cos(q_2 d_2) + \frac{i}{2}\left(\frac{n_1}{n_2} + \frac{n_2}{n_1}\right)\sin(q_2 d_2)\right], \\ T_{12} &= \mathrm{e}^{iq_1 d_1}\frac{i}{2}\left(\frac{n_1}{n_2} - \frac{n_2}{n_1}\right)\sin(q_2 d_2) \\ T_{21} &= T_{12}^*, \\ T_{22} &= T_{11}^*. \end{aligned} \tag{5.4.6}$$

Thus (5.4.3) is transformed into

$$\begin{pmatrix} A_{l+1} \\ B_{l+1} \end{pmatrix} = e^{ikd} \begin{pmatrix} A_l \\ B_l \end{pmatrix}. \tag{5.4.7}$$

Substituting this into (5.4.5), we have

$$(\boldsymbol{T} - e^{ikd} I) \begin{pmatrix} A_l \\ B_l \end{pmatrix} = 0, \tag{5.4.8}$$

where I is the unit matrix. If we consider the reverse process to get A_{l-1}, B_{l-1} from A_l, B_l, we can obtain

$$(\boldsymbol{T}^{-1} - e^{-ikd} I) = \begin{pmatrix} A_l \\ B_l \end{pmatrix}. \tag{5.4.9}$$

Combining these two expressions, we obtain

$$\cos kd = \frac{1}{2}(\boldsymbol{T} + \boldsymbol{T}^{-1}) = \frac{1}{2}\mathrm{Tr}\boldsymbol{T}. \tag{5.4.10}$$

Finally, we have the transcendental equation

$$\cos k(d_1 + d_2) = \cos\frac{n_1 \omega d_1}{c} \cos\frac{n_2 \omega d_2}{c} - \frac{1}{2}\left(\frac{n_1}{n_2} + \frac{n_2}{n_1}\right) \sin\frac{n_1 \omega d_1}{c} \sin\frac{n_2 \omega d_2}{c}, \tag{5.4.11}$$

from which the dispersion relation for electromagnetic wave propagation in the one-dimensional periodic dielectric structure is determined. The numerical result is shown in Fig. 5.4.1(b). It is clear that there are pass-bands and stop-bands for certain frequencies ω: The stop bands appear at the boundaries of the Brillouin zones.

5.4.2 Dynamical Theory of X-Ray Diffraction

As we have stated in §5.1.3, an early, systematic investigation of electromagnetic waves in periodic structures was the study of X-ray diffraction in three-dimensional crystals. A simple theory, known as kinematical theory, is often used. This theory is correct for small or imperfect crystals. For large and perfect crystals, multiple scattering cannot be neglected, because the crystal lattice is very regular over a large volume, the interaction between the incident and scattered waves is enhanced. The consequences of the interaction between the incident and the scattered waves depend mainly on the dynamical equilibrium between the resultant X-ray wavefield and the scattering atoms within the crystal. Therefore a dynamical theory of X-ray diffraction has been developed to account more precisely for this mechanism of diffraction. In contrast, the kinematical theory is not self-consistent and violates the principle of conservation of energy.

To describe the wavefield in a crystal, (5.1.13) is written in the form of conventional diffraction physics as

$$\nabla^2 \boldsymbol{D} + K^2 \boldsymbol{D} + \nabla \times \nabla \times [\chi(\boldsymbol{r})\boldsymbol{D}] = 0, \tag{5.4.12}$$

where ω and $\boldsymbol{K}$ are the frequency and wavevector, respectively, of the electromagnetic wave in vacuum, with $K = |\boldsymbol{K}| = \omega/c$. It is noted that $\chi(\boldsymbol{r})$ is the most important factor that determines the behavior of electromagnetic waves in a crystal. The crystal is now considered as a continuous distribution of electrons around the atomic sites. The scattering of an incident electromagnetic wave from any nucleus is negligibly small, due to its larger mass. Only scattering from the electronic distribution contributes to the resultant X-ray wavefield. Using a simple model, it can be shown that

$$\chi(\boldsymbol{r}) = -\frac{e^2}{m\omega^2}\rho(\boldsymbol{r}), \tag{5.4.13}$$

where $\rho(\boldsymbol{r})$ is the electron density, about 10^{23}–10^{25} cm^{-3}, so in the range of X-ray frequencies, we find that $\chi \approx 10^{-6}$–10^{-4}. This gives a dielectric constant $\epsilon \approx 1$. For such a small χ, or nearly spatially homogeneous dielectric constant, it is difficult to show the frequency gaps clearly. But we still expect that there is something arising from the periodicity of $\chi(\boldsymbol{r})$.

The quantity $\chi(\boldsymbol{r})$ is a three-dimensional periodic function with the same periods as those of the crystal lattice and can be be expressed as a Fourier series

$$\chi(\boldsymbol{r}) = \sum_{\boldsymbol{G}} \chi_{\boldsymbol{G}} \mathrm{e}^{-i\boldsymbol{G}\cdot\boldsymbol{r}}, \tag{5.4.14}$$

with

$$\chi_{\boldsymbol{G}} = \Omega_{\mathrm{c}}^{-1} \int_{\Omega_{\mathrm{c}}} \chi(\boldsymbol{r}) \mathrm{e}^{i\boldsymbol{G}\cdot\boldsymbol{r}} d\boldsymbol{r} = -\frac{4\pi e^2}{m\omega^2 \Omega_{\mathrm{c}}} F_{\boldsymbol{G}}, \tag{5.4.15}$$

where Ω_{c} is the volume of a unit cell, $F_{\boldsymbol{G}}$ the structure factor defined as

$$F_{\boldsymbol{G}} = \sum_j f_j \mathrm{e}^{i\boldsymbol{G}\cdot\boldsymbol{r}_j}, \tag{5.4.16}$$

and f_j is the atomic scattering factor of the atom at $\boldsymbol{r}_j$.

The solution of (5.4.12) is a Bloch function, which can be expressed as a superposition of plane waves

$$\boldsymbol{D}(\boldsymbol{r}) = \sum_{\boldsymbol{G}} \boldsymbol{D}_{\boldsymbol{G}} \mathrm{e}^{i\boldsymbol{k}_{\boldsymbol{G}}\cdot\boldsymbol{r}}, \tag{5.4.17}$$

where $\boldsymbol{k}_{\boldsymbol{G}}$ is the wavevector of the $\boldsymbol{G}$ reflection satisfying the Bragg equation, i.e.,

$$\boldsymbol{k}_{\boldsymbol{G}} = \boldsymbol{k}_0 + \boldsymbol{G}, \tag{5.4.18}$$

where $\boldsymbol{k}_0$ is the incident wavevector on the crystal. The coefficients $\boldsymbol{D}_{\boldsymbol{G}}$ are complex vectors. In this case, the Bloch wave describes an infinite number of plane waves with wavevector $\boldsymbol{k}_{\boldsymbol{G}}$. $\boldsymbol{D}_0$ is the electric displacement of the incident beam. In other words the $\boldsymbol{D}_{\boldsymbol{G}}$ represent the electric displacements of the diffracting beams. Substituting (5.4.14) and (5.4.17) into (5.4.12), we obtain the fundamental equation

$$\left(K^2 - k_{\boldsymbol{G}}^2\right) \boldsymbol{D}_{\boldsymbol{G}} - \sum_{\boldsymbol{G}'} \chi_{\boldsymbol{G}-\boldsymbol{G}'} \left[\boldsymbol{k}_{\boldsymbol{G}} \times (\boldsymbol{k}_{\boldsymbol{G}} \times \boldsymbol{D}_{\boldsymbol{G}'})\right] = 0. \tag{5.4.19}$$

As $\boldsymbol{G}$ can be an infinite number of reciprocal lattice vectors, (5.4.19) represents an infinite set of equations.

Usually, in X-ray diffraction, the two-wave approximation is adopted, because only in the case of $\boldsymbol{k}_{\boldsymbol{G}}$ very near to K, does the amplitude $\boldsymbol{D}_{\boldsymbol{G}}$ have a large contribution. Thus, in addition to the incident beam $\boldsymbol{D}_0$, there is only one diffracted wave $\boldsymbol{D}_{\boldsymbol{G}}$ to be taken into account, so

$$\begin{aligned}(k_0^2 - K^2)\boldsymbol{D}_0 &= k_0^2(\chi_0 \boldsymbol{D}_0 + \chi_{-\boldsymbol{G}} \boldsymbol{D}_{\boldsymbol{G}}),\\ (k_{\boldsymbol{G}}^2 - K^2)\boldsymbol{D}_{\boldsymbol{G}} &= k_{\boldsymbol{G}}^2(\chi_{\boldsymbol{G}} \boldsymbol{D}_0 + \chi_0 \boldsymbol{D}_{\boldsymbol{G}}).\end{aligned} \tag{5.4.20}$$

It is noted that $\chi_{-\boldsymbol{G}}$ also appears in the expression. This is easy to understand, because we should consider the interaction between the incident beam and the diffracted beam by dynamic diffraction theory, and the effect of the crystalline plane $-\boldsymbol{G}$ is also involved due to multiple scattering.

In general, there are two polarization states for an electric displacement vector $\boldsymbol{D}$: σ polarization is perpendicular to the plane of incidence, while π polarization is in the incident plane. Consequently equations (5.4.20) are transformed into scalar equations

$$\begin{aligned}(k_0^2 - K^2)D_0 &= k_0^2(\chi_0 D_0 + C\chi_{-\boldsymbol{G}} D_{\boldsymbol{G}}),\\ (k_{\boldsymbol{G}}^2 - K^2)D_{\boldsymbol{G}} &= k_{\boldsymbol{G}}^2(C\chi_{\boldsymbol{G}} D_0 + \chi_0 D_{\boldsymbol{G}}),\end{aligned} \tag{5.4.21}$$

where $C = 1$ for σ polarization, $C = \cos 2\theta$ for π polarization and 2θ is the angle between $\boldsymbol{D}_0^\pi$ and $\boldsymbol{D}_{\boldsymbol{G}}^\pi$.

We now introduce the index of refraction n, which equals the ratio of the wavenumber in the crystal to that in vacuo, i.e.,

$$n = \frac{k}{K} = 1 + \frac{1}{2}\chi_0 = \left(1 - \frac{e^2\langle\rho(\boldsymbol{r})\rangle}{m\omega^2}\right) < 1. \tag{5.4.22}$$

Again we introduce the approximations

$$k_0^2 - K^2 - k_0^2\chi_0 \simeq k_0^2 - K^2\left(1 + \frac{1}{2}\chi_0\right)^2 = k_0^2 - k^2,$$

$$k_G^2 - K^2 - k_G^2\chi_0 \simeq k_G^2 - k^2, \tag{5.4.23}$$

$$k_0 + k \simeq k_G + k \simeq 2K.$$

The condition for (5.4.21) having nonzero solution is the coefficient determinant takes zero, i.e.,

$$\begin{vmatrix} k_0 - k & -\frac{1}{2}CK\chi_{-G} \\ -\frac{1}{2}CK\chi_G & k_G - k \end{vmatrix} = 0. \tag{5.4.24}$$

Then we obtain

$$(k_0 - k)(k_G - k) - \frac{1}{4}C^2K^2\chi_{-G}\chi_G = 0. \tag{5.4.25}$$

This gives the important dispersion relation in X-ray dynamical diffraction theory and is a hyperbolic equation for the two variables k_0 and k_G. If the wavevectors are all real, i.e., the absorption of waves by the media is ignored, we can get the equal energy surface, called the dispersion surface, in reciprocal space. Figure 5.4.2 shows the dispersion surface in the two-wave approximation. In Fig. 5.4.2(a) the reciprocal wavevector points from O to G, where O and G are the end points of $\boldsymbol{k}_0$ and $\boldsymbol{k}_G$ respectively, so $\boldsymbol{k}_G = \boldsymbol{k}_0 + \boldsymbol{G}$ is satisfied. $\boldsymbol{k}_0$ and $\boldsymbol{k}_G$ have a common starting point called the tie point. The locus of the tie point, such as P in Fig. 5.4.2(b), is composed of two hyperbolic surfaces of rotation labelled as $\mathcal{S}^{(1)}$ and $\mathcal{S}^{(2)}$. Due to the difference of polarization factor C, each of these two dispersion surfaces is further divided into two branches, which are denoted by dashed line (σ polarization) and solid line (π polarization). After knowing the location of the tie point P on a dispersion surface, the directions and the amplitude ratio of $\boldsymbol{k}_0$ and $\boldsymbol{k}_G$ can be determined under dynamical diffraction. We must emphasize that the travelling waves under dynamical diffraction conditions are not plane waves, but Bloch waves modulated by periodic atomic planes, also the wavevector directions of incident waves and diffraction waves can deviate a little from the condition given by the Bragg equation. It should be noted that the dispersion surface in $\boldsymbol{k}$ space is divided into two parts, the gap between these two corresponds to the gap in the electronic energy band.

We are concerned with whether the gap can be tested by experiment. Because measurements for X-ray diffraction on crystals are always performed outside crystals, the properties of the crystal surfaces must influence the experimental results. We shall discuss surface-related problems in Chap. 7. Here we consider only the situation where the waves continuously penetrate through the surfaces,

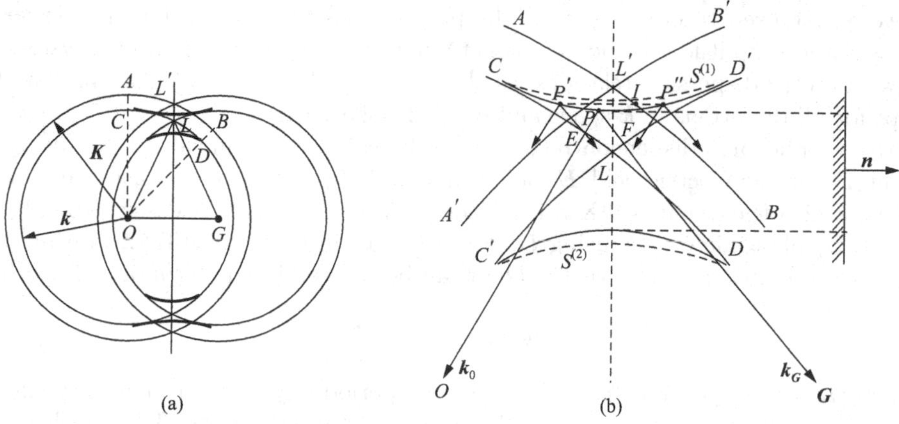

Figure 5.4.2 Dispersion surface in reciprocal space in two-wave approximation. (a) Globle picture, $(K - k = |\frac{1}{2}\chi_0|)$; (b) Local magnification.

and we discuss the continuity of wavevectors. If $\boldsymbol{\tau}$ is an any unit vector at a surface, the continuity condition at the interface for the waves on both sides can be expressed as

$$\exp(-i\boldsymbol{K}\cdot\boldsymbol{\tau}) = \exp(-i\boldsymbol{k}_0\cdot\boldsymbol{\tau}) = \exp(-i\boldsymbol{k}_{\boldsymbol{G}}\cdot\boldsymbol{\tau}). \tag{5.4.26}$$

That is, the tangential component of a wavevector is required to be continuous across the boundary, but the normal component permits a discontinuity, i.e.,

$$\boldsymbol{K} - \boldsymbol{k}_0 = \boldsymbol{K} - \boldsymbol{k}_{\boldsymbol{G}} = \delta k\boldsymbol{n}, \tag{5.4.27}$$

where $\boldsymbol{n}$ is the unit vector along the normal direction of the surface across the boundary, and $\delta k\boldsymbol{n}$ represents the difference of surface wavevectors outside and inside the surface.

If the normal to the crystal surface is parallel to the reciprocal lattice vector $\boldsymbol{G}$, it is called the symmetric Bragg case in diffraction physics. We can imagine that the end point of an incident wavevector $\boldsymbol{K}$ in vacuum moves along the curve AB in Fig. 5.4.2(b), corresponding to the variation of incident angle. When it enters the dynamical diffraction regime, we draw a straight line from a representative point on AB, such as I, parallel to the surface normal $\boldsymbol{n}$. There are two intersecting points, P' and P'', with one piece of dispersion surface, which are the two tie points connecting the initial points $\boldsymbol{k}_0$ and $\boldsymbol{k}_{\boldsymbol{G}}$ in the crystal, while the segments IP' and IP'' correspond to the two values of $\delta k\boldsymbol{n}$ satisfying (5.4.27). It is noted that as the Bragg condition is approached, this straight line may fall into the gap, and not intersect with any piece of the dispersion surface. Then there will be no excitation for $\boldsymbol{k}_0$ and $\boldsymbol{K}_{\boldsymbol{G}}$ in the crystal, so there is total reflection. The appearance of total reflection in the Bragg case verifies the existence of an energy gap, although this gap is very small and can be shown only under stringent experimental conditions.

5.4.3 Bandgaps in Three-Dimensional Photonic Crystals

As we have shown, it is very difficult for X-rays to show clear-cut bandgaps in real crystals. The problem arises from the fact that the variation of their dielectric constants for X-rays in all materials is very small. However, there may be some three-dimensional systems which can show bands and gaps for electromagnetic waves. These are artificial periodic structures. Due to the common characteristics of waves, there may be ranges of frequencies in which no allowed modes exist for electromagnetic wave propagation in these periodic structures. This type of frequency-wavevector relation for electromagnetic wave propagation is referred to as the photonic band structure. This relation depends on the details of lattice structure, i.e., the lattice constant, the shape of the embedded dielectric object, the dielectric constants of the constituent materials, and the filling fraction, which is the percentage of total crystal volume occupied by any one of the materials.

On the experimental side, artificial three-dimensional fcc dielectric structures, consisting of a regular array of spheres of one dielectric material embedded in a second one with a different dielectric constant, were first used to create gaps in the photon density of states for microwaves. Measurement of phases and amplitudes of electromagnetic waves through this photonic crystal map out the frequency-wavevector dispersion relations, as shown in Fig. 5.4.3.[b] Real photonic band gaps must display gaps for all directions in $\boldsymbol{k}$ space and in both polarizations.

Theoretically, photonic dispersion relations can be calculated equivalently by using the electric displacement $\boldsymbol{D}$, or the electric field $\boldsymbol{E}$, or the magnetic field $\boldsymbol{H}$, to establish the wave equations. For consistency with the dynamical X-ray diffraction theory, we shall discuss the bandgaps in three-dimensional (3D) photonic crystal using $\boldsymbol{D}$. We begin again from (5.1.13). According to the Bloch theorem, the electric displacement vector $\boldsymbol{D}(\boldsymbol{r})$ can be written in the form of a Bloch function

$$\boldsymbol{D}_{\boldsymbol{k}}(\boldsymbol{r}) = \boldsymbol{u}_{\boldsymbol{k}}(\boldsymbol{r})\mathrm{e}^{i\boldsymbol{k}\cdot\boldsymbol{r}}. \tag{5.4.28}$$

The functions $\boldsymbol{u}_{\boldsymbol{k}}(\boldsymbol{r})$ are periodic in $\boldsymbol{r}$ with the same periodicity as $\chi(\boldsymbol{r})$. The corresponding eigenvalues are $\omega^2(\boldsymbol{k})$, yielding the dispersion relation with band structure. The Bloch functions form the basis set for the crystal momentum representation. Since $\boldsymbol{u}_{\boldsymbol{k}}(\boldsymbol{r})$ is periodic, it can be expanded

[b] E. Yablonovitch and T. J. Gmitter, *Phys. Rev. Lett.* **63**, 1950 (1989).

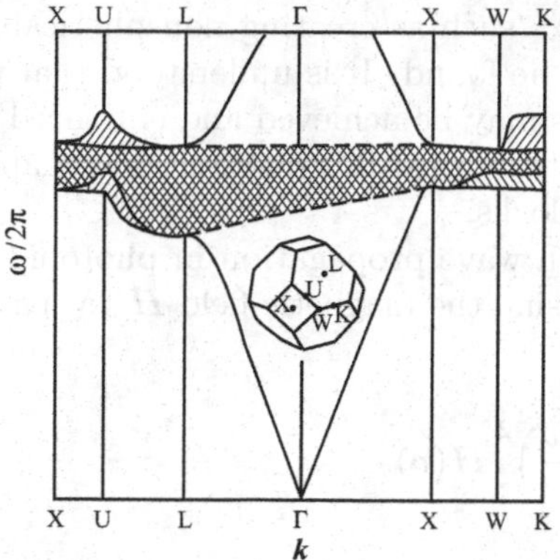

Figure 5.4.3 The experimentally observed photonic band structure in reciprocal space of spherical-air-atom crystal. From E. Yablonovitch and T. J. Gmitter, *Phys. Rev. Lett.* **63**, 1950 (1989).

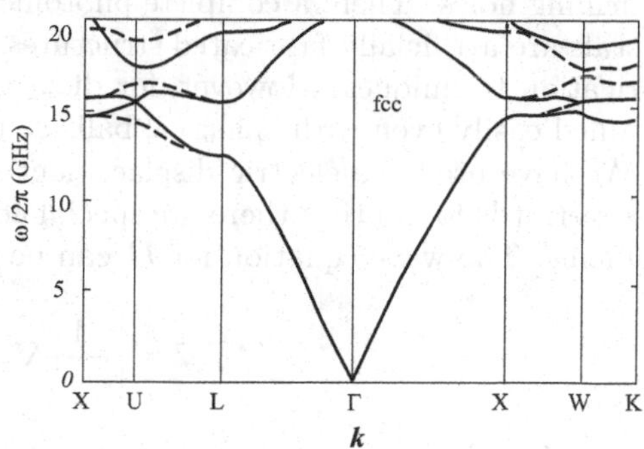

Figure 5.4.4 Theoretical photonic band structure of fcc dielectric structure. From K. M. Ho *et al.*, *Phys. Rev. Lett.* **65**, 3152 (1990).

in terms of plane waves with wavevectors $\boldsymbol{G}$ being reciprocal lattice vectors. Thus, $\boldsymbol{D_k}(\boldsymbol{r})$ can be expanded as

$$\boldsymbol{D_k}(\boldsymbol{r}) = \sum_{\boldsymbol{G}} \boldsymbol{D_G} \mathrm{e}^{i(\boldsymbol{k}+\boldsymbol{G})\cdot \boldsymbol{r}}, \tag{5.4.29}$$

where $\boldsymbol{k}$ is the Bloch momentum and lies within the first Brillouin zone (BZ). From $\nabla \cdot \boldsymbol{D} = 0$, (5.4.29) gives

$$\boldsymbol{D_G} \cdot (\boldsymbol{k} + \boldsymbol{G}) = 0. \tag{5.4.30}$$

Substituting (5.4.29) into (5.1.13), yields matrix equations for the Fourier coefficient $\boldsymbol{D_G}$,

$$Q_{\boldsymbol{G}} \boldsymbol{D_G} + \sum_{\boldsymbol{G}'} \chi_{\boldsymbol{G}-\boldsymbol{G}'} \left[(\boldsymbol{k}+\boldsymbol{G}) \cdot \boldsymbol{D}_{\boldsymbol{G}'} (\boldsymbol{k}+\boldsymbol{G}) - |\boldsymbol{k}+\boldsymbol{G}|^2 \boldsymbol{D}_{\boldsymbol{G}'} \right] = 0, \tag{5.4.31}$$

where

$$Q_{\boldsymbol{G}} = |\boldsymbol{k}+\boldsymbol{G}|^2 - \omega^2/c^2. \tag{5.4.32}$$

Expressing $\boldsymbol{D_G}$ in terms of Cartesian component, we get a secular determinant

$$\left|\left| Q_{\boldsymbol{G}} \delta_{\boldsymbol{G},\boldsymbol{G}'} \delta_{\alpha,\beta} \chi_{\boldsymbol{G}'-\boldsymbol{G}} \left[(\boldsymbol{k}+\boldsymbol{G}')_\alpha (\boldsymbol{k}+\boldsymbol{G}')_\beta - |\boldsymbol{k}+\boldsymbol{G}'|^2 \delta_{\alpha,\beta} \right] \right|\right| = 0, \tag{5.4.33}$$

$\alpha, \beta = x, y, z$. Equation (5.4.33) can be solved numerically, if we prescribe the dielectric constants ϵ_a inside the spheres and ϵ_b in the background. Taking N plane waves ($N \geq 300$ gives good convergence) it is a $3N \times 3N$ order determinant, which will give $3N$ eigenmodes. Because electromagnetic waves are transverse, there are two distinct helicity modes for each plane wave, i.e., $2N$ modes in total, so N unphysical solutions should be discarded.

Figure 5.4.4 shows the numerical results for the entire photonic band structure for $\boldsymbol{k}$ along the symmetry directions in the Brillouin zone.[c] This calculation was carried out with $\epsilon_a = 1$, $\epsilon_b = 12.25$. Comparing this to Fig. 5.4.3, we see that they are very similar. But in Fig. 5.4.4, we cannot find a true gap throughout the Brillouinn zone. Also, some discrepancies are found near the symmetry points W and U: At W, the second and third bands are degenerate, and at U, a crossing of the second and third band structures takes place. The reason for this discrepancy is simple: The higher symmetry of the unit cells assumed by the theoretical calculation leads to the degeneracy of the eigenfrequencies at W and U, but there are always imperfections in artificial photonic crystals measured in experiments. Based on these considerations, it is natural to change the degeneracy by

[c] K. M. Leung and Y. E. Liu, *Phys. Rev. Lett.* **65**, 2646 (1990).

decreasing the symmetry of the unit cells in the fcc lattice, such as creating nonspherical "atoms" by drilling holes. Then, a complete photonic bandgap can be found. It is understood that photonic crystals are artificially fabricated structures; full bandgaps may be achieved and enhanced by some fabrication techniques. However, for dielectric diamond structures, full photonic bandgaps can be obtained easily even with spherical balls as the building blocks.

We have used the electric displacement $\boldsymbol{D}$ to deal with wave propagation in photonic crystals. However, it is found that there are special advantages in using the magnetic field $\boldsymbol{H}$ to treat related problems. The wave equation for $\boldsymbol{H}$ can be expressed as

$$\nabla \times \left[\frac{1}{\epsilon(\boldsymbol{r})}\nabla \times \boldsymbol{H}(\boldsymbol{r})\right] = \left(\frac{\omega}{c}\right)^2 \boldsymbol{H}(\boldsymbol{r}). \tag{5.4.34}$$

Introducing the operator Θ,

$$\Theta \equiv \nabla \times \left[\frac{1}{\epsilon(\boldsymbol{r})}\nabla \times\right],$$

$$\Theta \boldsymbol{H}(\boldsymbol{r}) = \left(\frac{\omega}{c}\right)^2 \boldsymbol{H}(\boldsymbol{r}). \tag{5.4.35}$$

It is easy to verify that the operator Θ is Hermitian, so equation (5.4.35) is an eigenvalue problem for $\boldsymbol{H}$. To satisfy the series of operations to $\boldsymbol{H}(\boldsymbol{r})$ for the operator Θ, if $\boldsymbol{H}(\boldsymbol{r})$ is to really represent permitted electromagnetic modes, it must simply equals the original $\boldsymbol{H}(\boldsymbol{r})$ times a constant. It is not so simple when $\boldsymbol{D}(\boldsymbol{r})$ is used, so in theories for photonic crystals, it is more profitable to compute $\boldsymbol{H}(\boldsymbol{r})$ at first, then by using the relation

$$\boldsymbol{E}(\boldsymbol{r}) = \left[-\frac{ic}{\omega\epsilon(\boldsymbol{r})}\right] \nabla \times \boldsymbol{H}(\boldsymbol{r}), \tag{5.4.36}$$

one can obtain $\boldsymbol{E}(\boldsymbol{r})$. If $\epsilon(\boldsymbol{r})$ is periodically distributed, then $\boldsymbol{H}(\boldsymbol{r})$ has a Bloch form like

$$\boldsymbol{H}_{\boldsymbol{k}}(\boldsymbol{r}) = \boldsymbol{H}_0 \mathrm{e}^{i(\boldsymbol{k}\cdot\boldsymbol{r})}. \tag{5.4.37}$$

Furthermore, $\boldsymbol{E}(\boldsymbol{r})$ or $\boldsymbol{D}(\boldsymbol{r})$ can be obtained. As an eigenvalue problem, the study of photonic crystals is similar to band theory for electrons, except the Hamiltonian operator is replaced by Θ, i.e.,

$$\mathcal{H} \equiv -\frac{\hbar^2}{2m}\nabla^2 + V(\boldsymbol{r}) \to \Theta \equiv \nabla \times \left[\frac{1}{\epsilon(\boldsymbol{r})}\nabla \times\right].$$

The experimental study of photonic crystals began within the range of microwave frequencies, because the wavelengths of microwaves are longer, so the unit cell size of photonic crystals can be larger and easier to fabricate. In the 1990s, investigations were concentrated on the infrared and visible range. Due to the possibility of photonic crystals to manipulate light propagation, and applications in laser technology, especially optical communications, much is expected. From electromagnetic theory, there are no wavelength limitations in Maxwell equations to waves, so the behavior of photonic crystals has scale invariance, ranging from the size of centimeters for microwaves to the size of microns or submicrons for light waves, although the dispersion relations may be different for electromagnetic waves with various wavelengths propagating in media. The difficulty in studying photonic crystals lies in the experimental sample preparation. There are several preparation techniques, among which two methods are often adopted: One is microlithography in planar technology; the other is self assembly of equal spheres with small variance. The former is more suitable for two-dimensional (2D) structures, but the latter can be used both for two- or three dimensions. There is a colorful, natural precious stone, opal, formed in three dimensions by periodically arranged amorphous SiO_2 spheres with the size of microns. As introduced in §4.1.1, in recent years the photonic crystals fabricated by self-assembly are composed of either opal structure (micro spheres are the medium with a high index of refraction), or the inverse opal structure (micro spheres are the voids in a matrix with a high index of refraction).

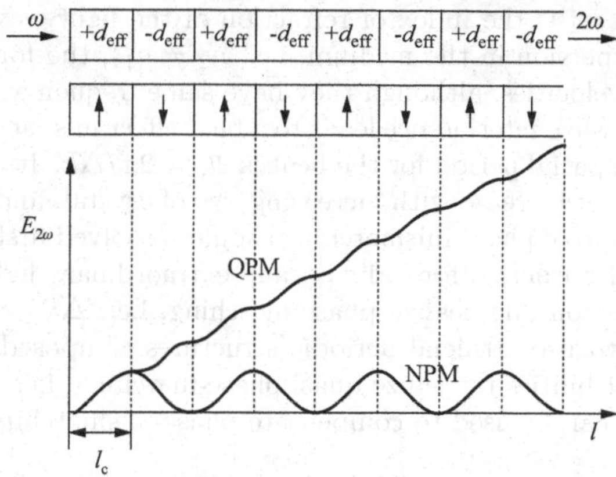

Figure 5.4.5 Amplitude of second harmonic waves versus optical path for a modulated structure satisfying the quasi phase matching.

5.4.4 Quasi-Phase-Matching in Nonlinear Optical Crystals

Photonic crystals are artificially modulated by the indices of refraction in linear optics, but with the advent of laser technology, a new era for nonlinear optics began. Under the strong electric field of lasers, the electric polarization in crystals has not only a linear term, but also nonlinear terms, such as

$$\boldsymbol{P} = \chi \boldsymbol{E} + \chi^{(2)} \boldsymbol{E}\boldsymbol{E} + \chi^{(3)} \boldsymbol{E}\boldsymbol{E}\boldsymbol{E} + \cdots , \tag{5.4.38}$$

where $\chi^{(2)}$ is a second-order nonlinear susceptibility, a third-order tensor. $\chi^{(3)}$ is third-order nonlinear susceptibility, a fourth-order tensor. Because the value of $\chi^{(3)}$ is smaller than that of $\chi^{(2)}$ by about five order of magnitudes, the term $\chi^{(3)}$ is ignored in the following discussion.

The existence of the nonlinear polarization term leads to frequency conversion of laser light, such as frequency doubling, sum frequency and difference frequency conversion. We will limit our discussion to the frequency doubling of laser light, that is the production of light wave with frequency 2ω from light wave with frequency ω. It is not difficult to extend the discussion to sum frequency and difference frequency conversion.

Consider a beam of monochromatic plane waves propagating in the x direction in a nonlinear optical crystal,

$$\boldsymbol{E}_1 = \boldsymbol{e}_1 E_0 \mathcal{E}_1 \sin(\omega_1 t - k_1 x),$$

where $\boldsymbol{e}_1$ is the unit vector of wave polarization, $k_1 = n_1\omega_1/c = 2\pi n_1/\lambda_1$. k_1, n_1, and λ_1 are the wavenumber, index of refraction, and wavelength of light waves at frequency ω_1. This kind of light will induce nonlinear polarization waves with frequency $2\omega_1$ in the crystal

$$\boldsymbol{P}(2\omega_1) = -\frac{1}{2}\chi^{(2)} \boldsymbol{e}_1 \boldsymbol{e}_1 \mathcal{E}_1^2 \cos(2\omega_1 t - 2k_1 x).$$

The propagation velocity of the polarization waves in the crystal is

$$v_{\mathrm{P}} = 2\omega_1/2k_1 = \omega_1/k_1 = v_1,$$

which is the same as the incident fundamental waves. The radiation of the induced dipoles at frequency 2ω gives second harmonics, with strength proportional to root $\chi^{(2)}$ as well as the square of the intensity of the incident waves.

The second harmonic is actually the result of the superposition of two waves. One of them is the forced wave

$$\boldsymbol{E}_2' = \boldsymbol{e}_2' \mathcal{E}_2' \cos(2\omega_1 t - 2k_1 x),$$

which is synchronized with the nonlinear polarization wave. The other is the free wave radiated by these dipoles with the wave number k_2,

$$\boldsymbol{E}_{2\omega} = \boldsymbol{e}_2 \mathcal{E}_2 \cos(2\omega_1 t - 2k_2 x),$$

where $k_2 = 2\omega_1 n_2/c$, and n_2 is the index of refraction of the light wave with frequency $\omega_2 = 2\omega_1$ in the crystal. Due to dispersion in the medium, i.e., $n_2 \neq n_1$), the forced wave and the free wave propagate with different velocities, although they have same frequency, and so phase mismatching, $\Delta = k_2 - 2k_1$ results. The wave interference leads to a beat effect in space, so the wave amplitude will be zero periodically. The spatial period for the beat is $2l_c = 2\pi/\Delta k$. It is obvious that the output of the second harmonic will not increase with increasing size of crystal, and, in order to obtain a highly efficient output, the problem of phase mismatching should be solved first. One effective method uses the difference in indices of refraction for ordinary and extraordinary light in birefringent crystals to compensate for the dispersion and realize phase-matching, i.e., $\Delta k = 0$. The other, proposed by Bloembergen in 1962, is to use artificial periodic structures composed of alternating positive and negative nonlinear susceptibilities to realize quasi-phase-matching, i.e., the reciprocal lattice vector $\boldsymbol{G}$ of a periodic structure can be used to compensate phase-mismatching Δk, so

$$\Delta k = k_2 - 2k_1 = |\boldsymbol{G}| = \frac{\pi}{l_c},$$

or more generally written in vector form

$$\Delta \boldsymbol{k} = \boldsymbol{G}.$$

Here $2l_c$ is the spatial period for positive and negative layers with alternating nonlinear susceptibilities, and l_c is called coherence length for nonlinear optical media satisfying

$$l_c = \frac{\lambda}{4(n_2 - n_1)},$$

which is about a micron for normal nonlinear optical crystals.

Figure 5.4.5 shows the variation of the amplitude of frequency-doubled light as a function of optical path in quasi-phase-matched and phase mismatched nonlinear optical crystals. The latter shows the spatial beat effect and the former increases monotonically with crystal size. The output of the second harmonic in the case of quasi-phase-matching can have the same order of magnitude as in the case of phase matching, or even stronger, because the large nonlinear optical coefficients of some crystals cannot be phase-matched at all. To realize quasi-phase matching, it is needed to fabricate the crystal with a periodically alternating positive and negative ferroelectric domain structure. The domain thickness is adjusted to l_c, or an odd multiple of l_c. The fabrication technique may involve using impurity striation in crystal growth to induce alternating positive and negative domain structures, or patterned electrodes to polarize a crystal.[d]

§5.5 Waves in Quasiperiodic Structures

Just as discussed above, electron waves, lattice waves and electromagnetic waves in periodic structures are all Bloch waves, characterized by an extended state. On the other hand, we shall see in Chap. 9 that waves in disordered structures exponentially decrease and are characterized by localized states. In this section we will discuss wave propagation in quasicrystal structures. It is clear that the structural features of a system will severely affect its physical properties. Because quasiperiodicity is intermediate between periodicity and disorder, it is somewhat more difficult to study. Loss of translation invariance means that the Bloch theory is ineffective, and the states may not be extended. On the other hand, long-range order in quasiperiodic structures leads the waves to be not strictly localized.

5.5.1 Electronic Spectra in a One-Dimensional Quasilattice

As a simple example, we investigate an electron moving in a Fibonacci lattice which is the prototype of one-dimensional quasicrystals.[e] To study the electronic properties of one-dimensional

[d] For the theoretical proposition for quasi-phase matching in nonlinear optical crystals, see J. A. Armstrong, N. Bloembergen *et al.*, *Phys. Rev.* **127**, 1918 (1962); The experimental verification is discussed in D. Feng, N. B. Ming *et al.*, *Appl. Phys. Lett.* **37**, 609 (1980); and recent progress is found in N. B. Ming, *Adv. Mater.* **11**, 1079 (1999).

[e] M. Kohmoto, B. Sutherland, and C. Tang, *Phys. Rev. B* **35**, 1020 (1987).

quasilattices, the tight-binding model is used. The basic equation is

$$t_i\psi_{i-1} + \varepsilon_i\psi_i + t_{i+1}\psi_{i+1} = E\psi_i, \tag{5.5.1}$$

where ψ_i and ε_i are, respectively, the probability amplitude and site potential of an electron being on the site i, while t_i is the transfer energy between sites i and $i-1$. Usually, one lets either t_i or ε_i take two values in the quasiperiodic lattice, and lets the other be independent of i. Here we consider the so called transfer model, in which the site potential $\varepsilon_i = 0$ for all i and two transfer energies t_A and t_B are arranged in Fibonacci sequence.

To obtain the solution of (5.1.1), we adopt the periodic boundary condition. Then the structure corresponds to a periodic chain whose unit cell is just the Fibonacci chain of the lth generation with the number of sites is $N = F_l$. The ideal Fibonacci chain can be understood as the limit of infinite l. Thus our treatment is based on (5.5.1) with $\varepsilon_i = 0$. If we define a transfer matrix

$$\boldsymbol{T}(t_{i+1}, t_i) = \begin{pmatrix} E/t_{i+1} & -t_i/t_{i+1} \\ 1 & 0 \end{pmatrix}, \tag{5.5.2}$$

and a column vector

$$\Psi_i = \begin{pmatrix} \psi_i \\ \psi_{i-1} \end{pmatrix}, \tag{5.5.3}$$

then equation (5.5.1) might be written in the form of a matrix

$$\Psi_{i+1} = \boldsymbol{T}(t_{i+1}, t_i)\Psi_i. \tag{5.5.4}$$

By using sequential products of the matrices, (5.5.4) is written as

$$\Psi_{i+1} = \boldsymbol{T}(i)\Psi_1, \tag{5.5.5}$$

where

$$\boldsymbol{T}(i) = \prod_{j=1}^{i} \boldsymbol{T}(t_{j+1}, t_j). \tag{5.5.6}$$

To solve the problem, the key point is therefore to calculate the transfer matrix $\boldsymbol{T}(i)$. The Fibonacci lattice permits an extremely effective method for doing this. When i is a Fibonacci number, $\boldsymbol{T}(i)$ can be obtained recursively. Define $\boldsymbol{T}_l \equiv \boldsymbol{T}(F_l)$, then

$$\boldsymbol{T}_{l+1} = \boldsymbol{T}_{l-1}\boldsymbol{T}_l, \tag{5.5.7}$$

with $\boldsymbol{T}_1 = \boldsymbol{T}(t_A, t_A)$ and $\boldsymbol{T}_2 = \boldsymbol{T}(t_A, t_B)\boldsymbol{T}(t_B, t_A)$. The transfer matrix for a general value i is given by

$$\boldsymbol{T}(i) = \boldsymbol{T}_{l_j} \cdots \boldsymbol{T}_{l_2}\boldsymbol{T}_{l_1}, \tag{5.5.8}$$

where $i = F_{l_1} + F_{l_2} + \cdots + F_{l_j}$ and $l_1 > l_2 > \cdots > l_j$.

The recursion relation (5.5.10) gives a powerful calculational scheme. However, the essential importance, rather, lies in the fact that it defines a nonlinear dynamical map and we can therefore use the theories and concepts of dynamical systems. By defining $\chi_l = (1/2)\mathrm{tr}T_l$, one can show that

$$\chi_{l+1} = 2\chi_l\chi_{l-1} - \chi_{l-2}, \tag{5.5.9}$$

which on successive iterations leads to a constant of the motion

$$I = \chi_{l+1}^2 + \chi_l^2 + \chi_{l-1}^2 - 2\chi_{l+1}\chi_l\chi_{l-1} - 1, \tag{5.5.10}$$

which is independent of l. It should be understood that a large I implies strong quasiperiodicity. It is easy to show for the transfer model that

$$I = \frac{1}{4}\left(\frac{t_B}{t_A} - \frac{t_A}{t_B}\right)^2.$$

For one-dimensional quasicrystals, the electronic energy spectra in general have the characteristics of the Cantor set with three branches, and the wavefunctions often display the self-similar amplitude distributions. It reminds us of the devil's staircase shown in Fig. 4.3.8. Figure 5.5.1(a) gives the integrated density of states (IDOS) of a Fibonacci lattice, the platforms corresponding to the energy gaps. Figure 5.5.1(b) shows the self-similar wavefunction at the center of the energy spectrum. This wavefunction is neither an extended state, nor a localized state, but a critical state.

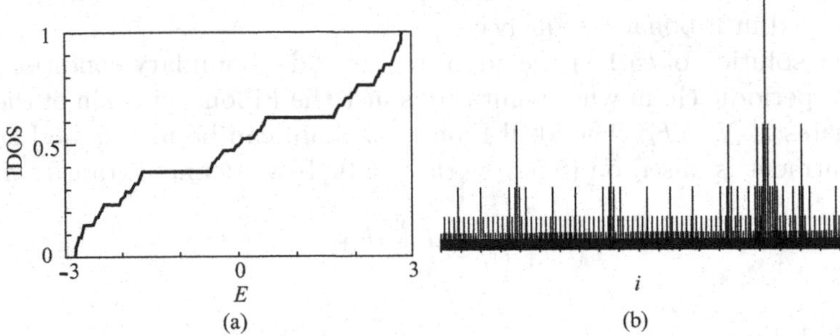

Figure 5.5.1 IDOS for one-dimensional Fibonacci lattice (a); a self-similar wave function (b).

In general,[f] there are three different types of energy spectrum: Absolute continuous, point-like and singular continuous. The absolute continuous spectrum is characterized by a smooth density of states $g(E)$, while the point-like spectrum is a set of δ functions defined on a countable number of points $\{E_i\}$. However, in the singular continuous spectrum the number of states whose energies are less than a specified energy E is continuously increasing but non-differentiable at any E, just like the Cantor set. Therefore, the density of states is not well-defined in the singular continuous spectrum. Nevertheless, the total number of states with energies less than E, i.e., the integrated density of states (IDOS), is always well-defined just as in the absolutely continuous spectrum. Wavefunctions are also classified into three kinds: Extended, localized, and critical. An extended state is defined by a wavefunction with asymptotical uniform amplitude as

$$\int_{|\boldsymbol{r}|<L} |\psi(\boldsymbol{r})|^2 d\boldsymbol{r} \sim L^d,$$

where L is the size of the sample and d the spatial dimension. A localized state is characterized by a square integrable wavefunction, i.e.,

$$\int_{|\boldsymbol{r}|<\infty} |\psi(\boldsymbol{r})|^2 d\boldsymbol{r} \sim L^0.$$

A critical state is, however, different from the former two kinds, a typical example of which is a power-law function $\psi(\boldsymbol{r}) \sim |\boldsymbol{r}|^{-\nu}$, with $\nu \leq d/2$, so

$$\int_{|\boldsymbol{r}|<L} |\psi(\boldsymbol{r})|^2 d\boldsymbol{r} \sim L^{-2\nu+d}, \qquad (0 < 2\nu < d).$$

These three different kinds of wavefunctions, namely extended state, localized state and critical state, may correspond to three different energy spectra, absolute continuous spectrum, point-like spectrum and singular continuous spectrum, respectively.

The transfer matrix method used here is not limited to the treatment of electronic structure, but can also be applied to that of lattice vibrations in one-dimensional quasiperiodic structures.

[f] H. Hiramoto and M. Kohmoto, *Int. J. Mod. Phys. B* **6**, 281 (1992).

5.5.2 Wave Transmission Through Artificial Fibonacci Structures

By molecular-beam epitaxy and lithography techniques, artificial Fibonacci structures can be fabricated.[g] These finite structures can be used to study the wave behavior in them experimentally as well as theoretically.[h]

Now we consider electromagnetic wave passing through a dielectric Fibonacci superlattice which is composed of two types of layers A and B with different dielectric constants. Consider an interface between two layers. The electric field for the light in layer A is given by

$$\boldsymbol{E}_{\mathrm{A}} = \boldsymbol{E}_{\mathrm{A}}^{(1)} \exp[i(\boldsymbol{k}_{\mathrm{A}}^{(1)} \cdot \boldsymbol{r} - \omega t)] + \boldsymbol{E}_{\mathrm{A}}^{(2)} \exp[i(\boldsymbol{k}_{\mathrm{A}}^{(2)} \cdot \boldsymbol{r} - \omega t)]. \tag{5.5.11}$$

The electric field in layer B is simply the same expression with subscript A replaced by B. Consider a polarization which is perpendicular to the plane of the light path (TE wave): The appropriate boundary conditions at the interface gives

$$\begin{gathered} \mathcal{E}_{\mathrm{A}}^{(1)} + \mathcal{E}_{\mathrm{A}}^{(2)} = \mathcal{E}_{\mathrm{B}}^{(1)} + \mathcal{E}_{\mathrm{B}}^{(2)}, \\ n_{\mathrm{A}} \cos\theta_{\mathrm{A}} \left(\mathcal{E}_{\mathrm{A}}^{(1)} - \mathcal{E}_{\mathrm{A}}^{(2)}\right) = n_{\mathrm{B}} \cos\theta_{\mathrm{B}} \left(\mathcal{E}_{\mathrm{B}}^{(1)} - \mathcal{E}_{\mathrm{B}}^{(2)}\right), \end{gathered} \tag{5.5.12}$$

where n_{A} and n_{B} are the indices of refraction of layers A and B, respectively, and the angles θ_{A} and θ_{B} are the incident and reflected angles.

It is convenient to choose two independent variables as

$$\mathcal{E}_{+} = \mathcal{E}^{(1)} + \mathcal{E}^{(2)}, \ \mathcal{E}_{-} = (\mathcal{E}^{(1)} - \mathcal{E}^{(2)})/i, \tag{5.5.13}$$

then from (5.5.17), we have the matrix equation

$$\begin{pmatrix} \mathcal{E}_{+} \\ \mathcal{E}_{-} \end{pmatrix}_{\mathrm{B}} = \boldsymbol{T}_{\mathrm{BA}} \begin{pmatrix} \mathcal{E}_{+} \\ \mathcal{E}_{-} \end{pmatrix}_{\mathrm{A}}, \tag{5.5.14}$$

where $\boldsymbol{T}_{\mathrm{BA}}$ is

$$\boldsymbol{T}_{\mathrm{BA}} = \begin{pmatrix} 1 & 0 \\ 0 & n_{\mathrm{A}} \cos\theta_{\mathrm{A}} / n_{\mathrm{B}} \cos\theta_{\mathrm{B}} \end{pmatrix}. \tag{5.5.15}$$

Now we show that $\boldsymbol{T}_{\mathrm{AB}} = \boldsymbol{T}_{\mathrm{BA}}^{-1}$.

The matrices $\boldsymbol{T}_{\mathrm{BA}}$ and $\boldsymbol{T}_{\mathrm{AB}}$ represent light propagation across interfaces B $\leftarrow$ A and A $\leftarrow$ B, respectively. The propagation within one layer of type A is represented by

$$\boldsymbol{T}_{\mathrm{A}} = \begin{pmatrix} \cos\delta_{\mathrm{A}} & -\sin\delta_{\mathrm{A}} \\ \sin\delta_{\mathrm{A}} & \cos\delta_{\mathrm{A}} \end{pmatrix}, \tag{5.5.16}$$

and the same expression for $\boldsymbol{T}_{\mathrm{B}}$ in which δ_{A} is replaced by δ_{B}. The phases are given by

$$\delta_{\mathrm{A}} = n_{\mathrm{A}} k d_{\mathrm{A}} / \cos\theta_{\mathrm{A}}, \ \delta_{\mathrm{B}} = n_{\mathrm{B}} k d_{\mathrm{B}} / \cos\theta_{\mathrm{B}}, \tag{5.5.17}$$

where k is the wave number in vacuum, and d_{A} and d_{B} are the thicknesses of the layers.

For one layer A, and two layers BA, the light propagation are given by

$$\boldsymbol{T}_1 = \boldsymbol{T}_{\mathrm{A}}, \ \boldsymbol{T}_2 = \boldsymbol{T}_{\mathrm{AB}} \boldsymbol{T}_{\mathrm{B}} \boldsymbol{T}_{\mathrm{BA}} \boldsymbol{T}_{\mathrm{A}}, \tag{5.5.18}$$

---

[g] One-dimensional quasiperiodic semiconductor superlattices were fabricated experimentally: Two units A and B composed of GaAs/AlAs were arranged in the Fibonacci lattice, X-ray and Raman scattering spectra show the structural features of this kind of quasiperiodic multilayers, as in R. Merlin, *et al. Phys. Rev. Lett.* **55**, 1768 (1985); quasiperiodic metallic superlattices can also be fabricated by sputtering technique, and even extended to k-component case, see A. Hu, S. S. Jiang *et al.*, *SPIE.* **22**, 2364 (1994).

[h] For wave transmission through Fibonacci structures, there are many theoretical studies, for electromagnetic wave transmission, see M. Kohmoto, B. Sutherland, and K. Iguchi, *Phys. Rev. Lett.* **58**, 2436 (1987); phonon transmission, see S. Tamura, J. P. Wolfe, *Phys. Rev. B* **36**, 3491 (1987); electron transmission, see G. Jin, Z. Wang, *et al. J Phys. Soc. Jpn.* **67**, 49 (1998).

and the recursion relation (5.5.10) is applicable. We can also get a constant of motion, defined in (5.5.13), as

$$I = \frac{1}{4}\sin^2\delta_{\rm A}\sin^2\delta_{\rm B}\left(\frac{n_{\rm A}\cos\theta_{\rm A}}{n_{\rm B}\cos\theta_{\rm B}} - \frac{n_{\rm B}\cos\theta_{\rm B}}{n_{\rm A}\cos\theta_{\rm A}}\right)^2. \tag{5.5.19}$$

For the case $n_{\rm A} = n_{\rm B}$ there is no quasiperiodicity and one has $I = 0$ as expected.

The transmission coefficient $\tilde{T}$ is given in terms of the matrix $\boldsymbol{T}_l$ as

$$\tilde{T} = 4/(|T_l|^2 + 2), \tag{5.5.20}$$

where $|T_l|^2$ is the sum of the squares of the four elements of $\boldsymbol{T}_l$. $\tilde{T}$ is a quantity that can be measured experimentally. Consider the simplest experimental setting: Taking the incident light to be normal, i.e., $\theta_A = \theta_B = 0$, and also choosing the thickness of the layers to give $\delta_A = \delta_B = \delta$, i.e., $n_A d_A = n_B d_B$. For $\delta = m\pi$, corresponding to a 1/2 wavelength layer, we have $I = 0$ and the transmission is perfect, while for $\delta = (m + \frac{1}{2})\pi$, corresponding to a 1/4 wavelength layer, I is a maximum and the quasiperiodicity is most effective. These cases are shown in Fig. 5.5.5. These characteristics have been verified experimentally.[i]

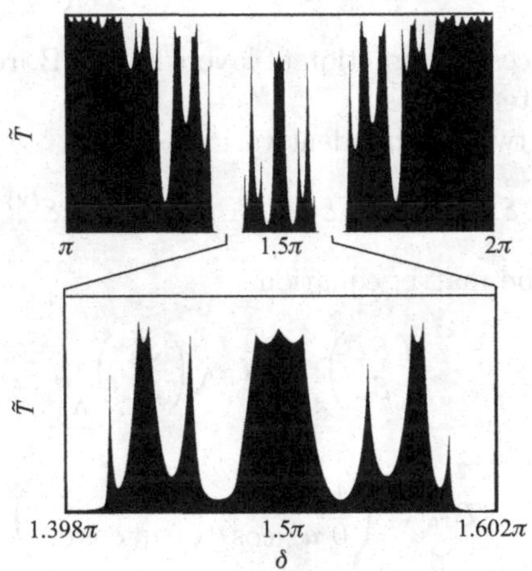

Figure 5.5.2 The transmission coefficient $\tilde{T}$ versus the optical phase length of a layer δ for a Fibonacci multilayer F_9 (55 layers). The indices of refraction are chosen as $n_A = 2$ and $n_B = 3$. From M. Kohmoto *et al.*, *Phys. Rev. Lett.* **58**, 2436 (1987).

As noted, there are different quasi-reciprocal vectors in a Fibonacci structure composed of positive and negative ferroelectric domains, so this can be used to extend the quasi-phase-matching of nonlinear optical crystals to the case of more than one reciprocal lattice vector, to compensate the phase mismatching in two nonlinear optical processes, to give rise to multi-wavelength frequency doubling and high efficiency third-order harmonic generation of light.[j]

5.5.3 Pseudogaps in Real Quasicrystals

In §5.5.1, we discussed the energy spectrum and wavefunctions of one-dimensional quasilattices in the tight-binding approximation and with the transfer matrix method. This approach can be extended to artificially layered quasiperiodic structures, the theoretical predictions can be in good agreement with experimental results, as is exemplified in §5.5.2. It is a pity that the approach which is so efficient in the treatment of one-dimensional quasiperiodic lattices cannot be applied to the study of the electronic behavior of real three- or two-dimensional quasicrystals. The tight-binding approximation has also been used in the treatment of the two- or three-dimensional Penrose lattice, but it is necessary to adopt the crystalline approximation to fulfill periodic boundary conditions,

[i] W. Gellermann *et al.*, *Phys. Rev. Lett.* **72**, 633 (1994).
[j] S. Zhu, Y. Zhu, *et al.*, *Phys. Rev. Lett.* **78**, 2752 (1997); S. Zhu, *et al. Science* **278**, 843 (1997).

then solve the wave equations by numerical methods. Theoretical results from these calculations still cannot be expected to illuminate experimental measurements. It is noted that real quasicrystals are all alloys, so if another approach, i.e., the nearly-free electron model, is adopted, we may get some intuitive insights which may help us to elucidate the electronic structures and properties of quasicrystals.

We have already seen that electrons in a perfect crystal will contribute strong Bragg diffraction peaks when the tuning condition is satisfied. This is the physical reason for the appearance of the energy gaps, and the peaks and valleys in the curves of $g(E)$ versus E. In a three-dimensional quasicrystal, the reciprocal lattice, as well as the Brillouin zone (BZ), are absent. However, sharp diffraction spots with icosahedral symmetry appear in the diffraction pattern. From these strong diffraction spots, a series of quasi-reciprocal vectors may be deduced and from the perpendicular bisecting planes of these quasi reciprocal vectors, a nearly spherical polyhedron can be constructed and taken as the quasi-Brillouin zone (QBZ).[k] Figure 5.5.3 shows QBZs for two icosahedral phases.

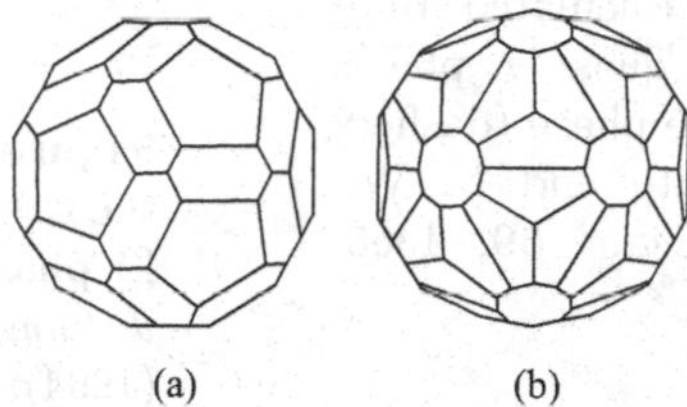

Figure 5.5.3 QBZ for two i phases. (a) Al-Mn alloys; (b)Al-Cu-Li alloys.

Why quasiperiodic order is stable is an interesting problem. There is sufficient evidence to indicate that the stability of quasicrystals is intimately related to their electronic structure. This means that the icosahedral structure may be established through energy considerations when the Fermi surface (FS) intersects the quasi Brillouin zone boundaries, where a pseudogap is produced in the DOS. The Fermi level situated near the minimum of the pseudogap tends to lower the electron energy. The results of theoretical calculation of the DOS curves are shown in Fig. 5.5.4 which exhibits very pronounced deviations from the free electron parabola and the singularities are much stronger than those found in the fcc Al calculation. The presence of peaks and valleys in the density of states suggest the stability of the quasicrystal relative to various competing crystalline phases varies rapidly with the average s-p electron concentration. For concentration at which the Fermi level is close to a minimum in the density of states, as observed in Fig. 5.5.4, the quasicrystal should be the most stable.

The origin of the pseudogap is attributed to strong electron scattering by the quasi-lattice and the touching of the FS with the QBZ boundaries, and the pseudogap causes an enhancement of cohesive energies. The existence of the pseudogap at the Fermi energy is generic in a system of alloys as crystalline approximants for quasicrystals. The pseudogap at the Fermi energy suggests the gap-opening mechanism by a touching of the Fermi surface at the effective Brillouin zone, which increases the cohesive energy. This band mechanism of enhancing the stability is known as the Hume–Rothery mechanism for alloys. It works more efficiently in quasicrystals because the effective Brillouin zone is almost spherical in a polyhedron shape.

The electronic structures of quasicrystals Al-Cu-Li is shown in Fig. 5.5.5. There is a small dip of the pseudogap at the Fermi surface. However, low values of the density of states at the Fermi level are observed in the crystalline as well as in the quasicrystalline phases. Hence, we conclude that although the pseudogap is a generic property of the quasicrystal, it is not a specific property distinguishing the quasiperiodic from the periodic or aperiodic phases. Crystalline, quasicrystalline, and amorphous alloys have to be considered as Hume–Rothery phases with a varying degree of bandgap stabilization.

The ordered icosahedral phase is the typical three-dimensional quasicrystal, which was discovered first. After this several two-dimensional quasicrystals were discovered in sequence, including the d

[k] In crystalline alloys, effective BZ known as Jones zone can be shown to be different from BZ. Refer to H. Jones, *The Theory of Brillouin Zones and Electronic States in Crystals*, North Holland, Amsterdam, 1960.

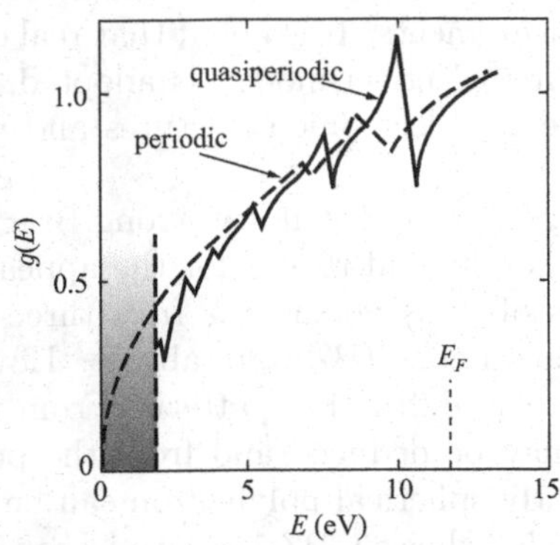

Figure 5.5.4 DOS curves calculated from the NFE model for Al quasi i phase (solid curve) and crystalline phase (dashed curve). From A. P. Smith and N. W. Ashcroft, *Phys. Rev. Lett.* **59**, 1365 (1987).

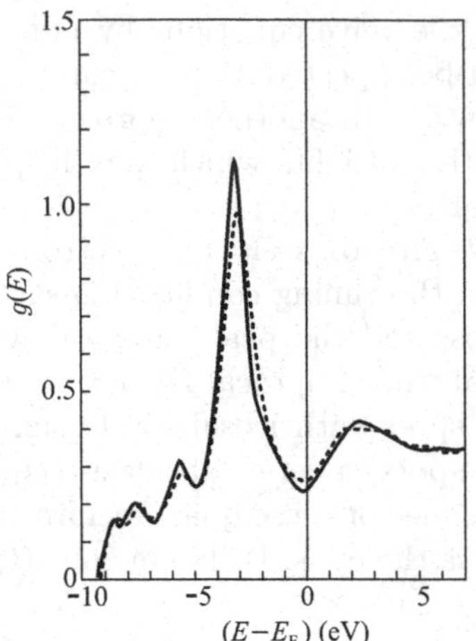

Figure 5.5.5 Pseudoenergy gap in the calculated DOS curve of i-Al-Cu-Li phase. From M. Windisch *et al.*, *J Phys: Condens. Matter* **6**, 6977 (1994).

phase quasicrystal with tenfold symmetry. For a two-dimensional quasicrystal, there is a two-dimensional quasiperiodic plane, but periodic direction perpendicular to it. It is clear that the QBZ can be constructed from the strong diffraction points, and then the Hume–Rothery mechanism can be used to demonstrate the stability of the two-dimensional quasicrystal in the same way.

Another major problem related to the electronic properties of quasicrystals concerns their unusual transport properties. Experimental measurements have shown that stable quasicrystals have semimetallic transport properties characterized by a high resistivity, a negative temperature coefficient of resistivity, and strong temperature and composition dependence of the Hall coefficient and thermopower. On the other hand, the experimental measurements of angular resolved photoelectronic spectra of two-dimensional Al-Ni-Co verified the NFE approximation;[l] some electronic properties of two-dimensional d phase quasicrystals, such as resistivity, Hall coefficient and thermopower are strongly anisotropic due to the presence of quasiperiodic order and periodic order in different directions.[m]

Bibliography

[1] Brillouin, L., *Wave Propagation in Periodic Structures*, Wiley, New York (1946).

[2] Ziman, J. M., *Principles of the Theory of Solids*, Cambridge University Press, Cambridge (1972).

[3] Pinsker, Z. G., *Dynamical Scattering of X-Ray in Crystals*, Springer-Verlag, Berlin (1978).

[4] Joannopoulos, J. D., Mead, R. D., and Winn, J. N., *Photonic Crystals*, Princeton University Press, Princeton (1995).

[5] Hui, P. M., and Johnson, H. F., Photonic band-gap materials, *Solid State Phys.* **49**, 151 (1996).

[6] Yeh, P., *Optical Waves in Layered Madia*, John Wiley & Sons, New York (1990).

[l] About the measurement of angle resolved photoelectronic spectra are referred to E. Rotenberg *et al.*, *Nature* **406**, 602 (2000).

[m] D. L. Zhang and Y. P. Wang, *Mat. Sci. Forum* **150–151**, 445 (1994); S. Y. Lin *et al.*, *Phys. Rev. Lett.* **77**, 1998 (1996).

[7] Fujiwara, T., and Tsunetsugu, H., *Quasicrystals: The State of the Art*, World Scientific, Singapore (1991).

[8] Quilichini, M., and Janssen, T., Phonon excitations in quasicrystals, *Rev. Mod. Phys.* **69**, 277 (1997).

[9] Stadnik, Z. M. (ed.), *Physical Properties of Quasicrystals*, Springer-Verlag, Berlin (1999).

[10] Poon, S. J., Electronic properties of quasicrystals: An experimental review, *Adv. Phys.* **41**, 303 (1992).

[11] Rolston, S., Optical Lattices, *Phys. World* **11**, Oct., 27 (1998).

Chapter 6

Dynamics of Bloch Electrons

The motion of Bloch electrons in periodic structures is an important topic in traditional solid state physics, especially in either electric or magnetic fields. This process is really a dynamical problem and it has brought about plentiful technical applications. In discussing this topic, wave-particle duality is stressed and semiclassical approach is mostly used.

§6.1 Basic Properties of Electrons in Bands

An electron has energy, momentum and mass, which are characteristics of a particle. However, in a crystal it behaves as Bloch wave with a dispersion relation that displays a band structure. The momentum becomes crystal momentum, or quasi-momentum, and its mass becomes the effective mass.

6.1.1 Electronic Velocity and Effective Mass

An electron in a Bloch state $\psi_{\boldsymbol{k}}(\boldsymbol{r})$ is characterized by its wavevector $\boldsymbol{k}$, so we may expect that its velocity moving through a periodic structure is a function of $\boldsymbol{k}$. To derive this function, it is useful to consider the electron as a wave packet of the Bloch state. The group velocity of the wave packet is given by $\boldsymbol{v} = \nabla_{\boldsymbol{k}}\omega(\boldsymbol{k})$, where ω is the frequency. Using de Broglie relation $\omega = E/\hbar$, we obtain the velocity of the Bloch electron

$$\boldsymbol{v} = \frac{1}{\hbar}\nabla_{\boldsymbol{k}}E(\boldsymbol{k}), \tag{6.1.1}$$

which states that the velocity of an electron in the state $\boldsymbol{k}$ is proportional to the gradient of the energy in $\boldsymbol{k}$-space. It follows that the velocity at every point in $\boldsymbol{k}$-space is normal to the energy contour passing through that point. In general, the contours are nonspherical, so the velocity is not necessary parallel to the wavevector $\boldsymbol{k}$, except the case of a free electron, whose velocity is given by $\boldsymbol{v} = \hbar\boldsymbol{k}/m$, which is obtained by combining (5.2.2) and (6.1.1). This velocity is proportional to and parallel to the wavevector $\boldsymbol{k}$.

It is known that in many cases near the center of the Brillouin zone, the parabolic dispersion relation $E = \hbar^2k^2/2m^*$ is almost satisfied and thus the velocity is

$$\boldsymbol{v} = \frac{\hbar\boldsymbol{k}}{m^*}, \tag{6.1.2}$$

which is of the same form as for a free electron, except using an effective mass m^* to replace the mass of the free electron m. This approximation is often very useful. It follows that near the center of the zone $\boldsymbol{v}$ is parallel to $\boldsymbol{k}$, and points radially outward. However, near the zone boundaries at which the energy contours are distorted as gaps arise, this simple relationship between $\boldsymbol{v}$ and $\boldsymbol{k}$ is not satisfied, and one must resort to the general result (6.1.1), which is quite different from that of a free electron.

When a force $\boldsymbol{F}$ is exerted on a Bloch electron in the crystal, the rate of energy absorption by the electron is

$$\frac{dE(\boldsymbol{k})}{dt} = \boldsymbol{F} \cdot \boldsymbol{v}, \tag{6.1.3}$$

where the term on the right is clearly the expression for the power absorbed by a moving object. The left hand side of (6.1.3) can be written as

$$\frac{dE(\boldsymbol{k})}{dt} = \nabla_{\boldsymbol{k}} E(\boldsymbol{k}) \cdot \frac{d\boldsymbol{k}}{dt} = \hbar \boldsymbol{v} \cdot \frac{d\boldsymbol{k}}{dt},$$

and so we obtain the simple relation

$$\hbar \frac{d\boldsymbol{k}}{dt} = \boldsymbol{F}, \tag{6.1.4}$$

where the vector $\hbar\boldsymbol{k}$ behaves like the momentum of the Bloch electron. Equation (6.1.4) simply states that the rate of the momentum change is equal to the force, which is Newton's second law.

When the wavevector $\boldsymbol{k}$ varies as (6.1.4), the electron undergoes an acceleration

$$\boldsymbol{a} = \frac{d\boldsymbol{v}}{dt}, \tag{6.1.5}$$

which can be combined with (6.1.1) and rewritten as

$$\boldsymbol{a} = \nabla_{\boldsymbol{k}} \boldsymbol{v} \cdot \frac{d\boldsymbol{k}}{dt} = \frac{1}{\hbar^2} \nabla_{\boldsymbol{k}} \nabla_{\boldsymbol{k}} E(\boldsymbol{k}) \cdot \boldsymbol{F}. \tag{6.1.6}$$

If we write this in Cartesian coordinates, we find its component are

$$a_i = \sum_j \frac{1}{\hbar^2} \frac{\partial^2 E}{\partial k_i \partial k_j} F_j, \; i, j = x, y, z, \tag{6.1.7}$$

which leads to the definition of dynamic effective mass as

$$\left(\frac{1}{m^*}\right)_{i,j} = \frac{1}{\hbar^2} \frac{\partial^2 E}{\partial k_i \partial k_j}, \; i, j = x, y, z. \tag{6.1.8}$$

The effective mass is now a second-order tensor, which has nine components. The concept of effective mass is very useful, in that it often enables us to treat the Bloch electron in a manner analogous to a free electron. Nevertheless, the Bloch electron exhibits many unusual properties which are different from those of the free electron.

In semiconductors, e.g., Si and Ge, the dispersion relation can often be written as

$$E(\boldsymbol{k}) = \alpha_1 k_x^2 + \alpha_2 k_y^2 + \alpha_3 k_z^2, \tag{6.1.9}$$

corresponding to an ellipsoidal contour, and the effective mass has three components: $m_{xx}^* = \hbar^2/2\alpha_1$, $m_{yy}^* = \hbar^2/2\alpha_2$, $m_{zz}^* = \hbar^2/2\alpha_3$. In this case the mass of the electron is anisotropic and depends on the direction of the external force. For example, when the force is along the k_x axis, the electron responds with a mass m_{xx}^*. Only when $\alpha_1 = \alpha_2 = \alpha_3$, is (6.1.9) reduced to a parabolic dispersion relation, and the free electron behavior is recovered.

As already stated, the quadratic dispersion relation is satisfied near the bottom of the band but it is no longer valid as $\boldsymbol{k}$ increases. Strictly m^* is not a constant, but a function of $\boldsymbol{k}$. There may exist an inflection point $\boldsymbol{k}_c$ beyond which the mass becomes negative, as the region is close to the top of the band. Negative mass means that the acceleration is negative, i.e., the velocity decreases for $\boldsymbol{k}$ beyond $\boldsymbol{k}_c$. This effect comes from the fact that in this region of $\boldsymbol{k}$-space the lattice exerts such a large retarding force on the electron that it overcomes the applied force and produces a negative acceleration. This picture is vastly different from the behavior of the free electron, and will bring in Bloch oscillations to be discussed in §6.2.1.

6.1.2 Metals and Nonmetals

One of the most successful applications of band theory is the treatment of the differences in the conducting behaviors of materials. Solids are divided into two major classes: Metals and nonmetals, the latter including insulators and semiconductors. A metal, or conductor, is a solid in which an electric current flows under an applied electric field. By contrast, application of an electric field produces no current in an insulator. There is a simple criterion for distinguishing between the two classes on the basis of the band theory. This criterion rests on the following statement: A band which is completely full of carriers gives no contribution to electric current, even in presence of an electric field. It follows therefore that a solid behaves as a metal only when some of the bands are partially occupied.

Take Na as an example: It has eleven electrons per atom. In an isolated atom these electrons have the configuration $[1s^2 2s^2 2p^6]3s^1$; the ten inner electrons form closed shells in the isolated atom, and form very narrow bands in the solid. Since the inner bands $1s$, $2s$, $2p$ are all fully occupied, they do not contribute to the current. We may therefore concern ourselves only with the uppermost occupied band, the valence band, which is the $3s$ band. This band can accommodate $2N$ electrons, where N is the total number of unit cells. Now in Na, of bcc structure, each cell has one atom, which contributes one valence electron, i.e., the $3s$ electron. Therefore the total number of valence electrons is N, and as these electrons occupy the band, only half of it is filled, as shown in Fig. 6.1.1(a). Thus sodium behaves like a metal. In a similar fashion, we conclude that the other alkalis, Li, K, etc., are also metals because their valence bands, the $2s$, $4s$, etc., respectively, are only partially occupied. The noble metals, Cu, Ag, Au, are likewise conductors for the same reason.

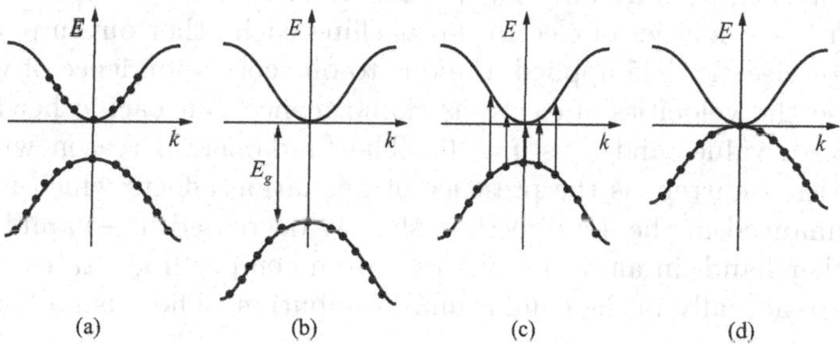

Figure 6.1.1 The distribution of electrons in the bands of (a) a metal, (b) an insulator, (c) a semiconductor, and (d) a semimetal.

As an example of a good insulator, we mention diamond which is an elemental solid of carbon. The electronic configuration of an isolated carbon atom is $1s^2 2s^2 2p^2$. For the diamond structure there is hybridization of the $2s$ and $2p$ atomic states, which gives rise to two bands split by an energy gap as shown in Fig. 6.1.1(b). Since these bands arise from $2s$ and $2p$ states and, since the unit cell here contains two atoms, these bands can accomodate $8N$ electrons. Now in diamond each atom contributes 4 electrons, resulting in 8 valence electrons per cell; thus the valence band is completely full, and the substance is an insulator.

There are substances which are intermediate between metals and insulators. If the gap between the valence band and the band immediately above it is small as shown in Fig. 6.1.1(c), then electrons are readily excited thermally from the former to latter band. Both bands become only partially filled and both contribute to the electric conduction; such a substance is known as a semiconductor. Typical examples are Si and Ge, which also form diamond structure just like C, but their gaps are narrower, about 1 and 0.7 eV, respectively; by contrast, the gap in diamond is about 7 eV. Roughly speaking, a substance behaves as a semiconductor at room temperature whenever its gap is less than 2 eV. In some substances the gap vanishes entirely, or the two bands even overlap slightly, and we call them semimetals as shown in Fig. 6.1.1(d). The best-known example is Bi; in addition, As, Sb, and white Sn also belong to this class of substances.

An interesting problem is presented in connection with the divalent elements, for example, Be, Mg, Ca, Zn, etc. For instance, the electronic configuration of the magnesium atom is

$[1s^2, 2s^2, 2p^6]3s^2$. Mg crystallizes in the hcp structure, with one atom per primitive cell. Since there are two valence electrons per cell, the $3s$ band should be completely filled up, resulting in an insulator. In fact, however, Mg is a metal. The reason for the apparent paradox is that the $3s$ and $3p$ bands in Mg overlap somewhat, so that electrons are transferred from the former to the latter, resulting in incompletely filled bands, and hence Mg is a metal. The same condition accounts for the metallicity of Be, Ca, Zn, and other divalent metals. We can conclude that a substance in which the number of valence electrons per unit cell is odd is necessarily a metal, since it takes an even number of electrons to fill a band completely. But when the number is even, the substance may be either an insulator or a metal, depending on whether the bands are separate or overlapping.

Metals and nonmetals show their characteristics in transport processes. We can derive the basic properties that a fully filled band carries no electric current, but a partially filled band does carry current, by noting that electrons with opposite Bloch wavevectors $\boldsymbol{k}$ and $-\boldsymbol{k}$ satisfy

$$\boldsymbol{v}(-\boldsymbol{k}) = -\boldsymbol{v}(\boldsymbol{k}), \tag{6.1.10}$$

according to (6.1.1). This equation follows from the symmetry relation $E(-\boldsymbol{k}) = E(\boldsymbol{k})$. The current density due to all electrons in the band is given by

$$\boldsymbol{j} = \frac{-e}{\Omega}\sum_{\boldsymbol{k}} \boldsymbol{v}(\boldsymbol{k}), \tag{6.1.11}$$

where Ω is the volume, $-e$ the electronic charge, and the sum is over all states in the band. As a consequence of (6.1.11), for any fully filled band, the sum over a whole band is seen to vanish, that is, $\boldsymbol{j} = 0$, with the velocities of electrons cancelling each other out in pairs; but for a partially filled band, when an electric field applied, the one-to-one correspondence of wavevectors $\boldsymbol{k}$ and $-\boldsymbol{k}$ will be destroyed, so the velocities of electrons cannot cancel out each other and the sum over this band gives a non-zero value, and thus $\boldsymbol{j} \neq 0$. The fundamental reason why an actual metal or semiconductor carries a current is the presence of relaxation effects which result in a distribution which is slightly enhanced in the $+v$ direction, slightly decreased in $-v$, and the end result is true current. On the other hand, in an insulator there are no empty final states available for scattering and the electrons are actually in the equilibrium distribution. There is no true current.

6.1.3 Hole

For a semiconductor, the forbidden gap between uppermost fully filled band and unfilled band is narrower, so that thermal excitation could empty a small fraction of the states in the uppermost occupied band, the valence band, simultaneously populating a few of the lower states in the next band, the conduction band. This thermal process leads to the appearance of vacant states in the valence band. Conduction by electrons in the conduction band and by the same number of positive holes in the valence band is the situation in an intrinsic semiconductor, and the energy separation between valence and conduction bands is the intrinsic gap. It is also suggested that at temperatures that are too low for appreciable intrinsic electron-hole generation, localized electron states within the forbidden gap associated with defects and impurities (to be discussed in §8.2.2) could still be thermally ionized to generate either free electrons or free holes. This second form of semiconducting behavior is termed extrinsic, since it depends on non-intrinsic properties of the conducting medium.

For further discussion, consider that a hole occurs in a band that is completely filled except for one vacant state (see Fig. 6.1.2). When we consider the dynamics of a hole in an external field, we find it far more convenient to focus on the motion of the vacant site than on the motion of the enormous number of electrons filling the band and the concept of the hole is an important one in band theory.

Suppose the hole is located at the wavevector $\boldsymbol{k}_1$, as shown in Fig. 6.1.2. The current density of the whole system is

$$\boldsymbol{j}_{\rm h} = \frac{-e}{\Omega}\sum_{\boldsymbol{k}}{}' \boldsymbol{v}_{\rm e}(\boldsymbol{k}), \tag{6.1.12}$$

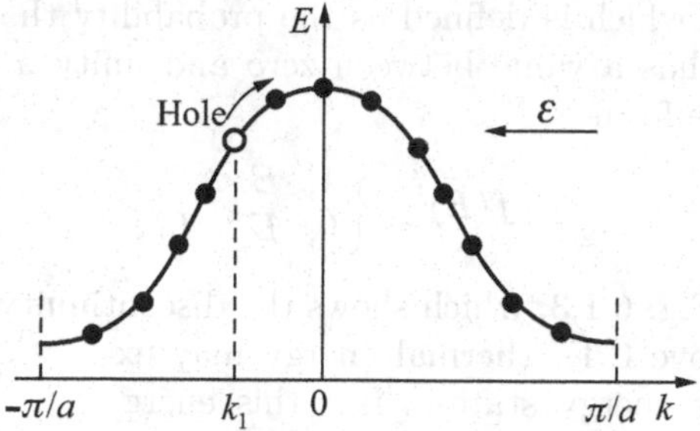

Figure 6.1.2 The hole and its motion in the presence of an electric field.

where the sum is over all the electrons in the band, with a prime over the summation indicating that the state $\boldsymbol{k}_1$ is to be excluded, as that state is vacant. Since the sum over the filled band is zero, the current density is equal to

$$j_{\mathrm{h}} = \frac{e}{\Omega}\boldsymbol{v}_{\mathrm{e}}(\boldsymbol{k}_1). \tag{6.1.13}$$

That is, the current is the same as if the band were empty, except for an electron of positive charge $+e$ located at $\boldsymbol{k}_1$. In practice a band contains not a single hole but a large number of holes, and in the absence of an electric field the net current of these holes is zero because of the mutual cancellation of the contributions of the various holes. When an electric field is applied, however, nonvanishing induced currents are created.

Assume the electric field is directed to the left in Fig. 6.1.2, so all the electrons move uniformly to the right in k_z-space at the same rate. Consequently the vacant site also moves to the right, together with the rest of the system in one dimension. The change in the hole current in a time interval δt can be found

$$\delta j_{\mathrm{h}} = \frac{e}{\Omega}\left(\frac{dv_{\mathrm{e}}}{dk}\right)_{k_1}\frac{dk}{dt}\delta t = \frac{e}{\Omega}\frac{1}{m^*(k_1)}F\delta t = \frac{-e^2}{\Omega}\frac{1}{m^*(k_1)}\mathcal{E}\delta t, \tag{6.1.14}$$

where $m^*(k_1)$ is the effective mass of an electron occupying state k_1. For simplicity, we take $m^*(k_1)$ as a scalar quantity. This equation gives the electric current of the hole induced by the electric field, which is the observed current. Since the hole usually occurs near the top of the band, due to thermal excitation of the electron to the next-higher band, where the mass $m^*(k_1)$ is negative, it is convenient to define the mass of a hole as

$$m_{\mathrm{h}}^* = -m^*(k_1), \tag{6.1.15}$$

which is a positive quantity, and (6.1.14) is now rewritten as

$$\delta j_{\mathrm{h}} = \frac{1}{\Omega}\frac{e^2}{m_{\mathrm{h}}^*}\mathcal{E}\delta t. \tag{6.1.16}$$

Note that the hole current, unlike the electron current, is in the same direction as the electric field.

By examing (6.1.13) and (6.1.16), we can see that the motion of the hole, both with and without an electric field, is the same as that of a particle with a positive charge e and a positive mass m_{h}^*. Viewing the hole in this manner results in a great simplification, in that the motion of all the electrons in the band has been reduced to that of a single "quasi-particle".

6.1.4 Electronic Specific Heat in Metals

The energy of electron in a metal is quantized: The electrons in the metal obey the Pauli exclusion principle and occupy each of the quantized levels with at most two electrons, one with spin up, and the other with spin down. All the electrons in the metal are accommodated from the lowest state up to the Fermi level. The distribution of electrons among the levels is usually described by the distribution function, $f(E)$, which is defined as the probability that the level E is occupied by an electron. In general, $f(E)$ has a value between zero and unity and the distribution function for electrons at $T = 0$ K has the form

$$f(E) = \begin{cases} 1, & E < E_\mathrm{F}, \\ 0, & E > E_\mathrm{F}. \end{cases} \tag{6.1.17}$$

This function is plotted in Fig. 6.1.3, which shows the discontinuity at the Fermi energy.

When the system is above 0 K, thermal energy may excite the electrons to higher energy states. But this energy is not shared equally by all the electrons, as would be the case in the classical treatment. It arises from the exclusion principle, because the energy which an electron may absorb thermally is of the order $k_\mathrm{B}T$ (0.025 eV at room temperature) which is much smaller than E_F (which is of the order of 5 eV). If we use $E_\mathrm{F} = k_\mathrm{B}T_\mathrm{F}$ to define a Fermi temperature, $T_\mathrm{F} \approx 60000$ K. Therefore only those electrons close to the Fermi level can be excited, because the levels above E_F are empty, and hence when those electrons move to a higher level there is no violation of the exclusion principle. Thus only these electrons, a small fraction of the total number, are capable of being thermally excited, and this explains the low electronic specific heat.

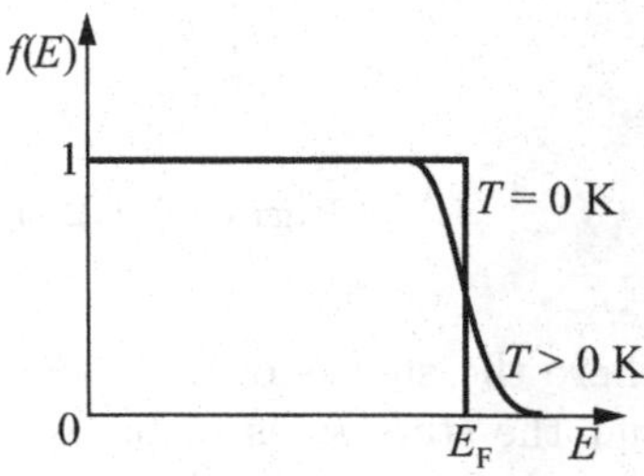

Figure 6.1.3 Fermi distribution function.

The distribution function $f(E)$ at temperature $T \neq 0$ K is given by

$$f(E) = \frac{1}{\mathrm{e}^{(E-E_\mathrm{F})/k_\mathrm{B}T} + 1}. \tag{6.1.18}$$

This is the Fermi-Dirac distribution which is also plotted in Fig. 6.1.3. It is substantially the same as the distribution at $T = 0$ K, except within $k_\mathrm{B}T$ of the Fermi level, where some of the electrons are thermally excited from below E_F to above it.

One can use the distribution function to evaluate the thermal energy and hence the heat capacity of the electrons, but this is a fairly tedious undertaking, so instead we shall attempt to obtain a good approximation with a minimum of mathematical effort. Since only electrons within the range $k_\mathrm{B}T$ of the Fermi level are excited, we conclude that only a fraction $k_\mathrm{B}T/E_\mathrm{F}$ of the electrons is affected. Therefore the number of electrons excited per mole is about $N(k_\mathrm{B}T/E_\mathrm{F})$, and since each electron absorbs an energy $k_\mathrm{B}T$, on the average, it follows that the thermal energy per mole is given approximately by

$$U = \frac{N(k_\mathrm{B}T)^2}{E_\mathrm{F}}, \tag{6.1.19}$$

and the specific heat is

$$C_e = \partial U/\partial T = 2R\frac{k_\mathrm{B}T}{E_\mathrm{F}}, \tag{6.1.20}$$

where $R = Nk_\mathrm{B}$. We see that the specific heat of the electrons is reduced from its classical value (of the order of R) by the factor $k_\mathrm{B}T/E_\mathrm{F}$. For $E_\mathrm{F} = 5$ eV and $T = 300$ K, this factor is equal to 1/200. This great reduction is in agreement with experiment. An more exact evaluation of the electronic specific heat yields the value

$$C_e = \frac{\pi^2}{2}R\frac{k_\mathrm{B}T}{E_\mathrm{F}}, \tag{6.1.21}$$

which is clearly of the same order of magnitude as the approximate expression (6.1.20), and both show the common feature that the electronic specific heat is linear in temperature.

§6.2 Electronic Motion in Electric Fields

Electronic states of motion will be changed when there is an applied electric field. These changes will take place in momentum space as well as in position space. The impurities or defects in otherwise perfect periodic structures affect the electronic transport substantially. The discussion below is based on the one-band model.

6.2.1 Bloch Oscillations

Figure 6.2.1(a) and (b) show a typical one-dimensional band structure and the corresponding velocity, respectively. As stated before, the dispersion curve is almost parabolic near the center of the zone, but distorted a lot near the boundaries. Now (6.1.1) is reduced to

$$v = \frac{1}{\hbar}\frac{\partial E}{\partial k}, \tag{6.2.1}$$

which says that the velocity is proportional to the slope of the energy curve. We see as k varies from the origin to the boundary of the Brillouin zone (BZ), the velocity increases at first, reaches a maximum, and then decreases to zero at the boundary. There is a inflection point corresponding to a maximum velocity.

When a static electric field $\mathcal{E}$ is applied to a crystalline solid, the electrons in the solid are accelerated. In absence of any scattering for a perfect periodic structure, their motions can be described in k-space,

$$\hbar\frac{dk}{dt} = F = -e\mathcal{E}. \tag{6.2.2}$$

This shows that the rate of change of k is proportional to, and lies in the same direction as, the electric force F, i.e., opposite to the field $\mathcal{E}$, so we can now consider the consequences of the acceleration of the electron in one dimension. Equation (6.2.2) shows that the wavevector k increases uniformly with time. Thus, as t increases, the electron traverses k-space at a uniform rate, as shown in Fig. 6.2.1(a). The electron, starting from $k = 0$, for example, moves up the band until reaches the top point A. Once the electron passes the zone edge at A, it immediately reappears at the equivalent point A', then continues to descend along the path $A'B'C'$. Note that the motion in k-space is periodic in the reduced-zone scheme, since after traversing the zone once, the electron repeats the motion. This process is known as Bloch oscillation. The period of the motion is readily found, on the basis of (6.2.2), to be

$$\tau_{\mathrm{B}} = \frac{2\pi\hbar}{e\mathcal{E}d}, \tag{6.2.3}$$

which corresponds to a frequency of the oscillation

$$\omega_{\mathrm{B}} = e\mathcal{E}d/\hbar. \tag{6.2.4}$$

Figure 6.2.1(b) shows the velocity of the electron as it traverses the k axis. Starting from $k = 0$, the velocity increases as time passes. After reaching a maximum, it decreases and then vanishes at the zone edge. Thus the electron turns around and acquires a negative velocity, and so on. The velocity we are discussing is the velocity in real space, i.e., the usual physical velocity. It follows that a Bloch electron, in the presence of a static electric field also executes an oscillatory periodic motion in real space.

The net displacement of the wave packet of a Bloch electron can be obtained by integrating the velocity with respect to time

$$\begin{aligned} z &= \int_0^t v(t)dt = \frac{1}{\hbar}\int_0^t \frac{\partial E}{\partial k}dt \\ &= \frac{1}{\hbar}\int_0^t \frac{\partial E}{\partial k}\frac{dt}{dk}dk = -\frac{1}{e\mathcal{E}}\left\{E[k(t)] - E(0))\right\}, \end{aligned} \tag{6.2.5}$$

from which it is evident that the maximum displacement in the z direction is $z_{\max} = B/e\mathcal{E}$, where B is the bandwidth, and after one cycle $t = \tau_{\mathrm{B}}$, the net displacement is $z = 0$. Thus after each period the electron returns to the original position.

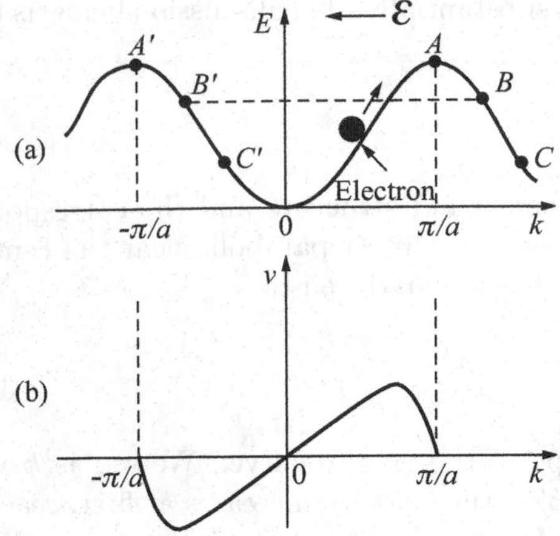

Figure 6.2.1 The motion of an electron in the presence of an electric field, (a) energy-wavevector relation, and correspondingly (b) velocity-wavevector relation.

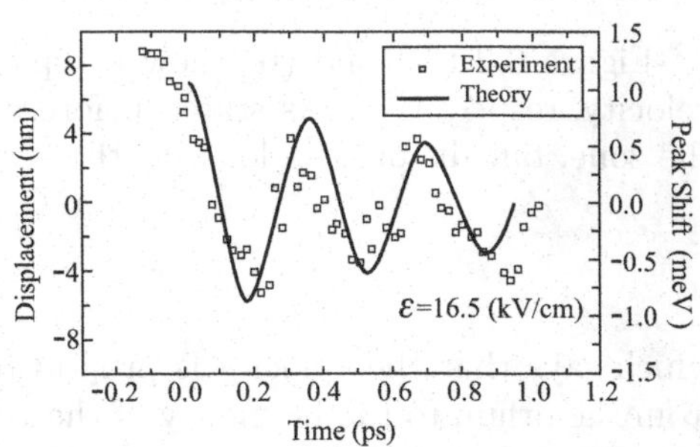

Figure 6.2.2 Spatially electronic oscillation. V. G. Lyssenko *et al.*, *Phys. Rev. Lett.* **79**, 301 (1997).

Although the description is reasonable for electron dynamics, the oscillatory motion is hard to observe in real crystals. The reason is simple: The crystal cannot be expected to be perfect. The period τ_B of (6.2.3) is about 10^{-5} s for usual values of the parameters, while a typical electron collision time $\tau = 10^{-14}$ s at room temperatures. Thus the electron undergoes an enormous number of collisions, about 10^9, in the time of one cycle. Consequently the oscillatory motion may be suppressed with the development of semiconductor superlattices, it is now possible to achieve $\tau_\mathrm{B} \leq \tau$ by growing superlattices of high purity.

For more perfect superlattices, on the other hand, the scattering time τ can be long enough so that the Bloch oscillations may be observed. Recently, Bloch oscillations have been observed in the time domain by using the four-wave mixing method. Figure 6.2.2 shows four-wave mixing peak shift (right scale) as a function of delay time for heavy-hole transition under an applied field. The peak shift can be related to the displacement shown on the left scale. The electron wave packet performs a sinusoidal oscillation with a total amplitude of about 140 Å, and the experimental results are in agreement with model calculations.

6.2.2 Negative Differential Resistance

Speaking of superlattices, an early successful application is the investigation of negative differential resistance, which is an interesting current transport phenomenon closely related to the scattering time τ. As in Fig. 6.2.1, under an external electric field $\mathcal{E}$, we can calculate the velocity increment of an electron in a time interval dt according to (6.1.6)

$$dv = \frac{e\mathcal{E}}{\hbar^2}\frac{\partial^2 E}{\partial k^2}dt. \tag{6.2.6}$$

The average drift velocity, taking into account the scattering time τ, is written as

$$v_\mathrm{d} = \int_0^\infty \mathrm{e}^{-t/\tau}dv = \frac{e\mathcal{E}}{\hbar^2}\int_0^\infty \frac{\partial^2 E}{\partial k^2}\mathrm{e}^{-t/\tau}dt. \tag{6.2.7}$$

The factor $\exp(-t/\tau)$ represents the probability of free acceleration for a time t by the electric field. If a sinusoidal approximation is used for the E-k relation, for example, $E = \alpha - 2\beta\cos kd$, then

$$v_\mathrm{d} = \frac{e\mathcal{E}\tau}{m^*(0)}\left[1 + \left(\frac{e\mathcal{E}\tau d}{\hbar}\right)^2\right]^{-1}, \tag{6.2.8}$$

and the current-field variation is given by

$$j = env_{\mathrm{d}}, \tag{6.2.9}$$

where $m^*(0)$ is determined by the curvature of $E(k)$ at $k = 0$ and n is the electron concentration. The current j, plotted as a function of $\mathcal{E}$ in Fig. 6.2.3(a), has a maximum at $e\mathcal{E}\tau d/\hbar = 1$ and thereafter decreases, giving rise to a negative differential resistance. This result indicates that, in an applied electric field, conduction electrons may gain enough energy to go beyond the inflection point in Fig. 6.2.1, whereupon they will be decelerated rather than accelerated by such an electric field.

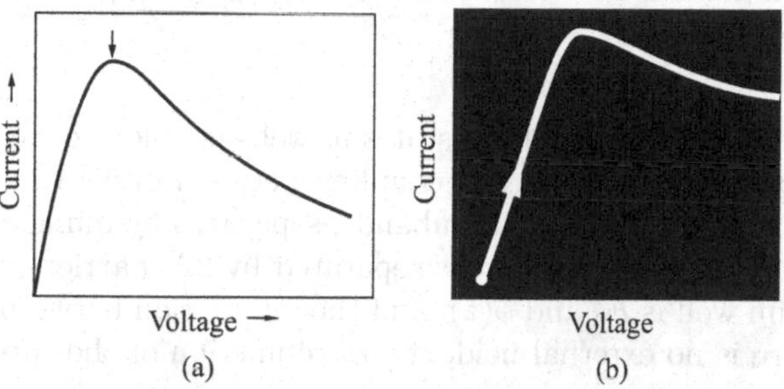

Figure 6.2.3 Current-voltage characteristic for negative differential resistance. (a) theoretical curve; (b) experimental result. From L. Esaki, in Synthetic Modulated Structures (eds. L. L. Chang and B. C. Giessen), Academic Press, Orlando (1985).

The obstacle for observing negative differential resistance in typical crystals arises from the fact that $\tau \ll \tau_{\mathrm{B}}$, so electrons have no chance to surmount the velocity inflection point. However, L. Esaki *et al.* (1972) found that the superlattice can be used to exhibit a negative differential resistance in its transport properties, which was, for the first time, interpreted on the basis of the predicted quantum effect. Figure 6.2.3(b) shows the current-voltage characteristic at room temperature for such a superlattice having 100 periods, with each period consisting of a GaAs well 60 Å thick and $Ga_{0.5}Al_{0.5}As$ barrier 10 Å thick.

6.2.3 Wannier–Stark Ladders

In addition to Bloch oscillations and negative differential resistance, there exists another way of describing electron motion in periodic structures in the presence of an applied electric field. It is evident that the periodic structure is actually destroyed by the static electric field, so electronic states are no longer given by Bloch-like solutions but instead are given by levels localized in space within a few periods of the lattice, and the energy spectrum is discrete with a level separation of $e\mathcal{E}d$. In this picture, localization occurs whenever the level broadening ($\hbar/\tau$) due to scattering is less than $e\mathcal{E}d$. These equally spaced energy levels are called Wannier–Stark ladders. The reason for this energy is obvious: A wavefunction displaced by just d still satisfies the same wave equation, but the energy displacement is $e\mathcal{E}d$.

Unfortunately, this effect is also impossible to observe in real crystals, but it may be observable in superlattices, because the larger lattice-modulated period leads to a larger energy splitting. For example, a semiconductor with $d \sim 2$ Å at $\mathcal{E} \sim 10^4$ V/cm will have $e\mathcal{E}d \sim 0.2$ meV. This splitting is extremely small; whereas for a superlattice with $d \simeq 100$ Å, $e\mathcal{E}d = 10$ meV, and this should make it easier to detect the Wannier–Stark levels in a superlattice. Evidence of this break-up of a superlattice miniband from extended Bloch states into discrete Stark levels has been observed by optical experiments.[a]

Here we will give an illuminated discussion on Wannier–Stark ladders in superlattices.[b] Beginning with the energy spectra of a superlattice, instead of the Kronig–Penney model in §7.2.4, we

[a]E. E. Mendez *et al.*, *Phys. Rev. Lett.* **60**, 2426 (1988).
[b]G. Bastard *et al.*, *Phys. Rev. Lett.* **60**, 220 (1988); *Superl & Micros.* **6**, 77 (1989).

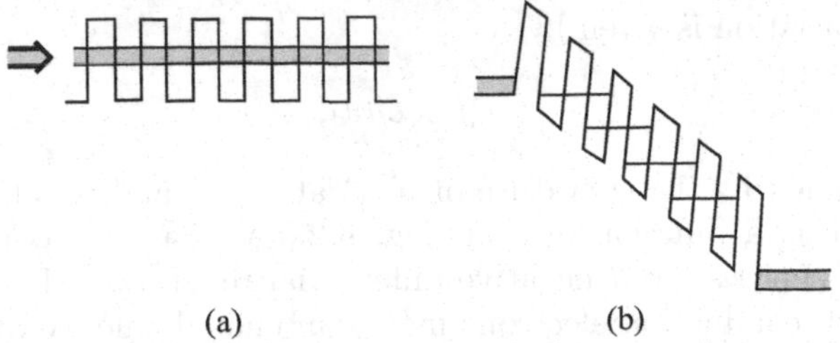

Figure 6.2.4 A Wannier–Stark ladder in a superlattice, (a) the perpendicular energies corresponding to the resonant transmission miniband states; (b) the miniband breaks up into discrete Wannier–Stark levels when a large electric field is applied.

consider a tight-binding picture of localized states in wells coupled to each other by the overlapping of wavefunctions through the barriers. The coupling between every two neighboring wells through the barrier will lead to the appearance of minibands, separated by minigaps. Assuming that a superlattice is composed of $2N+1$ quantum wells separated by $2N$ barriers, the level and wavefunction of each isolated quantum well is E_0 and $\phi(z)$, and the interaction between two neighboring quantum wells is $-\lambda$. When there is no external field, the wavefunction of the superlattice is

$$\psi_k(z) = \sum_{l=-N}^{N} c_k(l)\phi(z-ld), \; c_k(l) = \frac{1}{\sqrt{2N+1}} \mathrm{e}^{ikld}, \tag{6.2.10}$$

where l labels the quantum wells and k is the wavevector along z direction. The dispersion relation is

$$E(k) = E_0 - 2\lambda \cos kd. \tag{6.2.11}$$

These expressions describe the extended state and energy band. Figure 6.2.4(a) shows a miniband in the absence of an electric field.

When an electric field is applied in the modulated direction of the superlattice, the static electric energy is $-e\mathcal{E}z$, so the energy of lth quantum well is changed by $-e\mathcal{E}ld$. If the energy shift of neighboring wells $e\mathcal{E}d$ is larger than the half width λ of the miniband, the energy levels of the superlattice are determined by the levels of all quantum wells,

$$E_l = E_0 - e\mathcal{E}ld, \tag{6.2.12}$$

and a Wannier–Stark ladder appears in Fig. 6.2.4(b). In order to obtain the wavefunctions and energies of the system in the tight-binding approximation, we can write the coefficient equations recursively as

$$-\lambda c(l-1) + (E_0 - e\mathcal{E}ld - E)c(l) - \lambda c(l+1) = 0, \tag{6.2.13}$$

and the boundary conditions are assumed to be

$$c(-N-1) = c(N+1) = 0.$$

If $e\mathcal{E}d > \lambda$, in zeroth-order approximation, the eigenenergies can be written as

$$E = E_0 - \nu e\mathcal{E}d, \; -N \leq \nu < N. \tag{6.2.14}$$

For any eigenenergy, (6.2.13) is transformed to

$$c(l-1) + c(l+1) = \frac{2(l-\nu)}{(2\lambda/e\mathcal{E}d)} c(l). \tag{6.2.15}$$

This is the recursion relation of a Bessel function, so

$$c_\nu(l) = J_{l-\nu}\left(\frac{2\lambda}{e\mathcal{E}d}\right). \tag{6.2.16}$$

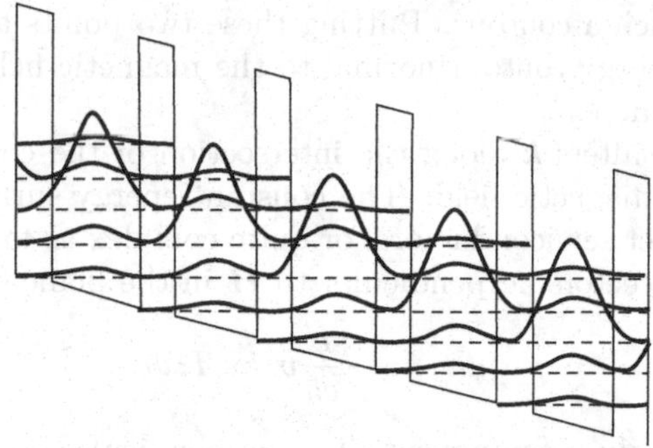

Figure 6.2.5 Schematic probability density of the Wannier–Stark ladder states in semiconductor superlattice.

To give insight into the wavefunction, we take $e\mathcal{E}d = 2\lambda$, then $J_{l-\nu}(1) = -0.0199$, 0.1150, −0.4401, 0.7652, 0.1150 and 0.0199 for $l - \nu = -3, -2, -1, 0, 1, 2, 3$. It is clear that the amplitude at the center ($l - \nu = 0$) is the largest, and decreases away from the center, so the Wannier–Stark state is basically localized. The absolute values of the wavefunction neighboring the center symmetrically decay. In the positive direction, the wavefunctions are always positive, but in the negative direction, they decay in an oscillatory fashion (see Fig. 6.2.5).

Very recently, theoretical and experimental studies on the quantum motion of ultracold atoms in an accelerating optical lattice have exhibited the same quantum behaviors as electrons, such as Bloch oscillations and Wannier–Stark ladders. This is easy to understand from the viewpoint of matter waves: The optical potential is spatially periodic yielding an energy spectrum of Bloch bands for the atoms, and the acceleration provides an inertial force in the moving frame, emulating an electric force on Bloch electrons.[c]

§6.3 Electronic Motion in Magnetic Fields

The investigation of electron dynamics in magnetic fields is fruitful: These have provided us with many results about the symmetries of electronic states in the Brillouin zone (BZ), carrier masses at band edges, and the shapes of the Fermi surfaces of crystalline solids, etc.

6.3.1 Cyclotron Resonance

In the presence of a magnetic field $\boldsymbol{H}$, it is assumed that the underlying picture of the electronic band structure remains intact and electrons are described by a wavevector $\boldsymbol{k}$. For nonmagnetic solids in a semiclassical treatment, the basic equation of motion describing an electron in a magnetic field is

$$\hbar\frac{d\boldsymbol{k}}{dt} = -\frac{e}{c}\boldsymbol{v} \times \boldsymbol{H}, \tag{6.3.1}$$

where the left hand side is the time derivative of the crystal momentum, and the right side is the well-known Lorentz force due to the magnetic field. According to this equation, the change in $\boldsymbol{k}$ in a time interval dt is given by

$$d\boldsymbol{k} = -\frac{e}{c\hbar}\boldsymbol{v} \times \boldsymbol{H}dt, \tag{6.3.2}$$

which shows that the electron moves in $\boldsymbol{k}$-space in such a manner that its displacement $d\boldsymbol{k}$ is perpendicular to the plane defined by $\boldsymbol{v}$ and $\boldsymbol{H}$. Since $d\boldsymbol{k}$ is perpendicular to $\boldsymbol{H}$, this means that the electron trajectory lies in a plane normal to the magnetic field. In addition, $d\boldsymbol{k}$ is perpendicular to the velocity $\boldsymbol{v}$, which is described by (6.1.1). As $\boldsymbol{v}$ is normal to the energy contour in $\boldsymbol{k}$-space, it

[c]For Bloch oscillations and Wannier–Stark ladders for ultracold atoms in optical potentials, see theoretical studies by Q. Niu, X. G. Zhao *et al.*, *Phys. Rev. Lett.* **76**, 4504 (1996). Experimental verifications can be seen at M. B. Daham, E. Peik *et al.*, *Phys. Rev. Lett.* **76**, 4508 (1996), and also S. R. Wilkinson, C. F. Bharucha *et al.*, *ibid.*, 4512.

means that $d\boldsymbol{k}$ lies along such a contour. Putting these two points together, we conclude that the electron rotates along an energy contour normal to the magnetic field, as shown in Fig. 6.3.1, and in a counterclockwise fashion.

The magnetic field thus alters $\boldsymbol{k}$ along the intersection of the constant energy surface and the plane perpendicular to the magnetic field. The constant energy surface can be quite simple as for the conduction band of direct semiconductors, or quite complex as for the valence bands. If $\boldsymbol{k}_\perp$ and $\boldsymbol{v}_\perp$ are the two component vectors perpendicular to $\boldsymbol{H}$ in the plane of the intersection, we have

$$d\boldsymbol{k}_\perp = -\frac{e}{c\hbar}\boldsymbol{v}_\perp \times \boldsymbol{H}dt, \tag{6.3.3}$$

which means that $d\boldsymbol{k}$ has only a component $d\boldsymbol{k}_\perp$, perpendicular to $\boldsymbol{H}$. On the other hand, $\boldsymbol{v}_\parallel$ is a constant if it is nonzero: The electron moves in a helical trajectory under the application of a magnetic field.

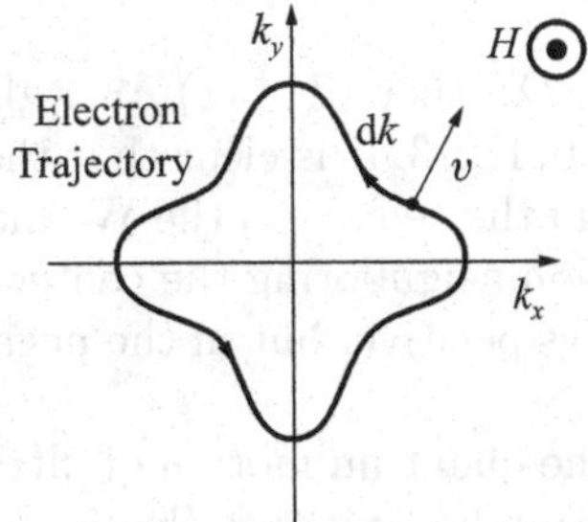

Figure 6.3.1 Trajectory of an electron in $\boldsymbol{k}$-space in the presence of a magnetic field $\boldsymbol{H}$.

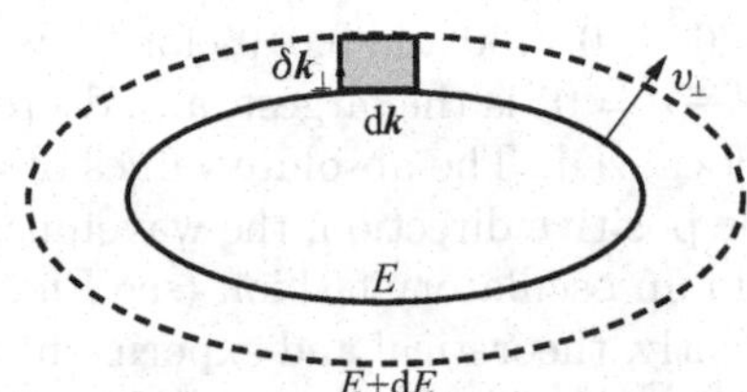

Figure 6.3.2 Two orbits of an electron at energies E and $E + dE$ in a magnetic field.

Let us now consider the frequency of the electron around the constant energy surface. We examine two orbits in the same $\boldsymbol{k}_\perp$-space with energies E and $E + dE$ as shown in Fig. 6.3.2. The $\boldsymbol{k}_\perp$-space separation of the orbits is

$$\delta k_\perp = \frac{dE}{|\nabla_{\boldsymbol{k}_\perp} E|} = \frac{dE}{\hbar|\boldsymbol{v}_\perp|}. \tag{6.3.4}$$

The rate at which an electron moving along one of the orbits sweeps out the annular area is given by

$$\left|\frac{d\boldsymbol{k}_\perp}{dt}\right| \delta k_\perp = \frac{e}{c\hbar^2}|\boldsymbol{v}_\perp \times \boldsymbol{H}|\frac{dE}{|\boldsymbol{v}_\perp|} = \frac{eH}{c\hbar^2}dE. \tag{6.3.5}$$

This rate is constant for constant dE, and if we define the time period of the orbit by T_c, then the annular area is

$$dS = T_c \cdot \frac{eH}{c\hbar^2}dE, \tag{6.3.6}$$

where S is the area in $\boldsymbol{k}$-space of the electronic orbit with energy less than E. Thus

$$T_c = \frac{c\hbar^2}{eH}\frac{dS}{dE}. \tag{6.3.7}$$

We can now introduce a cyclotron resonance frequency

$$\omega_c = \frac{2\pi}{T_c} = \frac{2\pi eH}{c\hbar^2}\frac{1}{dS/dE} = \frac{eH}{m_c c}, \tag{6.3.8}$$

where

$$m_c = \frac{\hbar^2}{2\pi}\frac{dS}{dE} \tag{6.3.9}$$

is defined as the cyclotron resonance mass, which is a property of the entire orbit and is not the same as the effective mass in general. However, for a parabolic band $E = \hbar^2 k^2/2m^*$ and we have $S = \pi k^2 = 2m^*\pi E/\hbar^2$, then $m_c = m^*$.

For a more complex band this relation will be appropriately modified. For example, the conduction band of indirect gap materials, such as Ge and Si, can be represented by an ellipsoidal constant energy surface coming from (6.1.9)

$$E(\boldsymbol{k}) = \hbar^2 \left(\frac{k_x^2 + k_y^2}{2m_t} + \frac{k_z^2}{2m_l} \right), \tag{6.3.10}$$

where m_t and m_l are the transverse and longitudinal effective masses, respectively. The velocity components are now

$$v_x = \frac{\hbar k_x}{m_t},\ v_y = \frac{\hbar k_y}{m_t},\ v_z = \frac{\hbar k_z}{m_l}. \tag{6.3.11}$$

If we assume that the magnetic field lies in the equatorial plane of the spheroid, and is parallel to the k_x axis, we get from the equation of motion

$$\frac{dk_x}{dt} = 0,\ \frac{dk_y}{dt} = -\omega_l k_z,\ \frac{dk_z}{dt} = \omega_t k_y, \tag{6.3.12}$$

with $\omega_l = eH/cm_l$ and $\omega_t = eH/cm_t$ Then from (6.3.11) and (6.3.12), we have

$$\frac{d^2 k_y}{dt^2} + \omega_l \omega_t k_y - 0, \tag{6.3.13}$$

which is the equation of motion of a classical harmonic oscillator with frequency

$$\omega_c = (\omega_l \omega_t)^{1/2} = \frac{eH}{c(m_l m_t)^{1/2}}. \tag{6.3.14}$$

It can be shown that if $\boldsymbol{H}$ is parallel to k_z, then the frequency is simply

$$\omega_c = \omega_t = \frac{eH}{cm_t}. \tag{6.3.15}$$

In general, if the magnetic field makes an angle θ with respect to the k_z direction, the cyclotron resonance mass

$$\left(\frac{1}{m_c}\right)^2 = \frac{\cos^2\theta}{m_t^2} + \frac{\sin^2\theta}{m_t m_l}. \tag{6.3.16}$$

Thus, by altering the magnetic field direction, one can probe various combinations of m_l and m_t. In a cyclotron resonance experiment, the cyclotron frequency can be measured directly and the carrier masses can be obtained.

In the above discussion of cyclotron motion, we have disregarded the effects of collisions. Of course, if this cyclotron motion is to be observed at all, the electron must complete a substantial fraction of its orbit during between two collisions, i.e., $\omega_c\tau \geq 1$. This necessitates the use of very pure samples at low temperatures under very strong magnetic fields.

6.3.2 Landau Quantization

In the quantum mechanical description of an electron in a magnetic field, the Hamiltonian of the system is

$$\mathcal{H} = \frac{1}{2m}\left(\boldsymbol{p} + \frac{e}{c}\boldsymbol{A}\right)^2 + V(\boldsymbol{r}), \tag{6.3.17}$$

where $\boldsymbol{A}$ is the vector potential and $V(\boldsymbol{r})$ is periodic potential. We will use the effective mass approximation to absorb the effect of the background crystal potential. This approach will be used in §7.2.2 when we addressed the problem of shallow impurities. The approach is general, and

(6.3.17) is now written as an effective mass equation for a band with effective mass m^*. The resulting Schrödinger equation is

$$\frac{1}{2m^*}\left(\frac{\hbar}{i}\nabla+\frac{e}{c}\boldsymbol{A}\right)^2\psi=E\psi, \tag{6.3.18}$$

where $\psi(\boldsymbol{r})$ is now to be considered as the envelope wavefunction which can reflect the main aspect of the magnetic field to the electron. It is noted that the interaction between the spin of the electron and the magnetic field is ignored at present: This interaction is $g_{\mathrm{L}}\mu_{\mathrm{B}}\boldsymbol{\sigma}\cdot\boldsymbol{H}$, where $\boldsymbol{\sigma}$ is the spin operator, $\mu_{\mathrm{B}}=e\hbar/2cm$ the Bohr magneton and g_{L} the Lande factor is related to the details of the state; for a free electron $g_{\mathrm{L}}=2$.

We write the vector potential in the gauge $\boldsymbol{A}=(0,Hx,0)$, which gives a magnetic field in the z direction, i.e., $\boldsymbol{H}=H\boldsymbol{z}$. The equation to be solved is

$$-\frac{\hbar^2}{2m^*}\left[\frac{\partial^2}{\partial x^2}+\left(\frac{\partial}{\partial y}+\frac{ieHx}{\hbar c}\right)^2+\frac{\partial^2}{\partial z^2}\right]\psi=E\psi, \tag{6.3.19}$$

where all energies are to be measured from the band edges. Since the Hamiltonian does not involve the coordinates y and z explicitly, the wavefunction can be written as

$$\psi(x,y,z)=\mathrm{e}^{i(k_y y+k_z z)}\phi(x). \tag{6.3.20}$$

Denoting

$$E'=E-\frac{\hbar^2}{2m^*}k_z^2, \tag{6.3.21}$$

we get the equation for $\phi(x)$ as

$$\left[-\frac{\hbar^2}{2m^*}\frac{d^2}{dx^2}+\frac{1}{2}m^*\omega_{\mathrm{c}}^2(x+l_{\mathrm{c}}^2k_y)^2\right]\phi(x)=E'\phi(x). \tag{6.3.22}$$

From (6.3.20) and (6.3.22), we can see that the motion of the electron along the magnetic field is unaffected, and the motion in the xy plane is given by a one-dimensional harmonic equation with frequency $\omega_{\mathrm{c}}=eH/cm^*$ and centered around the point

$$x_0=-l_{\mathrm{c}}^2k_y, \tag{6.3.23}$$

where

$$l_{\mathrm{c}}=\left(\frac{c\hbar}{eH}\right)^{1/2} \tag{6.3.24}$$

is called the cyclotron radius or magnetic length which is about 100 Å for $H=10^5$ Oe. The eigenfunctions of (6.3.22) are

$$\phi(x)\propto H_\nu(x-x_0)\mathrm{e}^{-(x-x_0)^2/2l_{\mathrm{c}}^2}, \tag{6.3.25}$$

where $H_\nu(x)$ are the Hermitian polynomials, and the eigenvalues

$$E'_\nu=\left(\nu+\frac{1}{2}\right)\hbar\omega_{\mathrm{c}},\ \nu=0,1,\ldots. \tag{6.3.26}$$

These discrete energies labelled by ν are called Landau levels. It is obvious that the states ψ are extended in the y and z directions, but localized in the x direction. The total energy is

$$E_\nu=\left(\nu+\frac{1}{2}\right)\hbar\omega_{\mathrm{c}}+\frac{\hbar^2}{2m^*}k_z^2. \tag{6.3.27}$$

The electron energy is quantized in the x-y plane and has continuous translational energy along the direction of the magnetic field. In (6.3.27), only the orbital quantization due to the magnetic field was accounted for. If the spin of electron is included, each Landau level will be split into two sublevels with additional energy $\pm g_L \mu_B \sigma H$. For semiconductors the contribution of $g_L \mu_B \sigma H$ is small if the magnetic field is not high.

Since the energy arising from the x-y plane motion is so drastically affected, it is important to ask what happens to the density of states of the system. Consider a box of sides L_x, L_y, and L_z. From the form of the wavefunction given by (6.3.20), it is clear that both k_z and k_y are quantized in units of $2\pi/L_z$ and $2\pi/L_y$ respectively, and in addition, the center x_0 in (6.3.23) must be inside the dimension of the system, i.e.,

$$0 \leq x_0 \leq L_x. \tag{6.3.28}$$

By using $\Delta k_y = 2\pi/L_y$, we have

$$\Delta x_0 = 2\pi l_c^2/L_y, \tag{6.3.29}$$

and the degeneracy of a level in two dimensions is

$$D = \frac{L_x}{\Delta x_0} = \frac{L_x L_y}{2\pi l_c^2}. \tag{6.3.30}$$

This degeneracy comes from linear oscillations with the same energy but different central positions. Equivalently, we note that the total magnetic flux through the x-y plane is $\Phi = HL_xL_y$, and the flux quantum $\phi_0 = hc/e$, (6.3.30) can also be expressed in the form $D = \Phi/\phi_0$, which addresses that the number of states equals the number of flux quanta.

One way to physically show the effect of magnetic field is to examine the distribution of states in $\boldsymbol{k}$-space as drawn in Fig. 6.3.3. Focusing on the k_x-k_y plane first, we can understand that, in absence of magnetic field, the k_x and k_y points are good quantum numbers and the points of allowed states are homogeneously distributed. However, after a magnetic field is applied, various (k_x, k_y) points condense into a series of circles which represent constant energy surfaces with energies $\hbar\omega_c/2$, $3\hbar\omega_c/2$, etc., as shown in Fig. 6.3.3(a). This rearrangement of states does not alter the total number of states in a macroscopic volume. It can be understood by examing the number of states in the presence of the magnetic field, per unit area, per unit energy, when the electron spin is not taken into account. This is the same as the two-dimensional density of states in the x-y plane without magnetic field

$$g_{2D}(E) = \frac{1}{L_x L_y} \frac{D}{\hbar\omega_c} = \frac{m^*}{2\pi\hbar^2}. \tag{6.3.31}$$

Extended to three dimensions by taking into account the k_z continuous component, each circle in Fig. 6.3.3(a) is transformed into a cylinder in Fig. 6.3.3(b).

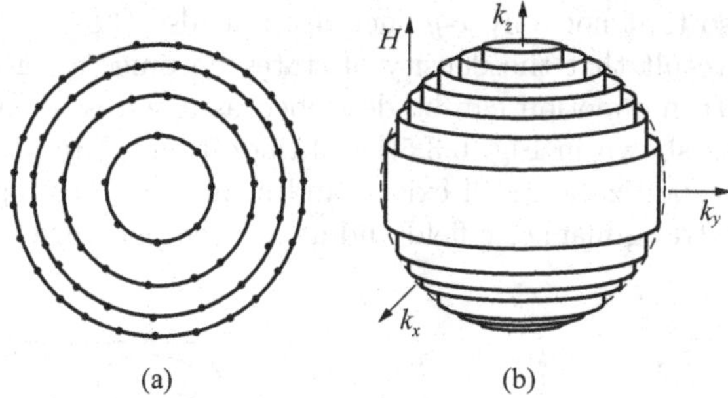

Figure 6.3.3 Quantization scheme for electrons when a magnetic field is applied. (a) In the k_x-k_y plane; (b) In whole $\boldsymbol{k}$-space.

The density of states of the three-dimensional system is essentially given by the one-dimensional density of states, weighted by the degeneracy factor D. Since k_z is still a good quantum number, the E-k_z relation gives the band structure, called Landau subbands. The various Landau levels are shown in Fig. 6.3.4(a). In the one-dimensional k_z-space the density of states for a particular Landau level with energy E_ν is

$$g_{1D}(E) = \frac{1}{4\pi}\left(\frac{2m^*}{\hbar^2}\right)^{1/2}(E - E_\nu)^{-1/2}. \tag{6.3.32}$$

Then by taking into account the two-dimensional density of states (6.3.31) and running over the contribution from all Landau levels with starting energies less than E, the total density of states is

$$g_{3D}(E) = \frac{1}{(4\pi)^2}\left(\frac{2m^*}{\hbar^2}\right)^{3/2} \hbar\omega_c \sum_{\nu}\left[E - \left(\nu + \frac{1}{2}\right)\hbar\omega_c\right]^{-1/2}, \tag{6.3.33}$$

which is shown in Fig. 6.3.4(b), where we see the van Hove singularities from the quantization of the states. In real systems, broadening due to impurities and thermal disturbance will wipe out these divergences, but the periodic variation of the density of states with magnetic field is retained. This variation has important effects on the physical properties of the system. It is obvious that as the magnetic field is altered, the separation of the Landau levels as well as the density of states (DOS) changes, and the Fermi level will gradually pass through the various sharp structures in the density of states. This leads to very interesting effects such as the de Haas–van Alphen (dHvA) effect and Shubnikov–de Haas (SdH) effect.

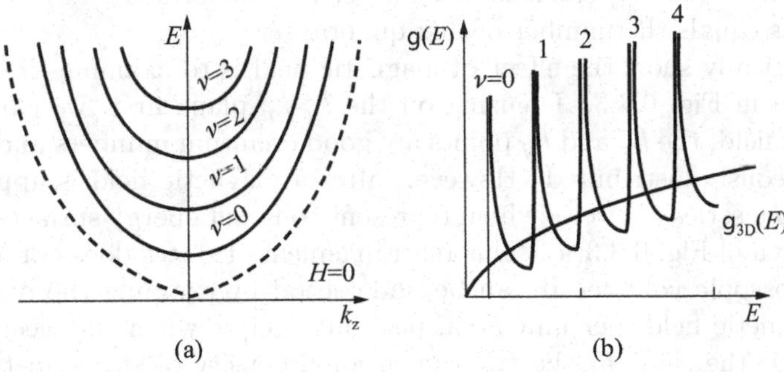

Figure 6.3.4 Effects of magnetic field in a three-dimensional electronic system on (a) the band structure; and (b) the density of states.

The treatment for electrons in three-dimensions can be easily extended to two-dimensions where the effect of the magnetic field becomes even more interesting. It is assumed that a two-dimensional electron gas is confined in z direction and a magnetic field is along the z axis. Now k_z is not a good quantum number, so that not only x-y energies, but also the z energies, are quantized. This leads to the remarkable result that the density of states becomes a series of δ-functions as shown in Fig. 6.3.5(a), and electronic motion can be described as a series of circles with cyclotron radius l_c covering the system, as shown in Fig. 6.3.5(b). In addition, if an electric field is applied along x direction, the Landau quantization still exists, and a more interesting effect, the quantum Hall effect may appear under strong magnetic field and at low temperature.

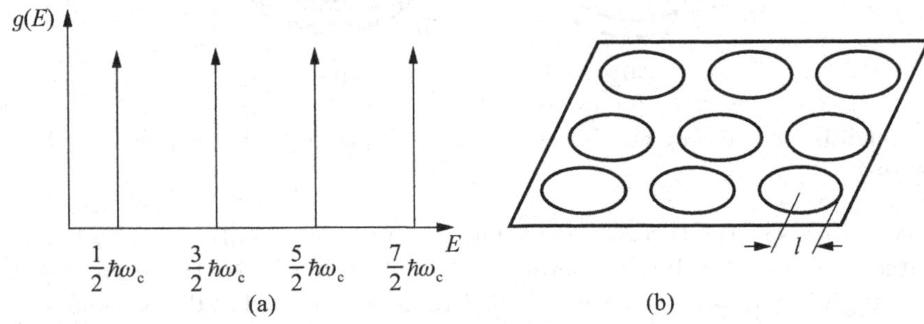

Figure 6.3.5 In a two-dimensional electronic system with a magnetic field perpendicular to it, (a) the density of states (DOS); (b) schematic trajectories of electronic cyclotron motion.

6.3.3 de Haas–van Alphen Effect

There are many magnetic oscillatory behaviors related to Landau quantization. The key point is the position of the Landau levels with respect to the Fermi energy. For a three-dimensional electron gas, the occupied states within the Fermi energy $E_{\rm F}$ are contained in a sphere of radius $k_{\rm F} = (2mE_{\rm F}/\hbar^2)^{1/2}$, if no magnetic field is applied. When a quantizing field exists, all the states situated within the sphere on the allowed cylinders in Fig. 6.3.3(b) are occupied. As H increases, the separation between the levels also increases, and hence the highest filled Landau level moves up in energy. Once this level is above the Fermi energy, it will start to be emptied into lower energy levels. This process repeats itself whenever

$$E'_{\rm F}/\hbar\omega_{\rm c} = m^* cE'_{\rm F}/\hbar eH = \nu + 1/2, \tag{6.3.34}$$

where $E'_{\rm F} = E_{\rm F} - \hbar^2 k_z^2/2m^*$. Thus, any physical quantities that can sense this oscillation should have a constant period in H^{-1} given by

$$\Delta(H^{-1}) = \frac{e\hbar}{m^* cE'_{\rm F}}. \tag{6.3.35}$$

This kind of oscillation with the period in H^{-1} can be seen in a variety of magnetic and galvanomagnetic properties of metals and semiconductors.

The de Haas–van Alphen (dHvA) effect is the oscillation of magnetic moment in solids as a function of the static magnetic field intensity. The effect can be observed in pure specimens at low temperatures, in strong magnetic fields satisfying $\hbar\omega_{\rm c} > k_{\rm B}T$, and $\omega_{\rm c}\tau > 1$. A thermodynamic calculation including temperature is sophisticated. So for clarity, in the following we shall consider the behavior at zero temperature, where all states below the Fermi energy $E_{\rm F}$ are filled, and all states above it are empty. It is reasonable to assume that the Fermi energy is unaltered by the application of a magnetic field parallel to the z axis. For simplicity, the electron spin is still neglected.

We imagine a planar slice cut in $\boldsymbol{k}$-space around k_z, of thickness δk_z (see Fig. 6.3.4(a)). Because there are $L_z\delta k_z/2\pi$ states in the k_z-space, the number of electronic states in this slice for a particular Landau level is

$$\delta N = D\frac{L_z\delta k_z}{2\pi} = \Omega\beta H, \tag{6.3.36}$$

where $\Omega = L_xL_yL_z$, and $\beta = e\delta k_z/4\pi^2 c\hbar$ is defined as the degeneracy per unit magnetic field per unit volume.

In the slice all Landau levels with energies less than $E'_{\rm F} = E_{\rm F} - \hbar^2k_z^2/2m^*$ are filled and all above are empty. If the highest filled level is ν, the total number of allowed electrons in the slice per unit volume, δn, is given by the expression

$$\delta n = (\nu + 1)\beta H, \tag{6.3.37}$$

which is proportional to H, so long as ν is constant. As H increases, δn also increases until the permitted level ν crosses the Fermi level $E'_{\rm F}$, and empties itself suddenly. Thus an infinitesimal change in magnetic field will spill electrons out of the ν level and put them into other regions on the Fermi surface. The movement of the levels is shown in Fig. 6.3.6, while the upper part of Fig. 6.3.7 shows the resulting variation of δn with H. In the latter, the magnetic field is designated by the quantum number $\nu + 1/2 = E'_{\rm F}/\hbar\omega_{\rm c} = m^*cE'_{\rm F}/e\hbar H$, which coincides with the Fermi level. This is equivalent to plotting against H^{-1}, and so the lines of the saw-tooth should not be quite straight; however in most real metals under experimental conditions ν is about 1000, and the difference is not noticeable. It is observable that δn oscillates about a mean value δn_0 which is the same as the electron content of the slice in zero field.

As a result of the change in δn the energy of the electrons contained in the slice also changes. The excess number, $\delta n - \delta n_0$, must be supplied by the rest of the Fermi surface, and these electrons are raised to a slightly higher energy to be put into the slice. Thus, the excess number would occupy an annulus of width $\delta k'$ satisfying

$$\delta n - \delta n_0 = 2\pi k'_{\rm F}\delta k'\delta k_z/(2\pi)^3, \tag{6.3.38}$$

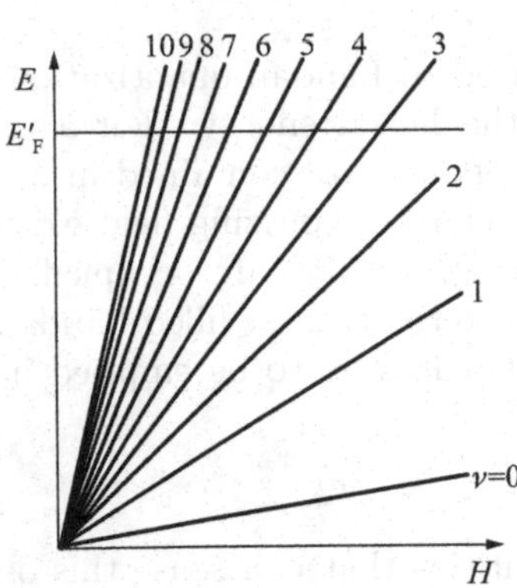

Figure 6.3.6 The spectrum of the Landau levels as a function of magnetic field H.

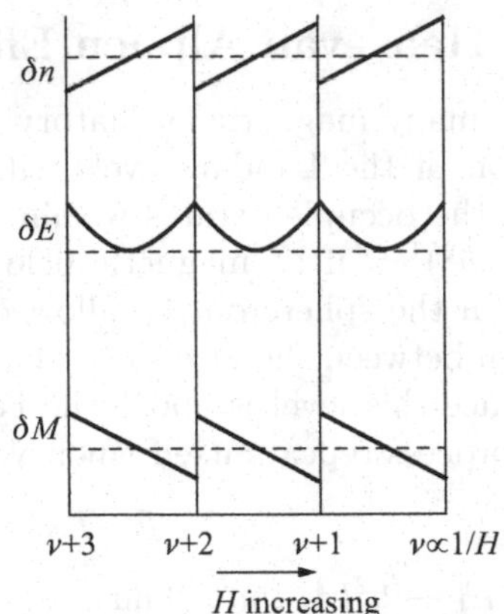

Figure 6.3.7 Variation of excess number δn, excess energy δE, and excess magnetic moment δM of slice δk_z, as the magnetic field is increased.

where k'_{F} is the radius of the Fermi surface in this slice, and a mean excess energy $\hbar^2 k'_{\mathrm{F}}\delta k'/2m^*$ can be obtained. The excess energy of the whole assembly due to the slice may thus be written

$$\delta E = \frac{\mu}{\beta}(\delta n - \delta n_0)^2, \tag{6.3.39}$$

where $\mu \equiv e\hbar/2m^*c$ is the effective number of Bohr magnetons. It is evident that δE is always positive and has a maximum at the point where a permitted orbit crosses the Fermi surface, as shown in the middle part of Fig. 6.3.7.

From this result, it is easy to derive the contribution of the slice to the magnetic moment by $dE/dH = -M$ at 0 K. Hence, from (6.3.39) and (6.3.37)

$$\delta M = -\frac{d}{dH}\left[\frac{\mu}{\beta}(\delta n - \delta n_0)^2\right] = -\frac{E'_{\mathrm{F}}}{H}(\delta n - \delta n_0). \tag{6.3.40}$$

Since $(\delta n - \delta n_0)$ oscillates between $\pm\beta H/2$, δM oscillates between $\mp\beta E'_{\mathrm{F}}/2$, as shown in the down part of Fig. 6.3.7.

To determine the response, M, of the entire system as a function of the magnetic field H, we have to add up the contributions δM from all slices of the electron distribution, having different E'_{F} and δn. Because δM is a periodic saw-tooth function of $1/H$ with period in (6.3.35) and overall amplitude $\beta E'_{\mathrm{F}}$, it may be expressed as a Fourier series

$$\delta M = \beta \sum_{p=1}^{\infty} A_p \sin px, \tag{6.3.41}$$

where

$$x = \frac{\pi E'_{\mathrm{F}}}{\mu H}. \tag{6.3.42}$$

Since for $-\pi < x < \pi$,

$$\delta M = -\frac{\beta E'_{\mathrm{F}}}{2\pi}x, \tag{6.3.43}$$

therefore

$$A_p = (-1)^p \frac{E'_{\mathrm{F}}}{p\pi}. \tag{6.3.44}$$

To sum over all slices, remembering that $E'_F = E_F - \hbar^2 k_z^2/2m^*$ and $\beta = e\delta k_z/4\pi^2 c\hbar$, the total magnetization is

$$M = \frac{e}{4\pi^3 c\hbar}\sum_{p=1}^{\infty}\frac{(-1)^p}{p}\int_{-k_F}^{k_F} E'_F \sin\left[\frac{p\pi}{\mu H}\left(E_F - \frac{\hbar^2 k_z^2}{2m^*}\right)\right] dk_z. \tag{6.3.45}$$

For a degenerate electron gas, in general, $E_F \gg \mu H$, the integrand oscillates very rapidly as a function of k_z and gives zero contribution unless it is at the stationary point $k_z \approx 0$. We thus replace E'_F by E_F and take it out of the integral. Then, by using the trigonometric function formula, we can expand the sin-expression and finish the integral for k_z, and finally obtain

$$M = \frac{eE_F(2m\mu H)^{1/2}}{4\pi^3 c\hbar}\sum_{p=1}^{\infty}\frac{(-1)^p}{p^{3/2}}\sin\left(\frac{p\pi E_F}{\mu H} - \frac{\pi}{4}\right), \tag{6.3.46}$$

with the approximation $k_F \to \infty$. In (6.3.46), the $p = 1$ term has the dominant contribution, and the periodic variation of M with $1/H$ is clearly obtained. This kind of oscillation with magnetic field is called the de Haas–van Alphen (dHvA) effect. In fact, many other physical properties, such as specific heat and thermoelectric power, which are all associated with the electronic density of states, also display the de Haas–van Alphen effect. The oscillation period in (6.3.46) is related to the Fermi energy E_F in the plane perpendicular to the magnetic field, and the measurement of the period can be used to determine the Fermi energy. However, there are a number of different periods if the Fermi surface is not spherical. By varying the crystal orientation with the field, the topological structure of a complex Fermi surface can thus be obtained.

6.3.4 Susceptibility of Conduction Electrons

There are many metals, such as the alkali metals, Li, Na, K, Rb and Cs, in which the atoms do not contain incomplete inner electron shells. There are weakly paramagnetic and show a susceptibility that varies little with temperature. We can give a simple explanation of this based upon the free-electron model in which the conduction electrons are assumed to move freely in metals.

At $T = 0$ K and when no field is applied there are two electrons per state at all energies up to Fermi energy E_F. When a uniform external field H is applied to the metal, the energy of electrons with spin direction parallel to the magnetic field decreases by $\mu_B H$, while the energy of electrons with antiparallel spin increases by the same amount. Thus, as shown in Fig. 6.3.8, a number $g(E_F)\mu_B H$ of electrons with antiparallel spin near the Fermi surface transfer to the parallel spin states; here $g(E_F)$ is the one-spin density of states at the Fermi surface. This change destroys the balance between the numbers of conduction electrons with spins parallel and antiparallel to the field so that the conduction electron system becomes magnetized. The magnetization is then

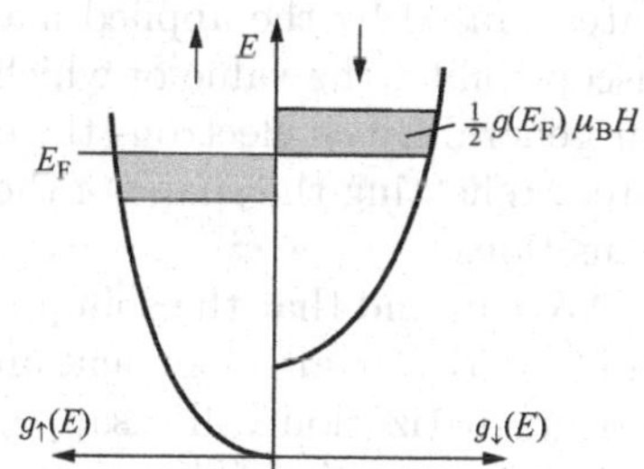

Figure 6.3.8 Energy distribution of electrons in the presence of a magnetic field.

$$M = \mu_B \int_0^{E_F} [g(E+\mu_B H) - g(E-\mu_B H)]dE = \mu_B \int_{E_F-\mu_B H}^{E_F+\mu_B H} g(E)dE. \tag{6.3.47}$$

Even in a very strong field, $\mu_B H/E_F$ will only be of the order of 10^{-3}, and the magnetization becomes

$$M = 2\mu_B^2 g(E_F)H. \tag{6.3.48}$$

The susceptibility due to such a process is called the Pauli paramagnetic susceptibility and is given by

$$\chi_P(0) = 2\mu_B^2 g(E_F). \tag{6.3.49}$$

It can be seen that the susceptibility is a measure of the electronic density of states (DOS) at the Fermi surface.

When the temperature is above zero, the distribution of electrons obeys the Fermi function $f(E_{k\sigma})$ and the magnetization is

$$M(T) = \mu_B \sum_{k} [f(E_{k\uparrow}) - f(E_{k\downarrow})], \tag{6.3.50}$$

where $E_{k\pm} = E_F \pm \mu_B H$. We can transform the summation to an integral and expand the Fermi function for weak field, then it becomes

$$M(T) = 2\mu_B^2 H \int_0^\infty \left[-\frac{\partial f(E)}{\partial E}\right] g(E) dE, \tag{6.3.51}$$

and the Pauli susceptibility at finite temperatures is

$$\chi_P(T) = 2\mu_B^2 \int_0^\infty \left[-\frac{\partial f(E)}{\partial E}\right] g(E) dE. \tag{6.3.52}$$

Furthermore, the susceptibility for the free-electron gas is

$$\chi_P(T) = \chi_P(0)\left[1 - \frac{\pi^2}{12}\left(\frac{k_B T}{E_F}\right)^2 + \cdots\right]. \tag{6.3.53}$$

Since $k_B T$ is always very much less than E_F, the susceptibility decreases a little with increasing temperature. When $T = 0$ K, $-\partial f/\partial E = \delta(E - E_F)$, (6.3.52) returns to (6.3.49).

In addition to the Pauli susceptibility related to electronic spin magnetic moments, a conduction electron system has a diamagnetic susceptibility; this originates from the change in the orbital states caused by the applied magnetic field. This portion is usually called the Landau diamagnetic susceptibility, the value of which is given by (6.3.52) multiplied by $-(1/3)$ for free electron system. For general Bloch electrons the expression for the diamagnetic susceptibility becomes rather complicated, reflecting the shape of the Fermi surface and also because of the contribution from interband transitions.

We can find that the spin polarization of electrons in Fig. 6.3.8 and the magnetization in (6.3.48) and (6.3.51) are all dependent on applied fields. If the applied field is removed, then spin polarization and magnetization will disappear. We can imagine that there exist interactions between electrons with different spins. These interactions are equivalent to an internal field, and each electronic spin is affected by the internal field. In this way, even there is no external field applied, spin polarization and magnetization will appear in the system. The formation of macroscopic magnetizations in magnetic materials like Fe, Co, and Ni is based on the existence of an internal field. The detailed discussion of this will be presented in Chap. 17.

It is interesting to consider the electron gas in a spatially varying external field $\boldsymbol{H}(\boldsymbol{r})$. We can decompose the field into Fourier components as

$$\boldsymbol{H}(\boldsymbol{r}) = \frac{1}{\Omega}\sum_{\boldsymbol{q}} \boldsymbol{H}_{\boldsymbol{q}} e^{i\boldsymbol{q}\cdot\boldsymbol{r}}, \tag{6.3.54}$$

with $\boldsymbol{H}_{\boldsymbol{q}} = \boldsymbol{H}^*_{-\boldsymbol{q}}$, and define the susceptibility, $\chi(\boldsymbol{q})$, in response to one of the components, $\boldsymbol{H}_{\boldsymbol{q}} \exp(i\boldsymbol{q}\cdot\boldsymbol{r})$. Then the magnetization is obtained as

$$\boldsymbol{M}(\boldsymbol{r}) = \frac{1}{\Omega}\sum_{\boldsymbol{q}} \chi_{\boldsymbol{q}} \boldsymbol{H}_{\boldsymbol{q}} e^{i\boldsymbol{q}\cdot\boldsymbol{r}}. \tag{6.3.55}$$

The key point is to get the susceptibility $\chi_{\boldsymbol{q}}$. For simplicity, the magnetic field is assumed to be applied along the z axis, s_{iz} is the z component of the spin of the i-electron, then the Zeeman energy is

$$\mathcal{H}' = -\mu_B \sum_i s_{iz}(H_{\boldsymbol{q}} e^{i\boldsymbol{q}\cdot\boldsymbol{r}_i} + H^*_{\boldsymbol{q}} e^{-i\boldsymbol{q}\cdot\boldsymbol{r}_i}), \tag{6.3.56}$$

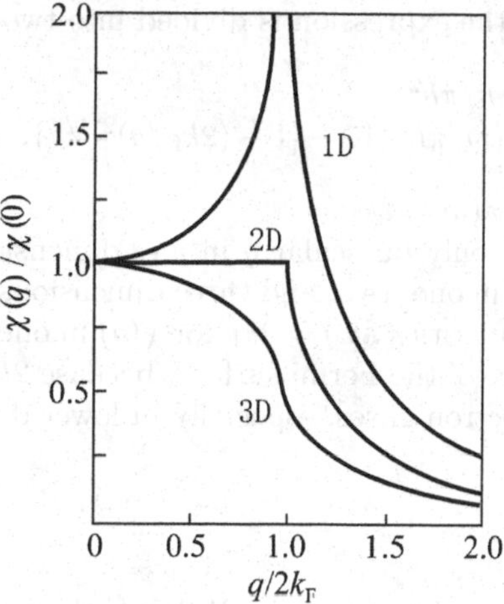

Figure 6.3.9 Normalized $\chi(q)$ versus q in one, two, and three dimensions. $\chi(0)$ is identified as the Pauli paramagnetic susceptibility at $q = 0$.

which will be taken as a perturbation. In first-order, the electron state described by a plane wave with wavevector $\boldsymbol{k}$ and spin + or − is modified to

$$\psi_{\boldsymbol{k}q\pm} = \frac{1}{\sqrt{\Omega}}e^{i\boldsymbol{k}\cdot\boldsymbol{r}}\left[1 \pm \frac{1}{2}\mu_{\mathrm{B}}\left(\frac{H_{\boldsymbol{q}}e^{i\boldsymbol{q}\cdot\boldsymbol{r}}}{\varepsilon_{\boldsymbol{k}+\boldsymbol{q}} - \varepsilon_{\boldsymbol{k}}} + \frac{H_{\boldsymbol{q}}^{*}e^{-i\boldsymbol{q}\cdot\boldsymbol{r}}}{\varepsilon_{\boldsymbol{k}-\boldsymbol{q}} - \varepsilon_{\boldsymbol{k}}}\right)\right]. \tag{6.3.57}$$

The square of the absolute value gives the number density of electrons with wavevector $\boldsymbol{k}$ and spin + or −,

$$\rho_{\boldsymbol{k}q\pm} = \frac{1}{\Omega}\left[1 \pm \frac{1}{2}\mu_{\mathrm{B}}\left(\frac{f_{\boldsymbol{k}}(1 - f_{\boldsymbol{k}+\boldsymbol{q}})}{\varepsilon_{\boldsymbol{k}+\boldsymbol{q}} - \varepsilon_{\boldsymbol{k}}} + \frac{f_{\boldsymbol{k}}(1 - f_{\boldsymbol{k}-\boldsymbol{q}})}{\varepsilon_{\boldsymbol{k}-\boldsymbol{q}} - \varepsilon_{\boldsymbol{k}}}\right)(H_{\boldsymbol{q}}e^{i\boldsymbol{q}\cdot\boldsymbol{r}} + H_{\boldsymbol{q}}^{*}e^{-i\boldsymbol{q}\cdot\boldsymbol{r}})\right], \tag{6.3.58}$$

where the Fermi distribution functions are added for non-zero temperature. The spatial density of the spin magnetic moment is obtained by multiplying (6.3.58) by $\pm\mu_{\mathrm{B}}$, adding both expressions for + and − spins, and then summing over $\boldsymbol{k}$ within the Fermi sphere, one finds

$$M_{\boldsymbol{q}}(\boldsymbol{r}) = \chi_{\boldsymbol{q}}\frac{1}{2}(H_{\boldsymbol{q}}e^{i\boldsymbol{q}\cdot\boldsymbol{r}} + H_{\boldsymbol{q}}^{*}e^{-i\boldsymbol{q}\cdot\boldsymbol{r}}), \tag{6.3.59}$$

where

$$\chi_{\boldsymbol{q}} = \mu_{\mathrm{B}}^{2}\sum_{\boldsymbol{k}}\frac{f_{\boldsymbol{k}} - f_{\boldsymbol{k}+\boldsymbol{q}}}{\varepsilon_{\boldsymbol{k}+\boldsymbol{q}} - \varepsilon_{\boldsymbol{k}}}. \tag{6.3.60}$$

It is usual to define the important function

$$F(\boldsymbol{q}) = \sum_{\boldsymbol{k}}\frac{f_{\boldsymbol{k}} - f_{\boldsymbol{k}+\boldsymbol{q}}}{\varepsilon_{\boldsymbol{k}+\boldsymbol{q}} - \varepsilon_{\boldsymbol{k}}}, \tag{6.3.61}$$

which is in fact dimension-dependent. Substituting the energy spectrum of the free electron gas into (6.3.61), and replacing the summation over $\boldsymbol{k}$ by an integral, one finds the analytic expressions in one and three dimensions

$$F_1(q) = \frac{2m}{\pi\hbar^2 q}\ln\left|\frac{2k_{\mathrm{F}} + q}{2k_{\mathrm{F}} - q}\right|, \tag{6.3.62}$$

$$F_3(q) = \frac{3N}{4\varepsilon_{\mathrm{F}}}\left[1 + \frac{4k_{\mathrm{F}}^2 - q^2}{4k_{\mathrm{F}}q}\ln\left|\frac{2k_{\mathrm{F}} + q}{2k_{\mathrm{F}} - q}\right|\right]. \tag{6.3.63}$$

In the case of two dimensions, the expression is divided into two parts

$$F_2(q) = \begin{cases} m/\pi\hbar^2, & \text{for } q < 2k_\mathrm{F}, \\ (m/\pi\hbar^2)\{1 - [1 - (2k_\mathrm{F}/q)^2]^{1/2}\}, & \text{for } q > 2k_\mathrm{F}. \end{cases} \tag{6.3.64}$$

where the two parts are joined at $q = 2k_\mathrm{F}$.

The function $F(\boldsymbol{q})$ depends only on scalar q in any dimension, so does the susceptibility from (6.3.60). Plots of $\chi(q)$ versus q in one, two, and three dimensions are given in Fig. 6.3.9 in which we see that there are different singularities at $q = 2k_\mathrm{F}$ for $\chi(q)$ in one, two, and three dimensions. These singularities reflect the existence of the Fermi surface, because $2k_\mathrm{F}$ is its diameter. The singularities lead to peculiar behavior in electron gases, especially in lower dimensions.

Bibliography

[1] Kittel, C., *Quantum Theory of Solids*, John Wiley & Sons, New York (1963).

[2] Ziman, J. M., *Principles of the Theory of Solids*, 2nd ed., Cambridge University Press, Cambridge (1972).

[3] Ziman, J. M., *Electrons in Metals*, Taylor & Francis, London (1963).

[4] Omar, M. A., *Elementary Solid State Physics: Principles and Applications*, Addison-Wesley, Reading (1975).

[5] Callaway, J., *Quantum Theory of the Solid State*, 2nd ed., Academic Press, New York (1991).

[6] Chang, L. L., and Giessen, B. C. (eds.), *Synthetic Modulated Structures*, Academic Press, New York (1985).

[7] Abrikosov, A. A., *Fundamentals of the Theory of Metals*, North-Holland, Amsterdam (1988).

[8] Chambers, R. G., *Electrons in Metals and Semiconductors*, Chapman and Hall, London (1990).

[9] Shoenberg, D., *Magnetic Oscillations in Metals*, Cambridge University Press, Cambridge (1984).

[10] Weisbuch, C., and Vinter, B., *Quantum Semiconductor Structures*, Academic Press, New York (1991).

[11] Singh, J., *Physics of Semiconductors and Their Heterostructures*, McGraw, New York (1993).

[12] Ogawa, T., and Kanemitsu, Y. (eds.), *Optical Properties of Low-Dimensional Materials*, World Scientific, Singapore (1995).

[13] Grahn, H. T. (ed.), *Semiconductor Superlattices*, World Scientific, Singapore (1995).

[14] Davis, J. H., *The Physics of Low-Dimensional Semiconductors*, Cambridge University Press, Cambridge (1998).

Chapter 7

Surface and Impurity Effects

The periodicity of perfect lattices has made it easy to study the propagation behavior of three different kinds of waves. But there are always imperfections in crystals, such as surfaces and impurities, that must be taken into account in treating the problem of real crystals. For the case of small imperfections, the band model is still effective, and only some slight modifications need to be introduced into it due to deviation from perfect lattice periodicity. We shall investigate the effects of surfaces and impurities on the band model in this chapter.

§7.1 Electronic Surface States

The surface of a solid obviously causes deviation from perfect periodicity. If the Born–von Karman cyclic boundary condition is abandoned in the direction normal to the surface, the real behavior of electrons will show some features not found in Chap. 5. The existence of the surface surely will introduce some modifications into the electronic structure for an infinite crystal. In general, the real structure of a solid surface is rather complex, involving rearrangements of atomic configuration as well as the segregation of chemical impurities. To simplify the discussion, we will adopt a model of an ideal surface in order to see the main effects that the surface brings.

7.1.1 Metal Surface

The work function and surface energy are two fundamental parameters that characterize a metallic surface. The work function can be understood as the energy difference between an electron located at the vacuum level outside a metal and at the Fermi level in the interior of the metal; in other words, it equals the work needed for an electron to be moved out through the surface of the metal to infinity (in practice 10 nm is enough). The surface energy is the energy needed to produce a unit surface area, and it equals the increase of electronic energy for a metal when all bonds across a planar surface are truncated. Both are involved in the electronic states related to the surface.

Here we use the nearly-free electron (NFE) approximation to study the electronic states for an ideal surface. Let the half-space for $z < 0$ be vacuum with constant potential, and the other half-space for $z > 0$ crystal with the periodic potential, as shown in Fig. 7.1.1. This can be simplified to a one-dimensional problem with a boundary at $z = 0$. The surface thus represents an abrupt transition at $z = 0$ between the vacuum and the periodic lattice. The problem now involves the solution of the Schrödinger equation of a semi-infinite periodic chain

$$\left[-\frac{\hbar^2}{2m}\frac{d^2}{dz^2} + V(z)\right]\psi(z) = E\psi(z), \tag{7.1.1}$$

with the potential

$$V(z) = \begin{cases} V_0, & \text{for } z < 0, \\ V(z + la), & \text{for } z > 0, \end{cases}$$

where a is the lattice constant, and l an integer. The solutions can be obtained for $z < 0$ and $z > 0$, respectively. For vacuum, the solutions of the Schrödinger equation must satisfy the requirement that they decrease with decreasing z, while within the periodic part, wavefunctions are travelling waves along the $\pm z$ directions, and so

$$\psi(z) = \begin{cases} A\exp\left[\frac{1}{\hbar}\sqrt{2m(V_0 - E)}z\right], & \text{for } z < 0, \\ Bu_k e^{ikz} + Cu_{-k}e^{-ikz}, & \text{for } z > 0, \end{cases} \tag{7.1.2}$$

where A, B, and C are constants. At $z = 0$ the wavefunctions and their derivatives must satisfy the continuity conditions.

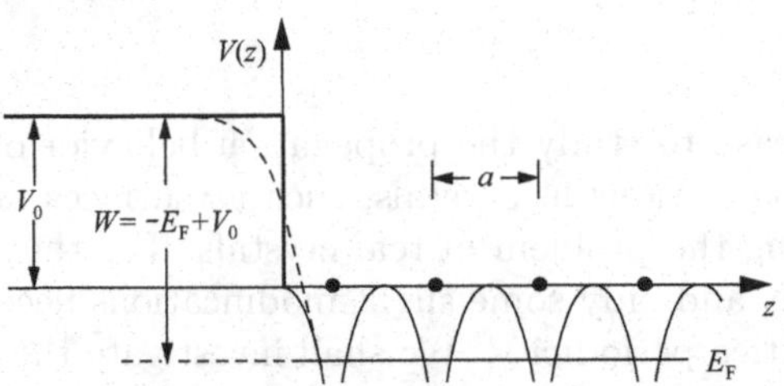

Figure 7.1.1 Schematic potential of an electron in a semi-infinite lattice.

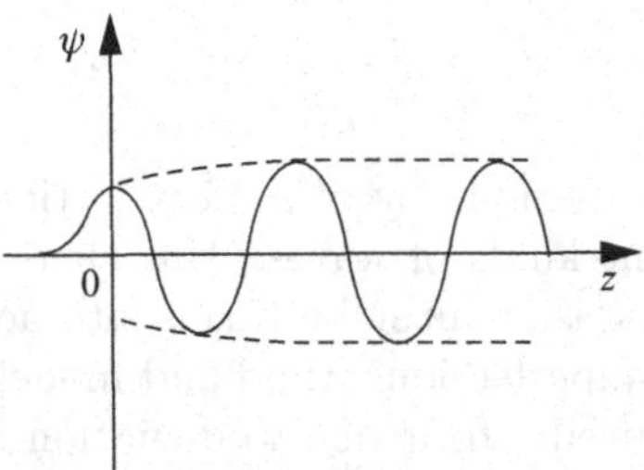

Figure 7.1.2 Extended wavefunction near metallic surface.

At the surfaces of metals, we are interested in the extended state for the energy bands which are not fully filled. The wavefunction obtained for such a state, as shown in Fig. 7.1.2, must be a Bloch wave in the crystal, and attenuate exponentially outside the surface. This shows that the wavefunction for an extended state will not terminate at the surface but will spill over the surface barrier into the vacuum like a tail, corresponding to a localized state near the surface. Thus, the one or two unit cells near the surface in a crystal are positively charged due to the deficiency of electrons that have spilled out, while the layer just outside the surface is negatively charged. Together these form a dipolar double layer across the surface. The existence of this dipolar double layer affects the potential profile across the crystal surface, which is changed from a step (from 0 to V_0) to a relatively smooth curve, like the dashed line in Fig. 7.1.1. The Fermi energy E_F in metals has the physical meaning of chemical potential, so the work function for an electron at a metallic surface is equal to

$$W = V(-\infty) - V(0) - E_F \approx V_0 - E_F.$$

The value of the electronic work function directly affects thermionic emission and field emission, which are important in technology.

To calculate the surface energy and electron density variation near a surface, we can make a further simplification from the nearly-free electron approximation, to the jellium model. In this model the positive charges of ions which give rise to the periodic potential are distributed homogeneously for a crystal, so in the interior of the crystal, the positive and negative charge densities cancel each other, satisfying $\rho^+ + \rho^- = 0$; but near the surface things may be different, such as appearance of a double layer as well as the oscillatory variation of the charge density, the so-called Friedel oscillation near the surface. Then the term for periodic potential in (7.1.1) vanishes. The problem now simplifies to solving the Schrödinger equation in a potential well of width L. Taking $L \to \infty$ in one direction gives the electronic surface state.

We consider first a one-dimensional potential well of width L. When the barriers are infinite, electrons cannot run over the surface. In this confined system the eigenfunctions are

$$\psi_k(z) = \left(\frac{2}{L}\right)^{1/2} \sin kz, \qquad k = \frac{n\pi}{L}, \qquad n = 1, 2, \ldots, n_F. \tag{7.1.3}$$

On the other hand, in an unconfined system, such as a normal crystal satisfying periodic boundary conditions with repeat width L, we have

$$\psi_{k'}(z) = \left(\frac{1}{L}\right)^{1/2} e^{ik'z}, \qquad k' = \frac{2n'\pi}{L}, \qquad n' = 0, \pm1, \pm2, \ldots, \pm n'_F. \tag{7.1.4}$$

Comparing the two cases leads to some interesting conclusions: The energy level $k = 0$ can only exist in the unconfined case and not in the confined case. In the confined case the separation of k values is $\Delta k = \pi/L$, while in the unconfined case it is $\Delta k' = 2\pi/L$, double the former. Also, in the confined case, k is always positive, but in the unconfined case it can be positive or negative, corresponding to waves travelling in opposite directions. Hence, for a definite energy level with $|k|$, there is one pair of electrons with spin up and down in the confined case, while there are two pairs of electrons in the unconfined case (see Fig. 7.1.3).

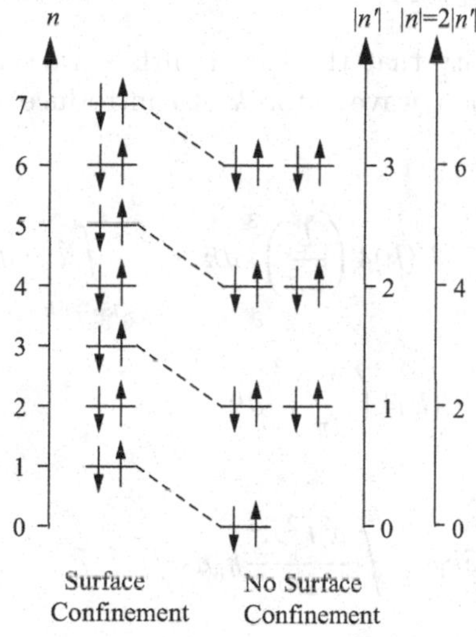

Figure 7.1.3 Filled energy levels of one-dimensional potential wells corresponding to the case with surface confinement and the case without surface confinement.

It is clear from (7.1.3) and (7.1.4) that the energy eigenvalues in the cases with and without confinement are

$$E_n = \frac{\hbar^2}{2m}\left(\frac{n\pi}{L}\right)^2, \qquad n = 1, 2, \ldots, n_F;$$

and

$$E_{n'} = \frac{\hbar^2}{2m}\left(\frac{2n'\pi}{L}\right)^2, \qquad n' = 0, \pm1, \pm2, \ldots, \pm n'_F. \tag{7.1.5}$$

The difference of two adjacent energy levels for the former is smaller than that of the latter. In the unconfined case the number of electrons accommodated in the state $|k|$ is double that of the confined case, so, using the same number of electrons to fill the energy level, the difference between the highest filled levels for both cases, is very small, and the two numbers satisfy $n_F - 2n'_F = 1$ or 0, the average is 1/2, corresponding to $k_F - k'_F = \Delta k_F = \pi/2L$, as can be seen from Fig. 7.1.3.

If we fill these two systems with N electrons, taking N even, there is an energy difference of

$$\Delta E = \sum_{n=1}^{n_F} E_n - \sum_{|n'|=0}^{n'_F} E_{n'}. \tag{7.1.6}$$

This difference arises from surface confinement. Compared to the unconfined system, surface confinement leads one half of the $(1/2)(N/2)$ levels to rise an energy $(\hbar^2/2m)(\pi/L)^2$, and so the total increment is

$$\Delta E = \frac{\hbar^2 k_{\mathrm{F}}^2}{2m}. \tag{7.1.7}$$

If we let $L \to \infty$, with N/L a constant, the difference between odd and even electron that are needed to fill levels will disappear, so the conclusion is independent of whether the number is odd or even. ΔE is equal to the energy to take an electron from $k = 0$ to the Fermi level $k = k_{\mathrm{F}}$. This is the principal origin of the metallic surface energy.

It is more exact to calculate the surface energy by using the jellium model for the surface of a three-dimensional solid. Assuming the surface energy is $\Delta E_{\mathrm{s}}^{\infty}$, which has arisen from cutting the bonds to form two surfaces, then

$$2L^2 \Delta E_{\mathrm{s}}^{\infty} = \int\limits_{k \le k_{\mathrm{F}}+\delta k_{\mathrm{F}},\ k_z \ge \pi/2L} E(\boldsymbol{k}) 2 \left(\frac{L}{2\pi}\right)^2 \frac{L}{\pi} d\boldsymbol{k} - \int\limits_{k \le k_{\mathrm{F}}} E(\boldsymbol{k}) 2 \left(\frac{L}{2\pi}\right)^3 d\boldsymbol{k}, \tag{7.1.8}$$

where $\delta k_{\mathrm{F}} = \pi/4L$ is the quantity that the Fermi surface raised due to the surface in the three-dimensional jellium model. For a wavevector $\boldsymbol{k}$, to introduce its component $k_{\parallel}$ parallel to the truncated plane, we can obtain

$$\begin{aligned}
2L^2 \Delta E_{\mathrm{s}}^{\infty} = &\int\limits_{k \le k_{\mathrm{F}}+\delta k_{\mathrm{F}},\ k_z \ge 0} E(\boldsymbol{k}) 4 \left(\frac{L}{2\pi}\right)^3 d\boldsymbol{k} - \int\limits_{k_{\parallel} \le k_{\mathrm{F}}+\delta k_{\mathrm{F}}} d^2 k_{\parallel} \int_0^{\pi/2L} E(\boldsymbol{k}) 4 \left(\frac{L}{2\pi}\right)^3 dk_z \\
&- \int\limits_{k \le k_{\mathrm{F}},\ k_z \ge 0} E(\boldsymbol{k}) 4 \left(\frac{L}{2\pi}\right)^3 d\boldsymbol{k} \\
= &\, E_{\mathrm{F}} \frac{L^3}{2\pi^3} 2\pi k_{\mathrm{F}}^2 \delta k_{\mathrm{F}} - \int_0^{k_{\mathrm{F}}} \frac{\hbar^2 k_{\parallel}^2 L^3}{2m} k_{\parallel} dk_{\parallel} \\
= &\int_0^{k_{\mathrm{F}}} \left(E_{\mathrm{F}} - \frac{\hbar^2 k_{\parallel}^2}{2m}\right) \frac{L^2}{4\pi^2} 2\pi k_{\parallel} dk_{\parallel}.
\end{aligned} \tag{7.1.9}$$

The last term in (7.1.9) shows that the physical reason for the surface energy is the energy increase for a half of electrons moved from the level $k = 0$ to the Fermi surface. To finish the integral, the result is

$$E_{\mathrm{s}}^{\infty} = \frac{\hbar^2 k_{\mathrm{F}}^4}{32\pi m}. \tag{7.1.10}$$

In the one-dimensional jellium model, for a system with two confined surfaces, it is easy for us to write the electron density distribution

$$\rho(z) = \sum_{n=1}^{N} \psi_n^*(z) \psi_n(z) = \frac{2}{L} \sum_{n=1}^{N} \sin^2 \frac{n\pi z}{L}. \tag{7.1.11}$$

For finite L, we carry out the summation of (7.1.11) associated with (7.1.4), the result is

$$\rho(z) = \frac{N+1/2}{L} - \frac{\sin[2\pi(N+1/2)z/L]}{2L \sin(\pi z/L)}. \tag{7.1.12}$$

In order to examine the effect of surface on the electron density near $z = 0$, we let L tend to infinity, and replace the summation by an integral

$$\rho(z) = \frac{2}{L} \int_0^N \sin^2 \frac{n\pi z}{L} dn. \tag{7.1.13}$$

It can be found that

$$\rho(z) = \rho_0 - \rho_0 \frac{\sin(2k_F z)}{2k_F z}, \tag{7.1.14}$$

where $\rho_0 = N/L$ is the average density, and k_F the Fermi momentum. Figure 7.1.4(a) displays the form of this distribution which is characterized by a density oscillation deep into the interior of the metal with wavelength π/k_F. This Friedel oscillation comes from a localized perturbation due to the wall at $z = 0$ in a free electron gas. The density rises from zero at the surface to its value ρ_0 when $\sin(2k_F z) = 0$, i.e., over a distance $\pi/2k_F$ which is one half of the de Broglie wavelength for an electron at the Fermi surface.

It should be pointed out that the calculated result for a one-dimensional system is somewhat different from the actual situation. One should use the three-dimensional jellium model with truncated surface using a finite barrier instead of an infinite one. Figure 7.1.4(b) shows a more correctly calculated result. The spilling of surface electrons and the density oscillation will decrease the surface energy a little. Certainly, a real calculation must take into account the modification due to the many-body effect from electronic interactions.[a]

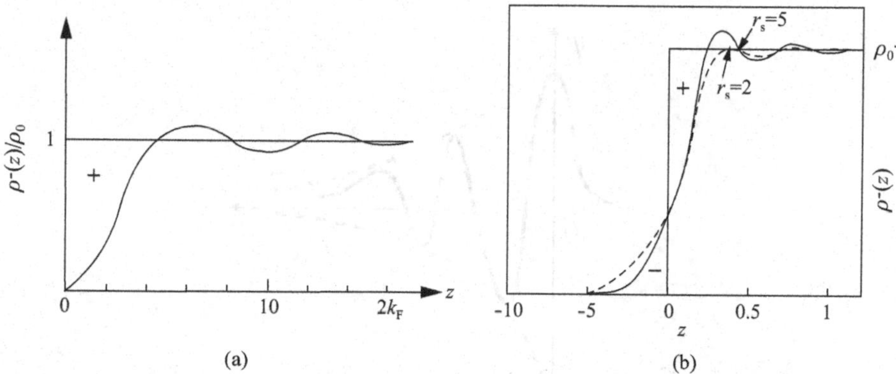

Figure 7.1.4 Electronic density oscillation near metal surface. (a) Infinite surface barrier, (b) finite surface barrier.

7.1.2 Semiconductor Surface States

For semiconductor surface states, we return to the nearly-free electron approximation, and concern ourselves with what happens in energy gaps. The half-infinite periodic chain with weak lattice potential is still used, and it is expected that the band structure will be almost the same as in Fig. 5.2.2. The first Brillouin zone extends from $k = -\pi/a$ to $+\pi/a$ and the $E(k)$ parabola for free electrons is distorted near the boundaries of the Brillouin zone (BZ) where an energy gap occurs. The solutions in the vicinity of $k = \pm\pi/a$ may be more interesting and we can write, for example, for k near $+\pi/a$

$$\psi_k(z) = \alpha e^{ikz} + \beta e^{i(k-2\pi/a)z}, \tag{7.1.15}$$

α and β can be found from

$$\left[\frac{\hbar^2}{2m}k^2 - E(k)\right]\alpha + V_1\beta = 0, \tag{7.1.16}$$

$$V_1^*\alpha + \left[\frac{\hbar^2}{2m}\left(k - \frac{2\pi}{a}\right)^2 - E(k)\right]\beta = 0, \tag{7.1.17}$$

where $V_1 = V_{2\pi/a}$ is the Fourier coefficient of the periodic potential.

[a]Experimental tests of Friedel oscillations are in general indirect, however, in recent years, Friedel oscillations on Be surfaces due to steps and defects have been observed directly by scanning tunnelling microscopy, see P. H. Hofmann *et al.*, *Phys. Rev. Lett.* **79**, 265 (1997); P. T. Sprunger *et al.*, *Science* **275**, 1764 (1997).

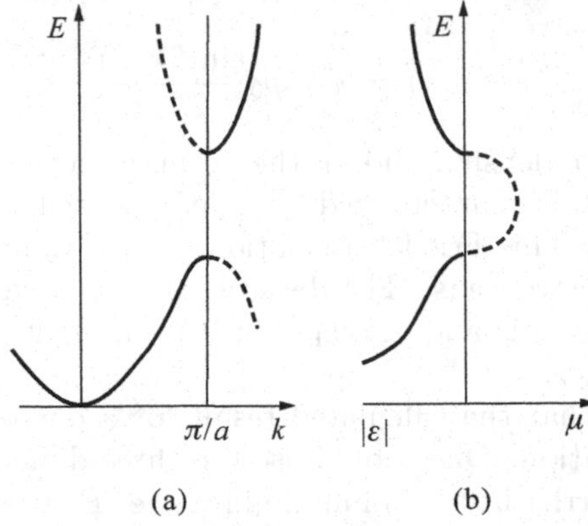

Figure 7.1.5 Dispersion relations when surface exists. (a) A little modified energy spectra for a nearly free electron model. (b) In the energy gap between the two bands, solutions with imaginary k can appear.

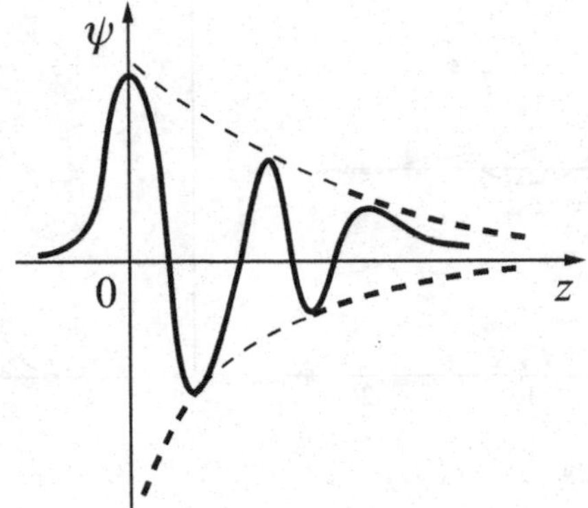

Figure 7.1.6 A localized wavefunction near semiconductor surface.

Defining $k = \pi/a + \varepsilon$ and $\gamma = (\hbar^2\pi/ma|V_1|)\varepsilon$, we find that, for real ε, the energies and wavefunctions are

$$E_\pm = \frac{\hbar^2}{2m}\left(\frac{\pi}{a}+\varepsilon\right)^2 + |V_1|(-\gamma \pm \sqrt{1+\gamma^2}), \tag{7.1.18}$$

and

$$\psi_\pm(z) = B\left[\mathrm{e}^{i\pi z/a} + \frac{|V_1|}{V_1}(-\gamma \pm \sqrt{1+\gamma^2})\mathrm{e}^{-i\pi z/a}\right]\mathrm{e}^{i\varepsilon z}, \text{ for } z > 0, \tag{7.1.19}$$

with B being an another constant. Equation (7.1.18) gives the bands shown in Fig. 7.1.5(a). For each energy, connecting the solutions in the vacuum and the periodic lattice, the two free parameters A in (7.1.9) and B in (7.1.14) are determined by the continuity of the wavefunction and its derivative at $z = 0$. In doing so, one needs to combine linearly the two solutions $\psi_k(z)$ and $\psi_{-k}(z)$. The result is the extended state shown in Fig. 7.1.2. The energy bands of an infinite lattice are only slightly modified.

On the other hand, for $\varepsilon = i\mu$ with real positive μ, solutions, that decrease exponentially into the crystal and are localized to the surface appear as shown in Fig. 7.1.5(b). Assuming $\gamma = i\sin(2\delta) = i(\hbar^2\pi/ma|V_1|)\mu$ we find that the energies are

$$E_\pm = \frac{\hbar^2}{2m}\left[\left(\frac{\pi}{a}\right)^2 - \mu^2\right] \pm |V_1|\left[1 - \left(\frac{\hbar^2\pi\mu}{ma|V_1|}\right)^2\right]^{1/2}, \tag{7.1.20}$$

and the corresponding solutions

$$\psi_\pm(z) = C\left[\mathrm{e}^{i(\pi z/a\pm\delta)} \pm \frac{|V_1|}{V_1}\mathrm{e}^{-i(\pi z/a\pm\delta)}\right]\mathrm{e}^{-\mu z}, \tag{7.1.21}$$

with C another constant. This represents a surface state. The energies in (7.1.20) are always real for $0 \le \mu \le \mu_{\max} = ma|V_1|/\hbar^2\pi$. For $\mu = 0$, (7.1.21) returns to the band edge resonant states (5.2.13).

The above result tells us that there are two kinds of solution of the Schrödinger equation (7.1.1). One with real k corresponds to the usual band solutions with a little modification, and the other with complex k decreases with increasing distance from the surface. The energy values associated with the complex wavevector k, according to (7.1.20), lie in the energy gap between the bands, as shown in Fig. 7.1.6. This solution represents a state in which the electron is localized to a narrow region at the surface: It is just the surface state we are seeking.

In the one-dimensional model a surface state has a discrete energy level, denoted as E_0, for fixed μ in the energy gap. Extending the model to three dimensions, we can regard the results as characterizing the component of $\boldsymbol{k}$ perpendicular to the surface. Since the crystal is still, presumably, periodic in directions parallel to the surface, the Bloch theorem must hold for translations in this plane. So the surface state is localized in one dimension, and may be delocalized in the other two dimensions, such as

$$\psi_k = \psi_0(z)\exp[i(k_x x + k_y y)], \tag{7.1.22}$$

where k_x and k_y are components of the wavevector $\boldsymbol{k}$ measured in the plane of the surface. In the nearly-free electron case, for example, we should expect this band to behave like

$$E(\boldsymbol{k}) = E_0 + \left(\frac{\hbar^2}{2m}\right)(k_x^2 + k_y^2). \tag{7.1.23}$$

Referring to §5.2.3, for transition metals, or covalent semiconductors like Si, Ge, and C, the tight-binding model is more suitable for study of the surface states. We can still assume atoms to form a one-dimensional semi-infinite periodic chain; an electron moves in the potential as shown in Fig. 7.1.1, so satisfying (7.1.1), but the wavefunction now is better constructed by a linear combination of atomic orbitals (LCAO) at various sites. Simple mathematical treatment shows that, when the transfer matrix elements satisfy some conditions, there are also electronic surface states that oscillate and are attenuated exponentially.

§7.2 Electronic Impurity States

The impurities in a solid also cause departures from perfect periodicity. In the case of dilute impurities, we may frame the problem as a single impurity in metals or semiconductors. The basic question is: What potential or interaction should be used to represent the effects of the impurity center on the electronic states and energy band structure?

7.2.1 Shielding Effect of Charged Center

We shall first deal with the simplest case, i.e., a static impurity charge Ze embedded in a metal described as a free electron gas. The original free electron gas has the plane wave solution $\Omega^{-1/2}e^{i\boldsymbol{k}\cdot\boldsymbol{r}}$ in (5.2.1), where Ω is the volume of the crystal. If the impurity charge provides a spherically symmetric scattering potential $U(r)$, then the waves of the conduction electrons are distorted into $\psi_{\boldsymbol{k}}(\boldsymbol{r})$, which is different from a plane wave. The wavevector $\boldsymbol{k}$ labels the unperturbed state before Ze is introduced.

Now the Schrödinger equation is

$$-\frac{\hbar^2}{2m}\nabla^2\psi_{\boldsymbol{k}} + U(\boldsymbol{r})\psi_{\boldsymbol{k}} = E_{\boldsymbol{k}}\psi_{\boldsymbol{k}}, \tag{7.2.1}$$

where $E_{\boldsymbol{k}} = \hbar^2k^2/2m$. To obtain the distorted wavefunction, (7.2.1) is written as

$$\nabla^2\psi_{\boldsymbol{k}} + k^2\psi_{\boldsymbol{k}} = \frac{2m}{\hbar^2}U(\boldsymbol{r})\psi_{\boldsymbol{k}}. \tag{7.2.2}$$

To solve this equation, we introduce the Green's function $G(\boldsymbol{r}\boldsymbol{r}')$ which satisfies

$$\nabla^2 G(\boldsymbol{r}\boldsymbol{r}') + k^2 G(\boldsymbol{r}\boldsymbol{r}') = -4\pi\delta(\boldsymbol{r}-\boldsymbol{r}'), \tag{7.2.3}$$

then the formal solution of (7.2.2) can be written as

$$\psi_{\boldsymbol{k}}(\boldsymbol{r}) = \Omega^{-1/2}\mathrm{e}^{i\boldsymbol{k}\cdot\boldsymbol{r}} - \frac{m}{2\pi\hbar^2}\int d\boldsymbol{r}' G(\boldsymbol{r}\boldsymbol{r}')U(\boldsymbol{r}')\psi_{\boldsymbol{k}}(\boldsymbol{r}'). \tag{7.2.4}$$

It should be noted that the Green's function has been widely used both in classical and quantum mechanics.[b] The physical meaning of the Green's function used here is just the well-known propagator, the wavefunction at $\boldsymbol{r}$ excited by the potential and associated wavefunction at $\boldsymbol{r}'$. In the mathematical treatment, it gives a simplified method of solving Schrödinger equation. $G(\boldsymbol{r}\boldsymbol{r}')$ in (7.2.3) is the so-called preliminary solution, provided (7.2.3) is satisfied. It is easy to verify that (7.2.4) is the solution of (7.2.2). Now the problem is transformed into finding the solution of (7.2.3) first. Actually the solution of (7.2.3) may be found simply by taking the Fourier transform of both sides of the equation. After some algebraic calculations and finishing with an integration, the result is

$$G(\boldsymbol{r}\boldsymbol{r}') = \frac{\mathrm{e}^{ik|\boldsymbol{r}-\boldsymbol{r}'|}}{|\boldsymbol{r}-\boldsymbol{r}'|}. \tag{7.2.5}$$

Since the Green's function is obtained, (7.2.4) is an integral equation related to the wavefunction, and its solution can be found by recursion methods to different orders. In the first-order Born approximation, we have

$$\psi_{\boldsymbol{k}}(\boldsymbol{r}) = \Omega^{-1/2}\left[\mathrm{e}^{i\boldsymbol{k}\cdot\boldsymbol{r}} - \frac{m}{2\pi\hbar^2}\int d\boldsymbol{r}' G(\boldsymbol{r}\boldsymbol{r}')U(\boldsymbol{r}')\mathrm{e}^{i\boldsymbol{k}\cdot\boldsymbol{r}'}\right]. \tag{7.2.6}$$

The electron density $\rho(\boldsymbol{r})$ can be obtained by summing $\psi^*_{\boldsymbol{k}}(\boldsymbol{r})\psi_{\boldsymbol{k}}(\boldsymbol{r})$ over all $\boldsymbol{k}$ up to the Fermi surface

$$\rho(\boldsymbol{r}) = \sum_{|\boldsymbol{k}|<k_{\mathrm{F}}} \psi^*_{\boldsymbol{k}}(\boldsymbol{r})\psi_{\boldsymbol{k}}(\boldsymbol{r}) = \Omega^{-1}\left\{\sum_{|\boldsymbol{k}|<k_{\mathrm{F}}} - \sum_{|\boldsymbol{k}|<k_{\mathrm{F}}} \frac{m}{2\pi\hbar^2}\int d\boldsymbol{r}' U(\boldsymbol{r}') \right.$$

$$\left. \times \left[G(\boldsymbol{r}\boldsymbol{r}')\mathrm{e}^{i\boldsymbol{k}\cdot(\boldsymbol{r}'-\boldsymbol{r})} + G^*(\boldsymbol{r}\boldsymbol{r}')\mathrm{e}^{-i\boldsymbol{k}\cdot(\boldsymbol{r}'-\boldsymbol{r})}\right]\right\}, \tag{7.2.7}$$

in which the 2nd-order terms are ignored. We change the summation over $\boldsymbol{k}$ into an integration by using (7.2.4), and so

$$\rho(\boldsymbol{r}) = \rho_0 - \frac{2m}{(2\pi)^4\hbar^2}\int d\boldsymbol{r}' U(\boldsymbol{r}')\int_{|\boldsymbol{k}|<k_{\mathrm{F}}} d\boldsymbol{k}[G(\boldsymbol{r}\boldsymbol{r}')\mathrm{e}^{i\boldsymbol{k}\cdot(\boldsymbol{r}'-\boldsymbol{r})}$$

$$+ G^*(\boldsymbol{r}\boldsymbol{r}')\mathrm{e}^{-i\boldsymbol{k}\cdot(\boldsymbol{r}'-\boldsymbol{r})}], \tag{7.2.8}$$

where $\rho_0 = N/\Omega$ is the average density. Now we substitute (7.2.5) into (7.2.8) and integrate over the directions of $\boldsymbol{k}$, we have

$$\rho(\boldsymbol{r}) = \rho_0 - \frac{2m}{(2\pi)^4\hbar^2}\int d\boldsymbol{r}' U(\boldsymbol{r}')\int_0^{k_{\mathrm{F}}} dk 4\pi k^2 \left[\frac{\sin k|\boldsymbol{r}-\boldsymbol{r}'|}{k|\boldsymbol{r}-\boldsymbol{r}'|}\cdot\frac{2\cos k|\boldsymbol{r}-\boldsymbol{r}'|}{|\boldsymbol{r}-\boldsymbol{r}'|}\right]. \tag{7.2.9}$$

After integrating over k, we get

$$\rho(\boldsymbol{r}) - \rho_0 = -\frac{mk_{\mathrm{F}}^2}{2\pi^3\hbar^2}\int d\boldsymbol{r}' U(\boldsymbol{r}')\frac{j_1(2k_{\mathrm{F}}|\boldsymbol{r}-\boldsymbol{r}'|)}{|\boldsymbol{r}-\boldsymbol{r}'|^2}, \tag{7.2.10}$$

where $j_1(x) = (\sin x - x\cos x)/x^2$ is the first-order spherical Bessel function.

[b]See, for example, F. W. Byron and R. W. Fuller, *Mathematics of Classical and Quantum Physics*, Addison-Wesley (1969).

From (7.2.10), it is clear that the final form of the density $\rho(\boldsymbol{r})$ depends on the impurity potential chosen. However, because the integrand includes the first-order spherical Bessel function which has the character of an oscillatory decay, we expect the density to be similar. For a simple example, which is however not self-consistent, let $U(\boldsymbol{r}) = U_0\delta(\boldsymbol{r})$, corresponding to an external short-range potential, the integral can be completed

$$\rho(\boldsymbol{r}) - \rho_0 = -\frac{mk_{\rm F}^2 U_0}{2\pi^3\hbar^2} \cdot \frac{j_1(2k_{\rm F}r)}{r^2} \sim \frac{\sin 2k_{\rm F}r - 2k_{\rm F}r\cos 2k_{\rm F}r}{r^4}, \tag{7.2.11}$$

and at large distance

$$\rho(\boldsymbol{r}) - \rho_0 \sim \frac{\cos 2k_{\rm F}r}{r^3}. \tag{7.2.12}$$

This shows the Friedel oscillation of electron density when an impurity atom is contained in the metal.

Conversely, we can derive the Friedel oscillation of the potential. Using the Poisson equation, (7.2.10) can be transformed into

$$\nabla^2 U = \frac{2me^2k_{\rm F}^2}{\pi^2\hbar^2}\int d\boldsymbol{r}' U(\boldsymbol{r}')\frac{j_1(2k_{\rm F}|\boldsymbol{r}-\boldsymbol{r}'|)}{|\boldsymbol{r}-\boldsymbol{r}'|^2}. \tag{7.2.13}$$

This is a self-consistent equation from which the asymptotic behavior of the potential is

$$U(\boldsymbol{r}) \sim \frac{\cos 2k_{\rm F}r}{r^3}, \tag{7.2.14}$$

as shown in Fig. 7.2.1. This oscillation of screened potential can be experimentally confirmed by the Knight shifts in dilute alloys. The electron density still retains the character of (7.2.12) in this self-consistently oscillating screened potential.

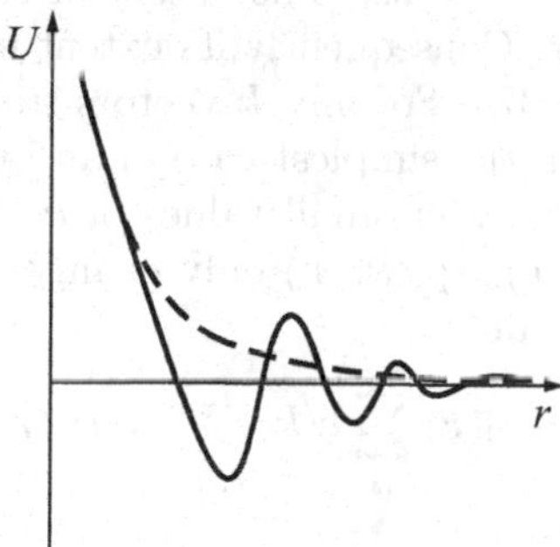

Figure 7.2.1 Screened potentials round a point charge in a sea of free electron gas.

7.2.2 Localized Modes of Electrons

For a single impurity charge in a semiconductor, the nearly-free electron model can be used. We shall only consider the effect of its electronic band structure. The Schrödinger equation (5.1.7) in a periodic potential $V(\boldsymbol{r})$, with Bloch solution $\psi_{\boldsymbol{k}}(\boldsymbol{r})$ and eigenenergy $E(\boldsymbol{k})$, is now transformed into

$$\left[-\frac{\hbar^2}{2m}\nabla^2 + V(\boldsymbol{r}) + U(\boldsymbol{r})\right]\psi = E\psi, \tag{7.2.15}$$

where $U(\mathbf{r})$ is the exotic potential introduced by the impurity charge which can be either negative or positive. Electrons can thus be bound to the impurity or repelled by it.

For further discussion, we consider a covalently-bonded elemental semiconductor of Z-valent atoms in which one of the original atoms was substituted by a $(Z+1)$-valent atom, a simple form of impurity potential $U(\boldsymbol{r})$ is provided by this donor atom. The donor atom introduces one electron and one additional positive charge to the nucleus. This electron is not required for the covalent bonds to the nearest neighbors. $U(\boldsymbol{r})$ in this case is the potential of the additional positive charge, in which the additional electron moves. For large separations of the electron from the positive charge, the crystal lattice screens the Coulomb potential just like a homogeneous medium with dielectric constant ϵ, and thus in (7.2.15) the potential is

$$U(\boldsymbol{r}) = -\frac{e^2}{\epsilon r}. \tag{7.2.16}$$

We have reduced the perturbation potential to the form of an hydrogen atom in a medium of dielectric constant ϵ.

For the motion of a conduction electron in the potential $U(\boldsymbol{r})$, its wavefunction can be expanded in terms of the Bloch functions

$$\psi = \sum_{\boldsymbol{k}} c(\boldsymbol{k})\psi_{\boldsymbol{k}}(\boldsymbol{r}). \tag{7.2.17}$$

By substitution into (7.2.15), we find

$$[E(-i\nabla) + U(\boldsymbol{r})]\psi = E\psi, \tag{7.2.18}$$

where $E(\boldsymbol{k})$ has been formally replaced by $E(-i\nabla)$. Because $E(\boldsymbol{k})$ is a periodic function in $\boldsymbol{k}$-space, it can be expanded as a Fourier series $E(\boldsymbol{k}) = \sum_l E_l \exp(i\boldsymbol{l}\cdot\boldsymbol{k})$, and

$$E(-i\nabla) = \sum_l E_l e^{\boldsymbol{l}\cdot\nabla}, \tag{7.2.19}$$

in which $\exp(\boldsymbol{l}\cdot\nabla)$ is a translational operator, and its role is to make any spatial function $f(\boldsymbol{r})$ equal to $f(\boldsymbol{r}+\boldsymbol{l})$.

In the following, we consider a shallow impurity with relatively weak binding potential for an electron; this condition is fulfilled for most donors. In this case, the orbital of the electron bound to the impurity traverses many lattice cells. The extent of the wave packet in space is thus large compared with the lattice constant. Consequently its extent in $\boldsymbol{k}$-space is small compared with the dimensions of the Brillouin zone (BZ). So only $\boldsymbol{k}$-vectors from a narrow region around the band minimum contribute to (7.2.17). In the simplest case of an isotropic parabolic minimum at $\boldsymbol{k} = 0$, the summation in (7.2.17) only runs over small values of $\boldsymbol{k}$. Since the lattice periodic part $u_{\boldsymbol{k}}(\boldsymbol{r})$ in the Bloch function $\psi_{\boldsymbol{k}}(\boldsymbol{r}) = u_{\boldsymbol{k}}(\boldsymbol{r})\exp(i\boldsymbol{k}\cdot\boldsymbol{r})$ only changes slowly with $\boldsymbol{k}$, we can replace $u_{\boldsymbol{k}}(\boldsymbol{r})$ directly by $u_0(\boldsymbol{r}) = \psi_0(\boldsymbol{r})$, and obtain

$$\psi = u_0(\boldsymbol{r})\sum_{\boldsymbol{k}} c(\boldsymbol{k})e^{i\boldsymbol{k}\cdot\boldsymbol{r}} = \psi_0(\boldsymbol{r})F(\boldsymbol{r}), \tag{7.2.20}$$

where $F(\boldsymbol{r}) = \sum_{\boldsymbol{k}} c(\boldsymbol{k})\exp(i\boldsymbol{k}\cdot\boldsymbol{r})$.

If one puts (7.2.20) into (7.2.18) and notes that the function $E(-i\nabla)$ appears like a function of the translation operator, then

$$E(-i\nabla)\psi_0(\boldsymbol{r})F(\boldsymbol{r}) = \psi_0(\boldsymbol{r})E(-i\nabla)F(\boldsymbol{r}), \tag{7.2.21}$$

it follows that

$$[E(-i\nabla) + U(\boldsymbol{r})]F(\boldsymbol{r}) = EF(\boldsymbol{r}). \tag{7.2.22}$$

(7.2.22) differs from (7.2.18) in that the rapidly changing function ψ has been replaced by the slowly changing envelope function $F(\boldsymbol{r})$. Then the operator $E(-i\nabla)$ can be expanded to the quadratic term

$$E(-i\nabla) = E_{\mathrm{c}} - \frac{\hbar^2}{2m^*}\nabla^2, \tag{7.2.23}$$

where E_{c} is the lower edge of the conduction band, and m^* is the effective mass. This leads to

$$\left(-\frac{\hbar^2}{2m^*}\nabla^2 - \frac{e^2}{\epsilon r}\right)F(\boldsymbol{r}) = (E - E_{\mathrm{c}})F(\boldsymbol{r}), \tag{7.2.24}$$

where we have inserted the explicit form (7.2.16) for $U(\boldsymbol{r})$.

Equation (7.2.24) is the Schrödinger equation for an electron with effective mass m^* in the potential of a positive charge in a medium of dielectric constant ϵ. We know the solution to this equation from the hydrogen atom. The eigenvalues are

$$E_n = E_c - \frac{e^4 m^*}{2\hbar^2\epsilon^2 n^2}, \qquad n = 1, 2, \ldots \tag{7.2.25}$$

which give discrete levels below the conduction band as shown in Fig. 7.2.2. The envelope function for the ground state is

$$F(r) = \frac{1}{(\pi a_0^{*3})^{1/2}} \exp\left(-\frac{r}{a_0^*}\right), \qquad a_0^* = \frac{\hbar^2}{me^2}\frac{m}{m^*}\epsilon, \tag{7.2.26}$$

which is known as a bound state.

In the case discussed so far, of a donor atom in a semiconductor with an isotropic, parabolic conduction band minimum, it should be noted, first of all, that the bound states of the electron form a hydrogen atom-like spectrum, which lies below the lower edge of the conduction band. The Bohr radius for the orbital of the ground state is increased by a factor $\epsilon m/m^*$ relative to that of the free H-atom (0.53 Å). For Si and Ge its value lies between 20 and 50 Å.

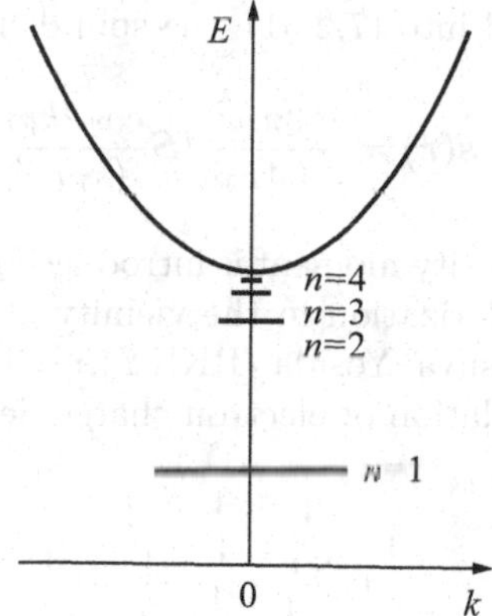

Figure 7.2.2 The E-k diagram for the localized impurity levels lie below the minimum of the conduction band.

7.2.3 Electron Spin Density Oscillation around a Magnetic Impurity

So far we have ignored the spin of electrons. However, if there is a magnetic impurity in a metal, the scattering of conduction electrons by the impurity moment can be spin-dependent. By taking spin into account, in the free electron approximation, the eigenstate of a conduction electron is

$$\psi_{\boldsymbol{k}\sigma}(\boldsymbol{r}) = \Omega^{-1/2} e^{i\boldsymbol{k}\cdot\boldsymbol{r}} |\sigma\rangle, \tag{7.2.27}$$

where spin index $\sigma = \uparrow$, or $\downarrow$ is denoted. Consider a local moment with spin $\boldsymbol{S}$ located at the site $\boldsymbol{R}$, which affects the spins of conduction electrons $\boldsymbol{s}_j$ described by a contact interaction

$$\mathcal{H}' = -J\sum_j \boldsymbol{s}_j \cdot \boldsymbol{S}\delta(\boldsymbol{r} - \boldsymbol{R}). \tag{7.2.28}$$

Then the spin of each conduction electron experiences an effective field coming from the impurity moment

$$\boldsymbol{H}_{\text{eff}}(\boldsymbol{r}) = -\frac{J}{g_L \mu_B}\boldsymbol{S}\delta(\boldsymbol{r} - \boldsymbol{R}), \tag{7.2.29}$$

where g_L and μ_B are the Landé factor and Bohr magneton, respectively.

The response of an electron gas to such a field is determined by its susceptibility $\chi(\boldsymbol{q})$. Since the Fourier transform of this field is

$$\boldsymbol{H}_{\text{eff}}(\boldsymbol{q}) = -\frac{J}{g_{\text{L}}\mu_{\text{B}}}\boldsymbol{S}, \tag{7.2.30}$$

we get the spin density at $\boldsymbol{r}$

$$\boldsymbol{s}(\boldsymbol{r}) = \frac{J}{g_{\text{L}}^2\mu_{\text{B}}^2\Omega}\sum_{\boldsymbol{q}}\chi(\boldsymbol{q})\text{e}^{i\boldsymbol{q}\cdot\boldsymbol{r}}\boldsymbol{S}, \tag{7.2.31}$$

by assuming $\boldsymbol{R} = 0$.

We have already derived the spin susceptibility of a free electron gas in §6.3.4, the result is

$$\chi(\boldsymbol{q}) = \frac{3g_{\text{L}}^2\mu_{\text{B}}^2}{16E_{\text{F}}}\frac{N}{\Omega}\left[1 + \frac{k_{\text{F}}}{q}\left(1 - \frac{q^2}{4k_{\text{F}}^2}\right)\ln\left|\frac{2k_{\text{F}} + q}{2k_{\text{F}} - q}\right|\right]. \tag{7.2.32}$$

The sum over $\boldsymbol{q}$ in (7.2.31) is evaluated by converting it into an integral, thus

$$\frac{1}{\Omega}\sum_{\boldsymbol{q}}\chi(\boldsymbol{q})\text{e}^{i\boldsymbol{q}\cdot\boldsymbol{r}} = \frac{3g_{\text{L}}^2\mu_{\text{B}}^2k_{\text{F}}}{128\pi E_{\text{F}}}\frac{N}{\Omega}\frac{\sin 2k_{\text{F}}r - 2k_{\text{F}}\cos 2k_{\text{F}}r}{(k_{\text{F}}r)^4}. \tag{7.2.33}$$

When this expression is substituted into (7.2.31), the spin density for $k_{\text{F}}r \gg 1$ is

$$\boldsymbol{s}(\boldsymbol{r}) = -\frac{3\pi n^2}{64E_{\text{F}}}J\boldsymbol{S}\frac{\cos 2k_{\text{F}}r}{(k_{\text{F}}r)^3}. \tag{7.2.34}$$

It is clear that when a localized impurity moment is introduced in a metal, the spins of the conduction electrons develop an oscillating polarization in the vicinity of the moment, as shown in Fig. 7.2.3. This is the Rudermann–Kittel–Kasuya–Yosida (RKKY) oscillation for electron spin density. It is quite analogous to the Friedel oscillation of electron charge density.

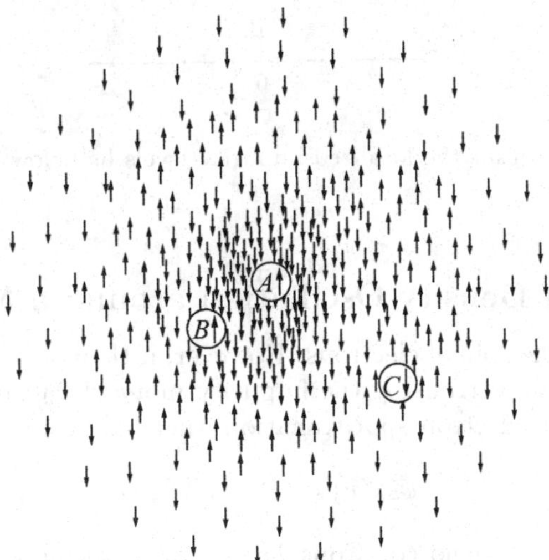

Figure 7.2.3 Electron spin density oscillation arising from an impurity moment at A.

§7.3 Vibrations Related to Surface and Impurity

As in the case of electrons in real crystals, when there is a surface or an impurity in an otherwise perfect periodic structure, the states related to the lattice waves or elastic waves will be slightly modified. In addition, localized modes appear near these imperfections.

7.3.1 Surface Vibrations

We consider a semi-infinite crystal in the $z > 0$ half space. There is a surface at $z = 0$ separating it from the vacuum of the $z < 0$ half space. If there is two-dimensional periodicity in the x-y plane, the Bloch theorem can still be applied. The problem of lattice vibrations may be formally reduced to the determination of the modes of a semi-infinite linear chain in z-direction. From the equations of motion, we can easily find two types of solutions: bulk modes, which can propagate to $z \to \infty$; and surface modes, which vanish rapidly as z increases.

Figure 7.3.1 A semi-infinite atomic chain used to surface vibrations.

Figure 7.3.1 shows a semi-infinite linear chain consisting of identical atoms of mass M with spacing a, the nearest-neighbor coupling constant is taken as β, except that $\beta_0 \neq \beta$ between the first two atoms. This deviation comes from effects of the surface. The equations of motion are

$$M\ddot{u}_0 = -\beta_0(u_0 - u_1), \tag{7.3.1}$$

$$M\ddot{u}_1 = -\beta_0(u_1 - u_0) - \beta(u_1 - u_2), \tag{7.3.2}$$

$$M\ddot{u}_l = -\beta(u_l - u_{l-1}) - \beta(u_l - u_{l+1}), \text{ for } l \geq 2. \tag{7.3.3}$$

Since there are two boundary conditions (7.3.1) and (7.3.2), then the solutions can be written as

$$u_0 = U_0 \mathrm{e}^{-i\omega t}, \tag{7.3.4}$$

and

$$u_l = U \mathrm{e}^{ikla - i\omega t}, \text{ for } l \geq 1, \tag{7.3.5}$$

where U_0 and U are two amplitude variables.

Substituting (7.3.5) into (7.3.3), we find the bulk modes for the infinite periodic chain satisfy the following dispersion relation

$$\omega_{\mathrm{b}} = \left(\frac{4\beta}{M}\right)^{1/2} \left|\sin\frac{ka}{2}\right|, \tag{7.3.6}$$

where ka is real. However, ka may be complex due to the existence of a surface, which represents a surface mode. To search for surface modes, we substitute (7.3.4), (7.3.5) into (7.3.1), (7.3.2), and keep (7.3.6) in mind, obtaining

$$[\beta_0 - 2\beta(1 - \cos ka)]U_0 - \beta_0 \mathrm{e}^{ika} U = 0,$$

$$-\beta_0 \mathrm{e}^{-ika} U_0 + [\beta_0 - \beta(1 - \mathrm{e}^{-ika})]U = 0.$$

These two equations have non-trivial solutions when the coefficient determinant is zero, thus

$$[\beta_0 - 2\beta(1 - \cos ka)][\beta_0 - \beta(1 - \mathrm{e}^{-ika})] - \beta_0^2 = 0, \tag{7.3.7}$$

from it one solution is $\exp(ika) = 1$, which corresponds to a rigid translation of the chain and has no physical interest. The other two solutions are

$$\mathrm{e}^{ika} = 1 \pm \sqrt{1 + \frac{1}{\varepsilon}}, \tag{7.3.8}$$

with

$$\varepsilon = \frac{\beta_0 - \beta}{\beta}. \tag{7.3.9}$$

Let us discuss the solutions in (7.3.8). Since β_0 and β are positive, we have $\varepsilon > -1$. Thus there are two possibilities: (1) $\beta_0 < \beta$, i.e., $-1 < \varepsilon < 0$, $(1 + 1/\varepsilon)$ is negative, and

$$e^{ika} = 1 \pm i\sqrt{-1 - \frac{1}{\varepsilon}},$$

which are not exponentially decaying solutions; (2) when $\beta_0 > \beta$, i.e., $\varepsilon > 0$, we have to consider the two solutions of (7.3.8) separately. It is easy to see that the solution with the plus sign is larger than 1 and thus corresponds to a negative imaginary k, which is contrary to the assumption, as can be seen from (7.3.5). The solution with minus sign corresponds to $ka = \pi + i\mu$, with

$$\mu = \ln(\varepsilon + \sqrt{\varepsilon^2 + \varepsilon}).$$

In this expression, if μ is positive, then

$$\varepsilon + \sqrt{\varepsilon^2 + \varepsilon} > 1,$$

or by using (7.3.9) it gives

$$\frac{\beta_0}{\beta} > \frac{4}{3}, \tag{7.3.10}$$

that is to say, when $\beta_0 > 4\beta/3$, we have a localized mode with a frequency

$$\omega_s = \left(\frac{2\beta}{M}\right)^{1/2} (1 + \cosh \mu) \tag{7.3.11}$$

above the bulk mode. The displacement of the lth atom is

$$u_l = U(-1)^l e^{-l\mu}, \text{ for } l \geq 1, \tag{7.3.12}$$

which is an oscillatory damped solution in the z direction, very similar to (7.1.21). By the way, U_0 can be expressed in terms of U from (7.3.1) and (7.3.11); however, if the condition (7.1.10) is not satisfied, there is no surface mode because μ is not a positive real number.

Alternatively, if the mass at site 0, M_0, also deviates from M (for sites $l \geq 1$), the condition for the existence of surface mode will be

$$\frac{\beta_0}{\beta} > \frac{4M_0}{2M_0 + M}. \tag{7.3.13}$$

7.3.2 Impurity Vibration Modes

As discussed in surface vibrations, there may exist two kinds of impurities to influence lattice vibrations: One introduces a mass difference, with the masses of a few atoms lighter or heavier than the atoms of the matrix; the other introduces a force constant defect, with the force constants changed between some atoms. In reality, these two kinds of defects may often be connected, but their detailed form is difficult to find. However, we only need to consider the simplest case to give an illustration. The most important results may be expected: slight influences of defects on the states in the branches of the phonon spectrum, and appearance of localized states between the acoustic and optical branches and above the optical branches.

For a perfect crystal with a single atom per primitive cell, the equations of motion (5.3.1) for the vibrations reduces to

$$\omega^2 \boldsymbol{u}_l - \frac{1}{M} \sum_{l'} \Phi_{ll'} \cdot \boldsymbol{u}_{l'} = 0 \tag{7.3.14}$$

in terms of the displacements $\boldsymbol{u}_l$ and the force constant $\Phi_{ll'}$. Correspondingly, (7.3.2) is transformed into

$$\boldsymbol{u}_l = \boldsymbol{U}_k e^{i\boldsymbol{k}\cdot\boldsymbol{l}}, \tag{7.3.15}$$

where $U_{\boldsymbol{k}}$ is normalized. Because l runs from 1 to N, (7.3.14) represents a set of N vector equations. We can define a $3N \times 3N$ matrix

$$\mathcal{D}(\omega^2) = \omega^2 I - (1/M)\Phi, \tag{7.3.16}$$

where Φ is a $3N \times 3N$ force constant matrix (not potential function as in Sec. 7.3.1). Then (7.3.14) is rewritten as

$$\mathcal{D}(\omega^2)u = 0, \tag{7.3.17}$$

where u is a $1 \times 3N$ column matrix standing for the set of displacements $\boldsymbol{u}_l$.

For simplicity, we will concentrate on the mass defect of a single atom, by supposing that the mass at the site $\boldsymbol{l} = 0$ is changed to M_0, and we will ignore the possible variation of force constant. Defining the mass difference $\delta M = M_0 - M$, the dynamic equations (7.3.14) becomes

$$\omega^2 \boldsymbol{u}_l - \frac{1}{M}\sum_{l'} \Phi_{ll'} \cdot \boldsymbol{u}_{l'} + \frac{\delta M}{M}\omega^2 \boldsymbol{u}_0 \delta_{l0} = 0. \tag{7.3.18}$$

This equation can be written as

$$\mathcal{D}(\omega^2)u + \delta\mathcal{D}(\omega^2)u = 0. \tag{7.3.19}$$

The mass defect leads to a perturbation of the dynamic matrix of the originally perfect crystal.

We now use the classical Green's function method and rewrite (7.3.18) in the algebraically equivalent form

$$(1 + \mathcal{G}\delta\mathcal{D})u = 0, \tag{7.3.20}$$

where $\mathcal{G}$, the Green function, is the matrix inverse of the dynamic matrix $\mathcal{D}$ satisfying

$$\mathcal{G}(\omega^2)\mathcal{D}(\omega^2) = 1. \tag{7.3.21}$$

After (7.3.17) is solved, the eigenfrequencies $\omega_{\boldsymbol{k}}$ and eigenvectors $u_{\boldsymbol{k}}$ are known, because they satisfy

$$\omega_{\boldsymbol{k}}^2 u_{\boldsymbol{k}} - \frac{1}{M}\Phi u_{\boldsymbol{k}} = 0. \tag{7.3.22}$$

We can now write an expression

$$\mathcal{G}^{-1} u_{\boldsymbol{k}} = \mathcal{D} u_{\boldsymbol{k}} = (\omega^2 - \omega_{\boldsymbol{k}}^2) u_{\boldsymbol{k}}, \tag{7.3.23}$$

or

$$\mathcal{G} u_{\boldsymbol{k}} = \frac{1}{\omega^2 - \omega_{\boldsymbol{k}}^2} u_{\boldsymbol{k}}. \tag{7.3.24}$$

Multiplying by $u_{\boldsymbol{k}}^*$ on both sides, and then summing over the wavevectors $\boldsymbol{k}$ to include all the vibrational modes, we have

$$\sum_{\boldsymbol{k}} \mathcal{G} u_{\boldsymbol{k}} u_{\boldsymbol{k}}^* = \sum_{\boldsymbol{k}} \frac{u_{\boldsymbol{k}} u_{\boldsymbol{k}}^*}{\omega^2 - \omega_{\boldsymbol{k}}^2}. \tag{7.3.25}$$

Note the orthogonality of $u_{\boldsymbol{k}}$. It is found that the matrix $\mathcal{G}(\omega^2)$ can be expressed in the reduced form

$$\mathcal{G}_{ll'}(\omega^2) = \frac{1}{N}\sum_{\boldsymbol{k}} \frac{\boldsymbol{U}_{\boldsymbol{k}} \boldsymbol{U}_{\boldsymbol{k}}^*}{\omega^2 - \omega_{\boldsymbol{k}}^2} \mathrm{e}^{i\boldsymbol{k}\cdot(\boldsymbol{l}-\boldsymbol{l}')}. \tag{7.3.26}$$

Because we assume that the mass defect at the site $\boldsymbol{l} = 0$ provides only a highly localized perturbation, substitution of (7.3.26) into (7.3.20) gives an equation involving only $\boldsymbol{u}_0$, the vector displacement on this site. We only need to consider $\mathcal{G}_{00}(\omega^2)$, i.e.,

$$\frac{\delta M}{NM}\sum_{\boldsymbol{k}} \boldsymbol{U}_{\boldsymbol{k}} \boldsymbol{U}_{\boldsymbol{k}}^* \frac{\omega^2}{\omega_{\boldsymbol{k}}^2 - \omega^2} = I. \tag{7.3.27}$$

This is in the form of a 3×3 matrix expression. We must now deal with the vector notation we have been using: The $\boldsymbol{U}_{\boldsymbol{k}}$ is a unit vector in the direction of polarization for the mode $\boldsymbol{k}$, $U_{\boldsymbol{k}\alpha}$ is the component of U_k in the αth direction. $U_k U_k^*$ is a matrix with elements in Cartesian coordinates of $U_{k\alpha}U_{k\beta}^*$. Written in components, (7.3.27) becomes

$$\frac{\delta M}{NM}\sum_k U_{k\alpha}U_{k\beta}^*\frac{\omega^2}{\omega_k^2-\omega^2}=\delta_{\alpha\beta}. \qquad (7.3.28)$$

Assuming that the frequency and polarization of each of the normal modes of the perfect lattice are known, all frequencies of the normal modes of the crystal with the defect can be found from (7.3.28).

If there is a defect with cubic symmetry in a cubic crystal, for all ω we can write a simplified relation $U_{k\alpha}U_{k\beta}^* = (1/3)\delta_{\alpha\beta}$, and (7.3.28) is reduced to

$$\frac{\delta M}{3NM}\sum_k \frac{\omega^2}{\omega_k^2-\omega^2}=1. \qquad (7.3.29)$$

The normal mode frequencies are the roots of this equation. They can be found graphically, as shown in Fig. 7.3.2. We look for points where the function

$$f(\omega^2)=\frac{1}{3N}\sum_k \frac{\omega^2}{\omega^2-\omega_k^2} \qquad (7.3.30)$$

intersects the horizontal line at $(-M/\delta M)$. If δM is positive, i.e. a heavy impurity, each root for (7.3.29) must lie below a pole ω_k^2 of $f(\omega^2)$, the normal modes of the perturbed system are interleaved in frequency between those of the perfect crystal. Since the values of ω_k form a dense band, whose spacing tends to zero like $1/N$, all of the new solutions are indistinguishable from the old. On the other hand, if δM is negative for light impurity, each root for (7.3.29) must lie above a pole ω_k^2 of $f(\omega^2)$, the normal modes of the perturbed system are also interleaved in frequency between those of the perfect crystal and form a dense band. However, we notice in Fig. 7.3.2 that the highest root is not constrained, but can move away from the top of the band, $\omega_{\max}$, by a finite amount. The frequency of the localized mode may be obtained from (7.3.29). The square amplitudes of the localized mode can also be obtained. They decay exponentially with distance. Localized lattice vibrations are often infrared active and thus can be detected in the absorption spectra of crystal.

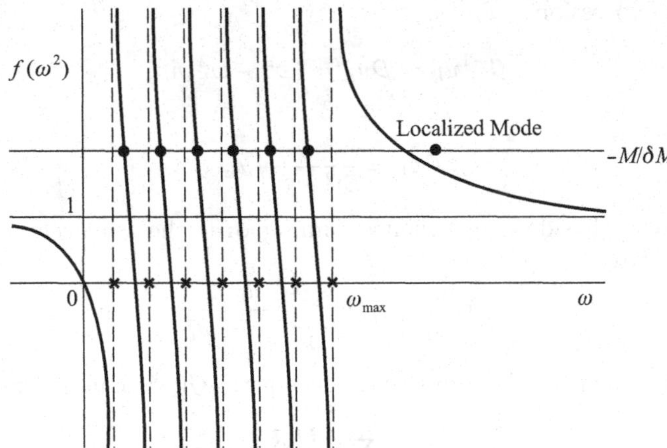

Figure 7.3.2 Graphical solution of lattice modes with a single mass defect.

§7.4 Defect Modes in Photonic Crystals

In Chap. 5 we introduced artificial periodic structures known as photonic crystals. These photonic crystals have photonic bandgaps, in which electromagnetic waves are forbidden to propagate along certain directions in space. As the result of the unified picture of wave propagation, we can also investigate surface and impurity modes in photonic crystals.

7.4.1 Electromagnetic Surface Modes in Layered Periodic Structures

In §5.4.1, we obtained the dispersion relation and band structure of electromagnetic waves in layered periodic dielectric structures. To investigate the surface modes in this kind of system, we consider a semi-infinite periodic dielectric medium consisting of alternating layers of different indices of refraction for the $z \geq 0$ half space, the parameters of structure are defined as in Fig. 7.4.1, and the other half space $z < 0$ is homogeneous medium with index of refraction n_0. There will be evanescent Bloch surface waves guided by the boundary of a semi-infinite periodic dielectric layered medium. These electromagnetic waves are propagating modes confined to the immediate vicinity of the interface between the two semi-infinite systems. One surface wave will create an eigenfrequency in the forbidden band.

For definiteness, we consider that the confined waves are propagating in the positive y direction and are transverse electric (TE) modes, where the electric field is polarized in the x direction. The electric field distribution for TE modes obeys the equation

$$\frac{d^2\mathcal{E}(y,z)}{dy^2} + \frac{d^2\mathcal{E}(y,z)}{dz^2} + \frac{\omega^2}{c^2}\epsilon(z)\mathcal{E}(y,z) = 0. \tag{7.4.1}$$

We write the solution as

$$\mathcal{E}(y,z) = \mathcal{E}(z)\mathrm{e}^{-ik_y y},$$

and substitute it into (7.4.1), then

$$\frac{d^2\mathcal{E}(z)}{dz^2} + \left(\frac{\omega^2}{c^2}\epsilon(z) - k_y^2\right)\mathcal{E}(z) = 0. \tag{7.4.2}$$

The solution of this equation can be divided into two parts

$$\mathcal{E}(z) = \begin{cases} C\mathrm{e}^{k_0 z}, & z \leq 0, \\ A_1\mathrm{e}^{iq_1 z} + B_1\mathrm{e}^{-iq_1 z}, & 0 \leq z < d_1, \end{cases} \tag{7.4.3}$$

where C, A_1, B_1 are constants, and $q_1 = n_1\omega/c$. The wavevector k_0 is given by

$$k_0 = \left[k_y^2 - \left(\frac{n_0\omega}{c}\right)^2\right]^{1/2}. \tag{7.4.4}$$

Actually, the part of the solution for $z > 0$ can be approximated as a Bloch wave $\mathcal{E}(z)\exp(ikz)$, as investigated in §5.4.1.

For a localized mode to exist, the wavevector k must be complex, so that the field decays as z deviates from $z = 0$. By using the continuity conditions of $\mathcal{E}(z)$ and $d\mathcal{E}(z)/dz$ at the interface $z = 0$,

$$\begin{aligned} C &= A_1 + B_1, \\ k_0 C &= iq_1(A_1 - B_1). \end{aligned} \tag{7.4.5}$$

Eliminating C from (7.4.5) and replacing the ratio A_1/B_1 by the matrix elements in (5.4.6), we obtain the mode condition for the surface waves

$$k_0 = iq_1\frac{T_{11} + T_{12} - \mathrm{e}^{ikd}}{T_{11} - T_{12} - \mathrm{e}^{ikd}}. \tag{7.4.6}$$

Combined with (7.4.4), this is a complicated transcendental equation to determine the dispersion relation $\omega(k_y, k)$. There are a lot of propagating wave solutions for the $z < 0$ or $z > 0$ half-spaces. To find a surface mode, we should first take $k_y > n_0\omega/c$ to assure that k_0 is real and positive. Because of the surface, k may become a complex number with a positive imaginary part, so the field amplitude decays exponentially as z increases. Actually, both sides of (7.4.6) are functions of k_y for a given ω. Values of k_y satisfying (7.4.6) for which k is complex, $k = \eta + i\mu$, correspond to surface

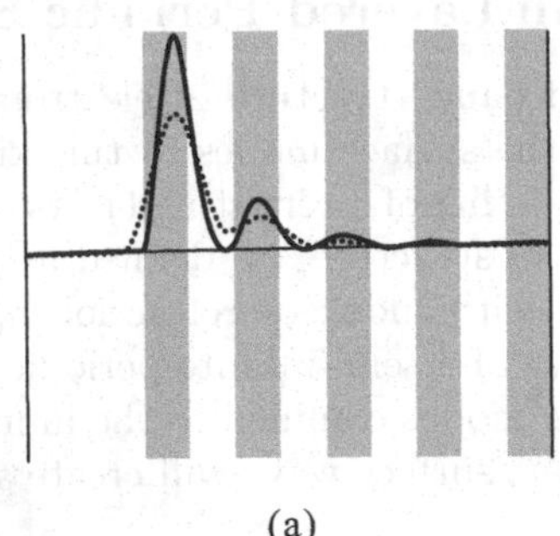

(a)

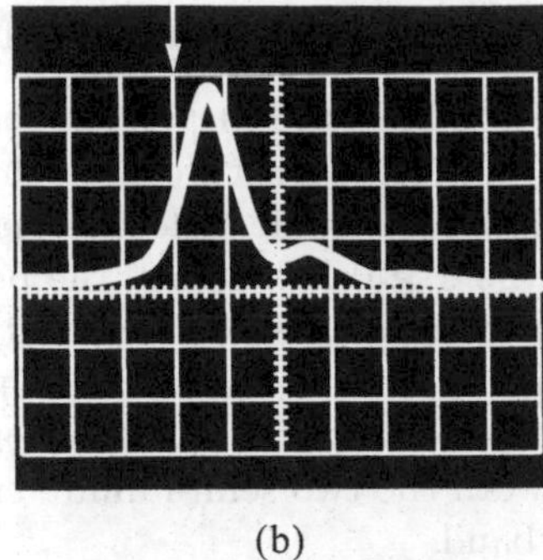

(b)

Figure 7.4.1 Transverse field distribution for typical surface modes guided by the surface of a semi-infinite periodic stratified media. (a) Theoretical result, (b) experimental measurement.

waves. The calculated transverse field distributions $E(z)$ of some typical surface waves are shown in Fig. 7.4.1.

It is evident that the energy is more or less concentrated in the first few periods of the semi-infinite periodic medium. It can easily be shown that the ratio of the energy in the first period to the energy in the whole semi-infinite periodic structure is $1 - \exp(-2\mu d)$. The electromagnetic surface wave can be observed experimentally by measuring its intensity distribution. The result is shown in Fig. 7.4.1(b).

Generally speaking, the fundamental surface wave has the highest μ and hence the highest degree of localization. The fundamental surface wave may happen to be in the zeroth or the first forbidden gap, depending on the magnitude of the index of refraction n_0.

7.4.2 Point Defect

A dielectric structure with defects can be described by the position-dependent dielectric function $\epsilon(\boldsymbol{r})$ as

$$\epsilon(\boldsymbol{r}) = \epsilon_0(\boldsymbol{r}) + \delta\epsilon(\boldsymbol{r}), \tag{7.4.7}$$

where $\epsilon_0(\boldsymbol{r})$ is a periodic function in real space, while $\delta\epsilon(\boldsymbol{r})$ represents the deviation at each site from $\epsilon_0(\boldsymbol{r})$. For a real defect, the deviation is actually concentrated in a certain region, for example, a unit cell, corresponding to a point defect; or a tube constructed with a sequence of unit cells, corresponding to a line defect. If defects are known, then the dielectric function (7.1.1) is determined and we can substitute it into the following equation

$$\nabla \times \left[\frac{1}{\epsilon(\boldsymbol{r})}\nabla \times \boldsymbol{H}(\boldsymbol{r})\right] = \frac{\omega^2}{c^2}\boldsymbol{H}(\boldsymbol{r}), \tag{7.4.8}$$

to obtain the eigenvalues and the localized eigenstates. The general method of solving the equation is to perform numerical calculations for supercells.

We take a two-dimensional photonic crystal with a point defect as an example. Consider a periodic structure composed of circular rods with same radius R and dielectric constant ϵ. Let the lattice constant be a and $R = 0.2a$. We choose a 7×7 supercell. The simplest point defect is obtained by inflating or deflating the radius of a dielectric rod in the middle of the structure. Certainly the real case for defects can be more sophisticated, such as filling materials with different ϵ, and involve more than one unit cell. We are concerned here about only the simplest case. Let the magnitude of R decrease from $0.2a$. In the beginning, the disturbance is not large enough to induce even one localized state. When the radius is decreased to $R = 0.15a$, there is a localized mode that appears not far from the top of the valence band. Afterwards it sweeps over the band gap until the rod disappears at $R = 0$, $\omega = 0.38c/a$. Conversely, we can let the radius increase gradually. When R reaches $0.25a$, there appears a pair of doubly degenerated localized modes, dipolar modes with nodes on the midplane, near the bottom of the conduction band. Continuously increasing

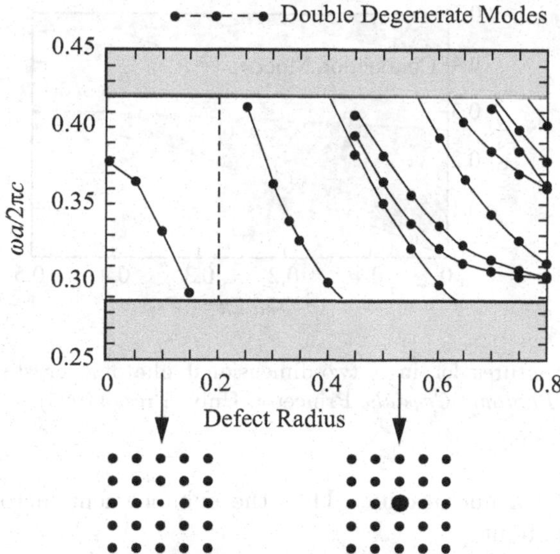

Figure 7.4.2 The relation of frequency and stick radius for local state in two-dimensional photonic crystals. From P. R. Villeneuve *et al.*, *Phys. Rev. B* **54**, 7837 (1996).

the radius causes the localized states with different symmetries sequentially to appear, and their eigenfrequencies sweep over the gap, as shown in Fig. 7.4.2. These localized states due to point defects correspond to the acceptor and donor states in doped semiconductors.

If there is an atom placed in an otherwise perfect photonic crystal, and the atomic transition frequency is just in the gap of the photonic crystal, then its spontaneous transition for radiation will be inhibited. But if this atom is placed in the point defect of a photonic crystal, the situation is quite different. When the atomic transition frequency matches the energy level of local modes of the point defect, the probability of the atomic spontaneous emission of radiation will be enhanced. Therefore, a void ($R = 0$) in a photonic crystal is like an optical resonant cavity, called a microcavity, surrounded by perfectly reflecting walls. The resonant frequency of this microcavity corresponds to the local mode of the point defect.

Usually the quality factor Q is used to characterize a resonant system, i.e.,

$$Q = \frac{\omega_0 E}{dE/dt} \simeq \frac{\omega_0}{\Delta\omega}, \tag{7.4.9}$$

where E is the energy stored by resonant cavity, ω_0 is the eigenfrequency of resonant cavity, and $\Delta\omega$ is the half width of the Lorentz profile of this frequency. The physical meaning of Q is the number of oscillating cycles in a resonant cavity after which its energy declines to a fraction $e^{-2\pi}$ ($\sim 0.2\%$) of its initial value. It is obvious that Q is closely related to the size of a photonic crystal. The more the number of unit cells, the larger the Q value. For example, when the number of unit cells is 9×9 for a two-dimensional rod periodic structure, the photonic crystal composed of parallel rods has $Q \simeq 10^4$.

Assume that the atom is coupled with a light field. Then the Einstein coefficient A_f, characterizing the possibility of the atomic spontaneous radiation in free space, is proportional to the photonic density of states (DOS) in a unit volume, that is

$$A_f \simeq \frac{1}{\omega_0 \lambda^3}, \tag{7.4.10}$$

while the corresponding coefficient in a microcavity is

$$A_c = \frac{1}{\Delta\omega\Omega}, \tag{7.4.11}$$

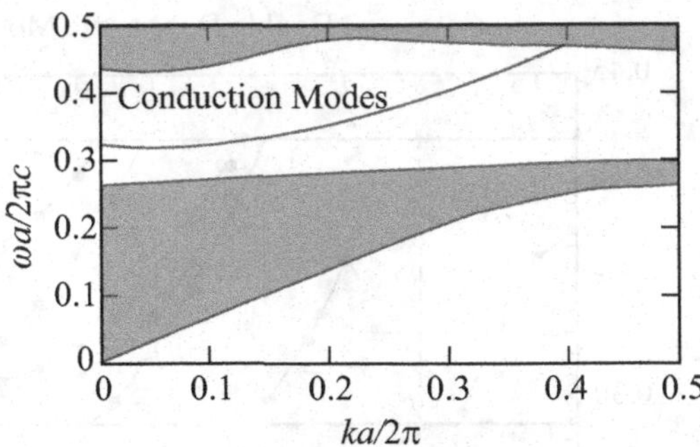

Figure 7.4.3 Energy structures for in a two-dimensional photonic crystal with a line defect. From J. D. Joannopoulos *et al.*, *Photonic Crystals*, Princeton Univ. Press (1995).

where Ω is the volume of a microcavity. Thus the enhancement factor in a microcavity for the spontaneous radiation is about

$$\frac{A_c}{A_f} \simeq \frac{\omega_0}{\Delta\omega}\frac{\lambda^3}{\Omega} = \frac{Q}{(\Omega/\lambda^3)}. \tag{7.4.12}$$

Since the size of a microcavity is $\Omega \sim \lambda^3$, the enhancement factor of a microcavity is almost equal to its Q value. A larger probability of spontaneous radiation is beneficial for the fabrication of highly efficient luminescent diodes and lasers. It is expected that there will be a large potential in application of point defects in photonic crystals as microcavities for lasers.

7.4.3 Line Defects

A point defect in a photonic crystal can be used to restrict electromagnetic waves in a local region; and a line defect can be used as a waveguide for electromagnetic waves; that is, it can guide electromagnetic waves from one place to other.

To illustrate this problem, we can still use the two-dimensional square array of dielectric circular rods, as in the case of point defects. The corresponding defect modes form a band of conduction waves, as shown in Fig. 7.4.3, just like the impurity band in a semiconductor. This band of conduction waves allows electromagnetic waves propagate freely along a narrow channel of a waveguide. Because it is impermissible for electromagnetic waves to penetrate the wall of a waveguide, even if the waveguide bends 90°, there is nearly no electromagnetic wave leaking out. Theoretical calculations have verified this conclusion.

In traditional waveguide techniques, at the microwave frequency range, metallic walls and coaxial cables are used to guide electromagnetic waves; while in the optical frequency range dielectric waveguides and optical fibres are used. The latter are based on the effects due to the gradient of indices of refraction and total reflection at boundary. From the viewpoint of applications, there are some shortfalls for optical fibre waveguides and dielectric waveguides, especially as the index of refraction has dispersion, i.e., there are differences for the velocities of light with different frequencies. The original very-short optical pulses (according to the uncertainty relation, very short in time regime means very wide in frequency regime) propagating in the dispersive media will widen, thus restricting the quality of information to be transmitted. In addition, if optical fibres are bent with large angles, the energy loss is considerable. The existence of these disadvantages provides opportunities for line defects of photonic crystals to be used as optical waveguides in applications for optical techniques of information transmission.[c]

[c]For line defects in photonic crystals as waveguides, see a review by J. C. Knight, *Nature* **428**, 847 (2003).

Bibliography

[1] Friedel, J., *Metallic Alloys*, Nuovo Cimento **7** Suppl, 287 (1958).

[2] Ziman, J. M., *Principles of the Theory of Solids*, Cambridge University Press, Cambridge (1972).

[3] Desjonqueres, M. C., and D. Spanjaard, *Concepts in Surface Physics*, Springer-Verlag, Berlin (1993, 1996).

[4] Joannopoulos, J. D., R. D. Mead, and J. N. Winn, *Photonic Crystals*, Princeton University Press, Princeton (1995).

[5] Yeh, P., *Optical Waves in Layered Madia*, John Wiley & Sons, New York (1990).

[6] Joannopoulos, J. D., P. R. Willeneuve, and S. Fan, *Photonic Crystals: Putting a New Twist on Light*, Nature **386**, 143 (1997).

[7] Davison, S. G., and M. Steslicka, *Basic Theory of Surface States*, Clarendon Press, Oxford (1996).

[8] Harrison, W., *Solid State Theory*, McGraw-Hill, New York (1970).

[9] Madelung, O., *Introduction to Solid State Theory*, Springer-Verlag, Berlin (1978).

[10] Kittel, C., *Quantum Theory of Solids*, John-Wiley & Sons, New York (1963, 1987).

[11] Hui, P. M., and N. F. Johnson, *Photonic Band-Gap Materials*, Solid State Physics **49**, 151 (1996).

Chapter 8

Transport Properties

Carriers in metals and semiconductors will move in a definite direction under applied electric fields and temperature gradients. At the same time, they will also be scattered by impurities, defects and lattice vibrations. If these two factors compete each other and finally arrive at equilibrium, we have stationary transport. In this chapter, we will at first use the semiclassical Boltzmann equation and its relaxation time approximation as the basis of a treatment of transport properties of solids, and we will illustrate this by looking at the conductivity of metals and semiconductors. Then we will discuss electronic transport in magnetic fields, involving charge transport as well as spin transport. Finally we will discuss electronic transport based on the tunneling effect, including the physical basis for tunneling magnetoresistance (TMR) and scanning tunneling microscopy (STM).

§8.1 Normal Transport

In this section we begin from the theoretical basis for transport properties in solids, including the Boltzmann equation and its relaxation time approximation, to discuss some problems in normal transport (i.e., without magnetic field), mainly concentrating on transport under electric fields or temperature gradients.

8.1.1 Boltzmann Equation

In a real solid, electrons are scattered by lattice imperfections and lattice vibrations. An electron in the state labelled by wavevector $\boldsymbol{k}$ may be scattered into another state $\boldsymbol{k}'$. Supposing the system is inhomogeneous on a macroscopic scale, and the scattering is weak, we may describe the motion of electrons by a semiclassical distribution function $f(\boldsymbol{k}, \boldsymbol{r}, t)$ which depends not only on wavevector $\boldsymbol{k}$, but also on position $\boldsymbol{r}$ and time t.

We now consider the change of the distribution function with time in the presence of an applied field. A particular state $(\boldsymbol{k}, \boldsymbol{r})$ may be occupied ($f = 1$) or not occupied ($f = 0$). If there is no scattering at all, this state will move through phase space satisfying the transport equation $df/dt = 0$. However, if there are scattering events, an electron will discontinuously change its momentum and therefore will make a discontinuous jump in phase space. Thus the transport equation is $df/dt = \partial f/\partial t|_{\text{coll}}$, where the term on the right hand side is the change in the distribution function due to collisions. The result is

$$\frac{\partial f}{\partial t} + \nabla_{\boldsymbol{k}} f \cdot \frac{d\boldsymbol{k}}{dt} + \nabla f \cdot \frac{d\boldsymbol{r}}{dt} = \left.\frac{\partial f}{\partial t}\right|_{\text{coll}}, \tag{8.1.1}$$

in which $d\boldsymbol{k}/dt$ is the rate of momentum variation and is proportional to the applied force $\boldsymbol{F}$ by (6.1.4); $d\boldsymbol{r}/dt$ is the rate of position variation and is equal to the velocity $\boldsymbol{v}$. Thus the transport equation is transformed into

$$\frac{\partial f}{\partial t} + \frac{1}{\hbar}\boldsymbol{F} \cdot \nabla_{\boldsymbol{k}} f + \boldsymbol{v} \cdot \nabla f = \left.\frac{\partial f}{\partial t}\right|_{\text{coll}}. \tag{8.1.2}$$

The rate of change of the distribution function contains three contributions: an acceleration term, in which an applied force causes electrons to change momentum states; a drift term, to account for the fact that electrons are leaving a region of space with velocity $\boldsymbol{v}$ if the distribution function varies in space; a collision term, which represents the scattering rate by defects or impurities.

In transport calculations, a relaxation time approximation for the collision term is usually used. The distribution function deviates from the equilibrium distribution function f_0, and it is expected that it will decay exponentially in time to the equilibrium value. It may be written as

$$\left.\frac{\partial f}{\partial t}\right|_{\text{coll}} = -\frac{f - f_0}{\tau}, \tag{8.1.3}$$

where τ is the relaxation time, and f_0 is the equilibrium distribution function. Using this relaxation time approximation, the Boltzmann equation is

$$\frac{\partial f}{\partial t} + \frac{1}{\hbar}\boldsymbol{F}\cdot\nabla_{\boldsymbol{k}} f + \boldsymbol{v}\cdot\nabla f = -\frac{f - f_0}{\tau}. \tag{8.1.4}$$

Frequently, we will be interested in applying fields to the system and seeking only the linear response, i.e., we can write the distribution function as $f = f_0 + f_1$, where f_1 is the deviation from the equilibrium distribution function f_0. This may be substituted in the Boltzmann equation in the collision approximation and only first-order terms in the applied fields retained. The result is the linearized Boltzmann equation

$$\frac{\partial f_1}{\partial t} + \frac{1}{\hbar}\boldsymbol{F}\cdot\nabla_{\boldsymbol{k}} f_0 + \boldsymbol{v}\cdot\nabla f_0 = -\frac{f_1}{\tau}. \tag{8.1.5}$$

The linearized Boltzmann equation can be used to study electric as well as heat transport, and the external fields include electric field, magnetic field, or a temperature gradient.

It should be noted that we have used a semiclassical theoretical framework, i.e., Boltzmann equation, to treat electronic transport which is in nature a behavior of a quantum many-particle system. So its limitations are apparent, and it is more proper to treat this kind of problem by quantum many-body theories, such as Kubo's quantum transport theory. The starting point of these theories is to take an external field as a perturbation for a many-particle system in equilibrium. This leads to a linear response and gives corresponding transport coefficients. For example, Kubo expresses the conductivity tensor $\sigma_{\mu\nu}$ as the time correlation function of the components of the current operators, and gives the formula for conductivity

$$\sigma_{\mu\nu} = \frac{1}{k_{\mathrm{B}}T}\int_0^{\infty} \langle j_\mu(t) j_\nu(0)\rangle dt, \tag{8.1.6}$$

where $j_\nu(0)$ is a component of current operator at $t = 0$; while $j_\mu(t)$ is another component after time interval t. In (8.1.6) an integral is taken over whole time range for their product in equilibrium. Although the time average of $j(t)$ is zero, the correlation of current fluctuations is in general nonzero and determined by scattering processes, which also reflects the effect of external fields. These linear response theories are the foundations of basic theorems involving various transport coefficients, for example the Onsager symmetry relations in irreversible processes. These theories are more complicated, generally begin from Green's functions,[a] and have only been carried out in a few simple standard cases. These confirm almost all the results obtained by the Boltzmann method.

8.1.2 DC and AC Conductivities

For a simple treatment, we consider the electrical conductivity under isothermal conditions. First let us consider a uniform, static electric field: There is an applied force given by $\boldsymbol{F} = -e\boldsymbol{E}$ for electrons or $+e\boldsymbol{E}$ for holes and we have $\partial f_1/\partial t = 0$. It is convenient to rewrite the left hand side as

$$\frac{1}{\hbar}\nabla_{\boldsymbol{k}} f_0 = \frac{1}{\hbar}\frac{\partial f_0}{\partial E}\nabla_{\boldsymbol{k}} E = \boldsymbol{v}\frac{\partial f_0}{\partial E},$$

[a]For Kubo's transport theory, one may consult any treatise of quantum many-body theory, such as G. D. Mahan, *Many-Particle Physics*, 3rd ed., Plenum Press, New York (1995).

and

$$\nabla f_0 = \frac{\partial f_0}{\partial E}\nabla E_{\mathrm{F}},$$

then the first-order term in the distribution function is

$$f_1 = e\tau\frac{\partial f_0}{\partial E}\boldsymbol{v}\cdot\boldsymbol{E}', \tag{8.1.7}$$

where $\boldsymbol{E}' = \boldsymbol{E} - (1/e)\nabla E_{\mathrm{F}}$ is the electromotive force.

Because there would be no contributions from the equilibrium distribution function f_0 to the current, we may simply use the first-order term, and adding up all the occupied states we get

$$\boldsymbol{j} = -\frac{2e}{(2\pi)^3}\int \boldsymbol{v}f_1 d\boldsymbol{k} = \sigma\boldsymbol{E}', \tag{8.1.8}$$

where

$$\sigma = -\frac{2e^2\tau}{(2\pi)^3}\int \boldsymbol{vv}\frac{\partial f_0}{\partial E}d\boldsymbol{k}. \tag{8.1.9}$$

In principle, σ is a tensor in crystals. For convenience, we now assume isotropic bands, and let the electric field $\boldsymbol{E}$ lie in the z direction. When we integrate over angle, the only surviving component of the current will lie in the z direction. We may therefore replace $\boldsymbol{v}$ by v_z, and average over angle to get $\langle v_z^2\rangle = v^2/3$. Then the conductivity is in the scalar form

$$\sigma = -\frac{2e^2\tau}{3(2\pi)^3}\int d\boldsymbol{k}v^2\frac{\partial f_0}{\partial E}. \tag{8.1.10}$$

For a simple case, we take a parabolic band with $E = \hbar^2k^2/2m^*$, and then $v(-\partial f_0/\partial E) = -(1/\hbar)\partial f_0/\partial k$. We can now perform a partial integration of (8.1.9)

$$\sigma = -\frac{2e^2\tau}{3(2\pi)^3m^*}\int_0^\infty dk4\pi k^3\frac{\partial f_0}{\partial k} = \frac{2e^2\tau}{3(2\pi)^3m^*}\left(-4\pi k^3 f_0\Big|_0^\infty + 3\int_0^\infty dk4\pi k^2 f_0\right).$$

The first term vanishes at the lower limit where $k = 0$ and at the upper limit where $f_0 = 0$. Thus we may immediately identify the conductivity

$$\sigma = \frac{ne^2\tau}{m^*}, \tag{8.1.11}$$

where $n = [2/(2\pi)^3]\int f_0 d\boldsymbol{k}$ is the number density of electrons. This equation can be understood intuitively: $e\mathcal{E}/m^*$ is the rate of acceleration of a particle with a charge e and mass m^* in a field $\mathcal{E}$; it will acquire a velocity $e\tau\mathcal{E}/m^*$ in the course of a scattering time τ and so it will carry a current $e^2\tau\mathcal{E}/m^*$. Finally, the total current per unit volume is obtained by multiplying by the number of electrons per unit volume.

In (8.1.8),

$$\boldsymbol{j} = \sigma\boldsymbol{E} - \frac{1}{e}\sigma\nabla E_{\mathrm{F}},$$

the first term is the current density due to the electric field, and the second term can be identified as the current density $-eD\nabla n$ due to diffusion, and by writing $\nabla n = \nabla E_{\mathrm{F}}(dn/dE_{\mathrm{F}})$, we obtain the Einstein relation

$$\sigma = e^2D\frac{dn}{dE_{\mathrm{F}}} = e^2Dg_{\mathrm{F}}, \tag{8.1.12}$$

where $g_{\mathrm{F}} = dn/dE_{\mathrm{F}}$ is the density of states (DOS) at the Fermi surface.

Now we extend Ohm's law in the relaxation time approximation to an alternating electric field. The magnitude of the electric field $\mathcal{E}$ is expressed in the complex form

$$\boldsymbol{\mathcal{E}} = \boldsymbol{\mathcal{E}}_0\mathrm{e}^{-i\omega t} \tag{8.1.13}$$

and the equation of motion for a quasi-free electron is

$$m^*\dot{v} = -e\mathcal{E} - m^*\frac{v}{\tau}. \tag{8.1.14}$$

Consider that v and $\mathcal{E}$ have the same alternating frequency, then

$$-i\omega m^* v = -e\mathcal{E} - m^*\frac{v}{\tau}. \tag{8.1.15}$$

The current density and conductivity can all be obtained. The results are

$$j = -nev = \frac{ne^2\tau}{m^*(1-i\omega\tau)}\mathcal{E}, \tag{8.1.16}$$

and

$$\sigma = \frac{ne^2\tau}{m^*(1-i\omega\tau)} = \frac{ne^2\tau(1+i\omega\tau)}{m^*(1+\omega^2\tau^2)}. \tag{8.1.17}$$

These formulas demonstrate that, under an alternating electric field, the conductivity becomes complex, with real and imaginary parts. Its real part varies with the electric field in phase, while the imaginary part is out of phase with a phase difference of $\pi/2$. Numerically, the conductivity decreases with increasing frequency and when $\omega\tau \gg 1$, τ disappears. Because τ is about 10^{-12} s, this limit is located in the infrared range. Due to the disturbance from reactance and skin effect at lower frequencies, this expression has been verified in the optical frequency range.

We can also get the displacement x from (8.1.14), and the electric polarization is

$$P = -nex = -\frac{\omega_\mathrm{p}^2}{\omega^2 + i\omega/\tau}\mathcal{E}. \tag{8.1.18}$$

Here we consider the damped plasma oscillation. There are positive and negative charges distributed in a medium with equal densities, but only one of them can move. In metals, the density of conduction electrons is just equal to the density of positively charged ions. ω_p represents the undamped plasma oscillation frequency

$$\omega_\mathrm{p}^2 = \frac{ne^2}{\epsilon_0 m^*}. \tag{8.1.19}$$

Therefore the dielectric function can be expressed as

$$\epsilon(\omega) = 1 - \frac{\omega_\mathrm{p}^2}{\omega(\omega + i/\tau)} = 1 - \frac{\omega_\mathrm{p}^2(\omega - i/\tau)}{\omega(\omega^2 + 1/\tau^2)}. \tag{8.1.20}$$

Its real and imaginary parts are reversed in phase compared with those of conductivity: Reactance contributes to the imaginary part of the dielectric function.

The frequency-dependent dielectric function (8.1.20) implies a specific physical meaning: When ω is large enough, but smaller than ω_p, the real part of the dielectric function is negative, so metals will have negative dielectric constants in the optical frequency range. Negative dielectric constants affect the optical properties of metals, making them opaque due to their strong reflection of optical waves. This illustrates that using metals as the matrix of photonic crystals is obviously different from using dielectric materials. It is well-known that the propagation velocity of electromagnetic waves in a medium is $c = (\epsilon\mu)^{-1/2}$, while the index of refraction of a medium is $n = (\epsilon\mu)^{1/2}$. For general dielectrics $\epsilon > 0$, but for metals $\epsilon < 0$ in the optical frequency range. It is different for permeability of media: Although there is diamagnetism in matter and its permeability is negative, its value is too small to transform μ from positive to negative, so all natural materials have positive permeability. In recent years, a kind of new artificial materials, with periodic structures, have been made; their basic units are split conducting rings of small size on a dielectric substrate. These artificial materials can be used to realize negative permeability in the microwave frequency range. We have already known that for frequencies near the plasma frequency ω_p, the dielectric constants of materials are negative. Scientists may design artificial materials to move the plasma frequency to the microwave range and then it will be possible to prepare materials with both ϵ and μ negative at the same frequency. The characteristic of these materials is that their indices of refraction are negative, i.e., $n = -(\epsilon\mu)^{1/2}$, and can show unusual propagation behavior for electromagnetic waves. These effects have already been verified experimentally.[b]

8.1.3 Microscopic Mechanism of Metallic Conductivity

We can understand from the discussion in the last subsection that the circumstances in which electrons are scattered in a material determines its conductivity or resistivity. One thing that should first be made clear is that the Bloch solutions for the Schrödinger equation of a perfect crystal correspond to the stationary state, so there are no contributions to the relaxation mechanism; only when a lattice deviates from perfection can there be transitions between electronic states.

There are two irregularities for a lattice, one is due to impurities and defects where the transition probability is independent of temperature; the other arises from lattice vibrations, which gives rise to a temperature-dependent transition probability. These two transition probabilities correspond, respectively, to the relaxation times from impurities and defects $\tau_{\rm d}$ and from lattice vibrations $\tau_{\rm l}$. If these two transitions are independent of each other, then the sum of their rates is just the total transition rate. Therefore the total relaxation time in the formula of conductivity $\tau(\boldsymbol{k})$ satisfies

$$\tau^{-1}(\boldsymbol{k}) = \tau_{\rm d}^{-1}(\boldsymbol{k}) + \tau_{\rm l}^{-1}(\boldsymbol{k}). \tag{8.1.21}$$

If $\tau_{\rm d}(\boldsymbol{k})$ and $\tau_{\rm l}(\boldsymbol{k})$ are both isotropic, then due to $\rho = \sigma^{-1}$, we find the Matthiesen rule for resistivity as

$$\rho = \rho_{\rm d} + \rho_{\rm l}. \tag{8.1.22}$$

This shows that resistivities due to impurities and defects (independent of temperatures) and to lattice vibrations (dependent on temperatures) can be simply added together. When $T \to 0$ K, $\rho_{\rm l} \to 0$, only $\rho_{\rm l}$ is left, this is the residual resistivity. The resistivity of dilute solid solutions can be calculated based on the electronic scattering by a single impurity as described in §7.2.

In the following we are mainly concerned with the influence of thermal vibrations of ions on resistivity. As stated in §5.3, lattice vibrations can be decomposed into a set of normal modes $\omega_i(\boldsymbol{q})$, where ω is angular frequency, $\boldsymbol{q}$ is wavevector, and i denotes polarization. Because energy is quantized, and expressed as $\hbar\omega_i(\boldsymbol{q})$, the normal modes of lattice vibration correspond to quasiparticles with definite energies and momentums, i.e., phonons. Phonons, like electrons, have energy band structures in wavevector space. If the Debye model is used, there exists a Debye cut-off frequency $\omega_{\rm D}$, related to the Debye temperature, $\hbar\omega_{\rm D} = k_{\rm B}\Theta_{\rm D}$, where $k_{\rm B}$ is the Boltzmann constant. When $T \gg \Theta_{\rm D}$, all the vibrational modes are excited; when $T < \Theta_{\rm D}$, only long waves (low frequency) vibrations are excited.

When an electron with wavevector $\boldsymbol{k}$ is scattered by a vibration mode $\boldsymbol{q}$, and makes a transition from state $\boldsymbol{k}$ to state $\boldsymbol{k}'$, the states before and after the transition must satisfy momentum conservation

$$\boldsymbol{k} + \boldsymbol{q} = \boldsymbol{k}',$$

and energy conservation

$$E(\boldsymbol{k}) + E(\boldsymbol{q}) = E(\boldsymbol{k}').$$

This is normal scattering process, called N process; however, if lattices $\boldsymbol{k}'$ is equivalent to $\boldsymbol{k} + \boldsymbol{G}$, where $\boldsymbol{G}$ is a reciprocal lattice vector, so it is possible to have scattering where

$$\boldsymbol{k} + \boldsymbol{q} = \boldsymbol{k}' + \boldsymbol{G}.$$

[b]In 1968, Veselago theoretically studied the propagation behavior of electromagnetic waves in materials with different combinations of negative and positive ϵ and μ, and pointed out that ϵ and μ are both positive for normal materials, which are called right-handed materials. The Poynting vector is in the same direction as the wavevector. If ϵ and μ are both negative, the material is called left-handed: the Poynting vector and the wavevector are in opposite directions, so their indices of refraction are negative. Readers may consult to V. G. Veselago, *Sov. Phys. Uspeshi.* **10**, 509 (1968) for earliest theoretical discussion; J. Pendry *et al.*, *IEEE* **47**, 11 (1999) for a detailed theoretical analysis; and J. B. Pendry and D. R. Smith, *Phys. Today.* **47**, 37 (2004) for a recent review of experimental results.

This is the Umklapp (U) scattering process, also called U process. Both scattering processes are shown in Fig. 8.1.1. Because the maximum value of a phonon is about $k_B\Theta_D$, the energy variation of an electron due to electron-phonon scattering, corresponding to emission or absorption of a phonon, is not large and $E(\boldsymbol{q})$ can be neglected. It can be seen from the figure that the main effect of collision of an electron and a phonon is to change the direction of the electron wavevector $\boldsymbol{k}$. It is more pronounced in a U process. But the value of phonon wavevector $\boldsymbol{q}$ is always smaller than $|\boldsymbol{G}/2|$, so only when the value of $\boldsymbol{q}$ is larger than $\overline{PQ}$ in the figure, can the U process be realized. At low temperatures, the value of $\boldsymbol{q}$ is very small, and the U process is almost absent.

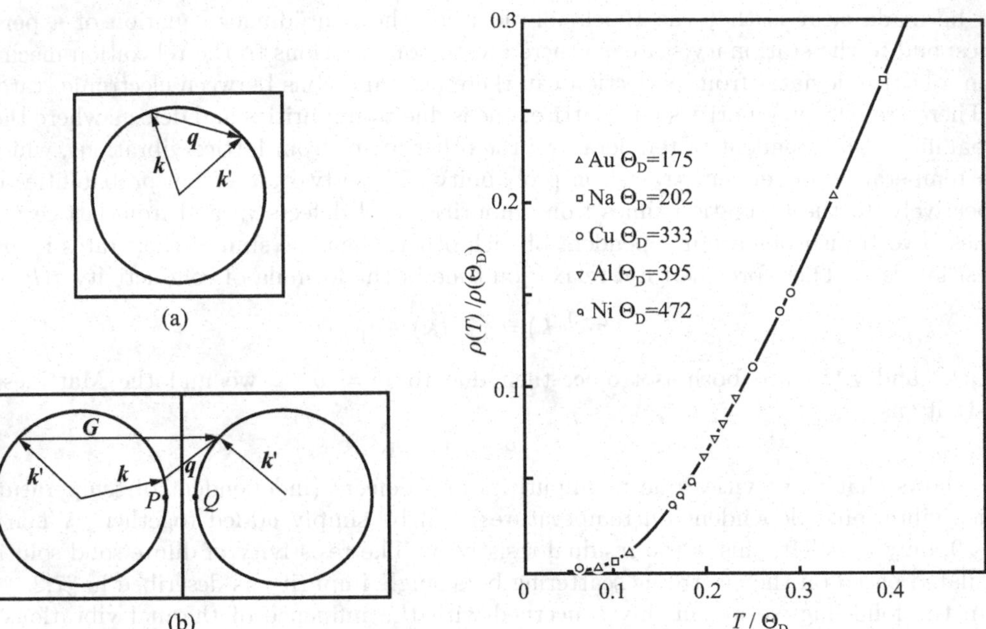

Figure 8.1.1 Electron-phonon scattering. (a) Normal process $\boldsymbol{k}+\boldsymbol{q}=\boldsymbol{k}'$; (b) Umklapp-process $\boldsymbol{k}+\boldsymbol{q}=\boldsymbol{k}'+\boldsymbol{G}$.

Figure 8.1.2 The relation between reduced resistivity $\rho(T)/\rho(\Theta_D)$ and reduced temperature T/Θ_D of several metals. Curves show the theoretical calculations.

During the scattering, the electronic wavevector is changed from $\boldsymbol{k}$ to $\boldsymbol{k}'$. We assume that the angle between them is θ and the scattering transition probability is $P(\theta)$ and then integrate the latter over the Fermi surface to get the total transition probability, which is equal to the reciprocal of the relaxation time τ, i.e.,

$$\tau^{-1} = 2\pi k^2 \int P(\theta)(1-\cos\theta)\sin\theta d\theta, \tag{8.1.23}$$

in which $(1-\cos\theta)$ represents the weighting factor due to large angle scattering. To average $P(\theta)$ and take it outside the integral, according to the Debye model for the specific heat of solids, we have

$$\langle P(\theta)\rangle \propto \frac{k_B T}{\Theta_D^2}, \tag{8.1.24}$$

then the integral can be transformed into

$$\int_0^x (1-\cos\theta)\sin\theta d\theta = \int_0^x 8\sin^3\frac{\theta}{2} d\left(\sin\frac{\theta}{2}\right). \tag{8.1.25}$$

At high temperatures ($T \gg \Theta_D$), and all lattice waves are excited. In the Einstein model, the upper limit of the integral in (8.1.23) is $x = \pi$; but in the Debye model, the value of x is determined by the Debye cut-off wavenumber q_D, i.e., $x = q_D/2k_F \sim 79°$. In both cases, x is always independent of T, so we obtain

$$\rho \propto \tau^{-1} \propto T, \tag{8.1.26}$$

which shows that the resistivity is proportional to T. At low temperatures ($T < \Theta_D$), only the lattice waves with low frequencies are excited (q_D is the cut-off wavenumber), the scattering angles of the electrons are very small and the cut-off angle is approximately $x = (q_D/2k_F)(T/\Theta_D)$, so

$$\rho_L \propto \tau^{-1} \propto T^5,$$

which denotes the resistivity is proportional to the fifth power of temperature. The combination of these relations for metallic resistivity with temperature at high and low temperatures is the Bloch–Grüneisen law, shown in Fig. 8.1.2.

The above estimates have not taken into account the energy change in electronic scattering, and are different from real situations. If the energy loss and addition from inelastic scattering are considered, a more sophisticated theoretical treatment can give a more rigorous expression (see [5] and [6] in the bibliography).

$$\rho_L = 4.225 \left(\frac{T}{\Theta_D}\right)^5 J_5\left(\frac{\Theta_D}{T}\right)\rho(\Theta_D), \tag{8.1.27}$$

where $\rho(\Theta_D)$ is the resistivity at $T = \Theta_D$ and

$$J_5(x) = \int_0^x \frac{z^5 dz}{(e^z - 1)(1 - e^{-z})}.$$

When $T \to 0$, $J_5(x) \to$ a constant; and when T is very large, $J_5 \to (\Theta_D/T)^4/4$, consistent with the Bloch–Grüneisen law.

It should be pointed out that these theories are not flawless, for example, the U process has not been taken into account, and the influence of a nonspherical Fermi surface has been neglected. But for alkali metals these factors give weak effects and experimental results have verified the Bloch–Grüneisen law.[c]

8.1.4 Electric Transport in Semiconductors

In semiconductors, as shown in Fig. 8.1.3, the Fermi level sits in the gap; its separation from the bottom of the conduction band or the top of the valence band is far larger than k_BT, i.e.,

$$E_c - E_F \gg k_BT, \qquad E_F - E_v \gg k_BT.$$

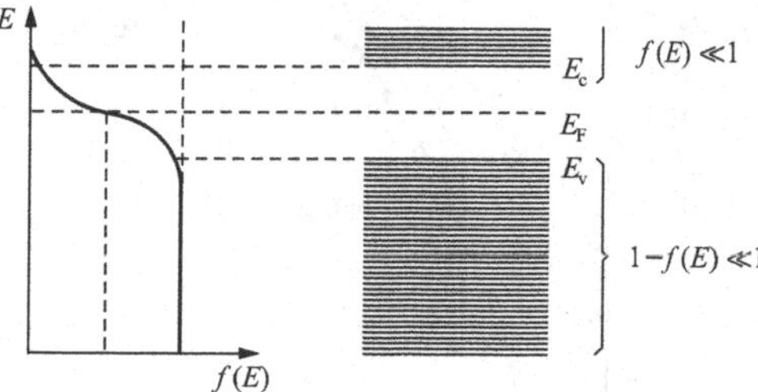

Figure 8.1.3 The distribution of energy levels of conduction band and valence band in semiconductors.

[c] A review of experimental results for alkali metals can be found in J. Bass, W. P. Pratt and P. A. Schroeder, *Rev. Mod. Phys.* **62**, 645 (1990). But at very low temperatures ($T < 2$ K), experimental results have shown the relationship $\rho \propto T^2$, due to the scattering between electrons. We shall discuss this topic later.

Although electrons are still distributed on conduction band levels, following the Fermi–Dirac statistics

$$f(E) = \frac{1}{e^{(E-E_F)/k_BT}+1},$$

the fact that

$$E - E_F > E_c - E_F \gg k_BT$$

leads the exponential term in the denominator of the distribution function to be much greater than one, so the electronic distribution in the conduction band satisfies

$$f(E) \approx e^{-(E-E_F)/k_BT}.$$

In the same way, the hole distribution in valence band is

$$1 - f(E) \approx e^{-(E_F-E)/k_BT}.$$

Both are approximately like the classical Boltzmann distribution, obviously different from the degenerate electronic gas in metals.

For intrinsic semiconductors, the thermal excitation of electrons from the valence band to the conduction band makes the electron number n_e and the hole number n_h equal. But, usually, semiconductors acquire carriers through doping: in the n-type semiconductors (donor doping, impurity levels near conduction band), the carriers are mainly electrons; while in the p-type semiconductors (acceptor doping, impurity levels near valence band), the carriers are mainly holes. Therefore, in general, the conductivity of semiconductors can be written as

$$\sigma = n|e|\mu_t = n_e|e|\mu_e + n_h|e|\mu_h,$$

where μ_t is the mobility, μ_e and μ_h are the mobilities of electrons and holes, respectively.

The mobility of a semiconductor is determined by the scattering from ionized impurities and lattice vibrations, with relaxation times τ_I and τ_L, respectively. Both of these are dependent on temperature in accordance with

$$\tau_I \propto T^{3/2}, \qquad \tau_L \propto T^{-3/2},$$

so

$$\tau^{-1} = aT^{-3/2} + bT^{3/2}.$$

This result was verified by experiment, see Fig. 8.1.4.

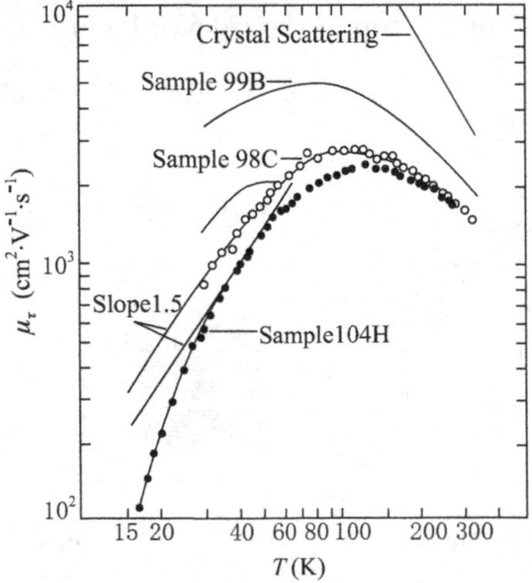

Figure 8.1.4 The relation of mobility and temperature in GaAs.

Carriers transporting current between two electrodes will cause Joule heating due to emission of phonons. The mean free path in the electron-phonon process in a metal is so short that specimen itself would be vaporized before this mechanism broke down. But in semiconductors, a modest electric field (≤ 1000 V/cm) can overload the energy transfer processes, so that the electronic energy distribution and lattice temperature are no longer in equilibrium. The corresponding electronic transport is called hot electron transport. The whole effect deviates from Ohm's law and makes the current of carriers tend to a saturated value irrespective of the electric field. Hot electron transport is a complicated phenomenon.[d]

8.1.5 Other Transport Coefficients

Now we shall have a brief discussion of the other transport coefficients besides conductivity, including the thermoelectric coefficient, thermal conductivity etc. The theoretical starting point is still the Boltzmann equation in the relaxation time approximation, but in addition to electric field, a temperature gradient ∇T is added. In this case, there is not only electric current density, but also thermal current density $\boldsymbol{j}_Q$. According to thermodynamics, the free energy is $F = U - TS$, where E is the internal energy and S is the entropy, the thermal transport quantities can be expressed as

$$dQ = TdS = dU - \mu dN, \tag{8.1.28}$$

where μ is the chemical potential and N the particle number. A temperature gradient also causes a chemical potential gradient $\nabla\mu$, so the electric field $\boldsymbol{E}$ acting on the electrons will be replaced by the effective field

$$\boldsymbol{E}' = \boldsymbol{E} + \frac{\nabla\mu}{e},$$

so the change of the Fermi distribution in (8.1.7) should be rewritten

$$f_1 = -\tau\frac{\partial f_0}{\partial E}\boldsymbol{v}\left[-e\left(\boldsymbol{E} + \frac{\nabla\mu}{e}\right) + \frac{E(\boldsymbol{k}) - \mu}{T}(-\nabla T)\right]. \tag{8.1.29}$$

Then from f_1 we can obtain the electric current density and thermal current density, respectively, as

$$\boldsymbol{j} = K_{11}\left(\boldsymbol{E} + \frac{\nabla\mu}{e}\right) + K_{12}(-\nabla T), \tag{8.1.30}$$

and

$$\boldsymbol{j}_Q = K_{21}\left(\boldsymbol{E} + \frac{\nabla\mu}{e}\right) + K_{22}(-\nabla T). \tag{8.1.31}$$

The coefficients K_{ij} can be obtained through the following expression

$$K_\alpha = e^2\int\frac{d\boldsymbol{k}}{4\pi}\left(-\frac{\partial f_0}{\partial E}\right)\tau E(\boldsymbol{k})\boldsymbol{v}(\boldsymbol{k})\boldsymbol{v}(\boldsymbol{k})[E(\boldsymbol{k}) - \mu]^\alpha, \tag{8.1.32}$$

where $K_{11} = K_0$, $K_{21} = TK_{12} = -(1/e)K_1$, $K_{22} = (1/e^2T)K_2$.

By using (8.1.9), we can establish a relation for K_α in terms of electrical conductivity, i.e.,

$$K_\alpha = \int dE\left(-\frac{\partial f_0}{\partial E}\right)(E - \mu)^\alpha\boldsymbol{\sigma}(E), \tag{8.1.33}$$

where there are approximate expressions $K_{11} \sim \sigma(E_F) \sim \sigma$, $K_{21} = TK_{12} \sim -(\pi^2/3e)(k_BT)^2\sigma'$, $K_{22} \sim (\pi k_BT/3e^2)\boldsymbol{\sigma}$, here $\sigma' = \partial\sigma(E)/\partial E|_{E=E_F}$. These expressions show how the influences of electric field and temperature gradient on electronic transport and thermal transport, form the physical basis of understanding the thermoelectric effect and thermal transport phenomena.

[d]The transport theory for hot electrons in semiconductors is discussed in C. S. Ting (ed.), *Physics of Hot Electron Transport*, World Scientific, Singapore (1992).

Usually, thermal conductivity is measured where the electric current is zero, so we let $\boldsymbol{j}$ in (8.1.30) be zero, then we have

$$\boldsymbol{E} + \frac{\nabla \mu}{e} = \frac{K_{12}}{K_{11}} \nabla T, \tag{8.1.34}$$

then we substitute this into the second part of (8.1.31) and find a formula for the variation of thermal current with the temperature gradient

$$j_Q = \kappa(-\nabla T), \tag{8.1.35}$$

where κ is the thermal conductivity and can be expressed as

$$\kappa = K_{22} - \frac{K_{21}K_{12}}{K_{11}}. \tag{8.1.36}$$

Because $\sigma' \sim \sigma/E_F$, the second term in (8.1.36) is about $(k_BT/E_F)^2$ times larger than the first term,

$$\kappa = K_{22} + O(k_BT/E_F)^2. \tag{8.1.37}$$

Then we deduce

$$\kappa = \frac{\pi^2}{3}\left(\frac{k_B}{e}\right)^2 T\sigma. \tag{8.1.38}$$

This is the well-known Wiedermann–Franz law, which relates thermal conductivity to electric conductivity in metals. The ratio $L = \kappa/\sigma T = (1/3)(\pi k_B/e)^2 \approx 2.45 \times 10^{-8}(\mathrm{V/K})^2$ is a universal constant. Certainly the influence of lattice thermal vibrations will constrain this relation to be valid only at $T > \Theta_D$.

The physical mechanism for thermal conductivity in nonmetals is thoroughly different. Because there are no freely moving electrons, temperature differences will cause a difference in phonon densities in different regions and cause phonon transport in these substances. This kind of thermal transport, arising purely from lattice vibrations, will not be discussed here in detail.

§8.2 Charge Transport and Spin Transport in Magnetic Fields

It has been discussed in §6.2 that an electron in a magnetic field moves cyclically. Here we will have a further discussion of the influence of this cyclic motion on electron transport properties, involving the classical Hall effect, Shubnikov–de Haas (SdH) effect, and normal magnetoresistance. An electron has not only charge, but also spin. The importance of spin transport phenomena has attracted much attention in recent years, so we will give a brief introduction to spin polarization and spin transport and discuss resistivity and magnetoresistivity in transition metals.

8.2.1 Classical Hall Effect

Before discussing the quantum theory of electron transport, it is instructive to look at the classical picture. Consider a solid with impurities in the presence of an electric field $\boldsymbol{E}$ and magnetic field $\boldsymbol{H}$. In the relaxation time approximation, the motion of an electron is determined by the Langevin equation

$$\frac{d\boldsymbol{v}}{dt} = -\frac{e}{m^*}\left(\boldsymbol{E} + \frac{1}{c}\boldsymbol{v} \times H\right) - \frac{\boldsymbol{v}}{\tau}, \tag{8.2.1}$$

where $\boldsymbol{v}$ is the average velocity and τ is the relaxation time. The current is given by $\boldsymbol{j} = -ne\boldsymbol{v}$, where n is the concentration of electrons. It is assumed that the magnetic field is always applied in the z-direction, i.e., $\boldsymbol{H} = H\boldsymbol{z}$; nevertheless, the electric field may be applied in any direction. We would like to give two examples: The first is the electric field parallel to the magnetic field, i.e., $\boldsymbol{E} = \mathcal{E}\boldsymbol{z}$; the track of an electron in this case is helical, as shown in Fig. 8.2.1(a). The electron is

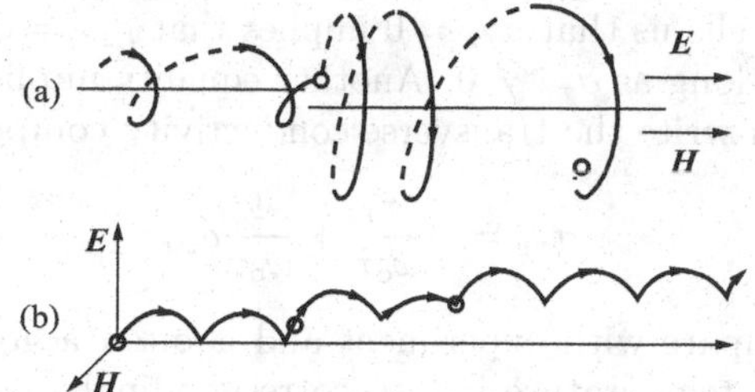

Figure 8.2.1 Classical motion of an electron in electric and magnetic fields. (a) $\boldsymbol{E} \parallel \boldsymbol{H}$, (b) $\boldsymbol{E} \perp \boldsymbol{H}$.

accelerated by the electric field, but scattered by impurities which limit the electronic velocity to a finite value. Its transverse motion in the x-y plane is attributed to the Lorentz force. The second is where the electric field is perpendicular to the magnetic field, for example, $\boldsymbol{E} = \mathcal{E}_x$. It is found that the trajectory of the electron is a series of arcs, as shown in Fig. 8.2.1(b). If there are no collisions, the electron moves in the direction perpendicular to $\boldsymbol{E}$ as well as $\boldsymbol{H}$, so the net current is zero in the direction of the electric field. By contrast, collision with impurities will lead to a drift current along the electric field. It is important to note the role of the relaxation term.

In general $\boldsymbol{E} = (\mathcal{E}_x, \mathcal{E}_y, \mathcal{E}_z)$, we can define the conductivity σ and resistivity ρ tensors by

$$\boldsymbol{j} = \sigma \cdot \boldsymbol{E}, \tag{8.2.2}$$

and inversely

$$\boldsymbol{E} = \rho \cdot \boldsymbol{j}. \tag{8.2.3}$$

In the steady state, $d\boldsymbol{v}/dt = 0$, which when combined with (8.2.1), (8.2.2) leads to

$$\sigma_0 \mathcal{E}_x = j_x + \omega_c \tau j_y, \qquad \sigma_0 \mathcal{E}_y = -\omega_c \tau j_x + j_y, \qquad \sigma_0 \mathcal{E}_z = j_z, \tag{8.2.4}$$

where σ_0 is the Drude conductivity given in (8.1.11), and $\omega_c = eH/m^*c$ is the cyclotron resonance frequency similar to (6.3.8). Because the third equality of (8.2.4) is a trivial result, we are now only concerned with the x and y directions. Then σ and ρ become 2nd rank tensors expressed as

$$\sigma = \begin{pmatrix} \sigma_{xx} & \sigma_{xy} \\ \sigma_{yx} & \sigma_{yy} \end{pmatrix}, \qquad \rho = \begin{pmatrix} \rho_{xx} & \rho_{xy} \\ \rho_{yx} & \rho_{yy} \end{pmatrix}, \tag{8.2.5}$$

where σ_{xx} and σ_{xy} are the longitudinal and the transverse components of the conductivity tensor, respectively. We arrive at the following conductivity formulas

$$\sigma_{xx} = \sigma_{yy} = \frac{\sigma_0}{1 + (\omega_c \tau)^2}, \qquad \sigma_{xy} = -\sigma_{yx} = -\frac{\sigma_0 \omega_c \tau}{1 + (\omega_c \tau)^2}, \tag{8.2.6}$$

and resistivity formulas

$$\rho_{xx} = \rho_{yy} = \frac{1}{\sigma_0}, \qquad \rho_{xy} = -\rho_{yx} = \frac{\omega_c \tau}{\sigma_0}. \tag{8.2.7}$$

It is actually equivalent to describe electric transport by either conductivity or resistivity. Theoretical investigation prefers the former and experimental measurement the latter. We may use either of them as is convenient. From (8.2.2) and (8.2.3), a general equation for conductivity and resistivity is

$$\sigma \cdot \rho = I, \tag{8.2.8}$$

where I is a 2×2 unit matrix, so we have the relationship between the resistivity and conductivity components as

$$\rho_{xx} = \frac{\sigma_{xx}}{\sigma_{xx}^2 + \sigma_{xy}^2}, \qquad \rho_{xy} = -\frac{\sigma_{xy}}{\sigma_{xx}^2 + \sigma_{xy}^2}. \tag{8.2.9}$$

The first equality of (8.2.9) tells us that $\sigma_{xx} = 0$ implies that $\rho_{xx} = 0$ at the same time, which may sound strange but is true as long as $\sigma_{xy} \neq 0$. Another equality implies $\rho_{xy} = -1/\sigma_{xy}$ where σ_{xx} or ρ_{xx} vanishes. In fact, we can write the transverse conductivity component in the form

$$\sigma_{xy} = -\frac{\sigma_0}{\omega_c \tau} + \frac{1}{\omega_c \tau}\sigma_{xx}, \tag{8.2.10}$$

which has been used to compare with experiment and is often a good approximation. When the magnetic field is strong and temperature is low, correspondingly, $\omega_c \tau \gg 1$, then the longitudinal conductivity approaches the limit $\sigma_{xx} = 0$, and we have the Hall conductivity

$$\sigma_H = \sigma_{xy} = -\frac{nec}{H}, \tag{8.2.11}$$

or the Hall resistivity

$$\rho_H = \rho_{xy} = -\frac{H}{nec}, \tag{8.2.12}$$

where $-1/nec$ is called the Hall coefficient. We find that the Hall resistivity changes continuously as the magnetic field and carrier density vary. This is purely a classical result. In fact, at low temperatures and under strong magnetic field, the quantum Hall effect can be observed, as will be discussed in Part VI.

8.2.2 Shubnikov–de Haas Effect

The Shubnikov–de Haas (SdH) effect is the oscillation of longitudinal conductivity, or resistivity, with magnetic fields, when the conditions $\hbar\omega_c > k_B T$ and $\omega_c \tau > 1$ are satisfied. This effect, just like the de Haas–van Alphen (dHvA) effect, comes from Landau quantization in magnetic fields. Although this effect has been observed in metals for many years, it has now become a very powerful tool to characterize transport in heterostructures. We will limit ourselves to this case.

Consider a magnetic field perpendicular to a two-dimensional electron gas causing the formation of discrete quantized orbits as shown in Fig. 6.3.5. The Landau levels increase in energy linearly with magnetic field, due to the cyclotron resonance frequency, $\omega_c = eH/cm^*$. As the magnetic field increases, these levels will sequentially pass through the Fermi surface. As we know, the conductivity is determined by the carrier concentration and scattering probability; the density of states at the Fermi surface influences both. The periodic variation of the Fermi surface with magnetic field must cause an oscillation of the conductivity. From (6.3.31) and (6.3.34) one can deduce the carrier density from the period of the SdH oscillations between two adjacent Landau levels $\Delta(1/H)$ as

$$n = \frac{e/ch}{\Delta(1/H)}. \tag{8.2.13}$$

These measured values are usually in excellent agreement with those determined by Hall measurements.

Alternatively, the oscillation can be realized by varying the carrier concentration in a constant magnetic field. We can understand this from (6.3.34) by taking $E'_F = E_F$, because the Fermi energy E_F is closely related to the electronic concentration. In a Si inversion layer (MOSFET), it was shown that the oscillation observed with electron number by varying the gate voltage has a constant period, which proves that each Landau level has the same number of states in two dimensions. This would not be the case in three dimensions because of k_z motion, and it can thus provide a signature for the two-dimensional character of the electron system.

Another interesting characteristic is the directional dependence of the SdH effect. It was proved that, for a magnetic field in any direction, only the perpendicular component of the field confines the x-y motion of carriers and determines the conductivity oscillation period. An experimental observation in a two-dimensional electron gas at a GaAs-AlGaAs:Si interface tested the theoretical prediction of directional dependence of the magnetoresistance. In Fig. 8.2.2 there are oscillations of magnetoresistance for perpendicular field before and after the sample was exposed to light. But for a parallel field there are no oscillations in either case.

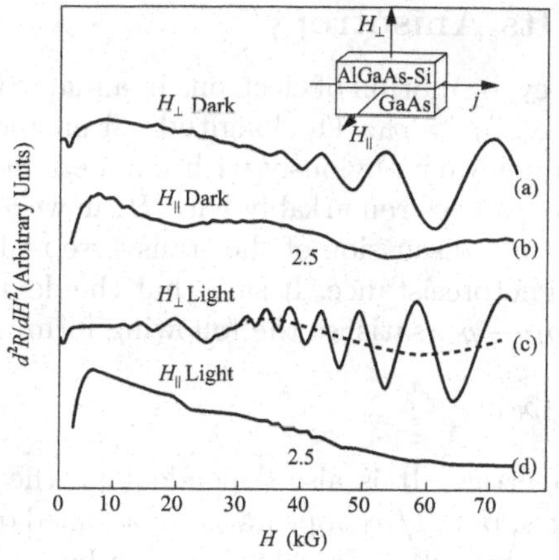

Figure 8.2.2 Angular dependence of the Shubnikov–de Haas oscillation. From H. Störmer and R. Dingle, *et al.*, *Solid State Commun.* **29**, 705 (1979).

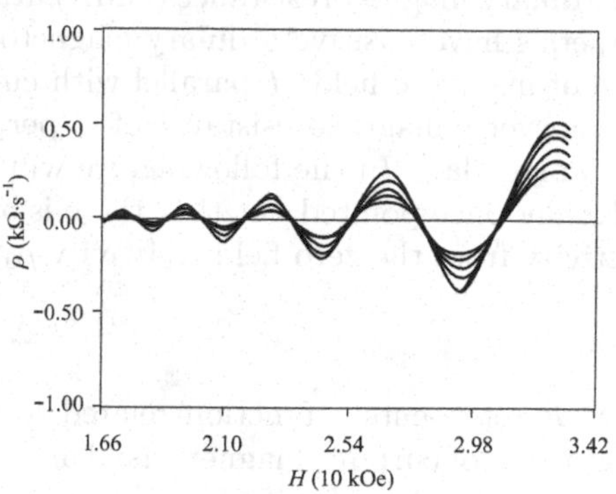

Figure 8.2.3 The oscillatory portion of the magnetoresistivity of a quantum well at six different temperatures between 1.7 K and 3.7 K. As the temperature increases, the amplitude of the oscillations decreases. From J. Singh, *Physics of Semiconductors and Their Heterostructures*, McGraw-Hill, New York (1993).

Theoretical treatments of the Shubnikov–de Haas (SdH) effect may involve different levels of sophistication. At finite temperatures, the typical analytic expression for magnetoconductivity can be derived as

$$\sigma_{xx} = \frac{ne^2\tau}{m^*}\frac{1}{1+(\omega_c\tau)^2}\left[1 - \frac{2(\omega_c\tau_c)^2}{1+(\omega_c\tau_c)^2}\frac{2\pi^2k_BT/\hbar\omega_c}{\sinh(2\pi^2k_BT/\hbar\omega_c)}\exp\left(-\frac{\pi}{\omega_c\tau_c}\right)\cos\left(\frac{2\pi E_F}{\hbar\omega_c}\right)\right], \quad (8.2.14)$$

where τ_c is the cyclic relaxation time corresponding to the dephasing of the Landau state, which is quite different from the relaxation time τ (τ_c may be an order of magnitude larger than τ). The theoretical formula (8.2.14) gives the conductivity, but the measured quantity is always the resistivity which can be transformed from (8.2.14) in an appropriate approximation

$$\rho_{xx} = \frac{\rho_{xx}(H=0)ne^2\tau}{m^*}\left[1 - \frac{2(\omega_c\tau_c)^2}{1+(\omega_c\tau_c)^2}\frac{2\pi^2k_BT/\hbar\omega_c}{\sinh(2\pi^2k_BT/\hbar\omega_c)}\exp\left(-\frac{\pi}{\omega_c\tau_c}\right)\cos\left(\frac{2\pi E_F}{\hbar\omega_c}\right)\right]. \quad (8.2.15)$$

It is clear that the oscillation effect is given by the cosine function in which $E_F/\hbar\omega_c$ determines the oscillation period as described in (6.3.34). From the temperature and magnetic field dependence of the oscillation amplitude, it is thus possible to extract m^* and τ_c.

Using typical values for the particle effective mass and scattering rates, we get $\omega_c\tau_c \leq 10000$. Therefore, for magnetic fields of interest, which are usually above 3 kG, we can approximate the oscillatory portion of the magnetoresistivity with sufficient accuracy as

$$\rho_{xx} \propto \frac{2\pi^2m^*ck_BT/\hbar eH}{\sinh(2\pi^2m^*ck_BT/\hbar eH)}\exp\left(-\frac{\pi}{\omega_c\tau_c}\right)\cos\left(\frac{2\pi^2c\hbar n}{eH}\right), \quad (8.2.16)$$

where $E_F = \hbar^2\pi n/m^*$ and $\omega_c = eH/cm^*$ are substituted. From this equation we may view the first factor as an amplitude that grows with increasing magnetic field and shrinks with increasing temperature and effective mass. The second factor represents scattering, and the last is a cosine term whose frequency is determined by the carrier concentration and the magnetic field. Figure 8.2.3 shows the oscillatory portion of the resistivity as a function of effective mass at six different temperatures between 1.7 K and 3.2 K. We can see that as the temperature increases, the amplitude of the oscillations decreases. Then the expression ρ_{xx} in (8.2.16) can be used to obtain the effective mass by examining its variation with temperature at fixed magnetic field.

8.2.3 Ordinary Magnetoresistance and Its Anisotropy

Ordinary magnetoresistance (OMR) arises from the cyclic motion of electrons in a magnetic field. All metals have positive ordinary magnetoresistance, i.e., $\rho_H > \rho_0$. The longitudinal magnetoresistance of magnetic field H parallel with current j does not vary obviously with magnetic field, but the transverse magnetoresistance of H perpendicular to j varies remarkably with H; however, there is no unique law. In the following, we will concentrate on a discussion of the transverse resistance.

Kohler once pointed out that there is a rule for magnetoresistance. It says that the deviation of resistivity from the zero field resistivity ρ_0, i.e. $\Delta\rho = \rho_H - \rho_0$, satisfies the following formula

$$\Delta\rho/\rho_0 = F(H/\rho_0), \tag{8.2.17}$$

where F represents a function related to metallic properties. It is also dependent on the relative orientations of current, magnetic field and crystalline axes, but H/ρ_0 appears as a combined quantity. Kohler rule has been verified experimentally for many metals, its physical reason can be understood qualitatively. A magnetic field causes electrons to move along a circular or helical orbit. The ratio of magnetic field to resistance depends on how many times electrons go around the orbit between collisions. This number is approximately the ratio of the electronic mean free path l to the orbit radius. For a free electron

$$l = mv/e^2 n\rho_0, \tag{8.2.18}$$

and

$$r = mv/eH, \tag{8.2.19}$$

so

$$\frac{l}{r} = \frac{H}{\rho_0}\frac{1}{ne}. \tag{8.2.20}$$

It is obvious that H/ρ_0 in the Kohler rule is actually the measurement of l/r. It should be noted that Kohler's rule has its restrictions.

The transverse magnetoresistance of crystals usually behaves in one of three ways. After a magnetic field is applied, (1) the resistance becomes saturated, with a value that may be several times the value at zero field, moreover the saturation appears along all measurement directions relative to the crystalline axes; (2) the resistance increases continuously for all crystalline axes; (3) the resistance saturates along some crystalline directions but not along others. That is there exists obvious anisotropy. All known crystals belong to one of these three types. The first type includes the crystals with closed Fermi surface, as In, Al, Na and Li; the second type includes crystals with the same number of electrons and holes; as Bi, Sb, W and Mo; and the third type is mainly seen in crystals with open orbits at the Fermi surface, as Cu, Ag, Au, Mg. Zn, Sn, Pb and Pt. For a Sn sample, the magnetic field dependence of $\Delta\rho/\rho_0$ in the directions with minimum and maximum magnetoresistance is shown in Fig. 8.2.4. Under high field, $\delta\rho_0$ in the minimum direction tends to saturation, but in the maximum direction it varies with H^2. Magnetoresistance can be used as a tool to investigate the Fermi surface to see whether it is closed or includes open orbits, and it can reveal the directions of those open orbits.

When we discussed cyclotron resonance in §6.3.1, only the situations in which the electron trajectories on the Fermi surface formed closed orbits was considered. But a real Fermi surface often has multi-connectivity, and taking one of its section, the boundary cannot be a closed curve; this is called an open orbit. Figure 8.2.5 gives schematic diagrams for several two-dimensional Fermi surfaces. When a Fermi surface does not contact with the Brillouin zone boundary, electrons move along closed orbits and the magnetoresistance saturates under high magnetic field. But the situation will be different when a Fermi surface contacts the Brillouin zone boundary. In some directions of magnetic fields, the orbits of electronic movement cross over the first Brillouin zone, even approach infinity. For these open orbits, irrespective of what value the magnetic field is, the cyclotron frequency is always zero. When the direction of a magnetic field is given and open orbits exist, saturation does not appear, the resistance will increase with the square of the applied field. Therefore, the investigation of magnetoresistance in single crystals no doubt can provide information for determining the topological structures of Fermi surfaces.

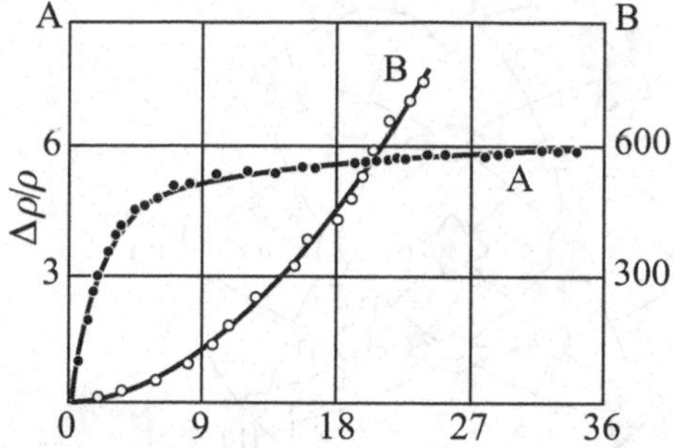

Figure 8.2.4 Variation of resistivity in two directions of magnetic field of a Sn single crystal. Curves A and B correspond to the minimum and maximum of resistivity variation. From J. L. Olsen, *Electron Transport in Metals*, John Wiley & Sons, New York (1962).

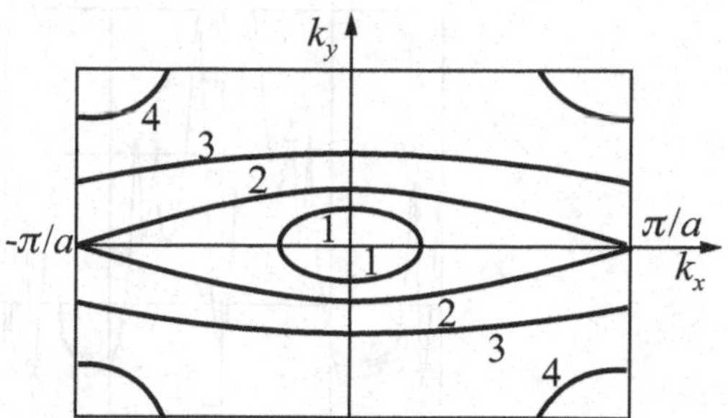

Figure 8.2.5 Schematic Fermi surfaces in two-dimensional Brillouin zone (BZ): 1. Closed orbit; 2. Self intersectional orbit; 3. Open orbit; 4. Hole orbit.

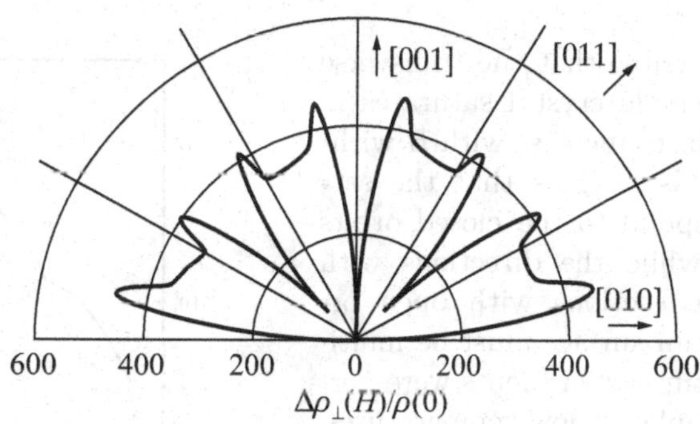

Figure 8.2.6 Anisotropic magnetoresistance in Cu single crystals. From J. R. Klauder and J. E. Kunzler, in *The Fermi Surface* (eds. W. A. Harrison and M. B. Webb', John Wiley & Sons, New York (1960).

When a perpendicular magnetic field is applied to determine the dc conductivity, it is necessary to take into account all the contributions from the Fermi surface. This means we must take note of all sections perpendicular to the field direction; maybe there are open orbits for some sections. It is not important to differentiate open or closed orbits under low fields. For a closed orbit, $\omega_c\tau \ll 1$, and electron scattering takes place before it finishes a cycle. So the magnetoresistance is only the average of local curvatures of the Fermi surface, i.e., the average of electronic velocities, no matter whether the trajectory of an electron before its scattering is a closed orbit or an open orbit, However, when a high field is applied, there is an important difference between a closed orbit and an open orbit. For a closed orbit, when $\omega_c\tau \gg 1$, an electron makes several orbits before its scattering. To calculate the conductivity, the electronic velocity should be the average velocity around the whole orbit. It can be seen that when H increases, the average velocity component perpendicular to $\boldsymbol{H}$ approaches zero. For open orbits, although the dependence of $\Delta\rho/\rho_0$ on magnetic direction is weak for low H, when H is high enough, the anisotropy of single crystals will be considerable.

The general characteristics of anisotropic magnetoresistance can be understood from considering the symmetry of crystals. For cubic crystals, assuming n is an integer, rotations around the [100] axis by $n\pi/2$, around the [111] axis by $2n\pi/3$, and around the [110] axis by $n\pi$, are all invariant. Figure 8.2.6 gives the anisotropic magnetoresistance for a Cu single crystal, at applied field $H = 18$ kOe, temperature $T = 4.2$ K. In these measurements, H is rotated in the plane perpendicular to j. In most directions, $\Delta\rho/\rho_0$ complies with the square law and shows large values, but in some directions, it saturates under low fields.

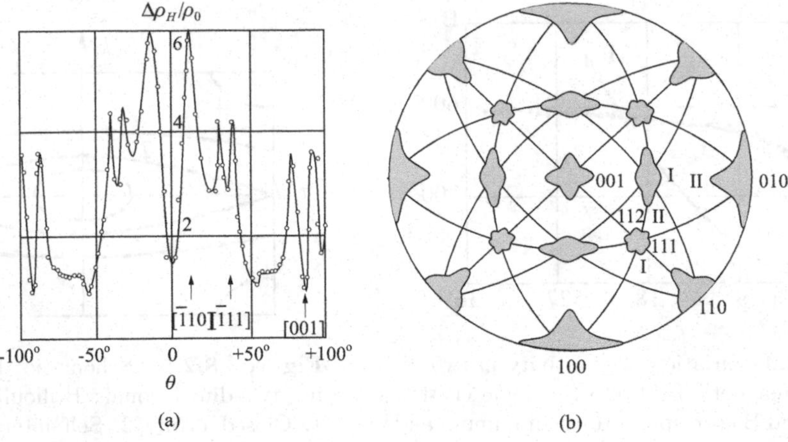

Figure 8.2.7 High field magnetoresistance in Au single crystals. (a) The relation of magnetoresistivity and angle of magnetic field. (b) Stereogram of distribution of magnetoresistance related to magnetic field direction. From J. M. Ziman, *Electron in the Metals*, Taylor & Francis, London (1963).

We have seen it is possible that the transverse magnetoresistance of a single crystal saturates in some directions, but seems to increase with H without limit in others. It is obvious that the saturation directions correspond to the closed orbits on the Fermi surface, while the directions with H^2 correspond to some sections with open orbits. In this case the Fermi surface must be multiconnected. Most interesting experiments were performed on very pure samples at low temperatures, under strong fields. For these experiments, the product of cyclotron frequency and relaxation time $\omega\tau \gg 1$; collision processes were restrained, but details of the Fermi surface were enhanced. As in Fig. 8.2.7(a), when the direction of the magnetic field is changed, the magnetoresistance of Au behaves in a very complicated fashion. One simple and intuitive method is to plot all the unsaturation directions of magnetoresistance into a stereogram like Fig. 8.2.7(b). By investigating the shape and size of the unsaturated regions in this stereogram, the characteristics of the Fermi surface can be obtained qualitatively. So high field magentoresistance is very useful, and can be applied to test and analyze topological structures of Fermi surfaces that have multi-connectivity.

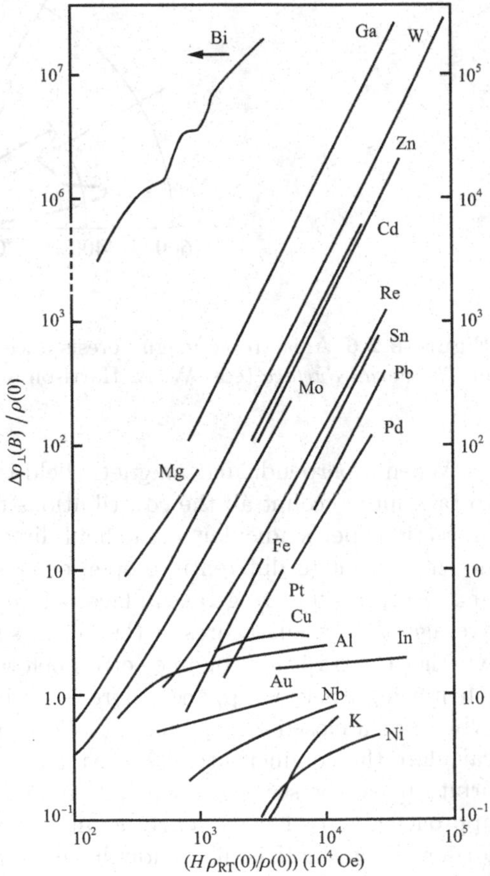

Figure 8.2.8 Transverse magnetoresistance of metals under high magnetic fields. From E. Fawcett, *Adv. Phys.* **13**, 139 (1964).

The calculation of magnetoresistance is a little more difficult than that for the Hall effect, because the magnetoresistance will be zero by using the simple model for electron having the same effective mass, velocity and relaxation time. It can be seen from Fig. 8.2.1 that metals have no magnetoresistance in the free electron model, i.e., ρ_{xx} is independent of H, the reason for this is that the Hall electric field E_y cancels the Lorentz force from magnetic field. In order to prevent such cancellation, we cannot expect to use only one drift velocity to describe the carrier's movement. One simple but important model for drift velocity is to introduce two kind of carriers. These two carriers can be electrons and holes, or s electrons and d electrons, or open orbits and closed orbits. This is the two-band model for magnetoresistance. This model gives a magnetoresistance is equal to zero under a longitudinal field, but under a transverse field the change of resistance can be written as

$$\frac{\Delta\rho}{\rho_0}=\frac{\sigma_1\sigma_2(\sigma_1/n_1+\sigma_2/n_2)^2(H/e)^2}{(\sigma_1+\sigma_2)^2+\sigma_1^2\sigma_2^2(1/n_1-1/n_2)^2(H/e)^2}, \tag{8.2.21}$$

where σ_1 and σ_2 are the conductivities for each band, while n_1 and n_2 are the corresponding carrier densities for each band. In general, this formula is not consistent with the Kohler rule, but when $\sigma_1=\lambda\sigma_2$, and λ is a constant, the above formula reduces to

$$\frac{\Delta\rho}{\rho_0}=\frac{A(H/\rho_0)^2}{1+C(H/\rho_0)^2}. \tag{8.2.22}$$

This represents a specific example satisfying the Kohler rule. It can be seen that the change of resistance under low field is in accordance with H^2; but approaches to saturation under high field except when $n_1=n_2$. The curve profile given by (8.2.22) is very consistent with the experimental results of some metals, such as In. When $n_1=n_2$, C is zero and the resistance increases with H^2 continuously and without limit. This phenomenon was observed in some metals. On the other hand, under high fields it was observed in polycrystalline Cu, Ag, Au, and many other metals that $\Delta\rho/\rho_0$ had a linear relation with H. Because under high fields there are only two possibilities for $\Delta\rho/\rho_0$ of single crystals varying with applied field, i.e., approaching saturation or in accordance with H^2. The appropriate combination of these two possibilities can be used to elucidate the linear dependence of magnetoresistance with H in polycrystalline samples.

One deficiency of the two band model is that it cannot be used to explain the anisotropy observed in single crystals. Theoretical computations of anisotropic magnetoresistance should take into account real shapes of Fermi surfaces, and the possible anisotropy of relaxation times.

Figure 8.2.8 shows the experimental data of several polycrystalline materials for their transverse magnetoresistance with magnetic field. It can be seen that the materials like the semimetal Bi have high ordinary magnetoresistances, and also for the single crystal Bi at low temperatures and under 1.2 T magnetic field, its OMR reaches as high as 10^2–10^3%. On the other hand, there are many semiconductors that have relatively large OMR, for example, for InSb-NiSb, under magnetic field 0.3 T and at room temperature the OMR $\approx$ 200%.

8.2.4 Spin Polarization and Spin Transport

An electron has not only charge, but also spin. It is expected that spin should also play an significant role in electron transport. The topic was involved for the first time in the study of transport in ferromagnetic alloys. After 1980, due to the discovery of the giant magnetoresistance effect, the transport problem of electrons with spins became the basis of a new discipline, spintronics.

The electrons of s and d bands in ferromagnetic metals and alloys all participate in conduction, and are involved in many scattering processes. Figure 8.2.9 schematically gives the densities of states of electronic spin subbands in ferromagnetic metals Fe, Co and Ni. The s bands are wide bands, but the d bands are narrow bands. The exchange splitting of d bands, simplified to an effective internal field, is the fundamental origin for spontaneous magnetization, causing a large difference between $g_\uparrow(E_F)$ and $g_\downarrow(E_F)$. Because the d bands are narrow, electrons have a large effective mass, but s bands are the opposite, so it is usually assumed that conduction is mainly due to the contribution from s electrons. Spin polarization requires $n_\uparrow\neq n_\downarrow$, so the spin susceptibility $(n_\uparrow-n_\downarrow)/(n_\uparrow+n_\downarrow)$ is not zero. It is usual to name $\uparrow$ as the majority spin, and $\downarrow$ as the minority spin.

To illustrate the resistance in ferromagnetic metals, N. F. Mott proposed a two-current model, i.e. electrons with different spins contributing to resistance correspond to two channels connected in parallel.[e] The basic assumption of this model is that electrons with different spins have different

[e]N. F. Mott, *Proc. Roy. Soc. London* **A153**, 699; **A156**, 368 (1936).

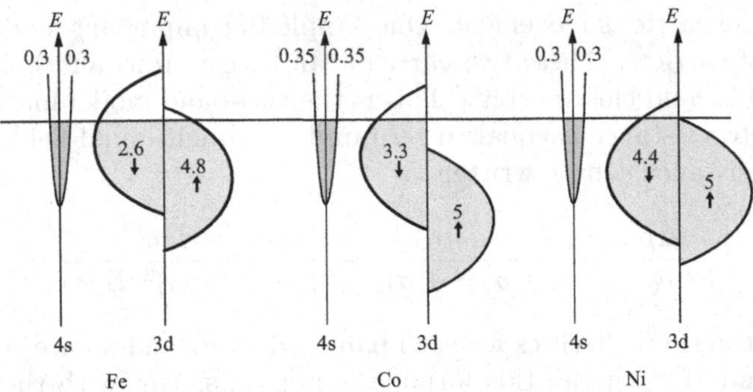

Figure 8.2.9 Schematic DOS curves in ferromagnetic metals.

distribution functions and relaxation times. If we consider that there exists a spin-flip scattering process, another relaxation time $\tau_{\uparrow\downarrow}$ can appear. Now the Boltzmann equation in §8.1.1 should be extended to a system including electrons with two kind of spins, then

$$e\boldsymbol{E}\cdot\boldsymbol{v}\frac{\partial f_0}{\partial E}=-\frac{f_\uparrow-f_0}{\tau_\uparrow}-\frac{f_\uparrow-f_\downarrow}{\tau_{\uparrow\downarrow}},\qquad e\boldsymbol{E}\cdot\boldsymbol{v}\frac{\partial f_0}{\partial E}=-\frac{f_\downarrow-f_0}{\tau_\downarrow}-\frac{f_\downarrow-f_\uparrow}{\tau_{\uparrow\downarrow}}. \tag{8.2.23}$$

I. A. Campbell and A. Fert solved these coupled equations,[f] and obtained an expression for the total resistance

$$\rho=\frac{\rho_\uparrow\rho_\downarrow+\rho_{\uparrow\downarrow}(\rho_\uparrow+\rho_\downarrow)}{\rho_\uparrow+\rho_\downarrow+4\rho_{\uparrow\downarrow}}, \tag{8.2.24}$$

where

$$\rho_\uparrow=\frac{m*}{ne^2\tau_\uparrow},\qquad \rho_\downarrow=\frac{m*}{ne^2\tau_\downarrow},\qquad \rho_{\uparrow\downarrow}=\frac{m*}{ne^2\tau_{\uparrow\downarrow}}.$$

If the effect of spin-flip scattering can be neglected, (8.2.4) will be reduced to the two-current model

$$\rho=\frac{\rho_\uparrow\rho_\downarrow}{\rho_\uparrow+\rho_\downarrow}. \tag{8.2.25}$$

The up and down spins form the current channels connected in parallel. The channel with small resistivity corresponds to a short mean free path. In the 1990s, the experimentally measured values of mean free path, proportional to relaxation time, of electrons with different spins are, cobalt: $l_\uparrow = 5.5$ nm, $l_\downarrow = 0.6$ nm; permalloy: $l_\uparrow = 4.6$ nm, $l_\downarrow = 0.6$ nm.

It is known that even in *s* metals, the electronic mean free path is not large, the magnitude is only about 10 nm. This means electrons in metals experience collisions very frequently, the momentum relaxation determines the value of resistivity. It may be asked if these frequent collisions cause an electron to lose its memory for previous spin orientation. Actually it does not, a spin flip can occur only through the exchange interaction or scattering by an impurity or defect with spin-orbit coupling. In nonmagnetic metals, where an electron experiences scattering many times, the original spin orientation can still retained. A spin memory effect in such large range can be described in thermodynamic language. If some spin-polarized electrons are injected into a nonmagnetic metal, the average time experienced (or average distance passed) with the original spin orientation in a transport process is the called spin mean time τ_s (or the spin diffusion length $L_s = \sqrt{2D\tau_s}$, D is the diffusion coefficient). In fact, in nonmagnetic metals the probability of spin-flip scattering only about 1/100 or 1/1000 that of scattering events, so the spin diffusion length is several hundred times of the electronic mean free path. At room temperature, the magnitude of L_s in Ag, Au and Cu is about 1–10 μm; at temperatures lower than 40 K, L_s in Al can even reach 0.1 mm, so it is expected that the transport of spin polarized electrons in nonmagnetic metals can also play a considerable role. Certainly one precondition is the injection of spin-polarized electrons

[f] I. A. Campbell *et al.*, *Phil. Mag.* **15**, 977 (1967); A. Fert *et al.*, *Phys. Rev. Lett.* **21**, 1190 (1968).

from ferromagnetic substances. Moreover, it is expected that once the spin-polarized electrons are injected, there may be spin accumulation within the spin diffusion length and this accumulation will lead to nonequilibrium magnetization in nonmagnetic metals. The same situation can also appear in nonmagnetic semiconductors. The long spin diffusion length in nonmagnetic metals provides the physical basis for development of spintronics.

8.2.5 Resistivity and Magnetoresistance of Ferromagnetic Metals

We now return to discuss the resistance and magnetoresitance of ferromagnetic metals. There are three sources of resistivity of ferromagnetic metals described by

$$\rho(T) = \rho_{\rm d} + \rho_{\rm l}(T) + \rho_{\rm m}(T), \tag{8.2.26}$$

where $\rho_{\rm d}$ is the residual resistivity, i.e., the resistivity at $T = 0$ K, arising from impurities and defects; $\rho_{\rm l}$ is from electronic scattering by lattice vibrations, which increases with temperature; $\rho_{\rm m}$ is the resistivity due to scattering related to magnetic order, it is also a function of temperature and limited to ferromagnetic metals. Take Ni as an example, its ρ-T relation is shown in Fig. 8.2.10. At $T \gg T_{\rm c}$ spins are all disordered, $\rho_{\rm m}$ approaches a temperature-independent saturated value; but when $T \leq T_{\rm c}$, the spin order causes the spontaneous magnetization to appear; within this region the variation of resistivity can be divided into lower and higher temperature parts for discussion.

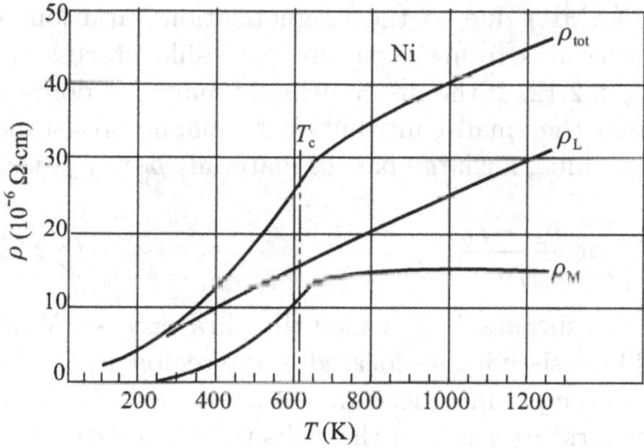

Figure 8.2.10 The relation of resistivity-temperature in Ni.

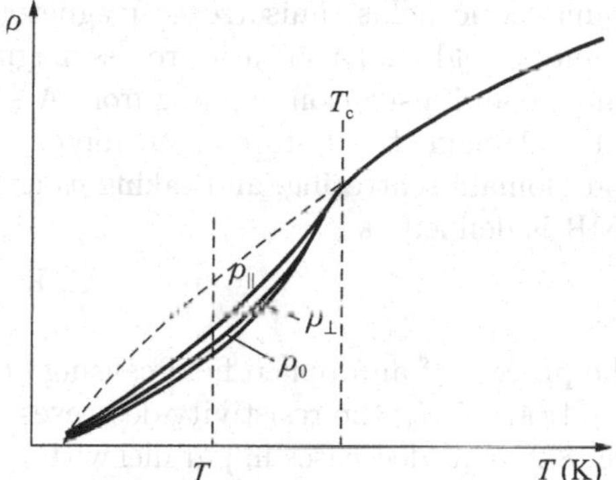

Figure 8.2.11 Schematic anisotropic resistivity in ferromagnets.

At lower temperatures, because $\rho_{\downarrow} > \rho_{\uparrow}$ and $\rho_{\uparrow\downarrow}$ can be neglected, according to the two-current model, it is possible to explain why $\rho_{\rm M}$ is so small (that is due to the short-circuit role played by $\rho_{\uparrow}$). As the temperature is raised, the spin-flip scattering gives rise to a spin mixing effect, and the short-circuit role of low resistance is reduced, so $\rho_{\rm M}$ rises, in accordance with the modified expression in the two-current model. In high temperatures, $\rho_{\uparrow}\rho_{\downarrow} \gg \rho_{\uparrow}$ (or $\rho_{\downarrow}$), and $\rho_{\rm M}$ approaches the saturation value

$$\rho_{\rm M}^{\rm s} = \frac{1}{4}(\rho_{\uparrow} + \rho_{\downarrow}).$$

In addition, below $T_{\rm c}$, the resistivity of ferromagnetic metals shows clear anisotropy, i.e., it varies with the relative orientation between current and spontaneous magnetization $M_{\rm s}$. It can be expressed as

$$\rho_{\parallel}(I \parallel M_{\rm s}) \neq \rho_{\perp}(I \perp M_{\rm s}),$$

and schematically sketched in Fig. 8.2.11. In the majority of materials $\rho_{\parallel} > \rho_{\perp}$. This is opposite to OMR stated in §8.2.1 and indicates that the principal mechanism of resistivity anisotropy in ferromagnetic metals cannot be ascribed to OMR arising from the internal field $M_{\rm s}$. Usually the structure of a ferromagnetic metal in the demagnetized state contains many domains, so the angles

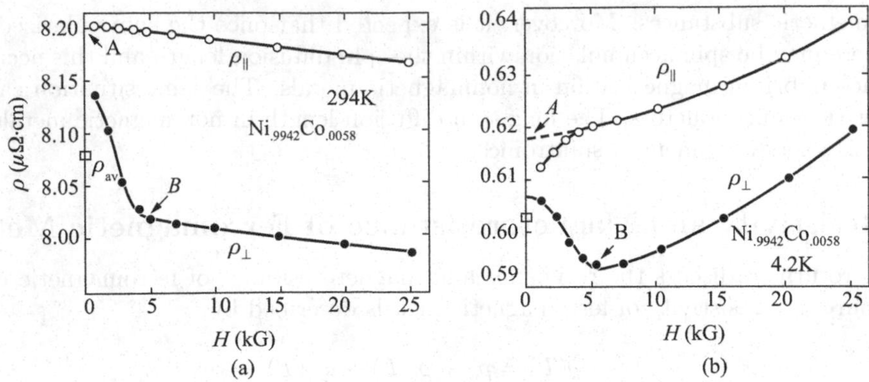

Figure 8.2.12 Dependence of resistivity with magnetic field in columnar Co-Ni alloys. (a) Room temperature, (b) low temperature (4.2 K).

between M_ss of various domains with current I have a certain distribution. If ρ_0 is the average value of resistivities of various domains, then in materials with $\rho_\parallel > \rho_\perp$, we have $\rho_\parallel > \rho_0 > \rho_\perp$, it is expected that ρ_0 includes the contribution due to scattering of electrons by domain walls.

The total magnetoresistance (MR) in a ferromagnetic metal includes OMR induced directly by magnetic fields, anisotropic magnetorisistance (AMR) due to the magnetization variation by magnetic field, and also para-process magnetoresistance arise from the paramagnetic-like process. We can give an illustration of going from A $\rightarrow$ B in Fig. 8.2.12. If the distribution of magnetic domains in the demagnetized state is isotropic, by neglecting the small contribution to magnetoresistance from domain scattering, and taking ρ_0 as its average value, for large part of materials $\rho_\parallel > \rho_\perp$, then AMR is defined as

$$\text{AMR} = \frac{\rho_\parallel - \rho_0}{\rho_0}, \text{ or } \frac{\rho_\perp - \rho_0}{\rho_0}. \tag{8.2.27}$$

The process of an applied field causing $(H + 4\pi M)$ to surpass M_s is called the para-process. When $H + 4\pi M > M_s$, the resistivity decreases further. The resistivity $\rho_\perp$ located in the region above B in Fig. 8.2.12(a) decreases in parallel with $\rho_\parallel$ belongs to the paramagnetic-like process. OMR is almost overshadowed. At low temperature, decreasing temperature leads to the reduction of resistivity, so OMR increases. In Fig. 8.2.12(b) OMR is dominant and paramagnetic-like process is overshadowed. $\rho_\parallel$ and $\rho_\perp$ all increase with H. For permalloy under weak fields (about several Oe), the AMR can reach a high value of 3–5%, and can also become negative. It has been used in read-out magnetic heads in computer magnetic disk drives.

§8.3 Tunneling Phenomena

Tunneling phenomena are based upon the wave nature of quantum particles. The study of them began in the earliest period of quantum mechanics. Different from the classical concept, a particle trapped in one region can tunnel through an energy barrier into another region. Two identical or different materials separated by a barrier layer can form a primary tunneling junction. Because the materials used to make a tunneling junction may be metals, semiconductors, or superconductors, and may be magnetic or nonmagnetic, tunneling phenomena are very rich, and are also important for modern nanoscience and technology.

8.3.1 Barrier Transmission

Consider a typical tunneling structure as shown in Fig. 8.3.1, where there is a potential barrier $V(z)$ in the region between z_1 and z_2, and the potential is zero outside this region. So outside the barrier, (in the regions to the left of the left boundary and to the right of the right boundary) the

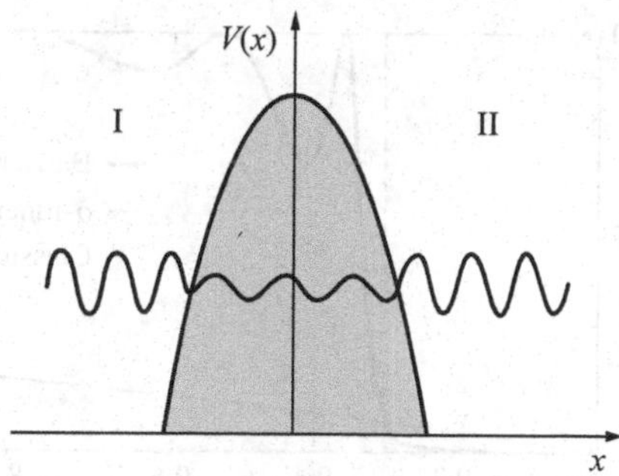

Figure 8.3.1 A particle with wave property and energy E tunneling from left to right across a barrier.

electronic wavefunction with energy $E = \hbar^2k^2/2m$ can be written as the linear combination of a forward travelling wave and a backward travelling wave, as

$$\psi_1(z) = A_1\mathrm{e}^{ikz} + B_1\mathrm{e}^{-ikz}, \qquad \psi_2(z) = A_2\mathrm{e}^{ikz} + B_2\mathrm{e}^{-ikz}, \tag{8.3.1}$$

the difference between them is that the coefficients may be different. We can also write the electronic wavefunction in the barrier, then using the continuity conditions for wavefunctions and their derivatives, we can obtain a transfer matrix $\boldsymbol{T}$

$$\begin{pmatrix} A_2 \\ B_2 \end{pmatrix} = \boldsymbol{T}\begin{pmatrix} A_1 \\ B_1 \end{pmatrix} = \begin{pmatrix} T_{11} & T_{12} \\ T_{21} & T_{22} \end{pmatrix}\begin{pmatrix} A_1 \\ B_1 \end{pmatrix}, \tag{8.3.2}$$

or a scattering matrix $\boldsymbol{S}$

$$\begin{pmatrix} A_2 \\ B_1 \end{pmatrix} = \boldsymbol{S}\begin{pmatrix} A_1 \\ B_2 \end{pmatrix} = \begin{pmatrix} t_1 & r_2 \\ r_1 & t_2 \end{pmatrix}\begin{pmatrix} A_1 \\ B_2 \end{pmatrix}, \tag{8.3.3}$$

in which $\tilde{T} = |t_1|^2$ represents the transmission probability of the particle from left-hand side of the barrier to the right-hand side. and $\tilde{R} = |r_1|^2$ is the reflection probability of the particle by the barrier.

From the fact that the conjugate ψ^* of one solution ψ of the Schrödinger equation is also a solution, it is easy to show that there are relations for the matrix elements in the $\boldsymbol{T}$ matrix

$$T_{11} = T_{22}^*, \qquad T_{12} = T_{21}^*, \tag{8.3.4}$$

and in the $\boldsymbol{S}$ matrix

$$t_1t_2^* = 1 - |r_1|^2, \qquad |r_1|^2 = |r_2|^2 = |\tilde{r}|^2, \qquad t_1r_2^* = -r_1t_2^*, \qquad t_2r_1^* = -r_2t_1^*. \tag{8.3.5}$$

It will be useful to express the transfer matrix as

$$\boldsymbol{T} = \begin{pmatrix} t_2^{*-1} & r_2t_2^{-1} \\ r_2^*t^{*-1} & t_2^{-1} \end{pmatrix}, \tag{8.3.6}$$

its determinant is det $\boldsymbol{T} = t_1/t_2 = t_1^*/t_2^*$. Finally using the conservation condition for probability flow

$$|t_1|^2 + |\tilde{r}|^2 = |t_2|^2 + |\tilde{r}|^2 = 1, \tag{8.3.7}$$

it is found that $t_1 = t_2 = \tilde{t}$, so det $\boldsymbol{T}$=1.

For a square barrier with height V_0, located between $z_1 = -a/2$ and $z_2 = a/2$, it can be shown that for a particle with energy $E < V_0$ the transmission amplitude is

$$\tilde{t} = |\tilde{t}|\mathrm{e}^{i\phi}, \tag{8.3.8}$$

where ϕ satisfies

$$\tan(\phi + ka) = \frac{1}{2}\left(\frac{k}{\kappa} - \frac{\kappa}{k}\right)\tanh \kappa a, \tag{8.3.9}$$

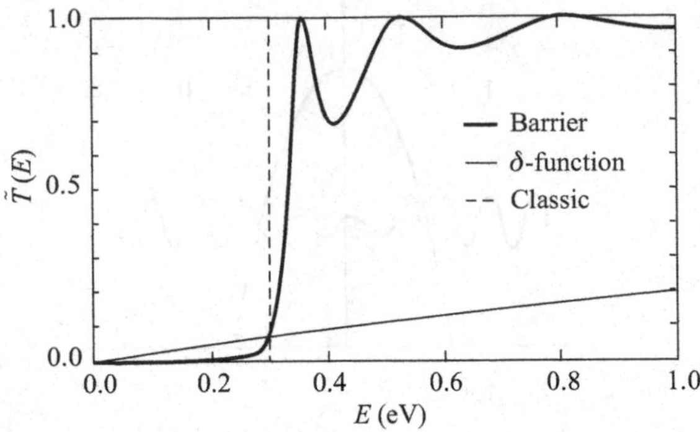

Figure 8.3.2 The relation of transmission coefficient and particle energy for a square barrier. From J. H. Davis, *The Physics of Low-Dimensional Semiconductors*, Cambridge University Press, Cambridge (1998).

in which $E = \hbar^2k^2/2m$ and $V_0 - E = \hbar^2\kappa^2/2m$. Then the transmission coefficient is

$$\tilde{T} = |\tilde{t}|^2 = \frac{4k^2\kappa^2}{4k^2\kappa^2 + (k^2+\kappa^2)^2\sinh^2\kappa a} = \left[1 + \frac{V_0^2}{4E(V_0-E)}\sinh^2\kappa a\right]^{-1}. \tag{8.3.10}$$

Figure 8.3.2 shows the relation of the transmission coefficient as a function of energy with $V_0 = 0.3$ eV and $a = 10$ nm. In the classical picture, when $E < V_0$, $\tilde{T} = 0$; and when $E > V_0$, $\tilde{T} = 1$, but quantum mechanics permits the particle to pass through the barrier when $E < V_0$. When κa is very large, (8.3.10) simplifies to

$$\tilde{T} = \frac{16E}{V_0}\exp(-2\kappa a). \tag{8.3.11}$$

and the tunneling probability is mainly determined by the exponential term. Often just $\exp(-2\kappa a)$ is used to give a simple estimate of the tunneling probability.

If in (8.3.1) z is replaced by $(z-d)$, which corresponds to the barrier being displaced by d, it is easy to verify that original $\boldsymbol{T}$ matrix is becomes

$$\boldsymbol{T}_d = \begin{pmatrix} e^{-ikd} & 0 \\ 0 & e^{ikd} \end{pmatrix} \boldsymbol{T} \begin{pmatrix} e^{ikd} & 0 \\ 0 & e^{-ikd} \end{pmatrix}. \tag{8.3.12}$$

For a double barrier structure with one barrier at $z = 0$ and the other at $z = d$, the total transfer matrix is

$$\boldsymbol{T}_t = \boldsymbol{T}_d\boldsymbol{T}. \tag{8.3.13}$$

It is not difficult to get its four matrix elements, and the reciprocal of the fourth matrix element is just the total transmission amplitude, the expression is

$$\tilde{t}_t = \frac{\tilde{t}^2}{1 + |\tilde{r}|^2 e^{2i(kd+\phi)}}. \tag{8.3.14}$$

Therefore the total transmission probability is

$$\tilde{T}_t = |\tilde{t}_t|^2 = \frac{\tilde{T}^2}{|1 + |\tilde{r}|^2 e^{2i(kd+\phi)}|^2} = \frac{(1-|\tilde{r}|^2)^2}{|1 + |\tilde{r}|^2 e^{2i(kd+\phi)}|^2}. \tag{8.3.15}$$

The interesting result is that for almost all energies the total transmission probability is approximately the square of the transmission probability of a single barrier, but for some specific values, which satisfy $2(kd+\phi) = (2n+1)\pi$, the double barrier is transparent, corresponding to a filter, which only permits the electrons with energies near the resonance values to be transmitted through this double barrier. This effect can be applied to fabricate the double barrier diode.

Using a barrier as a spacer and two similar or different materials as electrodes, various tunneling junctions can be constructed. When no voltage is applied, the Fermi surfaces for two electrodes are equal, so no current passes through the tunneling junction. When a bias voltage is applied, the Fermi surfaces of the two electrodes will have a relative displacement. Using the Fermi golden rule we can compute the number of electrons passing through the tunneling junction with a definite energy. This is proportional to the product of the density of states for an electron state to be occupied at one electrode, the other density of states which is empty at the other electrode and the tunneling probability. The total tunneling current is the sum of the various energy electrons which pass through the tunneling junction, i.e.

$$\begin{aligned} J(V) &\propto \int |M|^2 g_1(E-eV)g_2(E)\{f(E-eV)[1-f(E)]+[1-f(E-eV)]f(E)\}dE \\ &\propto \int |M|^2 g_1(E-eV)g_2(E)[f(E-eV)-f(E)]dE, \end{aligned} \tag{8.3.16}$$

where g_1 and g_2 are the densities of states at the Fermi surfaces, and $f(E)$ is the Fermi distribution function.

8.3.2 Resonant Tunneling through Semiconductor Superlattices

The above discussion about the electron transmission through single and double barriers can be extended to multi barrier, or even periodically repeated structures. In a periodic structure, a crystal or superlattice, the wave packet of an electron that is initially localized in space will spread by resonant tunneling, and eventually become delocalized. The quantized energy levels are broadened into energy bands, due to the tunneling process known as Bloch tunneling. In the following, we will be concerned with tunneling in artificial microstructures.

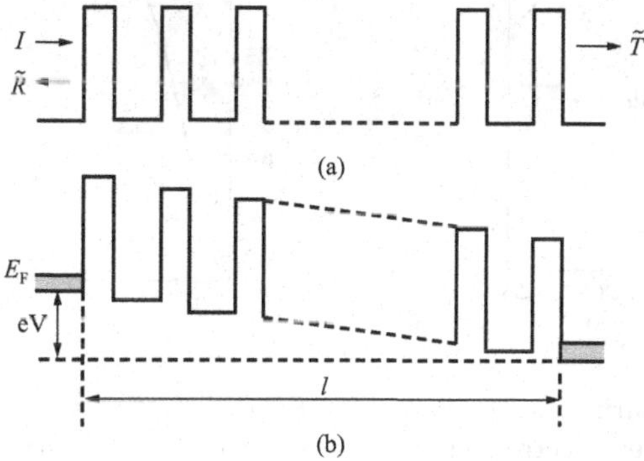

Figure 8.3.3 A finite superlattice of length l, (a) in equilibrium state with incidence, reflection, and transmission amplitudes, (b) after applying a voltage V.

We consider multiple barrier resonance tunneling.[g] In reality, the repetition period of a superlattice is always finite, so electron transport is related to an electron moving in multi-wells and barriers, as shown in Fig. 8.3.3. The energy of the incident electron includes two parts

$$E = E_l + \hbar^2 k_t^2/2m^*, \tag{8.3.17}$$

where E_l is the longitudinal energy, $\hbar^2 k_t^2/2m^*$ is the transverse energy, and the wavefunction is a product

$$\psi = \psi_l \psi_t. \tag{8.3.18}$$

At the left and right ends the wavefunctions are

$$\psi_l = \psi_t \cdot (e^{ik_1 z} + re^{-ik_1 z}), \tag{8.3.19}$$

[g] R. Tsu and L. Esaki, *Appl. Phys. Lett.* **22**, 562 (1973).

and

$$\psi_N = \psi_t \cdot t e^{ik_N z}, \tag{8.3.20}$$

respectively.

By matching the wavefunctions and their first derivatives at each interface, one can derive the reflection and transmission amplitudes, $\tilde{r}$ and $\tilde{t}$ and the transmission coefficient $|\tilde{t}|^2$ follows. In Fig. 8.3.4, $\ln \tilde{t}^*\tilde{t}$ is plotted as a function of electron energy for a double, a triple, and a quintuple barrier structure. In the calculation, the parameters are $m^* = 0.067m_e$, the height of barrier $V_0 = 0.5$ eV, the width of barriers $d_1 = 20$ Å and the width of wells $d_2 = 50$ Å. Note the splitting of the resonances. Each of the resonance peaks for a system with N repetitive periods has been split into $(N-1)$ small peaks. The resonance peaks represent the positions of electron energies. When the energy of the incident electron equals one of these energies, there is the largest transmission possibility. There is no doubt that, when N increases, the tunneling model is changing into band model.

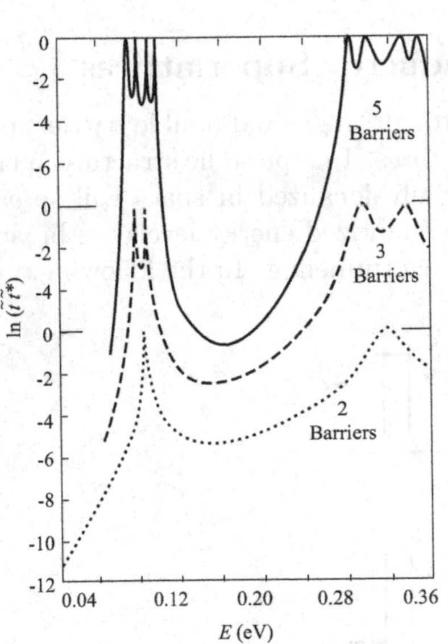

Figure 8.3.4 Plot of the logarithmic transmission coefficient $\ln t^*t$ versus electron energy showing peaks at the energies of the bound states in the potential well. From R. Tsu and L. Esaki, *Appl. Phys. Lett.* **22**, 562 (1972).

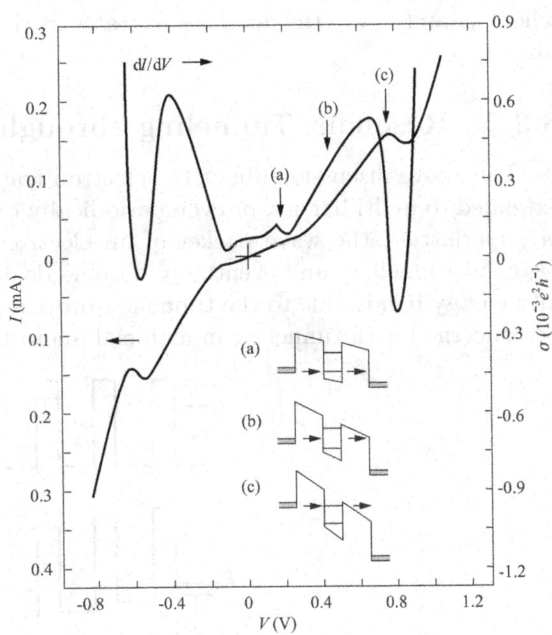

Figure 8.3.5 Current-voltage and conductance-voltage characteristics of a double-barrier structure at 77 K. Condition at resonance (a) and (c) and off-resonance (b) are indicated. From L. Esaki, in *Synthetic Modulated Structures* (eds. L. L. Chang and B. C. Giessen), Academic Press, Orlando (1985).

The net tunneling current J can be found as follows. At first, we need to define two energies, one is the energy of incident electron E, and the other is that of the transmitted electron E'.

$$J = \frac{e}{4\pi^3\hbar}\int_0^\infty dk_l \int_0^\infty dk_t [f(E) - f(E')]\tilde{t}^*\tilde{t}\frac{\partial E}{\partial k_l}. \tag{8.3.21}$$

Because of a separation of the variables, the transmission coefficient $\tilde{t}^*\tilde{t}$ is only a function of the longitudinal energy E_l. Together with the Fermi distribution function, the expression can be integrated over the transverse direction

$$J = \frac{em^*k_BT}{2\pi^2\hbar^3}\int_0^\infty \tilde{t}^*\tilde{t}\ln\left(\frac{1+\exp[(E_F - E_l)/k_BT]}{1+\exp[(E_F - E_l - eV)/k_BT]}\right) dE_l. \tag{8.3.22}$$

In the low temperature limit $T \to 0$, for $V \geq E_F$ we have

$$J = \frac{em^*}{2\pi^2\hbar^3}\int_0^{E_F}(E_F - E_l)t^*tdE_l, \tag{8.3.23}$$

and for $V < E_F$,

$$J = \frac{em^*}{2\pi^2\hbar^3}\left[V\int_0^{E_F-V}\tilde{t}^*\tilde{t}dE_l + \int_{E_F-V}^{E_F}(E_F - E_l)\tilde{t}^*\tilde{t}dE_l\right]. \tag{8.3.24}$$

We can see a resonant tunneling in double barriers in Fig. 8.3.5 where the well width of GaAs is 50 Å, the width of two barriers of $Ga_{0.3}Al_{0.7}As$ is 80 Å.

8.3.3 Zener Electric Breakdown and Magnetic Breakdown

In §6.2.1 when Bloch oscillations were discussed, only electron movement in one band was considered. This is reasonable if the electric field is not too strong. As shown in Fig. 6.2.1, according to the description for the Bloch oscillation, under a electric field, an electron moves from the point O to the point A. The point A is equivalent to the point A' with the difference of one reciprocal lattice vector, so the electron returns to A' to continue its movement. This is a repetitive process in reciprocal space and real space with acceleration and deceleration.

When an electric field is increased to a high value, it is possible that this picture of a single band needs modification. In contrast to Fig. 6.2.1, we see that, in Fig. 8.3.6, the second energy band is added to the dispersion relation and there is a energy gap E_g between its lowest point A'' and the point A. There is now a probability for an electron to acquire enough energy to jump from A to A'' which is in a different band. This transition is called the Zener effect. We can also refer to the energy diagram in real space (Fig. 8.3.7), where an electron acquires electric field energy $e\mathcal{E}z$, and at the point A it usually is reflected back to A'. However, if it can move a further distance $d = E_g/e\mathcal{E}$ to reach A'', then it acquires enough energy to cross the gap.

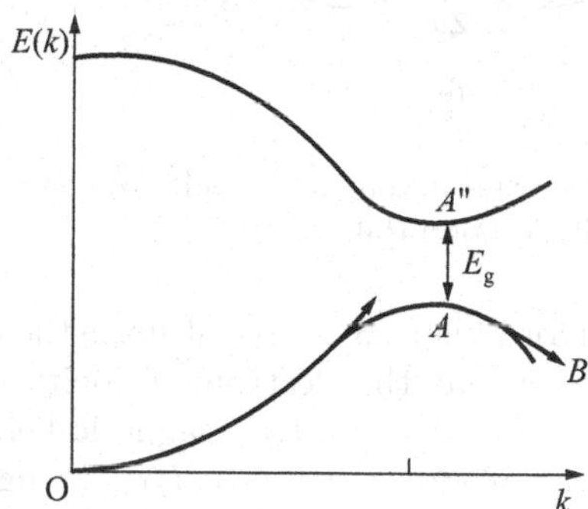

Figure 8.3.6 Electron transition between energy bands moving in Brillouin zone under electric fields.

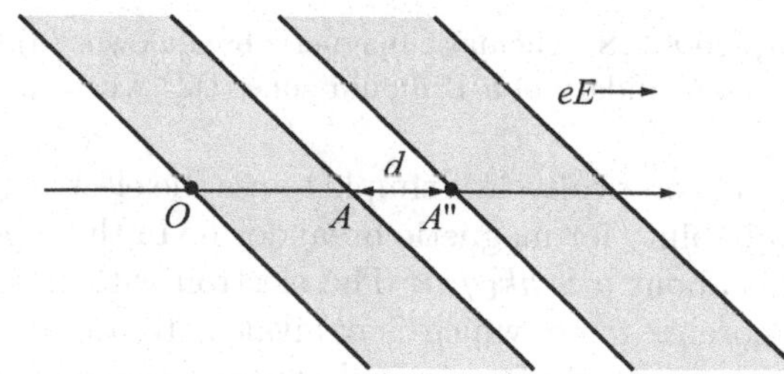

Figure 8.3.7 Tilted energy bands in real space under strong electric fields.

This is essentially a tunneling problem, because an electron must penetrate through the forbidden energy region and so, by combining the band theory in the near-free electron approximation and the tunneling theory from §8.3.1, we obtain the probability of electric breakdown:

$$\tilde{T} = \exp\left[-\frac{\pi^2}{4}\frac{E_g^2}{E_0 e\mathcal{E}a}\right], \tag{8.3.25}$$

where a is the lattice constant, $E_0 = (\hbar^2/2m)(G/2)^2$ is the electronic kinetic energy (equal to the Fermi energy) and $(G/2) = \pi/a$. For the energy gap E_g to ensure that there is electron tunneling, the electric field must satisfy

$$\frac{e\mathcal{E}aE_0}{E_g^2} > 1. \tag{8.3.26}$$

Similar to electric breakdown, a strong enough magnetic field can also cause electrons to transfer between different energy bands. This is magnetic breakdown which is also a tunneling problem. According to §6.3.2, if we consider a strong magnetic field along the z direction, then an electron describes a circular motion in the two-dimensional plane. We introduce a periodic potential for a crystal

$$V(\boldsymbol{r}) = \sum_{\boldsymbol{G}} V_{\boldsymbol{G}} e^{i\boldsymbol{G}\cdot\boldsymbol{r}}, \tag{8.3.27}$$

and take it as a perturbation. When the electronic orbit passes through the boundaries of the Brilbuin zone, i.e., its effective wavevector in the x direction k_x is $\pm G/2$, Bragg diffraction will take place. As shown in Fig. 8.3.8, the electronic orbit may be transferred from AB to AC. This comes from the possibility of orbit reconnection at the boundary of the Brillouin zone in a periodic lattice.

When the strength of perturbation is increased, the trajectory at point A will be split in energy, and the path AC will be preferred. Now the electron moves in open orbits according to the conventional scheme of repeat zones, so Sec. B in the original ring in Fig. 8.3.8(a) will be connected to form a separate branch. However, in a very strong magnetic field, the orbit may jump back to the free-electron path, i.e., electrons may return to circular motion. The electron does not move along AC normally, but an energy gap breakdown takes place, that is to say that there is tunneling between a region in reciprocal space to two separate orbits.

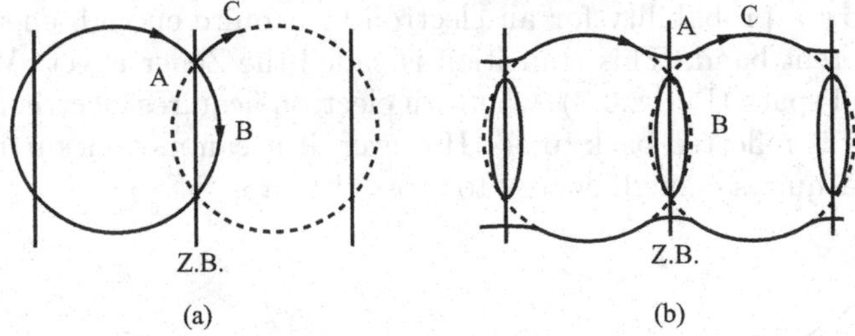

Figure 8.3.8 Schematic magnetic breakdown. (a) Free-electron orbit in magnetic field. (b) Orbit connected at the boundary of a Brillouin zone (BZ) when there is a periodic potential.

Theoretically, it is simple to use the electric breakdown formula (8.3.25) to estimate the transition probability for magnetic breakdown. In the repeated zone scheme, the electronic velocity near point A is about $v \simeq \hbar k_F/m$. The electron with this velocity moving through the magnetic field will feel a Lorentz force, which is equivalent to a electric field force of strength $\mathcal{E} \simeq vH/c$. Using (8.3.25), this equivalent electric field causes a tunneling process, if the parameters satisfy the inequality

$$\frac{e\hbar H}{m}\frac{k_F a E_0}{E_g^2} > 1. \tag{8.3.28}$$

By noting that $k_F a$ is of order unity, eH/m is the cyclotron resonance frequency ω_c, and $E_0 \approx E_F$, the condition of magnetic breakdown simplifies to

$$\frac{\hbar\omega_c E_F}{E_g^2} > 1. \tag{8.3.29}$$

8.3.4 Tunneling Magnetoresistance

Consider two ferromagnetic metals isolated by a thin nonmagnetic insulating layer. The electronic transport depends on the tunneling process through the insulating barrier and is related to the relative orientations of the magnetizations of the two metals; the latter depending mainly on their parallel or antiparallel alignment. The magnetization in a metal is determined by the spin polarization of its electrons, so this is a kind of spin-polarized transport.[h]

From this physical picture, only the electrons close to the Fermi level participate in the transport process, which is dominated by the s, p electrons with their more extended wavefunctions. The density of states of majority spins $n_\uparrow$ is higher than that of minority spins $n_\downarrow$. If the magnetization orientations in two ferromagnetic metal electrodes are in parallel, then the electrons of the electrode in the majority spin subband can enter the unoccupied states of the other electrode in the majority spin subband; the same situation occurs for the electrons in the minority spin subband. But if the magnetizations of two electrodes are in antiparallel, then the spin of electrons in the majority subband of one electrode is parallel to the spin of electrons in the minority subband of the other electrode. Consequently, in a tunneling process, the electrons in the majority subband of one electrode must enter into the unoccupied states in minority subband of the other electrode; *vice versa.* This is the reason why electrons with majority spin emitted into the junction meet a high density of empty states when the magnetization orientations are parallel. Therefore, the resistance is low; in the opposite case, the resistance is high.

In 1975, Jullière studied the transport properties of a Fe-Ge-Co tunneling junction, experimentally verifying that the resistance is related to the relative magnetization orientations of the two ferromagnetic layers and proposed a simple model.[i] This is a very short paper, but cited frequently in the literature of giant tunneling magnetoresistance (TMR). In this model, spin-flip is neglected in the tunneling process and the total conductivity $\sigma \propto 1/R$ is the sum of the two spin channels, which gives $\sigma_{\uparrow\uparrow} \sim n_\uparrow n_\uparrow + n_\downarrow n_\downarrow$ for parallel orientation and $\sigma_{\uparrow\downarrow} \propto n_\uparrow n_\downarrow + n_\downarrow n_\uparrow$ for antiparallel. Defining the spin polarization as

$$P = \frac{n_\uparrow - n_\downarrow}{n_\uparrow + n_\downarrow}, \tag{8.3.30}$$

and generalizing to two electrodes with different spin polarizations P_1 and P_2, one obtains a formula for reduced resistance $\Delta R/R$. At this point we have to mention that two definitions of $\Delta R/R$ exist in the literature, one is 'conservative', i.e., $(\Delta R/R)_\mathrm{c}$; the other is 'inflationary', i.e., $(\Delta R/R)_\mathrm{i}$. So we have

$$\left(\frac{\Delta R}{R}\right)_\mathrm{c} = \frac{R_{\uparrow\downarrow} - R_{\uparrow\uparrow}}{R_{\uparrow\downarrow}}, \qquad \left(\frac{\Delta R}{R}\right)_\mathrm{i} = \frac{R_{\uparrow\downarrow} - R_{\uparrow\uparrow}}{R_{\uparrow\uparrow}}, \tag{8.3.31}$$

while the former is always less than 100%, the latter can be infinite. Their relation is

$$\left(\frac{\Delta R}{R}\right)_\mathrm{i} = \left(\frac{\Delta R}{R}\right)_\mathrm{c} \Big/ \left[1 - \left(\frac{\Delta R}{R}\right)_\mathrm{c}\right]. \tag{8.3.32}$$

The resulting expressions for magnetoresistance are

$$\left(\frac{\Delta R}{R}\right)_\mathrm{c} = \frac{2P_1P_2}{1+P_1P_2}, \qquad \left(\frac{\Delta R}{R}\right)_\mathrm{i} = \frac{2P_1P_2}{1-P_1P_2}. \tag{8.3.33}$$

The Jullière model shows that the existence of spin polarization (or in other words, the difference of DOS at the Fermi level in spin subbands) in both electrodes is the dominant factor for TMR and this has been borne out in many subsequent experiments. However, it wholly neglects the influence of the momentum of tunneling electrons, as well as the physical characteristics of the barrier layer on TMR, so it is unsuitable for a quantitative comparison with experimental results. Slonczewski

[h] The earliest direct experimental verification of electronic spin polarization was by Tedrow and Meservery who measured the tunneling current of a superconductor-nonmagnetic isolating layer-ferromagnetic metal (S-I-FM) tunneling junction under applied magnetic fields. Their measurements confirmed that currents of magnetic metals, like Fe, Co, and Ni under applied electric fields, are spin-polarized. [P. M. Tedrow and R. Meservery, *Phys. Rev. Lett.* **26**, 192 (1971)]. Afterwards, Slonczewski suggested (1975) in an unpublished work that the superconductor in the tunneling junction could be replaced by another ferromagnetic metal, to form a tunneling junction composed of ferromagnetic metal-nonmagnetic isolating layer-ferromagnetic metal (FM-I-FM) tunneling junction.

[i] M. Jullière, *Phys. Lett.* **54A**, 225 (1975).

proposed an improved model for TMR based on a full quantum mechanical treatment of both electrodes and the barrier layer. He introduced the effective spin polarization P' satisfying

$$P' = \frac{(k_\uparrow - k_\downarrow)(\kappa^2 - k_\uparrow k_\downarrow)}{(k_\uparrow + k_\downarrow)(\kappa^2 + k_\uparrow k_\downarrow)}, \tag{8.3.34}$$

to replace the spin polarization P in (8.3.30). Here $k_\uparrow$ and $k_\downarrow$ are the Fermi wavenumbers in spin-up and spin-down subbands, and κ is the wavenumber in the barrier. So the formula for TMR is

$$\left(\frac{\Delta R}{R}\right)_{\text{c}} = \frac{2P_1'P_2'}{1+P_1'P_2'}, \qquad \left(\frac{\Delta R}{R}\right)_{\text{i}} = \frac{2P_1'P_2'}{1-P_1'P_2'}. \tag{8.3.35}$$

The Slonczewski model, and the subsequent extension of this model from rectangular to trapezoidal barrier, may explain more detailed experimental results of TMR showing the influence of bias voltage and physical characteristics of the barrier layer on TMR.[j]

Many applications are based on magnetic tunneling junctions: For example, two ferromagnetic layers with different coercivities. If at first the two ferromagnetic layers form a antiparallel configuration, the tunneling junction is in the high resistance state. Application of a magnetic field can transform it into the parallel configuration, and the tunneling junction is changed into a low resistance state. When the magnetic field is further reduced until it becomes negative, the magnetization of the ferromagnetic layer with lower coercivity will be reversed first, so that the two magnetic layers are antiparallel again, so on and so forth. It is clear that a magnetic field can be used to control the change of resistance from high to low or from low to high. Figure 8.3.9 shows the tunneling resistance of CoFe-Al_2O_3-Co at a temperature of 295 K under applied field, and also shows the resistances of CoFe and Co layers for comparison.

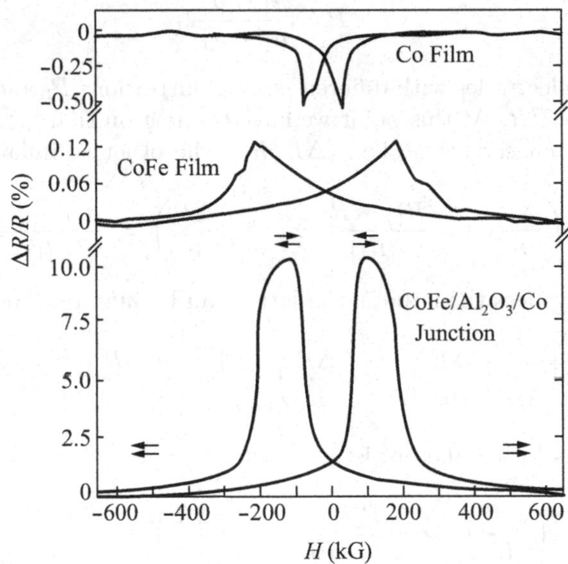

Figure 8.3.9 The relation of tunneling magnetoresistance (TMR) and magnetic field for CoFe-Al_2O_3-Co three layer structure. Arrows represent the magnetization directions in two magnetic layers. From J. S. Moodera *et al.*, *Phys. Rev. Lett.* **74**, 3273 (1995).

8.3.5 Scanning Tunneling Microscope

It has been known from §8.3.1 that a particle with mass m and energy E can penetrate a barrier by tunneling. If the barrier is a rectangular barrier with height U and width a, then (8.3.1) gives

[j] For theoretical models, see J. C. Slonczewski, *Phys. Rev. B* **39**, 6995 (1989); F. F. Li, Z. Z. Li *et al.*, *Phys. Rev. B* **69**, 054410 (2004). For experimental results on the decrease of TMR with increasing voltage, see J. S. Moodera and G. Mathon, *J. Magn. Magn. Mater.* **200**, 248 (1999); for TMR turning negative at high enough voltage, see M. Sarma *et al.*, *Phys. Rev. Lett.* **82**, 616 (1999).

the transmission probability

$$\tilde{T} \propto \exp(-2\kappa a), \tag{8.3.36}$$

where $\kappa = \sqrt{2m(U-E)/\hbar^2}$. In particular, this barrier structure can be considered as two metals separated by a thin insulating layer; this insulating layer is just the barrier for an electron.

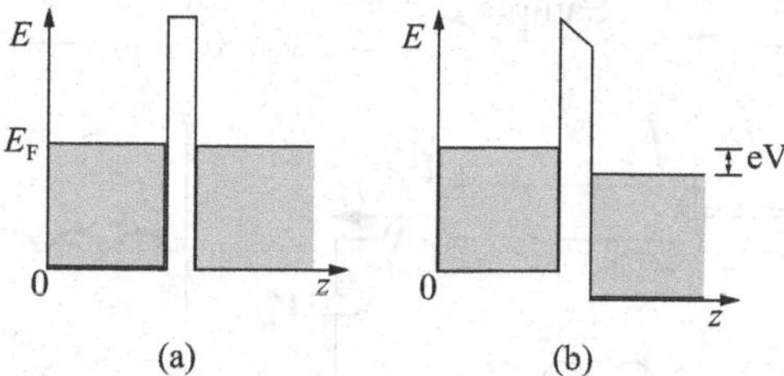

Figure 8.3.10 Potential as a function of position in a metal-insulator-metal tunnel junction, (a) the equilibrium state, (b) an applied potential V.

When the insulating layer is very thin, a tunneling current appears between the two metals. When a bias voltage is applied on the metal-insulator-metal structure shown in Fig. 8.3.10, the current contribution results from the tunneling probabilities of conduction electrons with energies larger than the Fermi energy. In the free electron model, the current density j can be written as

$$j = \frac{2e}{h}\sum_{k_z}\int_{E_z} \tilde{T}(E_z,V)[f(E) - f(E+eV)]dE_z, \tag{8.3.37}$$

where the tunneling probability $\tilde{T}(E_z,V)$ is related to an electron's total energy E, to its kinetic energy perpendicular to the barrier E_z, and to the applied bias voltage V. Summing over k_z gives the total contribution of electrons with the same energy E_z. Finishing the integral over $\boldsymbol{k}$ space, the total current density at 0 K is

$$j(V) = \frac{4\pi me^2V}{h^3}\int_0^{E_F-eV} \tilde{T}(E_z,V)dE_z + \frac{4\pi me}{h^3}\int_{E_F-eV}^{E_F}(E_F-E_z)\tilde{T}(E_z,V)dE_z. \tag{8.3.38}$$

When the bias is very small, the above expression gives a linear relation for I-V and $\tilde{T}(E_z,V)$ is taken as a constant; however, when the bias is increased to a value having the same order of magnitude as the barrier height U, the I-V relation becomes exponential. These theoretical predictions have been verified by many experiments.

The tunneling effect in metal-insulator-metal trilayer structures has an important application in the modern experimental technique known as the scanning tunneling microscope (STM), in which the insulating layer is vacuum or air.[k] Its essential component is a metal tip sharpened to a point with atomic dimensions, usually fabricated from narrow wires of W or Pt-Ir alloy. As shown in Fig. 8.3.11, the location of this detecting tip is controlled by three piezoelectric elements perpendicular to each other, the piezoelectric elements x and y lead the tip to scan in the xy plane, and the z piezoelectric element can be used to modulate the distance between the tip and sample below 1 nm. At this point the electronic wavefunction in the tip overlaps with the electronic wavefunction at the surface of a sample. Taking the tip and the surface of a sample as two electrodes, with an applied bias voltage, we see that current tunnels between the two electrodes.

Figure 8.3.11(a) shows the constant height mode of operation, where the tip scans over the surface of a sample from a fixed distance. Variation in the tunneling current are recorded to form an image of the surface. Due to the extreme sensitivity of the tunneling current to tip-surface separation and electron density, STM images routinely achieve atomic resolution. The tunneling current often changes by an order of magnitude even with the relatively small atomic scale variations across the

[k]The scanning tunneling microscopy was designed by Binnig and Rohrer (1982), and finished by Binnig, Rohrer, Gerber and Weibel (1982). G. Binnig and H. Rohrer, *Helv. Phys. Acta.* **55**, 726 (1982); G. Binnig, H. Rohrer, *et al.*, *Appl. Phys. Lett.* **40**, 178 (1982); *Phys. Rev. Lett.* **49**, 57 (1982).

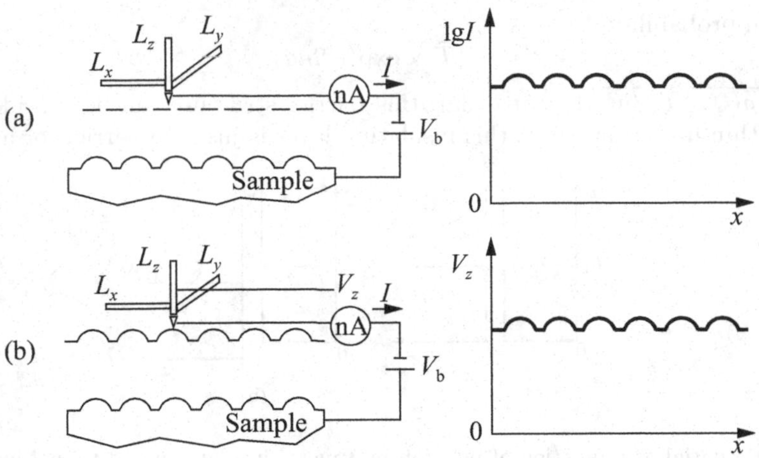

Figure 8.3.11 Two work modes for scanning tunneling microscope.

surface of a pure single crystal. The disadvantage of this mode is that if the fluctuations of the surface reach the nanometer scale, the tip may collide with the surface and be damaged. The constant current mode shown in Fig. 8.3.11(b) can avoid this problem. In the constant current mode, the tip is controlled by the z piezoelectric element. The tip moves up and down according to variations in the surface in order to maintain the same tunneling current. In this way, the separation between the tip and surface is kept constant. Variations in the z piezo voltage reflect the surface topography, although spatial variations in density of electron states must also be considered when interpreting the images obtained using STM. The constant current mode is the most common method of STM operation.

The tunneling current density j passing through the vacuum is a measure of the overlap of the electronic wavefunctions. It is related to the work functions of the tip and sample, ϕ_1 and ϕ_2, and also the separation between them. An approximate expression is

$$j = \frac{2e^2}{h}\left(\frac{\kappa}{4\pi^2 a}\right) V \exp(-2\kappa a), \tag{8.3.39}$$

where $\kappa = \hbar^{-1}[m(\phi_1 + \phi_2)]^{1/2}$. The tunneling current is very sensitive to the separation a between the tip and sample. If a is reduced by 1 nm, the tunneling current will increase by an order of magnitude. This sensitivity makes STM especially effective at measuring the fluctuations across the surface of a sample.

More detailed STM theories are mostly established on the basis of Bardeen's tunneling current theory. Its formula for tunneling current is

$$J = \frac{2\pi e}{\hbar} \sum_{\mu\nu} f(E_\mu)[1 - f(E_\nu + eV)]|M_{\mu\nu}|^2 \delta(E_\mu - E_\nu), \tag{8.3.40}$$

where $f(E)$ is the Fermi distribution function, V is the applied voltage, $M_{\mu\nu}$ is the transition matrix element between the tip state ψ_μ and the sample surface state ψ_ν and E_μ is the energy of ψ_μ in the absence of tunneling. Bardeen gave the expression for the matrix element

$$M_{\mu\nu} = \frac{\hbar^2}{2m} \int d\boldsymbol{S} \cdot (\psi_\mu^* \nabla \psi_\nu - \psi_\nu \nabla \psi_\mu^*), \tag{8.3.41}$$

where the integral is over the surface of the electrode located in the vacuum barrier. The key point to solving the last two expressions is to determine the eigenstates ψ_μ and ψ_ν for the tip and sample existing separately, to assume a reasonable barrier for the tunneling junction, and also to consider the influences of temperature and bias voltage.

STM is a very useful tool for condensed matter physics, chemistry, and biology. STM can be applied to ascertain local electronic structures at the atomic scale on solid surfaces, to infer their local atomic structures. It has been extended to other microscopic probes,[1] e.g., the atomic force microscope (AFM), can even be used to image local atomic structures on insulator surfaces. The imaging ability of STM and AFM in different environments, which involves very little disturbance to samples, has allowed an enormous range of applications. One typical, and most successful, example was the confirmation of the 7×7 structure on the Si (111) surface (see §12.4.5). In an investigation of organic molecular structures, monolayer absorption of benzene on the surface of Rh (111) was observed, and the Kekule ring structure was clearly shown. In biology, STM has been used to observe directly the structures of DNA, reconstructed DNA and HPI-protein by surface absorption on carriers.

Bibliography

[1] Mott, N. F., and H. Jones, *The Theory of the Properties of Metals and Alloys*, Oxford University Press, Oxford (1936).

[2] Ziman, J. M., *Principles of the Theory of Solids*, 2nd ed., Cambridge University Press, Cambridge (1972).

[3] Ziman, J. M., *Electrons in Metals*, Taylor & Francis, London (1963).

[4] Ziman, J. M., *Electrons and Phonons*, Oxford University Press, Oxford (1960).

[5] Wilson, A. H., *The Theory of Metals*, 2nd ed., Cambridge University Press, Cambridge (1958).

[6] Callaway, J., *Quantum Theory of the Solid State*, 2nd ed., Academic Press, New York (1991).

[7] Abrikosov, A. A., *Fundamentals of the Theory of Metals*, North-Holland, Amsterdam (1988).

[8] Olsen, J. L., *Electron Transport in Metals*, Interscience, New York (1962).

[9] Davis, J. H., *The Physics of Low-Dimensional Semiconductors*, Cambridge University Press, Cambridge (1998).

[10] Wolf, S. A. *et al.*, *Spintronics: Spin-based Electronics Vision for the Future*, Science **294**, 1488 (2001).

[11] Zutic, I., J. Fabian, and S. Das Sarma, *Rev. Mod. Phy.* **76**, 323 (2004).

[12] Duke, C. B., *Tunneling in Solids*, Academic Press, New York (1968).

[13] Wolf, E. L., *Principles of Electron Tunneling Spectroscopy*, Oxford University Press, New York (1985).

[14] Chen, C. J., *Introduction to Scanning Tunneling Microscopy*, Oxford University Press, New York (1993).

[15] Magonov, S. N., and M. H. Whangbo, *Surface Analysis with STM and AFM*, VCH, Weinheim (1996).

[1] Based on the principles of STM, a series of related techniques has been developed. They include, along with the atomic force microscope (AFM), the laser force microscope, electrostatic force microscope, magnetic force microscope, scanning thermal microscope, ballistic electron emitting microscope, scanning ionic conduction microscope, photon scanning tunneling microscope, and so on. The invention of these new types of microscopes provides us with powerful tools to study magnetic, electric, mechanical and thermal properties on surfaces or interfaces of matter.

Chapter 9

Wave Localization in Disordered Systems

We have shown in Chaps. 5 and 7 the formal analogy between wave propagation in structures with perfect or nearly perfect periodicity. Both de Broglie waves and classical waves have been treated, the latter including electromagnetic waves and lattice or elastic waves. In perfect periodic structures, wave behavior satisfies band theory, with all permitted modes belonging to extended states; while for those structures containing slight imperfections, such as impurities and surfaces, band theory needs to be modified a little. Almost all states are extended, but some localized modes emerge. When we turn to disordered systems, the picture of energy bands breaks down, and the wave behavior appears to be localized within finite regions. However, the formal analogy for three kinds of waves can still be demonstrated.

§9.1 Physical Picture of Localization

In a seminal paper entitled "Absence of diffusion in certain random lattices",[a] Anderson (1958) formulated the concept of disorder-induced electron localization which formed the basis of further investigations. In fact this concept may be applied to classical waves, as in John's extension to elastic waves and optical waves.[b]

9.1.1 A Simple Demonstration of Wave Localization

In Chap. 7, we introduced the wave equations for de Broglie waves, electromagnetic waves and lattice or elastic waves in parallel, and emphasized that the wave behavior will be determined by the spatially-dependent potential functions. It is expected that in disordered structures the random distribution of potential functions will play a crucial role: If the degree of disorder is strong enough, the wave will be localized. The eigenmodes of strongly disordered systems may be described by exponentially localized functions in space, as $\exp(-|\boldsymbol{r}-\boldsymbol{r}_0|/\xi)$, with $\boldsymbol{r}_0$ being a certain central position and ξ the localization length.

As a first step in understanding wave localization, we shall discuss a pedagogical example. For a general picture of wave behavior, the vector nature of electromagnetic or elastic waves, may be ignored for the time being, and the wave equations for electrons, elastic vibrations, and electromagnetic radiations can all be expressed in the same scalar form as

$$\nabla^2\psi + [k^2 - V(\boldsymbol{r})]\psi = 0. \tag{9.1.1}$$

[a] P. W. Anderson, *Phys. Rev.* **109**, 1492 (1958).

[b] S. John *et al.*, *Phys. Rev. B* **27**, 5592; **28**, 6358 (1983); *Phys. Rev. Lett.* **53**, 2169 (1984).

Here

$$k^2 = \begin{cases} 2mE/\hbar^2, & \text{for electrons;} \\ \omega^2/c^2, & \text{for classical waves,} \end{cases}$$

and the potential $V(\boldsymbol{r})$ was normalized with $2m/\hbar^2$ for electrons, or with the combination of a stiffness coefficient and mass density for elastic waves, or with dielectric constant for electromagnetic waves.

The advantage of (9.1.1) is evident: People have been more concerned with the problem of electron localization, but its clear indications are difficult to observe, because the Coulomb interaction between electrons (and other inelastic scattering processes of electrons) will smear the coherence of the eigenstates. In contrast, classical waves are cleaner systems, and it is easier to see wave localization. So we can use the experimental demonstration of classical waves to give an intuitive picture of the wave localization, for example with acoustic waves.

Consider a one-dimensional acoustic system made by a long steel wire in which a tension K is maintained. The periodic, or disordered, potential field V for the wire is provided by small masses along the wire, which may be equally separated or not. The masses are sufficiently small that the potential may be approximated as a series of δ functions with strength $m\omega^2/K$. The wave field ψ consists of transverse waves in the wire, generated by an electromechanical actuator at one end of the wire. Figure 9.1.1(a) and (b) show the frequency responses of both the periodic and the disordered system. For the periodic case, the frequency response shows distinct edges separating the pass band from the forbidden band on either side, while these features are lost in the random distribution case. The frequency response of a static disordered configuration in Fig. 9.1.1(b) illustrates the dramatic departure from the Bloch response in Fig. 9.1.1(a). On the other hand, some examples of the eigenstate amplitude distributions along the wire are shown in Fig. 9.1.2, among which the first two correspond to Bloch-wave-like eigenstates, and the other two to localized eigenstates.

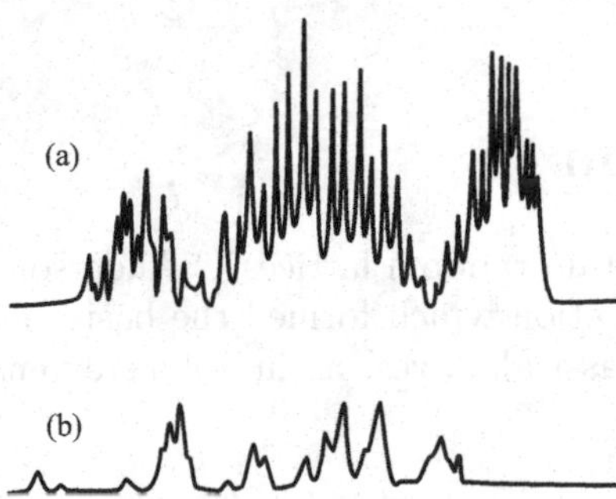

Figure 9.1.1 Frequency response of the wire as a one-dimensional acoustic system for (a) periodic potential and (b) random potential. From S. He and J. D. Maynard, *Phys. Rev. Lett.* **57**, 3171 (1987).

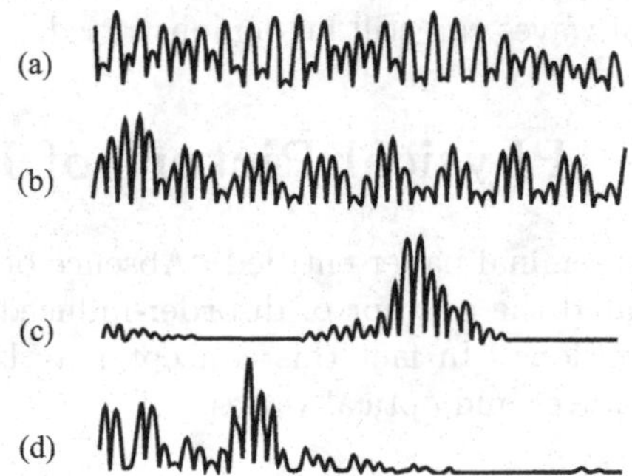

Figure 9.1.2 Eigenstate amplitude as a function of position along the wire, (a) and (b) are Bloch states corresponding to two eigenfrequencies in Fig. 9.1.1(a); (c) and (d) are eigenstates of the disordered system with frequencies corresponding to two peaks in Fig. 9.1.1(b). From S. He and J. D. Maynard, *ibid.*

Recognizing the analogy of classical waves and de Broglie waves, we can draw two main conclusions from the acoustic simulation, i.e. when sufficient disorder is introduced into a system, the bandgaps disappear and the wavefunctions are localized. In what follows, we shall mainly be concerned with electron waves and electromagnetic waves, although the investigation on elastic waves can also be generalized further.

9.1.2 Characteristic Lengths and Characteristic Times

From the above discussion, we have shown that the introduction of strong disorder will modify the wave behavior from extended states to localized ones. Localization comes from scattering processes.

Actually, in disordered systems, waves suffer multiple scattering among the randomly distributed scatterers.

For the investigation of a wave propagating through a non-dissipative disordered medium, three characteristic lengths need to be specified. The first is the wavelength λ, which is generally related to the eigenwavevector or eigenenergy. The second is the elastic mean free path l, which is a characteristic of the disorder, and can be estimated as $\sim (n\sigma^*)^{-1}$, where σ^* is a typical elastic scattering cross section of a scatterer, and n is the number of the scatterers per unit volume. The third is the size of the sample L.

The relative values for these three characteristic lengths are important in the consideration of wave localization. We can specify three regimes as follows:

(1) $l > L$, the medium acts as if it were homogeneous. It can be confirmed that in addition to the incident wave there are only reflected and transmitted waves, so this is the propagating case;

(2) $\lambda < l < L$, there is so much scattering that the wave loses memory of its initial direction and this is the diffusive case, described by multiple scattering. However, there is an additional coherent effect, weak localization, due to the interference of back-scattered waves;

(3) $l \leq \lambda < L$, the system satisfies the Ioffe–Regel criterion $l \leq \lambda$, and is strongly disordered: Coherent effects become so important that they lead to the vanishing of the diffusive constant. Nothing is transmitted through the system and all the energy is reflected, for large enough L. This is the case of formation of the localized state known as Anderson localization.

Among the three characteristic lengths, the mean free path deserves further discussion. Strictly speaking, non-dissipative media can be realized only at zero temperature. We can in turn define a collision time τ by

$$l = v\tau, \tag{9.1.2}$$

where v is the characteristic velocity, e.g., the Fermi velocity v_F in the case of electrons. τ denotes the average time interval of two consecutive elastic scatterings that are involved in the transition between eigenstates of different momenta and degenerate energy.

As the temperature rises, the medium may be thermally excited, so inelastic scattering arises and there are transitions between eigenstates of different energies. We can introduce another characteristic time τ_{in}, to denote the average time interval of two consecutive inelastic scatterings, where in general, $\tau_{\text{in}} > \tau$. This means that there are several elastic scattering events between two consecutive inelastic scatterings.

It should be noted that elastic scattering keeps phase coherence, while inelastic scattering breaks it. The phase coherent length is a very important quantity which can be defined as $l_{\text{in}} \equiv v\tau_{\text{in}}$. The inelastic scattering rate $1/\tau_{\text{in}}$ increases with temperature T as a power law T^p, where p is a constant. We give an estimate of this for electrons as an example. At room temperature, inelastic scattering by phonons occurs rapidly, $1/\tau_{\text{in}} \sim k_B T/\hbar \sim 10^{13}\ \text{s}^{-1}$, so $\tau_{\text{in}} \simeq 10^{-13}$ s. Phase information is always destroyed for a macroscopic sample, because $v_F \sim 10^8$ cm/s $= 10^{16}$ Å/s and $l_{\text{in}} = v_F\tau_{\text{in}} \approx$ 1000 Å, while the typical distance at which an electron remains phase coherent is the elastic mean free path $\sim$ 100 Å. So now the electrons may be treated as semiclassical particles in which wave behavior is not very apparent. However, there are two approaches to display the effects of phase coherence for electrons: One is to decrease sample size and temperature, whereupon we arrive at the mesoscopic electronic system discussed in Chap. 10. The other is to analyze physical effects in macroscopic samples that arise from multiple scattering at small scales, for example, the modification of conductivity to be discussed in next section.

9.1.3 Particle Diffusion and Localization

The concept of diffusion is of central importance in the theory of localization. For simplicity, we consider the classical diffusive behavior of a particle in d-dimensional disordered system first. At $t = 0$ the particle is assumed to be at the origin and to begin its random walk. After a time $t = \tau$,

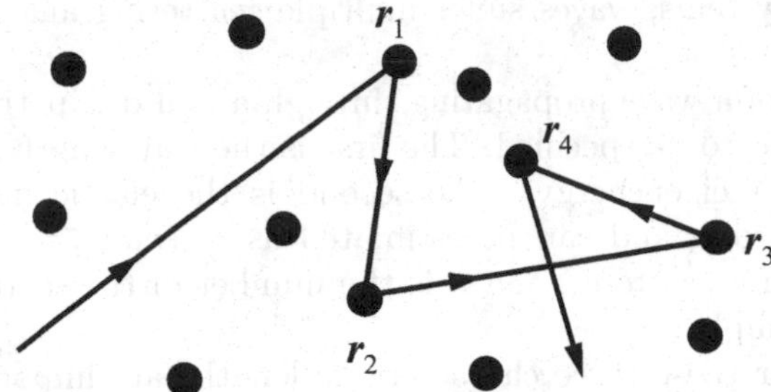

Figure 9.1.3 Diffusion path of a moving particle in the disordered system.

it experiences an elastic scattering at a certain place, and its direction of motion is changed. This process continues while it diffuses in the disordered medium from one impurity to another. The concept of localization may be defined through this classical diffusion process.

We can say that particle is localized if there is a nonzero probability for it to be around the origin as time approaches infinity; otherwise it is delocalized. The classical diffusion equation in d-dimensions for the probability density distribution $p(\boldsymbol{r}, t)$ is

$$\frac{\partial p}{\partial t} - D\nabla^2 p = 0, \tag{9.1.3}$$

where D is the diffusion coefficient. Its solution gives probability of the particle being at position $\boldsymbol{r}$ at time t

$$p(\boldsymbol{r}, t) = (4\pi Dt)^{-d/2} \exp(-r^2/4\pi Dt), \tag{9.1.4}$$

from which a diffusive volume V_{diff} can be defined. Because only for $r^2 \lesssim 4\pi Dt$ is $p(\boldsymbol{r}, t)$ of significance

$$V_{\text{d}} \approx (Dt)^{d/2}. \tag{9.1.5}$$

From (9.1.4), the chance of the particle returning to the origin at the time t is given by

$$p(0, t) = (4\pi Dt)^{-d/2}. \tag{9.1.6}$$

In Fig. 9.1.3 a possible path is drawn for the diffusion of a particle which might return to the origin.

It is important to calculate the integrated probability for the diffusive particle to return to the origin from its first scattering to some larger time t:

$$P(0, t) = \left(\frac{1}{4\pi D}\right)^{d/2} \int_\tau^t \frac{dt}{t^{d/2}} \propto \begin{cases} (t/\tau)^{1/2}, & d = 1; \\ \ln(t/\tau), & d = 2; \\ (t/\tau)^{-1/2}, & d = 3. \end{cases} \tag{9.1.7}$$

It can be seen from this expression that the integrated probability is closely related to the dimensionality of the system. The probability of revisiting the origin increases with time in a one-dimensional system, but decreases in a three-dimensional system. It is interesting to note that any disorder will lead to localization for $d = 1$, but the particle is always delocalized for $d = 3$. The two-dimensional system is the marginal case and shows a logarithmic dependence with time; the particle will be also localized as $t \to \infty$. A better discussion of localization as a function of dimensionality is to use scaling theory, which will be given in Part V. Moreover, it should be noted that the interference of scattered waves may modify this classical picture of diffusion, leading to enhanced backscattering as well as a size-dependent diffusion coefficient, to be treated in the next section.

§9.2 Weak Localization

We shall consider the effects of weak disorder, which is intermediate between the scattering from a single impurity and strong localization due to the scattering by a large number of random scatterers. Weak disorder means that the mean free path l is much greater than the characteristic wavelength λ and less than the size of the sample L. There are still a great many energy eigenstates that are extended although they are not periodic. However, we shall find that of weak localization is a precursor for strong localization.

9.2.1 Enhanced Backscattering

In classical diffusion there is an identical probability for a particle to propagate on the same path in the opposite direction. The two probabilities add up and contribute to the total probability. In reality, a microscopic particle like an electron has a wave-like character, instead of the classical random walk process, two partial waves can propagate in opposite directions on the same path. On their return to the origin, it is their amplitudes, instead of their intensities, that must be added together, and therefore the total probability or intensity is twice as large as in the classical diffusion problem expressed in (9.1.6). In addition, it is more instructive to study the wave behavior for the propagation of classical waves in disordered media, so we must examine wave diffusion in disordered systems more closely.

To understand the diffusive behavior of waves, we should look at the path of a wave diffusing from the origin O to some point O' in the medium, as shown in Fig. 9.2.1. Since the transport from O to O' can take place along different trajectories, there is a probability amplitude A_i connected to every path i. The total intensity I to reach point O' from O is then given by the square of the magnitude of the sum of all amplitudes

$$I = |\sum_i A_i|^2 = \sum_i |A_i|^2 + \sum_{i \neq j} A_i A_j^*. \tag{9.2.1}$$

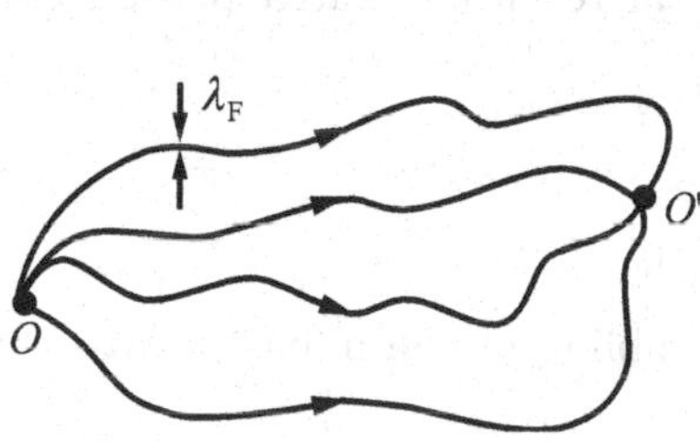

Figure 9.2.1 Various possible paths for a wave diffusing from O to O'.

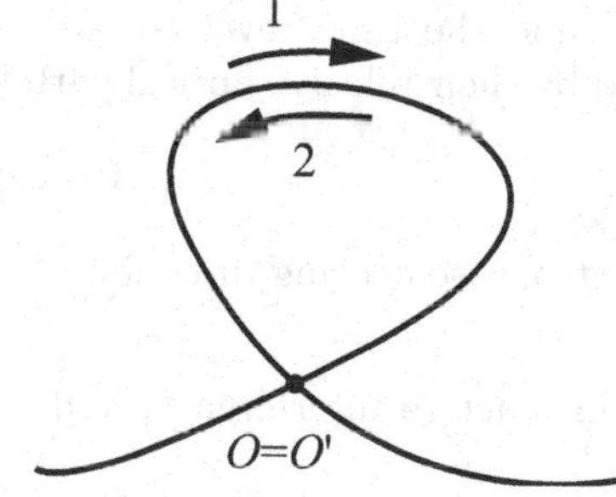

Figure 9.2.2 Self-crossing path of a diffusing particle.

The first term of the right hand of (9.2.1) describes the noninterference contribution corresponding to the classical case, while the second term represents the contribution due to interference of the paths, which comes from the nature of waves. In the conventional Boltzmann theory for electron transport these interference terms are neglected. In most cases this is justified, since the trajectories have different lengths and the amplitudes A_i carry different phases. On average, this leads to destructive interference. Hence, the quantum-mechanical interference terms suggested by Fig. 9.2.1 are generally unimportant.

There is, however, one important exception to this conclusion: namely, if the points O and O' coincide as in Fig. 9.2.2, i.e., if the path crosses itself. In this case, the starting point and the endpoint are identical, so that the path in between can be traversed in two opposite directions: forward and backward. The probability p to go from O to O' is thus the return probability to the starting point. Since paths 1 and 2 in Fig. 9.2.2 are identical, the amplitudes A_1 and A_2 are coherent in phase. This leads to constructive interference, so that the wave contribution to p becomes very important and (9.2.1) tells us that for $A_1 = A_2 = A$, the classical return probability is given by $2|A|^2$, while the wave character yields $2|A|^2 + 2A_1A_2^* = 4|A|^2$. The probability for a wave to return to some

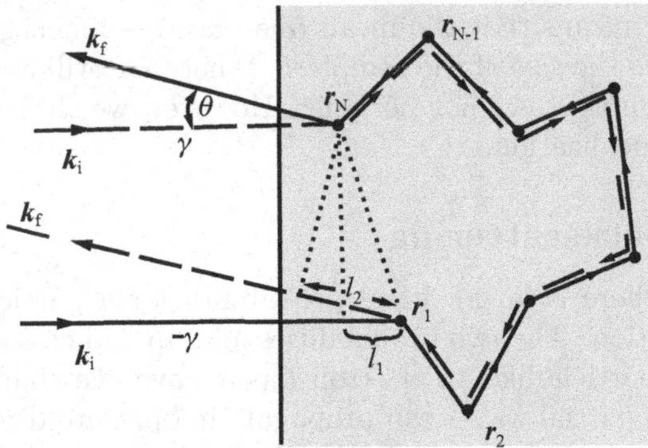

Figure 9.2.3 Schematic diagram of light backscattering.

starting point is thus seen to be twice that of a classical particle. One may therefore say that "wave diffusion" is slower than classical diffusion, due to the existence of a more effective backscattering effect in the former case. In other words, waves in a disordered medium are less mobile than classical particles.

In the case of optical waves, the backscattering interference effect has been vividly demonstrated in disordered dielectric media. This is well demonstrated by an experiment in which incident laser light of frequency ω enters a disordered dielectric half space or a slab, and the angular dependence of the backscattering intensity is measured. We consider the incident light with wavevector $\boldsymbol{k}_\mathrm{i} = \boldsymbol{k}_0$ entering a dielectric half space, as shown in Fig. 9.2.3. The light wave will be scattered at points $\boldsymbol{r}_1, \boldsymbol{r}_2, \ldots, \boldsymbol{r}_N$ into intermediate states with wavevectors $\boldsymbol{k}_1, \boldsymbol{k}_2, \ldots, \boldsymbol{k}_{N-1}$ and finally into the state $\boldsymbol{k}_N = \boldsymbol{k}_\mathrm{f}$. For scalar waves, the scattering amplitudes at the points $\boldsymbol{r}_1, \ldots, \boldsymbol{r}_N$ are the same for the path γ and the time reversed path $-\gamma$. The nature of interference between the two paths is determined by their relative optical path lengths and so the resulting relative phase factor is given by

$$A \sim \exp[i(\boldsymbol{k}_\mathrm{i} + \boldsymbol{k}_\mathrm{f}) \cdot (\boldsymbol{r}_N - \boldsymbol{r}_1)]. \tag{9.2.2}$$

In the exact backscattering direction

$$\boldsymbol{q} = \boldsymbol{k}_\mathrm{i} + \boldsymbol{k}_\mathrm{f} = 0$$

there is constructive interference, and a consequent doubling of the intensity over the incoherent background.

If the angle between $-\boldsymbol{k}_\mathrm{i}$ and $\boldsymbol{k}_\mathrm{f}$ is θ, the coherent condition for small θ is

$$\boldsymbol{q} \cdot (\boldsymbol{r}_N - \boldsymbol{r}_1) = 2\pi\theta|\boldsymbol{r}_N - \boldsymbol{r}_1|/\lambda < 1. \tag{9.2.3}$$

In the diffusion limit

$$|\boldsymbol{r}_N - \boldsymbol{r}_1|^2 \approx D(t_N - t_1) \approx lL^*/3,$$

where $D = lv_\mathrm{eff}/3$ is the photon diffusion coefficient, l is the elastic mean free path, v_eff the effective velocity, and L^* is the total length of path γ. Thus paths of lengths L^* contribute to the coherent intensity for angles less than

$$\theta_m = \lambda/(2\pi\sqrt{lL^*/3}). \tag{9.2.4}$$

The width and the enhancement factor of the contribution to the backscattering cone as a function of the depth in the sample has been studied experimentally by using a difference technique; the experimental results are depicted in Fig. 9.2.4(a), where the contribution of light that has seen the deeper part of the slab is shown. As the thickness L of the sample increases, it is reasonable to assume that $L^* \propto L$, so that the angle θ decreases in accordance with the theoretical analysis as $\theta \sim L^{-1/2}$. The theoretical intensity profiles, calculated from diffusion theory, are shown in Fig. 9.2.4(b). Both are in good agreement for the shape, width and relative intensity of the calculated and experimental cones. It is noted that the multi-scattering of light by fine particles can trap light which then moves back and forth in a finite region. This is, in some sense, similar to the case in which light is reflected

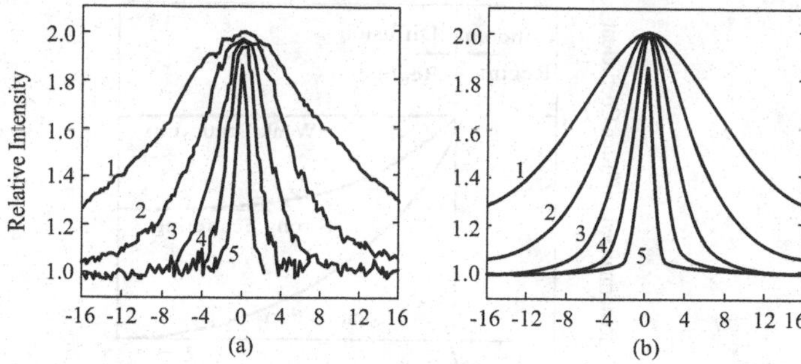

Figure 9.2.4 Light backscattering pattern, (a) experimental results; (b) calculated results. From M. B. van der Mark *et al.*, *Phys. Rev. B* **37**, 3575 (1988).

back and forth in a laser cavity. If the medium used has the property of light gain, then a random laser can be manufactured.[c]

9.2.2 Size-Dependent Diffusion Coefficient

It has been emphasized that in a disordered system, a diffusion process can be used to describe the motion of an electron as well as a photon. If the length scale considered is longer than the length of the elastic mean free path l, it is convenient to regard the particle going on a random walk, with the diffusion coefficient given by $D = vl/3$, where v is the effective speed of the particle. However, the classical random walk is not enough, because of the wave character of the electron and the photon, and their diffusion process must be described by an amplitude rather than a probability. Interference between all possible diffusion paths must be considered in evaluating the transport of waves. In this situation, a more precise definition of the diffusion coefficient for wave propagation is needed.

This new definition of diffusion coefficient is essential because the diffusion coefficient can no longer be looked upon as a local variable, but is determined by coherent wave interference throughout the entire disordered medium. When the scattering is very weak it can be reduced to the familiar diffusion coefficient given by the product $v_{\text{eff}}l/3$ of the effective medium speed v_{eff}, of wave propagation and the mean field path l. In the vicinity of incipient localization the diffusion coefficient D depends on the size of the entire sample. For clarity, but without loss of generality, we take the electromagnetic wave as an example.

The incorporation of wave interference and, in particular, of coherent backscattering into the theory of wave diffusion leads to a renormalized picture of transport, as shown in Fig. 9.2.5. In a situation where wave interference plays an important role in determining transport, the spread of wave energy is not diffusive at all, in the sense that a photon performs a classical random walk. Fortunately, there is a way of applying the concept of classical diffusion here: In a random medium it is reasonable to expect that scattering centers that are far apart do not, on average, cause large interference corrections to the classical diffusion picture. Because the fluctuations are significant, it follows that there exists a finite coherence length $\xi_{\text{coh}} \gtrsim l$ which represents a scale on which interference effects must be taken

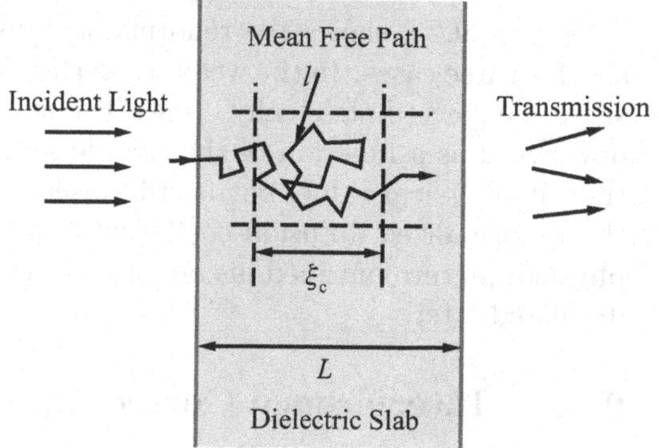

Figure 9.2.5 Physical picture of optical transport at incipient localization.

[c]For the theoretical suggestion of random lasers, see V. S. Letokhov, *Sov. Phys. JETP* **26**, 835 (1968); for experimental verification see C. Gouedard *et al.*, *J. Opt. Soc. Am. B* **10**, 2358 (1993); N. M. Lawandy *et al.*, *Nature* **368**, 436 (1994); D. S. Wiersma and A. Lagendijk, *Phys. Rev. E* **54**, 4256 (1996).

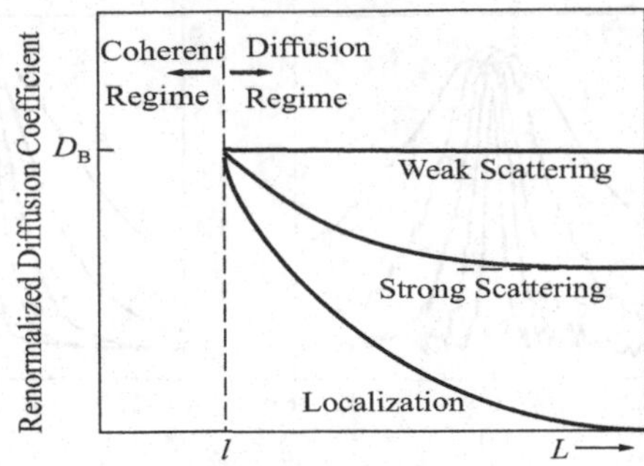

Figure 9.2.6 A schematic variation of the renormalized diffusion coefficient with sample size (which is larger than the mean free path before diffusion can be observed). Weak scattering, strong scattering, and localization are shown. D_B denotes the classical Boltzmann value of the diffusion coefficient. L and l are the sample size and mean free path, respectively. From P. Sheng, *Introduction to Wave Scattering, Localization, and Mesoscopic Phenomena*, Academic Press, New York (1995).

into account to determine the effective diffusion coefficient at any point within the coherence volume. In other words, the possible amplitudes for a wave to diffuse from O to O' within a coherence volume $\xi_{\rm coh}^d$ must interfere with each other. Depending on the distance between the point O at which the photon is injected into the medium and the point O' at which it is detected, the effective diffusion coefficient of the photon is strongly renormalized by wave interference. As a specific example, consider a finite size sample of linear size L. By changing the scale of the sample, the number of diffusion paths that can interfere changes, giving rise to an effective diffusion coefficient $D(L)$ at any point within the sample, which depends on the macroscopic scale L of the sample.

There are many formal mathematical ways of describing this physical picture. In incipient localization, there arises a new coherence length $\xi_{\rm coh} \geq l$ such that on a length scale L in the range $l < L < \xi_{\rm coh}$ the spread of the wave is sub-diffusive in nature as a result of coherent backscattering, which gives a significant interference correction to classical diffusion. In this range, the spread of wave energy may be interpreted in terms of a scale-dependent diffusion coefficient that behaves roughly as $D(L) \approx (vl/3)(l/L)$. On length scales long compared to $\xi_{\rm coh}$, the photon resumes its diffusive motion except with a lower, or renormalized, value $(vl/3)(l/\xi_{\rm coh})$ of the diffusive coefficient. We may finally combine them as

$$D(L) \approx \frac{vl}{3}\left(\frac{l}{\xi_{\rm coh}} + \frac{l}{L}\right). \tag{9.2.5}$$

Figure 9.2.6 shows the renormalized diffusion coefficient schematically as a function of sample size for three cases. In the weak scattering limit, the diffusion coefficient is independent of the sample size, as expected classically. When the scattering is strong, the diffusion coefficient is renormalized downward as a function of the sample size, with an asymptotic value that can be significantly less than its classical value, but is still observable at small sample sizes. When the asymptotic value of the renormalized diffusion coefficient vanishes, then by definition a localized state is created. The physical picture outlined is consistent with the scaling theory of electron localization which will be discussed later.

9.2.3 Interference Correction to Conductivity

It is well known that conductivity σ is related to the diffusion coefficient D by the Einstein relation

$$\sigma = e^2 D g_{\rm F}, \tag{9.2.6}$$

where e is the electronic charge and $g_{\rm F}$ the density of states (DOS) at the Fermi surface. Indeed, scattering of particles in a weakly-disordered medium leads to a diffusive motion of the electrons, and the semiclassical approach is applicable. In conventional Boltzmann transport theory, consecutive collisions of electrons are assumed to be uncorrelated; this implies that multiple scattering of an electron from a particular scattering center is not taken into account. However, if there is a finite probability for repeated occurrence of such multiple scatterings, the basic assumption of the independence of scattering events breaks down and the validity (9.2.6) for σ becomes questionable.

Actually, in contrast to conventional Boltzmann theory, we must take into account the interference between scattered waves. Then there is an effect called enhanced backscattering, which is a kind of special coherent superposition of scattered waves. This backscattering causes a decrease in conductance, due to the quantum mechanical effect, i.e., regarding electrons as de Broglie waves. As depicted in Fig. 9.2.7, we can consider that an electron with momentum $\boldsymbol{k}$ and wavefunction $\exp(i\boldsymbol{k}\cdot\boldsymbol{r})$ at $t=0$ penetrates into the medium and is scattered by scatterer $1,\ldots,N$. The momentum changes from $\boldsymbol{k}$ to $\boldsymbol{k}_1', \boldsymbol{k}_2', \ldots$, and finally $\boldsymbol{k}_N' = -\boldsymbol{k}$, the corresponding momentum transfer are $\boldsymbol{g}_1,\ldots,\boldsymbol{g}_N$. There is an equal probability for $\boldsymbol{k}$ to $\boldsymbol{k}_1'', \boldsymbol{k}_2'', \ldots, \boldsymbol{k}_N'' = -\boldsymbol{k}$; the momentum transfers are $\boldsymbol{g}_N,\ldots,\boldsymbol{g}_1$. These two complementary processes are time reversal symmetric, i.e., with time-dependent phase changes $(Et/\hbar)$ are identical, so the final amplitudes of A' and A'' are phase coherent and equal, $|A'| = |A''| = A$. Thus the backscattering intensity is

$$I = |A' + A''|^2 = |A'|^2 + |A''|^2 + A'^*A'' + A'A''^* = 4|A|^2 = 2\times 2|A|^2. \tag{9.2.7}$$

This leads to the decrease of the conductance σ.

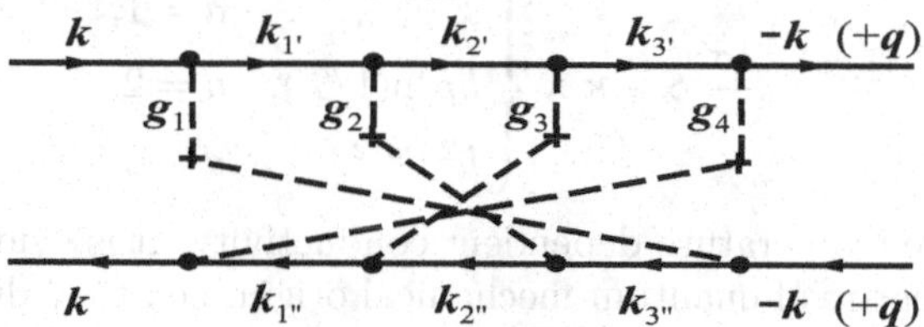

Figure 9.2.7 Diffusion path of a conduction electron.

The electronic characteristic wavelength is the Fermi wavelength $\lambda_F = 2\pi/k_F$, and weak disorder means that the mean free path l satisfies $l \gg \lambda_F$. Starting from the metallic regime, we shall examine how the precursor effects of localization, i.e., backscattering, will give rise to a correction $\delta\sigma$ to the metallic conductivity σ. Because of the expected decrease in conductivity, the sign of $\delta\sigma/\sigma$ must be negative. This change is proportional to the probability of occurrence of a closed path during diffusion. We consider a d-dimensional tube with diameter λ_F and cross section λ_F^{d-1}, as shown in Fig. 9.2.8. During the time interval dt the particle moves a distance $v_F dt$, so that the corresponding volume element of the tube is given by $dV = v_F dt \lambda_F^{d-1}$. On the other hand, the maximum possible volume for the diffusing particle is given by V_{diff} in (9.1.5). Thus, the probability for a particle to be in a closed tube is therefore given by the integral over the ratio of these two volumes

$$P = \int_\tau^{\tau_{\text{in}}} \frac{dV}{V_{\text{diff}}} = v_F\lambda_F^{d-1}\int_\tau^{\tau_{\text{in}}} \frac{dt}{(Dt)^{d/2}}. \tag{9.2.8}$$

The integration is taken over all times $\tau < t < \tau_{\text{in}}$, where τ is the time for a single elastic collision and τ_{in} is the inelastic relaxation time in the system. The latter determines the maximum time during which coherent interference of the path amplitudes is possible. Then the correction for conductivity can be expressed as

$$\frac{\delta\sigma}{\sigma} \propto -\kappa \times \begin{cases} (\tau_{\text{in}}/\tau)^{1/2}, & d=1, \\ \hbar\ln(\tau_{\text{in}}/\tau), & d=2, \\ \hbar^2(\tau_{\text{in}}/\tau)^{-1/2}, & d=3, \end{cases} \tag{9.2.9}$$

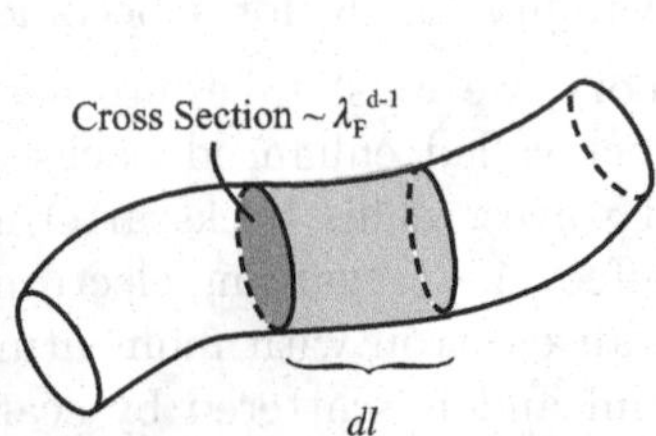

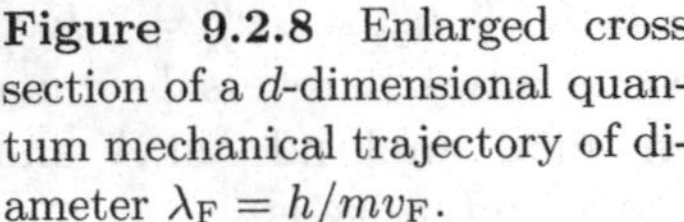

Figure 9.2.8 Enlarged cross section of a d-dimensional quantum mechanical trajectory of diameter $\lambda_F = h/mv_F$.

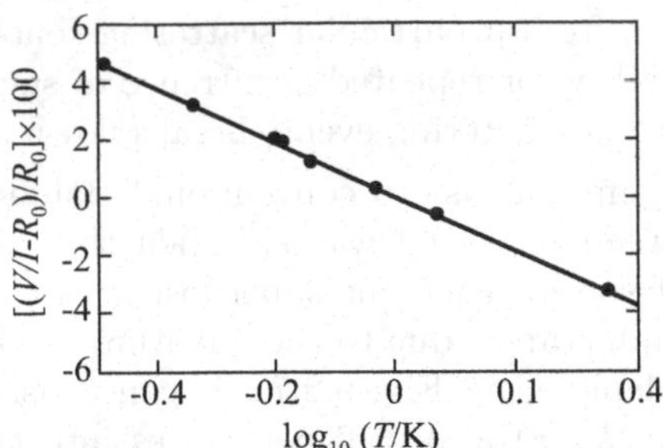

Figure 9.2.9 The dependence of resistance and temperature in disordered Au-Pd film. From G. J. Dolan and D. D. Osheroff, *Phys. Rev. Lett.* **43**, 721 (1979).

where κ is the disorder parameter. Assuming that the inelastic relaxation rate vanishes with some power of temperature T as $T \to 0$, i.e., $1/\tau_{\text{in}} \propto T^p$, where p is a positive constant, then (9.2.9) becomes

$$\frac{\delta\sigma}{\sigma} \propto -\kappa \times \begin{cases} T^{-p/2}, & d = 1, \\ \hbar p \ln(1/T), & d = 2, \\ \hbar^2 T^{p/2}, & d = 3. \end{cases} \tag{9.2.10}$$

This expression gives the temperature-dependent conductivity. It is evident that for the cases $d = 2$ and $d = 3$ the corrections are of quantum-mechanical origin, i.e., they disappear for $\hbar \to 0$. But the case $d = 1$ is the same as the classical case, for only forward and backward scattering are allowed, all paths are necessarily closed. Figure 9.2.9 shows the relationship of resistance and temperature in a disordered alloy thin film measured experimentally. It confirms the result for two dimensions in equation (9.2.10).

§9.3 Strong Localization

We have treated the localized electronic modes related to a single impurity in Chap. 7. Here we would like to consider the localization of one-electron states that occur in strongly disordered systems. In 1958 Anderson, for the first time, studied the diffusion of electrons in a random potential and found localization of electron waves if the randomness of the potential was sufficiently strong. Afterwards Mott applied Anderson's idea to investigate impurity conductance in doped semiconductors in which a disordered-induced metal-nonmetal transition may occur. We shall also discuss the strong localization of light.

9.3.1 Continuum Percolation Model

A very simple method to demonstrate the consequences of disorder for electrons is the percolation model. The concept of percolation was introduced in Chap. 4. Here, instead of site or bond percolation, we consider continuum percolation, which simulates the situation of an electron moving in a random mixture of conductors and insulators.

The problem is now looked as a classical particle with energy E moving in a space that can be divided in two parts, according to the irregular potential $V(\boldsymbol{r})$. One is the allowed region with $E > V(\boldsymbol{r})$ and the other the forbidden region with $E < V(\boldsymbol{r})$. A schematic diagram is described in Fig. 9.3.1 in terms of the motion of a particle moving on a water surface, which is equivalent to an energy surface of height E. At lower water levels, only lakes exist, so the particle is localized. As the water level rises, some channels form and connect the lakes. Further increasing the water

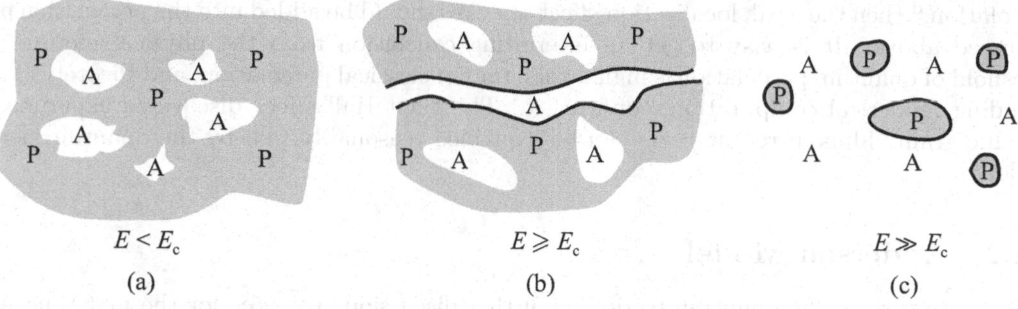

Figure 9.3.1 The motion of a classical particle in a random potential.

level, the particle becomes delocalized when an ocean has formed, i.e., when the channels extend to infinity. The peaks penetrating the ocean surface remain as scattering centers for the particle in the extended state.

Just like the occupation probability p for site percolation, there is a quantity $p(E)$ for continuum percolation. This quantity specifies the fraction of space allowed for particles of energy E; then we have

$$p(E) \equiv \int_{E>V(\boldsymbol{r})} d\boldsymbol{r} \bigg/ \int d\boldsymbol{r}. \tag{9.3.1}$$

It is clear that $p(E) \leq 1$, and increases monotonically with E. There must exist a percolation threshold p_c, such that when the energy-dependent allowed-space fraction $p(E)$ exceeds p_c, an infinitely extended allowed region appears. Corresponding to p_c, we can find a threshold energy E_c, that satisfies

$$p(E_c) = p_c. \tag{9.3.2}$$

In two dimensions, for a broad class of random potentials, $p_c = 1/2$, which comes from the fact that the lakes-ocean transition is the topological mirror image of the continent- island transition. In one dimension, it is evident that $p_c = 1$, since there is no way to go around the energy barrier for $E < V(\boldsymbol{r})$ as can be done in higher dimensions. In three dimensions, no rigorous result is available, but a reasonable estimate for many situations is $p_c \approx 0.16$.

Strictly speaking, electrons are not classical particles but obey quantum mechanics, so they can tunnel through regions in which $E < V(\boldsymbol{r})$. We can analyze some of the differences between classical continuum percolation and the quantum mechanical approach. The first concerns the nature of states with $E > E_c$. In the classical case, lakes can coexist with the infinite ocean. This suggests the possibility that localized states persist in the presence of extended states at the same energy but this is not correct in the quantum mechanical sense. A localized wavefunction cannot avoid mixing with an enveloping extended state present at the same energy. Thus localized and extended states are not cleanly separated by a threshold energy in the energy spectrum. The second concerns the dimension-dependence of the two pictures. Both percolation and quantum treatments have all states localized by disorder in one dimension. A percolation transition is present in two or higher dimensions, while a quantized localized-delocalized transition may occur only in dimensions larger than two, and so $d = 2$ is a borderline dimensionality. This problem will be discussed in Part V. In addition to these two pronounced differences, there is a third difference between Bloch states in a periodic structure and the extended state within the percolation model: In crystals, a Bloch state is an extended state with the same probability for an electron at any equivalent sites; while in a disordered system, an extended percolation state means that the wavefunction may extend to infinity, but with a marked fluctuation of $|\psi|^2$ in space.

The continuum percolation model gives us some intuitive ideas about extended states, localized states, and localization-delocalization transition in disordered systems. This approach might be applicable in certain situations, especially those in which the characteristic length of the topological features of $V(\boldsymbol{r})$ is much larger than a typical electron de Broglie wavelength, so its real usefulness is somewhat limited for electrons. On the other hand, if we extend classical percolation to quantum percolation,[d] then the weak localization effect of waves should be added into the percolation processes discussed above. It is easy to get an interesting conclusion from the physical picture that the threshold of quantum percolation is higher than that of classical percolation, and theoretical research, including numerical computation, confirms it. The giant Hall effect discovered experimentally in metallic grainy films in recent years can be explained reasonably well by the quantum percolation model.[e]

9.3.2 Anderson Model

Now we turn to the quantum model for further discussion. In 1958, for the first time Anderson studied electronic diffusion in random potentials and found that electron waves become localized when the random potentials are strong enough. Anderson's quantum theory of electrons in disordered solids is a tight-binding disordered one-electron model. We start with the Hamiltonian

$$\mathcal{H} = \sum_i \varepsilon_i |i\rangle\langle i| + \sum_{i\neq j} t_{ij} |i\rangle\langle j|, \tag{9.3.3}$$

where $|i\rangle$ is the state vector of the i-site using the Dirac symbol, ε_i is the energy level of an electron at site i, and t_{ij} is the transfer integral between sites i and j. In the Anderson model, nonzero transfer integrals are only for nearest neighbor sites and are assumed to be equal, i.e. $t_{ij} = \tilde{t}$, due to the fact that the separation for neighboring sites is almost the same, but ε_i is randomly chosen for each site i forming a spatial distribution with width W, as in Fig. 9.3.2, $(-W/2 \leq \varepsilon \leq W/2)$. This model is referred to as the diagonal disorder model, since disorder appears in the diagonal matrix elements. For further treatment, we need to know the probability distribution of energy. For mathematical convenience we may choose the uniform distribution expressed as

$$p(E) = \frac{1}{W}. \tag{9.3.4}$$

There are some other choices for the distribution of energy, for example a Lorentz distribution, but the fundamental physical implication is the same.

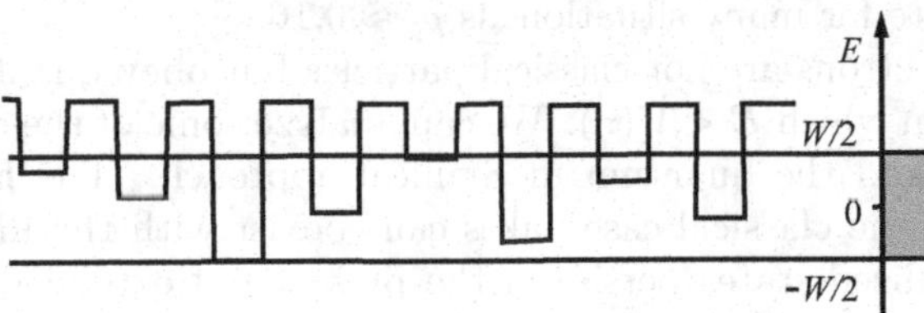

Figure 9.3.2 Disordered distribution of energy in the Anderson model.

The simplest way to discuss the model is in terms of a tight-binding expansion of the wavefunction

$$|\psi\rangle = \sum_i a_i |i\rangle, \tag{9.3.5}$$

where $|i\rangle$ is the atomic orbital centered at site i. If ψ is an eigenfunction that satisfies $\mathcal{H}\psi = E\psi$, we can get a matrix equation for the amplitudes a_i

$$Ea_i = \varepsilon_i a_i + \sum_j t_{ij} a_j. \tag{9.3.6}$$

[d] C. M. Soukoulis *et al.*, *Phys. Rev. B* **36**, 8649 (1987).
[e] X. X. Zhang *et al.*, *Phys. Rev. Lett.* **86**, 5562 (2001).

This leads to a solution for the stationary state. If we consider the time-dependent equation of motion, it is

$$\frac{\hbar}{i}\frac{da_i}{dt} = \varepsilon_i a_i + \sum_j t_{ij} a_j. \tag{9.3.7}$$

We can give a definition for localized states. Suppose, at time $t = 0$, an electron is placed at site i, so $a_i(t = 0) = 1$, but $a_j = 0$ for $j \neq i$. In principle, these initial conditions determine the subsequent evolution of the system by (9.3.7). We can examine $a_i(t)$ in the limit of t approaching infinity. If $a_i(t \to \infty) = 0$, the electron is in an extended state. But if $a_i(t \to \infty)$ is finite, the electron has only spread in a region near the neighborhood of site i, so it is in a localized state.

In the nearest neighbor approximation, (9.3.6) is transformed into

$$Ea_i = \varepsilon_i a_i + \tilde{t}\sum_{\alpha=1}^{z} a_{i+\alpha}, \tag{9.3.8}$$

where $\tilde{t}$ is the transfer integral between nearest neighbors and the sum extends over the z nearest neighbors of site i. It is instructive to examine some limiting cases. One limiting case is $W = 0$, which corresponds to all sites having the same energy, so there is no disorder. The bandwidth in this crystalline tight-binding approximation is proportional to $\tilde{t}$ and for a simple cubic structure

$$B = 2z\tilde{t}. \tag{9.3.9}$$

The opposite limit has the ε_i distribution in (9.3.4) left intact with W taking a finite value, and with $\tilde{t} = 0$. With the coupling removed, the bandwidth $B = 0$, the solutions are simply atomic orbitals at each site and the initial situation will be kept invariant, i.e., $a_i = 1$, and $a_j = 0$ for $j \neq i$.

Intermediate between these two limits, W and B are all finite values, it is evident that the magnitude of the ratio

$$\delta = W/B \tag{9.3.10}$$

shows the competition between disorder and order. It is expected that there is a critical value δ_c to determine the transition from delocalization to localization as the disorder increases, or *vice versa*. To determine this critical value, Anderson's full treatment with a Green's function based upon the perturbation method is rather difficult. We would like to give a qualitative discussion[f] here by approaching it from the strong disorder limit, i.e., $W \gg B$.

We begin with the localized limit $B = 0$, turn on the $\tilde{t}$, and take this coupling parameter as a perturbation. Consider at first $a_i = 1$, $a_j = 0$ for all $j \neq i$, first order perturbation mixes this state with neighboring sites by an amplitude of order $\tilde{t}/(\varepsilon_i - \varepsilon_j)$, higher orders of perturbation add terms containing higher powers of this quantity. Then it can be written

$$|\psi\rangle = |i\rangle + \frac{\tilde{t}}{\varepsilon_i - \varepsilon_j}|j\rangle + \cdots. \tag{9.3.11}$$

The question is: How big can $\tilde{t}/(\varepsilon_i - \varepsilon_j)$ be before localization is destroyed and an extended state arises? The site energies, ε_i and ε_j, are taken from a distribution of width W. Suppose ε_i is at the center of the energy distribution, and assume that the ε_j of the z nearest-neighbor sites are uniformly spaced over W/z. The smallest energy denominator in the perturbation parameter $\tilde{t}/(\varepsilon_i - \varepsilon_j)$ is $|\varepsilon_i - \varepsilon_j| = W/2z$, so the largest perturbation parameter is

$$\frac{\tilde{t}}{\varepsilon_i - \varepsilon_j} = \frac{2z\tilde{t}}{W} = \frac{B}{W}. \tag{9.3.12}$$

If the perturbation expansion converges, it is reasonable to require $(B/W) < 1$. When the convergence breaks down, delocalization appears and the localization-delocalization transition occurs

[f] D. J. Thouless, *Phys. Rep.* **13**, 93 (1974).

at $B = W$. When $W > B$, localization occurs at the center of the band, so we get the Anderson localization criterion

$$\delta_c = 1. \tag{9.3.13}$$

This is smaller than the value Anderson got originally but consistent with the result from computer simulation.

9.3.3 Mobility Edges

Based on the Anderson model, Mott put forward the useful idea of the mobility edge, which is schematically shown in Fig. 9.3.3. The mobility edge is a critical energy value E_c to distinguish the localized states from the delocalized ones. Its classical correspondence is the threshold energy of the percolation model. The concept of the mobility edge can be used to study the conductance of doped semiconductors when metal-insulator transitions takes place. Through changing the level of disorder, especially the impurity concentration in doped semiconductors, delocalized states near the Fermi surface can be localized. The conductance can be changed in this process, so there appears a metal-insulator transition in the material, called the Anderson transition.

For a small degree of disorder only the states in the band tails are localized. The energies of the localized states correspond to the tails at the top of the valence band and the bottom of the conduction band. The tails themselves stretch into the energy gap only because the energy ranges of additional states becomes available through the disordered potential distribution. Within the main body of each band, the states are extended. For an intermediate degree of disorder less than the critical value, there will be two energies which separate localized from extended states as shown in Fig. 9.3.3. The demarcation energies separating regions of localized and extended states are referred to as mobility edges. At either side of an energy gap in a disordered insulator or semiconductor, tails of localized states penetrate into the gap, as shown in Fig. 9.3.3; they may or may not overlap. The energy separation between two mobility edges is called the mobility gap. It is the direct extension of the concept of energy gap in a crystalline semiconductor.

Mott (1974) also emphasized the possibility of a metal-nonmetal transition, called the Anderson transition, if the degree of disorder is changed by doping, application of pressure, or electric field etc, the Fermi level E_F can be made to cross a mobility edge, then the delocalized states near the Fermi energy become localized. This process leads to a change in electric conduction, and a metal-insulator transition can occur. In Fig. 9.3.3, for example, if $-E_c < E_F < E_c$ the material is metallic and conducts at $T = 0$. If $E_F < -E_c$ or $E_F > E_c$ it will not conduct unless, at $T > 0$, the electrons can be thermally excited from one localized state to another or to an extended state. If $g(E_F)$ is high enough for the electron gas to be degenerate, and E_F lies in a mobility gap, the material is called a Fermi glass.

When the energy of the individual atomic states varies randomly over a range greater than the width of the band produced by overlap between adjacent atomic orbitals, the states at the center of

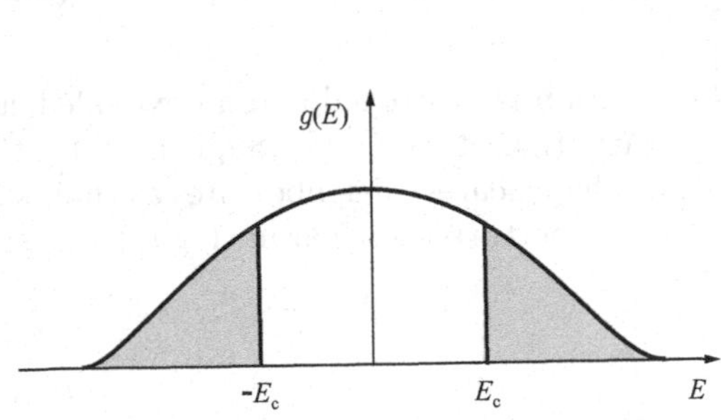

Figure 9.3.3 Mobility edges.

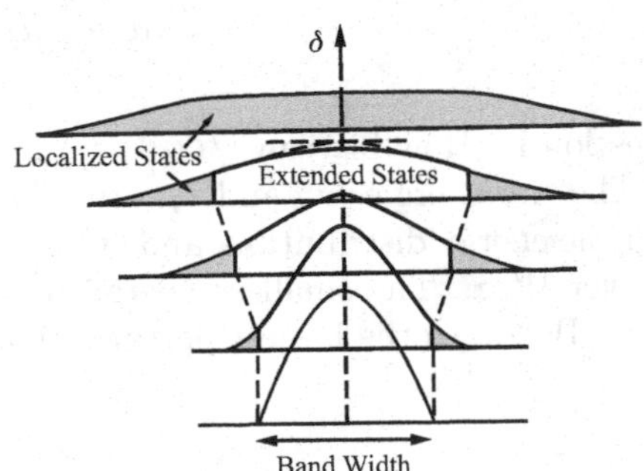

Figure 9.3.4 In the Anderson model, localized states increase as disorder parameter δ increases.

the band must be localized. Figure 9.3.4 shows how the effect of disorder produces localized states at the edges of the original band. As the disorder increases, these tails become longer, until eventually, the mobility edges move inward and coalesce at the center where all the states are localized.

The tail states are especially susceptible to localization, and Thouless (1974) provided a percolation argument for this. We may anticipate a specific feature of quantum transport among the disordered sites: An electron may move with relative ease from one site to a nearby one only if the energy levels of the two sites differ by an amount that is less than a small fraction of the bandwidth, an amount roughly given by B/z. Sites differing in energy by a greater amount are effectively decoupled. Suppose we slice the distribution W into discrete energy ranges of width B/z, and label (or "color") each site of the same energetically specified type (i.e., of the same color). A site of one color cannot communicate with any site of another type, and the situation now resembles a polychromatic percolation process. For a species of site corresponding to an energy slice taken from the fat central part of the density of states, there is an abundance of available sites; their spatial concentration is high, and percolation is easy. But, for a color corresponding to a slice taken from a thin tail of the distribution, the available sites are sparse and are spread far apart in space. Here a percolation path is absent and localization is likely. This conveys the reason that disorder most readily localizes electrons in the extremities of the density of states distribution and explains why Anderson localization most readily appears in a low density regime (distribution tail = low concentration in space) due to the random potential.

9.3.4 Edwards Model

Electrons in disordered materials with characteristic metallic behavior cannot be satisfactorily described in terms of tight-binding wavefunctions. So, for metals, it may not be correct to write the eigenstate as a linear combination of a small number of atomic orbitals as (9.3.5). In contrast to the Anderson model, Edwards (1958) considered the disordered distribution of scatterers in a nearly-free electron model.[g] We must now return to the one-electron Schrödinger equation

$$\left(-\frac{\hbar^2}{2m}\nabla^2 + V(\boldsymbol{r})\right)\psi(\boldsymbol{r}) = E\psi \tag{9.3.14}$$

to describe an electron in the conduction band of a disordered solid, where $V(\boldsymbol{r})$ is the random potential. This is called the Edwards model.

In this model, the random potential $V(\boldsymbol{r})$ can be chosen to have zero mean value

$$\langle V(\boldsymbol{r})\rangle = 0, \tag{9.3.15}$$

and a spatial autocorrelation function that satisfies

$$\langle V(\boldsymbol{r})V(0)\rangle = V_{\rm rms}^2 \mathrm{e}^{-r^2/\xi^2}, \tag{9.3.16}$$

where $\langle\cdots\rangle$ is an ensemble average, $V_{\rm rms}$ a root-mean-square amplitude, and ξ a length scale on which random fluctuations in the potential take place. The correlation length ξ for the disorder defines an energy scale $\varepsilon_\xi = \hbar^2/2m\xi^2$.

The one-electron density of states (DOS) in three dimensions is depicted in Fig. 9.3.5. The weak disorder corresponds to $V_{\rm rms} \ll \varepsilon_\xi$. For weak disorder there is a tail of strongly localized states, referred to as the Urbach tail, for $E < 0$ which is separated by a mobility edge $E_{\rm c}' \simeq -V_{\rm rms}^2/\varepsilon_\xi$. As the disorder parameter $V_{\rm rms}$ is increased, the mobility edge eventually moves into the positive continuum denoted as $E_{\rm c}$. Successive tunneling events allow an electron of energy greater than $E_{\rm c}$ to transverse the entire solid by a slow diffusive process and thereby conduct electricity, whereas electrons of energy lower than $E_{\rm c}$ are trapped and do not conduct electricity. The noteworthy point, however, for present consideration, is that only in the limit of very strong disorder, such that $V_{\rm rms} \gg \varepsilon_\xi$, do the states with $E > 0$ exhibit localization. Electrons with sufficiently negative energy E may get trapped in regions where the random potential is very deep. The rate for electrons tunneling out of

[g] S. F. Edwards, *Phil. Mag.* **3**, 1020 (1958).

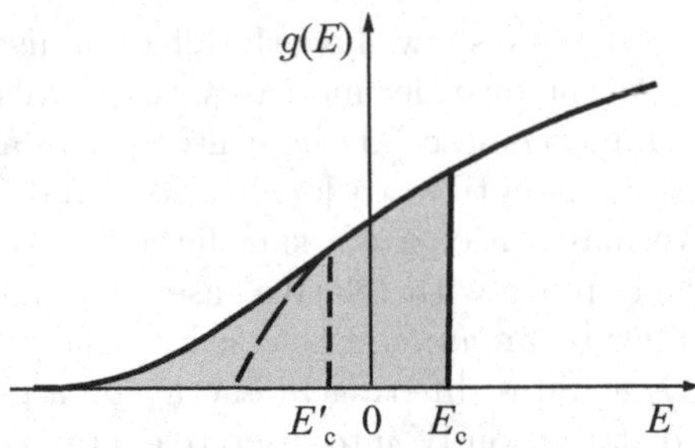

Figure 9.3.5 One electron density of states (DOS) in a correlated Gaussian random potential. For weak disorder, the mobility edge is denoted by E_c' below which states are localized. As the disorder is increased, the mobility edge moves into the positive energy regime denoted by E_c.

the deep potentials depends on the probability of finding nearby potential fluctuations into which the trapped electron can tunnel. This rate increases as the electron energy increases. For energy much greater than E_c, the scale on which scattering takes place grows larger than the electron's de Broglie wavelength, and the electron transverses the solid with relative ease.

9.3.5 Hopping Conductivity

Although we can discuss localization and mobility edges for disordered metals in the Edwards model, for real metallic materials, the most striking feature is the similarity between the phenomena exhibited by both crystals and glasses. There is a lack of experimental evidence for strong localization and the metal-nonmetal transition. In both the crystal and the glassy metals, conduction occurs via extended state electrons and is limited by disorder-induced scattering processes. But there are two significant differences of the electric properties of a metallic glass relative to the properties of its corresponding crystal: much higher resistivity and much lower sensitivity to temperature. The reason for both is the high degree of static disorder already present in the metallic glass; the additional dynamic disorder introduced by the presence of thermal phonons has little influence on the resistivity, which is already high as a result of the scattering caused by the intrinsic structural disorder. So the resistivity is high and flat for the glass.

We now turn our attention to more interesting conduction properties of semiconductors and insulators, which may be defined as materials whose conductivity vanishes at zero temperature. Referring to the schematic electron density of states (DOS) in Fig. 9.3.3, the Fermi energy E_F falls between the valence band and the conduction band, lying within the bandgap of the crystal or within the mobility gap of the glass. For the crystal, E_F lies at an energy devoid of states; for the glass, it lies within the region of localized states. In both cases, the conductivity is zero, since there is no electron motion unless thermal energy is supplied.

A localized electron cannot contribute to dc conductivity at zero temperature. At finite temperature, it may be thermally excited to an extended state, or to an empty localized state. For the latter case, an applied field will bias the motion and create a current down the field; called hopping conduction. The term "hopping" is an abbreviation for the phonon-assisted quantum mechanical tunneling of an electron from one localized state to another. If it can be recognized by some characteristic temperature dependence, it serves to reveal the presence of localized states. There is more than one regime for temperature-dependent hopping. One example is the hopping of an electron to sites with the smallest r and is therefore sometimes called nearest-neighboring hopping. The assumption here is that T is high enough to make hopping through energy difference $(\varepsilon_j - \varepsilon_i)$ to a nearest neighbor the most likely event. In the following we would like to discuss another example in which T is lower.

If T is lower enough, there is increased probability of a jump to a site j with energy nearer to ε_i but greater spatial distance. The idea was first proposed by Mott in 1968[h] and the process is known as 'variable-range hopping'. He gave an intuitive argument proceeding from defining the transition

[h]N. F. Mott, *Phil. Mag.* **17**, 1259 (1968).

probability p, which is the function of the hopping distance r, the energy separation W of the final and initial states, and the temperature T. For an upward energy hopping, we can write

$$p \sim \exp[-2\alpha r - W/k_B T]. \tag{9.3.17}$$

This expression is the product of two exponential factors. The first, $\exp(-2\alpha r)$ represents the probability of finding the electron at distance r from its initial site. (Here α is the inverse localization length that describes the exponential decay $\psi(\boldsymbol{r}) \sim \exp(-\alpha r)$ of the electron wavefunction at large distances.) From the viewpoint of multiple scattering, an electron of energy E is sufficiently scattered by the disorder: The scattered wavelets interfere destructively at distances larger than a characteristic length ξ and collectively reduce the amplitude $\psi_E(\boldsymbol{r})$ beneath the exponential envelope. The second factor, $\exp(-W/k_B T)$, denotes the contribution of phonon assistance in overcoming the energy mismatch W. In the low-temperature limit, the number of phonons of energy W is given by the Boltzmann factor $\exp(-W/k_B T)$.

The essential point emphasized by Mott is that there exists a competition between hopping distance r and mismatch energy W which implies that the nature of the dominant hops necessarily changes with temperature. There is an enhanced probability of encountering a smaller W by permitting the electron to choose among the larger range of finite-state sites contained within a larger neighborhood. A sphere of radius r surrounding the initial site is expressed by

$$(4\pi/3)r^3 W(r) g(E_F) = 1. \tag{9.3.18}$$

As is clear, $W(r)$ is proportional to r^{-3}, and is a reasonable measure of the minimum mismatch available for a hop of range r. Combining this expression with (9.3.16), then maximizing p, yields, for the most probable hopping distance

$$r = [\alpha k_B T g(E_F)]^{-1/4}. \tag{9.3.19}$$

Assuming that the most probable transition rate dominate the hopping conductivity, we obtain

$$\sigma \sim \exp(-A/T^{1/4}), \tag{9.3.20}$$

where

$$A = [\alpha^3/k_B g(E_F)]^{1/4}. \tag{9.3.21}$$

The $T^{1/4}$-law is characteristic of variable-range hopping and it can also be derived by percolation arguments.[i] In d space dimension the 1/4 is replaced by $1/(d+1)$. Mott's prediction of variable-range-hopping conductivity has been verified by experiments in a number of amorphous semiconductors.

9.3.6 Strong Localization of Light

The remarkable phenomenon of electron localization was first discussed by Anderson in 1958, but fully appreciated only in the 1980s. The reason is that the similarity of wave scattering and interference to photons and electrons has been realized over that period. In appropriately-arranged dielectric microstructures light can also be localized.

In the study of electron localization there is a complication due to the inevitable presence of electron-electron and electron-phonon interactions. However, the propagation of optical waves in nondissipative dielectric media provides the ideal realization of a single mode in a static random medium, even at room temperature. Light localization is an effect that arises entirely from coherent multiple scattering and interference and it may be understood purely from the point of view of classical electromagnetism. In traditional studies of the propagation of electromagnetic wave in dielectrics, scattering takes place on scales much longer than the wavelength of light. Localization of light, just the same as that of electrons, occurs when the scale of coherent multiple scattering is reduced to the wavelength itself.

[i] V. Ambegaokar, B. I. Halperin and J. S. Langer, *Phys. Rev.* **4**, 2612 (1974).

Consider monochromatic electromagnetic waves of frequency ω propagating in a disordered, nondissipative dielectric microstructure. It is proper to begin with Maxwell equations to describe the propagation of electromagnetic waves. The wave equation for the electric field $\boldsymbol{E}$ can be written as

$$-\nabla^2\boldsymbol{E}+\nabla(\nabla\cdot\boldsymbol{E})-\frac{\omega^2}{c^2}\epsilon_{\mathrm{f}}(\boldsymbol{r})\boldsymbol{E}=\epsilon_0\frac{\omega^2}{c^2}\boldsymbol{E}, \tag{9.3.22}$$

where the dielectric constant $\epsilon(\boldsymbol{r})$ written as

$$\epsilon(\boldsymbol{r})=\epsilon_0+\epsilon_{\mathrm{f}}(\boldsymbol{r}), \tag{9.3.23}$$

including two parts, i.e., ϵ_0 is an average value, and $\epsilon_{\mathrm{f}}(\boldsymbol{r})$ the random fluctuation satisfying

$$\langle\epsilon_{\mathrm{f}}(\boldsymbol{r})\rangle=0. \tag{9.3.24}$$

In a nondissipative material, the dielectric constant $\epsilon(\boldsymbol{r})$ is everywhere real and positive.

There are several important observations based on the analogy between the Schrödinger equation and Maxwell equations. First of all, the quantity $\epsilon_0\omega^2/c^2$ always positive, so the possibility of electromagnetic bounded states in deep negative potential wells is precluded. It is noteworthy that, unlike an electronic system, where localization is enhanced by lowering the electronic energy, lowering the photon energy, i.e., letting $\omega\to 0$, leads to a complete disappearance of scattering; the opposite limit, $\omega\to\infty$, geometric optics becomes operative and interference corrections to optical transport become less and less effective. In both limits the normal electromagnetic modes are extended, not localized. Therefore, unlike the familiar picture of electron localization, localized light must be in an intermediate frequency window, within the positive energy continuum. This is in agreement with the Ioffe–Regel rule, because the wavelength of electromagnetic waves ranges from γ-ray to radio waves, so localization only appears in the intermediate frequency window, depending on the filling ratio and arrangement of scatterers.

The propagation of electromagnetic waves in random nondissipative media exhibits three fundamentally different regimes. The underlying physics of the high and low frequency limits can be made more precise by considering scattering from a single dielectric sphere. If the wavelength of incident plane wave is λ, and the mean free path is l, then three different regimes can be distinguished.

The classical elastic mean free path l plays a central role in the physics of localization. Wave interference effects lead to large spatial fluctuations in the light intensity in disordered media. If $l\gg\lambda$, however, these fluctuation tend to average out to give a physical picture of essentially noninterfering, multiscattering paths for electromagnetic transport, it follows that the transport of wave energy is diffusive in nature on very long length scale, but that all states are extended.

When $\lambda\sim l$ in a strongly disordered medium, interference between multiply scattered paths drastically modifies the average transport properties, and a transition from extended to localized normal modes takes place. The condition for localization can be written as the Ioffe–Regel rule

$$2\pi l/\lambda\approx 1. \tag{9.3.25}$$

Figure 9.3.6 shows how the behavior of the elastic mean free path l varies with the wavelength λ. It follows that extended states are expected at both high and low frequencies: In the long wavelength Rayleigh scattering limit, $l\sim\lambda^4$; in the short wavelength limit $l\geq a$, where a is the correlation length. Also depicted in Fig. 9.3.6, for a strongly disordered medium (solid curve), there may exist a range of wavelengths for which $2\pi l/\lambda\approx 1$ where localization is exhibited, within a narrow frequency window when $\lambda/2\pi\approx a$. Our interest in light localization should be concentrated in this intermediate frequency regime.

The above discussion on independent scatterers is about free photon states that undergo multiple scattering. From the single scattering, or microscopic resonance, point of view, the free photon criterion for localization is very difficult to achieve. The dashed curve (Fig. 9.3.6) for dilute scatterers would not induce localization.

As described above, the free photon scattering among the independent scatterers may lead to strong localization provided that (9.3.25) can be satisfied. There is another approach that can be used to discuss light localization, i.e., beginning with the scatterers which are coherently arranged in the matrix. The most familiar example is photonic crystals introduced in Chap. 5.

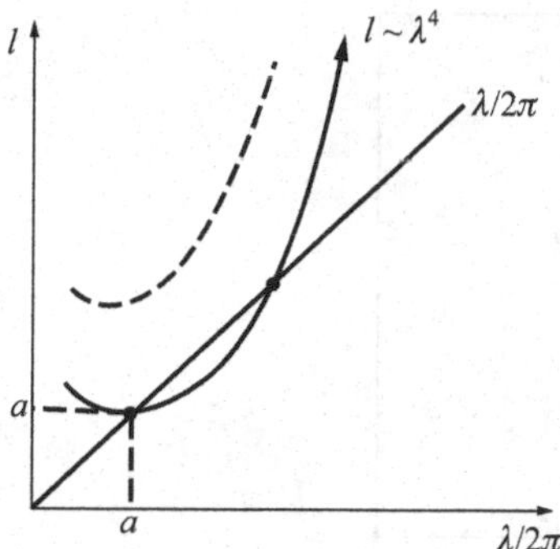

Figure 9.3.6 Behavior of the elastic mean free path l as a function of wavelength λ. From S. John, *Phys. Rev. Lett.* **53**, 2169 (1984).

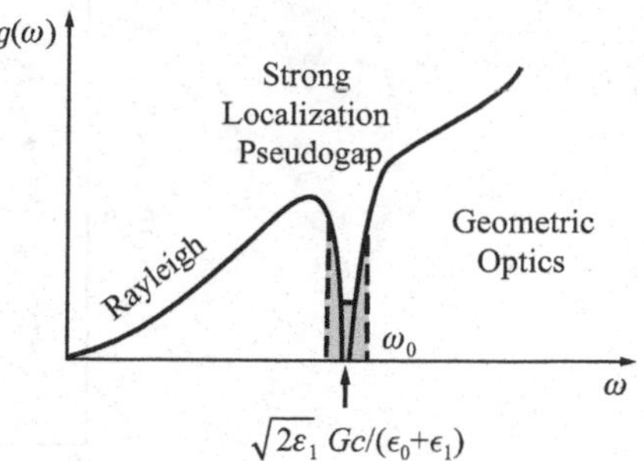

Figure 9.3.7 Photon density of states in a disordered coherent structure exhibiting low frequency Rayleigh scattering and high frequency geometric optics extended states separated by a pseudogap of strongly localized photons. From S. John, *Phys. Rev. Lett.* **58**, 2486 (1987).

Now consider the the dielectric constant $\epsilon_{\text{fluc}}(\boldsymbol{r})$ in (9.3.23) which includes two parts

$$\epsilon_{\text{f}}(\boldsymbol{r}) = \epsilon_1(\boldsymbol{r}) + V(\boldsymbol{r}), \tag{9.3.26}$$

where $\epsilon_1(\boldsymbol{r})$ is the spatially periodic function just like (7.1.6), so it can be Fourier expanded as

$$\epsilon_1(\boldsymbol{r}) = \epsilon_1 \sum_{\boldsymbol{G}} U_{\boldsymbol{G}} e^{i\boldsymbol{G}\cdot\boldsymbol{r}}, \tag{9.3.27}$$

while $V(\boldsymbol{r})$ is a perturbation to the originally periodic dielectric constant. Here $\boldsymbol{G}$ runs over the appropriate reciprocal lattice, and its value for the dominant Fourier component $U_{\boldsymbol{G}}$ is chosen so that the Bragg condition $\boldsymbol{k}\cdot\boldsymbol{G} = (1/2)G$ may be satisfied for a photon of wavevector $\boldsymbol{k}$. Setting $V(\boldsymbol{r}) = 0$ for the time being, the effect of the periodic dielectric structure gives photonic bands and gaps.

The existence of a gap in the photon density of states is of paramount importance in determining the transport properties and localization of light. The free photon Ioffe–Regel condition, discussed above, assumes an essential free photon density of states and completely overlooks the possibility of this gap and the concomitant modification in the character of propagating states. The electric field amplitude of a propagating wave of energy just below the edge of the forbidden gap is, to a good approximation, a linear superposition of the free-photon field of wavevector $\boldsymbol{k}$ and its Bragg reflected partner at $\boldsymbol{k}-\boldsymbol{G}$. As ω moves into the allowed band, this standing wave is modulated by an envelope function whose wavelength is given by $2\pi/q$, where q is the magnitude of the deviation of $\boldsymbol{k}$ from the Bragg plane. Under these circumstances the wavelength that must enter the localization criterion is that of the envelope. In the presence of even very weak disorder, the criterion $2\pi/\lambda_{\text{env}} \sim 1$ is easily satisfied as the photon frequency approaches the band edge frequency. In fact, near a band edge ω_{c}, $\lambda_{\text{env}} \sim |\omega - \omega_{\text{c}}|^{-1/2}$.

The perturbative introduction of disorder $V(\boldsymbol{r})$ now gives rise to localized states in the gap region, as depicted in Fig. 9.3.7. The existence of the band gap guarantees the existence of strongly localized photonic band tail states, at a certain frequency regime, analogous to the Urbach tail in the electron systems. This physical picture is entirely different from the free-photon Ioffe–Regel condition in that it is not the wavelength of the envelope wavefunction, rather that of the carrier wave which enters the localization criterion.

We have discussed in detail the two extreme limits: a structureless random medium for which the criterion $2\pi l/\lambda \approx 1$ applies and a medium with nearly sharp Bragg peaks and a band gap for which $2\pi l/\lambda_{\text{env}} \approx 1$ yields localization. Invariably a continuous crossover occurs between these

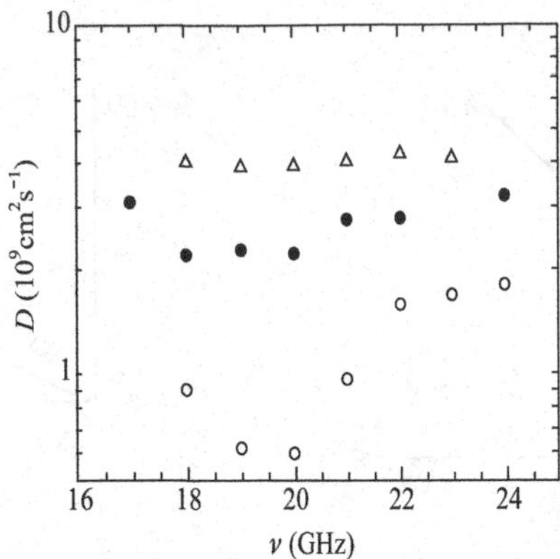

Figure 9.3.8 The experimental results for frequency dependence of diffusion coefficient D. The triangle, dot and circle are, respectively, for filling fractions $f = 0.20,\ 0.25,\ 0.35$. From A. Z. Genack and N. Garcia, *Phys. Rev. Lett.* **66**, 2064 (1991).

conditions as the structure factor of a high dielectric material evolves from one limit to the other. In real disordered systems it is likely that neither the band picture nor the single scattering approach will provide a complete description. In the limit of dilute uncorrelated scatterers, the microscopic resonance picture is entirely adequate, whereas in the high density limit of highly correlated scatterers a macroscopic resonance is essential. To investigate the classical wave localization which occurs in an intermediate regime of frequency, it might be more appropriate to take both of these conceptually different approaches into account.

There have been a lot of efforts to investigate photon localization experimentally, and some results have been reported. An experiment was performed in a random three-dimensional sample by randomly mixing metallic (aluminium) and dielectric (teflon) spheres. Localization of microwave radiation was found in a narrow frequency range. From the experimental results shown in Fig. 9.3.8, the variation of the diffusion coefficient D with filling fraction and frequency shows that $D \to 0$ at a metallic filling fraction $f = 0.35$, and frequencies around $\nu = 19$ GHz (f is the fraction of the total sample volume from which the wave is excluded by metallic surface). In the experiment, the filling fraction is associated with the sample length L. The results are consistent with the theoretical equation (9.2.5) for localization.

Bibliography

[1] Ziman, J. M., *Models of Disorder*, Cambridge University Press, Cambridge (1979).

[2] Edwards, S. F., *Localization via the Edwards Model*, Int. J. Mod. Phys. B **6**, 1583 (1992).

[3] Zallen, R., *The Physics of Amorphous Solids*, John Wiley & Sons, New York (1983).

[4] John, S., *The Localization of Light and Other Classical Waves in Disordered Media*, Comments Cond. Mat. Phys. **14**, 193 (1988).

[5] Mott, N. F., *Metal-Insulator Transitions*, Taylor and Francis, London (1974); 2nd ed. (1990).

[6] Sheng, P., *Introduction to Wave Scattering, Localization, and Mesoscopic Phenomena*, Academic Press, San Diego (1995).

[7] Lee, P. A., and T. V. Ramakrishnan, *Disordered Electron Systems*, Rev. Mod. Phys. **57**, 287 (1985).

[8] Vollhardt, D., and P. Wolfle, *Self-consistent Theory of Anderson Localization*, in *Electronic Phase Transition* (Hanke W, Kopaev Yu V. eds.), Elsevier Science Publisher B V, Amsterdam (1992).

[9] Cusack, N. E., *The Physics of Structurally Disordered Matter*, Adam Hilger, Bristle (1987).

Chapter 10

Mesoscopic Quantum Transport

The electron dynamics of a bulk material composed of large ensembles of particles can be calculated by averaging over many microscopic configurations. Although the quantum behavior of individual constituents of a macroscopic object are important over some length scale, typically a few lattice spacings, they are usually not correlated across the whole object. By contrast, electron transport in mesoscopic systems is dominated by wave behavior and the semiclassical approach can no longer be used. There are many quantum phenomena which are related to wave coherence.

§10.1 The Characteristics of Mesoscopic Systems

We can say that a mesoscopic system has a size between the microscopic and the macroscopic, but this statement is rather vague. The main characteristic of a mesoscopic electronic system is that an electron can keep its wavefunction phase-coherent throughout the sample. This puts a severe restriction on the size as well as the temperature of the sample.

10.1.1 Prescription of the Mesoscopic Structures

In a system of microscopic size, such as an atom or a small molecule with a length scale of several angstroms, the energy levels are all discrete, and so the physical properties are mainly controlled by quantum behavior.

On the other hand, for a macroscopic sample, larger than 1 mm in size, often a classical or semiclassical description can be used. For example, conductivity is determined by the average scattering rate. In the isotropic approximation, Boltzmann transport theory gives the conductivity $\sigma_0 = ne^2\tau(E_F)/m^*$ (§8.1.1). At room temperature, inelastic scattering by lattice vibrations occurs at a very high rate because the inelastic scattering time τ_{in} satisfies $1/\tau_{in} \approx k_BT/\hbar \approx 10^{13}$ s^{-1}, so $\tau_{in} \simeq 10^{-13}$ s. The phase information is always destroyed, because the Fermi velocity $v_F \approx 10^8$ cm/s $= 10^{16}$ Å/s, and the inelastic scattering mean free path $l_{in} = v_F\tau_{in} \approx 1000$ Å. On the other hand, the typical distance at which an electron remains phase coherent is the elastic mean free path, which is only about 100 Å. In this case the electrons can be treated as semiclassical particles; wave aspects are smeared out, and only the local interference correction in the conductivity needs to be considered, as discussed in §9.2.3.

With the development of microfabrication technology, such as molecular beam epitaxy and optical and electron-beam lithography, metal and semiconductor samples with size L less than 1 μm can be made. These are the so-called nanostructures; well-known examples include semiconductor superlattices that are nanosize in one direction, as discussed previously. In the following, however, we will be concerned with samples for which the size may be small in more than one direction. For these, if the temperature is also lowered to $T < 1$ K, the coherent lengths of waves will be larger than the sample size, and the Boltzmann equation is not appropriate to describe electronic transport. Their physical behaviors lie between the familiar macroscopic semiclassical picture and an atomic or molecular description. So it is obvious that there is a certain size range, which is intermediate between the microscopic and the macroscopic, at low temperatures where quantum coherence will be important. This is the characteristic size of mesoscopic structures. The effective length scale at low temperatures can be 10^2–10^4 times the characteristic microscopic scale, and correlations can involve more than 10^{11} particles.

The most important quantity is the phase coherence length defined as $L_\phi \equiv \sqrt{D\tau_{\rm in}}$, which is actually equivalent to $l_{\rm in}$. The inelastic scattering rate $1/\tau_{\rm in}$ decreases with the temperature T and, for a sample of size ~ 1 μm, the mesoscopic regime is reached when $T < 1$ K. In this situation, there is only elastic impurity scattering, which keeps the electron phase coherent. Then, a lot of quantum phenomena including quantized conductance, Aharonov–Bohm (AB) effect, weak localization, magneto-fingerprints, etc., become observable in these small metallic or semiconducting samples.

In fact, in various disordered macroscopic systems there is strong, as well as weak, localization of electronic systems caused by the coherence effect. It is reasonable to take these as mesoscopic effects in bulk materials. On the other hand, for propagation of classical waves in disordered composites, wave coherence may be easily achieved in conventional macroscopic samples, which can also be recognized as mesoscopic-like structures.

10.1.2 Different Transport Regimes

For electron transport, there are two kinds of descriptions. One is local, $\boldsymbol{j} = \sigma \boldsymbol{E}$, the conductivity σ relates the local current density $\boldsymbol{j}$ to the electric field $\boldsymbol{E}$; the other is global, $I = GV$, the conductance G relates the total current I to the voltage drop V. For a large homogeneous conductor, the difference between them is not important, and we can establish the relation

$$G = \sigma L^{d-2}. \tag{10.1.1}$$

Moreover, it is noted that the conductance G and conductivity σ have the same units in two dimensions. However, in mesoscopic structures, local quantities lose their meaning, so we will generally discuss conductance instead of conductivity.

There are several characteristic lengths for mesoscopic properties, such as the elastic mean free path l, the length L and width W of the sample, and the localization length ξ. Similar to the discussion in Chap. 9, we can define several regimes for electron transport in mesoscopic structures.

The first is the ballistic regime for which $W, L < l < \xi$, as shown in Fig. 10.1.1(a). The canonical example is the point contact; impurity scattering can be neglected, and electron scattering occurs only at the boundaries. Note that any local quantity has lost its meaning, so now only conductance plays a role, not conductivity.

The second case, $W < l < L < \xi$, for which boundary and internal impurity scatterings are of equal importance, can be called the quasi-ballistic regime, as shown in Fig. 10.1.1(b).

The third is the diffusive regime, for which $l < W, L < \xi$, as shown in Fig. 10.1.1(c). Conducting samples contain a significant amount of impurity atoms or structural disorder. The strength and concentration of impurities lead to $l \sim 100$ Å, which is the elastic scattering length, independent of temperature. In this case a simplified physical picture for electron transport is the random walk, with diffusion coefficient $D = v_{\rm F}l/3$ in three dimensions and $D = v_{\rm F}l/2$ in two dimensions.

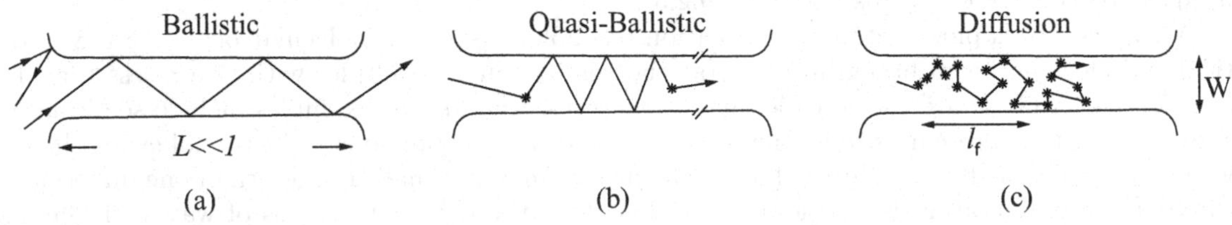

Figure 10.1.1 Narrow constriction in two-dimensional electron gas with width W and length L.

Finally, we may encounter the strongly localized case $L > \xi$, but we will not discuss that here.

10.1.3 Quantum Channels

Van Wees *et al.* (1988) presented a study on a two-dimensional electron gas in which a small constriction connected two half planes, as shown by the inset of Fig. 10.1.2. They found that the conductance, as a function of gate voltage, formed a ladder with step $= 2e^2/h$, when the width of the constriction varied between 0–360 nm. At $T < 1$ K, $l \geq 10$ μm, the transport is ballistic. When the width of the constriction was incremented by $\Delta W = \lambda_F/2$, there was a new channel added for an electron passing through the point contact.

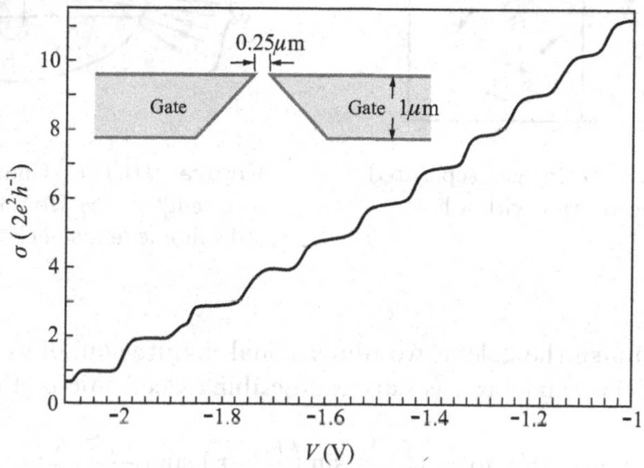

Figure 10.1.2 Experimental result for the quantum conductance as a function of gate voltage. Inset: Point contact layout. From B. J. van Wees *et al.*, *Phys. Rev. Lett.* **60**, 848 (1988).

An elementary explanation for the quantum conductance may be made as follows:[a] We can view the arrangement as two metals separated by an insulating barrier with a hole in it; at $T = 0$ K, the electrons fill energy states up to E_F. In classical thermodynamics, the number of particles which arrive at the hole of area of W^2 each second is

$$Q = \frac{1}{4} nvW^2, \tag{10.1.2}$$

where n is the number density. A difference of pressure can cause a difference of n, and like an effusing gas, electrons can flow through the hole.

In equilibrium, the Fermi energies of both sides are equal, i.e., $E_F^0(\mathrm{L}) = E_F^0(\mathrm{R})$, so there is no net flow. If an electric potential is applied which is $-V$ at the left and zero at the right, then

$$E_F(\mathrm{L}) = E_F^0(\mathrm{L}) + eV, \qquad E_F(\mathrm{R}) = E_F^0(\mathrm{R}). \tag{10.1.3}$$

The electrons near the Fermi energy can pass from the left to the right: For $eV \ll E_F$, the number density is $n = g(E_F)eV$, and the total number of electrons Q through the hole for each second is

$$Q = \frac{1}{4} g(E_F)eVv_F W^2. \tag{10.1.4}$$

In the free electron approximation, we have the Fermi energy $E_F = \hbar^2 k_F^2/2m$, and density of states at the Fermi surface $g(E_F) = 4mk_F/h^2$, and the conductance is

$$G = \frac{I}{V} = \frac{eQ}{V} = \frac{2e^2}{h}\frac{W^2}{4\pi}k_F^2. \tag{10.1.5}$$

[a] S. Kobayashi, *ASPAP News* **4**, 10 (1989).

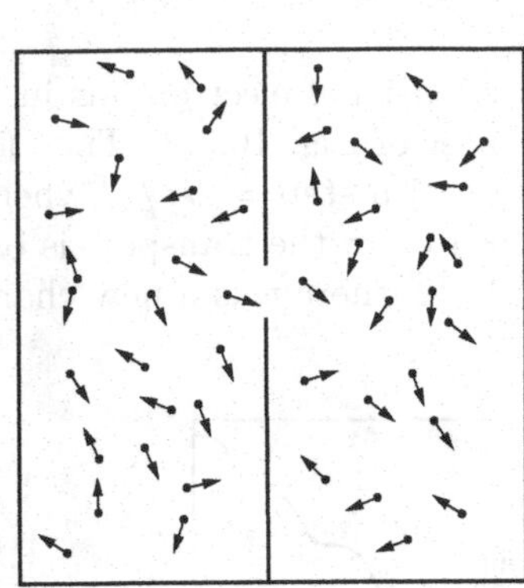

Figure 10.1.3 Fermi gas separated by a insulating barrier with a hole.

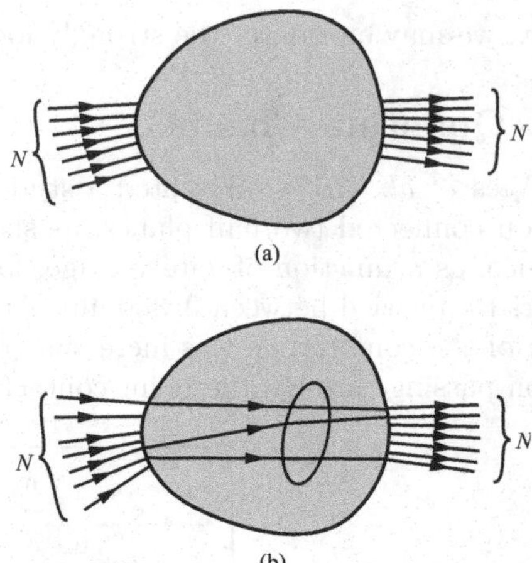

Figure 10.1.4 Channels with a finite length. (a) Homogeneous channels, (b) inhomogeneous channels.

For simplicity, we make the hole a two-dimensional infinite square well potential with width W and so the solution of the Schödinger equation describing electronic motion is

$$\psi(x, y) = \sqrt{\frac{4}{W^2}} \sin\left(\frac{n_x \pi}{W} x\right) \sin\left(\frac{n_y \pi}{W} y\right); \tag{10.1.6}$$

consequently the transverse quantized energies are

$$E_n = \frac{\hbar^2}{2m}\left(\frac{\pi}{W}\right)^2 (n_x^2 + n_y^2), \tag{10.1.7}$$

where n_x and n_y are quantum numbers.

In a degenerate electron gas, the quantum states labelled by the pairs (n_x, n_y) are occupied up to the Fermi wavenumber $k_{\rm F}$, and the corresponding occupied quantum states number is given by

$$N = \frac{W^2}{(2\pi)^2}\int d\mathbf{k} = \frac{W^2}{4\pi} k_{\rm F}^2. \tag{10.1.8}$$

By comparison of (10.1.5) and (10.1.8), we see

$$G = \frac{2e^2}{h} N, \tag{10.1.9}$$

where N is the channel number, the factor 2 coming from the electron spin, and e^2/h is the conductance of a single channel. $e^2/h \approx (25.8\ \text{k}\Omega)^{-1}$ sets the natural scale of conductance for all quantum transport. (10.1.9) thus interprets the experiments of van Wees *et al.*: The conductance is increased with step $2e^2/h$, as N is increased by one, in accordance with the size of the hole controlled by the gate voltage.

If the length of the hole is not zero, then it is like a tunnel, as schematically shown in Fig. 10.1.4. There are two possibilities: One is that the diameter of the tunnel is uniform, there is no scatterer within the tunnel, and the previous arguments still apply. The other is that the long tunnel is made of real small metals, the diameter may not be uniform, and there exist scatterers, such as defects, impurities etc. One to one correspondence of the channels is broken and channels may be mixed, so we can define an effective channel number $N_{\rm eff}$ satisfying

$$G = \frac{2e^2}{h} N_{\rm eff}. \tag{10.1.10}$$

It is evident that $N_{\rm eff} \leq N$.

§10.2 Landauer–Büttiker Conductance

Mesoscopic systems display a lot of interesting conduction properties. Because the configurations of these systems may be complex, it is not always easy to treat the problem by solving the Schrödinger equation. However, the phenomenological method proposed by Landauer, Büttiker and others has proved very powerful and can be used to study two-terminals with a single channel, two-terminals with multi-channels, and also multi-terminals with multi-channels. We will limit ourselves to the two-terminal cases.

10.2.1 Landauer Formula

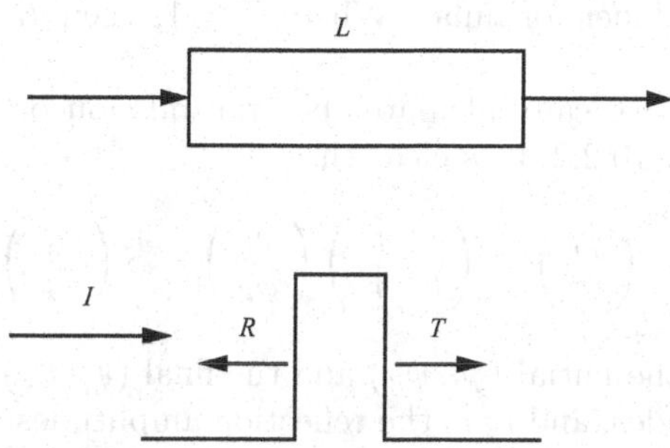

Figure 10.2.1 One-dimensional conductor and its barrier model.

Long ago, before the appearance of mesoscopic physics, Landauer proposed to study the conductance of a piece of one-dimensional disordered conductor,[b] as can be seen from Fig. 10.2.1. On the one hand, we can treat this problem in the barrier tunneling picture, where an incident electronic current comes into the sample with velocity v and density n. The net current density j for linear transport is

$$j = nev\tilde{T} = nev(1 - \tilde{R}), \tag{10.2.1}$$

where $\tilde{T}$ and $\tilde{R}$ are transmission and reflection probabilities, respectively. On the other hand, we may use the diffusion picture where the current density is related to the diffusion coefficient as

$$j = -eD\nabla n. \tag{10.2.2}$$

For an electron incident from the left, the relative particle densities on the left and right sides of the barrier will be $(1 + \tilde{R})$ and $(1 - \tilde{R})$, respectively. Then the densities at the two ends are given by $n(0) = (1 + \tilde{R})n$, $n(L) = \tilde{T}n = (1 - \tilde{R})n$, and the average density gradient can be written

$$\nabla n = -2\tilde{R}n/L. \tag{10.2.3}$$

From (10.2.1) and (10.2.2), we get the diffusion coefficient

$$D = \frac{vL}{2}\frac{1 - \tilde{R}}{\tilde{R}}. \tag{10.2.4}$$

Now, we know the Einstein relation for conductivity is

$$\sigma = e^2 D \frac{dn}{dE}, \tag{10.2.5}$$

[b] The original papers are R. Landauer, *IBM J. Res. Dev.* **1**, 223 (1957); *Phil. Mag.* **21**, 863 (1970). Landauer's approach is important in mesoscopic electronic transport, yet it is, in essence, a phenomenological theory. Recently, Das and Green analyzed the Landauer formula microscopically and suggested a straightforward quantum kinetic derivation for open systems. The details can be found in M. P. Das and F. Green, *J. Phys: Condens. Matter* **15**, L687 (2003).

where dn/dE is the density of states (DOS) at the Fermi surface. For one-dimensional free electron systems

$$\frac{dn}{dE} = \frac{2}{\pi\hbar v}, \tag{10.2.6}$$

so

$$\sigma = \frac{2e^2}{h} L \frac{1-\tilde{R}}{\tilde{R}}. \tag{10.2.7}$$

Combined with (10.1.1), we obtain the conductance

$$G = \frac{2e^2}{h} \frac{\tilde{T}}{\tilde{R}}. \tag{10.2.8}$$

This is the well-known Landauer formula. When $\tilde{T} = 1$, then $\tilde{R} = 0$ and $G \to \infty$ giving ideal conductance.

In a standard approach, we can calculate the transmission or reflection coefficients from the scattering matrix. From Fig. 10.2.2, it is clear that

$$\begin{pmatrix} \psi_{\tilde{r}} \\ \psi_{\tilde{t}} \end{pmatrix} = \begin{pmatrix} \tilde{r} & \tilde{t}' \\ \tilde{t} & \tilde{r}' \end{pmatrix} \begin{pmatrix} \psi_i \\ \psi_{i'} \end{pmatrix} = S \begin{pmatrix} \psi_i \\ \psi_{i'} \end{pmatrix}, \tag{10.2.9}$$

shows the relation between the initial $(\psi_i, \psi_{i'})$ and the final $(\psi_{\tilde{t}}, \psi_{\tilde{r}})$ states of scattering, where $\tilde{t}$, $\tilde{t}'$ are the transmission amplitudes, and $\tilde{r}$, $\tilde{r}'$ the reflection amplitudes. From current conservation and time reversal symmetry, one can prove

$$\tilde{t} = \tilde{t}', \qquad |\tilde{r}| = |\tilde{r}'|, \qquad \tilde{t}/\tilde{t}'^* = -\tilde{r}'/\tilde{r}^*,$$

obviously

$$\tilde{T} = |\tilde{t}|^2, \qquad \tilde{R} = |\tilde{r}|^2,$$

so the conductance can also be written as

$$G = \left(\frac{2e^2}{h}\right) \frac{|\tilde{t}|^2}{1 - |\tilde{t}|^2}. \tag{10.2.10}$$

10.2.2 Two-Terminal Single-Channel Conductance

The Landauer formula (10.2.8) is not very practical. A common way to drive current through a system is to connect ideal leads on its two sides to electron reservoirs of chemical potentials μ_1 and μ_2, assuming $\mu_1 > \mu_2$ for convenience, as shown in Fig. 10.2.3. The quantum point contact is

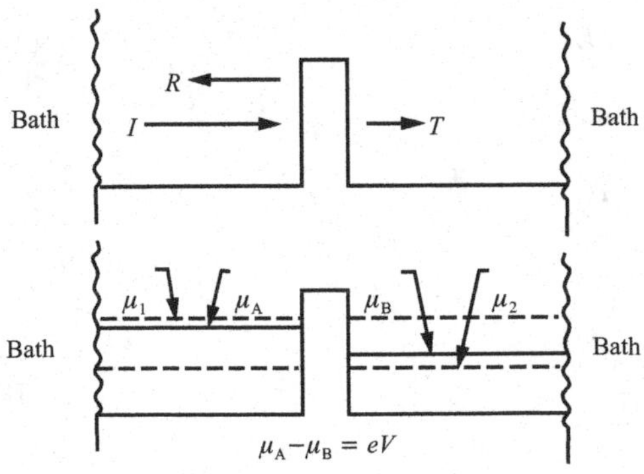

Figure 10.2.2 Two types of conductance.

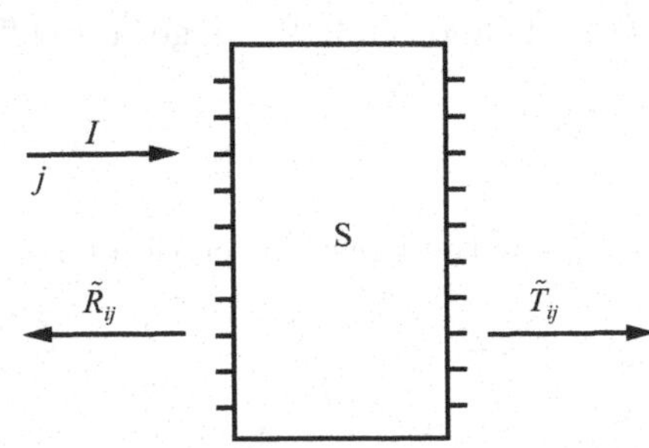

Figure 10.2.3 Transmission and reflection through a two terminal multichannel system.

an example of a short wire with a controllable width between zero and a few hundred nanometers. In semiconductors, these widths are comparable to the wavelengths of electrons, making the wire effectively one-dimensional. For a one channel system, we show in the following that, instead of the Landauer formula, the measurable conductance is[c]

$$G_c = \frac{eI}{\mu_1 - \mu_2} = \frac{2e^2}{h}\tilde{T}. \tag{10.2.11}$$

It is natural that even for $\tilde{T} = 1$, G_c is still a finite conductance.

For one channel connected with the ith end electrode, the incident current

$$I_i = \frac{2}{2\pi}e\int dk f(k,\mu_i)v = \frac{2e}{h}\int_0^{\mu_i} dE f(E,\mu_i), \tag{10.2.12}$$

where $v = \hbar^{-1}dE/dk$ is the electronic velocity, and f is the Fermi distribution function. At zero temperature, (10.2.10) gives

$$I_i = \frac{2e}{h}\mu_i. \tag{10.2.13}$$

If the difference of chemical potential between the two end electrodes $(\mu_1 - \mu_2)$ is very small, linear transport is satisfied

$$I = (I_1 - I_2)\tilde{T} = \frac{2e}{h}\tilde{T}(\mu_1 - \mu_2), \tag{10.2.14}$$

so (10.2.9) is proved.

Now we understand that the previously defined G in (10.2.8) is the conductance of the sample itself given by

$$G = \frac{eI}{\mu_A - \mu_B}, \tag{10.2.15}$$

where μ_A and μ_B are the chemical potentials on the left and right of the barrier. These two levels, which arose after the system was connected to the reservoirs with chemical potentials μ_1 and μ_2, are determined by making the number of occupied states (electrons) above μ_A (μ_B) equal to the number of empty states (holes) below μ_A (μ_B). Below energy μ_2 all states are fully occupied, so we need to consider the energy range from μ_2 to μ_1. Based on the discussion above, we can write

$$\tilde{T}\left(\frac{\partial n}{\partial E}\right)(\mu_1 - \mu_2) = 2\left(\frac{\partial n}{\partial E}\right)(\mu_B - \mu_2), \tag{10.2.16}$$

and

$$\tilde{T}\left(\frac{\partial n}{\partial E}\right)(\mu_1 - \mu_2) = 2\left(\frac{\partial n}{\partial E}\right)(\mu_1 - \mu_A), \tag{10.2.17}$$

then

$$\mu_A - \mu_B = \tilde{R}(\mu_1 - \mu_2). \tag{10.2.18}$$

Actually it is found from (10.2.8) and (10.2.11) that

$$G_c^{-1} = G^{-1} + \pi\hbar/e^2. \tag{10.2.19}$$

Here $\pi\hbar/e^2$ can be thought of as the two contact resistances, $\pi\hbar/2e^2$ each, i.e. the total resistance between the reservoirs is the sum of the barrier resistance and the two contact resistances due to the geometry of a narrow channel feeding into a large reservoir and to the electrons thermalizing in the baths by inelastic scattering.

At finite temperatures, we assume $(\mu_1 - \mu_2)$ is small, then

$$\begin{aligned} I &= \frac{2e}{h}\int dE\tilde{T}(E)[f(E-\mu_1) - f(E-\mu_2)] \\ &= \frac{2e}{h}\left[\int dE\tilde{T}(E)\left(-\frac{\partial f}{\partial E}\right)\right](\mu_1 - \mu_2), \end{aligned} \tag{10.2.20}$$

[c] M. Büttiker, Y. Imry, R. Landauer and S. Pinhas, *Phys. Rev. B* **31**, 6207 (1985).

so

$$G_c = \frac{2e^2}{h}\int dE\tilde{T}(E)\left(-\frac{\partial f}{\partial E}\right). \tag{10.2.21}$$

Similarly, we can determine the difference of chemical potentials $(\mu_A - \mu_B)$ by multiplying (10.2.16) and (10.2.17) with $-\partial f/\partial E$ and integrating them over the energy to get

$$\mu_A - \mu_B = \frac{\int dE(-\partial f/\partial E)\tilde{R}(E)(\partial n/\partial E)}{\int dE(-\partial f/\partial E)(\partial n/\partial E)}(\mu_1 - \mu_2). \tag{10.2.22}$$

Then the conductance of the sample is

$$G = \frac{2e^2}{h}\left[\int dE\tilde{T}(E)\left(-\frac{\partial f}{\partial E}\right)\right]\frac{\int dE(-\partial f/\partial E)v^{-1}(E)}{\int dE(-\partial f/\partial E)\tilde{R}(E)v^{-1}(E)}. \tag{10.2.23}$$

10.2.3 Two-Terminal Multichannel Conductance

If the sample has a finite cross section A, then the Landauer–Büttiker formulation must be generalized to the multichannel case, as illustrated in Fig. 10.2.3. There are discrete transverse energies E_i due to quantization in the transverse direction. $N_\perp$ conducting channels below the Fermi energy E_F, each at zero temperature characterized by longitudinal wavevectors k_i , so

$$E_i + \hbar^2 k_i^2/2m = E_F, \qquad i = 1, \ldots, N_\perp, \tag{10.2.24}$$

where $N_\perp \sim Ak_F^{d-1}$.

We can construct a $2N_\perp \times 2N_\perp$ $\boldsymbol{S}$ matrix

$$\boldsymbol{S} = \begin{pmatrix} \tilde{\boldsymbol{r}} & \tilde{\boldsymbol{t}}' \\ \tilde{\boldsymbol{t}} & \tilde{\boldsymbol{r}}' \end{pmatrix}, \tag{10.2.25}$$

to describe the relationships between the incoming and the outgoing waves for these $N_\perp$ channels. We define $\tilde{T}_{ij}$ and $\tilde{R}_{ij}$, which represent the probabilities of an incoming wave from the left jth channel being transmitted into the right-hand side ith channel and reflected into the left-hand side ith channel, respectively. We can also define $\tilde{T}'_{ij}$ and $\tilde{R}'_{ij}$ which represent the incoming waves from the right. The total transmission and reflection probabilities for the ith channel are

$$\begin{aligned} \tilde{T}_i &= \sum_j \tilde{T}_{ij}, \qquad \tilde{R}_i = \sum_j \tilde{R}_{ij}, \\ \tilde{T}'_i &= \sum_j \tilde{T}'_{ij}, \qquad \tilde{R}'_i = \sum_j \tilde{R}'_{ij}. \end{aligned} \tag{10.2.26}$$

The current conservation condition implies

$$\begin{aligned} \sum_i \tilde{T}_i &= \sum_i (1 - \tilde{R}_i) = N_\perp - \sum_i \tilde{R}_i, \\ \sum_i \tilde{T}'_i &= \sum_i (1 - \tilde{R}'_i) = N_\perp - \sum_i \tilde{R}'_i. \end{aligned} \tag{10.2.27}$$

If all incident channels on both sides of the barrier are fully occupied, all outgoing channels will also be fully occupied, and there are more detailed equations

$$\tilde{R}'_i + \tilde{T}_i = 1, \quad \tilde{R}_i + \tilde{T}'_i = 1. \tag{10.2.28}$$

The total current between two reservoirs is

$$I = \frac{2e}{h}\sum_i \int dE[f_1(E)\tilde{T}_i(E)(E) + f_2(E)\tilde{R}'_i(E)], \tag{10.2.29}$$

which is the extension of (10.2.14). Considering that $(\mu_1 - \mu_2)$ is small, we expand $f_1(E)$ at $\mu = \mu_2$, to obtain

$$I = (\mu_1 - \mu_2)\frac{2e}{h}\int dE(-\partial f/\partial E)\sum_i \tilde{T}_i(E), \tag{10.2.30}$$

which gives the conductance between the outside reservoirs

$$G_c = \frac{eI}{\mu_1 - \mu_2} = \begin{cases} (2e^2/h)\int dE(-\partial f/\partial E)\sum_i \tilde{T}_i(E), & \text{for } T \neq 0 \text{ K} \\ (2e^2/h)\sum_i \tilde{T}_i(E_F), & \text{for } T = 0 \text{ K}. \end{cases} \tag{10.2.31}$$

We can also study the conductance of the sample itself, defined by $eI/(\mu_A - \mu_B)$. As an extension of (10.2.18) for the one channel case,

$$\mu_A - \mu_B = \frac{\int dE(-\partial f/\partial E)\sum_i(1 + \tilde{R}_i - \tilde{T}_i)v_i^{-1}}{2\int dE(-\partial f/\partial E)\sum_i v_i^{-1}}(\mu_1 - \mu_2). \tag{10.2.32}$$

Finally by using (10.2.30) and (10.2.32), the conductance of the sample is

$$G = \frac{eI}{\mu_A - \mu_B} = \begin{cases} \dfrac{4e^2}{h}\dfrac{[\int dE(-\partial f/\partial E)\sum_i \tilde{T}_i][\int dE(-\partial f/\partial E)\sum_i v_i^{-1}]}{\int dE(-\partial f/\partial E)\sum_i(1 + \tilde{R}_i - \tilde{T}_i)v_i^{-1}}, & \text{for } T \neq 0 \text{ K}, \\ \dfrac{4e^2}{h}\dfrac{\sum_i \tilde{T}_i \sum_i v_i^{-1}}{\sum_i(1 + \tilde{R}_i - \tilde{T}_i)v_i^{-1}}, & \text{for } T = 0 \text{ K}. \end{cases} \tag{10.2.33}$$

When all $\tilde{T}_i \ll 1, 1 + \tilde{R}_i \simeq 2, G \simeq G_c$, that is, if the scattering is strong, the contact resistance may be neglected. This should be applicable for large $N_\perp$ whenever $G \ll 2e^2N_\perp/h$, or the sample length $L \gg l$.

§10.3 Conductance Oscillation in Circuits

Application of electromagnetic potentials to mesoscopic loop structures will introduce a few important effects in electron transport through the systems. The main characteristic of these effects is conductance oscillation.

10.3.1 Gauge Transformation of Electronic Wavefunctions

There is one type of symmetry, called gauge transformation invariance, related to electronic wavefunctions. To describe the electromagnetic field, there are two equivalent schemes, magnetic field $\boldsymbol{B}$ and electric field $\boldsymbol{E}$, or vector potential $\boldsymbol{A}$ and scalar potential φ. Both schemes are related by

$$\boldsymbol{B} = \nabla \times \boldsymbol{A}, \qquad \boldsymbol{E} = -\nabla\varphi - \frac{1}{c}\frac{\partial \boldsymbol{A}}{\partial t}, \tag{10.3.1}$$

but the relation is not a one to one correspondence. If an arbitrary scalar function $\Lambda(\boldsymbol{r}, t)$ is introduced to satisfy

$$\boldsymbol{A}' = \boldsymbol{A} + \nabla\Lambda, \qquad \varphi' = \varphi - \frac{1}{c}\frac{\partial \Lambda}{\partial t}, \tag{10.3.2}$$

then the new vector potential $\boldsymbol{A}'$ and scalar potential φ' give the same fields $\boldsymbol{B}$ and $\boldsymbol{E}$. (10.3.2) is called a gauge transformation. Any physical quantities must be invariant under a gauge transformation.

In classical electrodynamics, the fundamental equations of motion of charged particles can always be expressed directly in terms of the fields alone. The vector and scalar potentials were introduced as a convenient mathematical aid to obtain a classical canonical formalism. It is found that for the same $\boldsymbol{B}$ and $\boldsymbol{E}$, the potentials $(\boldsymbol{A}, \varphi)$ are not unique, so we can put an additional restriction on them; this restriction is called a gauge condition. There are two main gauges: The Coulomb gauge where $\nabla \cdot \boldsymbol{A} = 0$, and the Lorentz gauge which has $\nabla \cdot \boldsymbol{A} + \partial\varphi/\partial t = 0$.

In quantum theory, these potentials play a more significant role in the canonical formalism as we have seen in the study of Landau levels. We can go ahead further to discuss the deep meaning of electromagnetic potentials. The Hamiltonian for a single electron can be expressed in terms of $\boldsymbol{A}$ and φ as

$$\mathcal{H} = \frac{1}{2m}\left(\frac{\hbar}{i}\nabla + \frac{e}{c}\boldsymbol{A}\right)^2 - e\varphi + V(\boldsymbol{r}), \tag{10.3.3}$$

where $V(\boldsymbol{r})$ is the other scattering potential in addition to the electromagnetic potential. The concept of gauge invariance here comes from the fact that the wavefunction that characterizes the state of a system comprises two parts: amplitude and phase. Let us examine the time-dependent Schrödinger equation for an electron in external electromagnetic potentials

$$i\hbar\frac{\partial}{\partial t}\psi = \left[\frac{1}{2m}\left(\frac{\hbar}{i}\nabla + \frac{e}{c}\boldsymbol{A}\right)^2 - e\varphi + V(\boldsymbol{r})\right]\psi. \tag{10.3.4}$$

When the external potentials are transformed as in (10.3.2) we find that the wavefunction ψ must be correspondingly transformed as follows

$$\psi' = \psi e^{-ie\Lambda(\boldsymbol{r},t)/\hbar c}, \tag{10.3.5}$$

and the Schrödinger equation (10.3.4) stays invariant in its form, except for an additional phase factor in the wavefunction such as

$$\frac{e}{\hbar c}\Lambda(\boldsymbol{r},t) = \frac{e}{\hbar}\int\left(\varphi dt - \frac{\boldsymbol{A}}{c}\cdot d\boldsymbol{r}\right). \tag{10.3.6}$$

(10.3.5) is called the second, or local, gauge transformation, if Λ is a spatial function. When Λ is a constant, it is called the first, or global, gauge transformation. The electromagnetic potential $(\boldsymbol{A}, \varphi)$ is one of the gauge fields.

10.3.2 Aharonov–Bohm Effect in Metal Rings

Let us first consider that there is only a vector potential $\boldsymbol{A}$. It is straightforward to demonstrate that the wavefunction $\psi(\boldsymbol{r})$ of (10.3.4) with $\varphi = V(\boldsymbol{r}) = 0$ can be related to the wavefunction $\psi_0(\boldsymbol{r})$ with $\boldsymbol{A} = 0$ as

$$\psi(\boldsymbol{r}) = \psi_0(\boldsymbol{r})\exp\left(-\frac{ie}{\hbar c}\int \boldsymbol{A}\cdot d\boldsymbol{r}\right). \tag{10.3.7}$$

We can see when there is a vector potential $\boldsymbol{A}$, it may affect electronic behavior, even though there is no physical field $\boldsymbol{B}$ in the path electron passed. This is the well-known Aharonov–Bohm (AB) effect.[d]

To test their idea, Aharonov and Bohm put forward an experimental scheme: Figure 10.3.1 shows the geometry of a double slit experiment for an electron beam, but in addition there is a long solenoid enclosed by the electron paths. The wavefunctions for path γ_1 and path γ_2 are

$$\psi_1(\boldsymbol{r}) = \psi_0(\boldsymbol{r})\exp\left(\frac{ie}{\hbar c}\int_{\gamma_1} \boldsymbol{A}\cdot d\boldsymbol{r}\right), \ \psi_2(\boldsymbol{r}) = \psi_0(\boldsymbol{r})\exp\left(\frac{ie}{\hbar c}\int_{\gamma_2} \boldsymbol{A}\cdot d\boldsymbol{r}\right). \tag{10.3.8}$$

The electron density on the screen is

$$|\psi_1 + \psi_2|^2 = 2|\psi_0|^2 + 2|\psi_0|^2\cos\frac{2\pi\Phi}{\phi_0}, \tag{10.3.9}$$

where

$$\Phi = \int_{\gamma_1-\gamma_2} \boldsymbol{A}\cdot d\boldsymbol{r} = \oint \boldsymbol{A}\cdot d\boldsymbol{l} = \int\int \boldsymbol{B}\cdot d\mathbf{S}$$

[d]Y. Aharonov and D. Bohm, *Phys. Rev.* **115**, 485 (1959).

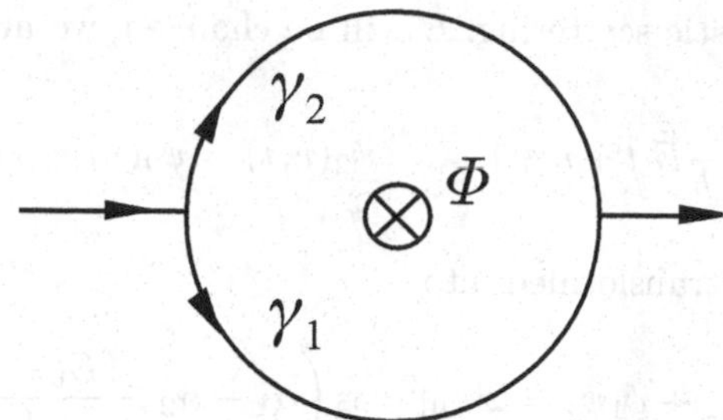

Figure 10.3.1 Schematic picture for the Aharonov–Bohm effect.

is the magnetic flux of the solenoid, and $\phi_0 = hc/e$ is a magnetic flux quantum for a system of single-electrons. When there is no magnetic field, i.e. $\Phi = 0$, (10.3.9) gives an electron interference pattern, which verifies the wave nature of electrons. Now when the flux increases, the interference fringes will move periodically, with a period of the magnetic flux quantum ϕ_0. This is a quantum effect arising from the modulation of the wavefunction phase factor by $\boldsymbol{A}$, and it is clear that at the quantum level $\boldsymbol{A}$ is a real entity. From the point of view of fundamental quantum mechanics, the AB effect is a good example that displays the geometric phase that comes from adiabatic cyclic evolution of microscopic particle.[e] The AB effect can be used to check that the frozen magnetic flux in a superconducting hollow cylinder is quantized, but the basic magnetic flux quantum Φ_0 is $hc/2e$, because of Cooper pairs.

In the original experiments which verified the AB effect, the electron waves propagated in free space.[f] What will happen in metals is a problem, for many people thought there would be no phase coherence in diffusion motion of electrons in metallic samples, and so perhaps the AB effect would not appear. However, as we have already discussed, elastic scattering does not destroy the phase coherence, and only inelastic scattering destroys it, so, in small samples such as mesoscopic metal rings, it is still possible to observe the AB effect.

Consider a metal ring: In classical mechanics, a particle entering such a ring must choose which of the two possible paths to take. The quantum mechanical wave nature of an electron, however, allows it to take both paths simultaneously. Part of the electron travels along one path, and part travels along the other. At the opposite end of the ring, the recombination of the two parts gives rise to interference. The resistance of the ring is low or high for constructive or destructive interference, respectively.

Let us give a more detailed discussion about this problem: There are two coherent electron beams in Fig. 10.3.1, but now in the medium they suffer multi-scatterings. In the beginning, the wavefunctions for the incident beams at origin are

$$\psi_1 = \psi_2 = \psi_0 \exp\left(-\frac{i}{\hbar}Et\right), \tag{10.3.10}$$

where E is the eigenenergy. In the elastic scattering case, E does not change, but collision phase shifts α_1, α_2 may appear, so the wavefunctions on the screen are

$$\psi_1(\boldsymbol{r},t) = \psi_0(\boldsymbol{r}) \exp\left(-\frac{i}{\hbar}Et + i\alpha_1\right), \qquad \psi_2(\boldsymbol{r},t) = \psi_0(\boldsymbol{r}) \exp\left(-\frac{i}{\hbar}Et + i\alpha_2\right), \tag{10.3.11}$$

and the interference term becomes

$$\psi_1^*\psi_2 + \psi_1\psi_2^* = 2|\psi_0|^2 \cos(\alpha_1 - \alpha_2), \tag{10.3.12}$$

where $(\alpha_1 - \alpha_2)$ is related to the distribution of elastic scatterers. In a mesoscopic system, these scatterers will give a fixed contribution to the interference.

[e] M. V. Berry, *Proc. Roy. Soc. London, Ser. A* **392**, 45 (1984).

[f] R. G. Chambers, *Phys. Rev. Lett.* **5**, 3 (1960); A. Tonomura *et al.*, *Phys. Rev. Lett.* **48**, 1443 (1982).

However, if there is inelastic scattering, E will be changed, we may write

$$\psi_1(\boldsymbol{r},t) = \psi_0(\boldsymbol{r})\exp\left(-\frac{i}{\hbar}E_1 t + i\alpha_1\right), \qquad \psi_2(\boldsymbol{r},t) = \psi_0(\boldsymbol{r})\exp\left(-\frac{i}{\hbar}E_2 t + i\alpha_2\right), \tag{10.3.13}$$

and the interference term is transformed into

$$\psi_1^*\psi_2 + \psi_1\psi_2^* = 2|\psi_0|^2\cos\left(\alpha_1 - \alpha_2 + \frac{E_1 - E_2}{\hbar}t\right). \tag{10.3.14}$$

Because the measurement in a normal metal sample is in the duration of microscopically long and macroscopically short times, averaging over microscopically long and macroscopically short times, (10.3.13) yields zero.

It is now clear that only elastic scattering can keep phase coherence. In this case in small metal samples, if there exists a flux Φ in the loop, (10.3.12) becomes

$$\psi_1^*\psi_2 + \psi_1\psi_2^* = 2|\psi_0|^2\cos\left(\frac{2\pi\Phi}{\phi_0} + \alpha_1 - \alpha_2\right). \tag{10.3.15}$$

In the AB effect in metal rings, electrons retain "phase memory" during the period of transport through the sample.

The inelastic scattering time in metal is typically around $\tau_{\text{in}} = 10^{-11}$ s at 1 K, $v_{\text{F}} = 10^8$ cm s^{-1}, then $l_{\text{in}} = v_{\text{F}} \cdot \tau_{\text{in}} \approx 10^5$ Å. But the elastic scattering length is $l \approx v_{\text{F}}\tau = 10^2$ Å, so the distance that an electron can cover without losing its phase memory is about 10 μm or 10^5 Å. For this reason, the sample must be shorter than the phase coherence length of order 10 μm to enable us to observe the interference effect in a metal. With the development of micro-fabrication technique, periodic oscillation of resistance has been observed in a gold ring enclosing a magnetic flux, as shown in Fig. 10.3.2. The inset shows a transmission electron photograph of the gold ring with an inside diameter of ~ 8000 Å and a width ~ 400 Å.

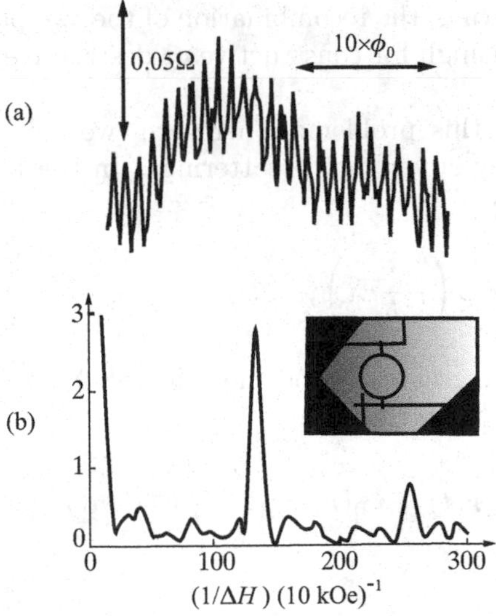

Figure 10.3.2 Aharonov–Bohm effect in a gold ring. (a) The oscillation in the magnetoresistance as a function of magnetic field. (b) The Fourier transform of the data in (a). From R. A. Webb *et al.*, *Phys. Rev. Lett.* **54**, 2696 (1985).

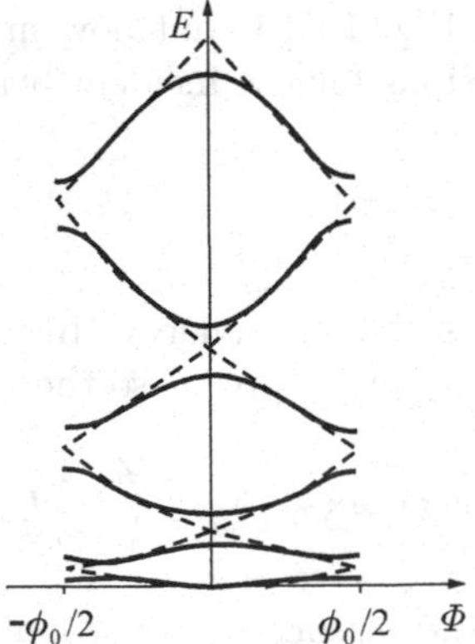

Figure 10.3.3 Energy diagram with dependence on flux in a metallic ring. From M. Büttiker *et al.*, *Phys. Lett. A* **96**, 365 (1983).

10.3.3 Persistent Currents

It is worthwhile to note that in (10.3.9) the phase change due to the magnetic flux in a ring can be written as

$$\Delta\theta = 2\pi\Delta\Phi/\phi_0, \tag{10.3.16}$$

thus the cases for flux Φ and $\Phi + n\phi_0$ are indistinguishable, where n is any integer. This issue was considered by using the simple model of a one-dimensional disordered ring.[g] Assuming the circumference is L, the boundary condition for (10.3.7) is similar to the case of the Bloch function ψ_k in a periodic potential with the unit cell of size L. Thus it is easy to identify that $2\pi\Phi/\phi_0$ and kL establish a one-to-one correspondence between the two problems. Here the circumference of the ring plays the role of the unit cell.

It is possible to estimate the flux dependence of the total energy E at low temperature. We rewrite (10.3.4) with $\varphi = 0$ as a stationary Schrödinger equation

$$\frac{1}{2m}\left(\frac{\hbar}{i}\frac{d}{dx} + \frac{e}{c}A_x\right)^2 \psi + V(x)\psi = E\psi, \tag{10.3.17}$$

where x denotes the direction along the ring and the magnetic field is in the direction z perpendicular to the plane of the ring. Because $\oint A_x dx = \int\int \boldsymbol{B}\cdot d\mathbf{S} = \Phi$, so $A_x = \Phi/L$. $V(x)$ is the scattering potential in the ring and has the period L. For an ideal ring, $V(x) = 0$, the Schrödinger equation is

$$\frac{1}{2m}\left(\frac{\hbar}{i}\frac{d}{dx} + \frac{e\Phi}{cL}\right)^2 \psi = E\psi. \tag{10.3.18}$$

By substituting a solution of plane wave form $\psi(x) = C\exp(ikx)$ into it, combining with a periodic condition $\psi(x + L) = \psi(x)$, the wavenumber is found to be $k = 2\pi n/L$, where n is any integer. The eigenenergy in (10.3.18) is then

$$E = \frac{\hbar^2}{2m}\left(\frac{2\pi}{L}\right)^2\left(n + \frac{\Phi}{\phi_0}\right)^2. \tag{10.3.19}$$

It is obvious from the geometry of a ring that E should be a periodic function of Φ. Taking into account $V(x)$ as a weak scattering potential, just as in the nearly-free electron approximation, we obtain a one-dimensional energy band structure with gaps. The first Brillouin zone satisfies $(-\phi_0/2) < \Phi < \phi_0/2$. In Fig. 10.3.3, the broken line is for $V(x) = 0$, and the solid line $V(x) \neq 0$. A scattering potential leads to the appearance of energy gaps at the center and edges of the Brillouin zone.

Due to quantum coherence, there will exist persistent currents in mesoscopic rings; we can write the flux-dependent currents as

$$I(\Phi) = -\frac{e}{L}\sum_n v_n(\Phi) = -c\sum_n \frac{\partial E_n(\Phi)}{\partial\Phi}. \tag{10.3.20}$$

The sum is over energy bands below the Fermi energy at the same Φ. Because $\partial E_n/\partial\Phi$ is larger for larger n, the dominant contribution to the current comes from the bands near the Fermi level. Stronger scattering leads to flatter bands, then smaller current. $I(\Phi)$ is also a periodic function of Φ with period ϕ_0.

The total current at $T = 0$ is obtained by adding all contributions from levels with energies less than the Fermi energy.[h] For an isolated ring with a fixed number of electrons N, the Fermi energy is

$$E_{\mathrm{F}} = \hbar^2(N\pi)^2/2mL^2.$$

The total persistent current is different for N odd and even. The result can be expressed as follows: for odd N

$$I(\Phi) = -I_0\frac{2\Phi}{\phi_0}, \qquad -0.5 \leq \frac{\Phi}{\phi_0} < 0.5, \tag{10.3.21}$$

[g] M. Büttiker, Y. Imry and R. Landauer, *Phys. Lett.* **96**A, 365 (1983).
[h] H. F. Cheung *et al.*, *Phys. Rev. B* **50**, 6050 (1988).

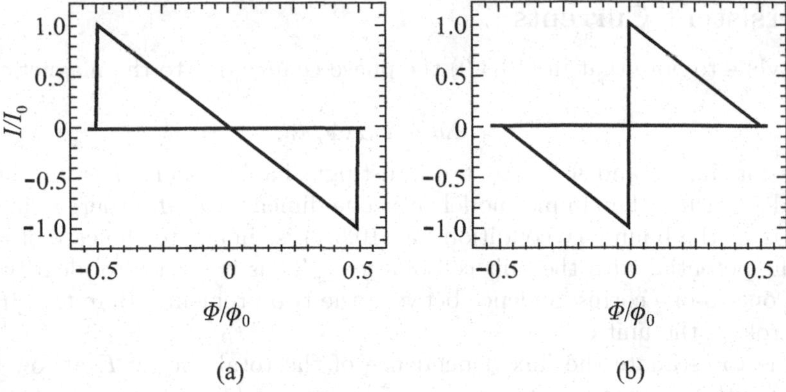

Figure 10.3.4 Persistent current over a period of the magnetic flux. The chemical potential is fixed such that the number of electrons in the ring is (a) odd and (b) even. From H. F. Cheung *et al.*, *Phys. Rev. B* **37**, 6050 (1988).

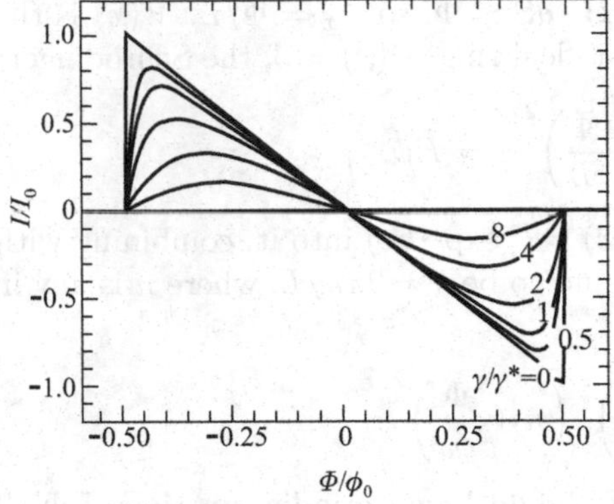

Figure 10.3.5 Effect on the persistent current of a single δ-function impurity of strength $\gamma/\gamma^* = 0, 0.5, 1, 2, 4,$ and 8 in a ring with an odd number of electrons. From H. F. Cheung *et al.*, *Phys. Rev. B* **37**, 6050 (1988).

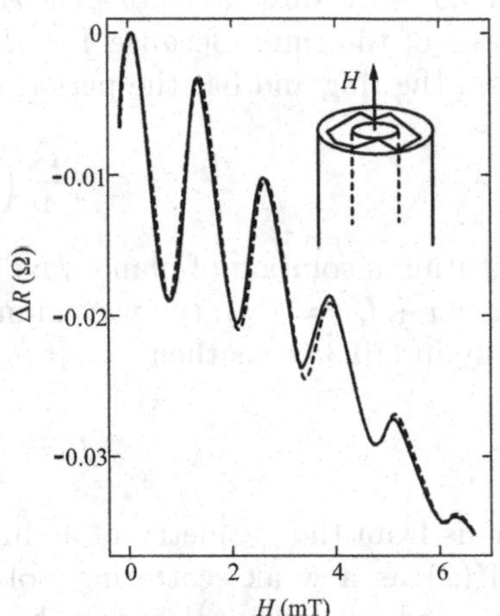

Figure 10.3.6 Magnetic field dependence of resistance measured in Li cylinder. From Yu V. Sharvin, *Physica B* **126**, 288 (1984).

and for even N

$$I(\Phi) = \begin{cases} -I_0\left(1 + \dfrac{2\Phi}{\phi_0}\right), & -0.5 \le \dfrac{\Phi}{\phi_0} < 0, \\ I_0\left(1 - \dfrac{2\Phi}{\phi_0}\right), & 0 \le \dfrac{\Phi}{\phi_0} < 0.5, \end{cases} \tag{10.3.22}$$

where $I_0 = heN/2mL^2$. These two cases are shown in Fig. 10.3.4.

Impurity scattering will affect the persistent currents. For simplicity, we consider the free electron model with an impurity characterized by a δ-potential, $V(x) = \gamma\delta(x)$, in (10.3.17). We can choose a reduced parameter

$$\gamma^* = \frac{\hbar v_{\mathrm{F}}}{\pi} = \frac{\hbar^2 N}{mL}, \tag{10.3.23}$$

then cases $\gamma \ll \gamma^*$ and $\gamma \gg \gamma^*$ are the weak- and strong-coupling regimes, respectively. Figure 10.3.5 exhibits the results of impurity potential on the persistent current for the I-Φ characteristics at $T = 0$ for a system with N odd.

Furthermore, if a time-dependent flux Φ is applied, there will be an induced electromotive force

$$U = -\frac{1}{c}\frac{d\Phi}{dt} \tag{10.3.24}$$

in the ring. For the case where U is pure d.c., this force will give rise to the Bloch oscillation for the electrons occupying the bands in Fig. 10.3.3. The equation of motion for electrons across the first Brillouin zone (BZ) is

$$-\frac{d\Phi}{dt} = cU, \tag{10.3.25}$$

and the oscillation frequency is

$$\omega = \frac{2\pi}{\phi_0}\left|\frac{d\Phi}{dt}\right| = \frac{eU}{\hbar}. \tag{10.3.26}$$

10.3.4 Altshuler–Aronov–Spivak Effect

In Fig. 10.3.2(b), besides the first peak representing the AB oscillation, there exists another oscillation with period corresponding to magnetic flux $hc/2e$. This comes from the Altshuler–Aronov–Spivak (AAS) effect, which is an interesting example confirming weak localization theory.[i] We have already met weak localization, which decreases the conductance remarkably in §10.2.3. Weak localization is an effect of the interference of electron waves. The physical picture here is very simple: When a magnetic flux Φ is enclosed by a loop, the phase of an electronic wavefunction changes by the vector potential along the path in the positive direction by $-\Delta\phi_1$, but the corresponding phase change along the same path in the negative direction is $\Delta\phi_2$, the phase difference of two partial waves is $\Delta\phi_1 - \Delta\phi_2$. After the paths go round the loop twice, as the electronic waves meet at the same point, $\Delta\phi_1 - \Delta\phi_2 = 2\Phi$, then there is a term

$$\psi_1^*\psi_2 + \psi_1\psi_2^* = 2|\psi_0|^2 \cos\frac{4\pi\Phi}{\phi_0}, \tag{10.3.27}$$

which is an interference effect with period $hc/2e$ instead of hc/e.

The AAS effect was theoretically predicted, and then confirmed by experiment.[j] The resistance of a hollow thin-walled metal cylinder, with a magnetic field in the cavity parallel to the cylindrical axis, was measured. A typical result is shown in Fig. 10.3.6.

Generally, the total electrical resistance $R(H)$ of a small two-lead metal loop threaded by a flux Φ due to an imposed magnetic field can be written as

$$R(H) = R_c + R_0 + R_1 \cos\left(\frac{2\pi\Phi}{\phi_0} + \alpha_1\right) + R_2 \cos\left(\frac{4\pi\Phi}{\phi_0} + \alpha_2\right) + \cdots, \tag{10.3.28}$$

where R_c is the classical resistance, which includes a magnetoresistance term proportional to H^2. Empirically, one finds that R_n and α_n are random functions of H.

10.3.5 Electrostatic Aharonov–Bohm Effect

So far we have not discussed the contribution of the scalar potential φ to the phase of the electronic wavefunction. When an electrostatic potential is exerted on an electron wave, the wave will accumulate a phase and this leads to the electrostatic Aharonov–Bohm effect. In this case the electron phase is proportional to the electric potential φ according to

$$\Delta\theta = (e/\hbar)\int \varphi dt. \tag{10.3.29}$$

The potential difference required to change the relative phase of the electron by 2π is $\Delta\varphi = h/e\tau$. Here τ is the upper bound of the integral and should be either the time the electron takes to travel through the device or the phase coherence time, $\tau_{\rm in}$, whichever is shorter.

[i] B. L. Al'tshuler, A. G. Aronov and B. Z. Spivak, *JETP Lett.* **33**, 94 (1981).
[j] D. Yu. Sharvin and Yu. V. Sharvin, *JETP Lett.* **34**, 272 (1981).

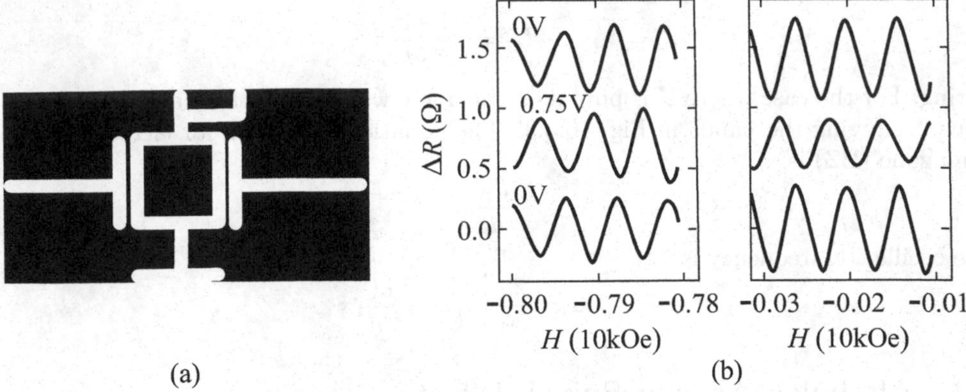

Figure 10.3.7 Electrostatic Aharonov–Bohm effect. (a) A square Sb loop placed between two metal electrodes. (b) When an electric field is established between the electrodes the phase of the magnetoresistance oscillations can be changed by 180°. From S. Washburn *et al.*, *Phys. Rev. Lett.* **59**, 1791 (1987).

The usual approach to a study of the electrostatic Aharonov–Bohm effect in metal rings is to explore the combined influence of electrostatic and magnetic potentials on electron waves. Washburn *et al.* investigated the effect of a transverse electric field on the magnetic Aharonov–Bohm oscillations. Their structure consisted of a metal Sb loop located between two electrodes of a parallel plate capacitor as shown schematically in Fig. 10.3.7(a). The loop is 0.82 μm on a side, and the capacitive probes are 0.16 μm from the arms of the loop. The Sb wires are approximately square in cross section being $d \simeq 0.075$ μm thick and wide. The magnetoresistance of the loop displayed Aharonov–Bohm oscillations which changed in phase by 180° when a bias of a 0.75 V was applied to the capacitor. The electric field between the plates of the capacitor appears to change the phase of the interfering electrons in much the same way as the magnetic vector potential. This can be regarded as the electrostatic Aharonov–Bohm effect.

Unfortunately there exists a little doubt about this, due to the applied voltage being much larger than that expected from the estimated phase coherence time of ~ 10 ps, possibly because of screening of the electric field within the metal. In practice the search for the electrostatic contribution to the electron phase in the metal rings has been hindered by the high conductance of metals, which makes it difficult to apply a well-defined voltage difference across the ring. Later, there was a measurement of quantum interference in a metal ring, that was interrupted by two small tunnel junctions, and this confirmed the electrostatic effect more convincingly.[k] It can be discerned that there are two periods in the measurement of transport properties: One is the period of the magnetic Aharonov–Bohm effect which is given by the ratio of the magnetic flux through the ring to the magnetic flux quantum hc/e. The other is related to the electrostatic Aharonov–Bohm effect which is given by the ratio of the so-called electrostatic flux VL^2/D to h/e, where V and L are the potential difference and the distance between the two tunnel barriers, and D is the diffusion coefficient.

§10.4 Conductance Fluctuations

We consider mesoscopic disordered metallic systems that contain a significant amount of impurity atoms. It has been shown both theoretically and experimentally that the mesoscopic conductance is sensitive to impurity. At low temperatures, they display interesting conductance fluctuation effects due to the coherence of electron waves transmitted through these systems.

[k] A. van Oudenaarden *et al.*, *Nature* **391**, 768 (1998).

10.4.1 Nonlocality of Conductance

Ohm's law, which is so successful in describing a linear current-voltage curve in any classical resistor, fails when the transport is phase coherent. Suppose a small conductor with a random impurity potential is biased between two reservoirs with different chemical potentials which will simply tilt the random potential distribution, as shown in Fig. 10.4.1.[1] The effective resistance of such a biased impurity potential is a random function of the bias voltage. The electron wavefunction extends all over a sample when $L < L_\phi$. Any disturbance at a given point modifies the wavefunction and, especially, its phase at any other point in the sample. So local conductivity has lost its meaning, Ohm's law is no longer applicable and the relation between I and V is nonlinear. Conductance fluctuates strongly when I changes. Define ΔG as the difference of measurement value and classical value of conductance, its amplitude of fluctuation is around the order of e^2/h.

Furthermore, one cannot even assume that $R(I)$, the resistance as a function of current, is equal to $R(-I)$. The resistance is not symmetric under exchange of the battery terminals. It can be illustrated in Fig. 10.4.1, that there is no inherent symmetry between positive and negative bias, because the potential has no mirror symmetry, or in a realistic three-dimensional wire it has no inversion center symmetry. This nonlinearity in the I-V curve has been measured: A representative example of non-ohmic behavior, found in an antimony wire at low temperatures, is shown in Fig. 10.4.2, where the deviation from Ohm's law is significant. One sees that quantum interference causes the conductance to fluctuate randomly and reproducibly throughout the range of current studied. Near zero current, the fluctuations are sharp, with an rms amplitude of e^2/h.

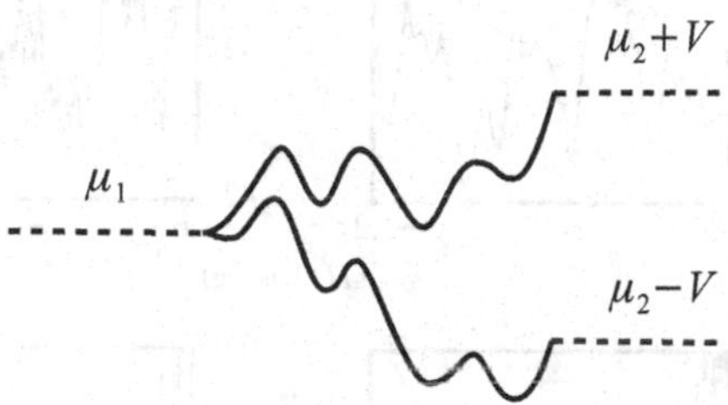

Figure 10.4.1 Illustration of the physics for the nonlinear response of a phase-coherent wire.

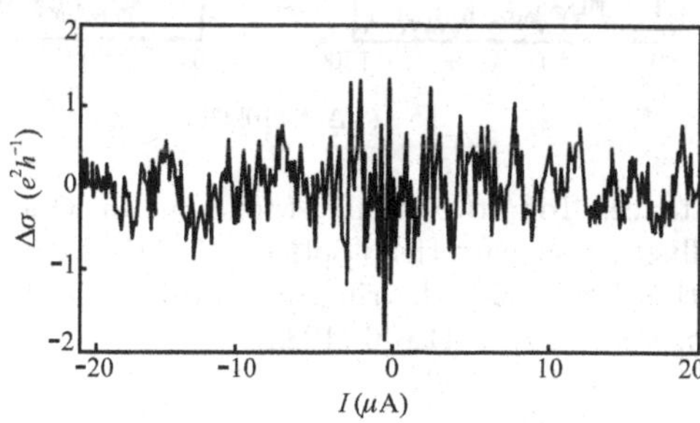

Figure 10.4.2 Conductance fluctuations measured as a function of current in a 0.6-μm length of Sb wire. From R. A. Webb and S. Washburn, *Phys. Today* **41**, 46 (1988).

One of the interesting experimental consequences is that the rms amplitude of the voltage fluctuations in a four-terminal sample becomes independent of length when the separation L between the voltage is less than L_ϕ. A four-terminal method, like the inset in Fig. 10.4.3, was used to measure conductance, where current is injected at lead 1 and removed at lead 4, and the voltage difference is measured between leads 2 and 3. Classically, the average resistance V_{23}/I_{14} depends linearly on L. The quantum-interference contributions to the voltage fluctuation with changing magnetic field behave very differently. In fact, when the distance L between 2 and 3 is less than L_ϕ, ΔV has no

[1] S. Washburn and R. A. Webb, *Rep. Prog. Phys.* **55**, 1311 (1992).

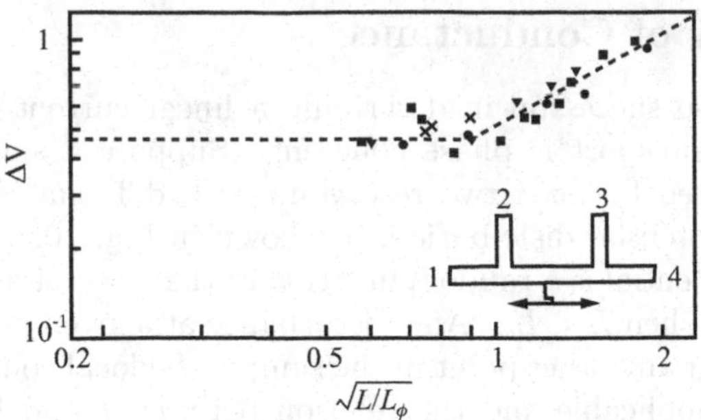

Figure 10.4.3 Length dependence of the voltage fluctuation amplitude. Inset is a four-lead sample. From R. A. Webb and S. Washburn, *Phys. Today* **41**, 46 (1988).

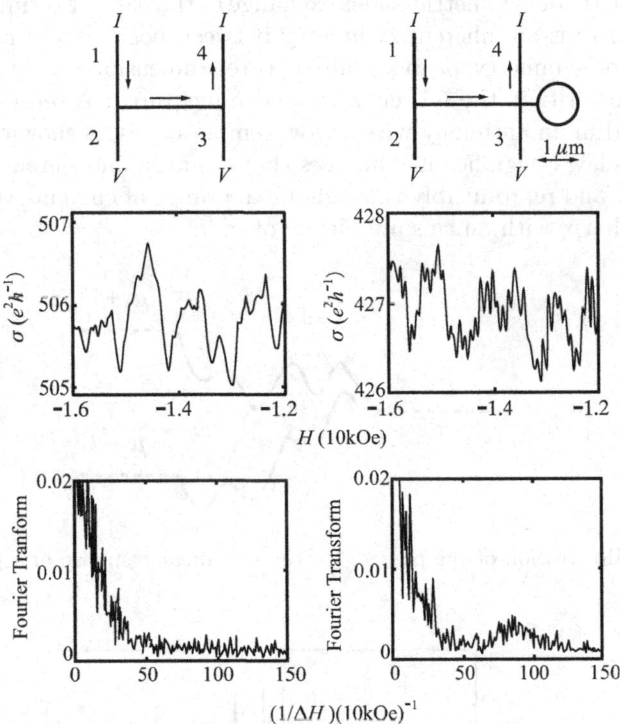

Figure 10.4.4 Magnetoresistance. Top: two similar wires, one of which has a ring dangling outside the path. Middle: conductance fluctuation patterns. Bottom: Fourier transforms of the conductance curves show an additional peak for the circuit with the ring, indicating hc/e Aharonov–Bohm oscillations. From R. A. Webb and S. Washburn, *Phys. Today* **41**, 46 (1988).

relation with length L. It is clear from Fig. 10.4.3 that for $L > L_\phi$, we find that ΔV is proportional to $(L/L_\phi)^{1/2}$ as expected, but for $L < L_\phi$, ΔV is a constant, even for $L \to 0$, and the voltage fluctuation ΔV does not vanish. The reason for this length independence arises from the fact that ΔV is not determined by the size of the sample, but rather by the correlation length. Here the voltage between 2 and 3 is not only determined by the electron path between 2 and 3, but also the paths between 1 and 3, 2 and 4. This is the nonlocal effect.

Since it is very difficult to observe the nonlocal property of the electric current by changing the positions of the voltage probes, the two devices shown in Fig. 10.4.4 are perhaps suitable for experimental measurement. Two identical four-probe wires were fabricated, but on the second sample a small ring was attached. The observation showed the dependence of the voltage drop between 2 and 3 upon the magnetic field applied in the direction perpendicular to the sample in the presence of the electric current between 1 and 4. The conductance of both wires exhibit random fluctuations with the same characteristic field scale, but the second sample shows additional high-frequency "noise". The Fourier transforms of both datasets clearly show that this high-frequency noise is in fact an hc/e oscillation. This is the AB effect which takes place in the small ring. Although this is impossible in classical conductors, the measured conductance as a function of the magnetic flux through the ring has a Fourier component corresponding to a period of hc/e. For this effect to be observable, some large fraction of the electrons must have enclosed the ring coherently.

10.4.2 Reciprocity in Reversed Magnetic Fields

Another counterintuitive discovery was that the magnetoresistance $R(H)$ measured in any four-terminal arrangement on a small wire does not remain the same when the magnetic field is reversed. Classically, of course, $R(H)$ should equal $R(-H)$. Figure 10.4.5 dramatically illustrates the asymmetric behavior found in a small gold wire, which is typical of all four-probe measurements made on small wires and rings. $R(-H)$ bears little resemblance to $R(+H)$. The second trace in Fig. 10.4.5(a) is the magnetoresistance obtained by interchanging the current and voltage leads. At first glance, it seems to bear little resemblance to the upper curve. Upon closer inspection, however, one sees that the positive-field half of the second trace is very similar to the negative-field half of the first trace. All the other permutations of leads yield different, asymmetric patterns of magnetoresistance fluctuation. These patterns are amazingly constant over time.

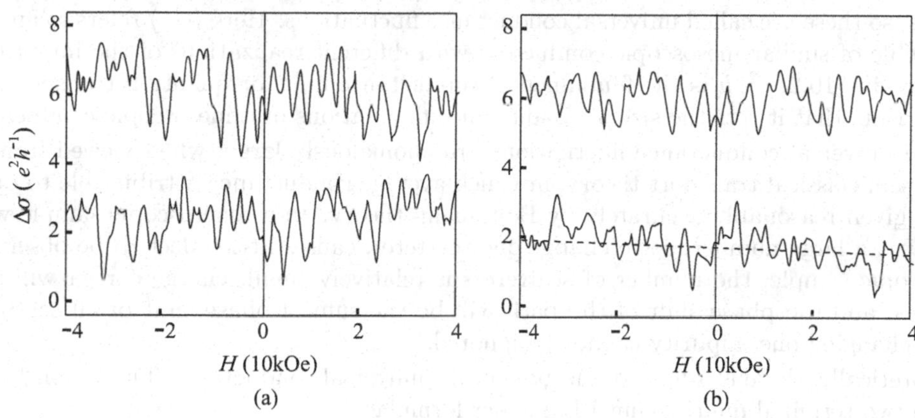

Figure 10.4.5 Asymmetric conductance fluctuation of a small gold wire. (a) Conductance fluctuation measured under magnetic field reversal. (b) Decomposing the measurements into symmetric and antisymmetric parts. From R. A. Webb and S. Washburn, *Phys. Today* **41**, 46 (1988).

None of these observations violates any fundamental symmetries. In fact, as Büttiker (1988) showed, the general properties displayed in Fig. 10.4.5(a) are consistent with the principle of reciprocity, which requires that the electric resistance in a given measurement configuration and magnetic field H be equal to the resistance at field $-H$ when the current and voltage leads are interchanged.

$$R_{14,23}(H) = R_{23,14}(-H), \tag{10.4.1}$$

where $R_{ij,kl} = V_{kl}/I_{ij}$. The Onsager symmetry relations for the local conductivity tensor may be more familiar, but they describe the *local* relationship between current density and electric field. As we have seen, conductance is a nonlocal quantity, and therefore the appropriate symmetries to consider are those involving the total resistance R.

To demonstrate reciprocity in the data, one determines the symmetric and antisymmetric parts of the resistance fluctuations as

$$\begin{aligned} R_S(H) &= \frac{1}{2}[R_{14,23}(H) + R_{23,14}(H)], \\ R_A(H) &= \frac{1}{2}[R_{14,23}(H) - R_{23,14}(H)]. \end{aligned} \tag{10.4.2}$$

Figure 10.4.5(b) shows this decomposition of the data. To within the noise level of the experiments, R_S is perfectly symmetric with respect to magnetic direction and R_A is perfectly antisymmetric. Applying this procedure to $R_{13,24}$ and $R_{24,13}$ results in a similar R_S but a different R_A. R_A is not simply due to the classical Hall resistance. The antisymmetric part fluctuates randomly with an amplitude similar to the symmetric fluctuations. Once again, the conductance fluctuation is of order e^2/h. The amplitude of the antisymmetric component is nearly constant, independent of L/L_ϕ, because it arises from nonlocal effects occurring within a distance L_ϕ of the junctions of the voltage probes. Excursions of electrons into the voltage probes account for all of the fluctuations in R_A.

10.4.3 Universal Conductance Fluctuations

Quantum coherence can lead to sample-specific and reproducible conductance fluctuations in mesoscopic systems. For mesoscopic conductors with $L \leq L_\phi$, the conductance varies from sample to sample, but its root-mean-square fluctuation

$$\Delta G \equiv \sqrt{\langle (G - \langle G \rangle)^2 \rangle}, \tag{10.4.3}$$

is roughly given by e^2/h, almost independent of impurity configuration, sample size, and spatial dimensions, so these are called universal conductance fluctuations. Here $\langle \cdots \rangle$ refers to an average over an ensemble of similar mesoscopic conductors with different realizations of the impurity positions. As we saw in §10.2, e^2/h is the fundamental conductance unit for quantum transport, so it is not too surprising that it sets the size of conductance fluctuations in a mesoscopic conductor.

These universal conductance fluctuations are anomalously large, when viewed from the standpoint of semiclassical transport theory, in which average conductance attributable to disorder scattering is given reasonably accurately by Boltzmann theory. In a macroscopic sample with N scatterers, N is a large number, so to change one scatterer, causes little effect to be observed. But for a mesoscopic sample, the number of scatterers is relatively small, charge carrier will visit each of impurities, and the phase shift of the path will be the sum of phase shift of all scatterers, so the effect of changing one impurity cannot be ignored.

Theoretically, there is a heuristic argument for universal conductance fluctuation[m] which begins with the two terminal multi-channel Landauer formula

$$G = \frac{e^2}{h}\tilde{T} = \frac{e^2}{h}\sum_{\alpha\beta}\left|\tilde{t}_{\alpha\beta}\right|^2, \tag{10.4.4}$$

where $\tilde{t}_{\alpha\beta}$ is the transmission amplitude from incoming channel β on the left to outgoing channel α on the right. Quantum mechanically, the transmission amplitude can be written as

$$\tilde{t}_{\alpha\beta} = \sum_{i=1}^{M} A_{\alpha\beta}(i), \tag{10.4.5}$$

where $A_{\alpha\beta}(i)$ represents the probability amplitude due to the ith Feynman path that connects channel β to α. Semiclassically, we can think of the Feynman path i as a classical random walk from the left of the sample to the right. Note that unlike the more familiar example of ballistic transport, where the important Feynman paths are restricted to a narrow tube of radius λ_F connecting the end points, here the presence of disorder means that the electron motion is diffusive, and the important Feynman paths are the random walks which cover much of the sample and M is very large. It is assumed that $A_{\alpha\beta}(i)$ are independent complex random variables. With this assumption, we can

[m] P. A. Lee, *Physica* **140A**, 169 (1986). More exact theoretical treatment can be found in the classical papers: P. A. Lee and A. D. Stone, *Phys. Rev. Lett.* **55**, 1622 (1985); P. A. Lee, A. D. Stone and H. Fukuyama, *Phys. Rev. B* **35**, 1039 (1987).

calculate the fluctuation in $|t_{\alpha\beta}|^2$, defined by $\Delta\langle|t_{\alpha\beta}|^2\rangle = [\langle|t_{\alpha\beta}|^4\rangle - \langle|t_{\alpha\beta}|^2\rangle^2]^{1/2}$. Ignoring terms of order unity compared with M,

$$\langle|\tilde{t}_{\alpha\beta}|^4\rangle = \sum_{ijkl}\langle A^*_{\alpha\beta}(i)A_{\alpha\beta}(j)A^*_{\alpha\beta}(k)A_{\alpha\beta}(l)\rangle = 2\left\langle\sum_i |A_{\alpha\beta}(i)|^2\right\rangle^2 = 2\langle|\tilde{t}_{\alpha\beta}|^2\rangle^2, \tag{10.4.6}$$

then we can immediately conclude that

$$\Delta\langle|\tilde{t}_{\alpha\beta}|^2\rangle = \langle|\tilde{t}_{\alpha\beta}|^2\rangle, \tag{10.4.7}$$

which states that the relative fluctuation of each transmission probability is of order unity.

With (10.4.4) and (10.4.7) we can attempt to estimate the fluctuation in G. The simplest assumption is that different channels are uncorrelated, i.e., there is no correlation between $|\tilde{t}_{\alpha\beta}|^2$ and $|\tilde{t}_{\alpha'\beta'}|^2$, but this gives too small conductance fluctuations compared to experimental results. We note that $t_{\alpha\beta}$ must involve multiple scattering in order to transverse the sample, so there may be stronger correlations among channels in the transmission, whereas the reflection amplitude is probably dominated by a few scattering events. Actually, the correct answer can be obtained by considering the reflection amplitude $\tilde{r}_{\alpha\beta}$ instead of $\tilde{t}_{\alpha\beta}$. Let us introduce the reflectance

$$\frac{e^2}{h}\tilde{R} = \frac{e^2}{h}\sum_{\alpha\beta}|r_{\alpha\beta}|^2. \tag{10.4.8}$$

By unitarity, $\tilde{T} + \tilde{R} = N$, so the fluctuation in $(e^2/h)\tilde{R}$ is the same as the fluctuation in G. Using similar reasoning that led to (10.4.7) we obtain

$$\Delta\langle|\tilde{r}_{\alpha\beta}|^2\rangle = \langle|\tilde{r}_{\alpha\beta}|^2\rangle. \tag{10.4.9}$$

If we assume that $|\tilde{r}_{\alpha\beta}|^2$ are uncorrelated from channel to channel, we obtain

$$\frac{e^2}{h}\Delta\tilde{R} \approx \frac{e^2}{h}N\langle|\tilde{r}_{\alpha\beta}|^2\rangle. \tag{10.4.10}$$

Using Ohm's law $G = \sigma L^{d-2}$, $\sigma = (e^2/h)k_{\mathrm{F}}^{d-1}l$ where l is the mean free path and $N = (Lk_{\mathrm{F}})^{d-1}$, we conclude that

$$\langle|\tilde{t}_{\alpha\beta}|^2\rangle \approx \frac{l}{NL}, \tag{10.4.11}$$

which when combined with unitarity leads to

$$\langle|\tilde{r}_{\alpha\beta}|^2\rangle \approx \frac{1}{N}\left(1 - \frac{l}{L}\right). \tag{10.4.12}$$

This simply expresses the fact as long as $l \ll L$, most of the incoming beam is reflected into N reflecting channels. Combining (10.4.10) and (10.4.12) we obtain the universal conductance fluctuation

$$\Delta G = \frac{e^2}{h}\Delta\tilde{R} \approx \frac{e^2}{h}. \tag{10.4.13}$$

Experimentally, these sample-to-sample conductance fluctuations are actually observed in one given sample, which is however subject to a varying applied magnetic field. As the magnetic field adds a phase factor to the electron partial waves between the various random scattering events, it has the same effect as changing the positions of the impurity atoms randomly from one sample to another. So we need to measure the magnetoresistance to verify universal conductance fluctuation in mesoscopic systems. For bulk metallic samples, the conductance changes smoothly with applied magnetic field. The physical properties of a bulk sample depend on the material, but various samples of the same material have the same physical properties. The resistance of a thin wire at low temperature is plotted against H in Fig. 10.4.6. This pattern of fluctuation is sample-specific, and for a given sample, the pattern is reproducible. But for another sample composed of the same material,

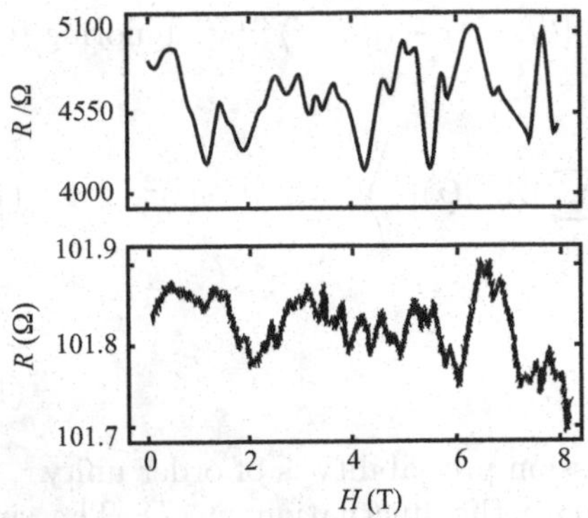

Figure 10.4.6 Experimental result for magnetoresistance fluctuation of a thin wire of AuPd and comparison with the numerical simulation. A. D. Stone, *Phys. Rev. Lett.* **54**, 2692 (1985).

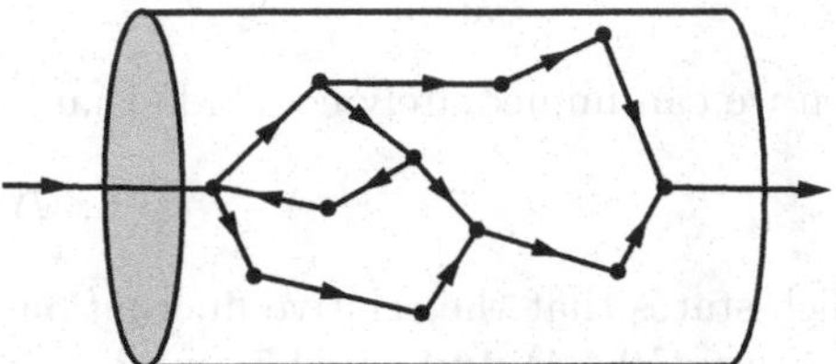

Figure 10.4.7 Multichannel conduction behavior with an applied magnetic field.

the magnetoresistance pattern will be different, because the number and distribution of scatterers in both samples are different. So the magnetoconductance fluctuations for a given mesoscopic conductor as a function of the applied field can be differentiated from noise. It is called a magneto-fingerprint. The numerical simulation in Fig. 10.3.6 gives very similar fluctuation pattern.

The magnetoresitance in a mesoscopic sample is closely related to AB effect. Consider the path of an electron that enters a conductor consisting of various scatterers as in Fig. 10.4.7. The path represents one of those N_{eff} paths, where N_{eff} is the number of effective channels. The path consists of various loops. When a magnetic field is applied, there will be a field dependent phase shift in each segment of loops according to the AB effect. The interference pattern at the next junction oscillates as a function of a magnetic field. The amplitude of the oscillation is of the order of e^2/h in each path, and its period is $\phi_0 = hc/e$ divided by the area A surrounded by the loops. As A may take any value from zero to the cross section of the sample, the conductance of the sample, which is a superposition of those oscillations, with loop-specific periods and phases, has an aperiodic fluctuation. The fluctuation in Fig. 10.4.6 observed in the experiment can, in this way, be interpreted.

Bibliography

[1] Beenakker, C. W. J., and H. van Houten, *Quantum Transport in Semiconductor Nanostructures*, Solid State Physics **44**, 1 (1991).

[2] Imry, Y., *Introduction to Mesoscopic Physics*, Oxford University Press, New York (1997).

[3] Washburn, S., and R. A. Webb, *Aharonov–Bohm Effect in Normal Metal, Quantum Coherence and Transport*, Adv. Phys. **35**, 375 (1986); *Quantum Transport in Small Disordered Samples from the Diffusive to the Ballistic Regime*, Rep. Prog. Phys. **55**, 1311 (1992).

[4] Sheng, P., *Introduction to Wave Scattering, Localization, and Mesoscopic Phenomena*, Academic Press, San Diego (1995).

[5] Thornton, T. J., *Mesoscopic devices*, Rep. Prog. Phys. **58**, 311 (1995).

[6] Datta, S., *Electronic Transport in Mesoscopic Systems*, Cambridge University Press (1995).

[7] Ferry, D. K., and S. M. Goodnick, *Transport in Nanostrucutures*, Cambridge University Press (1997).

Part III

Bonds and Bands with Things Between and Beyond

Only connect ⋯

— E. M. Forster

A genuine symbiosis may also emerge from complementary approaches. The typical chemist wants above all to understand why one substance behaves differently from another; the physicist usually wants to find principles that transcend any specific substance.

— Dudley Herschbach (1997)

Chapter 11

Bond Approach

In this part we are mainly concerned with the electronic structure of matter. We start from atoms and ions, pass through molecules, then concentrate on solids in the crystalline state. Both bond and band approaches are introduced, contrasted and sometimes used side by side. Single-electron methods such as the molecular orbital method and the band method are our major topics, fully discussed, justified and their inevitable deficiencies delineated. On the other hand, many-body effects are emphasized from the start, culminating in the chapter on correlated electronic states. A special chapter is devoted to the electronic properties of nanostructures in order to connect with contemporary research.

This chapter is devoted to the bond approach with a strong chemical flavor. We believe that students of condensed matter physics should know the basic concepts of quantum chemistry and acquire some chemical insights, especially those interested in more complex structures.

§11.1 Atoms and Ions

Atoms are fundamental building blocks for molecules and condensed matter. Here we review some basic ideas of atomic physics, especially those closely related to quantum chemistry and condensed matter physics.

11.1.1 A Hydrogen Atom

We begin with the simplest atom, hydrogen. It may be described as an electron moving in the Coulomb field of a stationary proton. The Schrödinger equation for this electron in spherical polar coordinates is written (see Fig. 11.1.1),

$$\left[\frac{\partial}{\partial r}\left(r^2\frac{\partial}{\partial r}\right)+\frac{1}{\sin\theta}\frac{\partial}{\partial\theta}\left(\sin\theta\frac{\partial}{\partial\theta}\right)+\frac{1}{\sin^2\theta}\frac{\partial^2}{\partial\phi^2}+\frac{2mr^2}{\hbar^2}\left(\frac{Ze^2}{r}+E\right)\right]\psi=0. \tag{11.1.1}$$

This equation can be solved analytically. Each solution is called the wavefunction or atomic orbital

$$\psi_{nlm}(r,\theta,\phi)=R_{nl}(r)\mathrm{Y}_{lm}(\theta,\phi), \tag{11.1.2}$$

where $R_{nl}(r)$ are the radial components and $\mathrm{Y}_{lm}(\theta,\varphi)$ are spherical harmonics which show the angular dependence of the wavefunctions. The wavefunctions are characterized by a set of quantum numbers, i.e., principal quantum number n, angular momentum quantum number l, and magnetic quantum number m. The energy eigenvalues are

$$E_n=-\frac{e^4}{2\hbar^2}\frac{1}{n^2},\ n=1,2,\ldots. \tag{11.1.3}$$

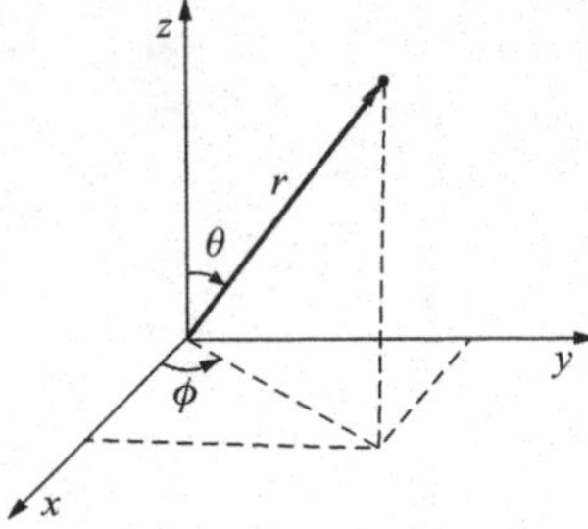

Figure 11.1.1 The coordinate system of a hydrogen atom.

With a given principal quantum number n, there are n^2 possible combinations, according to the following scheme

$$\sum_{l=0}^{n-1} \sum_{m=-l}^{l} 1 = \sum_{l=0}^{n-1} (2l+1) = n^2. \qquad (11.1.4)$$

The energy is the same, i.e., it is n^2-fold degenerate. This energy degeneracy will be partially lifted in atoms with many electrons and further reduced if the atom or the ion is situated in a non-spherical environment.

It is interesting to inspect the spatial distribution of different atomic orbitals (see Fig. 11.1.2): Only s-orbitals (with $l = 0$) are isotropic; others are anisotropic with nodes (i.e. loci where $\psi = 0$), such as p- (with $l = 1$), d- (with $l = 2$), and f- (with $l = 3$) orbitals. These geometrical shapes, which determine the spatial distribution of the electron clouds, have important consequences for chemical and physical properties. It is advisable that the reader should be able to visualize these shapes.

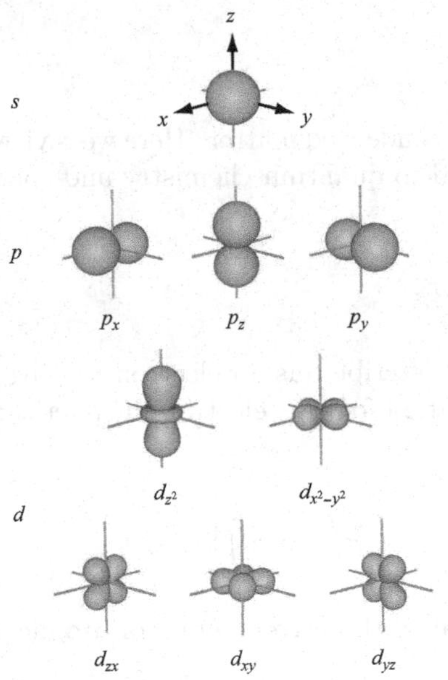

Figure 11.1.2 The three-dimensional graphs for s, p, d orbitals.

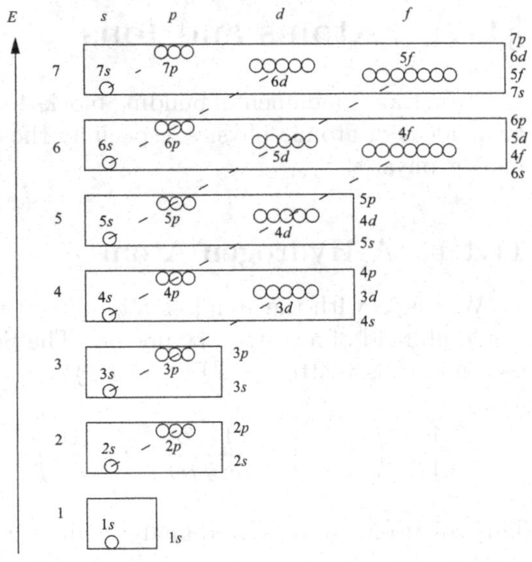

Figure 11.1.3 Schematic diagram for the energy levels of electron shells in many-electron atoms.

11.1.2 Single-Electron Approximation for Many-Electron Atoms

We consider an atom with Z electrons. The Hamiltonian of this system includes three parts: the kinetic energy; the attractive potential between every electron and nucleus; and the repulsive potential between every pair of electrons. The Schrödinger equation for this many-electron atom is

$$\left(-\frac{\hbar^2}{2m}\sum_{i=1}^{Z}\nabla_i^2-\sum_{i=1}^{Z}\frac{Ze^2}{r_i}+\frac{1}{2}\sum_{i\neq j}\frac{e^2}{|\boldsymbol{r}_i-\boldsymbol{r}_j|}\right)\Psi=E\Psi. \tag{11.1.5}$$

The many-electron atom problem is too complicated to be solved analytically, so some sort of approximation must be resorted to. The most useful approximation is the self-consistent field approximation introduced by D. R. Hartree and V. A. Fock, in which each electron is described by its one-electron Schrödinger equation,

$$\left[-\frac{\hbar^2}{2m}\nabla^2+v(\boldsymbol{r})\right]\psi=\varepsilon\psi, \tag{11.1.6}$$

where $v(\boldsymbol{r})$ includes both the ionic potential energy and the interaction energy from other electrons.

It can be assumed, with sufficient accuracy, that this potential is spherically symmetric, and therefore the type of solution is similar to that of hydrogen, i.e., the orbitals are still specified by s, p, d, and f states. However, one should also take into account the electron spin and the Pauli exclusion principle; each orbital may have two electrons with opposite spins. Therefore, in addition, a spin quantum number $m_s = \pm 1/2$, describing the spin coordinate, is introduced to the set of quantum numbers specifying the orbital. According to the Pauli exclusion principle and the principle of minimum energy, Z electrons occupy the different orbitals of a many-electron atom, from the lowest upward, forming the shell structure of electrons. So, the general outline of the periodic table of chemical elements is explained. Using a single-electron approximation, ground state electronic configurations may be assigned to all atoms, such as He ($1s^2$), Li ($1s^2$ $2s^1$), C ($1s^2$ $2s^2$ $2p^2$), O ($1s^2$ $2s^2$ $2p^4$), etc., and the details can be seen in the periodic table. It should be noted that spectroscopic evidence as well as detailed calculations show that the energy degeneracy for electrons with the same principal quantum number is lifted in many-electron atoms, and the general order for electron filling in valence shells and sub-shells is shown in Fig. 11.1.3. Actually the filling order is also dependent on atomic number Z (see Fig. 11.1.4), for instance, for light elements $3d$ is

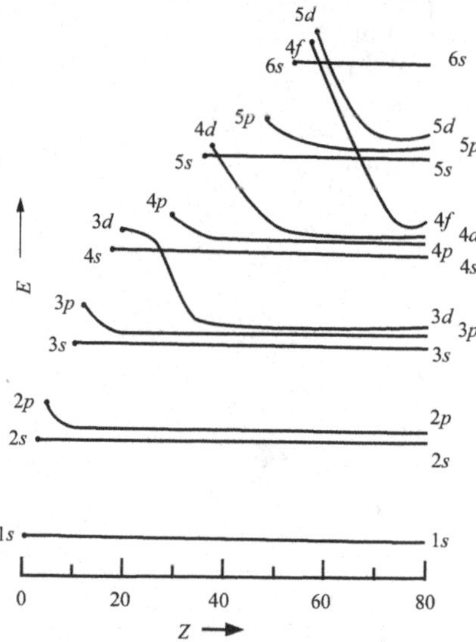

Figure 11.1.4 Schematic diagram for the ordering of the energy sub-shell in atoms as a function of atomic number. From F. Yang and J. H. Hamilton, *Modern Atomic and Nuclear Physics*, McGraw-Hill, New York (1996).

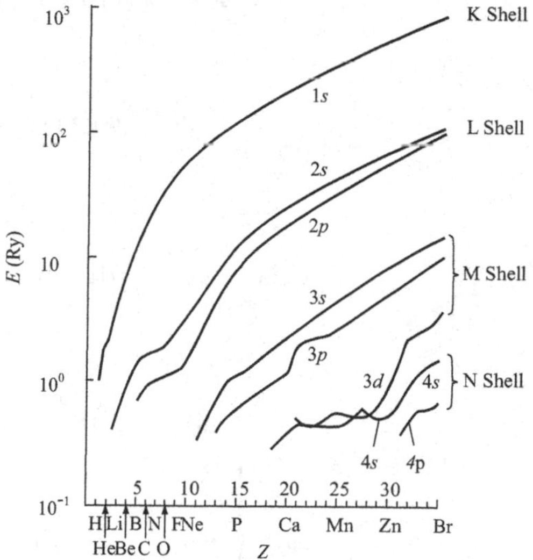

Figure 11.1.5 Electron binding energies as a function of atomic numbers for different shells. From F. Yang and J. H. Hamilton, *Modern Atomic and Nuclear Physics*, McGraw-Hill, New York (1996).

filled before 4s, while for heavier elements the reverse is true. In the crossing region ($z = 20 \sim 28$, i.e., 3d transition metal elements, from Ca to Ni), the situation is very complex (see Fig. 11.1.5), it is complicated by the many-body correlations of electrons. However the geometrical shapes of atomic orbitals are roughly the same as those of hydrogen, so we may still use the figures shown in Fig. 11.1.2 with some confidence.

11.1.3 Intraatomic Exchange

In spite of the enormous success achieved by the single-electron approximation, there are many problems that require a more rigorous treatment of the Coulomb interactions between the electrons in an atom. Let us consider the simplest case, a pair of electrons, for instance two electrons in a He atom, one in the ground state 1s, another in the first excited state 2s; or two electrons in an unfilled 3d shell of a transition metal atom, with the rest of electrons ignored. The state of each electron is described by a spin-orbital,

$$\psi(x, y, z, s_z) = \phi(x, y, z)\chi(s_z), \tag{11.1.7}$$

where $\phi(x, y, z)$ is the spatial part, and $\chi(s_z)$ is the spin part. The combination of the spins of two electrons may have a symmetric form, called the spin-triplet, $S = 1$; and antisymmetric form, called the spin-singlet, $S = 0$, as listed below (see Fig. 11.1.6)

$$\Psi_{\rm ss} = \begin{cases} \chi_{11} = \alpha(1)\alpha(2), \\ \chi_{10} = \dfrac{1}{\sqrt{2}}[\alpha(1)\beta(2) + \alpha(2)\beta(1)], \\ \chi_{1-1} = \beta(1)\beta(2), \end{cases} \tag{11.1.8}$$

$$\Psi_{\rm sa} = \chi_{00} = \frac{1}{\sqrt{2}}[\alpha(1)\beta(2) - \alpha(2)\beta(1)], \tag{11.1.9}$$

where α indicates spin up and β spin down.

Then we can express the total antisymmetric wave function as

$$\Psi = [\phi_0(\boldsymbol{r}_1)\phi_{nl}(\boldsymbol{r}_2) - \phi_0(\boldsymbol{r}_2)\phi_{nl}(\boldsymbol{r}_1)] \begin{pmatrix} \chi_{11} \\ \chi_{10} \\ \chi_{1-1} \end{pmatrix}, \tag{11.1.10}$$

or

$$\Psi = [\phi_0(\boldsymbol{r}_1)\phi_{nl}(\boldsymbol{r}_2) + \phi_0(\boldsymbol{r}_2)\phi_{nl}(\boldsymbol{r}_1)]\chi_{00}. \tag{11.1.11}$$

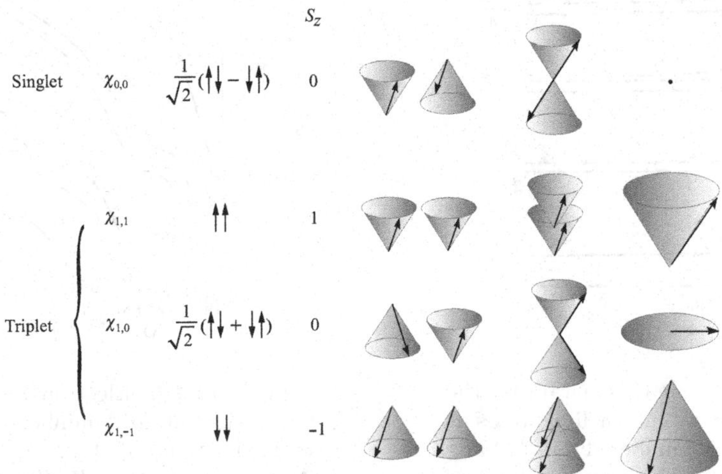

Figure 11.1.6 Schematic diagrams for spins of combined wave functions of two electrons.

Because the total wavefunction Ψ for a pair of electrons is antisymmetric (to satisfy Fermi–Dirac statistics), if the spin is symmetric, the spatial wave function will be antisymmetric and vice versa.

Now the question is: which state (spin-singlet or spin-triplet) has the lower energy? The answer may be obtained from the following argument: According to the Pauli principle, two electrons with parallel spins cannot be at the same point in space, because then they would occupy the same quantum mechanical state. So these two electrons tend to keep away in real space in order to minimize Coulomb repulsive energy. By contrast, two electrons with antiparallel spins have no restriction on being at the same point in space. Therefore, it is the Coulomb repulsive force that raises the energy of the singlet state; consequently parallel spins of the triplet state are energetically favored. This conclusion is reached for orthogonal orbitals in the same atom. This Coulomb interaction induced spin alignment is called the intraatomic exchange interaction, which is the physical origin of the magnetic moments of atoms or ions. Later, we shall see similar exchange interactions between electrons situated on different atoms, and delocalized into energy bands, play important roles in the ferro- and antiferromagnetism of molecules and solids. From the above argument, it is very clear that the "exchange interaction" is just a direct consequence of the Coulomb interaction between electrons; no extra '*mysterious*' exchange force is needed to explain it.

11.1.4 Hund's Rules and Magnetic Moments in Ions

From spectroscopic data, F. Hund formulated three empirical rules about the ground state of partially filled electronic shells of atoms or ions, known as Hund's rules.

1st rule: The ground state of an isolated atom or ion has the largest value of total spin S.

2nd rule: It also has largest value of the total orbital angular momentum L that is permitted by the first rule.

3rd rule: It has total angular momentum $J = \mid L - S \mid$ for less than half-filled shells, and $J = L + S$ for more than half-filled shells.

The first rule is a direct consequence of parallel spin alignment due to intra-atomic exchange interaction discussed in the previous section, while the third rule is related to spin-orbit coupling. From Hund's rules we can deduce the value of the effective magnetic moment for an atom or ion,

$$\mu_{\text{eff}} = g_{\text{L}}[J(J+1)]^{1/2}\mu_{\text{B}} = p\mu_{\text{B}}, \tag{11.1.12}$$

where g_{L} is the Landé factor, J is the total angular momentum, μ_{B} is Bohr magneton defined by

$$\mu_{\text{B}} = \frac{e\hbar}{2mc}. \tag{11.1.13}$$

This is the natural unit for magnetic moment in atoms or ions, and has a value $\mu_{\text{B}} = 9.27 \times 10^{-21}$ erg/Oe.

For rare earth elements, experimental values of effective magnetic moments agree reasonably well (with exceptions for Sm and Eu) with the predictions based on the three Hund's rules (see Table 11.1.1); while for $3d$ transition metal elements, no contribution from the orbital moments is found. The orbital moments are said to be quenched; only spins as predicted by the 1st Hund's rule contribute to experimental data (see Table 11.1.2). The reason for quenching of orbital moments of d-electrons will be discussed later (see §11.4).

§11.2 Diatomic Molecules

11.2.1 The Exact Solution for the Hydrogen Molecular Ion H_2^+

H_2^+ is the simplest molecule. This system may be described as one electron moving in the Coulomb field of two fixed protons. Assume the distance between the two protons A and B is R and the distances between the moving electron and fixed protons are r_A and r_B (see Fig. 11.2.1). Then the Schrödinger equation for the electron is

$$-\frac{\hbar^2}{2m}\nabla^2\psi - \left(\frac{e^2}{r_A} + \frac{e^2}{r_B}\right)\psi = E(R)\psi. \tag{11.2.1}$$

Table 11.1.1 Calculated and measured effective Bohr magneton numbers p of trivalent rare earth ions.

ment	Electronic state to Hund's rule							Calculated p	Measure
	f-shell ($l = 3$) n_d = d-electron number								
	n_d	$m_l = 3, 2, 1, 0, -1, -2, -3$	S	$L = \|\Sigma\|$		J	Ground state term		
e^{3+}	1	↑	1/2	3	5/2		$^2F_{5/2}$	2.54	2.4
r^{3+}	2	↑ ↑	1	5	4		$^2H_4$	3.58	3.4
d^{3+}	3	↑ ↑ ↑	3/2	6	9/2		$^4I_{9/2}$	3.62	3.5
n^{3+}	4	↑ ↑ ↑ ↑	2	6	4		$^5I_4$	2.68	—
n^{3+}	5	↑ ↑ ↑ ↑ ↑	5/2	5	5/2	$J = \|L - S\|$	$^6H_{5/2}$	0.84	1.6
u^{3+}	6	↑ ↑ ↑ ↑ ↑ ↑	3	3	0		$^7F_0$	0.00	3.4
d^{3+}	7	↑ ↑ ↑ ↑ ↑ ↑ ↑	7/2	0	7/2		$^8S_{7/2}$	7.94	7.9
b^{3+}	8	↓↑ ↑ ↑ ↑ ↑ ↑ ↑	3	3	6		$^7F_6$	9.72	9.5
y^{3+}	9	↓↑ ↓↑ ↑ ↑ ↑ ↑ ↑	5/2	5	5/2		$^6H_{15/2}$	11.63	11.4
o^{3+}	10	↓↑ ↓↑ ↓↑ ↑ ↑ ↑ ↑	2	6	8		$^5I_8$	11.60	11.4
r^{3+}	11	↓↑ ↓↑ ↓↑ ↓↑ ↑ ↑ ↑	3/2	6	15/2		$^4I_{15/2}$	9.59	9.4
m^{3+}	12	↓↑ ↓↑ ↓↑ ↓↑ ↓↑ ↑ ↑	1	5	6	$J = L + S$	$^3H_6$	7.57	7.1
b^{3+}	13	↓↑ ↓↑ ↓↑ ↓↑ ↓↑ ↓↑ ↑	1/2	3	7/2		$^2F_{7/2}$	4.54	4.9

Table 11.1.2 Calculated and measured effective number of Bohr magnetons p for $3d$ transition metal ions.

ment	Electronic state to Hund's rule							Calculated p		Measure
	d-electron number									
	n	$m_l = 2, 1, 0, -1, -2$	S	$L = \|\Sigma\|$	J		Ground state term	$J = S$	$J = \|L \pm S\|$	
i^{3+}	1	↑	1/2	2	3/2		$^2D_{3/2}$	1.73	1.55	—
$^{4+}$							$^2D_{3/2}$	1.73	1.55	1.8
$^{3+}$	2	↑ ↑	1	3	2		$^3F_2$	2.83	1.63	2.7
$^{2+}$	3	↑ ↑ ↑	3/2	3	3/2		$^4F_{3/2}$	3.87	0.77	3.8
r^{3+}						$J = \|L - S\|$	$^4F_{3/2}$	3.87	0.77	3.7
n^{4+}							$^4F_{3/2}$	3.87	0.77	4.0
r^{2+}	4	↑ ↑ ↑ ↑	2	2	0		$^5D_0$	4.90	0	4.8
n^{3+}							$^5D_0$	4.90	0	4.9
n^{2+}	5	↑ ↑ ↑ ↑ ↑	5/2	0	5/2		$^6S_{5/2}$	5.92	5.92	5.9
e^{3+}							$^6S_{5/2}$	5.92	5.92	5.9
e^{2+}	6	↓↑ ↑ ↑ ↑ ↑	2	2	4		$^5D_4$	4.90	6.70	5.3
o^{2+}	7	↓↑ ↓↑ ↑ ↑ ↑	3/2	3	9/2	J=L+S	$^4F_{9/2}$	3.87	6.54	4.0
i^{2+}	8	↓↑ ↓↑ ↓↑ ↑ ↑	1	3	4		$^3F_4$	2.83	5.59	2.9–3.5
u^{2+}	9	↓↑ ↓↑ ↓↑ ↓↑ ↑	1/2	2	5/2		$^2D_{3/2}$	1.73	3.55	1.7–1.9

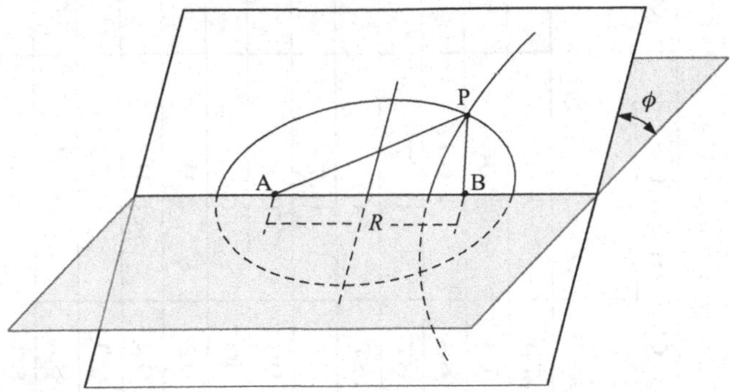

Figure 11.2.1 The coordinate system for hydrogen molecular ion.

Equation (11.2.1) can be solved in elliptic coordinates with the two protons as foci. Set

$$\xi = \frac{r_A + r_B}{R}(1 \le \xi < \infty),$$

$$\eta = \frac{r_A - r_B}{R}(-1 \le \eta \le +1). \qquad (11.2.2)$$

The third coordinate is defined by the azimuthal angle ϕ. The equation is transformed into

$$-\frac{2\hbar^2}{mR^2}\left[\frac{1}{\xi^2-\eta^2}\frac{\partial}{\partial\xi}\left\{(\xi^2-1)\frac{\partial\psi}{\partial\xi}\right\} + \frac{1}{\xi^2-\eta^2}\frac{\partial}{\partial\eta}\left\{(1-\eta)^2\frac{\partial\psi}{\partial\xi}\right\}\right.$$

$$\left. + \frac{1}{(\xi^2-1)(1-\eta^2)}\frac{\partial^2\psi}{\partial\phi^2}\right] - \frac{4e^2\xi}{R(\xi^2-\eta^2)}\psi = E(R)\psi, \qquad (11.2.3)$$

using the method of separation of variables, we set

$$\psi(\xi,\eta) = X(\xi)Y(\eta), \qquad (11.2.4)$$

and, by substitution into Eq. (11.2.3), we get two ordinary differential equations

$$\frac{d}{d\xi}\left\{(\xi^2-1)\frac{dX}{d\xi}\right\} + \left(\frac{mR^2E}{2\hbar^2}\xi^2 + \frac{mRe^2}{\hbar^2}\xi + A\right)X = 0. \qquad (11.2.5)$$

$$\frac{d}{d\eta}\left\{(\eta^2-1)\frac{dY}{d\eta}\right\} + \left(\frac{mR^2E}{2\hbar^2}\eta^2 + A\right)Y = 0. \qquad (11.2.6)$$

Energy eigenvalues E and the separation constant A for each value of R are determined by the boundary conditions from (11.2.2) at the end of the intervals. This set of equations were derived and solved numerically, for the first time by O. Burrau in 1927[a] and subsequently solved analytically under certain restrictions.[b] The result for E versus R is shown in Fig. 11.2.2. In order to find the equilibrium separation of two protons, the Coulomb repulsive potential energy e/R should be added to E:

$$U(R) = E(R) + \frac{e^2}{R}. \qquad (11.2.7)$$

Curves for $U(R)$ are shown in Fig. 11.2.3. The condition for a stable minimum is found to be $R = 2\,\text{AU} = 0.106\,\text{nm}$, with energy $U = -16.3$ eV, and the dissociation energy for this molecular ion is thus 2.8 eV.

This exact result gives a quantum mechanical picture for a molecular bond, i.e., a mobile electron acting as a 'glue' that binds two atoms together to form a molecule. Though, qualitatively speaking,

[a]O. Burrau, *Kgl. Denske Selskab. Mat. Fys. medd* **7**, 1 (1927).

[b]D. Bates, K. Ledsham and A. L. Stewart, *Phil. Trans. Roy. Soc. A* **246**, 215 (1953).

the reason for bond formation may be understood in classical physics, (for there is electrostatic attraction when a negative charge is placed between positive ones), an exact calculation (based on quantum mechanics) is required to establish it beyond doubt, to give a detailed quantitative description and to understand its real nature. The exact result agrees very well with experimental data for H_2^+.

11.2.2 The Molecular Orbital Method

In the exact solution of the H_2^+ problem discussed in the previous section, the electron is delocalized in space, without belonging to a particular atom. This is the spirit of the molecular orbital method, being somewhat similar to the band theory of solids. However, for molecular systems with two or more electrons, the Coulomb interaction between the electrons makes things difficult and some sort of approximation is needed. The molecular orbital method is a way to do this, based on single-electron approximation. The Coulomb interaction is either ignored or incorporated into a molecular field.

For a system of many-electron diatomic molecules, we may fill the orbitals of H_2^+ with electrons according to the Pauli principle and minimum energy requirement. The procedure is just like building up many-electron atoms. For the H_2 molecule, the obvious choice is to put the second electron in the $1s$ orbital to form an antiparallel pair. Indeed, experiment shows an extra energy gain (15.4 eV) from the addition of a second electron, as well as an increase of the dissociation energy to 4.7 eV. The increased stability of H_2 as compared with H_2^+ confirms the chemical intuition that the electron-pair bond is the most common bond in chemistry. Furthermore, if a third electron is introduced, from Fig. 11.2.2 and 11.2.3, the choice is $2p\sigma$ and no bound state is formed.

Two combinations are prescribed for us by the inversion symmetry of the H molecule about the midpoint between the two protons. The wavefunction has even parity, for it recovers itself after the inversion; while antiphase combination has odd parity, for it changes sign after inversion, i.e.

$$\psi(-x,-y,-z) = \pm\psi(x,y,z). \tag{11.2.8}$$

A commonly adopted approximation procedure is the linear combination of atomic orbitals (LCAO) to treat molecular problems, i.e.,

$$\psi(\boldsymbol{r}_i) = \psi_{\mathrm{A}}(\boldsymbol{r}_i) \pm \psi_{\mathrm{B}}(\boldsymbol{r}_i). \tag{11.2.9}$$

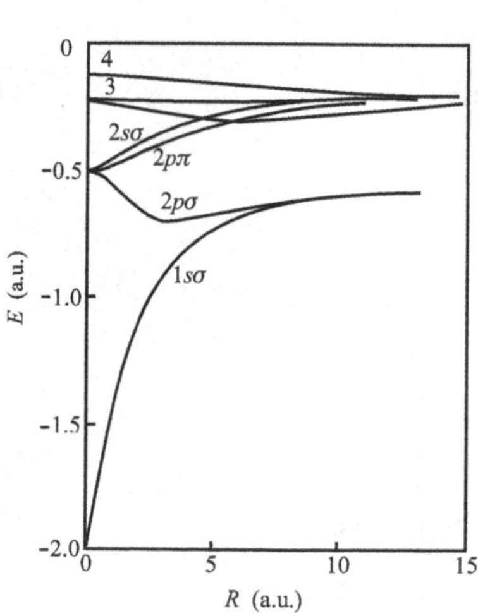

Figure 11.2.2 Plot of the electronic energy levels versus internuclear distance.

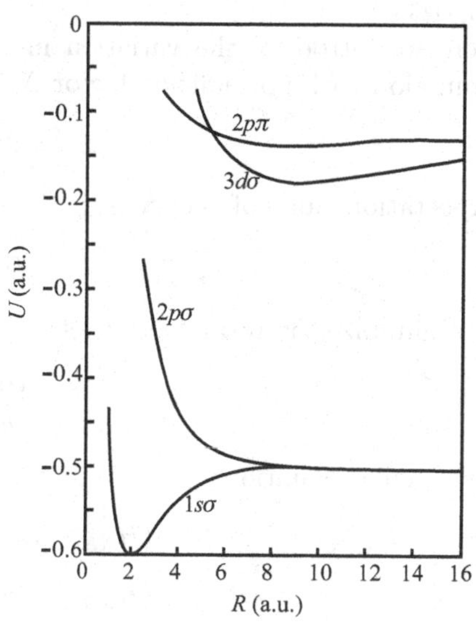

Figure 11.2.3 Plot of the total energy versus the internuclear distance.

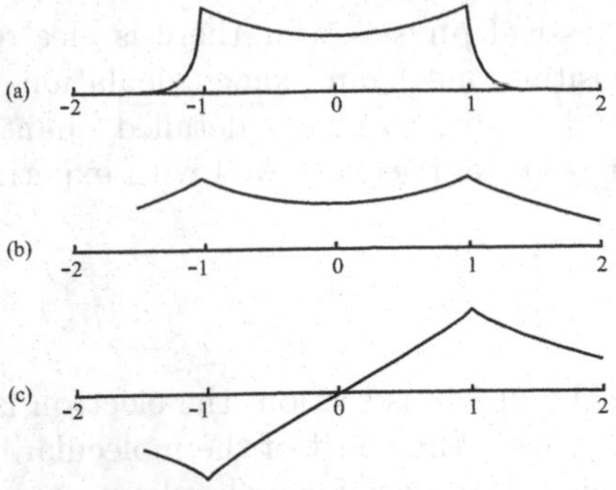

Figure 11.2.4 Ground-state wave functions of H_2^+ along the central line, (a) exact result, (b) bonding LCAO, (c) antibonding LCAO.

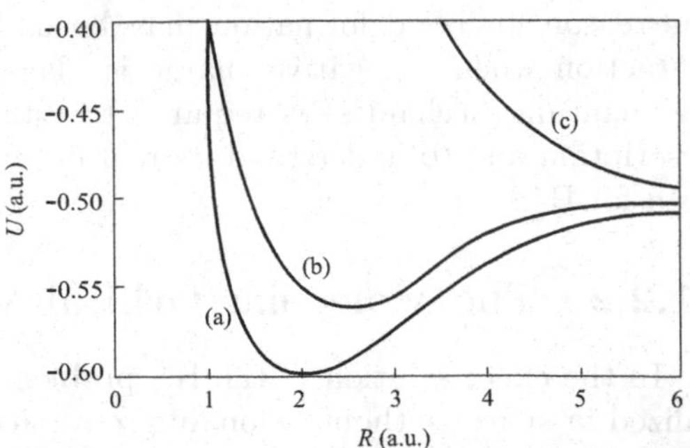

Figure 11.2.5 Bonding energy versus internuclear distance for H_2^+, (a) exact result, (b) bonding atomic orbital, (c) antibonding atomic orbital.

To judge whether this approximation can handle molecular problem with reasonable accuracy, we first apply it to H_2^+, where we have the exact results for comparison. Wave functions along the central line are plotted in Fig. 11.2.4, and the energies U versus the intermolecular distance R are plotted in Fig. 11.2.5. A bond is formed in the even parity state, but not in the odd one. This comparison shows that the LCAO scheme with its easy visualization and close connection to atomic entities works quite satisfactory to give a semi-quantitative picture of chemical bonds. These two combinations of AOs are called bonding and antibonding respectively because the enhancement of electronic density is located in the internuclear region for the former; located outside the internuclear region for the latter.

Historically, the first molecular orbital treatment of H_2 was proposed by F. Hund and R. S. Mulliken, they used just two orbitals of Eq. (11.2.9) as their single-electron wave functions satisfying the single-electron Schrödinger equation

$$\mathcal{H}_i\psi_i = \left(-\frac{\hbar^2}{2m}\nabla_i^2 - \frac{e^2}{r_{\mathrm{A}i}} - \frac{e^2}{r_{\mathrm{B}i}}\right) = \varepsilon_i\psi_i. \tag{11.2.10}$$

As well as the inversion symmetry of this molecular system, molecular orbitals (MOs) have even and odd parities.

Here we introduce the variation method to find the MOs through LCAO. Let us use a trial wavefunction with normalizing factor N_i as

$$\psi_i = N_i[c_{\mathrm{A}}\psi_{\mathrm{A}}(\boldsymbol{r}_i) + c_{\mathrm{B}}\psi_{\mathrm{B}}(\boldsymbol{r}_i)]. \tag{11.2.11}$$

Its expectation value of energy is

$$E(\psi) = \frac{\int \psi^* H\psi d\boldsymbol{r}_i}{\int \psi^*\psi d\boldsymbol{r}_i}, \tag{11.2.12}$$

and to minimize the total energy, let

$$\frac{\partial E}{\partial c_i} = 0 \quad (i = \mathrm{A},\ \mathrm{B}). \tag{11.2.13}$$

We get secular equations

$$(\mathcal{H}_{\mathrm{AA}} - ES_{\mathrm{AA}})c_{\mathrm{A}} + (\mathcal{H}_{\mathrm{AB}} - ES_{\mathrm{AB}}) = 0, \tag{11.2.14}$$

$$(\mathcal{H}_{\mathrm{AA}} - ES_{\mathrm{AA}})c_{\mathrm{A}} + (\mathcal{H}_{\mathrm{AB}} - ES_{\mathrm{AB}}) = 0, \tag{11.2.15}$$

with a corresponding determinant to be solved

$$\begin{vmatrix} \mathcal{H}_{\mathrm{AA}} - ES_{\mathrm{AA}} & \mathcal{H}_{\mathrm{AB}} - ES_{\mathrm{AB}} \\ \mathcal{H}_{\mathrm{AB}} - ES_{\mathrm{AB}} & \mathcal{H}_{\mathrm{BB}} - ES_{\mathrm{BB}} \end{vmatrix} = 0. \tag{11.2.16}$$

Here $\mathcal{H}_{AA}$, $\mathcal{H}_{BB}$, ... are the matrix elements of the Hamiltonian, and since A and B are chemically identical, let $H_{AA} = \mathcal{H}_{BB} = \alpha$, which is called the Coulomb integral

$$\alpha = \int \psi_A \mathcal{H}_i \psi_A d\boldsymbol{r}_i = \int \psi_B \mathcal{H}_i \psi_B d\boldsymbol{r}_i. \tag{11.2.17}$$

Now let $\mathcal{H}_{AB} = \mathcal{H}_{BA} = \beta$,

$$\beta = \int \psi_B \mathcal{H}_i \psi_A d\boldsymbol{r}_i = \int \psi_A \mathcal{H}_i \psi_B d\boldsymbol{r}_i. \tag{11.2.18}$$

This is the bond (or resonance) integral, for this term is crucial for molecular bonding. They are overlap integrals, obviously $S_{AA} = S_{BB} = 1$, and

$$S_{AB} = S_{BA} = \int \psi_A \psi_B d\boldsymbol{r}_i. \tag{11.2.19}$$

Solving the secular equations, two MOs are found,

$$\psi_\pm = \frac{1}{\sqrt{2(1 \pm S)}}(\psi_A \pm \psi_B) \tag{11.2.20}$$

with corresponding energies

$$E_\pm = \frac{\alpha \pm \beta}{1 \pm S_{AB}}. \tag{11.2.21}$$

This solution of the MO problem for H_2 is in agreement with our intuitive guess outlined above. Obviously the best choice is to put two electrons as an antiparallel pair into the even bonding orbital, and leave the odd antibonding orbital empty. This prescription for the ground state is quite similar to band theory, i.e., solving the one-electron problem, then filling-up the energy levels from the bottom up. The antibonding state is the 1st excited state for H_2 (see Fig. 11.2.6).

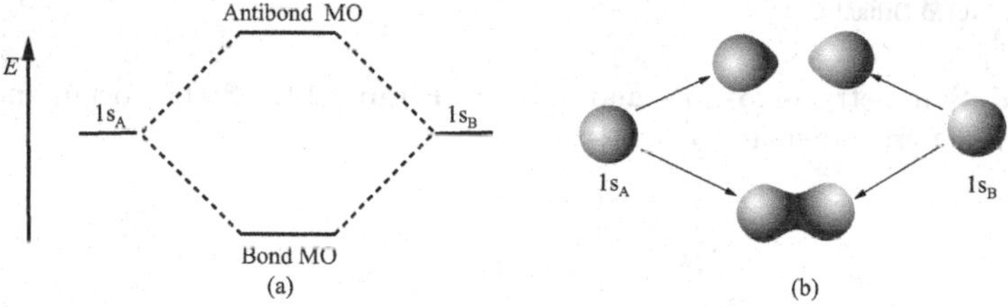

Figure 11.2.6 (a) Energy diagram showing MO formation with (b) the corresponding pictorial representation of orbitals.

This treatment is not very accurate for H_2; however, it is easy to generalize to more complex cases due to the flexibility of one particle wavefunctions and the adaptability of further refinements. Since the molecular orbital method (MO) has been improved with various techniques to increase its accuracy and can be fruitfully applied to various molecular problems, it is regarded as a mainstay of quantum chemistry.

Just like AOs, MOs may be displayed as geometrical shapes in real space (see Fig. 11.2.6) with their corresponding symmetries. If we view these orbitals along a line passing through two atoms, they appear to be cylindrically symmetrical. This is the molecular counterpart of the spherical symmetry of s atomic orbitals, so these orbitals are called σ-orbitals. The bonding state with even parity, is marked by the subscript 'g' (in German, gerade), while the antibonding state with odd parity is marked by 'u' (in German, ungerade). The two molecular orbitals shown in (11.2.20) and (11.2.21) are usually denoted by σ_g and σ_u.

MOs may be classified according to symmetry considerations. We can show some well-known bonding states such as σ, π, and δ orbitals: The σ orbitals have symmetry about the line joining the nuclei, if we take the z-axis as an axis of symmetry, there are bonds s–s, s–p_z, p_z–p_z, etc.; the π orbitals have a nodal plane containing the line between the nuclei, and there are bonds p_x–p_x, p_y–p_y, etc.; the δ orbitals possess two nodal planes, such as d_{xy}–d_{xy}, etc.. Note the following correspondence between AOs and MOs (see Fig. 11.2.7):

Atomic orbitals: $s, p, d, f\ldots$

Molecular orbitals: $\sigma, \pi, \delta, \phi\ldots$

The higher-energy MOs are of great importance for helping us to understand the more complex diatomic molecules, just like we use the atomic orbitals of the hydrogen atom to help us understand the electronic structures of many-electron atoms. It should be noted that the antibonding orbitals are an integral part of MOs, placed in the appropriate places in the energy level diagram, to be filled with antiparallel pairs of electrons just as MOs. Rules for filling MOs with electrons are just like that of AOs: the Pauli principle, Hund's rule and the minimum energy criterion. We may take the case of O_2 as an example, in which the triplet state ($S = 1$) with unpaired electrons in degenerate $1\pi_g$ levels appear according to Hund's rule (see Fig. 11.2.8), so O_2 is paramagnetic. At low temperature, solid β–O_2 is found to be an antiferromagnet.[c]

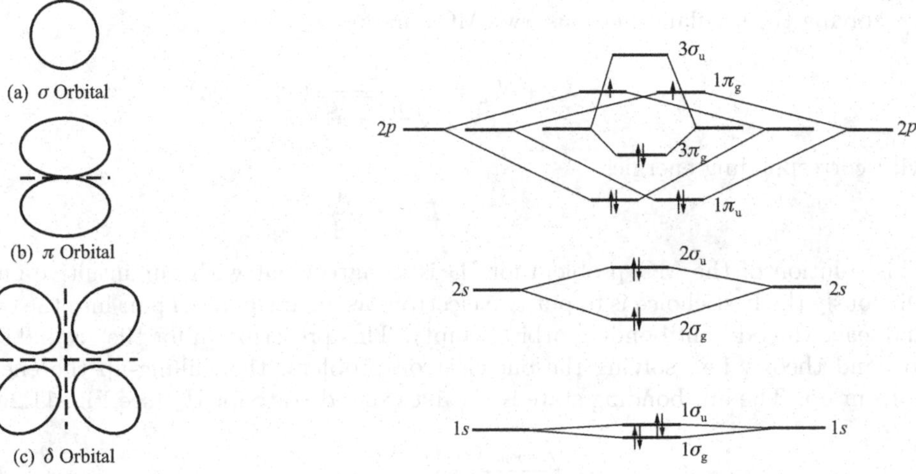

Figure 11.2.7 Symmetry of σ-, π- and δ-orbitals viewing perpendicular to bonding axis.

Figure 11.2.8 The occupancy of MOs in O_2.

11.2.3 Heitler and London's Treatment of Hydrogen Molecule

The starting point of Heitler and London's theory of the hydrogen molecule is different from that of the molecular orbital method, so instead of a one-electron approximation, it is based on an elementary treatment of the two-electron problem. Historically, it was the first theoretical treatment of the chemical bond, and it gave quite accurate values for the bonding energy of H_2. However, this model is difficult to generalize to more complex molecules, so as a quantum chemical method, it has been superseded by the more versatile molecular orbital method. On the other hand, this treatment had a strong impact on condensed matter physics, for it lead to the first explicit formulation of the magnetic exchange interaction in solids by Heisenberg. The reader may also be aware that we view this problem as an intermediate stage in our dealings with the many-body correlation of electrons from atomic physics to solid state physics.

This is a two-electron problem in the field of two protons, its Hamiltonian has several terms

$$\mathcal{H} = -\frac{\hbar^2}{2m}\nabla_1^2 - \frac{\hbar^2}{2m}\nabla_2^2 - \frac{e^2}{r_{A1}} - \frac{e^2}{r_{A2}} - \frac{e^2}{r_{B1}} - \frac{e^2}{r_{B2}} + \frac{e^2}{r_{12}} + \frac{e^2}{r_{AB}}. \tag{11.2.22}$$

[c] A. P. J. Jansen, *Phys. Rev. B* **33**, 6352 (1986).

The first two terms are due to isolated atoms, the next four terms are those which express the two electrons moving in the field of the two protons A and B, while final two terms are due to the electron-electron and the proton-proton interaction.

Let us consider two hydrogen atoms A and B widely separated, with $1s$ wave functions ψ_A and ψ_B respectively, then we bring them somewhat closer together to form a molecule. Two electrons sit in their own spatial orbitals with spins. According to the Heitler–London scheme, the two electrons reside in different atoms, so three spin-triplet states and one spin-singlet state are obtained as follows. The triplet ($S=1$):

$$|\Psi_1\rangle = \frac{1}{\sqrt{2(1-S_\mathrm{AB}^2)}}\alpha(s_1)\alpha(s_2)[\psi_\mathrm{A}(\boldsymbol{r}_1)\psi_\mathrm{B}(\boldsymbol{r}_2)+\psi_\mathrm{A}(\boldsymbol{r}_2)\psi_\mathrm{B}(\boldsymbol{r}_1)], \quad S^z=1. \tag{11.2.23}$$

$$|\Psi_2\rangle = \frac{1}{\sqrt{2(1-S_\mathrm{AB}^2)}}[\alpha(s_1)\beta(s_2)+\beta(s_1)\alpha(s_2)][\psi_\mathrm{A}(\boldsymbol{r}_1)\psi_\mathrm{B}(\boldsymbol{r}_2)-\psi_\mathrm{A}(\boldsymbol{r}_2)\psi_\mathrm{B}(\boldsymbol{r}_1)], \quad S^z=0. \tag{11.2.24}$$

$$|\Psi_3\rangle = \frac{1}{\sqrt{2(1-S_\mathrm{AB}^2)}}\beta(s_1)\beta(s_2)[\psi_\mathrm{A}(\boldsymbol{r}_1)\psi_\mathrm{B}(\boldsymbol{r}_2)-\psi_\mathrm{A}(\boldsymbol{r}_2)\psi_\mathrm{B}(\boldsymbol{r}_1)], \quad S^z=-1. \tag{11.2.25}$$

The singlet ($S=0$):

$$|\Psi_4\rangle = \frac{1}{2\sqrt{(1+l^2)}}[\alpha(\boldsymbol{s}_1)\beta(s_2)-\beta(\boldsymbol{s}_1)\alpha(\boldsymbol{s}_2)][\psi_\mathrm{A}(\boldsymbol{r}_1)\psi_\mathrm{B}(\boldsymbol{r}_2)+\psi_\mathrm{A}(\boldsymbol{r}_2)\psi_\mathrm{B}(\boldsymbol{r}_1)], \quad S^z=0. \tag{11.2.26}$$

The situation is somewhat similar to the case treated in §11.1.3, with the difference here that the two electrons are on non-orthogonal orbitals. For now, the wavefunctions of two electrons will overlap, described by the overlap integral

$$S_\mathrm{AB} = \int \psi_\mathrm{A}^*(\boldsymbol{r})\psi_\mathrm{B}(\boldsymbol{r})d\boldsymbol{r}. \tag{11.2.27}$$

The triplet energy and the singlet energy are

$$E_\mathrm{t} = \langle\Psi_1|H|\Psi_1\rangle = 2E_\mathrm{at}+\frac{C_\mathrm{AB}-I_\mathrm{AB}}{1-S_\mathrm{AB}^2}, \tag{11.2.28}$$

$$E_\mathrm{s} = \langle\Psi_4|H|\Psi_4\rangle = 2E_\mathrm{at}+\frac{C_\mathrm{AB}+I_\mathrm{AB}}{1+S_\mathrm{AB}^2}, \tag{11.2.29}$$

where the Coulomb and exchange integrals are

$$C_\mathrm{AB} = \int d\boldsymbol{r}_1\int d\boldsymbol{r}_2|\psi_\mathrm{A}(\boldsymbol{r}_1)|^2\frac{e^2}{|\boldsymbol{r}_1-\boldsymbol{r}_2|}|\psi_\mathrm{B}(\boldsymbol{r}_2)|^2-\int d\boldsymbol{r}_1\frac{e^2}{|\boldsymbol{r}_1-\boldsymbol{R}_\mathrm{B}|}|\psi_\mathrm{A}(\boldsymbol{r}_1)|^2 - \int d\boldsymbol{r}_2\frac{e^2}{|\boldsymbol{r}_2-\boldsymbol{R}_\mathrm{A}|}|\psi_\mathrm{B}(\boldsymbol{r}_2)|^2 \tag{11.2.30}$$

and

$$I_\mathrm{AB} = \int d\boldsymbol{r}_1\int d\boldsymbol{r}_2\psi_\mathrm{A}^*(\boldsymbol{r}_1)\psi_\mathrm{B}(\boldsymbol{r}_1)\frac{e^2}{|\boldsymbol{r}_1-\boldsymbol{r}_2|}\psi_\mathrm{B}^*(\boldsymbol{r}_1)\psi_\mathrm{A}(\boldsymbol{r}_2) - S_\mathrm{AB}\int d\boldsymbol{r}_1\frac{e^2}{|\boldsymbol{r}_1-\boldsymbol{R}_\mathrm{B}|}\psi_\mathrm{A}^*(\boldsymbol{r}_1)\psi_\mathrm{B}(\boldsymbol{r}_1) - S_\mathrm{AB}\int d\boldsymbol{r}_2\frac{e^2}{|\boldsymbol{r}_2-\boldsymbol{R}_\mathrm{A}|}\psi_\mathrm{B}^*(\boldsymbol{r}_2)\psi_\mathrm{B}(\boldsymbol{r}_2). \tag{11.2.31}$$

The difference in energy between the singlet and the triplet is

$$J_{12} = E_s - E_t = -2\frac{S_{AB}^2 C_{AB} - I_{AB}}{1 - S_{AB}^4}. \tag{11.2.32}$$

According to the original Heitler–London calculation, J_{12} is negative with a large overlap of wave functions, and the spins of the two electrons are antiparallel. This agrees with the result of the molecular orbital method and has been verified by experiment. For two atoms with infinite separation, J_{12} is close to zero. Whether there is a region in which J_{12} is positive when the interatomic distance is intermediate has attracted much theoretical attention. More accurate calculations of two hydrogen atoms show that J_{12} is always negative. These things have been critically reviewed by C. Herring (1963)[d] in detail, and the conclusion is that direct exchange interactions between atoms and ions always favors antiferromagnetic coupling. This conclusion has been amply confirmed by experimental findings; for example, in recent experiments on large clusters of Fe and Mn magnetic ions surrounded by nonmetallic ions and radicals, antiferromagnetic couplings are always found.[e] Some exceptions to this rule will be discussed in Chap. 13.

If we set $S_{AB} = 0$ in Eq. (11.2.27), the orbitals become orthogonal, and J_{12} is positive; this confirms our treatment of interatomic exchange in §11.1.3, which is the foundation of Hund's rule.

11.2.4 The Spin Hamiltonian and the Heisenberg Model

The Heitler–London model for H_2, outlined above, gives the energy of the molecule in terms of four states, three triplet states and one singlet state. For many problems higher molecular states may be ignored altogether, by representing the molecule simply as a four-state system. It is appropriate to introduce an operator, known as a spin Hamiltonian, whose eigenvalues are the same as those of the original Hamiltonian, and whose eigenfunctions give the spins of the corresponding states.

In order to construct the spin Hamiltonian some fundamental properties of the spin operator for the electron should be recalled: the spin operator for each electron satisfies the relationship

$$\boldsymbol{S}_i^2 = \frac{1}{2}\left(\frac{1}{2}+1\right) = \frac{3}{4}. \tag{11.2.33}$$

So the total spin satisfies

$$\boldsymbol{S}^2 = (\boldsymbol{S}_1 + \boldsymbol{S}_2)^2 = \frac{3}{2} + 2\boldsymbol{S}_1 \cdot \boldsymbol{S}_2. \tag{11.2.34}$$

Since $\boldsymbol{S}^2$ has the eigenvalue $S(S+1)$, then we know the operator $\boldsymbol{S}_1 \cdot \boldsymbol{S}_2$, has eigenvalue $+1/4$ in the triplet ($S = 1$) and $-3/4$ in the singlet ($S = 0$), so the spin operator can be written as

$$\mathcal{H}_s = \frac{1}{4}(E_s + 3E_t) - (E_s - E_t)\boldsymbol{S}_1 \cdot \boldsymbol{S}_2, \tag{11.2.35}$$

whose eigenvalue for each triplet state is E_t, for the singlet state is E_s just as we have desired. Omitting the constant term $(E_s + 3E_t)$ by redefining the zero of energy, the spin Hamiltonian becomes

$$\mathcal{H}_s = -J_{12}\boldsymbol{S}_1 \cdot \boldsymbol{S}_2. \tag{11.2.36}$$

This will favor parallel spins if J_{12} is positive and antiparallel spins if J_{12} is negative. This form of the spin Hamiltonian may be extended from the case of two atoms to the case of N atoms, and even to $N \to \infty$ as atoms in a crystalline lattice:

$$\mathcal{H}_s = -\sum_{i<j} J_{ij}\boldsymbol{S}_i \cdot \boldsymbol{S}_j, \tag{11.2.37}$$

[d]C. Herring, Direct exchange between well-separated atoms, in G. T. Rado and H. Suhl (eds.), *Magnetism*, Vol. IIB, Academic Press, New York (1963). For ions of transition metals, there is the added complication of the orbital degeneracy of d-electrons, this problem will be treated in §13.1.4.

[e]D. Gatteschi *et al.*, *Science* **265**, 1054 (1994).

in which the sum of all pairs of atoms (or ions) i, j is carried out, just as a straightforward extension of a two-atom molecule. This is the famous Heisenberg Hamiltonian, and the J_{ij} are coupling constants. Originally Heisenberg devised this Hamiltonian as a first attempt to understand ferromagnetism based on quantum mechanics, but the situation is not so simple as he imagined. This form of the Hamiltonian is formulated for localized spins, and this description is only suitable for insulators. However, the most important ferromagnets are metals and alloys in which the electrons are delocalized, while most insulators are antiferromagnetic. Now we have learned that simple direct exchange is not the basic ingredient of real ferromagnets: The physical origin of ferromagnetism must be sought elsewhere, such as in superexchange, double exchange, indirect exchange and the exchange interaction of itinerant electrons. This will, by no means, diminish the importance of the Heisenberg Hamiltonian for the study of the magnetic properties of matter, for this simple formulation with its easy visualization gives an effective framework to understand a lot of things related to magnetic order, both ground states and excited states.

§11.3 Polyatomic Molecules

11.3.1 The Molecular Orbital Method for Polyatomic Molecules

The main task of the molecular orbital method is to find the single-electron wavefunctions (i.e., MOs) as solutions of the wave equations for the polyatomic molecule and construct the corresponding energy level diagram. The principal method of solving these problems is to use the LCAO to find the MOs. The calculation begins by the choice of a basis, i.e., a set of AOs describing the electronic states of isolated atoms or ions. Furthermore, to simplify the problem, we may assume that the core electrons remain intact, so only electrons in outer shells are taken into account. Then single-electron MOs are formed as linear combinations of AOs, while the full wavefunctions $(1, 2, \ldots, N)$ form a Slater determinant:

$$\Psi = \frac{1}{\sqrt{N!}} \begin{vmatrix} \psi_1(1) & \psi_1(2) & \ldots & \psi_1(N) \\ \psi_2(1) & \psi_2(2) & \ldots & \psi_2(N) \\ \ldots & \ldots & \ldots & \ldots \\ \psi_N(1) & \psi_N(2) & \ldots & \psi_N(N) \end{vmatrix} \tag{11.3.1}$$

The numbers in the arguments of the orbital ψ_i stand for the four quantum numbers (orbital plus spin), and the function is antisymmetric. The larger the number of initial AOs in the basis, the more accurate the approximation of the MOs, but the more complicated calculations becomes, because the number of single-electron integrals of type (11.3.1) is then $\sim (1/2)N^2$, and that of two-electron integrals $\sim N^4$. Carrying out complete non-empirical, *ab initio* calculations takes account of all the electrons in the system, and further the positions of the nuclei are gradually determined by minimizing the total energy. Even for comparatively simple molecules such as CO_2 or H_2O, a basis with several tens of Slater or Gaussian functions $r^{n-1}\exp(-\beta r)Y_{lm}$ or $r^{n-1}\exp(-\alpha r^2)Y_{lm}$ is usually employed, and the number of integrals is $10^5 \sim 10^6$, which is a very time-consuming task. When N reaches several tens or more, and without symmetry to simplify the problem, it seems that accurate calculation is impossible even with the aid of modern electronic computers. This is just the exponential wall (i.e., the exponential increase of the number parameter for a satisfactory *ab initio* calculation of MOs versus the number of atoms) mentioned by W. Kohn in his Nobel lecture.[f]

Once the energy levels of MOs are determined, electrons fill up the energy ladder according to the building rules; thus ground states and excited states may be determined, just as we did in atomic physics. The electrons are delocalized in the molecular space and surely various MOs play important roles in the chemical bonds: it is often difficult to assign a particular bond to a definite orbital. Only when the total wavefunction is obtained can the picture of the electron distribution in the molecule be finally clarified.

[f]W. Kohn, *Rev. Mod. Phys.* **71**, 1253 (1999).

11.3.2 Valence Bond Orbitals

The localized behavior of chemical bonds is well established experimentally: Bond direction, bond length, bond dissociation energy and polarity are characteristic parameters for a chemical bond. They are almost equal in various molecules, and transferable from one to another. This evidence is the foundation of the concept of the electron-pair bond due to N. G. Lewis which was updated into the theory of valence-bonds based on hybridized atomic orbitals by L. Pauling.

We may construct valence bond orbitals by the hybridization of atomic orbitals of similar energies, either from the same atom or from different atoms. Let us consider the case of the hybridization of s and p orbitals from the same atom: In an isolated atom, the s orbital is isotropic in shape, while the p orbitals are directional along rectangular axes x, y, and z. For orbitals hybridized between s and p states, more pointed directional shapes may be formed, for instance, one s orbital may be hybridized with three p orbitals, i.e., four sp^3 orbitals along tetrahedral directions are formed (see Fig. 11.3.1(c)),

$$\sigma_1 = \frac{1}{2}(s + p_x + p_y + p_z), \quad \sigma_2 = \frac{1}{2}(s + p_x - p_y - p_z),$$

$$\sigma_3 = \frac{1}{2}(s - p_x + p_y - p_z), \quad \sigma_4 = \frac{1}{2}(s - p_x - p_y + p_z). \tag{11.3.2}$$

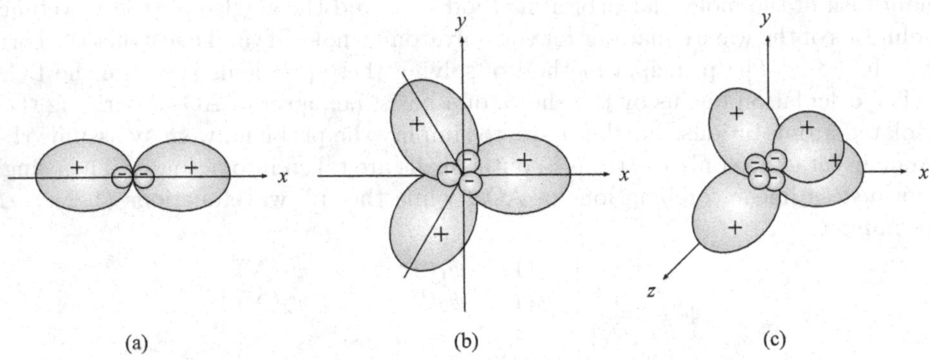

Figure 11.3.1 Hybridized orbitals for s, p electrons (a) sp orbitals, (b) sp^2 orbitals, (c) sp^3 orbitals.

Every hybridized orbital may accommodatc another electron from the other atom, so a spin-singlet pair is formed. This is the ideal bonding scheme for C, Si, Ge atoms, in which the valence shell lacks 4 electrons, so tetrahedral bonding with other atoms fill the complete octet set. This is the structure of the methane molecule, as well as the crystal structure for diamond, Si and Ge. One s-orbital may be hybridized with two p orbitals, i.e., three sp^2 orbitals along planar triangular directions are formed (see Fig. 11.3.1(b)). This type of bonding between one C atom and two other C atoms and one H atom forms the backbone of the benzene molecule; this bonding also appears in graphene (an atomic sheet of graphite). One s-orbital may hybridized with one p orbital, thus two sp orbitals along a line in opposite directions are formed (see Fig. 11.3.1(a)). In general, hybridization may cost some energy in an isolated atom, for it may raise some electrons from pure s state to mixed s and p states; however, this energy investment is amply returned with a profit once the electron-pair valence bonds are formed with neighboring atoms.

Also d-electrons may participate in hybridization, and hybridized orbitals such as d^3s with tetrahedral bonding, d^2sp^3 with octahedral bonding, dsp^2 with triangular-bi-pyramidal bonding, dsp^2 with planar square bonding can be formed. These may explain the prevalence of oxygen octahedra in many transition metal oxide crystals, such as the perovskites.

The valence bond orbital gives a clear picture of localized bonds in terms of the pairing of electrons described by a singlet wavefunctions as independent 'structures'. These valence bond orbitals obey rules of 'maximum overlap' and orthogonality. Then we can divide the electrons in a molecule into several independent parts: pairs in the inner shell, lone pairs and bonding pairs, denoted by A, B,

C, etc.. The total energy can be written as the sum of components,

$$E = E_A + E_B + E_C + \cdots + U_{AB} + U_{AC} + U_{BC} + \cdots, \tag{11.3.3}$$

where $E_A, E_B \ldots$ stands for electronic energy of A, B ... components in the field of nuclei and other electrons, while U_{AB}, $U_{AC} \cdots$ stands for the Coulomb repulsive energy between components A and B, A and C, ... In general, the minimum of $E_A + E_B + \cdots$ is attained by the optimization of bond lengths. However, for non-overlapping electron-pairs,

$$U \approx e^2 \int \frac{\rho_A(1)\rho_B(2)}{r_{12}} d\boldsymbol{r}_1 d\boldsymbol{r}_2. \tag{11.3.4}$$

This is just the Coulomb repulsive energy between electron density distributions $e\rho_A$ and $e\rho_B$, so the sum of these energy terms will be minimized by making various electron-pairs stay as far away as possible. We may visualize the situation as distribution of locations of electron-pairs on the spherical surface of a valence shell shared by the bonding atoms. The optimum solution to this problem is to place the electron-pairs on the vertices of a regular polyhedron circumscribed by this sphere. For 3-dimensional bonding: this ideal is realized for the case of 4 pairs in tetrahedral bonding; for 6 pairs in octahedral bonding; for 5 pairs, no ideal regular polyhedron is available and so the less symmetric trigonal bipyramid is adopted.

For two-dimensional bonding, for 3 pairs triangular bonding is ideal; for 4 pairs, square bonding; for 2 pairs, the diametric line. If there are lone pairs, things become more complicated as the Coulomb repulsive energy of lone pairs is somewhat larger than that of bonding pairs. If lone pairs are present in the valence shell, a compromise is reached for a less symmetrical bonding scheme.

The idea of electron pair bonding was proposed by G. N. Lewis and that of the octet structure was proposed by I. Langmuir before the advent of quantum mechanics; these ideas were synthesized to form valence shell electron pair repulsion theory, based on quantum mechanics. It is interesting to observe that the secret of this age-old chemical wisdom is the many-body correlation of electrons.

11.3.3 The Hückel Approximation for the Molecular Orbital Method

The Hückel approximation introduces a drastic simplification making the molecular orbital method possible to treat quite complicated organic molecules. Its basic assumption is that the carbon ion core and the localized σ-bonding electrons are to be regarded as a 'scaffold' or 'skeleton' for the electrons to move around; i.e., every π-electron moves in the effective field of the scaffold and other electrons. Then the Schrödinger equation for a single π-electron can be written as follows:

$$\left(-\frac{\hbar^2}{2m}\nabla^2 - \sum_i \frac{e^2}{r_i} \right) \psi = E\psi, \tag{11.3.5}$$

where ψ is the π-molecular orbital, and E is the energy eigenvalue. We may put p_z orbitals of all carbon atoms into LCAO for variation:

$$\psi = \sum_{i=1}^{N} c_i \phi_i \quad \cdot \tag{11.3.6}$$

and then we get a linear set of secular equations:

$$\begin{aligned} &(\mathcal{H}_{11} - E)C_1 + (\mathcal{H}_{12} - ES_{12})C_2 + \cdots + (\mathcal{H}_{1N} - ES_{12})C_N = 0, \\ &\cdots\cdots\cdots\cdots, \\ &(\mathcal{H}_{N1} - ES_{N1})C_1 + (\mathcal{H}_{N2} - ES_{N2})C_2 + \cdots + (\mathcal{H}_{NN} - E)C_N = 0, \end{aligned} \tag{11.3.7}$$

where the AOs $\phi_1, \phi_2, \ldots, \phi_N$ are supposed to be normalized; $\mathcal{H}_{rs}$ and $\mathcal{S}_{rs}$ denote the energy integral and overlap integral, respectively.

In Hückel's treatment, further simplifications are introduced by setting

(1) the Coulomb integral equal to $\mathcal{H}_{rr} = \alpha$,

(2) the resonance integral equal to $\mathcal{H}_{rs} = \beta$ if the r-atom connects with the s-atom, otherwise $\mathcal{H}_{rs} = 0$,

(3) the overlap between neighboring atoms is ignored $S_{rs} = 0$ when $r \neq s$.

Let energy E scale with ε, with α setting the zero point, it is found that

$$\frac{\alpha - E}{\beta} = -\varepsilon \quad \text{or} \quad E = \alpha + \beta\varepsilon. \tag{11.3.8}$$

Thus the secular equations are highly simplified.

The Hückel approximation is a powerful tool to treat problems of the electronic structures of complex organic molecules. Take conjugated molecules with rings such as benzene C_6H_6 as an example: It is a ring with $N = 6$ members. It is quite easy to treat the general case in which N equals any integer. The simplified secular equations may be written immediately as follows,

$$\begin{aligned} -\varepsilon c_1 + c_2 + c_N &= 0, \\ &\cdots\cdots\cdots\cdots, \\ c_{m-1} - \varepsilon c_m + c_{m+1} &= 0, \\ &\cdots\cdots\cdots\cdots, \\ c_1 + c_{N-1} - \varepsilon c_N &= 0. \end{aligned} \tag{11.3.9}$$

In fact, we only need to consider one general equation in Eq. (11.3.9),

$$c_{m-1} - \varepsilon c_m + c_{m+1} = 0,$$

which relates three neighboring coefficients. For the ring-like organic molecule shown in Fig. 11.3.2, the boundary condition must be periodic,

$$c_0 = c_N, \quad c_{N+1} = c_1, \tag{11.3.10}$$

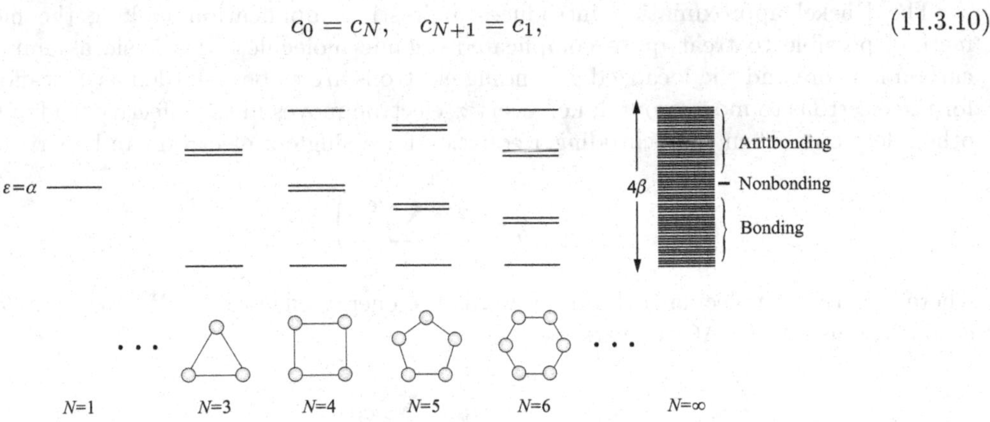

Figure 11.3.2 Conjugated rings and their energy level diagrams. Note when N increases, discrete levels are transformed into a continuous band.

Now we take the coefficient c_m with quantum number k as

$$c_m^k = c_k \exp\left(\frac{2\pi imk}{N}\right), \tag{11.3.11}$$

and substitute it into Eq. (11.3.9), so

$$\varepsilon_k = 2\cos\left(\frac{2k\pi}{N}\right), \tag{11.3.12}$$

and the eigenenergy is

$$E_k = \alpha + 2\beta \cos\left(\frac{2\pi k}{N}\right), \tag{11.3.13}$$

where

$$k = 0, \pm 1, \ldots, \begin{cases} \pm(N-1)/2, & \text{for} \quad N = \text{odd} \\ \pm N/2, & \text{for} \quad N = \text{even}. \end{cases} \tag{11.3.14}$$

A couple of k with the same value but opposite sign correspond to two different solutions with the same energy. So, except the first MO ($k = 0$) and the final MO ($k = N/2$) of an even number ring, all other MOs are travelling wave solutions corresponding to the electron moving in one direction (k positive) around the ring or opposite (k negative).

Energy levels are distributed about $E = \alpha$, with symmetry if N is even; with asymmetry if N is odd, as shown in Fig. 11.3.2. However, their energy levels have upper bound 2β and lower bound -2β, when $N \to \infty$, a densely populated continuous one-dimensional energy band with band width 4β is formed. This calculation may help us to trace how the bond approach is transformed into the band approach.

For long chain polymers such as $(CH)_n$, $[Pt(CN)_4^{2-}]$, ..., if N is sufficient large, end effects can be neglected. Their electronic structure may be mimicked by that of the N-ring with $N \to \infty$, i.e., a 1-D band structure with additional complexity due to chemical structure.

11.3.4 Electronic Structure of Some Molecules

In this section we shall discuss the electronic structure of some selected molecules to show the results of quantum chemical calculation.

(1) Water (H_2O)

Perhaps water is the most important substance both in the physical and in the life sciences. Water is also noted for its anomalous physical and chemical properties, so it is interesting to study its electronic structure. MO calculation gives the diagram of energy levels for H_2O shown in Fig. 11.3.3. The ground state configuration is (1a) (2a) (1b) (3a) (1b): here (1a) represents the inner 1s electron pair of oxygen atom; (2a) and (1b) are two pairs of bonding electrons, corresponding to two O-H bonds; (3a) and (1b) represent two pairs of non-bonding electrons of oxygen atoms, called lone-pairs.

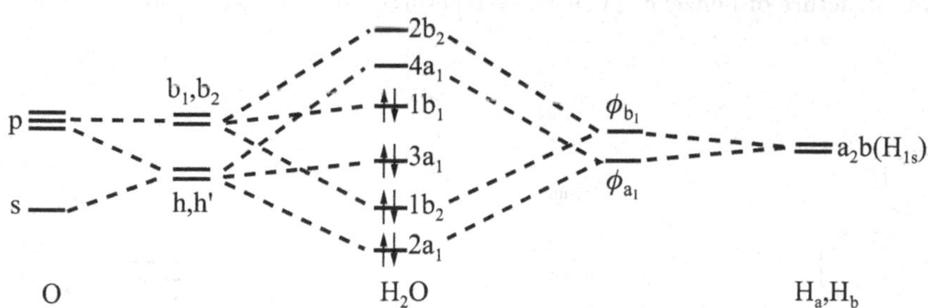

Figure 11.3.3 Energy level diagram of H_2O.

With the formation of the H_2O molecule, the valence shell of the oxygen atom is filled with the octet set of electrons, so it is very stable. According to the valence shell electron-pair repulsion rule, the spatial arrangement of bonding pairs and lone pairs should be nearly tetrahedral. Due to the fact that the repulsive force between the lone-pairs is slightly larger than that between bonding pairs, the angle between the two O-H bonds is 104.5°, somewhat less than the ideal tetrahedral angle of 109.5°. However, the incipient tetrahedral coordination should be noted. As partners of covalent bonding, oxygen and hydrogen atoms are actually unequal in strength, so polar covalent bonding results. H_2O is a strongly polar molecule, with the region near the protons positively charged and the region near the lone-pairs in the oxygen atom negatively charged. The lone-pair electrons may attract protons in nearby H_2O molecule to form hydrogen bonds. Hydrogen bonds, though much weaker, are also directional in real space like covalent bonds, but their origin is based on the attraction of unlike charges, somewhat like ionic bonds. This is why water molecules appear in the associated state. When NaCl is dissolved in water, each Na^+ or Cl^- ion is surrounded by a spherical shell of regularly arranged polar water molecules that are electrostatically attracted to it — a process called hydration. Hydration is the reason why water is such a good solvent for various salts and other substances. In some biochemical processes, the ions with their hydration spheres should be regarded as an entity somewhat like a molecule, say, when considering their passage through small pores in a membrane. When water is frozen into ice, it has a crystal structure like that of hexagonal diamond in which tetrahedral coordination is clearly displayed and hydrogen bonds and covalent bonds supplement each other. When ice is melted, this structure collapses, some of the hydrogen bonds are ruptured, so tighter packing of molecules is achieved making water denser than ice.

(2) Aromatics

Benzene and its derivatives are important organic molecules. The classical structural formula for benzene is characterized by an alternate arrangement of single and double bonds, known as Kekulé structures, shown in Fig. 11.3.4. Experimental results confirm that the six carbon bonds are actually identical in length and bond strength. So, according to molecular orbital theory, these π bonds are delocalized, shared by different carbon atoms; note the double-doughnut-shaped electron cloud (see Fig. 11.3.6) which Pauling called the 'resonance' between the two equivalent Kekulé structures. Nuclear magnetic resonance spectra show evidence of the existence of a 'circular current' of these delocalized π-electrons, induced by the magnetic field.

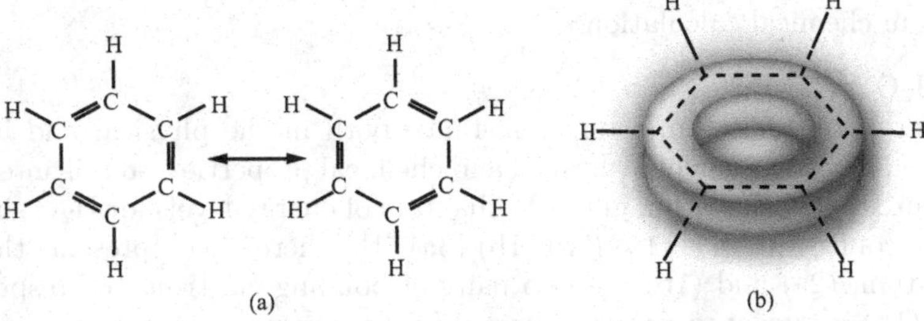

Figure 11.3.4 Structure of benzene. (a) Kekulé structures, (b) electron density for delocalized π electrons of benzene.

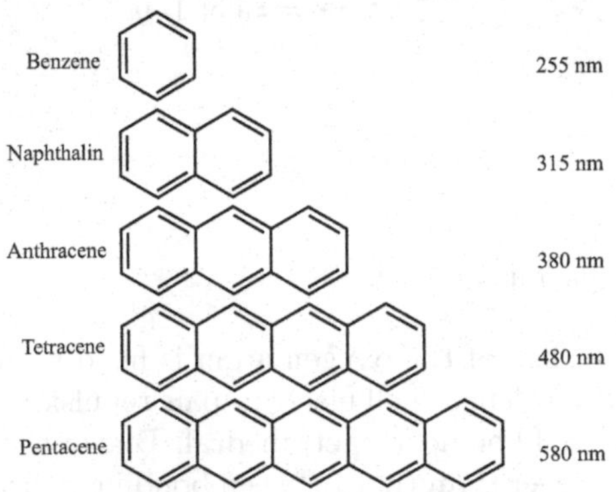

Figure 11.3.5 Structure and optical absorption of benzene and other molecules with planar region of fused rings (polyacenes).

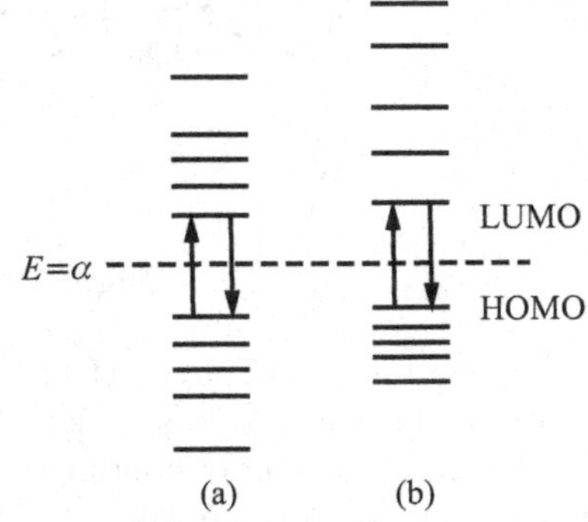

Figure 11.3.6 The molecular orbital energy levels of naphthalene. (a) Calculated by the Hückel approximation, (b) calculated by the Hückel approximation with additional overlap correction ($l = 0.25$).

Several planar benzene rings may share edges, and naphthalene, anthracene and other molecules are formed in this manner as shown in Fig. 11.3.5. Figure 11.3.6 shows the energy levels of naphthalene calculated according to the Hückel approximation and those with additional overlap correction. Note the change from a symmetrical arrangement of levels around $\varepsilon = \alpha$, to an asymmetrical arrangement; furthermore there is an energy gap between the lowest unoccupied molecular orbital (LUMO) and the highest occupied molecular orbital (HOMO) which plays a crucial role in optical transitions somewhat akin to band gaps in crystals. These molecules display interesting electronic and optical properties.

(3) Fullerene C

C_{60} is the soccer ball shaped molecule which consists of 60 carbon atoms arranged in a truncated icosahedron with icosahedral symmetry, the highest symmetry possible for a molecule. It is a great credit to quantum chemists to have predicted the existence of C_{60} and even published the MO and energy level sequence shown in Fig. 11.3.7 before the actual discovery of C_{60} molecule by Kroto, Smalley and Curl in 1985. It is easy to recognize that 60 electrons fill up 30 bonding MOs with a large gap between LUMO and HOMO, which testify to its stability.[g] Then other fullerene and carbon nanotubes were synthesized, and many novel properties were discovered. Their significance for chemistry, condensed matter physics and materials science cannot be overestimated.

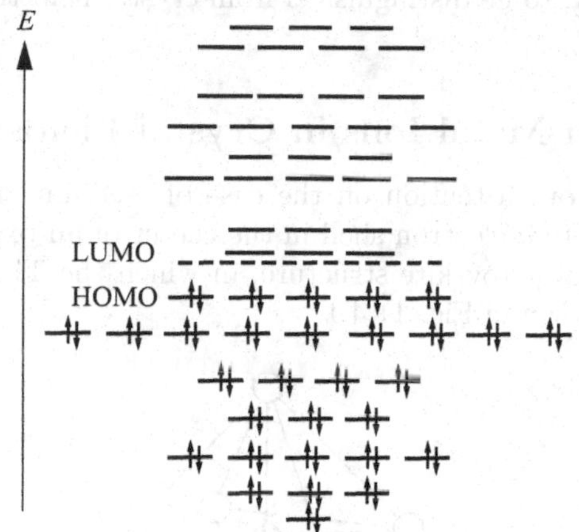

Figure 11.3.7 Energy levels of C_{60} from HMO calculation.

§11.4 Ions in Anisotropic Environments

11.4.1 Three Types of Crystal Fields

In §11.1 we were essentially concerned with free atoms or free ions. We have seen that for a hydrogen atom, orbitals with the same principal quantum number n are n^2-fold degenerate; however, in many-electron atoms, this degeneracy is partially lifted, the energy degeneracy in orbitals is now limited to the same sub-shell with the same angular momentum quantum number l, such as $3d$ or $4f$. If an atom or an ion is situated on a site in a crystal, the energy levels of this atom or ion certainly will be influenced by the anisotropic environment, i.e., the crystal fields. So further lifting of the energy degeneracy, or in other words further splitting of energy levels, will be expected. We have already seen that, in the experimentally measured magnetic moments of $3d$ transition metal ions, contribution from the orbital part is entirely missing. Presumably this is also the effect of crystal fields. To study the effect of crystal fields on atoms or ions is actually doing atomic physics in the

[g]Early in 1971, E. Osawa described C_{60} as the structure like a soccer ball in his book *Aromatics*; later the Russian scholar Bochvar discussed C_{60}'s structure by the Hückel approximation, see I. V. Stamkevich *et al.*, *Russ. Chem. Rev.* **53**, 604 (1984); then R. A. Davidson published his calculation, see R. A. Davidson, *Theor. Chem. Acta.* **58**, 193 (1981).

anisotropic environment. According to the strength of the crystal field, we may distinguish three types of crystal field effect:

1) The weak case (exchange splitting > spin-orbit coupling > crystal field)
This is the case for rare earth atoms in crystal fields: the $4f$ orbitals lie so deep within the ion core that other occupied shells of the same ion nearly screen out the electrostatic potential of neighboring ions. Thus, this situation is hardly different from the free ion. Even Hund's 3rd rule takes precedence over the crystal field effect.

2) The medium case (exchange splitting > crystal field > spin-orbit coupling)
This is the case for transition metal ions (particularly $3d$-ions). Since crystal fields now dominate the spin-orbit coupling, Hund's 3rd rule ceases to apply and J is no longer a good quantum number. This is why orbital magnetic moments are quenched by crystal fields.

3) The strong case (crystal field > exchange splitting > spin-orbit coupling)
Now the crystal field is comparable to (or even larger than) the exchange splitting, which gives rise to Hund's 1st and 2nd rule, so it mixes the states with those of other terms. This is the case for some $4d$ and $5d$ transition metal compounds, for which the mixing of the d-orbitals of the central cation with the p-orbitals of the surrounding anions should be taken into account. These are the subject of ligand field theory, to be distinguished from crystal field theory which uses a purely ionic description.

11.4.2 3*d* Transition Metal Ions in Crystal Fields

Initially, we shall focus our attention on the case of medium crystal fields, i.e., $3d$ transition metal ions with an incomplete d-electron shell in the center of an oxygen octahedron. Consider the case of $LaTiO_3$ with an ideal perovskite structure, in which the Ti ion is situated at the center of an oxygen octahedron as shown in Fig. 11.4.1.

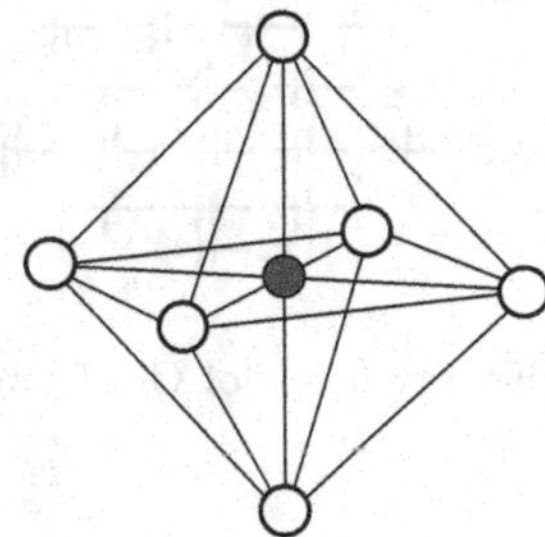

Figure 11.4.1 Local environment of a transition metal ion at the center of the oxygen octahedron.

The ionic state of $LaTiO_3$ can be summarized as La^{3+}, Ti^{3+}, O^{2-}, in which La^{3+} and O^{2-} have closed shells; the magnetic moment is carried by the single d-electron of the Ti^{3+} ion. Surely the motion of this d-electron is governed not only by the potential of the Ti ion core, but also by the electrostatic potential of the surrounding six O ions, so the Hamiltonian for the $3d$ electron is written as

$$\mathcal{H}_{3d}(\boldsymbol{r}) = \mathcal{H}^{(\mathrm{Ti})}(\boldsymbol{r}) + \sum_{j=1}^{6} V^{(\mathrm{O})}(\boldsymbol{r} - \boldsymbol{R}_j). \tag{11.4.1}$$

The crystal field effect in an octahedral field can be understood qualitatively by the symmetry argument shown in Fig. 11.4.2. In a free ion, the orbitals of the $3d$ electrons are 5-fold degenerate. In a crystal field, lobes of $d_{x^2-y^2}$ and d_{z^2} orbitals point directly to point charges of the O anions leading to a strong interaction. The Coulomb repulsive energy raises up their energy levels, while the lobes of the d_{xy}, d_{yz}, d_{zx} orbitals point in the diagonal directions, avoiding the O anions, and these energy levels will be lowered. So originally degenerate levels will split into two groups, the lower triplet one, called t_{2g}, the higher doublet one, called e_g. For La_2CuO_4, the parent compound of the first high T_c

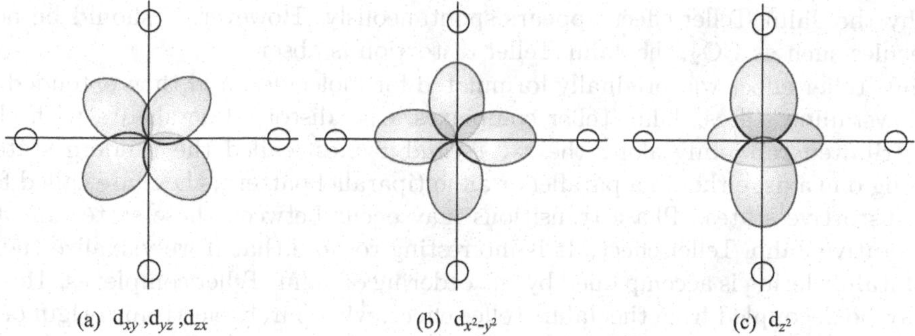

Figure 11.4.2 3*d* orbitals of a center ion in the octahedral environment of O anions.

superconductor, the local environment of the Cu ion is approximately octahedral. Its valence state is identified as La^{3+}, Cu^{2+}, O_4^{2-}, the ion with the partially filled shell is Cu in the 3*d* configuration. The problem is rather similar to Ti^{3+}: In the case of the Ti ion, we are concerned about only the *d*-state which is occupied; in the case of the Cu ion, the *d*-state has an unoccupied hole.

For tetragonal and cubic crystal fields, a similar argument may reverse the energy sequence for t_{2g} and e_g. More elaborate consideration of the point group symmetry of the crystal field will reach the same conclusion. The energy of this splitting is denoted by Δ; its value can be calculated theoretically through more elaborate manipulation using the Hamiltonian in Eq. (11.4.1).

11.4.3 Jahn–Teller Effect

Returning to our example of a Ti^{3+} ion at the center of a ideal oxygen octahedron, the ionic ground state is predicted to be orbitally degenerate, i.e., we cannot distinguish the orbitals in a doublet e_g or triplet t_{2g}. However, this is forbidden by the Jahn–Teller theorem which states: if the symmetry of the crystal field is so high that the ground state is predicted to be degenerate in energy, then spontaneous distortion should appear to lift the orbital degeneracy. This sort of distortion is called the Jahn–Teller effect. In the case of the Ti ion in the O octahedron, we assume that the octahedron is elongated along the *z*-axis by δ_z (see Fig. 11.4.3), this tetragonal distortion induces the distance between two ions in the *z* direction to become larger than the corresponding distances along the *x*- and *y*-axes. The difference in the Coulomb repulsion between the orbitals along the *z*-direction and those in the *x*-, or *y*-direction will lift the energy degeneracy (see Fig. 11.4.4).

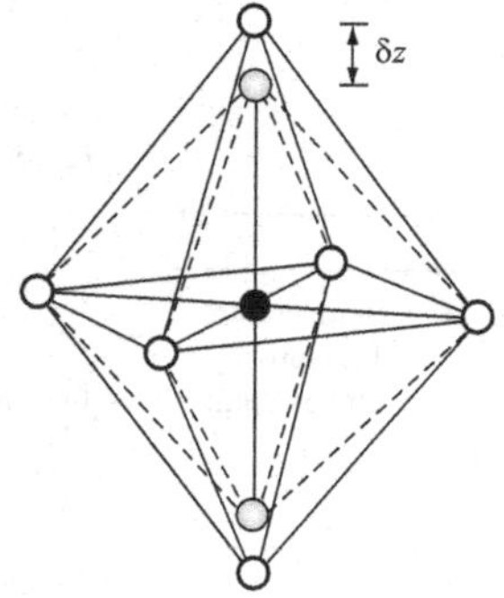

Figure 11.4.3 Jahn–Teller distortion of an oxide octahedron.

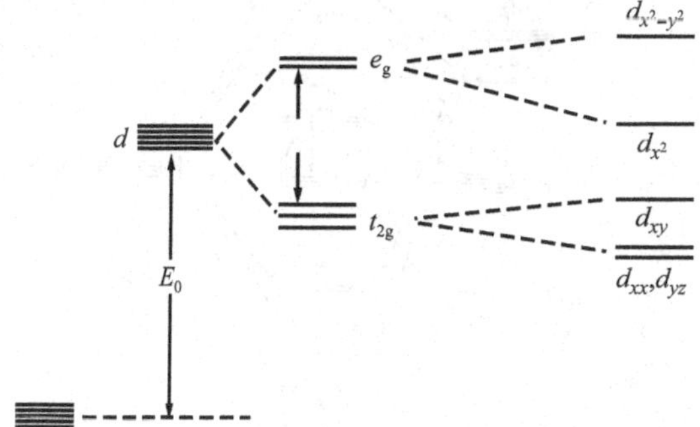

Figure 11.4.4 Energy levels in octahedral crystal field.

The energy relationship in the Jahn–Teller effect is summarized as follows: the distortion of the octahedron costs an elastic energy $\alpha\delta_z^2$; the splitting of electronic energy level is proportional to δ_z; two electrons drop to a lower level, but one electron rises to a higher level, and the net gain in energy is $-\beta\delta_z$. Thus, the total energy $\alpha\delta_z^2 - \beta\delta_z$ is minimized by finite distortion $\delta_z = \beta/2\alpha$. This neatly explains why the Jahn–Teller effect appears spontaneously. However, it should be noted that, for linear molecules such as CO_2, the Jahn–Teller distortion is absent.

The Jahn–Teller effect was originally formulated for molecules, and then extended to crystalline states. In crystalline states, Jahn–Teller complexes, e.g., distorted octahedra with elongated axes may be distributed randomly along the x-, y- and z-axes, called the random state; or coupled together along one axis, either in a parallel or an antiparallel pattern; these are called ferrodistortive or antiferrodistortive states. Phase transitions may occur between these states and this is known as the cooperative Jahn–Teller effect. It is interesting to note that if we visualize the ground state orbitals, orbital-ordering is accompanied by the ordering of Jahn–Teller complexes. However, orbital-ordering may be decoupled from the Jahn–Teller effect, with purely electronic origin or coupled with magnetic ordering.

In this section we have used z-axis elongation of oxygen octahedra as an example of Jahn–Teller effect mainly for its easy visualization; the actual distortion may be more complex and more difficult to imagine.

11.4.4 Ions in Ligand Fields

Let us consider the case where this is a strong crystal field in which covalent bonds may be formed between the central ion and its neighboring atoms, i.e., ligands. The same is true for the case of transition metal complexes, or coordination compounds, in this case, the ligands are nonmetallic atoms or molecules, such as TiF_6^{3-}, $Fe(CN)_6^{4-}$, ..., etc.

Now the situation is more complicated than ions in an ordinary crystal field. In addition, MOs must be introduced to account for the covalent bonds between the central metal ion M and the ligands, L, while the symmetry arguments of crystal field theory are still qualitatively valid.

Consider the case of complexes ML_6 with octahedral structure, with M at the origin of Cartesian coordinates, and 6 Ls at equal distances from the center. MOs may be divided into two groups: σ: s, p_x, p_y, p_z, $d_{x^2-y^2}$, d_{z^2}; π: d_{xy}, d_{yz}, d_{zx}. New orbitals are formed in the ligands to match the corresponding σ bonds and π bonds. Due to the fact that d_{xy}, d_{yz}, d_{zx} orbital lobes mismatch those of L (see Fig. 11.4.2), they become nonbonding orbitals. 6 orbitals of M and 6 orbitals of L form 12 delocalized MOs, one half are bonding orbitals, the another half are antibonding orbitals. The energy levels are shown schematically in Fig. 11.4.5.

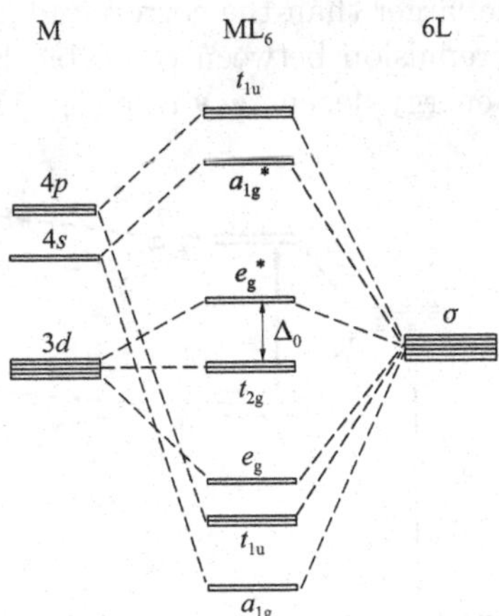

Figure 11.4.5 The energy levels for the MOs of octahedral transition metal complexes.

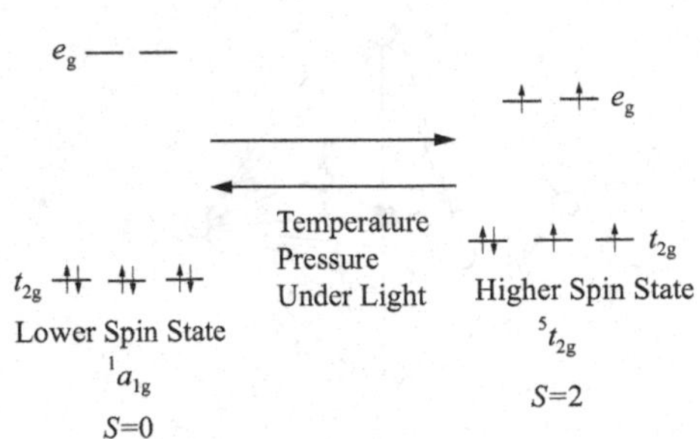

Figure 11.4.6 The electronic spin transition of Fe^{2+} ion in octahedral conjugated molecules.

Here the sequence of antibonding doublet e_g^* and nonbonding triplet t_{2g} just corresponds to that of e_g and t_{2g} in the case of ions in crystals, so the analysis, from point symmetry consideration of the octahedral field, is still valid for the case of the ligand field; however, the value of the splitting energy Δ is somewhat modified. To calculate the value of Δ directly from ligand field theory is quite difficult, so we may deduce these values from optical absorption spectra. Most of these complexes have optical absorption in the visible region. The simplest case is $[Ti(H_2O)_6]^{3+}$: It has an absorption band at 20 000 cm^{-1}, which corresponds to the transition from t_{2g} to e_g^*, so the approximate value is found to be 20 000 cm^{-1} or 2.4 eV. In general, the values lie between $1 \sim 4$ eV for most of these complexes.

Their magnetic properties are also very interesting. Take the vanadium complex $[V(H_2O)_6]^{3+}$ as an example: Its central ion is V^{3+} with two valence electrons. We must put these two electrons into the lowest triplet t_{2g} levels; according to Hund's rule, they should occupy two different degenerate levels with parallel spins.

For the chromium complex $[Cr(H_2O)_6]^{3+}$, we predict that three degenerate t_{2g} levels should be filled with parallel spins. However, for $[Cr(H_2O)_6]^{2+}$, with four electrons, there are two alternatives: either go into another t_{2g} orbital with antiparallel spin, or put into a higher e_g level with parallel spin and gain Coulomb energy. Which state is realized depends on the value of Δ. A low value of Δ favors the state with high spin (HS), while a high value of Δ favors the state of low spin (LS). In Table 11.4.1 we find that, when the number of d-electrons $n_d \leq 3$ or $n_d \geq 8$, the situation is clear, and the complex is in the HS state. For $4 \leq n_d \leq 7$, there are two alternatives: either the HS state or LS state. The equilibrium state at 0 K depends on the energy or enthalpy.

It is interesting to observe that, in some cases, the spin crossover from the LS state to HS state may be induced by change of temperature, pressure or irradiation by light. For instance, consider an Fe^{2+} ion surrounded by six ligands at the corners of an octahedron. The enthalpy of the LS state is slightly less than the HS state (see Fig. 11.4.6). At low temperature, the thermodynamically stable state is the LS state. Assume $T_{1/2}$ stands for the temperature for which there is coexistence of 50% of LS and 50% of HS complexes. When the temperature is higher than $T_{1/2}$, the IIS state becomes the thermodynamically stable state, because the entropy associated with the HS state is larger than that associated with the LS state, then the entropy gain term ΔS outweighs the energy (or the enthalpy) loss. This spin-crossover or transition is the physical basis of the new type of magnetic memory devices in the fast-developing field of molecular magnetism.[h] It should be remembered that the competition between high and low spin states may occur in the cases of ions in a crystal environment, such as ions with $3d^6$ configuration (realized by Fe^{2+} and Co^{3+} ions) in oxygen octahedra.

Table 11.4.1 High-spin and low-spin complexes with octahedral structures.

Number of d-electrons	High-spin state				Low-spin state			
	t_{2g}	e_g	n	p	l_{2g}	e_g	n	p
1	↑ — —	— —	1	1.73				
2	↑ ↑ —	— —	2	2.83				
3	↑ ↑ ↑	— —	3	3.87				
4	↑ ↑ ↑	↑ —	4	4.90	↑↓ ↑ ↑	— —	2	2.83
5	↑ ↑ ↑	↑ ↑	5	5.92	↑↓ ↑↓ ↑	— —	1	1.73
6	↑↓ ↑ ↑	↑ ↑	4	4.90	↑↓ ↑↓ ↑↓	— —	0	0
7	↑↓ ↑↓ ↑	↑ ↑	3	3.87	↑↓ ↑↓ ↑↓	↑ —	1	1.73
8	↑↓ ↑↓ ↑↓	↑ ↑	2	2.83				
9	↑↓ ↑↓ ↑↓	↑↓ ↑	1	1.73				

Previously, we have discussed the coordination compounds formed by σ-electron donation. Furthermore, organometallic coordination compounds may be formed by combining conjugated molecules with transition metal ions through π-electron donation. We may take ferrocene ($Fe(C_5H_5)_2$) as an example (see Fig. 11.4.7). The problem is how to combine the two π-MOs and the AO of Fe. The calculation by quantum chemistry gives the energy levels shown in Fig. 11.4.8: there

[h] O. Hahn and C. J. Marlinez, *Science* **279**, 44 (1998).

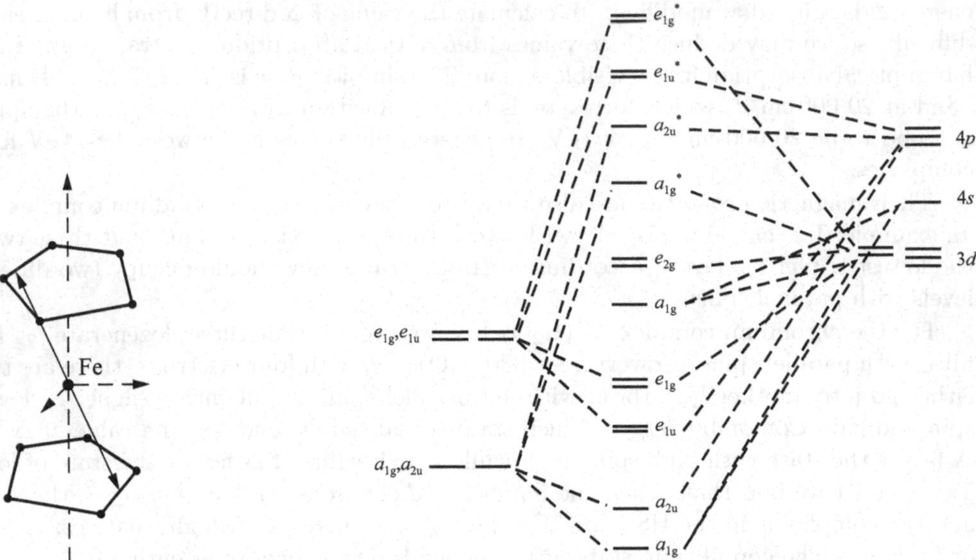

Figure 11.4.7 The structure of ferrocene. **Figure 11.4.8** The energy levels of ferrocene.

are six bonding MOs, three nonbonding MOs and a high-energy antibonding MO. Nine low-energy obitals contain 18 electrons (two C_5H_5 provide ten π-electrons and Fe atom (d^6s^2) provides eight electrons), forming the full electron shell of the inert element Kr ($d^{10}s^2p^6$). We can also find the magnetism of ferrocene from Fig. 11.4.8: It is antiferromagnetic because the nine MOs are occupied by electrons. If we replace an Fe atom by a C atom, an extra electron will take the first antibonding MO, so this would be ferromagnetic but not stable: If we substitute a Ni atom, two extra electrons will occupy a pair of antibonding MOs, and this is ferromagnetic according to the Hund's rule. The treatment of ferrocene above may be extended to other metallic coordination compounds with different rings. This kind of coordinate compounds provides a new opportunity for research on molecular magnets.

Bibliography

[1] Yang, F., and J. H. Hamilton, *Modern Atomic and Nuclear Physics*, Chaps. 5, 7 & 8, McGraw-Hill, New York (1996).

[2] Haken, H., and H. C. Wolf, *The Physics of Atom and Quanta*, 4th ed., Springer, Berlin (1994).

[3] Wannier, G. H., *Elements of Solid State Theory*, Chap. 8, Cambridge University Press, Cambridge (1959).

[4] Jiang Yuansheng, *Molecular Structural Theory*, Higher Education Press, Beijing (1999).

[5] Haken, H., and H. C. Wolf (translated by W. D. Brewer), *Molecular Physics and Elements of Quantum Chemistry*, Spinger, Berlin (1994).

[6] McWeeny, R., and S. Coulson, *Valence*, 3rd ed., Clarendon Press, Oxford (1979).

[7] Ashcroft, N. W., and N. D. Mermin, *Solid State Physics*, Chaps. 31 & 32, Holt, Rinehart and Winston, New York (1976).

[8] Fazekas, P., *Lecture Notes on Electron Correlation and Magnetism*, Chaps. 2 & 3, World Scientific, Singapore (1999).

[9] Marder, M. P., *Condensed Matter Physics*, Chaps. 25 & 26, Wiley, New York (2000).

Chapter 12

Band Approach

One of the most important aspects of condensed matter physics involves the electronic properties of many-particle systems. Energy band theory, based on the single-electron approximation, provides a conceptual framework for understanding a large part of the electronic properties of solids. As discussed in Chaps. 5 and 6, band theory is the basis for electrical classification of crystals as metals, semiconductors and insulators. Besides electrical properties, this theory is capable of explaining optical, magnetic and thermal properties. Next we introduce the many-particle Hamiltonian and the various approximations which give justification to, as well as show the limitations of, the single-electron approximation. Further, density functionals are introduced, which replace wavefunctions in the calculation of electronic structure; these procedures have proved to be effective and efficient. Finally, the results of band calculation are discussed and compared with the properties of various types of materials.

§12.1 Different Ways to Calculate the Energy Bands

As was stated in §5.2, in one-electron approximations, like the near-free electron model or the tight-binding electron model for crystalline materials, a set of independent electrons moving in a periodic potential leads to a band picture. However, nearly-free and tight-binding electron models are too crude to be useful in calculations that are to be compared with experimental results. More reliable band structure calculations in solids involve a lot of methods, such as the orthogonalized plane wave method, pseudopotential method, cellular method, augmented plane wave (APW) method, etc. In this section we shall introduce some of the common methods employed in calculations of real bands. The primary differences among the various methods of band calculations include two aspects: one is to choose a reasonable set of functions in which to expand the electronic wavefunctions, and the other is to approximate the real crystalline potential by an effective potential.

12.1.1 Orthogonized Plane Waves

The simplest complete orthogonal set of functions is plane waves,

$$|\boldsymbol{k}+\boldsymbol{G}\rangle = \Omega^{-1/2}\mathrm{e}^{i(\boldsymbol{k}+\boldsymbol{G})\cdot\boldsymbol{r}}, \tag{12.1.1}$$

where Ω is the volume of a crystal and $\boldsymbol{G}$ denotes the reciprocal lattice vectors. In principle, the wavefunction of an electron in a crystal can be expanded in plane waves,

$$\psi_{\boldsymbol{k}}(\boldsymbol{r}) = \Omega^{-1/2}\sum_{\boldsymbol{G}} c(\boldsymbol{G})\mathrm{e}^{i(\boldsymbol{k}+\boldsymbol{G})\cdot\boldsymbol{r}}. \tag{12.1.2}$$

We substitute this into the Schrödinger equation, and get the secular equations,

$$\sum_{\boldsymbol{G}'}\left\{\left[\frac{\hbar^2}{2m}(\boldsymbol{k}+\boldsymbol{G})^2 - E(\boldsymbol{k})\right]\delta_{\boldsymbol{G}\boldsymbol{G}'} + V(\boldsymbol{G}-\boldsymbol{G}')\right\}c(\boldsymbol{G}') = 0, \tag{12.1.3}$$

where the Fourier coefficients are,

$$V(\boldsymbol{G}-\boldsymbol{G}') = \frac{1}{\Omega}\int d\boldsymbol{r}V(\boldsymbol{r})\mathrm{e}^{-i(\boldsymbol{G}-\boldsymbol{G}')\cdot\boldsymbol{r}}. \tag{12.1.4}$$

The requirement of nontrivial solutions for (12.1.3) gives the following determinant equation,

$$\left|\left[\frac{\hbar^2}{2m}(\boldsymbol{k}+\boldsymbol{G})^2 - E(\boldsymbol{k})\right]\delta_{\boldsymbol{G}\boldsymbol{G}'} + V(\boldsymbol{G}-\boldsymbol{G}')\right| = 0. \tag{12.1.5}$$

Diagonalizing this determinant, the eigenvalues $E(\boldsymbol{k})$ and the coefficients $c(\boldsymbol{G})$ can all be obtained.

In deriving the secular equations (12.1.3), we did not have to assume that $V(\boldsymbol{r})$ was small, and in principle this set of equations enables us to find the energies $E(\boldsymbol{k})$ and the wavefunctions $\psi_{\boldsymbol{k}}(\boldsymbol{r})$ for an electron moving in any periodic potential $V(\boldsymbol{r})$. However, in practice, the equations can only be solved at all easily if most of the coefficients $V(\boldsymbol{G})$ are small (compared with the Fermi energy E_{F}). In the nearly-free electron approximation we assumed that they are all small, so that the solution wavefunctions contain only a few nonzero Fourier coefficients $c(\boldsymbol{G})$. In reality, this will not be so: $V(\boldsymbol{r})$ varies rapidly with $\boldsymbol{r}$ near the nuclei. Consequently the coefficients $V(\boldsymbol{G})$ are large, and only decrease in magnitude slowly as $|\boldsymbol{G}|$ increases. These equations then become a formidably large set of coupled equations. The solution wavefunctions contain many Fourier components, and within the ion core at least, they look nothing like plane waves. Physically, this is just what we should expect: within the ion cores, $\psi(\boldsymbol{r})$ must look something like an atomic wavefunction, and such a function will certainly need many Fourier components to represent it adequately. The energy bands of simple metals appear to be described well by the nearly-free electron (NFE) approximation; however, there are still some discrepancies. The reason for this is that the bands of simple metals originate from the s- and p-electrons. These states must be orthogonal to the s- and p-core functions, which oscillate with very short wavelengths.

Herring provided a modification in which the base functions include not only the plane waves with smaller momentum $\boldsymbol{k}+\boldsymbol{G}$, but also the isolated atomic wavefunctions with larger momentum, achieved by adding a sum of Bloch functions to the plane wave so that the orthogonalized plane wave (OPW) corresponding to $\boldsymbol{k}$ is

$$\phi_{\boldsymbol{k}}(\boldsymbol{r}) = \Omega^{-1/2}\mathrm{e}^{i\boldsymbol{k}\cdot\boldsymbol{r}} - \sum_j \mu_{j\boldsymbol{k}}\phi_{j\boldsymbol{k}}(\boldsymbol{r}), \tag{12.1.6}$$

where $\phi_{j\boldsymbol{k}}$ are Bloch functions formed from atomic orbitals $\chi_j(\boldsymbol{r})$ that describe core states

$$\phi_{j\boldsymbol{k}}(\boldsymbol{r}) = \frac{1}{\sqrt{N}}\sum_l \mathrm{e}^{i\boldsymbol{k}\cdot\boldsymbol{R}_l}\chi_j(\boldsymbol{r}-\boldsymbol{R}_l). \tag{12.1.7}$$

For example, to calculate the energy bands of potassium through to zinc, the Bloch functions included in the OPWs would be formed from the 1s-, 2s-, 2p-, 3s-, 3p-orbitals of the atom.

The conduction electron wavefunction should be orthogonal to the core states so the coefficents $\mu_{j\boldsymbol{k}}$ are then determined by $\mu_{j\boldsymbol{k}} = \Omega^{-1/2}\int\phi^*_{j\boldsymbol{k}}\mathrm{e}^{i\boldsymbol{k}\cdot\boldsymbol{r}}d\boldsymbol{r}$, which enables the wavefunction in (12.1.2) to be orthogonal to all the core states $\phi_{j\boldsymbol{k}}$. The OPW (12.1.6) has just such a character: within each ion core the $\phi_{j\boldsymbol{k}}$ terms are large; and outside it, they are small, as shown in Fig. 12.1.1. Now we use a set of OPWs to form a Bloch wave,

$$\psi_{\boldsymbol{k}}(\boldsymbol{r}) = \sum_{\boldsymbol{G}} c_{\boldsymbol{G}}\phi_{\boldsymbol{k}+\boldsymbol{G}}(\boldsymbol{r}). \tag{12.1.8}$$

This method leads the secular determinant for the eigenvalues to be identical with the NFE determinant, although, in addition to the Fourier component of the crystal potential $V(\boldsymbol{G})$, there is a repulsive contribution from core-orthogonality. This tends to cancel the attractive Coulomb potential term in the core region, which results in much smaller Fourier components and hence nearly-free-electron-like behavior of the band structure of simple metals.

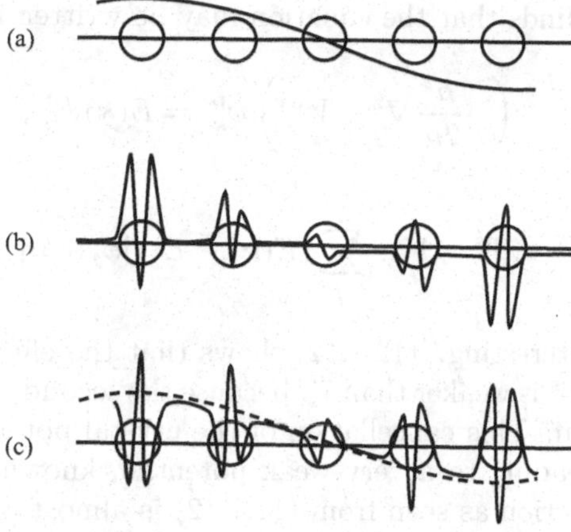

Figure 12.1.1 (a) Plane wave, (b) core state and (c) orthogonal plane wave state.

12.1.2 Pseudopotential

The OPW method leads in a natural way to the concept of pseudopotential. The true lattice potential $V(\boldsymbol{r})$ is replaced by a much weaker potential $V^{\mathrm{ps}}(\boldsymbol{r})$ which preserves the original eigenenergies $E(\boldsymbol{k})$.

Suppose we write the eigenfunction of the Schrödinger equation as

$$\psi_{\boldsymbol{k}}(\boldsymbol{r}) = \psi_{\boldsymbol{k}}^{\mathrm{ps}} - \sum_j \mu_j \phi_j, \tag{12.1.9}$$

where $\psi_{\boldsymbol{k}}^{\mathrm{ps}}$ is a plane-wave-like function and ϕ_j an atomic function. The sum over j extends over all the atomic shells which are occupied; for example, in Na, the sum extends over $1s$, $2s$, and $2p$ shells. The coefficients μ_j are chosen such that the function $\psi_{\boldsymbol{k}}(\boldsymbol{r})$, representing a $3s$-electron, is orthogonal to the core function ϕ_j. By requiring this orthogonality, we ensure that the $3s$-electron, when at the core, does not occupy the other atomic orbitals already occupied. The function $\psi_{\boldsymbol{k}}$ has the features we are seeking: away from the core, the atomic functions ϕ_j are negligible, and thus $\psi_{\boldsymbol{k}} = \psi_{\boldsymbol{k}}^{\mathrm{ps}}$, a plane wave-like function. At the core, the atomic functions are substantial and act so as to induce rapid oscillations, as shown in Fig. 12.1.2(a).

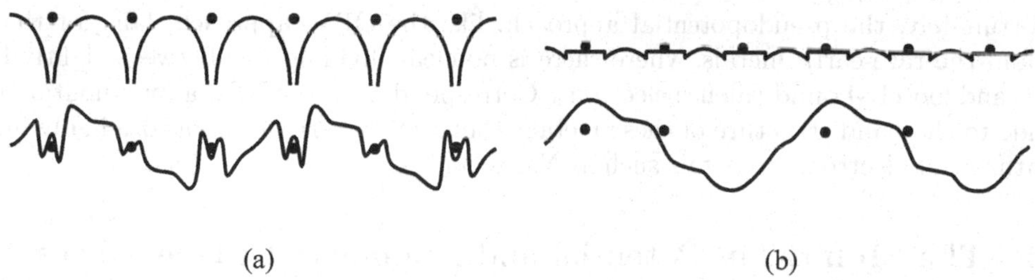

Figure 12.1.2 The pseudopotential concept: (a) The actual potential and corresponding wavefunction, (b) the pseudopotential and corresponding pseudofunction.

If we now substitute $\psi_{\boldsymbol{k}}$ into the Schrödinger equation, then

$$\left(-\frac{\hbar^2}{2m}\nabla^2 + V\right)\psi_{\boldsymbol{k}} = E(\boldsymbol{k})\psi_{\boldsymbol{k}}, \tag{12.1.10}$$

and we also, approximately, have

$$\left(-\frac{\hbar^2}{2m}\nabla^2 + V\right)\phi_j = E(j)\phi_j. \tag{12.1.11}$$

Rearranging the terms, one finds that the equation may be written in the form

$$\left(-\frac{\hbar^2}{2m}\nabla^2 + V^{\mathrm{ps}}\right)\psi_{\boldsymbol{k}}^{\mathrm{ps}} = E(\boldsymbol{k})\psi_{\boldsymbol{k}}^{\mathrm{ps}}, \tag{12.1.12}$$

where

$$V^{\mathrm{ps}} = V + \sum_j (E(\boldsymbol{k}) - E_j)\,|\phi_j\rangle\langle\phi_j|. \tag{12.1.13}$$

These results are very interesting. (12.1.12) shows that the effective potential is given by V^{ps}, while (12.1.13) shows that V^{ps} is weaker than V, because the second term on the right of the equation tends to cancel the first term. This cancellation of the crystal potential by the atomic functions is usually appreciable, often leading to a very weak potential, known as the pseudopotential. Since V^{ps} is so weak, the wavefunction as seen from (12.1.12) is almost a plane wave, given by $\psi_{\boldsymbol{k}}^{\mathrm{ps}}$, and is called a pseudofunction. The pseudopotential and pseudofunction are illustrated graphically in Fig. 12.1.2(b). Note that the potential is quite weak, and, in particular, the singularity at the core is entirely removed. Correspondingly, the rapid oscillations in the wavefunction have been erased, so that there is a smooth plane-wave-like function.

Now we can understand why the valence electron in Na, for instance, seems to behave as a free particle despite the fact that the crystal potential is very strong at the ionic cores. In fact, when the exclusion principle is properly taken into account, the effective potential is indeed quite weak. The free-particle behavior, long taken to be an empirical fact, is now borne out by quantum-mechanical calculations. The explanation of this basic paradox is one of the major achievement of the pseudopotential method. This method has also been used to calculate band structures in many metals and semiconductors, such as Be, Na, K, Ge, Si, etc., with considerable success.

It was realized that, except for the transition metals and the rare-earth metals which have unfilled d or f shells, the NFE model in fact works remarkably well: the band structure and the shape of the Fermi surface can be quite closely reproduced by choosing the right values for a few coefficiets $V(\boldsymbol{G})$ in the expansion and setting the rest equal to zero. These $V(\boldsymbol{G})$ are, in fact, the Fourier coefficients of a rather weak and smoothly-varying pseudopotential $V^{\mathrm{ps}}(\boldsymbol{r})$ for which the conduction electron band structure $E(\boldsymbol{k})$ happens to be much the same as for the (far stronger) potential $V(\boldsymbol{r})$. The corresponding wavefunctions

$$\psi^{\mathrm{ps}}(\boldsymbol{r}) = \sum c(\boldsymbol{G})\mathrm{e}^{i(\boldsymbol{k}+\boldsymbol{G})\cdot\boldsymbol{r}} \tag{12.1.14}$$

vary smoothly throughout the unit cell.

Unfortunately, the pseudopotential approach, like the OPW approach, fails for the transition metals and the rare-earth metals, where there is no clear dividing line between tightly-bound core electrons and loosely-bound valence electrons. Correspondingly, the NFE approximation is of no use as a guide to the band structure of these metals, though it is very useful, as originally intended, for the treatment of electrons in metals such as Na, K and Al.

12.1.3 The Muffin-Tin Potential and Augmented Plane Waves

From the start we have emphasized there are two different approaches for the energy bands of periodic solids: one approach is the NFE model, with pseudopotential theory as its improved form; another is the TBE model, in which the LCAO approximation is generally adopted. To combine the advantages, and to avoid the disadvantages, of both approaches is a crucial problem for band calculations. J. C. Slater (1937) proposed the concept of a muffin-tin potential to solve this problem: The periodic potential for the energy band is clearly divided into two parts: a spherically symmetric atomic potential for the region within ion cores; a constant potential (generally chosen to be zero) for the open space between ion cores. So the muffin-tin potential is expressed as

$$V(r) = \begin{cases} V_{\mathrm{a}}(r), & r < r_{\mathrm{c}}; \\ 0, & r \geq r_{\mathrm{c}}. \end{cases} \tag{12.1.15}$$

Here r_c is the radius of an ion core, surely smaller than the Wigner–Seitz (WS) radius r, so no contact or overlap of ion cores occurs (Fig. 12.1.3). The Muffin-tin potential may be easily generalized to the case of a lattice with a basis, in which ion cores with different radii are chosen.

There are many methods of calculating band structures utilizing the muffin-tin potential; we choose the augmented plane wave (APW) method as an illustrating example. The wavefunction for the wavevector $\boldsymbol{k}$ is now taken to be

$$w_{\boldsymbol{k}} = \begin{cases} \text{atomic function}, & r < r_c; \\ \Omega^{-1/2} e^{i\boldsymbol{k}\cdot\boldsymbol{r}}, & r \geq r_c. \end{cases} \qquad (12.1.16)$$

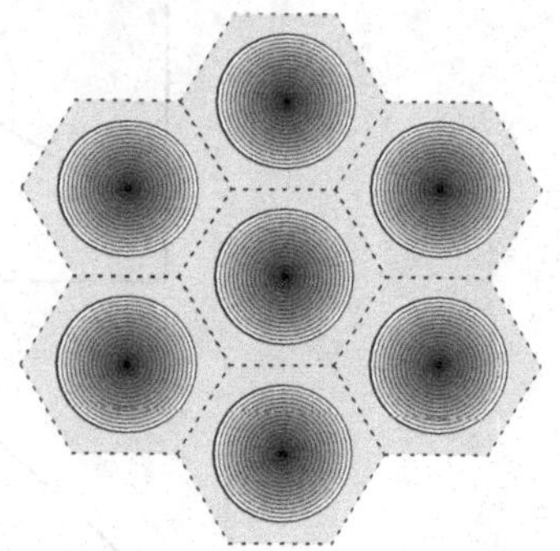

Figure 12.1.3 A schematic diagram for the muffin-tin potential.

Inside the core the function is atom-like, and is found by solving the appropriate free-atom Schrödinger equation. Outside the core the function is a plane wave because the potential is constant there. The atomic function in (12.1.16) is chosen such that it joins continuously to the plane wave at the surface of the sphere forming the core; this is the boundary condition here.

The function $w_{\boldsymbol{k}}$ does not have the Bloch form, but this can be remedied by forming the linear combination

$$\phi_{\boldsymbol{k}} = \sum_{\boldsymbol{G}} a(\boldsymbol{k}+\boldsymbol{G}) w_{\boldsymbol{k}+\boldsymbol{G}}, \qquad (12.1.17)$$

where the sum is over the reciprocal lattice vectors, which has the proper form. The coefficients $a(\boldsymbol{k}+\boldsymbol{G})$ are determined by requiring that $\phi_{\boldsymbol{k}}$ minimize the energy. An augmented plane wave consists of a plane wave $\Omega^{-1/2}\exp(i\boldsymbol{k}\cdot\boldsymbol{r})$ in the region between the spheres, and a linear combination of atomic-like wavefunctions satisfying the spherically symmetric potential within each sphere; the linear combination is chosen to match the plane wave at the surface of the sphere, so that the resultant APW wavefunction, $\phi_{\boldsymbol{k}}(\boldsymbol{r})$, is continuous across this surface. With composite wavefunctions, the APW method seems to attain the initial goal for combining both the NFE approach and the TBE approach.

The Bloch wave $\psi_{\boldsymbol{k}}(\boldsymbol{r})$ is now written as a sum of APWs

$$\psi_{\boldsymbol{k}}(\boldsymbol{r}) = \sum_{\boldsymbol{G}} c(\boldsymbol{G}) \phi_{\boldsymbol{k}+\boldsymbol{G}}(\boldsymbol{r}), \qquad (12.1.18)$$

and the coefficients $c(\boldsymbol{G})$ are chosen by a variational method to give the best solution. The equations determining these coefficients turn out to be almost identical in form to (12.1.3). But the coefficients $V(\boldsymbol{G})$ are no longer just the Fourier coefficients of $V(\boldsymbol{r})$; they are much more complicated objects. Nevertheless, relatively few of these coefficients are important, so the coupled set of equations is of manageable size and can be reasonably solved by computer. In practice, the series in (12.1.18) converges quite rapidly, and only four to seven terms (sometimes even less) suffice to give the desired accuracy. The APW is a sound and powerful method for calculating band structures in solids, especially for transition metals and rare earths.

There are many other methods for band-calculation using muffin-tin potentials, such as the Korringa–Kohn–Rostoker (KKR) method, muffin-tin orbitals (MTO) method, as well as linearized APW (LAPW) and linearized MTO (LMTO) but we shall not go into details of these here. For reference, the advanced textbook of J. Callaway (1996) is recommended.

12.1.4 The Symmetry of the Energy Bands and the $\boldsymbol{k}\cdot\boldsymbol{p}$ Method

The symmetry of crystal structures greatly simplifies band-calculations. The basic principle of symmetry of energy bands can be stated as follows: The energy function in the Brillouin zone has the full point group of the crystal; any symmetry operation, such as rotating the crystal around an axis that leaves it invariant, also transforms $\boldsymbol{k}$ into itself. Moreover, due to the fact the Hamiltonian for the one-electron Schrödinger equation is real, $E(\boldsymbol{k}) = E(-\boldsymbol{k})$. This is true for any symmetry of

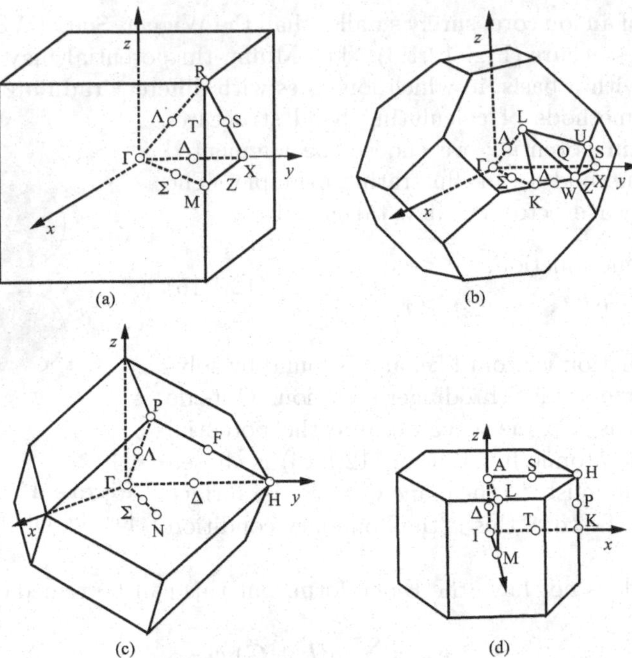

Figure 12.1.4 Symmetry points in the Brillouin zones of different real lattices: (a) sc, (b) fcc, (c) bcc, (d) sh.

the crystal, even for crystals without inversion symmetry. Actually, this arises from time reversal symmetry, and is called Kramers degeneracy.

Symmetry points and lines in the Brillouin zone for the sc, fcc, bcc, and sh structures are shown in Fig. 12.1.4; note the symbols for the symmetry points. We may obtain further information about the energy bands in the crystal by analogy with the crystal field splitting of atomic states.

We will select some special wavevectors corresponding to symmetry points in the Brillouin zone (BZ), such as the central point Γ, where $\boldsymbol{k} = 0$. We classify these states using group theory: as the wavevector moves away from this point, the group of wavevectors becomes smaller, and some degeneracies are lifted. A fair amount of information about the energy bands may be gathered from symmetry analysis alone, but to obtain the bands themselves requires detailed calculation. In general, the higher the symmetry of the point in the zone, the lower the amount of computational expenditure. So, group theory is an indispensable tool for actual calculation of the energy bands.

From the energy band calculation scheme outlined above, we can get the energy bands in the whole Brillouin zone. However, for some properties, such as the electrical properties of semiconductors, the important charge carriers (electrons and holes) are distributed in a small region around the bottom of the conduction band or near the top of the valence band. At room temperature, $k_{\rm B}T = 26$ meV, so only the band structure near the bottom of the conduction band or the top of the valence band may influence the electrical properties. Although conventional band calculations can give some information about the band structure in these regions, they are not very accurate and we need a good tool for analyzing problems and obtaining a quantitative relationship between energy E and wave vector $\boldsymbol{k}$.

The $\boldsymbol{k}\cdot\boldsymbol{p}$ perturbation method has been developed to fulfill this need. It takes the energy and wavefunction at $\boldsymbol{k} = 0$ as the zeroth approximation, and terms related to $\boldsymbol{k}$ in the Hamiltonian as the perturbation terms, and then gets the band structure near $\boldsymbol{k} = 0$ through the 1st order and 2nd order perturbations. Assuming the extrema of both the conduction and valence bands are situated at $\boldsymbol{k} = 0$, we can substitute the crystal wavefunction $e^{i\boldsymbol{k}\cdot\boldsymbol{r}}u_{n\boldsymbol{k}}(\boldsymbol{r})$ into the single-electron Schrödinger equation

$$\left[\frac{\boldsymbol{p}^2}{2m} + \frac{\hbar}{m}\boldsymbol{k}\cdot\boldsymbol{p} + \frac{\hbar^2k^2}{2m} + V(\boldsymbol{r})\right] u_{n\boldsymbol{k}}(\boldsymbol{r}) = E(\boldsymbol{k})u_{n\boldsymbol{k}}(\boldsymbol{r}), \tag{12.1.19}$$

and let $u_{n\boldsymbol{k}}(\boldsymbol{r})$ be expanded into wavefunctions of different bands at $\boldsymbol{k}=0$,

$$u_{n\boldsymbol{k}}(\boldsymbol{r})=\sum_{n'}C_{n'n}u_{n'0}(\boldsymbol{r}). \tag{12.1.20}$$

Equations for the coefficients $C_{n'n}$ can then be derived from

$$\sum_{n'}\left\{\left[E_n(0)+\frac{\hbar^2k^2}{2m}\right]\delta_{n'n}+\frac{\hbar}{m}\boldsymbol{k}\cdot\boldsymbol{p}_{n'n}\right\}C_{n'n}=E_n(\boldsymbol{k})C_{nn}, \tag{12.1.21}$$

where

$$\boldsymbol{p}_{n'n}=\int_\Omega u_{n0}^*(\boldsymbol{r})\boldsymbol{p}u_{n'0}(\boldsymbol{r})d\boldsymbol{r}. \tag{12.1.22}$$

The second term of the left hand side of (12.1.21) is the $\boldsymbol{k}\cdot\boldsymbol{p}$ perturbation term. From this analysis $E_n(\boldsymbol{k})$ can be obtained through rigorous diagonalization or perturbation.

§12.2 From Many-Particle Hamiltonian to Self-Consistent Field Approach

In spite of the enormous success of band theory in the elucidation of electronic structures of solids, the heart of band theory is the single-electron approximation which has been already used with much success in theories of many-electron atoms and polyatomic molecules. Now the problem is to place these single-electron approaches within the framework of more fundamental many-particle theory in order to seek justification and to show the limitations, as well as to hint at possible ways to improve these single-electron theories.

12.2.1 Many-Particle Hamiltonians

The Schrödinger equation for a system with more than one particle is

$$i\hbar\frac{\partial}{\partial t}\Psi=\mathcal{H}\Psi, \tag{12.2.1}$$

where $\mathcal{H}$ and Ψ are the many-particle Hamiltonian and wavefunction, respectively. For a set of noninteracting particles, the Hamiltonian of the system is

$$\mathcal{H}=\sum_i\mathcal{H}_i=\sum_i\left[-\frac{\hbar^2}{2m}\nabla_i^2+v(\boldsymbol{r}_i)\right], \tag{12.2.2}$$

where the independent-particle solution ψ_n for each $\mathcal{H}_i$ satisfies

$$\left[-\frac{\hbar^2}{2m}\nabla^2+v(\boldsymbol{r})\right]\psi_n=\varepsilon_n\psi_n. \tag{12.2.3}$$

These wavefunctions can be combined to form a total wavefunction for the system according to some prescribed rules.

Once we turn on the interaction terms, a piece of condensed matter is composed of an enormously large number (more than 10^{24}) of interacting nuclei and electrons, with equal positive and negative charge. The motion of any particle is now correlated with all other particles, and such an interacting many-particle system may be described by the following Hamiltonian

$$\mathcal{H}=\sum_i\frac{p_i^2}{2m}+\sum_\alpha\frac{P_\alpha^2}{2M_\alpha}+\frac{1}{2}\sum_{i\neq j}\frac{e^2}{|\boldsymbol{r}_i-\boldsymbol{r}_j|}+\frac{1}{2}\sum_{\alpha\neq\beta}\frac{Z_\alpha Z_\beta e^2}{|\boldsymbol{R}_\alpha-\boldsymbol{R}_\beta|}-\sum_{i\alpha}\frac{Z_\alpha e^2}{|\boldsymbol{r}_i-\boldsymbol{R}_\alpha|}, \tag{12.2.4}$$

where $\boldsymbol{r}_i$, $\boldsymbol{p}_i$, m and $-e$ are used to represent the coordinates, momenta, mass and charge of electrons, while $\boldsymbol{R}_\alpha$, $\boldsymbol{P}_\alpha$, M_α and $Z_\alpha e$ are the corresponding quantities of the nuclei. For the quantum mechanical treatment in coordinate representation, it is usual to take $\boldsymbol{p}_i\to -i\hbar\nabla_i$, and $\boldsymbol{P}_\alpha\to -i\hbar\nabla_\alpha$; here we have omitted the spin indices for brevity. In some circumstances, we should take the spins of the electrons into account, and then the magnetic properties of the system can be tackled.

12.2.2 Valence Electrons and the Adiabatic Approximations

In principle, we may say that all of condensed matter physics is contained in (12.2.4); however, it is impossible to solve the corresponding Schrödinger equation in order to extract any information. Since it is difficult to get results directly, various approximations and models were introduced with the purpose of simplifying the problem.

We consider a simple example for which there are N identical atoms in a solid, with a number of core electrons, which can be regarded as tightly-bound to the nuclei. These bound electrons are localized and generally make little contribution to the properties of the solid. However, the outer-orbital electrons can be delocalized. In fact, when the atoms are combined to form a solid, the configurations of these valence electrons vary greatly, but the core electrons do not change much. The electrical and optical properties are determined mainly by the valence electrons. Hence, we arrive at the valence electron approximation, in which each nucleus and its bounded electrons is looked upon as one ion and then the solid is considered to be composed of the valence electrons together with the ions. The valence electron approximation is correct for alkali metals, noble metals, and many other materials, but it is not always true. For instance, in the transition metals and rare earths, there is a mixed valence phenomena which will be discussed in Chap. 13; this makes the exact concept of the valence electron somewhat blurred. Often, however, the valence electron approximation is very effective.

If we further assume, on average, that there is only one valence electron for each ion in an identical particle system, then the Hamiltonian of the system can be written as

$$\mathcal{H} = -\frac{\hbar^2}{2m}\sum_i \nabla_i^2 - \frac{\hbar^2}{2M}\sum_\alpha \nabla_\alpha^2 + \frac{1}{2}\sum_{i\neq j}\frac{e^2}{|\boldsymbol{r}_i - \boldsymbol{r}_j|} + \sum_{i\alpha} v(\boldsymbol{r}_i, \boldsymbol{R}_\alpha) + \frac{1}{2}\sum_{\alpha\neq\beta} v(\boldsymbol{R}_\alpha, \boldsymbol{R}_\beta), \tag{12.2.5}$$

where $v(\boldsymbol{r}_i, \boldsymbol{R}_\alpha)$ is the shielded Coulomb potential, and $v(\boldsymbol{R}_\alpha, \boldsymbol{R}_\beta)$ the short range potential, respectively.

The concept of the adiabatic approximation comes from thermodynamics: We call a thermodynamic process an adiabatic one when the system remains in equilibrium, no heat transfer takes place and the entropy is constant. The quantum mechanical theory for such systems was developed by Born and Oppenheimer, so the adiabatic approximation is often called the Born–Oppenheimer approximation. For the electron system in a solid, the total electronic wavefunction depends on the instantaneous relative positions of the vibrating ions: The electrons move so rapidly that they adjust adiabatically to the much slower vibrations of the ions. On the other hand, as far as the ions are concerned, the rapidly moving electrons can be looked upon as a homogeneously smeared background. The direct result is, of course, that we may treat the electrons and vibrating ions as separate subsystems. It is instructive to assume that the total wavefunction can be written as a product of an electronic part and a ionic part, as implied by the adiabatic approximation. It can be shown that this wavefunction is an approximate solution to the total Hamiltonian with an error which depends on $(m/M)^{1/4}$ (m/M is the ratio between the masses of the electron and the ion), which is small.

Under the adiabatic approximation, the coupling between electrons and ions in (12.2.5) is ignored, and we only need to investigate the independent subsystems: one is the interacting ion system which is described by

$$\mathcal{H} = -\frac{\hbar^2}{2M}\sum_\alpha \nabla_\alpha^2 + \frac{1}{2}\sum_{\alpha\neq\beta} v(\boldsymbol{R}_\alpha, \boldsymbol{R}_\beta) + \sum_\alpha v_{\mathrm{e}}(\boldsymbol{R}_\alpha), \tag{12.2.6}$$

where $v_{\mathrm{e}}(\boldsymbol{R}_\alpha)$ is the contribution of electrons which may be looked as homogeneously distributed; the other is the interacting electron system with a static potential field,

$$\mathcal{H} = -\frac{\hbar^2}{2m}\sum_i \nabla_i^2 + \frac{1}{2}\sum_{i\neq j}\frac{e^2}{|\boldsymbol{r}_i - \boldsymbol{r}_j|} + \sum_i v(\boldsymbol{r}_i), \tag{12.2.7}$$

where $v(\boldsymbol{r}_i)$ is the potential provided by all ions to the ith electron. For a periodic structure, or a homogeneous structure, it is appropriate to assume that $v(\boldsymbol{r}_i)$ has the same form for all electrons.

As we can see, the electrons are assumed to be decoupled from the ionic vibrations. The rigid ionic potential field still exists and can be looked on as an external field for the electrons.

It should be noted that this decoupling may break down in some special cases, for instance the Jahn–Teller effect discussed in §11.4.3, and further examples that will be discussed in later chapters. In general, the wavefunction of a system is a one-electron function of the coordinates of all the electrons, $\Psi(\boldsymbol{r}_1, \ldots, \boldsymbol{r}_N)$. It is a well defined problem for a Schrödinger equation

$$\left[-\frac{\hbar^2}{2m}\sum_i \nabla_i^2 + \sum_i v(\boldsymbol{r}_i) + \frac{1}{2}\sum_{i\neq j}\frac{e^2}{|\boldsymbol{r}_i - \boldsymbol{r}_j|}\right]\Psi(\boldsymbol{r}_1, \ldots, \boldsymbol{r}_N) = E\Psi(\boldsymbol{r}_1, \ldots, \boldsymbol{r}_N). \tag{12.2.8}$$

This is easier than (12.2.4), but is not exactly solvable, either. The origin of the complexity arises from the Coulomb interaction, and we must adopt further approximations to simplify the interaction between electrons.

12.2.3 The Hartree Approximation

The single electron Schrödinger equation with external potential $v(\boldsymbol{r})$ is (12.2.3), where $v(\boldsymbol{r})$ can be a periodic potential. In other cases, the potential arises from different atoms in molecules or the central potential of the atomic structures. It must be justified why in some cases (12.2.8) may be reduced to (12.2.3), because there are actually many electrons interacting with each other.

Under one further approximation, it is possible to reduce the interacting many-electron system to an individual electron problem in an effective potential. This potential should be determined self-consistently by all other electrons in the system.

As a first step, neglecting its antisymmetric requirement, the total wavefunction for a system with N electrons could be written as the product of one-electron wavefunctions

$$\Psi(\boldsymbol{r}_1, \ldots, \boldsymbol{r}_N) = \prod_{i=1}^{N} \psi_i(\boldsymbol{r}_i), \tag{12.2.9}$$

for which Hartree suggested a variational calculation of

$$E = \frac{\langle\Psi|\mathcal{H}|\Psi\rangle}{\langle\Psi|\Psi\rangle}, \tag{12.2.10}$$

to minimize the energy. If Ψ were the exact ground state wavefunction of the system, then E would be the ground state energy. The variational principle states that E is stationary with respect to variation of Ψ, and is an upper bound to the ground state energy. From (12.2.9), this procedure leads to a set of Hartree equations

$$\left[-\frac{\hbar^2}{2m}\nabla^2 + v(\boldsymbol{r}) + \sum_j{}' e^2 \int \frac{\psi_j^*(\boldsymbol{r}')\psi_j(\boldsymbol{r}')d\boldsymbol{r}'}{|\boldsymbol{r} - \boldsymbol{r}'|}\right]\psi_i(\boldsymbol{r}) = \varepsilon_i \psi_i(\boldsymbol{r}), \tag{12.2.11}$$

where the prime is used to rule out the possibility of $j = i$, and ε_i are variational parameters, which look like the one-electron energy eigenvalues. We can now define an effective potential as

$$v_{\text{eff}} = v(\boldsymbol{r}) + \sum_j{}' e^2 \int \frac{\psi_j^*(\boldsymbol{r}')\psi_j(\boldsymbol{r}')d\boldsymbol{r}'}{|\boldsymbol{r} - \boldsymbol{r}'|}, \tag{12.2.12}$$

and so (12.2.11) is equivalent to (12.2.3). The potential seen by each electron is then determined from the average distribution $\sum_j \psi_j^*(\boldsymbol{r}')\psi_j(\boldsymbol{r}')$ of all the other electrons. We must notice that the ε_i are not truly one-electron energies: It is easy to illustrate this from

$$E = \frac{\langle\Psi|\mathcal{H}|\Psi\rangle}{\langle\Psi|\Psi\rangle} = \sum_i \varepsilon_i - \frac{1}{2}\sum_{i\neq j} e^2 \iint \frac{\psi_j^*(\boldsymbol{r}')\psi_j(\boldsymbol{r}')\psi_i^*(\boldsymbol{r})\psi_i(\boldsymbol{r})}{|\boldsymbol{r} - \boldsymbol{r}'|} d\boldsymbol{r}d\boldsymbol{r}'. \tag{12.2.13}$$

The Hartree equations (12.2.11) can be solved self-consistently by iteration. When we take this self-consistent field approximation, it includes a sum of terms, each of which depends on the coordinates of a single electron. A self-consistent calculation must be made, since the states themselves must be known in order to compute the interaction potential, which must, in turn, be known in order that we may compute the states. First we assume a particular set of approximate eigenstates, compute the effective potential and then recalculate the eigenstates repeatedly. This procedure leads to a set of states consistent with the potential.

Now we shall apply the Hartree approximation to the jellium solid, in which the positive ion cores are smeared out to form a continuum of positive charges just to neutralize the negative charges of valence electrons. Since everything is homogeneous, we can take a set of plane-wave states as the first approximation for the ground state

$$\psi_i(\boldsymbol{r}) = \frac{1}{\sqrt{\Omega}}\mathrm{e}^{i\boldsymbol{k}\cdot\boldsymbol{r}}, \tag{12.2.14}$$

normalized to the volume of this system. Each $\boldsymbol{k}$ is doubly occupied up to the Fermi wave vector and the Hartree potential is

$$v(\boldsymbol{r}) = \frac{2e^2}{\Omega}\sum_{k=0}^{k_\mathrm{F}}\int\frac{d\boldsymbol{r}'}{|\boldsymbol{r}-\boldsymbol{r}'|} - \frac{e^2}{\Omega}\int\frac{d\boldsymbol{r}'}{|\boldsymbol{r}-\boldsymbol{r}'|}. \tag{12.2.15}$$

The first term is cancelled by the positive background, because the system is neutral, while the second is the self-interaction term. If the point $\boldsymbol{r}$ is selected as the origin, then

$$\frac{e^2}{\Omega}\int\frac{d\boldsymbol{r}'}{r'} = \frac{4\pi e^2}{\Omega}\int r'dr'. \tag{12.2.16}$$

This integral scales linearly with the average of the total system and goes to zero in the limit of a large system. Thus the Hartree equation for the jellium solid is reduced to

$$\left[\frac{\hbar^2}{2m}\nabla^2 + E(\boldsymbol{k})\right]\psi_{\boldsymbol{k}}(\boldsymbol{r}) = 0. \tag{12.2.17}$$

This has plane wave solutions, and is described exactly by a complete set of single particle wavefunctions, just as for the noninteracting electron gas. This is simply the Sommerfeld model for free electrons in a solid, deduced with the help of the self-consistent field. The Hartree approximation also can be applied to real solids by replacing the original electron-ion potential by a new one-electron potential evaluated self-consistently. However, to obtain more realistic interaction effect we must go beyond it to introduce the Hartree–Fock approximation.

12.2.4 The Hartree–Fock Approximation

Because electrons are fermions, the Pauli principle must be considered. It is reasonable to take a linear combination of product wavefunctions to satisfy this antisymmetry condition and express it as a Slater determinant

$$\Psi(\{\boldsymbol{r}_i\}) = \frac{1}{\sqrt{N!}}\begin{vmatrix}\psi_1(\boldsymbol{r}_1) & \cdots & \psi_1(\boldsymbol{r}_N)\\ \cdots & \cdots & \cdots\\ \psi_N(\boldsymbol{r}_1) & \cdots & \psi_N(\boldsymbol{r}_N)\end{vmatrix}. \tag{12.2.18}$$

This is Hartree–Fock (HF) approximation. Variational calculation putting (12.2.18) into (12.2.10) leads to a set of Hartree–Fock equations:

$$\left[-\frac{\hbar^2}{2m}\nabla^2 + v(\boldsymbol{r}) + \sum_j{}' e^2\int\frac{\psi_j^*(\boldsymbol{r}')\psi_j(\boldsymbol{r}')}{|\boldsymbol{r}-\boldsymbol{r}'|}d\boldsymbol{r}'\right]\psi_i(\boldsymbol{r}) - \sum_j{}'\left[e^2\int\frac{\psi_j^*(\boldsymbol{r}')\psi_i(\boldsymbol{r}')}{|\boldsymbol{r}-\boldsymbol{r}'|}d\boldsymbol{r}'\right]\psi_j(\boldsymbol{r}) = \varepsilon_i\psi_i(\boldsymbol{r}). \tag{12.2.19}$$

The construction of an effective potential like (12.2.12) is trivial, but these equations should also be solved self-consistently.

We may obtain a value for the total energy in the Hartree–Fock approximation and this will again contain a correction to the simple sum over the parameters ε_i. The extra term is the exchange interaction

$$\frac{e^2}{2}\sum_{i\neq j}\iint\frac{\psi_i^*(\boldsymbol{r})\psi_j(\boldsymbol{r})\psi_j^*(\boldsymbol{r}')\psi_i(\boldsymbol{r}')}{|\boldsymbol{r}-\boldsymbol{r}'|}d\boldsymbol{r}d\boldsymbol{r}', \tag{12.2.20}$$

as distinguished from the direct interaction, which is also present in the Hartree approximation. The exchange interaction arises from every pair of parallel electrons.

In the application of HF equations, it is usually assumed that the spatial part of the wavefunction is the same for spin-up and spin-down electrons, i.e., every orbital is doubly occupied, and the wavefunctions of the Slater determinant are spin singlets. This is the so-called restricted Hartree–Fock (HF) method, and can be reasonably used in many problems not involving magnetism. In magnetic problems the HF equations are necessarily different. So the two sets of functions, for spin-up and spin-down, need not be identical or orthogonal; this is the unrestricted HF method. Certainly the solution of the unrestricted HF will be more laborious than the restricted HF.

Now we can also apply the Hartree–Fock (HF) approximation to the jellium solid. We can still assume plane waves as the starting wave functions. Just like the Sommerfeld solid, the other interaction terms cancel, and only the exchange interaction term remains. For a large volume Ω, the sum over $\boldsymbol{k}$ can be changed into an integral over the Fermi sphere, i.e., $(1/\Omega)\sum_{\boldsymbol{k}}=\int g(\boldsymbol{k})d\boldsymbol{k}$. Then by Fourier transform of $1/r$,

$$\int\frac{e^{i\boldsymbol{k}\cdot\boldsymbol{r}}}{r}d\boldsymbol{r}=\frac{4\pi}{k^2}. \tag{12.2.21}$$

Since the kinetic energy of the state $\boldsymbol{k}$ is $\hbar^2k^2/2m$, the exchange term may be evaluated, and we get

$$E(\boldsymbol{k})=\frac{\hbar^2k^2}{2m}-\frac{e^2}{2\pi^2}\int_0^{k_F}\frac{d\boldsymbol{k}'}{(\boldsymbol{k}'-\boldsymbol{k})^2}=\frac{\hbar^2k^2}{2m}-\frac{e^2}{\pi}k_F\left[1+\frac{(k_F^2-k^2)}{2k_Fk}\ln\left|\frac{k_F+k}{k_F-k}\right|\right]. \tag{12.2.22}$$

Because of the $1/r$ factor in the exchange integral, the total energy is $\sim n^{1/3}$, where n is the electron density. It seems that if the electron density varies slowly in space, the exchange energy can be calculated as a volume integral of this $n^{1/3}$ exactly, and the exchange term in the Hartree–Fock equations can be replaced by a potential proportional to $n^{1/3}(\boldsymbol{r})$.

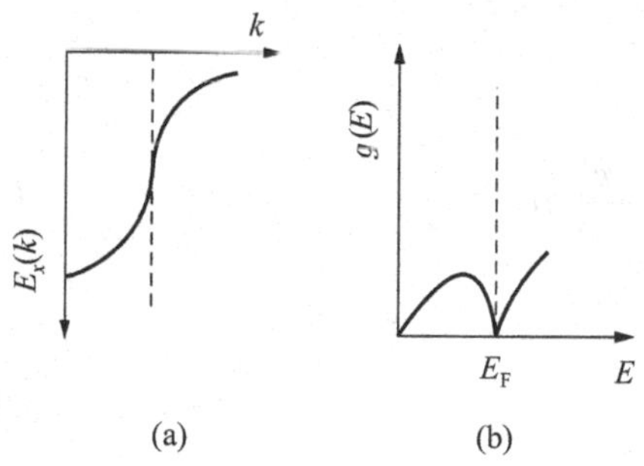

Figure 12.2.1 (a) E_x versus k and (b) $g(E)$ versus E of the jellium solid in the HF approximation.

Figure 12.2.1 shows the relationship of the exchange energy E_x versus k and density of states $g(E)$ versus E. This may be regarded as a smeared out Fermi distribution, with the infinite slope corresponding to the sharp drop at the Fermi surface, i.e., $(\partial E/\partial k)_{E_F}=\infty$, correspondingly, $g(E_F)\propto(\partial E/\partial k)_{E_F}^{-1}=0$, which makes the problem difficult. It should be noted that HF does not improve the Sommerfeld model for solids: the density of states at the Fermi level vanishes and metallic properties which depend on the density of states near the Fermi level are changed for the worse because the qualitative agreement of the Sommerfeld model is destroyed by the introduction of the exchange term. It also reflects the long-range nature of the Coulomb potential, for the Fourier transform of $1/r$ is $4\pi/k^2$, which diverges at $k=0$. But if we use the screened Coulomb potential $e^2\cdot e^{-k_0r}/r$ instead of the bare Coulomb potential, the corresponding Fourier transform is $4\pi e^2/(k^2+k_0^2)$, which eliminates the divergence at $k=0$. We can get a similar result to (12.2.22) by using the electronic screened effect, but k_F should be changed into $2k_F$. In this way the non-physical result on the Fermi surface is eliminated. When the HF approximation is used to calculate the electronic structure of molecules or solids, the potential is only concerned with the scale of the molecules or unit cells, so the difficulty will not arise. However the correction due to electron correlation should be retained.

The HF approximation has been widely used in the MO methods of quantum chemistry as well as in band calculations in periodic solids. In order to identify the one-electron equation in (12.2.3), we need a more direct relation between the parameters ε_i and the energies of interest in a crystal. It is instructive to calculate the total energy $E(N)$ and $E(N-1)$ using a Slater determinant by assuming that the individual one-electron functions are the same if the total number of the particles changes by one. Koopmans showed that the difference in these two total energy $\Delta E = E(N) - E(N-1)$ will simply be the parameter ε_i for the state that has been omitted. The conclusion is that the ionization energy of the crystal with respect to any given electron state is simply the Hartree–Fock parameter ε_i. Thus Koopmans theorem allows us to regard the calculated energy as one-electron energy eigenvalues. This kind of one-electron picture could be used in simple metals, but may break down in some cases such as transition or rare-earth metals.

§12.3 Electronic Structure via Density Functionals

12.3.1 From Wavefunctions to Density Functionals

Up to now we have utilized wavefunctions as basic variables which satisfy the Schrödinger equation, especially in the Hartree–Fock approximation using self-consistent field, in order to solve for the electronic structure of molecules and solids. From the wavefunction Ψ, we get the electron density distribution $n(\boldsymbol{r})$ in molecules and solids.

Is it possible to regard $n(\boldsymbol{r})$ as a basic variable in the calculation of electronic structure of solids and molecules? The answer is yes. Here we follow, essentially, the chain of reasoning in the illuminating account given in W. Kohn's Nobel lecture.[a]

In 1927, the Thomas–Fermi (TF) theory actually incorporated this concept to give a simplified treatment of the electronic structure of solids. It supposes electrons to be moving in an external potential $v_{\text{eff}}(\boldsymbol{r})$, and introduces a one-to-one implicit relation between $v_{\text{eff}}(\boldsymbol{r})$ and $n(\boldsymbol{r})$:

$$n(\boldsymbol{r}) = \gamma[\mu - v_{\text{eff}}(\boldsymbol{r})]^{3/2}, \tag{12.3.1}$$

with

$$\gamma = \frac{1}{3\pi^2}\left(\frac{2m}{\hbar^2}\right)^{3/2},$$

and

$$v_{\text{eff}}(\boldsymbol{r}) = v(\boldsymbol{r}) + e^2\int \frac{n(\boldsymbol{r}')}{|\boldsymbol{r}-\boldsymbol{r}'|}d\boldsymbol{r}'. \tag{12.3.2}$$

μ in (12.3.1) is the chemical potential, independent of $\boldsymbol{r}$. (12.3.1) is based on

$$n = \gamma(\mu - v)^{3/2}, \tag{12.3.3}$$

the density of a uniform degenerate electron gas in a constant external potential v, where the second term in (12.3.2) is just the classical electrostatic potential times (-1), generated by the electron density distribution $n(\boldsymbol{r})$.

However, TF theory is only useful for describing some qualitative trends such as total energies; it is too crude to account for chemical binding, so its application to the electronic structure of solids is quite limited. But the intimate connection of $n(\boldsymbol{r})$ and $v(\boldsymbol{r})$ in TF theory suggested that a knowledge of the ground-state electronic density $n(\boldsymbol{r})$ for any electronic system (with or without interactions) uniquely determines the system. If we can prove this, then a new framework for electronic structure calculation based on $n(\boldsymbol{r})$ may be built. This is the starting point for the formulation of density functional theory (DFT).

[a] W. Kohn, *Rev. Mod. Phys.* **71**, 1253 (1999).

12.3.2 Hohenberg–Kohn Theorems

Two basic theorems proved by Hohenberg and Kohn laid the foundation of density functional theory (DFT).

Theorem 1. The ground-state density $n(\boldsymbol{r})$ of a bound system of interacting electrons in some external potential $v(\boldsymbol{r})$ determines this potential uniquely. The term 'uniquely' means here: up to an uninteresting additive constant.

The proof: let $n(\boldsymbol{r})$ be the nondegenerate ground-state density of N electrons in the potential $v_1(\boldsymbol{r})$ and the ground state wavefunction be Ψ_1 with the energy E_1. We have

$$\begin{aligned} E_1 &= \langle\Psi_1|\mathcal{H}_1|\Psi_1\rangle \\ &= \int v_1(\boldsymbol{r})n(\boldsymbol{r})d\boldsymbol{r} + \langle\Psi_1|T+U|\Psi_1\rangle, \end{aligned} \tag{12.3.4}$$

where $\mathcal{H}_1$ is the total Hamiltonian corresponding to v_1, T and U are the kinetic and interaction energy operators. Now we assume that there is another $v_2(\boldsymbol{r})$, not equal to $v_1(\boldsymbol{r})$+constant, with ground state wavefunction Ψ_2, which is not equal to Ψ_1, but gives the same $n(\boldsymbol{r})$, then

$$E_2 = \int v_2(\boldsymbol{r})n(\boldsymbol{r})d\boldsymbol{r} + \langle\Psi_2|T+U|\Psi_2\rangle. \tag{12.3.5}$$

Since E is nondegenerate, the Rayleigh–Ritz variation principle gives two inequalities

$$\begin{aligned} E_1 &< \langle\Psi_2|\mathcal{H}_1|\Psi_2\rangle \\ &= \int v_1(\boldsymbol{r})n(\boldsymbol{r})d\boldsymbol{r} + \langle\Psi_2|T+U|\Psi_2\rangle \\ &= E_2 + \int [v_1(\boldsymbol{r}) - v_2(\boldsymbol{r})]n(\boldsymbol{r})d\boldsymbol{r}, \end{aligned} \tag{12.3.6}$$

and

$$E_2 < \langle\Psi_1|\mathcal{H}_2|\Psi_1\rangle = E_1 + \int [v_2(\boldsymbol{r}) - v_1(\boldsymbol{r})]n(\boldsymbol{r})d\boldsymbol{r}. \tag{12.3.7}$$

In the second inequality we use (12.3.4) since nondegeneracy is not assumed. Adding together these inequalities we get an absurd result

$$E_1 + E_2 < E_1 + E_2. \tag{12.3.8}$$

So we conclude that the assumption of the existence of a second potential $v_2(\boldsymbol{r}) \neq v_1(\boldsymbol{r})$ + constant, which gives the same $n(\boldsymbol{r})$, must be wrong. Now we can assert: since $n(\boldsymbol{r})$ uniquely determines both N and $v(\boldsymbol{r})$, it also determines implicitly all properties derivable from $\mathcal{H}$ through the solutions of the time-independent or time-dependent Schrödinger equations, including many-body effects.

Theorem 2. The ground-state energy E can be obtained through the variation of trial densities $\tilde{n}(\boldsymbol{r})$ instead of the trial wavefunctions $\tilde{\Psi}$. This is called the Hohenberg–Kohn variational principle which can be derived from the Raleigh–Ritz variational principle,

$$E = \min_{\tilde{\Psi}}\langle\tilde{\Psi}|\mathcal{H}|\tilde{\Psi}\rangle. \tag{12.3.9}$$

Here we follow the more simple derivation using the constrained search due to Levy (1982).[b] Every trial wavefunction $\tilde{\Psi}$ corresponds to a trial density $\tilde{n}(\boldsymbol{r})$ obtained by integrating over all variables, except the first, and multiplying by N. The minimization of (12.3.9) may be carried out

[b]M. Levy, *Phys. Rev. A* **26**, 1200 (1982).

in two stages. First we fix a trial $\tilde{n}(\boldsymbol{r})$ and denote the trial functions $\tilde{\Psi}^{\alpha}_{\tilde{n}(\boldsymbol{r})}$. With fixed $\tilde{n}(\boldsymbol{r})$, the constrained energy minimum is defined as

$$\begin{aligned} E_v[\tilde{n}(\boldsymbol{r})] &= \min_{\alpha}\langle\tilde{\Psi}^{\alpha}_{\tilde{n}}|\mathcal{H}|\tilde{\Psi}^{\alpha}_{\tilde{n}}\rangle \\ &= \int v(\boldsymbol{r})\tilde{n}(\boldsymbol{r})d\boldsymbol{r} + F[\tilde{n}(\boldsymbol{r})], \end{aligned} \tag{12.3.10}$$

where

$$F[\tilde{n}(\boldsymbol{r})] = \min_{\alpha}\langle\tilde{\Psi}^{\alpha}_{\tilde{n}(\boldsymbol{r})}|\mathcal{T}+U|\tilde{\Psi}^{\alpha}_{\tilde{n}(\boldsymbol{r})}\rangle, \tag{12.3.11}$$

$F[\tilde{n}(\boldsymbol{r})]$ is an universal functional of the density $\tilde{n}(\boldsymbol{r})$ which requires no explicit knowledge of $v(\boldsymbol{r})$. In the second step we minimize (12.3.10) over all n,

$$E = \min_{\tilde{n}(\boldsymbol{r})} E_v[\tilde{n}(\boldsymbol{r})] = \min_{\tilde{n}(\boldsymbol{r})}\left\{\int v(\boldsymbol{r})\tilde{n}(\boldsymbol{r})d\boldsymbol{r} + F[\tilde{n}(\boldsymbol{r})]\right\}. \tag{12.3.12}$$

For a nondegenerate ground state, the minimum is for the ground-state density $\tilde{n}(\boldsymbol{r})$; for degenerate ground states, it is one of the ground-state densities. As Kohn remarked, the HK minimum principle may be regarded as the formal proof of Thomas–Fermi theory. Thus the formidable problem of finding the minimum of $\langle\tilde{\Psi}|\mathcal{H}|\tilde{\Psi}\rangle$ with respect to the $3N$-dimensional trial function $\tilde{\Psi}$ has been transformed into the much easier problem of finding the minimum of $E_v[\tilde{n}(\boldsymbol{r})]$ with respect to the three-dimensional trial function $\tilde{n}(\boldsymbol{r})$.

12.3.3 The Self-Consistent Kohn–Sham Equations

The Hohenberg–Kohn theorem leads to a set of effective Schrödinger equations for single-particle functions.[c] The $F[n]$ can be separated into three parts

$$F[n] = \mathcal{T}[n] + \frac{e^2}{2}\iint\frac{n(\boldsymbol{r})n(\boldsymbol{r}')}{|\boldsymbol{r}-\boldsymbol{r}'|}d\boldsymbol{r}d\boldsymbol{r}' + E_{\mathrm{xc}}[n], \tag{12.3.13}$$

where the first term represents the kinetic energy, the second the ordinary Coulomb energy and the third the exchange and correlation energies.

We do not know exactly $\mathcal{T}[n]$ and $E_{\mathrm{xc}}[n]$, but we can bypass this problem for a moment. By varying the total energy, and adding the condition that the number of electrons must remain constant, i.e., $\int\delta n(\boldsymbol{r})d\boldsymbol{r} = 0$, we have

$$\int\delta n(\boldsymbol{r})\left[\frac{\delta\mathcal{T}[n]}{\delta n(\boldsymbol{r})} + v(\boldsymbol{r}) + e^2\int\frac{n(\boldsymbol{r}')}{|\boldsymbol{r}-\boldsymbol{r}'|}d\boldsymbol{r}' + \frac{\delta E_{\mathrm{xc}}[n]}{\delta n(\boldsymbol{r})} - \mu\right]d\boldsymbol{r} = 0, \tag{12.3.14}$$

where μ, coming in as a Lagrange multiplier, is a constant corresponding to the chemical potential. Then (12.3.14) gives

$$\frac{\delta\mathcal{T}[n]}{\delta n(\boldsymbol{r})} + v(\boldsymbol{r}) + e^2\int\frac{n(\boldsymbol{r}')}{|\boldsymbol{r}-\boldsymbol{r}'|}d\boldsymbol{r}' + \frac{\delta E_{\mathrm{xc}}[n]}{\delta n(\boldsymbol{r})} = \mu. \tag{12.3.15}$$

Thus, we can define an effective potential

$$v_{\mathrm{eff}}(\boldsymbol{r}) = v(\boldsymbol{r}) + e^2\int\frac{n(\boldsymbol{r}')}{|\boldsymbol{r}-\boldsymbol{r}'|}d\boldsymbol{r}' + v_{\mathrm{xc}}(\boldsymbol{r}), \tag{12.3.16}$$

in which the exchange-correlation potential is

$$v_{\mathrm{xc}}(\boldsymbol{r}) = \delta E_{\mathrm{xc}}[n]/\delta n(\boldsymbol{r}). \tag{12.3.17}$$

[c] W. Kohn and L. J. Sham, *Phys. Rev.* **A140**, 1133 (1965).

We now consider a collection of noninteracting electrons and treat the kinetic energy items by supposing that the system of N electrons has a set of N single-electron functions, so

$$n(\boldsymbol{r}) = \sum_{i=1}^{N} |\psi_i(\boldsymbol{r})|^2. \tag{12.3.18}$$

The kinetic energy functional can be written as

$$\mathcal{T}_{\rm s}[n] = \frac{\hbar^2}{2m}\sum_{i=1}^{N}\int \nabla\psi_i^*(\boldsymbol{r})\cdot\nabla\psi_i(\boldsymbol{r})d\boldsymbol{r} = \frac{\hbar^2}{2m}\sum_{i=1}^{N}\int \psi_i^*(\boldsymbol{r})\cdot(-\nabla^2)\psi_i(\boldsymbol{r})d\boldsymbol{r}. \tag{12.3.19}$$

It is believed that (12.3.19) is a proper approximation, although there has been no proof that $\mathcal{T}_{\rm s}[n]$ holds for exact $\mathcal{T}[n]$. Formally, this complication can be ignored, and the difference between $\mathcal{T}[n]$ and $\mathcal{T}_{\rm s}[n]$ can be absorbed into $E_{\rm xc}[n]$. The result is an eigenequation

$$\left[-\frac{\hbar^2}{2m}\nabla^2 + v_{\rm eff}(\boldsymbol{r})\right]\psi_i(\boldsymbol{r}) = \varepsilon_i\psi_i(\boldsymbol{r}), \tag{12.3.20}$$

with the effective potential defined in (12.3.16). This is a Hartree-like equation for the one-electron function $\psi_i(\boldsymbol{r})$, with $v_{\rm eff}$ and $n(\boldsymbol{r})$ defined by Eqs. (12.3.16) and (12.3.18). These self-consistent equations are called Kohn–Sham (KS) equations. The ground-state energy may be constructed from the solution of (12.3.20),

$$E_{\rm G} = \sum_{i=1}^{N}\varepsilon_i - \frac{e^2}{2}\int\frac{n(\boldsymbol{r})n(\boldsymbol{r}')}{|\boldsymbol{r}-\boldsymbol{r}'|}d\boldsymbol{r}d\boldsymbol{r}' + E_{\rm xc}[n] - \int n(\boldsymbol{r})v_{\rm xc}(\boldsymbol{r})d\boldsymbol{r}. \tag{12.3.21}$$

As Kohn also remarked, KS theory may be regarded as the formal demonstration of Hartree theory. In principle, all many-body effects are included in the exact $E_{\rm xc}$ and $v_{\rm xc}$. The practical usefulness of DFT depends entirely on whether approximations for the functional $E_{\rm xc}[\tilde{n}(\boldsymbol{r})]$ can be found with sufficient simplicity as well as accuracy. On the other hand, if the physical density $n(\boldsymbol{r})$ is independently known, either directly from experiment or derived theoretically from wavefunction-based accurate calculations for small systems, $v_{\rm eff}(\boldsymbol{r})$ and hence $v_{\rm xc}(\boldsymbol{r})$ can be obtained.[d]

12.3.4 Local Density Approximation and Beyond

Now the formal framework for DFT has been described, the next problem facing us is how to use it to solve physical problems. Some sort of approximation for $E_{\rm xc}[n(\boldsymbol{r})]$ must be adopted; the viewpoint of 'nearsightedness' is important here.

In 1996 Kohn proved the following principle: the local static physical properties of a many-electron system at $\boldsymbol{r}$ are dependent on the particles in the neighborhood of $\boldsymbol{r}$ (for example, within the sphere of radius $\sim\lambda_{\rm F}(\boldsymbol{r})$, the local Fermi wavelength $\lambda_{\rm F}(\boldsymbol{r}) \equiv [3\pi^2 n(\boldsymbol{r})]^{-1/3}$) and are insensitive to any change of potential outside this region. So approximations should be of a local, or quasi-local, nature.[e] This gives a delayed justification for the success of the simplest approximation, known as the local density approximation (LDA) proposed by Kohn and Sham.

Consider a system whose density varies slowly. Then a good approximation is to write $E_{\rm xc}[n]$ in the local form

$$E_{\rm xc}[n] \approx \int \varepsilon_{\rm xc}[n(\boldsymbol{r})]d\boldsymbol{r}. \tag{12.3.22}$$

This local form leads to

$$v_{\rm xc}[n(\boldsymbol{r})] \approx \frac{d\varepsilon_{\rm xc}[n(\boldsymbol{r})]}{dn(\boldsymbol{r})} \equiv \mu_{\rm xc}[n(\boldsymbol{r})], \tag{12.3.23}$$

---

[d]W. Wang and B. G. Parr, *Density Functional Theory of Atoms and Molecules*, Oxford University Press, Oxford, 1989.

[e]W. Kohn, *Phys. Rev. Lett.* **76**, 3168 (1996).

where $\mu_{\rm xc}[n(\boldsymbol{r})]$ is the exchange-correlation chemical potential of the uniform electron gas with its density equal to the local density $n(\boldsymbol{r})$. A particularly simple, albeit somewhat inaccurate, form for $\varepsilon_{\rm xc}[n(\boldsymbol{r})]$ is obtained by neglecting correlation at all in the uniform electron gas, as will be discussed in the next section. This produces the exchange energy

$$\varepsilon_{\rm xc}[n(\boldsymbol{r})] \approx -\frac{3e^2}{2\pi}(3\pi^2 n(\boldsymbol{r}))^{1/3} n(\boldsymbol{r}) \tag{12.3.24}$$

and the potential

$$v_{\rm xc}[n(\boldsymbol{r})] \approx -2e^2\left(\frac{3}{\pi}\right)^{1/3} n^{1/3}(\boldsymbol{r}). \tag{12.3.25}$$

Incorporating correlation into the problem can produce significant improvement.

The LDA, obviously exact for a uniform electron gas, is expected to be useful only for densities varying slowly on the scale of the local Fermi wavelength. This condition is rarely satisfied for atomic systems. However, the LDA has been found to be extremely useful for the solid state, especially for complex structures. It can also be extended to local spin density approximation (LSDA) in order to treat magnetic systems with unpaired spins.

Based on DFT (mostly LDA), in conjunction with band calculation methods, such as pseudopotential, APW, LMTO, etc., *ab initio* calculations for the energy bands have been developed with spectacular success, especially in tackling complex structures for energy bands in solid state physics and materials science. As for the treatment of valence bonds in chemistry, the superiority of DFT over methods based on wavefunctions, such as HF, becomes more apparent with the passage of time. The amount of computation depends crucially on the number of atoms N without any symmetry in the basic unit: in the case of HF, it depends exponentially on N, a rough estimate is about p^{3N} (with $p \sim 3$–10); in the case of standard methods of DFT, it scales with N^3, and new methods such as order-N or $\mathrm{O}(N)$ algorithm within the framework of DFT, which will scale linearly with N, are being vigorously investigated and developed. So DFT is now, by general consensus, the pillar of scientific computation in the physical sciences.

However, the original scheme of DFT was designed for the calculation of ground states, and modified schemes were needed to treat the excited states. So the quasiparticle (QP) approximation was developed, and further single-particle Green's function and dynamic screened Coulomb interaction (GW) approximations were also developed.

Although DFT has proved to be extremely successful in various applications, the way it is constructed precludes detailed insights into the electron-correlation problem contained in the pair-distribution function $g(\boldsymbol{r},\boldsymbol{r}')$ defined by

$$g(\boldsymbol{r},\boldsymbol{r}') = \frac{1}{n(\boldsymbol{r})n(\boldsymbol{r}')}\left\langle \Phi \left| \sum_{i\neq j} \delta(\boldsymbol{r}'-\boldsymbol{r}_i)\delta(\boldsymbol{r}-\boldsymbol{r}_j) \right| \Phi \right\rangle. \tag{12.3.26}$$

Here $|\Phi\rangle$ is the ground state wavefunction of a system, $\boldsymbol{r}$ and $\boldsymbol{r}'$ are coordinates of the electrons in the system, and $n(\boldsymbol{r})$ and $n(\boldsymbol{r}')$ are the corresponding electron densities. This function describes the change in the probability of finding an electron at point $\boldsymbol{r}'$ due to the presence of one at point $\boldsymbol{r}$, and gives a suitable description for electron-correlation due to Coulombic repulsion, i.e., the correlation hole. The pair-distribution function is partially incorporated into the exchange-correlation energy $E_{\rm xc}(n)$ in DFT, i.e.,

$$E_{\rm xc}(n) = \frac{e^2}{2}\iint d\boldsymbol{r}d\boldsymbol{r}' n(\boldsymbol{r})\frac{\tilde{g}(\boldsymbol{r},\boldsymbol{r}')-1}{|\boldsymbol{r}-\boldsymbol{r}'|}n(\boldsymbol{r}'), \tag{12.3.27}$$

where $\tilde{g}(\boldsymbol{r},\boldsymbol{r}')$ is related to $g(\boldsymbol{r},\boldsymbol{r}')$ via

$$\tilde{g}(\boldsymbol{r},\boldsymbol{r}') = \int_0^1 d\lambda g(\boldsymbol{r},\boldsymbol{r}',\lambda). \tag{12.3.28}$$

The parameter λ $(0 \le \lambda \le 1)$ modifies the Coulomb interaction to a fictitious one

$$\frac{e^2}{|\boldsymbol{r}-\boldsymbol{r}'|} \to \frac{\lambda e^2}{|\boldsymbol{r}-\boldsymbol{r}'|}, \quad v(\boldsymbol{r}) \to v_\lambda(\boldsymbol{r}). \tag{12.3.29}$$

The simplicity of DFT, or comparable approximations to DFT, is the result of making a reasonable assumption for $g(\boldsymbol{r}, \boldsymbol{r}')$ instead of calculating it from first principles.

Though, in principle, the many-body effects are included in DFT, in practice DFT has been employed with success only for systems with weakly or medium correlated electrons. For strongly correlated electronic systems, such as Mott insulators, doped Mott insulators and heavy electron metals, the results of DFT are still not coming out all correctly. So, in the next chapter, when we are concerned with systems with strongly correlated electrons, we will start our theoretical discussion by introducing some model Hamiltonians. How to combine the insights gained by model Hamiltonians with suitable modifications of DFT remains one of the crucial problems for current theoretical research, and we shall face this subject in Chap. 13.

12.3.5 Car–Parrinello Method

Molecular dynamics (MD) based on Newtonian equations of motion of classical mechanics is a very effective method for the computer simulation of equilibrium and nonequilibrium structures on the atomic scale. In general, this type of calculation utilizes atomic potentials, either empirically or theoretically derived. For systems of inert atoms, Lennard–Jones potentials are used; for metallic systems, the glue potential of embedded atoms are used; for covalent crystals, Stillinger–Weber potentials are used. These all-classical calculations may be carried out in comparatively large systems, so they have very important applications in computer simulations from the various physical processes occurring in clusters and crystalline lattices, including lattice dynamics. However, some subtle quantum mechanical effects may be lost in such simulations.

Since great success in the calculation of the electronic structures of solids has been achieved by DFT, an important problem facing the computation community is to connect MD with the DFT method. This goal was realized with the development of the Car–Parrinello method in 1985.[f]

Now we shall return to the many-particle Hamiltonian of mixed system of electrons and ions. Using the Born–Oppenheimer approximation, since they are more massive and moving slowly, the ions may be approximated as classical particles obeying Newtonian mechanics, and the electronic states may be described by KS equations. According to DFT, for a definite configuration of ions $\{R_i\}$, the ground-state total energy $E(\{R_i\})$ for a system of interacting electrons is a functional of the electron density $n(\boldsymbol{r})$. If the configuration of ions changes, then $E(\{R_i\})$ becomes a potential surface, called the Born–Oppenheimer potential surface. The ground state energy ε can be found by minimization of the energy functional with respect to the electronic degree of freedom E_i. The key point of the Car–Parrinello method is to introduce a fictitious system of electron dynamics into the real physical problem in which we are interested.

An electronic state can be expressed by a set of occupied orbitals φ_i $(i = 1, \ldots, n)$. The generalized classical Lagrangian of this fictitious system can be expressed as

$$L = \mathcal{T} - V, \tag{12.3.30}$$

where $\mathcal{T}$ is the kinetic energy and V the potential energy. The kinetic energy is given by

$$\mathcal{T} = \frac{1}{2}\sum_l M_l \dot{R}_l^2 + \frac{1}{2}\sum \mu_i |\dot{\varphi}_i|^2, \tag{12.3.31}$$

where the first term represents the real ionic kinetic energy ($\dot{R} = dR/dt$), and the second term is a fictitious kinetic energy associated with the 'velocities' of the electronic orbitals defined as

$$\dot{\varphi}_i = \frac{d\varphi_i}{dt}. \tag{12.3.32}$$

In (12.3.31), μ is a fictitious inertial mass assigned to the 'motion' of φ_i orbitals through the Born–Oppenheimer potential surface. The potential energy V includes both electronic energy and purely ionic contributions, and thus it is a function of both the ionic positions and electronic orbitals, $V = V(\{R_l\}, \{\varphi_i\})$.

[f] R. Car and M. Parrinello, *Phys. Rev. Lett.* **55**, 2471 (1985).

The equation of motion associated with the Lagrangian is

$$\frac{d}{dt}\frac{\partial L}{\partial \dot{\varphi}_i^*} - \frac{\partial L}{\partial \varphi_i^*} = 0. \tag{12.3.33}$$

The electronic orbitals must satisfy the constraint of orthonormality, i.e.,

$$\sigma_{ij} = \frac{1}{\Omega}\int_\Omega \varphi_i(r)\varphi_j(r)dr - \delta_{ij} = 0, \tag{12.3.34}$$

where Ω is the volume and δ_{ij} equals 1 (when $i = j$), or 0 (when $i \neq j$). Such constraint leads to additional constraint 'forces', and a Lagrange multiplier should be introduced. Car and Parrinello wrote the Lagrangian as,

$$L = \frac{1}{2}\sum_i \mu_i|\dot{\varphi}_i(\boldsymbol{r})|^2 + \frac{1}{2}\sum_l M_l\dot{R}_l^2 + V[\{\varphi_i\},\{R_l\}] + 2\sum_{ij}\lambda_{ij}\left(\frac{1}{\Omega}\int \varphi_i^*(\boldsymbol{r})\varphi_j(\boldsymbol{r})dr - \delta_{ij}\right). \tag{12.3.35}$$

Thus, two coupled equations of motion can be derived, a fictitious one for the orbitals which is equivalent to the KS equation

$$\mu\ddot{\varphi}_i = -\frac{\partial V}{\partial \varphi_i^*} - \sum \lambda_{ij}\varphi_j; \tag{12.3.36}$$

and another for the ions which is the classical equation of motion

$$M_l\ddot{R}_l = -\frac{\partial V}{\partial R_l}. \tag{12.3.37}$$

In this framework, the dynamics of electron parameters is a fictional process; however, it may be used to realize dynamical simulated annealing: when $\ddot{R}_l$, $\ddot{\varphi}_i$ become smaller, this is equivalent to lowering the temperature of the system, until at $T = 0$, the equilibrium state with minimum energy is reached. At the equilibrium state, this equation reduces to the Kohn–Sham equation. This method has proved to be very effective in the study of equilibrium and nonequilibrium structures on the atomic scale, dynamical properties of atomic systems, and properties of ions and electrons at finite temperatures.

§12.4 Electronic Structure of Selected Materials

12.4.1 Metals

(1) The Simple Metals. Simple metals are metals with valence electrons coming from the s and p shells. Their common characteristic is that the NFE model is a good approximation for them. The simplest among them are the monovalent alkali metals such as Na, K, Rb and Cs with the bcc structure, the measured Fermi surfaces are nearly spherical, with deviations only about 0.1%. Their transport behavior is free-electron-like.

Trivalent Al has the fcc crystal-structure and a Fermi surface that is free-electron-like, with two electrons to occupy the 1st Brillouin zone (BZ) completely, and another one to occupy the 2nd and 3rd ones. The net result is just like leaving a hole pocket of about $+e$ in the 2nd zone. This explains the high field Hall coefficient of Al which is about 1 hole per atom.

(2) The Noble Metals. The noble metals Cu, Ag and Au are also monovalent like alkali metals, with some differences: for instance, K has only a half-filled 4s band outside the argon core; while for Cu there is a d-shell filled with 10 electrons lying between the argon core and the half-filled 4s band. It is the overlap of the d-bands and s-band that makes things in Cu somewhat more complicated. The noble metals are fcc in structure, with a bcc reciprocal lattice. Experimental results show the Fermi surface of Cu is still sphere-like, but with 8 necks bulging out along the $\langle 111\rangle$ directions, touching the zone surfaces (Fig. 12.4.1). This induces more complexity in the transport phenomena in noble metals, such as the unsaturated high field magnetoresistance in Cu due to open orbits.

However, the Fermi surface still has only one branch, so, like the alkali metals, the noble metals may be regarded as one-band metals in the analysis of transport phenomena. The band structure of Cu is shown in Fig. 12.4.2. Some of the bands are clearly derived from the NFE model, shown as dashed lines, but there are also five almost horizontal bands running a little way below E_F, not accounted for at all on the NFE model, these are the d-bands. The filled d-bands near the Fermi surface in the noble metals may influence their physical properties, such as the optical properties exemplified by their reflectivity which is the origin of their metallic shine.

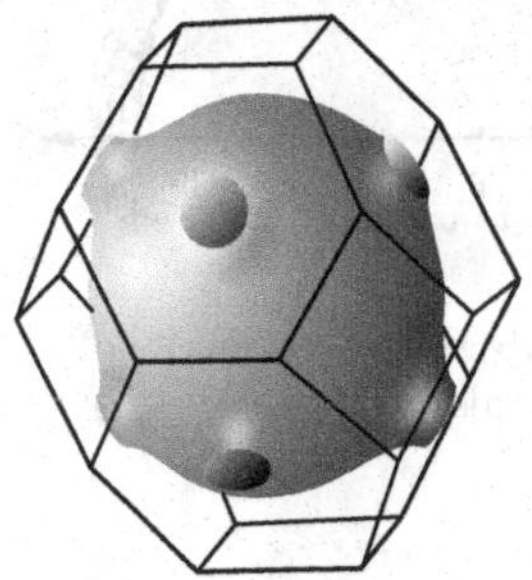

Figure 12.4.1 The Fermi surface of Cu.

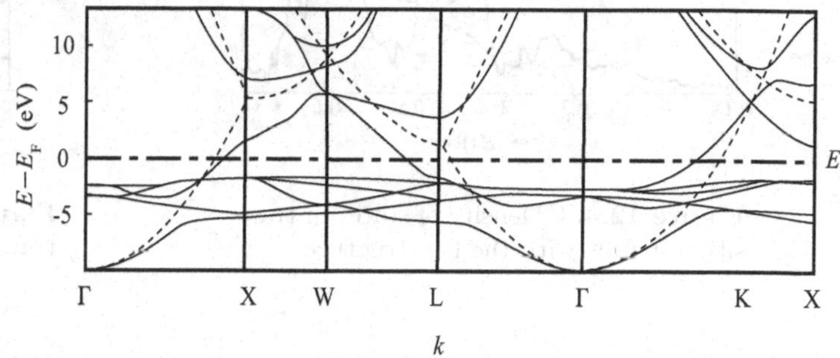

Figure 12.4.2 The band structure of Cu (full lines) compared with NFE approximation (broken lines).

(3) The Transition Metals. Transition metals are those with partially filled d bands, and this makes things more complicated. We should expect these to give rise to narrow bands fairly well-described by the tight-binding approximation. On the other hand, the s and p electrons should continue to give NFE bands. Take $3d$ transition metals as examples: Parts of the band structures of the first series of transition metals calculated by the APW method are shown in Fig. 12.4.3.

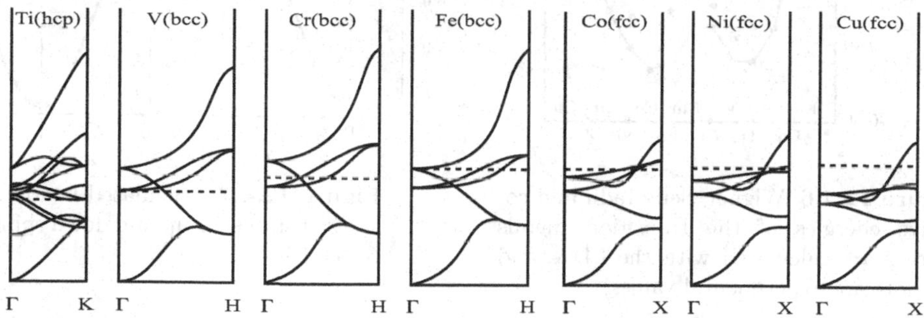

Figure 12.4.3 Parts of band structure of transition metals.

Some trends may be noted: with increasing atomic number, the d-shell gradually fills up, and becomes more compact, then the energy of the d-band becomes lower. Finally, when the d-shell is completely filled in Cu, the d-band drops below the Fermi surface. For transition metals, both the s-band and the d-band cut the Fermi surface, making transport phenomena more complicated.

We can see that almost the same curves for each crystal structure are obtained, except the position of the Fermi surface is different in Fig. 12.4.3. So we may derive a master curve for density of states (DOS) for the same crystal structure, the case for fcc is shown in Fig. 12.4.4. The density of states is much higher for d-bands than for s bands, and this effect can be observed in the electronic contribution to the low temperature specific heat.

The regular variation of cohesion energy with the filling of the d-band, peaking at half-filling for refractory metals, is well-known (Fig. 12.4.5). It is especially clear in the second and the third groups, where no magnetic complications are present. The LDA calculations of Wigner–Seitz (WS) radii and total energies of the transition metals agree quite well with the experimental data (Fig. 12.4.6). The

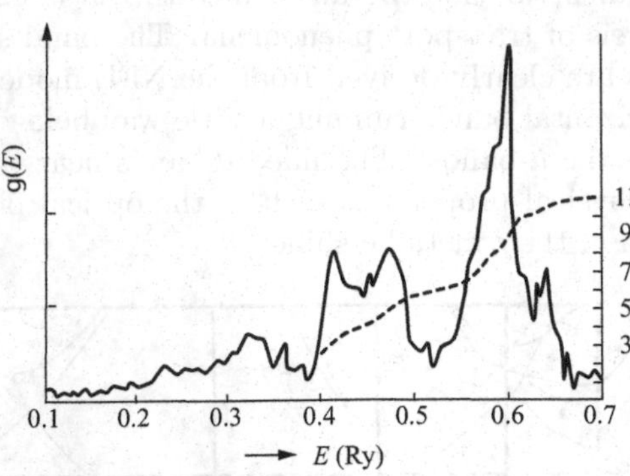

Figure 12.4.4 Density of states of transition metals with the fcc structure.

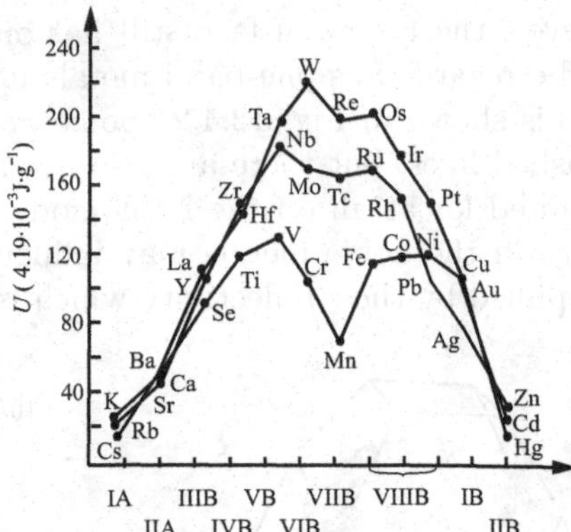

Figure 12.4.5 Cohesive energy of transition metals.

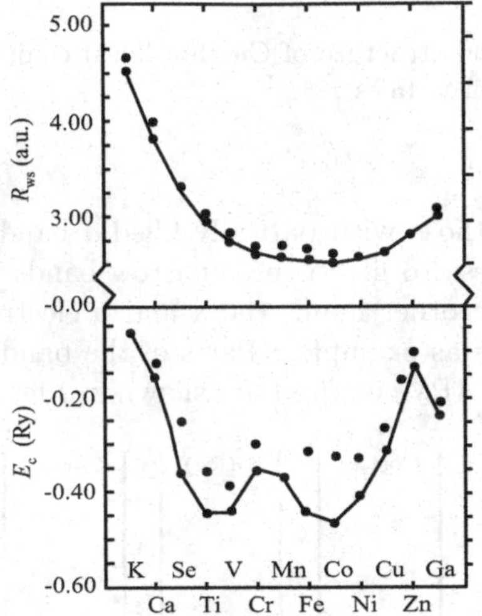

Figure 12.4.6 Wigner–Seitz radii and cohesive energies of the transition metals (curves are calculated with the LDA, and the dots are experimental values).

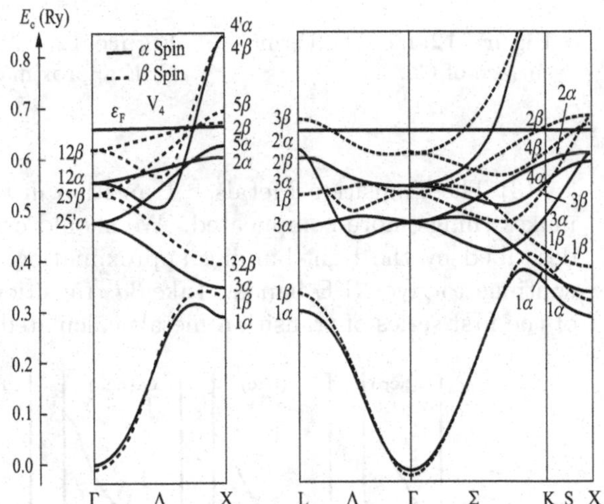

Figure 12.4.7 Calculated bands and densities of states for up and down spins (α and β) for Ni.

general tendency may be simply explained by the argument of J. Friedel:[g] A Hartree-like scheme is adopted, where the total energy is obtained by summing up the energies of the occupied one-electron states. The cohesive energy E_s per atom is then simply given by

$$E_s = -\frac{2}{N}\sum_k E_k + n_d E_0 = 2\int_0^{E_F} (E_0 - E)n(E)dE, \tag{12.4.1}$$

where n_d is the number of d-electrons per atom. From the analog to chemical bonding, in an incompletely filled d-band, more 'bonding' than 'antibonding' states are occupied. The effect should be a maximum when all the 'bonding' states, but no 'antibonding' states are occupied, i.e., for a

[g] J. Friedel, in *The Physics of Metals*, vol. 1 Electrons (ed. J. M. Ziman), Cambridge University Press, Cambridge, 1971, p. 340.

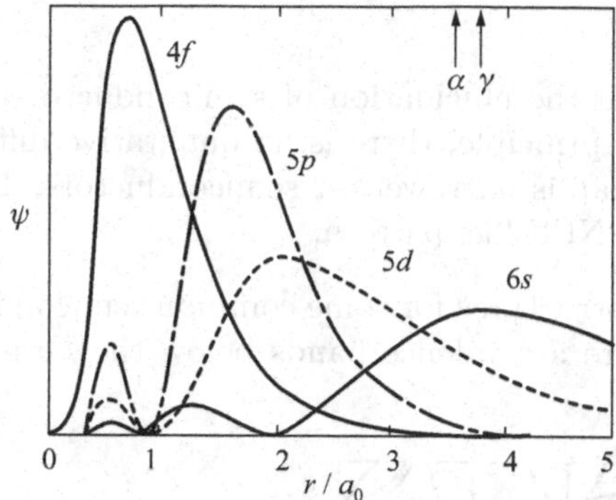

Figure 12.4.8 Schematic diagram for the radical distribution of orbits in Ce atom (the arrows point at the Wigner–Seitz radii of α- and γ-phase).

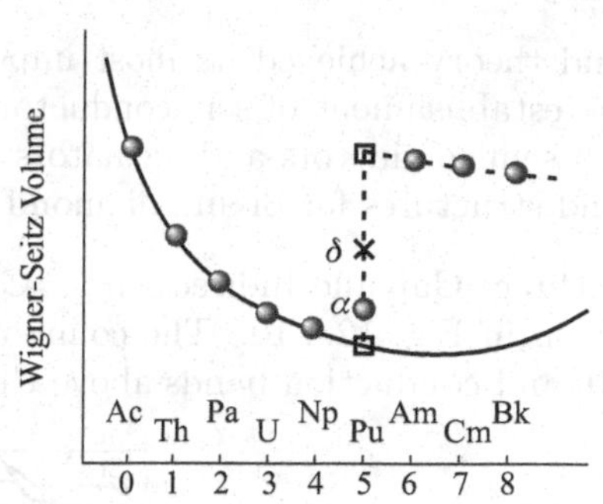

Figure 12.4.9 The variation of the Wigner–Seitz volumes of solid phases versus the number of f-electrons (from the data of α- and δ-phase of Pu).

Fermi level E_F equal to the energy level of isolated atoms, with

$$\rho = 2\int_0^{E_F} n(E)dE. \tag{12.4.2}$$

The experimental data can be roughly fitted to a shifted parabola

$$|E_s(n_d)| = [A(10 - n_d)n_d + Bn_d], \tag{12.4.3}$$

where A, B are constants.

If cohesion is interfered with by the Coulomb correlation of electrons, the ferromagnetism of Fe, Ni and Co and antiferromagnetism of Cr and Mn are the direct results of electron correlation. For treatment of metals with spin-polarized electronic density, LSDA was developed, and the density of states of up and down spins can be calculated separately; the result for Ni is shown in Fig. 12.4.7. Note the spin-polarized state below the Fermi surface of a ferromagnet. Though *ab initio* calculation can show the existence of ferromagnetism in Fe, Ni and Co, the mechanism for its formation is not clearly displayed and is all but buried in the enormous mass of calculations. We shall returned to this subject in Chap. 13.

(4) Lanthanides and Actinides. In the periodic table, there are two series of elements with unfilled f-shells. From Ce to Lu, the lanthanide series of the rare earth elements, the $4f$ shell is gradually filled; for the actinides running from Th to Lw, the $5f$ shell is gradually filled up.

The radial distribution of the atomic orbital of an rare earth atom (Fig. 12.4.8), shows that the radius of the $4f$ orbitals is only 1/5–1/4 of the atomic radius, which is mainly determined by that of $5d$-$6s$ orbitals. Thus, in crystalline solids, the $5d$ and $6s$ electrons are delocalized into broad unfilled conduction bands with 3 electrons showing metallic behavior, while the f-electrons remain in the ion core. In general, localized electrons in unfilled f-shells do not participate in the bonding of solids, but their spins are lined up, showing magnetic moments. The only exception for delocalized f-electrons participating in bonding occurs in the low temperature phase of Ce.

The situation of actinides is just in-between the $3d$ transition metals and the rare earth elements: for early actinides, from Ac to Pu, $5f$-shell filling induces the parabolic decrease of atomic volume, following Friedel's rule for the transition metals, that the bonding states are filled first. It shows that, for these metals, the f-electrons are delocalized and participate in bonding. However, just before the optimum bonding is reached, right after Pu and before Am, the atomic volume increases enormously (by 50%) and then the curve becomes quite flat (Fig. 12.4.9). Thus, the behavior of the later actinides resembles the rare-earth elements, in which the f-electrons are localized in the ion core and no longer participate in bonding. In order to describe this abrupt transition, we must go beyond the LDA, using a theory incorporating many-body effects.

12.4.2 Semiconductors

Band theory achieved its most important success in the elucidation of semiconductor physics and the establishment of semiconductor electronics. In principle, there is no qualitative difference between semiconductors and insulators, only the bandgap is narrower for semiconductors. In fact, the band structures for Si and diamond show a similar, NFE-like pattern.

(1) Direct-Gap and Indirect-Gap. Calculated band structures for some common semiconductors are shown in Fig. 12.4.10. The common feature is there are valence bands below the Fermi level $E_F = 0$, and conduction bands above the Fermi level.

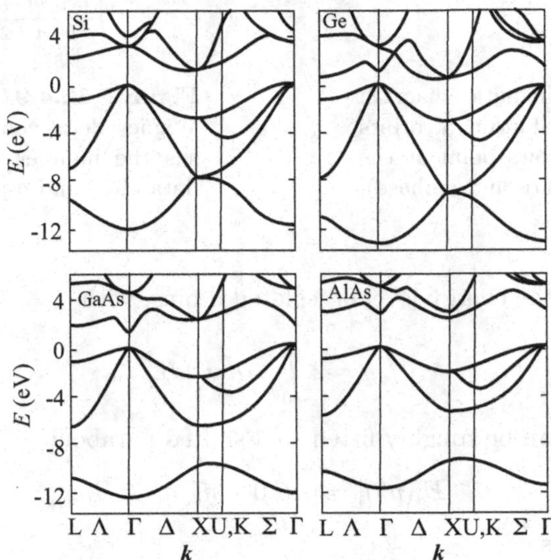

Figure 12.4.10 Band structure of four common semiconductors Si, Ge, GaAs and AlAs. The calculation excludes the spin-orbit interaction.

However, there is an important difference: for compound semiconductors (GaAs and AlAs), the tops of the valence bands and the bottoms of the conduction bands line up at the same values of k, the optical transitions between them may be accomplished by emission or absorption of a single photon; this is a direct gap semiconductor. For elemental semiconductors (like Si and Ge), the situation is different: the tops of the valence bands and the bottoms of the conduction bands do not line up at the same k, so optical transitions between them, besides having emission or absorption of photons are accompanied by a process which compensates the difference in k, such as emission or absorption of phonons. These are called indirect gap semiconductors. It is obvious that optical transitions are much more efficient in direct gap semiconductors, although Si is the most widely used semiconductor for electronics; compound semiconductors are mostly used for opto-electronic applications.

(2) The Band Gaps. It should be noted that, in general, band calculation results come out well and have been regarded as a useful tool for guiding practice; however, quantitatively speaking, calculated bandgaps are too small when compared with the experimental values. This means that, even in semiconductors, correlations between electrons in the excited state have been underestimated in the LDA; however more accurate calculations in the quasiparticle (QP) approximation have rectified this situation (Table 12.4.1).

Semiconductors are widely used as materials for light emitting diodes (LED) and laser diodes (LD). The bandgap is an important criterion in selecting materials for these applications: GaAs for near infrared, GaP for red, and the successful development by of GaN for blue and violet LEDs and LDs by Nakamura was a great technical achievement in the 1990s. In 1991, L. T. Canham discovered strong visible light emission from porous Si showing that not only is the bandgap widened in nanostructured Si, but also the emission efficiency is improved. This gives promise for nanostructured Si as a potential opto-electronic material.

Table 12.4.1 Bandgaps of semiconductors: experimental values versus theoretical ones.

Crystal	The type of the energy gap*	E_g (eV) (experimental values)		E_g (eV) (theoretical values)	
		300 K	0 K	LDA*	QP*
diamond	i		5.4	3.9	5.6
Si	i	1.11	1.17	0.5	1.29
Ge	i	0.66	0.744	< 0	0.75
αSn	d	0.00	0.00		
InSb	d	0.17	0.23		
InAs	d	0.36	0.43		
InP	d	1.27	1.42		
GaP	i	2.25	2.32		
GaAs	d	1.43	1.52	0.37	1.29
GaSb	d	0.68	0.81		
AlSb	i	1.6	1.65		
SiC (hex)	i	–	3.0		
GaN (W)*	d		3.5	2.3	3.5
GaN (ZB)*	d		3.3	2.1	3.1

Note: 1. i = indirect-gap, d = direct-gap.
2. (W)* = wurtzite structure; (ZB)* = zincblende structure.

(3) The Effective Masses. The electronic properties of semiconductors are completely determined by the comparatively small numbers of electrons excited into the conduction band, and holes left behind in the valence band. The effective masses of electrons and holes are important parameters for semiconductors. Based on the definition of the effective mass in §6.1, $E(k)$ may be Taylor-expanded into

$$E(k) = E(k_0) + \frac{\hbar^2(k_x - k_{0x})^2}{2m_x^*} + \frac{\hbar^2(k_y - k_{0y})^2}{2m_y^*} + \frac{\hbar^2(k_z - k_{0z})^2}{2m_z^*}. \tag{12.4.4}$$

This may be calculated using the $\boldsymbol{k} \cdot \boldsymbol{p}$ perturbation method introduced in §12.1.4. In direct-gap materials, the conduction band edge is spherical with effective mass m^*,

$$E_c = E_g + \hbar^2 k^2 / 2m^*. \tag{12.4.5}$$

referred to the valence band edge. The valence bands are degenerate near the edge, with heavy hole (hh) and light hole (lh) bands degenerate at the center, and a band due to splitting Δ' of the spin-orbit coupling. The values of effective masses can also be derived experimentally from cyclotron resonance in semiconductors (see Fig. 14.2.11).

In indirect-gap materials the situation is more complex. The valence band edges for Si and Ge are at $k = 0$, with light and heavy hole bands, as well as a split off one due to the spin-orbit interaction. However, the energy surfaces are not spherical but warped, making the effective masses harder to characterize.

The conduction band edges for Si and Ge are not situated at $k = 0$, but at points L for Ge; and on the lines labelled Δ, a little away from the boundary points X for Si. The energy surfaces are spherical there.

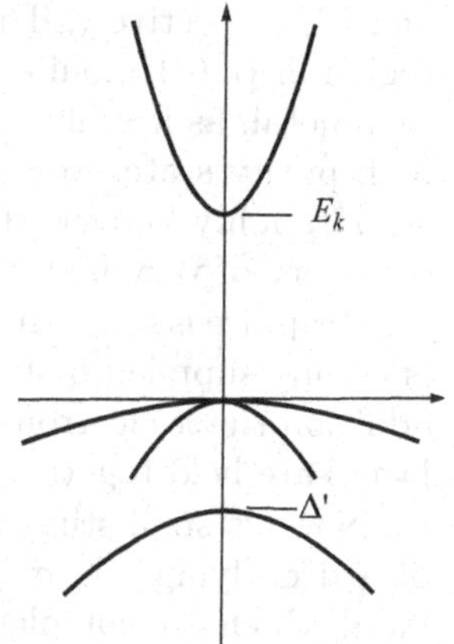

Figure 12.4.11 Schematic diagram for the band edges in a direct-gap semiconductor.

(4) Electron Density Distribution in Real Space. From band calculations we can compute the electron density distribution in real space; also this distribution may be measured directly from accurate X-ray diffraction data. Figure 12.4.12 shows the results for Si, and the agreement is quite

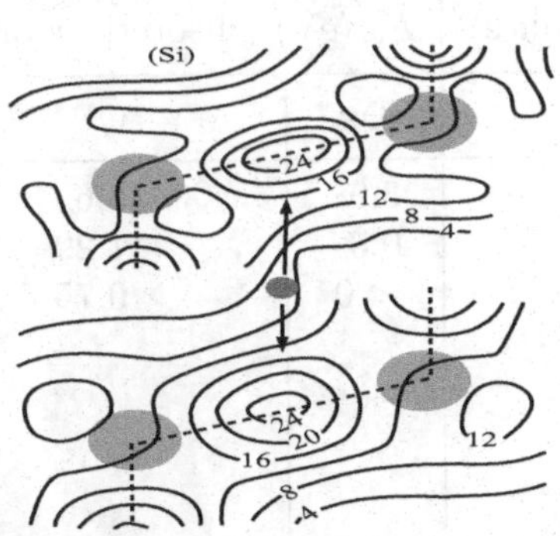

Figure 12.4.12 Valence electron density distribution in Si.

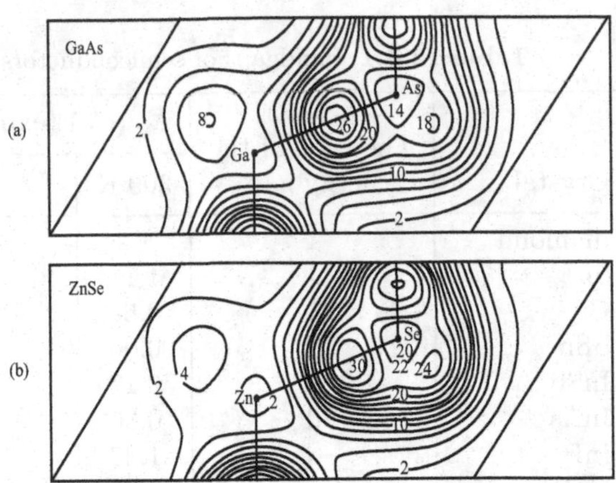

Figure 12.4.13 Valence electron density distribution: (a) GaAs, (b) ZnSe. From J. R. Chelikowsky and M. L. Cohen, *Phys. Rev. B* **14**, 556 (1976) .

good. Furthermore, the density of valence electrons is higher in the region around the mid-point of the line joining two neighboring atoms, and this gives clear evidence for covalent bonds. In the electron density distribution for GaAs, the valence electrons still concentrate on that line, but the center is shifted toward the As atom, showing the tendency of ionic bonding within the framework of covalent bonds. This tendency becomes more visible in the case of ZnSe (see Fig. 12.4.13).

12.4.3 Semimetals

Graphite, As, Sb and Bi are semimetals. For some directions in $\boldsymbol{k}$-space, their valence bands and conduction bands overlap, showing metallic behavior, while in other directions, the two kinds of bands do not overlap and show insulator behavior (see Fig. 12.4.14). The densities of charge carriers for transport are several orders of magnitude smaller than that of common metals, for instance, $n_{\rm e} = n_{\rm h} = 3 \times 10^{18}/\text{cm}^3$ for graphite, and $2 \times 10^{20}/\text{cm}^3$, $5 \times 10^{19}/\text{cm}^3$ and $3 \times 10^{17}/\text{cm}^3$ for As, Sb and Bi, respectively. The band structures of semimetals show overlap, which may occur in a special region of the Brillouin zone (see Fig. 12.4.14). For example, when the overlap is small we call it a semimetal, as it still shows metallic behavior. Bi single crystals, which can be easily fabricated in a high purity state, were much used in early experimental measurements, so the Fermi surface of Bi is thoroughly known. However its electronic properties still provide surprises, as exemplified by the discovery of very high magnetoresistance.[h]

Graphite is a very interesting material because of its electronic properties. The interlayer bonding is mainly supplied by bonds of sp^2 hybrids with a triangular arrangement in a honeycomb lattice; in addition the π-electrons on the p_z orbitals are delocalized throughout the layer. Individual graphite layers are held together by the much weaker van der Waals interactions.

Now we shall study the band structure of single atomic layer graphite, i.e. graphene: it consists of a deep-lying sp_x^2–σ bonding band, followed by a p_z–π bonding band at the top of the valence band which are completely filled; over them there are empty conduction bands, the antibonding π^* band and the higher σ^* band.

The Brillouin zone (BZ) is a hexagon rotated 60° with respect to the real lattice (Fig. 12.4.15). The gap is largest at Γ, becomes smaller at M and vanishes at K where the valence and conduction states are degenerate at the Fermi level. This is the ideal situation, in which the density of states is zero at the Fermi level, indicated in Fig. 12.4.16; but the residual interlayer interactions make the

[h]The magnetoresistance of Bi is just OMR which is produced by charge carriers executing cyclotronic motion in the magnetic field. Because the density of charge carriers in Bi is small, $\omega_{\rm c}\tau \gg 1$ (where $\omega_{\rm c}$ is cyclotron radius and τ is the scattering time of charge carries) and a large magnetoresistance at room temperature is observed in Bi single crystal thin films. See C. L. Chien *et al.*, *J. Appl. Phys.* **87**, 4659 (2000).

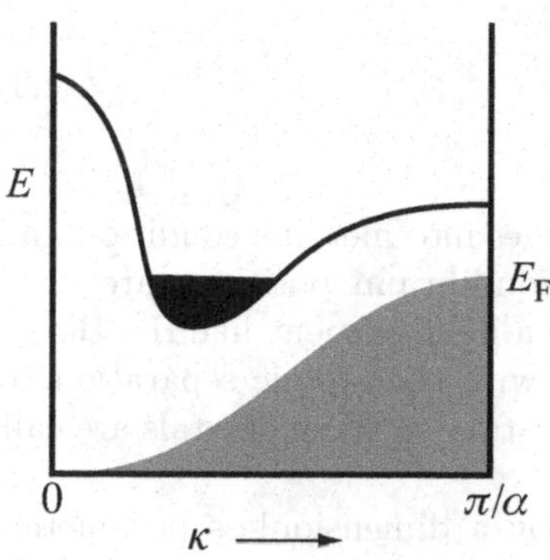

Figure 12.4.14 Schematic energy band picture for a semimetal.

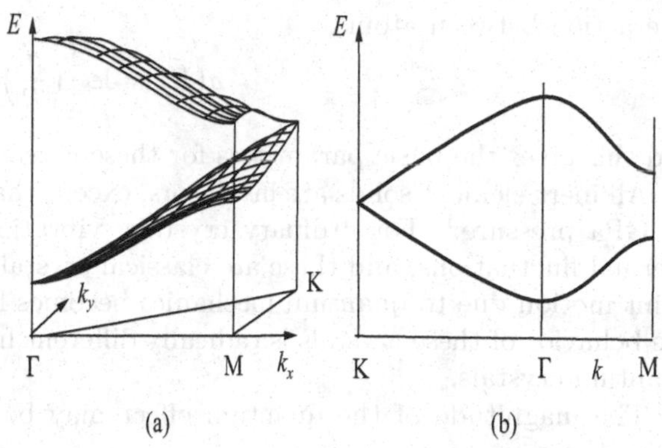

Figure 12.4.15 The band structure of a single graphite layer. (a) Band structure for the triangle MK in a 3D representation. (b) Band structure for the line KM in a 2D representation.

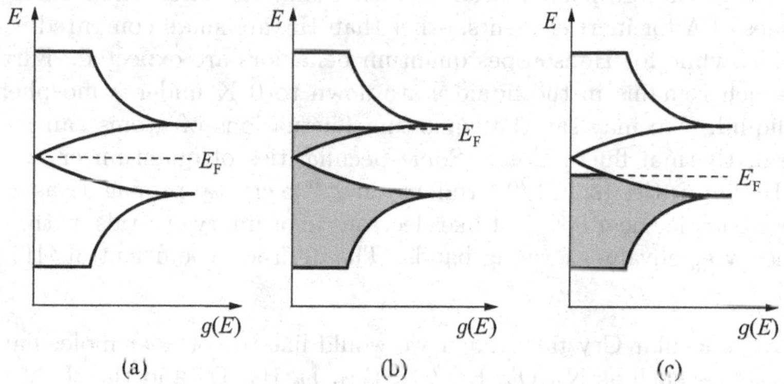

Figure 12.4.16 (a) The density of states versus energy of a single layer of graphite and (b), (c) its modification by intercalation.

density of states (DOS) small but finite; this is why graphite shows metallic behavior. The layered structure of graphite makes it easy to form intercalation compounds with intercalated layers of foreign atoms or molecules. Intercalation may induce charge transfer between the intercalated layer and the host: either donating electrons to graphite (e.g., alkali atoms as intercalants) or donating holes to graphite (in other words, accepting electrons, e.g., Br, AsF_5 and PtF_6 as intercalants); in both cases, intercalation makes the compound metallic.

12.4.4 Molecular Crystals

Molecular crystals are composed of molecules with saturated chemical bonds. In general, molecular crystals are insulators because all the electrons are localized in the molecules and participate in intra-molecular bonding. The cohesive forces forming the crystals come from weak attractive forces (e.g., van der Waals forces) between molecules. Therefore, molecular crystals are usually vapor or liquid phases at room temperature, and the melting points of their solid phases are very low.

(1) Monoatomic Crystals. Atoms in inert elements such as He, Ne, Ar, Kr and Xe with closed-shell electronic structure, are candidates for stable monoatomic molecular crystals at low temperature (for the crystallization of He, in addition, a pressure of several tens of atmospheres must be applied). The simplest model for these crystals uses the classical potential of the Lennard–Jones type for the

interaction between atoms,

$$u(R) = 4\varepsilon\left[\left(\frac{\sigma}{R}\right)^{12} - \left(\frac{\sigma}{R}\right)^{6}\right], \tag{12.4.6}$$

and this gives the basic parameters for these crystals.

All inert element solids are insulators, except that solid Xe changes into metallic condition under 50 GPa pressure.[i] For ordinary crystals, vibrations about the equilibrium positions are due to thermal fluctuations, and these are classical crystals. However, for a light element like He, the zero point motion due to quantum mechanics becomes large compared with the atomic separation, and the behavior of these crystals is radically different from classical crystals, so these crystals are called quantum crystals.

The magnitude of the quantum effect may be characterized by a dimensionless parameter Λ defined by

$$\Lambda = \frac{h}{\sigma\sqrt{m\varepsilon}}, \tag{12.4.7}$$

where σ and ε are the parameters from the Lennard–Jones potential (12.4.6), m is the mass of the atom and h is Planck's constant. The parameter Λ is a measure of the relative magnitude of the kinetic energy of zero-point motion of atoms and the interaction energy in the crystalline phase. The values of Λ for inert elements, other than He, are small compared with 1, so these show classical behavior; while for He isotopes quantum behaviors are expected. Furthermore, He is the only element which remains in the liquid state down to 0 K under atmospheric pressure, i.e., it is a quantum liquid. We may say that quantum fluctuations of atoms can melt the crystal even in the absence of thermal fluctuations. Some peculiarities of quantum crystals should be noted: For instance, ^{3}He impurities (spin 1/2 fermions) in a ^{4}He crystal propagate as waves in the periodic potential of the He lattice near 0 K, just like electrons in ordinary crystals, with k as a good quantum number and energy eigenvalues forming bands. The diffusion coefficient of ^{3}He impurities tends to infinity at 0 K.

(2) Diatomic Molecular Crystals. Next we would like to consider molecular crystals composed of diatomic molecules such as N_2, O_2, F_2, Cl_2, Be_2, I_2, H_2, D_2 and the slightly more complicated cases of CO, HCl, HBr, HI. For the case of N_2, the low temperature phase is fcc with a regular alignment of molecular axes. With a rise in temperature this alignment becomes more random due to thermal motion, and eventually the fcc phase becomes unstable and makes a transition to hcp at T_c. In the high temperature hcp phase, all the molecules perform nearly free rotations to appear spherical, and they close-pack into the hcp structure. Such a transition, in which molecular axes change from ordered to random configurations, is called a rotational phase transition, and this is a characteristic property of molecular crystals in general.

For a diatomic molecule with two atomic species, the molecule can have a permanent dipole moment, which contributes to the cohesive energy due to dipole-dipole interactions, and sometimes causes a dielectric anomaly at the rotational transition point. Some crystals such as DCl etc., even show ferroelectricity, with the permanent dipoles arranged in parallel.

The bonding of diatomic molecular crystals are of the van der Waals type, making them insulators. Under high pressure, some crystals undergo an insulator-metal transition; for instance, under 16–18 GPa, the bandgaps of I_2 disappear, and I_2 has metallic behavior; Br_2 changes from a diatomic molecular crystal to a monoatomic one under 80 GPa, coinciding with the insulator-metal transition; and a similar change is found in O_2 under 95 GPa.[j]

(3) H_2 Crystal and Metallic Hydrogen. Now we shall discuss problems related to the H_2 crystal in some detail, as well as in a wider perspective. H_2 molecules are placed on each lattice site, then we use a Lennard–Jones type of potential to determine their interaction energy and equilibrium separation. However, quantum mechanical effects manifest themselves in that the separation of

[i] R. Reichlin *et al.*, *Phys. Rev. Lett.* **62**, 669 (1989).

[j] There are some articles about the insulator-metal transition: I_2, B. M. Riggleman and H. G. Drickmer, *Chem. Phys.* **51**, 1117 (1963), K. Kasai *et al.*, *J. Phys. Soc. Jap.* **51**, 1811 (1982); Br_2, Y. Fujii *et al.*, *Phys. Rev. Lett.* **63**, 2998 (1991); O_2, K. Shimizu *et al.*, *Nature* **393**, 767 (1998).

nearest molecules is found to be 0.375 nm, which is much larger than the equilibrium separation of 0.32 nm determined classically, i.e. the actual crystal is expanded considerably due to zero point oscillations of protons, and the H_2 crystal is a kind of quantum crystal. Since the separation of molecules in the crystal is about 5 times larger than that of protons in the H_2 molecule, there is hardly any overlap between the wavefunctions of neighboring molecules, and it is a wide bandgap ($E_g = 15$ eV) insulator. However, hydrogen in the periodic table is in the same column as the alkali metals and it has been predicted that under high pressure, protons will occupy lattice sites, with the delocalized electrons as described by NFE model. In this way solid metallic hydrogen may be regarded as a special form of quasi-alkali metal and so metallic hydrogen remains the Holy Grail for some condensed matter physicists. The following are some research results in this direction.

In 1935, Wigner and Huntington predicted that H_2 molecule crystals would change into a monoatomic solid at 25 GPa; later this value was modified to 200 GPa. The H_2 molecular crystal can also change into the metallic phase by overlapping the bands; the transition pressure in this case is predicted to be lower than the value mentioned above. This transition pressure is related to the crystal structure, but the exact transition structure at 0 K is still unknown. If we extrapolate the pressure versus volume data with the pressure of hcp structure up to 20 GPa, we get 620 GPa as the transition pressure. N. W. Ashcroft *et al.* predicted that both monoatomic and diatomic metallic hydrogen are high T_c superconductors. Recently, static pressures of 290 GPa have been applied, due to the improvement of the diamond anvil technique, but solid metallic hydrogen has not yet been found, and only two kinds of orientationally ordered hydrogen molecular crystals, the II phase and the III phase, have been found.[k]

While research on solid metallic hydrogen was in something of a quandary, other experiments achieved conspicuous success: liquid metallic hydrogen was found. In 1996, Nellis *et al.* at Livermore Lab. observed the dc resistivity of liquid hydrogen drops to 0.5×10^4 Ω cm under a pressure of 140 GPa at 3000 K by a dynamic high pressure technique. This value coincides with those of the monoatomic liquids Cs and Rb, whose transitions to the metallic state occur at 2000 K. This provided direct proof of the existence of liquid metallic hydrogen. Figure 12.4.17 shows a schematic phase diagram, from which we see that the pressure for insulator-metal transition in the liquid state is much lower than in the solid state. There is lots of liquid hydrogen at the high pressures which exist in Jupiter and Saturn and the existence of liquid metallic hydrogen is of great significance for planetary science.

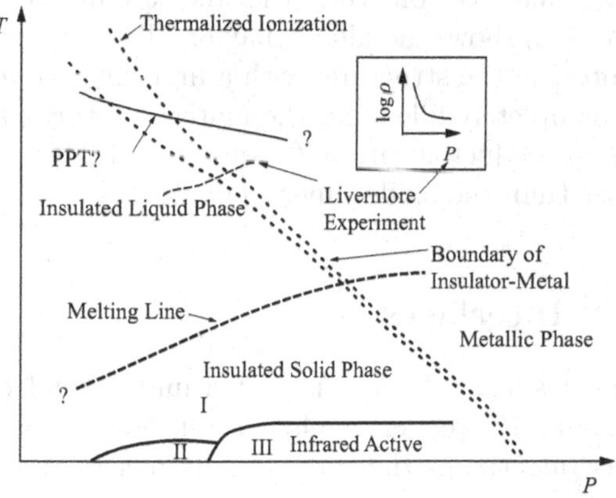

Figure 12.4.17 Schematic phase diagram of H_2. Solid state I is the disordered phase; II and III are orientational ordered phases, the former is quantum and the latter is classical. The solid line in high temperature regime denotes the transition to plasma phase (PPT). Inset gives the relation of resistance and pressure by theoretical prediction. From W. J. Nellis *et al.*, *Phil. Trans. Roy. Soc. Lond. A* **356**, 119 (1998).

[k]For a review of solid hydrogen, see H. K. Mao and R. J. Hemley, *Rev. Mod. Phys.* **66**, 671 (1994); for the explanation of different orientational ordered phases, see I. I. Mazin *et al.*, *Phys. Rev. Lett.* **78**, (1997); a review about research on liquid metallic hydrogen, is given by W. J. Nellis, A. A. Louis and N. W. Ashcroft, *Phil. Trans R. Soc. Lond. A* **356**, 119 (1998).

(4) C_{60} Solids. At room temperature, C_{60} molecules may be close-packed into an fcc structure, where the C_{60} molecules are held together by weak van der Waals interactions with inter-center spacings of about 1 nm and a minimum inter-cage separation of 0.3 nm. Part of the band structure of C_{60} solid is shown in Fig. 12.4.18(a), in which the energy bands expanded from HOMO and LUMO in Fig. 11.3.7 are shown. Pure crystalline C_{60} is a semiconductor with narrow band width of about 0.5 eV and a gap of 1.5 eV.

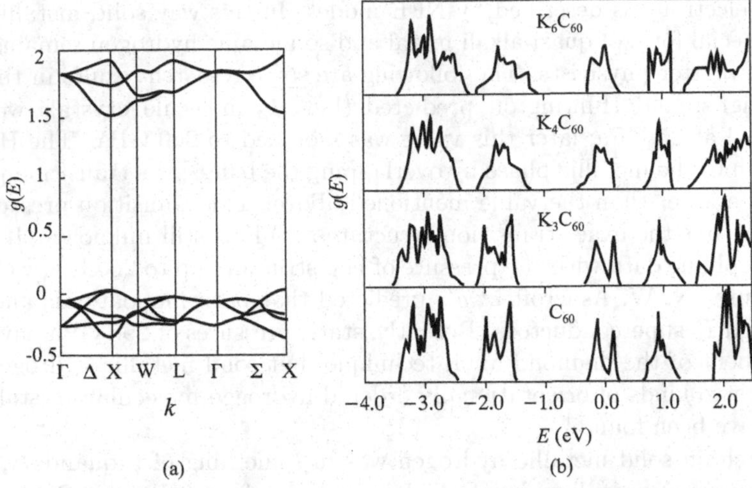

Figure 12.4.18 (a) Band structure of solid C_{60} showing the top of the valence band and the bottom of the conduction band. (b) Densities of states for solid C_{60} and some potassium fullerites.

Since the interstitial voids in solid C_{60} is quite large, foreign atoms, such as alkali atoms, may be inserted into voids by diffusion. Just like intercalation, this process is accompanied by charge transfer between the alkali atom M and the host C_{60}, forming fullerites. Donated electrons occupy the lowest one of the conduction bands. Thus at the composition M_3C_{60}, all interstitial voids (one octahedral and two tetrahedral voids per lattice site) of the fcc structure are filled with alkali atoms. If each alkali atom donates one electron, then the lowest band of the conduction states is half-filled with electrons, and M_3C_{60} shows metallic behavior. To accommodate further alkali atoms the structure is transformed into the bcc structure, with a limiting composition of M_6C_{60}, at this point the conduction bands are completely filled and the material returns to be a semiconductor. M_3C_{60} has been found to be a superconductor with a T_c somewhat higher than ordinary superconducting metals and alloys, but lower than the oxide superconductors.

12.4.5 Surfaces and Interfaces

In Chap. 7, we discussed surface electronic states under ideal conditions. Real surfaces and interfaces are very complex, and the reader should consult monographs on this subject. Here we shall cite some examples to illustrate things that are actually observed and their relation to theoretical calculations.

(1) 7×7 Cell of Reconstructed Si(111) Surface. One of the most stunning facts revealed by modern surface research techniques is the surface reconstruction of the Si(111) surface. The freshly cleaved (111) surface has been reconstructed with an enlarged cell of 2×1, and then, after annealing at high temperature, a reconstructed cell 7×7 appears. The STM image of a 7×7 reconstructed surface is shown in Fig. 12.4.19. An extremely complex dimerization adatom stacking fault model for a 7×7 cell of reconstructed Si(111) surface (DAS) model has been proposed to account for this image (Fig. 12.4.20). It is a credit to DFT that *ab initio* electronic calculation can substantiate this model, and the Car–Parrinello method may be used to explore its dynamical properties.

Figure 12.4.19 STM image of 7×7 cell of reconstructed Si(111) surface (provided by Prof. J. G. Hou).

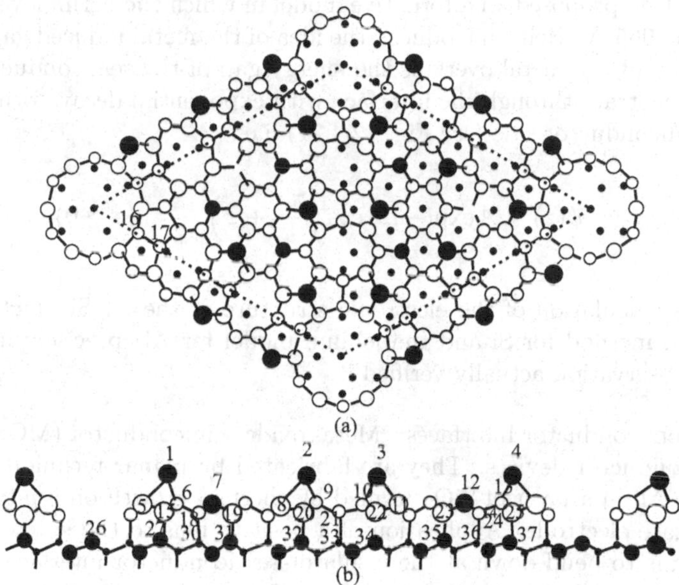

Figure 12.4.20 DAS model for 7×7 cell of reconstructed Si (111) surface. (a) The view perpendicular to the surface. (b) The view of cross-section along the long diagonal. Large full circles are atoms on the top; medium full circles are other atoms; hollow circles are dimers.

(2) Metal-Semiconductor Interfaces. The Schottky barrier at a metal-semiconductor interface is the physical origin of the rectifying effect of a semiconductor diode. This potential barrier may block one direction of electric current and inject current in the reverse direction. The bands of the semiconductor bend in the vicinity of the interface because, in order to equalize the chemical potentials (the levels of the Fermi surfaces) in both metal and semiconductor, electrical charges must flow from one to the other, to build up an electric field. In case of an n-type semiconductor in contact with a metal, electrons must flow from semiconductor to metal, leaving the semiconductor positively charged, with a depletion layer of thickness d, and the bands are bent upwards as a response.

In 1939 Schottky and Mott proposed a model to account for the Schottky barrier height as the difference of the work functions of the metal W_{m} and the affinity of the semiconductor ϕ_{s} (the difference of vacuum level and the bottom of conduction band), i.e., $E_{\mathrm{s}} = W_{\mathrm{m}} - \phi_{\mathrm{s}}$. According to this model, for a specific semiconductor, the Schottky potential barrier should scale linearly with the work functions of the metals. Experimentally it is found that this holds only for more ionic semiconductors such as ZnS, AlN, ZnO, ..., but is very weak, or almost non-existent, for standard semiconductors, like Si, Ge, GaAs, InSb, etc.

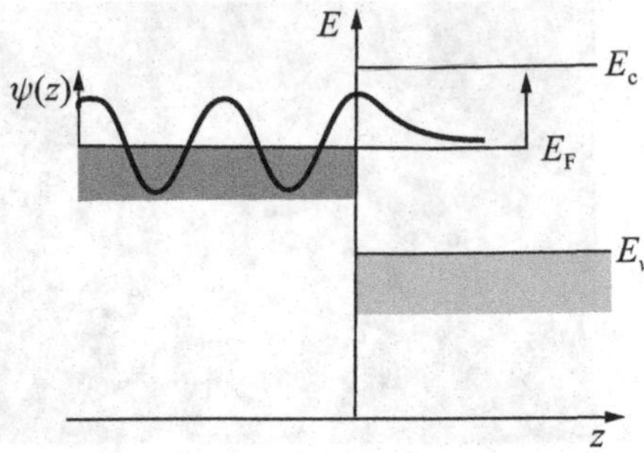

Figure 12.4.21 Schematic diagram for the metal induced gap states (MIGS).

In 1947, J. Bardeen proposed an alternative model in which the Fermi level is pinned at an interface state. Then in 1965, V. Heine introduced the idea of the metal induced gap states (MIGS): when the conduction band of the metal overlaps the energy gap of the semiconductor, the wavefunctions in the metal will penetrate through the interface with exponential decay, forming interface states in the gap on the semiconductor side (see Fig. 12.4.21), i.e.,

$$\psi(z) = A\exp(-qz)\cos\left(\frac{\pi z}{a} + \varphi\right), \qquad (z > 0). \tag{12.4.8}$$

More elaborate calculation of the electronic structure of the Al/Si junction (using the empirical pseudopotential method for Si and the jellium model for Al) precisely indicate the MIGS and subsequent STM observation actually verified it.

(3) Insulator-Semiconductor Interfaces. Metal-oxide-semiconductor (MOS) devices are the most widely used semiconductor devices. They are fabricated by planar techniques, involving the deposition of a layer of Al on a layer of SiO_2 formed by thermal growth on a p-type Si surface. The Al layer acts as the gate electrode. Application of a positive bias to the gate electrode will cause the semiconductor bands to bend down at the insulator-semiconductor interface (Fig. 12.4.22(a)). The majority carriers, i.e., holes, are repelled away from the interface, so a depletion layer about 100 nm thick is formed. As the gate voltage is increased over a critical value V_c, the bottom of the conduction band bends below the Fermi level and an inversion layer about 1 ~ 10 nm thick is formed, in which degenerate electrons are contained in the region, by an approximately triangular-shaped potential well ($U(z) = e\varepsilon_{\text{eff}}z$, for $z > 0$). We shall return to this in Chap. 14.

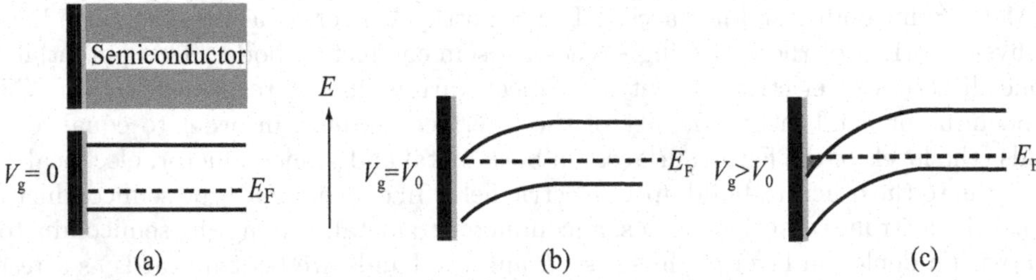

Figure 12.4.22 Metal-insulator-semiconductor interfaces in a MOS device: (a) $V_g = 0$, (b) $V_g = V_0$, depletion layer → inversion lager, (c) $V_g > V_0$, degenerate electrons in triangular-shaped potential well.

Field effect devices based on MOS technology not only play a leading role as the most important electronic devices in information technology, but also provide very powerful devices for basic research. For example, they led to the discovery of both the integral and fractional quantum Hall effects.

Bibliography

[1] Ziman, J. M., *Principles of the Theory of Solid*, Chaps. 3 & 4, 2nd ed., Cambridge University Press, Cambridge (1972).

[2] Harrison, W. A., *Solid State Theory*, McGraw-Hill, Chap. 2, New York (1970).

[3] Elliot, S. R., *The Physics and Chemistry of Solids*, Wiley, New York (1998).

[4] Callaway, J., *Quantum Theory of the Solid State*, 2nd ed., Academic Press (1999).

[5] Anderson, P. W., *Concepts in Solids*, Chap. 2, Benjamin, 1962; Addison-Wesley reprint (1991).

[6] Ashcroft, N. W., and N. D. Mermin, *Solid State Physics*, Chaps. 15 & 28, Rinehart and Winston, Holt (1976).

[7] Kittel, C., *Introduction to Solid State Physics*, 7th ed., Chaps. 8 & 19, Wiley, New York (1996).

[8] Matsubara, T. (ed.), *The Structure and Properties of Matter*, Chaps. 5, 6, & 7, Springer, Berlin (1982).

[9] Marder, M. P., *Condensed Matter Physics*, Part II, Wiley (2000).

[10] Kohn, W., *Nobel Lecture: Electronic Structure of Matter — Wave Functions and Density Functionals*, Rev. Mod. Phys. **71**, 1253 (1999).

Chapter 13

Correlated Electronic States

In this chapter we are mainly concerned with those electronic systems of correlated electrons in which band theory is found inadequate. The first section is devoted to Mott insulators, which should be metals according to band theory, but actually they are insulators. The Hubbard model is introduced to treat this kind of material. Doped Mott insulators were found to be strange conductors, such as the cuprate high T_c superconductors and manganites with colossal magnetoresistance. After this we will examine another set of phenomena due to the effects of magnetic impurities: The Anderson model is introduced, the Kondo problem is explained and a brief sketch of heavy electron metals and related problems is given. Finally, the outlook for this field is discussed both from the viewpoint of experiment and theory. It should be noted that some phenomena belonging to strongly correlated states, such as Wigner crystallization, fractional quantum Hall effect, will be discussed in Part VI.

§13.1 Mott Insulators

13.1.1 Idealized Mott Transition

It should be recognized that enormous success has been achieved by band theory in clarifying the electronic structures of the solid state; in explaining why some substances are metals, others semiconductors or insulators; especially in building up a theoretical framework for the quantitative understanding of transport properties of semiconductors and metals. However, in 1937 there was a serious setback because it gave incorrect ground states for some transition metal monoxides, such as CoO, MnO and NiO. We may take the case of CoO to illustrate this point: It has a slightly distorted rock salt (NaCl) structure with a unit cell which contains one Co atom and one O atom. The outer shell of the Co atom has electron configuration $3d4s$, and the O atom $2s2p$, so the number of electrons per unit cell is $9 + 6 = 15$, an odd number. According to band theory, crystals with a unit cell containing an odd number of electrons should be a metal, so the ground state of CoO was predicted to be metallic. However, this did not agree with experiment: CoO is actually an insulator with a large gap. What's wrong with band theory? Mott's verdict is that in spite of its sophisticated theoretical treatment it had neglected the correlation between electrons, so it must fail when it faces systems with strongly correlated electrons. These transition metal oxides are called Mott insulators.

For many years Mott insulators were neglected by most textbooks on solid state physics, only cultivated by a small number of scientists specializing in the metal-insulator transitions. The discovery of high T_c superconductors in cuprates and the rediscovery of colossal magnetoresistance in manganites have changed this situation. These materials have proved to be doped Mott insulators. So the physics of Mott insulators, doped Mott insulators, as well other systems of strongly correlated electrons have acquired an unprecedented importance in contemporary research in condensed matter physics. The central problem for Mott insulators and related materials is to find the correlation energy of the electrons. It is expressed as the Hubbard energy U for the double occupancy of the same orbital on the same atomic site by two electrons.

Mott considered the idealized metal-insulator transition for a Na crystal by changing the interatomic spacings. He predicted that at a critical interatomic spacing a_C, the electrical conductivity at zero temperature should jump abruptly from zero to a finite quantity; this first-order phase transition at which all valence electrons of a crystal are set free at once is now called the Mott transition (Fig. 13.1.1).

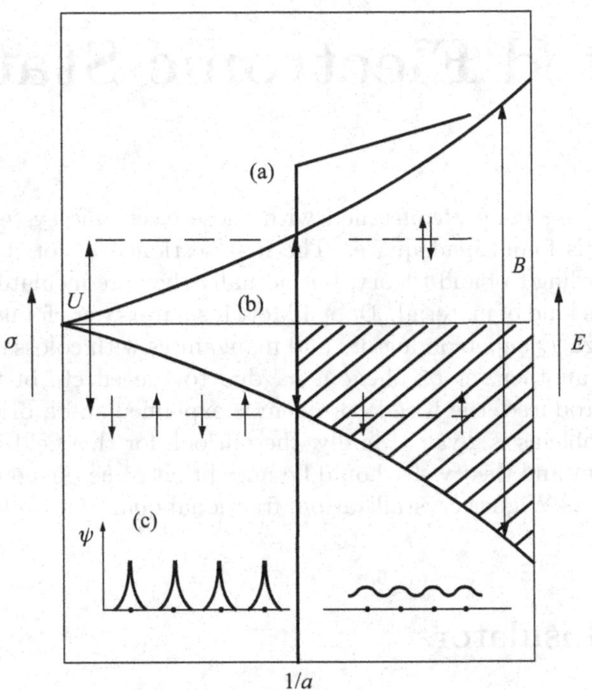

Figure 13.1.1 An illustration for the idealized Mott transition. (a) Electrical conductivity versus the inverse interatomic spacing; (b) bandwidth (half-filled 3s band of Na) versus the inverse interatomic spacing, also shown are Hubbard energy U (assumed to be independent of interatomic spacing) and site occupation on both sides of the Mott transition; (c) distribution of orbitals on both sides of the Mott transition.

From the tight-binding approximation of band theory, we may show that the bandwidth B should increase when the interatomic spacing decreases, due to the increasing overlap of neighboring atomic wavefunctions. On the other hand, electrical conductivity requires valence electrons to hop from one atom to another; this process may be expressed explicitly as a reaction involving charge fluctuation between neighboring atoms:

$$\mathrm{Na} + \mathrm{Na} \rightarrow \mathrm{Na}^+ + \mathrm{Na}^-.$$

So at a Na^- site, the double occupancy of an orbital at the same lattice site occurs. Since there is already an electron on the lattice site, if a second electron wants to occupy the same site, it must overcome the large Coulombic repulsive force. Thus the energy cost for double occupancy of an orbital on the same lattice site is called the Hubbard energy U. In the limit of isolated atoms,

$$U = I - A, \tag{13.1.1}$$

where I is the ionization energy and A is the electron affinity; for metallic elements, the values of U lie below 8–10 eV, but for nonmetallic elements, the values of U are higher.

Now we are ready to make an approximate estimate of the Mott transition. Since the band is half-filled, the average energy for electrons in a electron band of width B with N sites is about $(1/4)NB$. The ground state for a half-filled band is homogeneous: the probability that a site is occupied by either a spin-up or a spin-down electron is equal to 1/2; then the probability for a doubly occupied site (i.e., configuration Na^-) is equal to 1/4; and the same probability as for an unoccupied site (i.e., configuration Na^+). Thus the charge fluctuation for conduction requires an energy investment of about $(1/4)NU$. Hence the conclusion is that Mott transition occurs at

$$U = B. \tag{13.1.2}$$

For $U < B$, it is metallic; while for $U > B$, it is insulating. Since the bandwidth B depends on the atomic spacing but the Hubbard energy U is rather insensitive to atomic spacing, a change in the interatomic separation will cause a crossover from the metallic regime to an insulating one, i.e., a Mott transition can be described as an electron correlation-induced collective localization of all the free electrons, as shown in Fig. 13.1.1.

Now we can return to the subject of CoO: Its lattice constant is large enough to cause it to be on the insulating side, far from the Mott transition. It is a Mott insulator, i.e., it is in the highly correlated insulating state, in which electrons tend to minimize the Coulomb interaction energy by staying alone at each lattice site. In general, the ground state of most Mott insulators is also antiferromagnetic; however, it is not the magnetic interactions that drive these materials into the insulating state. Rather, it is Coulomb repulsions, which are still in operation even in the case of the paramagnetic state. We shall return to this topic in §13.1.3. Some substances, such as V_2O_3, are quite near the Mott transition, on the insulating side at ambient temperature and pressure but a change in temperature or pressure may bring them easily to the Mott transition. We shall discuss this subject further in Chap. 19.

13.1.2 Hubbard Model

Hubbard introduced a model Hamiltonian that included the possible competition of kinetic energy lowering by banding and Coulomb correlation energy lowering by localization. Hubbard considered that, for tight binding-like narrow band systems, the Coulomb interaction could be replaced by the on-site interaction U to the lowest order approximation, resulting in the widely studied Hamiltonian for an s band of the form

$$\mathcal{H} = \sum_{i,j,\sigma} t_{ij} c_{j\sigma}^{\dagger} c_{j\sigma} + U \sum_{i} n_{i\sigma} n_{i\bar{\sigma}}, \tag{13.1.3}$$

where $n_{i\sigma} = c_{i\sigma}^{\dagger} c_{j\sigma}$ is the number operator, $c_{i\sigma}^{\dagger}$ and $c_{j\sigma}$ are the creation and annihilation operators of electrons on atom i, and σ shows the spin. If we ignore the hopping integrals beyond nearest neighbors, the Hamiltonian contains three parameters: $\tilde{t}_0 = t_{ii}$, $\tilde{t} = t_{ij}$ (for i and j nearest neighbors), and U. It is easy to show that t_0 is the mean energy of the band, and t is equal to half the width of the band.

If $U = 0$ in (13.1.3), the problem becomes very simple: When we use the creation and annihilation operators of Bloch states given by

$$c_{i\sigma}^{\dagger} = \frac{1}{\sqrt{N}} \sum_{\boldsymbol{k}} \mathrm{e}^{i\boldsymbol{k}\cdot\boldsymbol{R}_i} c_{\boldsymbol{k}\sigma}^{\dagger}, \quad c_{i\sigma} = \frac{1}{\sqrt{N}} \sum_{\boldsymbol{k}} \mathrm{e}^{-i\boldsymbol{k}\cdot\boldsymbol{R}_i} c_{\boldsymbol{k}\sigma}, \tag{13.1.4}$$

the first term of (13.1.3) becomes

$$\sum_{\boldsymbol{k}\sigma} E_{\boldsymbol{k}\sigma} c_{\boldsymbol{k}\sigma}^{\dagger} c_{\boldsymbol{k}\sigma}, \quad E_{\boldsymbol{k}} = \sum_{ij} t_{ij} \mathrm{e}^{i\boldsymbol{k}\cdot(\boldsymbol{R}_i - \boldsymbol{R}_j)}, \tag{13.1.5}$$

where $E_{\boldsymbol{k}}$ is the energy of the Bloch state with the wave vector $\boldsymbol{k}$.

The meaning of the parameter U follows from a consideration of the limit of infinite lattice constant. Clearly when $\tilde{t}=0$, the Hamiltonian becomes diagonal and the energy becomes

$$E=\sum_i\left[\tilde{t}_0(n_{i\sigma}+n_{i\bar{\sigma}})+Un_{i\sigma}n_{i\bar{\sigma}}\right]=N_1\tilde{t}_0+N_2(2\tilde{t}_0+U), \tag{13.1.6}$$

where N_1 is the number of lattice sites occupied by one electron, and N_2 the number of sites occupied by two electrons. t_0 is therefore the energy needed to bind an electron on an isolated atom. t_0+U is the energy needed to attach the second electron with opposite spin, as shown in Fig. 13.1.1. Hence U is the Coulomb interaction energy of two electrons located in the same orbital of the same atomic site.

$$U=\int d\boldsymbol{r}_1\int d\boldsymbol{r}_2\,|\psi(\boldsymbol{r}_1-\boldsymbol{R})|^2\frac{e^2}{|\boldsymbol{r}_1-\boldsymbol{r}_2|}\,|\psi(\boldsymbol{r}_1-\boldsymbol{R})|^2\,. \tag{13.1.7}$$

This is the more precise definition of U in solids. In the ground state, one electron is accommodated at each atom, all electrons should have the same energy t_0, and $N_1=N$, $N_2=0$. In this limit we can find a strict localization of the electrons. The Hubbard approximation thus leads the band model to a local description.

The ground states for the Hubbard Hamiltonian are either itinerant states when $\tilde{t}\gg U$, or localized when $\tilde{t}\ll U$. We now look at the ground state of a system with finite lattice constant in which each lattice atom possesses one electron. The spin direction is taken to change from neighbor to neighbor (antiferromagnetic ground state). If we introduce into this system a further electron with a given spin, it can be accommodated at one of the $N/2$ atoms which already has an electron with opposite spin. The Pauli principle forbids its placement with any of the other $N/2$ atoms. The energy of this electron amounts to $\tilde{t}_0+U$ at any isolated atom. Due to the interaction between all $N/2$ states which can accept the electron, this energy splits up into a band centering around $\tilde{t}_0+U$ (now $\tilde{t}\neq 0$). The same arguments lead to the splitting up of the energy $\tilde{t}_0$ into a corresponding band. As long as the widths of the bands are smaller than the separation $(\tilde{t}_0+U)-\tilde{t}_0=U$, there will be a gap between the two bands. At a critical amount of splitting determined by the lattice constant the gap will disappear. The transition from the localized description to the band model then takes place.

This result can be derived quantitatively from the Hamiltonian in (13.1.3), but the calculations are too lengthy to reproduce here. In the example just considered, i.e., half-filled s-band and antiferromagnetic ground state, we can see the specific result in the figure for the density of states related to the ratio B/U. As shown in Figs. 13.1.2 and 13.1.3, for the large ratio, the curve only changes slightly; if the ratio is reduced to the critical value, the band splits into two separate subbands, and the gap of the subbands increases with reduced ratio B/U.

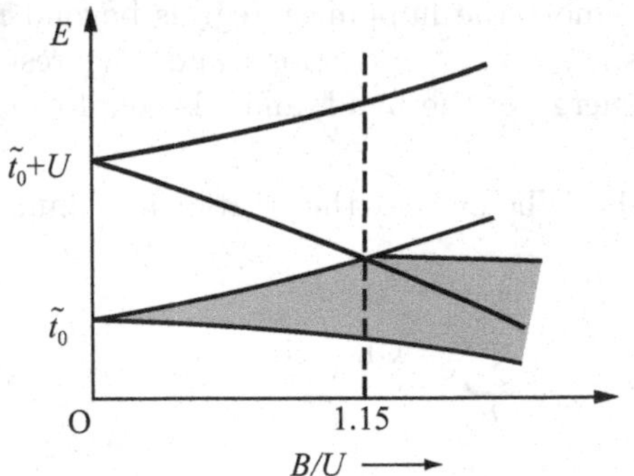

Figure 13.1.2 Transition from localized states to delocalized states in a half-filled energy band in the Hubbard model. t_0 is the energy needed to attach an electron to a free ion, t_0+U is the energy needed to attach a second electron to the same ion.

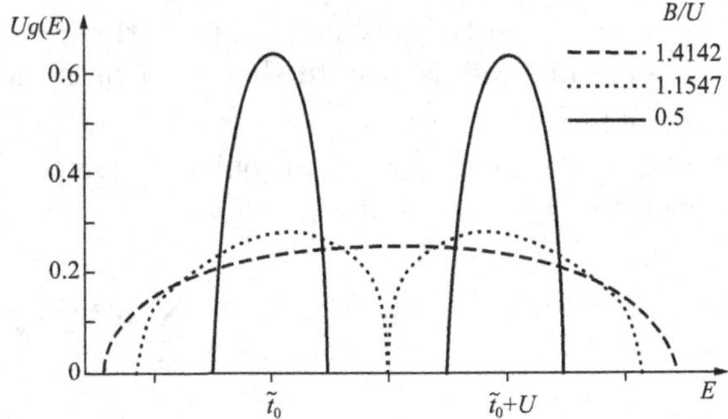

Figure 13.1.3 Density of states $g(E)$ showing band splitting by electron-electron interactions in the Hubbard model.

13.1.3 Kinetic Exchange and Superexchange

At exactly half-filling, for a large U, i.e., $\tilde{t}/U \ll 1$, the effective Hamiltonian is just the well-known antiferromagnetic Heisenberg model

$$\mathcal{H} = -J \sum_{ij} \boldsymbol{S}_i \cdot \boldsymbol{S}_j, \tag{13.1.8}$$

with $J = -4\tilde{t}^2/U$. This can be explained by the kinetic exchange process envisaged as follows: let us consider a pair of neighboring sites. If the spins are antiparallel, a virtual hopping process can create an intermediate pair state with one site empty and the other site doubly occupied with energy cost U. The associated energy gain according to second order perturbation theory is of order $-\tilde{t}^2/U$. If the spins are parallel, hopping is forbidden by the Pauli principle.

Kinetic exchange gives us the basic reason why most insulators are antiferromagnetic. In Fig. 13.1.4, the magnetic states of transition metal oxides and fluorides are tabulated: a great majority of these insulating crystals are indeed antiferromagnetic; while a few ionic compounds such as EuO, K_2CuF_4, CrO_2 are known to be ferromagnetic with low Curie temperatures. However, in most oxides and fluorides, this kind of direct kinetic hopping is almost impossible due to separation by the intervening oxygen or fluorine ions, so superexchange must be invoked.

Structures								
Rock Salt	TiO PM	VO PM		MnO 118 AF	FeO 198 AF	CoO 293 AF	NiO 520 AF	CuO Monoclinic 230 AF
Spinel	PM			Mn_2O_4 41 Ferri	Fe_3O_4 850 Ferri 119 (Verwey)	Co_3O_4 40 AF		
Corundum	Ti_2O_3 500 PM	V_2O_3 150 AF	α-Cr_2O_3 308 AF	α-Mn_2O_3 80 AF	α-Fe_2O_3 963 Slanted AF			
Rutile	TiO_2 DM	VO_2 340 PM	CrO_2 383 F	β-MnO_2 84 AF				
Rutile		VF_2 27-42 AF	CrF_2 53 AF	MnF_2 67 AF	FeF_2 78 AF	CoF_2 37 AF	NiF_2 73 AF	CuF_2 69 AF
Two Dimension				K_2MnF_4 42 Rb_2MnF_4 38 AF	Rb_2FeF_4 56 AF	K_2CoF_4 107 Rb_2CoF_4 101 AF	K_2NiF_4 97 Rb_2NiF_4 90 AF	K_2CuF_4 6 F

Insulator (or semiconductor) Metal

Figure 13.1.4 Summary of the magnetic states of various transition metal oxides and fluorides (compiled by J. W. Allen). From R. M. White and T. H. Geballe, *Long Range Order in Solids*, Academic Press, New York (1979).

Superexchange acquired its name because of the relatively large distances, occupied by diamagnetic ions, radicals or molecules, over which the exchange effect is realized. For example, in MnO, the Mn^{++} ions interact over a distance of 0.4 nm, and the overlap of their atomic d-orbitals is negligible. Surely the wave functions of these 'ligands' are modified by the presence of the magnetic ions, and they take part in a kind of exchange interaction. Kramers considered that this modification of wave functions had magnetic characteristics, which gave a kind of exchange interaction with other ions. Anderson further introduced the idea that there are covalently mixing d-states of magnetic ions and p-states of the ligand oxygen ion. In an antiferromagnet, such as MnO, which has the NaCl crystal structure, the spins in single (111) planes are parallel, while those in two adjacent (111) planes are antiparallel. The antiferromagnetic coupling is between second nearest-neighbor cations, such as

A = Mn^{++} and C = Mn^{++}, via an intervening (nearest neighbor) anion, B = O^{--}. The calculation of the exchange integral in this situation can be based on bonding and antibonding symmetrized LCAO. For example, if we assume that the configuration is perfectly ionic; each Mn ion would then have a single unpaired electron in a d-state, and the O ion would have, in the outermost occupied p states, two electrons with antiparallel spins as shown in Fig. 13.1.5.

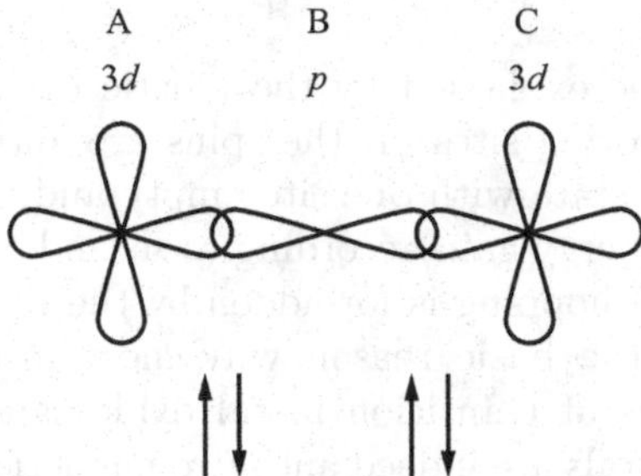

Figure 13.1.5 The spin configurations of four electrons in the 3d and p orbitals on sites Mn-O-Mn.

This situation may be generalized to cases with periodic structure, where the atomic orbitals should be replaced by Wannier functions, and the transfer integral between $\boldsymbol{R}_i$ and $\boldsymbol{R}_j$ sites becomes

$$b_{ij} = \langle w(\boldsymbol{r} - \boldsymbol{R}_i)|\, \mathcal{H}(\boldsymbol{R}_j)\, |w(\boldsymbol{r} - \boldsymbol{R}_j)\rangle\,, \tag{13.1.9}$$

where $w(\boldsymbol{r} - \boldsymbol{R}_i)$ is a Wannier function for an electron on an atom at $\boldsymbol{R}_i$. In fact, the superexchange may be regarded as a generalization of kinetic exchange to cases where electrons of the intervening oxygen atom participate in the virtual hopping processes.

According to Goodenough and Kanamori, exchange interactions between half-filled orbitals are described by the virtual hopping process as shown in Fig. 13.1.6(a). The antiferromagnetic coupling is derived to be

$$J_{ij}^{\text{kin}} = -2b_{ij}^2/4S^2U, \tag{13.1.10}$$

where S is the net spin of a magnetic atom and U is the Hubbard energy for the intermediate state when two electrons occupy the same orbital. This is known as the first Goodenough–Kanamori rule for superexchange.[a]

Furthermore, spin-up electrons may be transferred from a full orbital to an empty one [shown in Fig. 13.1.6(b)]; thus, the intraatomic exchange energy (Hund's rule coupling) Δ_{ex}/U^2 favors ferromagnetic exchange coupling:

$$J_{ij}^{\text{kin}} = +2b_{ij}^2\Delta_{ex}/4S^2U. \tag{13.1.11}$$

This is the second Goodenough–Kanamori rule for superexchange.

Goodenough also considered cases in which orbitals are in contact but without overlap, for instance, when the d_{xy} orbitals are at 90° with a p_x orbital as an intermediate link, where the overlap is zero by symmetry (see Fig. 13.1.7). This may be regarded as the realization of Hund's rules for parallel spins with orthogonal orbitals in an interatomic situation. The example demonstrates the importance of geometrical shapes of orbitals for the interatomic interactions. This aspect will be further discussed in the next subsection.

It should be noted that superexchange also plays an important role in ferrimagnetism, i.e., antiferromagnetic coupling of unequal spins on different atomic sites, so a net magnetization is realized. Thus, superexchange is also the physical foundation of other important classes of technical magnetic materials, such as ferrites and garnets.

[a] J. B. Goodenough, *Phys. Rev.* **100**, 564 (1955); *Phys. Chem. Solids* **6**, 287 (1958); J. Kanamori, *Phys. Chem. Solids* **10**, 87 (1959).

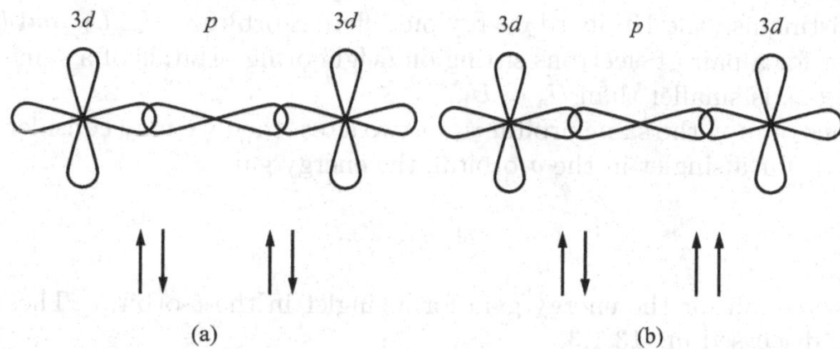

Figure 13.1.6 Virtual processes for the Goodenough–Kanamori rule of superexchange: (a) cation d-shell half filled (antiferromagnetic); (b) cation d-shell < half filled (ferromagnetic).

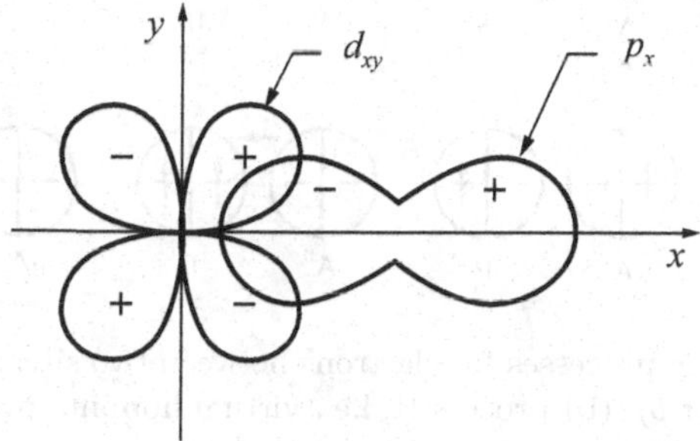

Figure 13.1.7 An illustration of the possibility for ferromagnetic alignment through superexchange.

13.1.4 Orbital Ordering versus Spin Ordering

The orbital degeneracy of d-electrons introduces further richness into the physics of Mott insulators, and the Hubbard model may be generalized to the case in which each site has n-fold degenerate orbitals.

First, we shall examine the simplest case, i.e., two electrons on two sites A and B, where each site has two-fold degenerate d orbitals, ψ_a and ψ_b. We may take two e_g orbitals in octahedral sites, i.e., $d_{x^2-y^2}$ and $d_{3z^2-r^2}$ as an example. Now, we consider a diatomic molecule in which each atom has two orbitals (see Fig. 13.1.8). We are facing the Heitler–London problem in a new context, two electrons on two sites with the added complexity of two orbitals. We may emulate the Heitler–London treatment of H_2; however, we should note the difference: instead of only 4 different low-energy states of H_2 classified into 1 singlet and 1 triplet; now we have 16 different states, to be classified into 4 singlets and 4 triplets. We shall not go into the theoretical details here but be content with making some general observations.

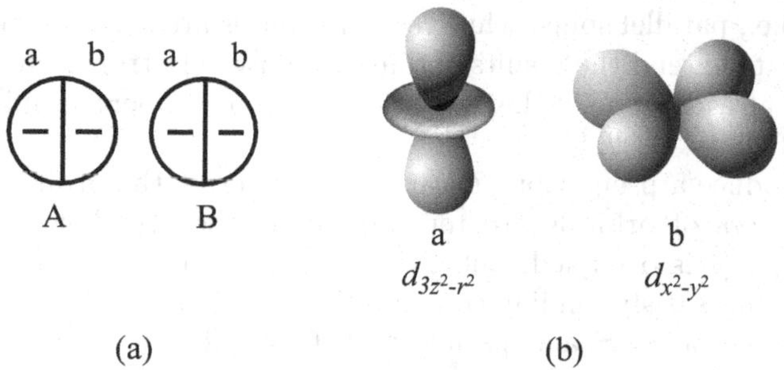

Figure 13.1.8 Schematic diagram of a e_g molecule for two orbitals a and b on two sites A and B.

We should distinguish the Hubbard energy on different orbitals, $U_{\rm a}$, $U_{\rm b}$ and $U_{\rm ab}$, where the last is the energy cost for a pair of electrons sitting on neighboring orbitals of a same site. Intuitively, it is plausible that $U_{\rm ab}$ is smaller than $U_{\rm a}$ or $U_{\rm b}$.

If two electrons occupy the same orbital ψ_a at two sites, then we may consider the virtual process I in Fig. 13.1.9(a): for a singlet in the a-orbital, the energy gain is

$$E_s \approx -\frac{4\tilde{t}_{\rm a}^2}{U_{\rm a}}, \tag{13.1.12}$$

and a similar expression for the energy gain for a singlet in the b-orbital. These are results of the kinetic exchange discussed in §13.1.3.

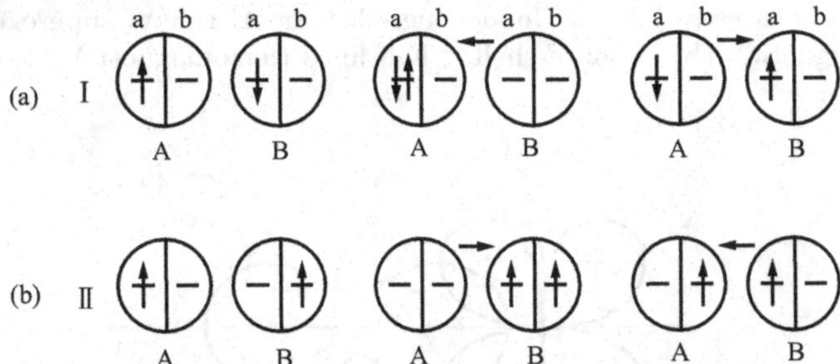

Figure 13.1.9 Virtual hopping processes for electrons between two sites: (a) process I, i.e., virtual hopping process in a single orbital (a or b); (b) process II, i.e., virtual hopping process in mixed orbitals (a and b).

For parallel spins on the same orbital at two sites, no hopping can take place; thus there is no energy gain. Other two singlets and two triplets occupy mixed a-b orbitals, so the triplets can have lower energy. Let us consider the virtual process II in Fig. 13.1.9(b): the intermediate state costs energy $U_{\rm ab}$-J where J is the energy gain due to Hund's rule coupling, so the energy gain for the triplet is about

$$E_t \approx -\frac{\tilde{t}_{\rm a}\tilde{t}_{\rm b}}{U_{\rm ab} - J}. \tag{13.1.13}$$

Since, roughly, $U_{\rm a} = U_{\rm b} = U > U_{\rm ab} = U - 2J$, so the triplets have lower energy than the singlets, and the ground state is the triplet with the lowest energy.

It is interesting to observe that intra-atomic exchange (Hund's rule coupling) may induce intersite ferromagnetic exchange by reducing the values of denominators in the energy expressions. This is a new type of ferromagnetic insulator that we have encountered. However, the ferromagnetic spin order in this case arises from the fact that electrons occupy different orbitals. We are already acquainted with the rule that the symmetric spin wave functions must be accompanied by the antisymmetric orbital wave functions, and *vice versa.* Thus, for degenerate orbitals, we may envisage a new kind of order, i.e., orbital order. Since the degenerate orbitals may have different shapes or orientations, so orbital order is related to the arrangement of orbital shapes or orientations on sites. Staggered orbitals have antisymmetric orbital wave functions which are accompanied by symmetric spin wavefunctions, i.e., parallel spins; while regular orbitals are accompanied by anti-parallel spins.

Now we are ready to extend the results obtained in a two-electron system to a lattice of localized electrons, as we have already done in the derivation of the Heisenberg Hamiltonian from the Heitler–London treatment of H_2 in §11.2.4.

Now we may introduce a pseudospin $\boldsymbol{T}$ variable to describe the orbital ordering on lattice sites: the two possible choices of orbitals are represented by the pseudospin $\boldsymbol{T}$, whose z component $T_z = +1/2$, when $d_{x^2-y^2}$ is occupied; and $T_z = -1/2$, when $d_{3z^2-r^2}$ is occupied. Three components of this pseudospin satisfy similar commutation relations to those of spin operators. There are interactions between spins $\boldsymbol{S}$ and pseudospins $\boldsymbol{T}$ of different ions. The following generalized Heisenberg Hamiltonian is used to account for these generalized interactions

$$\mathcal{H} = -\sum_{i<j} [J_S (\boldsymbol{S}_i \cdot \boldsymbol{S}_j) + J_T (\boldsymbol{T}_i \cdot \boldsymbol{T}_j) + J_{ST} (\boldsymbol{S}_i \cdot \boldsymbol{S}_j) (\boldsymbol{T}_i \cdot \boldsymbol{T}_j)], \tag{13.1.14}$$

where the exchange and pseudo-exchange coupling constants J_S and J_T originate from quantum mechanical processes with intermediate virtual states, and J_{ST} is the coupling constant between exchange and pseudoexchange. Rotational symmetry in the spin space leads to the inner product of the interactions between spins. Though its extension to the interaction between pseudospins may be somewhat questionable, it is commonly adopted for simplicity. When more than two orbitals are involved, depending on orbitals, many different situations can be realized, and spins and orbital pseudospins are intimately coupled. Furthermore, the transfer integral t_{ij} depends on the direction of bonds ij which are also dependent on the pairing of the orbitals $d_{x^2-y^2}$ and $d_{3z^2-r^2}$. This induces anisotropy of the Hamiltonian in the pseudospin space as well as in the real space.

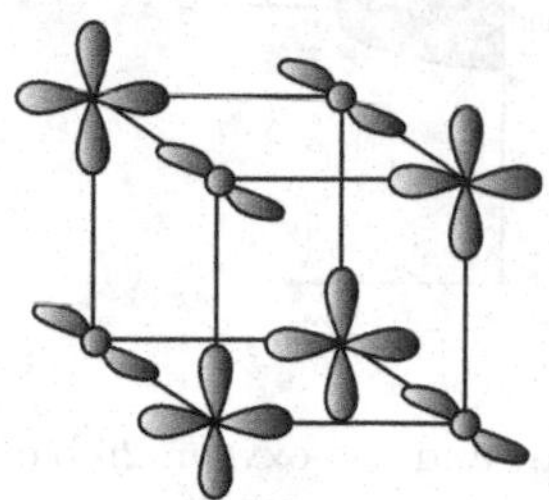

Figure 13.1.10 An example of the staggered orbital order (pseudospin antiferromagnetism): alternation of $d_{x^2-y^2}$ and $d_{3z^2-r^2}$ orbitals in a simple cubic lattice.

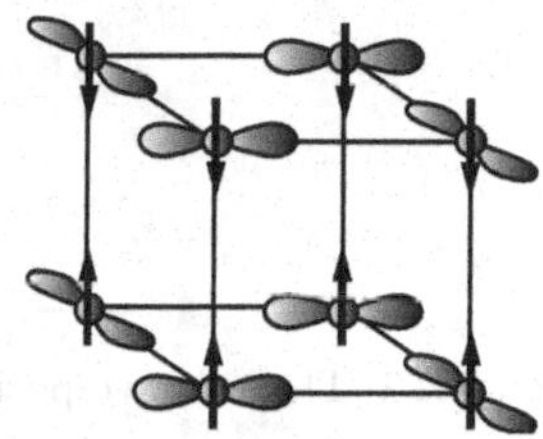

Figure 13.1.11 Orbital order and spin order of $LaMnO_3$.

Thus, for orbitals on different lattice sites, we can define the orbital order by its pseudospins. Using the pseudospin language, we may call the staggered orbital order as orbital antiferromagnetism; and the regular orbital order as orbital ferromagnetism. Figure 13.1.10 is an example of staggered orbital order, i.e., orbital antiferromagnetism. It should be emphasized that these terms are by no means connected with antiferromagnetism or ferromagnetism of orbital magnetic moments, but indicate ordered arrangement of pseudospins on the lattice, or in other words, staggered or regular orbital order. Since spins and orbital pseudospins are coupled together, in general, spin ferromagnetism favors pseudospin antiferromagnetism (staggered orbital order); while spin antiferromagnetism favors pseudospin ferromagnetism (regular orbital order).

Now we would like to inspect the real situation of Mott insulators: Let us consider K_2CuF_4, $CaMnO_3$ and $LaMnO_3$ as examples. Indeed K_2CuF_4 is found to be a ferromagnet, due to staggered orbital order; $CaMnO_3$ is found to be an antiferromagnet, due to regular orbital order; these examples seem correct. However, $LaMnO_3$ is found to be an antiferromagnet, why? Its arrangements of orbitals and spins on the lattice are shown in Fig. 13.1.11. We find that a planar ferromagnetic spin coupling is induced by the staggered orbital order in the planes, while antiferromagnetic spin coupling between planes is induced by the regular orbital order along the c-axis. Actually $LaMnO_3$ is a ferromagnet disguised as an antiferromagnet: its magnetic structure consists of antiferromagnetically coupled ferromagnetic layers, and 2/3 of the nearest neighbor sites of each magnetic ion are occupied by parallel spins. By the way, the Jahn–Teller effect couples the orbital order and the crystal lattice order. Surely orbital order, together with spin order, will play conspicuous roles in the physics of Mott insulators and doped Mott insulators.

13.1.5 Classification of Mott Insulators

Mott insulators are insulators with strong electron correlation effects: the partially filled d-band is split into a set of Hubbard subbands. However, these subbands are sandwiched between various $4s$ and $2p$ bands; their mutual arrangements will determine their insulating behavior. According to a classification scheme,[b] Mott insulators are classified into two types: the criterion is the relative values of the Hubbard energy, or gap, U and the charge transfer energy, or gap, Δ. The smaller gap

[b] J. Zanen, G. A. Sawatzty, and J. W. Allen, *Phys. Phys. Lett.* **55**, 418 (1985).

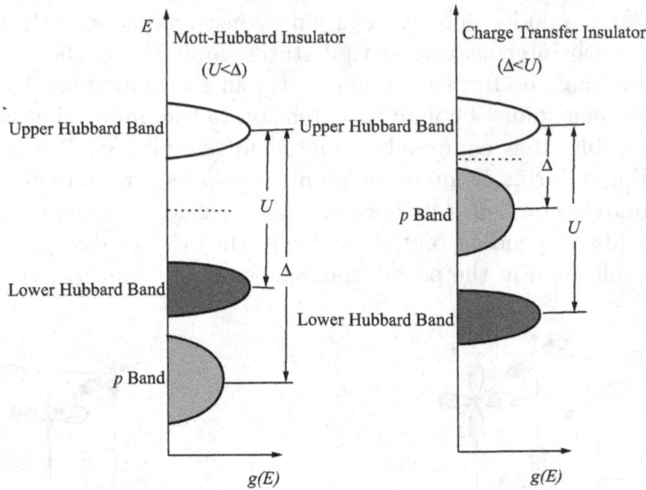

Figure 13.1.12 The relative positions of Hubbard subbands and the oxygen $2p$ band (schematic).

is of crucial importance for the transport behavior. If $\Delta > U$, the insulator is called Mott–Hubbard (MH) type; if $U > \Delta$, it is called charge transfer (CT) type, as shown in Fig. 13.1.12.

In MH insulators, the charge excitations for transport are d-electrons. But in CT insulators, there is overlap of the lower subband and $2p$ band, and the top of the $2p$ band is higher than that of the lower subband, so the lowest excitation is from the top of the $2p$ band to the bottom of the upper Hubbard subband. This excitation creates a d-like quasi-particle and a p-hole. Figure 13.1.13 shows the variation of experimentally determined gap energies for a series of compounds $LaMnO_3$ and YMO_3 (here M denotes some $3d$ transition metal element). A MH type of energy gap is observed for early members(indicated by full symbols), it increases in magnitude as the number of $3d$ electrons increases; by comparison, the CT type of energy gap between the O-$2p$ band and upper Hubbard band becomes smaller. The Mott–Hubbard gap around M = Cr decreases with further increase

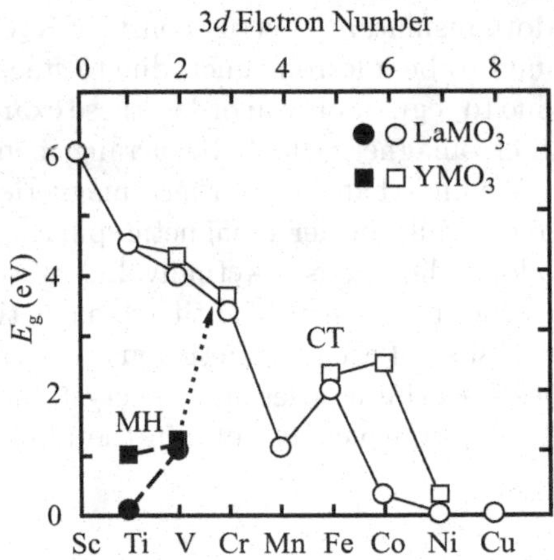

Figure 13.1.13 Optically determined gaps in Mott insulators in $LaMnO_3$ and YMO_3 (full symbols are MH gaps, open symbols are CT gaps). Taken from T. Arima *et al.*, *Phys. Rev. B* **48**, 17006 (1993).

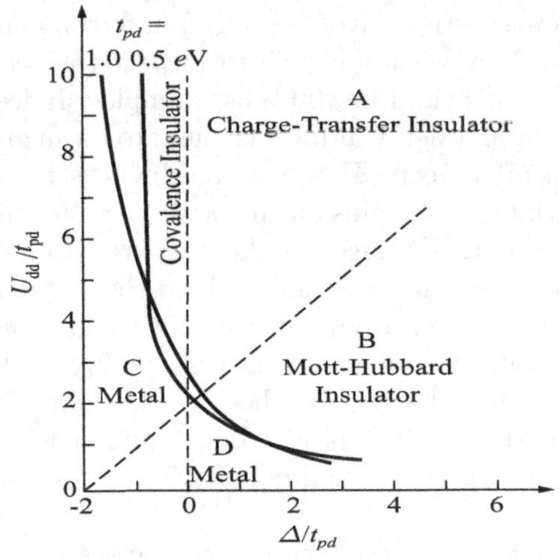

Figure 13.1.14 The Zaanen, Sawatsky and Allen diagram as modified by D. D. Sarma, here t_{pd} is the hybridization strength between d-electrons in transition metal ions and p-electrons in oxygen ions. Taken from C. N. R. Rao, *Chem. Comm.* **19**, 2217 (1996).

of d-electrons and eventually is closed for M = Ni and Cu. This tendency is equivalent to the increase of covalency between d-electrons of metal atoms and p-electrons of oxygen atoms in the first row transition metal perovskites. This is characterized by the metal oxygen hybridization strength t_{pd}. This parameter may be introduced into this classification scheme. Such a modified d-electron phase diagram $(U/t_{pd}, \Delta/t_{pd})$ is shown in Fig. 13.1.14. The dotted diagonal not only divides Mott insulators into region A (charge transfer insulator) and B (Mott–Hubbard insulator), but also divides the metallic region into C and D; further, a covalent insulator region E is marked, in which t_{pd} plays a crucial part.

§13.2 Doped Mott Insulators

13.2.1 Doping of Mott Insulators

Doping is an important means of modifying the physical properties of band insulators as well as semiconductors, the situation is just the same for Mott insulators. Mott insulators are characterized by an integer filling factor, i.e., the d-bands are filled with an integer number of electrons, or in other words they are stoichiometric. The filling factor may be increased or decreased by chemical doping. The former is called electron doping, the latter hole doping. The usual chemical doping process is the substitution of ions with different valence for cations (or anions) in the parent compound. For instance, we may substitute divalent Sr ions for trivalent La ions in the parent compounds $LaTiO_3$ (a Mott–Hubbard insulator) and $LaCuO_4$ (a charge transfer insulator) with $x = 0.125$, to produce $La_{1-x}Sr_xTiO_3$ and $La_{2-x}Sr_xCuO_4$. Though both parent compounds are antiferromagnetic insulators with $S = 1/2$, after doping both become metals with anomalous properties, the titanate is a nonmagnetic heavy Fermi liquid, while the cuprate is a high T_c superconductor.

It should be noted that the hole doping of Mott–Hubbard insulators is quite different from that of charge transfer insulators. For the former, the Fermi level lies above the top of the lower Hubbard subband, and doping creates holes in the d-bands. For the latter, the highest occupied band is the oxygen $2p$, the Fermi level lies above the top of the $2p$ band, so doping creates holes in the $2p$ band. However, for electron doping, the situations are quite similar for both types of Mott insulators, i.e., to put electrons in the upper Hubbard subband. Furthermore, hole doping and electron doping of charge transfer insulators are markedly different. While the majority of high T_c superconductors are hole-doped, only $Nd_{2-x}Ce_xCuO_4$ is electron-doped. It should be noted that a doped Mott insulator not only introduces extra charge carriers into the original insulator, but also introduces extra spins into the original ordered antiferromagnet, making the physics of doped Mott–Hubbard (MH) insulators complex and interesting.

Nonstoichiometry, or altered stoichiometry, of a compound's composition also modifies its d-band filling number. A well-known example of this is $YBa_2Cu_3O_7$(YBCO) with variable oxygen content. Figure 13.2.1 shows the range of filling control (control of d-band filling numbers) for some $3d$ transition-metal oxides with perovskite and perovskite-like structures. Black bars indicate the range of solid solution (mixed crystal) compounds so far successfully synthesized.

Substitution not only plays an important role in the control of filling but also brings distortions into structures, so it may indirectly reduce the band width B of compounds. So filling control, together with band width control, determines the schematic metal-insulator phase diagram with the relative electron correlation strength represented by U/B as ordinate and the band filling of $3d$ band as abscissa (Fig. 13.2.2). From this diagram we may see that YMO_3 shows stronger electron correlation than $LaMO_3$ due to the reduced B introduced by the distorted oxygen octahedra. The integer-filled $3d$ transition-metal oxides with the perovskite structure are mostly Mott insulators, apart from $LaCuO_3$ and $LaNiO_3$. A fractional valence or filling can drive the system metallic, yet in some cases the compounds remain insulating for large U/B.

13.2.2 Cuprates

Cuprate high T_c superconductors are doped charge transfer (CT) insulators; their crystal structures are mostly perovskite-like, such as K_2CuF_4 displayed in Fig. 13.2.3.

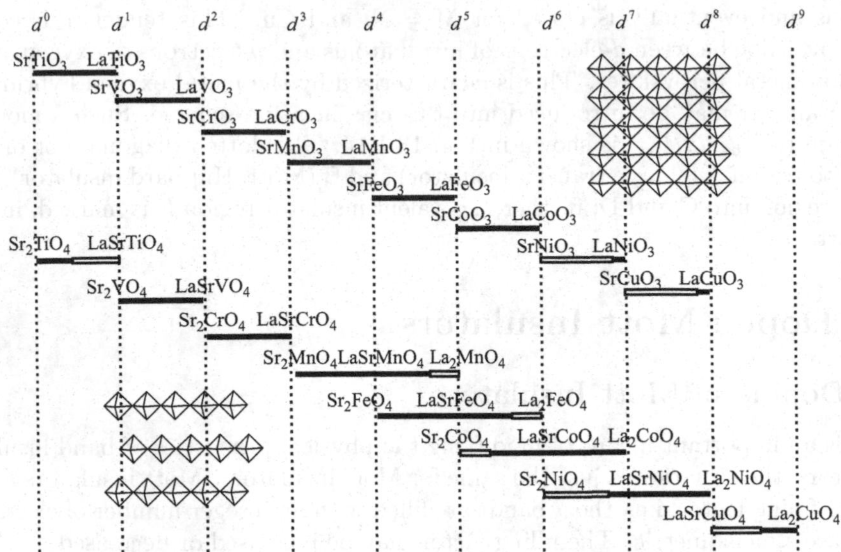

Figure 13.2.1 A guide map for the synthesis of filling-controlled $3d$ transition-metal oxides with perovskite and layered perovskite (K_2NiF_4-type) structures. Taken from M. Imada *et al.*, *Rev. Mod. Phys.* **70**, 1037 (1998).

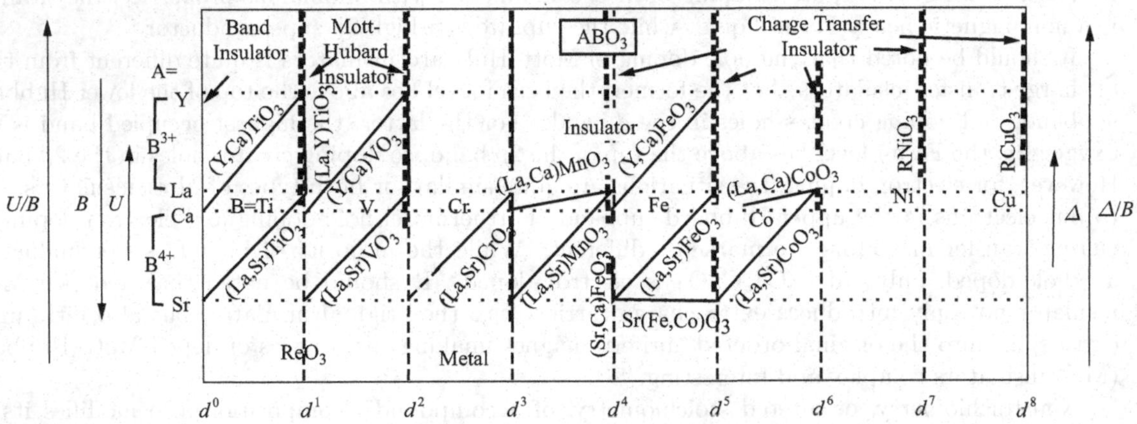

Figure 13.2.2 A schematic metal-insulator phase diagram for $3d$ transition-metal oxides with the perovskite structure. Taken from M. Imada *et al.*, *Rev. Mod. Phys.* **70**, 1037 (1998).

The crucial part played by CuO_2 layers in carrying current is universally recognized (Fig. 13.2.4). The CuO_2 active layers are sandwiched between block layers, which act as charge carrier reservoirs. The number of active layers n may vary from 1–4. For the La compound, $n = 1$; For Bi-Sr-Ca-Cu-O (BSCCO) compound, $n = 2$. For the Y-Ba-Cu-O (YBCO) compound, in addition, Cu-O chains are also important. All parent compounds are Mott insulators of the charge transfer type with antiferromagnetic order: for the La compound $T_N = 330$ K, for the Y compound $T_N = 500$ K.

We may take the comparatively simpler compound $La_{2-x}Sr_xCuO_4 (0 < x < 0.3)$ as an example to analyze its electronic structure. Cu atoms are situated in the centers of distorted oxygen octahedra, the Cu-O distance for O atoms that lie outside the CuO_2 plane is 0.23 nm; surely that is larger than 0.19 nm for O atoms in the plane. The chemical bonds may be described by the molecular orbital method: the stability of the structure of the CuO_2 plane can be ascribed to the σ bonds formed by Cu $3d_{x^2-y^2}$ and O $2p_{x,y}$ (see Fig. 13.2.5), while the much weaker π-bonds between Cu $3d_{x^2-y^2}$ and O $2p_z$ as well as other wavefunctions with slight overlap may be ignored. From energy band calculation,

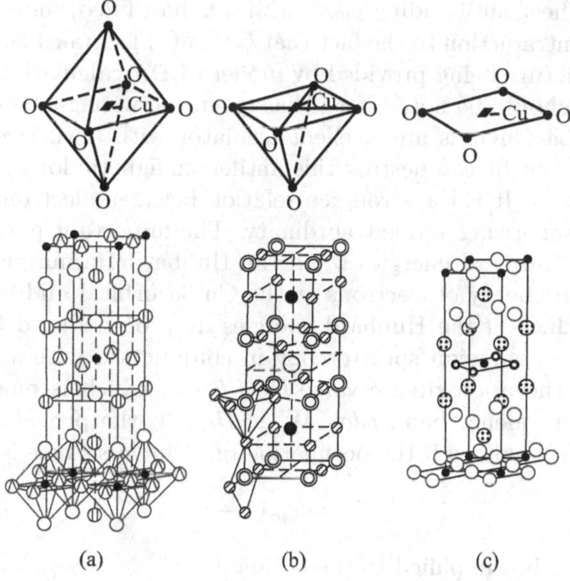

Figure 13.2.3 Crystal structures of some typical cuprates: (a) La_2CuO_4; (b) YBa_2CuO_7; (c) $NdCuO_4$.

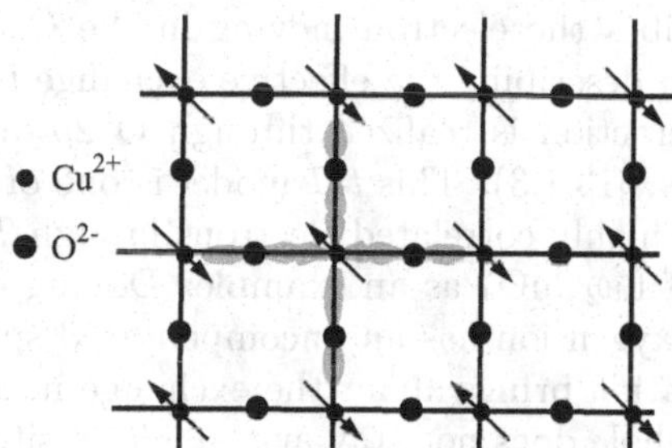

Figure 13.2.4 Schematic diagram of CuO_2 in cuprates in real space with arrows indicating the spin orientations and shaded areas the chemical bonds.

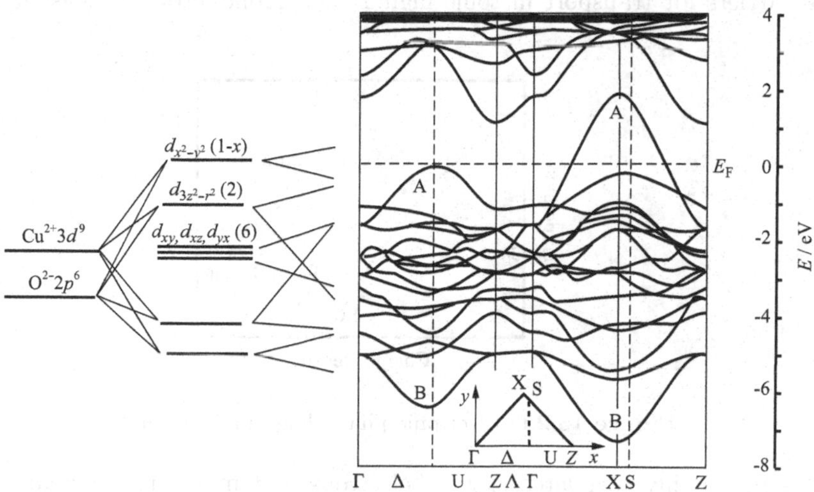

Figure 13.2.5 Electronic structure of the CuO_2 plane in cuprates.

it is found that the highest antibonding $pd\sigma^*$ orbital is half filled, indicating metallic character for La_2CuO_4. This is in contradiction to the fact that La_2CuO_4 is an insulator. In spite of many details of the energy band structure being provided by modern LDA calculations, the insufficiency of these single-particle theories should be noted, for it has given the wrong answer for the ground state.

Actually undoped La_2CuO_4 is an excellent insulator with antiferromagnetic long-range order. Only doping with sufficient Sr can destroy this antiferromagnetic long range order and turn it into a metallic superconductor. It is the strong correlation between electrons that makes the electronic properties of cuprates surprising and extraordinary. The important parameter to characterize electron correlation is the Hubbard energies U_i in the Hubbard Hamiltonian, here $U_i = U_{dd}$ or U_{pp}, denoting the correlation energy of electrons at the Cu $3d$ orbital and O $2p$ orbital respectively. It was shown that the values of the Hubbard energies may be deduced from experimental data on inner-shell photo-electron emission spectroscopy in conjunction with some theoretical model. For CuO_2 planes, U_{dd} has the approximate value $U = 7$ eV, which is much higher than the valence bandwidth $t \approx 2$ eV for valence band $pd\sigma$. When $U \gg t$, the probability of two $3d$ holes on the same atomic site is very small, so if the occupation of a single sd hole is expressed as

$$\tilde{c}_{i\sigma} = c_{i\sigma}(1 - n_{i\bar{\sigma}}), \tag{13.2.1}$$

the Hubbard model may be simplified to the t-J model

$$\mathcal{H} = t\sum_{ij\sigma} \tilde{c}^{\dagger}_{i\sigma}\tilde{c}_{j\sigma} - J\sum_{i<j} \boldsymbol{S}_i \cdot \boldsymbol{S}_j, \tag{13.2.2}$$

in which the first term describes the electron moving in the CuO_2 plane, while the second term is the Heisenberg Hamiltonian describing the effective exchange interaction of neighboring spins of $3d$ holes. This exchange interaction is realized through O $2p$ orbitals between Cu atoms by the superexchange mechanism (see §13.1.3). This t-J model is one of most popular models to describe the electronic behavior of the highly correlated electrons in high T_c superconductors.

We may take Sr-doping of La_2CuO_4 as an example: Doping creates p-holes in oxygen ions. A p-hole means a p-lobe of an oxygen ion has an uncompensated spin that is surrounded by spins on the Cu sites. So p-d hybridization brings about the exchange interaction between Cu and O spins. It may be envisaged that a p-hole does not stay at a single O site; instead it spreads out into four sites, occupying an orbital has the same symmetry as the d-orbital with the spin at the central Cu site. This p-state binds to the central d-spin, forming a pair of up and down spins called a Zhang–Rice singlet.[c] A dilute gas cloud of Zhang–Rice singlets may drive the system into the conducting state. There is some experimental evidence supporting the idea that it is Zhang–Rice singlets to act as charge carriers for transport in some high T_c superconductors; at least it is valid for light doping levels.

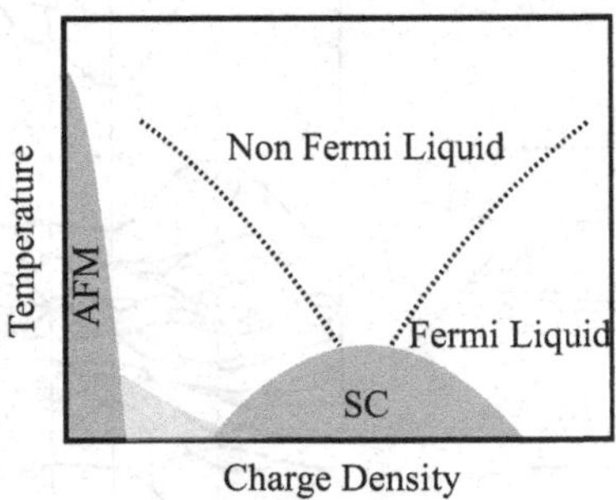

Figure 13.2.6 Electronic phase diagram for cuprates.

Surely, it is this highly correlated state of electrons that makes the normal properties of high T_c superconductors extremely anomalous and difficult to account for. Figure 13.2.6 is a typical electronic phase diagram for cuprates: Without doping, it is an antiferromagnetic insulator. At some

[c]F. C. Zhang and T. M. Rice, *Phys. Rev. B* **37**, 3759 (1988); **41**, 7243 (1990).

doping level ($x = 0.02$) the antiferromagnetic long range order collapses, and in its stead a short-range 2D antiferromagnetic order with metallic conductivity appears. Then, at about $x = 0.05$, superconductivity appears. Between $x = 0.02$–0.05, according to general consensus, it is a spin glass. At the optimum doping level of $x = 0.16$, T_c is highest. On the underdoped side ($x < 0.16$), there appears a pseudo-gap in the normal state above T_c; the true nature of this pseudo-gap is still unsettled (see §18.3.3). The physical properties of the underdoped normal state are extremely anomalous. The rich physics of cuprates appears to be still unexhausted after more than 15 years of worldwide intensive research. We shall discuss some of the problems related to the cuprate superconductivity in Chap. 18.

13.2.3 Manganites and Double Exchange

Manganite perovskites $T_{1-x}D_xMnO_3$, where T is a trivalent cation (e.g., La^{3+}) and D is a divalent cation (e.g., Ca^{2+}, Sr^{2+}, or Ba^{2+}), have been the focus of intense study since 1993 due to their colossal magnetoresistance (CMR) effects. The magnetic and orbital structure of the parent compound $LaMnO_3$ may be simply stated: it has a distorted perovskite structure in the orthorhombic system; it is a Mott insulator of charge transfer (CT) type and a peculiar antiferromagnet with in-plane ferromagnetism. With hole-doping by Sr in the composition range $x \approx 0.2$–0.4, charge balance is maintained by the creation of a fraction x of Mn^{4+} ions (d^3) distributed randomly throughout the crystal, while the $(1-x)$ manganese ions are in the Mn^{3+} (d^4) state. Resistivity versus temperature for different compositions of $La_{1-x}Sr_xMnO_3 (0 < x < 0.4)$ is shown in Fig. 13.2.7, in which resistivity drastically decreases when the temperature falls below the Curie temperature T_c. At temperatures above the ferromagnetic Curie temperature T_c, the resistivity behaves like an semiconductor with $(d\rho/dT) < 0$, but at temperatures lower than T_c, metallic behavior with $(d\rho/dT) > 0$ is indicated. It shows that the onset of ferromagnetism is coincident with the insulator-metal transition. This is explained by the double exchange mechanism for mixed valent oxides first proposed by C. Zener.[d]

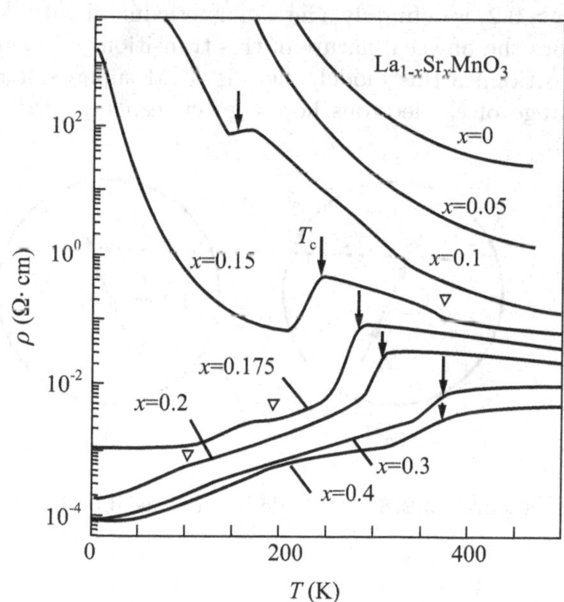

Figure 13.2.7 Resistivity versus temperature for $La_{1-x}Sr_xMnO_3 (0 < x < 0.4)$. Taken from A. Urshibara *et al.*, *Phys. Rev. B* **51**, 14103 (1995).

It should be noted that there is an important difference in electronic structure between cuprates and manganites. In cuprates, due to the Jahn–Teller distortion of the CuO_2 sheets, a large energy splitting is found between $d_{3z^2-r^2}$ and $d_{x^2-y^2}$ orbitals, and only the $d_{x^2-y^2}$ orbital is relevant, i.e.,

[d] C. Zener, *Phys. Rev.* **82**, 403 (1951), P. W. Anderson and H. Hasegawa, *Phys. Rev.* **118**, 675 (1955).

the orbital degree of freedom is quenched. However, for doped manganites, the orbital degree of freedom is still active, as exemplified by the double exchange which will be discussed below.

In the undoped state, $x = 0$, each ion on the lattice will remain Mn^{3+}; its ionic state is $t_{2g}^3\ e_g^1$, i.e., three d-electrons in t_{2g} orbitals plus one electron in e_g orbital, and the Hund's rule total spin is $S = 2$. Real hopping motion of electrons is forbidden by the Mott correlation effect. After doping with a finite value of x, an x-fraction of Mn^{3+} ions in random lattice sites are transformed into Mn^{4+} ions in the ionic state t_{2g} with the Hund's rule total spin $S = 3/2$, retaining only the core of Mn^{3+} ion, and making the e_g electrons free to move between lattice sites, which gives a large conductivity. The motion of e_g electrons on the background around the t_{2g} cores is described by the double exchange Hamiltonian

$$\mathcal{H} = -\tilde{t}\sum_{ij}(c_{i\sigma}^{\dagger}c_{j\sigma} + \text{h.c.}) - J_{\text{H}}\sum_{i}\boldsymbol{S}\cdot\boldsymbol{s}_i, \tag{13.2.3}$$

where the first term is the one-electron band term, and the second term describes the exchange interaction between the electron spin $\boldsymbol{s}_i$ of the itinerant electrons and the $\boldsymbol{S}$ of the t_{2g} orbitals of the localized electrons. If the hopping integral $\tilde{t}$ is much smaller than the intraatomic exchange J_{H} (Hund's rule coupling between itinerant e_g with localized t_{2g}), then the energy can be simplifies to

$$E = -J_{\text{H}}S \pm \tilde{t}\cos(\theta/2). \tag{13.2.4}$$

This is the energy associated with a modification of the transfer integral, where θ is the angle between neighboring spins. Figure 13.2.8 shows that two hopping states are not parallel. The bandwidth is, in fact, controlled by the directions of neighboring spins; it is wide when the spins are parallel, narrow when they are antiparallel. In contrast to superexchange, in which the exchange interaction is due to virtual hopping processes of electrons between neighboring atoms, real electron hopping processes between neighboring atoms is the foundation of double exchange (see Fig. 13.2.9).

Since $LaMnO_3$ is an insulator with planar ferromagnetic spins antiferromagnetically coupled, while $La_{1-x}Sr_xMnO_3(x > 0.2)$ is definitely a ferromagnetic metal with double exchange interactions, it is interesting to explore the physical nature of this transition. de Gennes used a pure spin model to account for this transition; in this model, there is AFM superexchange between fixed t_{2g} cores, as well as double exchange of e_g electrons hopping between t_{2g} sites. We may envisage that the

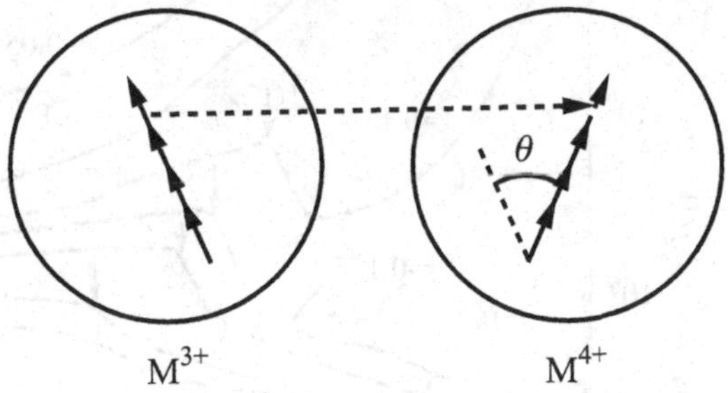

Figure 13.2.8 The angle between neighboring spins.

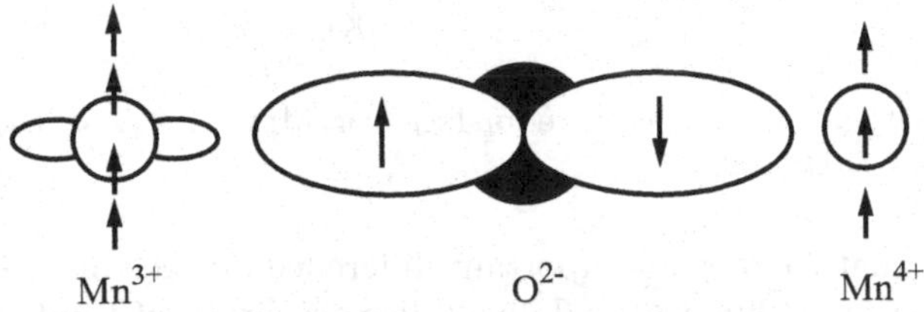

Figure 13.2.9 Mechanism of double exchange for a manganite.

sublattice magnetizations gradually tilt from antiparallel orientation toward parallel orientation. The spins on (100) Mn planes are uniformly polarized, but the angle θ subtended by the spins on neighboring planes are assumed to be dependent on the doping level x[1]. When $x = 0$, $\theta = \pi$; when x reaches some critical value x_c, then $\theta = 0$. The energy density is composed of two terms: the first term is due to superexchange between t_{2g} cores; the second is due to double exchange, i.e., the spin-orientation dependent hopping energy of e_g electrons,

$$E(\theta) = J\cos(\theta) - x[4\tilde{t} + 2\tilde{t}\cos(\theta/2)], \tag{13.2.5}$$

where, of course, classical spins are assumed and in the large J limit ($J/\tilde{t} \to \infty$), x is also assumed to be sufficiently small that the e_g electrons are situated near the band bottom. Minimizing $E(\theta)$ with respect to θ, we get

$$\cos(\theta/2) = \frac{\tilde{t}x}{2J}. \tag{13.2.6}$$

The solution gives a reasonable physical picture for the transition from an AFM insulator to a FM metal with doping level x: at $x = 0$, $\theta = \pi$, AFM state; at small x, $x \ll 1$, $\theta \approx \pi - (\tilde{t}x/J)$, the canted AFM state; at the critical value $x_c = 2J/\tilde{t}$, $\theta = 0$, FM state. So a gradual transition, from the canted phase for $0 < x < x_c$ to the full ferromagnetic alignment for $x > x_c$, is predicted.[e] However, it should be noted that the real situation is more complicated, for there is an orthorhombic-rhombohedral structural phase transition near $x \approx 0.175$.

13.2.4 Charge-Ordering and Electronic Phase Separation

Charge-ordering can occur in widely different systems, ranging from wholly localized systems such as alkali halide ionic crystals to a wholly delocalized one such as the electron crystallization envisaged by Wigner. For an ordinary delocalized system of electrons, the kinetic energy due to the Pauli exclusion principle is much more important than the potential energy due to Coulomb repulsion. However, the situation drastically changes when the electron concentration is very low. To explain it, we may use the jellium model, i.e., a system of homogeneous electron gas with a positively charged background spreading uniformly over it. Let r_0 denote the mean radius of the volume for a single electron, defined by $(4\pi/3)r_0^3 = \rho^{-1}$, where ρ is the electron density. The average kinetic energy of an electron due to the uncertainty principle is $\delta\varepsilon_{\text{kin}} = (\Delta p)^2/2m \approx 1/2mr_0^2$, but the average potential energy due to the Coulomb repulsion is $\delta\varepsilon_{\text{pot}} \approx e^2/r_0$. Since the former decreases with the inverse square power of r_0, while the latter decreases with inverse r_0, it is predicted that, for very low concentration, the potential energy becomes larger than the kinetic energy. If this situation happens, then the electrons will crystallize into a lattice to minimize the Coulomb repulsion, while the kinetic energy will maintain the zero-point motion of electrons around the equilibrium position. This change from a homogeneous to an inhomogeneous charge distribution will occur at a value of r_0 about 40–100 Bohr radius, is known as Wigner crystallization. We shall give a more detailed discussion of this problem in Chap. 19.

Here we shall focus our attention on the charge ordering of doped Mott insulators, especially manganites. Take $La_{1-x}Ca_xMnO_3$ as an example. Compared with $La_{1-x}Sr_xMnO_3$, the ionic radius of Ca is less than that of Sr, so replacement of Sr by Ca will introduce more distortion into the crystal structure, and its band width will be somewhat reduced. Instead of the typical double exchange behavior in the Sr doped manganite, we are faced with a more complicated situation, as indicated by its phase diagram (Fig. 13.2.10). At $x = 0$, there is no conduction electron for any Mn site; at $x = 1$, there is one conduction electron; between these limits, there is x electrons for each Mn site. Now the problem is how to distribute these electrons in space. We have already seen that spin-ordering and orbital-ordering play important roles in manganites, now charge ordering is added to make the physics of manganites more rich and complex.

In $La_{1-x}Ca_xMnO_3$, at $x = 0.5$, a stable charge-ordered AFM state below $T_c = 160$ K is found. This charge-ordered state can be explained by an ingenious model proposed by Goodenough (Fig. 13.2.11): Mn^{3+} and Mn^{4+} are arranged like a checkerboard, exhibiting the charge ordered,

[e] P. G. de Gennes, *Phys. Rev.* **118**, 114 (1960).

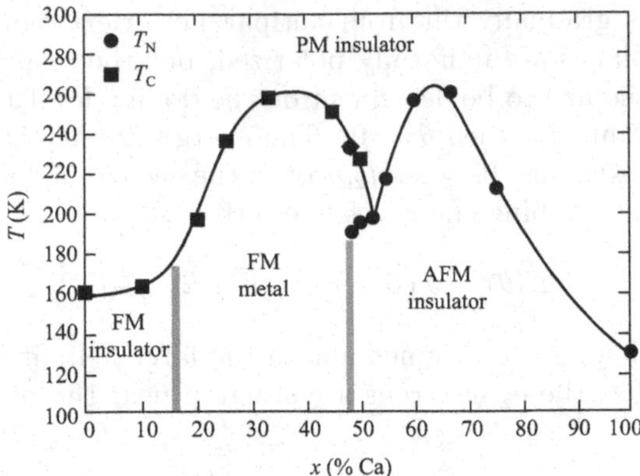

Figure 13.2.10 Phase diagram of $La_{1-x}Ca_xMnO_3$. From P. Schiffer *et al.*, *Phys. Rev. Lett.* **15**, 3336 (1995).

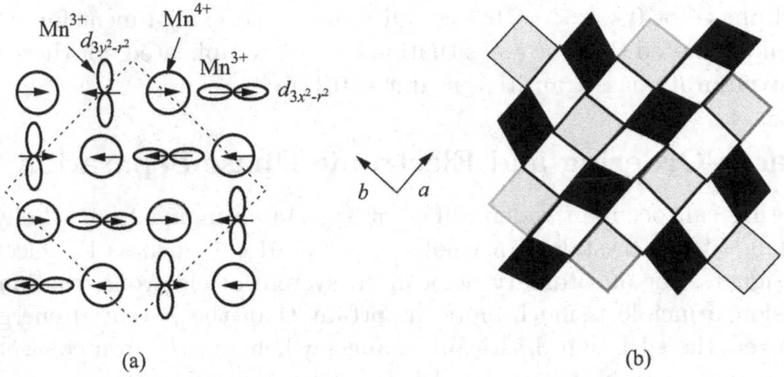

Figure 13.2.11 Spin, charge and orbital ordering in $La_{0.5}Ca_{0.5}MnO_3$ (a) schematic diagram (b) with corresponding JT distortions (the dark and light shaded squares show Mn^{3+} and Mn^{4+} sites) as the checkerboard pattern. Taken from T. Mizokawa and A. Fujimori, *Phys. Rev. B* **56**, R493 (1997).

spin ordered and orbital ordered states altogether. Since Mn^{3+} sites have a Jahn–Teller distortion, this periodic distribution of Mn ions reduces not only the Coulomb repulsive energy and exchange interaction energy, but also the Jahn–Teller distortion energy by the orbital ordering. So it seems an optimum solution to this problem; in fact, this model has been verified by X-ray and neutron diffraction measurements. The different types which describe charge, orbital and spin-ordering arrangements were classified by Wollan and Koehler in their neutron diffraction paper.[f]

It is interesting to note that manganites showing a large CMR effect are not those that exhibit pure double exchange behavior but are those involved with charge ordering. In Fig. 13.2.12, the magnetoresistances versus temperature for various compositions of $La_{0.7-x}Pr_xCaO_3$ are displayed, in which the inflationary expression $(\rho_0 - \rho_H)/\rho_H$ for magnetoresistance may reach 10^4–10^5, so it is fittingly called colossal. The theory for CMR is still unsettled: One type of theory stresses the importance of the Jahn–Teller effect in charge carriers; another type of theory invokes Anderson localization of electrons due to nonmagnetic impurities. The experimental evidence increasingly shows that the electronic phase separation into nanoclusters of FM metals and charge-ordered insulators may play an important role in CMR.[g]

[f] E. O. Wollan and W. C. Koehler, *Phys. Rev.* **100**, 545 (1955); for a theoretical explanation, see J. B. Goodenough, *Phys. Rev.* **100**, 564 (1955); for a recent review, see Y. Tokura and N. Naogosa, *Orbital physics in transition metal oxides*, *Science* **288**, 462 (2000).

[g] E. Dagotta, T. Hotta and A. Mareo, *CMR materials: the key role of phase separation*, *Phys. Rep.* **344**, 1 (2001).

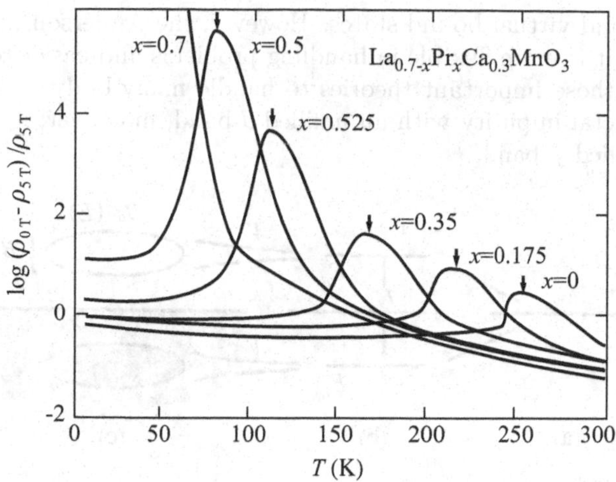

Figure 13.2.12 Magnetoresistance at 5 T versus temperature T(K) for compositions of $La_{0.7-x}Pr_xCa_{0.3}Mn$. Curie temperatures T_c are marked by arrows. Taken from H. Y. Hwang *et al.*, *Phys. Rev. Lett.* **75**, 914 (1995).

The electronic phase separation phenomenon is a general result of competition involving various types of ground states, such as metallic versus insulating states and various types of ordered states; also various types of interactions such as the Coulomb, exchange and Jahn–Teller interactions compete with each other. In the end, clusters or stripes of various electronic phases with nanometer sizes or larger may appear in specimens; their structures and physical consequences are still being hotly investigated. Phase separation also appears in other oxides; for instance, the 'stripe phases' in cuprates involving charge and spin ordering also form an important topic of current research. Formerly, the study of the electronic properties in condensed matter physics was most concerned with homogeneous states of materials; with electronic phase separation, a new chapter on the electronic properties of inhomogeneous materials is just emerging. Perhaps it will be some time before this field ripens.

§13.3 Magnetic Impurities, Kondo Effect and Related Problems

Another important area of correlated electronic states, developed in parallel with the Mott insulators, involves magnetic impurities in metals. The problem of magnetic moment formation and its influence on the resistivity of metals at low temperatures leads to the Kondo effect. This in turn has opened the field of the Kondo problem, as well as heavy electron metals and Kondo insulators. All these have attracted special attention from many distinguished theoreticians and experimentalists, and the main problem was solved in the 1970s and 1980s. Related problems such as heavy electron metals (or heavy fermions) and Kondo insulators are still among the important topics of strongly correlated electrons in current research. Full elucidation of the Kondo effect is beyond our scope; the reader could consult Hewson's monograph and related literature for that. Our attention will be focussed on the Anderson Hamiltonian of a magnetic impurity for its simplicity, as well as its importance in many-body physics. Then we will make a sketch of phenomena related to the Kondo problem.

13.3.1 Anderson Model and Local Magnetic Moment

In §7.2, we already treated impurity atoms in metals. Impurity atoms are considered as simple metallic ions with different charges situated in a gas of free electrons. This approach has been modified to treat the transition metal impurity problem by introducing the important concepts of resonance scattering and virtual bound states. However, the Anderson model can reach essentially the same results, and it is more flexible in handling problems and easier to get quantitative results, so it becomes one of those important theories to handle many-body effects. We shall discuss the case of a transition metal impurity with an unfilled d-band, moreover; it will be also correct for an impurity with an unfilled f-band.

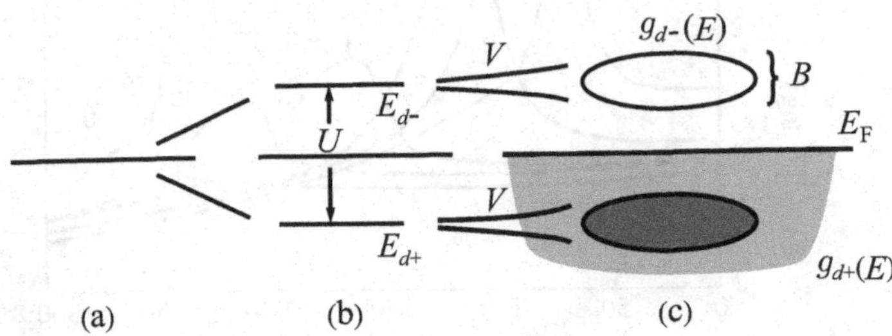

Figure 13.3.1 Pictorial representation of the Anderson model: (a) original atomic d-level; (b) it is split and polarized by on-site Coulomb repulsion; (c) it is further broadened by hybridization with s-electrons in the conduction band of the host.

The fundamental point of the Anderson model is the use of a localized description for the impurity and a delocalized description for the electrons in the host metal. Now consider the simplest case: the impurity atom has only a single orbital for a d-level occupied by a spin-up electron; double occupation will cost energy U, the Hubbard energy (Fig. 13.3.1). For d-electrons on the impurity site, the Hamiltonian may be written as

$$\mathcal{H}_d = \sum_{\sigma} \varepsilon_d n_{d\sigma} + U n_{d\uparrow} n_{d\downarrow}. \tag{13.3.1}$$

For s-electrons in the conduction band of the host, the Hamiltonian can be written as

$$\mathcal{H}_s = \sum_{\boldsymbol{k}\sigma} \varepsilon_{\boldsymbol{k}} n_{\boldsymbol{k}\sigma}. \tag{13.3.2}$$

Anderson further introduced an s-d hybridization interaction term with strength $V_{\boldsymbol{k}d}$

$$\mathcal{H}_{sd} = \sum_{\boldsymbol{k}\sigma} V_{\boldsymbol{k}\sigma}[c^\dagger_{\boldsymbol{k}\sigma} c_{d\sigma} + c^\dagger_{d\sigma} c_{\boldsymbol{k}\sigma}]. \tag{13.3.3}$$

When $V_{\boldsymbol{k}d}$ is not strong, it is equivalent to the s-d model first proposed by Zener (see Bib. [1]). The Anderson Hamiltonian is just the sum of the three terms introduced above,

$$\mathcal{H} = \mathcal{H}_d + \mathcal{H}_s + \mathcal{H}_{sd} \tag{13.3.4}$$

Now we shall examine the physical consequence of the Anderson model: the energy for single occupation of the d-level is ε_d, while that for double occupation is $\varepsilon_d + U$. For a magnetic moment to exist in the localized ion, it requires $\varepsilon_d < \varepsilon_{\mathrm{F}}$, where ε_{F} is the Fermi level of the s-band. So single occupation is energetically favorable; and $\varepsilon_d + U > \varepsilon_{\mathrm{F}}$, makes double occupation unfavorable. Due to the interaction between the localized d-state and the Bloch states of the conduction electrons of the host metal, s-d hybridization takes place as a perturbation of the d-level. This hybridization of the atomic d-level and the s-band of the host due to quantum mechanical tunneling can be interpreted as formation of a virtual bound state by resonant scattering of conduction electrons. The transition rate between the d-level and s-band can be described by the Fermi golden rule, in which the density of states in the s-band at the d-level $g(\varepsilon_{d\sigma})$ makes the d-level extend to some resonance width with

$$\Gamma_\sigma = \pi V^2 g(\varepsilon_{d\sigma});$$

thus, the sharp level at $\varepsilon_{d\sigma}$ will be replaced by a spectral density function of a Lorentzian form

$$\rho_{d\sigma}(\varepsilon) = \frac{1}{\pi} \frac{\Gamma_\sigma}{(\varepsilon - \varepsilon_{d\sigma})^2 + \Gamma_\sigma^2}. \tag{13.3.5}$$

A slight shift of the center can be added with no particular physical significance. The consequence of the broadening of the d-levels by resonance is that the occupancy of the $d\uparrow$ level $\langle n_{d\uparrow}\rangle$ becomes < 1, while $\langle n_{d\downarrow}\rangle$ becomes > 0, so whether the local moment is still retained on the impurity site demands further theoretical scrutiny. We may evaluate now

$$\langle n_{d\uparrow}\rangle = \frac{1}{\pi}\int_{-\infty}^{\varepsilon_{\rm F}} \frac{\Gamma d\varepsilon}{(\varepsilon-\varepsilon_d)^2+\Gamma^2} = \frac{1}{\pi}\text{arccot}\frac{\varepsilon_d-\varepsilon_{\rm F}+\langle n_{d\downarrow}\rangle U}{\Gamma}. \tag{13.3.6}$$

Of course, we have a comparable expression for $\langle n_{d\downarrow}\rangle$, so

$$\langle n_{d\downarrow}\rangle = \frac{1}{\pi}\text{arccot}\frac{\varepsilon_d-\varepsilon_{\rm F}+\langle n_{d\uparrow}\rangle U}{\Gamma}. \tag{13.3.7}$$

These equations must be solved self-consistently. Graphical solutions for chosen parameters are shown in Fig. 13.3.2. In case of small U, no magnetic moment remains. In case of large U, the lowest energy occurs at the intersections with unequal moments. In the lowest energy state, the local moment is given by $\langle n_{d\uparrow}\rangle - \langle n_{d\downarrow}\rangle = \pm 0.644\mu_{\rm B}$. We see that the formation of a local moment is a cooperative effect requiring an appropriate range of parameters. It arosc from the interaction term $Un_{d\uparrow}n_{d\downarrow}$ in the Hamiltonian. Through this analysis, Anderson explained why the iron group elements from V to Co show magnetic moments when dissolved in Cu, Ag and Cu but show no magnetic moment when dissolved in Al.

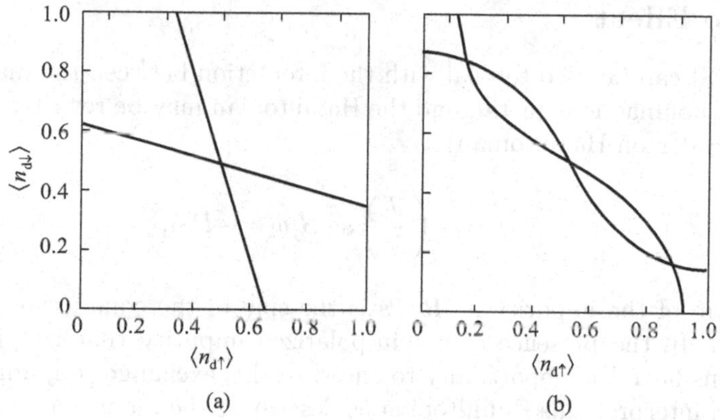

Figure 13.3.2 The self-consistency plot of $\langle n_{d\downarrow}\rangle$ against $\langle n_{d\uparrow}\rangle$: (a) $U = \Gamma$ and $\varepsilon_{\rm F} - \varepsilon_d = 0.5\Gamma$; the only solution is for $\langle n_{d\uparrow}\rangle = \langle n_{d\downarrow}\rangle$ and no local moment is formed. (b) $U = 5\Gamma$ and $\varepsilon_{\rm F} - \varepsilon_d = 2.5\Gamma$; there are three solutions with two corresponding to a local moment.

13.3.2 Indirect Exchange

If there is a local magnetic moment on the impurity, the next problem is to examine the spin polarization of the host electron gas around this magnetic moment. The situation is somewhat similar to the charge polarization treated in §7.2.3. If only contact action is considered, as in (7.2.28), we can get (7.2.34), i.e., the local magnetic moment of an impurity brings about an oscillation in the spin density of the free electron gas. This has been verified in Cu-Mn and Ag-Mn by NMR measurements.

Now further steps should be taken. Every impurity atom introduces spin density oscillation around it over quite a long range (Fig. 7.2.3), so it is expected that magnetic interaction may be set up between different impurity atoms. This is an indirect interaction called the Ruderman–Kittel–Kasaya–Yoshida (RKKY) interaction, after the theoreticians who first formulated it. Assuming that magnetic impurity 1 with spin $\boldsymbol{S}_1$ is situated at the origin, while impurity 2 with spin $\boldsymbol{S}_2$ is situated at $\boldsymbol{R}$. If the perturbation method is valid, then the interaction energy may be expressed as

$$E_{12} \approx J\boldsymbol{S}_2\cdot\boldsymbol{s}(r) \approx \frac{3\pi n^2}{64E_{\rm F}}\frac{\cos(2k_{\rm F}r)}{(k_{\rm F}r)^3}J^2\boldsymbol{S}_1\cdot\boldsymbol{S}_2. \tag{13.3.8}$$

Certainly this formula is used only to demonstrate the possibility of indirect exchange between localized spins via conduction electrons; more elaborate calculations should be made for this type of interaction in real materials. This type of interaction is used to explain the mechanism of ferromagnetism, antiferromagnetism and many other types of magnetism with exotic configurations such as the helix, cone etc., of the rare earth metals. It also explains the behavior of spin glasses, because the long-range oscillation of spin density may account for both the positive and negative, as well as more exotic, couplings of localized spins via the conduction electrons.

C. Zener proposed a theory of ferromagnetism of Fe based on the exchange interaction between localized spins via the conduction electrons. The physical picture is rather like the RKKY interaction described above but without invoking the oscillation of spin charge density of the electron gas. As a theory of ferromagnetism of Fe, Zener's theory fails to account for the itinerant character of the d-electrons, so it cannot be valid for Fe.[h] However, owing to the emergence of spintronics, research on ferromagnetic semiconductors has become a hot topic. GaAs-Mn with atomic concentration of Mn 5% has been fabricated with molecular beam epitaxy, and its Curie temperature has reached 110 K.[i] It is found that localized Mn spins couple through the charge carriers in GaAs, whose density is much lower than that of the electron gas in metals, so the oscillation of spin density is smeared out. Only ferromagnetic coupling remains possible, so it is unexpectedly found that Zener's theory is valid in this case.[j]

13.3.3 Kondo Effect

Equation (7.2.28) can be used to deal with the interaction between the magnetic impurities and band electrons in a nonmagnetic metal, and the Hamiltonian may be rewritten as (it can be directly derived from the Anderson Hamiltonian)

$$\mathcal{H} = -\left(\frac{J}{N}\right) \boldsymbol{s} \cdot \boldsymbol{S}_j \delta(\boldsymbol{r} - \boldsymbol{R}_j), \tag{13.3.9}$$

where $\boldsymbol{S}_j$ is the spin of the impurity at $\boldsymbol{R}_j$, $\boldsymbol{s}$ is the spin of the conduction electron, and J is the exchange coupling. In the presence of a spin-polarized impurity that is spin-up, only spin-down conduction electrons have the opportunity to enter, so this exchange coupling is antiferromagnetic, i.e., $J < 0$. We can interpret this Hamiltonian as describing the quantum mechanical exchange of a spin-up electron at the impurity site with a spin-down electron at the Fermi level of the conduction electrons (see Fig. 13.3.3). Many such events may produce an extra sharp resonance in the density of states at the Fermi energy, while a broader resonance is due to the coupling of the impurity level to the conduction band. Since transport properties of materials are determined by electrons with energies near the Fermi level, the resistivity of a material is dramatically influenced by this extra resonance, as exemplified by the Kondo effect.

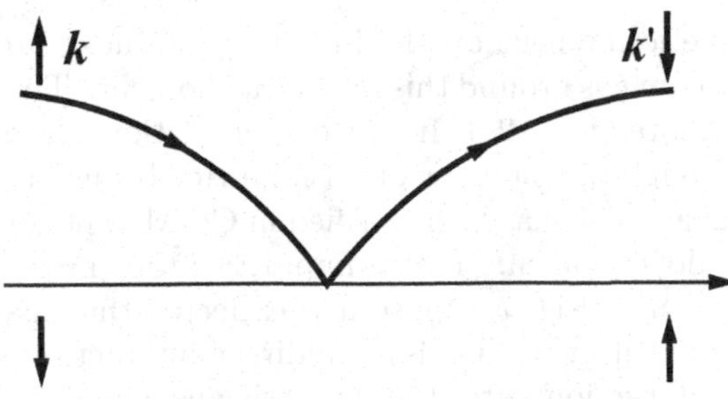

Figure 13.3.3 Spin flip scattering of a localized impurity moment by a conduction electron (the base line indicates the spin state of the impurity).

[h] For this theory, see C. Zener, *Phy. Rev.* **81**, 440 (1950); *Phys. Rev.* **83**, 299 (1950).
[i] H. Ohno *et al.*, *Phys. Rev. Lett.* **68**, 2664 (1992).
[j] T. Dietl *et al.*, *Science* **287**, 1019 (2000).

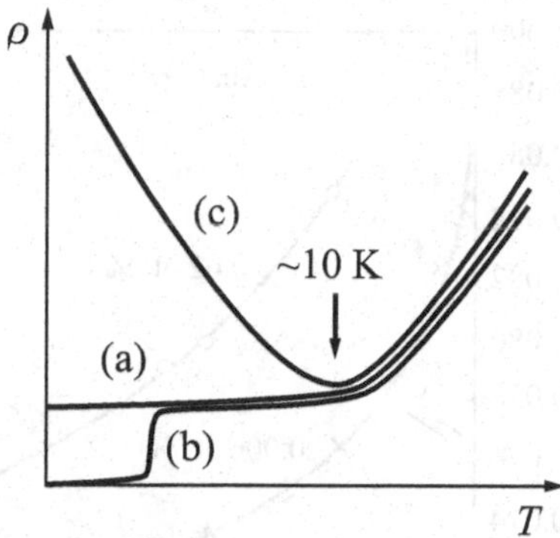

Figure 13.3.4 The resistance versus temperature curves for metals: curve (a) shows the residual resistance independent of temperature due to impurities and defects in the specimen; curve (b) shows the normal state to superconductor phase transition; curve (c) shows the Kondo effect.

The variation of electrical resistance with temperature is an important characteristic for materials. For metals, the resistance is mainly due to the scattering of electrons by lattice vibrations. Generally speaking, the value of resistance is reduced by a decrease of temperature. For nominally pure metals it is generally found that the resistance is gradually reduced to a certain value of residual resistance which is specimen-dependent and determined by impurities and defects (curve (a) in Fig. 13.3.4). Residual resistivities obey the Mathiessen's rule, i.e., they are additive and independent of temperature. For many metals such as Hg, Al, etc., the resistance is suddenly reduced to zero below a certain critical temperature T_c, they become superconductors (curve (b) in Fig. 13.3.4). Furthermore, there is another possibility, as shown in curve (c) in Fig. 13.3.4. The resistance of metal reaches a minimum then rises again when the temperature is further reduced. The puzzle of the resistance minimum of the noble metals was discovered in the early 1930s. It was suspected to be an impurity effect, but it was difficult to pin down. In the 1950s, the development of semiconductor technology brought with it tremendous progress in material purification and characterization, and experimentalists identified the resistance minimum as due to transition metal impurities dissolved in the nonmagnetic metal host. This effect was explained by J. Kondo theoretically in 1964, and subsequently became known as the Kondo effect. The Kondo effect is a many-electron effect in which localized electrons with spins interact with conduction electrons in the band. In this problem the interaction among conducting electrons is still ignored. This problem is complex but still soluble, and may be regarded as an archtype of many related problems.

Because a resistance anomaly is a common property in many alloys with dilute magnetic impurities, it is expected that the specific band structure does not have much influence, and the simplified density of states may be used, with B denoting the bandwidth. Kondo used this Hamiltonian to calculate the resistivity associated with the spin exchange scattering process in the second-order Born approximation. The resistivity due to impurities was found to be ρ_{imp} and that due to scattering of lattice vibrations to be aT^5, so the total resistivity may be expressed as

$$\rho = aT^5 + \rho_{\text{imp}} = aT^5 + c_{\text{imp}}\left(\rho_0 - \rho_1 \ln \frac{2k_{\text{B}}T}{B}\right), \tag{13.3.10}$$

where c_{imp} is the concentration of impurities and ρ_0 is the residual resistivity, $\rho_1 = 4|J|g_{\text{F}}$ is a constant. It should be noted that there is a logarithmic term which drastically increases when the temperature is lowered; the total resistivity will reach a minimum when $(d\rho/dT) = 0$, at the temperature

$$T_{\min} = \left(\frac{\rho_1}{5a}\right)^{\frac{1}{5}} c_{\text{imp}}^{\frac{1}{5}}. \tag{13.3.11}$$

This is in general agreement with experiment. Kondo's original paper presented a full comparison with the experimental results of Fe in Au, as shown in Fig. 13.3.5; the agreement was found to be very satisfactory. Other physical properties show similar behavior, e.g., specific heat and magnetic

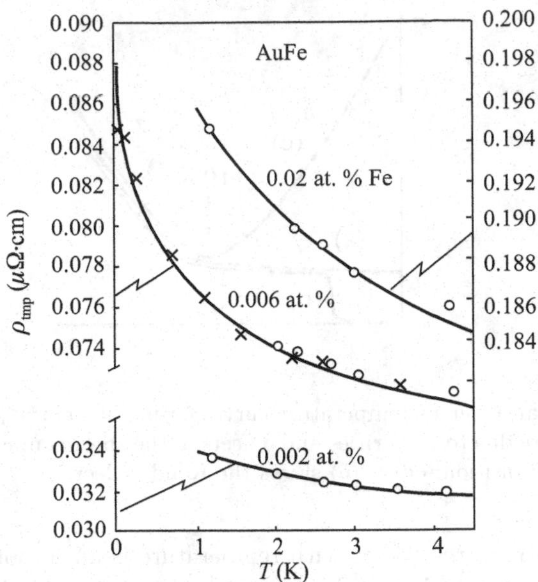

Figure 13.3.5 Comparison of experimental results for the resistivity of dilute Fe in Au at very low temperatures with the logarithmic form. From J. Kondo, *Prog. Theor. Phys.* **32**, 37 (1964).

susceptibility. But according to this theory: when $T \to 0$, $\rho \to \infty$, indicating that perturbation theory will lose its validity at low temperature. So physicists were greatly concerned with what happens with the Kondo resistance when $T \to 0$.

A. A. Abrikosov *et al.* dealt with this problem by introducing higher order perturbation terms (the so-called most divergent sum method). The magnetic susceptibility can be expressed as:

$$\chi_{\rm imp} = \frac{(g_{\rm L}\mu_{\rm B})^2 S(S+1)}{3k_{\rm B}T}\left\{1 + \frac{2Jg_{\rm F}}{1 - 2Jg_{\rm F}\ln(2k_{\rm B}T/B)} + c_2\,(2Jg_{\rm F})^2\right\}. \tag{13.3.12}$$

If we introduce the Kondo temperature

$$k_{\rm B}T_{\rm K} = \frac{1}{2}B\exp\left(-\frac{1}{2\,|J|\,g_{\rm F}}\right), \tag{13.3.13}$$

then at $T = T_{\rm K}$, we shall get the absurd result $\chi \to \infty$, indicating that near $T_{\rm K}$ the perturbation calculations become invalid. So though Kondo's original paper gave a correct physical explanation of the resistance minimum and other anomalous physical properties due to transition metal impurities; it raised new questions about the physics of systems near or below $T_{\rm K}$; the so-called Kondo problem.

In the late 1960s, Anderson introduced the idea of scaling, in which physical properties depend on temperature only in the form of $T/T_{\rm K}$, and identified $k_{\rm B}T_{\rm K}$ as the unique energy scale of the Kondo problem. At high temperature, i.e., $T \gg T_{\rm K}$, the spin interaction J is small but the bandwidth B is wide, so Kondo's results are correct for this regime and may be reformulated in the scaling form as a function of $\ln(T/T_{\rm K})$. At low temperature, $T \ll T_{\rm K}$, the opposite is true, J becomes strong, but B becomes small, so the physical relationship will be entirely different. In the early 1970s, K. G. Wilson applied the renormalization group method to the Kondo problem, with exemplary success. The low-temperature magnetic susceptibility is found to be

$$\chi_{\rm imp}(T) = \frac{\omega(g_{\rm L}\mu_{\rm B})^2}{4k_{\rm B}T_{\rm K}}\left\{1 - a_x\left(\frac{T}{T_{\rm K}}\right)^2 + O\left(\frac{T}{T_{\rm K}}\right)^4\right\}, \tag{13.3.14}$$

where $\omega = 0.5772$ is known as the Wilson number, and the low-temperature electrical resistance

$$\rho_{\rm imp}(T) = \rho_0\left\{1 - a_R\left(\frac{T}{T_{\rm K}}\right)^2 + O\left(\frac{T}{T_{\rm K}}\right)^4\right\}. \tag{13.3.15}$$

Now we can summarize the whole picture for the Kondo problem: At high temperature, $T \gg T_K$, the d-spin is essentially free; its contribution to magnetic susceptibility is Curie-like, like an independent magnetic moment. The scattering of conduction electrons is classified into two types: those of ordinary potential scattering and those involving a spin-flip by the magnetic impurity. As the temperature is lowered, in a wide crossover regime centered around T_K, the spin-flip scattering events become more frequent with a concurrent build-up of a spin compensating cloud. So numerical solutions for the whole temperature range were found by the renormalization group method,[k] furthermore, the exact solution of the s-d model and the Anderson model by the Bethe Ansatz substantiated these results.[l]

The physical nature of the resistance minimum is due to frequent spin-flip scattering events. At low temperature, $T \ll T_K$, spin-compensation is nearly complete; the impurity spin with its compensating electron spin cloud is strongly bound together, forming a Kondo singlet and appearing to be non-magnetic. Electrons are scattered by the big singlet with strong potential scattering, but spin-flip scattering disappears.

13.3.4 Heavy-Electron Metals and Related Materials

Heavy-electron metals or heavy fermions are rare earth or actinide compounds showing metallic behavior with a variety of anomalous thermodynamic, magnetic and transport properties. The prominent examples of heavy-electron metals are $CeAl_3$, $CeCu_2Si_2$, $CeCu_6$, UBe_{13}, UPt_3, UCd_{11}, U_2Zn_{17}, and $NpBe_{13}$, containing Ce, U and Np, respectively. At high temperature, these systems behave like weakly interacting collections of magnetic moments of f-electrons and conduction electrons with quite ordinary masses; at low temperatures, the f-electron magnetic moments become strongly coupled to the conduction electrons and to one another, and the effective masses of the conduction electrons becomes heavy, typically about 10 to 100 times the bare electron mass.

It is appropriate to give some basic experimental results for some typical heavy electron metals. Figure 13.3.6 shows the relation between the resistivity and temperature for some heavy electron metals, and the inset shows the linear relation between R and T^2 of $CeAl_3$ at low temperatures. For $T < 0.3$ K, $\rho(T)$ follows the T^2 law as shown in the inset of Fig. 13.3.6, and displays a broad maximum at $T_{\max} \approx 35$ K. Besides $\rho(T)$, $C_p(T)$ and $\chi(T)$ as shown in Fig. 13.3.7 also show the typical heavy electron behavior. In fact, below 0.2 K the linear specific heat term is enormous, with $\gamma = 1620 \times 10^4$ erg/mol K^2. In order to appreciate how huge this value is, let us recall that for ordinary metals, it is about 10^{-3} times smaller (for Cu, K and Ni: 0.695, 2.08 and 7.02 in units of 10^4 erg/mol K^2, respectively). Similarly, the magnetic susceptibility can be more than two orders of magnitude larger than the temperature-independent Pauli susceptibility observed in conventional conducting materials.

Magnetic susceptibility data show a Kondo-like behavior: at high temperatures the Curie-like behavior is due to the independent magnetic moments of f-atoms; at low temperatures, compensating electron-clouds about f-atoms are gradually build up to form a lattice of spin-singlets, so at $T = 0$, only enhanced Pauli susceptibility remains. From resistivity data, surely the Kondo-like scatterings play an important role. However, at low temperatures f-electrons are delocalized into the Fermi sea, thus the resistivity decreases again, so we see a resistivity maximum, instead of the minimum in Kondo effect.

For an explanation of the anomalous properties of heavy electrons, it is appropriate to generalize the Kondo Hamiltonian to the Kondo lattice model (KLM) as

$$\mathcal{H}_{\mathrm{KLM}} = -t\sum_{ij\sigma}(c_{i\sigma}^{\dagger}c_{j\sigma} + \mathrm{h.c.}) - J\sum_{i} \boldsymbol{s}_i \cdot \boldsymbol{S}_i, \tag{13.3.16}$$

[k] K. G. Wilson, *Rev. Mod. Phys.* **47**, 773 (1975).

[l] N. Andrei, *Phys. Rev. Lett.* **45**, 1379 (1979); P. B. Wiegmann, *Phys. Lett.* **81**A, 175 (1981).

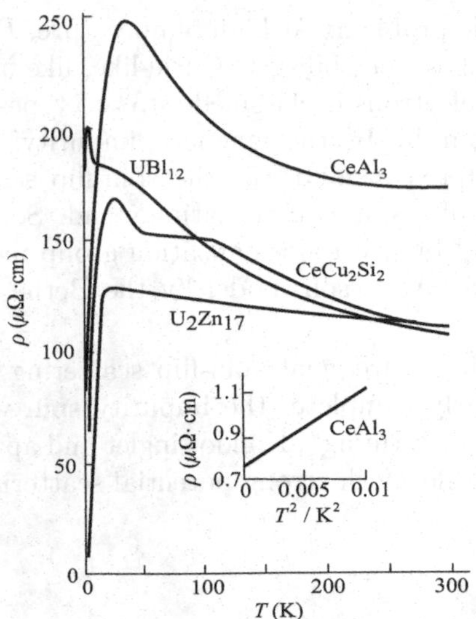

Figure 13.3.6 Temperature dependence of the electrical resistivity $R(T)$ for $CeAl_3$, UBe_{13}, $CeCu_2Si_2$ and U_2Zn_{11}. The inset shows the T^2 behavior for $CeAl_3$ for $T < 0.1$ K. From Z. Fisk *et al.*, *Nature* **320**, 124 (1986).

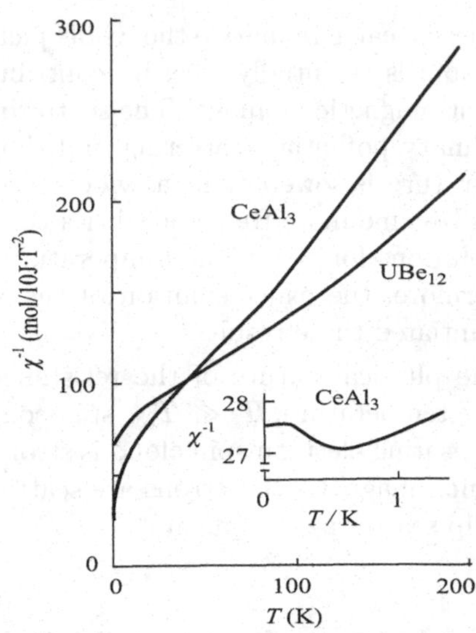

Figure 13.3.7 Inverse susceptibility $1/\chi$ versus T for $CeAl_3$ and UBe_{13}. The Curie-Weiss form above 100 K corresponds to local moment behavior of f electrons. The inset shows saturation indicating itinerant f electron behavior at low T. From Z. Fisk *et al.*, *Nature* **320**, 124 (1986).

where $\boldsymbol{s}_i = (1/2)\sum_{\sigma\sigma'} \tau_{\sigma\sigma'} c_{i\sigma}^\dagger c_{i\sigma'}$ are the spin-density operators of the conduction electrons and $\boldsymbol{S}_l = (1/2)\sum_{\sigma\sigma'} \tau_{\sigma\sigma} f_{l\sigma}^\dagger f_{l\sigma'}$ are localized spins, with τ being the Pauli matrices. The exchange interaction derived in this way is antiferromagnetic and inversely proportional to U. Therefore the limit of strong Coulomb interaction corresponds to the small-J limit of the Kondo lattice model.

The Kondo lattice model is one of the standard models for heavy electrons. In this model the charge fluctuations of the f electrons are completely omitted, and the lowest f-ion multiplet is taken into account as a localized spin. Thus the Kondo lattice model is an example of a coupled spin-fermion system. For this model an important question arises from the interplay between the Kondo screening and the RKKY interaction: Kondo screening favors a nonmagnetic singlet state, while the RKKY interaction tends to stabilize a magnetically-ordered phase, generally a spiral state. There may exist a transition between the large-J regime, where the localized spins are essentially screened by the conduction electrons, and the small-J regime, in which the effective interaction mediated by the conduction electrons leads to magnetic order among the local spins.

At higher temperatures, the heavy electron systems behave as nearly independent localized spins and conduction electrons. The two sets of degrees of freedom are coupled only weakly by the exchange coupling. As the temperature is lowered, the localized spins start to couple strongly with the conduction electrons. After anomalous behaviors in the crossover region, the heavy electron materials settle in various types of different ground states. An extensive Hartree–Fock treatment has been used on the three-dimensional Kondo lattice model with the usual electron density, and has given relative stability for three types of ground states — the magnetically disordered Kondo singlet, the ferromagnetic state, and the antiferromagnetic state. The mean-field phase diagram for the Kondo coupling J/t versus the conduction-electron number n_c is shown in Fig. 13.3.8.

From the above discussion, heavy electron metals may be theoretically interpreted by the Kondo lattice model: for $J = 0$, the ground state has the conduction electrons forming a Fermi sea surrounding localized spins of f-electrons, which are free to point either up or down. If an antiferromagnetic J is switched on, one possibility is the formation of a series of Kondo singlets, which then bind the f-spins to the conduction electrons; this may be identified as the non-magnetic heavy electron state. There is another possibility: to form a local singlet, a f-spin should polarize the surrounding

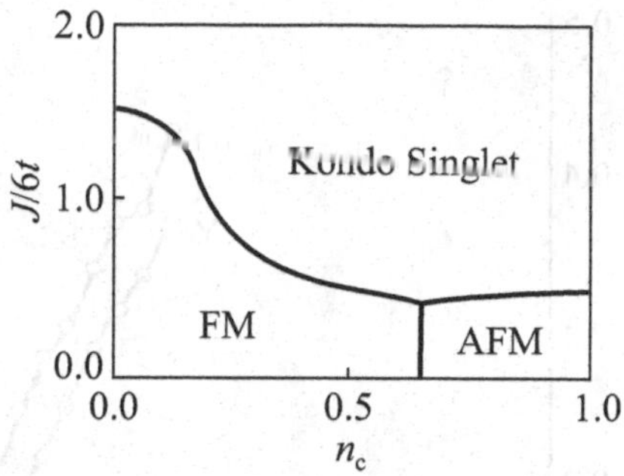

Figure 13.3.8 Schematic phase diagram, Kondo coupling versus electron density, obtained by a mean-field treatment of the three-dimensional Kondo lattice model. From H. Tsunetsugu *et al.*, *Rev. Mod. Phys.* **69**, 809 (1997).

Fermi sea. As a consequence the neighboring spins may feel this disturbance and line up into a spin ordered state, according to the RKKY interaction.

There are other possible ground states for Kondo-like materials: one is the Kondo insulator, a small gap of the order a few meV, in contrast to semiconductors with gaps of the order eV; another type of ground state is a superconductor. The heavy electrons form pairs which condense into the superconducting state, for example, URu_2Si_2, UNi_2Al_3, UPd_2Al_3, $CeCu_2Si_2$, UPt_3 and UBe_{13}. In general, these have rather small T_c ($T_c < 1$–2 K), but display unconventional properties and exotic pairing mechanisms; we shall not go into details here. Generally speaking, Kondo-like materials form another important class of materials with strongly correlated electrons in current research.

§13.4 Outlook

After a brief survey of correlated electronic states, we have seen some peaks of icebergs floating on the ocean surface, with an enormous unknown mass below. Now we would like to discuss the prospects of this field using an empirical view on the one hand and a theoretical approach on the other.

13.4.1 Some Empirical Rules

Most interesting materials for high-tech are those conductors on the verge of becoming insulators, or in other words bad metals. Semiconductors may be taken as a shining example, they are doped band insulators with their conductivity finely controlled. With advent of high T_c superconductors and colossal magnetoresistors, doped Mott insulators have become hot topics of current research and high-tech materials with great potential. Since the study of compounds with 4 or 5 elements is just beginning, the potential for further research in materials with strongly correlated electrons is, indeed, enormous.

Some empirical rules may be used as maps to guide us. Surely the periodic table of elements is a master map; however, to accentuate the elements with unfilled d and f shells which are closely related with narrow bands, the sequence of d and f shells should be suitably modified. The ratios of d or f shell volume to the volume of the Wigner–Seitz (WS) cell are shown in Fig. 13.4.1, the values of the ratio are all less than 0.5, showing these shells lie between inner core and bonding states. For lanthanides, the values are below 0.05, the f-shells are localized within the atoms. On the other hand, the d-electrons of the transition elements still participate in the bonding of metals. For actinides, the situation is in between: The early actinides are like transition metals, while the heavier actinides resemble the rare earths. With this in mind, the periodic table of elements may be redrawn according to the sequence of the spatial extent of f and d shell ($4f$-$5f$-$3d$-$4d$-$5d$) as proposed by Smith and Kmetko (Fig. 13.4.2). It should be noted that the elements with delocalized electrons lie on one side of the table, while those with localized electrons lie on the other. The transition region

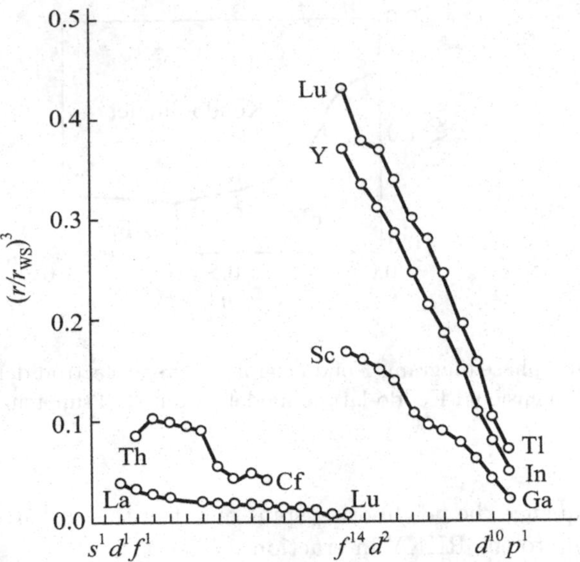

Figure 13.4.1 Ratio of l-shell volume to Wigner–Seitz volume in elements with unfilled f or d shells. From D. van der Marel and G. A. Sawatzky, *Phys. Rev. B* **37**, 10674 (1988).

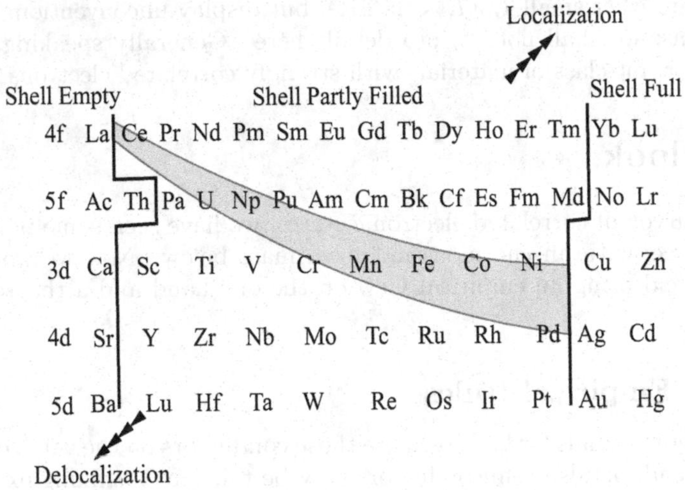

Figure 13.4.2 Modified periodic table with vertical column ($4f$-$5f$-$3d$-$4d$-$5d$). From J. L. Smith and E. A. Kmetko, *J. Less Comm. Met.* **90**, 83 (1983).

is drawn as an inclined band across the table, where strongly correlated electronic states occur. It is interesting to note that Ce and Pu are situated on the margin of localization-delocalization: Ce has two polymorphic phases, at room temperature the γ phase, with $4f$ electrons localized; while at high pressure or low temperature the α phase, with $4f$ electrons delocalized (see Fig. 12.4.8). Pu also has two polymorphic phases: at room temperature the α phase, with $5f$ electrons delocalized; while at high temperature the δ phase, with $5f$ electrons partly localized (see Fig 13.4.9). In Am, the $5f$ electrons are fully localized.

Though this kind of periodic table is a chart for elements, its usage may be extended to alloys and compounds whose basic ingredients occupy suitable places in this table. For instance, the basic ingredient for a heavy electron alloy is certainly a lanthanide or an actinide element; for the cuprates, the basic ingredient is a Cu^{2+} ion with one electron deficient in its d-shell, so it should be regarded as a transition metal to occupy the place of Ni in the table.

From the table, you may note the ferromagnetic metals Fe, Co and Ni with their strong ferromagnetism due to itinerant-electrons, though their narrow d-bands overlap with the wide s-band, showing typical metallic behavior. Ce and U based alloys are typical heavy electron metals; the cuprates are high-T_c superconductors; the manganites are colossal magnetoresistors; exotic superconductors with p-wave pairing have been found in UGe_2 and ruthenates. Most of these findings were discovered after this table was published.

From an historical point of view, intensive studies of the physical properties of compounds with many components began with the discovery of high T_c superconductors. Only a few have been studied in detail. How to prospect, explore and study the fertile field of complex compounds with superior or peculiar physical properties is one of the important topics of condensed matter physics in the 21st century.

13.4.2 Theoretical Methods

P. W. Anderson and J. R. Schrieffer are two distinguished physicists who have made important contributions to the many-body problems of condensed matter theory. Although they have different views on the theory of high T_c superconductors, they are unanimous about the present situation of condensed matter theory: There should be two volumes of condensed matter physics; volume one is about the physics of weakly or medium correlated electrons, already mature and well-established; volume two is about the physics of strongly-correlated electrons and is still waiting to be written.[m] It is expected that this will be a fertile field for theorists to introduce new ideas, new models and new methods. The effectiveness of these approaches has been testified to by the successes achieved in the fractional quantum Hall effect, a clean system of strongly correlated electrons.

However, for systems treated in this chapter there is added complexity due to chemistry. On the cross-disciplinary side, close synthesis with quantum chemistry is needed, as exemplified by the establishment of the orbital physics of transition metal oxides. Inhomogeneity in electronic structures such as stripe phases, phase separation and competing ground states in cuprates and manganites poses new problems for theorists. These are related to the physics of nonequilibrium states and nonlinear physics.

In the field of more traditional condensed matter theories, the problem is how to reconcile density functional theory with model Hamiltonians such as the Hubbard model and the Anderson model. Density functional theory (especially LDA) has achieved extraordinary success in the calculation of electronic structure of weakly and moderately correlated electronic structure of solids but is found to be inadequate in solving the electronic structure of strongly correlated systems. So we have emphasized model Hamiltonians, especially the Hubbard Hamiltonian and the Anderson Hamiltonian, in our account of the electronic structures of strongly correlated electronic states.

Though these model Hamiltonians proved their importance and enlightened our understanding of strongly correlated electronic systems, they are quite intractable and not able to give very detailed electronic structures for real materials. Real materials often have crystal structures that are quite complicated, with the number of atoms per unit cell usually reaching or exceeds 10–15, so the interaction between electrons and degrees of freedom of lattice demands more detailed study, which lies outside the model Hamiltonians. So an important direction of recent progress consists of the combination of these two different types of theoretical treatment.

One of the combination schemes is called the LDA+U method.[n] This method is to add a mean field Hubbard-like term to the LDA functional. The main idea of this method is similar to the Anderson impurity model: to separate electrons into two subsystems — localized d or f electrons for which the Coulomb interaction should be taken into account by a term containing U in a model Hamiltonian and delocalized s and p electrons that could be described by an orbital-independent one-electron potential (LDA). This method has been applied to the electronic structure and magnetism of Mott insulators with quite good results.

The shortcomings of the LDA+U method come from the mean field approximation; the physical properties related to self-energy such as the enhancement and decrement of mass are difficult to

[m] P. W. Andenson and J. R. Schrieffer, *Phys. Today* **44**, June 55 (1991).
[n] V. I. Anisimov, E. Aryasetiaswan and A. I. Lichtenstein, *J. Phys. Condens. Matter* **9**, 767 (1997).

account for. Further on the way, the LDA++ method and dynamical mean-field theory (DMFT) were developed.[o] As an example of the latter technique: it successfully solved the polymorphic transition from αPu to δPu at 600 K in which a $5f$ electron changes from a delocalized state to a partly localized state with a concurrent expansion of atomic volume by about 25%. It shows that the combination of two approaches can solve a difficult problem of strongly correlated electrons.[p]

Bibliography

[1] Fazekas, P., *Lecture Notes on Electron Correlation and Magnetism*, World Scientific, Singapore (1999).

[2] White, R. M., and T. H. Geballe, *Long Range Order in Solids*, Academic Press, New York (1978).

[3] Animalu, A. O. E., *Intermediate Quantum Theory of Crystalline Solids*, Prentice-Hall, Englewood Cliffs, New Jersey (1977).

[4] Mott, N. F., *Metal-Insulator Transitions*, Taylor and Francis, London (1990).

[5] Edwards, P. P., R. I. Johnston, F. Hansel, C. N. R. Rao, and D. P. Tunstall, *A Perspective on the Metal-Nonmetal Transition*, Solid State Physics **52**, 229 (1999).

[6] Imada, M., A. Fujimori, and Y. Tokura, *Metal-insulator Transitions*, Rev. Mod. Phys. **70**, 1039 (1998).

[7] Jaime, M. B., and M. Jaime, *The Physics of Maganites: Structure and Transport*, Rev. Mod. Phys. **73**, 583 (2001).

[8] Ziman, J. M., *Principles of Theory of Solids*, 2nd ed., Cambridge University Press, Cambridge (1972).

[9] Hewson, A. C., *The Kondo Problem to Heavy Fermions*, Cambridge University Press, Cambridge (1993).

[o] For more detail about LDA++, see A. I. Lichtenstein, M. I. Kastsnelson, *Phys. Rev. B* **57**, 6884 (1997). For DMFT, see A. Georges *et al.*, *Rev. Mod. Phys.* **68**, 13 (1996); G. Kotliar and D. Vollhardt, *Phys. Today* **57**, Mar. 53 (2004).

[p] S. Y. Sarasov, G. Kotliar, and Abrahams, *Nature* **410**, 793 (2001).

Chapter 14

Quantum Confined Nanostructures

In artificial nanostructures including quantum wells, wires and dots, the motion of electrons is governed by effective potentials, which confine the electrons in one, two, or three directions. These confinements bring about plentiful quantum effects, which are useful in designing electronic structures and tailoring physical properties. A fundamental problem is what characteristic size of nanostructure will make such a remarkable change to the electrons that it modifies their optical, transport and magnetic properties. Because a large part of the physical properties are determined by electrons at the Fermi surface, it is expected that the Fermi wavelength is this characteristic size. From §5.2.1, we obtain the Fermi wavelength for a free-electron gas; taking this simple case as an example, we find that Fermi wavelength decreases with electron density. According to the parameters of materials, it is easy to determine that this characteristic size is about 200 nm for semiconductors, and about 1 nm for metals. The former corresponds to the upper limit of nano-size, and the latter the lower one, so this difference is quite significant. An artificial structure fabricated at nano-size, at moderate temperature, will show quantum confinement effects.

§14.1 Semiconductor Quantum Wells

A semiconductor quantum well is a sandwich structure, in which a piece of narrow-gap material is placed between two pieces of wider-gap material. A heterostructure composed of a thin layer of GaAs embedded between two thick layers of AlGaAs, each with thickness much greater than the penetration length of the confined wavefunction, provides a simple picture of a quantum well. In this section we describe the electron states and optical properties of a quantum well in the GaAs-AlGaAs system. In this case, both types of carrier, electrons and holes, are all confined within the GaAs layer.

14.1.1 Electron Subbands

Consider an electron moving in a confined potential: Its energy levels can be calculated quite easily in the approximation of the envelope wavefunction, due to the modulation of the quickly oscillating Bloch function of the parent bulk materials. Roughly speaking, the Bloch function gives the solution of the bulk Hamiltonian, and the envelope function ensures that the boundary conditions at the surfaces of the film are met. The Bloch function wavelength is given by the atomic layer spacing via $\lambda_{\mathrm{Bloch}} = 2a$. The envelope function wavelength λ_{env} is determined by the thickness d of the film. Taking z as the growth direction, we can write the Schrödinger equation for the envelope function $\phi(z)$ as

$$\left[-\frac{\hbar^2}{2m^*(z)}\frac{d^2}{dz^2} + V_{\mathrm{c}}(z)\right]\phi(z) = E\phi(z), \tag{14.1.1}$$

where $m^*(z)$ is the electron effective mass, $V_{\mathrm{c}}(z)$ the energy level of the bottom of the conduction bands, and E the confined eigenenergy of the carriers. The continuity conditions at the interfaces

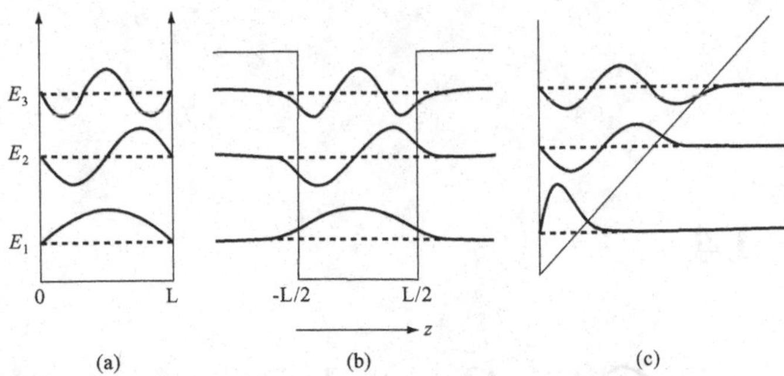

Figure 14.1.1 Schematic diagrams of quantum well energy levels and wavefunctions for three possible potentials. (a) Infinitely deep square well; (b) finite square well; and (c) triangular well.

are that ϕ and $(1/m^*)(d\phi/dz)$ are continuous; the latter is necessary for conservation of particle current. In the following, three different potential profiles are taken into account.

First consider the simple case of an infinitely deep square well, where the potential $V = 0$ within the well, and $V = \infty$ out of it. Here, the solution to (14.1.1) is simple, as the wavefunction must be zero outside the well region. Taking the origin of z at one interface, with the width of the well L, the eigenstate is $\phi_n = (2/L)^{1/2}\sin(n\pi z/L)$, with eigenenergy $E_n = n^2\hbar^2\pi^2/2m^*L^2$, where the quantum number $n = 1, 2, \ldots$. It is noted that the ϕ_n have even parity for odd n and odd parity for even n, about the center of the well.

In order to show clearly the symmetries of the wavefunctions and to extend the treatment to the general case of finite potential wells, it is more convenient to shift the origin to the center of the well. For the case $n = 1, 3, \ldots$, the solutions of (14.1.1) show even parity:

$$\phi_n(z) = \begin{cases} A\cos(kz), & |z| < L/2, \\ B\exp[\kappa(z + L/2)], & z < -L/2, \\ C\exp[-\kappa(z - L/2)], & z > L/2. \end{cases} \tag{14.1.2}$$

For the case $n = 2, 4, \ldots$, solutions of (14.1.1) show odd parity:

$$\phi_n(z) = \begin{cases} A\sin(kz), & |z| < L/2, \\ B\exp[\kappa(z + L/2)], & z < -L/2, \\ C\exp[-\kappa(z - L/2)], & z > L/2. \end{cases} \tag{14.1.3}$$

Figure 14.1.1(a) shows the solutions for an infinitely deep well, while Fig. 14.1.1(b) shows the solutions for a finite well. Note the difference: For the former the wavefunctions are limited within the well; but for the latter the wavefunctions are spilled out across the interfaces.

In the energy range $-V_0 < E < 0$, the eigenenergy

$$E_n = \frac{\hbar^2 k^2}{2m_{\mathrm{A}}^*} - V_0, \; E_n = -\frac{\hbar^2\kappa^2}{2m_{\mathrm{B}}^*}, \tag{14.1.4}$$

where the suffixes A and B denote the material of the well and barrier, respectively. By using the continuity conditions at $z = \pm L/2$, (14.1.2) and (14.1.3) yield implicit eigenvalue equations

$$(k/m_{\mathrm{A}}^*)\tan(kL/2) = \kappa/m_{\mathrm{B}}^*, \tag{14.1.5}$$

and

$$(k/m_{\mathrm{A}}^*)\cot(kL/2) = -\kappa/m_{\mathrm{B}}^*. \tag{14.1.6}$$

The equations can be solved numerically or graphically. As a simple example, when $m_{\mathrm{A}}^* = m_{\mathrm{B}}^*$, the above two equations can be transformed into

$$\cos(kL/2) = k/k_0, \; \text{for} \tan(kL/2) > 0, \tag{14.1.7}$$

and

$$\sin(kL/2) = k/k_0, \text{ for } \tan(kL/2) < 0, \tag{14.1.8}$$

where

$$k_0^2 = 2m^* V_0/\hbar^2. \tag{14.1.9}$$

The solution is shown in Fig. 14.1.2. There is always at least one bound state. The number of bound states is

$$1 + \left\lfloor \left(\frac{2m_A^* V_0 L^2}{\pi^2 \hbar^2} \right)^{1/2} \right\rfloor,$$

where $\lfloor x \rfloor$ indicates taking the integer part of x. It is evident that if $k_0 \to \infty$ the infinitely high barrier solutions can be found again.

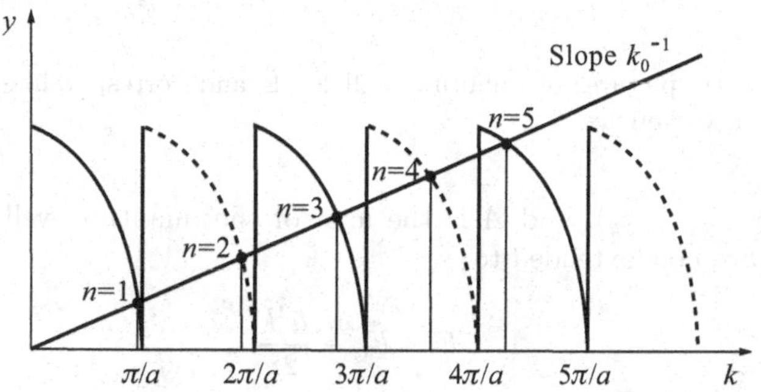

Figure 14.1.2 Graphical solutions for a finite square well potential. The intersections of the straight line $y = k/k_0$ with the curves $y = \cos(kL/2)$ (solid line) give the even solutions, and these with the curve $y = \sin(kL/2)$ (dot line) give the odd solutions.

Another often-encountered potential is the triangular quantum well, as shown in Fig. 14.1.1(c) for which the potential $V(z) = e\mathcal{E}_z$ for $z > 0$ and an infinite barrier for $z \leq 0$, where $\mathcal{E}_z$ is the electric field. The eigensolution of (14.1.1) is the Airy function

$$\phi_n = \mathrm{Ai}\left[\left(\frac{2m}{\hbar^2 e^2 \mathcal{E}_z^2} \right)^{1/3} (eF_z - E_n) \right]. \tag{14.1.10}$$

The boundary condition at $z = 0$ gives the eigenvalues

$$E_n = -\left(\frac{e^2 \mathcal{E}_z^2 \hbar^2}{2m} \right)^{1/3} a_n, \tag{14.1.11}$$

where a_n is the nth zero of $\mathrm{Ai}(z)$. Asymptotically, and also to a very good approximation for small n, one has

$$a_n \approx -\left[\frac{3\pi}{2} \left(n + \frac{3}{4} \right) \right]^{2/3}, \ n = 0, 1, \ldots,$$

so that

$$E_n \approx \left(\frac{\hbar^2}{2m} \right)^{1/3} \left[\frac{3\pi e \mathcal{E}_z}{2} \left(n + \frac{3}{4} \right) \right]^{2/3}. \tag{14.1.12}$$

In practice, the quantum well is a layered structure, so perpendicular to the growth direction z, an electron can execute two-dimensional movement. For simplicity, we shall assume a free electron approximation in the x-y plane for which the wavefunction is

$$\psi_{\boldsymbol{k}n}(\boldsymbol{r}, z) = A^{-1/2} e^{i\boldsymbol{k}\cdot\boldsymbol{r}} \phi_n(z), \tag{14.1.13}$$

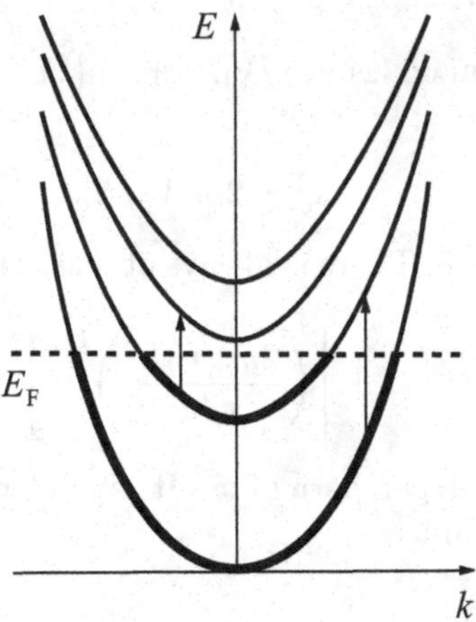

Figure 14.1.3 Schematic pictures of quantum well levels and corresponding subbands. Some possible optical transitions are also sketched.

where $\boldsymbol{r} = (x, y)$, $\boldsymbol{k} = (k_x, k_y)$, and A is the area of the quantum well. The eigenenergy levels previously obtained are now extended to

$$E = E_n + \frac{\hbar^2 k^2}{2m}, \tag{14.1.14}$$

leading to subbands shown in Fig. 14.1.3.

In this simple discussion for the quantum well structure, we have assumed that the conduction band state is a pure s-type state and we have used a simple effective mass theory to understand the quantum well band structure. A more sophisticated calculation can be performed, in which one retains the full description of the band structure of the individual components (e.g., an eight band model). When this is done, it turns out that the results for the conduction states are not affected very much.

14.1.2 Hole Subbands

The description presented above of the quantum well energy band is quite valid for electron states of the s-type, since the electron states, as noted earlier, in direct band gap materials are adequately described by a single s-type band. However, the valence band states are formed from p-type states, leading to heavy hole and light hole states, even in bulk semiconductors, due to strong anisotropy. This is unfortunate. While the heavy-hole $(3/2, \pm 3/2)$ and light-hole $(3/2, \pm 1/2)$ states are pure states at $k = 0$, they strongly mix away from $k = 0$. Thus, the dispersion relation for the hole states is much more complicated. Although, as far as subband level positions are concerned, the starting energies of the subbands can be solved just as for electrons, i.e., independently for the heavy-hole and light-hole states, the degeneracy of the hole states in the bulk valence bands must be taken into account. Most treatments of bulk valence band structure begin with simple models, to which ever-greater complications are added. The same is true for quantum well states.

According to the Kane model, which is an extension of the $\boldsymbol{k} \cdot \boldsymbol{p}$ approximation of §12.1.4 with the added complication of spin-orbit coupling, we take as a basis set the lowest conduction band which has s-type symmetry associated with the orbitals at each atomic site, and the three uppermost valence bands which have p-type symmetry at each atom, all at $\boldsymbol{k} = 0$. From these we form a fourfold degenerate set of orbitals, in the absence of interatomic interactions, but, once these are added in, the sharp energy levels broaden into bands. The two most important interaction terms are the effective mass parameter P, involving interband interactions, and a spin-orbit coupling parameter Q. The resulting energy bands are of the following form: The conduction band is given by

$$E = E_{\rm g} + \hbar^2 k^2/2m^* + [\sqrt{(E_{\rm g}^2 + 8P^2k^2/3)} - E_{\rm g}]/2, \tag{14.1.15}$$

and the valence band by

$$E = \begin{cases} -\hbar^2k^2/2m^*, & \text{(heavy holes)}, \\ -\hbar^2k^2/2m^* - [\sqrt{(E_g^2 + 8P^2k^2/3)} - E_g]/2, & \text{(light holes)}, \\ -Q - \hbar^2k^2/2m^* - P^2k^2/(3E_g + 3Q), & \text{(split off)}. \end{cases} \tag{14.1.16}$$

To these we must add the perturbations represented by the reduction of the dimensionality in a quantum well. This is exceedingly complicated, and requires numerical calculations. After appropriate simplification, we can get expressions for the bulk energy bands near $k = 0$. In the direction z

$$E = -\frac{\hbar^2 k_z^2}{2m}(\gamma_1 - 2\gamma_2), \text{ for } J_z = \pm 3/2, \tag{14.1.17}$$

thus the heavy-hole mass is $m/(\gamma_1 - 2\gamma_2)$, and

$$E = -\frac{\hbar^2 k_z^2}{2m}(\gamma_1 + 2\gamma_2), \text{ for } J_z = \pm 1/2, \tag{14.1.18}$$

with light-hole mass $m/(\gamma_1 + 2\gamma_2)$. For GaAs, we have $\gamma_1 = 6.790$ and $\gamma_2 = 1.924$.

For hole levels in a quantum well, a successive perturbation approach is used. After a first-order perturbation, the quantum well potential lifts the degeneracy between the $J_z = \pm 3/2$ and $\pm 1/2$ bands. The appropriate expressions for the in-plane dispersion at $k_z = 0$ in a quantum well are

$$E = \frac{\hbar^2 k^2}{2m}(\gamma_1 + \gamma_2), \text{ for } J_z = \pm 3/2, \tag{14.1.19}$$

$$E = \frac{\hbar^2 k^2}{2m}(\gamma_1 - \gamma_2), \text{ for } J_z = \pm 1/2. \tag{14.1.20}$$

It is noted that the transverse dispersion corresponding to $J_z = \pm 3/2$ (heavy-hole band along the z direction), now has a light mass $m/(\gamma_1 + \gamma_2)$, whereas the $J_z = \pm 1/2$ level now has a heavy mass $m/(\gamma_1 - \gamma_2)$.

In fact, the effects of quantum well perturbation on the Hamiltonian and the $\boldsymbol{k} \neq 0$ terms should be treated on the same footing. This is a numerical exercise with a simple solution when the well is infinitely deep, but otherwise it involves a complex interaction of the two phenomena. Typical results for the dispersion relations of valence bands in a quantum well are shown in Fig. 14.1.4. As can be seen quite clearly, the valence band is highly non-parabolic. It is also interesting to note that

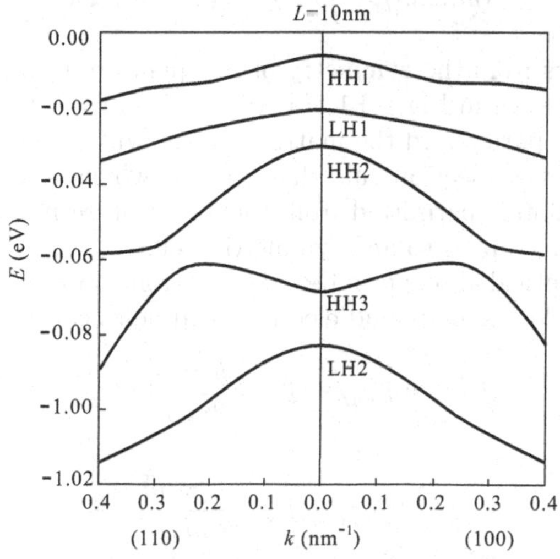

Figure 14.1.4 Dispersion curves for hole bands in a GaAs-$Al_{0.3}$ $Ga_{0.7}$As quantum well structure.

the first light hole band has a curvature opposite to the normal valence band structure, i.e., the hole state has a negative mass near the zone edge. The character of the hole states is represented by pure angular momentum states at the zone center, but there is a strong mixing of states as one proceeds away from there.

14.1.3 Optical Absorption

The most successful application of semiconductor quantum wells to date may be the exploitation of the optical properties of quantum wells. It is interesting to consider the optical absorption between bound states; in fact, we could measure the bound energy levels of a quantum well by shining light on the sample and determining which frequencies were absorbed. A photon is absorbed by exciting an electron from a lower level to a higher one, with the energy of the photon matching the difference in electronic energy levels. It is known that the states factorize into a product of a bound state in z and a transverse plane wave, as described by (14.1.13) and (14.1.14). Each state n for motion along z gives rise to a subband of energies.

The optical absorption is related to the matrix element between two such states $\langle m\boldsymbol{k}'|\boldsymbol{e}\cdot\boldsymbol{p}|n\boldsymbol{k}\rangle$, which depends strongly on $\boldsymbol{e}$, the polarization of the light. When $\boldsymbol{e} = (1,0,0)$ or $(0,1,0)$, i.e. polarization is in the x-y plane, the matrix element is zero, so no light is absorbed. Thus light that propagates normal to the layers, a convenient orientation for experiments, cannot be absorbed. On the other hand, when the electric field is normal to the quantum well, i.e. $\boldsymbol{e} = (0,0,1)$, which requires light to propagate in the plane of the well, the results are quite different: In this case $\boldsymbol{e}\cdot\boldsymbol{p} = -i\hbar\partial/\partial z$, which affects only the wavefunction of the bound states. Thus

$$\langle m\boldsymbol{k}'|\boldsymbol{e}\cdot\boldsymbol{p}|n\boldsymbol{k}\rangle = A^{-1}\int dz\int d^2\boldsymbol{r}\phi_m^*(z)e^{i(\boldsymbol{k}-\boldsymbol{k}')\cdot\boldsymbol{r}}p_z\phi_n(z). \tag{14.1.21}$$

The integral over $\boldsymbol{r}$ gives A if $\boldsymbol{k}' = \boldsymbol{k}$ and zero otherwise, so the two-dimensional wavevector is conserved. Thus, optical transitions are vertical in $\boldsymbol{k}$, as shown in Fig. 14.1.4. The remaining matrix element can be abbreviated to $\langle m|p_z|n\rangle$.

Because the allowed transitions are vertical in the transverse $\boldsymbol{k}$-plane, so the absorbed frequencies satisfy $\hbar\omega = E_m - E_n$, i.e., absorption is seen only at frequencies corresponding to the separation of the bound states of the well. Thus there are discrete lines, despite the continuous spectrum of states available, because of the restriction to vertical transitions. The lines may be broadened by any difference in effective mass between the subbands and transitions into the continuum above the quantum well occur at high energies.

The remaining task is to evaluate the matrix element

$$\langle m|p_z|n\rangle = -i\hbar\int\phi_m^*(z)\frac{d}{dz}\phi_n(z)dz. \tag{14.1.22}$$

An important result follows from the symmetry of the quantum well: The wavefunctions in a symmetric well, such as that shown in Fig. 14.1.5(a) where $V(-z) = V(z)$, are either even or odd in z. The derivative changes the parity, and the matrix element will be non-zero only if one state is even and the other odd. This is a selection rule that governs which transitions can be seen in optical absorption. Thus, absorption is permitted from the lowest state $n = 1$ to $n = 2, 4, \ldots$, but not to odd values of n. This result applies to any symmetric well.

Now, we consider the optical absorption between electron subbands and hole subbands, as shown in Fig. 14.1.5. The energy levels for bound electrons and holes can be rewritten as

$$E_{\mathrm{e}n_\mathrm{e}} = E_\mathrm{c} + \frac{\hbar^2\pi^2n_\mathrm{e}^2}{2m_\mathrm{e}L^2}; \tag{14.1.23}$$

and

$$E_{\mathrm{h}n_\mathrm{h}} = E_\mathrm{v} - \frac{\hbar^2\pi^2n_\mathrm{h}^2}{2m_\mathrm{h}L^2}, \tag{14.1.24}$$

respectively, where E_c is the bottom of the conduction band, and E_v is the top of the valence band. The conduction and valence bands are separated by the bandgap $E_\mathrm{g} = E_\mathrm{c} - E_\mathrm{v}$.

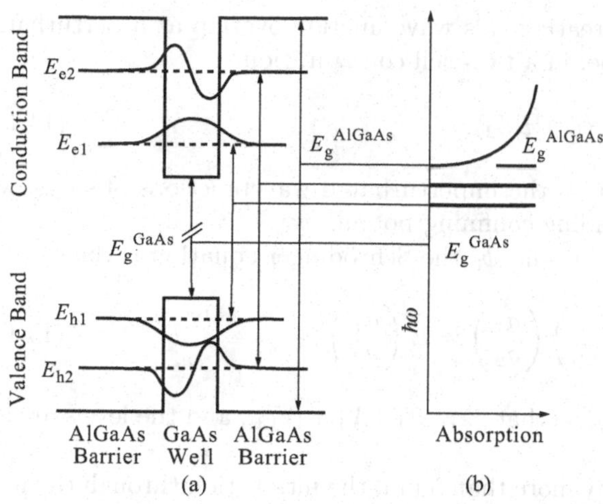

Figure 14.1.5 Optical absorption in a quantum well. (a) Potential wells in conduction and valence bands, showing two bound states in each; (b) Transition between states in the wells produce absorption lines between the bandgaps.

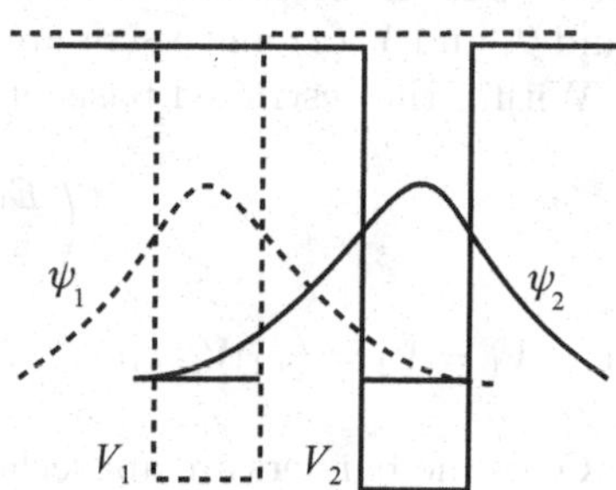

Figure 14.1.6 A double-well structure with the unperturbed potentials V_1, V_2 and the wavefunctions ψ_1, ψ_2 for the separate wells represented by dotted lines and solid lines, respectively.

At the start, the valence band is completely full and the conduction band completely empty (at zero temperature). Optical absorption may lift an electron from the valence band into the conduction band; this process leaves behind an empty state or hole in the valence band. In a bulk sample of GaAs, optical absorption can occur, provided that $\hbar\omega > E_g$. However, this is not so for a quantum well, because the states in the well are quantized due to confinement. The lowest energy at which absorption can occur is given by the difference in energy $E_{el} - E_{hl}$ between the lowest well state in the conduction band and the highest well state in the valence band. Absorption can occur at higher energies, by using other states. The strongest transitions occur between corresponding states in the two bands, so we set $n_e = n_h = n$. Therefore, strong absorption occurs at frequencies given by

$$\hbar\omega_n = E_{en} - E_{hn} = E_g + \frac{\hbar^2\pi^2 n^2}{2L^2}\left(\frac{1}{m_e} + \frac{1}{m_h}\right). \tag{14.1.25}$$

The energies look like those in a quantum well where the effective mass is m_{eh} given by $1/m_{eh} = 1/m_e + 1/m_h$. This is called the optical effective mass.

In practice there is a reverse process called photoluminescence. Light with $\hbar\omega > E_g$ is shone on the sample, which excites many electrons from the valence to the conduction band everywhere. Some of these electrons become trapped in the quantum well, and the same thing happens to the holes in the valence band. It is then possible for an electron to fall from the conduction band into a hole in the valence band and release the difference in energy as light. Experimentally, only the lowest levels are usually seen, so the photoluminescence spectrum often shows a line at $\hbar\omega_1$.

14.1.4 Coupled Quantum Wells

We have considered the optical absorption of a single quantum well. Actually, it is easy to fabricate multilayer structures with two types of semiconductor materials. As the barriers are thick enough, we can still treat the individual wells as independent. However, as the barriers become thinner (< 2 nm), there is significant interaction between adjacent wells, and so we should consider coupled multiple quantum wells.

Beyond the double-barrier single well structure, the simplest structure is the double-well configuration as shown in Fig. 14.1.6, which can be easily analyzed by the usual tight-binding perturbation model. As the barrier thickness is decreased, the exponentially-decaying wavefunction in the barrier will have some finite value in the next well. Treating this wavefunction overlap as a perturbation, one finds the perturbation matrix element to be, in a two-well configuration,

$$V_{12} = \langle \psi_1 | \mathcal{H} | \psi_2 \rangle, \tag{14.1.26}$$

where $\mathcal{H}$ is the electronic Hamiltonian, ψ_1 and ψ_2 the unperturbated wavefunctions of single wells 1 and 2, and $V_1(z)$ and $V_2(z)$ are the corresponding confining potentials.

Within the restricted basis of the functions ψ_1 and ψ_2 the Schrödinger equation is then

$$\begin{pmatrix} E_1 + V_1 & V_{12} \\ V_{12}^* & E_1 + V_1 \end{pmatrix} \begin{pmatrix} a_1 \\ a_2 \end{pmatrix} = E \begin{pmatrix} a_1 \\ a_2 \end{pmatrix}, \tag{14.1.27}$$

where $V_1 = V_2 = \langle \psi_1 | V_1(z) | \psi_1 \rangle = \langle \psi_2 | V_2(z) | \psi_2 \rangle$, so that $E = E_1 + V_1 \pm |V_{12}|$, and the levels are split by $2|V_{12}|$.

Once the barriers are appreciably thicker, say more than 5 nm, the interaction through them falls off sharply, and once the barriers exceed 10 nm there is, in effect, no interaction between adjacent wells. At this stage, the optical effects associated with adjacent wells simply add in parallel.

Introducing more wells leads to the creation of a continuous band of states. The transition from single wells to multiply-connected wells can be revealed by optical absorption. For N wells, the N-degenerate levels give rise to bands with $2N$ states.

§14.2 Magnetic Quantum Wells

A magnetic quantum well is also a sandwich structure composed of different materials, but at least one of them is a ferromagnetic metal. In writing this section we are much indebted to the excellent review on magnetic nanostructures by F. J. Himpsel *et al.* (Bib. [6]).

14.2.1 Spin Polarization in Metal Quantum Wells

Consider a single metal film confined by two interfaces, for simplicity with vacuum on each side. The energy dispersion relation $E(\boldsymbol{k}_\parallel)$, parallel to the surface of the film, is not affected by the confinement, so band theory can be used. Perpendicular to the surface, however, the energy spectrum $E(k_z)$ is discrete. We call these discrete thin film states quantum well states. The wavefunctions associated with the discrete states are characterized by a rapidly oscillating Bloch function that is modulated by a slowly-varying envelope wavefunction. The number of nodes in the envelope function determines the index of the discrete states (see §14.1.1).

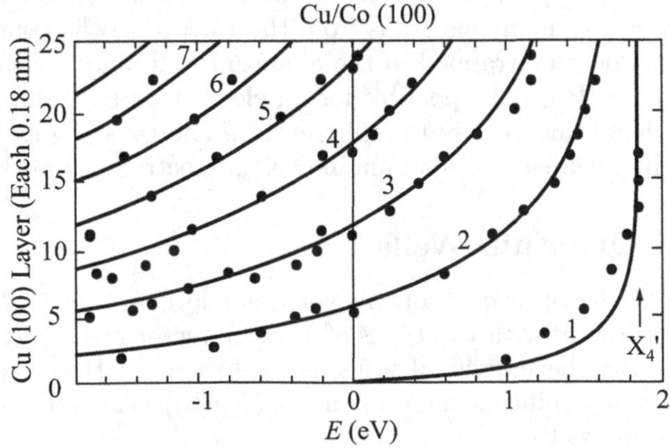

Figure 14.2.1 Energy versus thickness for quantum well states in Co-Cu structure. From F. J. Himpsel *et al.*, *Adv. Phys.* **47**, 511 (1998).

Extending this to metal multilayers, just like semiconductor quantum wells, a key feature of these quantum wells is still the reflection of electrons by the interfaces, which confine electrons within the layers with lower inner potential and quantized momentum and energy perpendicular to the layers. The discrete thin film states can be found using photoemission and inverse photoemission; the measurements give the density of states. Spectra of quantum well states in a simple magnetic structure are shown in Fig. 14.2.1, with epitaxial Cu(100) film of varying thickness grown on Co(100). It can be seen that the states become denser in thicker films and appear to converge towards the upper band edge at $X_{4'}$ (the curves are from a simple envelope function model, and the full circles come from photoemission and inverse photoemission data with $\boldsymbol{k}_{\|} = 0$). The energy positions of the quantized states with film thickness are described rather well by the envelope wavefunction model; the curves represent this model[a]

$$d_n(E) = \frac{n - 1 + \phi(E)}{1 - k(E)}, \tag{14.2.1}$$

where d_n is the thickness (in monolayers) at which the nth quantum well state appears at the energy E, $k(E)$ is the inverted bulk dispersion (with k in units of the Brillouin zone boundary), and $\phi(E)$ is the sum of the phase shifts for reflection at the two surfaces of the Cu film. Here, the phase function is obtained empirically by fitting a linear $\phi(E)$ relation to the $n = 2$ state; all the other states follow without adjustable parameters.

In magnetic metal quantum wells, an additional complication comes from the fact that the reflectivity is spin dependent, owing to the spin dependence of the inner potential in ferromagnets. Therefore, these quantized states may become spin-polarized. Actually, with ferromagnetic metals as barriers the quantum well states need to be spin-polarized, even in a noble metal film, such as Cu or Ag. This has been confirmed by experiment: Figure 14.2.2(a) shows a spin-polarized photoemission spectrum of a quantum well state in a Cu on fcc Co(100) near the Fermi level. It corresponds to state 2 in Fig. 14.2.1. The state has predominantly minority spin character. A first principles local density calculation arrives at the same conclusion, as shown in Fig. 14.2.2(b).

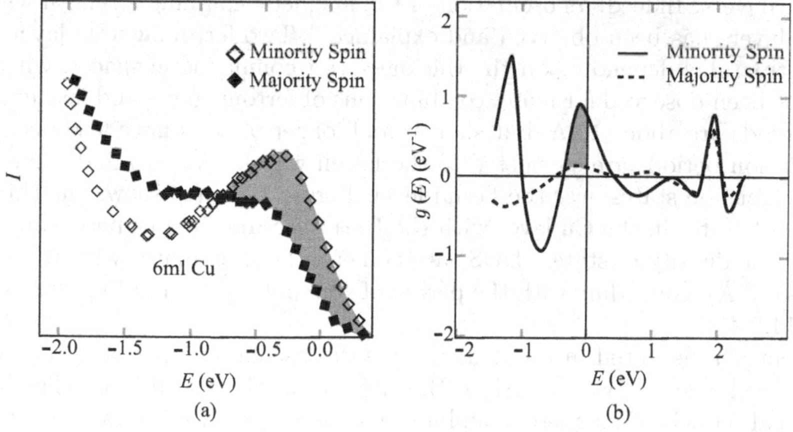

Figure 14.2.2 Spin polarization of quantum well states for Cu on Co(100). (a) Spin-polarized photoemission data; (b) First-principle calculation. From F. J. Himpsel *et al.*, *Adv. Phys.* **47**, 511 (1998).

There is a rather simple explanation for the spin polarization of quantum well states in magnetic multilayers: Generally, two magnetic layers are separated by a nonmagnetic spacer; quantized states exist for parallel magnetization only (bottom) and not for antiparallel magnetization. Quantum well states are formed by reflection of electrons at interfaces. The averaged inner potential of majority spin and minority spin states differs by the magnetic exchange splitting, leading to spin-dependent reflectivity, and only states with significant band offset to the nonmagnetic spacer band are confined. For metals to the right of the ferromagnets (particularly the noble metals) the majority bands are

[a] J. E. Ortega *et al.*, *Phys. Rev. B* **47**, 1540 (1993).

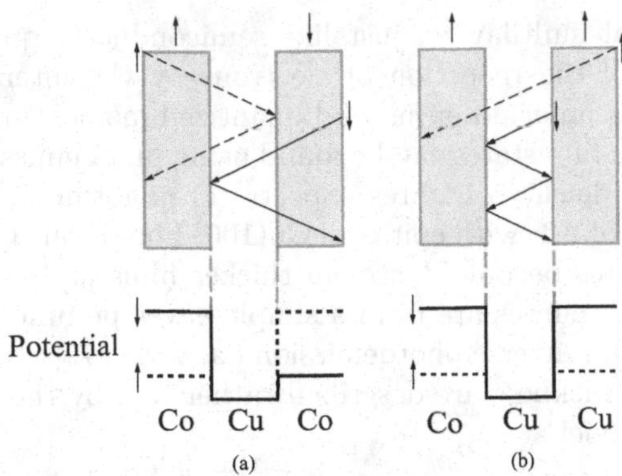

Figure 14.2.3 Schematic diagram of a magnetic quantum well in trilayers.

nearly lined up as shown in Fig. 14.2.3. Therefore, the minority spins experience a band offset and are confined in the quantum well states, but the majority spins behave like a continuous bulk band that extends throughout the noble metal and ferromagnet. For spacers to the left of the ferromagnets, for example Cr, the minority bands line up, and a majority spin polarization is expected for the quantum well states. This argument works independently of the specific band topology, but it does not guarantee full confinement, that is 100% reflectivity at the interface. Only in particular cases does one encounter a situation where the minority spins become totally Bragg reflected since they run into a bandgap in the ferromagnet.

14.2.2 Oscillatory Magnetic Coupling

Spin polarization in magnetic quantum well states can give rise to a fascinating characteristic of magnetic multilayers: Indeed, an oscillation of the magnetic coupling strength with thickness of the nonmagnetic layers has been observed and explained.[b] Two ferromagnetic layers line up, either in parallel or antiparallel depending on the thickness of a nonmagnetic spacer, with atomic precision. This effect has been observed for many combinations of ferromagnets and spacer materials. Typical oscillation periods are about 10 Å, but shorter and longer periods have also been observed.[c]

The oscillation periods are connected to the Fermi wavelength. To find the electronic origin for this effect we focus on states near the Fermi level. Figure 14.2.4(a) shows the thickness dependence of the density of states in the Cu layer with the Kerr effect on the magnetic coupling in the Co-Cu-Co trilayers. The density of states (DOS) at the Fermi level oscillates with the period of about six atomic layers (10 Å), coinciding with the period of spin polarization in Fig. 14.2.4(b) and saturation field in Fig. 14.2.4(c).

The wavelength measured in the density of states oscillations is that of the envelope function, and not the, much shorter, wavelength of Bloch states at the Fermi level. The latter is the length scale of classical models of magnetic coupling, such as the RKKY interaction. There is an another basic length scale in the system, that is, the lattice constant. The beat frequency between the Fermi wavelength and the lattice constant determines the longer period of magnetic oscillations. One has to determine the envelope function wavevector k_{env} by subtracting the Fermi wavevector $k_{\mathrm{F}} = 2\pi/\lambda_{\mathrm{F}}$ from the zone boundary wavevector k_{ZB}

$$k_{\mathrm{env}} = k_{\mathrm{ZB}} - k_{\mathrm{F}}. \tag{14.2.2}$$

The oscillation period is given simply by the inverse of k_{env}. The fact that the Fermi level crossing in Cu occurs at about one sixth of the Brillouin zone away from band maximum at X leads to a six-layer oscillation period. This result is identical with the prediction of RKKY theory after taking the discrete lattice into account.

[b] P. Grünberg *et al.*, *Phys. Rev. Lett.* **57**, 2442 (1986).
[c] J. Unguiris *et al.*, *Phys. Rev. Lett.* **67**, 140 (1991).

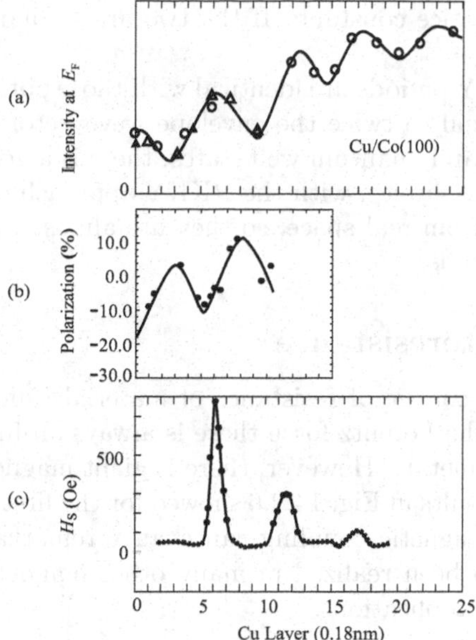

Figure 14.2.4 Simultaneous oscillations in (a) the density of states, (b) the spin polarization, and (c) the saturation field for the Cu/Co system. From F. J. Himpsel *et al.*, *Adv. Phys.* **47**, 511 (1998).

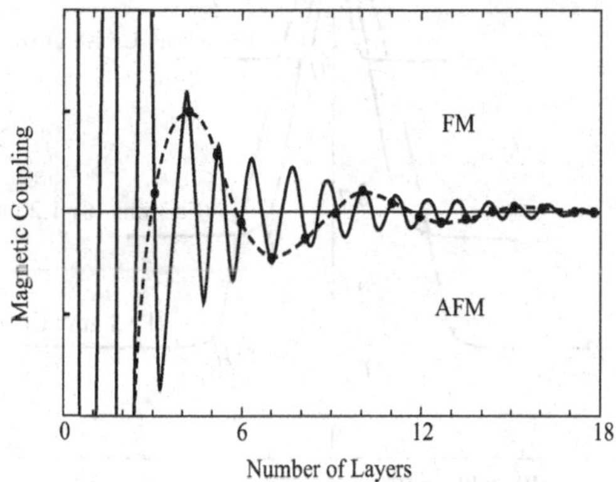

Figure 14.2.5 Formation of a beat frequency between the Fermi wavelength and the lattice periodicity for the RKKY coupling through Cu along the [100] direction. From F. J. Himpsel *et al.*, *Adv. Phys.* **47**, 511 (1998).

The RKKY model describes the coupling between two spin impurities via an intervening electron gas. The electron gas responds to the first spin by spin density oscillations, whose period is set by the Fermi wavevector $k_{\rm F}$. A second spin at a distance r from the first couples to the spin density wave. For a free electron gas, the interaction energy between the two spins takes the form

$$J(r) \propto \frac{\cos(2k_{\rm F}r)}{r^3}. \tag{14.2.3}$$

When summed over spins in two sheets, the coupling becomes

$$J_{\rm planar} \propto \frac{\cos(2k_{\rm F}z)}{z^2}, \tag{14.2.4}$$

where z is the spacing between the sheets. Actually, the quickly oscillating RKKY coupling can only be sampled at discrete crystal planes. Figure 14.2.5 shows the resulting beat frequency between the

Fermi wavelength and the lattice constant. If the two are similar, the resulting oscillation period can become quite long.

In most cases, these RKKY periods are identical with those obtained for the quantum well states, with the spanning vector equal to twice the envelope wevevector k_{env}. With the similarity in the periods predicted by RKKY and quantum well states, there is a good reason to believe that the two models are based on common physics, with the RKKY approach coming from reciprocal space and the quantum well approach from real space, so they are almost equivalent, although there may be some discrepancies in the details.

14.2.3 Giant Magnetoresistance

Magnetoresistance is the change of resistance of materials under applied magnetic fields. Just as discussed in §8.2, due to the Lorentz force there is always an intrinsic, but small, ordinary magnetoresistance (OMR) in all metals. However, there is giant magnetoresistance (GMR) in magnetic multilayers. Experimental results in Fig. 14.2.6 showed for the first time that the resistance of Fe/Cr superlattices with antiferromagnetic coupling can change remarkably under magnetic fields. This kind of GMR effect has also been realized in many other magnetic multilayer structures and has brought important technical applications.

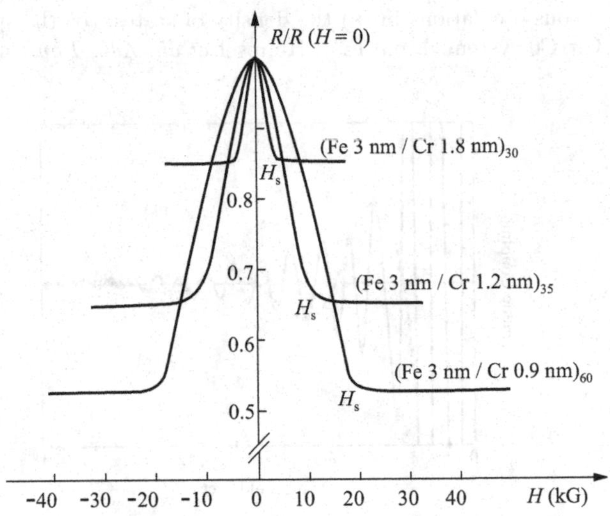

Figure 14.2.6 Magnetoresistance of three Fe/Cr superlattices at 4.2 K. From M. N. Baibich *et al.*, *Phys. Rev. Lett.* **61**, 2472 (1988).

The GMR effect is closely related to electronic spin polarization and oscillatory magnetic coupling. The basic structural unit is still a tri-layer, where two ferromagnetic layers are separated by a nonmagnetic layer, typically 1 nm thick. As we have seen, two magnetic configurations are possible by changing the thickness of the spacer: One with ferromagnetic layers oriented parallel, and the other antiparallel. The parallel configuration exhibits lower resistance than the antiparallel configuration. If one starts out with an antiparallel configuration and forces it into parallel alignment with an external field, the resistance decreases. Here the antiferromagnetic configuration as a starting point can be obtained by choosing the right spacer thickness. Figure 14.2.7 gives an example to illustrate the relationship of magnetoresistance to spacer thickness. It is consistent with Fig. 14.2.4, which shows the oscillatory magnetic coupling.

Experimental measurement of GMR can be done in two geometries: With the current in the plane (CIP) and the current perpendicular to the plane (CPP) of the layers. The CPP geometry is conceptually simpler and exhibits larger magnetoresistance, because there is no component through the normal metal layer; all current crossing over the layered structure must bear a spin-related

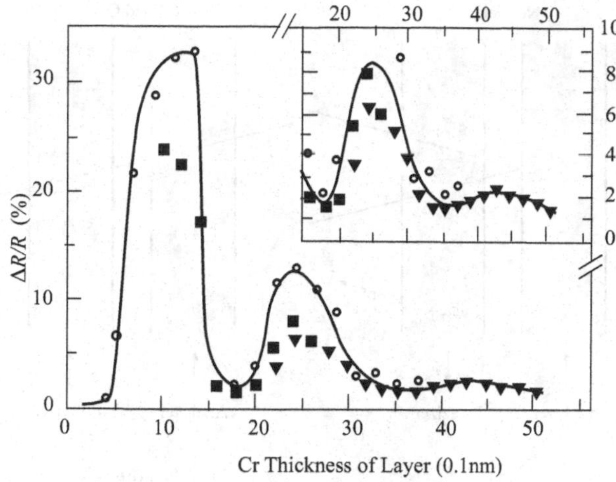

Figure 14.2.7 GMR of Fe/Cr magnetic multilayers versus thickness of the Cr layer. From S. S. P. Parkin *et al.*, *Phys. Rev. Lett.* **64**, 2304 (1990).

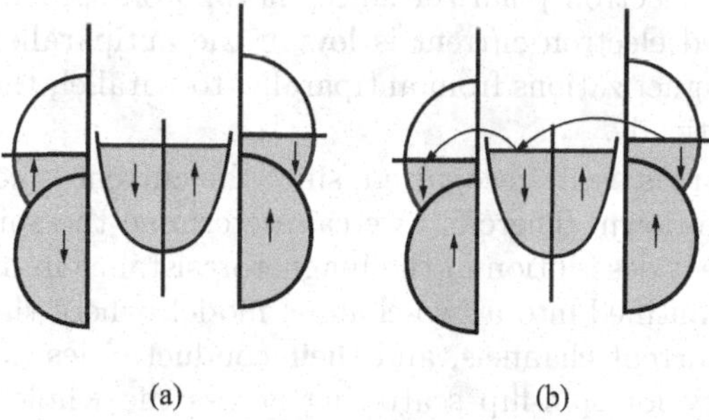

Figure 14.2.8 Schematic DOS and electronic transition in magnetic trilayers. (a) Antiparallel arrangement of magnetic moments; (b) Parallel arrangement of magnetic moments.

scattering at the interfaces. The GMR effect arises from spin-polarized electronic transport in magnetic multilayers; the spin-dependent scattering can take place at the interfaces and also in the bulk. However, a large number of experiments have confirmed that, for homogeneous multilayers, scattering mainly takes place at the interfaces. Certainly, defects in the interiors of layers, such as magnetic impurities in normal metallic layers and inhomogeneity in magnetic layers, can contribute to bulk spin scattering. Here we consider that the interfaces, including their roughnesses, play essential roles. Spin scattering at an interface must be deduced from the degree of matching of the conduction bands at the Fermi levels on both sides of the interface. Taking a Fe-Cr tri-layer shown in Fig. 14.2.8 as example, The paramagnetic d band of Cr matches the minority spin d band of Fe in energy, but does not match with its majority spin d band. This illustrates that the up and down spins at the interface are not equivalent. The minority spins can have antiparallel and parallel configurations, each giving rise to different effects on electronic transport. Arrows denote the electronic transport from one ferromagnetic layer through the nonmagnetic metallic layer to another magnetic layer.

Here we shall focus on the simpler CPP geometry to give a simple exposition of the key phenomena contributing to GMR. A good starting point is the optical polarizer analyzer analogy in Fig. 14.2.9. Conduction electrons become spin-polarized by spin-dependent scattering at the magnetic interfaces: Interfaces at the ferromagnetic layers act as spin-polarizers. The electron current perpendicular to the interfaces increases when the magnetic orientation is switched from antiparallel to parallel by an external field. Antiparallel orientation of the two ferromagnetic layers is equivalent to a crossed optical polarization filter: we note that a 90° rotation of an optical polarizer corresponds

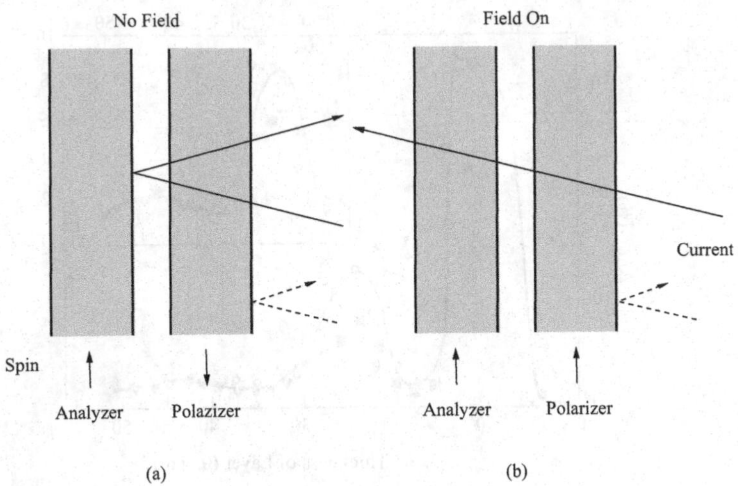

Figure 14.2.9 Simplified visualization of GMR via an optical polarizer-analyzer analogue.

to a 180° rotation for an electron polarizer since the photon spin is 1 and the electron spin 1/2. Therefore, the transmitted electron current is low in the antiparallel configuration. But when an applied field turns the magnetizations from antiparallel to parallel, the current perpendicular to the interface increases dramatically.

Perpendicular transport is easily recognized, since the current is composed of two unmixed components, spin-up and spin-down. Therefore we can determine the spin-dependent scattering coefficients and finish the correct description of the magnetoresistance in magnetic multilayer structures. Such a picture can be quantified into a two-channel model, where the majority and minority spins are treated as separate current channels, and their conductivities are just added up. This model assumes a low probability for spin-flip scattering across the whole stack. Each channel is then equivalent to a series of resistors when electrons are scattered at various interfaces and in the bulk. Figure 14.2.10 shows schematically this mechanism for magnetoresistance. Under a high magnetic field, the magnetizations of every layer are parallel. Because different spin orientations have different scattering probabilities, they have different resistivities $\rho_\uparrow$ and $\rho_\downarrow$, as shown in Fig. 14.2.10(a). The total resistivity of the system is

$$\rho_{\mathrm{F}} = \frac{\rho_\uparrow \rho_\downarrow}{\rho_\uparrow + \rho_\downarrow}. \tag{14.2.5}$$

There is a low resistance channel for electrons with one kind of spin. Current mainly passes through this channel, so the total resistance (14.2.5) has a lower value. Figure 14.2.10(b) represents the reversed situation, the neighboring magnetic layers are antiparallel, the electrons with lower resistance in one layer will show higher resistance in the neighboring layer. Therefore, each channel has the same resistivity $(\rho_\uparrow + \rho_\downarrow)/2$, and the total resistivity of the system is

$$\rho_{\mathrm{AF}} = \frac{1}{4}(\rho_\uparrow + \rho_\downarrow). \tag{14.2.6}$$

It is easy to see that in general $\rho_{\mathrm{AF}} \gg \rho_{\mathrm{F}}$, so the magnetoresistivity can be defined as

$$\mathrm{MR} = \frac{\rho_{\mathrm{AF}} - \rho_{\mathrm{F}}}{\rho_{\mathrm{F}}}. \tag{14.2.7}$$

For magnetic materials Fe, Co and Ni, the differences of resistivities $\rho_\uparrow$ and $\rho_\downarrow$ are very large and, by a simple argument, we can write the resistivity for spin σ as

$$\rho_\sigma = m_\sigma / n_\sigma e^2 \tau_\sigma. \tag{14.2.8}$$

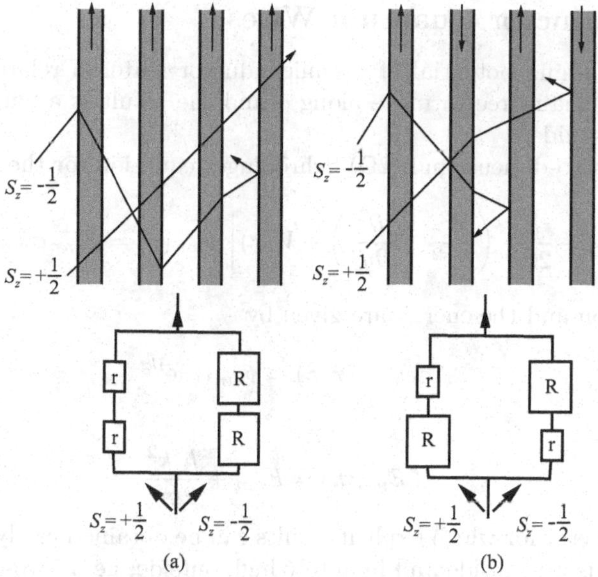

Figure 14.2.10 Two channel model for GMR. (a) Parallel magnetization under high magnetic fields; (b) Antiparallel magnetization under zero magnetic field and antiferromagnetic exchange interaction.

where m_σ and n_σ are the electronic mass and number density of spin σ. Taking V_σ as a scattering matrix element, in the Born approximation, the transition probability is

$$\tau_\sigma^{-1} \sim |V_\sigma|^2 n_\sigma(E_F). \tag{14.2.9}$$

Therefore, the spin-dependent resistivity ρ_σ can be calculated microscopically.

It should be noted that the spin relaxation length is usually greater than 10 nm, so an electron can cross through many layers before it loses its original orientation. Within this length, each magnetic interface represents a spin filter, so the more scattering interfaces there are, the more obvious the filtering effect is, i.e., magnetoresistance increases with the number of layers. A useful picture to show the interface scattering comes from the spin-polarized quantum well states is indicated in Fig. 14.2.3, i.e, there are spin-dependent steps in the inner potential which can even cause total reflection of the minority spins. Experiments have confirmed that GMR occurs only at a well-defined thickness of the spacer layer where the ferromagnetic layers are magnetized antiparallel. The resistance drops when the orientation is switched from antiparallel to parallel by an external field.

GMR occurs not only in well ordered multilayers but also in granular materials, such as ferromagnetic particles, e.g. Co segregated in noble metal Cu.[d] A random coupling between ferromagnetic particles is enough to produce GMR, because on average half of the particles are antiparallel to each other. If the sizes of particles are appropriate, the electronic spin-dependent scattering at the Co-Cu interfaces causes GMR. The resistance of a magnetic granular material will change considerably, when a magnetic field is applied. But if the sizes of particles are increased, the ratio of surface to volume decreases and GMR will diminish and disappear finally.

§14.3 Quantum Wires

In the last two sections, we investigated electrons that are confined in a bound state along one direction and behaved as though they were free in two dimensions. There is the possibility of confining electrons further and so reduce their effective dimensionality to one; this is the area of quantum wires.

[d] A. E. Berkowitz *et al.*, *Phys. Rev. Lett.* **68**, 3745 (1992); J. Q. Xiao *et al.*, *Phys. Rev. Lett.* **68**, 3749 (1992).

14.3.1 Semiconductor Quantum Wires

If we take the confining potential of a semiconductor material related to spatial position $\boldsymbol{r} = (x, y)$, the electrons remain free to move along z and the result is a wire, closely analogous to an electromagnetic wave guide.

Starting with the two-dimensional (2D) Schrödinger equation for the confining potential

$$\left[-\frac{\hbar^2}{2m^*}\left(\frac{\partial^2}{\partial x^2}+\frac{\partial^2}{\partial y^2}\right)+V(\boldsymbol{r})\right]\phi_{mn}(\boldsymbol{r}) = E_{mn}\phi_{mn}(\boldsymbol{r}). \tag{14.3.1}$$

The total wavefunction and the energy are given by

$$\psi_{mnk}(\boldsymbol{r}, z) = \phi_{mn}(\boldsymbol{r})\mathrm{e}^{ikz}, \tag{14.3.2}$$

and

$$E_{mn}(k) = E_{mn} + \frac{\hbar^2 k^2}{2m^*}. \tag{14.3.3}$$

The simplest geometry for which explicit results can be obtained easily is a rectangular geometry with a potential that is zero inside and infinitely high outside; i.e. a two-dimensional infinitely deep well. Assume that the respective effective masses for motions in the k_x, k_y, and k_z directions are m_x, m_y, and m_z; we can still use envelope functions, which are presumed to be slowly varying on an atomic scale. In the effective mass approximation, the envelope wavefunction of a state with quantum numbers m, n and wavevector k in a rectangular wire of dimensions a and b is taken to be

$$\psi_{mnk}(\boldsymbol{r}, z) = \left(\frac{4}{ab}\right)^{1/2} \sin\frac{\pi m x}{a} \sin\frac{\pi n y}{b}\mathrm{e}^{ikz}, \tag{14.3.4}$$

where $m, n = 1, 2, 3, \ldots$. The free electron-like motion along the wire leads to subbands of states with energies

$$E_{mn}(k) = \frac{m^2\pi^2\hbar^2}{2m_x a^2} + \frac{n^2\pi^2\hbar^2}{2m_y b^2} + \frac{\hbar^2 k^2}{2m_z}. \tag{14.3.5}$$

The subband energies for $k = 0$ are non-degenerate (not including spin) in general, but for a square with $a = b$ the states m, n and n, m will be degenerate when $m \neq n$. The density of states (DOS) per unit energy for the one-dimensional subband with quantum numbers m and n in a wire of length L is

$$g(E) = 2\cdot 2\frac{L}{2\pi}\frac{dk}{dE} = \left(\frac{L2m_z}{\pi\hbar^2}\right)^{1/2}(E - E_{mn})^{-1/2}, \tag{14.3.6}$$

where the first factor arises because there are states with both positive and negative values of k, and the second factor of 2 is the spin degeneracy. By the way, there is no simple closed form solution for the envelope wavefunctions and energy levels of a rectangular wire with a finite barrier.

Another simpler case may be the one with a circular cross section. To be a little more general, we consider an elliptical GaAs wire with semimajor axis a and semiminor axis b. It is assumed that the electrons are confined in this kind of wire by an infinite barrier. In elliptical coordinates $\boldsymbol{r} = (r, \theta, z)$ with r the radial coordinate, θ the angular coordinate, z the axial coordinate, the wavefunction can be separated into

$$\psi = AU(r)V(\theta)\mathrm{e}^{ikz}, \tag{14.3.7}$$

where k is the axial wavevector. The radial and angular components of the wavefunction satisfy the Mathieu equations

$$\frac{d^2U(r)}{du^2} - [\beta - 2\lambda\cosh(2r)]U(r) = 0, \tag{14.3.8}$$

$$\frac{d^2V(\theta)}{d\theta^2} + [\beta - 2\lambda\cos(2\theta)]V(\theta) = 0. \tag{14.3.9}$$

The angular functions are periodic $V(\theta+2\pi)=V(\theta)$ which implies that the separation constant β are quantized as β_m where m is the azimuthal quantum number. The solutions to the above equations which are regular to the origin are

$$\psi_{mnk}(\boldsymbol{r})=A_{mn}e^{ikz}\begin{cases}\mathrm{Ce}_m(r,\lambda_{mn})\mathrm{ce}_m(\theta,\lambda_{mn}), & \text{even},\\ \mathrm{Se}_m(r,\lambda_{mn})\mathrm{se}_m(\theta,\lambda_{mn}), & \text{odd},\end{cases} \tag{14.3.10}$$

where ce, se, Ce, and Se are the Mathieu functions, λ_{mn} is given by

$$\lambda_{mn}=\frac{1}{4}f^2k_{mn}^2 \tag{14.3.11}$$

with f the semifocal distance ($=ae'$ with e' the eccentricity) and k_{mn} is the confinement wavevector of the electron in the xy plane. The labels odd and even refer to the parities of the angular functions with respect to θ. The Mathieu functions with index m even have periodicity π, while those with m odd have periodicity 2π. The requirement that the above wavefunction vanishes at r_0 is given by

$$r=r_0=\cosh^{-1}(1/e') \tag{14.3.12}$$

and yields the quantum number n. The nth root of the radial component is labeled λ_{mn} and the normalized factor A_{mn} is given by

$$A_{mn}^{-2}=L\int_0^{2\pi}\int_0^{r_0}U_m^2(r,\lambda_{mn})V_m^2(\theta,\lambda_{mn})f^2(\sinh^2 r+\sin^2\theta)drd\theta, \tag{14.3.13}$$

where L is the length of the elliptical wire which is assumed to be effectively infinite. The total energy of the state (m,n,k) is given by

$$E_{mn}(k)=\frac{\hbar^2}{2m^*}\left(k_{mn}^2+k^2\right). \tag{14.3.14}$$

Figure 14.3.1 illustrates the confinement energy as a function of a/b for fixed $b=50$ Å. It is noted that both the cylindrical ($e'=0$) and the slab $e'\to 1$ limits are obtained, with the degeneracy in the cylindrical states lifted by the elliptical asymmetry. The number of electronic states, within the energy interval shown in Fig. 14.3.1, increases dramatically as e' approaches unity, and only a few of the states are shown in this figure. There is a tendency that, as the eccentricity increases, the ground state wavefunction tends to be localized near the region of lowest curvature, in other words towards the center of the ellipse.

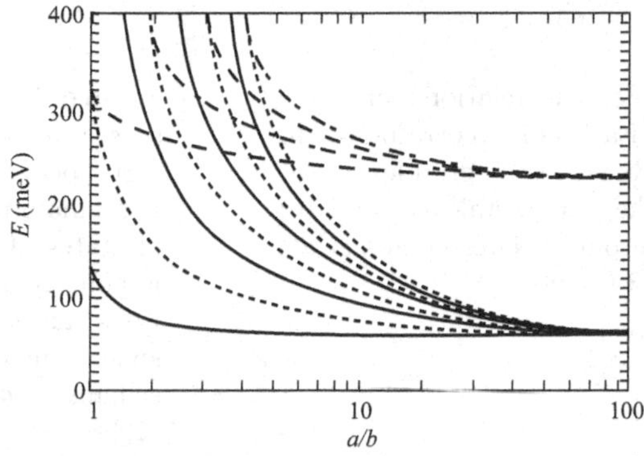

Figure 14.3.1 The subband energies of an elliptical wire. The solid curves correspond to even states with period π, the dotted curves are even states with period 2π, the dashed curves are odd states with period π and the dashed-dot curves are odd states with period 2π. From C. R. Bennett *et al.*, *J. Phys.: Condens. Matter* **7**, 9824 (1995).

14.3.2 Carbon Nanotubes

Carbon nanotubes have exotic electronic properties and can be considered as prototypes for a one-dimensional quantum wire. An ideal nanotube can be thought of as a hexagonal network of carbon atoms that has been rolled up to make a seamless cylinder (see §2.3.4). Just a nanometer across, the cylinder can be tens of microns long, and each end is "capped" with half a fullerene molecule. Single-wall nanotubes can be thought of as the fundamental cylindrical structure, and these form the building blocks of both multi-wall nanotubes as well as more complicated ropes. Many theoretical studies have been concerned with the properties of single-wall nanotubes.

The unique electronic properties of carbon nanotubes are due to the quantum confinement of electrons normal to the nanotube axis. In the radial direction, electrons are confined by the monolayer thickness of the graphene sheet. Around the circumference of the nanotube, periodic boundary conditions come into play: Going around the cylinder once introduces a phase difference of 2π.

Because of this quantum confinement, electrons can only propagate along the nanotube axis, and so their wavevectors point in this direction. The resulting number of one-dimensional conduction and valence bands effectively depends on the standing waves that are set up around the circumference of the nanotube. These simple ideas can be used to calculate the dispersion relations of the one-dimensional bands, which link wavevector to energy, from the well known dispersion relations in a graphene sheet.

The calculated dispersion relations for a small diameter nanotube show that about one-third of small-diameter nanotubes are metallic, while the rest are semiconducting, depending on their diameter and chiral angle. We see in Fig. 14.3.2 that an armchair (5, 5) nanotube and a zigzag (9, 0) nanotube are metallic, while a zigzag (10, 0) nanotube is a semiconductor. A small increase in diameter may have a major impact on the conduction properties of carbon nanotubes. Here we use the (m, n) notation for nanotubes introduced in §2.3.4.

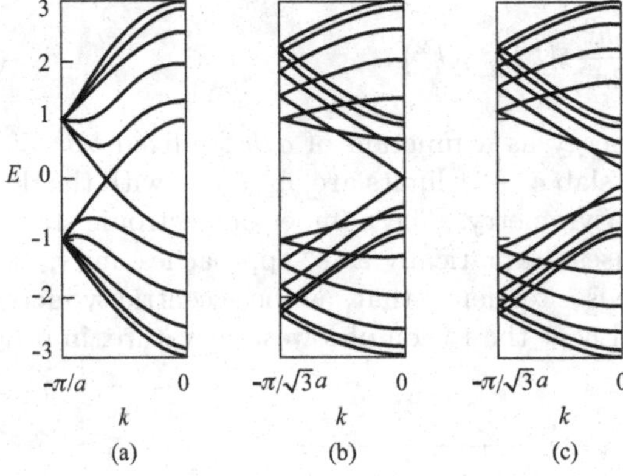

Figure 14.3.2 Dispersion relations of three types of nanotubes. Each curve corresponds to a single quantum subband. The Fermi level is at $E = 0$. (a) Metal-like; (b) Metal-like; and (c) Semiconductor-like. From M. Dresselhaus *et al.*, *Phys. World, Jan.*, 33 (1998).

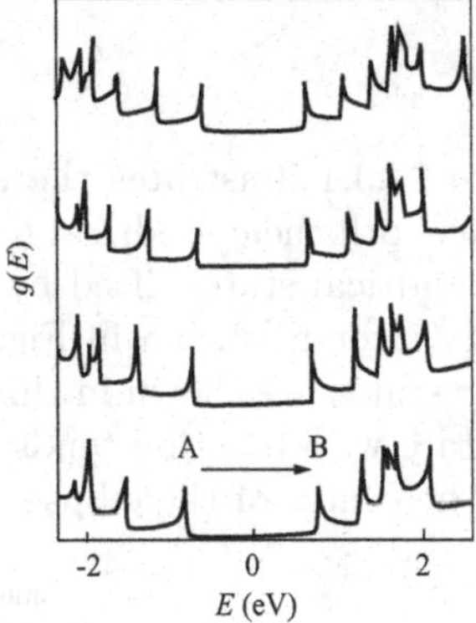

Figure 14.3.3 Density of states for several armchair nanotubes show discrete peaks at the positions of the band maxima or minima. The density of states of these metallic nanotubes is nonzero at $E = 0$. Optical transitions can occur between mirror-image spikes, such as A→B. From M. Dresselhaus *et al.*, *Phys. World, Jan.*, 33 (1998).

The electronic density of states has also been calculated for a variety of nanotubes. Figure 14.3.3 shows the calculated results for metallic (8, 8), (9, 9), (10, 10), and (11, 11) armchair nanotubes. While conventional metals have a smooth density of states, the DOS for these nanotubes are characterized by a number of singularities, where each peak corresponds to a single quantum subband.

These singularities arise from quasi-one-dimensionality. As the nanotube diameter increases, more wavevectors are allowed in the circumferential direction; since the bandgap in semiconducting nanotubes is inversely proportional to the tube diameter, the bandgap approaches zero at large diameters, just as for a graphene sheet. At a nanotube diameter of about 3 nm, the bandgap becomes comparable to thermal energies at room temperature.

A carbon nanotube is very much like a piece of graphene sheet with a hexagonal lattice that has been wrapped into a seamless cylinder. Since its discovery in 1991, the peculiar electronic properties of these structures have attracted much attention. Despite the difficulties, pioneering experimental work has confirmed the main theoretical predictions about the electronic structure of nanotubes. Their electronic conductivity, for example, has been predicted to depend sensitively on tube diameter and chiral angle (a measure of the helicity of the tube lattice), with only slight differences in these parameters causing a shift from a metallic to a semiconducting state. In other words, similarly shaped molecules consisting of only one element (carbon) may have very different electronic behavior. Scanning tunneling microscopy (STM) offers the potential to probe this prediction, as it can resolve simultaneously both the atomic structure and the electronic density of states. Scanning tunneling microscopy and spectroscopy on individual single-walled nanotubes have yielded atomically resolved images that allow us to examine the electronic properties as a function of tube diameter and chiral angle. Results show that there are both metallic and semiconducting carbon nanotubes, and that the electronic properties indeed depend sensitively on the chiral angle. The bandgaps of both tube types are consistent with theoretical predictions.

14.3.3 Metal Steps and Stripes

Most of the concepts related to the quantum wires discussed above can be extended to steps or stripes on metal surfaces, so we can call them metal wires. It is much more difficult to obtain truly confined states, however, since the bulk states are able to couple to the wire states easily. There are two confining directions: one perpendicular to the surface, and the other in-plane, but perpendicular to the wires. The crystal periodicity is kept only along the wire direction. A variety of step-related electronic and magnetic phenomena have been discovered, such as step states, lateral quantization and in-plane anisotropy.

The one-dimensional analog of a two-dimensional surface or interface is a single step. Electrons that are already confined perpendicular to the surface can become confined perpendicular to the step, too. Even if they are not totally confined, their wavefunctions are scattered elastically at the step edge, producing standing waves and ripples in the charge density. These oscillations have been sampled directly by scanning tunneling spectroscopy. The analogue of a surface state would be a wavefunction confined to a step edge.

A terrace confined by two parallel steps is the one-dimensional analog of a thin film. Such a structure produces lateral standing waves, similar to the standing waves induced by quantum well states in a thin film. On a terrace, the band structure perpendicular to the steps becomes quantized, Fig. 14.3.4(a); parallel to the step the continuum of energies remains. Scanning tunneling spectroscopy allows direct visualization of this effect. Figure 14.3.4(b) shows the first three quantized states on a Au(111) terrace 0.36 nm wide.[e] The dI/dV spectra are offset according to the sample bias. The arrows indicate maxima in the charge density when the bias voltage coincides with one of the quantized levels, and the step profile is shown as a broken curve. The results clearly demonstrate the ability to image the probability amplitudes and to obtain the spectra of confined states at metal surfaces, even at room temperature.

A simple theoretical description can be given: It is reasonable to assume that the reflection of electrons incident at a step from the side of the upper terrace is stronger than the reflection of electrons incident on the step from the lower terrace side. For certain step orientations, the barrier on the upper side of a step is found to be well represented by a hard-wall potential. It is likely that electrons incident on the lower side of the step can more easily be transmitted into the empty bulk states. In this simple model of a terrace of width a, we can place a hard-wall barrier at the upper side of the step leading to the lower terrace, which we take as the origin. That is, at $x = 0$, $V(0) = \infty$,

[e] P. Avouris and I. Lyo, *Science* **264**, 942 (1994).

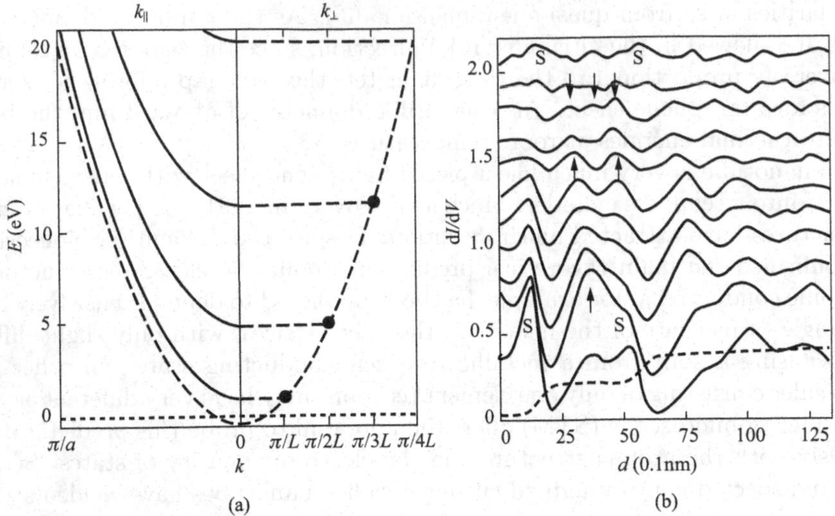

Figure 14.3.4 Band structure of electrons confined to a single stripe of length L. (a) Quantized levels and band dispersion along the directions perpendicular and parallel to the stripe, respectively; (b) Charge density probed by scanning tunneling spectroscopy. From J. F. Himpsel *et al.*, *Adv. Phys.* **47**, 511 (1998).

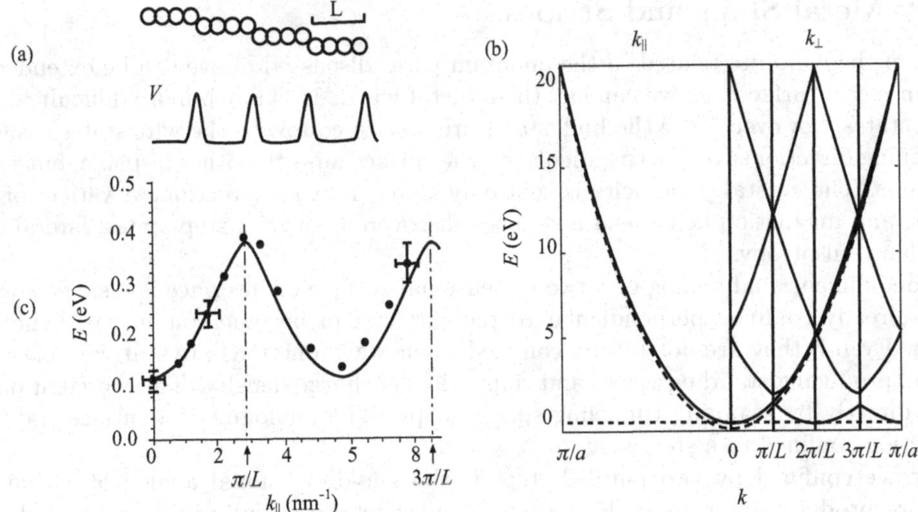

Figure 14.3.5 (a) Schematic diagrams of a lateral superlattice and its potential; (b) Dispersion relations along the directions parallel and perpendicular to the steps; (c) Band dispersion of an image state on a stepped Cu(100) surface measured by two-photon photoemission. From X. Y. Wang *et al.*, *Phys. Rev. B* **53**, 15738 (1996).

and at $x = L$, the bottom of the step, a delta function barrier may be used, i.e., $V(L) = V_0\delta(x - L)$ with V_0 determined by experiments. The wavefunction of the electrons in the narrow terrace was taken to be of the form

$$\psi(k_{\parallel}) = A\sin(k_x x)e^{ik_y y}, \qquad (14.3.15)$$

and the density of states can be calculated. The simulation based on this model can reproduce well the dI/dV scans of Fig. 14.3.4(b). On the other hand, it is observed that the energies of the confined states in terraces with widths in the range of 30 to 60 Å are proportional to L^{-2}.

A stepped surface having an infinite array of parallel steps forms a lateral superlattice, as schematically shown in Fig. 14.3.5(a) with L as the step spacing. A simple model potential is also shown for the steps; that is, a one-dimensional train of δ-functions will produce the nearly-free-electron-like band structure. For such a periodic structure we expect energy bands, but they are folded back into the small Brillouin zone of the reciprocal step lattice, as shown in Fig. 14.3.5(b). Its boundaries lie at $k = \pm\pi L$ in the direction perpendicular to the steps. Such folding-back of the bands has been observed for the $n = 1$ image state on Cu(100) shown in Fig. 14.3.5(c).

With respect to the flat surface, we notice two changes: the bottom of the band is shifted up in energy, since the average potential is higher after including the repulsive barriers, and the continuity is broken at the boundaries of the step-induced Brillouin zone, where small gaps open up. The shift of the surface band has been observed with STM and photoemission spectroscopy for the surfaces of Cu(111) and Au(111).

Steps have significant influences on magnetic properties, especially anisotropy. Using the stepped surface of a non-magnetic material, it is possible to create lateral magnetic superlattices by step decoration and step-flow growth of ferromagnets.

§14.4 Quantum Dots

It is possible to go one stage further and confine electrons or holes in all three dimensions. The typical approach is that they are first confined in one dimension by growth, a quantum well or doped heterojunction, and then restricted to a small area by etching or an electrostatic potential. The resulting zero-dimensional structure is a quantum dot, just like an artificial atom. Electrons in a quantum dot structure can occupy only discrete energy states, similar to the discrete states of atoms. The density of states is just a set of δ-functions, as there is no free motion in any direction. In general, quantum dots are small, with size less than 100 nm and with a number of electrons between one and a few thousand.

14.4.1 Magic Numbers in Metal Clusters

The term "clusters" is generally used to describe aggregates composed of several to several hundreds atoms. These clusters, with intermediate size, are often found to have hybrid properties characteristic of neither the molecular nor the bulk limits. For example, magnetism is a interesting problem. While a metal may be non-magnetic in the bulk phase, small clusters of its atoms may exhibit magnetism. One of the most important facts is that these clusters often do not have the same structure or atomic arrangement as a bulk solid, and they may change structure with the addition of just one or a few atoms. A striking phenomenon appears in the mass abundance spectra: Clusters consisting of certain numbers of atoms are more abundant than others; these favored numbers are called "magic numbers".

Clusters of simple metal atoms attract our attention: The particularly striking, and experimentally demonstrated, example of wave order is the valence electrons in a small drop of sodium, or some other simple metal, which form an ordered quantum state analogous to the ordered electron structures of atoms. An abundance distribution of sodium clusters is shown in Fig. 14.4.1. Clusters composed of 8, 20, 40 and 58 atoms, i.e. certain numbers of ions with the same number of valence electrons, are especially stable. It is the electrons, not the ions, that quantize and decide the variations in stability. As a consequence, small metal clusters form a periodic system, like the system of chemical elements — only much larger.

The jellium model, as a simple theory for electronic structure of metallic clusters, is borrowed from solid state physics, which ignores many-body interactions completely. Here, geometrical effects, like surface constraints, are particularly important in applying the jellium model to small metal clusters. Confinement of the jellium and the electrons to spherical or ellipsoidal regions leads to shell structure. Some simple quantum models illustrating electronic shell structure are depicted in Fig. 14.4.2. For a three-dimensional harmonic oscillator model, the energy level are equally spaced. When degeneracies are included in this model, there is a shell structure in the electronic energy-level occupation; i.e. degenerate levels are separated by wide gaps. A similar result is also found for a three-dimensional square-well potential, but with unevenly spaced energy levels. A model that

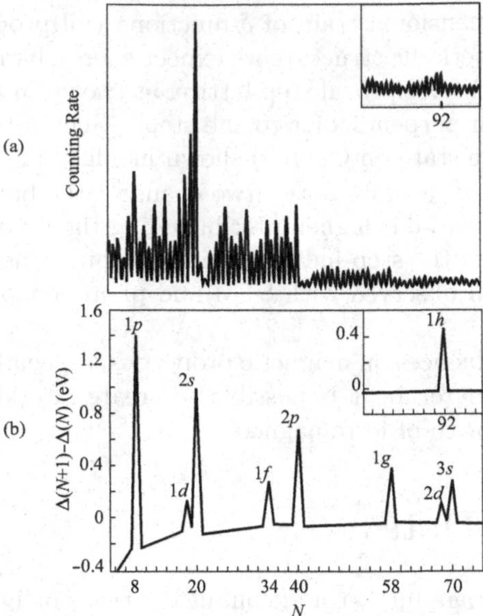

Figure 14.4.1 Abundance spectrum of sodium clusters. (a) Experimental results; (b) Second-order energy difference between neighboring clusters in a sequence. W. D. Knight *et al.*, *Phys. Rev. Lett.* **52**, 2141 (1984).

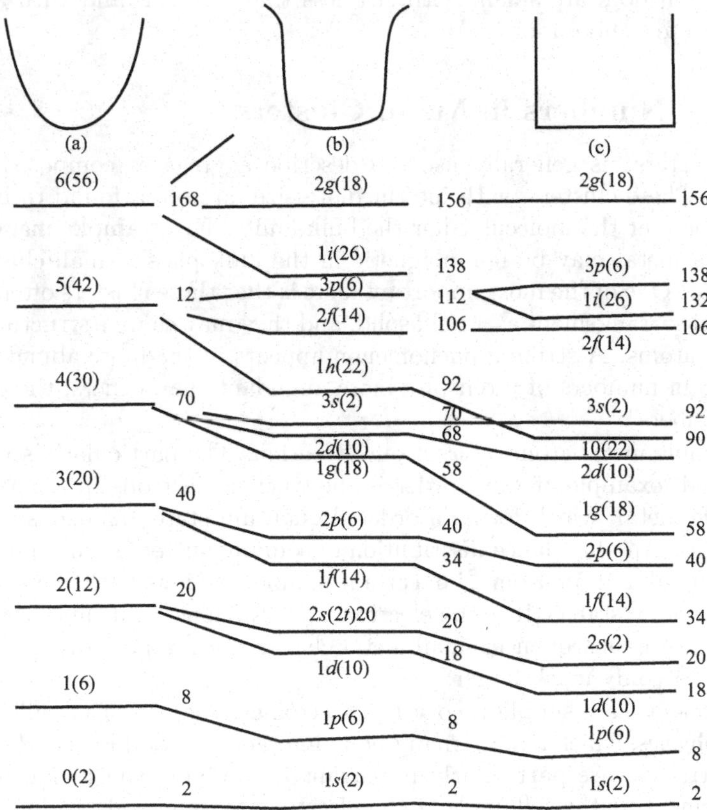

Figure 14.4.2 Energy-level spectra for a three-dimensional oscillator-well potential (a), a square-well potential (c), and a potential intermediate between the two (b). The energy-level labels with degeneracies and the total number of states are given. From M. L. Cohen and W. D. Knight, *Phys. Today* **43**, 42 (1990).

gives results similar to those found in self-consistent jellium calculations is intermediate between the harmonic oscillator and square-well models. In this model the energy levels are characterized by principal and angular momentum numbers (n, l). However, unlike electrons in atoms where l must be less than n, in this case there is no restriction on the relative values of l and n because the potential is not of the Coulomb form. The successive energy levels including their degeneracies for the intermediate model are 1s(2), 1p(6),1d(10), 2s(2), 1f(14), 2p(6), 1g(18), 2d(10), 3s(2), 1h(22), 2f(14), 3p(6), 1i(26), 2g(18) .... Hence, as electrons fill the shells, closings occur for total electron numbers 2, 8, 18, 20, 34, 40, 58, 68, 70, 92, 106, 112, 138, 156 and so on. In clusters of alkali or noble metals each atom contributes one electron, and shell closures occur for clusters containing the numbers of atoms in this series. Total energies should be low for clusters having these magic numbers, and hence clusters of these sizes are expected to be particularly stable.

We must note the characteristics of the effective potential, which is not the screened Coulomb potential around the positive charge of the point-like nucleus, but is more like a spherical cavity. The positive charge with the number of elementary charges is smeared out through the whole volume. The influence of the detailed positive charge configuration can be ignored to first order. Only the average positive charge density seems to matter. In this simple model the positive charge density distribution is chosen to be a homogeneously charged sphere as

$$\rho_+(r) = \rho_0\theta(r_0 - r), \tag{14.4.1}$$

where ρ_0 is the average density of the jellium, which is set equal to the valence electron density for the alkali metal, r_0 the cluster radius, and θ a step function. The cluster radius is related to the total number of electrons N as $r_0 = r_s N^{1/3}$, where r_s is the electron density parameter satisfying $\rho_0^{-1} = 4\pi r_s^3/3$.

As a reasonable assumption, we can adopt the effective one-electron potential inside the cluster, with a spherically symmetric rounded potential well of the form

$$U(r) = -\frac{U_0}{\exp[(r - r_0)/\epsilon] + 1}, \tag{14.4.2}$$

where U_0 is the sum of the Fermi energy (3.23 eV) and the work function (2.7 eV) of the bulk value. The parameter ϵ determines the variation of the potential at the edge of the sphere; $\epsilon = 1.5$ a.u. is suitable for this purpose. Based on the effective potential (14.4.2), The Schrödinger equation can be solved numerically for each N. It yields discrete electronic levels characterized by the angular momentum quantum number l with degeneracy $2(2l + 1)$. The electronic levels shift down slowly and continuously, as N increases, and the electronic energy for each cluster with N atoms, $E(N)$, is obtained by summing the eigenvalues of the occupied states. The difference in electronic energies between adjacent clusters, $E(N) - E(N - 1)$, is defined as $\Delta(N)$. The peaks result when $\Delta(N + 1)$ increases discontinuously, as an energy level is just filled at certain N and the next orbital starts to be occupied in the cluster with $N + 1$ atoms.

Numerical calculation shows that the total energy yield dips at the magic numbers. To make comparisons with measured abundance spectra, it is useful to calculate the second derivative of the total energy with respect to N,

$$\Delta_2(N) \equiv 2E(N) - E(N - 1) - E(N + 1). \tag{14.4.3}$$

It can be argued that if the clusters in the formation region are approximately in local thermal equilibrium, the observed abundances at temperature T can be expressed as

$$\ln \frac{I_N^2}{I_{N-1}I_{N+1}} \propto \frac{\Delta_2(N)}{k_B T}, \tag{14.4.4}$$

where I_N is the abundance intensity for an N-atom cluster. Figure 14.4.1 shows a comparison between the experimental abundance spectrum and $\Delta_2(N)$ for Na clusters. The peaks in $\Delta_2(N)$ coincide with the discontinuities in the mass spectra. So the main sequence N = 2, 8, 20, 40, 58, and 92 can be associated with an electronic shell structure. The shell structure is determined by large energy gaps between different energy levels.

Other peaks found in the calculation, at $N = 18, 34, 68$, and 70, are weaker than the observed ones and are more sensitive to the potential parameter. The good agreement between the experimental results and the model calculation suggests that there are no perturbations large enough to distort the main features of the level structure. A more sophisticated theoretical treatment of electronic structure is afforded by the density functional method, with the local density approximation for a chosen exchange-correlation potential. Its results are very similar to those obtained in a spherical pseudopotential method.

It is expected that closed-shell configurations will lead to spherical clusters. However, other configurations with some kind of distortion are also possible.[f] Ellipsoidal clusters are prevalent for open-shell configurations.

We would now like to give a little discussion about the problem of Coulomb explosion. Different from bulk materials, stability against charging is an experimentally important property of the clusters. When more and more electrons are stripped out from the cluster, the Coulomb repulsion of the positive charge distribution eventually beats the binding energy of the cluster, then the cluster explodes spontaneously. This fragmentation is a complicated dynamical process which we will not discuss here. It has been found that for alkali metals, the simple jellium model gives correctly the energetically most preferable fragmentation channels. This suggests that the jellium model can be used to study qualitatively the fragmentation of large clusters. The results show that the most preferable fragmentation channels are those for which the products are magic clusters. However the distribution of fragments from magic clusters are different from those where the parent is not magic. This is a result of the fact that, of charged clusters, the most stable are those where the number of electrons corresponds to a magic number. Then the clusters with one positive charge have magic numbers (number of atoms) 3, 9, 21, 41, etc. This has been seen experimentally in charged noble metal clusters.

14.4.2 Semiconductor Quantum Dots

At first, we consider the simple case where the semiconductor quantum dot has a spherical shape, where the radius R is on the order of a few nanometers. For example, semiconductor microcrystallites in glass matrices which have been extensively studied. One way to make such a quantum dot is to surround a small region of semiconductor with another semiconductor that has a larger band gap. Optically excited electrons and holes have are assumed to have effective masses m_e and m_h, the same as in the bulk material.

For the experimentally relevant case of spherical quantum dots, the single-particle Schrödinger equations for the electron and hole, in the absence of a Coulomb interaction, can be written as

$$-\frac{\hbar^2}{2m_\alpha}\psi_\alpha(\boldsymbol{r}) = E_\alpha\psi_\alpha(\boldsymbol{r}), \tag{14.4.5}$$

where $\alpha = \mathrm{e}$ or h. As an illustrating example, we consider the boundary condition of infinite confinement potential, so

$$\psi_\alpha(\boldsymbol{r}) = 0, \text{ for } r = R. \tag{14.4.6}$$

Then the eigenfunction and eigenenergy are

$$\psi_\alpha(\boldsymbol{r}) = \frac{\left(4\pi R^3\right)^{-1/2}}{\mathrm{j}_{l+1}(\kappa_{nl})}\mathrm{j}_l\left(\kappa_{nl}\frac{r}{R}\right)\mathrm{Y}_l^m(\theta,\phi), \tag{14.4.7}$$

and

$$E_\alpha = \frac{\hbar^2}{2m_\alpha}\left(\frac{\kappa_{nl}}{R}\right)^2, \tag{14.4.8}$$

where j_l is the lth order spherical Bessel function with κ_{nl} being its nth root, $\mathrm{Y}_l^m(\theta,\phi)$ are the spherical harmonics. Thus we can use quantum numbers n, l, and m to denote the quantum state

[f] K. Clemenger, *Phys. Rev. B* **32**, 1359 (1985).

of the system. The constant coefficient in (14.4.3) comes simply from the normalization of the wavefunction.

The boundary condition (14.4.6) is satisfied if $j_l(\kappa_{nl}) = 0$. Some of its solutions are $\kappa_{10} = \pi$, $\kappa_{11} = 4.4934$, $\kappa_{12} = 5.7635$, $\kappa_{20} = 6.2832$, $\kappa_{21} = 7.7253$, $\kappa_{22} = 9.0950$, $\kappa_{30} = 9.4248$, etc. It is customary to refer the nl eigenstates as ns, np, nd, etc., where s, p, d, etc. correspond to $l = 0, 1, 2, \ldots$, respectively, i.e. $\kappa_{10} = \kappa_{1s}$, $\kappa_{11} = \kappa_{1p}, \ldots$. It is different from the notation of atomic spectroscopy, where a $1s$ state would not be possible, because we are now treating a spherical confinement potential, not a Coulomb potential.

Consider the electrons and holes separately, and take the zero of the energy at the top of the valence band, then

$$E_e = E_g + \frac{\hbar^2}{2m_e}\left(\frac{\kappa_{n_e l_e}}{R}\right)^2, \tag{14.4.9}$$

and

$$E_h = -\frac{\hbar^2}{2m_h}\left(\frac{\kappa_{n_h l_h}}{R}\right)^2. \tag{14.4.10}$$

The lowest two energy levels are plotted schematically in Fig. 14.4.3. We see from this figure that the usual three-dimensional band structure is drastically modified and has become a series of quantized single-particle levels.

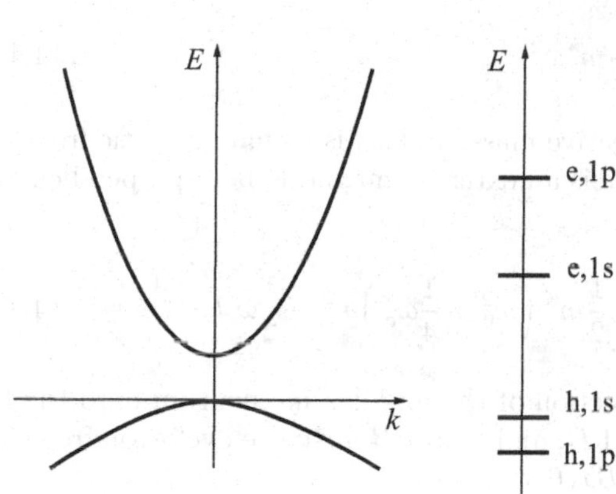

Figure 14.4.3 Schematic plot of the single-particle energy spectrum in bulk semiconductors (left) and in small quantum dots (right).

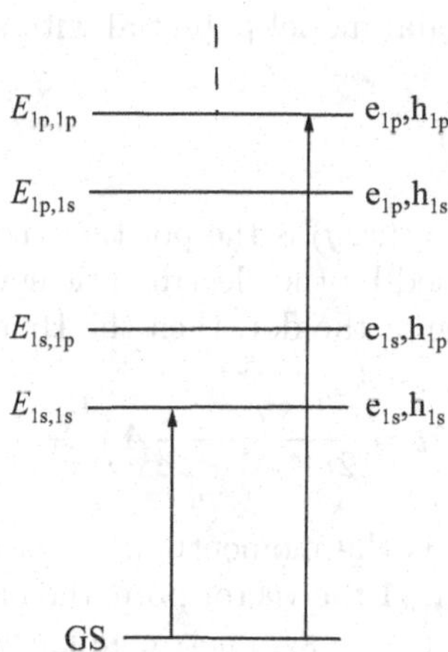

Figure 14.4.4 Schematic representation of the one-electron-hole-pair transitions in a semiconductor quantum dot.

The optical absorption spectrum is associated with electron-hole pairs. To describe an electron and a hole simultaneously, the Schrödinger equation of the system is written as

$$\left(-\frac{\hbar^2}{2m_e}\nabla_e^2 - \frac{\hbar^2}{2m_h}\nabla_h^2 + V_c\right)\Psi(\boldsymbol{r}) = E\Psi(\boldsymbol{r}), \tag{14.4.11}$$

still with the boundary condition (14.4.6). V_c is the Coulomb potential; if $V_c = 0$, there are analytical solutions for the eigenstates

$$\Psi(\boldsymbol{r}_e, \boldsymbol{r}_h) = \psi(\boldsymbol{r}_e)\psi(\boldsymbol{r}_h), \tag{14.4.12}$$

and eigenenergy

$$E = E_e + E_h = E_g + \frac{\hbar^2}{2m_e}\left(\frac{\kappa_{n_e l_e}}{R}\right)^2 + \frac{\hbar^2}{2m_h}\left(\frac{\kappa_{n_h l_h}}{R}\right)^2. \tag{14.4.13}$$

(14.4.13) shows that the absorption is blue shifted with respect to the bandgap E_g. The shift varies with the size R, like $1/R^2$, being larger for smaller sizes. Figure 14.4.4 exhibits the schematic representation of the one-electron-hole pair states. The notations e_{1s}, h_{1p}, etc., refer to the electron being in the $1s$ state, the hole being in $1p$, etc. The selection rules for the dipole-allowed interband transitions are $\Delta l = 0$ in the absence of Coulomb interaction. For example, the E_{1s-1s}-transition, where electron and hole are both of $1s$-type, is allowed.

When the Coulomb interaction is included, the problem can no longer be solved analytically and a numerical approach is necessary. The optical absorption may be weakly modified, since the kinetic energy terms dominate for quantum dots. The selection rules stated earlier are no longer valid, and transitions with $\Delta l \neq 0$ become weakly allowed.

Most realistic quantum dot systems contain dots of various radii, making it necessary to include the dot size distribution. Since the optical resonance energies strongly depend on the quantum dot radii, a radius distribution leads to a resonance distribution, which manifests itself as inhomogeneous broadening in the optical spectra.

14.4.3 Fock–Darwin Levels

A typical way to create a quantum dot is to produce a lateral confinement $V(x, y)$ that restricts the motion of the electrons, which are initially confined in a very narrow quantum well in the z direction. Then it forms a flat disk, with transverse size considerably exceeding its thickness. In the effective mass approximation, the confined electrons can be considered as moving in a two-dimensional model potential with a parabolic well; near the bottom

$$V(\boldsymbol{r}) = V_0 + \frac{1}{2}m^*\omega_0^2 r^2, \tag{14.4.14}$$

where $\boldsymbol{r} = (x, y)$ is the position vector, m^* is the effective mass, and ω_0 is a characteristic frequency determined by the electrostatic environment. If there is an external magnetic field perpendicular to the plane of the dot, then the Hamiltonian is

$$\mathcal{H} = \frac{1}{2m^*}\left(\boldsymbol{p} - \frac{e}{c}\boldsymbol{A}\right)^2 + \frac{1}{2}m^*\omega_0^2 r^2 = \frac{p^2}{2m^*} + \frac{1}{2}m^*\left(\omega_0^2 + \frac{1}{4}\omega_c^2\right)r^2 - \frac{1}{2}\omega_c l_z, \tag{14.4.15}$$

where $\boldsymbol{p}$ is the momentum, $l_z = xp_y - yp_x$ the projection of the angular momentum onto the field direction, $\boldsymbol{A}$ the vector potential of the magnetic field $\boldsymbol{B}$, and $\omega_c = eB/m^*c$ the cyclotron frequency. Here we take a symmetric gauge where $\boldsymbol{A} = (By, -Bx, 0)/2$.

To solve this problem, we can construct the complex variables

$$z = x + iy, \ z^* = x - iy, \tag{14.4.16}$$

and

$$\partial_z = \frac{1}{2}(\partial_x - i\partial_y), \ \partial_z^* = \frac{1}{2}(\partial_x + i\partial_y), \tag{14.4.17}$$

and also define the effective length

$$l_0 = \frac{l_B}{(1 + 4\omega_0^2/\omega_c^2)^{1/4}} \tag{14.4.18}$$

where

$$l_B = \left(\frac{\hbar c}{eB}\right)^{1/2} \tag{14.4.19}$$

is the magnetic length in the absence of a confining potential.

According to the above definitions, (14.4.15) might be transformed into the sum of two independent harmonic oscillators with the characteristic frequencies

$$\omega_\pm = \left(\omega_0^2 + \frac{1}{4}\omega_c^2\right) \pm \frac{1}{2}\omega_c, \tag{14.4.20}$$

and the corresponding eigenstates and eigenenergies

$$\psi_{n_+n_-}(z,z^*) = \frac{1}{\sqrt{2\pi}} \exp\left(\frac{zz^*}{4l_0}\right) \frac{(\partial_z)^{n_+}(\partial_z^*)^{n_-}}{(n_+!n_-!)^{1/2}} \exp\left(-\frac{zz^*}{2l_0^2}\right). \tag{14.4.21}$$

and

$$E(n_+n_-) = \hbar\omega_+ \left(n_+ + \frac{1}{2}\right) + \hbar\omega_- \left(n_- + \frac{1}{2}\right), \tag{14.4.22}$$

which are called Fock–Darwin states and levels, respectively. In the absence of a magnetic field, these levels are degenerate, $\omega_+ = \omega_- = \omega_0$, whereas a strong magnetic field leads to the formation of the structure of Landau energy levels, separated by the cyclotron energy $\hbar\omega_+ \approx \hbar\omega_c$. The system has circular symmetry, so the angular momentum operator component along the symmetry axis, $l_z = xp_y - yp_x = z\partial_z - z^*\partial_z^*$, commutes with $\mathcal{H}$. The operator l_z is diagonal on the basis of the Fock–Darwin states

$$l_z(n_+n_-) = n_+ - n_-. \tag{14.4.23}$$

Due to the circular symmetry of the Hamiltonian, the motion in the angular variable can be separated out, and the appropriate component of the angular momentum m is a good quantum number. The electron wavefunction can hence be written in the form

$$\psi_{nm}(r,\theta) = \phi_m(\theta) R_{nm}(r), \tag{14.4.24}$$

where the angle dependent function

$$\phi_m(\theta) = (2\pi)^{-1/2} e^{im\theta} \tag{14.4.25}$$

is the eigenfunction of the operator of the angular momentum projection, with eigenvalue m, while the radius-dependent function has the form

$$R_{nm}(r) = \frac{\sqrt{2}}{l_0} \left[\frac{n_r!}{(n_r + |m|)!}\right] \left(\frac{r}{l_0}\right)^{|m|} \exp\left(-\frac{r^2}{2l_0^2}\right) \mathcal{L}_{n_r}^{|m|}\left(\frac{r^2}{l_0}\right). \tag{14.4.26}$$

In the above expression $\mathcal{L}_{n_r}^{|m|}$ denotes the Laguerre polynomials

$$\mathcal{L}_{n_r}^{|m|}(z) = \frac{1}{m!} z^{-|m|} e^z \frac{d^{n_r}}{dz^{n_r}} \left(z^{n_r+|m|} e^{-z}\right), \tag{14.4.27}$$

$n = 0, 1, \ldots$ is the principal quantum number; $m = -n, -n+2, \ldots, n-2, n$ is the azimuthal quantum number; and $n_r = (n - |m|)/2$ is the radial quantum number. The pairs of quantum number (n, m) and (n_+, n_-) are related by

$$n = n_- + n_+, \quad m = n_- - n_+. \tag{14.4.28}$$

The eigenenergies expressed by the quantum numbers (n, m) are

$$E(n,m) = (n+1)\hbar \left(\omega_0^2 + \frac{1}{4}\omega_c^2\right)^{1/2} - \frac{1}{2} m\hbar\omega_c. \tag{14.4.29}$$

The evolution of the energy spectrum in an increasing magnetic field is presented in Fig. 14.4.3. The Zeeman splitting, very small for GaAs, is neglected here. The pairs of numbers on the vertical axis give the electron eigenstates (n_+, n_-), while the straight dashed lines show the energies of subsequent Landau levels (the levels in the absence of a parabolic potential in the plane of a dot, or $\omega_- = 0$)

$$E(n_+) = \hbar\omega_c \left(n_+ + \frac{1}{2}\right), \quad n_+ = 0, 1, \ldots. \tag{14.4.30}$$

The perpendicular arrows represent the allowed optical transitions. In the dipole approximation, the intraband optical transitions satisfy the following selection rules

$$n'_+ = n_+ \pm 1 \text{ or } n'_- = n_- \pm 1, \tag{14.4.31}$$

together with the allowed resonance frequencies $\omega_\pm$, indicated in Fig. 14.4.5.

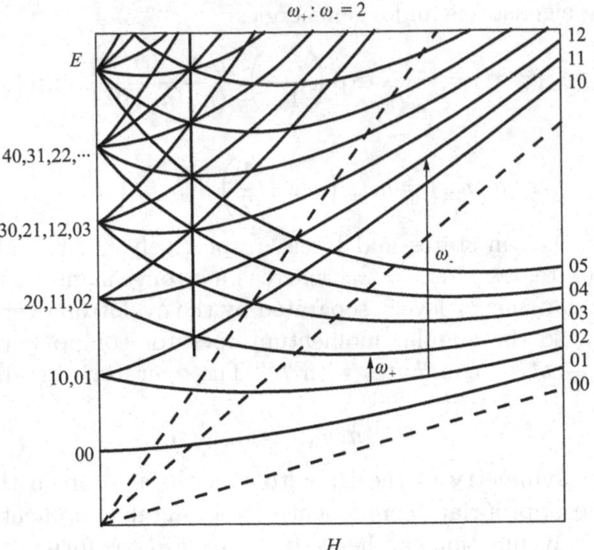

Figure 14.4.5 Evolution of the Fock–Darwin energy levels in a magnetic field. Dashed lines represent the Landau energy levels, and vertical arrows the allowed optical transitions. From L. Jacak, P. Hawrylak, and A. Wójs, *Quantum Dots*, Springer, Berlin (1998).

Besides the conduction band electrons, the valence band holes are the second type of carrier that can be bound in a quantum dot. Due to their opposite electric charge, and also due to the valence band building from atomic p-type orbitals, there is much complexity in treating holes.

The Fock–Darwin treatment is a single-electron approach, in the few-electron case it has reasonable agreement with experiment.[g] More exact theory must take into consideration of the Coulomb interactions.

14.4.4 Coulomb Blockade

It is not only the wave nature of electrons but also the discrete nature of charge, in units of e, that is important. The capacitance, C, of nanostructures can be so small that the charging energy, $e^2/2C$, for adding a single electron to, for instance, a quantum dot exceeds the thermal energy. A large charging energy can prevent the addition to, or removal from, a nanostructure of even one electron, resulting in transport effects such as the Coulomb blockade in tunneling. Single-electron transistors have been developed in which a switch from the "on" state to the "off" state is induced, as the name implies, by just one electron.

We have examined the electronic structures of quantum dots in the single-electron approximation. Another important aspect of nanostructures, including quantum dots, involves electronic correlation effects, especially in transport processes. Here we will discuss quantum dot situations in which the Coulomb interaction may be modelled by an effective capacitance.

Quantum dots may form small solid state devices in which the number of electrons can be made a well-defined integer N. In discussing transport, we can consider a small structure called a single-electron transistor, depicted in Fig. 14.4.6. There are several kinds of single-electron transistors. One of them can be made by depositing metal gates over a two-dimensional electron gas formed in a GaAs/AlGaAs heterostructure. Applying a negative voltage to these gates depletes the regions between them, creating a small dot, or an atomic-like box for electrons, coupled by tunneling to two separate two-dimensional electron gases acting as source and drain leads. For such a single-electron transistor, we denote I_{sd} as the source-to-drain current, V_{sd} the voltage between the leads, then the linear conductance is $G = I_{\text{sd}}/V_{\text{sd}}$, with V_{sd} kept very small. As the gate voltage V_{g} changes, the

[g] L. P. Kouwenhoven *et al.*, *Science* **278**, 1788 (1997).

result for conductance G is a series of periodically spaced peaks, each indicating a change in the number of electrons N in the dot by one, due to the Coulomb blockade.

The electronic states in the dot can be probed by transport when a small tunnel coupling is allowed between the dot and the nearby source and drain leads. This coupling is usually made as weak as possible, to prevent strong fluctuations in the number of confined electrons. The quantization of charge permits the use of a simple model in which all of the electron-electron interactions are captured in the single-electron charging energy $e^2/2C$, where C is the total capacitance between the dot and the rest of the system. This simple model has been successful in describing a transport phenomenon generally known as Coulomb blockade oscillations.

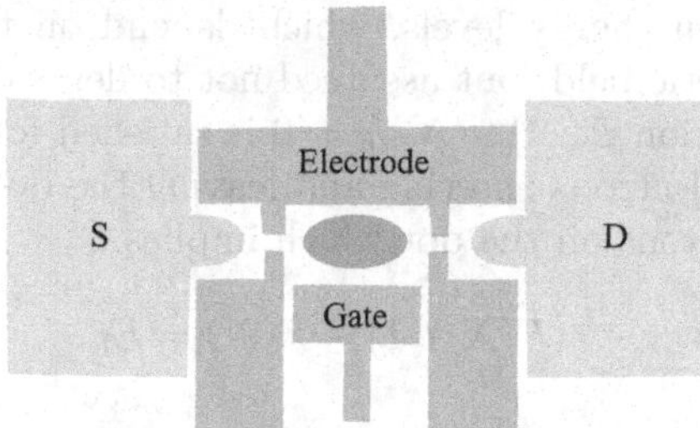

Figure 14.4.6 A schematic single-electron transistor with a quantum dot weakly coupled by tunnel barriers to two leads connecting to reservoirs. There is also a gate electrode.

If we consider a tunneling junction with area $S = 100\ \text{nm}^2$, thickness $d = 1$ nm, and dielectric constant $\epsilon = 10$, using the classical expression for the capacitance, then the capacitance is $C = \epsilon S/4\pi d = 10^{-15}$ F. The capacitance introduces an energy scale, the charging energy, corresponding to a single-electron charge $(-e)$, $E_C = e^2/2C \approx 10^{-4}$ eV, which corresponds to a temperature of $E_C/k_B \approx 1$ K. In a tunneling process the electrostatic energy changes by an amount of the order magnitude of E_C. Hence we expect, in the sub-Kelvin regime, electron transport to be affected by charging effects.

The Coulomb blockade oscillations of conductance are a manifestation of a single-electron tunneling through a quantum dot. In Fig. 14.4.6 the gate electrode is used to control the number of electrons in the dot. The dot is assumed to be small enough that the one-electron eigenenergies are well separated. A current I_{sd} can be passed through the dot by applying a voltage difference V_{sd} between the reservoirs. In the absence of charging effects, a conductance peak due to resonant tunneling occurs when the Fermi energy E_F in the reservoirs lines up with one of the energy levels in the dot. This condition is modified by the charging energy. The conductance oscillations occur as the voltage on the gate electrode is varied. The number N of electrons on the dot between two barriers is an integer, so that the charge $Q = -Ne$ on the dot can only change by a discrete amount e. In contrast, the electrostatic potential difference of the dot and the leads changes continuously as the electrostatic potential ϕ_{ext} due to the gate is varied. This gives rise to a net charge imbalance $C\phi_{ext} - Ne$ between the dot and the leads, which oscillates in a saw-tooth pattern with the gate voltage. Tunneling is blocked at low temperatures, except near the degeneracy points of the saw-tooth, where the charge imbalance jumps from $+e/2$ to $-e/2$. At these points the Coulomb blockade of tunneling is lifted and the conductance exhibits a peak.

In a simple model, the total electrostatic energy is written as

$$U(N) = (Ne)^2/2C - Ne\phi_{ext}, \tag{14.4.32}$$

where the external electrostatic potential ϕ_{ext} comes from external charges, in particular those on a nearby gate electrode. So it is reasonable to say that $Q_{ext} = C\phi_{ext}$ plays the role of an "external induced charge" on the dot, which can be varied continuously by means of an external gate voltage, in contrast to Q, which is restricted to integer multiples of $(-e)$. In terms of Q_{ext} one can write

$$U(N) = (Ne - Q_{ext})^2/2C - Q_{ext}^2/2C. \tag{14.4.33}$$

We emphasize that Q_{ext} is an externally controlled variable, via the gate voltage, regardless of the relative magnitude of the various capacitances in the system.

The probability of finding N electrons on the quantum dot in equilibrium with the reservoirs is given by

$$P(N) \propto \exp\{-[F(N) - NE_{\mathrm{F}}]/k_{\mathrm{B}}T\}, \tag{14.4.34}$$

where $F(N)$ is the free energy, which approaches the ground state energy $E(N)$ of the dot as $T \to 0$, for which we take the simplified form

$$E(N) = U(N) + \sum_{p=1}^{N} E_p, \tag{14.4.35}$$

where E_p are the one-electron energy levels, which depend on the size of the quantum dot, the gate voltages, and any magnetic field, but assumed not to depend on N. $P(N)$ is zero at very low temperature unless the condition $E(N) - NE_{\mathrm{F}} = 0$ is satisfied for some N. For current to flow we need a finite probability for electrons entering and leaving the dot, i.e. a finite probability of there being either N or $N+1$ electrons on the dot which implies

$$F(N+1) - F(N) = E_{\mathrm{F}}. \tag{14.4.36}$$

Combining with (14.4.35), it gives

$$E_N + U(N) - U(N-1) = E_{\mathrm{F}}. \tag{14.4.37}$$

Substitution of (14.4.32) into (14.4.37) gives a renormalized energy for the N-electron quantum dot

$$E_N^* = E_N + \left(N - \frac{1}{2}\right)\frac{e^2}{C} = E_{\mathrm{F}} + e\phi_{\mathrm{ext}}. \tag{14.4.38}$$

There will be peaks in the current as the bias is applied such that this condition is satisfied for successive N. The left-hand-side of (14.4.38) defines a renorlized energy level E_N^*. The renormalized level spacing

$$\Delta E^* = \Delta E + \frac{e^2}{C} \tag{14.4.39}$$

is enhanced above the bare level spacing by the charging energy. For a sufficiently large dot ($N \geq 100$), this equality is satisfied for successive values of N. If $E_N \gg e^2/C$, then the peaks in conductance will occur roughly periodically, with the period determined by $\Delta(e\phi_{\mathrm{ext}}) = e^2/C$, as seen in the conductance versus gate voltage in Fig. 14.4.7 where successive peaks going to the left arise from reducing one electron. The oscillation period of the conductance can be used to infer the

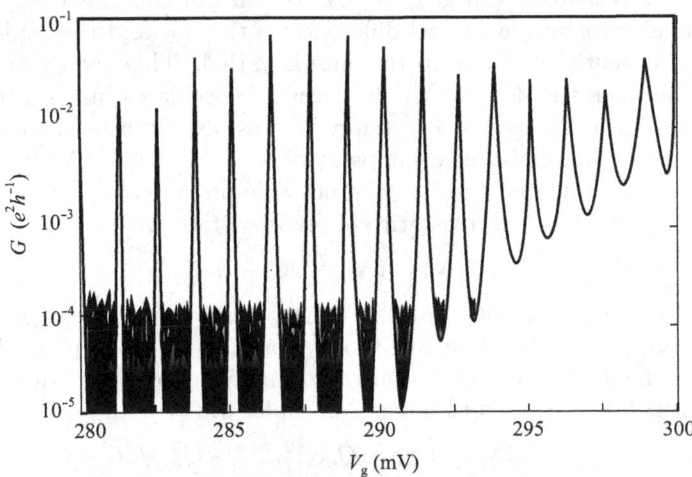

Figure 14.4.7 Coulomb blockade oscillations. From E. B. Foxman *et al.*, *Phys. Rev. B* **47**, 10020 (1993).

capacitance of the quantum dot. With the addition of a magnetic field, one can anticipate that the energy levels of the quantum dot due to a combination of electrostatic and magnetic confinement will show spin-splitting of the one-electron levels.

14.4.5 Kondo Effect

In §13.3, we had discussed the Kondo effect, i.e. for metals with dilute magnetic impurities there is a resistivity minimum as the temperature is lowered. Its physical essence is that conduction electrons interact with a single localized unpaired electron. At low temperatures a spin singlet state is formed between the unpaired localized electron and the delocalized electron at the Fermi energy.

Now, for quantum dot structures, it was predicted theoretically that a Kondo singlet state could also form.[h] This prediction has been verified experimentally in the low temperature transport of electrons using a single-electron transistor.[i] Because of this, a single-electron transistor could provide a means of investigating aspects of the Kondo effect under controlled circumstances that are not accessible in conventional systems. For examples, the number of electrons can be changed from odd to even, the energy difference between the localized state and the Fermi level can be tuned, the coupling to the leads can be adjusted, and a single localized state can be studied rather than a statistical distribution as in macroscopic materials.

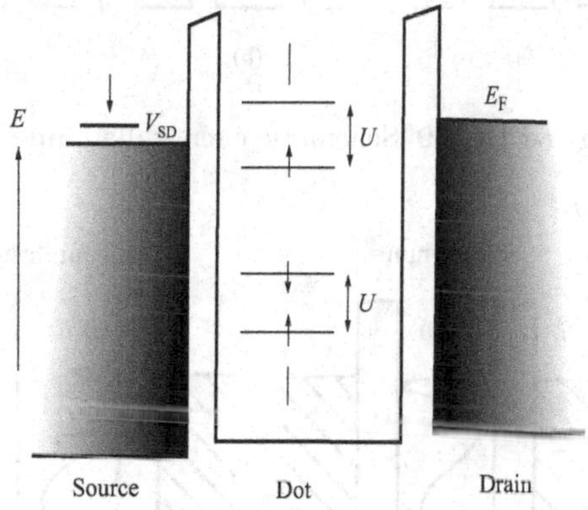

Figure 14.4.8 Energy diagram of a single-electron transistor in a semiconductor heterostructure.

Several important energy parameters and their relative sizes determine the behaviors of a single-electron transistor. At low temperature, the number of electrons N in the dot is an integer. This number could be changed by raising the voltage V_g of the gate electrode which lowers the energy of electrons in the dot relative to the Fermi level E_F in the leads. To analyze the Kondo effect, we must take the spins of the electrons into account, in addition to the electron-electron interaction. Just as in the appearance of Coulomb blockade, the energy required to add an electron to an empty dot is E_1 which is the energy of the lowest spatial state of the dot. A second electron, with the opposite spin, goes into the same spatial state, but its addition costs a larger energy $E_1 + U$, where U is the Coulomb repulsion energy between the two electrons and equals the charging energy in the capacitance model. Because a third electron can no longer enter the E_1 state, it enters the next available spatial state with energy E_2. Taking account of the Coulomb interaction with the first two electrons, its addition requires an energy $E_2 + 2U$. Continuing this procedure, the occupancy energy levels are at E_1, $E_1 + U$, $E_2 + 2U$, $E_2 + 3U$, and so forth, which occur in pairs as indicated in Fig. 14.4.8. The peak widths are finite because of a finite escape time onto the leads. Two peaks within a pair are separated by U, whereas the separation between different pairs, corresponding to different spatial states, is larger. It is worth to note that there is an essential difference depending on whether N is odd or even. If the number of electrons N in the dot is odd, the unpaired electron

[h] T. K. Ng and P. A. Lee, *Phys. Rev. Lett.* **61**, 1768 (1988).

[i] D. Goldhaber-Gordon *et al.*, *Nature* **391**, 156 (1998). For a review, see L. Kouwenhoven and L. Glazman, *Phys. World, Jan.*, 33 (2001).

causes the dot to have a local moment. At temperatures less than T_K, this local moment is screened by the nearby electrons in the leads, forming a singlet state of spin 0.

Another important energy Γ is the coupling of electronic states on the dot to those on the leads, resulting from tunneling. We have assumed Γ is small to ensure the Coulomb blockade. However, if Γ increases by tuning the tunnel barriers, the number of electrons on the dot becomes less well-defined. When the fluctuations in N become much greater than unity, the quantization of charge is completely lost. In this open regime, noninteracting electrons usually give a proper description of transport. It is more complicated in the intermediate regime where the tunnel coupling is relatively strong but the discrete nature of charge still plays an important role. Here, the transport description must incorporate higher order tunneling processes through virtual, intermediate states. In this case, spins should be taken into account, and the tunneling may be viewed as a magnetic-exchange coupling.

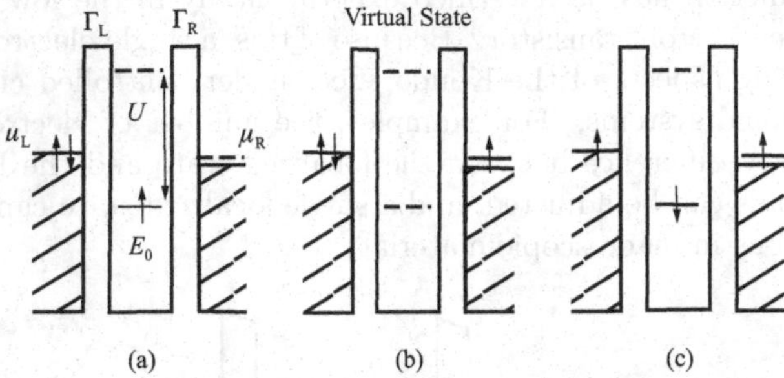

Figure 14.4.9 Schematic energy diagram of a dot.

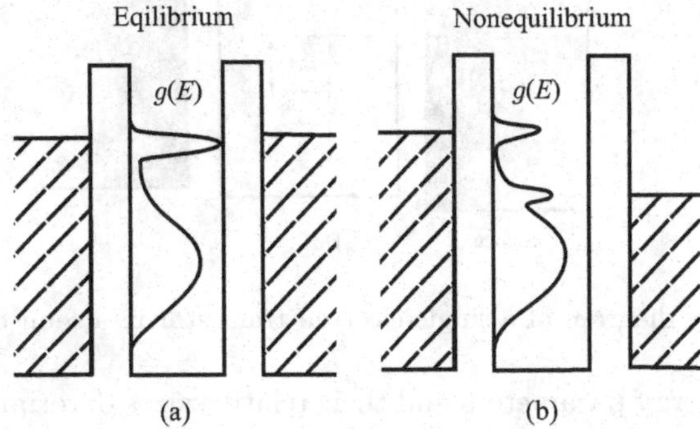

Figure 14.4.10 Kondo resonance in the density of states.

The energy that determines whether Kondo physics will be visible is $k_B T_K$. T_K is the Kondo temperature. As the samples are cooled to near the Kondo temperature, the inner shoulder of each pair of peaks in $G(V_g)$ broadens and is enhanced, whereas no broadening is seen outside of the pairs where the dot is non-magnetic. This is a signature of the Kondo effect in the dot systems, which is predicted theoretically. The Kondo effect is essentially a screening of the dot spin by nearby free electrons and so takes place only when the dot is magnetic. Below the Kondo temperature, the unpaired electrons in the dot hybridize with the conduction band states in the leads. Once this takes place, it is no longer appropriate to talk about an isolated dot. We should rather regard the dot plus the screening cloud as a new "quasi-dot" of spin zero. This hybridization produces in the local density of states of the quasi-dot a sharp peak (Kondo resonance) at $E = E_F$, which enhances G. For an appropriate gate voltage V_g, the first-order tunneling is blocked in the case shown in Fig. 14.4.9(a). An electron cannot tunnel onto the dot because the two electron energy $E_0 + U$ exceeds the Fermi energies of the leads μ_L and μ_R. Also, the electron on the dot cannot tunnel off because $E_0 < \mu_L, \mu_R$. This is the Coulomb blockade described before. In contrast to first-order tunneling, higher order processes, in which the intermediate state costs an energy of order U, are allowed for short time scales. In particular, we are interested in virtual tunneling events that effectively flip the spin on the dot. One such example is depicted in Fig. 14.4.9(b) and (c). Successive spin-flip processes effectively screen the local spin on the dot and form a spin-singlet state. This correlated state gives rise to the Kondo effect in a quantum dot which can be described as a narrow peak in the density of states at the Fermi levels of the leads, as shown in Fig. 14.4.10(a). This Kondo resonance gives rise to a temperature-sensitive enhancement of conductance through the dot. Out of equilibrium, when a bias voltage V_{sd} is applied between the source and drain, $eV = \mu_{\mathrm{L}} - \mu_{\mathrm{R}}$, the Kondo peak in the density of states splits into two peaks, each pinned to one chemical potential, as depicted in Fig. 14.4.10(b). This splitting will lead to new features in the differential conductance dI/dV.

From the above discussion on the Kondo phenomena in quantum dots, we are dealing with a spin system which allows one to study an individual, artificial magnetic impurity and tune, in situ, the parameters in Kondo problem. As N is even, there is no Kondo effect because the dot has no local moment. Sweeping the gate voltage V_{g} to change N, we can switch a quantum dot from a Kondo system to a non-Kondo system as the number of electrons on the dot is changed from odd to even.

§14.5 Coupled Quantum Dot Systems

Quantum dots are small conductive regions in which electrons are governed by the interplay of quantum mechanical and electrostatic effects. Because of confinement effects, electrons occupy well-defined, discrete quantum states, so an individual quantum dot is often referred to as an 'artificial atom'. Furthermore, it can be extended to consider two, three or lot of more quantum dots connected by electronic tunneling. This kind of coupled structure is then called an 'artificial molecule', or 'artificial solid' which can be made in many configurations with adjustable interdot tunneling rates.

14.5.1 Double Quantum Dots

Just as in double quantum well structures, when particles are allowed to tunnel back and forth between two quantum dots, the energy states of the individual dots mix and form new states that extended over both dots. The extended states are referred as the bonding or symmetric state, and the antibonding or antisymmetric state. There is an energy splitting between bonding and antibonding states. It is important for device application whether different dots can be coupled together in a quantum mechanically coherent way. Depending on the strength of the interdot coupling, the two dots can form 'ionic' or 'covalent' bonds. In the former case, the electrons are localized on individual dots, while in the latter, the electrons are delocalized over both dots. Covalent binding leads to bonding and antibonding states, whose energy difference is proportional to the degree of tunneling. Here we can study a transition from ionic bonding to covalent bonding in a double quantum dot by changing the interdot coupling.

Tunnel coupling between quantum dots can be continuously adjusted from the weak tunneling regime, in which the dots are well isolated, to the strong tunneling regime, in which the two dots effectively join into one. In the weak coupling regime, the number of electrons on each dot, N_1 and N_2, are quantized, and the Coulomb blockade theory applies to each dot individually. In the strong tunneling regime, N_1 and N_2 are not individually well defined, and the coupled dots system enters the interesting regime in which it acts as an artificial molecule. However, we assume the coupled dot system is well isolated from the leads and the total number of electron, $N = N_1 + N_2$, is quantized. We can still use the Coulomb blockade for the entire coupled dot system to probe its ground state energy.

We can expect that the dc current through a structure with double quantum dots in the presence of oscillating fields may display interesting phenomena not observable in single dots. Figure 14.5.1(a) gives a schematic picture of a double quantum dot structure. In this system, for the weak coupling regime, electrons are strongly localized on the individual dots when tunneling between the two dots is weak. Electron transport is then governed by single electron charging effects. The charging can be tuned away by means of the gate voltages. It is then energetically allowed for an electron to tunnel between dots when a discrete state in the left dot is aligned with a discrete state in the right

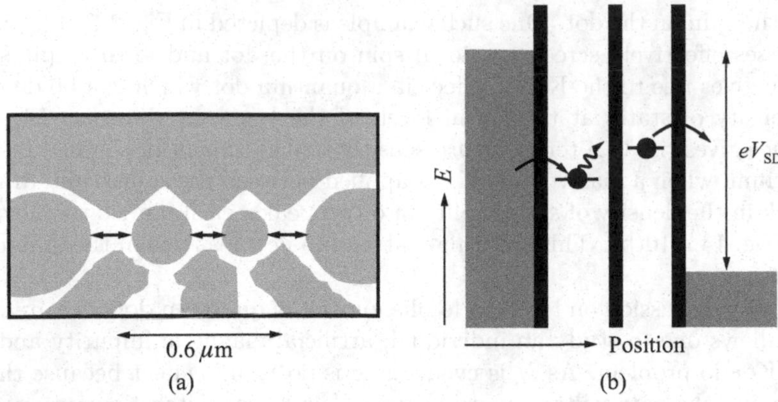

Figure 14.5.1 (a) A double quantum dot structure, and (b) diagram of its electron energies.

dot. External voltages also control the alignment of the discrete states. A current can flow when electrons tunnel, while conserving energy, from the left lead, through the left and right dots, to the right lead. There is another way the the energy is also conserved: when a photon of energy $\hbar\omega$, which matches the energy difference between the states of the two dots, is absorbed from a microwave field of frequency ω, as shown in Fig. 14.5.1(b).

It is found theoretically that the d.c. current is very sensitive to an oscillating field. A detailed description of photon-assisted tunneling in a double quantum dot has been given. The basic idea is that electrons can absorb quanta of fixed energy $\hbar\omega$ from a classical oscillating field. An a.c. voltage drop

$$V = V_{\text{ac}} \cos(\omega t) \tag{14.5.1}$$

across a tunnel barrier modifies the tunnel rate through the the barrier as

$$\tilde{T}'(E) = \sum_{n=-\infty}^{+\infty} J_n^2(\alpha)\tilde{T}(E + n\hbar\omega). \tag{14.5.2}$$

Here $\tilde{T}'(E)$ and $\tilde{T}(E)$ are the tunnel rates at energy E with and without an a.c. voltage, respectively; $J_n(\alpha)$ is the nth order Bessel function of the first kind, evaluated at $\alpha = eV_{\text{ac}}/\hbar\omega$, which describes the probability amplitude that an electron absorbs or emits n photons of energy $\hbar\omega$.

Microwave spectroscopy with frequencies in the range 0–50 GHz has been used to measure the energy difference between states in the two dots of the device shown in Fig. 14.5.1(a). The energy differences, including the bonding-antibonding splitting, are controlled by the gate voltage which tunes the coupling between the dots. The resonance in the lowest trace in Fig. 14.5.2 is due to an alignment of discrete states. The other traces are measured while applying a microwave signal. The satellites resonance are due to photon-assisted tunneling processes which involve the emission (left satellite) or absorption (right satellite) of a microwave photon. The satellite resonances induced by the external oscillating field can be of the same order of magnitude as the main static resonance with an even smaller width.

As the microwave power is increased, more satellite peaks appear in the current-gate-voltage plot corresponding to the absorption of multiple photons which are observed up to $n = 11$. At these high powers, the microwaves strongly perturb the tunneling. The separation of the satellite peaks from the main peak depends linearly on frequency from 1 to 50 GHz. This is the result of that the fact that the tunnel coupling is negligible. The electrons are thus localized on the individual dots, which have an ionic bonding. The coupling between the dots can be increased by changing the gate voltage on the center gate. In contrast to the weakly coupled dots, covalent bonding occurs when two discrete states that are spatially separated become strongly coupled. Electrons then tunnel quickly back and forth between the dots. In a quantum mechanical description this results in a bonding and antibonding state which are respectively lower and higher in energy than the original states.

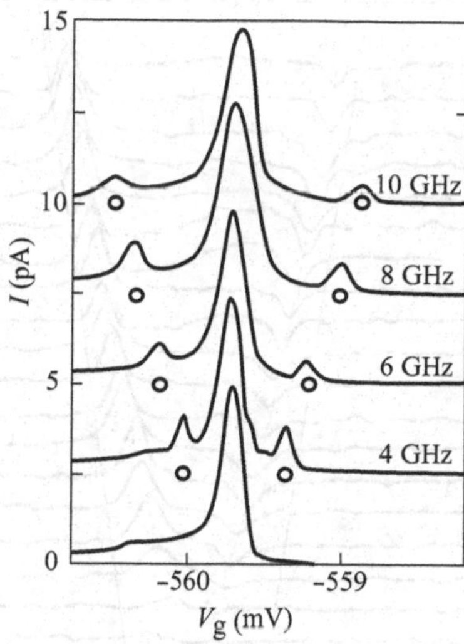

Figure 14.5.2 Current resonance through a double quantum dot. From T. H. Oosterkamp *et al.*, *Nature* **395**, 873 (1998).

To single out the current that is only due to the microwaves we can operate the device as an electron pump driven by photons in the way shown in Fig. 14.5.1(b). By sweeping the gate voltages we vary $\Delta E = E_{\mathrm{L}} - E_{\mathrm{R}}$, where E_{L} and E_{R} are the energies of the uncoupled states in the left and right dot. The bonding and antibonding states, that are a superposition of the wavefunctions corresponding to an electron in the left and right dot, have an energy splitting of $\Delta E^* = E_{\mathrm{A}} - E_{\mathrm{B}} = \left[(\Delta E)^2 + (2\Gamma)^2\right]^{1/2}$, where Γ is the tunnel coupling between the two dots. When the sample is irradiated, a photocurrent may result as illustrated in Fig. 14.5.3.

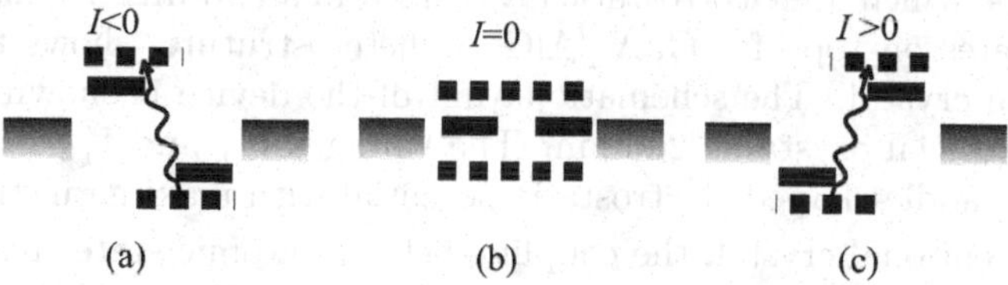

Figure 14.5.3 Measured pumped current through the strongly coupled double-dot. (a) $E_{\mathrm{L}} > E_{\mathrm{R}}$ which results in electron pumping from right to left corresponding to negative current; (b) The whole system is symmetric ($E_{\mathrm{L}} = E_{\mathrm{R}}$) and consequently the net electron flow must be zero; (c) $E_{\mathrm{L}} < E_{\mathrm{R}}$ which gives rise to pumping from left to right and a positive current.

A non-zero current indicates that an electron was excited from the bonding state to the antibonding state, thereby fulfilling the condition $\hbar\omega = \Delta E^*$, or conversely

$$\Delta E = \left[(\hbar\omega)^2 - (2\Gamma)^2\right]^2 . \tag{14.5.3}$$

Figure 14.5.4 shows measured current traces as a function of the uncoupled energy splitting ΔE, where from top to bottom the applied microwave frequency is decreased from 17 to 7.7 GHz in 0.5 GHz steps. The distance between the pumping peaks, which is proportional to $2\Delta E$, decreases as the frequency is lowered. However, the peak distance decreases faster than a straight line, and goes to zero as the frequency approaches the minimum energy gap between bonding and antibonding states, $\hbar\omega = 2\Gamma$. For $\hbar\omega < 2\Gamma$, the photon energy is too small to induce a transition from the bonding to the antibonding state.

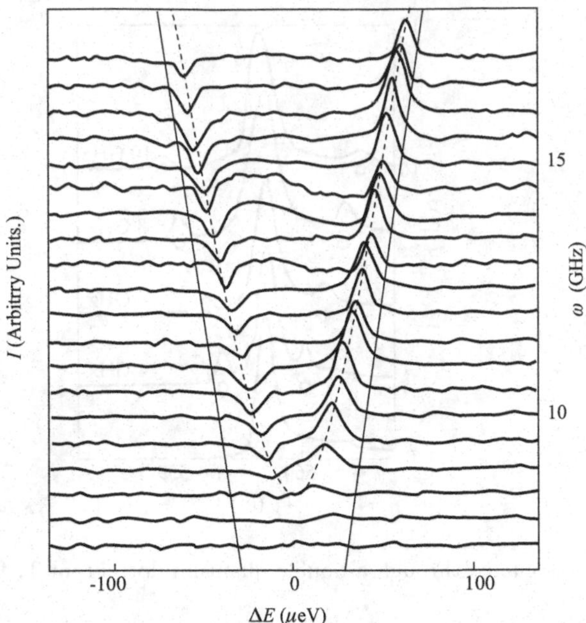

Figure 14.5.4 Measured pumped current through the strongly coupled double-dots. From T. H. Oosterkamp *et al.*, *Nature* **395**, 873 (1998).

14.5.2 Semiconductor Quantum Dot Superlattices

As we have seen, an individual quantum dot behaves like an artificial atom and a double quantum dot can be looked upon as an artificial molecule, so it is natural to think that a lot of quantum dots may be organized to form an artificial crystal. This is really true. Experimentally, a sequence of fifteen quantum dots, which are electrostatically defined in a two-dimensional electron gas by means of a two metallic gates on top of a GaAs/AlGaAs heterostruture, shows the transport properties of a one-dimensional crystal. The schematic layout of the device is shown in Fig. 14.5.5, in which the period of the artificial crystal is 200 nm. The voltages V_{g1} and V_{g2} make the depletion region resemble a periodic saddle-shaped electrostatic potential with maxima in the narrow regions.

In a perfect conventional crystal, the coupling between atomic states results in a collective state characterized by energy bands separated by energy gaps. The conducting properties of a solid strongly depend on the location of the Fermi energy in the band structure. These basic concepts are still effective in artificial crystals. In the one-dimensional quantum dot array shown in the inset of Fig. 14.5.5, spatial quantization is realized in all three directions. The transport properties of single quantum dots fabricated with the same split-gate technique demonstrates the formation of zero-dimensional states through oscillating conductance. In a sequence of equal quantum dots with equal coupling to nearest neighbors, the levels develop into minibands. The number of states within a miniband is equal to the number of dots, and the energy gap between consecutive bands is determined by the coupling between dots. Weak coupling yields a narrow band and a large gap, while strong coupling will result in a wide band and a small gap.

The theoretical consideration has been confirmed by conductance measurements, as shown in Fig. 14.5.5. The application of a magnetic field establishes adiabatic transport. The only scattering now takes place within a single subband at the potential maxima defined by the fingers. As can be seen, the two deep peaks enclose fifteen oscillations, which corresponds exactly with the number of quantum dots in the one-dimensional crystal. The deep peaks can be associated with energy gaps, and the small oscillations with the discrete states in the miniband. The effect of lowering the gate voltage V_{g2} here is mainly the decrease in Fermi energy and the reduction in dot area. Note that the reduction in area results in large energy separations that increase the bandwidth. Both effects move the Fermi energy through the miniband structure. A maximum in the conductance occurs

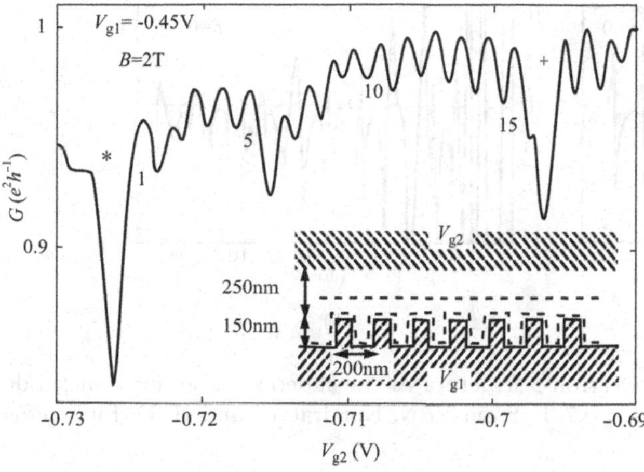

Figure 14.5.5 Conductance versus gate voltage V_{g2} on the second gate at magnetic field 2T and $V_{g1} = -0.45$ V on the first gate. The inset schematically shows the gate geometry; the dashed lines indicate the depletion regions in the two-dimensional electron gas. The upper depletion is moved towards the fingers when V_{g2} is made more negative. From L. P. Kouwenhoven *et al.*, *Phys. Rev. Lett.* **65**, 361 (1990).

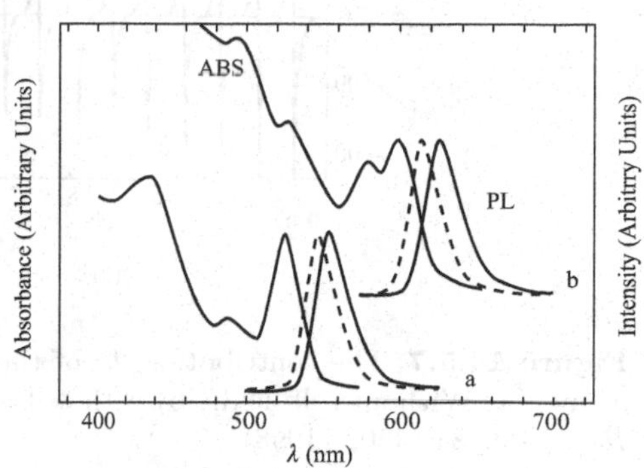

Figure 14.5.6 Optical absorption (ABS) and photoluminescence (PL) spectra at 10 K for a close-packed solid of CdSe quantum dots that are 38.5 Å (curve a) and 62 Å (curve b) in diameter. Dotted lines are photoluminescence spectra of the same dots but in a dilute form dispersed in a frozen solution. From C. B. Murray *et al.*, *Science* **270**, 1335 (1995).

when the Fermi energy coincides with the energy of a discrete state within the miniband. A simple one-dimensional model of resonant transmission can be used to illustrate the experimental results.

One of the important applications for quantum confinement effects is semiconductor lasers. There has been development of semiconductor lasers from quantum well lasers, to quantum wire lasers, and now to quantum dot lasers. To make a quantum dot laser, we need a dense array of equal-sized dots within the active region.

Artificial crystals can be extended from one-dimension to higher dimensions. Semiconductor nanocrystallites of CdSe have been self-organized into three-dimensional quantum dot superlattices (colloidal crystals). The size and spacing of the dots within superlattice are controlled with near-atomic precision. This kind of well-defined ordered artificial solid provides opportunities for optimizing properties of materials and offer possibilities for observing interesting and potentially useful new collective physical phenomena. Optical spectra of closed-packed CdSe quantum dots show the effects of quantum confinement on the individual dots as well as evidence of interdot interactions. The solid curves in Figure 14.5.6 show 10 K optical absorption and photoluminescence spectra of a thin solid film of close-packed CdSe quantum dots that are 38.5 Å (curve a) and 62 Å (curve b) in diameter. The discrete and size-dependent optical absorption features and the band edge emission are characteristics of the quantized electronic transitions of individual quantum dots. Comparison of optical spectra for dots closed-packed in the solid with dots in a dilute matrix reveals that, although the absorption spectra are essentially identical, the emission line shape of the dots in the solid is modified and red-shifted, an indication of interdot coupling.

14.5.3 Metal Quantum Dot Arrays

We should now like to discuss the magnetic properties of transition metal quantum dot arrays in which the dots are separated by non-magnetic tunnel barriers. Recent experimental work on one- and two-dimensional self-organized arrays of nanosize transition metal dots show magnetic ordering. For example, in the case of a two-dimensional system of Fe dots on an insulator substrate, a long

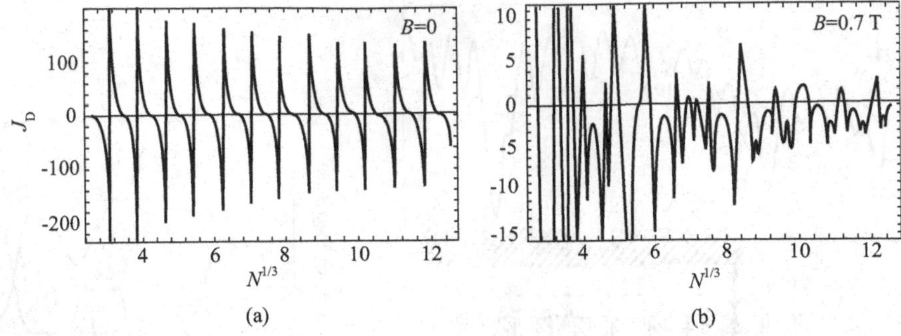

Figure 14.5.7 The contribution J_D of the dot electronic structure to the superexchange for a nickel dot array. (a) Without self-field; (b) with self-field $B = 0.7$ T. From V. N. Kondratyev and H. O. Lutz, *Phys. Rev. Lett.* **81**, 4508 (1998).

range order has been found, which has been attributed to a contribution of the exchange coupling between the dot supermoments.[j]

When the transition metal dots are sufficiently densely packed, the electronic structure will be modified due to exchange coupling. For instance, if ferromagnets are separated by a non-magnetic insulator, the tunnel exchange spin current results in Anderson-type superexchange coupling, nonoscillatory with separation distance. A similar coupling can be expected in a regular dot array with a coherent state of the dot supermoments. There are two geometrical parameters, one is dot size, and the other is the distance between any two dots. Variations of these two parameters that are too large will prevent the formation of a coherent state. The limiting condition can be expressed, within Anderson localization theory, as $\Gamma/B < 2$, with Γ the level broadening due to variations, and B the miniband splitting. For sufficiently small Γ, after some mathematics, the superexchange coupling constant J_0 for zero temperature can be approximately written as

$$J_0 = J_D J_B, \tag{14.5.4}$$

where J_D and J_B are determined by the dot electronic structure and the barrier properties, respectively.

Expression (14.5.4) is the Anderson-type superexchange coupling originating from tunneling between the superparamagnetic dots. The sign of the coupling constant is determined by the dot electronic structure and remains unchanged with interdot separation, similar to what is obtained for ferromagnetic layers abutted by an insulator. Numerical results for the dot-size dependence contained in J_D for the Ni system is shown in Fig. 14.5.7, which displays regular oscillations with varying dot diameter. The sign of J_D indicates a ferromagnetic type of exchange when the gross shells are more than half filled, and it changes smoothly to an antiferromagnetic type for a gross-shell occupation below one-half. This behavior remains also if the self-field is taken into account in the case of small dot diameters (~ 1 nm), while for large sizes the exchange coupling of supermoments is preferentially ferromagnetic. J_B in equation (14.5.4) exponentially decreases: This arises from the exponentially decaying overlap of superparamagnetic dot wavefunctions extending their tail into the barrier. This restricts the interdot separation at which exchange can contribute to the magnetic ordering, in agreement with experiments.

Bibliography

[1] Weisbuch, C., and B. Vinter, *Quantum Semiconductor Structures*, Academic Press, Boston (1991).

[2] Davies J. H., *The Physics of Low-Dimensional Semiconductors*, Cambridge University Press, Cambridge (1998).

[j] M. R. Scheinfein *et al.*, *Phys. Rev. Lett.* **76**, 1541 (1996).

[3] Singh, J., *Physics of Semiconductors and Their Heterostructures*, McGraw-Hill, New York (1993).

[4] Kelly, M. J., *Low-Dimensional Semiconductors*, Clarendon Press, Oxford (1995).

[5] Levy, P. M., *Giant Magnetoresistance in Magnetic Layaered and Granular Materials*, Solid State Physics **47** (1994).

[6] Himpsel, F. J., J. E. Ortega, G. J. Mankey, and R. F. Willis, *Magnetic Nanostructures*, Adv. Phys. **47**, 511 (1998).

[7] Bastard, G., J. A. Brum, and R. Ferreira, *Electronic States in Semiconductor Heterostructures*, Solid State Physics **44**, 229 (1991).

[8] Butcher, P., N. H. March, and M. P. Tosi (eds.), *Physics of Low-Dimensional Semiconductor Structures*, Plenum Press, New York (1993).

[9] Ridley, B. K., *Electrons and Phonons in Semiconductor Multilayers*, Cambridge University Press, New York (1997).

[10] Dresselhaus, M., G. Dresselhaus, and P. C. Eklund, *Science of Fullerenes and Carbon Nanotubes*, Academic Press, San Diego (1996).

[11] Peyghambarian, N., S. W. Koch, and A. Mysyrowicz, *Introduction to Semiconductor Optics*, Prentice-Hall, New Jersey (1993).

[12] de Heer, W. A., W. D. Knight, M. Y. Chou, and M. L. Cohen, *Electronic Shell Structure and Metal Clusters*, Solid State Physics **40**, 93 (1987).

[13] Jacak, L., P. Hawrylak, and A. Wójs, *Quantum Dots*, Springer, Berlin (1998).

[14] Kastner, M. A., *Artificial Atoms*, Phys. Today **46**, Jan., 24 (1993).

[15] Ashoori, R. C., *Electrons in Artificial Atoms*, Nature **379**, 413 (1996).

[16] Yoffe, A. D., *Semiconductor Quantum Dots and Related Systems: Electronic, Optical, Luminescence and Related Properties of Low-dimensional Systems*, Adv. Phys. **50**, 1 (2001).

Part IV

Broken Symmetry and Ordered Phases

Order is heav'n's first law ⋯

— Alexander Pope

Symmetry cannot change continuously: what I have called the first theorem of condensed matter physics.

— P. W. Anderson (1981)

Chapter 15

Landau Theory of Phase Transitions

Phase transitions are cooperative phenomena involving a global change of structure and physical properties of a system when a certain external variable, in most cases temperature or pressure, is changed continuously. Mean-field theory has been successfully used in many kinds of phase transitions: e.g., the theories of van der Waals for the vapor-liquid transition (1873), Weiss for the paramagnetism-ferromagnetism transition (1907), and Bragg–Williams for the order-disorder transition in alloys (1934). The well-known superconductivity theory of Bardeen–Cooper–Schrieffer (1957) is also a mean-field theory. The theory of second-order phase transitions posed by Landau (1937), has attracted much attention because of its simplicity of formalism and universality of application. It can be used to illustrate ferroelectric, structural, magnetic and even the superconducting and superfluid phase transitions. The Landau theory of phase transitions is a phenomenological theory based on thermodynamic principles; it unifies various mean-field theories.

§15.1 Two Important Concepts

In the Landau theory of phase transitions, we emphasize two closely related important, general concepts: Broken symmetry and the order parameter.

15.1.1 Broken Symmetry

A phase transition is usually accompanied by some breakdown of symmetry. What is symmetry? We have touched this subject in Chap. 1. Symmetry is the invariance of some physical quantities under some kind of operations, all of which may form a closed set called a symmetric group. Table 15.1.1 enumerates some continuous symmetric operations of several different systems. Take the liquid state for example: Its physical properties are invariant to arbitrary translation and rotation, that is to say, invariant, under the Euclidean group E(3) transformation.

In general, a physical system is described by a Hamiltonian, so the symmetry possessed by the system is closely related to the invariance of the Hamiltonian under the transformations. For the liquid state, the Hamiltonian $\mathcal{H}$ of the system is invariant under group E(3). We might take $\mathcal{G}_0$ to denote the symmetric group of a system which is described by a Hamiltonian $\mathcal{H}$, and let g be an element of the group, then the symmetry requires

$$g^{-1}\mathcal{H}g = \mathcal{H}. \tag{15.1.1}$$

At sufficiently high temperature, when all microscopic states of the system become more or less equally accessible, the symmetry of $\mathcal{H}$ becomes that of Gibbs free energy G.

When macroscopic conditions are changed (the temperature is decreased, or the pressure is increased, or an external field is applied) one or more symmetric elements may disappear: This is the phenomenon of *broken symmetry*. Broken symmetry refers to the situation in which the state of a system does not have the full symmetry possessed by the Hamiltonian used to describe the system. A magnetic system is a well-known example: At temperatures above its Curie temperature the system has zero magnetization in zero field; it is symmetric, i.e., does not have any preferred direction of the magnetization. As the temperature is lowered below the Curie temperature, however, a spontaneous magnetization develops in a specific direction; consequently the full symmetry of directions for the magnetization breaks down.

Phase transitions occur in systems with large numbers of particles, and many-body interactions or correlations play an important role, so we need to go from an independent particle model to a many-body problem. Different kinds of interactions lead to different ordered phases through symmetry breaking, when the temperature is decreased or the pressure increased. Now, interactions between particles are not a weak correction but the dominant factor which determines various ordered phases. For many-particle systems, due to interactions, qualitatively different ordered phases appear, as shown in Table 15.1.2. Broken symmetry leads to the appearance of ordered phases; at zero temperature, the ground state has broken symmetry. At low temperature, a system may display many different ordered states.

Let us consider a structural phase transition, as temperature decreases the symmetry may be transformed from the liquid state to a crystalline state. In this process, continuous translation and rotation symmetries are broken. The resulting symmetry group of the crystal is one of the 230 possible space groups. In many cases, especially of second-order phase transitions, the symmetry group $\mathcal{G}$ of the broken symmetric state is a subgroup of the initial group $\mathcal{G}_0$, i.e., $\mathcal{G} \subset \mathcal{G}_0$. Assume $|\mathrm{G}\rangle$ is the ground state of a system, then for an operation $h \in \mathcal{G}$, broken symmetry leads to $h|\mathrm{G}\rangle = |\mathrm{G}\rangle$. However, for an operation $g \notin \mathcal{G}$, we have $g|\mathrm{G}\rangle \neq |\mathrm{G}\rangle$, even though $g \in \mathcal{G}_0$.

In some systems, there may be degenerate ground states. In these cases, the real ground state is only one of these possible ground states. The physical phenomena taking place in this special ground state will not or only partly display the original symmetries of the physical laws. Here the symmetries are not broken by external factors; the breaking is thoroughly spontaneous. In reality, the symmetries of the physical laws have not been destroyed but cannot be displayed in the special background. So we can distinguish two types of broken symmetries, one is spontaneous breaking for which $\mathcal{H}$ remains invariant, and the other is externally disturbed as $\mathcal{H} \to \mathcal{H} + \mathcal{H}'$, where $\mathcal{H}'$ is an additional perturbation to the original Hamiltonian $\mathcal{H}$.

Landau emphasized the importance of broken symmetry: A given symmetry element is either there or it is not. In each state there is either one symmetry or the other, the situation is never ambiguous. When the symmetry is broken, there is concurrent ordering. It should be noted that the transition between phases with different symmetries, like liquid and crystal or different crystalline states, cannot occur in a continuous manner, that is to say, it is impossible to change symmetry gradually.

Most phase transitions are associated with a sudden change of symmetry. We may ask if there are phase transitions without change of symmetry? The answer is yes! The vapor-liquid transition, as will be discussed in §19.2.1, is an example that has no symmetry change. The states above and below the transition temperature are all isotropic; only the density changes drastically. Other examples such as liquid-glass transition, paramagnetic-spin glass transition, and metal-insulator transition are also unrelated to broken symmetry. We shall discuss these phase transitions in §19.2, §19.3, and §19.4, respectively. They are involved in a more general concept called broken ergodicity.

15.1.2 Order Parameter

We shall give a quantitative description of the phase transition in a system. According to the spirit of broken symmetry, phase transitions are characterized by the loss or gain of some symmetry elements when the macroscopic variables of the system are changed. When a system is transformed from a high symmetry phase to a low symmetry phase, there is a physical quantity η, called the order parameter, which varies in such a way that it is zero in the high symmetry phase and takes nonzero values in the low symmetry phase. For instance, in a structural phase transition where the atoms are displaced from their equilibrium positions in the high symmetry phase, η may be taken

Table 15.1.1 Transformations associated with various continuous symmetries.

Symmetric group	Infinitesimal generator	Finite transformation	Characteristics of transformation
translation	$\boldsymbol{k}$	$e^{i\boldsymbol{k}\cdot\boldsymbol{r}}$	translation $\boldsymbol{r}$ in wavevector $\boldsymbol{k}$ direction
rotation	$\boldsymbol{L}$	$e^{i\phi\boldsymbol{n}\cdot\boldsymbol{L}}$	rotation ϕ along $\boldsymbol{n}$ axis with angular momentum $\boldsymbol{L}$
spin precession	$\boldsymbol{S}$	$e^{i\chi\boldsymbol{n}\cdot\boldsymbol{S}}$	rotation χ along $\boldsymbol{n}$ axis when spin $\boldsymbol{S}$ precesses
gauge	N	$e^{i\theta N}$	phase change θ caused by the action of number operator N

as the amount of the displacement. For a magnetic transition, η may be taken as the macroscopic magnetic moment per unit volume of a ferromagnet, or the magnetic moment of the sublattice for an antiferromagnet.

Since the order parameter is closely connected with the symmetry of a system, we can call the high symmetry phase the disordered phase, and the low symmetry phase the ordered phase. According to the Landau theory of phase transitions, there may exist a macroscopic order parameter η which measures the ordered phase below the transition temperature T_c. η is a thermodynamic variable, as it is the ensemble average of some microscopic variables σ_i. Such variables σ_i are functions of the space-time coordinates around the site i, so that the time variation, as well as the spatial distribution, is significant for the averaging of distributed variables. In the disordered phase above T_c, those variables σ_i are usually in fast random motion so that their time average $\langle\sigma_i\rangle_t$ vanishes at each lattice point and hence is independent of the site i. In contrast, at temperatures below T_c, they are correlated and in slow motion, so that the ordered phase is dominated by their spatial distribution.

It must be emphasized again that the symmetry of a system is changed only when η becomes nonzero, any nonzero value of the order parameter, no matter how small, brings about a lowering of the symmetry. Therefore, change of the symmetry is always abrupt, but the order parameter can vary in two different ways.

The manner in which the symmetry is broken enables us to define the order of the transition. There are two kinds of phase transitions in general. One is the first-order phase transition, for which the order parameter appears discontinuously below the transition temperature T_c. The two symmetry groups for the high symmetry phase and low symmetry phase may or may not have any group-subgroup relationship to each other. The other is the second-order phase transition, also known as the continuous phase transition, for which the order parameter appears gradually. The symmetries on the two sides of the transition are related, the symmetry group on the lower symmetry side must be a subgroup of the higher symmetry side.

For instance, in ferroelectric systems, the order parameter is the electric polarization. As the temperature decreases through T_c, the polarization can go from zero to a finite value continuously or discontinuously. Figure 15.1.1 shows the first-order or discontinuous phase transition in $BaTiO_3$. At high temperatures, $BaTiO_3$ has a cubic lattice whose unit cell is composed with the barium atoms at the vertices, the titanium atoms at the centers, and the oxygen atoms at the centers of the faces. As the temperature decreases below T_c, the titanium and oxygen atoms begin to move relative to the barium atoms, parallel to an edge of the cube. Once this happens, the symmetry of $BaTiO_3$ is changed, it becomes tetragonal instead of cubic. In this process, the electric polarization (now taken as the order parameter) jumps from zero to a finite value at $T_c = 120°C$. In contrast, Fig. 15.1.2 shows the second or continuous phase transition at $T_c = 110$ K. The structural symmetry change also from cubic to tetragonal by a tilt of the neighboring oxygen octahedra. The tilting angle can be taken as the order parameter; we see the order parameter grows gradually from zero.

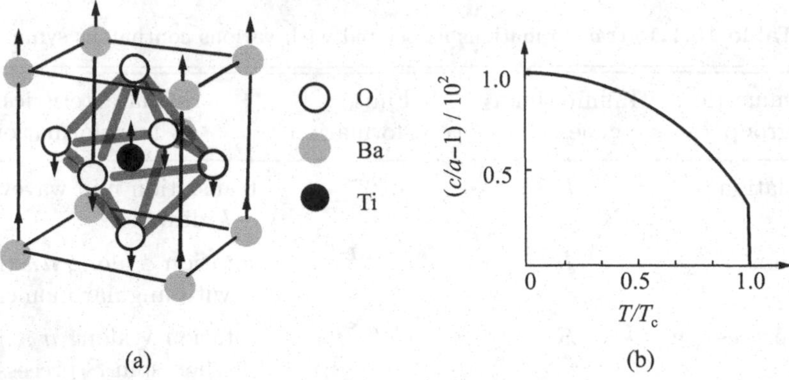

Figure 15.1.1 First-order phase transition in $BaTiO_3$. c/a is the ratio of lattice constants.

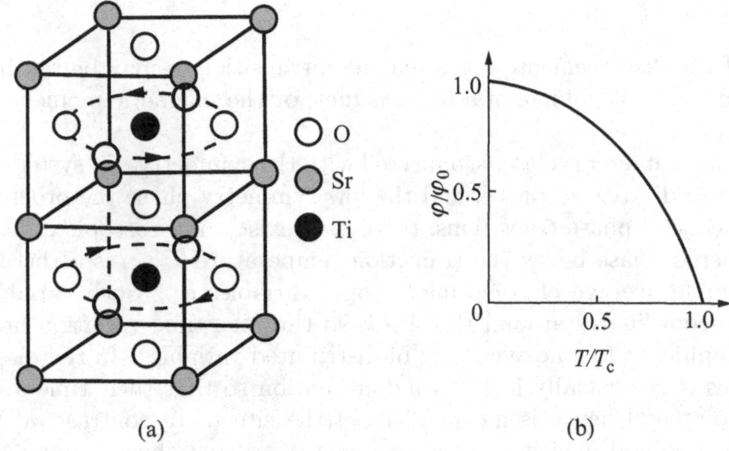

Figure 15.1.2 Second-order phase transition of $SrTiO_3$. φ is the tilting angle of the oxygen octahedra. $\varphi_0 = 1.3°$ is the maximum value of the tilting angle.

In some cases it is possible to use a variable external force to change the nature of the transition from first-order to second-order. If we can change the variable external force in arbitrary small steps we can go from first-order to second-order passing through a threshold point between the two cases which is called a "tricritical point".

Any order parameter, as a physical quantity, can be a scalar, or a vector, or a tensor. Generally speaking, the order parameter may have multicomponents. In the simplest case, the order parameter is a scalar. In this case the number of components of the order parameter $n = 1$. For example, we have chosen $(c/a - 1)$ and φ/φ_0 as the order parameters for $BaTiO_3$ and $SrTiO_3$ respectively. They are all scalar order parameters. The order parameter can also be a vector with a number of components $n = 3$. A well-known example is a bulk ferromagnet: Below the Curie temperature, the material has macroscopic magnetization $\boldsymbol{M}$, which is a vector. For a two-dimensional isotropic ferromagnet, the magnetization is limited to a plane, so its number of components is $n = 2$. Similarly, for superfluids and superconductors, their macroscopic wave functions are chosen as order parameters. Because a macroscopic wavefunction, written as $\psi = \psi_0 \exp(i\theta)$, is complex with modulus ψ_0 and phase θ, so $n = 2$. For the vapor-liquid phase transition, we cannot distinguish the symmetry of the vapor and liquid, so there is no change of symmetry at all, but at the transition temperature, gas and liquid are separated, we can take the density difference $\rho_l - \rho_g$ as the order parameter, as shown in Fig. 15.1.3. For some strong first-order reconstruction phase transitions, although there is no group and subgroup relation between their high-temperature phase and low-temperature phase, it is still possible to define suitable order parameters to treat them in the framework of Landau theory.

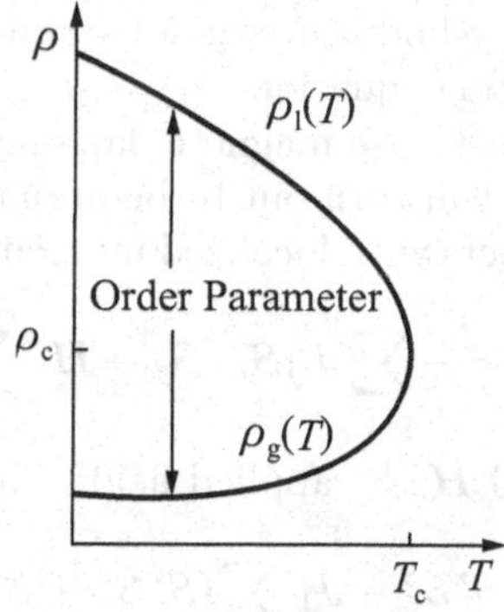

Figure 15.1.3 The ρ-T phase diagram of gas-liquid.

Table 15.1.2 Broken symmetry and ordered phases.

Phase	Broken symmetry	Order parameter
crystal	translation and rotation	$\rho = \sum_G \rho_G e^{i\boldsymbol{G}\cdot\boldsymbol{r}}$
nematic	rotation	$\eta_{ij} = \frac{1}{2}(3\eta_i\eta_j - \delta_{ij})$
smectic	rotation and 1D translation	$\eta_{ij} = A\lvert\psi\rvert \cos(qz - \phi)$
ferroelastic	inversion	$\boldsymbol{P}$
antiferroelastic	inversion	$\sum \boldsymbol{p}$ (sublattice)
ferromagnetic	time reversal	$\boldsymbol{M}$
antiferromagnetic	time reversal	$\sum \boldsymbol{m}$ (sublattice)
superfluid ^{4}He	gauge (U(1) group)	$\psi = \lvert\psi\rvert e^{i\theta}$
superconductivity	gauge (U(1) group)	$\psi = \lvert\psi\rvert e^{i\theta}$

15.1.3 Statistical Models

We would like to consider the physical realization of the order parameters at the microscopic level. In most cases, internal interaction is the main reason for spontaneous symmetry breaking when the temperature is lowered below T_c, because internal interactions will suppress thermal fluctuations and lead to an internal field conjugate to the order parameter, which in turn helps to drive the entire system into an ordered state.

Every phase transition is accompanied by the appearance, at the phase transition point T_c, of a set of physical quantities that are absent in the initial phase. These quantities may be divided into two major groups: microscopic and macroscopic parameters. Examples of micro parameters are atomic displacements or atomic spins arising at the phase transition point T_c, and also variations of the probability of finding an atom of a given species on a given site.

In addition, a variety of physical properties of the substance are described by macroscopic variables, such as electric polarization, magnetization, strain tensor, etc. Measurement of these quantities underlies the various experimental techniques of investigating phase transitions in materials.

Phase transitions are induced by the mutual interactions of many particles, and are cooperative in nature. In order to understand the nature of cooperative transitions, it is necessary to employ more powerful microscopic theories which take into account the details of atomic interactions beyond simple thermodynamic theories; this leads to statistical models.

There are some fundamental models that describe cooperative behavior in condensed matter systems. Although these models may be too simple to imitate real physical systems, they still include enough information of many-body interactions and can give qualitative prediction of the behavior, by solving the corresponding equations.

It is conventional and convenient to use magnetic language and write the model Hamiltonian in terms of spin variables, although it will turn out to be applicable to many non-magnetic systems.

A realistic model for many magnets with localized moments is provided by the Heisenberg Hamiltonian

$$\mathcal{H} = -\sum_{ij} J_{ij}\boldsymbol{S}_i \cdot \boldsymbol{S}_j - \boldsymbol{H} \cdot \sum_i \boldsymbol{S}_i, \tag{15.1.2}$$

where J is the exchange energy and $\boldsymbol{H}$ the applied field. This Hamiltonian can be rewritten in the form

$$\mathcal{H} = -J_z \sum_{ij} S_i^z S_j^z - J_\perp \sum_{ij} (S_i^x S_j^x + S_i^y S_j^y) - H \sum_i S_i^z, \tag{15.1.3}$$

where x, y, z label Cartesian axes in spin space, and the applied field is assumed along the z-axis. For $J_\perp = 0$, it reduces to the Ising model, while for $J_z = 0$, it is the XY model.

In some systems the combined cooperative interaction and local crystal field interaction forces the spins to point up or down in a specific direction, which implies a one-dimensional order parameter, i.e., the Ising case. In some other systems the spins can only rotate within a single plane, implying a two-dimensional order parameter, which is the XY model. Still in other magnetic systems the allowed direction for the spins is not restricted to a line or to a plane but may be in any spatial direction, so there is a three-dimensional order parameter: the Heisenberg case. In all these three cases, the transition from the paramagnetic to magnetically-ordered state can be characterized by the occurrence of mean magnetic moment vectors on the sites.

The spin-1/2 Ising model is a remarkably successful model for an interacting system. A classical spin variable, which is allowed to take the values ± 1, is placed on each lattice site. By taking $J_\perp = 0$ and omitting the subscript z in (15.1.3), we write the Ising Hamiltonian as

$$\mathcal{H} = -J\sum_{ij} S_i S_j - H \sum_i S_i. \tag{15.1.4}$$

It is clear that a positive J favors parallel, and negative J antiparallel, alignment of the spins. The main restriction of the Ising model is that the spin vector can only lie parallel to the direction of quantization introduced by the magnetic field. This means that the Ising Hamiltonian can only prove useful in describing a magnet which is highly anisotropic in spin space. There are some physical systems, MnF_2 for example, which to a good approximation obey this criterion. However, despite its simplicity the Ising model is widely applicable, because it can describe any interacting two-state system, such as order-disorder transitions in binary alloys, which will be discussed in §20.2.1.

We take the macroscopic mean value of the Ising system, i.e. the magnetization, as the order parameter η. At high temperature, the Ising spin system has the symmetries of the Z_2 group. The elements of the Z_2 group are

$$Z_2 = \{E, I\},$$

here E is the identity transformation and I is the inversion transformation. If a system has the symmetry of Z_2, then under the symmetry operation of Z_2, the free energy of the system is invariant for $\eta \to \eta, \eta \to -\eta$. For an Ising spin system, at $T > T_c$, $\eta = 0$, so $E\eta = \eta$, and $I\eta = \eta$, satisfying Z_2 symmetry. But at $T < T_c$, $\eta \neq 0$, the transformations are

$$E\eta = \eta, \qquad I\eta = -\eta.$$

and Z_2 symmetry is broken. The thermodynamic state below T_c lacks the full symmetry of $\mathcal{H}$, i.e., broken symmetry!

For the XY model, any spin i is a two-dimensional vector, which can written as

$$\boldsymbol{S}_i = \boldsymbol{i}S_{ix} + \boldsymbol{j}S_{iy} = S(\boldsymbol{i}\cos\theta_i + \boldsymbol{j}\sin\theta_i).$$

If we take the average of $\boldsymbol{S}_i$ as the order parameter

$$\boldsymbol{\eta} = \langle \boldsymbol{S}_i \rangle,$$

then in the high-temperature phase, $T > T_c$, the spin vectors at sites are randomly distributed, satisfying the symmetry of the $\mathcal{O}(2)$ group, corresponding to the case in which the orientational angle in the two-dimensional plane can take any value, so

$$\eta = 0,$$

but in the low-temperature phase, $T < T_c$, the symmetry of the $\mathcal{O}(2)$ group is broken. If $J_\perp > 0$, the spin vectors will predominantly lie along a certain direction in the xy plane; the O_2 symmetry is broken simultaneously and the average θ_i takes a definite value, i.e.,

$$\eta = \langle S\rangle(\boldsymbol{i}\cos\theta + \boldsymbol{j}\sin\theta).$$

Broken symmetry comes from the interactions of spins at different sites, $J_\perp$, and there exists the following relation for spins at any sites i and j

$$\boldsymbol{S}_i \cdot \boldsymbol{S}_j = S_{ix}S_{jx} + S_{iy}S_{jy} = S^2\cos\theta_{ij}. \tag{15.1.5}$$

It is obvious $\langle\theta_{ij}\rangle \to 0$ when the symmetry is broken.

For an isotropic XY system, the ordered states are infinitely degenerate; a different realization of the ordered state corresponds to different values of the phase for a definite free energy from a reference direction in the xy plane. It is known that the symmetry groups $\mathcal{O}(2)$ and $U(1)$ are isomorphic, so the order parameter can also expressed as a complex number

$$\eta = \langle S\rangle e^{i\theta}.$$

The components of the complex number are its amplitude $\langle S\rangle$ and phase θ. The XY model is the simplest model for the study of continuous symmetry breaking. A complex parameter, such as a macroscopic wavefunction $\psi = |\psi|e^{i\theta}$, can also be used to describe superfluid and superconducting phase transitions with the breaking of gauge symmetry.

§15.2 Second-Order Phase Transitions

The order of a phase transition is defined in the Ehrenfest scheme in which the order of the lowest derivatives of the free energy shows a discontinuity at the transition point. Landau formulated the central principles of the phenomenological theory of second-order phase transition based on the idea of spontaneous symmetry breaking at the phase transition. By means of this approach, it has been possible to treat phase transitions of a different nature in distinct systems from a unified viewpoint. In what follows, we will usually take a scalar order parameter to give an illustration of the principles.

15.2.1 Series Expansion of Free Energy

The quantitative theory of second-order phase transitions can be started from the Gibbs free energy G of the system, which is a function of pressure P, temperature T and order parameter η. It must be kept in mind that, in the function $G(P,T,\eta)$, the variable η is not on the same footing as the variables P and T; whereas the pressure and temperature can be specified arbitrarily, the value of η must itself be determined from the condition of thermal equilibrium, i.e. the condition that G is minimized for given P and T.

The continuity of the change of state in a second-order phase transition implies that the quantity η takes arbitrarily small values near the transition point. In the vicinity of the phase transition point the Gibbs free energy G can be expanded in a power series of η. For the case of scalar order parameter η, the free energy is written as

$$G(P,T,\eta) = G_0 + \alpha\eta + A\eta^2 + C\eta^3 + B\eta^4 + \cdots, \tag{15.2.1}$$

where G_0 is the Gibbs free energy of the high symmetry phase and is unrelated to the phase transition but α, A, C, B are certain parameters of the system that are dependent on P and T. We will use temperature as the macroscopic variable to induce a phase transition in what follows. Actually, it is also possible to substitute another variable for the temperature, for instance, a pressure sufficient to trigger the transition in the case of a ferroelectric, where the temperature is fixed, while the smectic-nematic liquid crystal transition can be driven by an external magnetic field.

The stability condition requires that G, as a function of η, should be a minimum and thus satisfy

$$\left(\frac{\partial G}{\partial \eta}\right) = 0, \qquad \left(\frac{\partial^2 G}{\partial \eta^2}\right) > 0, \tag{15.2.2}$$

from which it follows that the linear terms in (15.2.1) cancel out. The equilibrium value of order parameter η is found by combining (15.2.1) and (15.2.2). For the high symmetry phase, $T > T_c$, the equilibrium value $\eta = 0$, it is necessary that $A > 0$; on the other hand, for the low symmetry phase, $T < T_c$, η takes a non-zero value, and it is required that $A < 0$. Therefore, at the transition point $T = T_c$, $A = 0$. In the vicinity of T_c the coefficient of the quadratic term A may be expected to be a linear function of temperature

$$A(P,T) = a(P)(T - T_c), \tag{15.2.3}$$

with $a(P) > 0$.

If the phase transition point, $T = T_c$, itself is stable, the conditions

$$\left(\frac{\partial^2 G}{\partial \eta^2}\right)_{\eta=0} = 0, \qquad \left(\frac{\partial^3 G}{\partial \eta^3}\right)_{\eta=0} = 0, \qquad \left(\frac{\partial^4 G}{\partial \eta^4}\right)_{\eta=0} > 0, \tag{15.2.4}$$

must also fulfilled, then

$$A(P,T_c) = 0, \qquad C(P,T_c) = 0, \qquad B(P,T_c) > 0. \tag{15.2.5}$$

Assuming that the two possibilities of broken symmetry for η and $-\eta$ are equivalent, the coefficient C is identically equal to zero. Generally, B is weakly temperature dependent; we take it as a positive constant here. By neglecting the high order terms, the free energy is in the form

$$G(P,T,\eta) = G_0 + A(P,T)\eta^2 + B\eta^4. \tag{15.2.6}$$

From $\partial G/\partial \eta = 0$, we obtain

$$\eta(A + 2B\eta^2) = 0, \tag{15.2.7}$$

which can be called the equation of state, because it governs the relationship of P and T in the system. From it there are two solutions,

$$\eta = 0, \tag{15.2.8}$$

and

$$\eta = \pm\left(-\frac{A}{2B}\right)^{1/2} = \pm\left[\frac{a(T_c - T)}{2B}\right]^{1/2}. \tag{15.2.9}$$

For $T \geq T_c$, $\eta = 0$ is stable, but for $T < T_c, \eta = 0$ corresponds to the free energy being a maximum, and thus only the nonzero solution is stable, which corresponds to the appearance of the ordered phase. We can see the situation in Fig. 15.2.1. Here, for simplicity, we take the free energy of the high temperature phase G_0 as the zero point of energy.

The dependence of the order parameter on temperature in (15.2.9) shows that the transition is continuous at the transition point. This characteristic is displayed in Fig. 15.1.2.

15.2.2 Thermodynamic Quantities

Phase transitions may bring many extraordinary physical properties to systems. The thermodynamic quantities may change drastically; examples showing anomalies are thermal expansion coefficients, elastic constants, refractive indices etc. Even transport coefficients such as the thermal and electric conductivity often present pronounced anomalies in the vicinity of the phase transition. For instance, the dielectric constant of ferroelectrics diverges as T_c is approached from both sides.

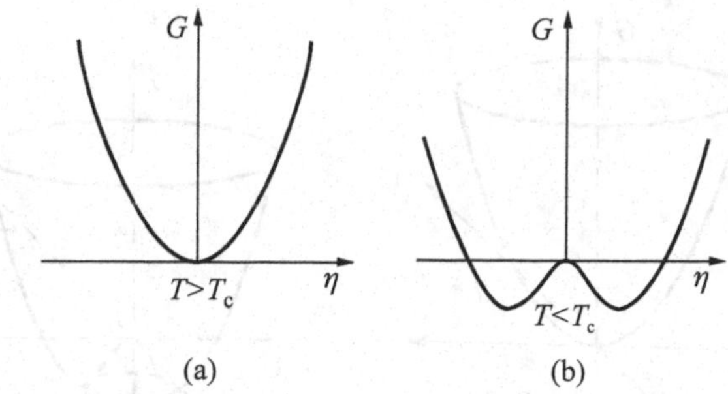

Figure 15.2.1 Free energy as a function of the scalar order parameter in the vicinity of the second-order phase transition. (a) $T > T_c$; (b) $T < T_c$.

For a second-order phase transition, the absence of any discontinuous change of state at the phase transition point has the result that the thermodynamic functions of the system, including its entropy, energy, volume, etc., vary continuously as the transition point is passed. Hence a second-order phase transition, unlike the first-order, is not accompanied by emission or absorption of heat. In fact, it is the derivatives of the these thermodynamic quantities, i.e. the specific heat, the thermal expansion coefficient, the compressibility, etc., that are discontinuous at a transition point of the second-order.

Now we discuss the temperature dependence of entropy and specific heat at the transition point. The entropy is given by $S = -\partial G/\partial T$. For $T > T_c$, in the high symmetry phase, $\eta = 0$, so

$$S = -\frac{\partial G_0}{\partial T} = S_0, \tag{15.2.10}$$

however, when $T < T_c$, $\eta \neq 0$, and

$$S = S_0 + \frac{a^2}{2B}(T - T_c). \tag{15.2.11}$$

It is clear that at $T = T_c$, $S = S_0$. Thus the entropy is continuous at the transition point. This continuity of first-order derivatives of G indicates that the phase transition is second-order.

The specific heat at constant pressure is evaluated from $C_P = T(\partial S/\partial T)_P$. For the high symmetry phase,

$$C_P = T\left(\frac{\partial S_0}{\partial T}\right)_P, \tag{15.2.12}$$

but for the low symmetry phase,

$$C_P = T\left(\frac{\partial S_0}{\partial T}\right)_P + \frac{a^2 T_c}{2B}, \tag{15.2.13}$$

just at T_c, there is no divergence, but a discontinuous jump of C_P between T_{c-} and T_{c+}. The size of the discontinuity is

$$\Delta C_P = \frac{a^2 T_c}{2B}. \tag{15.2.14}$$

Other quantities besides C_P, such as the thermal expansion coefficient, compressibility, etc., are also discontinuous. There is no difficulty in deriving relations between the discontinuities of all these quantities.

15.2.3 System with a Complex Order Parameter

The symmetry that is broken in the formation of the superconducting or the superfluid state is the gauge symmetry, as will be discussed in Chap. 18. A macroscopic wavefunction emerges below the transition temperature and can be introduced to denote the broken gauge symmetry. From

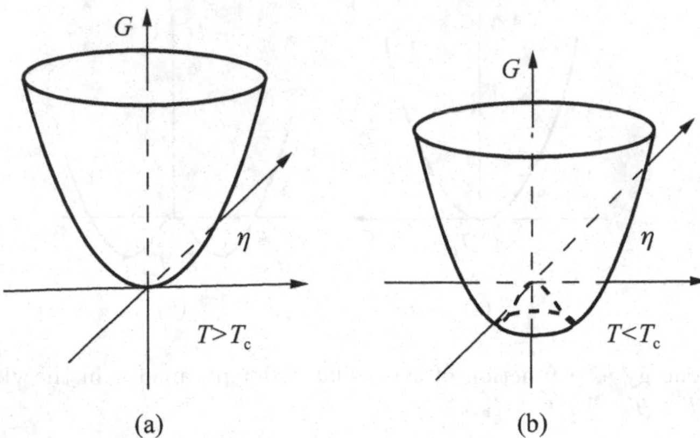

Figure 15.2.2 Free energy surfaces for complex order parameter corresponding to two cases for (a) $T > T_c$ and (b) $T < T_c$.

the thermodynamic standpoint, the macroscopic wavefunction can be taken as a complex order parameter.

$$\eta = \eta_0 e^{i\theta}.$$

There are two real components, the amplitude η_0 and phase angle θ.

These two components should be spatially homogeneous for a system, in the absence of applied field. According to Landau theory, we expand the free energy to fourth order

$$G = G_0 + A|\eta|^2 + B|\eta|^4, \tag{15.2.15}$$

where $A = a(T - T_c)$ and $B > 0$, as usual. The minimum of free energy in (15.2.15) will be given by $\partial G/\partial \eta_0 = 0$, so we have

$$(A + 2B\eta_0^2)\eta_0 = 0. \tag{15.2.16}$$

The solutions are $\eta_0 = 0$, for $T > T_c$; or $\eta_0 = (A/2B)^{1/2} = [a(T_c - T)/2B]^{1/2}$ for $T < T_c$.

In the normal state for $T > T_c$, $\eta_0 = 0$ and $G = G_0$, we could say that θ takes any value, so gauge symmetry is intact. When $T < T_c$, $\eta_0 \neq 0$ and θ takes a definite value. The gauge symmetry is broken. Figure 15.2.2 gives the free energy surface in these two cases. However, we note that in (15.2.15) the phase factor does not appear unambiguously. If we perform the transformation

$$\eta \to \eta' = e^{i\theta'}\eta, \tag{15.2.17}$$

where θ' is an additional phase difference, the form of (15.2.15) does not change. This transformation corresponds to a rotation by the angle θ' in the complex plane. The set of all such rotation constitutes the continuous group U(1). It is clear that G is invariant under the transformation of group U(1). So it seems that there still exists gauge symmetry in the free energy. In fact, a gauge transformation applied to the ordered state changes θ to some other value corresponding to a different ordered state with the same free energy. These ordered states are degenerate, analogous to ferromagnetic states. In the low temperature phase, the minimum of G lies on a circle. Every point on the circle represents a possible ordered state. When the symmetry is spontaneously broken, the system is in a state represented by one of the points of the circle, see Fig. 15.2.2(b).

The phase of the macroscopic wavefunction arises in the condensation process. Just below the transition temperature, only a small fraction of the particles are in the condensate and participate in superflow. As the temperature is lowered, the thermal effects, which tend to destroy the condensate, become less important and a large fraction of particles condense into the condensate. At $T = 0$ K, the maximum number of particles have entered the condensate. We can expect that the phase and the particle density are a couple of conjugate variables. In general, it is impossible to specify the density and the phase simultaneously. Careful analysis shows that there is an uncertainty relation for large N

$$\Delta N \Delta\theta \sim 1. \tag{15.2.18}$$

More intricate broken gauge symmetries are involved in which appropriate "internal" degrees of freedom are associated with the ordering process. Examples of this kind include liquid crystalline phases and the phase of superfluid ^{3}He.

§15.3 Weak First-Order Phase Transitions

Landau theory which has been successfully used for second-order phase transitions can be extended to treat some weak first-order phase transitions. For these first-order phase transitions, the concept of order parameter is still effective.

15.3.1 Influence of External Field

In a number of systems, phase transitions are involved in a pair of conjugated variables whose product is often an energy. For example, pressure P and volume V in vapor-liquid transitions, magnetic field $\boldsymbol{H}$ and magnetization $\boldsymbol{M}$ in paramagnetic-ferromagnetic transitions, electric field $\boldsymbol{E}$ and polarization $\boldsymbol{P}$ in paraelectric-ferroelectric transitions, and stress σ and strain ε in paraelastic-ferroelastic transitions.

We are interested in the contribution of conjugated fields of order parameters to phase transitions. For simplicity, consider a conjugate field h of the scalar order parameter η which leads the free energy to add a term of $-\eta h$, then the free energy takes the form

$$G_h(P,T,\eta) = G_0 + a(T-T_c)\eta^2 + B\eta^4 - \eta h. \tag{15.3.1}$$

Figure 15.3.1 shows that the free energy is asymmetric about order parameter η. Note that the minimum of free energy above T_c is not at $\eta = 0$, and below T_c, the two minimum values are not equal again.

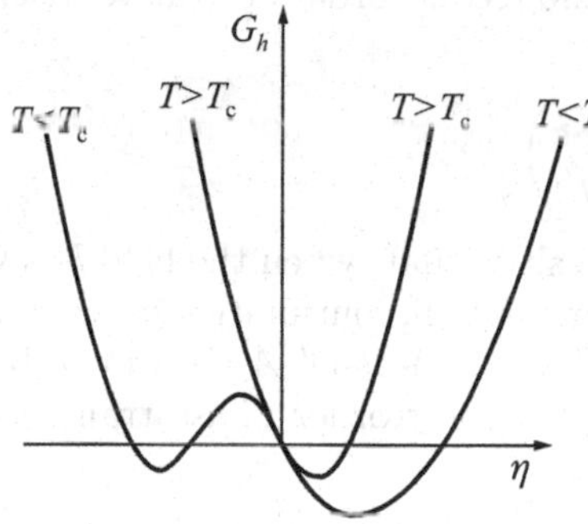

Figure 15.3.1 Asymmetric free energy under external field.

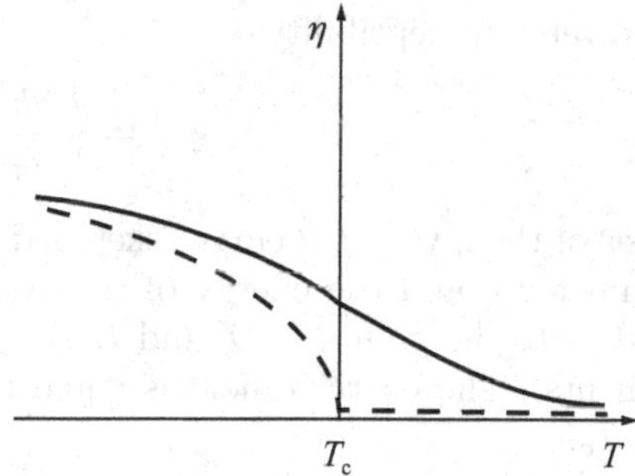

Figure 15.3.2 Phase diagram of η versus T under the fixed external field h, dot line corresponds to $h = 0$.

By using the equilibrium condition $\partial G_h/\partial\eta = 0$, we have the equation of state

$$2a(T-T_c)\eta + 4B\eta^3 - h = 0, \tag{15.3.2}$$

and under a fixed external field h, we can plot η as a function of T in Fig. 15.3.2.

We can evaluate the susceptibility $\chi = (\partial\eta/\partial h)_{T,h\to 0}$; the result is

$$\chi = \frac{1}{2a(T-T_c) + 12B\eta^2}, \tag{15.3.3}$$

then at $T > T_c$,

$$\chi = \frac{1}{2a(T-T_c)}, \tag{15.3.4}$$

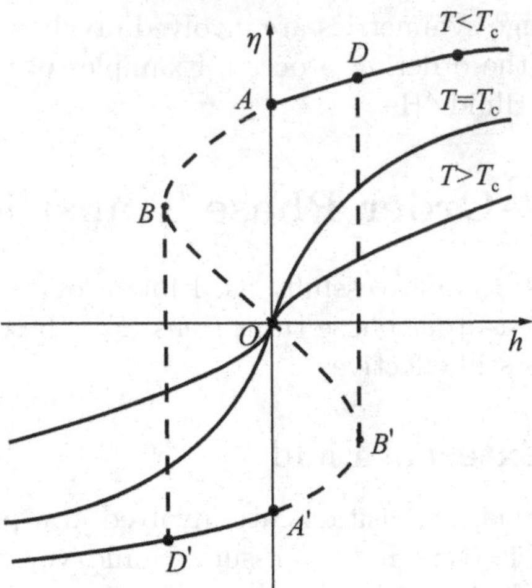

Figure 15.3.3 Phase diagram of η versus T under the fixed external field h, dotted line corresponds to $h = 0$.

and at $T < T_c$,

$$\chi = \frac{1}{4a(T_c - T)}. \tag{15.3.5}$$

When $T \to T_c$, $\chi \to \infty$. This is the Curie–Weiss law.

Figure 15.3.3 gives η as a function of h at different temperatures for $T > T_c$ and $T < T_c$. The η-h solid lines refer to stable states of the system; the dashed lines refer to unstable states. The segments A-B and A'-B' of the η versus h curve correspond to metastable states. The segments B-O and B'-O refer to unstable states with negative values of the second-order derivative, $\partial^2 G/\partial h^2 < 0$, or with a inverse susceptibility

$$\chi^{-1} = \left(\frac{\partial h}{\partial \eta}\right)_{\eta=0} = \left(\frac{\partial^2 G}{\partial \eta^2}\right)_{\eta=0}. \tag{15.3.6}$$

Analysis of the η versus h curves sketched in Fig. 15.3.3 show that, when the field h is varied, the order parameter η and the energy of the system should exhibit discontinuities between the states corresponding to the points B-D' and D-B'. A hysteresis loop D-A-B-D'-A'-B' should be observed in experiments. The coercive field is equal to $(h_{B'} - h_B)/2$. First-order phase transition appears when $T < T_c$.

15.3.2 Landau–Devonshire Model

In many ferroelectrics, weak first-order phase transitions have been observed experimentally, even without an external electric field, but the transformation properties of the order parameters exclude a cubic term. In describing such transitions, the free energy should be expanded to higher terms (Devonshire, 1949).

We assume that the spontaneous polarization in a ferroelectric is along a fixed direction, so the polarization is taken as a scalar order parameter. Suppose $B < 0$, but for stability of the low temperature phase, we expand the free energy to the 6th power

$$G(P, T, \eta) = G_0 + a(T - T_c)\eta^2 + B\eta^4 + D\eta^6, \tag{15.3.7}$$

where $D > 0$. It is noted that the coefficient $A = a(T - T_c)$ is kept invariant, because we assume (15.3.7) is only a small modification of (15.2.6). Now T_c is not a transition temperature. Its meaning will be derived later.

The equilibrium condition $\partial G/\partial \eta = 0$ gives the equation of state

$$2a(T - T_c)\eta + 4B\eta^3 + 6D\eta^5 = 0. \tag{15.3.8}$$

There are solutions

$$\eta = 0, \tag{15.3.9}$$

$$\eta^2 = \frac{-B + [B^2 - 3aD(T - T_c)]^{1/2}}{3D}, \tag{15.3.10}$$

and

$$\eta^2 = \frac{-B - [B^2 - 3aD(T - T_c)]^{1/2}}{3D}. \tag{15.3.11}$$

The condition for (15.3.10) and (15.3.11) having real roots gives an upper limit of temperature T_+

$$T_+ = T_c + \frac{B^2}{3aD} > T_c. \tag{15.3.12}$$

For $T < T_+$, it can be verified that (15.3.10) is a solution that minimizes the free energy, but to form an ordered state, (15.3.11) is unstable or meaningless.

It must be emphasized that T_+ is not a transition temperature, though (15.3.10) can represent a metastable polarized state. We should see if G is larger or less than G_0 after (15.3.10) is substituted into (15.3.7). As a matter of fact, the real transition temperature $T = T_t$ is determined from the condition $G - G_0 = 0$; this gives

$$a(T - T_c)\eta^2 + B\eta^4 + D\eta^6 = 0. \tag{15.3.13}$$

Then, from the condition of real root, we have

$$T_t = T_c + \frac{B^2}{4aD}, \tag{15.3.14}$$

which is less than T_+. We now have three characteristic temperatures which are arranged in a sequence $T_+ > T_t > T_c$. T_t is the temperature of the phase transition. At $T = T_t$, there are three minima of G: $\eta = 0$, and $\eta = \pm(-B/2D)^{1/2}$.

We depict the curves of free energy versus polarization η at different temperatures T in Fig. 15.3.4. When $T > T_+$, only $\eta = 0$ corresponds to the minimum of the free energy, so the disordered phase is stable; for $T_+ > T > T_t$, there are $\eta = 0$ and $\eta \neq 0$ as equilibrium values for G, but still the disordered phase is more stable, and the ordered phases are metastable. At $T = T_t$, for which

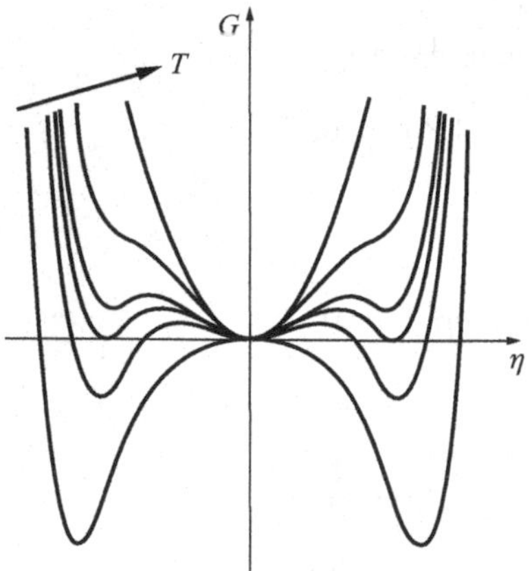

Figure 15.3.4 Free energy versus order parameter in the Landau–Devonshire theory.

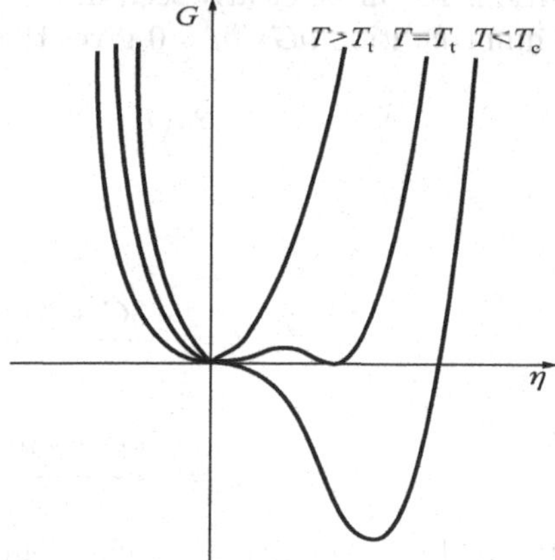

Figure 15.3.5 Free energy versus order parameter in Landau–de Gennes theory.

$G - G_0 = 0$, a first-order phase transition takes place. The polarization changes discontinuously from zero to a finite value

$$\eta^2 = \frac{B}{2D}. \tag{15.3.15}$$

The change of entropy can be computed

$$\Delta S = \frac{\partial G}{\partial T} - \frac{\partial G_0}{\partial T} = \frac{aB}{2D}, \tag{15.3.16}$$

which also changes discontinuously. If T is lower than T_t, the disordered phase becomes unstable, and the ordered phase stable. Finally, at $T = T_c$, for $\eta = 0$, $\partial G/\partial \eta = 0$ and $\partial^2 G/\partial \eta^2 = 0$, so $\eta = 0$ is a spinodal point. T_c corresponds to the absolutely unstable limit of the disordered phase, and $\eta = \pm(-2B/3D)^{1/2}$ are perfectly stable.

We can also adopt (15.3.7), in the case of $B > 0$, to study ferroelectric phase transitions. There is no doubt that this is related to second-order phase transitions. As in the last subsection, we can add the influence of an external field into consideration; we need only to add a term $-h\eta$ into (15.3.7). Now G is not symmetric about the η axis, and first-order phase transitions related to T or h may take place.

15.3.3 Landau–de Gennes Model

We shall discuss the microscopic theory of the isotropic-nematic phase transition of liquid crystals in §16.3. For nematic liquid crystals, η denotes the orientational order parameter. If this phase transition is first-order, some anomalies appear. This is due to the fact that, in the isotropic phase, there is no long-range order in the direction of the alignment of the molecules, or in other words, the order parameter vanishes, on average. However, small nematic droplets can exist in the isotropic phase, even though the orientations of successive droplets are uncorrelated.

In 1971, de Gennes proposed a phenomenological description of these effects on the basis of the Landau theory of phase transitions. The point is that the free energy should include the cubic term

$$G(P, T, \eta) = G_0 + a(T - T_c)\eta^2 + C\eta^3 + B\eta^4, \tag{15.3.17}$$

where $C < 0$, still $B > 0$. T_c represents the temperature of a phase transition of second-order if $C = 0$. Now the free energy G contains a nonzero term η^3. This odd function of η ensures that the states, with some nonvanishing value of η due to some alignment of molecules, will have different free energy values depending on the direction of the alignment. A state with an order parameter η is not the same as the state with $-\eta$. We shall find such a free energy predicts a discontinuous phase transition. It can be clearly seen in Fig. 15.3.5.

Equilibrium condition $\partial G/\partial \eta = 0$ gives the equation of state

$$2a(T - T_c)\eta + 3C\eta^2 + 4B\eta^3 = 0, \tag{15.3.18}$$

the solutions are

$$\eta = 0, \tag{15.3.19}$$

$$\eta = \frac{-3C + [9C^2 - 32aB(T - T_c)]^{1/2}}{8B}, \tag{15.3.20}$$

and

$$\eta = \frac{-3C - [9C^2 - 32aB(T - T_c)]^{1/2}}{8B}. \tag{15.3.21}$$

To satisfy the real root condition, we can define a temperature limit

$$T_+ = T_c + \frac{9C^2}{32aB}. \tag{15.3.22}$$

When $T > T_+$, only $\eta = 0$ is stable. As $T < T_+$, there is a metastable minimum for $\eta \neq 0$.

The first-order phase transition point can be obtained from $G - G_0 = 0$

$$T_t = T_c + \frac{C^2}{4aB} < T_+. \tag{15.3.23}$$

The system has two stable minima at $T = T_t$; corresponding to $\eta = 0$ and $\eta \neq 0$. A true phase transition occurs at temperature T_t and there is a jump of order parameter at T_t of amount

$$\Delta\eta = -\frac{C}{2B}. \tag{15.3.24}$$

Absolute instability appears at $T \leq T_c$. Here $T = T_c$ is a spinodal point for $\eta = 0$, because $\partial^2 G/\partial\eta^2 = 0$. T_c is the absolutely unstable limit for the high symmetry phase. Taking the equilibrium condition from (15.3.18), we can find $\eta = -3C/4B$ from (15.3.20).

We have arrived at the conclusion that the presence of a cubic term in the expansion of G makes the phase transition first-order.

15.3.4 Coupling of Order Parameter with Strain

In structural phase transitions, there may appear an interplay of the strain ε with the order parameter η. An interaction of the type $\eta^2\varepsilon$ is a reasonable choice in some simple cases. We could add this and also a term representing the elastic energy to the free energy

$$G = G_0 + a(T - T_c)\eta^2 + B\eta^4 + J\eta^2\varepsilon + \frac{1}{2}K\varepsilon^2, \tag{15.3.25}$$

where J is the coupling constant and K is the elastic constant, all assumed to be independent of temperature, near T_c.

The condition for a minimum of the free energy $\partial G/\partial\eta = 0$ gives

$$a(T - T_c) + 2B\eta^2 + J\varepsilon = 0. \tag{15.3.26}$$

In addition, the equation of state for the variable ε can be obtained as

$$\sigma = \left(\frac{\partial G}{\partial \varepsilon}\right)_{\eta,T} = J\eta^2 + K\varepsilon. \tag{15.3.27}$$

In the case of no external stress, i.e. $\sigma = 0$, we get spontaneous strain below the transition temperature T_c

$$\varepsilon = -\frac{J\eta^2}{K}, \tag{15.3.28}$$

and

$$\eta^2 = -\frac{a(T - T_c)}{2B*}, \tag{15.3.29}$$

and

$$B^* = B - \frac{J^2}{2K}. \tag{15.3.30}$$

It follows that the equilibrium value ε depends linearly on temperature.

The inverse susceptibility is easily found from (15.3.27)

$$\chi^{-1} = \left(\frac{\partial\sigma}{\partial\varepsilon}\right)_{\sigma=0} = K + 2J\eta\left(\frac{\partial\eta}{\partial\varepsilon}\right)_{\sigma=0}. \tag{15.3.31}$$

From it we find that the model (15.3.25) gives a discontinuous change at the transition point

$$\chi^{-1} = K, \text{ for } T > T_c; \tag{15.3.32}$$

and

$$\chi^{-1} = K - \frac{J^2}{2B}, \text{ for } T < T_c. \tag{15.3.33}$$

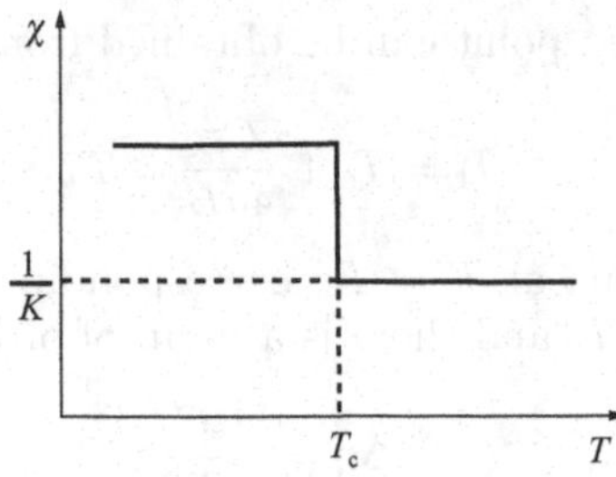

Figure 15.3.6 Susceptibility versus temperature for the model with coupling between strain and order parameter.

Figure 15.3.6 shows the temperature dependence of the susceptibility for second-order transitions described by the free energy (15.3.25). It is clear that there a finite jump at T_c.

We introduce a conjugate field h in the order parameter η, and the field is defined by

$$h = \left(\frac{\partial G}{\partial \eta}\right)_{\varepsilon,T} = 2a(T - T_c)\eta + 4B\eta^3 + 2J\eta\varepsilon. \tag{15.3.34}$$

The inverse susceptibility is

$$\chi_\eta^{-1} = 2a(T - T_c) + 12B\eta^2 + 2J\varepsilon. \tag{15.3.35}$$

Above T_c, $\eta = 0$, $\varepsilon = 0$, so

$$\chi_\eta^{-1} = 2a(T - T_c). \tag{15.3.36}$$

Below T_c, η and ε take the equilibrium values in (15.3.28) and (15.3.29), respectively, then

$$\chi_\eta^{-1} = 4a\frac{B}{B*}(T_c - T). \tag{15.3.37}$$

We see that the susceptibility satisfies the Curie–Weiss law. From the results obtained, it can be understood that the susceptibility corresponding to the parameter η diverges at the transition point T_c, whereas the susceptibility corresponding to the parameter ε remains finite within the framework of the model (15.3.25).

The statement above corresponds to the situation in which the coupling between the order parameter and strain is weak. (15.3.25) gives a description of a second-order phase transition. If the coupling is strong the situation will be different. Actually, we can substitute (15.3.28) into (15.3.25), the free energy is

$$G = G_0 + a(T - T_c)\eta^2 + B^*\eta^4. \tag{15.3.38}$$

This expression is just like the one component free energy in equation (15.2.6), but there is a substitution of $B \to B^*$. There is no doubt that if $B > B^* > 0$, the phase transition is still second-order. However, if the coupling is strong enough to lead to $B^* < B$, the high symmetry state is unstable, and a higher degree term, such as $D\eta^6$, needs to be included into the free energy. Then the coupling of η-ε may drive the phase transition from second-order to first-order.

§15.4 Change of Symmetry in Structural Phase Transitions

The symmetry consideration of Landau theory is perhaps most significant in analyzing structural phase transitions in crystals. A prerequisite for this study is the theory of space groups of crystals. As temperature decreases, a crystal can undergo a series of structural phase transitions. When T is lowered through a certain T_c, the symmetry of the crystal can be changed from high to low. A generalized order parameter with multicomponents will be used to characterized the broken symmetry in structural phase transitions.

15.4.1 Density Function and Representation Theory

Crystal symmetry is lowered by a reduction in the number of symmetry elements of both rotations and translations when passing through a structural phase transition. This reduction gives rise to a new crystal structure. The theoretical analysis consists of enumerating all the possible structural types that may be obtained from a parent crystal as a result of phase transitions and determining how low symmetry space groups are contained in the space group of the initial phase.

We can begin from the density function $\rho(\boldsymbol{r})$ to describe crystal structures and to ascertain the symmetries of crystals. For concreteness, $\rho(\boldsymbol{r})d\boldsymbol{r}$ is the probability to find the number of electrons in the volume element $d\boldsymbol{r}$ in the neighborhood of the point $\boldsymbol{r}$.

Let the initial high symmetry phase be specified by the symmetry group $\mathcal{G}_0$, which leaves the density function $\rho_0(\boldsymbol{r})$ of the system invariant. Below, but near, T_c, the density function for the low symmetry phase becomes

$$\rho(\boldsymbol{r}) = \rho_0(\boldsymbol{r}) + \delta\rho(\boldsymbol{r}), \tag{15.4.1}$$

where $\delta\rho$ is the change of density function to form the low symmetry phase. Since the state changes continuously at a second-order phase transition, the symmetry of the new phase may become lower only due to the loss of part of the symmetry elements and will be described by a group $\mathcal{G}$ that is a subgroup of the initial group $\mathcal{G}_0$, i.e., $\mathcal{G} \subset \mathcal{G}_0$.

The method of analyzing symmetry variation at a second-order phase transition, proposed by Landau, is based on expanding the density function $\rho(\boldsymbol{r})$ or $\delta\rho(\boldsymbol{r})$ in a complete set of the basis functions ψ_i^ν of the irreducible representations of the initial group $\mathcal{G}_0$,

$$\delta\rho(\boldsymbol{r}) = \sum_\nu{}' \sum_i \eta_i^\nu \psi_i^\nu, \tag{15.4.2}$$

where ν denotes the different irreducible representations (IR), and i the basis functions of the same IR. There is no identity representation in $\sum_\nu'$, because it keeps the symmetry of $\mathcal{G}_0$. In general, each second-order phase transition is related to only one IR, and the density function can be reduced to

$$\delta\rho(\boldsymbol{r}) = \sum_{i=1}^{d} \eta_i \psi_i(\boldsymbol{r}), \tag{15.4.3}$$

where d denotes the dimension of the IR. η_i is an expansion coefficient which is independent of coordinate but varies with pressure P and temperature T. It is reasonable to view the set $\{\eta_1, \ldots, \eta_d\}$ as a vector order parameter, $\boldsymbol{\eta}$, transforming according to the IR, with the same transforming properties as the basis functions. In the high symmetry phase, at $T > T_c$, all the $\eta_i = 0$, but when $T < T_c$ at least some of these coefficients should become non-zero. Since the density function varies continuously at the phase transition point as $T \to T_c$, the coefficients η_i tend to zero and may be considered small in the vicinity of T_c. The physical meaning of (15.4.3) is that the ordered phase is formed by the freezing of a particular density fluctuation of individual structure characterized by one IR of $\mathcal{G}$.

In dealing with structural phase transitions in crystals, the problem has focused on finding the IR of space group. The successful method adopts the modulated wave vector. The space group IRs are specified by wavevectors defined with the help of reciprocal lattice vectors. Using the symmetry of the reciprocal lattice, it is possible to classify all the points that belong to a Brillouin zone (BZ).

Some specific points in a Brillouin zone are called Lifshitz points and have particularly high symmetry, that is some set of symmetry elements that leave a given point fixed or transform it into an equivalent one. The construction of the IRs of space group $\mathcal{G}$ is associated with the action of its elements g on a given wavevector $\boldsymbol{k}$. Some of the elements of group $\mathcal{G}$ leave the wavevector $\boldsymbol{k}$ invariant or transform it into an equivalent one, which differs by an arbitrary reciprocal lattice vector $\boldsymbol{G}$. The point-group symmetry operation g (rotation and reflection) of Group $\mathcal{G}_0$ leaves $\boldsymbol{k}$ invariant, i.e.,

$$g\boldsymbol{k} = \boldsymbol{k}, \tag{15.4.4}$$

while the non-symmorphic operation (glide-reflection or screw rotation) changes $\boldsymbol{k}$ to an equivalent wavevector, different by a reciprocal lattice vector, i.e.,

$$g\boldsymbol{k} = \boldsymbol{k} + \boldsymbol{G}. \tag{15.4.5}$$

The totality of all such elements satisfying the above condition forms a subgroup $\mathcal{G}_{\boldsymbol{k}}$ of the group $\mathcal{G}_0$ and is called the wavevector group.

The IRs of the wavevector group $\mathcal{G}_{\boldsymbol{k}}$ are characterized by the wavevector $\boldsymbol{k}$ and the representation number ν. The IRs of the entire space group $\mathcal{G}_0$ are characterized by a wavevector star, which is labelled $\{\boldsymbol{k}\}$. A star is a set of nonequivalent wavevectors that are obtained from a given wavevector by the action of all the space group elements, and the individual vectors comprised in this set are said to be the star arms, $\boldsymbol{k}_L$. The latter can be obtained by the action of g_L elements on a given wavevector $\boldsymbol{k}$

$$\boldsymbol{k}_L = g_L \boldsymbol{k}. \tag{15.4.6}$$

These elements are representative elements of coset decomposition of group $\mathcal{G}_0$ relative to its subgroup $\mathcal{G}_{\boldsymbol{k}}$

$$\mathcal{G}_0 = \sum_{L=1}^{l_{\boldsymbol{k}}} g_L \mathcal{G}_{\boldsymbol{k}}. \tag{15.4.7}$$

The number of star arms $l_{\boldsymbol{k}}$ is evidently equal to the index of the subgroup $\mathcal{G}_{\boldsymbol{k}}$ in the group $\mathcal{G}_0$.

The basis functions of the group $\mathcal{G}_{\boldsymbol{k}}$ IR are Bloch functions of the form

$$\psi^{\nu}_{\boldsymbol{k}\lambda}(\boldsymbol{r}) = u^{\nu}_{\boldsymbol{k}\lambda}(\boldsymbol{r}) \mathrm{e}^{i\boldsymbol{k}\cdot\boldsymbol{r}}, \tag{15.4.8}$$

where $u^{\nu}_{\boldsymbol{k}\lambda}(\boldsymbol{r})$ have the periodicity of the lattice. Under the action of a group $\mathcal{G}_{\boldsymbol{k}}$ element, the set of $\psi^{\nu}_{\boldsymbol{k}\lambda}$ $(\lambda = 1, 2, \ldots, d_\nu)$ transforms according to the equation

$$g\psi^{\nu}_{\boldsymbol{k}\lambda}(\boldsymbol{r}) = \sum_{\mu=1}^{d_\nu} \Gamma^{\nu}_{\boldsymbol{k}\mu\lambda}(g) \psi^{\nu}_{\boldsymbol{k}\mu}. \tag{15.4.9}$$

A basis of the IR of the entire group $\mathcal{G}$ is generated by the set of Bloch functions $\{\psi^{\nu}_{\boldsymbol{k}_1\lambda}\}$, $\{\psi^{\nu}_{\boldsymbol{k}_2\lambda}\}, \ldots,$ $\{\psi^{\nu}_{\boldsymbol{k}_{l_{\boldsymbol{k}}}\lambda}\}$ prescribed on all star arms $\boldsymbol{k}_1, \boldsymbol{k}_2, \ldots, \boldsymbol{k}_{l_{\boldsymbol{k}}}$. Here we note that $\boldsymbol{k}$ determines the translational symmetry of ψ_i, and also $\delta\rho$, i.e., the property of the lattice of the new phase.

15.4.2 Free Energy Functional

For structural phase transition based on the density function description, the free energy functional of a crystal is written as

$$G = G(P, T, \rho(\boldsymbol{r})). \tag{15.4.10}$$

This functional form of the free energy can be transformed according to an IR as in (15.4.3). Here we fix the ψ_i and let $\{\eta_i\}$ transform under the operations of $\mathcal{G}$, then

$$G = G(P, T, \{\eta_i\}), \tag{15.4.11}$$

where η_i can be found by the equilibrium condition. Because the coefficients η_i of the basis functions of the responsible IR can be defined as a multicomponent order parameter, the number of these components is equal to the dimensionality of the responsible IR; the theoretical scheme discussed in §15.2 will be useful. It is clear that for $T \geq T_c$, $\delta\rho = 0$, then all $\eta_i = 0$. This is the high symmetry phase. However, at $T < T_c$, $\delta\rho \neq 0$, there must be at least one $\eta_i \neq 0$, and the low symmetry phase appears. As T approaches T_c, $\delta\rho \to 0, \eta_i \to 0$.

G may be expanded in powers of $\{\eta_i\}$ near the critical temperature. Since the free energy of a crystal must obviously be independent of the choice of coordinates, it must be invariant under any transformation of the coordinate system and in particular under the transformation of the group $\mathcal{G}$. Thus the expansion of G in powers of the η_i can contain in each term only an invariant combination of the η_i that is of the appropriate power. This corresponds to constructing polynomial expansions of the free energy in powers of a multicomponent order parameter. No linear invariant can be formed from quantities that are transformed according to a non-unit IR of a group, for otherwise that representation would contain the unit representation and would be reducible. The structure of the quadratic terms is determined by the fact that only one second-order invariant $\sum_i \eta_i^2$ exists for each irreducible representation according to which the quantities η_i transform under the action of the group $\mathcal{G}$ elements. Higher order, such as cubic, fourth order, etc., expansion terms are invariant polynomials of the corresponding order. If we introduce the normalization definition

$$\eta_i = \eta\gamma_i, \qquad \sum_i \gamma_i^2 = 1, \tag{15.4.12}$$

then

$$\eta^2 = \sum_i \eta_i^2. \tag{15.4.13}$$

Now $\{\gamma_i\}$ describes the symmetry of the ordered states, while the scale η is a measure of the degree of order. Above T_c, η is zero, and it increases continuously from zero when T is lowered below T_c. We expand the free energy to fouth order

$$G = G_0(P,T) + \eta^2 A(P,T) + \eta^3 \sum_\alpha C_\alpha(P,T) I_\alpha^{(3)}(\gamma_i) + \eta^4 \sum_\alpha B_\alpha(P,T) I_\alpha^{(4)}(\gamma_i), \tag{15.4.14}$$

$I_\alpha^{(3)}, I_\alpha^{(4)}$ are polynomials of the 3rd and 4th orders formed from quantities γ_i, the sum over α indicating the number of independent invariant formed by γ_i.

One may apply the same lines of argument as those used in the elementary Landau analysis enunciated in §15.2. For insight into this, we write only the leading terms in the expansion of G, which are of the form

$$G = G_0 + A(P,T) \sum_i \eta_i^2 = G_0 + A(P,T)\eta^2. \tag{15.4.15}$$

From the minimization of the free energy $\partial G/\partial \eta = 0$, and $\partial^2 G/\partial \eta^2 > 0$, we see that at $T > T_c$, the coefficient of the second-order term A should be positive, so that the equilibrium value of the parameters η_i is equal to zero; at $T < T_c$, A becomes negative, and an ordered state occurs with at least one of the η_i taking a non-zero value.

15.4.3 Landau Criteria

The general scheme of the Landau theory enables us to find all allowed ordered phases arising from a given initial phase via second-order phase transitions. The corresponding analysis reduces to the construction of an expansion of the free energy in powers of the order parameter transformed according to an IR of some group G, followed by minimization of the free energy to find the stable phase.

Landau himself proposed and solved in general form the problem that the initial group IRs cannot give rise to a second-order phase transition. As was implicit in §15.3.3, the presence in the free energy of cubic terms leads inevitably to a first-order phase transition. Therefore, the condition which restricts the list of IRs describing the second-order phase transition consists of the requirement that the IR allows no third-order invariants constituted by the coefficients η_i for the corresponding IR. In fact, from the condition of transition point itself should be stable, we have the Landau criterion

$$I_\alpha^{(3)}(\gamma_i) = 0, \tag{15.4.16}$$

or to say that it is impossible to construct third-order invariants, and the fourth-order term must be positive in (15.4.15).

We give some details of this: At $T = T_c^-$, $A(P,T_c) = 0$, so the second-order term vanishes. The free energy is

$$G = G_0 + \eta^3 C_\alpha(P,T) I_\alpha^{(3)}(\gamma_i) + \eta^4 B_\alpha(P,T) I_\alpha^{(4)}(\gamma_i). \tag{15.4.17}$$

Here, for brevity, only one third order term and one fourth order term are presented. From the equilibrium condition $\partial G/\partial\eta = 0$,

$$3I_\alpha^{(3)}(\gamma_i)C_\alpha(P,T)\eta^2 + 4I_\alpha^{(4)}(\gamma_i)B_\alpha(P,T)\eta^3 = 0, \tag{15.4.18}$$

so there are solutions

$$\eta = 0, \rightarrow G = G_0, \tag{15.4.19}$$

related to the high symmetry phase; and

$$\eta = -\frac{3I_\alpha^{(3)}(\gamma_i)C_\alpha(P,T)}{4I_\alpha^{(4)}(\gamma_i)B_\alpha(P,T)} \rightarrow G = G_0 - \frac{3^2}{4^3}\frac{[I_\alpha^{(3)}(\gamma_i)C_\alpha(P,T)]^4}{[I_\alpha^{(4)}(\gamma_i)B_\alpha(P,T)]^3} \tag{15.4.20}$$

represents the low symmetry phase. We must assume

$$B_\alpha(P,T)I_\alpha^{(4)}(\gamma_i) > 0, \tag{15.4.21}$$

otherwise for

$$B_\alpha(P,T)I_\alpha^{(4)}(\gamma_i) < 0 \ \rightarrow G > G_0,$$

or

$$B_\alpha(P,T)I_\alpha^{(4)}(\gamma_i) = 0 \ \rightarrow G \rightarrow -\infty,$$

both of these are unreasonable, so there is no stable solution. Thus, it is clear at $T = T_c^-, \eta \neq 0$, the order parameter changes from 0 to $-3I_\alpha^{(3)}C_\alpha/4I_\alpha^{(4)}B_\alpha$ discontinuously. This does not suggest a second-order phase transition, unless $3I_\alpha^{(3)}C_\alpha = 0$. Because $C_\alpha(P,T) = 0$ is not a general case, we require that (15.4.16) is satisfied, and

$$G = G_0 + \eta^2 A(P,T) + \eta^4 \sum_\alpha B_\alpha(P,T)I_\alpha^{(4)}(\gamma_i). \tag{15.4.22}$$

We have arrived at two Landau criteria for second-order phase transitions. The first is that the group $\mathcal{G}$ for the low symmetry phase is the subgroup of the initial group $\mathcal{G}_0$ for the high symmetric phase. The second is that there is no third-order invariant in the free energy functional.

15.4.4 Lifshitz Criterion

The original Landau theory assumed that the ordered phase arising from the phase transition be homogeneous. Lifshitz demonstrated that there may be spatially inhomogeneous phases occurring, if the free energy involves terms containing order parameter derivatives with respect to the coordinates. Linear invariants in derivatives have come to be called Lifshitz invariants.

We have so far studied the situation where the ordering is uniform throughout the medium, in which η has the same value everywhere. When we consider the situation where the thermodynamic fluctuations play an important role, we need to introduce the density of Gibbs free energy g,

$$g = g(P,T,\eta_i(\boldsymbol{r}),\nabla\eta_i(\boldsymbol{r})). \tag{15.4.23}$$

Here for simplicity, only the first derivative term is included, $\nabla\eta_i(\boldsymbol{r})$. Note that $\eta_i(\boldsymbol{r})$ is now a local quantity. The free energy of the system is

$$G = \int g(P,T,\eta_i(\boldsymbol{r}),\nabla\eta_i(\boldsymbol{r}))d\boldsymbol{r}. \tag{15.4.24}$$

We can still use the picture of a modulated wavevector. In the case of a solid that has lost spatial homogeneity, there will be a response of the phase transition to continuous change of wavevector $\boldsymbol{k}$. For a homogeneous phase transition, at the transition point T_c there is only one characteristic wavevector, $\boldsymbol{k} = \boldsymbol{k}_0$ to satisfy $A(\boldsymbol{k}_0) = 0$, and G will be a minimum for stability of the new phase.

The free energy of the new phase with periodicity corresponding to $\boldsymbol{k}_0$ should be a minimum around k_0. However, if inhomogeneity is to appear, we may consider

$$\boldsymbol{k} = \boldsymbol{k}_0 + \boldsymbol{\kappa}, \tag{15.4.25}$$

$\boldsymbol{\kappa}$ is a small quantity, and $1/\boldsymbol{\kappa}$ offers a spatial modulation, i.e., a kind of macroscopic inhomogeneity. Then the order parameter η becomes a slowly varying function spatially, so the free energy will contain terms composed of $\partial\eta_i/\partial x_p$, and $\eta_j\partial\eta_i/\partial x_p$, where $i, j = 1, 2, \ldots, d$ and $p = 1, 2, 3$ denote the components of the order and space, respectively. In this first-order approximation, the density of free energy takes the form

$$\begin{aligned} g\left(P, T, \eta_i, \frac{\partial \eta_i}{\partial x_p}\right) = {} & g_0(P, T, \eta_i) + \sum_{ip} U_{i,p}(P,T)\frac{\partial \eta_i}{\partial x_p} \\ & + \frac{1}{2}\sum_{ijp} V_{ijp}(P,T)\left[\eta_i \frac{\partial \eta_j}{\partial x_p} + \eta_j \frac{\partial \eta_i}{\partial x_p}\right] \\ & + \frac{1}{2}\sum_{ijp} V_{ijp}(P,T)\left[\eta_i \frac{\partial \eta_j}{\partial x_p} - \eta_j \frac{\partial \eta_i}{\partial x_p}\right] + \cdots \end{aligned} \tag{15.4.26}$$

where g_0 does not include the derivatives of the order parameter. The expansion coefficients are defined as

$$U_{ip}(P,T) = \frac{\partial G}{\partial(\partial \eta_i/\partial x_p)}, \tag{15.4.27}$$

which equals zero, due to the equilibrium condition, and

$$V_{ijp}(P,T) = \frac{\partial^2 G}{\partial \eta_i \partial(\partial \eta_i/\partial x_p)}. \tag{15.4.28}$$

It is clear that

$$\int \left(\eta_i \frac{\partial \eta_j}{\partial x_p} + \eta_j \frac{\partial \eta_i}{\partial x_p}\right) dx_p \sim \eta_i \eta_j,$$

which can be included into the first term of (15.4.26). So the total free energy is

$$G = \int g d\boldsymbol{r} = \int g_0 d\boldsymbol{r} + \frac{1}{2}\sum_{ijp} V_{ijp}(P,T)\int \left(\eta_i \frac{\partial \eta_j}{\partial x_p} - \eta_j \frac{\partial \eta_i}{\partial x_p}\right) d\boldsymbol{r}. \tag{15.4.29}$$

Here the term $(\eta_i\partial\eta_j/\partial x_p - \eta_j\partial\eta_i/\partial x_p)$ cannot be transformed into $\eta_i\eta_j$ after integration and will play an important role in the form of inhomogeneous structures.

From the stability condition of $\partial\eta_i/\partial x_p$ as an independent variable,

$$\frac{\delta G}{\delta(\partial \eta_j/\partial x_p)} = \sum_i V_{ijp}\eta_i(\boldsymbol{r}) = 0 \ (i, j = 1, \ldots, d). \tag{15.4.30}$$

This is a set of linear equations: In the low symmetry phase, $\eta_i(\boldsymbol{r})$ are not all zero, so for fixed p $(p = 1, 2, 3)$, the coefficient matrix $V^{(p)} = \{V_{ijp}\}$ should satisfy

$$\det[V^{(p)}] = 0. \tag{15.4.31}$$

Because V_{ijp} is a function of P and T, it is accidental that $\det[V^{(p)}] = 0$. In general, we may require

$$\sum_{ijp} V_{ijp}\int dx_p \left[\eta_i \frac{\partial \eta_j}{\partial x_p} - \eta_j \frac{\partial \eta_i}{\partial x_p}\right] = 0. \tag{15.4.32}$$

This is the Lifshitz criterion, which means that the nonexistence of the Lifshitz invariant is the condition for a phase transition between two homogeneous phases to be possible.

The Lifshitz condition is also called the homogeneous condition; it eliminates the possibility of transition from high temperature homogeneous phase to lower temperature inhomogeneous phase. Lifshitz also proved that only the Γ point, and the end points with high symmetry of $\boldsymbol{k}$ at the boundary of the Brillouin zone, relate to second-order phase transitions. According to his argument, the wavevector $\boldsymbol{k}$ of the new phase is a simple fraction of the wavevector of the initial phase. We can write

$$\boldsymbol{a}_i' = \sum_{ij} l_{ij}\boldsymbol{a}_j \ (i,j=1,2,3), \tag{15.4.33}$$

where $\boldsymbol{a}_i$ is a basis vector of the initial structure, and $\boldsymbol{a}_j'$ that of the new phase and l_{ij} is an integer.

However, when the Lifshitz condition is unfulfilled, then $G < \int g_0 d\boldsymbol{r}$, and the inhomogeneous phase may have a lower free energy than the homogeneous phase. In this case, the state of the crystal should be stable with respect to loss of macrohomogeneity, and an incommensurate phase may appear. We shall discuss the commensurate-incommensurate phase transition in §16.2.3.

By the way, we note that (15.4.29) is related to the fact that G is only expanded to first-order derivatives $\partial\eta/\partial x_p$, so the Lifshitz condition is not very strict. Higher order expansion may lead to domain structure. One of the examples is the Ginzburg–Landau free energy density

$$g(P,T,\eta,\nabla\eta) = g_0(P,T,\eta) + K(\nabla\eta)^2, \tag{15.4.34}$$

where $K > 0$. The total free energy is

$$G = \int g(P,T,\eta,\nabla\eta)d\boldsymbol{r}, \tag{15.4.35}$$

from its minimum, the distribution of order parameter, and domain structure can be determined.

However, besides the criteria due to Landau and Lifshitz discussed above, there may be other criteria that will be neglected here.[a] On the other hand, the Landau theory of phase transitions also has been fruitfully extended to another type of structural phase transitions — reconstructive phase transitions, in which the group-subgroup relationship is entirely missing by retaining only the concept of the order parameter.[b]

Bibliography

[1] Landau, L. D., and E. M. Lifshitz, *Statistical Physics I*, Pergomon Press, Oxford (1980). §15.

[2] Tolédano, J. C., and P. Tolédano, *The Landau Theory of Phase Transitions*, World Scientific, Singapore (1987).

[3] Izyumov, Y. A., and V. N. Syromyatnikov, *Phase Transitions and Crystal Symmetry*, Kluwer Academic, Dordrecht (1990).

[4] Yeomans, J. M., *Statistical Mechanics of Phase Transitions*, Clarendon Press, Oxford (1992).

[5] Fujimoto, M., *The Physics of Structural Phase Transitions*, Springer-Verlag, New York (1997).

[6] Blinc, R., and B. Zeks, *Soft Modes in Ferroelectrics and Antiferroelectrics*, North-Holland, Amsterdam (1974).

[a] Readers may consult the monograph by Y. A. Izyumov and V. N. Syromyatnikov, *Phase Transitions and Crystal Symmetry*, see Bib. [3].

[b] Readers may consult the monograph, P. Tolédano and V. Dmitriev, *Reconstructive Phase Transitions*, World Scientific, Singapore (1996).

Chapter 16

Crystals, Quasicrystals and Liquid Crystals

In Part I we discussed the structures of crystals, quasicrystals, and liquid crystals. They are all the result of broken spatial translational and orientational symmetries. In this chapter we would like to go beyond geometry, in order to clarity the physical reason for their formation. Landau theory may be the first step in this direction, but more microscopic theories are needed for further elucidation.

§16.1 Liquid-Solid Transitions

Starting from a homogeneous and isotropic liquid and lowering the temperature gradually, mass or compositional density waves will arise in the liquid. Below a certain temperature, some density wave modes are locked in and an ordered solid is formed. One would like to explain the structure from first-principle calculations, taking into account the actual electronic properties of the constituent atoms. However, such a calculation of crystal stability is very elaborate, and some understanding of the relative stability of solids and liquids can be obtained by using postulated periodic density waves within the framework of the phenomenological Landau theory of phase transitions.

16.1.1 Free Energy Expansion Based on Density Waves

Consider a two- or three-dimensional liquid which has full translational and rotational symmetries corresponding to the Euclidean group. The liquid phase may condense into the solid phase, and we would ask what are the possible ordered structures that arise at low temperatures. To simplify this problem, we may ignore the difference in mean density between liquid and solid, so the Gibbs free energy may be replaced by the Helmholtz free energy F.

In the homogeneous and isotropic liquid phase, the density function ρ_0 is a constant. When temperature decreases, the original higher symmetry is broken. At the point of transition, $\rho_0 \rightarrow \rho_0 + \delta\rho = \rho$, where $\delta\rho$ and ρ have the symmetry of the ordered solid. According to Landau theory, the condensed phase is described by a symmetry-breaking order parameter which transforms as an irreducible representation of the symmetry group of the liquid phase. Due to the translational symmetry, the irreducible representation is labeled by the wavevectors $\boldsymbol{q}$, and the density of the low-temperature ordered phase is written

$$\rho(\boldsymbol{r}) = \rho_0 + \sum_{\boldsymbol{q}} \rho_{\boldsymbol{q}} e^{i\boldsymbol{q}\cdot\boldsymbol{r}}. \tag{16.1.1}$$

The complex constants $\rho_{\boldsymbol{q}}$, labelled by the wavevectors $\boldsymbol{q}$, are the order parameters of the phase transition. $\rho(\boldsymbol{r})$ is real, so

$$\rho_{\boldsymbol{q}} = \rho^*_{-\boldsymbol{q}}, \tag{16.1.2}$$

(where the symbol * denotes complex conjugate). To determine which structure may actually become stable, the free-energy of the system F is expanded in terms of the possible order parameters $\rho_{\boldsymbol{q}}$. Because of the rotational symmetry, the free energy depends only on the magnitude $|\boldsymbol{q}|$ and not on its direction. In general, the order parameters corresponding to wavevectors with a single length are important. It is reasonable to fix $\boldsymbol{q}$ to $\boldsymbol{G}$, where $\boldsymbol{G}$ are the reciprocal lattice vectors of the solid, and $\rho_{\boldsymbol{G}}$ is the Fourier component of the density.

The free energy of the solid is a functional of $\rho(\boldsymbol{r})$, i.e., $F = F(P, T, \rho(\boldsymbol{r}))$. Near the transition point, F can be expanded in powers of $\rho_{\boldsymbol{G}}$, such as

$$F = F_0 + F_1 + F_2 + F_3 + F_4 + \cdots, \tag{16.1.3}$$

where F_0 is the free energy of the liquid phase and F_n for $n \neq 0$ contains terms with $\rho_{\boldsymbol{G}_1}\rho_{\boldsymbol{G}_2}\cdots\rho_{\boldsymbol{G}_n}$. It is easy to see that the permissible F_n can only include terms satisfying

$$\boldsymbol{G}_1 + \boldsymbol{G}_2 + \cdots + \boldsymbol{G}_n = 0. \tag{16.1.4}$$

Actually, F should not change under any translation of the origin of the coordinates, i.e. under coordinate transformation $\boldsymbol{r} \to \boldsymbol{r} + \boldsymbol{R}$,

$$\rho_{\boldsymbol{G}_1}\rho_{\boldsymbol{G}_2}\cdots\rho_{\boldsymbol{G}_n} = \rho_{\boldsymbol{G}_1}\rho_{\boldsymbol{G}_2}\cdots\rho_{\boldsymbol{G}_n}\exp\{i(\boldsymbol{G}_1 + \boldsymbol{G}_2 + \cdots + \boldsymbol{G}_n)\cdot\boldsymbol{R}\}.$$

Because $\boldsymbol{R}$ is chosen as an arbitrary constant vector, (16.1.4) must be fulfilled.

(16.1.4) gives a basic relation to constrain the possible wavevectors. Taking $n = 1$, we have for the first-order term $\boldsymbol{G}_1 = 0$, so $F_1 = 0$. This is consistent with the minimum of free energy given in last chapter. For the second-order terms, $n = 2$, $\boldsymbol{G}_1 = -\boldsymbol{G}_2$, so F_2 satisfies

$$F_2 = \sum_{\boldsymbol{G}} A_{\boldsymbol{G}}|\rho_{\boldsymbol{G}}|^2, \tag{16.1.5}$$

where $A_{\boldsymbol{G}}$ are constants depending on pressure P and temperature T as well as $\boldsymbol{G}$. Because of the isotropy of the liquid, the quantities $A_{\boldsymbol{G}}$ depend only on the magnitude, but not the direction, of the vector $\boldsymbol{G}$. On the other hand, near the transition point, we can expect that density waves arise that correspond only to plane waves with one definite wavelength, and $A_{\boldsymbol{G}}$ will have a minimum. Designating coefficient $A_{\boldsymbol{G}}$ simply by A, we have

$$F_2 = A\sum_{\boldsymbol{G}} |\rho_{\boldsymbol{G}}|^2, \tag{16.1.6}$$

where the summation is over $\boldsymbol{G}$ with only different direction.

The third-order terms have the form

$$F_3 = \sum_{\boldsymbol{G}_1\boldsymbol{G}_2\boldsymbol{G}_3} C_{\boldsymbol{G}_1\boldsymbol{G}_2\boldsymbol{G}_3}\rho_{\boldsymbol{G}_1}\rho_{\boldsymbol{G}_2}\rho_{\boldsymbol{G}_3}, \tag{16.1.7}$$

where in every term

$$\boldsymbol{G}_1 + \boldsymbol{G}_2 + \boldsymbol{G}_3 = 0. \tag{16.1.8}$$

But, as has just been pointed out, near the transition point, density waves should have the same period. Therefore, in the third-order terms only those $\boldsymbol{G}_1, \boldsymbol{G}_2, \boldsymbol{G}_3$ which have the same absolute magnitude and differ only in direction take part. (16.1.8) means therefore that $\boldsymbol{G}_1, \boldsymbol{G}_2, \boldsymbol{G}_3$ should form an equilateral triangle. In all third-order terms these triangles have equal size, because the quantity $\boldsymbol{G}$ is determined by the second-order term, and differ only in their orientation in space. Because of the isotropy of the liquid the coefficients $C_{\boldsymbol{G}_1\boldsymbol{G}_2\boldsymbol{G}_3}$ can depend only on the sizes, but not

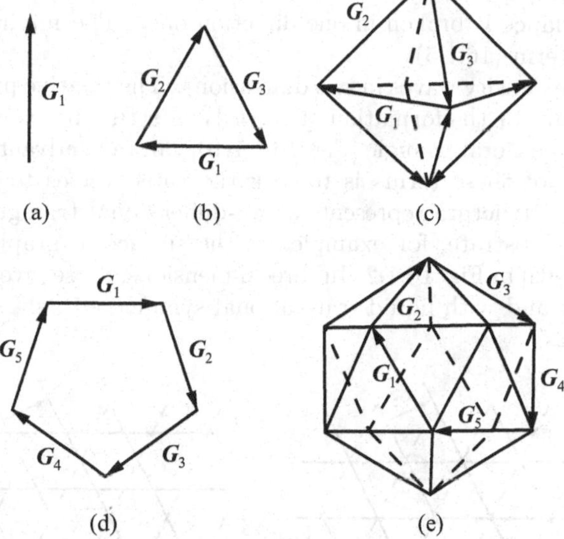

Figure 16.1.1 Wavevector combinations representing (a) smectic structure, (b) rodlike triangular structures or triangular atomic monolayers, (c) bcc structures, (d) two-dimensional Penrose structures or three-dimensional rodlike lyotropic structures, and (e) icosahedral quasicrystals.

on the orientations, of these triangles. Therefore all $C_{G_1G_2G_3}$ in the third-order terms are equal, their common value is denoted by C here. In this way, we write

$$F_3 = C \sum_{G_1G_2G_3} \rho_{G_1}\rho_{G_2}\rho_{G_3}, \tag{16.1.9}$$

where the summation is over $\boldsymbol{G}_1, \boldsymbol{G}_2, \boldsymbol{G}_3$.

In the same way F_4 and F_5, etc. can also be written out. Finally, the free energy expanded to fifth order takes the form as follows

$$\begin{aligned} F = F_0 + A\sum_{\boldsymbol{G}} |\rho_{\boldsymbol{G}}|^2 + C \sum_{|\boldsymbol{G}_i|=G} \rho_{G_1}\rho_{G_2}\rho_{G_3} \\ + B \sum_{|\boldsymbol{G}_i|=G} \rho_{G_1}\rho_{G_2}\rho_{G_3}\rho_{G_4} \\ + E \sum_{|\boldsymbol{G}_i|=G} \rho_{G_1}\rho_{G_2}\rho_{G_3}\rho_{G_4}\rho_{G_5}. \end{aligned} \tag{16.1.10}$$

From this we can discuss the stability of a variety of structures. The wavevector combinations of some possible structures are shown in Fig. 16.1.1 and will be analyzed in the following two subsections. We shall see that the third-order terms in (16.1.10) are essential for some liquid-solid transitions. The third-order terms violate the Landau criteria for continuous phase transitions, so liquid-solid transitions are first order. However, in these first-order phase transitions, the Landau theory is still valid.

16.1.2 Crystallization

As the simplest example, we first consider a single density wave

$$\rho(\boldsymbol{r}) = \frac{1}{\sqrt{2}}\rho\cos(\boldsymbol{G}\cdot\boldsymbol{r}), \tag{16.1.11}$$

which describes a smectic liquid crystal with wavevector $\boldsymbol{G}$ as well as $-\boldsymbol{G}$, as shown in Fig. 16.1.1(a). The translational invariance is broken in one direction only. The minimum of free energy must be fulfilled by the second term (16.1.5).

Next we consider the density waves in two dimensions. The relative phases of the different density waves are very important for the formation of crystals. A structure composed by superposing three waves which form an equilateral triangle [Fig. 16.1.1(b)] can take advantage of the third-order terms in (16.1.10). The role of these terms is to lock the three waves together. In two dimensions, the resulting "triple-$\boldsymbol{G}$" structure represents a two-dimensional triangular (or honeycomb) crystal absorbed on a smooth substrate, for example, on the surface of graphite, xenon atoms can form triangular lattice, as shown in Fig. 16.1.2. In three dimensions, these give rodlike structures with two-dimensional periodicity and with liquid translational symmetry in the third direction, as observed for lyotropic mesophases.

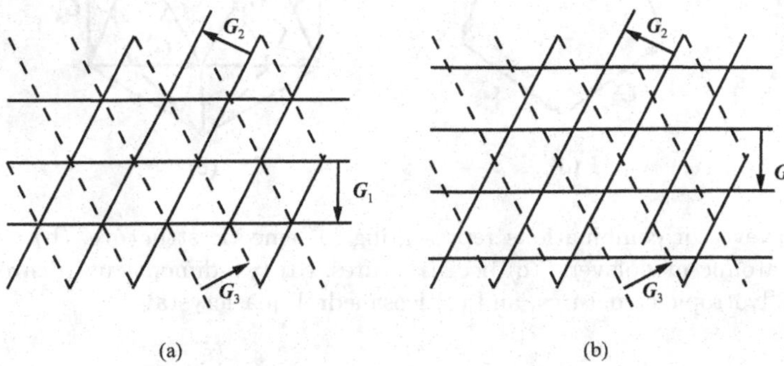

Figure 16.1.2 Density waves for two-dimensional crystals.

For isotropic and homogeneous media, there are many choices for reciprocal lattice vectors, so many kind of three-dimensional crystals can be formed. The actual lattice is dependent on the combination of the coefficients in the free energy F. It has been known that the metallic elements on the left side of the periodic table of the elements, i.e., the elements of groups IA, IIA, IIIB-VIB, except Mg, and almost all the lanthanides and actinides, when near or lower than the melting curve, are all bcc structure. At lower temperature, a large number of them change into other structures. Among the high temperature phases, more than 40 elements have the bcc structure. Thus there must be a general factor which controls the formation of the bcc structure. These results can be understood by considering the symmetry when a liquid is transformed into a solid. To generalize the Landau theory, one can directly conclude that the bcc structure will appear first when the first-order behavior of the phase transition near the melting curve is not pronounced. This is possible for many metals, because when a phase transition takes place in a metal, its volume changes very little, and the latent heat is less than $k_{\rm B}T$, so it is a weak first-order transition. This provides an illustration for the formation of bcc structure.

S. Alexander and J. McTaque pointed out that,[a] if six pairs of $\pm\boldsymbol{G}_i$ form an octahedron, as shown in Fig. 16.1.1(c), the free energy, in general, decreases, and this leads to the formation of a three-dimensional body-centered cubic structure. Here $\rho_{\boldsymbol{G}_i}$ could be written as $\rho_{\boldsymbol{G}_i} = (1/2\sqrt{6})\rho\exp(i\theta_i)$, and the density is

$$\rho(\boldsymbol{r}) = \sum_{\text{octa}} \frac{\rho}{\sqrt{6}} \cos(\boldsymbol{G}_i \cdot \boldsymbol{r} + \theta_i), \tag{16.1.12}$$

where the sum includes each pair of vectors in the octahedron, and the higher order terms are omitted. An octahedron has four pairs of triangular faces, each of which gives a contribution to the free energy, so one expects, in general, the bcc structure to have lower free energy than the rodlike lyotropic structure. Not all six pairs of vectors $\boldsymbol{G}_i$ are linearly independent; they can all be formed by linear combinations of three vectors. The third-order term of the free energy takes the form

$$F_3 = \sum \frac{C}{6\sqrt{6}} \rho^3 \cos(\theta_i + \theta_j + \theta_k), \tag{16.1.13}$$

[a] S. Alexander and J. McTaque, *Phys. Rev. Lett.* **41**, 702 (1978).

and the free energy can be minimized by choosing

$$\theta_i + \theta_j + \theta_k = \pi p,$$

where p is an integer. Only three of the four constraints are linearly independent, so there are only three degrees of freedom leaving the free energy invariant

$$F_3 = -\frac{2C}{3\sqrt{6}}\rho^3. \tag{16.1.14}$$

From this theory, we can understand why the high temperature solid phase of almost all metallic elements is bcc.

By the way, it should be noted that there is a kind of mean-field theory for liquid-solid phase transitions — density functional theory in which the difference in density between liquid and solid is taken seriously.[b]

16.1.3 Quasicrystals

As stated in Chap. 2, the experiment on $Al_{86}Mn_{14}$ alloy showed a diffraction spectrum with two- three- and five-fold symmetry, the apparent point group symmetry is icosahedral. The icosahedral ordering can be described as a 6-$\boldsymbol{G}$ structure corresponding to a six-dimensional space group formed by superposition of compositional density waves. All 6D space-group operations describe actual symmetry operations in the real 3D crystals. The diffraction spots can be labelled by six Miller indices $(n_1, \ldots, n_6)$, and the pattern is spanned by six linearly independent reciprocal lattice vectors $\boldsymbol{G}_1, \ldots, \boldsymbol{G}_6$. The actual atomic or electronic densities can be thought of as superpositions of six density waves with wavevectors $\boldsymbol{G}_i$, and higher harmonics. Only one length scale is involved since $|\boldsymbol{G}_1| = |\boldsymbol{G}_2| = \cdots = |\boldsymbol{G}_6|$. Comparison with experiments indicates that the actual 6D space group of the Mn-Al alloy is the simple-cubic version.

The Landau theory allows for the existence and complete stability of systems with icosahedral symmetry; S. Alexander and J. McTague (1978), in fact, predicted the existence of icosahedral structure. Anyhow, the melting transition of icosahedral structures is first order.

In addition to the three-dimensional icosahedral case, a slightly simpler two-dimensional structure is formed by superposition of five density waves with wavevectors $\boldsymbol{G}_1, \ldots, \boldsymbol{G}_5$ to form a regular pentagon. The resulting ordered structure has five-fold rotational symmetry, but no discrete translational invariance. We call this crystal a generalized Penrose structure. The Penrose structure is characterized by a five-dimensional space group which gives a description of fivefold rotation in 5D space. An example of a Penrose structure is the decagonal phase of 2D quasicrystals.[c]

The fifth-order term in (16.1.10) favors a two-dimensional structure composed of five density waves with wavevectors forming regular pentagon as shown in Fig. 16.1.1(d). Writing $\rho_i = (1/2\sqrt{5})\rho\exp(i\theta_i)$, the density becomes

$$\rho(\boldsymbol{r}) = \sum_{i=1}^{5} \frac{\rho}{\sqrt{5}}\cos(\boldsymbol{G}_i \cdot \boldsymbol{r} + \theta_i), \tag{16.1.15}$$

and the fifth-order term of the free energy takes the form

$$F_5 = \frac{E}{25\sqrt{5}}\rho^5\cos(\theta_1 + \theta_2 + \theta_3 + \theta_4 + \theta_5). \tag{16.1.16}$$

If E is positive the minimum of the free energy is

$$\left(F_5^{\text{pent}}\right)_{\min} = -\frac{E}{25\sqrt{5}}\rho^5. \tag{16.1.17}$$

---

[b]T. V. Ramakrishnan and M. Yussouff, *Phys. Rev. B* **19**, 2775 (1979).

[c]For a detailed discussion of the stability of Penrose and icosahedral structures in terms of Landau theory see P. Bak, *Phys. Rev. Lett.* **54**, 1517 (1985); *Phys. Rev. B* **32**, 5764 (1985).

In contrast to the situations for the triangular 2D case and the bcc 3D case, these operations cannot be represented by two-dimensional translations. This is related to the fact that the five vectors $\boldsymbol{G}_i$ cannot be formed as linear combinations of two vectors spanning a regular reciprocal lattice. Four of the vectors are linearly independent. The resulting $\rho(\boldsymbol{r})$ for $\theta_i = 0$ has fivefold symmetry, but does not form a regular space-filling Bravais lattice. Such structures can be called generalized Penrose structures as an extension of original Penrose tilings. For $\theta_i = 0$ the structure actually has tenfold symmetry, since $\theta_i \to \theta_i + \pi$ leaves it invariant. Figure 2.4.6 shows the symmetry. The straight lines represent maxima of the individual density waves, so that at the center $\boldsymbol{r} = 0$ the density is maximized since all the waves have maxima at this point, which could represent actual atoms.

AlMn quasicrystals were first produced during crystallization of a melt via a first-order phase transition when the melt was subjected to sufficiently rapid quenching. The quasicrystalline state was shown to be a sufficiently stable metastable state. We must heat this quasicrystal to 400°C for more than an hour to make it transform to the usual crystal phase Al_6Mn_4. Subsequently mm-scale AlLiCu quasicrystals were discovered by traditional melt-casting; this is a thermodynamic equilibrium phase with defects. A new generation of icosahedral quasicrystals, such as AlFeCu, AlRuCu, and AlPdMn has equilibrium phases with fully ordered structures.

It is interesting to note that in (16.1.10) the fifth order terms in combination with the third order terms favor more complicated structures in three dimensions composed of wavevectors forming regular icosahedra as shown in Fig. 16.1.1(e). An icosahedron has twenty regular triangular faces, twelve corners, and thirty edges. The 15 pairs of edge vectors $\pm\boldsymbol{G}_i$ define a structure

$$\rho(\boldsymbol{r}) = \sum_i \frac{\rho}{\sqrt{15}} \cos(\boldsymbol{G}_i \cdot \boldsymbol{r}_i + \theta_i), \tag{16.1.18}$$

and the third- and fifth-order terms of free energy become

$$F_3 = \frac{\rho^3 C}{15\sqrt{15}} \sum_{10 \text{ triangles}} \cos(\theta_i + \theta_j + \theta_k), \tag{16.1.19}$$

and

$$F_5 = \frac{\rho^5 E}{225\sqrt{15}} \sum_{6 \text{ pentagons}} \cos(\theta_i + \theta_j + \theta_k + \theta_l + \theta_m). \tag{16.1.20}$$

If the signs of C and E are the same, the minimum of $F_3 + F_5$ is located at $\theta_i = 0$ or $\theta_i = \pi$. For instance, when C and E are all positive, the resulting free energy becomes

$$(F_3 + F_5)_{\min} = -\frac{2\rho^3 C}{3\sqrt{15}} - \frac{2\rho^5 E}{75\sqrt{15}}, \tag{16.1.21}$$

which, for small C/E, can become favorable compared with both the free energy of the bcc structure in (16.1.12) and the free energy of the Penrose structure in (16.1.15), which in three dimensions is an icosahedral structure.

The discussion above is purely phenomenological, and cannot be used to predict the existence of icosahedral structures in any given material. However, the Landau theory allowed us to show that the icosahedral structures may, in principle, be stable in some circumstances. By the way, the fourth-order term in the free energy has been ignored so far, or that is to say, it was assumed that this term has the same value for the bcc phase and the icosahedral phase. In practice, this term will contribute to the stability of the related phases.

Up to now, we have considered a set of vectors all having the same modulus corresponding to the edges of a regular icosahedron. Alternatively, we can adapt the 12 vectors joining the center of the icosahedron to its 12 vertices. Their modulus is a little different from the one of the edge vectors. Geometrically, each edge vector is an integral combination of two vertex vectors. Several models have been proposed, for the stabilization of the icosahedral phase by considering both the vertex and the edge vectors, in order to give a more rigorous illustration of the diffraction patterns.

This analysis of quasicrystals by Landau's thermodynamic theory is based on the postulate that, over a certain temperature interval, the quasicrystalline state may have a lower free energy as compared with the usual crystalline state with a bcc structure. The icosahedral symmetry of quasicrystal structure thus obtained is the unique representative of quasi-crystallographic symmetry of three-dimensional quasicrystals, including simple and body-centered two space groups, for example, AlMn and AlLiCu belong to the former, while AlFeCu and AlPdMn belong to the latter. They are somewhat different in structure.

§16.2 Phase Transitions in Solids

The ordered structures formed under the melting line are unstable as the temperature is lowered continuously. The symmetry will be broken further. So phase transitions may take place from solid to solid, with change of symmetry. There are abundant phase transitions in solids: We will discuss three typical examples in this section.

16.2.1 Order-Disorder Transition

We have already discussed in §3.1 the order-disorder transition in substitutional binary alloys which have periodic lattices.

The mutual solubility of two metals capable of forming an alloy can be described in terms of a simple model which assumes that the cohesive energy is the sum of interactions between nearest-neighboring sites, such as $\varepsilon_{\mathrm{AB}}$, $\varepsilon_{\mathrm{AA}}$ and $\varepsilon_{\mathrm{BB}}$. For $T = 0$ K, (a) if $\varepsilon_{\mathrm{AB}} > (\varepsilon_{\mathrm{AA}} + \varepsilon_{\mathrm{BB}})/2$, then the case where all A atoms are separated from the B atoms is energetically more favorable; (b) if $\varepsilon_{\mathrm{AB}} < (\varepsilon_{\mathrm{AA}} + \varepsilon_{\mathrm{BB}})/2$, the case of A atoms mixed with B atoms is dominant. These are the ordered states. At higher temperatures, entropy will play a prominent role in mixing the two types of atoms on lattice sites, so a disordered state is established. At some critical temperature T_{c}, a order-disorder phase transition will take place.

Bragg and Williams (1934) proposed the first satisfactory theoretical model which described the order-disorder transition in alloys, based upon a mean-field approach.

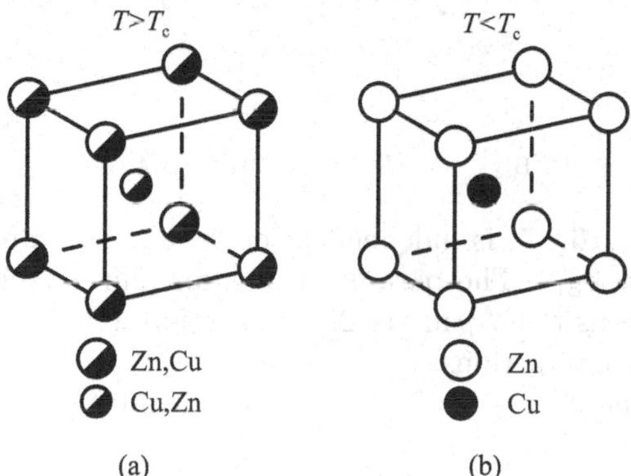

Figure 16.2.1 Unit cell of CuZn alloy in the ordered and disordered phases.

Consider a simple case such as that of β-brass (CuZn), the unit cell of which has been depicted in Fig. 16.2.1. In the disordered phase, each atomic position can be occupied by either A or B atoms with equal probability. In the ordered phase, there are two interpenetrating simple cubic lattice in which each atom A is surrounded by eight nearest neighbor atoms B and *vice versa.*

In binary alloys AB such as CuZn (β brass), atomic ordering takes place below the transition temperature T_{c}, arising from a diffusive rearrangement of atoms among lattice sites. Since the process is slow (often quasistatic), the rearrangement can be described by a variable σ_i defined by the difference

$$\sigma_i = p_i(\mathrm{A}) - p_i(\mathrm{B}), \tag{16.2.1}$$

where $p_i(\mathrm{A})$ and $p_i(\mathrm{B})$ are local probabilities for the site i to be occupied by an atom A and by an atom B, respectively, and satisfying

$$p_i(\mathrm{A}) + p_i(\mathrm{B}) = 1. \tag{16.2.2}$$

The macroscopic order parameter η is given by the spatial average

$$\eta = \langle \sigma_i \rangle = \frac{1}{N} \sum_i \sigma_i, \tag{16.2.3}$$

where the summation is taken over the whole subsystem.

We can formulate the pseudospin model to represent probabilities for ordering in binary systems. It is reasonable to assume that the correlation energy in a correlated system is generally expressed by the Hamiltonian

$$\mathcal{H} = -\sum_{ij} J_{ij} \sigma_i \sigma_j, \tag{16.2.4}$$

where J_{ij} is a parameter for the magnitude of the correlation between σ_i and σ_j, and the negative sign is attached for convenience. In the following, we shall show that (16.2.4) can be derived from a physical description of the short-range interactions in crystals.

The short-range correlation energy E_i, arising from those interactions between the site i and the neighboring sites j, can be expressed in terms of local probabilities, $p_i(\mathrm{A})$, $p_i(\mathrm{B})$, $p_j(\mathrm{A})$, $p_j(\mathrm{B})$. Namely,

$$E_i = \sum_j p_i(\mathrm{A})p_j(\mathrm{A})\varepsilon_{\mathrm{AA}} + p_i(\mathrm{B})p_j(\mathrm{B})\varepsilon_{\mathrm{BB}} + p_j(\mathrm{A})p_i(\mathrm{B})\varepsilon_{\mathrm{AB}} + p_j(\mathrm{B})p_i(\mathrm{A})\varepsilon_{\mathrm{BA}}. \tag{16.2.5}$$

According to (16.2.1) and (16.2.2),

$$p_i(\mathrm{A}) = \frac{1}{2}(1+\sigma_i), \qquad p_i(\mathrm{B}) = \frac{1}{2}(1-\sigma_i). \tag{16.2.6}$$

Substituting these into (16.2.5), the energy E_i can be expressed in terms of the order variables σ_i and σ_j, i.e.,

$$E_i = \sum_j \mathrm{const.} - K(\sigma_i + \sigma_j) - J\sigma_i\sigma_j, \tag{16.2.7}$$

where

$$\mathrm{const.} = \frac{1}{2}(2\varepsilon_{\mathrm{AB}} + \varepsilon_{\mathrm{AA}} + \varepsilon_{\mathrm{BB}}), \qquad K = \frac{1}{4}(\varepsilon_{\mathrm{BB}} - \varepsilon_{\mathrm{AA}}), \qquad J = \frac{1}{4}(\varepsilon_{\mathrm{AA}} + \varepsilon_{\mathrm{BB}} - 2\varepsilon_{\mathrm{AB}}).$$

The first constant term in (16.2.7) is independent of the ordering process, while K is zero for most binary alloys where $\varepsilon_{\mathrm{AA}} \approx \varepsilon_{\mathrm{BB}}$. The parameter J is for pair correlations with nearest neighbors, and is essentially the same as the J_{ij} in (16.2.4). Therefore (16.2.4) and (16.2.7) are considered as expressions for pseudospin interactions.

Considering z nearest neighbors $i = 1, \ldots, z$ in the vicinity of σ_i, the short-range E_i is given by

$$E_i = -J\sigma_i \sum_j \sigma_j. \tag{16.2.8}$$

In this expression, the quantity $J\sum_j \sigma_j$ may be interpreted as the local field F_i at the site i due to the nearest group of σ_j. In the mean-field approximation the average $\langle \sum_j \sigma_j \rangle$ taken over the group of z neighbors may be replaced by $z\eta$ applied to the whole subsystem, and hence $F = \langle F_i \rangle = Jz\eta$, which is analogous to the Weiss field in a ferromagnet which will be discussed in the next chapter. As remarked, the ordered phase of a binary system consists of two subsystems characterized by $\pm\eta$, which are however thermodynamically indistinguishable because of the invariant Gibbs free energy under inversion $\eta \to -\eta$.

Alternatively, the local probabilities averaged over all lattice sites in the subsystems can be written as

$$p(\mathrm{A}) = \langle p_i(\mathrm{A}) \rangle, \qquad p(\mathrm{B}) = \langle p_i(\mathrm{B}) \rangle,$$

where

$$p(\mathrm{A}) + p(\mathrm{B}) = 1,$$

the order parameter can be defined as

$$\eta = p(\mathrm{A}) - p(\mathrm{B}). \tag{16.2.9}$$

Here the average probabilities $p(\mathrm{A})$ and $p(\mathrm{B})$ can take values in the continuous range between 1 and 0, and they can be expressed as

$$p(\mathrm{A}) = \frac{1}{2}(1+\eta), \qquad p(\mathrm{B}) = \frac{1}{2}(1-\eta).$$

For complete disorder, $p(\mathrm{A}) = p(\mathrm{B}) = 1/2$, and hence $\eta = 0$. On the other hand, ordered states $\eta = \pm 1$ correspond to $p(\mathrm{A}) = 1$, $p(\mathrm{B}) = 0$ and $p(\mathrm{B}) = 1$, $p(\mathrm{A}) = 0$, respectively.

The mean-field $F = zJ\eta$ gives a self-consistent equation for the order parameter

$$\eta = \tanh \frac{zJ}{2k_{\mathrm{B}}T}\eta. \tag{16.2.10}$$

The solution of (16.2.10) can be obtained graphically from the intersection of the straight line $y = (zJ/2k_{\mathrm{B}}T)\eta$ and the hyperbolic curve $\eta = \tanh y$, as illustrated in Fig. 16.2.2. It is noticed that for $2k_{\mathrm{B}}T/zJ \geq 1$ the intersection is only at $\eta = 0$, whereas for $2k_{\mathrm{B}}T/zJ < 1$ there is another intersection at which the nonzero η represents a partially ordered state. The transition temperature is given by

$$T_{\mathrm{c}} = \frac{zJ}{2k_{\mathrm{B}}}. \tag{16.2.11}$$

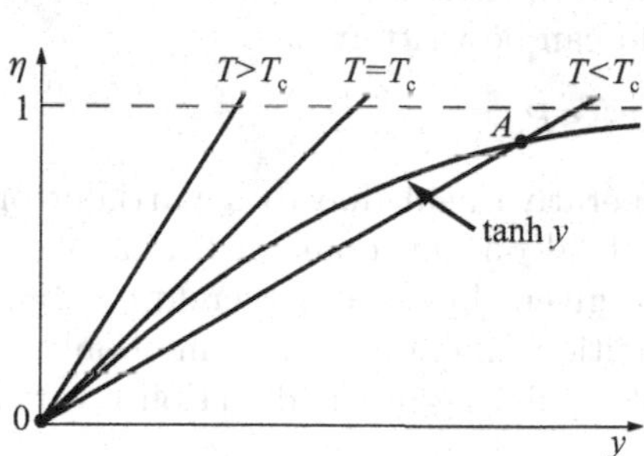

Figure 16.2.2 Graphically solutions for the order parameter η.

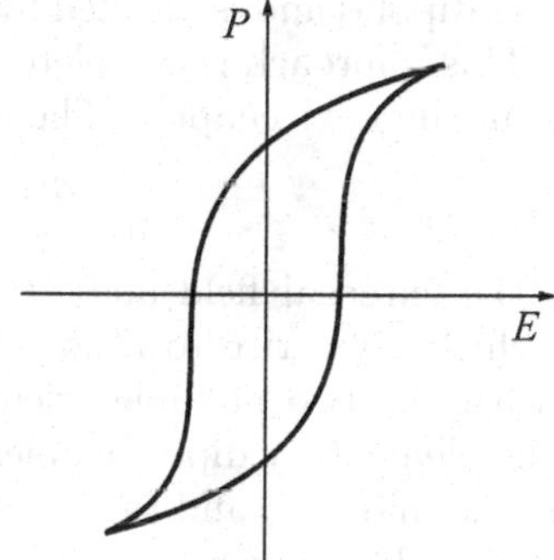

Figure 16.2.3 P versus E curves and $P_s(T)$ for a typical second order transition ferroelectric crystal.

16.2.2 Paraelectric-Ferroelectric Transition

The phenomenon of ferroelectricity is closely related to piezoelectricity and pyroelectricity. A ferroelectric crystal can be defined as a piezoelectric possessing a spontaneous electric polarization which is reversible under the action of an external electric field. Pyroelectrics possess a temperature dependent spontaneous polarization. The polarization can switched back and forth along the polar axis under the action of an external electric field.

The paraelectric-ferroelecric phase transition is in general a structural transition in which a change in the crystal structure is accompanied by the appearance of a spontaneous electric polarization with anomalous dielectric properties, such as where the dielectric constant has a sharp peak at the transition temperature, and a plot of electric polarization $\boldsymbol{P}$ versus electric field $\boldsymbol{E}$ shows a hysteresis loop, as in Fig. 16.2.3.

Ferroelectric phase transitions can be displacive phase transition or disorder-order type. If the ferroelectric phase is realized by the minute displacement of atoms or molecules in the paraelectric phase, then such a transition is said to be displacive. In the order-disorder transition, the ferroelectric phase results from the ordering of certain atoms or molecules in the paraelectric phase. It is possible for a transition to have characteristics of both.

Ferroelectric phase transitions can be either continuous or discontinuous. In the case of continuous phase transitions, the spontaneous polarization varies continuously with temperature and tends to zero at the transition temperature (see Fig. 16.2.4(a)). A discontinuous phase transition is characterized by an abrupt change in the spontaneous polarization at the transition temperature (see Fig. 16.2.4(b)).

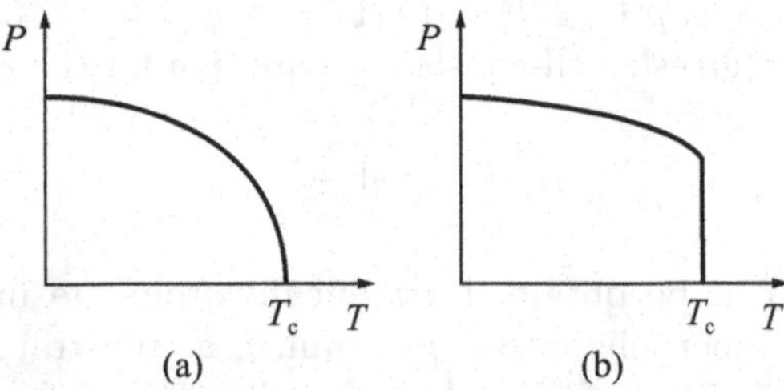

Figure 16.2.4 Temperature dependence of the spontaneous polarization in (a) continuous and (b) discontinuous ferroelectric phase transitions.

Here we shall give a treatment of the simple case of a uniaxial system with rigid elementary dipoles that can reorient in either one of two opposite directions. Because any dipole is surrounded by many other dipoles and is affected by them, we can use the effective field approach to ferroelectric transitions. This approach is completely analogous to the Weiss theory for ferromagnets which will be discussed in the next chapter. The effective field can be written as

$$E_{\text{eff}} = E + \gamma P, \tag{16.2.12}$$

where E is the external field and γP is the cooperative field due to a partially ordered system of dipoles, which gives rise to a non-zero dipolar field on any point of the lattice. The energies associated with the two possible orientations of a given dipole are, therefore $w = \pm(E + \gamma P)\mu$, where μ is the elementary dipole moment. The partition function is the sum of only two Boltzmann factors with $+w$ and $-w$, and the number of dipoles pointing in the direction favored and opposed by the effective field is given by

$$N_+ = \frac{N}{Z}\mathrm{e}^{w/k_BT}, \qquad N_- = \frac{N}{Z}\mathrm{e}^{-w/k_BT}, \tag{16.2.13}$$

where N is the total number of the dipoles, and Z the partition function

$$Z = \mathrm{e}^{w/k_BT} + \mathrm{e}^{-w/k_BT}. \tag{16.2.14}$$

The polarization is then given by a self-consistent equation

$$P = (N_+ - N_-)\mu = N\mu \tanh\frac{(E + \gamma P)\mu}{k_BT}. \tag{16.2.15}$$

To study spontaneous polarization, we simply let $E = 0$. As T approaches T_c from below, P approaches zero, one gets the critical temperature

$$T_c = \gamma N\mu^2/k_B. \tag{16.2.16}$$

Then, it is easy to obtain the spontaneous polarization

$$P_s = N\mu\sqrt{3}[1 - (T/T_c)]^{1/2}. \tag{16.2.17}$$

The dielectric constant, which is defined as $\epsilon = 4\pi dP/dE$, shows a strong temperature dependence in the vicinity of the transition point. It is easy to prove that the dielectric constant at T near T_c obeys the Curie–Weiss law

$$\epsilon(T) = \frac{C}{T - T_c}, \qquad T \geq T_c, \tag{16.2.18}$$

and

$$\epsilon(T) = \frac{C}{2(T - T_c)}, \qquad T \leq T_c \tag{16.2.19}$$

where $C = 4\pi T_c/\gamma$ is the Curie constant.

The specific heat and other thermal properties of the ferroelectric system can be calculated from the temperature dependence of the internal energy associated with the ordering of the dipole system

$$U = -\frac{1}{2}E_{\text{eff}}P = -\frac{1}{2}\gamma P^2, \tag{16.2.20}$$

with the following results for transition heat, transition entropy, and specific heat discontinuity at T_c

$$\Delta U = U(T_c) - U(0) = \frac{1}{2}Nk_B T_c, \tag{16.2.21}$$

$$\Delta S = \int_0^{T_c} d\left[-\frac{1}{2}\gamma P_s^2(T)\right] \Big/ T, \tag{16.2.22}$$

and

$$\Delta C_p = T_c \frac{dS}{dT} = \frac{3}{2}Nk_B. \tag{16.2.23}$$

Ferroelectric transitions are usually accompanied by pronounced anomalies near T_c in many other physical properties: structural properties (unit cell dimensions, atomic position), thermal properties (specific heat, thermal conductivity), elastic properties (sound velocity and attenuation, elastic constants), optical properties (refractive indices, birefringence, optical activity), etc. These facts make ferroelectric crystals useful in a variety of applications.

16.2.3 Incommensurate-Commensurate Transitions

Incommensurate phases occur in various materials when broken symmetry develops a spatially periodic variation of structure, or composition, or charge density, or spin density with a period which is not a simple multiple of that in prototypic phase. Usually, the incommensurate phase is stable only in a limited temperature range, and there its lattice period becomes longer with decreasing temperature. At some temperature a commensurate structure becomes more stable than the incommensurate one, and so an incommensurate-commensurate transition takes place. A succession of prototypic-incommensurate-commensurate phase transitions is experimentally observed with a decrease in temperature. We can take a ferroelectric to show this process. Molecular crystal thiourea is paraelectric above 202 K and ferroelectric below 169 K. Between 169 K and 202 K, it exhibits an incommensurate phase. In this temperature range, there is a polarization wave of dipole moments in the crystal whose wavelength is incommensurate with the underlying lattice periodicity.

It is natural to adapt the density wave description. The spatial variation of the density can be expressed in terms of the basis functions of the symmetry group for the high-temperature phase

$$\delta\rho_{\boldsymbol{k}}(\boldsymbol{r}) = \sum_i \eta_{i\boldsymbol{k}}\psi_{i\boldsymbol{k}}(\boldsymbol{r}). \tag{16.2.24}$$

The modulated wavevector $\boldsymbol{k}$ changes with temperature and denotes two transitions: At $T = T_I$, the transition is from high-temperature prototypic phase to incommensurate phase; and at $T = T_L$, the transition from incommensurate phase to low-temperature commensurate phase.

We shall give a simple example related to the quantitative derivation of an incommensurate transition. In this example. the order parameter has two components denoted by η_1 and η_2 corresponding to a two-dimensional representation of the symmetry group of the prototypic phase. According to the transformation properties of the symmetry group, the free energy density should be in the form of a combination of invariants. Since, in the incommensurate phase, the order parameter depends on spatial coordinates it is necessary to include the gradient invariants in the free energy. In the case of a two-component order parameter, it is sufficient to consider the dependence on only one coordinate, for example, on x. Here we write the free energy density by taking into account two fourth-order invariants and gradient terms including a Lifshitz invariant and a Ginzburg term

$$g = g_0 + A(\eta_1^2 + \eta_2^2) + B_1(\eta_1^2 + \eta_2^2)^2 + B_2\eta_1^2\eta_2^2 + \delta\left[\eta_1\frac{\partial\eta_2}{\partial x} - \eta_2\frac{\partial\eta_1}{\partial x}\right] + \frac{\kappa}{2}\left[\left(\frac{\partial\eta_1}{\partial x}\right)^2 + \left(\frac{\partial\eta_2}{\partial x}\right)^2\right]. \tag{16.2.25}$$

We note that the case of two-component order parameter with the Lifshitz invariant is realized in numerous crystals in which incommensurate phases are observed. For a structural phase transition of the displacive type, the normal coordinates of the soft mode are taken to be the order parameter. The incommensurate phase corresponds to a soft mode wavevector located in a general point in the Brillouin zone (BZ). The soft mode eigenvectors Q, Q^* can be taken as the complex order parameter. This order parameter (Q, Q^*) is actually equivalent to (η_1, η_2) by noting the replacement $Q = \eta_1 + i\eta_2$ and $Q^* = \eta_1 - i\eta_2$.

It is convenient to introduce the transformation

$$\eta_1 = \eta\sin\theta, \qquad \eta_2 = \eta\cos\theta, \tag{16.2.26}$$

and define $\alpha = 2A$, $\beta_1 = 4[B_1 + B_2/8]$, $\beta_2 = -B_2/2$, then the free energy density has the form

$$g = g_0 + \frac{\alpha}{2}\eta^2 + \frac{\beta_1}{4}\eta^4 + \frac{\beta_2}{4}\eta^4\cos 4\theta - \delta\eta^2\frac{\partial\theta}{\partial x} + \frac{\kappa}{2}\left[\left(\frac{\partial\eta}{\partial x}\right)^2 + \eta^2\left(\frac{\partial\theta}{\partial x}\right)^2\right], \tag{16.2.27}$$

where $\eta(x)$ and $\theta(x)$ are modulated along the x-direction. The free energy density defined in (16.2.27) has been successfully used to describe the successive prototypic-incommensurate-commensurate phase transitions and anomalies of physical properties in ferroelectrics, e.g., in ammonium fluoroberyllate, $(NH_4)_2BeF_4$.

In order to ensure the stability of the commensurate phase in a certain temperature interval, without expanding to higher degree terms, we must have $\beta_1 > \beta_2$. On the other hand, a positive wavenumber k implies $\delta > 0$ and $\kappa > 0$. The free energy is then

$$G = \int_L g\left(\eta, \theta, \frac{\partial\eta}{\partial x}, \frac{\partial\theta}{\partial x}\right) dx, \tag{16.2.28}$$

where L is the length of the crystal in the x-direction. From the equilibrium conditions, $\partial F/\partial\eta = 0, \partial F/\partial\theta = 0$, we obtain a set of coupled nonlinear differential equations

$$\alpha\eta + \beta_1\eta^3 + \beta_2\eta^3\cos 4\theta - 2\delta\eta\frac{\partial\theta}{\partial x} + \kappa\eta\left(\frac{\partial\theta}{\partial x}\right)^2 - \kappa\frac{\partial^2\eta}{\partial x^2} = 0, \tag{16.2.29}$$

$$\beta_2\eta^4\sin 4\theta + 2\kappa\eta\frac{\partial\eta}{\partial x}\left(\frac{\partial\theta}{\partial x} - \frac{\delta}{K}\right) + \kappa\eta^2\frac{\partial^2\theta}{\partial x^2} = 0. \tag{16.2.30}$$

For general values of the coefficients, the solutions of this set of equations can only be obtained by numerical methods. Analytic treatments can be carried out for the constant amplitude approximation in which η is taken as a constant, and only $\theta(x)$ is spatial modulated. In this simplifying assumption, the thermodynamic quantities and temperature dependence of the modulated wavevector can be calculated. However, the mathematical analysis involved is still complex. We prefer to limit ourselves to a discussion of some special cases from (16.2.29) and (16.2.30), and we hope to account for the principal characteristics of incommensurate transitions.

We consider the solutions for commensurate phases. If it is required that $\partial\eta/\partial x = 0, \partial\theta/\partial x = 0$, then (16.2.29) and (16.2.30) are simplified to

$$\eta(\alpha + \beta_1\eta^2) + \beta_2\eta^3\cos 4\theta = 0, \tag{16.2.31}$$

and

$$\beta_2\eta^4\sin 4\theta = 0. \tag{16.2.32}$$

Obviously, there are two sets of solutions. The first set of solutions is $\eta = 0$, and θ can take an arbitrary value; this case is the high-temperature prototypic phase. The second set of solutions is related to $\eta \neq 0$, and so the structure is ordered. Here $\sin 4\theta = 0$ determines eight θ values defining eight directions in the plane $(\eta_1\eta_2)$. Only four directions correspond to the minima of free energy, which depends on the sign of β_2. The results are: for $\beta_2 > 0$, $\theta = \pm\pi/4, \pm 3\pi/4$, and for $\beta_2 < 0$, $\theta = 0, \pm\pi/2, \pi$. The amplitude of the order parameter is in the same form

$$\eta^2 = -\frac{\alpha}{\beta_1 - |\beta_2|}, \tag{16.2.33}$$

as usual, we set $\alpha = \alpha_0(T - T_{\rm L})$. When $T < T_{\rm L}, \eta^2 > 0$, the low-temperature phases are commensurate and ordered. Figure 16.2.5(a) shows the low-temperature commensurate phases denoted by some isolated dots on the (η_1, η_2) plane with amplitude η_e, and argument θ_e.

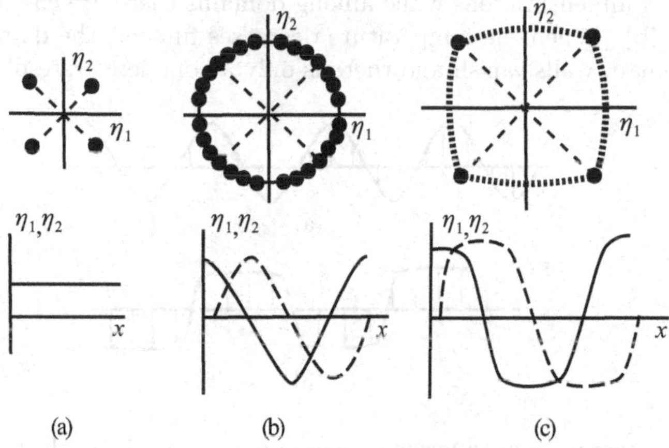

Figure 16.2.5 The thermodynamic stable solutions in the order parameter plane with $\beta_2 > 0$. (a) Low-temperature commensurate phases; (b) ignoring anisotropic energy; (c) numerical results.

Because of the existence of the Lifshitz invariant, it is in fact prohibited for the second-order transition to go directly to the commensurate phase from the prototypic phase. We should discuss the solution for incommensurate phase: Near the transition point of the prototypic phase to incommensurate phase, the order parameter may be considered as a small quantity, and has a form of plane wave

$$\eta_1 = \eta\sin(k_{\rm I}x), \qquad \eta_2 = \eta\cos(k_{\rm I}x), \tag{16.2.34}$$

i.e., $\eta \simeq 0$ and $\theta = k_{\rm I}x$. The modulated wavevector can be obtained by ignoring the higher order terms in (16.2.30), then

$$K\eta\frac{\partial\eta}{\partial x}\left(\frac{\partial\theta}{\partial x} - \frac{\delta}{K}\right) = 0. \tag{16.2.35}$$

We find that

$$k_{\rm I} = \delta/\kappa. \tag{16.2.36}$$

It is seen that modulated wavevector is determined by the coefficients of the Lifshitz term and Ginzburg term. Supposing $\beta_1 \gg \beta_2$ and neglecting anisotropic terms, (16.2.29) can be transformed into

$$\alpha_0\left(T - T_{\rm L} - \frac{\delta^2}{\alpha_0\kappa}\right) + \beta_1\eta^2 = 0. \tag{16.2.37}$$

From this expression, we can define the transition temperature from the prototypic phase to the incommensurate phase

$$T_{\rm I} = T_{\rm L} + \frac{\delta^2}{\alpha_0 \kappa}. \tag{16.2.38}$$

When $T_{\rm L} < T < T_{\rm I}$, the amplitude of the order parameter is

$$\eta_{\rm I}^2 = -\frac{\alpha_0(T - T_{\rm I})}{\beta_1} > 0. \tag{16.2.39}$$

As $T \to T_{\rm I}$, $\eta_{\rm I}^2 \to 0$, so the transition is continuous. In the vicinity of $T_{\rm I}$ the modulation wavelength $\sim 1/k_{\rm I} = \kappa/\delta$ is irrational with respect to the lattice period.

It can be seen that on the order parameter plane, the stable phase is represented by any point on the circle with radius $\eta_{\rm I}$, η_1 and η_2 are changing along the x-axis sinusoidally with amplitude $\eta_{\rm I}$ and wavenumber $k_{\rm I}$, as shown in Fig. 16.2.5(b). This is called the single plane wave form of incommensurate phase, as shown in Fig. 16.2.6(a). As the temperature decreases, the anisotropic energy increases, and the representative points on the (η_1, η_2) plane are not distributed homogeneously but will be dense near the low-temperature commensurate phases. Numerical solutions of (16.2.29) and (16.2.30) show that near $T_{\rm L}$ the incommensurate modulated wave becomes a square wave, as shown in Fig. 16.2.5(c). This square wave is composed of a lot of domain structure. The domain walls are discommensurations while among domains there are commensurate structures, as shown in Fig. 16.2.6(b). When the temperature decreases further, the domain walls diminish, and finally at $T_{\rm L}$, the domain walls vanish and there is only a commensurate phase.

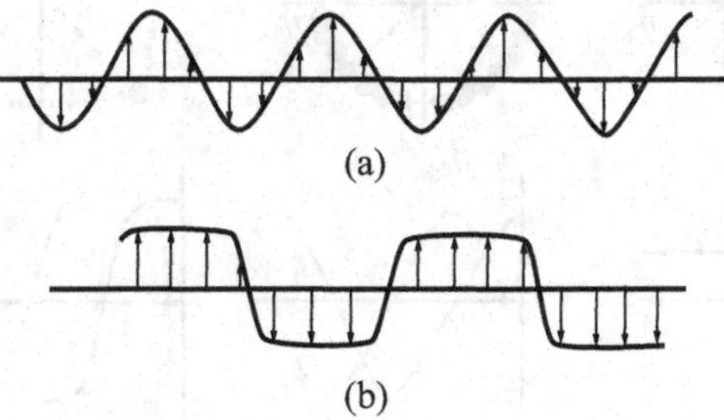

Figure 16.2.6 Modulation wave in an incommensurate phase. (a) Single plane modulation for T near $T_{\rm I}$. (b) Domain and wall structure for $T_{\rm c} < T < T_{\rm I}$.

§16.3 Phase Transitions in Soft Matter

Soft matter includes liquid crystals, polymers, colloids, etc. Their polymorphic configurations lead to a large number of interesting phenomena, especially as phase transitions can be driven by entropy as well as energy. In this section we will discuss first isotropic-nematic transition in thermotropic liquid crystals, and then briefly introduce the phase separation in hard-sphere packing.

16.3.1 Maier–Saupe Theory for Isotropic-Nematic Transition

As introduced in §3.3, the thermotropic liquid crystal is supposed to be composed of rod-like molecules. As the temperature decreases, it will experience an isotropic-nematic transition. The nematic phase differs from ordinary liquids in its anisotropy. Its symmetry is cylindrical, that is, there exists a unique axis along which some of the properties are quite different from those perpendicular to this axis. The symmetry axis, denoted as $\bar{\boldsymbol{n}}$, is referred to as the director. The anisotropy of nematics arises from the tendency of the rod-like molecules in the fluid to align their long axis parallel to the director. At finite temperature, the thermal motion prevents perfect alignment with $\bar{\boldsymbol{n}}$, the orientations of the molecules are in fact distributed over angle θ, as shown in Fig. 16.3.1, where ϕ is the azimuthal angle. If there is no preference for a particular θ, then all such angles become equally probable and complete isotropy results; this is the isotropic normal liquid phase. Ordering in the polar angle θ distinguishes the nematic structure from the isotropic liquid.

We should define the long-range orientational order parameter in the nematic phase. One may expect projection of the molecules along $\bar{\boldsymbol{n}}$, $\cos\theta$, would be a natural order parameter, but this is not right, because the direction $\bar{\boldsymbol{n}}$ and $-\bar{\boldsymbol{n}}$ are fully equivalent, i.e., the preferred axis is non-polar. So we need to consider the term $\cos^2\theta$ rather than $\cos\theta$ to describe the molecules. Furthermore, we desire the average value, $\langle\cos^2\theta\rangle$, averaged over all molecules in the liquid. When all the molecules are fully aligned with $\bar{\boldsymbol{n}}$, all $\theta = 0$ and $\langle\cos^2\theta\rangle = 1$. On the other hand, if the molecules are randomly distributed in direction, all values of θ are equally likely and $\langle\cos^2\theta\rangle = 1/3$. It is natural to choose

$$\eta = \langle P_2\rangle = \frac{1}{2}\langle 3\cos^2\theta - 1\rangle, \tag{16.3.1}$$

as a scalar order parameter for the nematic phase. P_2 is just the second-order Legendre function.

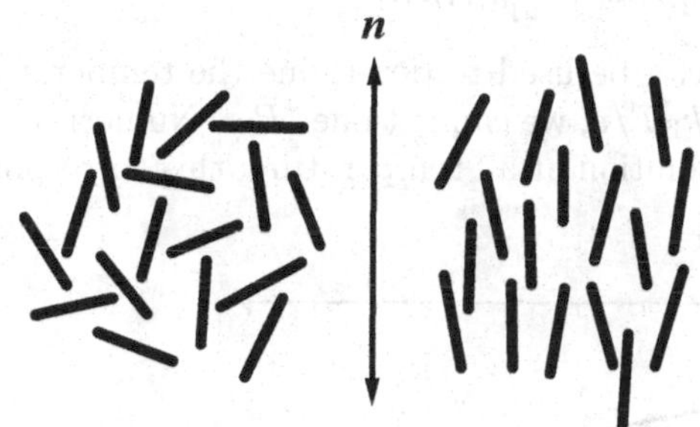

Figure 16.3.1 Schematic diagrams of the structure of a nematic liquid crystal and single rod-like molecule.

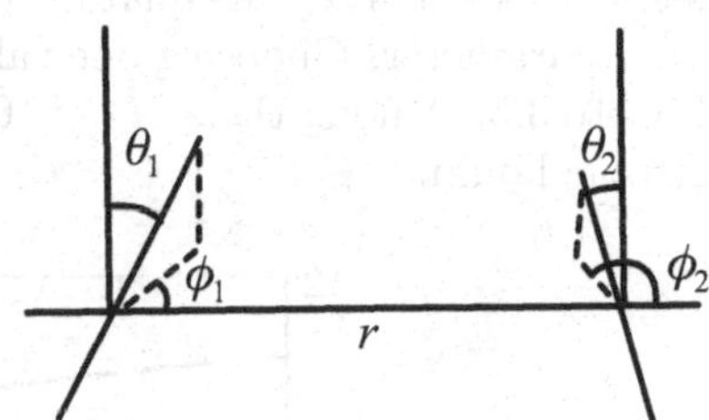

Figure 16.3.2 Schematic diagram of the interaction between two rod-like molecules.

The stability of the nematics results from interactions between the constituent molecules. The pair potential between two rod-like molecules can generally be expressed as

$$V_{12} = V_{12}(r, \theta_1, \phi_1, \theta_2, \phi_2), \tag{16.3.2}$$

where r is the distance between the centers of mass, θ_i and ϕ_i are orientational and azimuthal angles, respectively. However, it is very difficult to get the exact form of (16.3.2).

An approach has proved to be extremely useful in developing a theory of spontaneous long-range orientational order and the related properties is the Maier–Saupe molecular field method (1958).[d] We should get a single-molecule potential, then a molecule is in the mean field of all other molecules, such as

$$V(\cos\theta) = -vP_2(\cos\theta)\langle P_2\rangle, \tag{16.3.3}$$

where the contribution of all other molecules is characterized by the degree of order $\langle P_2\rangle$, $-P_2(\cos\theta)$ describes the angular-dependence of potential which is a minimum when the molecule $\parallel \boldsymbol{n}$, and maximum when $\perp \boldsymbol{n}$, and v is the strength of the intermolecular interaction, $v > 0$.

We now need an orientational distribution function, which describes how the molecules are distributed among the possible directions about the director. It gives the probability of finding a molecule at some prescribed angle θ from $\bar{\boldsymbol{n}}$. With this function we can calculate the average values of various quantities of interest pertaining to the nematic phase. From the classical statistical

[d] In addition to being applied to isotropic-nematic transitions, Maier–Saupe theory can be extended to other thermotropic phase transition in liquid crystals, for example, one from isotropic phase to smectic phase, which has orientational order and one-dimensional translational order simultaneously, refer to W. L. McMillan, *Phys. Rev. A* **4**, 1238 (1971).

mechanics, the orientational distribution function is

$$f(\cos\theta) = Z^{-1}\exp[-\beta V(\cos\theta)], \tag{16.3.4}$$

and the single molecule partition function is

$$Z = \int_0^1 \exp[-\beta V(\cos\theta)]d(\cos\theta), \tag{16.3.5}$$

where $\beta = 1/k_B T$.

Now the order parameter, just like the average of the second-order Legendre function, can be calculated from

$$\begin{aligned}\langle P_2\rangle = \eta &= \int_0^1 P_2(\cos\theta)f(\cos\theta)d(\cos\theta)\\ &= \frac{\int_0^1 P_2(\cos\theta)\exp[\beta v P_2(\cos\theta)\cdot\eta]d(\cos\theta)}{\int_0^1 \exp[\beta v P_2(\cos\theta)\cdot\eta]d(\cos\theta)}.\end{aligned} \tag{16.3.6}$$

This is a self-consistent integral equation which can be used to determine the temperature dependence of order parameter. Choosing one value of $k_B T/v$, we can get one $\langle P_2\rangle$. Numerical results are shown in Fig. 16.3.3. Among them $\langle P_2\rangle = 0$ is a solution at all temperature; this corresponds to the normal isotropic liquid.

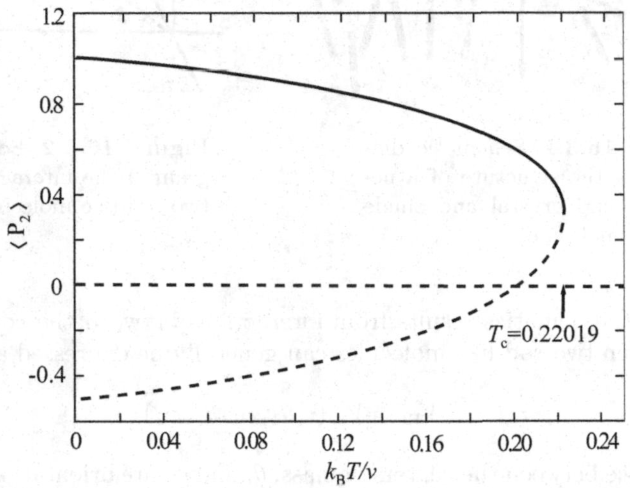

Figure 16.3.3 Phase diagram of Maier–Saupe transition. The stable equilibrium solutions are shown as the solid lines.

Here the transition temperature is in fact $T_c = 0.22019v/k_B$. For temperatures T below T_c, two other solutions appear. The upper branch tends to unity at absolute zero and represents the nematic phase. The lower branch tends to $-1/2$ at absolute zero and represents a phase in which the molecules to line up perpendicular to the director without azimuthal order. We can judge which one of the three solutions is stable by minimizing the free energy.

The internal energy is the average of the potential

$$U = \frac{1}{2}N\langle V\rangle = \frac{1}{2}N\int_0^1 V(\cos\theta)f(\cos\theta)d(\cos\theta), \tag{16.3.7}$$

where N is the number of molecules, and the factor 1/2 is required to avoid counting the intermolecular interactions twice. The entropy is calculated by taking the average of the logarithm of the partition function

$$S = -Nk_B\langle\ln f\rangle = \frac{N}{T}\langle V\rangle + Nk_B\ln Z. \tag{16.3.8}$$

Combining (16.3.7) and (16.3.8), the Helmholtz free energy

$$F = -Nk_{\rm B}T\ln Z - \frac{1}{2}N\langle V\rangle. \tag{16.3.9}$$

The reason for the appearance of the second term is the replacement of pair interactions by a temperature-dependent single molecular potential. We can verify its correctness by setting $\partial F/\partial\langle P_2\rangle = 0$, and see that the self-consistent equation (16.3.6) is regained. Thus, as required by thermodynamics, the self-consistent solutions to the problem must be those that represent the extrema of the free energy. From the minimum of F, we can show that, when $T < T_{\rm c}$, the nematic phase is stable.

Numerical calculation shows that the order parameter decreases from unity to a minimum value of 0.4289 at $T = T_{\rm c}$. For temperatures above $T_{\rm c}$ the isotropic phase with vanishing order parameter is stable. The stable phases are shown by the solid line in Fig. 16.3.3. The phase transition is first-order, because the order parameter discontinuously changes from 0.4289 to 0. The general trend of the temperature dependence of $\langle P_2\rangle$ displayed in Fig. 16.3.3 is in agreement with experimental results.

16.3.2 Onsager Theory for Isotropic-Nematic Transition

In the Maier–Saupe theory we have seen how an anisotropic, attractive interaction (16.3.3) can give rise to a first-order isotropic-nematic transition. The origin of the anisotropy lies in the fact that the molecules are rod-like and quite rigid. Then one expects that, besides the anisotropic attractive interaction, there must also be an anisotropic steric interaction which is due to the impenetrability of the molecules. Taking into account only the steric interaction, Onsager established his theory for the transition of a system of hard rods from the isotropic phase to anisotropic phase as the density is increased.[e]

To understand Onsager theory, we should consider two kinds of entropy in a gas of hard rods. One is the entropy due to the translational degrees of freedom, and the other is the orientational entropy. More important is that there is a coupling between these two kinds of entropy through the effect of excluded volume. The excluded volume is the volume into which the center of mass of one molecule cannot move due to the impenetrability of the other molecule. The excluded volume is always larger when two hard rods lie at an angle with each other than they are parallel. It is clear that the translational entropy favors parallel alignment of the hard rods because this arrangement gives less excluded volume and, therefore, more free space for the molecules to jostle around. However, parallel alignment represents a state of low orientational entropy. Therefore, a competition exists between the tendencies to maximize the translational entropy and to maximize the orientational entropy. In the limit of zero density the tendency to maximize the orientational entropy always wins because each molecule rarely collides with another molecules, and the gain in excluded volume due to parallel alignment would only be minimal addition to the already large volume of space within which each molecule can move about. When the density is increased, however, the excluded volume effect becomes more and more important. In the limit of tight-packing density, the hard rods must be parallel. A transition between the isotropic and anisotropic states therefore must occur at some intermediate density.

Simply put, at sufficiently low densities the rods can assume all possible orientations and the fluid will be isotropic. As the density increases, it becomes much more difficult for the rods to point in random directions and intuitively one may expect the fluid to undergo a transition to a more ordered anisotropic phase with uniaxial symmetry. This was first proved by Onsager. Onsager's approach is based on an exact density expansion for free energy.[f]

We consider a fluid of long thin hard-rod molecules with well defined length L and diameter D, satisfying $L \gg D$. The only forces of importance correspond to steric repulsion, i.e., the rods cannot

[e] It is noted that, just as Maier–Saupe theory, Onsager theory was also successfully applied to the formation of smectic phase. See A. Stroobants, H. N. W. Lekkerkerker and D. Frenkel, *Phys. Rev. A* **36**, 2929 (1987); X. Wen, R. B. Meyer and D. L. D. Caspar, *Phys. Rev. Lett.* **63**, 2760 (1989).

[f] For the Onsager theory for isotropic-nematic transition and its extension, see G. J. Vroege and H. N. W. Lekkerkerker, *Rep. Prog. Phys.* **55**, 1241 (1992).

interpenetrate each other and the volume fraction $\nu = (1/4)\rho\pi LD^2$, where ρ is the concentration of the rods, is much smaller than unity.

To study this system of hard rods, we must specify not only the overall concentration ρ, but also the angular distribution of the rods, so we may define $f(\Omega)$ as the number of rods per unit volume pointing in a solid angle Ω. It is clear that the sum over all the solid angles must satisfy the condition of normalization, so

$$\int f(\Omega)d\Omega = 1. \tag{16.3.10}$$

The free energy of the system expanded to first order in density is

$$F = F_0 + k_BT\left\{\int f(\Omega)\ln[4\pi f(\Omega)]d\Omega + \frac{1}{2}\rho\int\int f(\Omega)f(\Omega')u(\Omega\Omega')d\Omega d\Omega'\right\}. \tag{16.3.11}$$

The first term on the right side of the above expression can be taken as a constant, so will be neglected in the following discussion; the second term describes the entropy contribution associated with molecular alignment; the third term describes the excluded volume effects, $u(\Omega\Omega')$ is the volume excluded by one rod in direction Ω as seen by one rod in direction Ω'. The calculation of u is simple for the long rods, where end effects are ignored and it is expressed as

$$u = 2L^2D|\sin\gamma|, \tag{16.3.12}$$

where γ is the angle between Ω and Ω'.

We can obtain a self-consistent equation for the distribution function $f(\Omega)$ by specifying that the free energy (16.3.11) is a minimum for all variations of $f(\Omega)$ that satisfy the constraint (16.3.10). Taking λ as the Lagrange multiplier, we can write

$$\delta F = k_BT\lambda\int \delta f(\Omega)d\Omega, \tag{16.3.13}$$

and give the self-consistent equation

$$\ln[4\pi f(\Omega)] = \lambda - 1 - \rho\int u(\Omega\Omega')f(\Omega')d\Omega'. \tag{16.3.14}$$

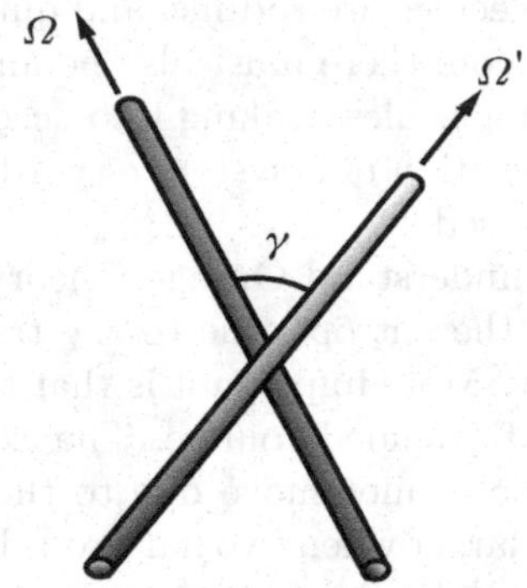

Figure 16.3.4 Excluded volume of two hard rods with angle γ.

λ is then determined by the normalization condition (16.3.10). (16.3.12) and (16.3.14) show that the concentration ρ enters the problem only through the combination $\rho L^2D \propto \times\nu L/D$.

Equation (16.3.14) always has an isotropic solution, $f(\Omega) = 1/4\pi$, independent of $\boldsymbol{a}$, but if $\nu L/D$ is large enough, it may also have anisotropic solutions describing a nematic phase. To solve the nonlinear integral equation (16.3.15), Onsager adopted a variational method, based on a trial function of the form

$$f(\Omega) = A\cosh(\alpha\cos\theta), \tag{16.3.15}$$

where α is a variational parameter, θ the angle between $\boldsymbol{a}$ and the nematic axis, and A is a constant which should be chosen to normalize f according to (16.3.10). In the region of interest, α turns out to be large (~ 20) and the function f is strongly peaked around $\theta = 0$ and $\theta = \pi$. The order parameter is

$$\eta = \frac{1}{2}\int f(\Omega)(3\cos^2\theta - 1)\sin\theta d\theta =\simeq 1 - 3/\alpha, \tag{16.3.16}$$

for $\alpha \gg 1$. Minimizing the energy F in (16.3.11) with respect to α, one obtains a function $F(c)$ which shows a first-order phase transition from isotropic ($\alpha = 0$) to nematic ($\alpha \geq 18.6$). The volume fraction ν occupied by the rods in the nematic phase, just at the transition point, is $\nu_c^n = 4.5D/L$. At the same point, the value of ν for the isotropic phase, in equilibrium with the nematic phase, is smaller: $\nu_c^i = 3.3D/L$.

Note that ν_c^{n} and ν_c^{i} are independent of T in this model. This means that the hard rods are an 'athermal' system, and this phase transition is driven by entropy. Of particular interest is the value of the order parameter η_c in the nematic phase just at the transition: This turns out to be quite high ($\eta_c \simeq 0.84$). Thus the Onsager solution leads to a rather abrupt transition between a strongly-ordered nematic and a completely disordered, isotropic phase.

16.3.3 Phase Separation in Hard-Sphere Systems

In recent years, the concept of the entropy-driven phase transitions has been extensively applied to illuminate the phase behavior of soft matter. Here we will give a discussion of the phase separation of hard sphere systems.[g]

For phases of condensed matter in equilibrium, the condition of free energy minimum must be satisfied. In the free energy

$$F = U - TS, \tag{16.3.17}$$

U is the internal energy, T the temperature and S the entropy. In conventional solids, which may be designated as the hard matter, the contribution of internal energy is larger than the entropy. That is to say, the internal energy determines the structure of the equilibrium phase. Taking the crystallization discussed in §16.1.2 as example, when temperature decreases, a phase transition takes place from disordered liquid to ordered crystal for the system. In this process the reduction of the entropy will increase the free energy, so the appearance of the ordered phase is due to the decrease of internal energy, in order to ensure the free energy is a minimum. This kind of phase transition has arisen from energy, so it can be called an energy-driven phase transition. In the case of soft matter, the situation is just the opposite: compared with TS, the contribution from U is too small to have an influence on the configuration of the system. Now the decrease of the free energy is mainly due to the increase of the entropy; the equilibrium state is determined by the entropy maximum instead of the internal energy minimum. The key point is that the increase of microscopic disorder is found to be beneficial to the appearance of macroscopic order. Formally, the entropy deviation from an equilibrium value will give rise to an entropic force; its effect is just like the gradient of a potential in practice. Entropic force will drive a system to develop into a new phase with a minimum free energy. This is the driving force for the entropy-driven phase transition.

The simplest model to describe the entropy-driven phase transitions is the Alder–Wainright's computer model of hard spheres for fluids developed in the 1950s. Actually there are weak attractive interactions between atoms. Just as will be discussed in §19.2.1, these interactions lead to a gas-liquid phase transition. To investigate the entropy-driven phase transition, we may assume that the internal energy of a system is only a function of temperature and not density. If the temperature is fixed, corresponding to a fixed internal energy, but the sphere density is varied, it is possible for us to observe the entropy-driven phase transition directly.

In a hard sphere system with a single size for the sphere radius, the internal energy is always zero for different configurations. The forces between particles and the free energy of the system are thoroughly determined by the entropy. It is clear that the entropy of the system is only related to the total volume fraction occupied by hard spheres ν. When ν is small, the chances of collision between particles is less, the system is looked as a ideal gas. As ν increases, the restriction of the movement of a particle by collisions with neighboring particles is also increased. In the case of close packing, all particles are trapped. The pronounced characteristic for hard sphere systems is that there are two close packing densities, the hexagonal close-packing density $\nu_{\mathrm{h}} = 0.7405$ and the random close packing density $\nu_{\mathrm{r}} = 0.638$. In the case of random close packing, particles are arranged randomly, but each particle contacts with other particles, so its movement is inhibited. It is noted that $\nu_{\mathrm{r}} < \nu_{\mathrm{h}}$. We can imagine magnifying the lattice of the hexagonal close packing structure, while the crystalline structure is kept invariant. It is obvious as the lattice is magnified, ν decreases, and each particle can move freely around a site in the magnified lattice. The result of this free movement for particles leads the entropy to increase. It should be pointed out that for random close packing,

[g] A concise introduction to the role of entropy in soft condensed matter is found in T. C. Lubensky, *Solid St. Commun.* **102**, 187 (1997).

there are a lot of configurations which form residual entropy for glasses. But just as will be discussed in §19.2.3, there is less chance for different configurations to access each other, so it is reasonable to take the entropy of each configuration as zero. Therefore, the magnified lattice with $\nu = nu_{\mathrm{r}}$ has higher entropy compared to the random close packing structure with the same volume fraction. This means for a liquid phase in the quasi-equilibrium process, when the volume fraction increases, the system is favored to form a periodic crystal, and not to be trapped into a random liquid structure. This is an entropy-driven first-order liquid-solid phase transition. To realize the glass transition for a liquid, it is necessary to use quenching, a nonequilibrium process.

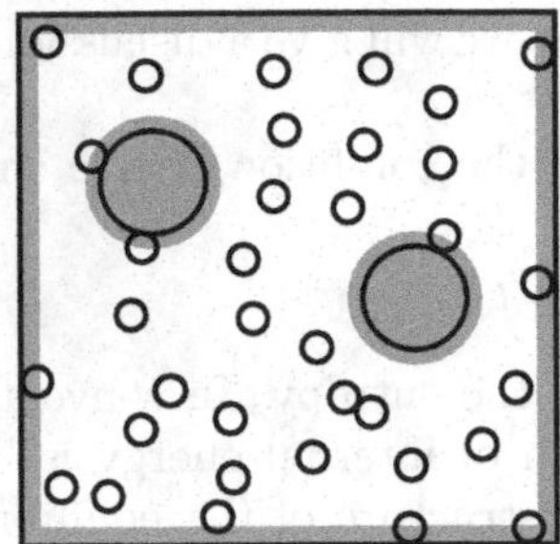
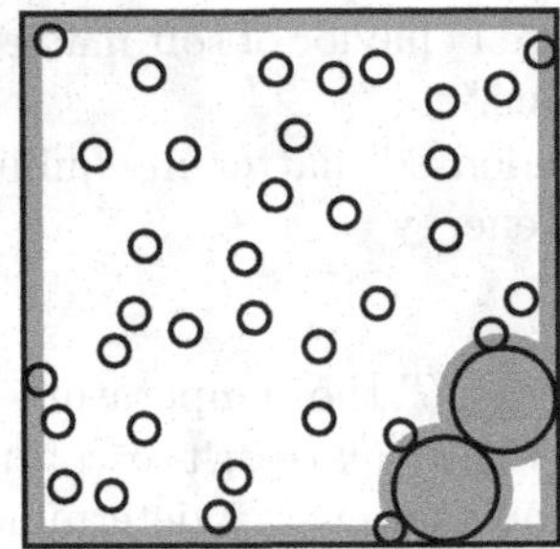

Figure 16.3.5 Schematic diagram of exclusive volume between large sphere, small sphere, and wall of container.

Mixtures of hard spheres with different sizes provide interesting examples showing the role of entropy. For a simple discussion, we consider a dispersion system composed of colloidal spheres with two different diameters. The diameter of a large sphere d_{L} is much larger than that of a small one d_{S}. If both the volume fractions are about the same, then the number of small spheres exceeds that of large ones, so the entropy of small spheres play a principal role in determining the structure of the system. The configuration adopted by large spheres must let the entropy of the small spheres be a maximum. It can be seen from Fig. 16.3.5 that two large spheres will provide more free space for small spheres when they are in contact and near the wall of the container, due to the excluded volume effect. In Fig. 16.3.5 solid lines represent the profiles of spheres and the walls of the container, Shadow corresponds the impenetrable regions. So between large spheres there is an entropy-induced attractive force, called the depletion force. The concept of the depletion force is very important in colloidal, latex, and biological systems. Particles involved are not only hard spheres, but can also be hard rods, or sphere-rod mixtures. Contact of large particles with the walls of a container can contribute more free volume to small particles, so surfaces provide also a kind of attractive force. Experiments verify that large particles can not only separate on surfaces but also form crystals if their density is high enough. Using the jargon of modern statistical physics, the entropic force gives rise to a crystalline phase of large particles, which wet a solid surface.

Based on the discussion for attractive force between large spheres, we can expect that a phase separation will appear between large and small particles by increasing volume fractions. The entropic mechanism for this effect is very obvious. If the large spheres form a close-packed crystal, all voids between large spheres cannot contain large spheres, so more free space is left to small spheres. This two-component hard-sphere mixture shows very rich phase behaviors. These phase behaviors are dependent on three parameters, i.e., the volume fractions of large and small spheres ν_{L} and ν_{S}, and the ratio of radii of large and small spheres $\alpha = r_{\mathrm{L}}/r_{\mathrm{S}}$. The ideal hard-sphere gas attracts much attention currently. The colloidal dispersion phase composed of polystyrene spheres of diameters from 0.06 μm to 8 μm has been used to verify experimentally the theoretical predictions from the hard sphere model. Figure 16.3.6 shows the phase diagram from the experimental measurements and theoretical analysis for a colloidal system with large and small spheres of diameters 0.825 μm and 0.069 μm. The horizontal and vertical axes represent the volume fractions of large and small spheres, respectively. It is found that the phase boundary determined from the experiments is consistent with the theoretical calculation.

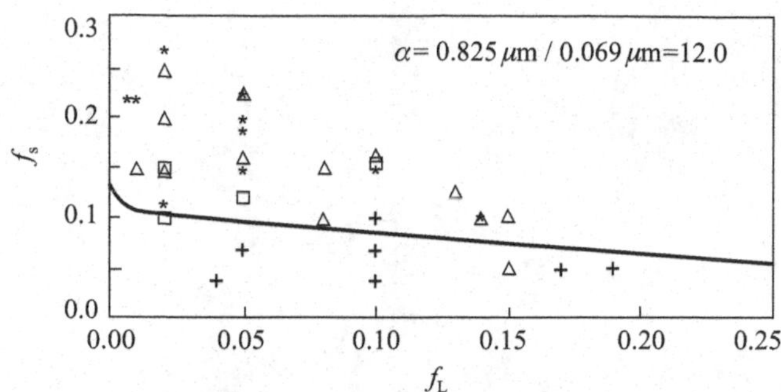

Figure 16.3.6 Phase diagram of mixture with two components. The experimental result is denoted by: +, no phase separation; □, solid phase appears on the surface of container; △, phase separations appear on surface as well as in bulk; ∗, liquid-solid phase separation appears only in bulk. The solid line denotes a phase boundary by theoretical calculations. From A. D. Dinsmore *et al.*, *Phys. Rev. E* **52**, 4045 (1995).

Bibliography

[1] Landau, L. D., Phys. Z. Sowjet **11**, 26; 545 (1937); or in *Collected Papers of L. D. Landau*, ed. D. ter Haar, Pergamon, Oxford (1965), p. 193.

[2] Izyumov, Y. A., and V. N. Syromyatnikov, *Phase Transitions and Crystal Symmetry*, Kluwer Academic, Dordrecht (1990).

[3] Tolédano, J. C., and P. Tolédano, *The Landau Theory of Phase Transitions*, World Scientific, Singapore (1987).

[4] Fujimoto, M., *The Physics of Structural Phase Transitions*, Springer-Verlag, New York (1997).

[5] Gonzaro, J. A., *Effective Field Approach to Phase Transitions and Some Applications to Ferroelectrics*, World Scientific, Singapore (1991).

[6] Sannikov, D. G., *Phenomenological Theory of the Incommensurate-Commensurate Phase Transition*, in *Incommensurate Phases in Dielectrics: Fundamentals* (eds. R. Blinc and A. P. Levanyuk), North-Hollan, Amsterdam (1986).

[7] de Gennes, P. G., and J. Prost, *The Physics of Liquid Crystals*, 2nd ed., Clarendon Press, Oxford (1993).

[8] Chandrasekhar, S., *Liquid Crystals*, 2nd ed., Cambridge University Press, Cambridge (1992).

[9] Priestley, E. B., P. J. Wojtowicz and P. Shen, *Introduction to Liquid Crystals*, Plenum, New York (1975).

[10] Daoud, M., and C. E. Williams (eds.), *Soft Matter Physics*, Springer, Berlin (1999).

[11] Klemen, M., and O. D. Lavrentovitch, *Introduction to Soft Matter Physics*, Springer, Berlin (2003).

Chapter 17

Ferromagnets, Antiferromagnets and Ferrimagnets

Magnetic ordering, such as ferromagnetism, antiferromagnetism and ferrimagnetism, arises from the broken symmetry of time-reversal or spin-rotation. There are two physical models to describe the magnetism of condensed matter: One is the localized; the other, the itinerant. The former has been used successfully in magnetic insulators, and the latter for magnetic metals. These two models are opposed but complementary to each other and illustrate the intrinsic properties of magnetism of materials, however, they are still in the stage of development with increasing sophistication. Certainly, in many cases, these two models cannot be separated from each other. There has been a trend to combine both to develop a unified theory to understand magnetism. In this chapter we are mainly concerned with the formation of magnetically-ordered phases and intrinsic magnetic properties. The problems of technical magnetization and micromagnetics will be discussed in Part VII.

§17.1 Basic Features of Magnetism

Magnetic properties of materials are closely related to the spin of microscopic particles. Spin, as a new degree of freedom, is purely quantum mechanical, although sometimes it can be considered as a classical vector, such as in Langevin's treatment of paramagnetism, and also in many statistical models, like the classical Ising, XY, and Heisenberg models.

17.1.1 Main Types of Magnetism

Magnetic behavior in solids is, in general, involved in the orientations of magnetic dipoles. Each of these dipoles, or magnetic moments, is composed of the electronic spins, electronic orbitals, and nuclear magnetic moments. Because the nuclear magnetic moment is so much smaller than the electronic magnetic moment (by three orders of magnitude), when macroscopic magnetism in solids is investigated, the nuclear moment can be neglected and only ionic and electronic moments are considered.[a] Furthermore, in magnetic compounds of transition metals, the orbital moment is always quenched, so the actual moment is mainly provided by the electronic spin. Magnetism in solids, according to its magnitude and sign, principally includes five types: Diamagnetism, paramagnetism,

[a]It should be noted that there are obvious interactions between nuclear spins and electronic spins. This is the basis of using nuclear magnetic resonance and the Mössbauer spectra to study magnetism of matter. By adiabatic demagnetization, substances can be cooled to ultra low temperature, to temperatures in the range of μ K. Experimentally, nuclear magnetic ordered phases have been observed at 10^{-3}–10^{-7} K, for example, the nuclear ferromagnetism (T_C=0.40 mK) of ^{141}Pr in $PrNi_5$ and the nuclear antiferromagnetism (T_N=1.03 mK) in ^{3}He crystals, see A. Abragam and M. Goldman, *Nuclear Magnetism: Order and Disorder*, Clarendon Press, Oxford (1982). Recent experimental results show that nuclear spin polarization may appear in a semiconductor adjacent to a ferromagnet; its internal field gives considerable influence on the transport of electrons with spins, so it has practical significance in spintronics. Refer to R. Kawakami *et al.*, *Science* **294**, 131 (2001).

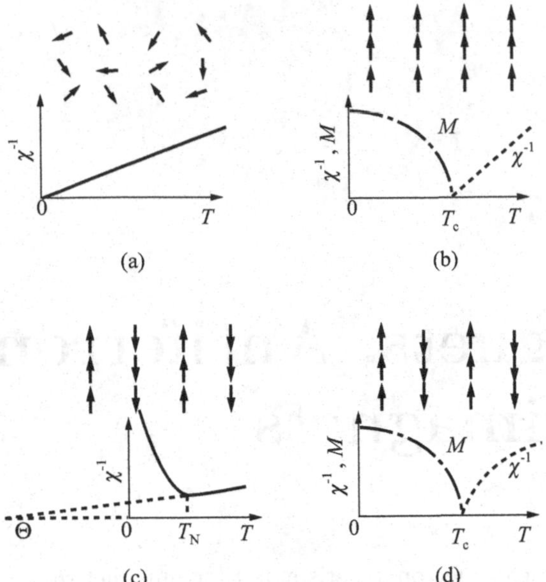

Figure 17.1.1 Several kind of magnetic behavior. (a) Paramagnetism; (b) ferromagnetism; (c) antiferromagnetism; and (d) ferrimagnetism.

ferromagnetism, antiferromagnetism, and ferrimagentism. The first two types only represent the properties of the independent moment ensemble, but the latter three types reflect cooperative phenomena of a large number of moments. Figure 17.1.1 shows these important types. The upper part displays the moment distribution, the magnetization $\boldsymbol{M}$ of a sample is the vector sum of these moments; the lower part represents the susceptibility $\chi = M/H$ versus temperature, H is the applied magnetic field.

(1) Diamagnetism. Diamagnetic substances have negative susceptibility. Actually, all substances, such Cu, Zn, Au, H_2O have basic diamagnetism, but it is very weak and often overshadowed by the positive paramagnetic susceptibility, usually larger by one or two orders of magnitude. Basic diamagnetism is independent of temperature and arises from the effect of applied fields on the inner shell electrons of atoms. Electronic orbitals around a nucleus can sometimes be looked upon as a current. When a magnetic field is applied, the electronic motion is disturbed, corresponding to the moment being modified and an induced moment appears. According to the Lenz law for electromagnetic induction, this induced moment is opposite to the applied field, therefore its susceptibility χ_d is negative. χ_d is not only small, but also independent of temperature and external field.

(2) Paramagnetism. Many solids, such as Na, Al, V, Pd have paramagnetism. In Fig. 17.1.1(a), the magnetic moments are oriented randomly. Under the application of an applied field $\boldsymbol{H}$, the number of moments will increase along the $+\boldsymbol{H}$ direction but decrease along $-\boldsymbol{H}$ direction. This process leads to a small magnetization $\boldsymbol{M}$, which is linearly dependent on the applied field; moreover, once the applied field is removed, the magnetization disappears instantly. It is easy to show that the relation between susceptibility and temperature is $\chi \propto T^{-1}$.

(3) Ferromagnetism. Typical ferromagnetic substances are Fe, Co, Ni. As shown in Fig. 17.1.1(b), the ferromagnetism appears for them when temperature is below a transition temperature (Curie point) T_c. That is to say, below T_c spins tend to take parallel orientation spontaneously. Above T_c, ferromagnets have paramagnetism; the spins are oriented randomly. The relation between susceptibility and temperature satisfies the Curie–Weiss law $\chi \propto (T - T_c)^{-1}$.

(4) Antiferromagnetism. Typical antiferromagnetic substances are Cr and Mn, and also oxides like MnO, CrO, CoO; their behavior is displayed in Fig. 17.1.1(c). Above a transition temperature (Néel point) T_N, spins distribute randomly and have paramagnetic behavior. However, below T_N, one half of the spins are antiparallel to the another half. So the resultant magnetization is zero. Above T_N the relation between susceptibility and temperature satisfies $\chi \propto (T + \Theta_A)^{-1}$.

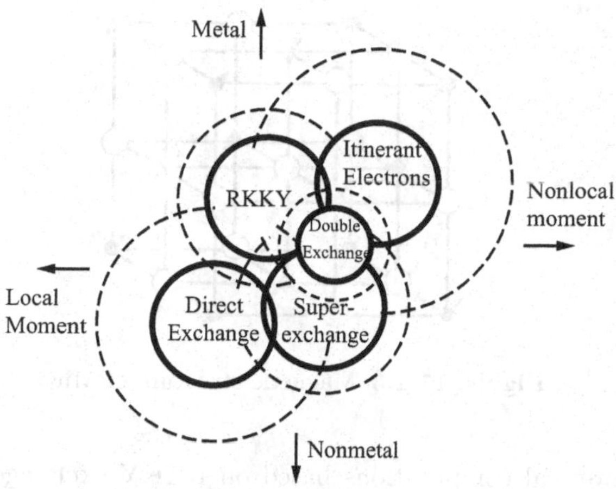

Figure 17.1.2 The relationship among five exchange interactions.

(5) Ferrimagnetism. Figure 17.1.1(d) shows ferrimagnetism. A ferrite, like Fe_3O_4, is a typical ferrimagnet. Above T_c the spins are oriented randomly, but below T_c they are arranged antiparallel. But it is different from an antiferromagnet, the magnetic moments are not equal on different sublattices, and a net magnetization appears. This is similar to a ferromagnet, but usually ferromagnets are metals and ferrimagnets are nonmetals. Another important difference from ferromagnets is that for most ferrimagnets their susceptibility-temperature relation does not follow the Curie–Weiss law in a large temperature range above T_c. Only after $T > 2T_c$, does the temperature dependence of χ^{-1} asymptotically approaches linearity.

There are other spin configurations, like helical, canted, spiral and umbrella-like, but the five types discussed above are the main ones. Among them ferromagnets attracts the most attention: Ferromagnets display complicated magnetic hysteresis; this is the basic problem for technical magnetization and we will discuss it in Part VII.

The magnetically ordered structures are the results of direct or indirect interactions between ionic moments on sites or delocalized electronic moments in crystals. Previously, we have discussed several kinds of magnetic interactions, such as the direct exchange (including kinetic exchange), superexchange and double exchange, and also the indirect, or RKKY, exchange between localized moments mediated by conduction electrons. In addition, there are exchange interactions between itinerant electrons. This situation is more complicated, due to that 3*d* electrons are partly delocalized into the Fermi sea, and partly localized around atomic sites; moreover, these two aspects cannot be distinguished completely, and this is related to the complex many-body problems of electrons.

All these exchange interactions contribute to the formation of diverse magnetically ordered structures through cooperative phenomena for macroscopic magnetism. For example, direct exchange and superexchange will be used in the magnetization theory of local moments in §17.2; the exchange of itinerant electrons will be discussed in §17.3, and the RKKY interaction will be introduced to discuss spin glasses in §19.3.

Although these five kinds of exchange interactions were proposed for different cases and have been applied to various circumstances appropriately, there are no clear borderlines between them. They are related to each other, and there are overlaps of their regions of application. Figure 17.1.2 is a schematic diagram which shows the relationships between these five exchange interactions. The solid circles represent the main region of application for each type of interaction, while the dashed circles indicate enlarged regions of application. Strictly speaking, in real substances it is possible there are several exchange interactions which coexist and are mixed together. For transition metals, it is most suitable, in principle, to use the exchange interactions between the itinerant electrons. The RKKY exchange has been used mainly to illustrate the magnetic ordered phases in rare earth metals, but in some circumstances, it is also effective for some transition metals and alloys. In fact, the RKKY theory was first proposed to explain experiments in CuMn. In recent years, for the interlayer coupling

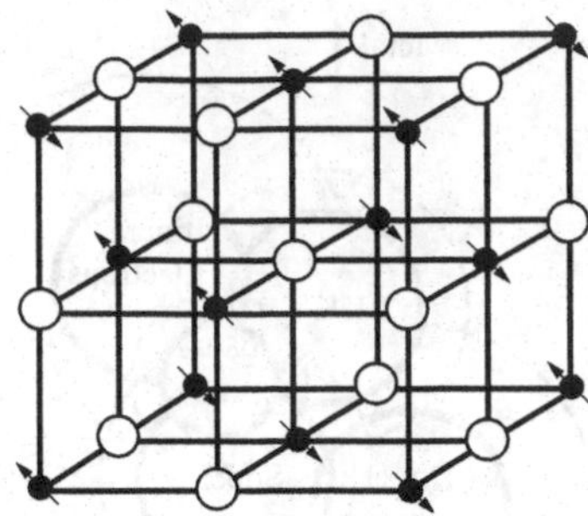

Figure 17.1.3 Magnetic structure of MnO.

of multilayers, the theoretical computations based on RKKY exchange interactions have become consistent with experimental results. This means there are localized moments on the transition group atoms in metals or alloys. Neutron scattering experiments can provide information about localized spin densities in transition metals, but it only gives the spin density distribution probability. From the theoretical point of view, general Bloch functions with the tight-binding approximation can give indications about itinerancy as well as localization of electrons. We take into account the *s-d* exchange interaction and strong correlations, itinerant *s* and *d* electrons always have large probability and time to locate around nuclei of the transition elements, so there exist localized moments, in the probabilistic sense.

17.1.2 Spatial Pictures of Magnetic Structures

Neutron diffraction has played a pivotal role in illuminating magnetic ordered structures. X-ray diffraction can only be used to locate atoms or ions in a crystal; neutron diffraction can further determine the distribution of magnetic moments in the crystal. The theory of antiferromagnets, proposed by Néel, was verified by neutron diffraction. In the following, we will begin our discussion of magnetic structures of the oxides, representative of insulators.

The magnetic structure of MnO is typical: Its crystalline structure is of the NaCl type, and Mn is a magnetic ion. In MnO the spins of the Mn ions are arranged alternately positive and negative, as shown in Fig. 17.1.3, so the whole structure is antiferromagnetic. The magnetic structures of FeO, CoO, NiO show some similarities, but are more complicated. The trivalent ions Ti^3, V^{3+}, Cr^{3+}, Fe^{3+} which can be used to form M_2O_3 oxides with the Al_2O_3 type crystalline structure, are also antiferromagnets generally. Among them, $\alpha-Fe_2O_3$ is a somewhat special, at temperatures 950 K–260 K, the magnetic moments are located in the basal plane, perpendicular to the body diagonal of the rhombohedron, the Fe ions in adjacent atomic layers are arranged antiparallel with a small tilt. The small tilt angle leads to the neighboring moments being unable to cancel each other, so a weak ferromagnetism appears, see Fig. 17.1.4; but its magnetic behavior approaches that of an antiferromagnet.[b] Below the temperature $T = 260$ K, called the Morin temperature, moments are turned perpendicular to the basal plane, then $\alpha-Fe_2O_3$ becomes a normal antiferromagnet. The oxides with the perovskite structure, like $LaCrO_3$, $LaMnO_3$, often appear antiferromagnetic: Neutron diffraction experiments have shown very complicated magnetic structures, some of which have been discussed in Chap. 13. The magnetic data of some important antiferromagnetic substances are compiled in Table 17.1.1.

[b]The weak ferromagnetism arising from the moment tilt of an antiferromagnet has attracted the interest of theorists. First, I. Dzyaloshinsky, *J. Phys. Chem. Solids* **4**, 241 (1958) gave a phenomenological explanation; then by the perturbation method of the Anderson superexchange interaction, T. Moriya, *Phys. Rev.* **120**, 91 (1960) gave it a microscopic theoretical foundation. In recent years, scientists have adopted *ab initio* calculation of electronic structure to give further explanations. Similar, but with magnetic configurations being completely different, is the weak ferromagnetism of Mn_3Ga and Mn_3Sn in which magnetic ions are arranged in a triangle which is not closed, so there are residual moments which show weak ferromagnetism. It should be emphasized that the physical nature of this weak ferromagnetism due to the tilt of localized moments is completely different from the weak ferromagnetism in itinerant electrons, which will be discussed later in §17.3.3.

Table 17.1.1 Intrinsic magnetic properties and crystalline structures of several antiferromagnets.

Compounds	Crystalline structures	$T_N(°K)$	$\Theta(°K)$	Θ/T_N	C_{mole}	$\chi_p(0)/\chi_p(T_N)$
MnO	fcc	122	610	5.0	4.40	0.69
FeO	fcc	185	570	3.1	6.24	0.77
CoO	fcc	291	280	0.96	3.0	—
NiO	fcc	515	—	—	—	0.67
MnS	fcc	165	528	3.2	4.30	0.82
MnF_2	rutile	74	113	1.5	4.08	0.75
FeF_2	rutile	85	117	1.4	3.9	0.72
CoF_2	rutile	40	53	1.3	3.3	—
NiF_2	rutile	78	116	1.5	1.5	—
MnO_2	corundum	86	—	—	—	0.93
Cr_2O_3	corundum	307	1070	3.5	2.56	0.76
α-Fe_2O_3	corundum	950	2000	2.1	4.4	—
FeS	layered hexagonal	613	857	1.4	3.44	—
$FeCl_2$	layered hexagonal	24	−48	−2.0	3.59	< 0.2
$CoCl_2$	layered hexagonal	25	−38.1	−1.5	3.46	~ 0.2
$NiCl_2$	layered hexagonal	50	−68.2	−1.4	1.36	—
$FeCO_3$	complex	57	—	—	—	0.25
$CuCl_2{\cdot}2H_2O$	rhombohedronal	4.3	5	1.16	—	—
$FeCl_2{\cdot}4H_2O$	rhombohedronal	1.6	2	1.2	3.61	—
$NiCl_2{\cdot}6H_2O$	rhombohedronal	5.3	—	—	—	—

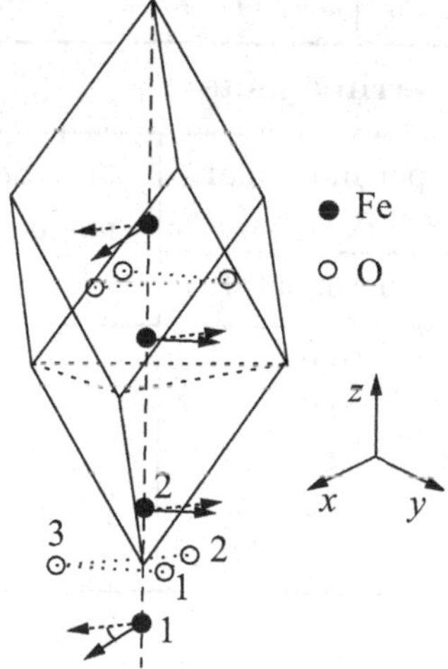

Figure 17.1.4 Magnetic structure of αFe_2O_3.

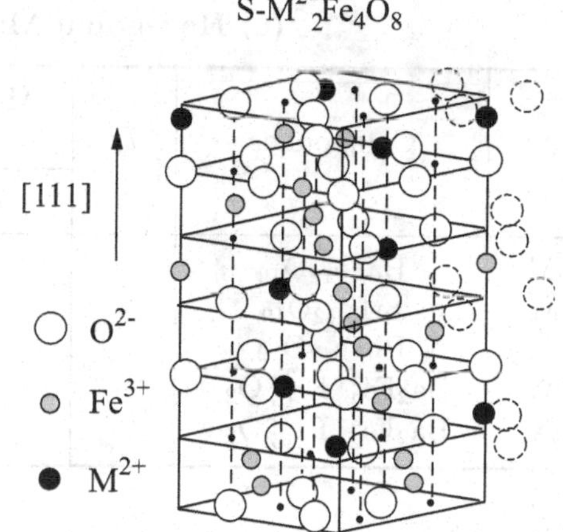

Figure 17.1.5 Magnetic structure of Fe_3O_4.

The typical oxides with ferrimagnetic structure are ferrites: Magnetic ions on two, or more, sublattices have unequal moments. Although they are arranged antiparallel, the net moments are not zero. The typical crystalline structures of ferrites include the cubic spinel, garnet, and magnetoplumbite structures. We take a ferrite with the spinel structure an example: The chemical formula of a spinel ferrite is MFe_2O_4, where M is M^{2+} or M^{3+} metallic ions. Its crystalline structure, is shown in Fig. 17.1.5: Oxygen ions are arranged according to face centered cubic close packing, each unit cell contains eight molecular formula units, there are 64 tetrahedral voids and 32 octahedral voids;

Table 17.1.2 Intrinsic magnetic properties and crystalline structures of several ferrimagnets.

(a) Spinel Ferrimagnets

Compounds	Structures types	Curie point T_c/K	Ionic moments (μ_B)				Saturation magnetization
			A	B	Net value		
			position	position	Theoretical	Experimental	M_s/kG
$MnFe_2O_4$	I	575	$-(1+4)$	$1+9$	5	$46 \sim 5$	0.40
Fe_3O_4	I	860	-5	$4+5$	4	4.1	0.50
$CoFe_2O_4$	I	790	-5	$3+5$	3	3.7	0.45
$NiFe_2O_4$	I	865	-5	$2+5$	2	2.3	0.33
$CuFe_2O_4$	I	728	-5	$1+5$	1	1.3	
$Li_{0.5}Fe_{2.5}O_4$	I	943	-5	$0+7.5$	2.5	$2.5 \sim 3$	0.033
$MgFe_2O_4$	I	700	-5	$0+5$	0	1.1	0.0092

(b) Garnet Ferrimagnets

Compounds	Curie points T_c/K	Compensation temperature T^*/K	Magnetic moment per molecular formula at 0 K			Saturation magnetization at RT
			Theoretical		Experimental	
			$\lvert 3(2S)-5 \rvert$	$\lvert 3(L+2S)-5 \rvert$		M_s/kG
$Y_3Fe_5O_{12}$	560	—	5	5	4.96	0.14
$GdFe_5O_{12}$	564	290	16	16	15.2	0.010
$DyFe_5O_{12}$	563	220	10	25	17.2	0.032

(c) Hexagonal Magnetic Lead Ferrimagnets

Type	Compounds	T_c	Magnetic moment per molecular formula at 0 K		Saturation magnetization at RT
			Theoretical	Experimental	M_s/kG
BaM	$BaFe_{12}O_{19}$		20	19.9	0.37
SrM	$SrFe_{12}O_{19}$		20	20.2	0.37
PbM	$PbFe_{12}O_{19}$		20	19.6	0.33
BaW	$Ba_2Fe_2^{2+}Fe_{16}^{3+}O_{27}$		28	27.6	0.40
BaX	$Ba_2Fe_2^{2+}Fe_{16}^{3+}O_{27}$		48	47.5	0.20

but cations actually occupy eight tetrahedral voids (A sites) and 16 octahedral voids (B sites). The spinel structure can be divided into two types: Normal (N) and inverse (I). In the normal type, M^{2+} occupy the A sites, M^{3+} occupy the B sites; but in the inverse type, M^{3+} occupy the A sites, equal numbers of M^{2+} and M^{3+} occupy the B sites. Magnetite was the earliest magnet discovered by humans, but it is not a ferromagnet, rather a ferrimagnet. Neutron diffraction verified that its structure is of the inverse spinel type, i.e., its magnetic moments are distributed as $(Fe^{3+})_A(Fe^{3+}Fe^{2+})_BO_4$. Because the spins on sites A and B are arranged antiparallel, the moments of Fe^{3+} on sites A and B cancel each other, and the net moments are provided by Fe^{2+} on sites B. Some data on ferrimagnetic substances are shown in Table 17.1.2.

Neutron diffraction studies of the magnetic structures of the transition metals have also been very enlightening. Unlike insulators, the three typical ferromagnetic metals, Fe, Co and Ni have lattice site magnetic moments which are smaller than those of the isolated ions; moreover, they

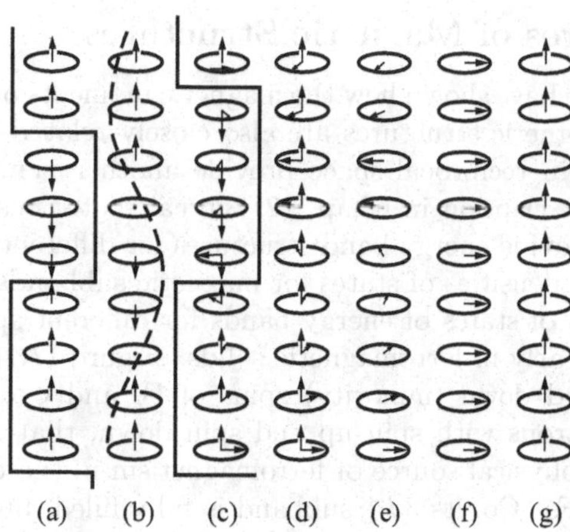

Figure 17.1.6 Different types of magnetic ordered structures for rare earth metals.

Table 17.1.3 Magnetic structures of several rare earth metals and alloys.

Element	Gd	Tb	Dy	Ho	Er	Tm	
	293 K	PM 230 K helical χ=18–20°(*e*) 220 K	PM	PM	PM	PM	300 K
	FM (along c axis, orientantional angles of moments are variable)	FM moment along base plane is 9.0 μ_B(*f*)	176 K helical AFM χ=26–43°(*e*)	130 K helical AFM			200 K
			88 K FM moment along base plane is 9.5 μ_B(*f*)	χ=30–50°(*e*) 20 K	85 K c axis modulated AFM 7(*f*) 52 K 20 K	57 K 32 K	100 K
				cone helical $\mu_\perp$=9.5 μ_B(*d*) $\mu_\parallel$=1.7 μ_B	cone helical $\mu_\perp$=4.3 μ_B(*d*) $\mu_\parallel$=7.6 μ_B		0 K

are all non-integers, showing that their *d* electrons partly participate the itinerant process. On the other hand, experimental results from neutron scattering show that the spin density in Cr fluctuates sinusoidally, providing an example for the existence of spin density waves.

Neutron diffraction has also obtained very rich results from the study of rare earth metals and alloys, especially showing the non-collinear structure of the magnetic order, see Fig. 17.1.6. Table 17.1.3 lists the various complicated cases and very large ionic moments for the magnetic ordered structures of the magnetic rare earth metals.

17.1.3 Band Pictures of Magnetic Structures

In the last subsection we have shown how the magnetic moments of various materials are arranged in real space. Because magnetic structures are also closely related to the electronic spins, energy band structures displayed in reciprocal space provide another viewpoint to understand magnetic structures. Just like the description in Chap. 12, the calculated results based on the spin density functional approach can provide energy band structures for different spin orientations, up or down, and also the corresponding densities of states for magnetic subbands. For nonmagnetic solids, such as Si and Na, the densities of states of energy bands for different spins are completely symmetric, but asymmetry appears clearly in ferromagnetic solids. Figure 17.1.7 shows the densities of states for up (majority) spins and down (minority) spins of Fe and Co. Their asymmetry leads to a number difference for electrons with spin up and spin down, that is, $N_\uparrow > N_\downarrow$. This is the spin polarization, which is the physical source of ferromagnetism in these metals. It is noted that Fe is clearly different from Co. For Co, its $3d\uparrow$ subband is fully filled, but for Fe it is not fully filled, so their spin polarizations $\rho = (n_\uparrow - n_\downarrow)/n_\downarrow$ are different. The $3d\uparrow$ subbands of Co and Ni are fully filled, so ρ is very high; but the $3d\uparrow$ of Fe is not filled and ρ is lower. According to the filling of the two subbands by $3d$ electrons, we can understand ferromagnetism, antiferromagnetism, and the corresponding results of magnetic moment values in a series of transition metals, see Table 17.1.4.[c]

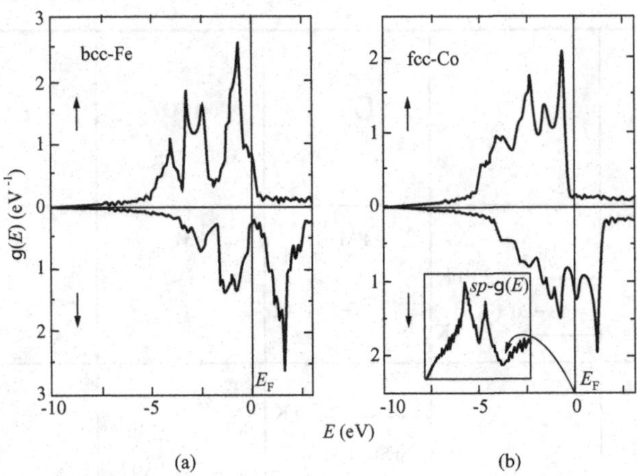

Figure 17.1.7 DOS curves. (a) Fe; (b) Co. From J. Kübler, *Theory of Itinerant Electron Magnetism*, Oxford University Press, Oxford (2000).

Moments of highly spin-polarized ferromagnetic metals can be obtained by counting the d electrons. The electronic configuration of most $3d$ metals is $[\mathrm{Ar}]3d^n4s^2$. For the elements with $n > 5$, there are about 1.35 $4s$ electrons (not spin polarized on the whole) entering the $3d$ band. The $3d$ band can accomodate ten electrons, if the $3d\uparrow$ subband is fully filled, then the number of electrons with down spin is $n_\downarrow = n + 1.35 - 5$; the net moment is

$$m = (n_\uparrow - n_\downarrow)\mu_\mathrm{B} = [5 - (n + 1.35 - 5)]\mu_\mathrm{B} = (8.65 - n)\mu_\mathrm{B}. \tag{17.1.1}$$

From this formula it is found that $m(\mathrm{Co}) = 1.65\mu_\mathrm{B}$, $m(\mathrm{Ni}) \approx 0.65\mu_\mathrm{B}$, which are almost consistent with the experimentally measured values. In iron, $m(\mathrm{Fe}) \approx 2.65\mu_\mathrm{B}$, but the experimentally measured value is only $m(\mathrm{Fe}) = 2.2\mu_\mathrm{B}$; this is ascribed to the $3d\uparrow$ subband not being fully filled (weakly spin polarized), so the magnetic moment is decreased.

The measured results for moments in some two-component alloys of transition metals can be fitted by the Slater–Pauling curve as shown in Fig. 17.1.8, in which the sloping line on the right-hand side

[c] In some references the high spin polarization Co and Ni is called strong ferromagnetism, while the low spin polarization in Fe is called weak ferromagnetism. These terms are easily misunderstood, because there are other criteria for the magnitude of ferromagnetism. For example, the strength of exchange interactions, which determines the value of T_c, and the number of net magnetic moments gives saturated magnetizations. From these criteria, the ferromagnetism of Fe is not weak. The net magnetic moment of Fe is larger than that of Co and Ni, while its T_c is in between them.

Table 17.1.4 Electron distribution in energy bands and intrinsic magnetic properties of ferromagnetic metals.

Element	Electronic configuration of isolated atoms	Distribution of band electrons				Hole number		Spin number	E value of moments (μ_B)			
		$3d\uparrow$	$3d\downarrow$	$4s\uparrow$	$4s\downarrow$	$3d\uparrow$	$3d\downarrow$		Magnetic measure	Neutron diffraction 3d	4s value	Net
Cr	$3d^44s^2$	2.7	2.7	0.3	0.3	2.3	2.3	0	0			0
Mn	$3d^54s^2$	3.2	3.2	0.3	0.3	1.8	1.8	0	0			0
Fe	$3d^64s^2$	4.8	2.6	0.3	0.3	0.2	2.4	2.2	2.216	2.39	−0.21	2.18
Co	$3d^74s^2$	5.0	3.3	0.35	0.35	0	1.7	1.7	1.715	1.99	−0.28	1.71
Ni	$3d^84s^2$	5.0	4.4	0.3	0.3	0	0.6	0.6	0.616	0.620	−0.105	0.515

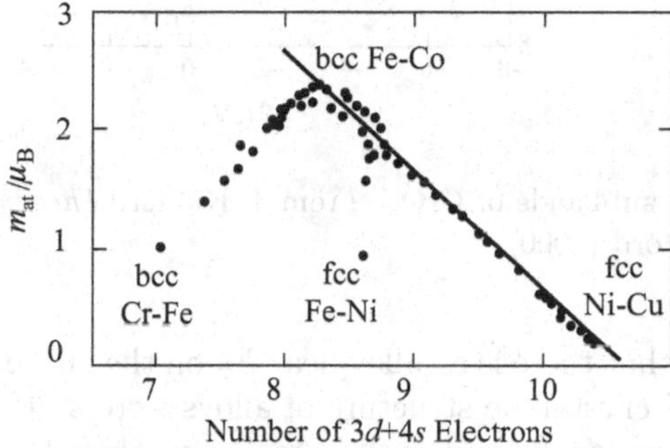

Figure 17.1.8 Slater–Pauling curve for two component $3d$ magnetic alloys.

represents the relation in (17.1.1). This accounts for the role the number of d electrons modified by alloying. For two-component alloys, we can write

$$m/\mu_B = (1-x)m'_A + xm'_B, \tag{17.1.2}$$

where m'_A and m'_B are the moments of A and B, respectively. For alloys satisfying the right side of the Slater–Pauling curve, $m' = 8.65-n$; but for metalloid, $m' = 0.65-n_v$ (n_v is the number of valence electrons), for example, for H, $m' = -0.15$. It is simple and convenient to use this rule to treat the magnetic structures of alloys, moreover its validity even surpasses that of the simple rigid band model. $Fe_{6.5}Co_{3.5}$ alloy gives the highest value of moment ($2.5\mu_B$), which is an important constituent of the strong magnetic alloy Alnico (an alloy of Fe, Co, Ni, Al, etc.). It should be noted that the Fe-Ni alloy deviates from the Slater–Pauling curve, showing an anomaly. Some of the Fe-Ni alloys have anomalous physical properties. For example, $Fe_{32}Ni_{66}$ is called as permalloy, which has excellent soft magnetic properties. $Fe_{64}Ni_{36}$, called as invar, has its zero thermal expansion coefficient, which is related to its anomalous electronic structure. There are also a series of ferromagnetic alloys without ferromagnetic elements, such as the binary alloys of Mn with N, P, As, Sb, Bi; Cr with S, Te, Pb; and also three-component alloys called Heusler alloys, like Cu_2MnAl, Cu_2MnSn, Cu_2MnIn, Cu_2MnGa, etc. In addition, there are low T_c ferromagnets without magnetic ions, like $ZrZn_2$ with $T_c \sim 22$ K; Sc_3In with $T_c \sim 6.1$ K. Both are purer itinerant electron systems than Fe, Co, Ni, and will be discussed in §17.3.3. These kinds of alloys also have a very weak ferromagnetism, but its physical origin is entirely different from the weak ferromagnetism due to canted antiferromagnetism as stated in §17.1.2.

For transition alloys containing rare earth elements, the dominant role is often played by the sublattice of transition metals; some alloys with excellent permanent magnet properties, such as $Nd_2Fe_{14}B$ and $SmCo_5$, belong to this category.

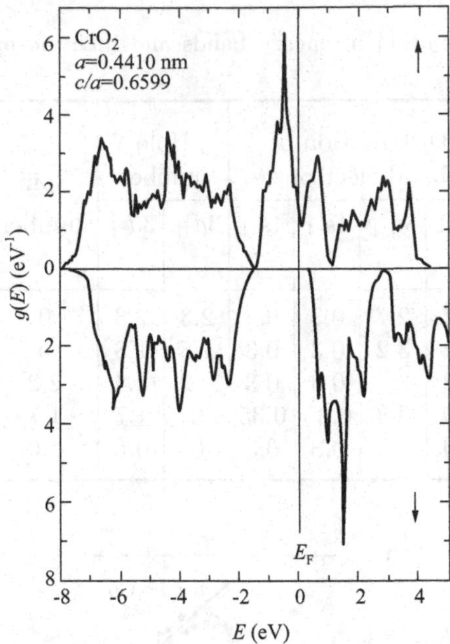

Figure 17.1.9 DOS of spin subbands of CrO_2. From J. Kübler, *Theory of Itinerant Electron Magnetism*, Oxford University Press, Oxford (2000).

Figure 17.1.7(a) shows that there is a valley near E_F on the curve of the spin $\downarrow$ density of state for Fe. If the components and crystalline structure of alloys were suitably designed, this energy valley could developed into a pseudogap, and even a real gap, then become completely spin polarized ($\rho = 100\%$) ferromagnets, i.e., the half-metallic ferromagnet, could be obtained. To arrive at this objective, it is necessary to have complicated components and structures. The Heusler alloy X_2MnY or half Heusler alloy XMnY then became the focus of theoretical study, here X = Co, Ni, Cu, Pd, Pt, etc., Y = Al, Sn, In, Sb; the crystalline structures are complicated ordered structures based on the bcc structure with the unit cell enlarged by a factor of eight. Kübler *et al.* were first to find, in the energy band calculations for Heusler alloys, that there is a valley going to zero in the density of states of the spin $\downarrow$ subband, similar to a pseudo-gap; afterwards, de Groot *et al.* discovered clear energy gaps in the half Heusler alloys NiMnSb, PdMnSb and PtMnSb.[d] The characteristics of this kind of electronic structure can be summarized as follows: In two spin subbands, one shows an energy gap and is insulator (or semiconductor)-like, while the other is metal-like without an energy gap, de Groot called them half-metallic ferromagnets. The most typical, and also simplest in structure among half-metallic ferromagnets, is CrO_2. Its crystalline structure is of the rutile type. The density of states (DOS) is shown in Fig. 17.1.9; the gap width of the spin down subband is $E_g \sim 1.88$ eV, while the conduction band is located about 0.38 eV above E_F. The reason why CrO_2 becomes a semi-metallic ferromagnet is simply due to the exchange splitting, i.e., the difference between the spin-up subband and spin down subband is larger than the bandwidth occupied by spin-up electrons. Therefore all the valence electrons are spin up; none are spin down. That is, in CrO_2, the spins are completely polarized; all the valence electrons are spin-oriented in the same direction. The ferromagnets $La_{1-x}Ca_xMnO_3$, etc., introduced in §12.2.3, due to the double exchange interaction also belong to this type; their metallic behavior comes from the spins of doped holes being aligned parallel to each other.

Another type of ferromagnets is insulators or semiconductors. There are energy gaps in all electronic subbands including spin-up and spin-down. EuO with NaCl-like structure is ferromagnetic with a $T_c \sim 70$ K; furthermore, the O can be substituted by S, Se or Te. With the emergence of spintronics, research on semiconducting ferromagnets has become a hot topic. In conventional

[d]The theoretical problems about the half metallic ferromagnets are referred to Bib. [11]. A review article on this field is W. E. Picket and J. S. Moodera, *Phys. Today* **54** (5), 39 (2001).

compound semiconductors, such as GaAs, InSb, by doping with excessive Mn, Co, Fe, they become ferromagnetic. For example, in GaAs the equilibrium solubility of Mn is very low ($< 0.1\%$), but, by the molecular beam epitaxy technique, doped Mn in GaAs film can reach 5%, and its $T_c \sim 110$ K.[e] In the anatase structure doped with 8% Co, its T_c is higher than room temperature.[f] Also by the molecular beam epitaxy technique the digital alloy or δ doping can be fabricated; for example, an atomic layer of MnAs can be inserted every few atomic layers of GaAs. In this way a ferromagnet with a high T_c was obtained[g] The formation of this type of ferromagnet can be explained by the RKKY exchange interactions coupled by the charges in semiconductors. Because the concentration of charge carriers in semiconductors is far less than the concentration of conduction electrons in metals, instead of RKKY oscillation, only ferromagnetic coupling appears. Its mechanism is somewhat similar to the theory of ferromagnetism based on *s*-*d* interaction originally proposed by Zener. In ferromagnetic semiconductors, the Zeeman splitting by internal fields leads to complete spin polarization for parabolic energy bands, so they are possible candidate materials for spintronics.

Organic magnets have also attracted much attention. In §11.4, we discussed ferrocene, which is an example of coordination compounds with 3*d* magnetic ions. If we combine it with other organic molecules to form charge-transfer salts, then connecting them to form net-like solids, among them there may be some magnetically-ordered phases, which might be antiferromagnetic, or ferrimagnetic, or ferromagnetic. A few of them have T_c higher than room temperature, for example, the ferrimagnet $[V(TCNE)_2]\cdot 12CH_2Cl_2$.[h] It is noted that in organic compounds without 3*d* ions, completely composed of light elements C, N, H, O, magnetically-ordered phases have also been observed. For example, in 1991, $C_{13}H_{16}N_3O_4$ was found to have ferromagnetism,[i] its $T_c \sim 0.65$ K. Afterwards, the solid formed by grafting C_{60} molecules to TADE has also found to be ferromagnet at low temperatures with T_c equal to 16 K.[j] These results tell us that there is potential for molecular solids to form magnetically ordered structures.

17.1.4 Hamiltonians with Time Reversal Symmetry

One of the important symmetry operations is time reversal; this operation can be described by the time reversal operator $\mathcal{T}$. For objects in condensed matter physics, time reversal is related to complex conjugation; this connection can be proved from the time-dependent Schrödinger equation

$$\mathcal{H}\psi = i\hbar\frac{\partial\psi}{\partial t}. \tag{17.1.3}$$

By performing operation $\mathcal{T}$ and noting $\mathcal{H}$ is Hermitian, we have

$$\mathcal{H}\psi^* = -i\hbar\frac{\partial\psi^*}{\partial t} = i\hbar\frac{\partial\psi^*}{\partial(-t)}. \tag{17.1.4}$$

Comparing (17.1.4) with (17.1.3), we can see the only variation is that t changes its sign and ψ becomes ψ^*, while the Schrödinger equation is still satisfied. If we take an another complex conjugate operation for (17.1.4), then (17.1.3) is obtained again. That is for two time inversions, the equation of motion is invariant.

It is essential and effective to consider the commutativity between some physical quantities and the time inversion operator. For the position vector $\boldsymbol{r}$, it is trivial to write

$$\mathcal{T}\boldsymbol{r}\mathcal{T}^{-1} = \boldsymbol{r}, \tag{17.1.5}$$

while for the momentum operator $\boldsymbol{p}$, in classical mechanics $\boldsymbol{p} = d\boldsymbol{r}/dt$, or in quantum mechanics, $\boldsymbol{p} = (\hbar/i)\nabla$, and we obtain

$$\mathcal{T}\boldsymbol{p}\mathcal{T}^{-1} = -\boldsymbol{p}. \tag{17.1.6}$$

[e] H. Ohno *et al.*, *Appl. Phys. Lett.* **69**, 363 (1996); H. Ohno, *Science* **281**, 951 (1998).
[f] R. Kawakami, *Appl. Phys. Lett.* **77**, 2379 (2000).
[g] Y. Matsumoto *et al.*, *Science* **291**, 854 (2001).
[h] G. Du *et al.*, *Appl. Phys.* **73**, 6556 (1993).
[i] M. Kahasi *et al.*, *Phys. Rev. Lett.* **67**, 746 (1991).
[j] P. M. Allemend *et al.*, *Science* **253**, 302 (1991).

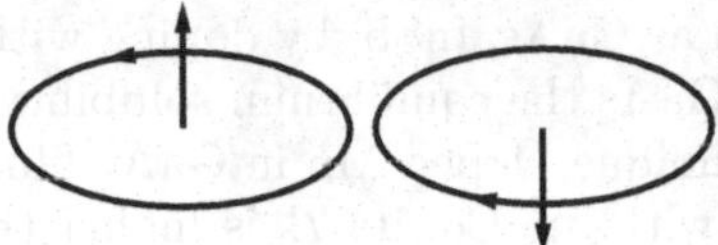

Figure 17.1.10 Schematic diagram for magnetic moments given by currents.

By combining the last two equations, the orbital angular momentum $\boldsymbol{L}$ should be satisfied by

$$\mathcal{T}\boldsymbol{L}\mathcal{T}^{-1} = -\boldsymbol{L}. \tag{17.1.7}$$

When spin is taken into account, because it is generalized from orbital angular momentum in the quantum sense, the commutativity should be in the same way as in (17.1.7)

$$\mathcal{T}\boldsymbol{S}\mathcal{T}^{-1} = -\boldsymbol{S}, \tag{17.1.8}$$

and the operation $\mathcal{T}$ thus reverses a magnetic moment. Alternatively, we can also understand the magnetic moment in a semiclassical way by thinking of $\mathcal{T}$ as reversing the direction of an electric current, because an orbiting charge constitutes a current as shown in Fig. 17.1.10. Since $I = dQ/dt$, we may write $-I = dQ/(-dt)$ and therefore we could also associate $\mathcal{T}$ as a time inversion operator.

Suppose that the Hamiltonian of a physical system is a function of coordinates, momenta and spins described by

$$\mathcal{H}(\boldsymbol{r}, \boldsymbol{p}, \boldsymbol{S}),$$

then according to (17.1.5), (17.1.6) and (17.1.8), it will be transformed as

$$\mathcal{T}\mathcal{H}(\boldsymbol{r}, \boldsymbol{p}, \boldsymbol{S})\mathcal{T}^{-1} = \mathcal{H}(\boldsymbol{r}, -\boldsymbol{p}, -\boldsymbol{S}). \tag{17.1.9}$$

Except when an external magnetic field is present, in general, the Hamiltonian will be invariant under the simultaneous transformation $\boldsymbol{r} \to \boldsymbol{r}$, $\boldsymbol{p} \to -\boldsymbol{p}$, $\boldsymbol{S} \to -\boldsymbol{S}$, such that

$$\mathcal{T}\mathcal{H}(\boldsymbol{r}, \boldsymbol{p}, \boldsymbol{S})\mathcal{T}^{-1} = \mathcal{H}(\boldsymbol{r}, \boldsymbol{p}, \boldsymbol{S}). \tag{17.1.10}$$

Therefore we can say that the Hamiltonian has time-reversal symmetry, even if it includes spin operators.

In most band theories, spin is irrelevant, for example, for a single electron in the periodic potential, the Hamiltonian is

$$\mathcal{H} = \frac{\boldsymbol{p}^2}{2m} + V(\boldsymbol{r}), \tag{17.1.11}$$

which has time reversal symmetry. Its eigenstates are the Bloch functions in space, and for each $\psi_{\boldsymbol{k}}(\boldsymbol{r})$ we can put two electrons with opposite spins into the two states with this wavefunction. Then time-reversal symmetry always leaves a Kramers degeneracy between any Bloch state $\psi_{\boldsymbol{k}}(\boldsymbol{r})$ and its complex conjugate $\psi^*_{\boldsymbol{k}}(\boldsymbol{r})$, which describes a state in which both the wavevector and the spin of the electron have been reversed. This implies $E(\boldsymbol{k}) = E(-\boldsymbol{k})$, regardless of point group symmetries of the lattice.

Another simple example for which this is true is a single-particle Hamiltonian containing spin-orbit coupling

$$\mathcal{H} = \frac{1}{2m}\boldsymbol{p}^2 + V(\boldsymbol{r}) + \frac{\hbar}{4m^2c^2}\sigma \cdot (\nabla V \times \boldsymbol{p}). \tag{17.1.12}$$

which is obviously invariant under time reversal operation.

In the rest of this chapter, we will be concerned with magnetic ordering. There are two theoretical models to treat the ordering process, one begins with the Heisenberg Hamiltonian (§11.2.4), and the other with the Hubbard Hamiltonian (§13.1.2). The former recognizes the existence of local magnetic moments, while the latter favors band electrons by taking their spins into account.

The Heisenberg Hamiltonian is based on the exchange interaction of local electrons introduced first in §14.4. If there are magnetic ions with spin vector $\boldsymbol{S}_i$ at lattice sites $\boldsymbol{R}_i$, we can construct the Heisenberg Hamiltonian

$$\mathcal{H} = -\sum_{i>j} J_{ij}\boldsymbol{S}_i \cdot \boldsymbol{S}_j, \tag{17.1.13}$$

by considering the interaction of near neighbor ion pairs; then J_{ij} is the exchange integral, as a function of the distance between ions i and j, i.e.,

$$J_{ij} = J_{ij}(|\boldsymbol{R}_i - \boldsymbol{R}_j|), \tag{17.1.14}$$

In most magnetic insulators, the nearest neighboring interaction is dominant, and can be taken as a constant J. The Hamiltonian is now written as

$$\mathcal{H} = -J\sum_{i>j} \boldsymbol{S}_i \cdot \boldsymbol{S}_j. \tag{17.1.15}$$

It is not difficult to note that the Heisenberg Hamiltonian (17.1.9) or (17.1.10) is invariant under time-reversal. However, the addition of a term $-\boldsymbol{H}\cdot\boldsymbol{S}$, where $\boldsymbol{H}$ is a fixed external magnetic field, destroys the invariance. This is broken time-reversal symmetry, broken by an applied field. But, even though there is no external field, as temperature decreases, there will be a magnetization which arises in systems described by the Heisenberg Hamiltonian. We shall find later that the materials with $J > 0$ are ferromagnets, and $J < 0$ antiferromagnets. In these cases, the time-reversal symmetry is spontaneously broken.

In contrast to the Heisenberg model for local magnetic moments, to investigate the itinerant electrons with interactions can start from the Hubbard Hamiltonian

$$\mathcal{H} = \sum_{ij\sigma} T_{ij}c^{\dagger}_{i\sigma}c_{i\sigma} + \frac{1}{2}U\sum_{i\sigma} n_{i\sigma}n_{i\bar{\sigma}}, \tag{17.1.16}$$

where i and j describe the atomic positions, T_{ij} is a matrix element between two sites, and the interaction is

$$U = \int\int |\psi(\boldsymbol{r}_1)|^2 \frac{e^2}{|\boldsymbol{r}_1 - \boldsymbol{r}_2|}\psi(\boldsymbol{r}_2)|^2 d\boldsymbol{r}_1 d\boldsymbol{r}_2. \tag{17.1.17}$$

It is easy to see that if the spins in (17.1.16) are all reversed, its form is the same, so the Hubbard Hamiltonian is also invariant under time-inversion. It only remains to apply an external field or to decrease the temperature to break the time-reversal symmetry.

Time-reversal is associated with inversion of the spin orientations. In the case of a crystal without magnetic moments, or loosely speaking in the case of a paramagnetic substance in which the local moments is randomly distributed, the crystal Hamiltonian will be invariant under time-reversal. However, broken time-reversal invariance leads to the appearance of a rich variety of magnetic structures. As introduced in Chap. 1, taking the spin operator related to black and white operations, there are 1191 magnetic space groups, instead of 230 colorless space groups (see §1.5.2). There are different magnetic states, e.g., paramagnetic, ferromagnetic, antiferromagnetic, and ferrimagnetic, as shown in Fig. 17.1.1, or even more complicated magnetic ordering structures like helical, canted, spiral, umbrella-like, etc. All these show the diversity of magnetic structures arising from spins.

§17.2 Theory Based on Local Magnetic Moments

The physical picture of magnetism in solids was first established on the basis of local moments. Weiss (1907) was the first to introduce the concept of molecular field to study ferromagnetism. It was later used by Néel (1936, 1948) to study antiferromagnetism and ferrimagnetism. The essential idea is that each spin is acted upon by an effective magnetic field proportional to the magnetization of the crystal. Weiss's theory was proposed before Heisenberg's quantum theory. Now in this section we shall use Heisenberg Hamiltonian to derive Weiss's effective field, then to investigate the ferromagnetic, antiferromagnetic and ferrimagnetic transitions, finally we go beyond the framework of the molecular field to discuss the problem of magnetic ground state. The ferromagnetic ground state is very simple, the exact result is same as that derived using molecular field; but the antiferromagnetic ground state is more sophisticated, so far a perfect solution is still lacking.

17.2.1 Mean-Field Approximation for Heisenberg Hamiltonian

When we consider coupling only between atoms and their nearest neighbors, from Heisenberg Hamiltonian (17.1.15), we can write the Hamiltonian for a cluster around a single atom at lattice point i

$$\mathcal{H}_i = -J\boldsymbol{S}_i \cdot \sum_{j=1}^{z} \boldsymbol{S}_j, \tag{17.2.1}$$

where z is the number of nearest neighbors. An effective field $\boldsymbol{H}_{\text{eff}}$ can be defined as

$$g_{\text{L}}\mu_{\text{B}}\boldsymbol{H}_{\text{eff}} = J\sum_{j=1}^{z}\boldsymbol{S}_j = J\sum_{j=1}^{z}\langle\boldsymbol{S}_j\rangle = zJ\langle\boldsymbol{S}_j\rangle, \tag{17.2.2}$$

where g_{L} is the Landé factor, and μ_{B} the Bohr magneton equal to $e\hbar/2mc$. The essential point is that we have taken the average $\langle\boldsymbol{S}_j\rangle$ to replace $\boldsymbol{S}_j$. The single atom Hamiltonian now becomes

$$\mathcal{H}_i = -g_{\text{L}}\mu_{\text{B}}\boldsymbol{S}_i \cdot \boldsymbol{H}_{\text{eff}}. \tag{17.2.3}$$

Because the magnetization is

$$\boldsymbol{M} = Ng_{\text{L}}\mu_{\text{B}}\langle\boldsymbol{S}_j\rangle, \tag{17.2.4}$$

the effective field is

$$\boldsymbol{H}_{\text{eff}} = \frac{zJ}{g_{\text{L}}\mu_{\text{B}}}\langle\boldsymbol{S}_j\rangle = \frac{zJ}{Ng_{\text{L}}^2\mu_{\text{B}}^2}\boldsymbol{M} = \gamma\boldsymbol{M}, \tag{17.2.5}$$

with $\gamma = zJ/Ng_{\text{L}}^2\mu_{\text{B}}^2$.

If there is an external field $\boldsymbol{H}$ added, then total field is

$$\boldsymbol{H}_{\text{t}} = \boldsymbol{H} + \boldsymbol{H}_{\text{eff}} = \boldsymbol{H} + \gamma\boldsymbol{M}. \tag{17.2.6}$$

Assume that $\boldsymbol{H}$ is along the z axis; due to the isotropy of (17.1.16), broken symmetry leads $\boldsymbol{H}_{\text{eff}}$ to also be along the z axis, then the field may be treated as scalars. Instead of (17.2.1), the single atom Hamiltonian is

$$\mathcal{H}_i = -g_{\text{L}}\mu_{\text{B}}S_{iz}H_{\text{t}}. \tag{17.2.7}$$

The eigenvalues of $\mathcal{H}_i$ are

$$E_\nu = -g_{\text{L}}\mu_{\text{B}}\nu H_{\text{T}},\ \nu = -S, \ldots, S. \tag{17.2.8}$$

For studying the thermodynamic behaviors of the system, the partition function is necessary. It can be written according to (17.2.8)

$$Z = \sum_{\nu=-S}^{S} \mathrm{e}^{-E_\nu/k_{\text{B}}T} = \sum_{\nu=-S}^{S} \mathrm{e}^{\nu g_{\text{L}}\mu_{\text{B}}H_{\text{t}}/k_{\text{B}}T}, \tag{17.2.9}$$

and after summation

$$Z = \frac{\sinh[g_{\text{L}}\mu_{\text{B}}H_{\text{t}}(2S+1)/2k_{\text{B}}T]}{\sinh[g_{\text{L}}\mu_{\text{B}}H_t/2k_{\text{B}}T]}. \tag{17.2.10}$$

The magnetization in (17.2.4) is

$$M = Nk_{\text{B}}T\frac{\partial \ln Z}{\partial H_{\text{t}}} = Ng_{\text{L}}\mu_{\text{B}}S\mathrm{B}_S(x), \tag{17.2.11}$$

where B_S is the so-called Brillouin function, defined as

$$\mathrm{B}_S(x) = \frac{2S+1}{2S}\coth\left(\frac{2S+1}{2S}x\right) - \frac{1}{2S}\coth\left(\frac{1}{2S}x\right), \tag{17.2.12}$$

and

$$x = \frac{g_L \mu_B S H_t}{k_B T}. \tag{17.2.13}$$

By defining the maximum of magnetization

$$M_0 = N g_L \mu_B S, \tag{17.2.14}$$

we can get the reduced magnetization as

$$m = M/M_0 = B_S(x), \tag{17.2.15}$$

where $B_S(x)$ can take values from 0 to 1 as x changes from infinity to zero. Let $x_0 = g_L \mu_B S H / k_B T$, then

$$x = x_0 + (zJS^2/k_B T)m. \tag{17.2.16}$$

We see that (17.2.15) and (17.2.16) are two coupled equations and can be solved graphically. The result is shown in Fig. 17.2.1.

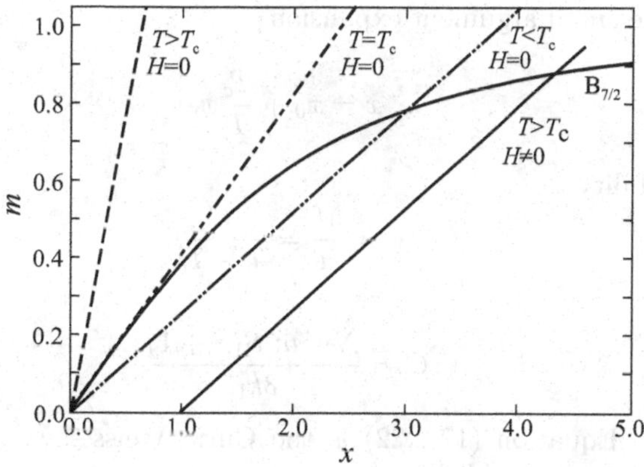

Figure 17.2.1 Graphical solution of magnetization.

17.2.2 Ferromagnetic Transition

Ferromagnets must experience a spontaneous breaking of time-reversal symmetry from the paramagnetic phase to the ferromagnetic phase, after the temperature decreases through the transition temperature T_c.

First consider that the external field $H = 0$. For small values of x, the Brillouin function (17.2.12) is expanded as

$$B_S(x) = \frac{S+1}{3S}x - \frac{[(S+1)^2 + S^2](S+1)}{90S^3}x^3 + \cdots. \tag{17.2.17}$$

A spontaneous internal magnetic field develops at temperature T_c, where it is first possible to satisfy (17.2.15) and (17.2.16). Substituting (17.2.17) to (17.2.15), we get

$$T_c = \frac{2JzS(S+1)}{3k_B}. \tag{17.2.18}$$

For $T > T_c$, a magnetic solution exists only if an external field is present. For $T < T_c$, the spontaneous magnetization in zero field satisfies the transcendental equation

$$m = B_S\left(\frac{3S}{S+1}\frac{T_c}{T}m\right). \tag{17.2.19}$$

In the vicinity of T_c, m is given by

$$m^2 = \frac{10}{3}\frac{(S+1)^2}{(S+1)^2+S^2}\frac{T_c - T}{T_c}, \tag{17.2.20}$$

if the expansion (17.2.17) is used. The relationship between the magnetization and temperature is

$$m \sim (T_c - T)^{1/2}.$$

We note that above T_c, the magnetic dipoles in a solid are randomly oriented, so in the absence of an applied magnetic field there is no net magnetic moment. Below T_c, there is a partial alignment of the dipoles and a spontaneous magnetization. It is interesting to note that the magnetization $\boldsymbol{M}$ is a vector whose direction is not unique because the basic Hamiltonian (17.1.16) is isotropic. Thus upon heating a magnet above T_c and recooling, the same magnitude of magnetization will occur, but the direction may be different. If the magnetic order is influenced by crystal structures, an anisotropic term should be added into the Hamiltonian; then this degeneracy will be lifted and lowered.

We will investigate some other thermodynamic quantities related to phase transitions. We find from (17.2.16) using the small argument expansion

$$x = x_0 + \frac{T_c}{T}x. \tag{17.2.21}$$

This gives the susceptibility

$$\chi = \frac{M}{H} = \frac{C}{T - T_c}, \tag{17.2.22}$$

where

$$C = \frac{N g_L^2 \mu_B^2 S(S+1)}{3k_B} \tag{17.2.23}$$

is the Curie constant. Equation (17.2.22) is the Curie–Weiss law. When $T \to T_c, \chi \to \infty$, a spontaneous magnetization is established.

The specific heat can be deduced from the internal energy

$$U = -M\left(\frac{1}{2}\gamma M + H\right) = -\frac{1}{2}\gamma M_0^2 m\left(m + \frac{2H}{\gamma M_0}\right),$$

that is

$$C_M = \frac{\partial U}{\partial T} = -\gamma M_0^2\left(m + \frac{H}{\gamma M_0}\right)\frac{\partial m}{\partial T}. \tag{17.2.24}$$

When $H = 0$, for $T > T_c$, $m = 0$, so

$$C_M = 0, \tag{17.2.25}$$

and for $T < T_c$, $m \neq 0$, then

$$C_M = -\frac{1}{2}\gamma M_0 \frac{dm^2}{dT} = 5Nk_B\frac{S(S+1)}{(S+1)^2+S^2}. \tag{17.2.26}$$

We can see that the specific heat has a finite discontinuity at the Curie temperature. This indicates a second-order phase transition. In fact the results from the molecular field, including magnetization, susceptibility, and specific heat, are consistent with Landau theory. If we take $m = M/M_0$ as the order parameter to construct the free energy

$$G = G_0 + A(T)m^2 + Bm^4 - Hm, \tag{17.2.27}$$

we can confirm these results easily by using the treatment in Chap. 15.

17.2.3 Antiferromagnetic Transition

A negative value of the exchange integral J in the Heisenberg Hamiltonian (17.1.16) may lead to the appearance of antiferromagnetism. For simplicity, we take two sublattice system as an example. Alternatively, we consider the next nearest neighbor model in which there are two interaction constants between two kinds of spins. $J_{12} = J_{21}$ couples the different sublattices, but $J_{11} = J_{22}$ couples the same sublattice. The basic treatment stated below is not difficult to generalize to the case of n sublattices.

It is natural to use the effective field method discussed before, but for two sublattices i, j (i, $j =$ 1, 2), there are two effective fields

$$\boldsymbol{H}_i = \boldsymbol{H} + \sum_j \gamma_{ij} \boldsymbol{M}_j, \tag{17.2.28}$$

in which each effective field includes the contribution of magnetizations from two sublattices through the coupling constants

$$\gamma_{ij} = \frac{2 z_{ij} J_{ij}}{N g^2 \mu_{\mathrm{B}}^2}, \tag{17.2.29}$$

where z_{ij} is the number of neighbors of an atom on sublattice i which are on sublattice j and are connected with i by the exchange parameter J_{ij}. If we suppose that the nearest neighbors of an atom on sublattice 1 are all on sublattice 2, and *vice versa*, then $z_{12} = z_1$ is the total number of nearest neighbors. We suppose that the next nearest neighbors of a given atom are on the same sublattice, so we put $z_{11} = z_{22} = z_2$ is the total number of next nearest neighbors. We also relabel $J_1 = J_{12}$, the exchange parameter for nearest neighbors, and $J_2 = J_{11}$ the exchange parameter for next nearest neighbors. Then

$$\gamma_{12} = \gamma_{21} = \frac{2 z_1 J_1}{N g_{\mathrm{L}}^2 \mu_{\mathrm{B}}^2}, \gamma_{11} = \gamma_{22} = \frac{2 z_2 J_2}{N g_{\mathrm{L}}^2 \mu_{\mathrm{B}}^2}.$$

Repeating the arguments of the ferromagnetic case, reduced magnetization on sublattice i is

$$m_i = \mathrm{B}_S(x_i), \tag{17.2.30}$$

where $x_i = g_{\mathrm{L}} \mu_{\mathrm{B}} H_i S / k_{\mathrm{B}} T$, and $M_{0i} = N g_{\mathrm{L}} \mu_{\mathrm{B}} S/2$. At higher temperatures, the material is in the paramagnetic phase; the material is only magnetized in the direction of applied field H. We expand the Brillouin function (17.2.30) and retain only the first term, then

$$m_i = \mathrm{B}_S(x_i) = \frac{S+1}{3S} x_i = \frac{(S+1) g_{\mathrm{L}} \mu_{\mathrm{B}} H_i}{3 k_{\mathrm{B}} T}.$$

Substitute it into (17.2.28), it is found

$$M_i - \frac{C}{2T} \sum_j \gamma_{ij} M_j = \frac{C}{2T} H, \tag{17.2.31}$$

where C is the Curie constant defined by (17.2.23). We sum (17.2.31) with respect to i and make use of the symmetry properties of the coefficients γ: $\gamma_{ij} = \gamma_{ji}$, for $i \neq j$, and $\gamma_{ii} = \gamma_{jj}$. The result is

$$M \left(1 - \frac{C}{2T} \sum_j \gamma_{ij} \right) = \frac{C}{T} H. \tag{17.2.32}$$

Now the paramagnetic susceptibility is derived directly from (17.2.32)

$$\chi = \frac{M}{H} = \frac{C}{T + \Theta_A}, \tag{17.2.33}$$

where

$$\Theta_A = -\frac{C}{2}\sum \gamma_{ij} = -\frac{2S(S+1)}{3k_B}\sum_i z_i J_i. \tag{17.2.34}$$

Antiferromagnetism is characterized by the dominance of terms with negative J, and Θ_A will be a positive temperature. Thus, the susceptibility of an antiferromagnet in the paramagnetic region obeys a Curie–Weiss law with a negative Curie temperature $T_c = -\Theta_A$.

To investigate the possibility of a transition to an antiferromagnetic ordered state in zero external field, we set $H = 0$ in (17.2.31) and look for a nontrivial solution by

$$\begin{vmatrix} 2T/C - \gamma_{11} & \gamma_{12} \\ \gamma_{12} & 2T/C - \gamma_{11} \end{vmatrix} = 0. \tag{17.2.35}$$

The roots of this equation give a set of two possible transition temperatures, the eigenvectors give the ratios of the M_i, that is, the relative magnetization of the two sublattices. The solutions for the transition temperatures are

$$T = \frac{C}{2}(\gamma_{11} \pm \gamma_{12}) = \frac{2S(S+1)}{3k_B}(z_2 J_2 \pm z_1 J_1). \tag{17.2.36}$$

The solution with the plus (+) sign has $T = -\Theta_A$. This describes a transition at negative temperature, not observable. In fact, this solution corresponds to ferromagnetism as is seen from the eigenvector ($M_1 = M_2$), and therefore is an unstable state of the system. The other root has an eigenvector which gives $M_1 = -M_2$, so that the two sublattices are magnetized oppositely. This is the antiferromagnetic state, and the transition temperature, called the Néel temperature, is

$$T_N = \frac{2S(S+1)}{3k_B}(z_2 J_2 - z_1 J_1). \tag{17.2.37}$$

If there is only nearest neighbor interaction, then $J_1 = -|J|,\ J_2 = 0$, thus

$$T_N = \Theta_A = \frac{2S(S+1)}{3k_B} z_1 |J|.$$

17.2.4 Ferrimagnetic Transition

Ferrimagnets are the extension of antiferromagnets, but unlike antiferromagnets, their magnetic moments on different sublattices cannot fully cancel each other. This means that the antiparallel configuration of the magnetic moments of two sublattices is the lower-energy state of the system, but the two types of magnetic moments are different. Néel's mean-field treatment to ferrimagnetic transition can give a reasonable explanation of many magnetic properties of ferrites.

Using similar reasoning for antiferromagnets in §17.2.3, we still obtain the expressions (17.2.28) and (17.2.29) in effective fields, in which the magnetizations of the two sublattices are M_1 and M_2, respectively, and γ_{ij} represent the coupling constants among the same sublattice or between two sublattices. For ferrimagnets, $\gamma_{12} = \gamma_{21}$ is still satisfied, and both are negative, but $M_1 \neq M_2$, and also $\gamma_{11} \neq \gamma_{22}$. In general, γ_{11} and γ_{22} can be positive or negative, but for most of the ferrimagnetic materials, they are negative; moreover, $|\gamma_{11}|$ and $|\gamma_{22}|$ are far less than $|\gamma_{12}|$. Taking the z axis as the symmetry-breaking direction and denoting the effective fields and magnetizations by scalar quantities, the magnetizations in equilibrium are

$$M_i = N_i g_L \mu_B S_i \mathrm{B}_{S_i}(x_i), \tag{17.2.38}$$

where N_i is the number of atoms with spin quantum number S_i in unit volume, $\mathrm{B}_{S_i}(x)$ and x_i are defined similar to (17.2.12) and (17.2.13), only the spins and total field are changed.

According to (17.2.17), near the transition temperature expasion of the Brillouin function to the first order can give

$$M_i = \frac{C_i}{T} H_i, \tag{17.2.39}$$

where the Curie constant C_i is obtained by adding superscript and subscript i to N and S in (17.2.23). The next step is to substitute the effective fields with the scalar expressions in (17.2.28). We can write the following explicit equations

$$(T - C_1\gamma_{11})M_1 - C_1\gamma_{12}M_2 = C_1H,$$
$$-C_2\gamma_{12}M_1 + (T - C_2\gamma_{22}M_2) = C_2H. \tag{17.2.40}$$

From these two equations, the magnetizations of the sublattices are obtained, respectively, as

$$M_1 = \frac{C_1(T - C_2\gamma_{22}) + C_1C_2\gamma_{12}}{(T - C_1\gamma_{11})(T - C_2\gamma_{22}) - C_1C_2\gamma_{12}^2}H,$$
$$M_2 = \frac{C_2(T - C_1\gamma_{11}) + C_1C_2\gamma_{12}}{(T - C_1\gamma_{11})(T - C_2\gamma_{22}) - C_1C_2\gamma_{12}^2}H. \tag{17.2.41}$$

The total magnetization is $M = M_1 + M_2$, and the susceptibility is determined by $\chi = M/H$. In order to show the relation of susceptibility and temperature, it is more convenient to write its reciprocal form

$$\frac{1}{\chi} = \frac{T}{C} - \frac{1}{\chi_0} - \frac{\sigma}{T - \Theta}, \tag{17.2.42}$$

where $C = C_1 + C_2$, the other constants χ_0, σ and Θ can all be conveniently determined by the material parameters C_i and γ_{ij}. The relation $1/\chi - T$ in (17.2.42) is a hyperbola, as shown in Fig. 17.1.1(d). Its asymptotic behavior at high temperatures ($T \to \infty$) can be simplified to

$$\frac{1}{\chi} = \frac{T}{C} - \frac{1}{\chi_0}. \tag{17.2.43}$$

For general ferrimagnetic materials, the intercept $-1/\chi_0$ is positive, then one positive temperature can be defined as

$$\Theta = -C/\chi_0, \tag{17.2.44}$$

so (17.2.43) becomes

$$\chi = \frac{C}{T + \Theta}. \tag{17.2.45}$$

This expression is very similar in form to the paramagnetic susceptibility of antiferromagnets (17.2.33). That is to say, at high enough temperatures, the relation between $1/\chi$ and T for ferrimagnets is essentially linear, similar to the mean-field results of antiferromagnetic materials.

Taking $H = 0$ and getting the roots from the coefficient determinant in (17.2.40), we can find the Néel temperature for a ferrimagnetic transition

$$T_N = \frac{1}{2}\{C_1\gamma_{11} + C_2\gamma_{22} + [(C_1\gamma_{11} - C_2\gamma_{22})^2 + 4C_1C_2\gamma_{12}^2]^{1/2}\}. \tag{17.2.46}$$

Below this Néel temperature, the two sublattices tend to be antiparallel, and spontaneous magnetizations can appear. The magnetizations of the sublattices are determined by the two transcendental equations, also the external field $H = 0$; then the composite spontaneous magnetization is

$$M(T) = M_1(T) + M_2(T). \tag{17.2.47}$$

Due to the difference of material parameters in ferrimagnets, there are some interesting characteristics in their magnetic properties which are absent in ferromagnets and antiferromagnets. Generally speaking, the maximum of magnetization of each sublattice is located at zero temperature, corresponding to $M_1(0)$ and $M_2(0)$. As the temperature rises, the magnetization decreases and becomes zero at the Néel temperature. It is possible that the relation of the composite magnetization and temperature is similar to the standard ferromagnetic curve in Fig. 17.1.1(b), but some unconventional curves can often appear; one example is shown in Fig. 17.2.2. It can be seen from the figure that the direction of the composite magnetization can be reversed in some temperature range, denoted

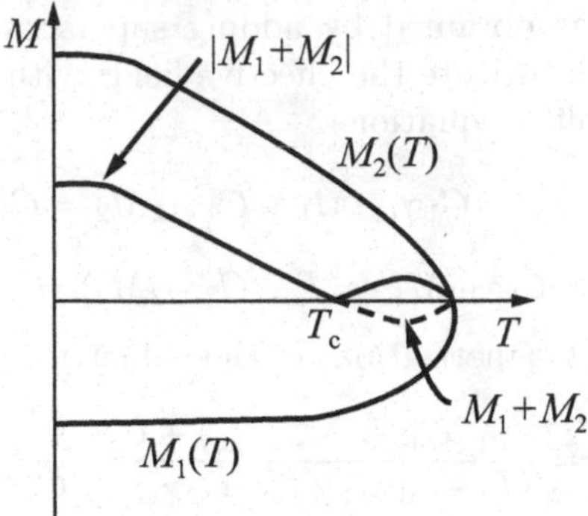

Figure 17.2.2 Reversal of magnetization direction in ferrimagnets.

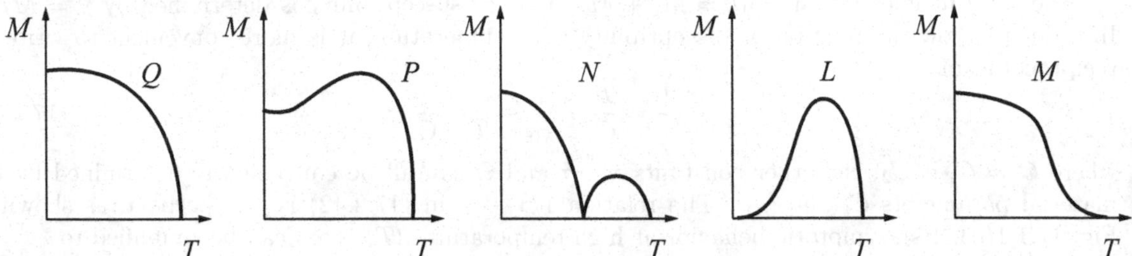

Figure 17.2.3 Relation of magnetization-temperature in ferromagnets.

by the dashed line. However, the physical quantity actually measured is $|M_1(T) + M_2(T)|$, i.e., the solid line in the figure. Then there is a temperature corresponding to the composite magnetization equal to zero, represented in the figure by T_c, which is called the compensated point. According to the investigation by Néel, all the $M(T)$-T curves in Fig. 17.2.3, denoted by Q, P, N, L and M, respectively, can appear.

17.2.5 Ferromagnetic and Antiferromagnetic Ground States

In §17.2.2 and §17.2.3, we have seen the appearance of ferromagnetic and antiferromagnetic ordered phases when the temperature decreases. We are interested here in their ground states.

The Heisenberg Hamiltonian (17.1.15) can be rewritten in the form of raising and lowering operators,

$$\mathcal{H} = -\sum_{i>j} J_{ij} \boldsymbol{S}_i \cdot \boldsymbol{S}_j = -\sum_{i>j} J_{ij} \left(S_i^z S_j^z + \frac{1}{2} S_i^+ S_j^- + \frac{1}{2} S_i^- S_j^+ \right), \tag{17.2.48}$$

here the raising and lowering operators S_i^+, S_i^- are defined as

$$S_i^{\pm} = S_i^x \pm i S_i^y, \tag{17.2.49}$$

and the commutation relations are

$$[S_i^z, S_j^{\pm}] = \pm \delta_{ij} S_i^{\pm}, \ [S_i^+, S_j^-] = 2\delta_{ij} S_i^z. \tag{17.2.50}$$

It is noted that spin operators for different sites commute.

We can write the difference of the operators

$$\frac{1}{2} S_i^- S_j^+ - \frac{1}{2} S_i^+ S_j^- = i(S_i^x S_j^y - S_i^y S_j^x), \tag{17.2.51}$$

which is antisymmetric about i and j, so vanishes on summation. Now the Heisenberg Hamiltonian is

$$\mathcal{H} = -\sum_{i,j} J_{ij}(S_i^z S_j^z + S_i^- S_j^+). \tag{17.2.52}$$

It is convenient to define $J_{ii} = 0$, and thus extend the sum over all i, j, without restriction.

Let us consider the state of the coupled spins in which each S_i^z attains its maximum value S. We call this state $|0\rangle$, which represents the completely ferromagnetic state

$$|0\rangle = \prod_i |S\rangle_i. \tag{17.2.53}$$

This state has the property that

$$S_i^z|S\rangle_i = S|S\rangle_i,\ S_i^+|S\rangle_i = 0, \tag{17.2.54}$$

in which the left equality means any spin takes its maximum quantum number on the z axis, while the right hand equality says that no spin can be raised. The eigenenergy of this state for Hamiltonian (17.2.42) is

$$\mathcal{H}|0\rangle = E_0|0\rangle,\ E_0 = -S^2\sum_{ij} J_{ij}.$$

In the case in which J_{ij} differs from zero only when i and j are nearest neighbors, the sum over j gives N, the total number of atoms in the crystal. Taking into account the number of nearest neighbors z, the energy

$$E_0 = -NzS^2J. \tag{17.2.55}$$

The state $|0\rangle$ can easily be seen to be the ground state of the system for positive J. If any spin has less than its maximum z component, then the eigenvalue of (17.2.42) will be less. So, in the case of ferromagnetic spin systems ($J > 0$), the state with all spins are aligned along one direction, becomes the ground state of the Heisenberg Hamiltonian.

It is relatively simple to find out the ferromagnetic ground state. In contrast, the antiferromagnetic ordered state described by the molecular field is not the ground state of the Heisenberg Hamiltonian for quantum spins. In the case of an antiferromagnetic spin system ($J < 0$) investigated above, the ordered state in the molecular field approximation consists of two sublattices, in which all spins on each of the sublattices are aligned anti-parallel along the same direction. But this state is not the ground state of the Heisenberg Hamiltonian; it is not even an eigenstate.

In the molecular field approximation for the antiferromagnetic spin system, as Anderson (1951) pointed out,[k] the Néel state energy is determined solely by the longitudinal part of the exchange interaction. Therefore the molecular field value

$$E = NzS^2J \tag{17.2.56}$$

is an upper bound for the ground state energy of antiferromagnetic spin systems. We consider a cluster consisting of a spin $\boldsymbol{S}_i$ on one sub-lattice and its z neighboring spins $\boldsymbol{S}_{i+\delta}$ on the other sub-lattice, which are interacting with the former. The Hamiltonian for this cluster is given by

$$\mathcal{H}_i = -2J\boldsymbol{S}_i \cdot \sum_\delta \boldsymbol{S}_{i+\delta}. \tag{17.2.57}$$

If we write $\boldsymbol{S}_\mathrm{t} = \sum_\delta \boldsymbol{S}_{j+\delta}$, the energy of this spin cluster is given by

$$\begin{aligned} -2J\boldsymbol{S}_i \cdot \boldsymbol{S}_\mathrm{t} &= -J[(\boldsymbol{S}_i + \boldsymbol{S}_\mathrm{t})^2 - \boldsymbol{S}_i^2 - \boldsymbol{S}_\mathrm{t}^2] \\ &= -J[(\boldsymbol{S}_i + \boldsymbol{S}_\mathrm{t})^2 - S(S+1) - S_\mathrm{t}(S_\mathrm{t}+1)]. \end{aligned}$$

[k] P. W. Anderson, *Phys. Rev.* **86**, 694 (1951).

For $J < 0$, this energy is the lowest when the magnitude of $\boldsymbol{S}_i + \boldsymbol{S}_t$ is $S_t - S$. The energy of the cluster for this case is

$$-J[(S_t - S)(S_t - S + 1) - S(S+1) - S_t(S_t + 1)] = 2JS(S_t + 1),$$

which is lowest when S_t has its maximum value zS. Thus the lowest energy is given by

$$2JS(zS + 1). \tag{17.2.58}$$

Since the total Hamiltonian is the sum of (17.2.47) over i, the ground state energy is clearly larger than $N/2$ times (17.2.48), so that this value gives a lower bound to the ground state energy E_G. Form (17.2.48), we find the following inequalities

$$NzJS^2 > E_G > NzJS^2\left(1 + \frac{1}{zS}\right). \tag{17.2.59}$$

The right-hand inequality suggests that the energy is lower than the molecular field approximation, when the quantum effect increases as zS becomes smaller. The results suggest, in particular, that in the one-dimensional lattice with spin 1/2 (giving the smallest zS), the ground state energy may become as low as twice the value of the molecular field approximation.

The ground state of antiferromagnets is an unsolved problem, except in the special case of a one-dimensional array of spin 1/2 ions with coupling only between nearest neighbors. The rigorous treatment of the one-dimensional antiferromagnetic ground state is based on the Bethe ansatz. Its mathematics are quite complicated and we shall not discuss them here. But we will give a simplified analysis, following Anderson (1972).[1] For a one-dimensional antiferromagnetic chain with $S = 1/2$, we can construct the Néel state

$$\psi_N = \alpha(1)\beta(2)\alpha(3)\beta(4)\cdots, \tag{17.2.60}$$

where α and β denote two opposite spin states. The energy of this system is

$$E = \langle\psi_N|\mathcal{H}|\psi_N\rangle = NJzS^2/2 = NJ/4. \tag{17.2.61}$$

For an alternating chain of paired atoms, the state is

$$\psi = \frac{\alpha(1)\beta(2) - \alpha(2)\beta(1)}{2}(34)(56)\cdots, \tag{17.2.62}$$

and the energy is

$$E = \frac{NJ}{2}S(S+1) = 0.75NJ, \tag{17.2.63}$$

which is lower than the energy in (17.2.61), but much closer to the correct energy $E_0 = 0.886NJ$.

These solutions of the chain are obtained by writing the wavefunction as a linear combination of products of pair-bond wave functions of the type of (17.2.62). Anderson called it the resonating valence bond (RVB) model, schematically shown in Fig. 17.2.4. Furthermore, Anderson applied this treatment to a two-dimensional case of a triangular lattice divided into three sublattices with spins of the three sublattices at 120° to each other. The energy of this state is

$$E_N^{\triangle} = NJ\left(3 \times \frac{1}{2}S^2\right) = 0.75NJ, \tag{17.2.64}$$

with each spin parallel to its local field. This energy as is

$$E_N^{\triangle} = 2 \times (0.463 \pm 0.007)NJ. \tag{17.2.65}$$

The two-dimensional case also permits a pair-bond trial wave function. The ground state may be that in which these many different bond configurations are linearly combined in the same wavefunction to lower the energy. Further result shows

$$E^{\triangle} = 2 \times (0.54 \pm 0.01)NJ. \tag{17.2.66}$$

[1] P. W. Anderson, *Mat. Rev. Bull.* **8**, 153 (1973); P. Fazekas and P. W. Anderson, *Phil. Mag.* **30**, 432 (1974).

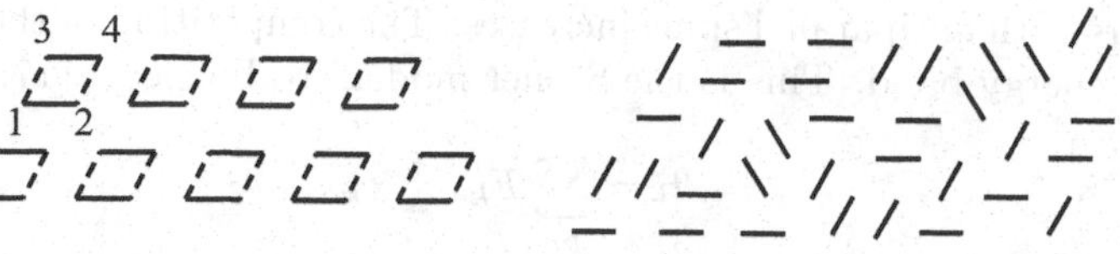

(a) (b)

Figure 17.2.4 Two-dimensional resonating valence bonds.

At absolute zero, the Néel state can be consistently postulated to be a locally stable minimum in the total energy, but not the absolute minimum. The relation between the Néel state and the real ground state is that of a quantum solid versus a quantum liquid. The Néel state is a solid: it has condensed into a spin lattice, whereas the real ground state is a fluid of mobile valence bonds, i.e., pairs of spins correlated together into singlets. The two states are so far apart in phase space as to be unavailable quantum-mechanically to each other. $LiNiO_2$, with a triangular lattice, has been recognized as a possible candidate for RVB as antiferromagnetic ground state. But actually, due to the interaction between the spin order and orbital order, it is more complicated than expected.[m]

§17.3 Theory Based on Itinerant Electrons

We have considered the magnetism that may be treated as though the electrons are contained within the cores of the atoms in solids. However, there are magnetic metals in which magnetism arises from conduction electrons. In the following, we will discuss the problem in the band picture. For conduction electrons with exchange interactions, we can understand why certain metals are ferromagnetic (Fe, Co, Ni), antiferromagnetic (Cr, Mn, γ-Fe) or nonmagnetic (Sc, V, Ti, etc.).

17.3.1 Mean-Field Approximation of Hubbard Hamiltonian

There is 5-fold orbital degeneracy for d electron, but for simplicity, we only consider the one-band model. Since the d-orbitals in the transition metals are much more localized than those of the s-electrons, the overlap of d-wavefunctions is much less than the s-wavefunctions. So we treat a conduction-electron system in one band with short-range interaction, i.e., an electron system described by the Hubbard Hamiltonian (17.1.16). Under the mean-field approximation, (17.1.16) can be rewritten in the form

$$\mathcal{H} = \sum_{ij\sigma} T_{ij} c_{i\sigma}^\dagger c_{i\sigma} + U \sum_{i\sigma} n_{i\sigma} \langle n_{i\bar{\sigma}} \rangle. \tag{17.3.1}$$

In a homogeneous system, $\langle n_{i\bar{\sigma}} \rangle$ has no dependence on position, so

$$\langle n_{i\bar{\sigma}} \rangle = \langle n_{\bar{\sigma}} \rangle.$$

Now (17.3.1) becomes

$$\mathcal{H} = \sum_{ij\sigma} T_{ij} c_{i\sigma}^\dagger c_{i\sigma} + U \sum_{i\sigma} \langle n_{\bar{\sigma}} \rangle c_{i\sigma}^\dagger c_{i\sigma}. \tag{17.3.2}$$

We can transform this into the Bloch representation,

$$\mathcal{H} = \sum_{\boldsymbol{k}\sigma} \varepsilon_{\boldsymbol{k}\sigma} c_{\boldsymbol{k}\sigma}^\dagger c_{\boldsymbol{k}\sigma} + \sum_{\boldsymbol{k}\sigma} U \langle n_{\bar{\sigma}} \rangle c_{\boldsymbol{k}\sigma}^\dagger c_{\boldsymbol{k}\sigma}. \tag{17.3.3}$$

The first term is the total energy of band electrons; its value is a minimum for $n_\sigma = n_{\bar{\sigma}}$. The second term represents the Coulomb energy for anti-parallel electrons; its value is lower, when the difference

[m]For the magnetic order of $LiNiO_2$, the experimental investigation can be seen in R. Reynaud *et al.*, *Phys. Rev. Lett.* **86**, 3638 (2001); and theoretical study in Y. Q. Li *et al.*, *Phys. Rev. Lett.* **81**, 3537 (1998).

of electron numbers with antiparallel spins increases. The competition of these two terms might lead to the splitting of energy band. This is the Stoner model. To be clear, (17.3.3) is rewritten as

$$\mathcal{H} = \sum_{\boldsymbol{k}\sigma} E_{\boldsymbol{k}\sigma} c_{\boldsymbol{k}\sigma}^{\dagger} c_{\boldsymbol{k}\sigma}, \tag{17.3.4}$$

with

$$E_{\boldsymbol{k}\sigma} = \varepsilon_{\boldsymbol{k}\sigma} + U\langle n_{\bar{\sigma}}\rangle, \tag{17.3.5}$$

to represent the energy of an itinerant electron. Although (17.3.4) is analogous to a Hamiltonian describing noninteracting electrons, the single electron energy $E_{\boldsymbol{k}\sigma}$ is correlated to the number of antiparallel electrons.

Actually U provides a molecular field for the energy band electrons. Define

$$n = \langle n_{\uparrow}\rangle + \langle n_{\downarrow}\rangle, \ m = \langle n_{\uparrow}\rangle - \langle n_{\downarrow}\rangle, \tag{17.3.6}$$

in which n is the number of itinerant electrons for each atom, and m is the relative magnetization per atom. If N is the number of atoms in unit volume, the magnetization is

$$M = N\mu_{\mathrm{B}}m. \tag{17.3.7}$$

From (17.3.6) and (17.3.7), we have

$$\langle n_{\sigma}\rangle = \frac{1}{2}(n + \sigma m), \ \sigma = \pm 1, \tag{17.3.8}$$

so

$$E_{\boldsymbol{k}\sigma} = \left(\varepsilon_{\boldsymbol{k}\sigma} + \frac{1}{2}nU\right) - \sigma\mu_{\mathrm{B}}\left(\frac{U}{2N\mu_{\mathrm{B}}^2}M\right). \tag{17.3.9}$$

The second term corresponds to a molecular field. If $\langle n_{\uparrow}\rangle \neq \langle n_{\downarrow}\rangle$, the energy band splitting is

$$E_{k\downarrow} - E_{k\uparrow} = U(\langle n_{\uparrow}\rangle - \langle n_{\downarrow}\rangle) \equiv 2\Delta, \tag{17.3.10}$$

as schematically shown in Fig. 17.3.1, there is spontaneous magnetization in the system.

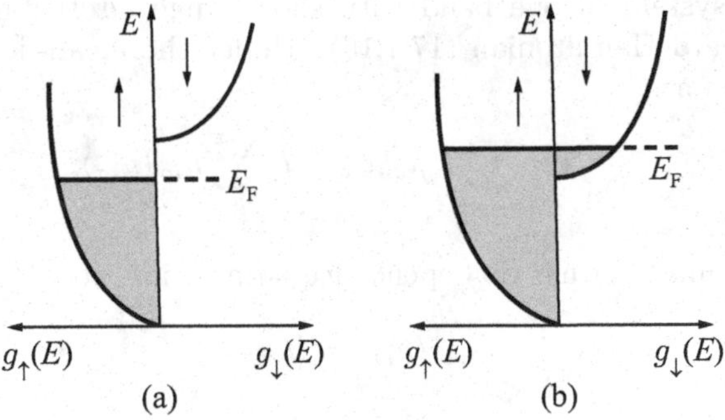

Figure 17.3.1 Band splitting of ferromagnetic phases in transition metals. $g_{\uparrow}(E)$ is the density of states with spin up, and $g_{\downarrow}(E)$ is the density of states with spin down. (a) Half-metallic ferromagnet with full spin polarization; (b) ordinary ferromagnet with partial spin polarization.

17.3.2 Stoner Theory of Ferromagnetism

First we should like to introduce the Stoner criterion for ferromagnetism. According to (17.3.6), the population of itinerant electrons in the Bloch state $(\boldsymbol{k}\sigma)$ satisfies Fermi distribution

$$\langle c_{\boldsymbol{k}\sigma}^{\dagger} c_{\boldsymbol{k}\sigma}\rangle = \langle n_{\boldsymbol{k}\sigma}\rangle = f(E_{\boldsymbol{k}\sigma}).$$

If there is an applied field H, we can rewrite (17.3.9) as

$$E_{k\sigma} = \varepsilon_{k\sigma} - \sigma\mu_{\rm B}\left(H + \frac{U}{2N\mu_{\rm B}^2}M\right), \tag{17.3.11}$$

where the constant term $(1/2)nU$ is omitted. By translation invariance, the magnetization is

$$M(T) = N\mu_{\rm B}(\langle n_\uparrow\rangle - \langle n_\downarrow\rangle) = \mu_{\rm B}\sum_k [f(E_{k\uparrow}) - f(E_{k\downarrow})]. \tag{17.3.12}$$

This is a self-consistent equation in combination with (17.3.11). The detailed result will depend on the band structure. For weak magnetic field and weak mean field, $f(E_{k\sigma})$ can be expanded and

$$M(T) \propto 2\mu_{\rm B}^2\left(H + \frac{U}{2N\mu_{\rm B}^2}M\right)\int_0^\infty \left[-\frac{\partial f(E)}{\partial E}\right] g(E)dE, \tag{17.3.13}$$

where $g(E)$ is the density of states. This is the generalization of (6.3.51) by adding the molecular field to the external field.

For itinerant electrons, $U \neq 0$, by using (6.3.52), we have

$$M(T) = \frac{\chi_{\rm P}(T)}{1 - 2(U/N)[\chi_{\rm P}(T)/4\mu_{\rm B}^2]}H \equiv \chi(T)H, \tag{17.3.14}$$

where $\chi_{\rm P}(T)$ is the paramagnetic susceptibility of itinerant electrons. Taking $4\mu_{\rm B}^2$ as the unit of magnetic susceptibility, we can see from (17.3.14) that $\chi_0(T)$ becomes infinite when $2(U/N)\chi_{\rm P}(T) = 1$. This shows that the paramagnetic phase is unstable and a spontaneous magnetization appears. The condition of stability for the ferromagnetic phase is

$$2\frac{U}{N}\chi_{\rm P}(T) > 1. \tag{17.3.15}$$

In terms of (6.3.52), $\chi_{\rm P}(T=0)/4\mu_{\rm B}^2 = g(E_{\rm F})/2$, the condition of ferromagnetism at zero temperature is

$$\frac{U}{N}g(E_{\rm F}) > 1. \tag{17.3.16}$$

This is called the Stoner criterion. Now the ferromagnetic ground state is energetically more favorable, and the total numbers for electrons with spin up and spin down are not the equilibrium values, as indicated in Fig. 17.3.1. Moreover, according as the interaction is strong or weak, one can have total or partial ferromagnetism.

For the paramagnetic phase at zero temperature, we find that

$$\chi(T=0) = \frac{\chi_{\rm P}(T=0)}{1 - (U/N)g(E_{\rm F})}. \tag{17.3.17}$$

It is clear that, due to the interaction U, the paramagnetic susceptibility at $T = 0$ K should be multiplied by the Stoner factor S

$$S = \frac{1}{1 - (U/N)g(E_{\rm F})}. \tag{17.3.18}$$

It can be seen from the Stoner criterion (17.3.16) that whether there is ferromagnetism from the itinerant electrons depends on the product of the interaction constant and the DOS at Fermi surface. Figure 17.3.2 gives the results from a theoretical computation of the exchange integral J and the DOS at Fermi surface $g(E_{\rm F})$ for some metals in the periodic table. It is to be noted that the exchange integrals J of some simple metals, like Li, Be, Na, Mg, K, are obviously larger than those of Fe, Co, Ni, but for $g(E_{\rm F})$ the former is much smaller than the latter. The reason for Fe, Co, Ni becoming metallic ferromagnets is that their J and $g(E_{\rm F})$ are all relatively larger, so the Stoner criterion is satisfied. In the periodic table, the whole trends of variation for J and $g(E_{\rm F})$ may be summarized as: in general, J decreases with Z, but in the transition metal group, J increases with

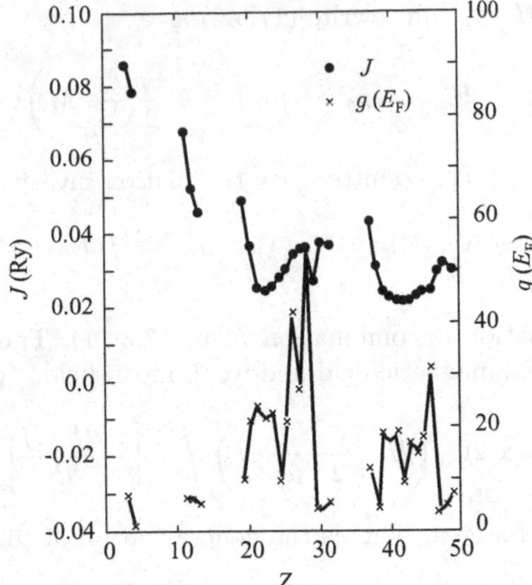

Figure 17.3.2 Exchange integral J and DOS at Fermi surface $g(E_{\mathrm{F}})$ with atomic number.

Z. The s band and p band of simple metals are wide, so their $g(E_{\mathrm{F}})$ are small, while the d bands of transition metals are narrow, so $g(E_{\mathrm{F}})$ is larger. In transition metals, the tendency for localization of electronic orbitals near the top of d bands makes J and $g(E_{\mathrm{F}})$ larger. These characteristics may explain why Fe, Co, Ni of 3d group are ferromagnets.

Now we will discuss the spontaneous magnetization. For a transition metal satisfying the condition (17.3.16), the spin polarization may be strong or weak, determined by its d electron band structure. To study the spontaneous magnetization, we consider the situation at $T \neq 0$ K without an applied field. The electronic energy from (17.3.11) is

$$E_{\boldsymbol{k}\sigma} = \varepsilon_{\boldsymbol{k}\sigma} - \frac{1}{2}\sigma U m, \tag{17.3.19}$$

and the Fermi distribution function is

$$f(E_{\boldsymbol{k}\sigma}) = \left[\exp\frac{E_{\boldsymbol{k}\sigma} - \frac{1}{2}\sigma U m - \mu}{k_{\mathrm{B}}T} + 1\right]^{-1}, \tag{17.3.20}$$

where μ is the chemical potential and equal to the Fermi energy E_{F} at $T = 0$ K.

The total number of electrons in the spin-up band N_+ is

$$N_+ = \int_0^\infty \left[\exp\frac{E - \frac{1}{2}Um - \mu}{k_{\mathrm{B}}T} + 1\right]^{-1} g(E)dE, \tag{17.3.21}$$

where the density of states can be different for different metals. We take the free electron form for further calculation,

$$g(E) = \frac{3}{4}NE_{\mathrm{F}}^{-3/2}E^{1/2}. \tag{17.3.22}$$

For convenience, we use the following abbreviations

$$x = \frac{E}{k_{\mathrm{B}}T}, \quad \eta = \frac{\mu}{k_{\mathrm{B}}T}, \quad \beta = \frac{Um}{2k_{\mathrm{B}}T}, \tag{17.3.23}$$

and define the function

$$F_{1/2}(\alpha) = \int_0^\infty \frac{x^{1/2}dx}{\mathrm{e}^{x-\alpha} + 1}. \tag{17.3.24}$$

Then

$$N_+ = \frac{3}{4}N\left(\frac{k_B T}{E_F}\right)^{3/2} F_{1/2}(\eta+\beta). \tag{17.3.25}$$

Similarly the total number of electrons in the spin-down band is

$$N_- = \frac{3}{4}N\left(\frac{k_B T}{E_F}\right)^{3/2} F_{1/2}(\eta-\beta). \tag{17.3.26}$$

The total number of electrons in two spin directions is $N = N_+ + N_-$, and the magnetization becomes $M = \mu_B(N_+ - N_-)$. We write them as

$$N = \frac{3}{4}N\left(\frac{k_B T}{E_F}\right)^{3/2} \left[F_{1/2}(\eta+\beta) + F_{1/2}(\eta-\beta)\right], \tag{17.3.27}$$

and

$$M = \frac{3}{4}N\mu_B\left(\frac{k_B T}{E_F}\right)^{3/2} \left[F_{1/2}(\eta+\beta) - F_{1/2}(\eta-\beta)\right]. \tag{17.3.28}$$

For ferromagnetic metals, μ is usually several eV, but $k_B T \sim 2.5\times10^{-3}$ eV for $T = 300$ K, so $\eta \gg 1$. Starting from (17.3.27) and (17.3.28), after a little mathematics, we get the following equation

$$\frac{U}{E_F} = \frac{1}{m}\left[(1+m)^{2/3} - (1-m)^{2/3}\right]\left[1 + \frac{\pi^2}{12}\left(\frac{k_B T}{E_F}\right)^2 (1-m^2)^{-2/3}\right], \tag{17.3.29}$$

which we will now use to determine the spontaneous magnetization.

When $T = 0$ K, (17.3.29) is reduced into

$$\frac{U}{E_F} = \frac{1}{m}\left[(1+m)^{2/3} - (1-m)^{2/3}\right]. \tag{17.3.30}$$

We can write the condition for spontaneous magnetization

$$\frac{4}{3} < \frac{U}{E_F} < 2^{2/3}. \tag{17.3.31}$$

When $T = T_c$, $m = 0$, so the Curie temperature can obtained from (17.3.31)

$$\frac{\pi^2}{12}\left(\frac{k_B T_c}{E_F}\right)^2 = \frac{3}{4}\frac{U}{E_F} - 1. \tag{17.3.32}$$

Near the Curie point, $m \ll 1$, and also noting that $E_F \gg k_B T$, (17.3.29) can be approximated to

$$m^2 = \frac{9\pi^2}{8}\left(\frac{k_B T_c}{E_F}\right)^2 \left[1 - \left(\frac{T}{T_c}\right)^2\right], \tag{17.3.33}$$

or

$$M = M_0 \frac{3\pi}{2\sqrt{2}}\left(\frac{k_B T_c}{E_F}\right) \left[1 - \left(\frac{T}{T_c}\right)^2\right]^{1/2}, \tag{17.3.34}$$

which differs from the result $M(T) \sim (T_c - T)^{1/2}$ from molecular field theory. When $T > T_c$, the system is in the paramagnetic phase; in a reasonable approximation, from (17.3.14), the paramagnetic susceptibility is derived

$$\chi(T) \sim \frac{T_F^2}{T^2 - T_c^2}, \tag{17.3.35}$$

which is different from the Curie–Weiss law.

The Stoner band model is in agreement with almost all experimental measurements related to the d-bands, such as electronic specific heat, Fermi surface, transport properties, etc., and also explains the non-integer values of magnetization at $T = 0$. Anyway, the band model gives a satisfactory explanation for many properties of metals and alloys. But it should be noted that the Stoner theory is a mean-field theory for itinerant ferromagnets and there are still some discrepancies between experiment and theory. For a better understanding of these properties, it is necessary to have detailed knowledge of the band structure; it is also necessary to treat the interactions between the electrons in a less simplistic way, i.e., to include the fluctuations.

17.3.3 Weak Itinerant Ferromagnetism

We have seen above that the itinerant electron model description of the ferromagnetism of transition metals is very successful, however, its shortcomings are also obvious, especially the temperature dependence of the magnetization, as well as the paramagnetic susceptibility, is inconsistent with Curie–Weiss law. On the other hand, the T_c values estimated from this model for transition metals are often too high. The reason is that spin fluctuation has not been taken sufficiently into account; this is a common fault for mean-field theory. However we may note here that there is an interesting class of materials known as very weak itinerant ferromagnets which have very low Curie temperatures, i.e., ~ 10 K, and the energy splitting between the up and down spin subbands is very small. In these materials the temperature dependence of magnetization and susceptibility are in consistent with (17.3.34) and (17.3.35).

Theoretical consideration can similarly begin from (17.3.21), but now we allow an applied magnetic field, so the number of electrons with positive or negative spin can be written as

$$N_{\pm} = \int_0^{\infty} \left[\exp \frac{E - \mu \mp \frac{1}{2}Um \mp \mu_{\mathrm{B}} H}{k_{\mathrm{B}} T} + 1 \right]^{-1} g(E) dE. \tag{17.3.36}$$

The reduced magnetization is a function of temperature, as well as magnetic field, i.e., $m = m(H,T)$. Weak itinerant ferromagnetism means $m_0 = m(0,0) \ll 1$. If the applied field is not very strong, it is reasonable to assume $m(H,T) \ll 1$. Therefore, (17.3.36) can be expanded for temperatures satisfying $k_{\mathrm{B}}T/E_{\mathrm{F}} \ll 1$, we have

$$\frac{2}{N} g(E_{\mathrm{F}}) \left(\frac{mU}{2} + \mu_{\mathrm{B}} H \right) = m \left[1 + \alpha \left(\frac{T}{T_{\mathrm{c}}} \right)^2 \right] + \gamma m^3 \cdots , \tag{17.3.37}$$

with

$$\alpha = \frac{1}{6} \pi^2 (k_{\mathrm{B}} T_{\mathrm{c}})^2 (D_1^2 - D_2), \ \gamma = \frac{1}{8} \left[\frac{N}{g(E_{\mathrm{F}})} \right]^2 \left(D_1^2 - \frac{D_2}{3} \right), \ D_{\nu} = g^{(\nu)}(E_{\mathrm{F}})/g(E_{\mathrm{F}}),$$

where T_{c} is the Curie temperature and $g^{(\nu)}(E_{\mathrm{F}})$ is the ν-order derivative of g.

It is found from (17.3.37) that, at zero temperature and zero applied field, there is an expression

$$\frac{1}{N} g(E_{\mathrm{F}}) U = 1 + \gamma m_0^2, \tag{17.3.38}$$

which returns to the Stoner criterion when its right-hand side is ≥ 1. So $\gamma \gg 0$ is required. This is the necessary, but not sufficient, condition for the appearance of weak ferromagnetism.

Defining a zero-field dynamic susceptibility as

$$\chi_0 = \chi(0,0) = \left(\frac{\partial M(H,T)}{\partial H} \right)_{0,0}, \tag{17.3.39}$$

it can be deduced that

$$\chi_0 = g(E_{\mathrm{F}}) \mu_{\mathrm{B}}^2 \mu_0 / \gamma m_0^2. \tag{17.3.40}$$

This formula tells us that the susceptibility increases with decreasing m_0 and this has physical meaning. From (17.3.37) and (17.3.38), and T_c as the divergent temperature for $(\partial m/\partial H)_{H=0}$, i.e., $(\partial H/\partial m)_{m=m(0,T_c)} = 0$, we have

$$\alpha = \gamma m_0^2. \tag{17.3.41}$$

Because α is proportional to T_c^2, so above expression can be used to determine T_c.

From (17.3.38) to (17.3.41), (17.3.37) can be written into

$$\left[\frac{M(H,T)}{M(0,0)}\right]^3 - \left[\frac{M(H,T)}{M(0,0)}\right]\left[1-\left(\frac{T}{T_c}\right)^2\right] = \frac{2\chi_0 H}{M(0,0)}, \tag{17.3.42}$$

or further into

$$M^2(H,T) = M^2(0,0)[1-(T/T_c)^2 + 2\chi_0 H/M(H,T)]. \tag{17.3.43}$$

So the theory predicts that, at different temperatures, the relation between M^2 and H/M gives a series of parallel straight lines, and the line for $T = T_c$ passes through the origin. This type of plot is called an 'Arrott chart'. The parallel straight lines in Fig. 17.3.3 describe this variation, and are the results of one typical weak itinerant ferromagnet, $ZrZn_2$.

From (17.3.42), the temperature dependence of zero-field dynamic susceptibility $\chi = \chi(0,T)$ can be given above or below T_c

$$\chi = \begin{cases} \chi_0\left[1-\left(\dfrac{T}{T_c}\right)^2\right]^{-1}, & T < T_c, \\ 2\chi_0\left[\left(\dfrac{T}{T_c}\right)^2 - 1\right]^{-1}, & T > T_c. \end{cases}$$

From (17.3.43), we find that the temperature dependence of spontaneous magnetization is

$$M^2(0,T) = M^2(0,0)\left[1-\left(\frac{T}{T_c}\right)^2\right], \tag{17.3.44}$$

that is to say, if the relation of M^2 and T^2 is plotted, it should be a straight line. This was verified in the investigation of the compound $ZrZn_2$, as shown in Fig. 17.3.4.

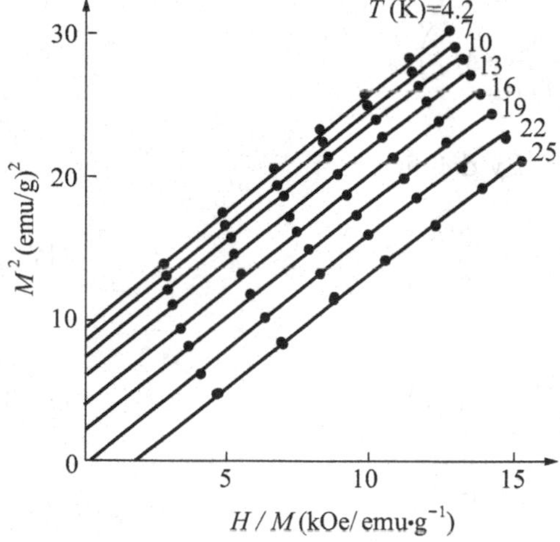

Figure 17.3.3 Square magnetization of $ZrZn_2$ compounds $M^2(H,T)$ versus $H/M(H,T)$. From B. Barbara, D. Gignoux, and C. Vettier, *Lectures on Modern Magnetism*, Science Press and Springer-Verlag, Beijing (1988).

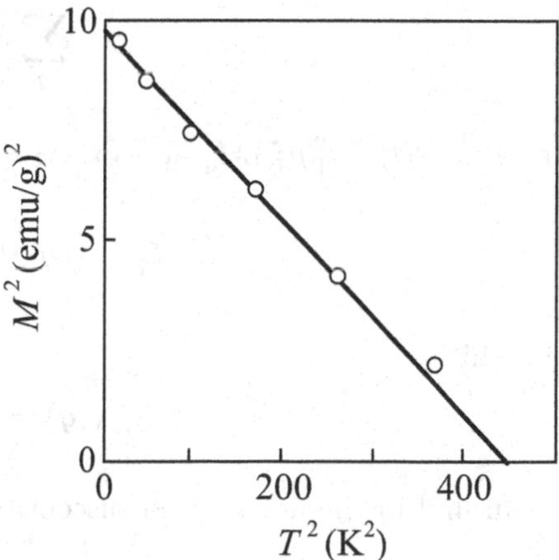

Figure 17.3.4 Square spontaneous magnetization $M^2(0,T)$ versus square temperature. From B. Barbara, D. Gignoux, and C. Vettier, *Lectures on Modern Magnetism*, Science Press and Springer-Verlag, Beijing (1988).

17.3.4 Spin Density Waves and Antiferromagnetism

Antiferromagnetism in metals can also be investigated in the band approach when the electron-electron interaction is considered. Its ground state is characterized by a periodic modulation of spin density. This type of ground state was first proposed by Overhauser (1960, 1962) for isotropic metals;[n] by common consensus, the antiferromagnetism of chromium is due to spin density waves[o] and it has been confirmed that spin density waves appear in highly anisotropic, so-called quasi-one-dimensional, metals.[p]

Just as in §6.3.4, we introduce a spatially varying external magnetic field along the z axis

$$H(\boldsymbol{r}) = \sum_{\boldsymbol{q}} H_{\boldsymbol{q}} e^{i\boldsymbol{q}\cdot\boldsymbol{r}}. \tag{17.3.45}$$

The coupling of the electron system to this field is described by an extra term

$$\mathcal{H}' = -\sum_{\boldsymbol{q}} M_{\boldsymbol{q}} H_{-\boldsymbol{q}}, \tag{17.3.46}$$

where $M_{\boldsymbol{q}}$ is the $\boldsymbol{q}$-th component of magnetization along z and can be related to the spin operator by

$$M_{\boldsymbol{q}} = g_{\rm L}\mu_{\rm B} S_{\boldsymbol{q}}. \tag{17.3.47}$$

Now we can write the total Hamitonian as

$$\mathcal{H} = \sum_{\boldsymbol{k}\sigma} \varepsilon_{\boldsymbol{k}} c^\dagger_{\boldsymbol{k}\sigma} c_{\boldsymbol{k}\sigma} + U\sum_i n_{i\uparrow} n_{i\downarrow} - \sum_{\boldsymbol{q}} M_{\boldsymbol{q}} H_{-\boldsymbol{q}}. \tag{17.3.48}$$

This Hamiltonian can be rewritten as

$$\mathcal{H} = \sum_{\boldsymbol{k}\sigma} E_{\boldsymbol{k}} c^\dagger_{\boldsymbol{k}\sigma} c_{\boldsymbol{k}\sigma} - \frac{2U}{g_{\rm L}^2\mu_{\rm B}^2} \sum_{\boldsymbol{q}} M_{\boldsymbol{q}} M_{-\boldsymbol{q}} - \sum_{\boldsymbol{q}} M_{\boldsymbol{q}} H_{-\boldsymbol{q}}, \tag{17.3.49}$$

where $E_{\boldsymbol{k}} = \varepsilon_{\boldsymbol{k}} + nU/2$. The first term describes the non-magnetic behavior, and the last two terms are related to magnetism. We can write these last two terms as

$$-\sum_{\boldsymbol{q}} M_{-\boldsymbol{q}} \left(H_{\boldsymbol{q}} + \frac{2U}{N g_{\rm L}^2 \mu_{\rm B}^2} M_{\boldsymbol{q}} \right).$$

The term $(2U/Ng_{\rm L}^2\mu_{\rm B}^2)M_{\boldsymbol{q}}$ acts exactly as a mean field. By definition,

$$M_{\boldsymbol{q}} = \chi(\boldsymbol{q}) H_{\boldsymbol{q}} = \chi_0(\boldsymbol{q}) \left(H_{\boldsymbol{q}} + \frac{2U}{N g_{\rm L}^2 \mu_{\rm B}^2} M_{\boldsymbol{q}} \right), \tag{17.3.50}$$

from which

$$\chi(\boldsymbol{q}) = \chi_0(\boldsymbol{q}) \left(1 + \frac{2U}{N g_{\rm L}^2 \mu_{\rm B}^2} \chi(\boldsymbol{q}) \right). \tag{17.3.51}$$

We can find in the normalized susceptibility, in unit of $g_{\rm L}^2\mu_{\rm B}^2$, that

$$\chi(\boldsymbol{q}) = \frac{\chi_0(\boldsymbol{q})}{1 - (2U/N)\chi_0(\boldsymbol{q})}. \tag{17.3.52}$$

[n] A. W. Overhauser, *Phys. Rev. Lett.* **4**, 462 (1960); *Phys. Rev.* **128**, 1437 (1962); *Phys. Rev.* **167**, 691 (1968).
[o] E. Fawcett, *Rev. Mod. Phys.* **60**, 209 (1988).
[p] G. Grüner, *Rev. Mod. Phys.* **66** 1 (1994).

The normalized susceptibility of independent particles was given in §6.2, i.e.,

$$\chi_0(\boldsymbol{q}) = \sum_{\boldsymbol{k}} \frac{f(E_{\boldsymbol{k}+\boldsymbol{q}}) - f(E_{\boldsymbol{k}})}{E_{\boldsymbol{k}} - E_{\boldsymbol{k}+\boldsymbol{q}}}. \tag{17.3.53}$$

For a given $\boldsymbol{q}$, $\chi(\boldsymbol{q})$ may become infinite at the temperature $T_c(\boldsymbol{q})$ when

$$\frac{2U}{N}\chi_0(\boldsymbol{q}) = 1. \tag{17.3.54}$$

If, at $T = 0$, this condition is not satisfied for any value of the $\boldsymbol{q}$ the system is non-magnetic at all temperatures. If it is satisfied for various values of $\boldsymbol{q}$, a system which is non-magnetic at high temperature will be ordered at the temperature $T_c = \max\,[T_c(\boldsymbol{q})]$.

We obtain ferromagnetic order if $\boldsymbol{q}_0 = 0$ and antiferromagnetic order if $\boldsymbol{q}_0 = \pi/a$ for a simple cubic, for example. If $\boldsymbol{q}_0$ has an arbitrary value, there may be two types of order (the z axis being arbitrary), i.e., one is a sinusoidal spin density wave with

$$M(\boldsymbol{R}_i) = M_0 \cos(\boldsymbol{q}_0 \cdot \boldsymbol{R}_i), \tag{17.3.55}$$

and the other is a helicoidal spin density wave with

$$M_x(\boldsymbol{R}_i) = M_0 \cos(\boldsymbol{q}_0 \cdot \boldsymbol{R}_i),\ M_y(\boldsymbol{R}_i) = \pm M_0 \sin(\boldsymbol{q}_0 \cdot \boldsymbol{R}_i). \tag{17.3.56}$$

The choice of plus or minus corresponds to two possible rotations of the magnetization when $\boldsymbol{q}_0 \cdot \boldsymbol{R}_i$ increases. Starting with these three spin density waves, one may construct spin density waves of any polarization. At T_c, one of the three spin density waves builds up with an infinitesimal amplitude. The existence of static spin density waves depends crucially on the zero order susceptibility $\chi_0(\boldsymbol{q})$. We have seen the form of $\chi_0(\boldsymbol{q})$ for free particles: In this case $\boldsymbol{q}_0 = 0$ and the system becomes ferromagnetic. In general $\chi_0(\boldsymbol{q})$ depends in a detailed fashion on the band structure, but one can give some general guidelines: $\chi_0(\boldsymbol{q})$ is important if, for a large number of $\boldsymbol{k}$ values, $(E_{\boldsymbol{k}} - E_{\boldsymbol{k}+\boldsymbol{q}})$ is small. This situation occurs if two portions of the Fermi surface coincide (nearly) over a large area for a translation of wavevector $\boldsymbol{q}_0$ or $\boldsymbol{q}_0 + \boldsymbol{K}$. This condition is just the one which gives Kohn anomalies.

In one dimension $\chi_0(2k_F)$ is infinite, and there is an instability for any infinitesimal value of the interaction for $\boldsymbol{q}_0 = 2k_F$. In three dimensions, $\chi_0(\boldsymbol{q})$ is finite. But with a Coulomb interaction, the Hartree–Fock theory gives an instability for $\boldsymbol{q}_0 = 2k_F$ whatever the strength of the interaction may be.

Bibliography

[1] Morrish, A. H., *The Physical Principles of Magnetism*, John Wiley & Sons, New York (1965); IEEE Press (1980); (2001).

[2] Callaway, J., *Quantum Theory of the Solid State*, 2nd ed., Academic Press, New York (1991).

[3] Blandin, A., *Band Magnetism*, in *Magnetism: Selected Topics* (ed., S. Foner), Gorden and Breach Science Publishers, New York (1976).

[4] Wohlfarth, E. P., *Band Magnetism and Applications*, in *Magnetism: Selected Topics*, (ed. S. Foner), Gorden and Breach Science Publishers, New York (1976).

[5] White, R. M., *Quantum Theory of Magnetism*, Springer-Verlag, Berlin (1983).

[6] Barbara, B., D. Gignoux, and C. Vettier, *Lectures on Modern Magnetism*, Science Press and Springer-Verlag, Beijing (1988).

[7] Yosida, K., *Theory of Magnetism*, Springer-Verlag, Berlin (1996).

[8] Skomski, R. and J. M. D. Coey, *Permanent Magnetism*, Institute Physics Publishing, Bristal (1999).

[9] Fazekas, P., *Lecture Notes on Electron Correlation and Magnetism*, World Scientific, Singapore (1999).

[10] Kübler, J., *Theory of Itinerant Electron Magnetism*, Oxford University Press, Oxford (2000).

[11] Bacon, G. E., *Neutron Diffraction*, 3rd ed., Clarendon Press, Oxford (1975).

Chapter 18

Superconductors and Superfluids

§18.1 Macroscopic Quantum Phenomena

The key to the theoretical understanding of macroscopic quantum phenomena was provided by Einstein (1924) in his prediction of the Bose–Einstein condensation (BEC) of an ideal gas composed of identical bosons. Although the first experimental confirmation of the BEC of dilute gases with very weak interactions (close to an ideal gas) came much later (1995), the postulate of BEC in quantum liquids has enabled the study of superconductivity and superfluidity and has stood the test of time.[a] Here, we will give a brief account of BEC and subsequent experimental verification, before a general introduction of superfluids and superconductors.

18.1.1 The Concept of Bose–Einstein Condensation

Quantum statistical mechanics predicts that there is a phase transition in an ideal gas of identical bosons when the thermal de Broglie wavelength $\lambda_T = (2\pi\hbar/k_{\mathrm{B}}T)^{1/2}$ exceeds the mean spacing between particles. Under such conditions, bosons are stimulated by the presence of other bosons in the lowest energy state to occupy that state as well, resulting in the macroscopic occupation of a single quantum state i with energy ε_i satisfying the distribution

$$n_i(T) = \frac{1}{\exp\left[\beta\left(\varepsilon_i - \mu\right)\right] - 1}, \tag{18.1.1}$$

where $\beta = 1/k_{\mathrm{B}}T$ and μ is the chemical potential. The total number of particles, N, will be given by summing over the quantum states i

$$N = \sum_i \frac{1}{\exp\left[\beta\left(\varepsilon_i - \mu\right)\right] - 1} \tag{18.1.2}$$

at any temperatures. Because the particle number is now conserved, the chemical potential is determined from (18.1.2). When T is low enough, all particles are in the state with lowest energy ε_0, so it is reasonable to assume

$$n_0(T) \equiv N_0(T) = \frac{1}{\exp\left[\beta\left(\varepsilon_0 - \mu\right)\right] - 1} \approx N. \tag{18.1.3}$$

Because N is a macroscopic number ($\sim 10^{23}$ say), the exponent in (18.1.3) is very small, with the result that $k_{\mathrm{B}}T/\left(\varepsilon_0 - \mu\right) \approx N$. It follows that μ is very close to ε_0, but just below it. Also, since the number of particles in the next higher level $n_1(T) \ll N$, $(\varepsilon_1 - \mu) \gg (\varepsilon_0 - \mu)$. Therefore

[a] For the original theoretical prediction, see A. Einstein, Sitzgber Press Acad Wissen 3 (1925); the first observation on BEC of dilute gases, see M. H. Anderson *et al.*, *Science* **269**, 198 (1995); the early explanation for superconductivity and superfluidity as macroscopic quantum phenomena, see F. London and H. London, *Proc. Roy. Soc.* (Lond.) **A149**, 71 (1935), and F. London, Superfluids: Vols. I, II., Wiley, New York, 1950, 1964.

the gap between ε_1 and ε_0 is much larger than that between ε_0 and μ. At finite temperatures μ is further below ε_0.

Under cyclic boundary conditions, the quantum states of free atoms in a box can be characterized by wavevectors q, and their energy levels are $\varepsilon_q = \hbar^2 q^2/2M$. In an independent system, ε can be treated as a continuous variable, and we can introduce the density of states

$$g(\varepsilon) = \frac{V}{4\pi^2}\left(\frac{2M}{\hbar^2}\right)^{3/2} \varepsilon^{1/2}. \tag{18.1.4}$$

The summation over states in (18.1.2) is now replaced by an integral over ε, but it should be noted that the integral does not include the particles in the ground state, due to $g(0) = 0$. The omission of the particles with $\varepsilon = 0$ is serious since $N_0(T)$ can be of order N at low temperatures. Thus, we must write

$$N = N_0(T) + N'(T), \tag{18.1.5}$$

where

$$N'(T) = \frac{V}{4\pi^2}\left(\frac{2M}{\hbar^2}\right)^{3/2} \int_0^\infty \frac{\varepsilon^{1/2} d\varepsilon}{\exp[\beta(\varepsilon-\mu)]-1}. \tag{18.1.6}$$

Because $\mu \leqslant 0$, at a given temperature this expression takes its maximum value for $\mu = 0$, so putting $\mu = 0$ gives us an upper bound for $N'(T)$. With $x = \beta\varepsilon$, we have

$$N'(T) \leqslant \frac{V}{4\pi^2}\left(\frac{2Mk_{\mathrm{B}}T}{\hbar^2}\right)^{3/2} \int_0^\infty \frac{x^{1/2} dx}{\mathrm{e}^x - 1}. \tag{18.1.7}$$

The definite integral in (18.1.7) can be evaluated and is $2.612 \times \sqrt{\pi}/2$, and the maximum number of particles in excited states is therefore

$$N'_{\mathrm{m}}(T) = 2.612V\left(\frac{Mk_{\mathrm{B}}T}{2\pi\hbar^2}\right)^{3/2}. \tag{18.1.8}$$

At sufficiently high temperatures, $N'_m(T)$ is large enough for all the particles to be accommodated in excited levels. However, as the temperature is reduced, a critical temperature T_{c} is reached below which $N'_{\mathrm{m}}(T)$ is less than N. In other words, below T_{c} particles start moving into the lowest energy level, and do so in increasing number as the temperature is lowered further. The critical temperature T_{c} is thus defined by setting $N'_{\mathrm{m}}(T_{\mathrm{c}}) = N$, which yields

$$T_{\mathrm{c}} = \frac{2\pi\hbar^2}{Mk_{\mathrm{B}}}\left(\frac{N}{2.612V}\right)^{2/3}. \tag{18.1.9}$$

Combination of the last two equations shows that the number of particles in excited states is

$$N'(T) = N\left(\frac{T}{T_{\mathrm{c}}}\right)^{3/2}. \tag{18.1.10}$$

The remaining particles are in the ground state, and from (18.1.10), we find

$$N_0(T) = N\left[1 - \left(\frac{T}{T_{\mathrm{c}}}\right)^{3/2}\right]. \tag{18.1.11}$$

At absolute zero all the particles are in the lowest energy level; above T_{c} almost all the particles are in excited levels. Between absolute zero and T_{c} the particles are divided into two groups, some in the lowest level and some in excited levels. In a system of macroscopic size, the single particle quantum state with lowest energy remains occupied by a macroscopically large number of particles up to a finite temperature (see Fig. 18.1.1). This is the BEC: The particles in the lowest level comprising what is called the "condensate". However, this condensation is quite different from what

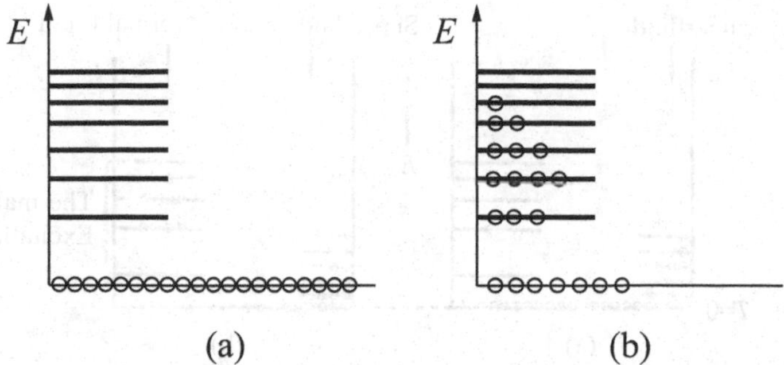

Figure 18.1.1 Schematic diagram showing occupation of energy levels for a BE condensate of ideal gas: (a) $T = 0$; (b) $0 < T < T_c$.

occurs, for example, when a gas is liquified. In the latter, the particles form two phases separated by a well defined boundary in position space. In contrast, the BEC can be regarded as a separation in momentum space, but there is no physical boundary between the condensate and the excited particles. Nevertheless, the particles are ordered according to their momenta, and from this point of view, the BEC is an example of an order-disorder transition. BEC provides a basic concept for macroscopic quantum phenomena, such as superfluidity and superconductivity. Though real bosons have interactions, BEC happens in the case of real quantum gases and liquids. Since entropy is carried by the thermally excited particles, these particles form a normal fluid, while the condensate, in which a macroscopic number of particles have zero momentum, forms a kind of superfluid.

Since the theory of BEC was originally proposed for an ideal gas, a crucial problem has been how to generalize this to a gas of interacting bosons. This problem was first tackled by Bogoliubov (1947) in his theory of superfluidity for a system of weakly repulsive bosons. Then Penrose and Onsager (1956) gave a more general theory showing that the BE condensate may survive, in the case of interacting bosons, as a state with long range order in momentum space or, in the more precise way introduced by C. N. Yang, with off-diagonal long range order. It was suggested that this state may be characterized by an off-diagonal component of a density matrix.[b] We may write the density matrix in terms of the creation and annihilation operators $\hat{\Psi}^\dagger(\boldsymbol{r})$ and $\hat{\Psi}(\boldsymbol{r})$

$$\rho(\boldsymbol{r}, \boldsymbol{r}') = \langle \hat{\Psi}(\boldsymbol{r}) \hat{\Psi}^\dagger(\boldsymbol{r}') \rangle = N_{0c} \Psi_0(\boldsymbol{r}) \Psi_0^*(\boldsymbol{r}') + g(\boldsymbol{r} - \boldsymbol{r}'), \tag{18.1.12}$$

where $\hat{}$ is introduced to distinguish operators from one-particle states, $\langle \cdots \rangle$ denotes the statistical average, and $g(\boldsymbol{r}-\boldsymbol{r}')$ is the short-range correlation within the de Broglie wavelength at a temperature T. It is assumed that the wavefunction $\Psi_0(\boldsymbol{r})$ of the condensate changes over a macroscopic scale. There is only short range order in the high temperature phase. However, $|\boldsymbol{r} - \boldsymbol{r}'|$ reaches infinity in the Bose–Einstein condensation (BEC), and $\rho(\boldsymbol{r}, \boldsymbol{r}')$ remains finite and of the order of particle density, which is called off-diagonal long range order (ODLRO). In comparison, crystalline order is characterized by periodic variation of density $n(\boldsymbol{r})$ in space. It should be emphasized that the ODLRO can only arise in a system of particles governed by quantum statistics, while crystalline order readily appears in a system of particles obeying classical statistics though their interactions may have a quantum nature. Although a system of fermions does not condense into a single-particle state due to the Pauli principle, an ODLRO in a two-particle density matrix associated with the pair formation of particles may appear.

Now we may show schematically the occupation of energy levels for the BE condensate of interacting bosons in Fig. 18.1.2. Compared with the ideal gas, two differences should be noted: First, the number of particles condensing into the lowest level is reduced, i.e., at 0 K, not all particles are in the lowest level, which means that the condensate is depleted by interactions. Second, the nature of the excited levels is altered, as shown by the excited levels in superfluid at 0 K and the additional excitations above 0 K. The normal fluid fraction is related to these thermal excitations. It should

[b]see N. N. Bogoliubov, *J. Phys. USSR* **11**, 1923 (1947); R. Penrose and L. Onsager, *Phys. Rev.* **104**, 412 (1956); C. N. Yang, *Rev. Mod. Phys.* **34**, 694 (1962).

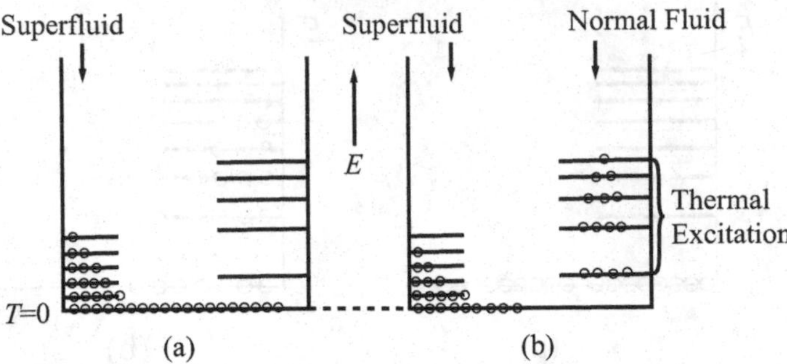

Figure 18.1.2 Schematic diagram showing occupation of energy levels for a BE condensate of interacting bosons: (a) $T = 0$; (b) $0 < T < T_c$.

be noted that, due to interactions, the condensate fraction is no longer identified with the superfluid fraction; the latter is 100% at 0 K, while the former may be much smaller.

18.1.2 Bose–Einstein Condensation of Dilute Gases

In the real world there is no ideal gas to give a direct test of the theory of BEC. The closest approximation to an ideal gas is a dilute gas with weak interactions. In 1995, the first successful BEC of a trapped dilute gas of Rb atoms, under ultra low temperature with laser cooling plus evaporation cooling, was achieved by E. Cornell and C. Wieman. Afterwards BEC for other dilute gases such as Na, Li, ^{1}H as well as the metastablc first excited state of ^{4}He have been also realized. So a new family of condensed matter (condensed in momentum space) have been added to our repertoire. The challenge was to cool the gases to temperatures below 1 μK, while preventing the atoms from condensing into a solid or liquid. These neutral atoms are cooled by using a combination of laser and evaporative cooling to the very low temperatures (under 100 nK). The laser cooling can produce confined gases at relatively low densities around 10^{12} atoms cm^{-3} and the number of atoms in the trap is typically about 10^3–10^7. These alkalis are effectively weakly interacting gases at sufficiently low temperatures, so that the interaction does not overwhelm the nature of quantum statistics in the BEC process. Then the quantum statistics of particles, rather than the interactions between them, dominates the transition.

The fact that these gases are highly inhomogeneous has one important consequence: At the critical temperature, a sharp peak was observed, centered at the zero velocity in velocity distribution. This shows the obvious character of BEC, i.e., the BEC shows up not only in momentum space, but also in real space. This double possibility of investigating the effects of condensation is very interesting from both theoretical and experimental viewpoints. Preceeding and parallel to this dramatic experimental progress, a large body of theoretical work has been done on trapped dilute-gas BE condensates. An impressive agreement between theory and experiment has already been achieved for such things as condensate size, shape, interaction energy and excitation frequencies.

The confining potential for alkali atoms in the magnetic traps now used is in the quadratic form

$$V(\boldsymbol{r}) = \frac{M}{2}(\omega_x^2 x^2 + \omega_y^2 y^2 + \omega_z^2 z^2). \tag{18.1.13}$$

As a first step, we consider an ideal Bose atomic gase, i.e., ignoring atom-atom interactions. The Hamiltonian of the system is the sum of single-particle Hamiltonians whose eigenvalues have the form

$$\varepsilon_{n_x n_y n_z} = \left(n_x + \frac{1}{2}\right)\hbar\omega_x + \left(n_y + \frac{1}{2}\right)\hbar\omega_y + \left(n_z + \frac{1}{2}\right)\hbar\omega_z, \tag{18.1.14}$$

here n_x, n_y, n_z are non-negative integers. The ground state is described by the wavefunction

$$\Psi(\boldsymbol{r}_1, \ldots, \boldsymbol{r}_N) = \prod_i \Psi_0(\boldsymbol{r}_i), \tag{18.1.15}$$

in which

$$\Psi_0(\boldsymbol{r}) = \left(\frac{M\omega_0}{\pi\hbar}\right)^{1/3} \exp\left[-\frac{M}{2\hbar}(\omega_x x^2 + \omega_y y^2 + \omega_z z^2)\right], \tag{18.1.16}$$

where ω_0 represents the geometric average of the oscillator frequencies

$$\omega_0 = (\omega_x \omega_y \omega_z)^{1/3}. \tag{18.1.17}$$

The density distribution then becomes $\rho(\boldsymbol{r}) = N\,|\Psi_0(\boldsymbol{r})|^2$, and its value grows with N. The size of the cloud instead is independent of N and is given by the harmonic oscillator length

$$a_0 = \left(\frac{\hbar}{M\omega_0}\right)^{1/2}. \tag{18.1.18}$$

which corresponds to the average width of the Gaussian in (18.1.16). This is an important length scale, which is typically of the order of 1 μm in experiments. At finite temperatures only a fraction of the atoms occupy the lowest state, the others being thermally distributed over the excited states. The radius of the thermal cloud is larger than a_0 and, with a rough estimate, is $a_T = a_0(k_B T/\hbar\omega_0)^{1/2}$, for $k_B T \gg \hbar\omega_0$.

One can derive from quantum statistics of an ideal gas in a harmonic trap the T dependence of the condensate fraction for $T < T_c$

$$\frac{N_0}{N} = 1 - \left(\frac{T}{T_c}\right)^3, \tag{18.1.19}$$

where T_c is the predicted critical temperature for an ideal gas in a harmonic trap in the thermodynamic limit. It is different from that for a uniform ideal gas in (18.1.11), with a T^3 dependence instead of a $T^{3/2}$ one. The experimental curve slightly deviates from the simple theoretical one, but the agreement can be improved by including a correction for the finite number of atoms ($\sim$ 4000) in the condensate (Fig. 18.1.3). The condensate fraction at 0 K is about 99%, showing that the depletion due to the interaction is small.

The appearance of the condensate as a narrow peak in both real space and momentum space is a peculiar feature of trapped Bose gases which has important consequences in both experiments and theory. The spatial variation of the condensate density has been observed by the absorption of an

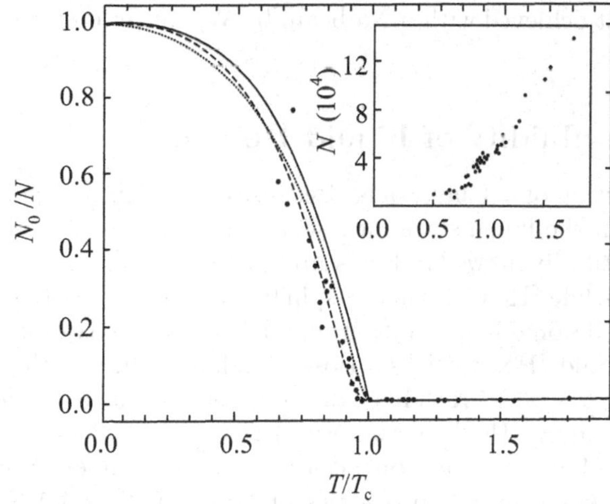

Figure 18.1.3 The condensate fraction number N versus the scaled temperature for a BE condensate of Rb gas of 4000 atoms: the solid curve is the theoretical curve for ideal gas in harmonic trap in the thermodynamic limit; the dotted curve includes a correction for the finite number of atoms; the dashed curve is the best fit for the experimental data, shown as points. From J. R. Ensher *et al.*, *Phys. Rev. Lett.* **77**, 4984 (1996).

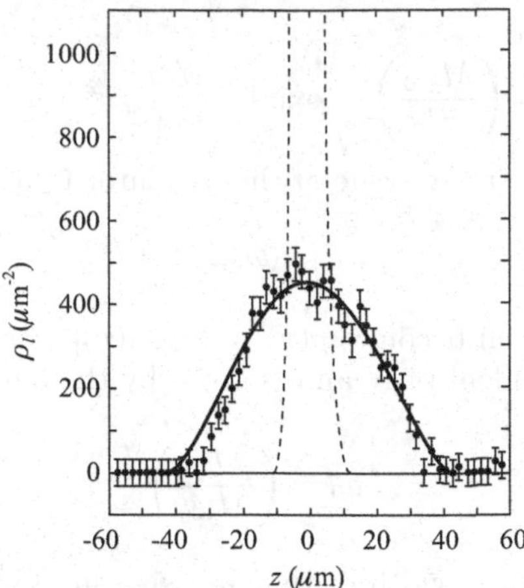

Figure 18.1.4 The condensate column density versus the radial coordinate: the dashed line is the theoretical prediction for non-interacting cloud; solid line is theoretical curve with inclusions of two-body interactions verified by experimental data shown as points. From F. Dalfovo *et al.*, *Rev. Mod. Phys.* **71**, 463 (1999).

appropriate laser light: The observed peaks are much modified by the interactions between atoms as shown in Fig. 18.1.4. After all, weak interactions between atoms have important roles to play in many delicate experiments on BE condensates.

Due to the low occupation of the ground state, the superfluidity of Bose–Einstein condensation (BEC) is rather difficult to observe. There is reliable, but indirect, experimental evidence for the superfluidity of BE condensates of dilute gases, one of which is the vortex formation to be discussed in the following section. The fact that atoms in a BE condensate occupy the ground state is somewhat similar to photons occupying a single quantum state, with identical momentum and energy, in a laser beam. However, photons are massless, and their number is not a conserved quantity, so lowering the temperature will not induce BEC of photons. A laser beam can be produced by a nonequilibrium process: Pumping electrons to a higher energy level will achieve the population inversion, then stimulated emission will outstrip absorption in a laser cavity. A BE condensate of atoms in a trap may be coupled out of the trap as a beam of atoms with definite momentum and energy. This beam of atoms has the same monochromatic and parallel qualities as a laser beam, and is called the atom laser. This feat was first achieved with a Na beam by W. Ketterle *et al.* who confirmed its coherent nature.[c]

18.1.3 The Superfluidity of Liquid Helium

We consider a collection of helium atoms. Because of its high zero point energy and small atomic mass, helium remains in the liquid state for a wide range of pressures all the way down to absolute zero. Helium occurs naturally in two kinds of stable isotopes: ^{3}He and ^{4}He. ^{3}He with nuclear spin 1/2 obeys Fermi statistics, while ^{4}He with nuclear spin 0 obeys Bose statistics. At very low temperatures, where quantum effects become important, ^{3}He and ^{4}He provide two of the few examples in nature of quantum liquids. Liquid ^{4}He, which is a Bose liquid, exhibits a rather straightforward transition to a superfluid state at $T = 2.17$ K. This can be understood as a condensation of particles into a single quantum state. Liquid ^{3}He, being a Fermi liquid, also undergoes a transition to a superfluid state by the formation of atomic pairs, but at a much lower temperature $T < 2.7 \times 10^{-3}$ K.

There are some striking physical properties of ^{4}He. At $T > 4.2$ K, it forms a normal gas or vapor, and when the temperature is lowered to $T = 4.2$ K at a pressure of 1 atm, a normal gas-liquid transition takes place. It remains a normal liquid, the so-called He I state, until $T_c = 2.17$ K, which is called the λ point because of the shape of its $C_P(T)$ versus T variation, as shown in Fig. 18.1.5. The sharp specific heat peak is the signature of the transition from the normal to the

[c]M. O. Mews *et al.*, *Phys. Rev. Lett.* **78**, 582 (1997); M. R. Andrews *et al.*, *Science* **275**, 639 (1997).

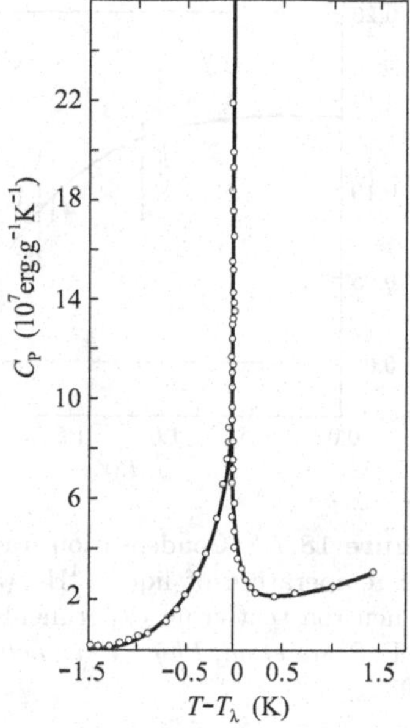

Figure 18.1.5 The specific heat anomaly in the superfluid transition of ^{4}He.

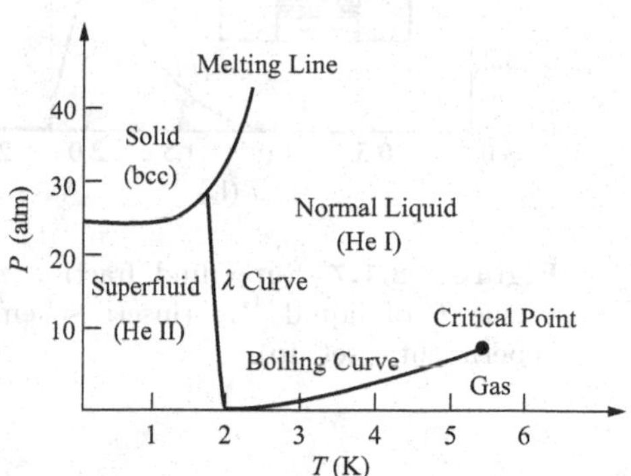

Figure 18.1.6 Phase diagram of ^{4}He.

superfluid state. At $T \leqslant T_c$, liquid ^{4}He moves collectively without viscosity, i.e., without energy dissipation, and it can even appear to defy gravity by flowing upwards in the form of a thin film over the walls of a container.

The phase diagram for ^{4}He is shown in Fig. 18.1.6. At low temperatures ^{4}He has four phases. The solid phase only appears for pressures above 2.5 MPa, and the phase transition continues down to $T = 0$ K. However there are, in fact, two liquid phases separated by a line of λ-point occurring at about $T = 2$ K, the exact temperature depending on the pressure. There is a triple point at each end of the line.

The characteristic for superfluid ^{4}He is flow without viscosity in a capillary tube with a diameter of the order of 10^{-7} m. However, a nonvanishing viscosity is observed if it is measured at finite temperatures through the decay of rotational vibration of a disk suspended in a fluid, as is commonly used to measure the viscosity of a liquid. The two-component model has been proposed based upon this observation and the so-called fountain effect. The fluid in this model is made of a superfluid component with number density n_s, which flows without viscosity, and the normal component with number density n_n, which has a finite viscosity and carries entropy. The component n_s takes a finite value at $T < T_c$ and $n_s = n$ (total number density) at $T = 0$ K as shown in Fig. 18.1.7. The relation was verified by macroscopic experiment by measuring the damping of oscillations of metallic disks immersed in liquid ^{4}He.

London conjectured that the BEC of an ideal gas is related to the superfluidity of liquid ^{4}He. He substituted the value of the density of liquid ^{4}He into the N/V in (18.1.9), and obtained $T = 3.1$ K which is quite close to 2.17 K, the experimental value of T_c. So, London made a bold assertion that T_c marks the onset of BEC of liquid ^{4}He. However, liquid ^{4}He is obviously a system in which the attractive interactions between atoms play an essential role. Experimentally, neutron scattering and direct momentum measurement from thermal evaporation of surface atoms have determined how the condensate fraction varies with temperature, as shown in Fig. 18.1.8. This curve may be fitted to an empirical relation

$$n_0(T) = n_0 \left[1 - \left(\frac{T}{T_\lambda}\right)^\alpha\right], \tag{18.1.20}$$

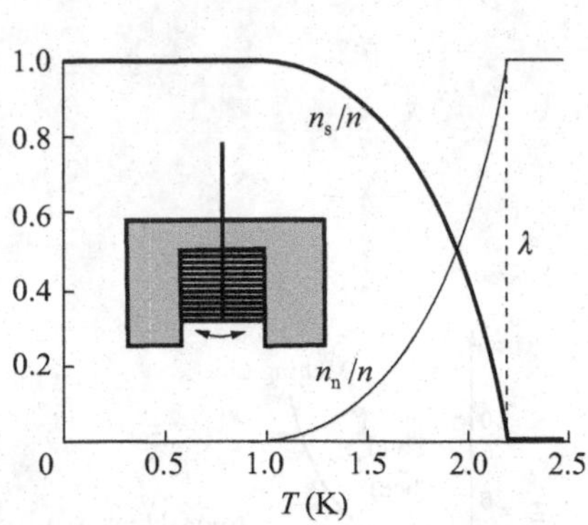

Figure 18.1.7 Superfluid fraction n_s/n versus T of liquid ^{4}He (inset: schematic experimental set-up).

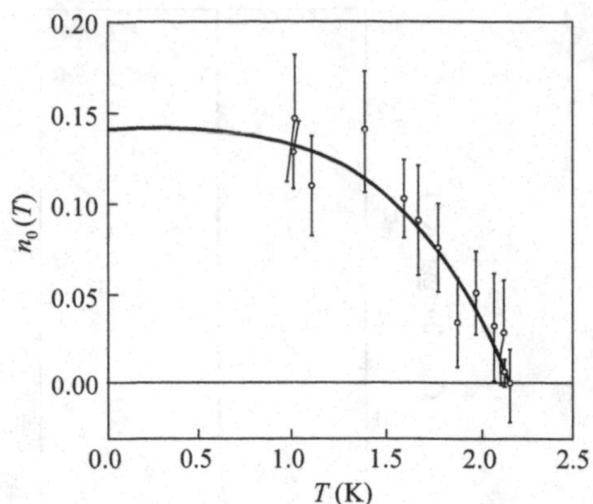

Figure 18.1.8 Condensation fraction versus temperature of liquid ^{4}He (according to neutron scattering experiments). From V. F. Sears *et al.*, *Phys. Rev. Lett.* **49**, 279 (1982).

where the exponent $\alpha = 3.6$ (which is not equal to 3/2 as for the ideal Bose gas) and $n_0(0) = 13.9\%$. The reduced condensate fraction is surely due to interaction-depletion, which makes the condensate fraction smaller than the superfluid fraction as already discussed in §18.1.1.

Based on the discussion above, to describe a macroscopic quantum state we can use a macroscopic wavefunction

$$\Psi(\boldsymbol{r}) = \Psi_0(\boldsymbol{r})e^{i\theta(\boldsymbol{r})}, \tag{18.1.21}$$

where the modulus Ψ_0 and phase $\theta(\boldsymbol{r})$ are all real functions of the position $\boldsymbol{r}$. A reasonable and successful application of the macroscopic wavefunction is to illustrate the normalized $\Psi(\boldsymbol{r})$ as the probability amplitude, so that $\Psi^*(\boldsymbol{r})\Psi(\boldsymbol{r})$ is equal to the average number of superflowing particles per unit volume, that is,

$$\Psi^*(\boldsymbol{r})\Psi(\boldsymbol{r}) = \Psi_0^2(\boldsymbol{r}) = n_s. \tag{18.1.22}$$

It will be found later that the phase $\theta(\boldsymbol{r})$ is the key to understand macroscopic quantum phenomena. Now we introduce the condensate momentum $\boldsymbol{p}$, which satisfies

$$\boldsymbol{p}\Psi = \frac{h}{i}\nabla\Psi. \tag{18.1.23}$$

For a homogeneous system, Ψ_0 is independent of $\boldsymbol{r}$, then

$$\boldsymbol{p} = \hbar\nabla\theta. \tag{18.1.24}$$

It is convenient to interpret the canonical momentum of one particle of the superfluid ^{4}He as

$$\boldsymbol{p} = M_4\boldsymbol{v}_s. \tag{18.1.25}$$

If the superfluid velocity is $\boldsymbol{v}_s$, this, in turn, can be written as

$$\boldsymbol{v}_s = \frac{\hbar}{M_4}\nabla\theta. \tag{18.1.26}$$

This equation means that the superfluid velocity is proportional to the gradient of the phase of the condensate wavefunction. Thus when the superfluid velocity is zero, the phase has the same value throughout, and when the superfluid velocity is finite and constant, the phase varies uniformly in the direction of $\boldsymbol{v}_s$. The effect of phase coherence is to lock the condensate particles together in a state of uniform motion. This may provoke the idea of a rigid structure moving as a whole, but it must be remembered that the 'rigidity' exists in momentum space rather than in position space. The phase coherence gives us a qualitative understanding of how a constant superfluid velocity can be maintained over long times. A sudden change of $\boldsymbol{v}_s$ would necessarily involve a simultaneous identical alteration in the velocity for all members of a macroscopically large number of particles, an event so unlikely that it can be discounted.

Another manifestation of the macroscopic quantum phenomena is the quantization of circulation. Consider superfluid ^{4}He in an annular region, such as the space between two concentric cylinders as shown in Fig. 18.1.9. The circulation of a closed loop can be written as

$$\Gamma = \oint_L \boldsymbol{v}_s \cdot d\boldsymbol{l}, \tag{18.1.27}$$

and, due to (18.1.26), it may be expressed in terms of the phase of the wavefunction.

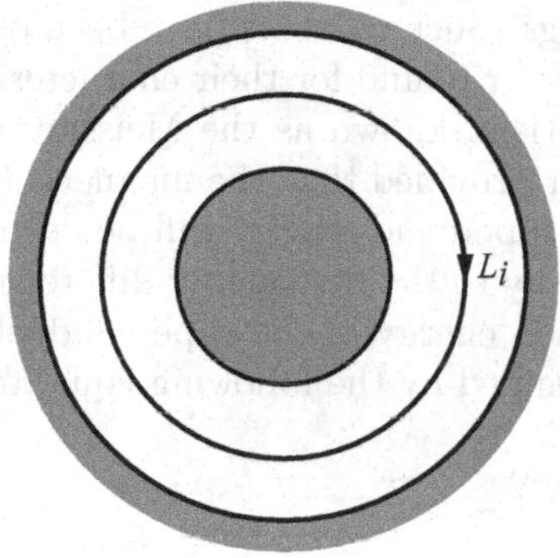

Figure 18.1.9 Superfluid in annular (multiply connected) region with a closed contour L.

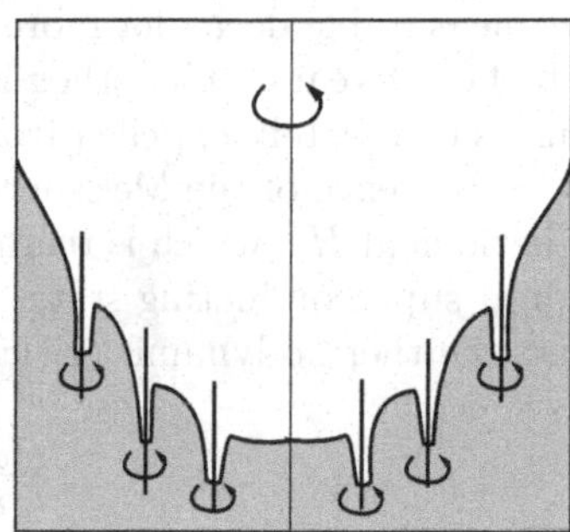

Figure 18.1.10 Array of vortex lines in rotating superfluid ^{4}He.

Since the superfluid wavefunction is single-valued, a trip around a closed loop must leave it unchanged, which means that the change in θ can only be zero or an integral multiple of 2π. Thus the circulation is quantized with values

$$\Gamma = \nu \frac{h}{M_4}, \quad \nu = 0, \pm 1, \pm 2, \ldots, \tag{18.1.28}$$

where M_4 is the mass of the ^{4}He atom, h/M_4 is called the quantum of circulation which has a value of 9.97×10^{-8} cm^{2}s^{-1}. It should be noted that the annulus in Fig. 18.1.9 is an example of a multiply connected region, it contains a 'hole' (i.e., a non-superfluid region) in the superfluid so that closed loops, such as L, may be drawn.

By taking the curl of (18.1.26), we can show that

$$\nabla \times \boldsymbol{v}_s = 0. \tag{18.1.29}$$

This condition establishes the irrotational nature of superfluid flow. However, the irrotationality of superflows does not mean no rotation at all, instead it means that the angular momentum of a rotating superfluid should be carried by an array of vortex lines, as schematically shown in Fig. 18.1.10. The cores of the vortex lines are actually normal regions in which $\nabla \times \boldsymbol{v}_s \neq 0$, while the contour enclosing a vortex core yields a quantized circulation. Thus, an apparently singly-connected superfluid region, as shown in Fig. 18.1.10, is actually multiply connected. This explains the mystery of rotation of the superfluids by formation of vortex lines, and this type of experimental observation in rotating BE condensates of dilute gases gave credence to their superfluidity.

The liquid ^{3}He isotope is composed of fermions, however, below 2.7 mK, it was found to be in a superfluid state. The explanation of this lies in BEC of bosons consisting of pairs of ^{3}He atoms.

18.1.4 Superconductivity of Various Substances

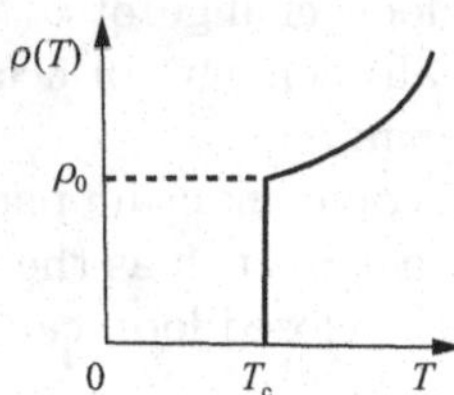

Figure 18.1.11 Temperature dependence of the resistivity of a superconductor.

The superconducting state has two fundamental features. One is zero electrical resistivity below T_c (Fig. 18.1.11). The complete disappearance of resistance is most sensitively demonstrated by experiments with persistent currents in superconducing rings. Such currents have been observed to flow without measurable decay for more than a year, so a lower bound for their characteristic decay time is set to be 10 years. The other is perfect diamagnetism, known as the Meissner effect; the magnetic flux is completely expelled from a superconductor provided that the magnetic field is not too large. The existence of the Meissner effect implies that superconductivity will be destroyed by a critical magnetic field H_c, which is related thermodynamically to the free energy difference between the normal and superconducting state, i.e., the condensation energy of the superconducting state. It means that the thermodynamic critical field H_c is determined by the following equation

$$\frac{H_c^2(T)}{8\pi} = f_n(T) - f_s(T), \tag{18.1.30}$$

where $f_n(T)$ and $f_s(T)$ are the Helmholtz free energies per unit volume in the respective states in zero field. It was found empirically that $H_c(T)$ is described by a parabolic law (Fig. 18.1.12)

$$H_c(T) \approx H_c(0)\left[1 - \left(\frac{T}{T_c}\right)^2\right], \tag{18.1.31}$$

and there is a specific heat anomaly in the superconducting transition (Fig. 18.1.13).

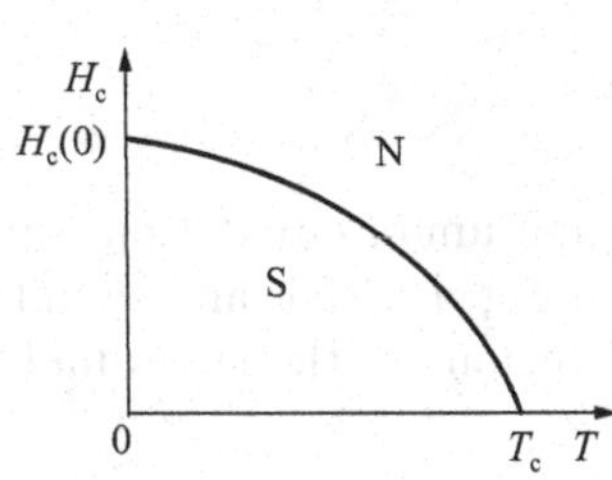

Figure 18.1.12 Phase diagram of a type I superconductor.

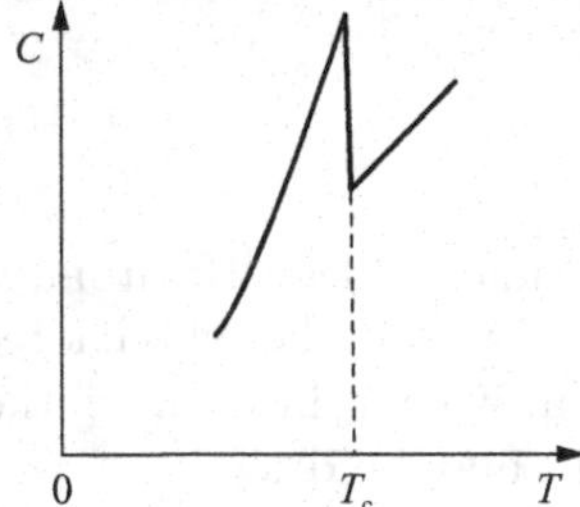

Figure 18.1.13 Specific heat anomaly in the superconducting transition.

Those superconductors with a complete Meissner effect (sharp H_c) are called type I superconductors (Fig. 18.1.14); but there are also some superconductors with an incomplete Meissner effect, the sharp H_c is replaced by the gradual decrease from H_{c1} to H_{c2} (Fig. 18.1.15); these are called type II superconductors.

In the equilibrium state, a superconductor is characterized by $\boldsymbol{B} = 0$ and $\boldsymbol{E} = 0$. In a superconductor the condensate 'particles' are Cooper pairs of electrons with mass $2m$ and charge $2e$. In an applied magnetic field $\boldsymbol{B} = \nabla \times \boldsymbol{A}$, the canonical momentum of a Cooper pair is given by

$$\boldsymbol{p} = 2m\boldsymbol{v} + 2e\boldsymbol{A}, \tag{18.1.32}$$

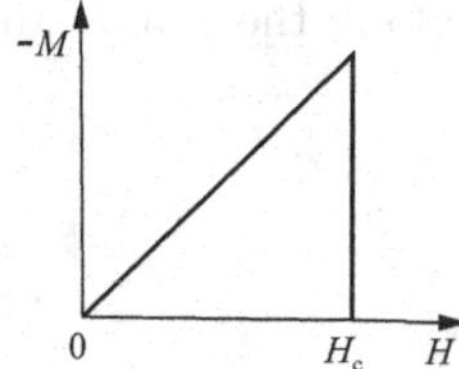

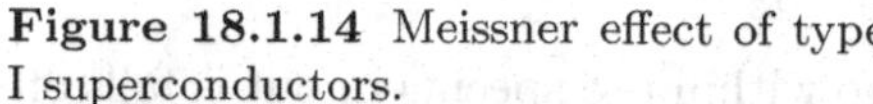

Figure 18.1.14 Meissner effect of type I superconductors.

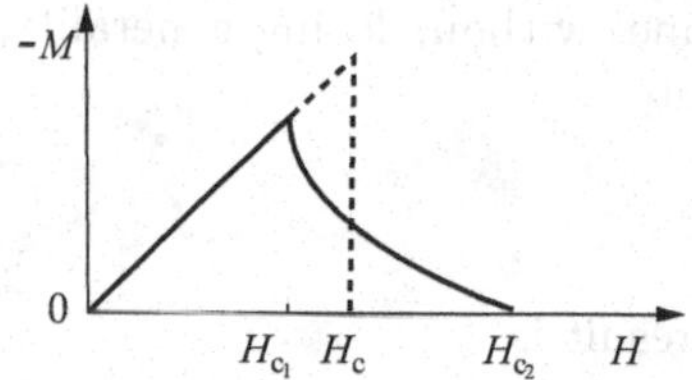

Figure 18.1.15 Meissner effect of type II superconductors.

where $\boldsymbol{v}$ is the supercurrent velocity. If $\boldsymbol{p}$ is eliminated from (18.1.32) by (18.1.24), we find

$$\boldsymbol{v}_s = \frac{\hbar}{2m}\nabla\theta - \frac{e\boldsymbol{A}}{mc}. \tag{18.1.33}$$

This equation corresponds to (18.1.26) of a superfluid; it means we may use a macroscopic wave-function to describe the behavior of superconductors, as already had been pointed out in 1935 by the brothers F. and H. London.

Introducing the superconducting electron density n_s, the supercurrent density is

$$\boldsymbol{j}_s = n_s e \boldsymbol{v}_s. \tag{18.1.34}$$

Combining (18.1.33) and (18.1.34), we find

$$\boldsymbol{j}_s = \frac{n_s e\hbar}{2m}\nabla\theta - \frac{n_s e^2}{mc}\boldsymbol{A}. \tag{18.1.35}$$

Taking the curl of (18.1.35), assuming that the superconductor is nonmagnetic, so $\boldsymbol{B} = \boldsymbol{H}$, we can write $\boldsymbol{H} = \nabla \times \boldsymbol{A}$, then

$$\nabla \times \boldsymbol{j}_s + \frac{n_s e^2}{mc}\boldsymbol{H} = 0. \tag{18.1.36}$$

In the time-independent case, Maxwell equations give

$$\nabla \times \boldsymbol{H} = \frac{4\pi}{c}\boldsymbol{j}_s. \tag{18.1.37}$$

Substituting this into (18.1.36), we can get the London equation

$$\boldsymbol{H} + \lambda_L^2 \nabla \times \nabla \times \boldsymbol{H} = 0, \tag{18.1.38}$$

with London penetration depth

$$\lambda_L = (mc^2/4\pi n_s e^2)^{1/2}. \tag{18.1.39}$$

When $T_c = 0$ K, $n_s = n$, the estimated valued λ_L is about 50–200 nm.

The London equation can be used to illustrate the Meissner effect, and in addition, shows a magnetic field penetration which, together with the shielding supercurrents existing in the surface of a superconductor, is about the length scale of λ_L. Let us now investigate the situation for a semi-infinite specimen. The surface is in the xy plane, the region $z < 0$ being empty space. From (18.1.38) and with $\nabla \cdot \boldsymbol{H} = 0$ from Maxwell equations, we can get

$$\nabla^2 \boldsymbol{H} - \frac{1}{\lambda_L^2}\boldsymbol{H} = 0. \tag{18.1.40}$$

Due to the symmetry of the problem, $\boldsymbol{H}$ is a function of z only. In the direction parallel to the z axis, according to $\nabla \cdot \boldsymbol{H} = 0$, we can get $dH/dz = 0$, so H is a constant and $j_s = 0$. If $\boldsymbol{H}$ is in the xy plane, without losing generality, we can set $\boldsymbol{H}$ to be along the x axis and (18.1.40) can be changed into

$$\frac{d^2H}{dz^2} = \frac{H}{\lambda_{\mathrm{L}}^2}. \tag{18.1.41}$$

The final result is

$$H(z) = H(0)\mathrm{e}^{-z/\lambda_{\mathrm{L}}}, \ j_{\mathrm{s}} = \frac{c}{4\pi}\frac{dH}{dz} = j_{\mathrm{s}}(0)\mathrm{e}^{-z/\lambda_{\mathrm{L}}}. \tag{18.1.42}$$

Hence the magnetic field does not fall abruptly to zero within the specimen, but is falls off exponentially inside the superconductor as shown in Fig. 18.1.16. The corresponding characteristic length is just the London penetration depth λ_{L}. Thus, the Meissner effect is complete in most of the bulk region but remains incomplete within the surface region. Figure 18.1.17 gives an illustration of the Meissner effect of a superconductor.

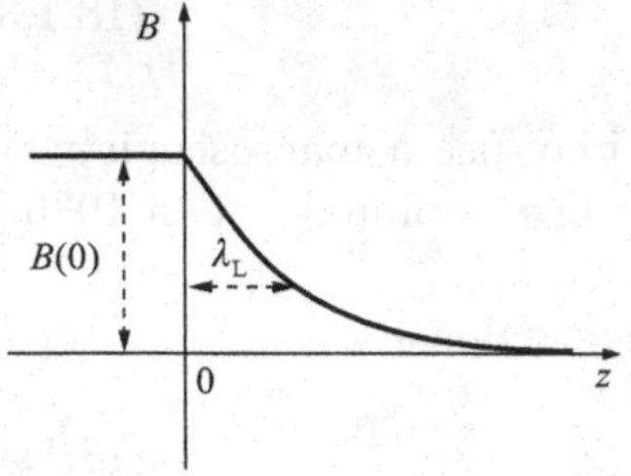

Figure 18.1.16 Penetration of the magnetic field inside the surface layer of a superconductor according to the London equation.

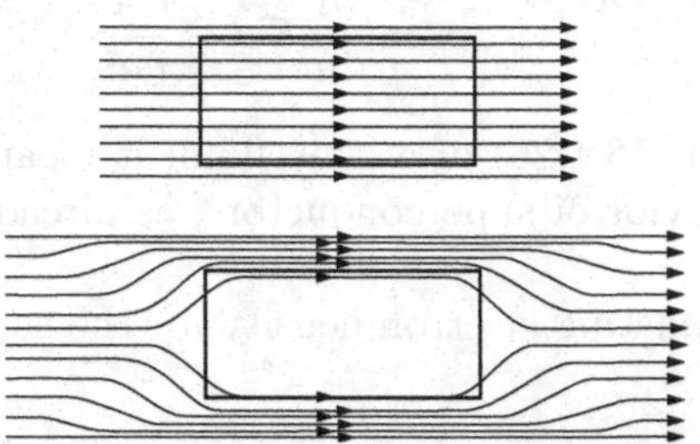

Figure 18.1.17 An illustration of the Meissner effect with an additional consideration on the penetration of a magnetic field into the surface layer.

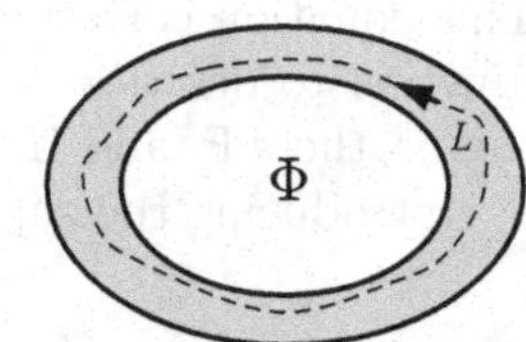

Figure 18.1.18 A ring-shaped superconductor.

Furthermore, in multiply-connected superconductors, such as a superconducting ring (Fig. 18.1.18), we may demonstrate flux quantization just like the quantization of circulation in a multiply-connected superfluid. Since $H = 0$ and $\langle j_{\mathrm{s}} \rangle = 0$ along the path L, provided that the cross-sectional radius of the ring $d \gg \lambda_{\mathrm{L}}$, then we may derive from (18.1.35)

$$\oint_{\mathrm{L}} \nabla\theta \cdot d\boldsymbol{l} = \oint_{\mathrm{L}} \frac{2e}{\hbar c}\boldsymbol{A} \cdot d\boldsymbol{l}. \tag{18.1.43}$$

By Stokes' theorem, the right-hand side of above equation is equal to $2e\Phi/\hbar$, where Φ is the total magnetic flux trapped inside the ring. However, on the left hand side of it is the phase of the superconductor; its total variation along a closed loop must be an integral multiple of 2π. Thus the quantization of the magnetic flux is derived

$$\Phi = 2\pi\nu\frac{\hbar c}{2e} = \nu\frac{hc}{2e}, \tag{18.1.44}$$

where ν is an integer. The magnetic flux passing through the loop can only take a discrete set of values. Experimental measurement of flux quantization in superconductors shows that the effective charge $e^* = 2e$, this fact shows that superconductivity is due to paired electrons. Thus the flux quantum or fluxoid is defined as

$$\Phi_0 = \frac{hc}{2e} = 2.07 \times 10^{-7}\ \mathrm{G} \cdot \mathrm{cm}^2. \tag{18.1.45}$$

Flux quantization has been observed experimentally, and its value gives direct confirmation of the pairing of electrons in superconductors.

Table 18.1.1 The periodic table showing superconducting elements.

s								*s-d*						*s-p*			
H ?								Elements T_c(K) (T_c) Metastable									He
Li	Be 0.026 (9)											B (11)	C ?	N	O	F	Ne
Na	Mg											Al 1.18	Si (7.1)	P (5.5)	S (17)	Cl	Ar
K	Ca	Sc	Ti	V 5.4	Cr	Mn	Fe (2)	Co	Ni	Cu	Zn 0.85	Ga 1.08	Ge (5.3)	As (0.5)	Se (5.9)	Br	Kr
Rb	Sr	Y (2.5)	Zr 0.61	Nb 9.25	Mo 0.92	Tc 7.7	Ru 0.49	Rh .0003	Pd	Ag	Cd 0.52	In 3.41	Sn 3.72	Sb (3.5)	Te (4.3)	I	Xe
Cs (1.5)	Ba (5.4)	La (6.0)	Hf 0.12	Ta 4.47	W 0.01	Re 1.7	Os 0.66	Ir 0.11	Pt	Au	Hg 4.15	Ti 2.38	Pb 7.2	Bi (8.5)	Po	At	Rn
Fr	Ra	Ac	Th 1.4														
e/a=1	2	3	4	5	6	7	8	9	10	1	2	3	4	5	6	7	8

Ce (1.8)	Pr- Yb	Lu (2.8)
	Pa (1.4)	U (2.4)

After the discovery of superconductivity, the T_c's of various elements were measured, and the results are listed in Table 18.1.1.[d] There has also been enormous research activity in quest of materials with higher T_c or H_{c2} as well as the search for exotic pairing mechanisms besides the *s*-wave singlet pairing of BCS theory. Table 18.1.2 tabulates the achievements in this field.[e] A few highlights in the quest for high T_c materials: Nb_3Ge with $T_c \approx 23$ K held the T_c record for nearly 13 years after 1973; Bednorz and Müller's discovery of superconductivity of LBCO in 1986 pushed T_c over the 30 K barrier; the discovery of YBCO, by several research groups independently in 1987, pushed T_c into liquid nitrogen regime ($\sim$ 93 K); then BSSCO pushed T_c still higher; the present record of high T_c is still held by Hg cuprate ($T_c \approx 134$ K, and 165 K under high pressure). By the end of 2000, the T_c record for metallic alloys was raised to 39 K by the unexpected discovery of superconductivity in the intermetallic compound MgB_2. Here we also mention some highlights in search of exotic pairings: The singlet *d*-wave pairing for high T_c cuprates has received general consensus, though the mechanism for its pairing is still uncertain; there is some evidence that some heavy electron superconductors, as well as some ruthenates, may have triplet *p*-wave pairing signaling that a magnetic mechanism is involved; while the pairing mechanism in organic superconductors is being hotly investigated, both from the viewpoint of conventional and exotic mechanisms; there is also a recent report on the experimental coexistence of ferromagnetism and superconductivity in the metallic alloy $ZrZn_2$, suggesting that exotic pairing may even infiltrate into the classic domain of the superconductivity of metallic alloys. We reserve a special section §18.3 on pairing states to discuss these problems.

[d]This is a slightly modified version of the chart by T. H. Geballe published in Science **293**, 223 (2002). The data for Fe are taken from K. Shimitzu *et al.*, *Nature* **412**, 316 (2001); for B from M. I. Erements *et al.*, *Science* **293**, 272 (2001); for S from V. V. Struzhkin *et al.*, *Nature* **360**, 282 (1999); for Pt from R. Konig, *Phys. Rev. Lett.* **82**, 4528 (2001).

[e]The values of T_c in brackets are the metastable ones. Data for $ZrZn_2$ taken from C. Pfleiderer *et al.*, *Nature* **412**, 58 (2001); for MgB_2 from J. Magamatzu *et al.*, *Nature* **410**, 63 (2001) and D. C. Larbalestier *et al.*, *Nature* **410**, 186 (2001), for a review of MgB_2, see C. Day, *Phys. Today* **54**(4), 17 (2002); for UGe_2 from S. S. Saxena *et al.*, *Nature* **406**, 587 (2000); for $SrRuO_4$ from S. Nishizaki *et al.*, *J. Phys. Soc. Jpn.* **65**, 1875 (1996), for a review of $SrRuO_4$, see A. P. Mackensi and Y. Maeno, *Rev. Mod. Phys.* **75**, 657 (2003); for $RuSr_2GdCu_2O_8$ from W. E. Pickett *et al.*, *Phys. Rev. Lett.* **83**, 3713 (1999); for SW, from Z. K. Tang *et al.*, *Science* **292**, 2462 (2001); for a review on superconducting phases of UPt_3, see R. Joynt and L. Taillefer, *Rev. Mod. Phys.* **74**, 235 (2002).

Table 18.1.2 Some superconducting alloys and compounds.

type	material	T_c (K)	H_{c2} (kOe)	remark
alloys (solid solutions)	$Mo_{0.5}Re_{0.5}$	12.5	115	
	$Nb_{0.4}Ti_{0.6}$	9.3	124	practical superconducting
	$Nb_{0.26}Ti_{0.7}Ta_{0.04}$	9.9	91	materials for cables
	$Nb_{0.75}Zr_{0.25}$	11	~26	
	$Pb_{0.75}Bi_{0.25}$	8.7		
alloys (compounds)	V_3Ga	16.8	240	A15 structure, practical superconducting materials for cables
	Nb_3Al	18.8	300	ditto
	Nb_3Sn	18.1	245	ditto
	$Nb_3Al_{0.75}Ge_{0.25}$	21.0	420	ditto
	Nb_3Ge	23.2		A15 structure, the highest T_c
	NbN	17		B1 structure
	Pb_3Mo_6S	14.7	600	Chevrel-phase
	$Ga_{0.25}Eu_{0.3}Pb_{0.7}Mo_6S_8$	14.3	700	Chevrel-phase with highest H_{c2} losing
	$ErRh_4B_4$	8.5		superconducting at 0.9K, changing into
	$ErMo_6S_8$	2.2		ferromagnet <0.2K, coexistance of SC with AFM
	$ZrZn_2$	0.29		coexistence of SC and weak FM
	MgB_2	39	16~18	The highest T_c for alloys
heavy electron intermetallic compounds	$CeCu_2Si_2$	0.5~0.6		
	UPt_3	0.5		having three superconducting phases,
	UBe_3			p wave pairing(?)
	UGe_2	<(1)		coexistence of SC and weak FM
oxides	$SiTiO_{3-x}$	0.05~0.5		perovskite type structure
	$LiTiO_4$	13		perovskite type structure
	$Ba(Pb_{1-x}B_{ix})O_3$	13		x~0.25.
	$Ba_{1-2}K_xBiO_3$	>30		x>0.35 insulator x~0.4
	$SrRuO_4$	1.2		layered perovskite type structure, p-wave pairing(?)
cuprates	$La_{2-x}Ba_xCuO_4$	>30		HTSC, layered perovskite type structure
	$YBa_2Cu_3O_{7-x}$	93	150(77K)	discovered first superconductor with T_c in liguid nitrogen
	$Bi_2Sr_2CaCu_2O_8$	110	500(4.2K)	
	$Tl_2Ba_2Ca_2Cu_3O_8$	125		
	$HgBa_2Ca_2Cu_3O_{8-x}$	135		very unstable,
		164 (at high pressure)		coexistence of SC and WFM in different layers
	$RuSr_2GdCu_2O_8$	15~40		
organic compounds and molecular materials	$(TMTSF)_2ClO_4$	1.2~1.4		1D charge transfer salt
	$[BEDT-TTF]_2Cu(MCS)_2$	11.4		1D charge transfer salt,
	K_3C_{60}	~18		having the highest T_c for organics
	Rb_3C_{60}	~28~15		
	single wall nanocarbon-tube			abnormal Meissner effect, and 1D SC uctuations

§18.2 Ginzburg–Landau Theory

The phenomenological theory developed by Ginzburg and Landau has proved to be one of the most fertile approaches to superconductivity. It gives a comprehensive description of thermodynamic and electrodynamic properties of superconductors near T_c; sometimes it was even found to be useful outside the realm of its proven validity. It can be applied to both homogeneous and inhomogeneous superconductors. The crucial insight of Ginzburg–Landau (GL) theory is to identify the order parameter Ψ of superconductors, somewhat like the macroscopic wavefunction Ψ: This means the order parameter must be complex and can vary in space. Once the free energy has been written down as a function of Ψ and the vector potential $\boldsymbol{A}$, from the minimum free energy condition, an equation of motion for Ψ, as well as an equation for the supercurrent in terms of $\boldsymbol{A}$, may be derived. The latter has form of the London equation, so it may be regarded as a generalization of the London equation to the case of a spatially varying Ψ. The GL theory was developed before the microscopic BCS theory, but later it was shown by Gor'kov that it is a rigorous consequence of the microscopic theory, within a certain domain of temperature and magnetic field.

18.2.1 Ginzburg–Landau Equations and Broken Gauge Symmetry

The crucial wavefunction for a spin-singlet s-wave superconductor is the pair wavefunction $\Psi(\boldsymbol{r}) = \langle\psi_\uparrow(\boldsymbol{r})\psi_\downarrow(\boldsymbol{r})\rangle$. The phase transition from normal to superconducting phase is a transition in which gauge symmetry is broken. The condensed phase in a superconductor corresponds to a macroscopically occupied quantum state and is therefore described by a macroscopic wavefunction. We may interpret $|\Psi|^2 = \Psi^*\Psi$ as n_s, the local pair density. The Ginzburg–Landau order parameter Ψ is closely modeled on this macroscopic wavefunction Ψ, though it is not identical to it.

Here, for brevity, we will introduce the Ginzburg–Landau theory in accord with the phenomenological approach following Chap. 15. Based on the postulate that Ψ is small and varies slowly in space, the density functional of free energy can be written as

$$f_s(\boldsymbol{r}) = f_n + \alpha|\Psi(r)|^2 + \frac{\beta}{2}|\Psi(\boldsymbol{r})|^4 + \frac{1}{2m^*}\left|\left(-i\hbar\nabla + \frac{e^*}{c}A\right)\Psi\right|^2 + \frac{H^2(\boldsymbol{r})}{8\pi}, \tag{18.2.1}$$

where the pair charge and mass are $e^* = 2e$ and $m^* = 2m$; f_0 is the free energy density for the normal phase; $[-i\hbar\nabla + (e^*/c)\boldsymbol{A}]\Psi$ the canonical momentum associated with the condensate and $\boldsymbol{H}(\boldsymbol{r})$ the magnetic field.

After taking the minimum of the free energy with respect to Ψ

$$\delta F = \delta\int f(\boldsymbol{r})dr = 0, \tag{18.2.2}$$

and using the gauge condition $\nabla\cdot\boldsymbol{A} = 0$, we have the GL equation

$$\frac{1}{2m^*}\left(-i\hbar\nabla + \frac{e^*}{c}A\right)^2\Psi + \alpha\Psi + \beta|\Psi|^2\Psi = 0. \tag{18.2.3}$$

This differential equation will give us the variation of Ψ within a specimen, once we know $\boldsymbol{A}$. We have assumed that, on the boundary of the sample, the normal component is zero; that is,

$$\boldsymbol{n}\cdot\left(-i\hbar\nabla + \frac{e^*}{c}\boldsymbol{A}\right)\Psi = 0. \tag{18.2.4}$$

If on the other side we take the minimum of the free energy with respect to $\boldsymbol{A}$, the supercurrent equation is

$$\boldsymbol{j}_s = -\frac{ie^*\hbar}{2m^*}(\Psi^*\nabla\Psi - \Psi\nabla\Psi^*) - \frac{e^{*2}}{m^*c}\Psi^*\Psi\boldsymbol{A}. \tag{18.2.5}$$

In general, $\Psi = |\Psi|\exp(i\theta)$ is a complex function, so if its amplitude is constant, the supercurrent equation is

$$\boldsymbol{j}_s = \frac{e^*}{m^*}|\Psi|^2\left(\hbar\nabla\theta - \frac{e^*}{c}\boldsymbol{A}\right) = e^*|\Psi|^2\boldsymbol{v}_s, \tag{18.2.6}$$

which is consistent with (18.1.34). This is just the formula for current in quantum mechanics. The analogy of $\Psi(\boldsymbol{r})$ to a wavefunction of quantum mechanics cannot be taken too literally: $\Psi(\boldsymbol{r})$ is not a wavefunction in the usual sense. The GL equation is nonlinear in the term $\beta|\Psi(\boldsymbol{r})|^2\Psi(\boldsymbol{r})$. Neither the quantum-mechanical principle of superposition holds for $\Psi(\boldsymbol{r})$, because of this nonlinearity, nor does $\Psi(\boldsymbol{r})$ have to be normalized because $|\Psi(\boldsymbol{r})|^2$ is not a probability distribution. Really $\Psi(\boldsymbol{r})$ in the GL equations is just the order parameter of the superconducting phase transition. However, since the superconducting state is a manifestation of macroscopic quantum phenomena, this order parameter acquires some wavefunction-like behavior.

The physics of symmetry changes at the superconducting transition temperature T_c is most simply understood using GL theory. For any $\Psi(\boldsymbol{r})$ we can readily find another with exactly the same free energy, i.e., $\exp(i\theta)\Psi(\boldsymbol{r})$, where θ is the phase angle. This transformation is called a global gauge transformation.[f] When θ takes different constants, since the free energy f in (18.2.1) is invariant to the transformation, all these transformations form a gauge symmetry group of f. We may visualize the global gauge symmetry by the Argand diagram shown in Fig. 18.2.1. The symmetry operation which changes $\Psi(\boldsymbol{r})$ into $\exp(i\theta)\Psi(\boldsymbol{r})$ is simply a rotatation to any point on the diagram by an angle ϕ about the origin. The symmetry group of these 2D rotations is known mathematically as $U(1)$, the group of unitary transformations in one dimension. A matrix M is unitary if its Hermitian conjugate is equal to its inverse, i.e., $M^* = M^{-1}$, a complex number $\exp(i\theta)$ satisfies this condition for its complex conjugate $\exp(-i\theta)$ equals its inverse $1/\exp(i\theta)$.

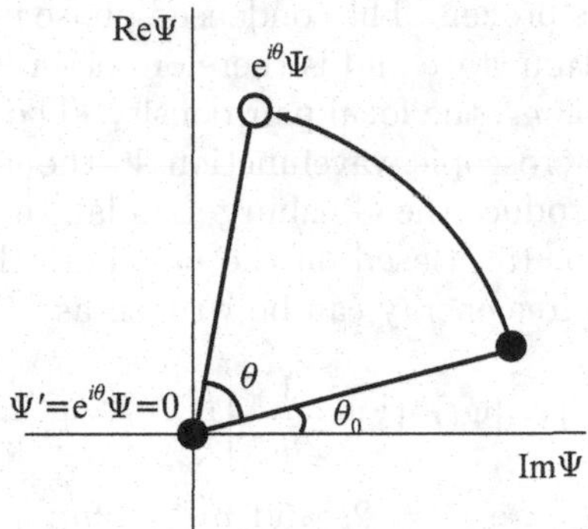

Figure 18.2.1 Schematic diagram showing global gauge transformation of the GL order parameter.

Now the crucial point is that, though the free energy has gauge symmetry, its minima need not possess this symmetry. The symmetry can be broken spontaneously on cooling through T_c. The GL equations give a simple illustration for superconductors. For simplicity, we assume that the size of the superconductor is infinite (so surface effects may be neglected) and the magnetic field is absent ($\boldsymbol{A} = 0$). Then the minimum of free energy in (18.2.1) corresponds to a spatially constant $\Psi(\boldsymbol{r}) = \Psi_0$. The GL equation (18.2.3) can be simplified to

$$\alpha\Psi_0 + \beta|\Psi_0|^2\Psi_0 = 0. \tag{18.2.7}$$

This has two solutions $\Psi_0 = 0$ and $|\Psi_0| = (-\alpha/\beta)^{1/2}$, for $\alpha/\beta < 0$. If we assume that $\beta > 0$, and α changes sign as $\alpha = a(T - T_c)$ at T_c, the solutions for $T > T_c$ and $T < T_c$ are found to be

$$|\Psi_0| = \begin{cases} 0, & T > T_c \\ [a(T-T_c)/\beta]^{1/2}, & T < T_c \end{cases}. \tag{18.2.8}$$

[f]It should be noted that the global gauge transformations are quite different from the local gauge transformations which are commonly encountered in particle physics and electromagnetism. In a local gauge transformation $\Psi(\boldsymbol{r}) \to \exp[i\theta(\boldsymbol{r})]\Psi(\boldsymbol{r})$ in which $\theta(\boldsymbol{r})$ can be a function of position. The gauge transformations in electromagnetism $A = A + \nabla\chi$ are also local and not global.

Consider the normal state $T > T_c$, $\Psi_0 = 0$: Applying the global gauge transformation gives the trivial identity $\Psi_0 \exp(i\theta) = 0$, showing it is invariant after the gauge transformation. Then consider the superconducting state $T < T_c$: The global gauge transformation gives $\Psi_0 \to \Psi_0 \exp(i\theta) \neq 0$, it is not invariant after the gauge transformation. Thus for the state above T_c, the global gauge symmetry is intact, but it is spontaneously broken below T_c. This situation is clearly shown in Fig. 18.2.1.

The GL-like theory was formulated for the treatment of ^{4}He superfluid as well as the BE condensate of dilute gases in trap. We now consider the Ginzburg–Pitaevskii equation for ^{4}He, which is identical to (18.2.3) when $\boldsymbol{A} = 0$,

$$\left(-\frac{\hbar}{2M_4}\nabla^2 + \alpha + \beta|\Psi(\boldsymbol{r})|^2\right)\Psi(\boldsymbol{r}) = 0; \tag{18.2.9}$$

this is an extension to the case of a normal fluid which is not stationary, the kinetic energy of the normal fluid must be included in the free energy term

$$f(\boldsymbol{r}) = \frac{1}{2}M_4|(-i\hbar\nabla - \boldsymbol{v}_{\rm n}))\Psi(\boldsymbol{r})|^2 + f_0(|\Psi(\boldsymbol{r})|^2), \tag{18.2.10}$$

where f_0 is the free energy density of the stationary fluid in equilibrium. The Gross–Pitaevskii equation for a BE condensate in the trap is

$$\left(-\frac{\hbar^2}{2M}\nabla^2 + V(\boldsymbol{r}) + g|\Psi(\boldsymbol{r})|^2\right)\Psi(\boldsymbol{r}) = \mu\Psi(\boldsymbol{r}), \tag{18.2.11}$$

where g and μ denote the interaction strength and chemical potential, respectively. The absence of interactions ($g = 0$) will reduce it to the Schrödinger equation for single particles confined in a harmonic trap.

18.2.2 Penetration Depth and Coherence Length

In the interior of a superconductor, no magnetic field exists and the gradient term vanishes, the free energy density in (18.2.1) is reduced to

$$f_{\rm s} - f_{\rm n} = \alpha|\Psi|^2 + \frac{1}{2}\beta|\Psi|^4, \tag{18.2.12}$$

where the temperature dependence of parameters are assumed $\alpha = a(T - T_c)$, and β is a constant larger than zero determined from the free energy minimum. In this case, the first term in (18.2.3) vanishes, and we can write

$$|\Psi|^2 = |\Psi_\infty|^2 = -\frac{\alpha}{\beta} \propto (1-t), \tag{18.2.13}$$

where $t = T/T_c$. Combined with (18.1.30), the thermodynamic critical field is determined from

$$f_{\rm s} - f_{\rm n} = \frac{H_c^2}{8\pi} = -\frac{\alpha^2}{2\beta}, \tag{18.2.14}$$

i.e.,

$$H_c = \left(\frac{4\pi}{\beta}\right)^{1/2}|\alpha| \propto (1-t), \tag{18.2.15}$$

from which when $T \to T_c$, $H_c \to 0$.

From (18.2.6), the current is simply given by

$$\boldsymbol{j}_{\rm s} = \frac{e}{m}\frac{|\alpha|}{\beta}\left(\hbar\nabla\theta - \frac{2e}{c}\boldsymbol{A}\right). \tag{18.2.16}$$

If we notice that $\boldsymbol{B} = \nabla \times \boldsymbol{A}$ and $\boldsymbol{j}_s = (c/4\pi)\nabla \times \boldsymbol{B}$, we find the usual London penetration depth is temperature dependent

$$\lambda_L = \left(\frac{mc^2\beta}{8\pi e^2|\alpha|}\right)^{1/2} \sim (1-t)^{-1/2}. \tag{18.2.17}$$

The supercurrent is confined within this distance λ_L from the surface.

To study the variation of the wavefunction near the surface, we consider the GL equations (18.2.3) in the absence of field, i.e., $\boldsymbol{A} = 0$. Since all of the coefficients are real, Ψ must be real, and there is no supercurrent, $\boldsymbol{j}_s = 0$. Introducing a dimensionless wavefunction $\Psi' = \Psi/\Psi_\infty$, and assuming that it varies only in the z direction, (18.2.3) becomes

$$\xi^2(T)\frac{d^2\Psi'}{dz^2} + \Psi' - \Psi'^3 = 0 \tag{18.2.18}$$

where a coherence length $\xi(T)$ can be defined as

$$\xi(T) = \left(\frac{\hbar^2}{2m^*|\alpha|}\right)^{1/2} \propto (1-t)^{-1/2}. \tag{18.2.19}$$

ξ is the characteristic length for variation of the order parameter, which diverges at $T = T_c$. To see the significance of $\xi(T)$ more clearly, we seek an approximate solution for (18.2.18). Let $\Psi' = 1+\Delta\Psi'$, with $\Delta\Psi' \ll 1$, then the first-order of (18.2.18) becomes

$$\xi^2\frac{d^2}{dz^2}\Delta\Psi' + (1+\Delta\Psi') - (1+3\Delta\Psi' + \cdots) = 0, \tag{18.2.20}$$

or

$$\frac{d^2}{dz^2}\Delta\Psi' = \frac{2}{\xi^2}\Delta\Psi', \tag{18.2.21}$$

and its solution is

$$\Delta\Psi'(z) \sim e^{\pm\sqrt{2}z/\xi(T)}. \tag{18.2.22}$$

This means that Ψ will decay to Ψ_∞ within the characteristic length $\xi(T)$.

There is an important GL parameter defined as a ratio of above two length parameters

$$\kappa = \frac{\lambda_L}{\xi}. \tag{18.2.23}$$

In type I superconductors, $\lambda_L \sim 50$ nm, $\xi \sim 400$ nm, so $\kappa \ll 1$ for these materials. Figure 18.2.2 shows the crossover region of the normal-superconductor boundary, how the magnetic field penetrates

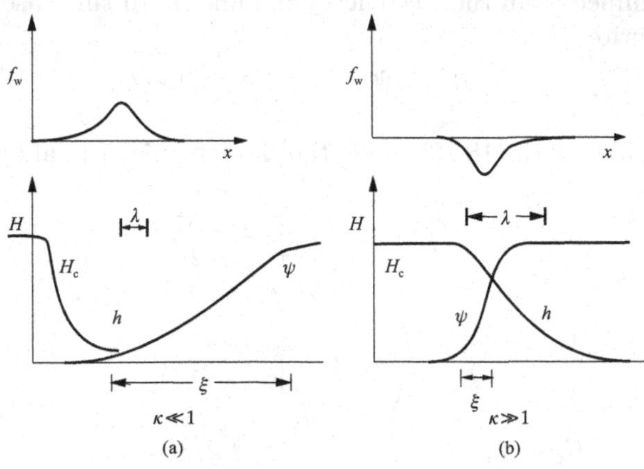

Figure 18.2.2 Schematic diagrams showing how the free energy of the wall, local magnetic field and order parameter vary with the distance from a normal-superconductor interface. (a) Type 1 and (b) type II superconductors.

the superconductor to a depth λ_L, and how Ψ increases in the superconductor to its value at infinity Ψ_∞ in a distance ξ. Abrikosov (1957) shows the importance of the $\kappa \gg 1$ case which leads to type II superconductors. Essentially all superconducting compounds and all high-T_c materials have $\kappa \gg 1$. Usually, the value $\kappa = 1/\sqrt{2}$ is used as a criterion to separate superconductors of types I and II.

18.2.3 Magnetic Properties of Vortex States

As stated in §18.1.4, flux quantization is an important concept related to the macroscopic wavefunction. It should be noted that there is a profound difference in the magnetic behavior of type I and type II superconductors, as deduced from L. V. Shubnikov's experimental results in the 1930s. In contrast to type I, the magnetization of type II material begins to decrease at $H > H_{c1}$, then decreases more gradually, and goes to zero at $H > H_{c2}$ see Figs. 18.1.14 and 18.1.15. This peculiar magnetic behavior is instrumental in making the applications of superconducting magnets and cables possible. Abrikosov in 1957 first explained the magnetic properties of type II superconductors theoretically from the solutions of the GL equations. The essential point is that, in a type II material, magnetic flux can penetrate into the bulk specimen as an array of vortex lines, between H_{c1} and H_{c2}. The Meissner effect is not so complete as in type I materials; this mixture of a superconductor with normal state in the cores of a vortex lattice is called the vortex state or the mixed state. This bold hypothesis was experimentally confirmed by neutron diffraction in 1965 and direct observations after that.

With the GL theory, we can determine the value of H_{c2} even without an explicit knowledge of the structure of a vortex state. Since $H \to H_{c2}$, $\Psi \to 0$; then at magnetic field slightly less than H_{c2}, Ψ is very small, so the nonlinear term in GL equation may be dropped and instead the simplified linear GL equation

$$\frac{1}{4m}\left(-i\hbar\nabla + \frac{2e}{c}\boldsymbol{A}\right)^2 \Psi = |\alpha|\Psi \tag{18.2.24}$$

may be used. Here, $\boldsymbol{A}$ may be regarded as the vector potential for the homogeneous field H at $\Psi = 0$; in this situation, the magnetic field permeates the normal state. Formally (18.2.24) is just a Schrödinger equation which describes the motion of a particle with charge $2e$ and mass $2m$ in the magnetic field, and this leads to Landau levels (as discussed in §7.3.2), where $|\alpha|$ is the spacing of the energy levels. Furthermore, the boundary condition $\Psi = 0$ at infinity is the same for both problems. We have already solved the problem of a charged particle in a magnetic field; the minimum value for energy is $E_0 = \hbar\omega_H/2$, where $\omega_H = eH/2mc$. Due to the similarity of these two problems, so we may conclude that the superconducting phase appears only when the condition

$$|\alpha| > \frac{e\hbar}{2mc}H \tag{18.2.25}$$

is satisfied. Then we arrive at

$$H_{c2} = \frac{2mc|\alpha|}{e\hbar} = \frac{\Phi_0}{2\pi\xi(T)^2} = \sqrt{2}\kappa H_c. \tag{18.2.26}$$

This is the highest field at which superconductivity can nucleate in a bulk sample in a decreasing external magnetic field.

H_{c1} marks the magnetic field at which the first vortex in the form of a flux quantum penetrates into the specimen. By definition, when $H = H_{c1}$, the Gibbs free energy must have the same value whether the first vortex is in or out of the sample. Thus, H_{c1} is determined by the condition

$$G_s^{(0)} = G_s^{(1)}, \tag{18.2.27}$$

$G_s^{(0)}$ and $G_s^{(1)}$ are the Gibbs free energies with no magnetic flux trapped and with the first isolated vortex line entering the sample. Since

$$G = F - \frac{\boldsymbol{H}}{4\pi} \cdot \int \boldsymbol{h} d\boldsymbol{r}, \ G_s = F_s, \tag{18.2.28}$$

we may introduce ε_i (or in other words, the line tension of the vortex) and L the length of vortex line, whereupon the condition becomes

$$F_s = F_s + \varepsilon_i L - \frac{H_{c1}\Phi_0 L}{4\pi}. \tag{18.2.29}$$

Thus we get

$$H_{c1} = \frac{4\pi\varepsilon_i}{\Phi_0}. \tag{18.2.30}$$

In the extreme type II limit $\kappa = \lambda/\xi \gg 1$, at the center of an isolated vortex, $\Psi = 0$, then it gradually rises to a limiting value at radius ξ. This defines a vortex core region within which the behavior is like the normal metal, while outside the core region the behavior is just like an ordinary London superconductor (see Fig. 18.2.3).

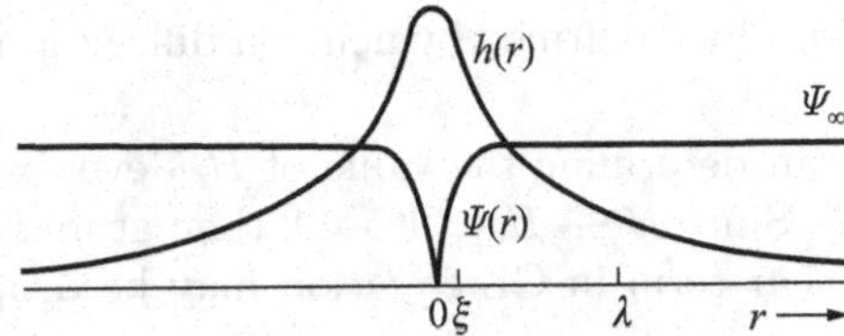

Figure 18.2.3 Magnetic field and superconducting order parameter of an isolated vortex line (schematic).

18.2.4 Anisotropic Behavior of Superconductors

After the discovery of high T_c superconductors, it was found that the GL equation is still valid; however, the parameters become very anisotropic. Now (18.2.3) can be rewritten as

$$-\frac{\hbar^2}{2}\left(\nabla - i\frac{2e}{\hbar c}\boldsymbol{A}\right) \cdot \left(\frac{1}{m}\right) \cdot \left(\nabla - i\frac{2e}{\hbar c}\boldsymbol{A}\right)\Psi + \alpha\Psi + \beta\left|\Psi\right|^2\Psi = 0, \tag{18.2.31}$$

where $(1/m)$ is the reciprocal mass tensor with principal values $1/m_{ab}$ and $1/m_c$. Due to the fact that the interlayer coupling is weak, $m_c \gg m_{ab}$. The anisotropy of mass causes the coherence length ξ to be very anisotropic. We may generalize (18.2.19) to the anisotropic case and get

$$\xi_i^2(T) = \frac{\hbar^2}{2m_i|\alpha(T)|}, \tag{18.2.32}$$

where the subscript i refers a particular principal axis. Since $\alpha(T)$ is isotropic and proportional to $(T - T_c)$, ξ_i scales with $1/\sqrt{m_i}$ and diverges as $|T - T_c|^{-1/2}$ when $T \to T_c$. The penetration depth λ is also anisotropic by the relation

$$2\sqrt{2}\pi H_c(T)\xi_i(T)\lambda_i(T) = \Phi_0. \tag{18.2.33}$$

This shows that the anisotropy of the penetration depth λ_i will be the inverse of that of ξ_i since H_c is isotropic, and since λ_i describes the screening by supercurrents flowing along the ith axis, not the screening of a magnetic field along the ith axis. We may take an Abrikosov vortex line in a sample with the magnetic field along the a axis. In an isotropic superconductor, the vortex will be circularly symmetric; however, in an anisotropic superconductor, the core radius along the plane

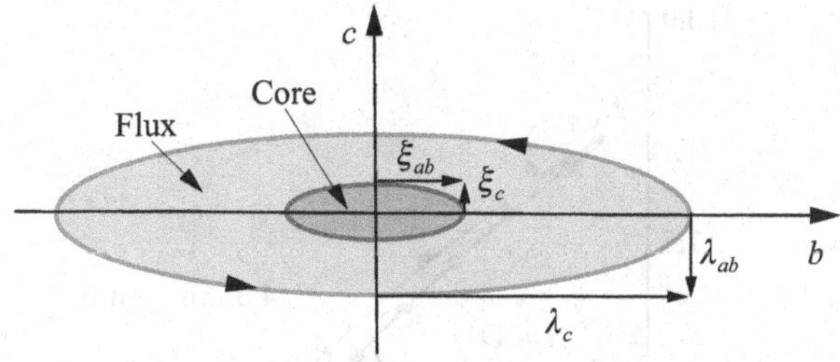

Figure 18.2.4 Cross-section of a vortex line along the a axis in an anisotropic superconductor (schematic).

direction will be ξ_{ab}, but the core radius along the c direction will be $\xi_c \ll \xi_{ab}$. On the other hand, the flux-penetration depth will be λ_c along the plane direction, and will be much smaller, λ_{ab} along the c direction. Thus, both the core and the current streamlines confining the flux are flattened into ellipses with their long axes parallel to the planes (b axis), and an aspect ratio $(m_c/m_{ab})^{1/2}$, as shown in Fig. 18.2.4.

Furthermore, the anisotropy of the upper critical field along the two distinct principal axes may be derived,

$$H_{c2\|c} = \frac{\Phi_0}{2\pi}\xi_{ab}^2, \; H_{c2\|ab} = \frac{\Phi_0}{2\pi}\xi_{ab}\xi_c. \tag{18.2.34}$$

Since $\xi_{ab} \gg \xi_c$, $H_{c2\|ab} \gg H_{c2\|c}$. Because $H_{c1} \sim 1/\lambda^2$, which is inversely related to H_{c2}, the anisotropy in H_{c1} will be inverse to that for H_{c2}, i.e., $H_{c1\|ab} \ll H_{c1\|c}$.

Now we introduce the dimensionless anisotropy parameter

$$\gamma = \left(\frac{m_c}{m_{ab}}\right)^{1/2} = \frac{\xi_{ab}}{\xi_c} = \frac{\lambda_c}{\lambda_{ab}} = \left(\frac{H_{c2\|ab}}{H_{c2\|c}}\right) = \left(\frac{H_{c1\|c}}{H_{c1\|ab}}\right). \tag{18.2.35}$$

The mass ratio m_c/m_{ab} and γ for YBCO are about 50 and 7 respectively, while for BSCCO, these are 20 000 and 150 respectively. This large anisotropy is one of the decisive factors which make the high T_c superconductors act so differently from the conventional ones. Near T_c, $\xi_c(T) \approx \xi_c(0)$ $(1-t)^{-1/2}$ will always be large enough to justify the GL approximation discussed above. But when the temperature is lowered, $\xi_c(T)$ shrinks toward a limiting value. If this value is smaller than the interplanar spacing, it is obvious that the smooth variation assumed in the GL equations will break down at some intermediate temperature T. At temperatures below T_c, it is expected that the 3D continuum approximation will be replaced by the 2D behavior of a stack of individual layers. This may be described by a model proposed by Lawrence and Doniach (LD)[g] In this model, the free energy may be expressed as

$$F = \sum_n \int \left[\alpha|\Psi_n|^2 + \frac{1}{2}\beta|\Psi_n|^4 + \frac{\hbar^2}{2m_{ab}}\left(\left|\frac{\partial\Psi_n}{\partial x}\right|^2 + \left|\frac{\partial\Psi_n}{\partial y}\right|^2\right) + \frac{\hbar^2}{2m_c s^2}|\Psi_n - \Psi_{n-1}|^2\right] dxdy, \tag{18.2.36}$$

where the z is along c axis, x, y are the coordinates in the plane, s is the distance between the layers, the sum runs over layers and the integral is over the area of each layer. Note that if we write $\Psi_n = |\Psi_n|e^{i\theta_n}$, and assume that all $|\Psi_n|$ are equal, the last term of (18.2.36) becomes

$$\frac{\hbar^2}{m_c s^2}|\Psi_n|^2[1 - \cos(\Psi_n - \Psi_{n-1})]. \tag{18.2.37}$$

This term is equivalent to a Josephson coupling energy $(1/m_c)$ between adjacent planes (see §18.4). This crossover from 3D behavior $[H_{c2} \propto (T_c - T)]$ to a 2D one $[H_{c2} \propto (T_c - T)^{1/2}]$ in superconductors has been verified experimentally by the artificial layered composites Nb/Ge in certain thickness ranges (Fig. 18.2.5). We can expect that high T_c superconductors would have an analogous behavior.

[g] W. E. Lawrence and S. Doniach, *Proc 12th Int. Cont. Low Temp. Phys. (Kyoto, 1970)*; E. Kanda (ed.), Keigaku, Tokyo, 1971.

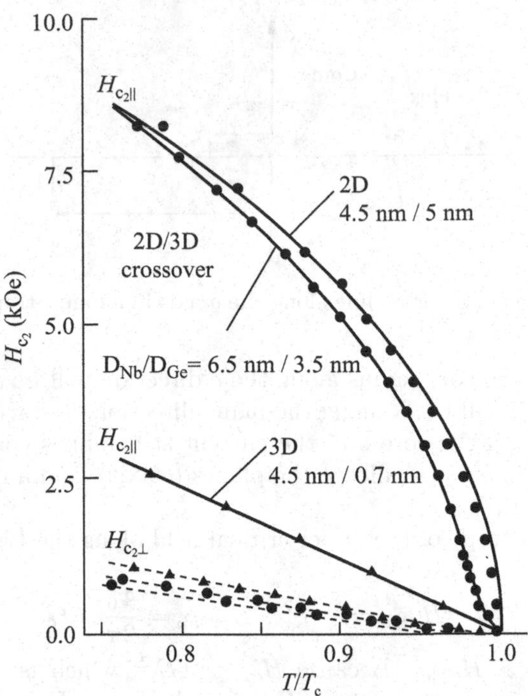

Figure 18.2.5 Upper critical fields of Nb/Ge composites with layer thickness $D_{\rm Nb}$ for Nb layer and $D_{\rm Ge}$ for Ge layer. From S. T. Ruggiero *et al.*, *Phys. Rev. Lett.* **45**, 1299 (1980).

§18.3 Pairing States

At the heart of the microscopic theory of superconductivity, as well as that of ^{3}He superfluidity, is the problem of the symmetry of the paired electrons, or atoms, and the mechanism of the corresponding pair formation. The famous Bardeen–Cooper–Schrieffer (BCS) theory solved these problems for conventional superconductors with exemplary success; however, we have no intention to develop a complete description of BCS theory here. This has been the subject of several books listed in the bibliography. In this section, our task is limited to giving an elementary sketch of the generalized Cooper pairs and then applying this to various cases: The spin-singlet s-wave pairing of conventional superconductors, the spin-singlet d-wave pairing of cuprate superconductors, and the spin-triplet p-wave pairing of ^{3}He superfluid and some exotic superconductors. In writing of this section, authors are much indebted to the excellent introduction of this subject in Mineev and Somokhin's monograph (Bib. [9]).

18.3.1 Generalized Cooper Pairs

The Cooper problem, as one of the fundamental ideas of Bardeen–Cooper–Schrieffer (BCS) theory, is that the normal metallic state is unstable with respect to the Fermi sea in the presence of an arbitrarily small attractive interaction between electrons; thus Cooper pairs are formed. Let us consider a simple model in which two electrons, located at $\boldsymbol{r}_1$ and $\boldsymbol{r}_2$, are added to a Fermi sea of electrons at $T = 0$: Suppose that the two electrons interact via the potential $V(\boldsymbol{r}_1 - \boldsymbol{r}_2)$ and the presence of other electrons manifests itself only through Pauli exclusion principle (Fig. 18.3.1). The wavefunction for this pair of electrons satisfies the Schrödinger equation

$$-\frac{\hbar^2}{2m}(\nabla_1^2 + \nabla_2^2)\Psi(\boldsymbol{r}_1, \boldsymbol{r}_2) + V(\boldsymbol{r}_1, \boldsymbol{r}_2)\Psi(\boldsymbol{r}_1, \boldsymbol{r}_2) = (E + 2E_{\rm F})\Psi(\boldsymbol{r}_1, \boldsymbol{r}_2), \tag{18.3.1}$$

where the energy E is measured from the state where two electrons are at the Fermi level. We may assume the center of mass of these two electrons $\boldsymbol{R} = (\boldsymbol{r}_1 + \boldsymbol{r}_2)/2$ is at rest, the wavefunction of the

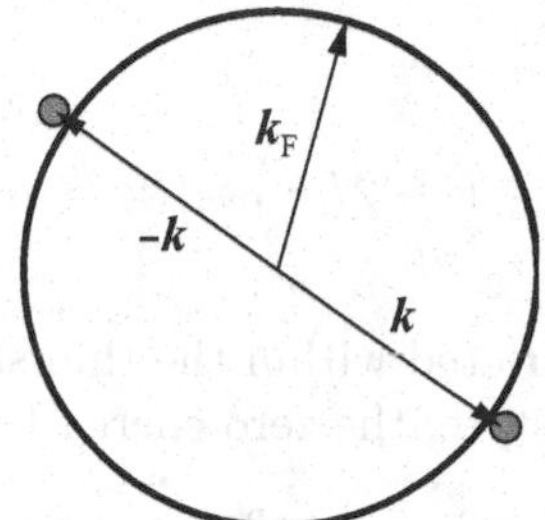

Figure 18.3.1 Two electrons with oppositely directed momenta just outside the Fermi sphere.

electron pair depends only on $\boldsymbol{r} = \boldsymbol{r}_1 - \boldsymbol{r}_2$. Since the condensate in the ground state responsible for superconductivity must have zero momentum, only a pair of electrons with opposite momenta $\boldsymbol{k}$ and $-\boldsymbol{k}$ is considered, the coordinate $\boldsymbol{R}$ in the wavefunction can be omitted, and we have a simplified equation

$$-\frac{\hbar^2}{m}\nabla^2\Psi(\boldsymbol{r}) + V(\boldsymbol{r})\Psi(\boldsymbol{r}) = (E + 2E_\mathrm{F})\Psi(\boldsymbol{r}). \tag{18.3.2}$$

In the momentum representation the wavefunction is written as

$$\phi(\boldsymbol{k}) = \int d\boldsymbol{r}\, \mathrm{e}^{-i\boldsymbol{k}\cdot\boldsymbol{r}}\Psi(\boldsymbol{r}). \tag{18.3.3}$$

In general, the wavefunction which has a definite angular momentum quantum number l and associated magnetic quantum number m can be expanded by spherical harmonics Y_{lm}, and this is true for the wavefunction in real space as well as in k-space. Let $\boldsymbol{\kappa} = \boldsymbol{k}/k_\mathrm{F}$, for a definite l, the wavefunction in momentum space is

$$\phi_l(\boldsymbol{k}) = \sum_{m=-l}^{l} a_{lm}(k)\mathrm{Y}_{lm}(\boldsymbol{\kappa}). \tag{18.3.4}$$

Since the anisotropy of wavefunction leads to the anisotropy of interaction potential which is also l-dependent and a function of the direction of $\boldsymbol{k}$, the potential can be expanded by Fourier transform

$$V(\boldsymbol{k}) = \int d\boldsymbol{r}\, \mathrm{e}^{-i\boldsymbol{k}\cdot\boldsymbol{r}}V(\boldsymbol{r}), \tag{18.3.5}$$

and it can also expanded by spherical harmonics

$$V_l(\boldsymbol{k}) = \sum_{m=-l}^{l} V_{lm}(k)\mathrm{Y}_{lm}(\boldsymbol{\kappa}). \tag{18.3.6}$$

The equation for $\phi_l(\boldsymbol{k})$ is easily found to be

$$\frac{\hbar^2k^2}{m}\phi_l(\boldsymbol{k}) + \sum_{\boldsymbol{k}'}\phi_l(\boldsymbol{k}')V_l(\boldsymbol{k}-\boldsymbol{k}') = (E + 2E_\mathrm{F})\phi_l(\boldsymbol{k}). \tag{18.3.7}$$

When $k < k_\mathrm{F}$,

$$\phi(\boldsymbol{k}) = 0. \tag{18.3.8}$$

This means simply that all states below the Fermi level are occupied. This equation (Bethe–Goldstone equation) has a continuous spectrum of solutions for $E > 0$ corresponds to the collision of two electrons $\pm\hbar k$. However, Cooper had the insight to point out that the bound states exist for $E < 0$ provided that V is attractive.

A drastic simplification is introduced by the assumption that this interaction is attractive and constant within a thin shell over Fermi surface with the thickness ε_l, while otherwise is zero

$$V_l(\boldsymbol{k}-\boldsymbol{k}') = \begin{cases} -V_l, & E_\mathrm{F} < \hbar^2k^2/2m,\ \hbar^2k'^2/2m < E_\mathrm{F} + \varepsilon_l, \\ 0, & \text{other cases.} \end{cases} \tag{18.3.9}$$

Then (18.3.7) may be rewritten as

$$\left(\frac{\hbar^2 k^2}{m} - E - 2E_{\mathrm{F}}\right)\phi_l(\boldsymbol{k}) = V_l \sum_{\boldsymbol{k}'} \phi_l(\boldsymbol{k}'), \tag{18.3.10}$$

where the summation over $\boldsymbol{k}'$ is restricted within the thin shell defined above.

In order to calculate the bound state, the zero energy level is redefined at the Fermi level, thus

$$\xi_{\boldsymbol{k}} = \frac{\hbar^2 k^2}{2m} - E_{\mathrm{F}}, \tag{18.3.11}$$

and (18.3.10) can be rewritten as

$$1 = V_l \sum_{\boldsymbol{k}} \frac{1}{2\xi_{\boldsymbol{k}} - E}. \tag{18.3.12}$$

According to the usual method, the summation of $\boldsymbol{k}$ can be replaced by an integration over energy, so we can get

$$1 = V_l \int_0^{\varepsilon_l} N(\xi) \frac{d\xi}{2\xi - E}. \tag{18.3.13}$$

If $\varepsilon_l \ll E$, $N(\xi)$ may be replaced by the value $N(0)$ at Fermi level, then

$$1 = \frac{1}{2} N(0) V_l \ln \frac{E - 2\varepsilon_l}{E}. \tag{18.3.14}$$

Finally, we obtain the equation satisfied by the binding energy E, and define the gap function at 0 K as $\Delta = -E/2$. For $N(0)V \ll 1$, the weak-coupling approximation is introduced, and the solution is

$$E_l = -2\varepsilon_l \exp\left(-\frac{2}{N(0)V_l}\right), \quad \Delta_l = \varepsilon_l \exp\left(-\frac{2}{N(0)V_l}\right). \tag{18.3.15}$$

Then an allowed energy state exists with $E < 0$. Thus a bound-pair of electrons with a finite binding energy $2\Delta_l$ is called a generalized Cooper pair.

According to the analysis above, we shall make several remarks:

1) The instability exists even for a very weak interaction V_l; the only requirement is that V_l is attractive. However, what type of interaction responsible for attractive pairing remains unspecified.

2) In the preceding derivation, the Pauli exclusion principle is taken into account between a pair of electrons at $\boldsymbol{r}_1$, $\boldsymbol{r}_2$ and the electrons at the Fermi surface. The total wavefunction of the Cooper pair must be antisymmetric with respect to the exchange of $\boldsymbol{r}_1$ and $\boldsymbol{r}_2$. The orbital wavefunction $\phi_l(\boldsymbol{k})$ is even or odd according to the value of its angular momentum quantum number l is even or odd: $\phi_l(-\boldsymbol{k}) = (-1)^l \phi_l(\boldsymbol{k})$. On the other hand, the spin function for paired fermions, each with spin 1/2, is odd for a spin-singlet $(\alpha_1\beta_2 - \alpha_2\beta_1)$ with $S = 0$, and even for the spin-triplets $(\alpha_1\alpha_2, \alpha_1\beta_2 + \alpha_2\beta_1, \beta_1\beta_2)$ with $S = 1$, here α_i $(i = 1, 2)$ means the spin-up state for particle i, and β_i for its spin-down state. Since the total wavefunction for a pair of fermions must be antisymmetric with the exchange of two particles, we may conclude: The pairing states for even values of l, such as s-wave $(l = 0)$, d-wave $(l = 2)$, g-wave $(l = 4), \ldots$, from spin-singlets; while the pairing states for odd values of l, such as p-wave $(l = 1)$, f-wave $(l = 3), \ldots$, form spin-triplets.

3) It should be noted that the particular form of E in (18.3.18) cannot be expanded in powers of V_l when V_l approaches zero. This explains why the microscopic theories of superconductivity via conventional perturbation methods must fail.

4) According to the uncertainty principle, we may estimate the spatial extent of a generalized Cooper pair by requiring $\xi_0 \delta k \geqslant 1$. The uncertainty in $\boldsymbol{k}$ may be estimated from the kinetic energy of a Cooper pair

$$2\Delta \sim \delta E \sim \frac{\hbar^2}{m} k_{\mathrm{F}} \delta k \sim \hbar v_{\mathrm{F}} \delta k, \tag{18.3.16}$$

then we obtain

$$\xi_0 \sim \frac{\hbar v_{\rm F}}{2\Delta}. \tag{18.3.17}$$

So the coherence length ξ_0 is deduced from the microscopic formation of Cooper pairs. The instability of a normal Fermi gas against pair formation may be estimated from the critical temperature, i.e., $k_{\rm B}T_{\rm c} \sim 2\Delta$. The Fermi velocity may be estimated from the relationship $mv_{\rm F} \sim \hbar n^{1/3}$, here n denotes the electron density and $n^{-1/3}$ the mean separation of the fermions. In conventional superconductors, $T_{\rm c}$ is several Kelvins, and the electron mass is roughly the free electron mass, hence $\xi_0 \sim 10^{-4}$–10^{-5} cm; this is the same order of magnitude as the coherence length from the GL theory. The mass of a ^{3}He is about several thousand times larger than that of an electron, while $T_{\rm c}$ is three orders of magnitude lower, then $\xi_0 \sim 10^{-6}$ cm. In heavy electron superconductors, the mass is about 100–1000 times the free electron mass, but the $T_{\rm c}$ is about 1 K or lower; in cuprate superconductors, the $T_{\rm c}$ reaches 40–100 K, so ξ_0 of these materials is about 10^{-6}–10^{-7} cm. In conventional superconductors, ξ_0 is much larger than $n^{-1/3} \sim 10^{-8}$ cm. From this we may conclude: There are many electron-pairs in a conventional superconductor which are distributed crisscrossing and overlapping with each other in the same spatial range. This is the original picture of Cooper pairs in momentum space as described by the Bardeen–Cooper–Schrieffer (BCS) theory. Schafroth postulated a theory for a pair of electrons in real space just like a two-electron molecule to explain conventional superconductivity without success. However, for cuprates and heavy fermions, though ξ_0 is still slightly larger than $n^{-1/3}$, it is not much larger. So some theories use the concept of bipolarons, which bear some resemblance to the Schafroth pair.

5) The Cooper problem is a simplistic version of the many-body problem: Its solution gave a decisive clue in the complex problem of the microscopic theory of superconductivity. For realistic calculations of the physical properties of conventional superconductivity; however, we must go outside the Cooper problem. This was accomplished admirably by the BCS theory (for more details about the BCS theory, see Bibs. [4, 5]).

18.3.2 Conventional Pairing of Spin-Singlet *s*-Wave

When the interaction V_l is applied to a gas of free electrons considered as s-wave, the electrons will form spin-singlet pairs with the release of energy. The physical mechanism for the attractive potential between electrons arises from coupling of electrons with lattice vibrations (electron-phonon mechanism) which correctly explains the isotope effect of $T_{\rm c}$ found experimentally by substituting atoms in the materials by isotopes with different masses, i.e.,

$$T_{\rm c}M^{\alpha} = \text{const.}, \tag{18.3.18}$$

where $\alpha \sim 1/2$ for most elements.

More accurate calculation of $\Delta(0)$ by Bardeen–Cooper–Schrieffer (BCS) theory with many-body formulation obtained a formula in weak-coupling approximation, i.e.,

$$\Delta(0) = 2\hbar\omega_{\rm D}{\rm e}^{-2/N(0)V}. \tag{18.3.19}$$

It should be noted that the form is just the same as (18.3.15), except the value of Δ is twice as large. It signifies that the simple derivation with a two-electron approximation in the treatment of the Cooper problem has captured the essential physics of conventional superconductors.

Now we shall try to explain the physical meaning of the gap function $\Delta(0)$. The Cooper pairs with their pair wavefunctions of zero momentum form a sort of Bose–Einstein condensation (BEC) in the ground state responsible for superconductivity. The excitations of the superconducting state show some peculiarities: The excitation of quasiparticles including electrons and holes is formed by the breaking of the Cooper pairs; this break needs to absorb an energy of 2Δ, so for these quasiparticles, the density of states between $\pm\Delta$ measured from the Fermi level is zero. From the BCS theory we can get

$$N(E) = \begin{cases} 0, & \text{when } |E| < \Delta; \\ \dfrac{N(0)}{\sqrt{E^2-\Delta^2}}, & \text{when} |E| > \Delta, \end{cases} \tag{18.3.20}$$

where we let the Fermi level be the zero point of E, and $N(0)$ is the density of states at the Fermi level for a normal state metal. This result can be seen in Fig. 18.3.2. The density of states between $\pm\Delta$ measured from the Fermi level for a normal state metal is transferred away and accumulated in the range out of the gap of 2Δ. For this reason, Δ is known as BCS gap function. In Fig. 18.3.3, the Fermi sphere, the energy gap and some quasiparticles are shown. The analogy to semiconductors should be noted: A definite energy just like the gap in semiconductors is required to produce the excited states, and it influences some physical properties such as absorption of electromagnetic waves and single-electron tunneling.

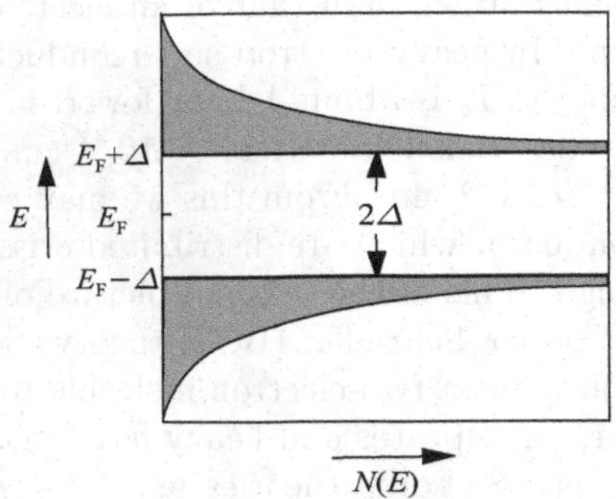

Figure 18.3.2 A schematic of energy gap in superconductor.

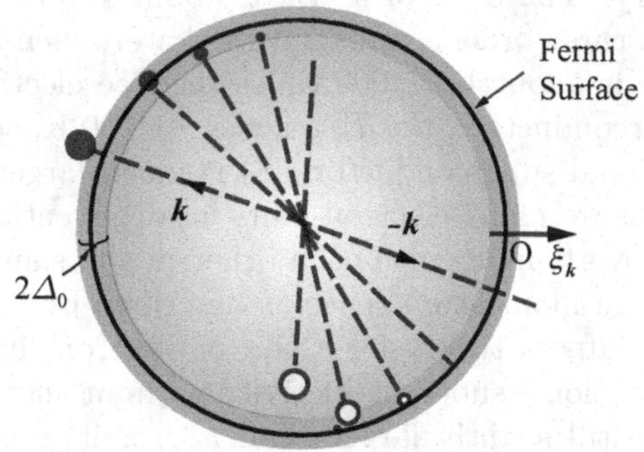

Figure 18.3.3 The energy gap and associated quasiparticles.

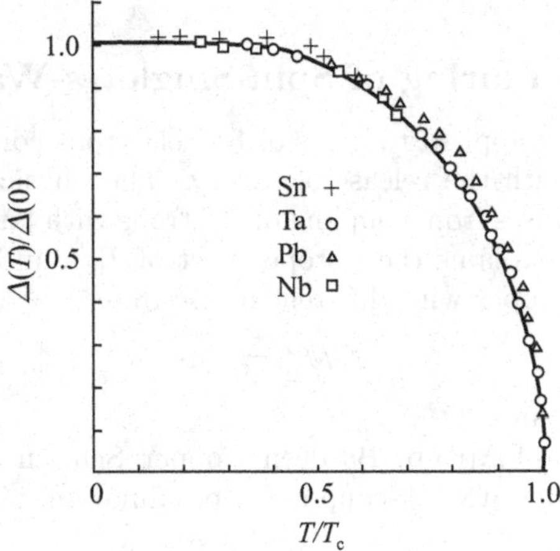

Figure 18.3.4 Temperature dependence for BCS energy gap function.

Furthermore, the temperature dependence of the gap function is derived from the BCS theory and its relationship to T_c is clarified. At $T = T_c$, $\Delta = 0$, so a formula for T_c may be derived as

$$\Delta(0) = 3.52k_BT_c. \tag{18.3.21}$$

Variation of Δ versus T is shown graphically in Fig. 18.3.4. In the vicinity of T_c, a simple approximation is obtained

$$\Delta(T) = 1.74\left(1 - \frac{T}{T_c}\right)^{1/2}. \tag{18.3.22}$$

From the values of T_c of conventional superconductors, we estimate the energy gap is actually very small, i.e., has the order of 10^{-4} eV. The gap can be measured experimentally by several different method such as tunneling, absorption and reflection of infrared light and nuclear magnetic resonance (NMR). In general, excellent agreement was found with the BCS theory. In some cases, the slight deviations are due to the weak coupling approximation, which can be rectified by using strong coupling theory instead.

The BCS gap function also provides the microscopic interpretation of the GL order parameter, first derived theoretically by Gor'kov in 1959. The proportionality of the GL order parameter $\Psi(\boldsymbol{r})$ to the BCS gap function $\Delta(\boldsymbol{r})$ is not unexpected, since both Ψ and Δ are complex quantities with magnitude and phase. They are both zero in the normal state, when $T > T_c$; at $T < T_c$, the gap opens at the Fermi surface and the order parameter becomes non-zero. These may vary in space or with applied magnetic field or both. Furthermore, the microscopic theory restricts the range of validity for GL equation near T_c and to gradual spatial variations of Ψ and $\boldsymbol{A}$. However, the concept of the order parameter identified as the gap function can be used without restraint even at 0 K.

Though, in general, sharp energy gaps are found for BCS superconductors, there are a few exceptional cases in which the fields, currents and gradients may act as 'pair-breakers' so the energy gaps become blurred. For instance, magnetic impurities which break the time-reversal symmetry lead to strong depression of T_c and modification of BCS states. Also excitation spectra of superconductors are modified if they carry currents. Strong current may add a common drift momentum $\boldsymbol{K}$ to the paired electrons lifting the degeneracy of $\boldsymbol{k}$ and $-\boldsymbol{k}$. So superconductors with concentration of magnetic impurities in a certain range become gapless for a finite current range before superconductivity is destroyed.

18.3.3 Exotic Pairing for Spin-Singlet *d*-Wave

Symmetry-breaking in conventional superconductors is only the breaking of gauge symmetry $U(1)$. However, for unconventional superconductors, other symmetries may also be broken at the phase transition: For instance, the point symmetry group of a crystal ($\mathcal{G}_c$) and the time-reversal symmetry group $\mathcal{T}$. So the symmetry group of a normal state may be expressed as

$$\mathcal{G} = U(1) \times \mathcal{T} \times \mathcal{G}_c. \tag{18.3.23}$$

Transition into unconventional superconducting, or superfluid, states involves additional spontaneously breaking of the symmetry group $\mathcal{G}_c$ or $\mathcal{T}$, besides that of $U(1)$. Formerly, we use names, such as s-wave, p-wave, d-wave and g-wave for the angular momentum of the Cooper pairs, but, strictly speaking, the presence of the crystal lattice makes the angular momentum l no longer a good quantum number. A crystal lattice will still have a definite symmetry, though Cooper pairs have a quantum mixture of states with different momenta. However, point symmetry groups of crystals may be described by writing down the simplest orbital wavefunctions or spherical harmonics. So s, $d_{x^2-y^2}$, d_{xy}, ... with different l may be used to characterize these symmetry states.

Now we consider the energy gap of unconventional superconductivity, taking the d-wave pairing of cuprates as an example. Since the gap function Δ must have the same symmetry as the GL order parameter Ψ, in conventional superconductors the order parameter is unchanged under any symmetry operation of the crystal group. This means the gap function Δ must have the full symmetry of the crystal, however, for unconventional superconductors, the situation is different: The order parameter Ψ as well as the gap function Δ cannot have the full symmetry of the crystal. For example, in the $d_{x^2-y^2}$ superconductor, the order parameter and the gap function change sign under rotation of 90° and reflections in the body diagonals (see Fig. 18.3.5). Since $\boldsymbol{k}$ is a wavevector in the Brillouin zone, the simplest function which has the right symmetries is

$$\Delta(\boldsymbol{k}) = \Delta_d[\cos(k_x a) - \cos(k_y a)]. \tag{18.3.24}$$

For small k we may expand the cosines and find, approximately, $\Delta(\boldsymbol{k}) \propto k_x^2 - k_y^2$ indicating $d_{x^2-y^2}$ symmetry. The magnitude of this function $\Delta(\boldsymbol{k})$ is the energy gap for the superconductor, it vanishes at four special points, where the Fermi surface crosses the square diagonals($k_x = \pm k_y$). Experimental

Symbol of Group Theory	A_{1g}	A_{2g}	B_{1g}	B_{2g}
Base Function for Order Parameter	Constant	$xy(x^2-y^2)$	x^2-y^2	xy
Wave Function	s	g	$d_{x^2-y^2}$	d_{xy}
Sketch for $\Delta(\boldsymbol{k})$ in B.Z.	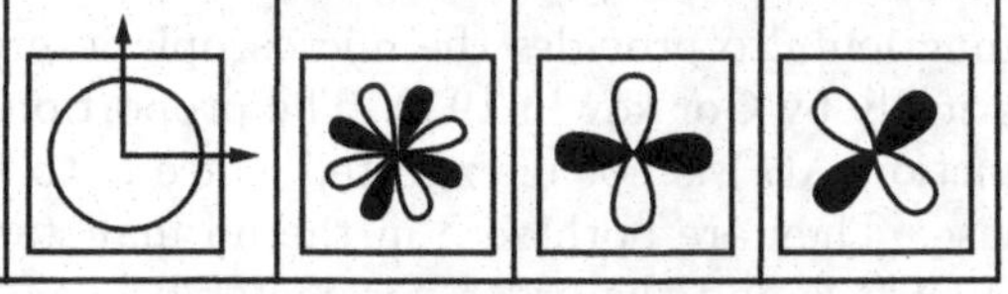			

Figure 18.3.5 $\boldsymbol{k}$-space representation of allowed symmetry basis functions for tetragonal symmetry appropriate for the CuO_2 planes in high T_c cuprates. From C. C. Tsuei and J. R. Kirtley, *Rev. Mod. Phys.* **72** 969 (2000).

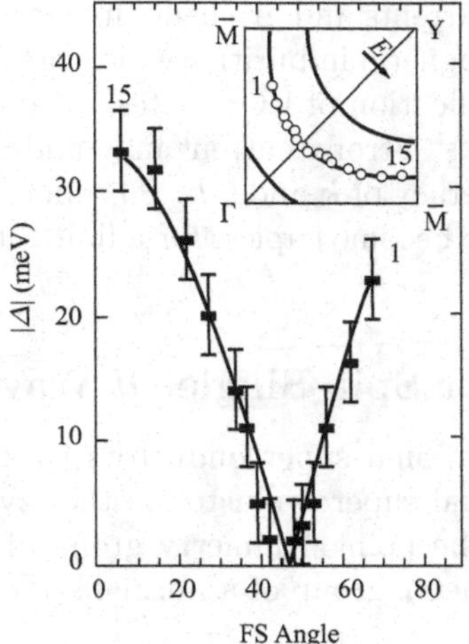

Figure 18.3.6 Energy gap in Bi-2212: full circles indicates the values measured with ARPES as a function of angle on the Fermi surface; the solid curve is a fit to the data using the d-wave order parameter; the inset indicates the locations of the data points in the Brillouin zone. From H. Ding *et al.*, *Phys. Rev. B* **54**, 9678 (1996).

determination of energy gap in Bi-2212 with the angle resolved photoemission spectroscopy (ARPES) roughly confirmed $d_{x^2-y^2}$ pairing in high T_c superconductors (see Fig. 18.3.6). But it still cannot settle this problem, further work with phase-sensitive techniques are needed, and we shall discuss this in §18.4.3.

18.3.4 Pseudogaps and Associated Symmetry

In the underdoped cuprates, $p < p_0$ (here p is the density of charge-carrier per Cu atom in CuO_2 plane, and p_0 is its optimum value at optimum doping), the pseudogap in the normal state above T_c was observed by various experimental methods, such as nuclear magnetic resonance (NMR), neutron scattering, specific heat measurement, photoemission spectra and infrared optical spectra. A result from specific heat measurements is shown in Fig. 18.3.7.

The most unusual property of the pseudogaps is, with decreasing value of p, the density of states is depleted near the Fermi level as shown schematically in Fig. 18.3.8. An approximately linear relation may be fitted to these experimental results,

$$E_{g0} = J\left(1 - \frac{p}{p_0}\right). \tag{18.3.25}$$

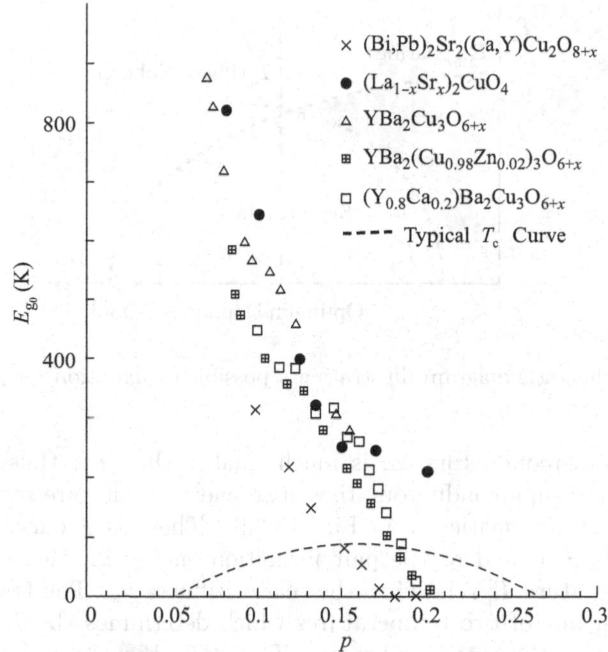

Figure 18.3.7 Magnitude of pseudogaps E_{g0} versus charge-carrier density p for various cuprates deduced from the specific heat measurements. From Prof. W. Y. Liang.

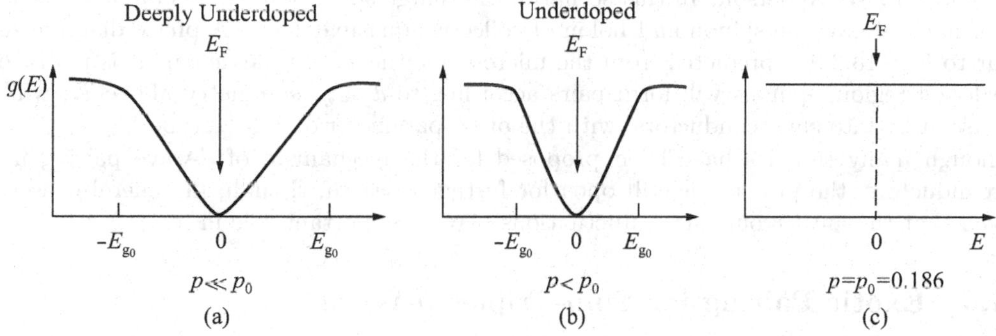

Figure 18.3.8 Schematic diagram for the density of states versus energy for pseudogaps in samples with different charge carrier densities: (a) deeply underdoped sample; (b) moderately underdoped sample; (c) optimum doped sample.

For $p \geqslant p_0$, $E_{g0} = 0$; and the maximum value of E_{g0} is J which is about 1200–1500 K. Furthermore, ARPES studies by Z. X. Shen and associates show that the symmetry of the pseudogap in $\boldsymbol{k}$-space is $d_{x^2-y^2}$ just like that of the superconducting gap discussed above.

The pseudogaps in cuprate superconductors pose an important theoretical problem: What is the physical nature of the pseudogaps?[h] This problem is still not definitely resolved but there are two different theoretical approaches. One is the phenomenal approach exemplified by V. J. Emery, and S. A. Kivelson; their argument may be summarized as follows: There are two energy scales for conventional superconductors: One is the Bardeen–Cooper–Schrieffer (BCS) gap Δ which measures the binding energy of Cooper pairs; and the other is the phase stiffness n_s, the superfluid density at $T = 0$ K, which measures the energy needed to maintain the macroscopic coherence for supercurrents.

[h] For the theory of the pseudogap, see V. J. Emery and S. A. Kivelson, *Nature* **347**, 434 (1995); Y. I. Uemara *et al.*, *Phys. Rev. Lett.* **62**, 2317 (1989); P. A. Lee and X. G. Wen, *Phys. Rev. Lett.* **78**, 4111 (1997); P. W. Anderson, *Science* **235**, 1196 (1987).

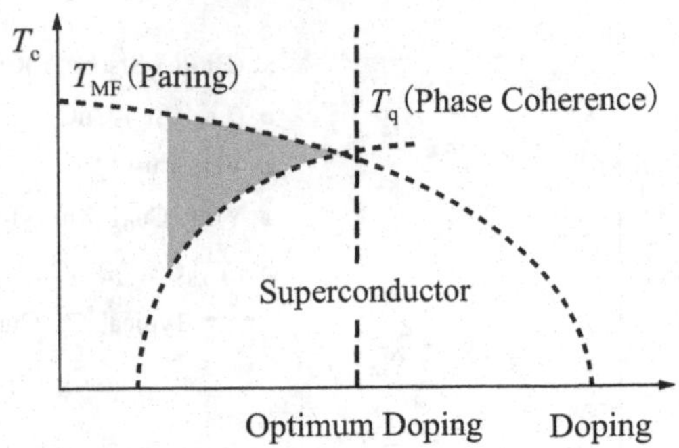

Figure 18.3.9 A schematic diagram illustrating a possible explanation for pseudogaps in cuprates.

For conventional superconductors, Δ is much smaller than n_s; thus it is Δ that determines T_c. However, for cuprate superconductors, these two energy scales are more closely balanced. The situation may be shown schematically in Fig. 18.3.9. There is a curve showing the mean field critical temperature T_{MF} related to the pair formation energy Δ; there is another curve for the phase coherence temperature T_q related to the phase stiffness n_s. The two curves cross at optimal doping. It is the lower one of two temperatures which determines the T_c for the superconductive phase transition. Thus, in the overdoped region, $T_{MF} < T_q$, then $T_c = T_{MF}$; in underdoped region, $T_q < T_{MF}$, $T_c = T_q$, the temperature below which the phase coherence becomes macroscopic, and T_{MF} interpreted as the pseudogap temperature at which the pair formation begins. Another approach for the pseudogaps is based on the newer development of RVB (resonating valence bond) theory first proposed by P. W. Anderson. In this scenario, the charge-spin separation is already taken place in the normal state, as the spinon and holon of collective quasiparticles. A phase diagram somewhat similar to Fig. 18.3.9 is predicted from the microscopic theory. In particular, it is predicted in the underdoped region, spinons will form pairs according to $d_{x^2-y^2}$ symmetry above T_c; while holons condense at T_c into superconductors, with the order parameter of $d_{x^2-y^2}$ type.

Though many theories have been proposed for the mechanism of d-wave pairing in cuprate superconductors, the problem is still open for further research; though, in general consensus, it is believed that the antiferromagnetic fluctuations play an important role in it.

18.3.5 Exotic Pairing for Spin-Triplet *p*-Wave

The pairing of p-wave into a spin-triplet not only breaks the gauge symmetry, it may also break time-reversal symmetry. The most thoroughly studied case of p-wave ($l = 1$) pairing into a spin-triplet is the superfluid state of ^{3}He. The interaction responsible for pairing of ^{3}He atoms is due to the exchange of magnetic excitations of the surrounding atomic sea. The phase diagram for liquid ^{3}He is shown in Fig. 18.3.10. It should be noted that there are three different superfluid phases, a phenomenon typical of p-wave pairing.

For p-wave pairing, $l = 1$ (an odd value), so the associated spin-pairing state is symmetrical for particle-exchange, i.e., spin-triplet (see §11.1.3). So the total pairing wavefunction Ψ_{pair} is a linear combination of the components of the product of the orbital wavefunction $\phi_\alpha(\boldsymbol{k})$ (expressed by momentum) and the spin function,

$$\Psi_{\text{pair}} = \phi_1(\boldsymbol{k})\,|\uparrow\uparrow\rangle + \phi_2(\boldsymbol{k})\,(|\uparrow\downarrow\rangle + |\downarrow\uparrow\rangle) + g_3(\boldsymbol{k})\,|\downarrow\downarrow\rangle, \tag{18.3.26}$$

where the orbital wavefunction can be expressed by spherical harmonics Y_{lm}, i.e.,

$$\phi_\alpha(\boldsymbol{k}) = \sum_{m=-1}^{1} a_{lm}^{\alpha}\mathrm{Y}_{lm}(\boldsymbol{\kappa}), \tag{18.3.27}$$

they are amplitudes of spin states for $S_z = -1,\ 0,\ 1$.

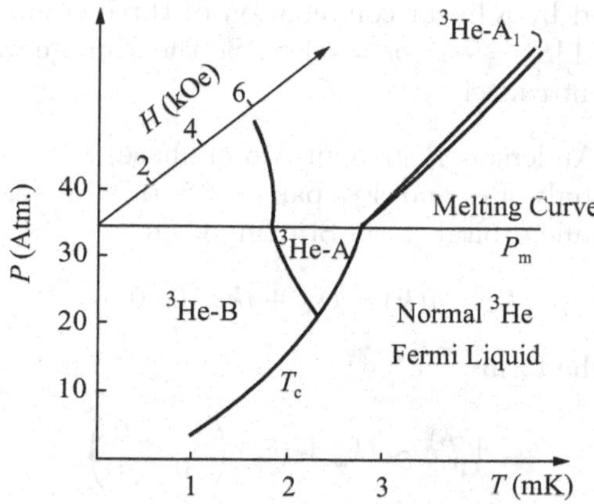

Figure 18.3.10 Phase diagram of liquid ^{3}He in a 3D plot.

We now introduce the spin-vector $\boldsymbol{d}(\boldsymbol{k})$ and Pauli matrix which have components as follows,

$$\sigma_x = \begin{pmatrix} 0 & 1 \\ 1 & 0 \end{pmatrix}, \sigma_y = \begin{pmatrix} 0 & -i \\ i & 0 \end{pmatrix}, \sigma_z = \begin{pmatrix} 1 & 0 \\ 0 & -1 \end{pmatrix}.$$

We can express the pairing wavefunction by the components of $\boldsymbol{d}(\boldsymbol{k})$ using the symmetrical matrix $i\boldsymbol{\sigma}\sigma_y = (i\sigma_x\sigma_y,\ i\sigma_y\sigma_y,\ i\sigma_z\sigma_y)$, and get another expression of Ψ_{pair},

$$\begin{aligned}\Psi_{\text{pair}} &= i\left(\boldsymbol{d}(\boldsymbol{k})\cdot\boldsymbol{\sigma}\right)\sigma_y \\ &= \begin{pmatrix} -d_x(\boldsymbol{k}) + id_y(\boldsymbol{k}) & d_z(\boldsymbol{k}) \\ d_z(\boldsymbol{k}) & d_x(\boldsymbol{k}) + id_y(\boldsymbol{k}) \end{pmatrix}.\end{aligned} \tag{18.3.28}$$

Since the spherical harmonics $\mathrm{Y}_{lm}(\boldsymbol{\kappa})$ with $l = 1$ can be expressed as linear function of the vector $\boldsymbol{k}$,

$$\mathrm{Y}_{11}(\boldsymbol{\kappa}) \sim k_x + ik_y,\ \mathrm{Y}_{1-1}(\boldsymbol{\kappa}) \sim k_x - ik_y,\ \mathrm{Y}_{10}(\boldsymbol{\kappa}) \sim k_z. \tag{18.3.29}$$

We may write the vector $\boldsymbol{d}$ of the order parameter as

$$d_\alpha(\boldsymbol{k}) = A_{\alpha i}k_i. \tag{18.3.30}$$

The complex 3×3 matrix $A_{\alpha i}$ may be identified as the order parameter for the superfluid ^{3}He. And it transforms as a vector against rotation in both the spin space and orbital space. Given a state specified by a set of $A_{\alpha i}$, we can rotate it in both spaces with rotation operators $R^{(S)}$ and $R^{(O)}$ into a new state. As the Hamiltonian of this system is invariant against the rotations, all the states generated in this way are degenerate in energy. It should be noted that these states are physically different, just as ideal ferromagnets with spontaneous magnetization in different directions are different. This is the essential idea of broken symmetry.

Below we shall give a brief sketch of the order parameters for various superfluid phases distinguished by the configurations of the spin vector $\boldsymbol{d}$ (or matrix $A_{\alpha i}$).

1) B-phase (also called the Balian–Werthamer phase)

Consider the states of ^{3}He in which the total angular momentum is zero. Since $J = 0$, the pair state of ^{3}He must be invariant against the simultaneous rotation in spin and orbital space, so

$$\boldsymbol{d}(\boldsymbol{k}) \sim \boldsymbol{k}. \tag{18.3.31}$$

The pair wavefunction now takes the form

$$\begin{aligned}\Psi^{\mathrm{B}}_{\text{pair}} &\sim \begin{pmatrix} -k_x + ik_y & k_z \\ k_z & k_x + ik_y \end{pmatrix} \\ &= (-k_x + ik_y)\,|\uparrow\uparrow\rangle + k_z\left(|\uparrow\downarrow\rangle + |\downarrow\uparrow\rangle\right) + (k_x + ik_y)\,|\downarrow\downarrow\rangle.\end{aligned} \tag{18.3.32}$$

The B phase is described by a linear combination of three equiprobable states $|S_z = +1,\ m = -1>$; $|S_z = 0,\ m = 0>$ and $|S_z = -1,\ m = +1>$. So the average values of spin magnetic moment and orbital magnetic moment cancel.

2) A-phase (also called Anderson–Brinkman–Morel phase)

In this phase there are only $\Psi_{\uparrow\uparrow}$ and $\Psi_{\downarrow\downarrow}$ pairs if we take the axis of spin space appropriately and both pairs are in the same orbital state. Starting from

$$\boldsymbol{d}(\boldsymbol{k}) \sim (k_x + ik_y,\ 0,\ 0), \tag{18.3.33}$$

the pair wavefunction has the form

$$\begin{aligned}\Psi^{(\mathrm{A})}_{\mathrm{pair}} &\sim (k_x + ik_y)\begin{pmatrix} -1 & 0 \\ 0 & 1 \end{pmatrix} \\ &= (k_x + ik_y)(|\uparrow\uparrow\rangle - |\downarrow\downarrow\rangle).\end{aligned} \tag{18.3.34}$$

Thus the A-phase is a linear combination of two equiprobable states $|S_z = +1,\ m = 1>$ and $|S_z = -1,\ m = 1>$. The total spin in this phase is zero; this is a magnetic state with orbital magnetism.

In the presence of a magnetic field, the A_1 state appears, in which the spins are not compensated, and it has both orbital and spin magnetic moments. In general, the different states of ^{3}He superfluid can be explained by p-wave pairing. It means that the spin-triplet p-wave pairing is more complex than spin-singlet pairing. Some indications of exotic pairing of p-wave electrons have been found in some unconventional superconductors, for instance, the heavy electron superconductor UPt_3 has three superconducting phases, one below 0.5 K, the other two below 0.44 K; while $U_{1-x}Th_xBe_{13}$ has four distinct superconducting phases, in which one phase is considered to be magnetic (see Fig. 18.3.11).

Also superconductivity has been found in materials which are weakly or marginally ferromagnetic, such as Sr_2RuO_4, UGe_2, $ZrZn_2$. Even pure Fe under high pressure, after losing ferromagnetism, becomes a superconductor. In conventional superconductors, magnetic impurities are pair-breakers, now there are clear indications which show the coexistence (or near coexistence) of superconductivity and ferromagnetism in these materials (see Fig. 18.3.12). Thus, evidence of spin-triplet p-wave pairing in these exotic superconductors is mounting and continues to be intensely studied both

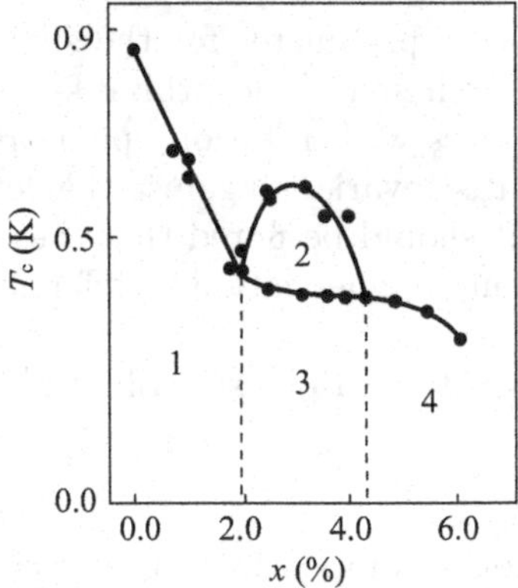

Figure 18.3.11 Phase diagram for $U_{1-x}Th_xBe_{13}$. The four superconducting regions (numbered) can be described by two pair wavefunctions. In phase there is evidence from muon spin rotation for an anomalous magnetic moment of order 0.001 – 0.01 μ_B. From R. H. Heffner *et al.*, *Phys. Rev. Lett.* **65**, 2816 (1991).

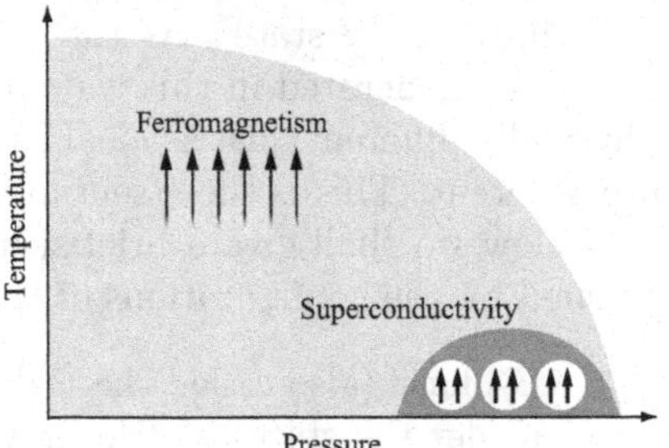

Figure 18.3.12 Schematic temperature-pressure phase diagram for UGe_2. Showing the coexisting region of superconductivity and ferromagnetic state.

experimentally and theoretically. References for these discoveries are listed in the footnote under Table 18.1.2.

§18.4 Josephson Effects

There are many peculiar phenomena involved in the phases of two weakly-coupled macroscopic quantum systems. Josephson explored problems related to superconducting weak-links which are called Josephson junctions. The effect named after him is typical example of broken gauge symmetry. It clearly embodies Heisenburg uncertainty relation for phase θ and number of particles N, i.e., $\Delta\theta\Delta N \simeq 1$.

18.4.1 Josephson Equations

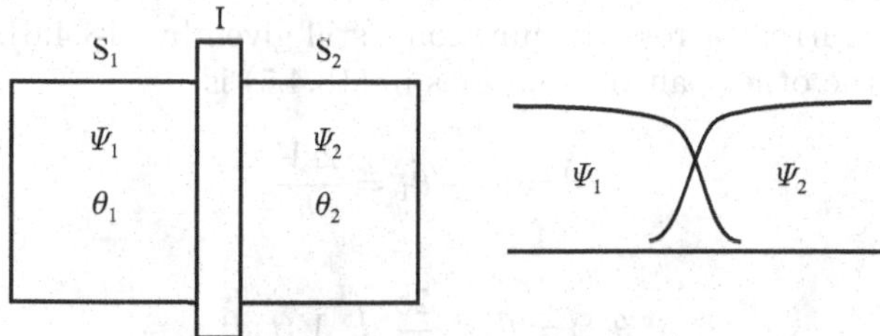

Figure 18.4.1 Two superconductors separated by a thin insulator.

We consider two superconductors separated by an insulating layer as shown in Fig. 18.4.1. The insulating layer acts as a potential barrier for electrons. If it is thick, the electrons cannot get through it, but for the case of a thin layer, the electrons can tunnel across. We choose Ψ_1 and Ψ_2 to represent the wavefunctions of electron pairs, i.e., the macroscopic wavefunctions of left and right superconductors. Because the two superconductors are weakly coupled, the wavefunctions should satisfy the following equations

$$i\hbar\frac{\partial\Psi_1}{\partial t} = E_1\Psi_1 + K\Psi_2, \; i\hbar\frac{\partial\Psi_2}{\partial t} = E_2\Psi_2 + K\Psi_1, \tag{18.4.1}$$

where K is the coupling constant corresponding to the transition amplitude for an electron pair between two superconductors, and E_1, E_2 are the ground state energies when $K = 0$.

Suppose the two superconducting regions are connected to the two terminals of a battery, so that there is a potential difference V across the junction, then $E_1 - E_2 = 2$ eV. If we define the zero of energy at $(E_1 + E_2)/2$, the two weak link equations are

$$i\hbar\frac{\partial\Psi_1}{\partial t} = eV\Psi_1 + K\Psi_2, \; i\hbar\frac{\partial\Psi_2}{\partial t} = -eV\Psi_2 + K\Psi_1. \tag{18.4.2}$$

Actually, when the coupling exists, no matter how weak it is, the two superconductors become one system and should be described by a single condensate wavefunction.

The densities of Cooper pairs in the two superconductors are

$$|\Psi_1|^2 = \rho_1, \; |\Psi_2|^2 = \rho_2. \tag{18.4.3}$$

so the two macroscopic wavefunctions can be written as

$$\Psi_1 = \sqrt{\rho_1}e^{i\theta_1}, \; \Psi_2 = \sqrt{\rho_2}e^{i\theta_2}. \tag{18.4.4}$$

Substituting these into (18.4.2), we get four equations by equating the real and imaginary parts in each case. The result is

$$\dot{\rho}_1 = \frac{2K}{\hbar}\sqrt{\rho_2\rho_1}\sin(\theta_2 - \theta_1),\ \dot{\rho}_2 = -\frac{2K}{\hbar}\sqrt{\rho_2\rho_1}\sin(\theta_2 - \theta_1),$$

$$\dot{\theta}_1 = \frac{K}{\hbar}\sqrt{\frac{\rho_2}{\rho_1}}\cos(\theta_2 - \theta_1) - \frac{eV}{\hbar},\ \dot{\theta}_2 = \frac{K}{\hbar}\sqrt{\frac{\rho_1}{\rho_2}}\cos(\theta_2 - \theta_1) + \frac{eV}{\hbar}. \tag{18.4.5}$$

The first two equations tell us how the densities would change, and therefore describe the kind of current that would begin to flow, even though there were no extra electric forces due to an imbalance of potential. This current from sides 1 to 2 would be just $2e\dot{\rho}_1$(or $-2e\dot{\rho}_2$), and

$$J = \frac{4eK}{\hbar}\sqrt{\rho_2\rho_1}\sin\theta = J_0\sin\theta, \tag{18.4.6}$$

where the phase difference $\theta = \theta_2 - \theta_1$ and the maximum current density $J_0 = 4eK\sqrt{\rho_2\rho_1}/\hbar$ are defined. We remember that the two sides are connected by wires to the battery, ρ_1 and ρ_2 do not in fact change, but the current across the junction s still given by (18.4.6).

What we get from the other pair of equations in (18.4.5) is

$$\dot{\theta} = \dot{\theta}_2 - \dot{\theta}_1 = \frac{2eV}{\hbar}, \tag{18.4.7}$$

or

$$\theta(t) = \theta_0 + \frac{2e}{\hbar}\int_0^t V(t)dt, \tag{18.4.8}$$

where θ_0 is the value of θ at $t = 0$. Equations (18.4.6) and (18.4.8) are called the Josephson equations.

18.4.2 The Josephson Effects in Superconductors

The theoretical prediction for (18.4.6) and (18.4.8) were quickly confirmed by experiments, which showed furthermore that suitable weak links are not limited to pure tunnel junctions, but can be made in a large variety of ways, such as point contacts, microbridges, etc. There are a lot of quantum effects related to weak-links which show the importance of the phase in superconductors. We shall enumerate some of the main results in the following.

If there is no voltage applied, i.e., take $V = V_0 = 0$, then

$$J = J_0\sin\theta_0. \tag{18.4.9}$$

This is the dc Josephson effect, which was experimentally observed by Anderson and Rowell in 1963. It means the current in a weak link with no applied voltage can be any amount between J_0 and $-J_0$, depending on the value of θ_0. This effect confirms the meaning of the phase difference among the superconductors.

If there is a dc voltage across the junction, that is, $V = V_0 \neq 0$, $2eV_0$ should be less than the gap width to avoid pair breaking, then

$$J = J_0\sin\left(\theta_0 + \frac{2e}{\hbar}V_0 t\right). \tag{18.4.10}$$

This is the ac Josephson effect, which means a dc voltage leads to an ac Josephson current. The alternating frequency is $\omega = 2\pi\nu = 2eV_0/\hbar$, which was detected experimentally through the electromagnetic radiation of the oscillatory current. Actually, there is a constant related to the flux quantum defined by $\nu/V_0 = 2e/h = 4.8 \times 10^{14}$ $\mathrm{HzV^{-1}}$.

Further, we can apply an ac voltage at microwave frequency in addition to the dc voltage. The total voltage is $V = V_0 + v\cos\omega t$, but the condition $v \ll V_0$ is required. Now the Josephson current is

$$J = J_0\sin\left(\theta_0 + \frac{2e}{\hbar}V_0 t + \frac{2ev}{\hbar\omega}\sin\omega t\right). \tag{18.4.11}$$

Using the Fourier expansion, $\sin(x + \Delta x) \approx \sin x + \Delta x \cos x$, the last expression can be approximated by

$$J = J_0 \left[\sin\left(\theta_0 + \frac{2e}{\hbar} V_0 t\right) + \frac{2ev}{\hbar\omega} \sin\omega t \cos\left(\theta_0 + \frac{2e}{\hbar} V_0 t\right)\right], \tag{18.4.12}$$

in which the first term is zero on the time average, but the second term gives a dc current if $V_0 = \hbar\omega/2e$. More exactly, using the Fourier–Bessel expansion to higher ranks, we can find that there are dc components in the frequency modulated current, as the dc voltage satisfies

$$V_0 = \frac{\hbar}{2e} n\omega, \tag{18.4.13}$$

where n is any integer, and its amplitude is

$$J = J_0 J_n \left(\frac{2ev}{\hbar\omega}\right) \sin\theta_0, \tag{18.4.14}$$

where $J_n(2ev/\hbar\omega)$ is the n-rank Bessel function. The physical implication for this dc current can be easily understood. As a matter of fact, a dc voltage V_0 leads to an energy difference $2eV_0$ for a Cooper pair located in one superconductor from the other superconductor of the junction. The tunneling process will be facilitated by applied radiation, especially (18.4.12) is fulfilled. Such a resonance effect has been called Shapiro steps,[i] which can be demonstrated by experiments like in Fig. 18.4.2.

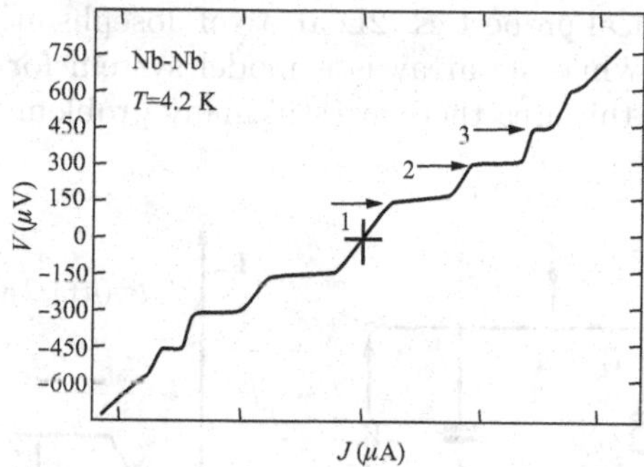

Figure 18.4.2 Current versus voltage of Josephson junction, showing Shapiro steps. From C. C. Grimes and S. Shapiro, *Phys. Rev.* **169**, 397 (1968).

It is very important to understand the effect of an applied magnetic field on the Josephson junctions from both the viewpoints of basic physics and of applied superconducting electronics. Now, for simplicity, we focus our attention on the effect of magnetic field through the loop. The vector potential $\boldsymbol{A}$ will influence the phase of the macro-wavefunction. So θ in (18.4.6) and (18.4.8) can be changed into

$$\gamma = \theta - \frac{2\pi}{\Phi_0} \int \boldsymbol{A} \cdot d\boldsymbol{l}, \tag{18.4.15}$$

where Φ_0 is the magnetic flux quantum.

We shall consider a special case for a pair of Josephson junctions (denoted by 1 and 2) between a pair of superconducting electrodes as shown in Fig. 18.4.3. Since $\boldsymbol{B} = \nabla \times \boldsymbol{A}$, the enclosed magnetic flux can be found as the line integral of $\boldsymbol{A}$ around the contour. If the electrodes are thicker than λ, we can take the integration contour wholly in the region where $\boldsymbol{v}_s$ vanishes. From (18.1.33), we may infer that $A = (\Phi_0/2\pi)\nabla\theta$ in the electrodes, so

$$\Phi = \oint \boldsymbol{A} \cdot d\boldsymbol{l} = \frac{\Phi_0}{2\pi} \oint \nabla\theta \cdot d\boldsymbol{l}. \tag{18.4.16}$$

[i] S. Shapiro, *Phys. Rev. Lett.* **11**, 80 (1963).

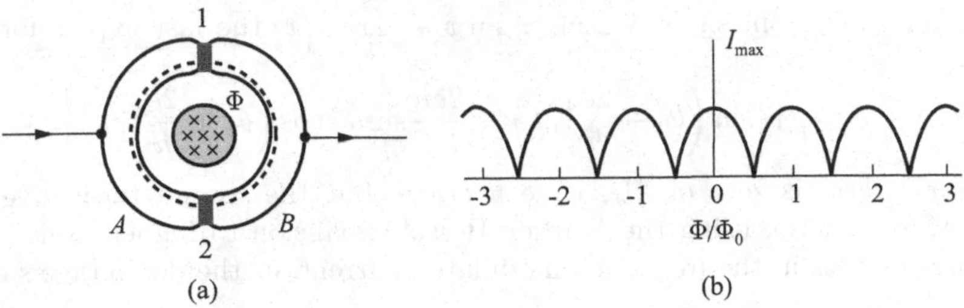

Figure 18.4.3 (a) Schematic diagram for two Josephson junctions in magnetic field; (b) corresponding quantum interference pattern.

Because the phase must be single-valued, $\oint \Delta\theta \cdot \boldsymbol{dl}$ should be divided exactly by 2π. So we find that the phase difference is $2\pi\Phi/\phi_0$, and the maximum supercurrent should be

$$I_{\text{max}} = 2I_c \cos(2\pi\Phi/\Phi_0). \tag{18.4.17}$$

The relation plotted in Fig. 18.3.4 is reminiscent of the two slit interference pattern in optics. This is the physical basis of the dc-SQUID (superconducting quantum interference device) magnetometer, the most sensitive device for the measurement of magnetic flux, to the accuracy of a extremely small fraction of Φ_0. Just as a system of two slits in optics may be generalized to gratings (1D, 2D or 3D), the ideal for junction SQUID in a magnetic field may be generalized to an array of junctions (1D, 2D or 3D) with interesting physical properties. 2D array of Josephson junctions may be used to study quantum phase transitions; while 3D array is a model system for granular superconductivity for ceramic superconductors. In this area there are still many problems worthy of further experimental or theoretical study.

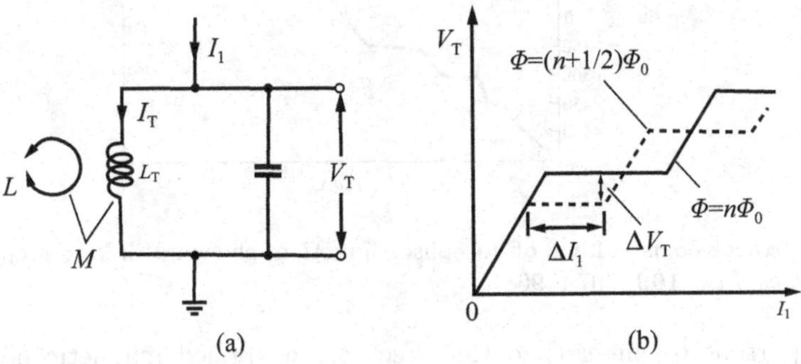

Figure 18.4.4 (a) Schematic diagram for the circuit of rf-SQUID; (b) corresponding V_T versus I_l relations for integral and half-integral numbers of flux quanta.

Another form of SQUID is the rf-SQUID in which only a single Josephson junction is used. A constant rf ($\approx$ 20–30 MHz) current I_l is supplied via a coil resonantly coupled to the SQUID loop as shown in Fig. 18.4.4(a). The rf loss as a function of the magnitude of the rf voltage V_T across it, is shown in Fig. 18.4.4(b).

18.4.3 Phase-Sensitive Tests of Pairing Symmetry

More definite tests of pairing symmetry of high T_c cuprates can be made using phase-sensitive techniques with Josephson devices. In a square-shaped single crystal of a cuprate with d-wave pairing we will have different phases on orthogonal sides. This fact may be used to discriminate d-wave pairing from s-wave pairing by a suitable design of Josephson junctions or SQUID devices, e.g., the corner-type junction shown in Fig. 18.4.5. The results of these experimental tests all favor d-wave pairing for high T_c cuprates, though they still contain some complications to be clarified.

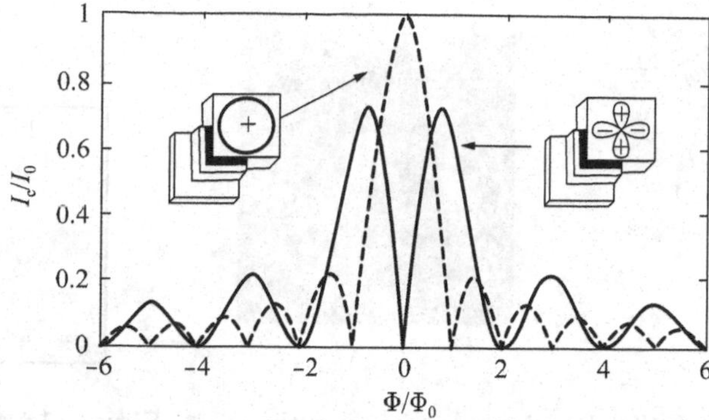

Figure 18.4.5 Josephson current of a high T_c superconductor corner-type junctions. The dashed line is the normal superconductor junctions. From D. A. Wollman *et al., Phys. Rev. Lett.* **74**, 797 (1995).

The principle for another set of tests is related to a half-integer flux effect in superconducting rings, first suggested theoretically in connection with p-wave pairing in heavy electron superconductors: C. C. Tsuei *et al.* used the ring geometry of tricrystal grain boundary junctions with controlled orientations. Based on the fundamental requirement of a single-valued macroscopic pair wavefunction, the flux quantization of a superconducting ring with self-inductance L can be expressed by

$$\Phi_a + I_s L + \frac{\Phi_0}{2\pi}\sum_{ij}\gamma_{ij} = n\Phi_0. \tag{18.4.18}$$

The supercurrent circulation in the ring is

$$I_s = I_c^{ij}(\theta_i, \theta_j)\sin\gamma_{ij}, \tag{18.4.19}$$

where $I_c(\theta_i, \theta_j)$ is the critical current of the junction between superconducting electrodes i and j, while θ_i and θ_j are corresponding angles of the crystallographic axes with respect to the junction interface. Flux quantization of a multiply-connected superconductor must be valid for a superconducting ring with any pairing symmetry. For a ring with an odd number of sign changes in the circulating supercurrent I_s, it is sufficient to consider the case in which only one critical current is negative (say $I_c^{12} = -|I_c^{12}|$), then $I_s = |I_c^{12}|\sin(\gamma_{12} + \pi)$. In absence of an external field, $\Phi_a = 0$, $n = 0$ for the ground state, the combined conditions of (18.4.18) and (18.4.19) lead to

$$I_s = \frac{\pi}{2\pi\left(\dfrac{L}{\phi_0}\right) + \dfrac{1}{|I_c^{12}|} + \dfrac{1}{|I_c^{23}|} + \cdots} \approx \frac{\Phi_0}{2L}, \tag{18.4.20}$$

provided that $|I_c^{12}|L \gg \Phi_0, \ldots, |I_c^{ij}|L \gg \Phi_0$. It is interesting to note that, in the ground state of a superconducting ring with an odd number of sign changes: π ring, when the external magnetic field is zero, half integer flux quantum will appear; with even number (including 0) of sign changes (0 ring), $I_s = 0$, no half integer flux quantum will appear. An series of ingenious experiments by C. C. Tsuei *et al.* using a YBCO tricrystal with controlled orientations and photolithographically patterned rings detected half integer flux in π rings, but not in 0 rings. In consequence it gave an unambiguous confirmation of the d-wave pairing in high T_c cuprates (see Fig. 18.4.6).

18.4.4 Josephson Effect in Superfluids

Since the Josephson effect is a manifestation of macroscopic quantum phenomena, it is expected that what happens in superconductors can also occur in superfluids. The Josephson equations are still valid in this case, except the electric current should be replaced by the mass transport of neutral atoms.

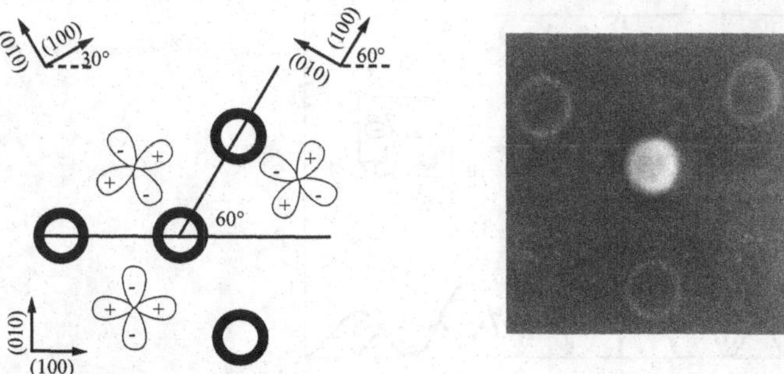

Figure 18.4.6 Scanning SQUID microscope image with three different geometrical configurations to allow an unambiguous test of the symmetry of the order parameter of YBCO. The central bright spot is π ring, and the others are 0 rings. From C. C. Tsuei and J. R. Kirtley, *Rev. Mod. Phys.* **72**, 969 (2000).

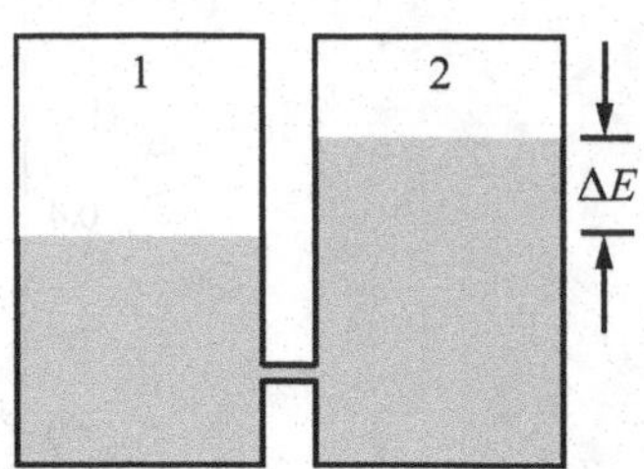

Figure 18.4.7 Two reservoir of liquid helium connected by a small aperture.

We can schematically construct a superfluid weak link as in Fig. 18.4.7 where two volumes of superfluid helium are connected by a small aperture. In this arrangement there is a level difference Δz between the two reservoirs for which the temperatures are assumed to be equal; the weak link is a very small aperture. This level difference will give chemical potential difference

$$\mu_2 - \mu_1 = Mg\Delta z, \tag{18.4.21}$$

where M is the mass of one ^{4}He or two ^{3}He atoms. Then, we have

$$\dot{\theta} = \dot{\theta}_2 - \dot{\theta}_1 = -\frac{1}{\hbar}(Mg\Delta z), \tag{18.4.22}$$

and a characteristic frequency could be defined as

$$\omega = mg\Delta z/\hbar. \tag{18.4.23}$$

It corresponds to the ac Josephson frequency $2eV_0/\hbar$ for the tunneling of Cooper pairs. It should be noted that ac Josephson effects in superconductors are electrodynamic effects, while in superfluid helium they are hydrodynamic effects due to the fact that helium atoms are neutral. However, the coherence lengths, or healing lengths, ξ_c of superfluids He at low temperature are very small: For ^{4}He, $\xi_c \sim 0.1$ nm; for ^{3}He, $\xi_c \sim 50$ nm. It is the small coherence length that prevents easy realization of the Josephson effect in liquid He.

In spite of the more severe requirement for low temperature, the successful realizations of Josephson effect in superfluid ^{3}He with the larger coherence length came first.[j] The crucial technique is to fabricate an array of apertures with diameters less than the coherence length as the weak links between the two reservoirs of superfluids by lithography, a modern microfabrication technique. Further along this path, the interference effect of dc SQUID-like devices of superfluid ^{3}He have been reported, in which the rotation of earth acts like a magnetic field, with the prospect of developing a sensitive rotation sensor.

Now return to ^{4}He, surely this is a more difficult problem. However, the coherence length diverges in the critical region near T_c in the following way

$$\xi = \xi_0 \left(1 - \frac{T}{T_\lambda}\right)^{-\gamma}, \quad \gamma = 0.672. \tag{18.4.24}$$

[j]For Josephson effect in superfluid ^{3}He, see E. Varoquax and O. Avenel, *Phys. Rev. Lett.* **60**, 416 (1988); S. V. Pereversev *et al.*, *Nature* **388**, 449 (1997); R. W. Simmonds *et al.*, *Nature* **412**, 55 (2001).

When $t = T_\lambda - T$ is sufficiently small, the coherence length may reach the scale of 10 nm, which is accessible to modern microfabrication; however, in the critical region, fluctuation effects may smear out the Josephson signal. An ingenious experiment using $t = 3.72$ mK actually found the ac Josephson signal above the noise, and accomplished the goal of realization of Josephson effect in ^{4}He superfluid.[k]

Bose–Einstein condensation (BEC) of dilute gases form another type of superfluid. One experiment that was carried out first used a laser beam to cut a cigar-shaped BE condensate into two separated parts. Then the confining potential and the laser were switched off, the two independent parts of the BE condensates then expand and eventually overlap. In this way clean interference patterns have been observed in the overlapping region.[l] Moreover, it opens the door for the study of quantum interference and the Josephson effect in more complicated situations: For instance, a 1D array of Josephson junctions has been realized by loading BE condensates into the optical lattice potential generated by a standing-wave laser field.[m]

Bibliography

[1] Dalfovo, F., S. Giogini, and L. P. Pitaevskii, *Theory of Bose–Einstein condensation in trapped gases*, Rev. Mod. Phys. **71**, 463 (1999).

[2] Tilley, D. R., and J. Tilley, *Superfluidity and Superconductivity*, 2nd ed., Adam Hilger, Bristol (1990).

[3] Tsuneto, T., *Superconductivity and Superfluidity*, Cambridge University Press, Cambridge, (1998).

[4] Tinkham, M., *Introduction to Superconductivity*, 2nd. ed., McGraw-Hill, New York (1996).

[5] Schrieffer, J. R., *Theory of Superconductivity*, revised ed., Addison-Wesley, Reading (1983).

[6] Matsubara, T. (ed.), *The Structure and Properties of Matter*, North-Holland, Amsterdam (1988).

[7] de Gennes, P. G., *Superconductivity of Metals and Alloys*, Benjamin, New York (1966); reprint, Addison-Wesley, Reading (1989).

[8] Saint-James, D., G. Sarma, and E. Thomas, *Type II Superconductors*, Clarendon Press, Oxford (1969).

[9] Mineev, V. P., and K. V. Somokhin, *Introduction to Unconventional Superconductivity*, Gordon & Breach, Australia (1999).

[10] Tsuei, C. C., and J. R. Kirtley, *Pairing symmetry in cuprate superconductors*, Rev. Mod. Phys. **72**, 969 (2000).

[11] Damascelli, A., Z. Hussain, and Z. X. Shen, *Angle-resolved photoemission studies of cuprate superconductors*, Rev. Mod. Phys. **75**, 473 (2003).

[k] For the experiment of Josephson effect in ^{4}He superfluid, see K. Sikhatme *et al.*, *Nature* **411**, 280 (2001).
[l] For the interference experiment with BE condensate, see M. R. Andrews *et al.*, *Science* **275**, 637 (1997).
[m] For SQUID arrays of BE condensates. F. S. Cataliotti *et al.*, *Science* **293**, 843 (2001).

Chapter 19

Broken Ergodicity

A lot of phase transitions in condensed matter are associated with broken symmetries. However, there are many other phase transitions in which symmetry breaking cannot be seen, but there is what we call broken ergodicity. Broken symmetry is seen to be a part of broken ergodicity.

§19.1 Implication of Ergodicity

In the late 19 century, Boltzmann introduce the hypothesis of ergodicity as the basis of statistical mechanics.[a] Afterwards Gibbs introduced ensemble theory to substitute for the ergodic hypothesis, but the difference is not large. We will use the weaker expression, i.e., quasi-ergodicity. This is the statistical assumption that for a many-particle system in thermal equilibrium, if the experimental time is long enough, the phase space trajectory which describes the time evolution of the system will come arbitrarily close to any specified point in the phase space accessible to the system. As a result, observed quantities are given by the average taken over all of the allowed phase space. In essence, ergodicity means that the ensemble average, i.e., the phase space average, can be used to replace the time average of the variables evolving from any single initial condition.

19.1.1 Ergodicity Hypothesis

A condensed material includes a great number of particles composed of electrons and ions. From a dynamical point of view, we can define a microscopic state by specifying all of the dynamical variables of the system. However, only a few physical quantities, say the temperature, the pressure and the density, are usually taken to specify macroscopic state of the system.

Although the atomic world must obey quantum statistical mechanics, often classical treatments are still reasonable and instructive. So, here we start from classical phase space. Let $(q_1, q_2, \ldots, q_N)$ be the generalized coordinates of a system with N degrees of freedom and $(p_1, p_2, \ldots, p_N)$ their conjugate momentum. A microscopic state of the system is specified by the values of $(q_1, q_2, \cdots, q_N, p_1, p_2, \ldots, p_N)$. The $2N$-dimensional space constructed from these $2N$ variables are the coordinates of the phase space of the system. Each phase point corresponds to a microscopic state and the microscopic states in classical statistical mechanics make a continuous set of points in phase space. The evolution of the system is determined by the canonical equations of motion

$$\frac{\partial p_i}{\partial t} = -\frac{\partial \mathcal{H}}{\partial q_i}, \ \frac{\partial q_i}{\partial t} = \frac{\partial \mathcal{H}}{\partial p_i}, \ (i = 1, 2, \ldots, N). \tag{19.1.1}$$

[a] Ergodicity has been a long-standing mathematical problem. The original form of ergodicity has been proved incorrect by mathematicians. Even the quasi-ergodicty stated here is still a hypothesis and cannot be proved mathematically. But the concept of ergodicity has proved useful and important in physics, especially in elucidating thermodynamic properties as well as nonequilibrium properties. We shall use this concept, in spite of the fact that its mathematical rigor is questionable.

These describe the motion of the phase point, $\boldsymbol{x}^N(t)$, denoting the dynamical state of the system at time t. $\boldsymbol{x}^N$ are the dynamical degrees of freedom as a function of time t, position and momentum for a liquid, for example. The trajectory of the phase point is called a phase orbit. For a conservative system, the energy is constant, i.e.,

$$\mathcal{H}(\{q_i, p_i\}) = E. \tag{19.1.2}$$

Therefore the phase orbit must be constrained to lie on a constant energy surface $S(E)$ with $(2N-1)$-dimensions.

Ergodicity means that every allowed point in phase space will have been visited by the system after a sufficiently long time. Allowed points of phase space are those which satisfy macroscopic constraints, such as that of constant energy. A dynamical quantity A is represented by a time dependent quantity $A(t) = A(\boldsymbol{x}^N(t))$ which changes in time according to the motion of the phase point. An observed value of A is therefore to be considered as a time average

$$\langle A \rangle_t = \lim_{t\to\infty} \frac{1}{t} \int_0^t A(\boldsymbol{x}^N(t'))dt'. \tag{19.1.3}$$

Ergodicity implies that observed quantities are given by averages over all of allowed phase space. We can write this as

$$\langle A \rangle_s = \frac{1}{s(E)} \int_{s_E} A(\boldsymbol{x}^N) ds_E, \tag{19.1.4}$$

where

$$s(E) = \int_{s_E} ds_E = \int_\Gamma \delta(\mathcal{H}(\boldsymbol{x}^N) - E) d\boldsymbol{x}^N \tag{19.1.5}$$

is the area of the energy surface. It appears possible to justify the principle of equal weight for time average and ensemble average as

$$\langle A \rangle_t = \langle A \rangle_s, \tag{19.1.6}$$

which is called the ergodic hypothesis. This prescription means that thermodynamic quantities can be obtained by the computation of a partition function Z and then its differentiation.

An isolated system, which has reached thermal equilibrium, will have the normalized probability density

$$\rho(\boldsymbol{x}^N, s_E) = \frac{1}{s(E)}, \tag{19.1.7}$$

which shows equal probability of states on the energy surface. This is called the micro-canonical ensemble and forms the base of statistical mechanics of the equilibrium state, as well as a large part of the nonequilibrium state. A system satisfying the micro-canonical ensemble is ergodic.

Alternatively, we consider a quantum system in which q and p cannot be specified simultaneously due to the uncertainty principle, so that classical phase space loses its rigorous meaning. In quantum statistical mechanics, a microscopic state of a stationary dynamical system must be one of the quantum states determined by the equation

$$\mathcal{H}\Psi_l = E_l \Psi_l, \ (l = 1, 2, \ldots).$$

Here $\mathcal{H}$ is the Hamiltonian of the system, E_l and Ψ_l are the energy and wavefunction, respectively, of the lth quantum state. In fact, a quantum state, denoted by the quantum number l, corresponds to a phase orbit in classical mechanics, so for a dynamical quantity A, its observed value in the macroscopic sense must be the ensemble average

$$\langle A \rangle = \sum_m \langle l|A|l \rangle, \tag{19.1.8}$$

where m denotes all possible microscopic states which can be realized by the system under a certain macroscopic condition. A statistical ensemble is defined by the distribution function that characterizes it. The most fundamental ensemble is still the micro-canonical ensemble, which is defined by the principle of equal weight.

We now understand that ergodicity involves an important assumption, that is, the phase space average of an system is equal to its infinite time average, and then the infinite time average of an observable gives a good estimate of the observed value. In various systems related to phase transitions, this assumption is always impossible or unrealistic, especially as things are related to time scale.

19.1.2 Involvement of Time Scale

We discuss an example in which broken translational symmetry leads to broken ergodicity. Assume a N-particle system that is translational invariant. Its Hamiltonian is $\mathcal{H}(\boldsymbol{r}_i)$, where $i = 1, \ldots, N$. By varying the temperature, the system may undergo a phase transition. The perfect translational symmetry means that for an arbitrary vector $\boldsymbol{a}$ we have

$$\mathcal{H}(\boldsymbol{r}_i) = \mathcal{H}(\boldsymbol{r}_i + \boldsymbol{a}). \tag{19.1.9}$$

We can write the particle density of the system

$$\rho(\boldsymbol{r}) = \sum_i \delta(\boldsymbol{r} - \boldsymbol{r}_i), \tag{19.1.10}$$

and its Fourier component

$$\rho_i(\boldsymbol{q}) = \mathrm{e}^{i\boldsymbol{q}\cdot\boldsymbol{r}_i}. \tag{19.1.11}$$

If we use m to denotes the microstates, the expectation value is

$$\langle \rho_i(\boldsymbol{q}) \rangle_m = \frac{1}{Z} \sum_m \rho_i(\boldsymbol{q}) \mathrm{e}^{-\beta\mathcal{H}(\boldsymbol{r}_i)}, \tag{19.1.12}$$

where Z is the partition function

$$Z = \sum_m \mathrm{e}^{-\beta\mathcal{H}(\boldsymbol{r}_i)}. \tag{19.1.13}$$

Just as described in the liquid-solid transition, for $T > T_\mathrm{c}$, $\langle \rho_i(\boldsymbol{q}) \rangle = 0$, while for $T < T_\mathrm{c}$, $\langle \rho_i(\boldsymbol{q}) \rangle \neq 0$, and at $T = T_\mathrm{c}$, a density wave arises with certain wavevector $\boldsymbol{q}$, and $\langle \rho_i(\boldsymbol{q}) \rangle_m$ becomes nonzero. Thus the transition is associated with some ordering of the particles in space. This is the spontaneous breaking of the symmetries associated with translational and rotational invariance.

To proceed, we must first specify which microstates should be included in m. If all possible microstates are included, then because of the translational invariance of the Hamiltonian expressed in (19.1.9), we can prove from (19.1.11) and (19.1.12) that

$$\langle \rho_i(\boldsymbol{q}) \rangle_m = \mathrm{e}^{i\boldsymbol{q}\cdot\boldsymbol{a}} \langle \rho_i(\boldsymbol{q}) \rangle_m, \tag{19.1.14}$$

and hence $\langle \rho_i(\boldsymbol{q}) \rangle_m = 0$ regardless of the value of T. The only way in which $\langle \rho_i(\boldsymbol{q}) \rangle_m$ can be nonzero for $T < T_\mathrm{c}$ is to assume that m does not include all the allowed microstates. However, this statement is not precise enough: If some microstates are omitted from m, but all the residual microstates remain in m under the action of the translation operation, then $\langle \rho_i(\boldsymbol{q}) \rangle_m$ will still vanish. Therefore, it is only if the residual set m is not invariant under the symmetry group of $\mathcal{H}(\boldsymbol{r}_i)$ that $\langle \rho_i(\boldsymbol{q}) \rangle_m$ can acquire a non-zero value. The residual set of microstates represents the system subject to the constraint that the center of mass is at certain position in space. It is not very difficult to see that there are infinitely many other residual sets m' in phase space, each corresponding to a different center of mass position.

Any broken symmetry also breaks ergodicity, but the converse is far from true. The requirement that the set of microstates included in the summation be restricted for $T < T_\mathrm{c}$ is, once again, ergodicity breaking. In this example, it is associated with spontaneous symmetry breaking, although this need not be the case: On the co-existence line of the liquid-vapor transition, for example, phase space is fragmented into two distinct sets, which are not distinct in terms of their symmetry properties.

Although one usually presupposes ergodic behavior in statistical mechanics, many systems are not ergodic in practice. Some systems may technically be ergodic in an infinite time limit but far from ergodic on reasonable physical timescales, so that a phase space or ensemble average does not give a measurable time average. Broken ergodicity is equivalent to a restricted ensemble: The phase space trajectory remains restricted to certain subsets of the allowed phase space for all reasonable time scales of observation, then a physical quantity A satisfies

$$\langle A\rangle_t = \langle A\rangle_{s(\text{partly})} \neq \langle A\rangle_s.$$

Time scale plays a critical role and merits further attention. There are two types of time scales, one is τ_0 related to observation, and the other is τ related to the slowest dynamical processes.

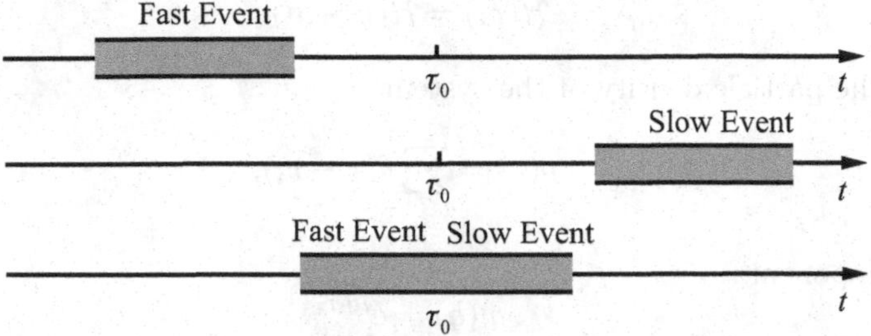

Figure 19.1.1 Time scales and ergodicity.

We can define a thermal equilibrium state in which all the fast processes have happened and all the slow ones have not. It is quite clear that the concept of thermal equilibrium depends crucially on the observational timescale τ_0, which itself determines the meaning of fast and slow, as shown in Fig. 19.1.1.

Let us take the Ising ferromagnet as an example to discuss the role of the time scale. Assuming $H = 0$ in (19.1.4), the spontaneous magnetization is expressed as $M = \sum_i S_i$. At $T < T_c$, and the system has macroscopic states of positive (up) or negative (down) magnetization.

If at the beginning, the system is denoted by $\langle \sum S_i^z \rangle = M$, through fluctuations there will be a spin-down cluster which appears with probability

$$P \propto \mathrm{e}^{-\beta\Delta F}, \tag{19.1.15}$$

where $\beta = 1/k_B T$ and ΔF is the free energy of formation of the spin cluster. Only after the cluster is larger than a critical size is it possible for the system to grow and arrive at $\langle \sum S_i^z \rangle = -M$. Because the free energy is proportional to the surface area of the cluster

$$\Delta F \propto N^{(d-1)/d} J, \tag{19.1.16}$$

where d is the spatial dimensionality,

$$P \propto \exp[-\beta J N^{(d-1)/d}]. \tag{19.1.17}$$

The lifetime of the cluster of given size is

$$\tau \propto 1/P \propto \exp[\beta J N^{(d-1)/d}]. \tag{19.1.18}$$

So, for $d > 1$, in the thermodynamic limit $\tau \gg 1$ the macroscopic magnetization is determined by initial conditions, and ergodicity for positive or negative magnetization is broken.

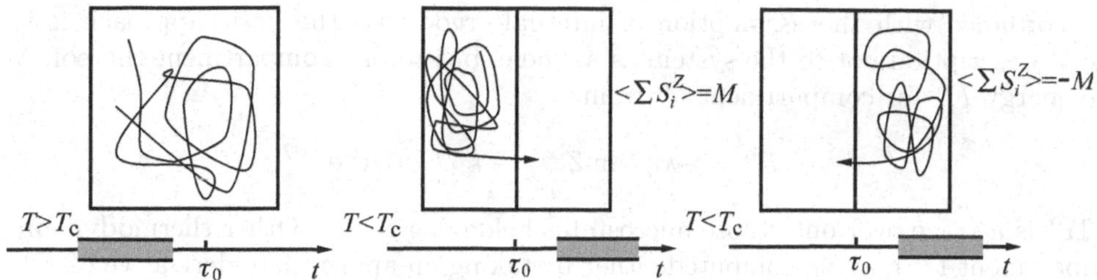

Figure 19.1.2 Phase transition of Ising spin system and ergodicity.

Figure 19.1.2 shows schematically the phase space trajectories in the disordered and the ordered states, as well as the relevant time scales. Γ represents the phase space. When $T > T_c$, the system is in the disordered state, $\langle\sum S_i^z\rangle = 0$, and $\tau_0 \gg \tau$, the system explores the whole of Γ and ergodicity is restored. The point, however, is that this would require τ_0 to be astronomically large when $T \ll T_c$.

The existence of a time scale τ longer than τ_0 leads to broken ergodicity and necessitates redefining the Ising Hamiltonian or the calculation procedure if a physical result is to be obtained. Indeed the Hamiltonian (19.1.4) with $H = 0$ gives $\langle S_i\rangle = 0$ by symmetry in the canonical prescription. This is incorrect for the physical case, $\tau_0 < \tau$. The main effect of $\tau_0 < \tau$ is the effective decomposition of phase space Γ into two parts, Γ^+ and Γ^-, for up and down states. So

$$\Gamma = \Gamma^+ \bigcup \Gamma^-. \tag{19.1.19}$$

We now understand that, for the ordered state, $\langle\sum S_i^z\rangle = M$, or $\langle\sum S_i^z\rangle = -M$. There is essentially no chance of the phase point moving from one to the other within a time τ_0. A proper redefinition of the problem for $\tau_0 < \tau_{\text{flip}}$ must treat a system with phase space either Γ^+ or Γ^-, not Γ. The slow process must not be allowed to happen. This example displays 'broken symmetry' because the compartments do not have the inversion symmetry of the Hamiltonian (1.1) at $h = 0$. The inversion symmetry does apply to the union of all compartments and thus implies that the up and down compartments must be congruent. There are many known examples of broken symmetry in condensed matter physics, all characterized by the decomposition of phase space into a number of congruent compartments, disconnected on physical timescales. But the phenomenon of ergodicity breaking is not limited to broken symmetry.

19.1.3 Internal Ergodicity

In a system that is non-ergodic on physical time scales, the phase point is effectively confined in one compartment of phase space. In essence, there may be free energy barriers surrounding a compartment Γ^α in phase space, which prevent escape. Because of these barriers between compartments, the energy surface breaks into several disconnected parts due to the existence of hills and valleys in phase space. Theoretical treatments of such systems should calculate thermal averages over one compartment at a time. The probability distribution of physical properties can then be obtained from the probability of occurrence for each compartment, and moments of these distributions may be used to predict the results of typical measurements.

The assumption of ergodicity, particularly in the broad sense that a phase average gives a time average, is appropriate within a particular compartment. That is, the usual apparatus of equilibrium statistical mechanics is effective when applied to one compartment at a time. Each compartment thus behaves as a 'system' in its own right, but the real physical system can have several such compartments, all effectively isolated from one another. Within Γ^+ or Γ^-, there is internal ergodicity.

$$\langle A_+\rangle_s = \langle A_+\rangle_t, \ \langle A_-\rangle_s = \langle A_-\rangle_t. \tag{19.1.20}$$

Equilibrium statistical mechanics can be effectively applied to one compartment, but not between compartments.

In accordance with the assumption of internal ergodicity, the basic approach is to apply the canonical prescription not to the system as a whole, but to one compartment thereof. We compute the free energy F^α for compartment Γ^α from

$$F^\alpha = -k_\mathrm{B}T\ln Z^\alpha = -k_\mathrm{B}T\ln \mathrm{Tr}^\alpha \mathrm{e}^{-\beta\mathcal{H}}, \tag{19.1.21}$$

where Tr^α is a trace over only those microstates belonging to Γ^α. Other thermodynamic quantities for compartment Γ^α may be computed either by taking an appropriate derivative of F^α, or directly by averaging an observable A over Γ^α

$$\langle A\rangle_\alpha = \frac{1}{Z}\mathrm{Tr}^\alpha A\mathrm{e}^{-\beta\mathcal{H}}. \tag{19.1.22}$$

This is the consequence of the modified prescription — the restricted ensemble. The restricted ensemble is assumed to be internally ergodic within each compartment; then the infinite time average can be replaced by an average over τ_0. Restrictions may either be replaced explicitly on the trace itself, as in (19.1.21) and (19.1.22), or may be realized implicitly by modifying the Hamiltonian $\mathcal{H}$.

Actually many approximation methods, such as mean-field theory, do take account of non-ergodicity in some ways. They add information about the system and modify the prescription. A common case is the specification of an order parameter and its use in a Landau expansion.

The trace is frequently restricted by regarding some variables as fixed or quenched and others as dynamical. Another is the use of periodic potentials in the electronic theory of crystals, associating atoms or ions with particular lattice points instead of allowing them to be anywhere in space. It has not yet been rigorously proved that the latter leads to the former, but we can observe that crystals exist and construct our theories — or restrict our traces — accordingly.

Instead of deriving macroscopic properties from a specified microscopic Hamiltonian, a different approach, empirical and macroscopic, is to construct trial functions for the free energy $F^\alpha(T,H)$ in compartment Γ^α. How well this can be done depends on the system and one's knowledge of its compartment structure. Among examples of this approach is the Landau theory, which has been successfully applied to broken symmetry as before and can also be used to generalize broken ergodicity. In the general case of broken ergodicity, we still try to construct $F^\alpha(T,H)$; stability criteria continue to apply, and we may make expansions in T or H about any point desired.

Usually, and particularly in a system with a large number of compartments, we are more interested in the distribution or average of measurable quantities than in their values for one specific compartment. Thus some parameters in F^α may be variables dependent on α, and we need to know the frequency of occurrence of different parameter values. Let us write $F(T,H,K)$ for the dependence of a single free energy function F on some parameters K, which take the value K^α in compartment Γ^α. Then $F^\alpha(T,H) = F(T,H,K^\alpha)$. The parameters K could be regarded as frozen order parameters. Suppose we wish to compute an average, say of F^α,

$$\langle F\rangle = \sum_\alpha p^\alpha F^\alpha(T,h), \tag{19.1.23}$$

using probability p^α for compartment Γ^α.

§19.2 From Vapor to Amorphous Solid

Two classical examples showing broken ergodicity without broken symmetry are transitions from vapor to liquid and from liquid to amorphous solid by avoiding crystallization. An interesting example involving the behavior of a liquid drop on a solid surface is also related to broken ergodicity.

19.2.1 Vapor-Liquid Transition

The vapor-liquid transition is of first order, as is the liquid-solid transition discussed in §16.1, because both have a latent heat and change of volume. Experimentally, a sufficiently dense vapor will

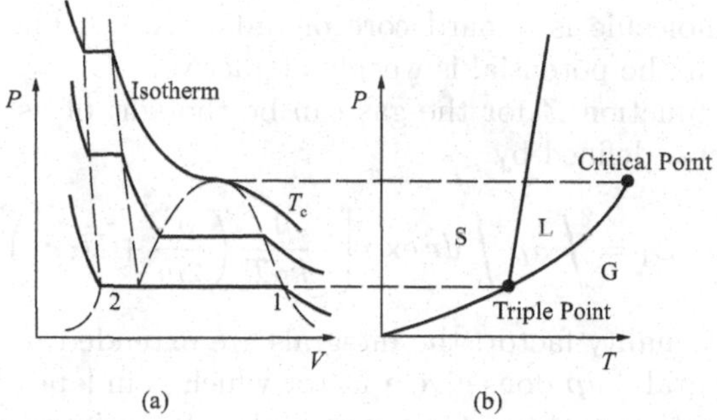

Figure 19.2.1 P-V and P-T diagram of a typical substance.

condense into a liquid if the pressure is increased or the temperature is lowered. Theoretically, van der Waals modified the ideal gas equation of state which relates pressure, volume and temperature for the constituent molecules with no interaction among them, by introducing two correction terms, one to account for the finite volume of the molecules and another to account for their mutual attraction.

Let us consider the transition between the gas phase and the liquid phase. The P-V and P-T diagrams are shown in Fig. 19.2.1. The transition takes place at a constant temperature and pressure. The pressure $P(T)$ is called the vapor pressure at the temperature T. If the system is initially in state 1 in one isotherm, where it is all gas, as heat is subtracted from the system, some of the gas will be converted into liquid, and so on until we reach state 2, where the system is all liquid. The isotherm in the P-V diagram is horizontal during the phase transition, because the gas phase has a smaller density than the liquid phase. Consequently, when a certain mass of gas is converted into liquid, the total volume of the system shrinks, although P and T remain unchanged. Such a transition is of first order.

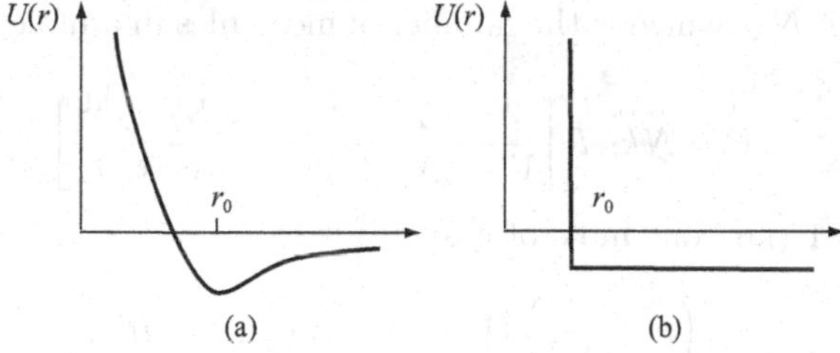

Figure 19.2.2 Effective interaction potential of an individual molecule in a gas. (a) Qualitative form; (b) Simplified form.

In most substances the potential energy between two molecules as a function of the intermolecular separation has the qualitative shape shown in Fig. 19.2.2(a). The attractive part of the potential energy originates from the mutual electric polarization of the two molecules and the repulsive part from the Coulomb repulsion and the Pauli principle of the overlapping electronic clouds of the molecules. The situation can be idealized by approximating the repulsive part by an infinite hard-sphere repulsion and the attractive part by a infinite range shallow well, so that the potential energy looks like that in Fig. 19.2.2(b). The main effect of the hard core would be to forbid the presence of any other molecule in a certain volume around a molecule. The qualitative effect of the attractive part of the potential energy is a tendency for the system to form a bound state.

The van der Waals theory can be viewed as an example of the mean-field approach to a phase transition. Suppose each molecule in a gas feels an effective interaction potential due to all other molecules in the gas which is of the form

$$U(r) = \begin{cases} \infty, & r \le r_0 \\ u < 0, & r > r_0. \end{cases} \tag{19.2.1}$$

This means that each molecule is a hard core of radius r_0; within it the potential is infinitely repulsive, but outside of it the potential is weakly attractive.

The global partition function Z for the gas can be thought of as the product of the N single-particle partition functions, defined by

$$Z_1 = \int d\boldsymbol{p} \int d\boldsymbol{r} \exp\left[-\frac{1}{k_B T}\left(\frac{p^2}{2m} + U(\boldsymbol{r})\right)\right], \tag{19.2.2}$$

where we omit a proportionality factor; the integrals are extended in momentum $\boldsymbol{p}$ and position $\boldsymbol{r}$ space, as usual. The integral $\int d\boldsymbol{p}$ does give a factor which is independent of V, and therefore does not contribute to the equation of state. On the other hand, the integral $\int d\boldsymbol{r}$ can be performed and gives

$$(V - V_e)e^{-u/k_B T},$$

where V_e is the 'excluded volume' corresponding to the hard core of all molecules in the gas interacting with a single molecule. The pressure is given by

$$P = -\left(\frac{\partial F}{\partial V}\right)_T = k_B T \frac{\partial \ln Z}{\partial V}. \tag{19.2.3}$$

Noticing that $Z = Z_1^N$, we have

$$P = N k_B T \frac{\partial}{\partial V}\left[\ln(V - V_e) - u/k_B T\right]. \tag{19.2.4}$$

It is reasonable to assume that V_e is proportional to the total number of molecules in the gas, and u, which corresponds to an attractive interaction, is proportional to the density of molecules in the gas. Therefore,

$$V_e = (b/N_A)N, \quad u = -(a/N_A^2)(N/V), \tag{19.2.5}$$

where b and a are constants whose physical meaning will become apparent later on, defined in terms of the Avogadro constant N_A, which is the number of molecules in a mole of gas. Combining (19.2.4) with (19.2.5), we get

$$P = N k_B T\left[\frac{1}{V - (N/N_A)b} - \frac{a(N/N_A)^2}{N V^2 k_B T}\right], \tag{19.2.6}$$

which, for $n = N/N_A = 1$ (i.e., one mole of gas), leads to

$$\left(P + \frac{a}{V^2}\right)(V - b) = N_A k_B T = RT, \tag{19.2.7}$$

which is the van der Waals equation, where $V$is the volume per mole and R the gas constant. It may be noted that for $P \gg a/V^2$ and $V \gg b$, i.e., a mole of a gas occupying a sufficiently large volume at a moderately high temperature, (19.2.7) reduces to the ideal gas equation $PV = RT$.

To investigate in more detail the van der Waals' equation, we write (19.2.7) as

$$(PV^2 + a)(V - b) = RTV^2, \tag{19.2.8}$$

or equivalently,

$$V^3 - \left(b + \frac{RT}{P}\right)V^2 + \frac{a}{P}V - \frac{ab}{P} = 0, \tag{19.2.9}$$

which reflects in a visible way the cubic character of the van der Waals equation. To each pair of values (T, P) there are, in general, three solutions for V. We may note that for small T the three solutions are real; for a certain $T = T_c$, the three solutions become a single solution with $V = V_c$, $P = P_c$; finally, for large T a pair of roots become complex and a single real solution remains. It implies that there is a critical point, characterized by P_c, V_c, T_c which can be used as a point of reference to describe the behavior of the gas. Let us write down

$$(V - V_c)^3 = V^3 - 3V_c V^2 + 3V_c^2 V - V_c^3 = 0, \tag{19.2.10}$$

at $V = V_c$. We can compare (19.2.10) with (19.2.9) and equate coefficients of identical powers of V. Thus

$$\left(b + \frac{RT_c}{P_c}\right) = 3V_c, \; a/P_c = 3V_c^2, \; ab/P_c = V_c^3. \quad (19.2.11)$$

Finally we get

$$V_c = 3b, \; P_c = a/27b^2, \; T_c = 8a/27bR. \quad (19.2.12)$$

These results give a direct physical meaning to the constants a and b, which are the characteristic parameters for a mole of a specific gas in terms of its volume V_c and its pressure P_c at the critical temperature T_c for the continuous phase transition from the vapor state to the liquid state.

19.2.2 Wetting Transition

Wetting is a process related to the spreading of liquid on a substrate; for convenience in discussion, we take a solid as the substrate. When a liquid drop is put in contact with a flat solid surface, two distinct equilibrium regimes may be found: partial wetting with a finite contact angle θ, or complete wetting $\theta = 0$, just as described in Fig. 19.2.3. In cases of partial wetting, the wetted portion of the surface is limited by a contact line, for example, a circle.

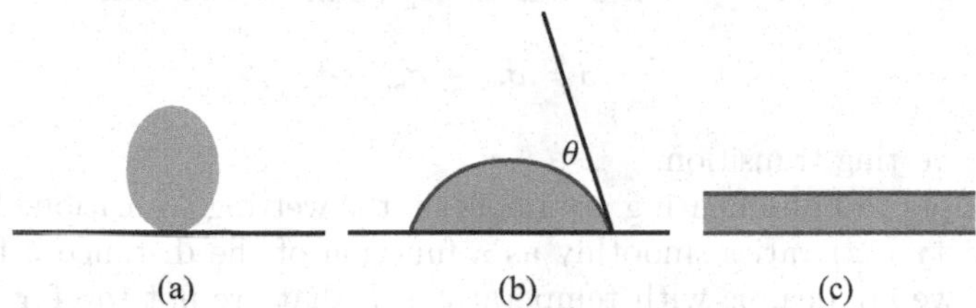

Figure 19.2.3 A liquid drop on a surface of a solid with contact angle θ. (a) Partial wetting; (b) Stronger partial wetting; (c) Complete wetting, $\theta = 0$.

For clarity, we will deal with a macroscopic wedge, and the line $\mathcal{L}$ is normal to the plane of Fig. 19.2.4. Three phases are in contact at the line: The solid S, the liquid L, and the corresponding equilibrium vapor V. Each interface has a certain free energy per unit area such as σ_{sl}, σ_{sv}, and σ_{lv}; the latter, as usual, will simply be called σ. The condition of force equilibrium gives Young's

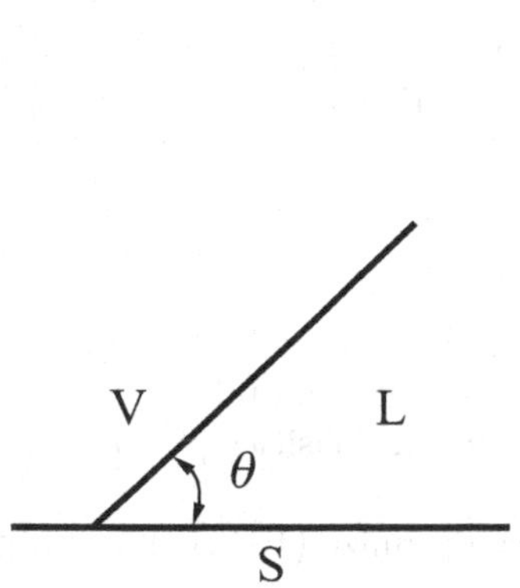

Figure 19.2.4 Definition of contact angle between vapor, liquid and solid.

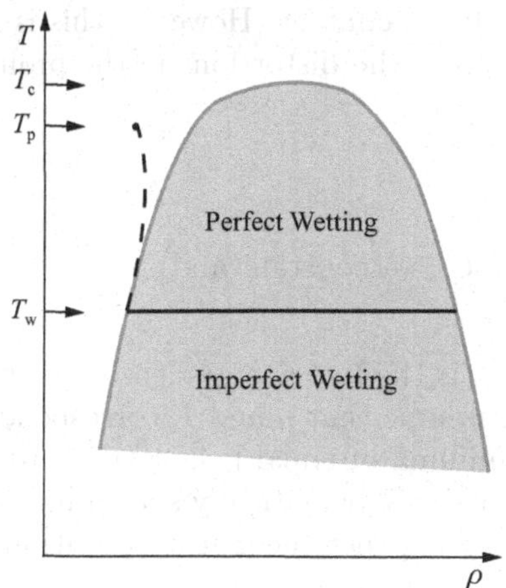

Figure 19.2.5 Phase diagram for the existence of a wetting transition.

equation

$$\sigma \cos\theta = \sigma_{\rm sv} - \sigma_{\rm sl}, \tag{19.2.13}$$

which shows that the contact angle θ is entirely defined in terms of thermodynamic parameters. The shape of a drop on a substrate is given by the minimum of the total free energy of the system.

A liquid-vapor interface in the vicinity of a solid S may exhibit either a finite equilibrium contact angle θ (partial wetting) or a strictly vanishing contact angle (complete wetting). There may exist a particular temperature $T_{\rm w}$ at which we switch from one regime to the other. This is called the wetting transition temperature. The phase diagram for vapor and liquid, related temperature and concentration, can be shown in Fig. 19.2.5. There are two regions of co-existence separated by a horizontal line denoted as the wetting transition temperature $T_{\rm w}$. There is a pre-wetting phase transition line for the system.

The occurrence of a wetting transition can be argued from the Young equation (19.2.13). The surface tension depends not only on the concentration but also on the temperature along the coexistence line in the phase diagram. If the system approaches the critical point $T_{\rm c}$ of the gas-liquid coexistence line, the difference between the liquid and gas phase vanishes and so does σ. It was analyzed by Cahn[b] that in this process the difference $(\sigma_{\rm sv} - \sigma_{\rm sl})$ also vanishes, but more slowly than σ. A related result is that $\cos\theta$ in (19.2.13) will diverge as $T \to T_{\rm c}$. This is in contradiction to $\cos\theta \leq 1$. The conclusion is that there is always a wetting transition at $T_{\rm w} < T_{\rm c}$ so that $\theta > 0$ for $T < T_{\rm w}$ and $\theta = 0$ for $T \geq T_{\rm w}$. One can define the spreading coefficient

$$S = \sigma_{\rm sv} - \sigma_{\rm sl} - \sigma \tag{19.2.14}$$

to characterize the wetting transition.

Cahn gave a simple and illuminating argument on the wetting transition. He considered that the liquid number density $\rho(z)$ varies smoothly as a function of the distance z from the solid surface. This is adequate if we are dealing with temperatures T that are not too far from the critical point $T_{\rm c}$. There is an important assumption that the forces between solid and liquid are of short range ($\sim a$), and can, in fact, be described simply by adding a special energy $\sigma_{\rm c}(\rho_{\rm s})$ at the solid surface. Here $\rho_{\rm s} = \rho(z=0)$ is the liquid density at the surface and $\sigma_{\rm c}$ is a certain functional

$$\sigma_{\rm c} = \sigma_0 - \sigma_1\rho_{\rm s} + \frac{1}{2}\sigma_2\rho_{\rm s}^2 + \cdots, \tag{19.2.15}$$

where σ_0, σ_1, and σ_2 are constants. The σ_1 term, favoring large $\rho_{\rm s}$, describes an attraction of the liquid by the solid. The σ_2 term represents a certain reduction of the liquid/liquid attraction interactions near the surface. The parameters σ_1 and σ_2 describe the essential features at the interface. We may say that $\sigma_{\rm c}$ is the contribution to the solid-liquid interfacial energy that comes from direct contact. However, this is not all of the interfacial energy. Another contribution $\sigma_{\rm d}$ will come from the distortions in the profile $\rho(z)$, which can be written as a classical "gradient square" functional

$$\sigma_{\rm d} = \int dz \left[\frac{1}{2}C\left(\frac{d\rho}{dz}\right)^2 + W(\rho)\right], \tag{19.2.16}$$

where C is a constant and

$$W(\rho) = f(\rho) - \rho\mu - P. \tag{19.2.17}$$

Here f is the free energy density of the bulk liquid, μ its chemical potential and P its pressure. We shall assume that μ and P correspond to the exact coexistence of liquid and vapor. Then $W(\rho)$ has two minima of equal height ($W = 0$) for the two equilibrium densities $\rho = \rho_{\rm L}$ (liquid) and $\rho = \rho_{\rm v}$ (vapor), as schematically shown in Fig. 19.2.6.

To construct the density profile in the liquid $\rho(z)$ we optimize (19.2.16) and obtain

$$-C\frac{d^2\rho}{dz^2} + \frac{dW(\rho)}{d\rho} = 0, \tag{19.2.18}$$

[b] J. W. Cahn, *J. Chem. Phys.* **66**, 3677 (1977).

from which a first integral

$$\frac{1}{2}C\left(\frac{d\rho}{dz}\right)^2 = W(\rho) \tag{19.2.19}$$

is derived. There is no integration constant in (19.2.19). If we consider a point deep in the bulk, where $\rho = \rho_b$ (ρ_b being either ρ_l or ρ_v), we must have $d\rho/dz = 0$ and $W(\rho_b) = 0$ as explained in Fig. 19.2.6. By using (19.2.19), the distortion energy σ_d is

$$\sigma_d = \int_{\rho_b}^{\rho_s} [2CW(\rho)]^{1/2} d\rho. \tag{19.2.20}$$

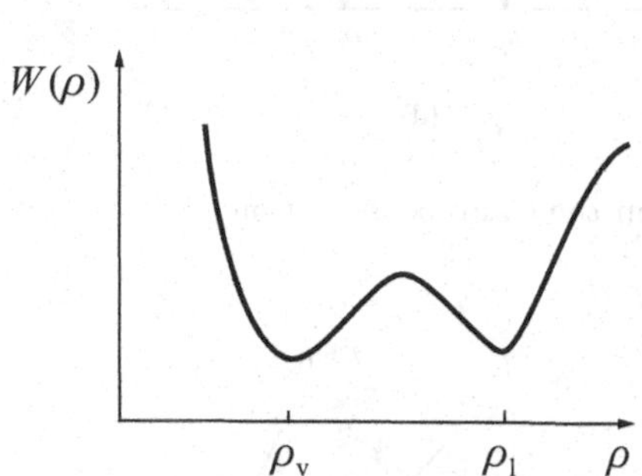

Figure 19.2.6 The "effective free energy" $W(\rho)$ as a function of the density ρ.

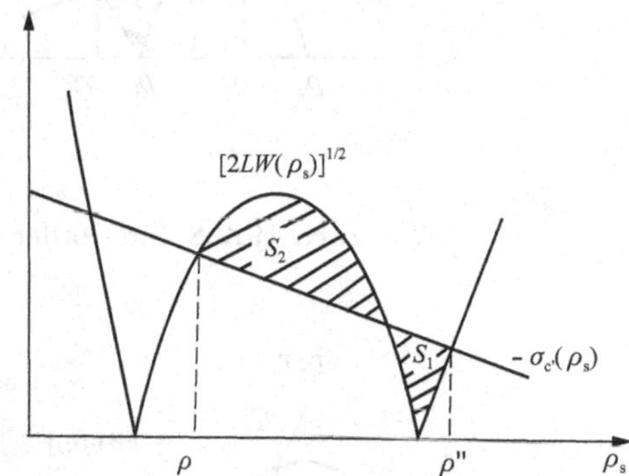

Figure 19.2.7 The Cahn construction determining the density at the surface ρ_s. There are two locally stable roots (ρ', ρ''). The other two roots are unstable.

The last step is to determine the surface density ρ_s by minimization of the total energy $\sigma_d + \sigma_c(\rho_s)$. The resulting condition is

$$-\sigma_c'(\rho_s) = [2CW(\rho_s)]^{1/2}, \tag{19.2.21}$$

where $\sigma_c'(\rho) = d\sigma_c/d\rho$. This leads to the graphical construction of Fig. 19.2.7. Here, for the form of $\sigma_c(\rho_s)$ proposed in (19.2.15), it gives a linear plot for $-\sigma_c'(\rho_s) = \sigma_1 - \sigma_2\rho_s$.

If the slope σ_2 is small, the condition (19.2.21) may give four roots for ρ_s. Two of these are locally stable, while the others correspond to a maximum of the free energy and are unstable. In this regime we find a competition between a state of low ρ_s ($\rho_s = \rho'$) describing a nearly "dry" solid in contact with the vapor ($\rho_b = \rho_v$) and a state of high ρ_s ($\rho_s = \rho'' > \rho_l$) describing a wet solid in contact with the liquid ($\rho_b = \rho_l$). The energies of these two states are

$$\sigma_{sv} = \sigma_d(\rho_v, \rho') + \sigma_c(\rho'), \quad \sigma_{sl} = \sigma_d(\rho_l, \rho'') + \sigma_c(\rho''), \tag{19.2.22}$$

and the liquid/vapor interfacial energy σ can be derived from the same analysis

$$\sigma = \sigma_d(\rho_v, \rho_l). \tag{19.2.23}$$

Using (19.2.22) and (19.2.23), one can check that S in (19.2.14) has a simple graphical interpretation: In Fig. 19.2.7, $S = S_1 - S_2$, which is the difference of the two shaded areas.

Let us now vary the temperature, as indicated in Fig. 19.2.8. (1) At $T \ll T_c$ the difference $\rho_l - \rho_v$ is large, and S_2 is larger than S_1. This gives $S < 0$, i.e., $\cos\theta$ is finite, leading to partial wetting. (2) If T is raised, the difference $S_1 - S_2$ decreases and vanishes at a special temperature $T = T_w$. Here $S = 0$ and $\theta = 0$. (3) At temperatures $T > T_w$, $S_2 < S_1$, and S is positive. This regime cannot be observed in thermal equilibrium. Instead of building up a liquid/vapor interface with $\rho_s = \rho'$, the system prefers to achieve it in two steps, through a macroscopic film of L wetting the surface and giving a total surface energy $\sigma_{sl} + \sigma$. Thus, here, we keep $\theta = 0$: The case of complete wetting.

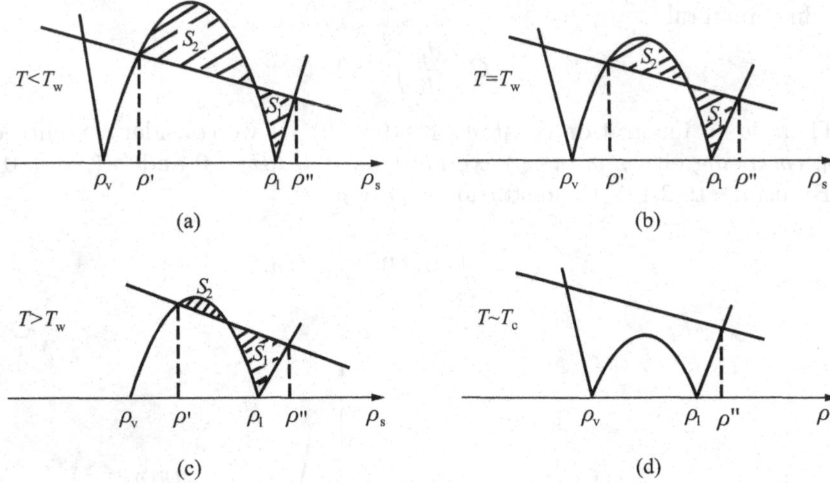

Figure 19.2.8 First-order transition from the Cahn construction.

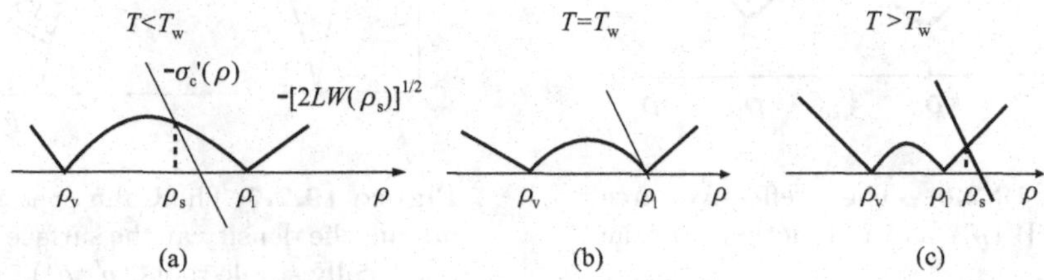

Figure 19.2.9 Second-order transitions from the Cahn construction.

Ultimately, at high temperature ($T \sim T_{\rm c}$), only one stable root is left, corresponding to a solid-liquid interface.

In this scenario the transition at $T_{\rm w}$ involves a jump from one energy minimum ($\rho_{\rm s} = \rho'$) to a distinct minimum ($\rho_{\rm s} = \rho''$) and is clearly of first order. The plot of $\cos\theta$ versus temperature in the partial wetting regime has a finite slope, and intersects $\cos\theta = 1$ at $T = T_{\rm w}$.

If the slope σ_2 of $-\sigma_{\rm c}'(\rho_{\rm s})$ is large, at all temperatures T we find only one root $\rho_{\rm s}$ from the construction illustrated in Fig. 19.2.9. (1) At $T < T_{\rm w}$, $\rho_{\rm s} < \rho_{\rm l}$, and we can construct two density profiles corresponding to two physical situations: One profile where $\rho(z)$ decreases from $\rho_{\rm s}$ to $\rho_{\rm v}$ (describing S/V) and one profile where $\rho(z)$ increases from $\rho_{\rm s}$ to $\rho_{\rm l}$ (describing S/L). Again a discussion of areas allows one to compare the surface energies. One finds a negative spreading coefficients, $S < 0$ corresponding to partial wetting. (2) At high temperatures ($T > T_{\rm w}$), the surface density $\rho_{\rm s}$ is higher than $\rho_{\rm l}$; there is only one profile associated with $\rho_{\rm s}$, where $\rho(z)$ decreases from $\rho_{\rm s}$ to $\rho_{\rm l}$ (S/L interface). The S/V interface must then involve a macroscopic film of L, and we have complete wetting. Clearly this scenario corresponds to a continuous, or second-order transition. At $T = T_{\rm w}$, $\rho_{\rm s} = \rho_{\rm l}$.

19.2.3 Glass Transition

A glass is an amorphous solid that lacks the periodicity characteristic of a crystal. Many liquids, inorganic or organic, metallic or insulating, can form glasses upon cooling. When a liquid is cooled, there may appear either crystalline or glass phases. The crystallization may take place at the melting point $T_{\rm m}$, if the temperature has been lowered sufficiently slowly so that the system has remained in a state of quasi-thermal-equilibrium. The liquid will become 'supercooled' for temperatures below $T_{\rm m}$ through rapid quenching of the system and becomes more viscous with decreasing temperature, and may ultimately form a glass. These changes can be observed readily by monitoring the volume

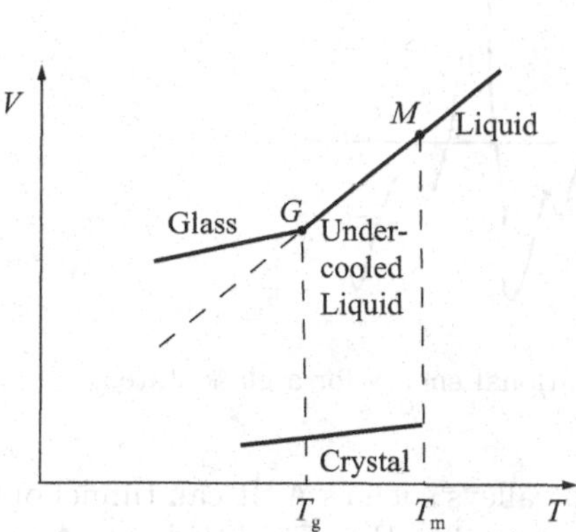

Figure 19.2.10 Schematic drawing of the glass transition.

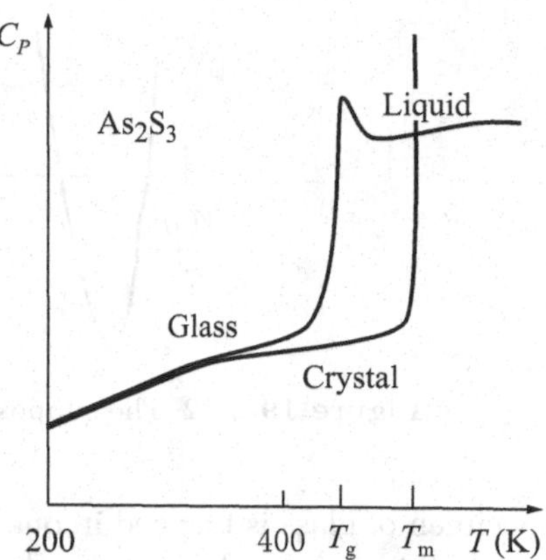

Figure 19.2.11 The specific heat of As_2S_3 near the glass transition.

as a function of temperature, and a typical result is shown schematically in Fig. 19.2.10. The crystallization process is manifested by an abrupt change in volume at $T_{\rm m}$, whereas glass formation is characterized by a gradual change of slope; the region over which this change of slope occurs is called the glass transition temperature $T_{\rm g}$. Since the transition to the glassy state is continuous, the glass transition temperature is not well defined. It depends on the rate of cooling of the supercooled liquid. It is found that the slower the rate of cooling, the larger is the region for which the liquid may be supercooled, and hence the lower is the glass transition temperature. Thus, the glass transition temperature of a particular material depends on its thermal history, and is not an intrinsic property of the substance. The actual value of the glass transition temperature may vary by as much as 10 to 20% for widely differing cooling rates. As an example, for silica-base glasses, the change in $T_{\rm g}$ for different cooling rates may be as much as 100–200 K for values of $T_{\rm g}$ in the range of 600–900 K.

The nature of the glass transition is very complex and even now is poorly understood. It seems that the glass transition is a second-order phase transition, due to the fact that certain thermodynamic variables, e.g., V and S, are continuous, yet the derivatives

$$\alpha_T = (\partial \ln V/\partial T)_P, \ \kappa_T = -(\partial \ln V/\partial P)_T, \ C_P = T(\partial S/\partial T)_P$$

are discontinuous at the transition.

From the behavior observed near $T_{\rm g}$ in Figs. 19.2.10 and 19.2.11, it may be seen that characteristic properties of the glass transition closely resemble a second-order thermodynamic transition. While $V(T)$ is continuous through the vicinity of $T_{\rm g}$, $C_p(T)$ near $T_{\rm g}$ has a step in a narrow temperature interval. However, these changes are not as sharp as they should be in a true second-order transition.

The liquid-glass transition is an old problem and there has been no reliable microscopic theory until now. One well-known theoretical treatment, based on an equilibrium thermodynamic viewpoint, is the free-volume model. The basic idea is to approach the delocalization-localization transition for the atoms and molecules by considering the volume available to each molecule, and asking if there is adequate room for molecular motion. The glass transition occurs when the free volume is sufficiently squeezed out of the system; this theory has been partially successful to illustrate the glass transition.

It must be emphasized that glass transition is not related to a symmetry change; in fact we can take glass as the classical example of broken ergodicity: There is no unique structure, i.e., configuration. For glass, the structure is whatever the liquid finds itself trapped in. We plot in Fig. 19.2.12, the configuration energy or potential energy $V(X)$, where X is the configuration coordinates for all degrees of freedom. So, V is a multidimensional surface with ridges and valleys.

The phase space can be separated into compartments A, B, C, etc.,

$$\Gamma = \Gamma^{\rm A} \bigcup \Gamma^{\rm B} \bigcup \Gamma^{\rm C} \cdots . \qquad (19.2.24)$$

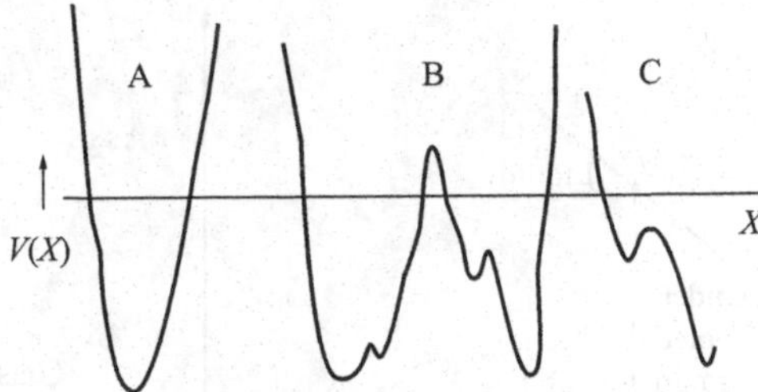

Figure 19.2.12 The supposed configurational energy for a glass state.

A given specimen of glass is trapped in one of the deep valleys such as A. It can tunnel or thermally fluctuate to neighboring low-lying states, but access to states like B or C is ruled out. A macroscopic specimen of glass has in it compartments representative of a sufficient number of the different realizable configurations like A, B, C, . . . , and each of these compartments is itself sufficiently macroscopic for thermodynamics to be applicable. One can thus average over configurations before the results of theory are compared with experiment. This "self-average" property of systems exhibiting quenched disorder is valid for most properties of physical interest.

From kinetic aspects, the transition observed near T_g differs from a strict second-order phase transition. The important aspect for a glass transition concerns the relaxation processes that occur when a supercooled liquid cools. We have already seen that the experimentally measured value of T_g is not unique; it depends on the time scale of the experiment used to observe it.

The configurational changes that cause the relaxation of the supercooled liquid become increasingly slow with decreasing temperature until, at the glass transition temperature, the material behaves as a solid. The rapid increase of the viscosity or characteristic relaxation time τ with decreasing temperature is one of the most important features of the relaxation process. For a time of observation, τ_0, long compared with the structural relaxation time τ the material appears 'liquid-like', whereas for $\tau_0 > \tau$ the material behaves as if it were 'solid-like', which can be specified by a certain viscosity, say $10^{14.6}$ poise. A 'transition' will appear to take place when $t_0 = t_r$. T_g can thus be defined.

The main unsolved problem for the glass transition may be summarized as follows: Whether this is a special kind of phase transition or just a purely dynamical phenomenon. Both viewpoints have adherents and have gathered evidence to support their theories. In recent years, there has been significant progress in the glass transition, strongly stimulated by achievements of the mode-coupling theory. This theory is formulated in terms of non-linear coupling between density fluctuation modes and predicts an ideal kinetic glass transition. The most important finding is the existence of a crossover temperature T_{cr} above the glass transition temperature T_g where significant changes in the dynamics occur. Although the theory still has to overcome several problems, it highlights a few essential features in the dynamics of supercooled liquids and has stimulated many experimental efforts.[c]

The description of broken ergodicity by using the uneven energy landscape, as shown in Fig. 19.2.12, will also be effective in discussing the spin glass transition in next section. In recent years, it has played an important role in treating the problem of protein folding.

§19.3 Spin Glass Transition

Spin glass is a typical example of a condensed matter system characterized by the absence of long-range order, but exhibiting broken ergodicity. The spin glass is a fundamental form of magnetism besides ferromagnetism and antiferromagnetism. There are many phenomena analogous to the spin glass. In condensed matter physics, the analogs are electric dipolar and quadrupolar glasses, granular superconductors, etc. We can also enumerate many non-physical analogs to the spin glasses, such as combinational optimizations, neural networks and biological evolution, . . . , etc.

[c]W. Gotze and L. Sjogren, *Rep. Prog. Phys.* **55**, 241 (1992).

19.3.1 Spin Glass State

The spin glass is a kind of magnetically disordered structure in which the underlying perfect crystalline lattice may be preserved. This situation occurs in some dilute magnetic alloys, such as CuMn or AuFe, with a magnetic component from 0.1 to 10 atomic percent. The local moments in a metallic host are randomly frozen, as indicated in Fig. 19.3.1. The formation of a spin glass state is due to the fact that the exchange interaction between moments in a metal has an oscillatory spatial dependence.

According to the concentration of magnetic impurities in a magnetic alloy, the magnetic state of the alloy falls into one of several regimes, as schematically represented in Fig. 19.3.2. When the concentration is very low, the interaction between the spins is negligible; this is the Kondo regime in which the phenomenon of resistivity minimum may appear. When the concentration is large, this is the inhomogeneous long-range magnetic ordered regime. In between these two, for intermediate concentrations, is the spin-glass regime, including cluster spin glass.

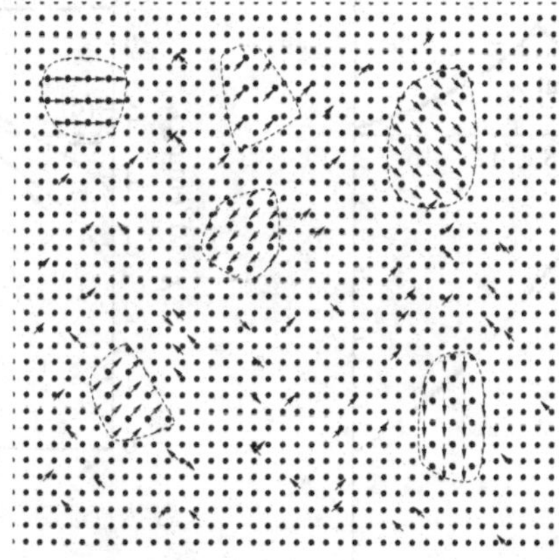

Figure 19.3.1 Randomly frozen magnetic moments on a nonmagnetic metal lattice.

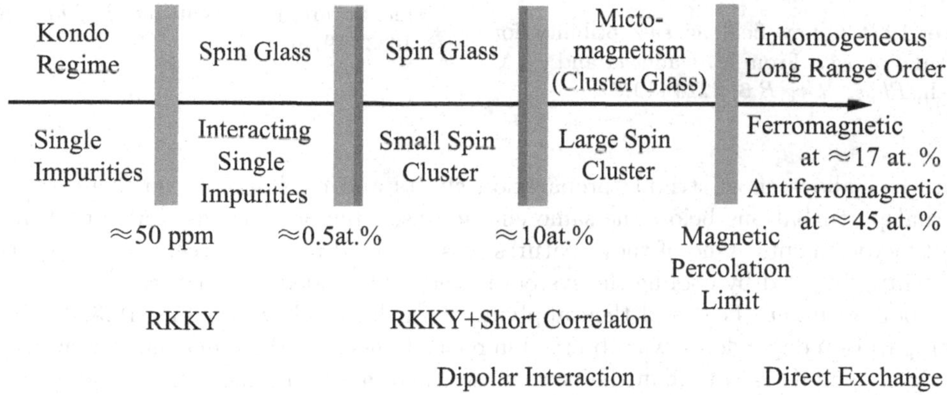

Figure 19.3.2 Various concentration regimes for a magnetic alloy illustrating the different types of magnetic behavior. From J. A. Mydosh, *Spin glasses: An Experimental Introduction*, Taylor & Francis, London (1993).

Spin glass systems are characterized by a random competition between ferromagnetic and antiferromagnetic interactions. As a result, conventional magnetic long-range order is not possible. This constitutes a new state of magnetism, distinctly different from the long-range ordered ferro- and antiferromagnetic phases. Yet similar to these magnets, the spin glass also has a cooperative or collective nature in the frozen state. For the dilute magnetic alloys, which are the canonical example of spin glass systems, the competition arises from the oscillatory RKKY interactions between the magnetic ions, mediated by the conduction electrons of the host metal.

Experiments have shown that there are rather sharp cusps in the temperature dependence of the low frequency susceptibility at low field, in the dilute metallic alloys AuFe and CuMn. Figure 19.3.3 reproduces the first measurements on AuFe which coined the term spin glass, showing the sharp cusps; the frequency was in the low audio range (50–155 Hz) and the driving field was 5 Gauss. Notice how the peak height increases in magnitude and shifts to higher temperatures as the concentration of Fe is increased. At higher temperatures, the magnetic alloy is in the paramagnetic state; however, as the temperature decreases and approaches a characteristic temperature T_f, relaxation times become extremely long and the system begins to exhibit hysteresis. A paramagnet-spin glass transition is assumed to take place at T_f, which is often called the freezing temperature.

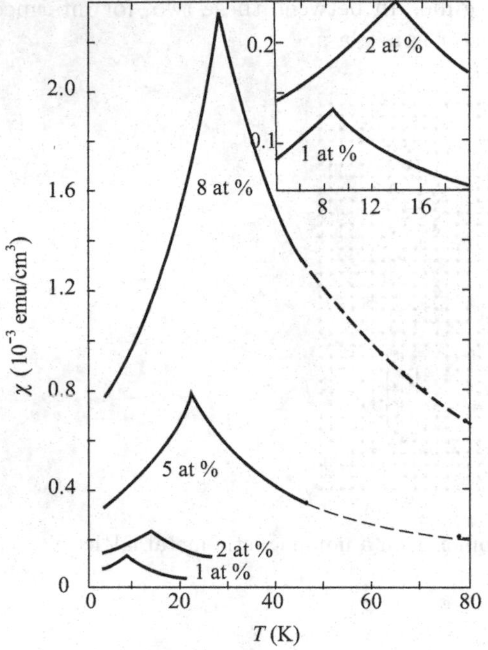

Figure 19.3.3 Low-field ac susceptibility for four AuFe alloys. From V. Cannella and J. A. Mydosh, *Phys. Rev. B* **6**, 4220 (1972).

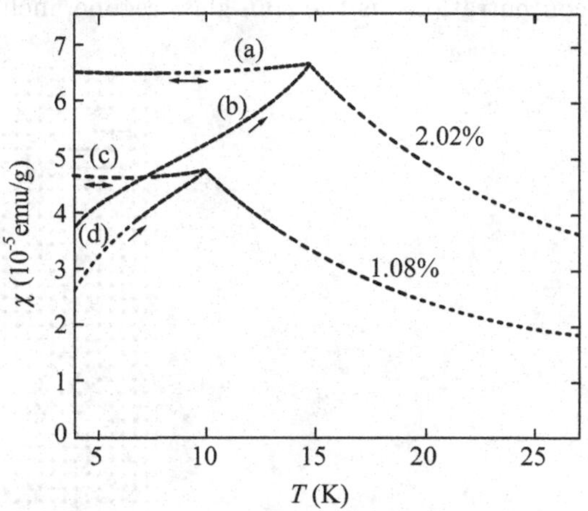

Figure 19.3.4 Dc susceptibility measurements for two CuMn alloys with 1.08 and 2.02 at.% Mn. Curves (a) and (c) were obtained by cooling in the measurement field (FC), (b) and (d) are the results of zero-field-cooled (ZFC) experiments. From S. Nagata *et al.*, *Phys. Rev. B* **19**, 1633 (1979).

Spin glass alloys show striking preparation effects and considerable slowing-down of response to external perturbations below the same characteristic temperature as that found in the ac susceptibility experiments. One of these features was observed in a dc susceptibility experiment: The susceptibility obtained by cooling the system in the field yielded a higher value than that obtained by first cooling in zero field and then applying the field, as shown in Fig. 19.3.4. Similarly, dramatic preparation dependence was observed in measurements of the remanent magnetization. These observations demonstrate that in the new phase, there are many metastable states whose relative free energies vary in different ways with external perturbations and which have significant energy barriers impeding motion from one state to another.

Just as in Fig. 19.2.12, for the spin glass we can also construct a picture of the energy surface in phase space. For all $T < T_f$ there exist an infinite number of distinct equilibrium states, or valleys, in the free energy landscape, into which the system may fall when cooled below T_f. Each such state is separated by infinite energy barriers from all the others. An explanation of the onset of preparation dependence between field-cooled and zero-field-cooled measurements comes from different modifications with field of the relative energies of different minima, so that the lowest minimum at zero field is no longer the lowest in even a small field.

The lowest-lying minima represent the pure equilibrium state; at higher energies there are the many metastable states. Upon cooling, a spin glass may become stuck in one of these states. If the system cannot explore all of phase space, then we call it non-ergodic. When the barriers between the valleys become infinite, ergodicity is lost. The time it takes to go from one valley to another is the exponential of the height of the barrier between these two valleys. It is not surprising that there are many different relaxation times present. Theoretical treatment really shows that below the freezing temperature T_{f}, the system condenses into non-ergodic state characterized by an infinite number of pure phases. In comparison, the simple Ising ferromagnets exhibits only two pure phases, with positive and negative magnetizations. Below the Curie temperature, the broken ergodicity is associated with spontaneous symmetry breaking, the system will stay in one pure phase for infinite time in the thermodynamic limit.

Although we have taken a few very specific materials, e.g., CuMn and AuFe, as the examples for spin glasses, the spin glass is actually a very general phenomenon. Over 500 different systems, including transition metal alloys, rare-earth binary or pseudo-binary alloys, amorphous magnetic alloys, and even magnetic semiconductors and magnetic insulators, can display the behavior of spin glasses, if there is sufficient magnetic coupling, even if the interaction is not RKKY-like.

19.3.2 Frustration and Order Parameter

Edwards and Anderson (1975) attributed the salient phenomena of spin glasses to the competition between random ferromagnetic and antiferromagnetic interactions for local magnetic ions.[d] Our discussion can begin from a model Hamiltonian

$$\mathcal{H} = -\sum_{i,j} J_{ij}\boldsymbol{S}_i \cdot \boldsymbol{S}_j, \tag{19.3.1}$$

where the i, j label the magnetic ions, the $\boldsymbol{S}_i$ are the corresponding spin orientation vectors, and J_{ij} the exchange interaction between the pair of spin (ij). The spin locations $\boldsymbol{R}_i$ are randomly located and J_{ij} is a function of $(\boldsymbol{R}_i - \boldsymbol{R}_j)$ that oscillates in sign with separation. In magnetic alloys the spins are coupled through the RKKY interaction

$$J(R) = J_0 \frac{\cos(2k_{\mathrm{F}}R)}{(2k_{\mathrm{F}}R)^3}, \tag{19.3.2}$$

which is a long-range indirect exchange interaction mediated by conduction electrons with its oscillation period π/k_{F}, where k_{F} is the Fermi wavevector. This interaction is either ferromagnetic or antiferromagnetic as a function of distance.

Competing interactions lead to frustration, which refers to conflicts between interactions that contradict each other and cannot be obeyed simultaneously. More clearly, frustration arises as pairs of spins get different ordering instructions through the various paths that link i and j, either directly or via intermediate spins. Frustration is a far-reaching concept, so let us look a bit closer at it. Suppose we have a triangle of three Ising spins with three bonds of $+J$ or $-J$ as shown in Fig. 19.3.5. For the configuration (a) three bonds are all positive, so all the bond energies are satisfied and there will only be a two-fold degenerate ordered state. Here we say that the triangle is unfrustrated. However, for the configuration (b), two bonds are positive and one bond is negative, all the bond energies cannot simultaneously be satisfied. One spin remains frustrated no matter what we do. The degeneracy of the ground states is six-fold. We may write a frustration function as $\Phi = \prod \text{sign}\ (J_{ij}) = \pm 1$, where the minus sign denotes a frustrated system. For the case in Fig. 19.3.5, $\Phi = +1$ for the unfrustrated (a) and -1 for the frustrated (b).

Generally, frustration increases the energy and the degeneracy of the lowest energy states. Frustration is a necessary ingredient for spin glasses. Antiferromagnetic interactions alone can provide frustration. A classical demonstration is given by the nearest neighbor Ising model on a triangular lattice. If all the interactions are ferromagnetic, it exhibits the usual paramagnet-ferromagnet

[d] The first theoretical work on spin glass was made by Edwards and Anderson; refer to S. F. Edwards and P. W. Anderson, *J. Phys. F* **5**, 965 (1975).

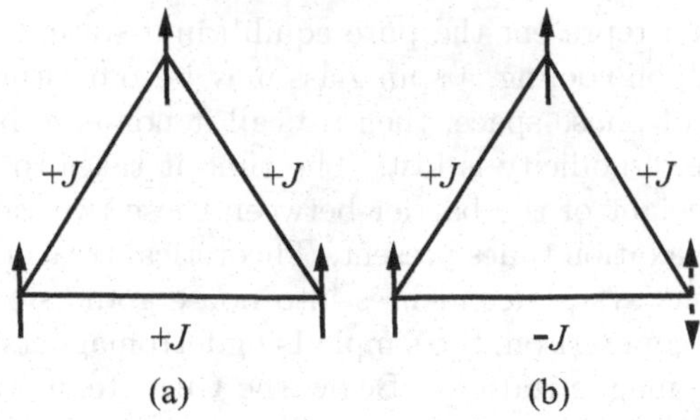

Figure 19.3.5 Triangle lattice with mixed interaction. (a) Unfrustrated plaquettes and (b) frustrated plaquettes.

transition at finite temperature, but if the interactions are all antiferromagnetic and equal then the resultant frustration around each plaquette suppresses cooperative order. The frustration combined with randomness creates a multidegenerate, metastable, frozen ground state for the spin glass. In the dilute magnetic alloys, when temperature is lowered through a temperature T_f, RKKY interaction leads to magnetic frozen state. Randomly frozen magnetic moments gives rise the frustration phenomenon. Here the free energy of the system is at one stable point, but not at the absolute minimum. The mixed ferromagnetic and antiferromagnetic interactions are essential to install the competition and ensure the cooperativeness of the freezing process.

Edwards and Anderson hypothesized that the cusp of ac susceptibility and other anomalous behaviors at T_f were associated with a true phase transition with spontaneously broken spin-flip symmetry to produce the low temperature spin glass phase. Each spin S_i develops a nonzero expectation value, $m_i = \langle S_i \rangle_T$ in this phase, but the sign and magnitude of m_i's vary from site to site because of the random competing interactions. The spatial average of the m_i's, that is, the average magnetization density, thus vanishes.

To describe the spin glass transition, Edwards and Anderson defined the order parameter in the time domain

$$q = \lim_{t\to\infty} \langle \langle \mathbf{S}_i(0) \boldsymbol{S}_i(t) \rangle_\mathrm{T} \rangle_\mathrm{C}, \tag{19.3.3}$$

where subscripts T and C are the thermal and configurational averages, respectively. When $T > T_\mathrm{f}$, $q = 0$, it is in the paramagnetic state, but when $T < T_\mathrm{f}$, $q \neq 0$, it is in the spin glass state. It is clear that in the paramagnetic phase we have $\langle\langle S_i \rangle_\mathrm{T}\rangle_\mathrm{c} = 0$ and $q = 0$, in the ferromagnetic phase $\langle\langle S_i \rangle_\mathrm{T}\rangle_\mathrm{c} \neq 0$ and $q \neq 0$ and in the spin glass phase $\langle\langle S_i \rangle_\mathrm{T}\rangle_\mathrm{c} = 0$ and $q \neq 0$.

The recognition of the combination of disorder and frustration led Edwards and Anderson to propose a more convenient model in which there are spins $\boldsymbol{S}_i$ on all the sites of a lattice but the J_{ij} are chosen randomly from a Gaussian distribution centered at zero. It follows that the time average for the order parameter can be replaced by the ensemble average, and then we can write

$$q = \langle \langle S_i \rangle_T^2 \rangle_\mathrm{c}, \tag{19.3.4}$$

where the inner brackets an represent ensemble, or thermal, average while the outer brackets are for a configurational average. q is called the Edwards–Anderson spin glass order parameter. Edwards and Anderson went on to construct an approximate mean-field solution for their model, finding that q did indeed develop a non-zero value at a continuous phase transition at finite temperature, with an associated cusp in the susceptibility, consistent with the measurements.

We now understand the freezing process as follows: At high temperature, there will be a collection of paramagnetic spins. As T approaches T_f the various spin components begin to interact with each other over a longer range, because the temperature induced disorder is being removed. The system seeks its ground-state ($T = 0$) configuration for the particular distribution of spins and exchange interactions. This means the original random alignment of spins can be locked into a preferred direction due to the local anisotropy. Since there is a spectrum of energy differences between the frozen states, the system may become trapped in a metastable configuration of higher energy.

19.3.3 Theoretical Models

The mechanism of the spin-glass transition, similar to that of the glass transition, is not completely understood. We will give an outline of this problem in what follows. A successful theoretical approach to the spin-glass transition was initiated by Edwards and Anderson (1975), then developed by Sherrington and Kirkpatrick (1975), and further modified by Parisi (1979).

Edwards and Anderson (EA) introduced a simple Heisenberg-version model, which focuses on the bond competition, with a Hamiltonian (19.3.1), in which the sum is over nearest-neighbor pairs of sites located on a three-dimensional cabic lattice. The exchange interaction J_{ij} is an independent random variable, satisfying a Gaussian distribution

$$P(J_{ij}) = \frac{1}{\sqrt{2\pi}\Delta} \exp\left(-\frac{J_{ij}}{2\Delta^2}\right), \tag{19.3.5}$$

where Δ is the variance. The EA model and its mean field approximation was the first step to understand the spin-glass behavior and possible phase transitions. The model is really very simple, yet elegant: Replace the site disorder and RKKY interaction by a random set of bonds which satisfy a Gaussian distribution. What still remains is to establish the true mean-field theory for this model.

In order to derive a mean-field theory for the EA model, Sherrington and Kirkpatrick (SK) proposed a modification in which the interactions J_{ij} are not restricted to nearest-neighbor pairs of sites, but couple all pairs of sites, i.e., the interactions are of infinite-range. In addition they took the spins to be Ising variables, $S_i = \pm 1$, for simplification. The model is still defined by (19.3.1), except that a random exchange bond J_{ij} connects every pair of sites (i, j), not just the nearest-neighbor pairs. Such infinite range model provides a formulation of mean field theory, which can be solved exactly. Various models distinguish the probability distribution of the exchange interactions.

The SK model gives an ordinary paramagnet at high temperatures: As the critical temperature $T_{\rm f}$ is approached from above, there is a continuous spin glass transition. Allowing for the lack of spatial symmetry, a simple extension of a conventional mean-field approximation yields a set of self-consistent equations

$$\langle S_i \rangle_T = \tanh\left(\sum_j J_{ij}\beta\langle S_j\rangle_T\right), \tag{19.3.6}$$

where $\langle\cdots\rangle_T$ is a thermodynamic average. Near the transition temperature $T_{\rm f}$, by linearizing the tanh, we have

$$\langle S_i\rangle_T = \sum_j J_{ij}\langle S_j\rangle_T / k_{\rm B}T_{\rm f}. \tag{19.3.7}$$

Averaging over sites and bonds, and neglecting correlations between those averages, it is found

$$\langle\langle S_i\rangle_T\rangle_{\rm c} = \sum_j \langle J_{ij}\rangle_{\rm c}\langle\langle S_j\rangle_T\rangle_{\rm c}/k_{\rm B}T_{\rm f}. \tag{19.3.8}$$

For a symmetric distribution of exchange interactions $\langle J_{ij}\rangle_{\rm c} = 0$, no non-trivial result can be obtained. Now we consider (19.3.7) first squared on each side and then averaged. The result is

$$\langle\langle S_i\rangle_T^2\rangle_{\rm c} = \sum_j \langle J_{ij}^2\rangle_{\rm c}\langle\langle S_j\rangle_T^2 S^2\rangle_{\rm c}/(k_{\rm B}T_{\rm f})^2, \tag{19.3.9}$$

which gives a non-trivial solution

$$k_{\rm B}T_{\rm f} = \left(\sum_j J_{ij}^2\right)^{1/2}. \tag{19.3.10}$$

To understand and quantify the spin-glass problem further, we should study the average free energy $\langle F\rangle_{\rm c} = -k_{\rm B}T\langle \ln Z\rangle_{\rm c}$ of the magnetically disordered system, with partition function

$Z = \mathrm{Tr}\, \mathrm{e}^{-\beta\mathcal{H}}$. A new statistical method, known as replica theory, was adopted to aid in the analysis of physical averages over quenched disorder, by introducing the mathematical identity

$$\ln Z = \lim_{n\to 0} \frac{1}{n} (Z^n - 1) . \tag{19.3.11}$$

Z^n may be interpreted as the partition function of n identical replicas of the original system, and written as

$$Z^n = \mathrm{Tr}_n \exp\left(-\sum_{\alpha=1}^{n} \beta\mathcal{H}^\alpha \right) , \tag{19.3.12}$$

where α denotes the different replicas with corresponding Hamiltonian $\mathcal{H}^\alpha$, and trace Tr_n is the trace over all the variables. The average over Z^n can be done in terms of cumulants. For the case of J_{ij} satisfying a Gaussian distribution, the average free energy is

$$\langle F \rangle_{\mathrm{c}} = -k_{\mathrm{B}} \lim_{n\to 0} \frac{1}{n} \left\{ \mathrm{Tr}_n \exp \left[\sum_{ij} \left(\beta \tilde{J}_0 \sum_{\alpha=1}^{n} S_i^\alpha S_j^\alpha + \beta^2 \tilde{J}^2 \sum_{\alpha,\beta=1}^{n} S_i^\alpha S_i^\beta S_j^\alpha S_j^\beta \right) \right] - 1 \right\} . \tag{19.3.13}$$

where $\tilde{J}_0$ is the mean and $\tilde{J}$ the standard deviation of the nearest-neighbor $P(J_{ij})$ and the sum (ij) now refers to nearest neighbors. We thus effectively replaced the original disordered system of (19.3.1) by one with a periodic effective temperature-dependent Hamiltonian

$$\mathcal{H}_{\mathrm{eff}} = -\sum_{ij} \left(\tilde{J}_0 \sum_{\alpha}^{n} S_i^\alpha S_j^\alpha + \beta \tilde{J}^2 \sum_{\alpha,\beta=1}^{n} S_i^\alpha S_i^\beta S_j^\alpha S_j^\beta \right) , \tag{19.3.14}$$

involving more complicated interactions and requiring analysis in the limit $n \to 0$.

By analogy with conventional magnetism, we can use a replica mean-field approximation by replacements

$$\sum_{ij} S_i^\alpha S_j^\alpha \to \sum_{ij} (2S_i^\alpha m_j^\alpha - m_i^\alpha m_j^\alpha) ,$$

where $m_i^\alpha = \langle \sigma_i^\alpha \rangle_T$, and

$$\sum_{ij} S_i^\alpha S_i^\beta S_j^\alpha S_j^\beta \to \sum_{ij} \left(2S_i^\alpha S_i^\beta q_j^{\alpha\beta} - q_i^{\alpha\beta} q_j^{\alpha\beta} \right) ,$$

where $q_i^{\alpha\beta} = \langle \sigma_i^\alpha \sigma_i^\beta \rangle_T$, for $\alpha \neq \beta$. The thermodynamic average $\langle \cdots \rangle_T$ is taken for the effective Hamiltonian.

Further treatment can invoke the replica-symmetric ansatz by letting the order parameters $m^\alpha = m$ for all α, and $q^{\alpha\beta} = q$ for all $\alpha \neq \beta$. After some mathematics and taking the limit $n \to 0$, the free energy is

$$\langle F \rangle_{\mathrm{c}} = N \left[\frac{J_0 m^2}{2} - \frac{\beta J^2}{4} (1-q)^2 - k_{\mathrm{B}} T \int dh P(h) \ln(2\cosh\beta h) \right] , \tag{19.3.15}$$

where $J_0 = \tilde{J}_0 z$, $J = \tilde{J} z^{1/2}$, z is the coordination number. The order parameters satisfy the self-consistent equations

$$m = \int dx P(x) \tanh\beta x, \tag{19.3.16}$$

and

$$q = \int dx P(x) (\tanh\beta x)^2, \tag{19.3.17}$$

where

$$P(x) = (2\pi J q^2)^{-1/2} \exp\left[-\frac{(x - J_0 m)^2}{2Jq^2} \right] . \tag{19.3.18}$$

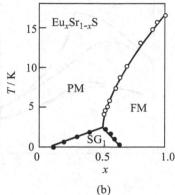

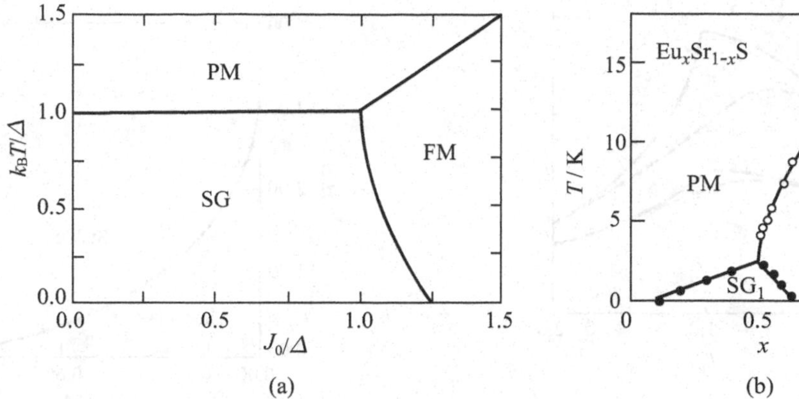

Figure 19.3.6 Phase diagram of magnetic alloy. (a) Theoretical the result obtained in replica-symmetric mean-field method for a random-bond Ising model. From D. Sherrington and S. Kirkpatrick, *Phys. Rev. Lett.* **35**, 1792 (1975). (b) Experimental result of $Eu_xSr_{1-x}S$. From H. Maletta and P. Convert, *Phys. Rev. Lett.* **42**, 108 (1979).

We can get a theoretical phase diagram by using (19.3.16–18) in numerical calculations and confirm that for the paramagnetic phase, $m = q = 0$, for the ferromagnetic phase, $m \neq 0$ and $q \neq 0$, but for spin glass phase, $m = 0$ and $q \neq 0$. The calculated theoretical phase diagram in Fig. 19.3.6(a) qualitatively agrees with the experimental phase diagram.

The ac susceptibility in zero field can be expressed in the fluctuation correlation form as

$$\chi(T) = \frac{1}{k_BT} \sum_{ij} (\langle S_i S_j \rangle - \langle S_i \rangle \langle S_j \rangle) . \tag{19.3.19}$$

SK calculation gives it in the $q(T)$ function as

$$\chi(T) = \frac{\chi_0(T)}{1 - J_0 \chi_0(T)}, \tag{19.3.20}$$

with $\chi_0(T) = (1 - q(T))/k_BT$. By including an applied field H in the Hamiltonian (19.3.1), the SK calculation can give the field dependence of $\chi(T)$. Figure 19.3.7 exhibits the susceptibility behavior for $J_0/\Delta = 0$ and 0.5 with and without a field $H = 0.1\Delta$. There is a sharp cusp in the zero field and the cusp becomes rounded and shifted down in a dc field. These features are well confirmed by experiments.

The SK model offers a reasonable first basis for comparison with experiment. The predicted phase diagram can be nicely mimicked by real spin-glass materials, and the calculated susceptibility is in qualitative agreement with measurement. However, there is something wrong with the SK model. A severe drawback is that the entropy goes to a negative value at $T = 0$: This is closely unphysical. Other difficulties became apparent with the free energy which turns out to be a maximum with respect to q for the solutions $q = 0$, $T > T_f$ and $q \neq 0$, $T < T_f$. Moreover, the $q = 0$ solution, if analytically continued below T_f, has lower free energy than the spin glass state $q \neq 0$. These results contradict to the conventional second-order phase transition. A detailed analysis of the SK solution (de Almeida and Thouless, 1978) showed it to be unstable at low temperature both in the spin-glass and ferromagnetic phases. Such behavior is shown in Fig. 19.3.8 where an H-T line gives the stability limit of the SK solution. The cause of the instability of the SK model lies in the replica-symmetric ansatz which treats all the replicas as indistinguishable, i.e., $q_{\alpha\beta} = q$. Such an assumption leads to an invalid solution of the mean-field EA model. Despite these theoretical difficulties, the SK solution seems correct above the instability line in Fig. 19.3.8. However, in many other measurements, such as specific heat, magnetization, etc., this instability induces the SK model no longer valid, and a more refined model is needed to describe the subtleties of the frozen spin-glass.

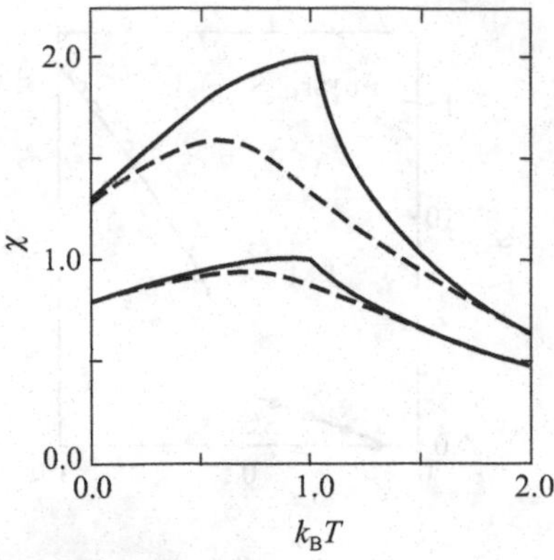

Figure 19.3.7 Differential susceptibility as a function of temperature from the SK model. The solid curves are for $H = 0$, dashed curves for $H = 0.1J_0$; lower curves $J_0/\Delta = 0$, upper curves $J_0/\Delta = 0.5$. From D. Sherrington and S. Kirkpatrick, *Phys. Rev. Lett.* **35**, 1792 (1975).

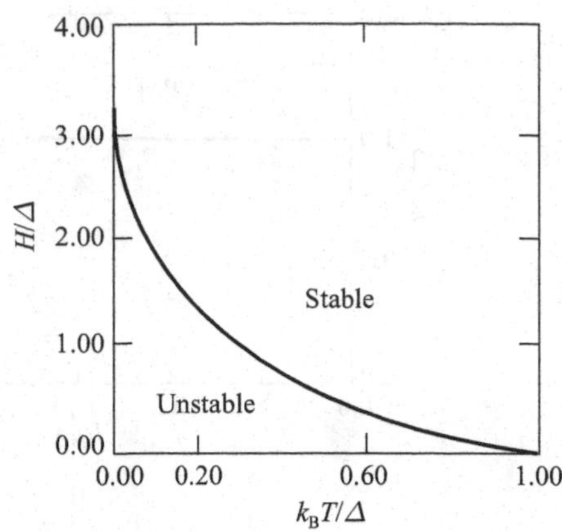

Figure 19.3.8 H-T phase diagram illustrating the stability limits of the SK solution for the case $J_0 = 0$. From J. R. L. de Almeida and D. J. Thouless, *J. Phys. A* **11**, 983 (1978).

A reasonable approach is a replica-symmetry-breaking scheme to be incorporated into the theory. In this scheme, the symmetric requirement $q^{\alpha\beta} = q$ is abandoned. After a number of attempts, a correct treatment was finally proposed by Parisi (1979).[e]

§19.4 Metal-Nonmetal Phase Transitions

Previously we have investigated the disorder-induced metal-nonmetal transition (Anderson transition) and electronic correlation-induced transition (Mott transition). Here we will continue our discussion on electronic transition between extended states and localized states, mainly concerned with interacting electron systems. The broken ergodicity in electron systems is more interesting. Because electrons are quantum particles, the uncertainty relation $\Delta x \cdot \Delta k \sim 1$ will play important role. It can be argued that a localized-delocalized transition in configuration space must be accompanied by an ergodic-nonergodic transition in momentum space, and *vice versa.* Metal-nonmetal phase transitions are involved in electrons, so they will be controlled by the rules in quantum mechanics. There are two kind of phase transitions: One is the normal thermodynamic phase transition, taking place at nonzero temperatures, although interactions should be described according to quantum mechanics, but the realization of phase transitions depends on classical thermal fluctuations. The other is the quantum phase transition, expected to take place at zero temperature. The transition point is approached through changing pressure, composition or magnetic field, and the realization of phase transition depends on quantum zero point fluctuation related to the uncertainty principle. Metal-nonmetal transition can be realized at finite temperature, this is the normal case; but it can also take place at zero temperature, such as the idealized Mott transition described in §13.1.1, which is a quantum phase transition. Although exactly zero temperature cannot be reached in laboratories, it can be extrapolated from experimental results taken at temperatures near 0 K; moreover theoretical study can give many instructive results. The study of the quantum phase transition is important for condensed matter physics, its conclusions are not only applicable to the cases near zero temperature, but often give enlightened insights on physical properties in a wide temperature range. In this section we will discuss first the empirical rules of the metal-nonmetal transition, then the Wigner crystallization, next the Mott transition, and finally the influence of electronic correlation on the Anderson transition will be discussed.

[e] G. Parisi, *Phys. Lett. A* **73**, 203 (1979); *J. Phys. A* **13**, L115; 1101; 1887 (1980).

19.4.1 Semi-Empirical Criteria

Here we will first introduce two important semi-empirical criteria for metal-nonmetal phase transitions, the Goldhammer–Herzfeld (GH) criterion (1927) and Mott criterion (1952), and then explain their relationship. Goldhammer–Herzfeld criterion is used to distinguish various elements in the periodic table; whether they belong to metals or nonmetals. Their point of view concentrated on that the density variation can lead to the variation of electric polarization (per mole) α_{m}. When the density of a substance arrives at a critical value, there will be $\alpha_{\mathrm{m}} \to \infty$, or equivalently $\epsilon \to \infty$, then all the valence electrons bound previously to interior of atoms will be delocalized at once, so the insulator is transformed to a metal.

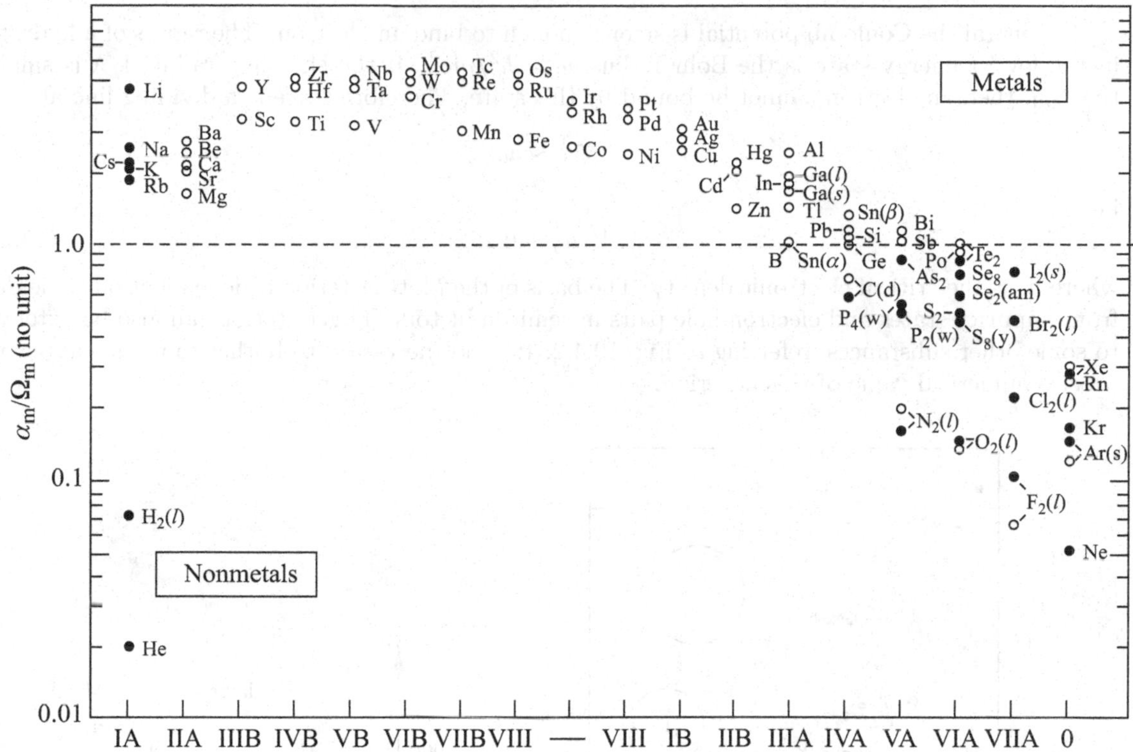

Figure 19.4.1 The $\alpha_{\mathrm{m}}/\Omega_{\mathrm{m}}$ values for several elements in the Periodic Table. Black dots represent the elements in which α_{m} and Ω_{m} all have experimental values; circles represent the elements that have experimental values for Ω_{m}, but only theoretical values for α_{m}. From P. P. Edwards *et al.*, A perspective on the metal-nonmetal transition, *Solid State Phys.* **52**, 229 (1999).

GH adopted the Clausius–Mossotti relation in dielectrics to understand the physical essence of this transition, i.e.,

$$\frac{n^2-1}{n^2+2} = \frac{\alpha_{\mathrm{m}}}{\Omega_{\mathrm{m}}}, \tag{19.4.1}$$

where n is the index of refraction ($n = \sqrt{\epsilon}$ if $\mu = 1$), $\alpha_{\mathrm{m}} = \frac{4}{3}\pi N_{\mathrm{A}}\alpha$ is the molar polarization (N_{A} is the Avogadro constant, and Ω_{m} is the molar volume.

According to this criterion, if $\alpha_{\mathrm{m}}/\Omega_{\mathrm{m}} = 1$, then $n^2 - 1 = n^2 + 2$. This means that the high frequency dielectric constant $\varepsilon \to \infty$, and the valence electrons previously quasi-elastically bound to the interior of atoms are all delocalized to form a sea of free electrons; correspondingly there is a transition from the nonmetal to the metal. Figure 19.4.1 shows the related data for some elements, which illustrates that the criterion can be used to distinguish approximately metallic and nonmetallic elements.

The Mott criterion begins from the metallic state. If there is a positively charged impurity atom in a Fermi sea, free electrons affected by the Coulomb interactions give rise to a shielding effect to the positive charge, then the interaction between the impurity atom and an electron changes from the Coulomb potential to the Yukawa potential

$$-\frac{e^2}{r}\exp(-\lambda r).$$

If the density of free electrons is n, then according to semi-classical Thomas–Fermi theory, we have

$$\lambda^2 \simeq \frac{4me^2n^{1/3}}{\hbar^2}.$$

In general the Coulomb potential is strong enough to bind an electron. The radius of a hydrogen in the lowest energy state is the Bohr radius $a_0 = \hbar^2/me^2$. If the shielding radius $1/\lambda$ is smaller than a_0, then an electron cannot be bound by this state, Therefore there is a dividing line at

$$\lambda^{-1} \sim a_0,$$

i.e.,

$$n_c^{1/3}a_0 \sim 0.25, \tag{19.4.2}$$

where n_c is the critical electronic density. The basis of the Mott criterion is dependent on some data from impurity states and electron-hole pairs in semiconductors. This criterion can also be extended to some other substances, referring to Fig. 19.4.2. It is not necessary to bother too much about the precise numerical value of this criterion.

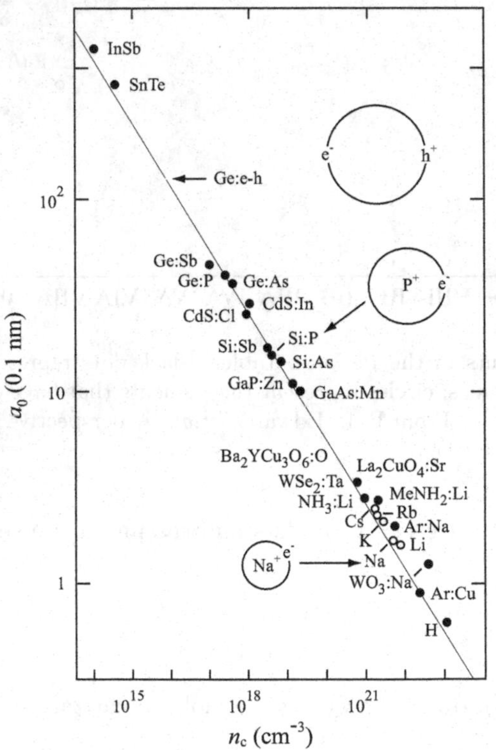

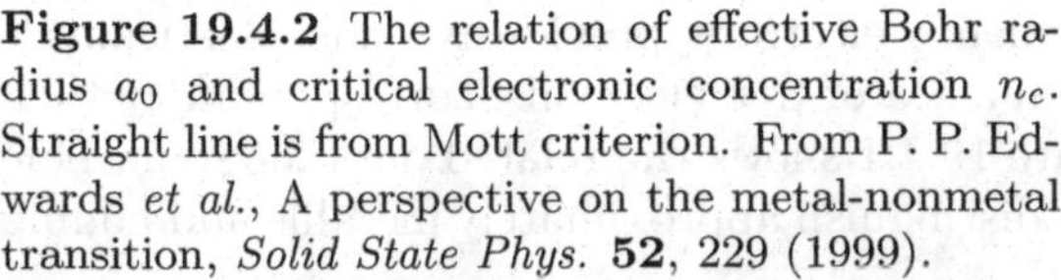
Figure 19.4.2 The relation of effective Bohr radius a_0 and critical electronic concentration n_c. Straight line is from Mott criterion. From P. P. Edwards *et al.*, A perspective on the metal-nonmetal transition, *Solid State Phys.* **52**, 229 (1999).

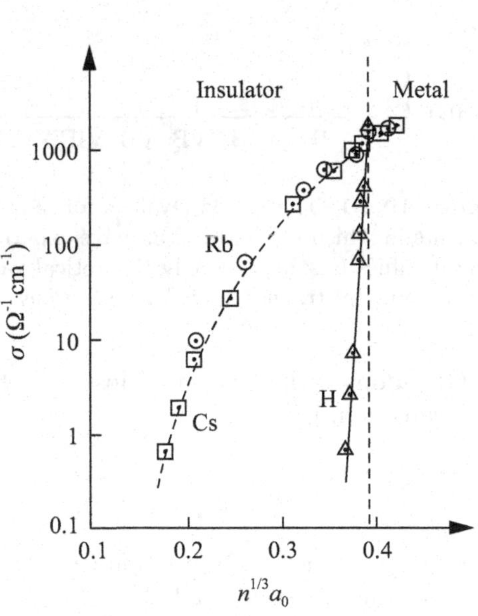

Figure 19.4.3 The relation of conductivity and Mott parameter ($n^{1/3}a_0$). From P. P. Edwards *et al.*, A perspective on the metal-nonmetal transition, *Solid State Phys.* **52**, 229 (1999).

We can also consider the Mott criterion from another aspect. Beginning from an insulator, the Coulomb potential in a dielectric is

$$V(r) = -\frac{e^2}{\epsilon r}, \tag{19.4.3}$$

where ϵ is the effective dielectric constant of the medium. When $n = n_c$, $\epsilon \to \infty$, certainly the effective electronic density is determined by the separation between neighboring atoms. In this way the Mott criterion is related to the GH criterion. If one has calculated n_c from the latter, we only need to modify the value by

$$n_c^{1/3} a_0 = 0.38. \tag{19.4.4}$$

The metal-nonmetal transitions in liquids of alkali metals, like Ce, Rb, etc., were confirmed experimentally many years ago. In recent years, it has been found that there is a similar transition in liquid hydrogen under high temperature and high pressure; the experimental data are displayed in Fig. 19.4.3. It is easy to see that transitions all appear at $n_c^{1/3} a_0 = 0.38$ and this indicates the effectiveness of both criteria and the similarity of condensed hydrogen with alkali metals, because they are all located in the group IA in the periodic table.

19.4.2 Wigner Crystallization

Wigner (1934, 1938) pointed out, for a low density electronic system, the Coulomb interaction dominates the kinetic energy, and the electrons may localize into a crystalline lattice. Electrons in such a phase do not move freely through the available space; they are confined to the sites of a lattice and their quantum mechanical zero-point motions are limited to small oscillations about their mean positions. We can expect a Wigner electron crystal has a insulating ground state, and long-range magnetic order, most probably Néel-like antiferromagnetism, may appear.

The electronic density $n = 3/4\pi {r_0}^3$ (r_0 is half of the average electronic separation; it can be written as $r_0 = r_s a_0$, a_0 is the Bohr radius) is an important parameter to determine the behavior of electronic systems in metals. Because the kinetic energy is proportional to ${r_0}^{-2}$, while the Coulomb repulsive potential is proportional to ${r_0}^{-1}$, it can be deduced that: In high densities, $r_0 \sim a_0$, the electronic kinetic energy dominates, Coulomb potential can be neglected, the behavior of the electronic system is like a free gas: In low densities, $r_0 \gg a_0$, the Coulomb repulsive potential surpasses the kinetic energy to play the principal role, and to reduce the Coulomb repulsive interactions, electrons will be separated from each other as far as possible; therefore they will be localized to form a Wigner crystal.

A simple theoretical treatment can begin from the jellium model, in which the homogeneous electron gas moves in a fixed uniform background of neutralized positive charges. We consider a Wigner–Seitz (WS) cell, approximately taken as spherical, and the effect from neighboring cells is neglected. In this way the computation of electric potential distribution at r from the center of the unit cell is a simple electrostatic problem. It is easy to get the result

$$\varphi(r) = \frac{3e}{2r_0} - \frac{er^2}{2{r_0}^3}. \tag{19.4.5}$$

The potential in the cell E_p includes two terms: One is the self interaction energy due to the homogeneously distributed positive charge E_s, expressed as

$$E_s = \frac{e}{2}\int_{r<r_0} 4\pi r^2 ne\varphi(r)dr = \frac{3}{5}\cdot\frac{e^2}{r_0}; \tag{19.4.6}$$

the other is the interaction energy of an electron at $r = 0$ with the background of positive charges, expressed as

$$E_i = -e\varphi(0) = -\frac{3}{2}\cdot\frac{e^2}{r_0}. \tag{19.4.7}$$

Because the electron is located in a parabolic potential well, its kinetic energy (zero point energy) compels it to oscillate around the center of the sphere to form an isotropic harmonic oscillator. Its oscillation frequency is

$$\omega = \left(\frac{e^2}{m{r_0}^3}\right)^{1/2}, \tag{19.4.8}$$

so the electronic zero point energy in the unit cell in three dimensions is

$$E_{\rm k} = \frac{3}{2}\hbar\omega = \frac{3}{2}\frac{e\hbar}{(m{r_0}^3)^{1/2}}. \tag{19.4.9}$$

It is clear when $r_0 \to \infty$ the kinetic energy for N electrons $NE_{\rm k}$ is proportional to ${r_0}^{-3/2}$, and not proportional to ${r_0}^{-2}$, as in high densities. This result means that the picture of the Fermi sphere in momentum space is no longer valid, and it also means that the abrupt jump of density of states (DOS) at the Fermi surface disappears.

Introducing the Rydberg constant $\text{Ry} = e^2/2a_0 = 13.61$ eV, the total energy of a Wigner crystal composed of N electrons is

$$N(E_{\rm i} + E_{\rm s} + E_{\rm k}) = N\left(-\frac{1.8}{r_{\rm s}} + \frac{3}{{r_{\rm s}}^{3/2}}\right) \text{Ry}. \tag{19.4.10}$$

According to the competition between kinetic and potential energy, and also the empirical data from ordinary crystal melting, the critical density for the instability of Wigner lattices is about $r_{\rm s} \sim 20$–100.

It is fairly obvious that the Wigner crystal is a charge density wave state with minimum potential energy. The Wigner lattice breaks the continuous translational symmetry of the underlying Hamiltonian. Hence there will be a critical temperature for the melting of the Wigner lattice into an 'electronic liquid' which, since the electrons are mobile again, will be metallic. Thus we should observe a thermodynamic phase transition from the Wigner insulator into a conducting state. When we consider the Wigner transition at finite temperature, it is convenient to define a ratio of the average Coulomb potential to the average thermal kinetic energy as

$$\Gamma \equiv \frac{e^2/r_s}{k_{\rm B}T}. \tag{19.4.11}$$

It was suggested when $\Gamma > 155$, electron crystallization would occur at sufficiently low density.

Although Wigner predicted the bcc electron crystal in three dimensions, it was first observed experimentally that a two-dimensional electron crystal appeared on the surface of liquid helium at electronic density $\rho = 4.4 \times 10^8$ cm^{-2}, $T = 0.42$–0.46 K. Changing the temperature, the melting curve was obtained. The phase boundary in Fig. 19.4.4 is determined by

$$\Gamma_m = \frac{\pi^{1/2}e^2\rho^{1/2}}{k_{\rm B}T} \sim 131 \pm 7. \tag{19.4.12}$$

In fact, low-dimensional systems are more effective to give rise to a Wigner transition, due to stronger localization in lower dimensions since the average kinetic energy as a function of density vanishes more rapidly. Hence, the Coulomb interaction is effectively stronger in low dimensions because of the restricted phase space. We may further reduce the effective dimension in a two-dimensional electron gas by applying a strong magnetic field, which restricts the motion of the electrons to the lowest Landau level such that we may hope to observe the Wigner insulator experimentally. The evidence for a Wigner crystal in high magnetic fields was obtained in a two-dimensional electron gas. Its existence has been confirmed by various methods, such as measurement of the conductivity and of time-resolved photoluminescence. Near fractional filling factors, the problem for its observation is the proximity of the Wigner crystal to a correlated quantum liquid, the fractional quantum Hall state. The experimental phase diagram of the two-dimensional electron gas in

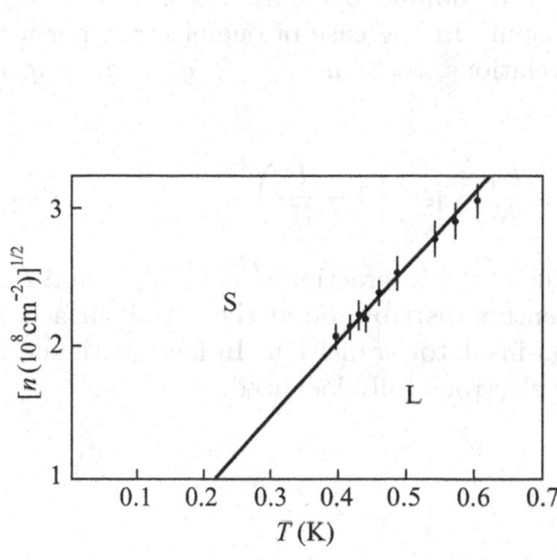

Figure 19.4.4 Melting curve for the two-dimensional electronic system on the surface of liquid helium. From C. C. Grimes and G. Adams, *Phys. Rev. Lett.* **42**, 795 (1979).

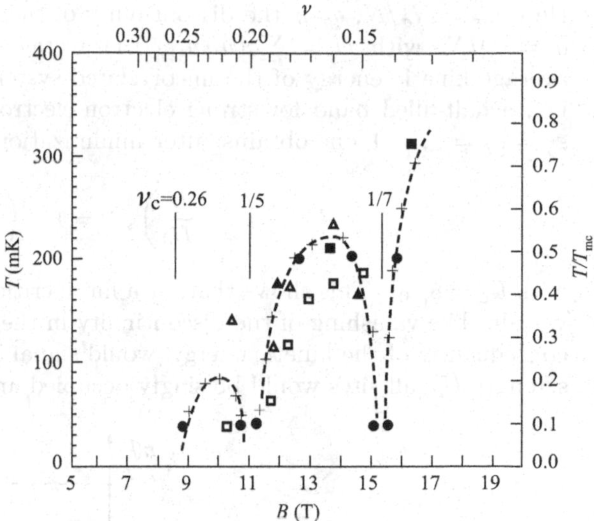

Figure 19.4.5 Phase diagram of a two-dimensional electron gas. The filling factor is the number of electrons per magnetic flux line. From I. V. Kukushkin *et al.*, *Europhys. Lett.* **23**, 211 (1993).

GaAs/AlGaAs heterostructures in a strong magnetic field is shown in Fig. 19.4.5. The system is a correlated electron liquid above the dotted curves, and a Wigner crystal below the curves.

19.4.3 Gutzwiller Variation Method and Phenomenological Treatment of Mott Transition

As already discussed in §13.1, the Hubbard energy U disfavors double occupancy of the same lattice site. In Gutzwiller's variational approach, the ground state wavefunction can be approximately written as

$$|\Psi\rangle = \prod_i (1 - g n_{i\uparrow} n_{i\downarrow})|\Psi_0\rangle, \tag{19.4.13}$$

where $|\Psi_0\rangle$ is the uncorrelated free electron ground state wavefunction, such as a Slater determinant, and g is a variational parameter which takes into account the reduction of double occupancy of electrons at the same site. Clearly, in the wavefunction Ψ, the compartments containing doubly-occupied sites are reduced by a fractional amount g $(0 \le g \le 1)$ with respect to their value in the uncorrelated wavefunction Ψ_0. It is straightforward to show that for $g = 0$ one simply regains the uncorrelated state, which is of course the exact ground state for zero on-site repulsion, $U = 0$. On the other hand, $g = 1$ corresponds to a fully correlated wavefunction in which the compartments containing doubly occupied sites are suppressed for $U = \infty$. For finite repulsion, one will have $0 < g < 1$, since the effect of correlation is precisely to reduce the number of doubly occupied sites present in the uncorrelated state.

The ground state energy

$$E_0(g) = \frac{\langle\Psi|\mathcal{H}|\Psi\rangle}{\langle\Psi|\Psi\rangle} \tag{19.4.14}$$

is estimated by optimization of the parameter g. This estimate satisfies the variational principle and hence gives an upper bound on the ground-state energy.

Within Gutzwiller's approximation, the average of (19.4.10) with the Hubbard Hamiltonian can be written as

$$\frac{E_0}{N} = n_\uparrow q_\uparrow \varepsilon_\uparrow + n_\downarrow q_\downarrow \varepsilon_\downarrow + Ud. \tag{19.4.15}$$

Here $n_\sigma = N_\sigma/N$, q_σ is the discontinuity of the momentum distribution $n_{\boldsymbol{k}\sigma}$ at the Fermi surface, $d = D/N$, with $D = \sum_i \langle n_{i\uparrow} n_{i\downarrow} \rangle$ the average number of doubly occupied sites, and ε_σ is the average kinetic energy of the uncorrelated system per spin. In the case of one electron per atom, i.e., a half-filled band for strong electron-electron correlations, $n_\uparrow = n_\downarrow = 1/2$, $q_\uparrow = q_\downarrow = q$, and $\varepsilon_\uparrow = \varepsilon_\downarrow = \varepsilon_0 < 0$, one obtains, after minimization

$$d = \frac{1}{4}\left(1 - \frac{U}{U_c}\right), \quad q = 1 - \left(\frac{U}{U_c}\right)^2, \quad \frac{E_0}{N} = |\varepsilon_0| \left(1 - \frac{U}{U_0}\right)^2, \tag{19.4.16}$$

with $U_c = 8|\varepsilon_0|$. This shows that, at a finite critical value of the interaction $U = U_c$, d, q, and E_0 all vanish. The vanishing of the discontinuity in the momentum distribution at the Fermi surface, and consequently of the kinetic energy, would signal a metal-insulator transition. In fact, at the critical strength U_c, all sites would be singly occupied and the electrons fully localized.

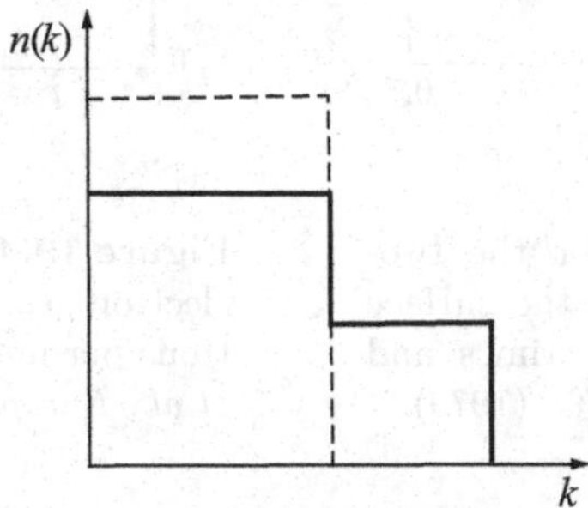

Figure 19.4.6 Ground state momentum distribution. The dotted lines denote the free electron gas, while the solid lines are in the Gutzwiller correlation wavefunction.

Gutzwiller's calculation gave a step function for the momentum distribution, as shown in Fig. 19.4.6; actually, it can be proved that the discontinuity at k_F decreases with increasing correlation. A phenomenological treatment proposed by March *et al.*[f] parallels some features of the Gutzwiller theory: We can take the discontinuity q at the Fermi surface of the single particle occupation number $n(k)$ as a generalized order parameter. In the metallic phase, q has a finite value, while $q = 0$ represents an insulating state. U is a tuning parameter; when U approaches the critical strength U_c, $q = 0$, a metal-insulator transition takes place.

The ground-state energy $E(q)$ can be expanded as a series of q about $q = 0$,

$$E(q) = E_0 + E_1 q + E_2 q^2 + \cdots. \tag{19.4.17}$$

The form of E_1 is assumed

$$E_1(U) = \alpha(U - U_c), \text{ for } \alpha > 0, \tag{19.4.18}$$

i.e., $E_1(U_c) = 0$, while $E_2(U_c) > 0$. The energy minimum is determined by $\partial E/\partial q = 0$; it evidently yields

$$E_1 + 2E_2 q + \cdots = 0. \tag{19.4.19}$$

Near the metal-insulator transition for very small q,

$$q = -\frac{E_1}{2E_2} = Q\left(1 - \frac{U}{U_c}\right), \quad Q = \frac{\alpha U_c}{2E_2}. \tag{19.4.20}$$

Substituting q into $E(q)$, we have

$$E_{\min} = E_0 + (-\alpha U_c Q + E_2 Q^2)\left(1 - \frac{U}{U_c}\right)^2. \tag{19.4.21}$$

[f] N. H. March, M. Suzuki, and M. Parrinello, *Phys. Rev. B* **19**, 2027 (1979).

The difference in energy between metal and insulator states is

$$\Delta E = (-\alpha U_c Q + E_2 Q^2)\left(1 - \frac{U}{U_c}\right)^2 \tag{19.4.22}$$

and

$$d = \frac{d(\Delta E)}{dU} = -\frac{2}{U_c}(-\alpha U_c Q + E_2 Q^2)\left(1 - \frac{U}{U_c}\right). \tag{19.4.23}$$

These dependences of q, ΔE, and d on $1 - U/U_c$ are equivalent to those given by Gutzwiller's variational calculation.

Furthermore, the spin susceptibility near the metal-insulator transition can be obtained. Take h as an applied magnetic field; then we may write an energy function as

$$E(m,q) = E_0(m) + am^2 + E_1 q + E_2 q^2 + bqm^2 - hm. \tag{19.4.24}$$

Taking the energy minimum $\partial E(m,q)/\partial m = 0$, we have

$$2am + 2bqm = h, \tag{19.4.25}$$

which gives

$$\chi = \frac{m}{h} = \frac{2}{a(U) + bq}, \tag{19.4.26}$$

as $U \to U_c$, $q \to 0$, if $a(U) \to 0$ as $1 - U/U_c$ or faster, then

$$\chi \propto \frac{1}{1 - U/U_c}.$$

The susceptibility diverges at the critical strength.

19.4.4 Electron Glass

The behavior of non-interacting electrons moving in a random potential and the related phenomena of localization have been studied in Chap. 9. However, the spatial distribution of localized electrons in a disordered solid may be strongly influenced by the long-ranged Coulomb repulsion between them. This will cause a depletion of the single particle density of states (DOS) near the Fermi energy. At zero temperature the density of states vanishes at the Fermi energy but is non-zero elsewhere. This is known as the Coulomb gap, and at low temperature it leads to deviations from Mott's $T^{1/4}$ law for electric conduction by variable-range hopping, since this law was derived by assuming a constant density of states (DOS) near the Fermi level, and neglecting electron-electron interaction.

The interplay of the electron-electron interactions and disorder is particularly evident deep on the insulating side of metal-insulator transition. Both experimental and theoretical studies have demonstrated that this leads to the formation of a soft 'Coulomb gap', a phenomena that is believed to be related to the glassy behavior of electrons. The classic work of A. L. Efros and B. I. Shklovskii in 1975 has clarified some basic aspects of this behavior.[g] A simplified spinless Hamiltonian for a system of localized electrons can be adopted in the form

$$\mathcal{H} = \sum_i \varepsilon_i n_i + \frac{1}{2}\sum_{i \neq j} \frac{e^2}{\epsilon r_{ij}} n_i n_j, \tag{19.4.27}$$

where n_i, equal to zero or one, is the occupation number, ε_i the on-site energy, ϵ the dielectric constant, and r_{ij} the distance between the states i and j. An excellent practical realization of such a system is the impurity band of a lightly doped, compensated semiconductor, where the disorder arises from the random distribution of impurities over the host lattice sites.

[g] A. L. Efros, B. I. Shklovskii, *J. Phys. C* **8**, L49, (1975); A. L. Efros, *J. Phys. C* **9**, 2021 (1976).

It is reasonable to introduce the energies of single-site excitations

$$E_i = \varepsilon_i + \sum_j \frac{e^2}{r_{ij}} n_j. \tag{19.4.28}$$

At zero temperature, $n_i = 1$ for $E_i < E_{\rm F}$ and $n_i = 0$ for $E_i > E_{\rm F}$, where $E_{\rm F}$ is the Fermi level. The ground state of the system should also satisfy another condition: For two states i and j, which in the ground state are occupied and unoccupied respectively, the transfer of an electron from state i to j should increase the energy of the system. The energy increment is

$$\Delta E_{ji} = E_i - E_j - \frac{e^2}{\epsilon r_{ij}} > 0, \tag{19.4.29}$$

where the last term describes the attraction of the created electron-hole pair, and its presence causes the Coulomb gap. So in the ground state any two energies E_i and E_j separated by the Fermi level should satisfy the inequality (19.4.25).

Now we can show that the density of states (DOS) $g(\varepsilon)$ should vanish at the Fermi level. We assume $g(E_{\rm F}) = g_0$ and consider an energy interval of small width ε centered at the Fermi level. For this interval, an average distance R between the states is determined by the condition $g_0 R^3 \varepsilon \approx 1$ and equals $(g_0\varepsilon)^{-1/3}$. If $\varepsilon \ll \Delta = e^3 g_0^{1/2}/\epsilon^{3/2}$ the interaction energy of the states $e^2/\epsilon R = (e^2/\epsilon)(g_0\varepsilon)^{1/3}$ exceeds ε, and the inequality (19.4.25) inevitably breaks down. Thus a constant density of states contradicts the inequality (19.4.25), and $g(\varepsilon)$ at $|E - E_{\rm F}| < \Delta$ should decrease with $E - E_{\rm F}$ and vanish at the Fermi level.

A self-consistent density of states near the Fermi level may be found from the condition that for any $\varepsilon < \Delta$, the mean interaction energy $e^2/\epsilon r_{ij}$ of the states within the ε interval has to be of the order of ε. In other words, the average distance between the states in the ε interval has to be of the order of $e^2/\epsilon\varepsilon$, i.e., $g(\varepsilon)(e^2/\epsilon\varepsilon)^3\varepsilon \approx 1$, so $g(\varepsilon) \approx (\epsilon^3/e^6)\varepsilon^2$. For the two-dimensional case the same arguments give $g(\varepsilon) \approx (\epsilon^2/e^4)|\varepsilon|$. We can finally write the density of states (DOS) for a d-dimensional system in a common formula

$$g(E) \propto |E - E_{\rm F}|^{d-1}, \tag{19.4.30}$$

which really characterizes the Coulomb gap. The Coulomb gap is a "soft" gap, meaning that it vanishes only at $E = E_{\rm F}$. It is produced by the long-range Coulomb forces and is therefore to be distinguished from a Hubbard gap, which is due to short-range forces. Only in the case of one electron per site does a Hubbard gap separate a filled and an empty band. It is different from the present model, because of the restriction $n_i = 0$ or 1, so the chemical potential always lies in the lower Hubbard band.

Numerical calculations based on the time averaging Monte Carlo method were carried out for the sites put on a square or simple cubic lattice.[h] The disorder is introduced by selecting the site energies ε_i from a rectangular distribution of width 2. The number of electrons was one-half the number of sites, and to maintain electrical neutrality each site had charge $+1/2$. The single particle density of states is shown in Fig. 19.4.7 for two- and three-dimensional systems. The Coulomb gap can be clearly seen. It is noted that at low temperature, single particle excitations might not be bare, like those considered above, but rather might be electron polarons. If an electron is added to a site, the system can relax, if nearby electrons move away thereby creating a polarization cloud. The Coulomb gap for polarons is clearly much narrower than that for bare excitations.

The Coulomb gap is a property of the ground state and therefore exists strictly only at zero temperature. In a non-interacting system, the occupation of states changes as the temperature rises, but not their energies. In the presence of interactions, a change in the occupation of one state alters the energies of all the others; consequently the form of the density of states, as well as its occupation, is altered. The finite temperature must have an effect on $g(E)$; the numerical results for two dimensions are shown in Fig. 19.4.8. At the lowest values of $T = 0.05$ K, $g(E)$ has a well-defined

[h] For a detailed discussion, see J. H. Davies, P. A. Lee, T. M. Rice, *Phys. Rev. Lett.* **49**, 758 (1982); *Phys. Rev. B* **29**, 4260 (1984).

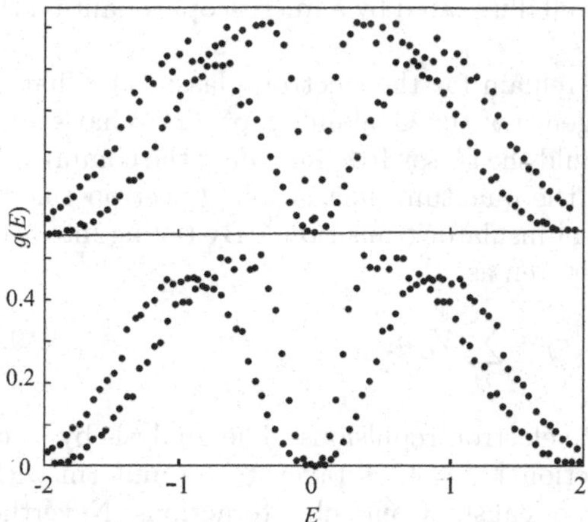

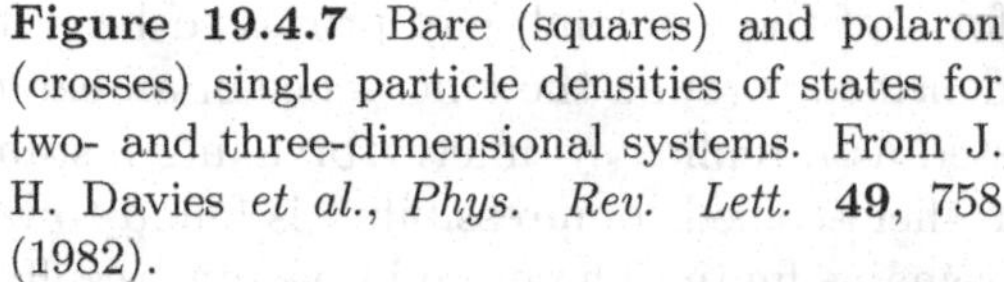
Figure 19.4.7 Bare (squares) and polaron (crosses) single particle densities of states for two- and three-dimensional systems. From J. H. Davies *et al.*, *Phys. Rev. Lett.* **49**, 758 (1982).

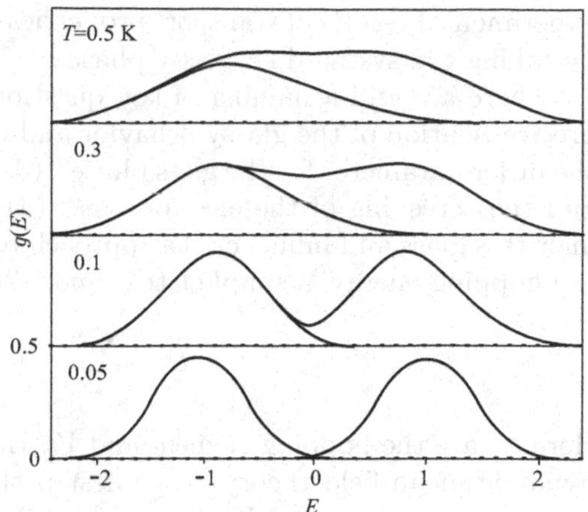

Figure 19.4.8 Bare single particle density of states for a two-dimensional system at various temperatures. The lower curve in each cases shows the density of occupied states. From J. H. Davies *et al.*, *Phys. Rev. Lett.* **49**, 758 (1982).

gap; this gap closes gradually as the temperature is raised. At $T = 0.5$ K there is no gap at all, and the Coulomb gap is washed out at $T = 0.3$ K. This is something like a transition temperature. As an example, in silicon doped with donors at about 1% of the concentration at the metal-insulator transition, a typical Coulomb energy between nearby impurities is a few meV, so the Coulomb gap should be important at temperatures below about 10 K.

The ground state shows behavior typical of a glass. It is instructive to rewrite the lattice-gas Hamiltonian (19.4.23) in an Ising form. Setting $\sigma_i = n_i - 1/2$, we get for the half-filled band

$$\mathcal{H} = \sum_i \varepsilon_i \sigma_i + \frac{1}{2} \sum_{i \neq j} \frac{e^2}{\epsilon r_{ij}} \sigma_i \sigma_j, \tag{19.4.31}$$

where the site energies are now distributed about zero energy. In this form, the Hamiltonian is that of an Ising model in a random field, but with long-range antiferromagnetic interactions. This is analogous to an Ising spin glass model, so the effective methods for spin glass may be useful here. For example, a modified order parameter from the spin glass can be defined as

$$q = 2\langle [\langle \sigma_i \rangle_t - f(\varepsilon_i)]^2 \rangle_S, \tag{19.4.32}$$

where the t denotes the time average and S an average over samples, $f(\varepsilon_i)$ is the average value of the spin at site i in the absence of interaction. The non-zero value of the order parameter q at low temperature suggests a glass transition occurs.

A mean field approach was used to investigate the electronic properties of a three-dimensional system of localized Coulomb-interacting electrons.[i] There were a large number of metastable states found by solving the mean-field equation. After 'heating up' the system and subsequently slowly cooling down, lower metastable states are reached. Akin to spin glasses, the calculation results for specific heat, susceptibility and order parameter strongly support the existence of an electron glass transition.

In a series of experiments on high-mobility two-dimensional electron gases, a metal-insulator transition was found in strong magnetic fields. Depending on the degree of disorder, three different mechanisms for the electron localization are possible: First, the formation of an electron crystal which is expected for the clean system; second, the formation of a 'glassy' electron solid at intermediate disorder, and third, an Anderson-type single particle localization at dominant disorder. The

[i] M. Grünewald *et al.*, *J. Phys. C* **15**, L1153 (1982).

experimental results of transport properties may be well illustrated by a microscopic quantum theory by taking the system in a glassy phase.[j]

There are still a number of key questions which remain for the electron glass: (1) What is the precise relation of the glassy behavior and the emergence of the Coulomb gap? (2) What should be the order parameter for the glass phase? (3) How should the glassy freezing affect the compressibility and the screening of the electron gas? (4) How do the quantum fluctuations (electron tunneling) melt this glass and influence the approach to the metal-insulator transition.[k] By taking into account the hopping energy, a simple lattice model can be written as

$$\mathcal{H} = \sum_i \varepsilon_i n_i - \sum_{ij} t_{ij} c_i^\dagger c_j + \sum_{ij} V_{ij} n_i n_j. \tag{19.4.33}$$

Here t_{ij} are the hopping element and V_{ij} the intersite electron repulsions. The analysis by so called dynamic mean-field theory is simplest if the interaction $V_{ij} = V$ is taken to be uniform within a volume containing z neighbors, as opposed to the more realistic Coulomb interactions. Nevertheless, most of the qualitative features of the Coulomb glass are still captured if we consider the coordination number $z \gg 1$. Within this model, the universal form of the Coulomb gap proves to be a direct consequence of glassy freezing. The glass phase is identified through the emergence of an extensive number of metastable states, which can be manifested as a replica symmetry breaking instability. As a consequence of this ergodicity breaking, the zero-field-cooled compressibility is found to vanish at $T = 0$, in contrast to the field-cooled one, which remains finite. The more important result from this model is that quantum fluctuations can melt this glass at $T = 0$. If the hopping elements are taken as $t_{ij} = t$, the disorder strength is W, the theoretical phase diagram is shown in Fig. 19.4.9.

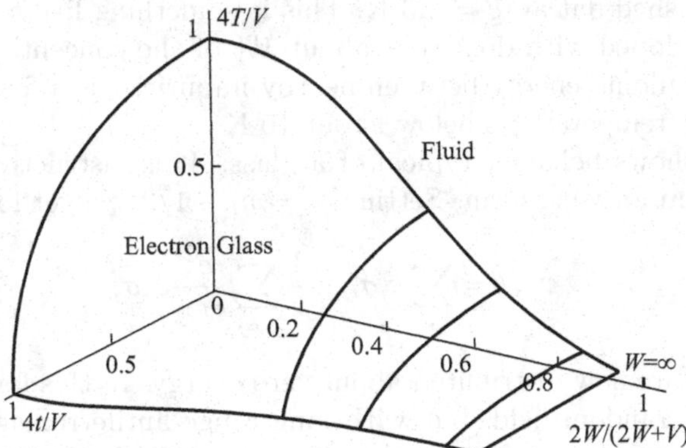

Figure 19.4.9 Phase diagram as a function of temperature T, disorder strength W, and hopping element t. From A. A. Pastor and V. Dobrosavljevic, *Phys. Rev. Lett.* **83**, 4642 (1999).

Bibliography

[1] Anderson, P. W., *Amorphous Systems*, in *Ill-condensed Matter*, Les Houches, Sessions XXXI (eds. Balian, R. *et al.*), North-Holland, Amsterdam (1979).

[2] Palmer, R. G., *Broken ergodicity*, Adv. Phys. **31**, 669 (1982).

[3] Kubo, K. *et al.*, *Statistical Mechanics*, North-Holland, Amsterdam (1965).

[4] Reichl, L. E., *A Modern Course in Statistical Physics*, University of Texas Press (1980).

[5] Gonzalo, J. A., *Effective Field Approach to Phase Transitions and Some Applications to Ferroelectrics*, World Scientific, Singapore (1991).

[j] U. Wulf, J. Kucera, and E. Sigmund, *Phys. Rev. Lett.* **77**, 2993 (1996).

[k] A. A. Pastor and V. Dobrosavljevic, *Phys. Rev. Lett.* **83**, 4642 (1999).

[6] de Gennes, P. G., *Wetting: Statics and dynamics*, Rev. Mod. Phys. **57**, 827 (1985).

[7] Dietrich, S., *Wetting phenomena*, Phase Transitions and Critical Phenomena **12**, 1 (1988).

[8] Zallen, R., *The Physics of Amorphous Solids*, John Wiley & Sons, New York (1983).

[9] Mydosh, J. A., *Spin Glasses: An Experimental Introduction*, Taylor & Francis, London (1993).

[10] D. Sherrington, *Spin Glasses*, in *Electronic Phase Transitions* (eds. W. Hanke and Y. V. Kopaev) (Modern Problems in Condensed Matter Sciences **32**), North-Holland, Amsterdam (1992).

[11] S. Sondhi *et al.*, *Continuous quantum phase transition*, Rev. Mod. Phys. **69**, 315 (1997).

[12] Fulde, F., *Electron Correlations in Molecules and Solids*, Spinger-Verlag, Berlin,.

[13] March, N. H., *Electron Correlation in Molecules and Condensed Phases* (1996).

[14] Edwards, P. P., R. L. Johnston, F. Hensel, C. N. R. Rao, and D. P. Tunstall, *A perspective on the metal-nonmetal transition*, Solid State Phys. **52**, 229 (1999).

[15] Gebhard, F., *The Mott Metal-Insulator Transition*, Springer-Verlag, Berlin (1997).

[16] Rice, T. M., *Electron Glass*, in *Highlights of Condensed Matter Theory* (eds. F. Bassani, F. Fumi, and M. P. Tosi) North-Holland, Amsterdam (1985).

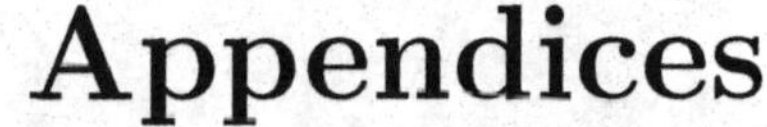

Appendices

Appendix A

Units and Their Conversion

Measured physical quantities are expressed in units. There are three systems of units commonly used in physics and technology. The Gaussian cgs system traditionally used in treatises and textbooks of physics; the international SI system, established by the 11th General Conference on Weights and Measures in 1960, now widely used in physical measurements and technological literature; besides, there is the atomic system used sometimes in some theoretical texts on microscopic physics. A system of units is chosen for convenience and clarity. A system of units once chosen affects not only the values of physical quantities obtained from experimental measurements, but also the forms of formulas relating various physical quantities. We may choose Maxwell equations with related formulas and Schrödinger equation for an atom with Z electrons to illustrate this point (Tables A.1 and A.2).

A glance at the tables listed below may convince us that the simplest formulation in use is the atomic system of units. This system of units by setting many universal constants, such as e (electron charge), m (electron mass), $\hbar$ ($\hbar = h/2\pi$, h Planck constant), and c (velocity of light in vacuo) to unity; thus it achieves simplicity at the expense of physical clarity. So it seems to be a reserved domain for the initiated, unsuitable for an introductory text on condensed matter physics. In comparison, the Gaussian formulation appears to be simpler in the case of Schrödinger equation for an atom, while the SI formulation appears to be simpler in the case of Maxwell equations. But by further scrutiny the latter simplicity is somewhat deceptive due to the fact that the constitutive relations become more complicated with the introduction of two artificially contrived universal constants, i.e., the dielectric constant in vacuo ϵ_0 and magnetic permeability in vacuo μ_0:

$$\epsilon_0 = 10^7/4\pi c^2 = 8.854188 \text{ Fm}^{-1}, \; \mu_0 = 4\pi \times 10^7 = 12.56637 \text{ NA}^{-2}.$$

These two universal constants with artificial dimensionalities are introduced without any justification from microscopic physics, for in the vacuum, there is actually no difference between $\boldsymbol{E}$ and $\boldsymbol{D}$, or $\boldsymbol{B}$ and $\boldsymbol{H}$. The only pretext for their introduction is to integrate this system of units with those practical units already widely used for electromagnetic measurements in laboratories. Considering that the Schrödinger equation and Coulomb potential as well as the constitutive relations for the electromagnetic media occupy important places in condensed matter physics, we naturally choose the Gaussian system of units to formulate our equations. This just follows the tradition of most textbooks and treatises on solid state physics and condensed matter physics. However, we are not purists in our use of units; in our text we do not shy away from using some familiar SI units when listing experimental results, as well as some mixed units already widely used in scientific literature, for instance, to specify electrical resistivity of materials in units of ohm-cm, or using eV as an unit for energy,

$$1 \text{ eV} = 1.60219 \times 10^{-12} = 1.60219 \times 10^{-19} \text{ J}, \; 1 \text{ eV}/h = 2.41797 \times 10^{14} \text{ Hz},$$
$$1 \text{ eV}/hc = 8.06546 \times 10^3 \text{ cm}^{-1} = 8.06546 \times 10^5 \text{ m}^{-1}, \; 1 \text{ eV}/k_B = 1.16048 \times 10^4 \text{ K}.$$

Also SI prefixes (Table A.3) have been widely used. As to conversion of units from one system to another, readers may consult Table A.4. Table A.5 gives the values of some fundamental physical constants in both Gaussian and SI units. The main difficulty in conversion of units lies in magnetic units. Our recommendation is to always use $\boldsymbol{B}$-field in SI units, i.e., using $\boldsymbol{B}_0 = \mu_0 \boldsymbol{H}$ in free space instead of $\boldsymbol{H}$, so we may easily convert units with the relation: 1 tesla (T) = 10^4 gauss (G). For this problem, one may consult J. Crangle and M. Gibbs, *Phys. World* **6**, November, 31 (1994).

§A.1 Maxwell Equations and Related Formulas in Three Systems of Units

	Maxwell equation	Constitutive relations	Lorentz force
Gaussian	$\nabla\times\boldsymbol{H} = \frac{4\pi}{c}\boldsymbol{j} + \frac{1}{c}\frac{\partial \boldsymbol{D}}{\partial t}, \nabla\cdot\boldsymbol{D} = 4\pi\rho$	$\boldsymbol{D} = \boldsymbol{E} + 4\pi\boldsymbol{P} = \epsilon\boldsymbol{E}$	$\boldsymbol{F} = q\left(\boldsymbol{E} + \frac{\boldsymbol{v}}{c}\times\boldsymbol{B}\right)$
	$\nabla\times\boldsymbol{E} + \frac{1}{c}\frac{\partial \boldsymbol{B}}{\partial t} = 0, \nabla\cdot\boldsymbol{B} = 0$	$\boldsymbol{H} = \boldsymbol{B} - 4\pi\boldsymbol{M} = \frac{\boldsymbol{B}}{\mu}$	
SI	$\nabla\times\boldsymbol{H} = \boldsymbol{j} + \frac{\partial \boldsymbol{D}}{\partial t}, \nabla\cdot\boldsymbol{D} = \rho$	$\boldsymbol{D} = \epsilon_0\boldsymbol{E} + \boldsymbol{P} = \epsilon\epsilon_0\boldsymbol{E}$	$\boldsymbol{F} = q(\boldsymbol{E} + \boldsymbol{v}\times\boldsymbol{B})$
	$\nabla\times\boldsymbol{E} + \frac{\partial \boldsymbol{B}}{\partial t} = 0, \nabla\cdot\boldsymbol{B} = 0$	$\boldsymbol{H} = \frac{1}{\mu_0}\boldsymbol{B} - \boldsymbol{M} = \frac{\boldsymbol{B}}{\mu}\mu_0$	
Atomic	$\nabla\times\boldsymbol{H} = 4\pi\boldsymbol{j} + \frac{\partial \boldsymbol{D}}{\partial t}, \nabla\cdot\boldsymbol{D} = 4\pi\rho$	$\boldsymbol{D} = \boldsymbol{E} + 4\pi\boldsymbol{P} = \epsilon\boldsymbol{E}$	$\boldsymbol{F} = q(\boldsymbol{E} + \boldsymbol{v}\times\boldsymbol{B})$
	$\nabla\times\boldsymbol{E} + \frac{\partial \boldsymbol{B}}{\partial t} = 0, \nabla\cdot\boldsymbol{B} = 0$	$\boldsymbol{H} = \boldsymbol{B} - 4\pi\boldsymbol{M} = \frac{\boldsymbol{B}}{\mu}$	

§A.2 Schrödinger Equation for a Many-Electron Atom in Three Systems of Units

Gaussian	$\left(-\frac{\hbar}{2m}\sum_{i=1}^{Z}\nabla_i^2 + \sum_{i=1}^{Z}\frac{Ze^2}{\boldsymbol{r}_i} + \frac{1}{2}\sum_{i\neq j}^{z}\frac{e^2}{\boldsymbol{r}_i - \boldsymbol{r}_j}\right)\Psi = E\Psi$
SI	$\left(-\frac{\hbar}{2m}\sum_{i=1}^{Z}\nabla_i^2 + \sum_{i=1}^{Z}\frac{Ze^2}{4\pi\epsilon_0\boldsymbol{r}_i} + \frac{1}{2}\sum_{i\neq j}\frac{e^2}{4\pi\epsilon_0(\boldsymbol{r}_i - \boldsymbol{r}_j)}\right)\Psi = E\Psi$
Atomic	$\left(-\frac{1}{2}\sum_{i=1}^{Z}\nabla_i^2 + \sum_{i=1}^{Z}\frac{Z}{\boldsymbol{r}_i} + \frac{1}{2}\sum_{i\neq j}\frac{1}{\boldsymbol{r}_i - \boldsymbol{r}_j}\right)\Psi = E\Psi$

§A.3 SI Prefixes

Factor	10^{-18}	10^{-15}	10^{-12}	10^{-9}	10^{-6}	10^{-3}	10^3	10^6	10^9	10^{12}	10^{15}	10^{18}
Prefix	atto	femto	pico	nano	micro	milli	kilo	mega	giga	tera	peta	exa
Symbol	a	f	p	n	μ	m	k	M	G	T	P	E

§A.4 Conversion of Units Between SI and Gaussian Systems

Physical Quantity	Symbol	SI	Gaussian
Length	l	1 meter (m)	10^2 centimeter (cm)
Mass	m	1 kilogram (kg)	10^3 gram (g)
Time	t	1 second (s)	1 second (s)
Force	F	1 newton (N)	10^5 dyne
Pressure (Stress)	p	1 pascal (Pa)	0.9868×10^{-5} atm
Work	W	1 joule (J)	10^7 erg
Energy	U	1 joule	10^7 erg
Power	P	1 watt (W)	10^7 erg s^{-1}
Charge	q	1 Coulomb (C)	3×10^9 esu
Charge density	ρ	1 coul m^{-3}	3×10^3 esu cm^{-3}
Current	I	1 ampere (A)	3×10^9 esu
Current density	j	1 A m^{-2}	3×10^5 esu cm^{-2}
Electric field	$\boldsymbol{E}$	1 V m^{-1}	$\frac{1}{3} \times 10^{-4}$ esu cm^{-1}
Potential	φ V	1 volt(V)	$\frac{1}{300}$ esu
Electric polarization	$\boldsymbol{P}$	1 C m^{-2}	3×10^5 esu cm^{-2}
Electric displacement	$\boldsymbol{D}$	1 C m^{-2}	$12\pi \times 10^5$ esu cm^{-2}
Conductivity	σ	1 ohm $m^{-1}(\Omega^{-1}m^{-1})$	9×10^9 s^{-1}
Resistance	R	1 ohm (Ω)	$\frac{1}{9} \times 10^{-11}$ s^{-1}
Capacitance	C	1 farad (F)	9×10^{11} cm
Magnetic flux	Φ	1 weber (Wb)	10^8 gauss cm^2(G cm^2)
Magnetic induction	$\boldsymbol{B}$	1 tesla (T)	10^4 gauss (G)
Magnetic field	$\boldsymbol{H}$	1 ampere-turn m^{-1}(Am^{-1})	$4\pi \times 10^{-3}$ oersted (Oe)
Magnetic polarization	$\boldsymbol{J} = \mu_0 \boldsymbol{M}$	1 tesla (T)	$\frac{1}{4\pi} \times 10^4$ gauss (G)
Magnetization per unit volume	$\boldsymbol{M}$	1 JT^{-1} m^{-3}	10^{-3} erg Oe^{-1} cm^{-3}
Magnetic susceptibility per unit volume	χ	1 JT^{-2} m^{-3}	10^{-4} Oe^{-2} cm^{-3}
Inductancc	L	1 henry (H)	$\frac{1}{9} \times 10^{-11}$ Gaussian unit

Note: All factors of 3 which occurs in conversion factors (such as 3, 9, 12, 1/3, 1/9/, 1/300) except those in the exponentials should be replaced by 2.997825 for accurate work.

§A.5 Fundamental Physical Constants

Quantity	Gaussian	SI
Electron charge e	$4.803207 \cdot 10^{-10}$ esu	$1.602176 \cdot 10^{-19}$ C
Speed of light c	$2.997925 \cdot 10^{10}$ cm s^{-1}	$2.997925 \cdot 10^{8}$ m s^{-1}
Planck constant $\hbar = h/2\pi$	$1.054572 \cdot 10^{-27}$ erg	$1.054572 \cdot 10^{-34}$ J
	$6.582122 \cdot 10^{-16}$ eV	$6.582122 \cdot 10^{-16}$ eV
h	$6.626068 \cdot 10^{-27}$ erg	$6.626068 \cdot 10^{-34}$ J
	$4.135663 \cdot 10^{-15}$ eV	$4.135663 \cdot 10^{-15}$ eV
Electron mass m	$9.109382 \cdot 10^{-28}$ g	$9.109382 \cdot 10^{-31}$ kg
Mass ratio(proton/electron) M_p/m	$1.836153 \cdot 10^{3}$	$1.836153 \cdot 10^{3}$
Mass ratio(neutron/electron) M_n/m	$1.838684 \cdot 10^{3}$	$1.838684 \cdot 10^{3}$
Boltzmann constant k_{B}	$1.380650 \cdot 10^{-16}$ erg K^{-1}	$1.380650 \cdot 10^{-23}$ J K^{-1}
	$8.617385 \cdot 10^{-5}$ eV K^{-1}	$8.617385 \cdot 10^{-5}$ eV K^{-1}
Avogadro constant N_{A}	$6.022142 \cdot 10^{23}$ mol^{-1}	$6.022136 \cdot 10^{23}$ mol^{-1}
Rydberg constant Ry	13.605698 eV	13.605 698 eV
Bohr radius $a_0 = \hbar^2/me^2$	$5.291772 \cdot 10^{-9}$ cm	$5.291772 \cdot 10^{-11}$
Bohr magneton $\mu_{\mathrm{B}} = e\hbar/2mc$	$9.274009 \cdot 10^{-21}$ erg G^{-1}	$9.274009 \cdot 10^{-24}$ J T^{-1}
Magnetic moment ratio μ_p/μ_{B} (proton/electron)	$1.521032 \cdot 10^{-3}$	$1.521032 \cdot 10^{-3}$
Fine structure constant $\alpha = e^2/\hbar c$	$7.297353 \cdot 10^{-3}$	$7.297353 \cdot 10^{-3}$
Inverse fine structure constant α^{-1}	137.0360	137.0360
Magnetic flux quantum $\Phi_0 = hc/e$	$4.135667 \cdot 10^{-7}$ G cm^2	$4.135667 \cdot 10^{-15}$ T m^2
Conductance quantum $2e^2/h$	$6.963638 \cdot 10^{7}$ esu	$7.748092 \cdot 10^{-5}$ Ω^{-1}
Quantized Hall resistance $R_{\mathrm{H}} = h/e^2$	$2.872062 \cdot 10^{-8}$ esu	$2.581281 \cdot 10^{4}$ Ω
Josephson constant $2e/h$	$1.449794 \cdot 10^{17}$ esu erg^{-1} s^{-1}	$4.835979 \cdot 10^{14}$ Hz V^{-1}

Source: P. J. Mohr and B. N. Taylor, The fundamental physical constants, *Phys. Today* **55**, August BG 6–16 (2002).

Appendix B

List of Notations and Symbols

$\boldsymbol{A}$	vector potential
A	mass number
$\mathrm{Ai}(y)$	Airy function
$\boldsymbol{a}$	acceleration
$\boldsymbol{a}, \boldsymbol{b}, \boldsymbol{c}$	basic vectors of the lattice
$\boldsymbol{a}^*, \boldsymbol{b}^*, \boldsymbol{c}^*$	basic vectors of the reciprocal lattice
a, b, c	lattice constants
a_0	Bohr radius
$\boldsymbol{B}$	magnetic induction
B	bandwidth
$\mathrm{B}_S(x)$	Brillouin function
C	cyclic group
C_e	electronic specific heat capacity
C_P	specific heat capacity at constant pressure
c	speed of light in vacuum
$\boldsymbol{D}$	electric displacement
D	diffusion coefficient
D	fractal (or Hausdorff) dimension
D	dihedral group
d	lattice spacing
d	Euclidean dimension
d_t	topological dimension
$\boldsymbol{E} = \mathcal{E}\boldsymbol{e}$	electric field
E	energy
$E(\boldsymbol{k})$	dispersion relation
E_c	mobility edge
E_F	Fermi energy
E_g	band gap energy
E_k	kinetic energy
E_ex	exchange energy
E_xc	exchange-correlation energy
$\boldsymbol{e}$	unit vector
e	electronic charge
$\mathrm{e} = 2.71828$	base for natural logarithm
e'	eccentricity
$\boldsymbol{F}$	force
F	Helmholtz free energy
f	Helmholz free energy density
$f(E)$	Fermi distribution function
f_c	covering density
f_p	packing density
$\mathcal{G}$	symmetry group
$\boldsymbol{G}$	reciprocal lattice vector
$G(\boldsymbol{r}, \boldsymbol{r}')$	Green's function
G	conductance
G	Gibbs free energy
g	Gibbs free energy density
$g(\boldsymbol{r}, \boldsymbol{r}')$	correlation function
g_L	Landé factor
g	symmetry operation (or element)
$g(E)$	density of states
$g(\omega)$	frequency distribution function
g_F	density of states at Fermi level
g_t	genus
$\mathcal{H}$	Hamiltonian
$\boldsymbol{H}$	magnetic field
h	height
$\hbar = h/2\pi$	Planck constant
(h, k, l) or h, k, l	index for lattice planes
I	electric current
I_AB	exchange integral
$\boldsymbol{i}, \boldsymbol{j}, \boldsymbol{k}$	unit vectors along the Cartesian axes
$\boldsymbol{J}$	angular momentum
J	total angular momentum quantum number
J_{ij} or J	exchange coupling constant

J_n	the nth rank Bessel function
$\boldsymbol{j}$	electric current density
j	angular momentum quantum number
$\boldsymbol{K}_0$	incident wavevector outside the surface
$\boldsymbol{K}_{\mathrm{G}}$	diffracted wavevector outside the surface
$\boldsymbol{k}$	wavevector
$\boldsymbol{k}_{\mathrm{F}}$	Fermi wavevector
k_{F}	Fermi wavenumber at Fermi level
k_{B}	Boltzmann constant
L	Lagrangian
L	total azimuthal quantum number
L	diffusion length
L	sample size
L_x, L_y, L_z	depolarization factors
$\boldsymbol{l}$	lattice vector
l	mean free path
l	azimuthal quantum number
$\boldsymbol{M}$	matrix for group representation
$\boldsymbol{M}$	magnetization
M	atomic (or ionic) mass
m	electronic mass
m^* or m_{e}^*, m_{h}^*	effective mass of an electron (or hole) in the energy band
m	magnetic quantum number
N_{A}	Avogadro constant
N	total number of particle (or cell)
$\boldsymbol{n}$	unit normal
$\tilde{\boldsymbol{n}}$	director of liquid crystal
n	number density of electrons
O	octahedral group
$\boldsymbol{P}$	electric polarization
P	probability
P_l	the lth order Legendre polynomial
$\boldsymbol{p}$	momentum
p	effective number of Bohr magneton
p_{c}	percolation threshold
$\{p, q\}$ or $\{p, q, r\}$	Schläffli symbol for regular polyhedra or space filing
Q	quality factor for resonant cavity
$\boldsymbol{q}$	wavevector for lattice vibration
$\boldsymbol{R}$	atomic (or ionic) position vector
R	electric resistance
R	radius of gyration of polymers
R_{H}	Hall resistance
R_{nl}	radial wave function
$\tilde{R}$	reflection coefficient
$\tilde{r}$	reflection amplitude
$\boldsymbol{r}, \boldsymbol{r}'$	position vector
$\boldsymbol{S}, \boldsymbol{s}$	spin
S	entropy
S	total spin quantum number
S_{AB}	overlap integral
s	spin quantum number
$\mathcal{T}$	time reversal symmetry group or operator
$\mathcal{T}$	kinetic energy
$\mathcal{T}[n]$	kinetic energy functional
$\boldsymbol{T}$	transfer matrix
T	thermodynamic temperature
T	tetrahedral group
T_{c}	critical temperature or Curie temperature
T_{F}	Fermi temperature
T_{g}	glass transition temperature
T_{m}	melting temperature
T_{N}	Néel temperature
T_0	quantum degeneracy temperature
$\tilde{T}$	transmission coefficient
$\tilde{t}$	transmission amplitude
$t = (T - T_{\mathrm{c}})/T_{\mathrm{c}}$	reduced temperature
t	time
U	internal energy
U	Hubbard energy
$\boldsymbol{u}$	displacement vector from the lattice site
$[u, v, w]$ or $\langle u, v, w\rangle$	indices for lattice direction
V	potential, voltage
$\boldsymbol{v}$	velocity
W	work function
B	bandwidth
$w(\boldsymbol{r})$	Wannier funtion
x, y, z	position component along the Cartesian coordinates
Y	icosahedral group
Y_{lm}	spherical harmonics
$\langle y\rangle$	statistical average of y
$\lfloor y\rfloor$	the integer part of y

Z	atomic number
Z	partition function
z	coordination number
α	Coulomb integral
β	bond (or resonance) integral
$\Gamma(\boldsymbol{r}, \boldsymbol{r}')$	correlation function
γ	critical exponent
$\Delta(T)$	energy gap of superconductor
$\delta(\boldsymbol{r})$	delta function
$\mathcal{E}$	the magnitude of electric field
ε	single-particle energy
ϵ	dielectric constant
$\epsilon(\omega)$	dielectric function
η	order parameter
η, τ, σ	order parameter for smectic liquid crystal
Θ_A	Curie–Weiss temperature
Θ_D	Debye temperature
$\boldsymbol{\kappa} = \boldsymbol{k}/\boldsymbol{k}_F$	reduced wave vector
κ	curvature
κ	thermal conductivity
κ_m	mean curvature
κ_G	Gaussian curvature
λ	wavelength
λ_L	London penetration depth
μ	chemical potential
μ_B	Bohr magneton
$\mu_t = \mu_e + \mu_h$	mobility (electron mobility + hole mobility)
μ_{eff}	effective magnetic moment
ν	frequency
ξ_c	correlation length
ξ_0	coherence length of superconductor
ξ_s	screening length
Π	product
$\pi = 3.14159$	ratio of circumference of circle to diameter
$\sum$	sum
σ	electrical conductivity
σ_g	molecular orbital with even parity
σ_u	molecular orbital with odd parity
$\sigma_x, \sigma_y, \sigma_z$	spin components of Pauli matrix
τ	relaxation time
$\tau = (1+\sqrt{5})/2 = 1.61803$	golden mean
Φ	magnetic flux
Φ_0	magnetic flux quantum for paired-electrons
ϕ_0	magnetic flux quantum for single-electrons
ϕ or χ	orbital wave function
χ_P	Pauli susceptibility
χ	spin wave function
χ	magnetic susceptibility
Ω	solid angle
Ω	volume
Ω_0	volume of unit cell
ω	angular frequency
ω_c	cyclotron frequency
ω_p	plasma frequency

Index

$3d$ transition metal, 279, 295, 321
α-helix chain, 94
β-brass, 433
β-sheet, 94
δ function, 64, 148
δ-potential, 264
π orbital, 285
π polarization, 141
π ring, 517
π-electron, 291
σ orbital, 285
σ polarization, 140, 141
σ-bonding electron, 291
d-band, 318, 319
d-wave, 507
d-wave pairing, 516, 517
g-wave, 507
n-grid, 67
n-type semiconductors, 206
p-wave, 507
p-wave pairing, 510
s-d hybridization, 352
s-band, 318, 319
s-point grain correlation, 105
s-wave, 507
sp orbital, 290
sp^2 orbital, 290
sp^3 orbital, 290
0 ring, 517

Abelian group, 31
Abrikosov vortex lattice, 20
Abrikosov, A. A., 7, 16, 356, 499
absolute continuous spectrum, 148
absolutely unstable limit, 418, 419
absorption, 507
absorption spectra, 192
ac conductivity, 202
ac Josephson effect, 514, 518
ac Josephson frequency, 518
ac susceptibility, 541
acceptor, 195
acoustic, 123
acoustic mode, 134
acoustic wave, 136, 232
actinide, 321
adiabatic approximation, 308
adiabatic continuity, 17
Aharonov, Y., 260
Aharonov–Bohm (AB) effect, 252, 260
Airy function, 365
Alexander, S., 430, 431
alkali metals, 545
Allen, J. F., 16
alloy, 73, 537
alternating electric field, 201
alternating frequency, 514
Altshuler–Aronov–Spivak (AAS) effect, 265
amino-acid, 95
Ammann quasilattice, 68
amorphous magnetic alloy, 537
amorphous semiconductor, 14, 247
amorphous solid, 532
amphiphilic interface, 93
amphiphilic molecule, 83
amplitude, 138, 411, 414, 515
Anderson Hamiltonian, 16, 351, 352, 354, 361
Anderson localization, 14, 15, 124, 233, 245, 350, 400
Anderson localization criterion, 244
Anderson model, 16, 242, 244, 333, 352, 361
Anderson transition, 244, 542
Anderson, P. W., 5, 14, 16, 17, 19, 124, 231, 240, 337, 361, 469, 470, 510, 514, 537–539
Anderson-type single particle localization, 551
Andrews, T., 16
angles of crystallographic axis, 517
angular momentum, 279, 507
angular momentum quantum number, 275, 295, 503, 504
angular momentum state, 368
anisotropic, 7, 82, 103, 276, 500
anisotropic environment, 296
anisotropic magnetorisistance (AMR), 218
anisotropic molecule, 11
anisotropic solution, 444
anisotropic term, 464
anisotropy, 212, 217, 383, 503
anomalous dielectric property, 435

anomalous dispersion, 41
anthracene, 295
anti-ferromagnetic interaction, 535
antibonding orbital, 298
antibonding state, 285
antiferrodistortive, 298
antiferroelectrics, 18
antiferromagnet, 19, 279, 286, 337, 341, 347, 407, 452, 461, 467
antiferromagnetic, 289, 300, 335, 337, 452, 459, 537
antiferromagnetic chain, 470
antiferromagnetic coupling, 288, 338
antiferromagnetic fluctuation, 510
antiferromagnetic ground state, 19, 470
antiferromagnetic Heisenberg model, 337
antiferromagnetic insulator, 343, 346
antiferromagnetic long-range order, 346, 469, 479
antiferromagnetic phase, 535
antiferromagnetic state, 358
antiferromagnetism, 16, 321, 341, 449, 456, 465, 478, 545
antiparallel, 288, 338, 467
antiparallel alignment, 225
antisymmetric, 278, 504
antisymmetric wave function, 278
aperiodic fluctuation, 272
aperiodic structure, 124
applied field, 410, 473
approximate mean-field solution, 538
Archimedean solids (or polyhedra), 37
argon core, 318
Arrott chart, 477
artificial atom, 395
artificial layered composite, 501
artificial molecule, 395
artificial periodic structure, 142
artificial solid (crastal), 395, 398
Ashcroft, N. W., 327
associated state, 294
associativity rule, 30
atom laser, 486
atomic correlation, 75
atomic displacement, 122
atomic distribution function, 77
atomic orbital (AO), 245, 275, 276, 278, 321
atomic physics, 275, 286, 289, 295
atomic position, 437
atomic scale, 229
atomic scattering factor, 140
atomic spacing, 133, 335
attenuation, 437
augmented plane wave (APW) method, 301, 305
average free energy, 540
average scattering rate, 251
average t-matrix (ATM), 14
average velocity, 208
Avogadro constant, 543
azimuthal angle, 282
azimuthal quantum number, 389

Büttiker, M., 255, 257, 262
Bak, P., 22
ballistic regime, 252
ballistic transport, 15
band approach, 15, 16, 275, 293
band calculation, 323
band edge, 165, 168, 249
band index, 125
band insulator, 343
band method, 275
band model, 177, 222
band picture, 250
band structure, 131, 142, 366
band tail, 244
band tail state, 124
band theory, 15, 124, 157, 158, 231, 333
band-calculation, 305
bandgap, 142, 192, 232, 246, 368, 372, 381
bandgap stabilization, 151
bandgaps, 142
bandwidth, 245, 335, 398
Bardeen, J., 6, 22, 330
Bardeen–Cooper–Schrieffer (BCS) gap, 509
Bardeen–Cooper–Schrieffer (BCS) theory, 6, 7, 16, 19, 502, 505
barrier, 178, 364, 371
basic vector, 33
basis, 289
basis function, 437
bcc electron crystal, 546
Bednorz, J. G., 7, 493
benzene, 290, 292, 294
Bernal, J. D., 79
Bessel function, 164, 386, 515
Bethe ansatz, 19, 470
bicontinuous, 52
binary alloy, 7
biological evolution, 534
biology, 229
biomembrane, 87
biopolymer, 8, 10, 82
bipolarons, 505
birefringence, 437
black-white group, 45

Bloch electron, 155, 161, 165, 174
Bloch function, 138, 140, 142, 186, 302, 422
Bloch momentum, 143
Bloch oscillation, 156, 162, 163, 223, 265
Bloch representation, 471
Bloch solution, 203
Bloch state, 335, 472
Bloch theorem, 123, 130, 132, 136, 142, 183
Bloch tunneling, 221
Bloch wave, 123, 140, 141, 146, 178, 193
Bloch wavevector, 158
Bloch, F., 6, 14
Bloch–Grüneisen law, 205
Bloch-wave-like, 232
body-centered-cubic (bcc) structure, 125, 318, 328, 430
Bogoliubov, N. N., 6
Bohm and Pines theory of plasmon, 16
Bohr magneton, 168, 187, 279, 462
Bohr radius, 187, 349
Boltzmann constant, 9, 97, 203
Boltzmann distribution, 206
Boltzmann equation, 199, 207, 216, 251
Boltzmann factor, 247, 436
Boltzmann transport theory, 235, 238, 251, 270
Boltzmann, L., 21, 521
bond (or resonance) integral, 285
bond approach, 15, 16, 293
bond direction, 290
bond dissociation energy, 290
bond length, 290
bond percolation, 106
bonding of solid, 321
bonding orbital, 298
bonding pair, 290, 291
bonding state, 285
bonding-antibonding splitting, 396
Born approximation, 377
Born, M., 14, 308
Born–Oppenheimer approximation, 308, 317
Born–Oppenheimer potential surface, 317
Born–von Karman cyclic boundary condition, 123, 177
Bose–Einstein condensation (BEC), 10, 12, 481, 483, 486, 505, 519
Bose–Einstein statistics, 10
boson, 10
bottom of the conduction band, 368
bound state, 527
boundary conditions, 22, 164
Bragg condition, 128, 142, 249
Bragg diffraction, 127, 151
Bragg equation, 141
Bragg, W. H., 142
Bragg, W. L., 142
Bravais lattice, 39, 41, 63
Brillouin function, 462, 463, 465, 466
Brillouin zone, 212
Brillouin zone (BZ), 123, 125, 131, 136, 143, 151, 161, 165, 181, 186, 213, 224, 265, 306, 318, 324, 383, 421, 438
Brillouin zone boundary, 126, 212
Brillouin's theory, 97
Brillouin, L., 14
broken ergodic phase, 22
broken ergodicity, 22, 406, 521, 523, 526
broken gauge symmetry, 413
broken spin-flip symmetry, 538
broken symmetry, 7, 16, 17, 19, 21, 22, 405, 406, 410, 420, 437, 462, 521, 523, 525, 526
broken translational symmetry, 523
Brownian motion, 114
buckminsterfullerene, 37
building block, 144
bulk material, 363
bulk semiconductor, 366
Burrau, O., 282

Cambell, I. A., 216
Canham, L. T., 322
canonical equations of motion, 521
canonical example, 535
canonical momentum, 490, 495
canted, 451, 461
canted antiferromagnetism, 457
Cantor bar (or set), 110, 148
Cantor, G., 110
capacitance, 390
Car–Parrinello method, 9, 317
carbon nanotube, 60, 61, 295, 380
carrier mass, 165
carrier wave, 249
Carroll, L., 29
Cartesian axis, 410
Cartesian coordinate system, 28
catenoid, 52
Cauchy, A. L., 13
cellular method, 301
centered lattice, 39, 41
chain-folding model for polymer crystallization, 92
change of volume, 526
channels, 254
characteristic frequency, 518
characteristic length, 100, 233
characteristic size, 363

characteristic temperature, 536
charge density, 122
charge density wave (CDW), 11, 20, 546
charge ordering, 349
charge transfer (CT) type, 342, 343, 347
charge transport, 199
charge-spin separation, 510
charge-transfer salt, 7, 459
chemical potential, 257, 258, 329, 474, 481, 497, 518, 530, 550
chemistry, 229
chiral angle, 62, 380, 381
chiral molecules, 29
chiral vector, 61, 62
cholesteric phase, 82, 84
circle of best fit, 50
circulation, 489
classical crystallography, 67, 72
classical Hall effect, 208
classical harmonic oscillator, 167
classical physics, 283
classical potential, 325
classical statistical mechanics, 521
classical thermal fluctuation, 542
classical waves, 14, 121, 124, 231, 232
Clausius–Mossotti relation, 543
close-packing, 54, 55
closed Fermi surface, 212
closed orbit, 213
Closure rule, 30
clusters, 383
co-precipitation, 100
coden, 95
coherence, 266
coherence length of nonlinear optical media, 146, 237
coherence length of superconductor, 500, 505, 518
coherent backscattering, 237
coherent effect, 252
coherent multiple scattering, 247
coherent nature, 486
coherent potential approximation (CPA), 14
cohesion energy, 319
collective quasiparticles, 510
collision, 199, 209
collision term, 200
color group, 45
colossal magnetoresistance (CMR), 20, 62
column matrix, 133
columnar phase, 84
combination rule for symmetric axes, 34
combination rules, 35
combinational optimization, 534
commensurate, 64
commensurate phase, 437, 439
commensurate-incommensurate phase transition, 426
commutative, 31
commutativity, 459
compartment, 525, 526, 533
compensated point, 468
complete wetting, 529, 530, 532
complex order parameter, 438
complex plane, 414
complex point group, 47
complex wavevector, 183
composite spontaneous magnetization, 467
composites, 102
compound metallic, 325
compound semiconductor, 322
compressibility, 413, 552
computer simulation, 317
condensate, 495, 503
condensate fraction, 485, 487
condensate wavefunction, 488
condensation, 486
condensation energy, 490
condensed matter, 8, 12, 22, 327, 409, 521
condensed matter physics, 6, 8, 9, 12, 13, 15, 17, 22, 27, 100, 110, 229, 275, 308, 534, 542
conductance, 252, 254, 258, 391
conductance fluctuation, 266
conductance measurement, 398
conductance oscillation, 259
conduction band, 158, 244, 246, 324, 366, 369
conduction band edge, 323
conductivity, 106, 200–202, 206, 209, 233, 238, 246, 251, 252, 255, 546
conductor, 157, 240
configuration space, 542
confining potential, 378
conjugate momentum, 521
conjugated field, 415
connectivity, 104, 106
conservation of energy, 139
constant energy surface, 166, 522
constructive interference, 236
contact angle, 530
continued fraction, 63
continuity condition, 178, 364
continuous group, 33
continuous phase transition, 407, 436
continuous spin glass transition, 539
continuous translational symmetry, 546
continuum, 109, 135
continuum percolation, 240

conventional superconductors, 502, 505, 509, 512
convolution, 44, 66
convolution theorem, 44
cooling speed, 76
Cooper pair, 12, 19, 261, 490, 502, 504, 505, 507, 518
Cooper problem, 502, 505
Cooper, L., 6
cooperative Jahn–Teller effect, 298
cooperative phenomena, 16, 17, 450, 451
coordination compound, 459
coordination number, 54, 56, 58, 59, 77, 540
core region, 500
core state, 302
Cornell, E., 484
correlated electron liquid, 547
correlated electronic state, 12, 275
correlated electrons, 20, 322, 333
correlated quantum liquid, 546
correlation, 75, 102, 105
correlation between electrons, 333
correlation function, 75, 76, 78, 91
correlation length, 21, 75, 245, 248
coset decomposition, 422
Coulomb blockadc, 393
Coulomb blockade oscillation, 391
Coulomb energy, 299
Coulomb explosion, 386
Coulomb gap, 549, 550, 552
Coulomb gauge, 260
Coulomb integral, 285, 287
Coulomb interaction, 20, 278, 279, 283, 309, 479, 545
Coulomb potential, 302, 311, 387, 544, 545
Coulomb repulsion, 349
Coulomb repulsive energy, 291
coupled multiple quantum well, 369
covalent binding, 395
covalent bond, 294
covalent semiconductor, 183
covering density, 54, 56, 68
covering lattice, 107
covering model, 68
criterion for localization, 248
critical, 75
critical current of type II superconductor, 20
critical density, 546
critical electronic density, 544
critical field, 497
critical index, 8
critical indices, 16
critical phenomena, 5, 8, 16, 20, 92
critical point, 528, 530
critical region, 20, 92
critical state, 148
critical strength, 548, 549
critical temperature, 17, 482, 485, 539
crossover temperature, 534
crystal, 246, 427, 532
crystal field, 295, 296
crystal field theory, 296
crystal growth, 100, 146
crystal lattice, 27, 134
crystal momentum, 121
crystal plasticity, 20
crystal structure, 42
crystal structure analysis, 45
crystal surfaces, 141
crystalline approximants for quasicrystal, 151
crystalline order, 483
crystalline solid, 11, 123, 165
crystalline state, 13, 73, 275
crystallization, 12, 77, 532
crystallography, 27, 37
cube, 35
cubic, 37, 39, 62
cubic terms, 423
cumulants, 540
cuprate, 346, 347, 505
cuprate superconductors, 509
Curie constant, 437
Curie temperature, 406, 537
Curie–Weiss law, 416, 420, 437, 464, 466, 475, 476
Curie-like behavior, 357
current in the plane (CIP), 374
current operators, 200
current perpendicular to the plane (CPP), 374
curvature, 50, 51
curve, 50, 111
curved surface, 111
cut and projection, 65
cut-off frequency, 121
cyclic groups, 35, 37
cyclic relaxation time, 211
cyclotron energy, 389
cyclotron frequency, 167, 212, 214
cyclotron radius, 170
cyclotron resonance, 165, 167, 212, 323
cyclotron resonance frequency, 166, 209, 224
cyclotron resonance mass, 167
cylinder, 52
cylindrical state, 379

D surface, 52
damped plasma oscillation, 202
damping of oscillations, 487

Darwin, C. G., 14
dc conductivity, 213
dc Josephson effect, 514
dc-SQUID, 516
de Broglie relation, 155
de Broglie wavelength, 8, 101, 121, 483
de Broglie waves, 9, 14, 15, 121, 231
de Gennes, P. G., 17, 92, 348
de Haas–van Alphen (dHvA) effect, 170, 171, 173, 210
Debye characteristic temperature, 136
Debye cut-off wavenumber, 205
Debye frequency, 136
Debye model for the specific heat of solids, 135, 204
decagonal phase, 68, 431
defect-mediated phase transitions, 20
defects, 160, 199
deflation, 65
degeneracy, 143, 276, 464, 537
degeneracy per unit magnetic field per unit volume, 171
degenerate electron gases, 11
degenerate orbitals, 339
degenerate perturbation theory, 126, 127
degree of order, 423
delocalization-localization, 533
delocalized, 183, 234, 289
delocalized states, 129, 244
density, 521, 523, 543
density correlations, 112
density distribution, 323
density fluctuation, 421
density function, 421
density functional, 301
density functional theory (DFT), 312, 313, 361, 386
density of states (DOS), 131, 132, 170, 173, 195, 201, 238, 245, 246, 256, 319, 325, 372, 378, 382, 456, 458, 546, 549, 550
density wave, 427, 437, 523
dephasing, 211
depletion, 485, 549
depletion force, 446
depletion layer, 329, 330
depolarization factors, 102
designing electronic structure, 363
determinant, 134, 284
devil's staircase, 115, 148
diagonal disorder, 242
diagonal matrix element, 242
diamagnet, 6
diamagnetism, 449, 450
diametric line, 291
diamond, 144, 290
diatomic molecule, 326
dielectric anomaly, 326
dielectric constant, 105, 122, 149, 412, 435, 437, 549
dielectric function, 202
dielectrics, 247, 543
difference frequency conversion, 145
differential geometry, 50
diffraction pattern, 42, 70, 71, 151
diffraction physics, 142
diffusion, 76, 201, 233
diffusion coefficient, 234, 236–238, 252, 255, 326
diffusion equation, 234
diffusive case, 233
diffusive regime, 252
dihedral groups, 35, 37
dimerization adatom stacking fault model for a 7×7 cell of reconstructed Si(111) surface, 328
dipolar double layer, 178
dipolar modes, 194
Dirac symbol, 242
Dirac's delta function, 43, 79
Dirac, P. A. M., 5
direct band gap material, 366
direct exchange, 289, 451
direct lattice, 123
direct-gap, 323
director, 83, 84, 440
disclinations, 16
discommensurations, 440
discontinuous phase transition, 407, 418, 436
discrete energy state, 383
discrete lattice, 135
dislocations, 20
disorder, 146, 233
disorder-induced electron localization, 231
disordered phase, 418
disordered state, 433
disordered systems, 231
dispersion, 14
dispersion curve, 132, 133
dispersion in the medium, 146
dispersion relation, 15, 124–126, 132, 135, 137, 139, 141, 142, 144, 155, 182, 193, 380
dispersion surface, 141
displacement, 133
displacement type, 62
displacement vector, 136
displacive phase transition, 436
distorted perovskite structure, 347

distortion energy, 531
distribution function, 75, 160, 199, 201, 522
DNA, 6, 29, 94
dodecahedron (dodecahedra), 35, 69
domain structure, 440
domain theory for ferromagnets, 16
domain walls, 440
donated electron, 328
donor, 185
donor doping, 206
donor states, 195
doped charge transfer (CT) insulator, 343
doped Mott insulator, 16, 317, 333, 341, 349
doped Mott–Hubbard (MH) insulator, 343
doped semiconductors, 195
doping, 244
dot supermoments, 400
double barrier, 220
double barrier diode, 220
double exchange, 16, 347, 348, 451
double occupancy, 334
double occupancy of the same, 547
double quantum dot, 395
doublet, 107
doubly degenerated, 194
Drude conductivity, 209
dual, 55
dual lattice, 68
duality, 52
duplication, 94
Duwez, P., 76
dynamic high pressure technique, 327
dynamic matrix, 133, 191
dynamic mean-field theory, 552
dynamic structure, 79
dynamic structure factor, 79
dynamical mean-field theory (DMFT), 362
dynamical quantity, 522
dynamical simulated annealing, 318
dynamical X-ray diffraction theory, 139, 141, 142
dynamics of a hole, 158

eccentricity, 379
Edwards model, 245, 246
Edwards, P. P., 537–539
Edwards, S. F., 17, 245
effective channel number, 254
effective diffusion coefficient, 238
effective field, 6, 436, 461, 462, 465, 466
effective interaction potential, 527
effective magnetic moment, 279
effective mass, 155, 168, 186, 323
effective mass approximation, 388
effective mass parameter, 366
effective medium approximation (EMA), 106
effective number of Bohr magnetons, 172
effective potential, 385
effective temperature-dependent Hamiltonian, 540
Efros, A. L., 549
eigenenergy, 123, 164, 233
eigenfrequency, 123, 133, 143
eigenfunction, 288, 303
eigenstate amplitude, 232
eigenvalue, 288
eigenvector, 133
eigenwavevector, 233
Einstein model, 135, 205
Einstein relation, 201, 238, 255
Einstein, A., 10
elastic, 124, 136
elastic constants, 437
elastic mean free path, 236, 248, 252
elastic scattering cross section, 233
elastic scatterings, 233
elastic wave, 13, 121, 136, 231
electric conductivity, 208, 412
electric current, 517
electric dipolar, 534
electric displacement, 122, 124, 140, 142, 144
electric field, 138, 142, 199–201, 207, 208, 224, 244, 248, 249, 252, 415
electric force, 165
electric polarization, 145, 409, 543
electric susceptibility, 124
electrical conductivity, 106
electrodynamic effects, 518
electrodynamic properties, 495
electromagnetic, 121
electromagnetic modes, 248
electromagnetic potential, 260
electromagnetic theory, 144
electromagnetic wave, 121, 122, 124, 137, 139, 142, 146, 193, 196, 202, 231, 237, 247
electromagnetic wave guide, 378
electromagnetism, 247
electromechanical actuator, 232
electromotive force, 201
electron, 206, 257, 323, 545
electron correlation, 15, 16, 311, 321
electron crystal, 551
electron crystallization, 546
electron density, 139, 185, 363
electron density distribution, 312
electron doping, 343
electron dynamics, 317
electron energy, 125

electron glass transition, 551
electron localization, 240, 247
electron microscopy, 99
electron pair, 503
electron polarons, 550
electron probability distribution, 127
electron spin, 171
electron state, 363
electron subband, 368
electron trajectory, 165
electron transport, 208, 215, 235, 251
electron wavefunction, 247
electron waves, 127, 146
electron-electron interaction, 19, 549
electron-hole generation, 158
electron-hole pairs, 544
electron-lattice, 20
electron-pair bond, 283, 290
electron-pair valence bond, 290
electron-phonon interaction, 19, 247
electron-phonon mechanism, 505
electronic band structure, 127, 165, 185
electronic charge, 158
electronic conductivity, 381
electronic cyclotron motion, 170
electronic energy band, 141
electronic motion, 170
electronic orbital, 449
electronic phase diagram, 346
electronic phase separation phenomenon, 351
electronic properties, 301, 324
electronic specific heat, 160, 476
electronic spectra, 146
electronic spin, 449
electronic spin magnetic moments, 174
electronic states, 183, 203
electronic structure, 6, 148, 229
electronic surface state, 178, 183
electronic system, 252, 333
electronic velocity, 257
electrostatic Aharonov–Bohm effect, 265, 266
electrostatic field, 106
elemental semiconductor, 185
elementary excitations, 16, 17, 20
Eliashberg, G.M., 6
ellipses, 501
ellipsoid, 102, 103
elliptic, 51, 52
elliptic coordinates, 282
elliptical asymmetry, 379
emergent phenomena, 7
Emery, V. J., 509
emission or absorption of phonon, 322
emission or absorption of photon, 322
empirical rules, 542
enantiomorphic, 41
energy band, 6, 127, 129, 178, 366
energy band structure, 183, 263, 456
energy band theory, 301
energy barrier, 218, 536
energy conservation, 203
energy eigenvalue, 179, 275
energy gap, 128, 142, 151, 181, 223, 244, 295, 507
energy increment, 550
energy integral, 291
energy ladder, 289
energy level, 179, 363
energy spectra, 148
energy spectrum, 148, 150, 370
energy surface, 536
energy-driven phase transition, 445
enhanced backscattering, 234, 239
enhanced Pauli susceptibility, 357
enhancement factor, 196
ensemble average, 521, 522, 524
ensemble theory, 521
entropic force, 445
entropy, 11, 413, 443, 445, 487, 541
entropy-driven ordering, 11
entropy-driven phase transition, 445
envelope function, 186
envelope wavefunction, 168, 249, 363, 370
equation of motion, 129, 133
equation of state, 415, 528
equatorial plane, 167
equilibrium condition, 419
equilibrium phases, 432
equilibrium state, 522, 536, 537
equilibrium statistical mechanics, 525
equilibrium value, 412
ergodic hypothesis, 521, 522
ergodic-nonergodic transition, 542
ergodicity, 21, 521, 523, 537
ergodicity breaking, 552
Esaki, L., 163
Euclidean dimension, 111
Euclidean plane, 60
Euclidean space, 45
Euler formula, 52, 60
Euler theorem, 30, 34, 36
Euler–Poincaré characteristic, 50
evanescent Bloch surface waves, 193
evaporation cooling, 10
even parity, 283, 285
exact ground state, 547
exchange coupling, 400
exchange energy, 410

exchange integral, 287
exchange interaction, 350, 451, 539
exchange parameter, 465
exchange splitting, 296
exchange-correlation energy, 316
exchange-correlation potential, 386
excited state, 322
exciton, 7, 19
excluded volume, 443, 528
exotic pairing mechanisms, 359
expectation value, 523
exponential decay, 247
exponential term, 220
extended state, 146, 148, 178, 231, 241, 246
extended-zone scheme, 125, 127
external field, 462
external perturbations, 536
extrinsic, 158

face-centered-cubic (fcc) structure, 55, 75, 125, 318, 319, 326
far from equilibrium, 22
fcc, 125
femtoseconds, 8
Fermi distribution, 472
Fermi distribution function, 175, 221, 222, 228, 257
Fermi energy, 125, 151, 160, 171, 178, 210, 246, 258, 263, 302, 385, 394, 398, 474, 549
Fermi gas, 20
Fermi glass, 244
Fermi golden rule, 352
Fermi level, 101, 151, 159, 160, 171, 177, 225, 244, 263, 322, 324, 330, 372, 393, 502, 506, 549, 550
Fermi liquid, 17, 20
Fermi sea, 502
Fermi sphere, 175, 546
Fermi surface, 11, 151, 171–173, 176, 201, 204, 212–214, 238, 244, 256, 311, 318, 321, 329, 363, 473, 476, 546
Fermi temperature, 160
Fermi velocity, 233
Fermi wavelength, 239, 363, 372, 373
Fermi wavevector, 125
Fermi–Dirac statistics, 10, 279
fermion, 10, 326
ferrimagnet, 454, 467
ferrimagnetic, 459
ferrimagnetic structure, 453
ferrimagnetic transition, 466, 467
ferrimagnetism, 338, 449, 451
ferrite, 338, 453, 466
ferrocene, 299, 459
ferrodistortive, 298
ferroelectric crystal, 435
ferroelectric phase transitions, 418, 436
ferroelectricity, 326, 435
ferroelectrics, 4, 18
ferromagnet, 4, 19, 279, 289, 321, 341, 372, 407, 434, 451, 454, 461, 474
ferromagnetic, 300, 371, 459, 512, 535, 537
ferromagnetic coupling, 459
ferromagnetic ground state, 19, 469
ferromagnetic insulator, 340
ferromagnetic metal, 217
ferromagnetic order, 479
ferromagnetic phase transition, 7
ferromagnetic phases, 541
ferromagnetic semiconductor, 354
ferromagnetic state, 358, 414
ferromagnetic substances, 217
ferromagnetism, 15, 321, 341, 354, 449, 450, 456, 472, 476, 512
Fert, A., 216
Fibonacci, 64
Fibonacci lattice, 65, 146, 148
Fibonacci sequence, 65, 68, 147
Fibonacci superlattice, 149
field emission, 178
filter, 121
first-order Born approximation, 184
first-order phase transition, 17, 407, 423, 444
fluctuation, 21, 237, 241, 476
fluctuation correlation form, 541
fluctuation effect, 519
fluidity, 84
flux-penetration depth, 501
fluxoid, 492
Fock, V. A., 277
Fock–Darwin state, 389
folding-back, 383
forbidden band, 121, 193, 232
force constant, 133, 190
fountain effect, 487
four-probe measurements, 269
four-wave mixing method, 162
Fourier coefficient, 143, 302
Fourier component, 249, 428, 523
Fourier expansion, 133, 515
Fourier series, 137, 140, 186
Fourier transform, 43, 44, 64, 66, 112, 184, 188, 311, 503
Fourier transformation, 123
Fourier transforming, 75
Fourier–Bessel expansion, 515
Fröhlich, H., 6

fractal, 99
fractal dimension, 112, 113
fractal structure, 14, 110
fractals, 110
fractional quantum Hall effect, 20, 333
fractional quantum Hall state, 546
fracton, 14
Frank–Kaspar phase, 57
free electron, 155, 544
free electron gas model, 126
free electron model, 264
free electron parabola, 151
free electron systems, 256
free electrons, 543
free energy, 495
free energy density, 530
free energy functional, 424
free energy landscape, 536
free energy per unit area, 529
free-electron approximation, 125
free-photon field, 249
free-volume model, 533
freezing temperature, 536, 537
frequency conversion, 145
frequency distribution function, 134, 136
frequency doubling, 145
frequency gap, 134, 139
frequency spectrum, 134
frequency-wavevector relation, 124
Friedel oscillation, 178, 181, 185
Friedel's rule, 321
Friedel, J., 320
frozen order parameter, 526
frozen state, 535
frustrated system, 537
frustration, 19, 537
fullerene, 37, 60, 295
fullerites, 61
fully ordered structures, 432

Gamov, R. I., 95
gap, 127, 249, 263, 506
gap function, 504–507
garnet, 338, 453
gas constant, 528
gas-liquid coexistence line, 530
gas-liquid transition, 16, 486
gases, 9
gate voltage, 253, 391, 396
gauge fields, 260
gauge invariance, 260
gauge symmetry, 18, 19, 413, 414, 496, 510
Gauss–Bonnet formula, 51
Gaussian curvature, 51, 86
Gaussian distribution, 539, 540
Gaussian function, 289
Gaussian random potential, 246
gels, 10
generalized classical Lagrangian, 317
generalized Cooper pair, 502, 504
generalized coordinates, 521
generalized crystallography, 63
generalized order parameter, 548
generalized Penrose structure, 431
generalized rigidity, 17, 19
generalized symmetry, 45
genetic information, 95
genus, 86
geometric phase, 261
giant Hall effect, 242
giant magnetoresistance (GMR) effect, 215, 374
Gibbs free energy, 411, 434, 499
Gibbs free energy[G], 500
Gibbs, J. W., 521
Ginzburg term, 439
Ginzburg, V. I., 6, 495
Ginzburg–Landau (GL) equation, 7
Ginzburg–Landau (GL) free energy density, 426
Ginzburg–Landau (GL) theory, 495
Ginzburg–Pitaevskii equation, 497
glass, 11, 22, 73, 76, 246, 532, 534
glass transition, 12, 77, 533, 539, 551
glass transition temperature, 92, 533, 534
glide-reflection, 39, 422
global, 252
global gauge transformation, 496, 497
global partition function, 528
Goldhammer–Herzfeld (GH) criterion, 543
good quantum number, 169, 170
Goodenough, J. B., 338, 349
Goodenough–Kanamori rule, 338
Gor'kov, L. P., 7, 507
granular superconductor, 516, 534
graphene, 61, 290, 324, 380
graphite, 59, 290
Green's function, 183, 184, 191, 200, 243
Gross–Pitaevskii equation, 497
ground state, 10, 293, 316, 468, 469, 478, 484, 486, 503, 505, 517, 547, 550
ground state energy, 309, 317, 469
ground state wavefunction, 309
group theory, 306
group velocity, 155
group-subgroup relationship, 407
growth morphology, 114
Gutzwiller's variational approach, 547, 549

Hölder exponent, 115, 116
Hückel approximation, 291, 295
Haken, H., 22
half integer flux, 517
half-filled band, 551
half-metallic ferromagnet, 458
Hall coefficient, 210
Hall conductivity, 210
Hamiltonian, 7, 167, 168, 242, 260, 285, 297, 306, 388, 405, 434, 460, 484, 522, 523, 525, 526, 539–541, 546
handedness, 29
hard core, 528
hard sphere systems, 445
harmonic approximation, 122
harmonic oscillator, 135
harmonic oscillator length, 485
harmonic trap, 497
Hartree approximation, 310, 311
Hartree equation, 309, 310
Hartree theory, 315
Hartree, D. R., 277
Hartree–Fock (HF) approximation, 310, 311
Hartree–Fock equation, 310, 311
Hartree–Fock theory, 479
Hartree-like equation, 315
Hausdorff, F., 111
healing lengths, 518
heat transport, 200
heavy electron, 16, 20
heavy electron metal, 317, 333, 351, 357
heavy electron superconductor, 505, 512, 517
heavy fermions, 505
heavy hole, 323, 366
Heine, V., 330
Heisenberg Hamiltonian, 289, 346, 410, 461, 462, 465, 468, 469
Heisenberg model, 15, 449
Heisenberg uncertainty principle, 12
Heisenberg's theory of ferromagnetism, 16, 462
Heisenberg, W., 286, 289
Heitler and London's theory of the hydrogen molecule, 286
Heitler, W., 15
Heitler–London problem, 339
helical, 451, 461
helical trajectory, 166
helicoidal spin density wave, 479
helix, 50
helix-coil transition, 8
Helmholtz free energy, 490
Hermitian, 133, 144, 459, 496
Hermitian polynomials, 168
Herring, C., 288
Hertz, J. A., 17
heterogeneous material, 102, 105
Heusler alloy, 457
Hewson, A. C., 351
hexagon, 54, 60, 67
hexagonal, 39
hexagonal close-packed (hcp) structure, 55
hexagonal close-packing, 55, 445
hexagonal diamond, 59
hexagonal honeycombs, 54
hexagrid, 67
high T_c cuprates, 516, 517
high T_c superconductor, 16, 20, 62, 343, 346, 500, 501
high frequency cutoff, 134
high index of refraction, 144
high pressure, 545
high spin, 299
high symmetric phase, 424
high symmetry phase, 406, 407, 412, 413, 419
high temperature, 545
high temperature phases, 430
highest occupied molecular orbital (HOMO), 295, 328
Himpsel, F. J., 370
Hohenberg, P. C., 313
Hohenberg–Kohn variational principle, 313
hole, 158, 206, 257, 323
hole doping, 343
hole state, 366
hole subband, 368
holon, 510
homogeneous condition, 426
honeycomb, 55
honeycomb lattice, 107
hopping conduction, 246
hopping energy, 552
hot electron transport, 207
Hubbard energy, 15, 334, 338, 340, 352, 547
Hubbard gap, 550
Hubbard Hamiltonian, 336, 461, 471, 547
Hubbard model, 333, 339, 361
Hubbard subbands, 341
Hubbard, J., 15, 335
Hume–Rothery mechanism, 151, 152
Hume–Rothery phases, 151
Hund's rule, 279, 286, 288, 299, 300, 338
Hund's rule coupling, 338, 340, 348
Hund, F., 279, 284
Huntington, H. B., 327
Hurst exponent, 114
hybridization, 290, 394
hybridized, 290

hydration, 294
hydrodynamic effects, 518
hydrogen, 275
hydrogen atom, 276
hydrogen bond, 293
hyperbolic, 51, 52
hyperbolic curve, 435
hyperbolic equation, 141
hyperbolic plane, 53
hyperbolic surface, 52, 60, 141
hyperspace, 14
hysteresis, 436, 536

icosahedral phase, 432
icosahedral structure, 151
icosahedral symmetry, 35, 63, 69, 70, 151, 295, 433
icosahedron (icosahedra), 35, 36, 59, 69, 432
ideal surface, 177
identical replicas, 540
identity element, 31
imperfections in crystals, 177
impurity, 160, 177, 183, 199, 264
impurity band, 549
impurity center, 183
impurity levels, 206
impurity modes, 192
impurity states, 544
in-plane anisotropy, 381
incident fundamental waves, 145
incident wavevector, 140
incipient localization, 238
incommensurate, 64, 71, 437
incommensurate phase, 11, 437, 439
incommensurate-commensurate transition, 437
incommensurately modulated structure, 14, 71
independent electron model, 125
index of refraction, 138, 140, 145, 202, 203, 543
index space, 47
indirect gap, 322, 323
indirect interaction, 353
inelastic scattering, 233, 251
inelastic scattering mean free path, 251
inertial force, 165
infinite cluster, 107, 108
infinite energy barrier, 536
infinite hard-sphere repulsion, 527
infinite periodic minimum surfaces (IPMS), 52, 61
infinite range shallow well, 527
infinite time average, 523
infinite time limit, 524
inflation, 65
inflection point, 156, 161, 163
information, 96, 97
infrared active, 192
infrared light, 507
infrared optical spectra, 508
inhomogeneous broadening, 388
inhomogeneous long-range magnetic ordered, 535
inhomogeneous media, 136
inhomogeneous phase, 426
inhomogeneous structure, 99
inhomogeneous superconductors, 495
injection, 216
instability, 546
insulating ground state, 545
insulating layer, 513
insulating state, 548
insulator, 157, 240, 244, 246, 322, 325–327, 337, 452
insulator-metal, 327
insulator-metal transition, 326, 347
insulator-semiconductor interface, 330
integer, 515
integral equation, 184
integral multiple, 492
integrated density of states (IDOS), 148
interaction constant, 473
interaction energy, 545
interaction of itinerant electrons, 289
interaction potential, 503
interaction strength, 497
interaction-depletion, 488
interatomic exchange, 288
intercalant, 325
intercalation, 325
interface state, 330
interference, 236, 239, 247, 261
interference correction, 238
interference effect, 265
interference of back-scattered waves, 233
interference term, 235, 261
interlayer bond, 324
intermetallic compound, 493
intermolecular separation, 527
internal energy, 11, 437, 445
International notation, 35, 37
interstitial void, 328
intrinsic magnetic properties, 449
intrinsic semiconductors, 206
invar, 457
invariance, 405
inverse (I) type of spinel structure, 454

inverse element, 31
inverse Fourier transform, 43
inverse localization length, 247
inverse photoemission, 371
inverse susceptibility, 416, 419
inversion, 29, 434
inversion center, 41
inversion layer, 210, 330
inversion rotation, 30
inversion symmetry, 283, 306
Ioffe–Regel rule, 233, 248, 249
ion, 275
ionic crystal, 56
irrational number, 63–65, 71, 440
irrational slope, 66
irreducible representation (IR), 33, 421, 427
irregular fractal, 112
Ising ferromagnet, 524, 537
Ising Hamiltonian, 410, 525
Ising model, 7, 8, 19, 410, 449, 551
Ising spin glass model, 537, 551
isometric transformation, 52
isometry, 28, 47
isotherm, 527
isotope effect, 505
isotropic, 77, 85, 103, 105, 113, 276
isotropic approximation, 251
isotropic parabolic minimum, 186
isotropic phase, 444
isotropic-nematic transition, 418, 440, 443
itinerancy, 452
itinerant electron, 451, 472
itinerant electron model, 476
itinerant ferromagnet, 476, 477
itinerant state, 336

Jahn–Teller complex, 298
Jahn–Teller effect, 297, 309, 350
jellium model, 178, 180, 349, 383, 545
Jerome, D., 7
John, S., 124, 231
Jones, H., 151
Josephson coupling energy, 501
Josephson devices, 516
Josephson effect, 7, 517
Josephson equations, 514
Josephson junction, 7, 513, 515, 516, 519
Josephson signal, 519
Josephson, B., 7
Jullière model, 225
Jullière, M., 225
Jupiter, 327

Kadanoff, P. L., 5, 7, 8
Kagomé lattice, 107
kaleidoscope theorem, 30
Kanamori, J., 338
Kane model, 366
Kapitza, P. L., 16
Kekulé structures, 294
Kelvin, W. J., 13
Kepler's conjecture, 56
Kepler, J., 27, 55
Kerr effect, 372
Ketterle, W., 486
Kevlar, 93
kinematical theory, 139
kinetic energy, 121, 317, 497, 545
kinetic exchange, 337, 338, 340, 451
Kirkpatrick, S., 539
kissing number, 54
Kivelson, S. A., 509
Klémen, M., 16
Knight shifts, 185
Koehler, W. G., 350
Kohler rule, 212
Kohler, M., 212
Kohn anomalies, 479
Kohn, W., 289, 313, 315
Kohn–Sham (KS) equation, 315, 318
Kondo coupling, 358
Kondo effect, 16, 354, 355, 357, 393
Kondo Hamiltonian, 16, 357
Kondo insulator, 16, 351, 359
Kondo lattice model, 357, 358
Kondo problem, 333, 351
Kondo regime, 535
Kondo resonance, 394, 395
Kondo singlet, 357, 358
Kondo temperature, 394
Kondo, J., 351, 355
Kondo-like behavior, 357
Koopmans theorem, 312
Korringa–Kohn–Rostoker (KKR) method, 305
Kramers degeneracy, 306
Kramers, H. A., 337
Kronig–Penney model, 129, 163
Kubo's quantum transport theory, 200
Kubo, R., 200
Kuhn, T. S., 13

Lagrange multiplier, 444
Laguerre polynomial, 389
Lamé coefficients, 136
Landé factor, 168, 187, 279
Landau criterion, 423, 429
Landau diamagnetic susceptibility, 174

Landau energy level, 389
Landau level, 168, 170, 171, 210, 260, 499, 546
Landau quantization, 167, 210
Landau state, 211
Landau subbands, 169
Landau theory, 17, 20, 415, 464, 526
Landau theory of Fermi liquids, 16
Landau theory of phase transitions, 405, 427
Landau, L. D., 6, 17, 495
Landau–de Gennes theory, 417
Landau–Devonshire theory, 417
Landauer formula, 255, 256, 270
Landauer, R., 255
Landauer–Büttiker formulation, 258
Langevin's treatment, 449
Langmuir, I., 291
lanthanide series, 321
large angle scattering, 204
laser, 196, 486
laser cavity, 486
laser cooling, 10
laser diodes (LD), 322
laser vitrification, 77
latent heat, 526
lateral quantization, 381
lateral superlattice, 382
lattice, 9, 13, 43, 75, 422, 460
lattice constant, 372
lattice modes, 192
lattice periodicity, 437
lattice site, 461
lattice vector, 121, 122
lattice vibration, 9, 121, 123, 132, 148, 199, 251, 505
lattice wave, 14, 121, 132, 134, 146, 231
lattice-gas Hamiltonian, 551
Lawrence and Doniach (LD) model for layered superconductor, 501
layered periodic media, 137
layered perovskite, 62
LCAO, 338
LDA++ method, 362
LDA+U method, 361
Lee, D. M., 16
left coset, 33
Legendre function, 441, 442
Legendre polynomial, 84
Legget, A. J., 16
length scale, 485
Lennard–Jones type, 326
Lesbeque measure, 115
level, 389
Levy, M., 313
Lewis, N. G., 290, 291
Lifshitz condition, 426
Lifshitz invariant, 424, 425, 439
Lifshitz point, 421
Lifshitz term, 439
Lifshitz, E. M., 16
ligand field theory, 296
ligands, 298
light emitting diodes (LED), 322
light hole, 323, 366
light localization, 247
light waves, 124
lightly doped, compensated semiconductor, 549
line defect, 194, 196
linear combination of atomic orbitals (LCAO), 183, 283
linear optics, 145
linear superposition, 249
linear transport, 257
linearized APW (LAPW), 305
linearized Boltzmann equation, 200
linearized MTO (LMTO), 305
lines, 111
linkage structures, 58
liquid, 22, 76
liquid crystal, 10, 11, 16, 17, 73, 82, 418, 427, 440
liquid crystalline phase, 415
liquid He, 9
liquid hydrogen, 327, 545
liquid metallic hydrogen, 327
liquid state, 327
liquid-glass transition, 406
liquids, 8, 9, 11, 73, 532, 545
Little, W., 7
local, 252
local conductivity tensor, 269
local current density, 252
local density approximation (LDA), 315, 361, 386
local density calculation, 371
local density of state, 394
local magnetic ions, 537
local moments, 535
local spin density approximation (LSDA), 316, 321
localization, 14, 77, 124, 163, 194, 233, 235, 245, 246, 248, 452
localization criterion, 249
localization length, 252
localization-delocalization transition, 241
localized, 73, 183, 231, 232, 234, 244
localized atomic wavefunction, 128
localized electron states, 158

localized function, 128
localized impurity moment, 188
localized lattice vibrations, 192
localized mode, 188, 190, 194, 231
localized spins, 289
localized state, 148, 164, 178, 233, 238, 246
localized wavefunction, 182
localized-delocalized transition, 542
logarithmic normal distribution, 102
London, F., 6, 10, 487, 491
London, H., 287
London equation, 491
London penetration depth, 491, 498
lone pair, 290, 291
long chain molecules, 87
long-range, 73, 106
long-range antiferromagnetic interactions, 551
long-range correlation, 92
long-range Coulomb forces, 550
long-range magnetic order, 545
long-range order, 11, 74, 75, 78, 146
long-range orientational order parameter, 441
long-range positional order, 84
long-ranged Coulomb repulsion, 549
longitudinal, 135
longitudinal conductivity, 210
longitudinal effective mass, 167
longitudinal energy, 221
lonsdalite, 59
Lorentz distribution, 242
Lorentz force, 165, 224, 374
Lorentz gauge, 260
Lorentz profile, 195
low density electronic system, 545
low spin (LS) state, 299
low symmetry phase, 407, 412, 413, 424
low temperature phase, 321
low-dimensional physics, 14
lowering operator, 468
lowest unoccupied molecular orbital (LUMO), 295, 328
Luttinger liquid, 20
lyotropic mesophases, 430
lyotropics, 82, 93

Müller, K. A., 7, 493
Mackay icosahedron (icosahedra), 57, 58
macromolecules, 87
macroscopic inhomogeneities, 100
macroscopic magnetization, 524
macroscopic occupation, 481
macroscopic pair wavefunction, 517
macroscopic quantum phenomena, 483, 489, 496, 517
macroscopic scale, 238
macroscopic state, 521
macroscopic wavefunction, 6, 488, 491, 513
magic clusters, 386
magic number, 383, 385
magnetic Aharonov–Bohm effect, 266
magnetic alloys, 537
magnetic breakdown, 224
magnetic coercivity, 20
magnetic dipoles, 449
magnetic domain walls, 20
magnetic exchange interaction, 286
magnetic field, 122, 142, 144, 165, 167, 170, 200, 208, 210, 212, 224, 226, 263, 269, 271, 294, 393, 415, 490, 491, 495, 500, 517, 549
magnetic flux, 492, 516
magnetic flux quantum for single electron, 266
magnetic frozen state, 538
magnetic group, 46
magnetic hysteresis, 451
magnetic impurities in metal, 351
magnetic impurity, 333, 357, 512
magnetic induction, 122
magnetic insulators, 537
magnetic long-range order, 535
magnetic moment, 295, 321, 357, 407, 449, 450, 452, 456, 461, 466
magnetic moment formation, 351
magnetic multilayer, 372
magnetic ordering, 298, 399, 460
magnetic ordering structure, 461
magnetic permeability, 122
magnetic quantum number, 275, 503
magnetic semiconductor, 537
magnetic space group, 461
magnetic state, 512
magnetic structure, 461
magnetic subband, 456
magnetic susceptibility, 122
magnetically-ordered phase, 358, 449, 459
magnetism, 534, 540
magnetite, 454
magnetization, 102, 174, 409, 415, 464, 472, 473, 475, 479
magneto-fingerprint, 252, 272
magnetoconductivity, 211
magnetometer, 516
magnetoplumbite structure, 453
magnetoresistance, 208, 212, 269, 318, 324, 374
magnetoresistance fluctuation, 269
magnetoresistance pattern, 272
magnetoresistivity, 208, 211

magnons, 16, 19
Maier–Saupe theory, 443
majority band, 371
majority spin subband, 225
Mandelbrot, B. B., 114
manganites, 16, 349
many-body correlation, 291
many-body correlation of electrons, 278, 286
many-body effect, 275, 321
many-body formulation, 505
many-body interactions, 410
many-electron atom, 277
many-particle Hamiltonian, 301, 307
mass abundance spectra, 383
mass of a hole, 159
mass spectrum, 58
mass transport, 517
matrix, 28, 496
matrix element, 193, 368, 461
matrix equation, 143, 149
matrix form, 133
Matthiesen rule, 203
maximum current density, 514
maximum overlap, 290
Maxwell equations, 121, 122, 144, 248, 491
McTaque, J., 430, 431
mean curvature, 51
mean free path, 239, 248
mean-field approximation, 7, 471, 527, 539
mean-field critical temperature, 510
mean-field theory, 16, 20, 212, 233, 476, 539
measure, 116
Meissner effect, 490, 491, 499
Meissner, W., 6, 490
melt spinning, 77
melting, 76
melting point, 92
melting temperature, 76
membrane, 83
mesophases, 10, 82
mesoscopic, 93, 99
mesoscopic physics, 15, 255
mesoscopic systems, 251, 271
metal induced gap states (MIGS), 330
metal-insulator transition, 15, 22, 106, 244, 333, 334, 406, 548, 549, 552
metal-nonmetal transition, 240, 244, 246, 543, 545
Metal-oxide-semiconductor (MOS), 330
metal-semiconductor interface, 329
metallic, 56, 380, 543
metallic behavior, 324, 328
metallic conductivity, 239, 347
metallic phase, 548
metallic superconductor, 346
metallic surface energy, 180
metals, 157, 199, 218, 363
metastable state, 416, 536, 537
methane, 290
method of separation of variables, 282
metric, 51
micro-canonical ensemble, 522
microbridge, 514
microcavity, 195, 196
micrograph, 105
microlithography, 144
micromagnetics, 449
micron structure, 101
microscopic, 252
microscopic inhomogeneities, 100
microscopic state, 521
microstates, 523
microstructure, 100, 102, 103
microwave frequencies, 144
microwave frequency range, 202
microwave spectroscopy, 396
mineral crystals, 27
minibands, 164
minigaps, 164
minimum energy criterion, 286
minimum in the density of states, 151
minimum surface, 52
minizones, 131
minority spin, 371
minority spin subband, 225
mirror reflection, 29
mirror rotation, 30
mixed valence phenomena, 308
mobilities, 206
mobility edge, 124, 244–246
mobility gap, 244, 246
model Hamiltonian, 317, 335, 361, 410, 537
modulated wavevector, 437, 439
modulation, 261
modulus, 488
molar polarization, 543
molar volume, 543
molecular bond, 285
molecular crystal, 325, 326
molecular dynamics (MD), 317
molecular field approximation, 470
molecular magnets, 300
molecular orbital, 284
molecular orbital method, 275
molecular orbital method (MO), 283, 285, 286, 288, 291
molecule, 275
molecules with saturated chemical bond, 325

momentum, 8, 121, 460
momentum conservation, 203
momentum order, 12
momentum space, 10, 11, 22, 160, 505, 542
monochromatic electromagnetic waves, 248
monochromatic plane waves, 145
monoclinic, 39, 62
Monte Carlo method , 550
Morin temperature, 452
most divergent sum method, 356
Mott criterion, 543, 544
Mott insulator, 15, 16, 317, 333, 335, 341, 343, 347, 351
Mott transition, 15, 334, 335, 542
Mott's $T^{1/4}$ law, 247, 549
Mott, N. F., 14, 15, 215, 240, 244, 247, 329
Mott–Hubbard (MH) insulator, 342, 343
muffin-tin orbitals (MTO) method, 304, 305
muffin-tin potential, 305
Mulliken, R. S., 284
multi-colored groups, 46
multi-columns, 101
multi-connected, 214
multi-connectivity, 212
multi-layers, 101
multi-phased, 102
multi-wavelength frequency doubling, 150
multicomponent order parameter, 422, 423
multicomponents, 408
multidimensional surface, 533
multiple scattering, 14, 233
multiplication table, 32
multiplication table for point group, 32
multiply-connected, 492
myoglobin, 94

Néel state, 471
Néel temperature, 466, 467
Néel, L., 468
Nakamura, S., 322
Nambu, Y., 6
nanostructure, 15, 100, 275
nanostructured Si, 322
nanotechnology, 101
naphthalene, 295
narrow band, 132
narrow-gap, 126
narrow-gap material, 363
near-free electron model, 301
nearest neighbors, 465
nearest-neighboring hopping, 246
nearly-free electron (NFE) approximation, 177, 302
nearly-free electron model, 126, 127, 151, 185, 245
negative Curie temperature, 466
negative dielectric constant, 202
negative differential resistance, 163
negative permeability, 202
Nellis, W. J., 327
nematic liquid crystal, 18
nematic phase, 82, 441, 443, 444
net moment, 456
neural networks, 534
neutron diffraction, 7, 452
neutron inelastic scattering, 79
neutron scattering, 508
neutron stars, 9
neutrons, 9
Newton equation, 122
next nearest neighbors, 465
noble metal, 318
non-Abelian, 31
non-dissipative media, 233
non-ergodic, 537
non-ergodic state, 537
non-Euclidean, 51
non-magnetic, 357
non-orthogonal orbitals, 287
non-trivial solutions, 189
noncentered lattice, 39
noncrystalline state, 73
nondissipative dielectric microstructure, 248
nonequilibrium state, 522
nonequilibrium states, 361
nonequivalent wavevectors, 422
nonlinear optical crystals, 145, 146
nonlinear optics, 145
nonlinear phenomena, 22
nonlinear physics, 361
nonlinear polarization, 122, 145
nonmagnetic heavy Fermi liquid, 343
nonmetallic elements, 543
nonmetals, 157
nonspherical Fermi surface, 205
nonsymmorphic groups, 41
nonsymmorphic symmetry operations, 41
normal component, 142
normal curvature, 50, 51
normal fluid fraction, 483
normal liquid, 486
normal metal, 500
normal mode, 192
normal state, 497, 508
normal state conductivity, 16
normal thermodynamic phase transition, 542
normal-superconductor boundary, 498

normalization, 387
normalized susceptibility, 478
nuclear magnetic moment, 449
nuclear magnetic resonance (NMR), 294, 507, 508
nucleic acids, 95
number density of electrons, 201

oblate, 70
oblate spheroid, 102
observational timescale, 524
occupation number, 549
octagrid, 67
octahedral, 37
octahedral bonding, 290
octahedral void, 56
octahedron (octahedra), 35, 36, 62
odd parity, 283, 285
off-diagonal long range order (ODLRO), 483
Ohm's law, 201, 207, 267
on-site energy, 549
one-dimensional, 127, 256, 470
one-dimensional periodic lattice, 133
one-dimensional quasilattice, 146
one-dimensional quasiperiodic structures, 148
one-electron Schrödinger equation, 305
Onnes, K., 6
Onsager symmetry relations, 200, 269
Onsager theory, 443
Onsager, L., 8, 483
onset of ferromagnetism, 347
opal, 100
open orbit, 212, 213, 318
Oppenheimer, J.P., 308
optical absorption, 299, 368
optical activity, 437
optical branches, 123, 190
optical frequency range, 202
optical lattice potential, 519
optical microscopy, 99
optical mode, 134
optical transition, 322
optical waves, 231
optimization, 547
opto-electronic material, 322
orbit radius, 212
orbit reconnection, 224
orbital, 186
orbital antiferromagnetism, 341
orbital degree of freedom, 348
orbital ferromagnetism, 341
orbital magnetism, 512
orbital moment, 279
orbital order, 340
orbital physics, 361
orbital wavefunction, 504, 507, 510
orbital-ordering, 298, 349
orbitals, 317
order parameter, 6, 17, 74, 75, 405, 407, 408, 411, 412, 415, 419, 438–440, 495, 496, 498, 507, 526, 538, 540, 551
order-disorder transition, 16, 73, 96, 433, 483
ordered phase, 16, 21, 407, 412, 421, 433
ordered state, 414, 417, 423, 433
ordering, 441
ordinary differential equations, 282
ordinary London superconductor, 500
ordinary magnetoresistance (OMR), 374
organic molecules, 291
organic superconductors, 493
organometallic coordination compound, 299
orientational distribution function, 442
orientational entropy, 443
orientational order, 11, 73, 82
Orowan, E., 16
orthogonal orbitals, 279
orthogonality, 290
orthogonalized plane wave (OPW), 301, 302
orthorhombic, 39, 62, 347
oscillation frequency, 265
oscillation period, 210
oscillatory magnetic coupling, 374
oscillatory spatial dependence, 535
Osheroff, D. D., 16
Overhauser, A. W., 478
overlap, 129
overlap correction, 295
overlap integral, 287, 291
oxide, 452
oxygen octahedron, 296, 297

P surface, 52
p-type semiconductors, 206
packing density, 54, 56, 57
packing structures., 56
pair correlation function, 75, 112
pair formation, 510
pair-breakers, 512
pair-distribution function, 316
paired electrons, 502
pairing state of superconductors and superfluids, 504
pairing symmetry, 516, 517
parabolic, 52
parabolic dispersion relation, 131, 155, 156
parabolic well, 388
paradigm, 13–15
paraelastic-ferroelastic transition, 415

parallel, 225
paramagnet, 539
paramagnet-spin glass transition, 536
paramagnetic phase, 465
paramagnetic susceptibility, 467, 476
paramagnetic-ferromagnetic transition, 415
paramagnetic-spin glass transition, 406
paramagnetism, 449, 450
parent compound, 343
Parisi, G., 539, 542
partial wetting, 529, 530, 532
partition function, 436, 462, 522, 523, 540
pass band, 121, 134, 139, 232
Pauli exclusion principle, 10, 159, 277, 504
Pauli matrix, 511
Pauli paramagnetic susceptibility, 173
Pauli principle, 337
Pauli susceptibility, 174
Pauli, W., 6
Pauling formula, 60
Pauling model for ice, 60
Pauling's model for tcp structure, 59
Pauling, L., 57, 290, 294
peak shift, 162
Penrose structure, 431
Penrose tiling, 68
Penrose, R., 483
pentagon, 60, 67
pentagrid, 67, 68
percolation, 99, 103, 106, 245
percolation model, 241
percolation theory, 14, 107
percolation threshold, 107, 113, 241
perfect diamagnetism, 490
perfect disorder, 11
perfect order, 11
perfect periodicity, 177
perimeter polynomial, 108
perimeter sites, 108
periodic, 68, 121, 124, 151
periodic Anderson model, 16
periodic approximants, 71
periodic boundary condition, 76, 131, 147
periodic composite, 132, 136, 137
periodic density waves, 427
periodic function, 64, 136, 186
periodic materials, 121
periodic modulation, 478
periodic potential, 123, 127, 167, 177, 178, 224, 301
periodic structure, 14, 37, 122, 124, 137, 142, 146, 155, 161
periodic table, 277, 473, 543
periodic table of elements, 359
periodicity, 11, 14, 37, 67, 146, 532
permalloy, 457
permanent dipole moment, 326
perovskite structure, 296, 452
perovskites, 61, 290, 343
persistent current, 19, 264, 490
perturbation method, 6, 243, 504
perturbation potential, 127
perturbation theory, 126
phase, 43, 411, 414, 488
phase boundary, 546
phase coherence, 233, 488
phase coherence length, 252
phase coherence temperature, 510
phase diagram, 349, 358, 541
phase difference, 380, 514
phase factor, 123, 128, 261
phase function, 47
phase mismatching, 146, 150
phase orbit, 522
phase point, 521
phase problem of X-ray analysis, 45
phase separation, 361, 445
phase space, 199, 471, 525, 533, 536
phase space average, 521, 523
phase transition, 12, 16, 27, 92, 405, 412, 415, 421, 424, 481, 521, 523, 527, 538, 539, 542
phase-sensitive techniques, 516
phenomenological theory, 405, 495
phonon transport, 208
phonon-assisted, 246
phononic crystals, 136, 137
phonons, 16, 19, 203
photoemission, 371
photoemission spectra, 508
photoluminescence, 369
photonic band structure, 142
photonic band tail states, 249
photonic band theory, 125
photonic bandgap, 124, 137, 144
photonic bands, 249
photonic crystal, 14, 100, 125, 142, 144, 195
physical field, 260
physical properties, 525
piezoelectricity, 41, 435
planar square bonding, 290
planar technology, 144
Planck's constant, 8
plane of incidence, 140
plane wave, 140, 301
plane wave solution, 125
plane-wave state, 310
planetary science, 327

plasmons, 19
plastic crystals, 11
Platonic solids (or polyhedra), 35, 36
point contacts, 514
point defect, 194, 196
point group, 34, 35, 41, 305, 460
point symmetry, 59, 63
point-group symmetry operation, 421
point-like spectrum, 148
Poisson equation, 185
Polanyi, M., 16
polar covalent bonding, 293
polaritons, 19
polarity, 290
polarization, 102, 415
polarization cloud, 550
polarization of the light, 368
polarons, 19, 550
polyaramide, 93
polyatomic molecule, 289
polychromatic percolation, 245
polyethylene, 87
polygons, 60
polymer, 10, 82, 87
polymorphic phases, 360
polynomials, 64, 108
population inversion, 486
porous material, 113
porous Si, 322
position space, 160
position vector, 136
positional order, 11, 73, 82
positive hole, 158
potential barrier, 218, 513
potential energy, 122, 133, 317, 527
potential fluctuation, 246
potential profile, 178
potential well, 135, 178
precipitation, 92
precipitation-hardened, 57
preferred orientation, 92
preparation effects, 536
pressure, 244, 405, 411, 415, 521, 527, 528, 530
Prigogine, I., 22
primary structure, 95
primitive cell, 133
principal quantum number, 275, 389
principle of reciprocity, 269
probability, 74, 421
probability amplitude, 381
probability density, 165
probability distribution, 525
prolate spheroid, 102
propagating modes, 193
propagating states, 249
propagation velocity, 202
protein, 29, 94
protein folding, 534
proton, 9
prototypic phase, 437, 439
pseudogap, 508
pseudogap for alloys and quasicrystals, 150, 151
pseudogap or underdoped cuprate superconductor, 347, 510
pseudopotential method, 301, 303, 304, 316
pseudospin, 340
pseudospin interactions, 434
pyroelectricity, 435
pyroelectrics, 435

quadrupolar, 534
quality factor of resonant cavity, 195
quantization of circulation, 489
quantized conductance, 252
quantized single-particle level, 387
quantum chemistry, 15, 275, 299
quantum coherence, 252
quantum conductance, 253
quantum confinement effect, 363
quantum crystal, 326, 327
quantum degeneracy temperature, 9
quantum dot, 15, 383, 390–392
quantum effect, 470
quantum fluctuations, 552
quantum Hall effect, 170, 210, 330
quantum interference, 267
quantum liquid, 471, 481
quantum mechanical description, 167
quantum mechanical effect, 239, 326
quantum mechanical tunneling, 246
quantum mechanics, 8, 10, 218, 283, 291, 496
quantum number, 275, 295
quantum phase transition, 8, 12, 516, 542
quantum phenomena, 251, 252
quantum solid, 471
quantum statistical mechanics, 481
quantum well, 15, 363, 368
quantum well potential, 367
quantum well state, 374
quantum wire, 15, 377, 381
quantum zero point fluctuation, 542
quartz, 41, 71
quasi Brillouin zone boundaries, 151
quasi-ballistic regime, 252
quasi-Brillouin zone (QBZ), 151
quasi-electrons, 19
quasi-ergodicity, 521

quasi-momentum, 155
quasi-one-dimensional (quasi-2D), 14
quasi-particle, 159
quasi-phase-matching, 145, 146, 150
quasi-two-dimensional (quasi-2D), 14, 131
quasicrystal, 11, 37, 63, 68–70, 121, 151, 427
quasicrystallography, 37, 63, 70
quasimomentum, 121
quasiparticle (QP) approximation, 316, 322
quasiparticles, 203, 506
quasiperiodic, 68, 70
quasiperiodic function, 63, 64
quasiperiodic lattice, 65
quasiperiodic structures, 63, 121, 146
quasiperiodicity, 11, 67, 146
quenching, 279

radial distribution function (RDF), 77
radial quantum number, 389
raising, 468
random Brownian motion, 113
random close packing density, 445
random competition, 535
random field, 551
random fluctuation, 248
random laser, 237
random potential, 245
random state, 298
random walk, 91, 92, 237, 252
randomly frozen, 535
rare earth atom, 296
rare earth element, 279, 321
rare-earth binary, 537
ratio of the average Coulomb potential to the average thermal kinetic energy, 546
rational fractions, 58
rationals, 63
Rayleigh number, 22
Rayleigh scattering limit, 248
Rayleigh–Ritz variation principle, 313
reactance, 202
real crystals, 177
real lattice, 324
real space, 294, 323, 374
real structure of a solid surface, 177
reciprocal lattice, 43, 66, 123, 125, 137, 151, 249
reciprocal lattice vector, 43, 44, 137, 146, 150, 301, 421, 428
reciprocal space, 43, 131, 141, 374, 456
reduced-zone scheme, 125, 127
reducible representation, 33
reflection, 28, 507
reflection coefficients, 256
reflection probabilities, 255
refractive index, 437
refractory metal, 319
regular fractals, 112
regular icosahedron, 432
regular orbital order, 341
regular orbitals, 340
regular polyhedron, 36, 37, 291
regular tiling, 53
relative free energies, 536
relaxation processes, 534
relaxation term, 209
relaxation time, 203, 208, 211, 214, 216, 534, 536, 537
relaxation time approximation, 199, 200, 207, 208
remanent magnetization, 536
renormalization group, 5, 17, 21, 92
renormalized diffusion coefficient, 238
repeated zone scheme, 224
replica mean-field approximation, 540
replica symmetry breaking instability, 552
replica-symmetric ansatz, 540, 541
replica-symmetry-breaking, 542
representations of symmetry groups, 32
reservoir, 257
residual resistivity, 203
resistance, 261
resistivity, 203, 208, 209, 246, 347
resistivity maximum, 357
resistivity minimum, 535
resonant frequency, 195
resonant states, 183
resonant transmission, 399
resonant tunneling, 221
resonating valence bond (RVB) model, 19, 470
response to external perturbations, 536
restricted ensemble, 526
restricted Hartree–Fock (HF) method, 311
rf-SQUID, 516
rhomb, 37, 68
rhombic, 69
rhombic dodecahedron, 55, 125
rhombohedral, 39, 62
Richardson, R. C., 16
ridges, 533
right coset, 33
rigid band model, 457
room temperature, 251
root-mean-square amplitude, 245
rotation, 28, 29
rotational invariance, 523
rotational phase transition, 326
rotational symmetry, 17

rotational vibration, 487
Rowell, J. M., 514
Ruderman–Kittel–Kasaya–Yoshida (RKKY) oscillation, 188, 353
RKKY interaction, 358, 372, 451, 459, 536–538
RKKY theory, 372, 374

sample-to-sample conductance fluctuations, 271
satellites resonance, 396
saturated value, 207
saturation, 212
Saturn, 327
scalar, 408
scalar equations, 140
scalar form, 201
scalar order parameter, 415
scalar potential, 265
scalar wave, 122, 124, 236
scale invariance, 14, 28, 112, 113
scaling, 5, 8, 17
scaling laws, 20
scaling theory of localization, 14
scaling transformation, 113
scanning tunneling microscopy (STM), 199, 381
scattering matrix, 200, 219
scattering potential, 260
scattering process, 232, 246
scattering transition probability, 204
Schönflies notation, 35, 37
Schafroth pair, 12, 505
Schafroth, M. R., 505
Schläffli symbol, 80
Schottky barrier, 329
Schottky potential, 329
Schottky, W., 329
Schrödinger equation, 121, 168, 177, 178, 183, 185, 187, 248, 263, 275, 277, 279, 291, 301, 303, 305, 308, 309, 312, 313, 363, 370, 378, 385, 387, 497, 499, 502
Schrödinger, E., 96
Schrieffer, J. R., 6, 361
Schwarz, H., 52, 61
screened Coulomb potential, 311
screening, 358
screening of the electron gas, 552
screw axes, 39, 84
second harmonic generation, 41
second-order Born approximation, 355
second-order nonlinear susceptibility, 145
second-order phase transition, 8, 16, 17, 20, 407, 411, 415, 421, 464, 533, 534, 541
secondary structure, 94
secular determinant, 143, 302
secular equation, 133, 134, 284, 285, 291, 301
segregation, 177
Seitz, F., 13
selection rule, 389
self assembly, 144
self interaction energy, 545
self similarity, 111
self-affine, 114
self-affine fractal, 114
self-affine transformation, 113
self-assembled, 10
self-assembled membranes, 82
self-avoiding walk, 90, 91
self-consistent equation, 436, 444, 539, 540
self-consistent field approximation, 277, 310
self-duality, 52
self-inductance, 517
self-similar transformation, 113
self-similar wavefunction, 148
self-similarity, 65, 113
semi-empirical criteria, 543
semi-infinite crystal, 189
semi-infinite periodic chain, 177
semi-regular polyhedron, 37
semiclassical approach, 155, 238, 251
semiclassical particles, 233
semiconducting, 380
semiconducting ferromagnet, 458
semiconductor, 158, 244, 322, 323, 328, 347, 363, 506
semiconductor superlattice, 165
semiconductor diode, 329
semiconductor electronics, 322
semiconductor physics, 322
semiconductor quantum well, 363
semiconductor superlattice, 129
semiconductor surface, 181, 182
semiconductors, 4, 157, 199, 217, 218, 246, 544
semimetal, 157, 324
semiregular, 53
series expansion, 108
Sham, L. J., 315
Shannon, C., 96
shape of Fermi surface, 174
shapes of the Fermi surface, 165
Shapiro steps, 515
Shechtman, D., 37
Sherrington, D., 539
Sherrington–Kirkpatrick (SK) model, 541
shielding supercurrents, 491

Shklovskii, B. I., 549
short-range correlation, 483
short-range forces, 550
short-range interactions, 434
short-range order, 11, 75
short-range order parameter, 75
Shubnikov, L. V., 7
Shubnikov–de Haas (SdH) effect, 170, 208, 210, 211
simple cubic (SC) lattice, 550
simple metal, 318
single mass defect, 192
single quantum state, 486
single scattering approach, 250
single-electron approximation, 275, 277, 283, 301, 307
single-electron Schrödinger equation, 284, 306
single-electron tunneling, 391
single-particle Green's function and dynamic screened Coulomb interaction (GW) approximation, 316
single-particle Hamiltonian, 460, 484
single-particle partition function, 528
single-particle Schrödinger equation, 386
single-site excitations, 550
single-wall nanotube, 380
singlet energy, 287
singly-connected, 489
singular continuous spectrum, 148
sinusoidal spin density wave, 479
site percolation, 106, 107
size-dependent diffusion coefficient, 234
Slater determinant, 289, 312, 547
Slater, J. C., 289
Slater–Pauling curve, 456
slowest dynamical processes, 524
slowing-down, 536
small imperfection, 177
smectic phase, 84
soft 'Coulomb gap', 549
soft mode eigenvectors, 438
solid, 8, 9, 275
solid state, 327
solid state physics, 12, 13, 15, 17, 121, 155, 286
solid surfaces, 229
Sommerfeld model, 310
sound velocity, 437
space group, 27, 37, 41, 421, 461
space partition, 52
spatial dimensionality, 524
specific heat, 413, 437, 464, 551
specific heat measurement, 508
specific heat peak, 486
spectra of confined state, 381
spectra of lattice modes, 135
sphere, 52
spherical cavity, 385
spherical clusters, 386
spherical harmonics, 275, 386, 503, 510
spherical polar coordinates, 275
spherical pseudopotential method, 386
spherical surface, 52
spheroid, 167
spin, 460
spin crossover, 299
spin density, 188, 478
spin density functional approach, 456
spin density wave (SDW), 11, 20, 478
spin fluctuation, 476
spin function, 504
spin glass, 22, 534, 536–538, 551
spin glass state, 535
spin glass transition, 538
spin glasses, 8
spin Hamiltonian, 288
spin memory effect, 216
spin operator, 288, 468
spin orientation, 456
spin polarization, 174, 208, 225, 372, 374
spin quantum number, 277, 466
spin relaxation length, 377
spin singlet state, 393
spin subband, 476
spin susceptibility, 549
spin transport, 199, 208
spin variable, 410
spin vector, 461
spin waves, 16
spin-dependent resistivity, 377
spin-down, 376
spin-down band, 475
spin-flip, 357
spin-flip scattering, 216, 357
spin-glass, 535, 541
spin-orbit coupling, 279, 296, 323, 366, 460
spin-ordering, 349
spin-polarized electronic density, 321
spin-polarized electrons, 10, 216
spin-polarized photoemission spectrum, 371
spin-polarized state, 321
spin-polarized transport, 225
spin-singlet, 278, 287, 357, 395, 504
spin-singlet d-wave pairing, 502
spin-singlet s-wave pairing, 502
spin-singlet pairing, 512
spin-triplet, 278, 287, 504, 510
spin-triplet p-wave pairing, 502, 512

spin-up, 376
spin-up band, 474
spinel structure, 453
spinless Hamiltonian, 549
spinodal decomposition, 100
spinodal point, 418, 419
spinon, 510
spintronics, 215, 217, 458
spiral, 451, 461
spiral state, 358
splat quenching, 76
splitting of energy band, 472
spontaneous breaking of the symmetries, 461, 523
spontaneous distortion, 297
spontaneous magnetization, 215, 464, 467, 472, 524
spontaneous polarization, 435, 436
spontaneous radiation, 196
spontaneous strain, 419
sputtering, 100
square, 54, 550
square bonding, 291
square lattice, 107
square matrix, 133
squared moduli, 127
staggered orbital order, 341
standing wave, 127, 249
standing-wave laser field, 519
star arms, 422
states of up and down spins, 321
static electric field, 200
stationary states, 132
statistical mechanics, 21, 521, 522, 524
statistical model, 409, 449
statistical physics, 8, 10
statistical symmetry, 47
step, 381
step state, 381
steric interaction, 90
stimulated emission, 486
stoichiometric, 343
Stokes' theorem, 492
Stoner band model, 472, 476
Stoner criterion, 472, 473, 476
Stoner factor, 473
stop-bands, 134, 139
strain, 419
stripe, 381
stripe phase, 351, 361
strong anisotropy, 366
strong coupling theory, 398, 507
strong diffraction spots, 151
strong electron-electron correlations, 548
strong localization, 246
strong magnetic field, 546
strongly correlated electrons, 317, 333, 351, 359, 361, 362
strongly correlated state, 333
strongly disordered systems, 240
strongly localized case, 253
structural phase transition, 62, 349, 420, 421, 435
structure factor, 74, 140
subband, 366
subgroup, 32, 63, 407, 421, 422, 424
sublattice, 465
substitution rule, 64
substitutional binary alloys, 433
substitutional disorder, 73
substitutionally-ordered alloys, 11
successive iterations, 147
sulfuric acid, 93
sum frequency, 145
super-cooled liquids, 73
super-microstructure, 100
superconducting cables, 7
superconducting electronics, 7, 515
superconducting phase transition, 411
superconducting quantum interference device, 516
superconducting ring, 517
superconducting state, 20, 490
superconducting transition temperature, 496
superconductivity, 6, 15, 16, 347, 481, 483, 490, 499, 502, 512
superconductor, 4, 6, 19, 218, 328, 359, 490, 492, 495, 512, 513, 518
supercooled liquid, 76, 534
supercurrent, 495, 498, 500
supercurrent circulation, 517
supercurrent velocity, 491
superexchange, 16, 337, 338, 346, 451
superexchange coupling, 400
superfluid, 4, 411, 483, 486, 489, 492, 518
superfluid ^{4}He, 487
superfluid ^{3}He, 415, 510
superfluid fraction, 488
superfluid state, 413, 487, 489, 507
superfluid velocity, 488
superfluid wavefunction, 489
superfluid weak link, 518
superfluidity, 16, 481, 483, 486, 502
superlattice, 15, 101, 131, 132, 163, 164
superparamagnetic, 400
superspaces, 111
superstructure, 74
supramolecular structures, 73, 82

surface, 10, 50, 114, 177, 192, 491
surface band, 383
surface confinement, 179
surface density, 531
surface electronic state, 328
surface energy, 177, 178
surface mode, 189, 193
surface of a solid, 177
surface physics, 14
surface reconstruction, 328
surface state, 183
surface wave, 193
surfaces of metals, 178
susceptibility, 176, 415, 464, 479, 536, 551
swollen chain, 91
symmetric Bragg case, 142
symmetric properties, 133
symmetric well, 368
symmetry, 27, 305, 405, 423, 502
symmetry change, 533
symmetry element, 421
symmetry group, 30, 405, 437, 523
symmetry of reciprocal space, 47
symmetry operation, 28, 47
symmetry states, 507
symmorphic groups, 41
systematic extinction, 41, 47, 75

tail states, 245
tailoring physical properties, 363
tangential component, 142
Taylor, G. I., 16
technical magnetization, 449, 451
temperature, 170, 405, 411, 445, 521, 523, 527
temperature gradient, 199, 200, 207, 208
temperature-dependent conductivity, 240
temperature-dependent hopping, 246
tensor, 122, 201
termination coden, 95
terrace, 381
tertiary structure, 94
tetragonal, 39, 62
tetrahedral, 37, 59
tetrahedral bonding, 290, 291
tetrahedral coordination, 80, 293, 294
tetrahedral void, 56
tetrahedrally close-packed (tcp) structure, 57, 58
tetrahedron, 35
tetrahedron (tetrahedra), 58
Theatatus, 36
theory of chaos, 22
theory of measure, 116
thermal conductivity, 207, 208, 412, 437
thermal current, 208
thermal equilibrium, 8, 10, 524
thermal excitations, 483
thermal expansion coefficient, 413
thermal fluctuation, 76, 122
thermal transport phenomena, 207
thermally excited, 233
thermionic emission, 178
thermodynamic average, 539
thermodynamic behaviors, 462
thermodynamic limit, 485, 524
thermodynamic parameters, 530
thermodynamic quantities, 413
thermodynamic temperature, 11
thermoelectric coefficient, 207
thermoelectric effect, 207
thermoplastic, 92
thermotropics, 82, 440
thiolate, 101
thiourea, 437
third-order harmonic generation, 150
third-order invariant, 423, 424
third-order nonlinear susceptibility, 145
Thom, R., 22
Thomas–Fermi (TF) theory, 312, 544
three-body distribution function, 77
three-dimensional (3D), 132, 142
three-dimensional crystals, 139
threshold energy, 241, 244
tie point, 141
tight-binding approximation, 128, 147, 150, 164, 301, 319, 334, 452
tight-binding disordered one-electron model, 242
tight-binding wavefunctions, 245
tight-packing density, 443
time average, 521, 522, 524
time correlation function, 200
time reversal, 459, 460
time reversal operator, 459
time reversal symmetry, 306
time scale, 523, 524
time-dependent Schrödinger equation, 260, 459
time-resolved photoluminescence, 546
time-reversal symmetry, 461, 510
top of the valence band, 368
topological classification of defects, 16
topological defects, 16, 17, 20
topological dimension, 111
topological structures, 212, 214
topology, 111
total current, 252
total free energy, 530

total number of modes, 136
total pairing wavefunction, 510
total reflection, 142
total resistance, 257
total resistivity, 355
totally Bragg reflected, 372
Toulouse, G., 16
trace, 540
trajectories, 170
transfer energy, 147
transfer matrix, 138, 147, 148, 150, 219
transfer matrix element, 183
transfer model, 147
transition, 527, 532
transition entropy, 437
transition heat, 437
transition metal, 183, 208, 319, 449, 476
transition metal alloys, 537
transition point, 412, 413
transition probability, 203, 377
transition temperature, 433, 435, 436, 466, 539
transitions, 203
translation invariance, 473
translation symmetry, 38
translational, 523
translational entropy, 443
translational invariance, 14
translational operator, 186
translational symmetry, 14, 17, 73, 422
transmission, 255, 256
transmission amplitude, 219, 270
transmission coefficient, 220
transparent insulators, 15
transport coefficient, 412
transport equation, 199
transport phenomena, 15, 318
transport properties, 152, 199, 476, 552
transversal modes, 135
transverse, 167
transverse energy, 221
travelling wave, 127
triacontahedron, 36
trial solutions, 133
triangular bonding, 291
triangular quantum well, 365
triangular-bi-pyramidal bonding, 290
triclinic, 39
tricontahedron, 59, 69
tricritical point, 408
trigonal bipyramid, 291
triple point, 487
triplet, 107, 339
triplet code, 95
triplet energy, 287
truncated icosahedron, 36, 37, 59
truncated octahedron, 55, 125
Tsuei, C. C., 517
tube diameter, 381
tuning parameter, 548
tunneling, 15, 507, 518
tunneling current, 222
tunneling junction, 218, 221
tunneling magnetoresistance (TMR), 199, 225, 226
tunneling model, 222
tunneling phenomena, 218
turbulence, 22, 114
Turnbull, D., 13
two sublattices, 467
two-band model, 215
two-body (or pair) distribution function, 77
two-component order parameter, 438
two-current model, 215, 216
two-dimensional (2D), 378
two-dimensional (2D) electron gas, 131
two-dimensional (2D) structures, 144
two-electron problem, 286
two-phase alloy, 103
two-terminal multichannel conductance, 258
two-wave approximation, 140, 141
type I superconductors, 490, 498
type II superconductors, 7, 490, 499

ultrasonic wave, 137
umbrella-like, 451, 461
Umklapp (U) scattering process, 204, 205
uncertainty principle, 504, 522, 542
unconventional properties, 359
unconventional superconductors, 19, 507, 512
undamped plasma oscillation frequency, 202
underdoped cuprate, 508
uniform distribution, 242
uniform motion, 488
unit cell, 125, 128, 138, 437
unit matrix, 133
unit vector, 138
unitary transformation, 496
universal conductance fluctuation, 270, 271
universality, 5, 8, 17, 21
unrestricted HF method, 311
unstable state, 416
upper bound, 469
upper cut-off frequency, 136
Urbach tail, 245, 249

valence band, 158, 244, 246, 322–324, 368
valence bond orbital, 290
valence electron approximation, 308

valence shell electron pair repulsion theory, 291
valleys, 533, 536
van der Waals bond, 59
van der Waals equation, 528
van der Waals force, 325
van der Waals interaction, 324
van der Waals theory, 527
van der Waals, J. D., 16
van Hove correlation function, 79
van Hove singularity, 135, 170
vapor deposition, 100
vapor pressure, 527
vapor-liquid transition, 415, 526
variable-range hopping, 246, 247, 549
vector, 84, 408
vector order parameter, 421
vector potential, 167, 260, 515
vector-wave, 124
vibrational state, 133
virtual hopping process, 337, 338
viscosity, 487, 534
voltage drop, 252
volume, 415
volume per mole, 528
von Laue, M., 14
Voronoi cells, 77
Voronoi polyhedra, 77
vortex, 500
vortex formation, 486

Wannier functions, 128
Wannier–Stark ladder, 163–165
Wannier–Stark state, 165
water, 293
wave amplitude, 132
wave coherence, 251
wave diffusion, 236
wave equation, 121–123, 138, 231, 248
wave packet, 155
wave propagation, 121, 122, 132, 231
wave-particle duality, 155
wavefunction, 125–128, 148, 150, 164, 168, 178, 183, 232, 242, 260, 261, 275, 301, 307, 312, 382, 387, 502, 503, 513
wavefunction phase, 251
wavefunctions of electron pairs, 513
waveguide, 196
wavelength, 100, 233, 247, 248
wavevector, 121, 123, 140, 142, 165, 183, 249, 421, 424, 428, 460, 523
wavevector group, 422
wavevector star, 422
weak coupling approximation, 398, 507
weak disorder, 239, 245
weak ferromagnetism, 452, 457, 476
weak first-order phase transition, 415, 416
weak localization, 15, 233, 235, 252, 265
weak scattering, 238
weak scattering potential, 263
weakly attractive, 528
weakly coupled, 513
weighting factor, 204
Weiss field, 434
Weiss theory for ferromagnets, 436
Weiss, P. E., 16
wetting transition, 530
Weyl, H., 27
Widom, B., 5
Wiedermann–Franz law, 208
Wieman, C., 484
Wigner crystal, 11, 20, 545–547
Wigner crystallization, 12, 333, 349, 542
Wigner transition, 546
Wigner, E. P., 327, 545, 546
Wigner–Seitz (WS) cell, 54, 123, 359, 545
Wigner–Seitz (WS) radii, 319
Wigner–Seitz (WS) radius, 305
Wilson, A. C., 14
Wilson, K. G., 5, 356
window function, 66
Wollan, E. O., 350
work function, 177, 178, 329, 385
wüstite, 71
Wyckoff symbol, 42

X-ray diffraction, 42, 86, 124, 140
XY model, 410, 411, 449

Yabolonovitch, E., 124
Young's equation, 529
Yukawa potential, 544

Zeeman splitting, 389
Zener electric breakdown, 223
Zener's theory of ferromagnetism, 354
Zener, C., 16, 347
zero on-site repulsion, 547
zero point energy, 546
zero point oscillation, 327
zero temperature, 542
zero thermal expansion coefficient, 457
zero-point fluctuation, 122
zeroth-order approximation, 164
Zhang–Rice singlet, 346
zigzag type of carbon nanotube, 61, 380